W9-BXI-907

The University of Chicago School Mathematics Project

Algebra

Second Edition

About the Cover The art on the cover was generated by a computer. The planes, grid, and intersecting lines suggest the integrated approach of *UCSMP Algebra.* This course uses geometry and statistics as a setting for work with linear expressions and sentences, and much work is done with graphing.

Authors

John W. McConnell Susan Brown
Zalman Usiskin Sharon L. Senk Ted Widerski Scott Anderson
Susan Eddins Cathy Hynes Feldman James Flanders Margaret Hackworth
Daniel Hirschhorn Lydia Polonsky Leroy Sachs Ernest Woodward

ScottForesman

A Division of HarperCollins*Publishers*

Editorial Offices: Glenview, Illinois
Regional Offices: Sunnyvale, California • Tucker, Georgia
Glenview, Illinois • Oakland, New Jersey • Dallas, Texas

ACKNOWLEDGMENTS

Authors

John W. McConnell
Instructional Supervisor of Mathematics, Glenbrook South High School, Glenview, IL

Susan Brown
Mathematics Department Chair, York High School, Elmhurst, IL

Zalman Usiskin
Professor of Education, The University of Chicago

Sharon L. Senk
Associate Professor of Mathematics, Michigan State University, East Lansing, MI (Second Edition only)

Ted Widerski
Mathematics Teacher, Waterloo High School, Waterloo, WI (Second Edition only)

Scott Anderson
UCSMP (Second Edition only)

Susan Eddins
Mathematics Teacher, Illinois Mathematics and Science Academy, Aurora, IL (First Edition only)

Cathy Hynes Feldman
Mathematics Teacher, The University of Chicago Laboratory Schools (First Edition only)

James Flanders
UCSMP (First Edition only)

Margaret Hackworth
Mathematics Supervisor, Pinellas County Schools, Largo, FL (First Edition only)

Daniel Hirschhorn
UCSMP (First Edition only)

Lydia Polonsky
UCSMP (First Edition only)

Leroy Sachs
Mathematics Teacher (retired), Clayton High School, Clayton, MO (First Edition only)

Ernest Woodward
Professor of Mathematics, Austin Peay State University, Clarksville, TN (First Edition only)

Design Development

Curtis Design

UCSMP Production and Evaluation

Series Editors: Zalman Usiskin, Sharon L. Senk
Directors of First Edition Studies: Sandra Mathison (director); Assistants to the Directors: Penelope Flores, Catherine Sarther
Directors of Second Edition Studies: Gurcharn Kaeley, Geraldine Macsai
Technical Coordinator: Susan Chang
Second Edition Teacher's Edition Editor: David Witonsky
Second Edition Consultants: Amy Hackenberg, Mary Lappan
First Edition Managing Editor: Natalie Jakucyn

We wish to acknowledge the generous support of the Amoco Foundation and the Carnegie Corporation of New York in helping to make it possible for the First Edition of these materials to be developed, tested, and distributed, and the continuing support of the Amoco Foundation for the Second Edition.

We wish to thank the many editors, production personnel, and design personnel at ScottForesman for their magnificent assistance.

Multicultural Reviewers for ScottForesman

Winifred Deavens
St. Louis Public Schools, St. Louis, MO
Seree Weroha
Kansas City Public Schools, Kansas City, KS
Efraín Meléndez
Dakota School, Los Angeles, CA
Linda Skinner
Educator, Edmond, OK

ISBN: 0-673-45765-6

5 6 7 8 9 10 - DR - 01 00 99 98 97

It is impossible to thank everyone who has helped create and test this book. We wish particularly to thank Carol Siegel, who coordinated the use of the test materials in schools; Tina Klawinski and Lynn Libby of our editorial staff; Sara Benson, Anil Gurnarney, Dae S. Lee, Jee Yoon Lee, and Sara Zimmerman of our technical staff; and Eileen Fernandez, Rochelle Gutiérrez, Suzanne Levin, Nancy Miller, and Gerald Pillsbury of our evaluation staff.

A first draft of *Algebra* was written and piloted during the 1985–86 school year. After a major revision, a field trial edition was tested in 1986–87 at these schools:

Clearwater High School
Clearwater, Florida

Aptakisic Junior High School
Buffalo Grove, Illinois

Washington High School
Von Steuben Metropolitan Science Center
Disney Magnet School
Austin Academy
Chicago, Illinois

Morton East High School
Cicero, Illinois

O'Neill Middle School
Downers Grove, Illinois

Glenbrook South High School
Glenview, Illinois

Elk Grove High School
Elk Grove Village, Illinois

McClure Junior High School
Western Springs, Illinois

Hubble Middle School
Wheaton, Illinois

Parkway West Middle School
Chesterfield, Missouri

Northeast High School
Clarksville, Tennessee

A second revision underwent a comprehensive nationwide test in 1987–88. The following schools participated in those studies:

Rancho San Joaquin Middle School
Lakeside Middle School
Irvine High School
Irvine, California

Mendocino High School
Mendocino, California

Lincoln Junior High School
Lesher Junior High School
Blevins Junior High School
Fort Collins, Colorado

Bacon Academy
Colchester, Connecticut

Rogers Park Junior High School
Danbury, Connecticut

Hyde Park Career Academy
Bogan High School
Chicago, Illinois

Morton East High School
Cicero, Illinois

John H. Springman School
Glenview, Illinois

Carl Sandburg Junior High School
Winston Park Junior High School
Palatine, Illinois

Fruitport High School
Fruitport, Michigan

Taylor Middle School
Van Buren Middle School
Albuquerque, New Mexico

Crest Hills Middle School
Shroder Paideia Middle School
Walnut Hills High School
Cincinnati, Ohio

Easley Junior High School
Easley, South Carolina

R.C. Edwards Junior High School
Central, South Carolina

Liberty Middle School
Liberty, South Carolina

Glen Hills Middle School
Glendale, Wisconsin

Robinson Middle School
Maple Dale Middle School
Fox Point, Wisconsin

Since the ScottForesman publication of the First Edition of *Algebra* in 1990, thousands of teachers and schools have used the materials and have made additional suggestions for improvements. The materials were again revised, and the following teachers and schools participated in field studies in 1992–1993:

Dallas Russell
D.W. Griffith Jr. High School
Los Angeles, California

Michael Mueller
Mendota High School
Mendota, Illinois

Pat Carlson
Fruitport High School
Fruitport, Michigan

Sally Jackman
Hanks High School
El Paso, Texas

Claire V. Giambalvo
Chaffey High School
Ontario, California

Sally Cadagin
Grant Middle School
Springfield, Illinois

Jerry Johnson
Sauk Rapids-Rice Schools
Sauk Rapids, Minnesota

Bonnie L. Buehler
John H. Springman School
Glenview, Illinois

Marilyn Morse
Eagleview Middle School
Colorado Springs, Colorado

Brian Anderson
Central Junior High School
Lawrence, Kansas

Michael R. Casey
Lake Oswego Sr. High School
Lake Oswego, Oregon

Sidney Caldwell
Safety Harbor Middle School
Safety Harbor, Florida

Melanie Kellum
Old Rochester High School
Mattapoisett, Massachusetts

Stephen Mazurek
Springfield High School
Springfield, Pennsylvania

We wish also to acknowledge the contribution of the text *Algebra Through Applications with Probability and Statistics,* by Zalman Usiskin (NCTM, 1979), developed with funds from the National Science Foundation, to some of the conceptualizations and problems used in this book.

THE UNIVERSITY OF CHICAGO SCHOOL MATHEMATICS PROJECT

The University of Chicago School Mathematics Project (UCSMP) is a long-term project designed to improve school mathematics in grades K–12. UCSMP began in 1983 with a 6-year grant from the Amoco Foundation. Additional funding has come from the National Science Foundation, the Ford Motor Company, the Carnegie Corporation of New York, the General Electric Foundation, GTE, Citicorp/Citibank, and the Exxon Education Foundation.

UCSMP is centered in the Departments of Education and Mathematics of the University of Chicago. The project has translated dozens of mathematics textbooks from other countries, held three international conferences, developed curricular materials for elementary and secondary schools, formulated models of teacher training and retraining, conducted a large number of large and small conferences, engaged in evaluations of many of its activities, and through its royalties has supported a wide variety of research projects in mathematics education at the University. UCSMP currently has the following components and directors:

Resources	Izaak Wirszup, Professor Emeritus of Mathematics
Elementary Materials	Max Bell, Professor of Education
Elementary Teacher Development	Sheila Sconiers, Research Associate in Education
Secondary	Sharon L. Senk, Associate Professor of Mathematics, Michigan State University Zalman Usiskin, Professor of Education
Evaluation Consultant	Larry Hedges, Professor of Education

From 1983 to 1987, the director of UCSMP was Paul Sally, Professor of Mathematics. Since 1987, the director has been Zalman Usiskin.

Algebra

The text *Algebra* has been developed by the Secondary Component of the project, and constitutes the core of the second year in a six-year mathematics curriculum devised by that component. The names of the six texts around which these years are built are:

Transition Mathematics
Algebra
Geometry
Advanced Algebra
Functions, Statistics, and Trigonometry
Precalculus and Discrete Mathematics

The content and questions of this book integrate geometry, probability, and statistics together with algebra. Pure and applied mathematics are also integrated throughout. It is for these reasons that the book is deemed to be a part of an integrated series. However, algebra is the trunk from which the various branches of mathematics studied in this book emanate. It is for this reason that we call this book simply *Algebra.*

The first edition of *Algebra* introduced many features that have been retained in this edition. There is **wider scope,** including significant amounts of geometry and statistics, and some combinatorics and probability. These topics are not isolated as separate units of study or enrichment. They are employed to motivate, justify, extend, and otherwise enhance important concepts of algebra. The geometry is particularly important because many students have in the past finished algebra without the content prerequisites for success in geometry. A **real-world orientation** has guided both the selection of content and the approaches allowed the student in working out exercises and problems. This is because being able to do mathematics is of little use to an individual unless he or she can apply that content. We require **reading mathematics,** because students must read to understand mathematics in later courses and must learn to read technical matter in the world at large. The use of **up-to-date technology** is integrated throughout, with *scientific calculators* assumed and *graphics calculators* strongly recommended.

Four dimensions of understanding are emphasized to maximize performance: skill in carrying out various algorithms; developing and using mathematics properties and relationships; applying mathematics in realistic situations; and representing or picturing mathematical concepts. We call this the SPUR approach: **S**kills, **P**roperties, **U**ses, **R**epresentations.

The **book organization** is designed to maximize the acquisition of both skills and concepts. Ideas introduced in a lesson are reinforced through Review questions in the immediately succeeding lessons. This daily review feature allows students several nights to learn and practice important concepts and skills. Then, at the end of each chapter, a carefully focused Progress Self-Test and a Chapter Review, each keyed to objectives in all the dimensions of understanding, are used to solidify performance of skills and concepts from the chapter so that they may be applied later with confidence. Finally, to increase retention, important ideas are reviewed in later chapters.

Since the ScottForesman publication of the first edition of *Algebra* in 1990, the entire UCSMP secondary series has been completed and published. Thousands of teachers and schools have used the first edition and some have made suggestions for improvements. There have been advances in technology and in thinking about how students learn. We have attempted to utilize these ideas in the development of the second edition.

Those familiar with the first edition will note a rather significant reorganization of the content in the second edition, a restructuring of the beginning of the course so that some ideas are introduced one or two months earlier than before. We were encouraged to do this because a high percentage of *Algebra* students enter this course with a better background than could have expected when we wrote the first edition. Many of these have benefited from *Transition Mathematics.*

There are many other changes. We have reorganized the material on square roots, absolute value, and quadratics somewhat. Technology that graphs functions, introduced in Chapter 12 in the first edition, is now introduced in Chapter 5 and applied thereafter.

There are also a number of features new to this edition, including the following: **Activities** have been incorporated into many of the lessons to help students develop concepts before or as they read. There are **projects** at the end of each chapter because in the real world much of the mathematics that is done requires a longer period of time than is customarily available to students in daily assignments. There are many more questions requiring **writing,** because writing helps students clarify their own thinking, and writing is an important aspect of communicating mathematical ideas to others. In the area of technology, **spreadsheets** are introduced early as a pattern-finding and a problem-solving tool.

Comments about these materials are welcomed. Please send them to: UCSMP, The University of Chicago, 5835 S. Kimbark, Chicago, IL 60637.

CONTENTS

CHAPTER 1 4

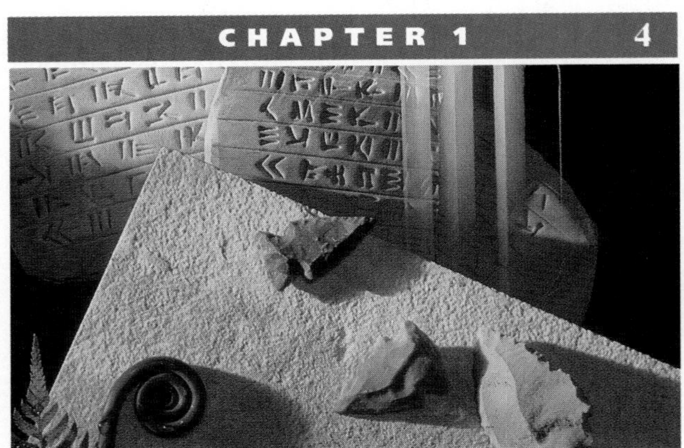

USES OF VARIABLES

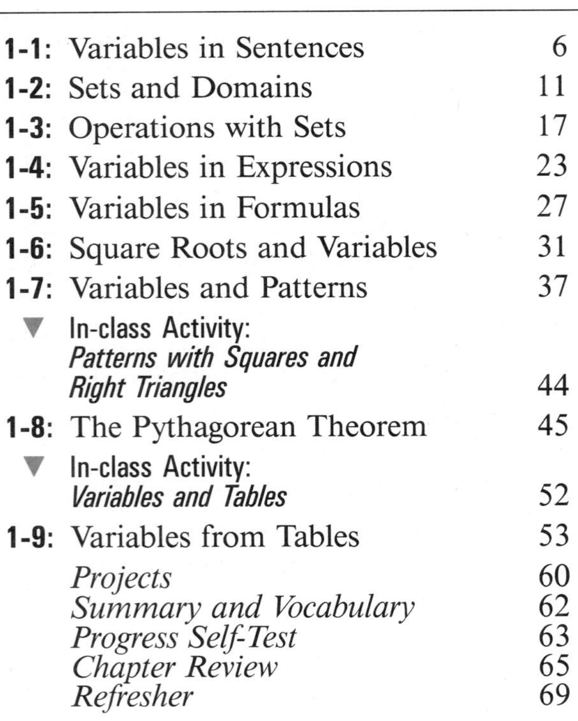

CHAPTER 2 70

MULTIPLICATION IN ALGEBRA

CHAPTER 12 718

FACTORING

CHAPTER 13 772

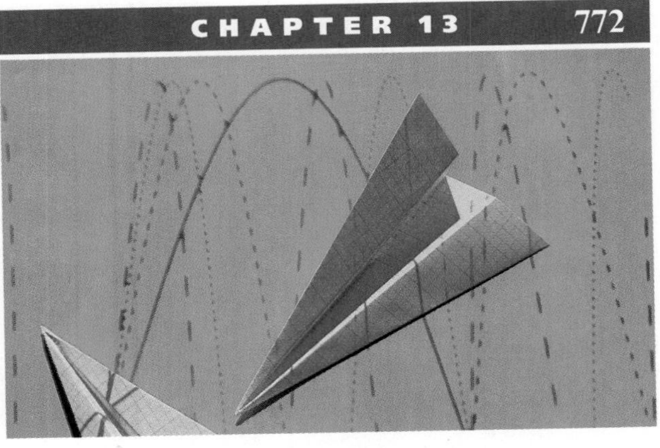

FUNCTIONS

To the Student

Getting Started

Welcome to Algebra.
We hope you enjoy this book; it was written for you.

Studying Mathematics

This book has several goals. It will introduce you to the language of algebra. It will help you prepare for geometry and other mathematics. It will help you learn about the many uses of algebra in the real world and deal with the mathematics around you.

ANOTHER GOAL OF this book is for you to continue to develop your study skills in mathematics. To accomplish this goal, you should take advantage of all the resources you have. The authors, who are experienced teachers, offer the following advice on studying algebra.

1 You can watch basketball hundreds of times on television. Still, to learn how to play basketball, you must actually dribble, shoot, and pass a ball. Mathematics is no different. You cannot learn much just by watching other people do it. You must participate. Some teachers have a slogan: *Mathematics is not a spectator sport.*

Mathematics is not a spectator sport.

2 You are expected to read each lesson, and it is vital for you to understand what you have read. Here are some ways to improve your reading comprehension.

Read slowly, paying attention to each word and symbol.

Look up the meaning of any word you do not understand.

Work examples yourself as you follow the steps in the text.

Reread sections that are unclear to you.

Discuss difficult ideas with a fellow student or your teacher.

3 Writing is a tool for communicating your solutions and thoughts to others. It can help you understand mathematics, too. So you will sometimes be asked to explain your solution to a problem, to justify an answer, or to write down information that can help you study. Writing good explanations takes practice. You can look at the solutions to the examples in each lesson as a guide for your own writing.

4 If you cannot answer a question immediately, don't give up! Read the lesson again. Read the question again. Look for examples. If you can, go away from the problem and return to it a little later. Ask questions and talk to others when you do not understand something.

Equipment Needed for This Book

You need to have some tools to do any mathematics. The most basic tools are paper, pencil, and erasers. For this book, you will also need the following equipment:

a ruler with both centimeter and inch markings,

a protractor,

graph paper,

and **a scientific calculator**.

The calculator should have the following keys:

x^y or y^x or ^ (powering),

$\sqrt{x}$ (square root), x! (factorial),

±, +/−, or (−) (for negative numbers),

π (pi), and 1/x (reciprocals).

Your calculator should also display very large or very small numbers in scientific notation.

Getting Acquainted with UCSMP *Algebra*

If you have never used a UCSMP book before, spend some time getting acquainted with this book. The questions that follow are designed to help you become familiar with *Algebra.*

We hope you join the hundreds of thousands of students who have enjoyed UCSMP *Algebra.* We wish you much success.

QUESTIONS

Covering the Reading

1. What are the goals of *Algebra?*
2. List some materials you will need for your work in *Algebra.*
3. Explain the meaning of the statement, "Mathematics is not a spectator sport."
4. Of the five things listed that you can do to improve reading comprehension, list the three you think are most helpful.
5. Where can you look for a model to help you write an explanation justifying your answer to a problem?

Knowing Your Textbook

For 6–13, answer the questions by looking at the Table of Contents, the chapters, the appendices, or the index at the end of the book.

6. Refer to the Table of Contents beginning on page *vi.* What lesson would you read to learn about the factorial symbol?
7. The appendices contain information about topics you may need to review.
   a. How many appendices does *Algebra* have?
   b. What topics are reviewed?

In 8 and 9, refer to Lesson 3-4.

8. What are the four categories of questions in Lesson 3-4?
9. Suppose you just finished the questions in Lesson 3-4. On what page can you find answers to check your work? For which questions are answers given?
10. Refer to a Progress Self-Test at the end of a chapter. When you finish the test, what is recommended that you do?
11. What kinds of questions are in the Chapter Review at the end of each chapter?
12. Use the index in the back of your book to find the pages where Mount Rushmore is mentioned. What are the page numbers? What is the title of that lesson? Which presidents are depicted on Mount Rushmore?

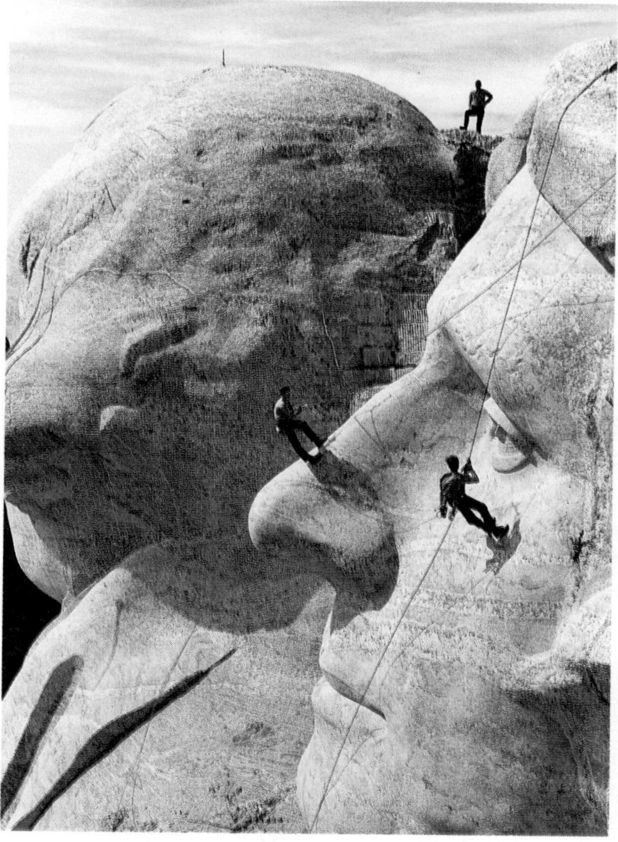

13. Each chapter is introduced with an application of a major idea from the chapter or a brief history of a mathematical concept presented in one of the lessons. Read the opener for Chapter 1. Who was François Viète, and what was his contribution to mathematics?

CHAPTER 1

USES OF VARIABLES

About 3500 years ago, an Egyptian scribe, Ahmose, copied the following hieroglyphics from a source then about 200 years old. The hieroglypics tell how to find the area of a rectangle with length 10 units and width 2 units.

Today's description in English can be shorter.

The area of a rectangle equals its length times its width.

The statement can be shortened more by using symbols for "equals" and "times."

$$\begin{aligned}\text{Area of a rectangle} &= \text{length} \times \text{width}\\ &= 10 \text{ units} \times 2 \text{ units}\\ &= 20 \text{ square units}\end{aligned}$$

The statement can be abbreviated even more by using *variables.*

$$A = \ell w$$

Today we call letters such as A, ℓ, and w variables because their values can vary. Beginning algebra is the study of variables like those found in formulas. For many people, formulas are clearer and easier to use than words. Formulas are one way in which algebra makes relationships between quantities easier to understand and use.

The first person to use letters to describe general arithmetic patterns as we do today was François Viète (Fran swah Vee yet), in 1591. He was also one of the greatest mathematicians of the 16th century. He believed that his invention was so powerful that with it "there is no problem that cannot be solved."

Viète was both right and wrong. Despite the power of algebra, there remain problems that algebra cannot solve.

In this chapter, you will see how variables appear in sentences, expressions, formulas, and tables.

LESSON

1-1

Variables in Sentences

Electrifying facts. *Lightning travels at speeds up to 100,000 miles per second! A flash of lightning between a cloud and the ground could be 9 miles long.*

What Is a Variable?

There is an old *rule of thumb* for estimating your distance from a flash of lightning. First, determine the number of seconds between the flash and the sound of thunder. Divide this number by 5. The result is the approximate distance in miles. For example, if you count 10 seconds between the flash and the sound of thunder, you are about $\frac{10}{5}$, or 2, miles from the lightning.

In the language of algebra, if s is the number of seconds between the flash and the thunder, then $\frac{s}{5}$ is your approximate distance (in miles) from the lightning. The letter s is a *variable.* A **variable** is a letter or other symbol that can be replaced by any number (or other object) from some set.

Variables may be capital or lower-case letters. In this situation the letter s was chosen because s is the first letter of "seconds." If m is your distance in miles from a flash of lightning, then in the language of algebra,

$$m = \frac{s}{5}.$$

The sentence "$m = \frac{s}{5}$" is read "m is equal to s divided by 5."

Types of Sentences Used in Algebra

A **sentence** in algebra is a grammatically correct set of numbers, variables, or operations that contains a verb.

The sentence $m = \frac{s}{5}$ uses the verb "is equal to," denoted by the symbol =. Any sentence using the verb = is called an **equation.** Symbols for other mathematical verbs are shown on the following page.

$\neq$ is not equal to	$\approx$ is approximately equal to
$<$ is less than	$>$ is greater than
$\leq$ is less than or equal to	$\geq$ is greater than or equal to

A sentence with one of the verbs listed above is called an **inequality.** For instance, the sentence $\frac{1}{2} < \frac{5}{8}$ is an inequality.

In Example 1 and throughout this book we show what you might write using this special writing font.

Example 1

Write an inequality that compares $\frac{1}{3}$ and $\frac{3}{8}$.

Solution 1

Convert $\frac{1}{3}$ and $\frac{3}{8}$ to equivalent fractions with a common denominator. The least common multiple of 3 and 8 is 24, so we use 24 as the denominator.

$$\frac{1}{3} = \frac{8}{24} \text{ and } \frac{3}{8} = \frac{9}{24}.$$

$$\text{Since } \frac{8}{24} < \frac{9}{24},$$

$$\frac{1}{3} < \frac{3}{8}.$$

Solution 2

A fraction indicates division. Perform the division to convert the fraction to a decimal.

$$\frac{1}{3} = 1 \div 3 = 0.\overline{3} \approx 0.333$$

$$\frac{3}{8} = 3 \div 8 = 0.375$$

In decimal form, $0.\overline{3}$ is seen to be smaller than 0.375.

$$\text{Since } 0.\overline{3} < 0.375,$$

$$\frac{1}{3} < \frac{3}{8}.$$

A correct answer using the "is greater than" sign is $\frac{3}{8} > \frac{1}{3}$.

Open Sentences and Solutions

A sentence with a variable is called an **open sentence.** The sentence $m = \frac{s}{5}$ is an open sentence with two variables. It is called "open" because its truth cannot be determined until the variables are replaced by values. The sentence $\frac{1}{3} < \frac{3}{8}$ is not an open sentence. There are no variables. A **solution** to an open sentence is a replacement for the variable that makes the statement true.

In later chapters you will learn methods for finding solutions to equations and inequalities. For now, you may have to solve equations and inequalities by using trial and error or what you know about numbers.

Example 2

Which of the numbers 7, 8, or 9 is a solution to the following open sentence?

$$3 \cdot x + 15 = 4 \cdot x + 6$$

(Recall that the dot means multiplication.)

Solution

Try 7.	Does $3 \cdot 7 + 15 = 4 \cdot 7 + 6$?	No, $36 \neq 34$.
Try 8.	Does $3 \cdot 8 + 15 = 4 \cdot 8 + 6$?	No, $39 \neq 38$.
Try 9.	Does $3 \cdot 9 + 15 = 4 \cdot 9 + 6$?	Yes, $42 = 42$.

So 9 is a solution.

Example 3

Find all solutions to $p^2 = 36$.

Solution

Recall that p^2 means $p \cdot p$. Thus the equation above means $p \cdot p = 36$. To find p, ask yourself, "What number multiplied by itself equals 36?" You know that $6 \cdot 6 = 36$, so one solution is 6. Now, is any other number multiplied by itself equal to 36? Yes: $-6 \cdot -6 = 36$. So there are two solutions: 6 and -6. This may also be written:

$$p = 6 \text{ or } p = -6.$$

Example 4

Water will remain ice for all temperatures less than 0° Celsius.

a. Write an open sentence describing this situation.

b. Give three numbers that make the sentence true.

Solution

a. You may write: Let T be the Celsius temperature of the water. Then water is ice if $T < 0°$. ($0° > T$ would also be correct.)

b. The open sentence is $T < 0°$. Thus any number less than 0 is a solution. For instance,

-3°, -10°, and -20° are three solutions.

If at first you don't succeed. *After three unsuccessful attempts in previous Olympic games, U.S. world champion speedskater Dan Jansen won an Olympic gold medal in 1994. He set a new world's record in the 1,000-meter race with a time of 1:12.43.*

QUESTIONS

Covering the Reading

These questions are designed to check your understanding of the reading. If you cannot answer a question, you should go back to the reading for help in obtaining the answer.

1. You clocked the time between lightning and thunder as 8 seconds. About how far away was the lightning?

2. What is a *variable?*

In 3–6, identify each as a sentence, an open sentence, an equation, an inequality, or none of these. (More than one answer may apply.)

3. $5 \cdot x + 3 < 2$

4. $5 \cdot x + 3$

5. $-5 = r$

6. $-5 \leq 0$

In 7 and 8, write a symbol for each phrase.

7. is approximately equal to

8. is greater than

In 9 and 10, two numbers are given.
a. Write an inequality to compare the two numbers.
b. Explain how you got your answer.

9. $\frac{5}{8}, \frac{4}{7}$

10. $\frac{5}{6}, \frac{17}{20}$

11. *Multiple choice.* $z \geq 100$ means the same as which sentence?
(a) $z \leq 100$ (b) $100 \geq z$ (c) $100 \leq z$

12. Which of the numbers 5, 6, or 7 is a solution of $2 \cdot y + 3 = 4 \cdot y - 9$?

13. **a.** Which of the numbers 12, 72, or 142 is a solution to $x^2 = 144$?
b. Find the solution to $x^2 = 144$ that is not mentioned in part **a.**

14. Find all solutions to $n^2 = 100$.

In 15 and 16, list three solutions to the sentence.

15. $p \geq 2$

16. $y < \frac{2}{3}$

17. The temperature was above 10°F all day.
a. Write an open sentence to describe the situation.
b. List three solutions to the sentence in part **a.**

Applying the Mathematics

These questions extend your understanding of the content of the lesson. Study the examples and explanations if you cannot get an answer. For some questions, you can check your answers with those in the back of the book.

18. List all of the symbols $=, \neq, \approx, >, \geq, <,$ or $\leq$ that make a true statement when written in the blank.

$$8.999 \ \_\_\_ \ 9$$

If I had a hammer. Home Improvement, *a popular television series, features actor Tim Allen as a carpenter who hosts his own television show, "Tool Time." At left is Bob Vila, who hosts his own home improvement show,* Home Again.

19. Order from smallest to largest: $\frac{7}{10}, \frac{2}{3}, \frac{3}{4}$.

20. Carpenters often measure lengths to the nearest fourth, eighth, or sixteenth of an inch.
a. A carpenter cut a piece of wood $15\frac{5}{16}''$ long. Write this measure in decimal form.
b. Will the piece of wood in part **a** fit in a space $15\frac{1}{4}''$ long? Why or why not?

21. Which of the values 2, 4, 8, and 16, are solutions to $5 \cdot x < 40$?

22. *Multiple choice.* Let m = the number of miles per gallon your neighbor's car gets. Which of the following sentences means that your neighbor's car gets over 20 miles per gallon?
(a) $20 > m$ (b) $20 < m$ (c) $m \geq 20$ (d) $20 \leq m$

23. The following sentence contains two inequalities: Before 1990, households in the United States contained an average of more than 2.1 persons.
a. Use the variable y to write an inequality for "before 1990."
b. Use the variable p to write "an average of more than 2.1 persons."

Review

Every lesson contains review questions to give you practice on ideas you have studied earlier.

In 24 and 25, what addition problem is pictured on the number line? *(Previous course)*

24. 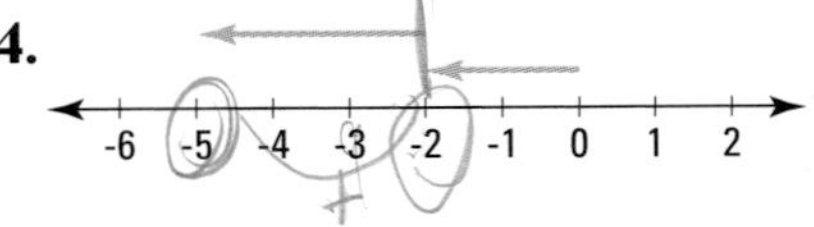

25. 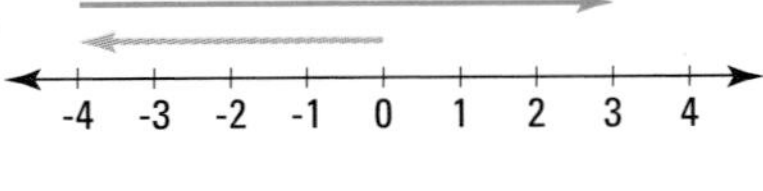

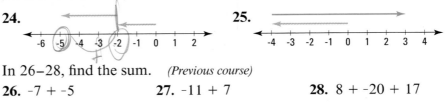

In 26–28, find the sum. *(Previous course)*

26. -7 + -5 **27.** -11 + 7 **28.** 8 + -20 + 17

29. In Example 1, the statement is made that $\frac{1}{3} \approx 0.333$. Give a closer decimal approximation to $\frac{1}{3}$. *(Previous course)*

Often it is quicker and more convenient to do problems in your head. Pressing calculator keys for simple problems is time consuming and may lead to careless mistakes. Do not use your calculator or work with paper and pencil on these problems. Just write an answer.

In 30–32, compute in your head. *(Previous course)*

30. $10 \cdot 3.7$ **31.** $1\frac{1}{2} \cdot 2$ **32.** $4 \cdot \$2.25$

Exploration

These questions ask you to explore mathematics topics related to the chapter. Sometimes these questions require that you use dictionaries, encyclopedias, and other sources of information.

33. **a.** Find a fraction between $\frac{11}{20}$ and $\frac{14}{25}$. Describe the method you used.
b. Find some other fractions between $\frac{11}{20}$ and $\frac{14}{25}$.
c. How many fractions are between $\frac{11}{20}$ and $\frac{14}{25}$? Explain your reasoning.

34. Explain the origin of the word *hieroglyphics*.

LESSON

1-2

Sets and Domains

One for the records. *In 1992, Carol Moseley-Braun of Chicago, Illinois, became the first African-American woman to be elected to the United States Senate. She was sworn in by Vice President Dan Quayle in the presence of Senator George Mitchell.*

What Is a Set?

A **set** is a collection of objects called **elements** or **members.** Usually the elements are grouped for a purpose.

Name of set	Name of member
herd of dairy cattle	cow
team	player
committee	member
the U.S. Senate	senator
class	student
deck (of cards)	card

A set often has properties different from those of its members. For example, a team in baseball can win the World Series, but a player cannot. The Senate can pass legislation, but a senator cannot. A card might be a king, but a deck is not.

The standard symbols used for a set are braces { . . . }, with commas used to separate the elements. Sets are often named with letters. For instance, when Lulu, Mike, Nell, Oscar, Paula, and Quincy are the six members of a committee, you could call the committee C and write

$$C = \{\text{Lulu, Mike, Nell, Oscar, Paula, Quincy}\}.$$

The order of naming elements in a set makes no difference. {Oscar, Mike, Lulu, Nell, Paula, Quincy} is the same committee C. Two sets are **equal** if and only if they have the same elements.

Frequently Used Sets of Numbers

The following sets are frequently used in arithmetic and algebra.

Name of Set	Description	Examples of Elements
whole numbers	{0, 1, 2, 3, . . . }	five, $\frac{16}{2}$, 1995, 7 million
integers	{0, 1, −1, 2, −2, . . . }, the whole numbers and their opposites	$\frac{21}{3}$, −17.00, negative one thousand
real numbers	the set of all numbers that can be represented as terminating or non-terminating decimals	5, 0, π, −0.0042, $-3\frac{1}{3}$, $0.\overline{13}$, $\sqrt{2}$, one hundred thousand

Notice that all whole numbers are also integers, and all integers are also real numbers.

Domains of Variables

All the values that *may* be meaningfully substituted for a variable make up the **domain** of the variable. The three sets mentioned above are often used as domains for variables. The **solution set** of an open sentence is the set of numbers from the domain that actually are solutions.

Solution sets for inequalities are often pictured on a number line. Notice how the domain can affect a solution set.

Example 1

Graph all solutions to $x < 9$ when the domain of x is the indicated set.
a. whole numbers **b.** integers **c.** real numbers

Solution

a. The set of whole-number solutions is {0, 1, 2, 3, 4, 5, 6, 7, 8}. Draw a number line, label it with the variable x, and plot these points.

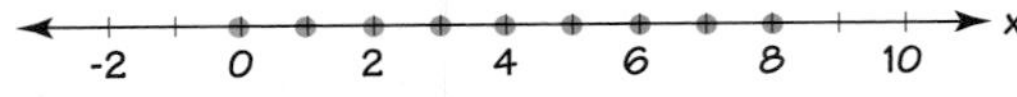

b. The solutions include those in part **a**, and all negative integers as well.

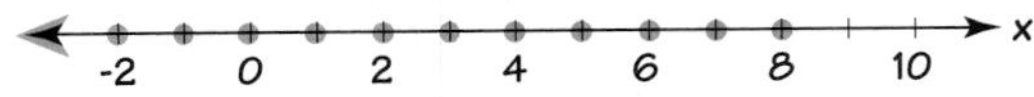

The larger arrow to the left of −2 means "and so on."

c. The solutions include all those in part **b**, and all other decimals less than 9. For instance, 7.3, 8.9, and 8.99 are also solutions. To show all solutions, draw a solid line through all points to the left of 9. Draw an open circle at 9 to indicate that 9 is not a solution.

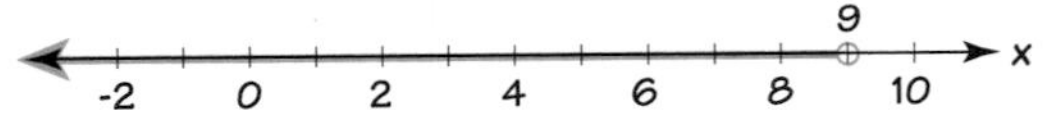

Choosing a Reasonable Domain

When solving real problems, you often must decide what domain makes sense for a situation.

Preparing tomorrow's leaders. *These students are participating in a mock legislative session held in the state senate chamber in Austin, Texas.*

Example 2

Let x = the number of people at a meeting.

a. Name a reasonable domain for x.

b. Assume there are more than 15 people at the meeting, and write an algebraic sentence using x.

c. Graph the solution set to part **b**.

Solution

a. Since x is a count of the number of people at the meeting, it does not make sense to have x be a fraction or a negative number. It *does* make sense for x to be any positive whole number or 0. The domain for x is the set of whole numbers.

b. $x > 15$

c. You must show all the whole numbers greater than 15. Each element in the solution set is marked with a dot on the number line. The dots are not connected because the numbers between these whole numbers are not in the domain of the variable.

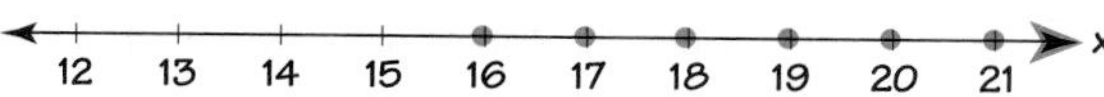

In the next example, the variable can be any real number between two whole numbers. The graph is no longer a set of separate or *discrete* dots; it is connected.

Example 3

Let w = the weight of a meat roast.

a. Give a reasonable domain for w.

b. The roast weighs less than six pounds. Write this as an inequality using w.

c. Graph the solution set to part **b**.

Solution

a. The roast can weigh a fraction of a pound, but not a negative number. So, The domain could be the set of positive real numbers.

b. Here are two conditions: $w > 0$ (the weight must be positive), and $w < 6$. The two inequalities can be written as one: $0 < w < 6$. This is read "w is greater than 0 and less than 6."

c. You need to show all real numbers between 0 and 6, but not including 0 or 6. The graph is shown below.

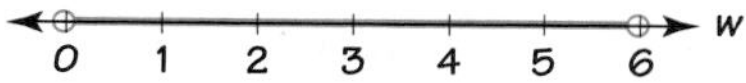

Intervals

The solution set in Example 3 is called an *interval.* An **interval** is a set of numbers between two numbers a and b, possibly including a, or b, or both a and b. The numbers a and b are called the **endpoints** of the interval. The interval $0 < w < 6$ is called an **open interval** because it does not include the endpoints. This is pictured on the graph by the open circles at 0 and 6. The interval $0 \le w \le 6$ is called a **closed interval** because its endpoints are included. An interval that includes one endpoint and not the other is neither open nor closed.

The table below shows four types of intervals. Notice how for each interval the inequality, graph, and verbal description describe the same set.

Interval	Inequality	Graph	Verbal Description
Open	$0 < w < 6$	0 1 2 3 4 5 6 w	all real numbers between 0 and 6
Closed	$0 \le w \le 6$	0 1 2 3 4 5 6 w	all real numbers from 0 to 6
Neither open nor closed	$0 < w \le 6$	0 1 2 3 4 5 6 w	all positive real numbers less than or equal to 6
Neither open nor closed	$0 \le w < 6$	0 1 2 3 4 5 6 w	all non-negative real numbers less than 6

QUESTIONS

Covering the Reading

1. What are the objects in a set called?

In 2–4, a set and an element are given. Name another element in the set.

2. set: family; element: mother

3. set: {2, 11, -6}; element: 11

4. set of real numbers; element: 2

5. Which of the following sets are equal?
$A = \{2, 0, -5\}$ $B = \{-5, 2, 0\}$ $C = \{-5, 2\}$

In 6–11, a number is given.
a. Is the number a whole number?
b. Is the number an integer?
c. Is the number a real number?

6. -10 **7.** $\frac{6}{2}$ **8.** 0.5

9. 0 **10.** 3.6×10^9 **11.** 47.3928

In 12–14, graph the solutions to the sentence for each of the following domains: **a.** set of whole numbers; **b.** set of integers; **c.** set of real numbers.

12. $n < 3$ **13.** $n \geq 3$ **14.** $5 < x$

Multiple choice. In 15 and 16, which is the most reasonable domain for the variable?

(a) set of whole numbers (b) set of integers
(c) set of real numbers (d) set of positive real numbers

15. n = the number of performers at a piano recital

16. t = the time in minutes to prepare a meal

17. Let ℓ = the length of a sauropod.
a. Give a reasonable domain for ℓ.
b. Lengths of fully-grown sauropods (dinosaurs) ranged from about 23 m to 46 m. Write this fact as an algebraic sentence involving ℓ.
c. Graph the solution set to part **b.**

In 18 and 19, a sentence and a domain are given.
a. Graph the solution set on a number line.
b. Tell whether the interval graphed is open, closed, or neither.

18. $2 < x < 9$, where x is a real number.

19. $-7 < x \leq -3$, where x is a real number.

The super sauropods. *Called the "largest dinosaur ever found," sauropods were huge plant eaters during the late Jurrasic period, 152 million years ago. Shown above is archeologist, Jim Jensen, standing next to a reconstruction of a sauropod leg that is 6 meters tall.*

Applying the Mathematics

20. Let S be the solution set for the sentence $-8 < y < 8$. Graph S if y has the indicated domain.
a. the set of whole numbers **b.** the set of integers

In 21–23, an interval is graphed.
a. Tell whether the interval is open, closed, or neither.
b. Describe the set in words.
c. Write an inequality to describe the interval.

21.
22.
23.

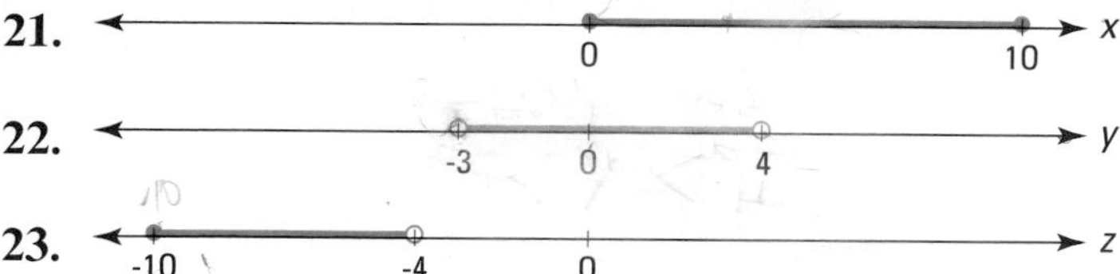

24. Let E = the elevation of a place in the United States.
a. Give a reasonable domain for E.
b. Elevations in the U.S. range from 86 meters below sea level (in Death Valley) to 6194 meters above sea level (at the top of Mt. McKinley). Write a sentence involving E.
c. Graph the solution set to part **b.**

25. Using {3, 4, 6, 9} as the domain for t, find the solution set of $5t + 2 > 23$.

Review

26. At sea level, water is steam at all temperatures greater than 100°C. Write an inequality describing this situation. *(Lesson 1-1)*

27. Write a different inequality with the same meaning as $-15 < y$. *(Lesson 1-1)*

In 28 and 29, find all solutions. *(Lesson 1-1)*

28. $x^2 = 16$

29. $16^2 = x$

30. Find the sum without using a calculator. *(Previous course)*

a. $-5 + -9$ **b.** $-5 + 9$ **c.** $5 + -9$

In 31–34, evaluate each of the following without a calculator. *(Previous course)*

31. $3 \cdot -5$ **32.** $4 \cdot -10$ **33.** $5 \cdot -70$ **34.** $-8 \cdot -25$

35. For its grand opening, Harold's Electronics gave away pens to customers. The cost of the pens was $70.00 per day. What was the cost of pens for five days? *(Previous course)*

Exploration

36. Collections of animals frequently are given special names. Match the group name with the correct animal as in "school of fish."

DENNIS THE MENACE

"THIS WEEK WE'RE STUDYING REAL DEEP STUFF, LIKE: A BUNCH OF SHEEP IS A FLOCK ... AND A FLOCK OF FLOWERS IS A BUNCH."

Group name	Animal
cloud	ants
colony	bees
exaltation	crows
gaggle	fish
hive	foxes
leap	geese
mob	gnats
nest	kangaroos
pride	larks
school	leopards
skulk	lions
watch	nightingales
yoke	oxen

37. Name three sets outside of mathematics that are not mentioned in this lesson. For each, indicate what an element is usually called.

LESSON

1-3

Operations with Sets

Safety first! *Safety Village, in Clearwater, Florida, is a permanent miniature town. Children in primary grades take field trips there to learn safety skills. For example, children are taught how to walk safely as they cross an intersection.*

Intersection of Sets

A police report of an accident stated that "the car was in the intersection of Main and Oak Streets when it was struck by a truck." The intersection, the region where the streets overlap, is shaded in the map at the right. Since the car was in the intersection, it was in both Main Street and Oak Street.

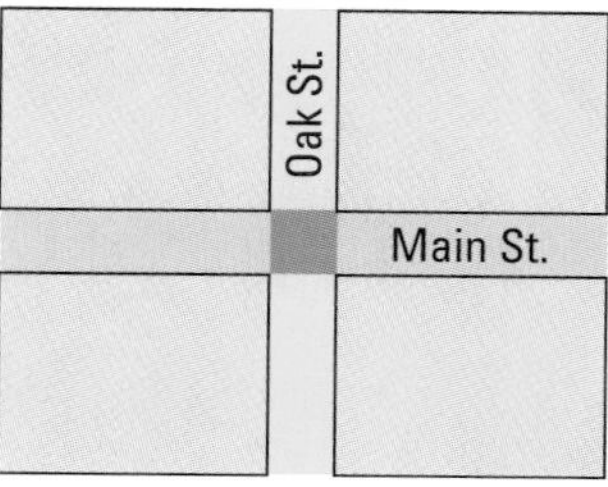

The term *intersection* has a similar meaning when used with sets.

Intersection of Sets
The **intersection** of sets A and B, written $A \cap B$, is the set of elements that are in both A and B.

Example 1

Let $A = \{1, 3, 5, 7, 9, 11\}$ and $B = \{1, 4, 7, 10\}$. Give the intersection of A and B.

Solution

The elements that are in both A and B are 1 and 7.

$A \cap B = \{1, 7\}$.

Relations between sets can be illustrated by *Venn diagrams,* named after John Venn (1834–1923), an English mathematician who used these diagrams in his work. In a Venn diagram, each set is represented by a circle or other closed figure. In the Venn diagram below, the elements of set A from Example 1 are in the circle labeled A. The elements of set B are in the circle labeled B. Overlapping parts of figures represent the intersection of sets. $A \cap B$ is represented by the green overlap.

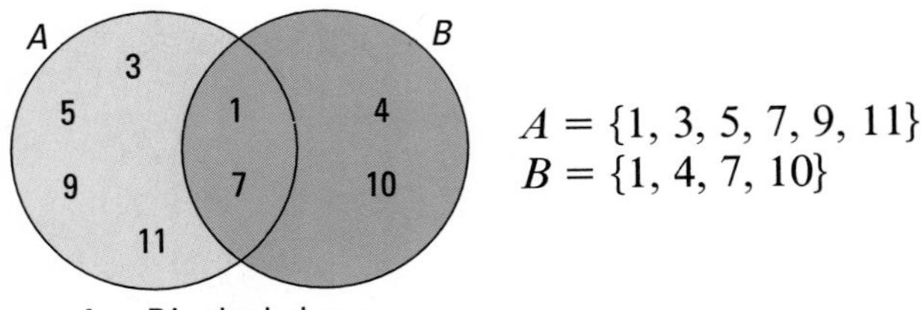

$A \cap B$ is shaded green.

Union of Sets

A second operation used with sets is *union.* The symbol for union looks like the letter U.

> **Union of Sets**
> The **union** of sets A and B, written $A \cup B$, is the set of elements in either A or B (or in both).

Contrast the definition of union with that of intersection. The key word for intersection is "and"; the key word for union is "or." Notice that the union of two sets is not just the result of "putting them together." Elements are not repeated if they are in both sets.

Example 2

Let $A = \{1, 3, 5, 7, 9, 11\}$ and $B = \{1, 4, 7, 10\}$ as in Example 1.

a. Find $A \cup B$.

b. Make a Venn diagram of the two sets and shade the union.

Solution

a. The union is the set of elements in one set or the other or both.
$\{1, 3, 5, 7, 9, 11\} \cup \{1, 4, 7, 10\} = \{1, 3, 5, 7, 9, 11, 4, 10\}$.
Notice that the elements 1 and 7 are in both sets, but they are written only once in the union.

b.

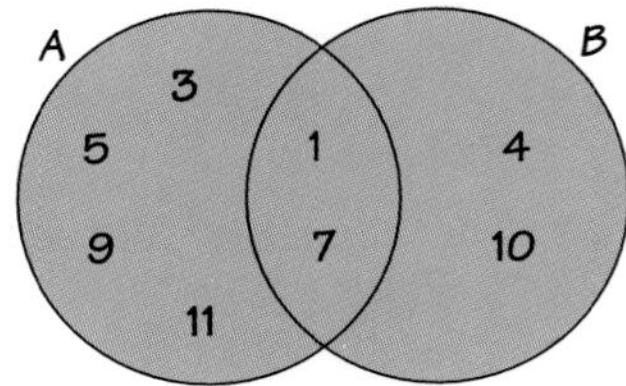

$A \cup B$ is shaded.

There is a set that has no elements in it. It is called the **empty set,** or **null set.** The symbol { } or the Danish letter ø can be used to refer to this set. The empty set might refer to many things, including,

the set of whole numbers between 11 and 12;
the set of U.S. presidents under 35 years of age;
the set of all living dinosaurs.

Example 3

Let E = the set of even integers and
O = the set of odd integers. Find $E \cap O$.

Solution

E = { . . . , -6, -4, -2, 0, 2, 4, 6, . . . } and
O = { . . . , -5, -3, -1, 1, 3, 5, . . . }.
No integer is both odd and even.
$E \cap O = \{\}$.

Graphs of Intersections and Unions

Parts of the number line can be described as the graph of the intersection or union of two sets.

Example 4

Graph the set of all numbers s such that $s > -2$ or $s \leq -10$.

Solution

The word "or" in the above sentence indicates that you must find the union of the solution set to $s > -2$ and the solution set to $s \leq -10$. Draw the graph of $s > -2$.

Draw the graph of $s \leq -10$.

The union includes all points that satisfy either sentence or both sentences. It is:

Example 5

A family wants to bake a casserole and muffins at the same time. The casserole recipe calls for an oven temperature of 325° to 375°. The muffins can bake at any temperature from 350° to 400°.

a. Describe each of the two intervals with an inequality.

b. Graph the intervals in part **a**, and describe their intersection with an inequality.

c. What temperature settings are right for both the casserole and muffins?

Solution

a. Let t be the oven temperature. Then for the casserole, $325 \le t \le 375$, and for the muffins, $350 \le t \le 400$.

b. First graph the solution to each inequality separately. It helps to line up the scales on the two number lines.

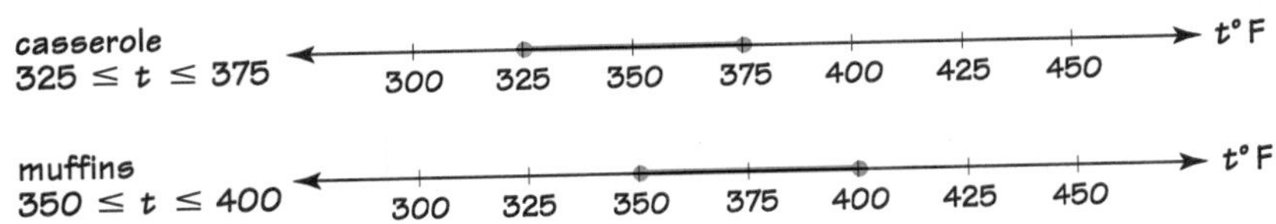

The intersection is that part of the number line where the two graphs overlap.

right for both
$350 \le t \le 375$
t° F
300 325 350 375 400 425 450

c. Temperature settings from 350° to 375° are right for both the casserole and muffins.

QUESTIONS

Covering the Reading

1. Define: *intersection* of two sets.

2. Define: *union* of two sets.

In 3 and 4, let $A = \{2, 4, 6, 8, 10\}$ and $B = \{3, 6, 9, 12, 15\}$.

3. Give the intersection of A and B.

4. Find the union of A and B.

In 5 and 6, suppose $R = \{2, 5, 6, 11, 13\}$ and $S = \{3, 4, 5, 6, 7\}$. Draw a Venn diagram to illustrate the set.

5. $R \cap S$

6. $R \cup S$

7. a. What does ø represent?

b. Give an example of a set equal to ø.

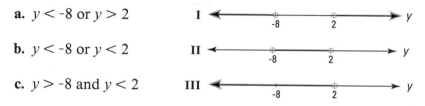

8. Suppose W = the set of whole numbers less than 5, and X = the set of whole numbers greater than 5 and less than 10. List the elements of each set.
   a. W **b.** X
   c. $W \cup X$ **d.** $W \cap X$

9. Graph the solution sets. Use the set of real numbers for the domain.
   a. $x \geq -3$ **b.** $x \leq 7$
   c. $x \geq -3$ and $x \leq 7$ **d.** $x \geq -3$ or $x \leq 7$

10. Match the sentence at the left with its graph at the right.
   a. $y < -8$ or $y > 2$

   I 

   b. $y < -8$ or $y < 2$

   II 

   c. $y > -8$ and $y < 2$

   III 

11. Graph the set of all numbers z such that $z < -2$ or $z \leq 4$.

12. During the winter months, Mrs. King is comfortable with indoor temperatures of 68°F to 75°F. Mr. King is comfortable at temperatures of 65°F to 70°F.
   a. Write each person's comfort zone as an inequality.
   b. Graph the temperatures that are comfortable for both Mr. and Mrs. King.
   c. Write the set in part **b** as an inequality, and tell how it is related to the sets in part **a.**

A glimpse of Africa. *The National Museum of African Art in Washington, D.C., is the first museum in the U.S. devoted to the collection and exhibition of African art. As part of the Smithsonian Institution, admission is always free.*

Applying the Mathematics

13. Let E = the set of odd numbers from 1 to 10.
   Let F = the set of all multiples of 3 between 1 and 10.
   Describe each set.
   a. $E \cap F$ **b.** $E \cup F$

14. Let S = the solution set for $x^2 = 16$,
   and T = the solution set for $x + 2 = 16$.
   List the elements of each set.
   a. $S \cup T$ **b.** $S \cap T$

15. Graph the ages for which admission to a museum is free under the following rule: You will get in free if you are younger than 5 or a senior citizen (62 or older).

16. An outing club is planning a bicycle trip. The graphs below show the number of kilometers three people want to cycle per day.

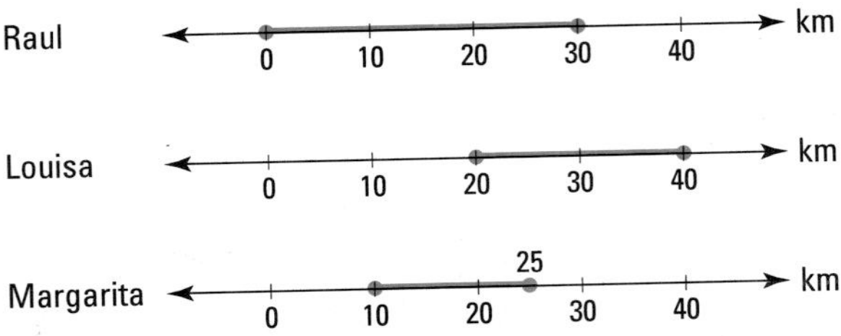

a. Draw a graph showing the distances that would be acceptable to all three.

b. How is the graph in part **a** related to the graphs above?

Review

17. Consider $Z < \frac{8}{3}$. Graph the solution set if Z has the given domain. *(Lesson 1-2)*

a. set of whole numbers

b. set of real numbers

18. Name an integer that

a. is a whole number;

b. is not a whole number. *(Lesson 1-2)*

In 19–21, *true or false.* *(Lessons 1-1, 1-2)*

19. The number 13.23×10^5 is a whole number.

20. $-0.125 < -0.1$

21. The number 7 is a solution to $7 \geq n$.

In 22 and 23, fill in the blanks. Describe the patterns you see. *(Previous course)*

22.
$5 \cdot 3 = 15$
$5 \cdot 2 = 10$
$5 \cdot 1 = 5$
$5 \cdot 0 = 0$
$5 \cdot -1 = \underline{\ ?\ }$
$5 \cdot -2 = \underline{\ ?\ }$
$5 \cdot -3 = \underline{\ ?\ }$

23.
$-4 \cdot 3 = -12$
$-4 \cdot 2 = -8$
$-4 \cdot 1 = -4$
$-4 \cdot 0 = 0$
$-4 \cdot -1 = \underline{\ ?\ }$
$-4 \cdot -2 = \underline{\ ?\ }$
$-4 \cdot -3 = \underline{\ ?\ }$

Exploration

24. Is it possible to have sets A and B with $A \cap B$ having more elements than $A \cup B$? Explain why or why not.

LESSON

1-4

Variables in Expressions

Algebra uses the same symbols for numerical operations as arithmetic does, but with one exception. In algebra, multiplication is signified by either a raised dot or by putting two expressions next to each other with no symbol between. So "3 times x" is written either as $3 \cdot x$ or $3x$, and "5 times $(A + B)$" is written $5 \cdot (A + B)$ or $5(A + B)$.

Evaluating Expressions

Numerical expressions, like $6 + 3^2 - 1$, combine numbers only. An expression, such as $4 + 3x$, that includes one or more variables is called an **algebraic expression.** Expressions are not sentences because they do not contain verbs, such as equal or inequality signs. Finding the numerical value of an expression is called **evaluating** the expression. To evaluate an algebraic expression, you must have values to substitute for its variables. When evaluating any expression, you must do operations in the following order.

Order of Operations in Evaluating Expressions

1. Do operations within parentheses or other grouping symbols.
2. Within grouping symbols, or if there are no grouping symbols:
   a. Do all powers from left to right.
   b. Do all multiplications and divisions from left to right.
   c. Do all additions and subtractions from left to right.

Example 1

a. Evaluate $4 + 3x$ when $x = 9$. **b.** Evaluate $4 + 3x$ when $x = -1$.

Solution

a. Let $x = 9$. Then $4 + 3x = 4 + 3 \cdot 9$ — Substitute 9 for x.
$= 4 + 27$ — Multiply first.
$= 31$ — Add.

b. Let $x = -1$. Then $4 + 3x = 4 + 3 \cdot -1$ — Substitute -1 for x.
$= 4 + -3$
$= 1$

Example 2

a. If $n = 2$, find $7n^3$.

b. If $n = 2$, calculate $(7n)^3$.

Solution

a. Substitute 2 for n. There are no grouping symbols, so do the power before the multiplication.

$$7 \cdot 2^3 = 7 \cdot 8 = 56$$

b. Substitute 2 for n. Do the operation within the parentheses first.

$$(7 \cdot 2)^3 = 14^3 = 2744$$

Most scientific calculators use the same order of operations as algebra. Examples 1 and 2 could have been done with a calculator. Check Examples 1 and 2 using a calculator. (If you have never used a scientific calculator, see Appendix A.)

Example 1a: 4 [+] 3 [×] 9 [=] [31]

Example 2a: 7 [×] 2 [y^x] 3 [=] [56]

You can count on them. *The first mechanical calculating device was the abacus. The first true mechanical calculator was an adding machine invented by Blaise Pascal in 1642. Today's advanced, scientific calculators perform many functions that are useful to people in all walks of life.*

Example 3

Evaluate $\frac{5(A + B)}{2}$, when $A = 3.4$ and $B = 7.2$.

Solution

Substitute 3.4 for A and 7.2 for B.

Then $\frac{5(A + B)}{2} = \frac{5(3.4 + 7.2)}{2}$ Work inside the parentheses.

$= \frac{5(10.6)}{2}$ The fraction bar is a grouping symbol, so work with the numerator and denominator separately. 5(10.6) means $5 \cdot 10.6$. Multiply.

$= \frac{53}{2}$

$= 26.5$

Computer programs and graphing calculators also use algebraic expressions, but the expressions must be in a language the computer understands. Most computer languages use the order of operations listed on page 23. In **BASIC** (Beginner's All-purpose Symbolic Instruction Code), + is used for addition, − for subtraction, * for multiplication, / for division, ^ for powering, and ( ) for grouping. Usually the symbol for *every* operation must be shown. So, in BASIC, 5(A + B) is written 5 * (A + B). The power 2^3 is written 2 ^ 3.

Example 4

Write $3(4x - 5) + y$ in BASIC.

Solution

Use * for the multiplications and capital letters for the variables.

```
3 * (4 * X - 5) + Y
```

With most calculators and computer software, you must input fractions on one line. Then you must include grouping symbols. For instance, to evaluate $\frac{10 + 4}{2 + 5}$, you must enter (10 + 4) / (2 + 5) to get the correct answer, 2. If you enter 10 + 4 / 2 + 5, the machine will follow order of operations, do the division first, and get 17.

Example 5

What would you input in order to use a calculator or computer to evaluate $\left(\frac{x + 10.7}{y + 4}\right)^7$ when $x = 3.1$ and $y = 0.6$?

Solution

The computer expression is written ((X + 10.7) / (Y + 4)) ^ 7. Substitute for each variable and input

((3.1 + 10.7) / (0.6 + 4)) ^ 7.

PCs. *In 1975, the world's first personal computer was introduced. PCs became popular around 1977 when smaller size and lower-cost models became available. Tomorrow's computers will undoubtedly be smaller, faster, and more powerful than today's models.*

QUESTIONS

Covering the Reading

In 1–3, identify each expression as numerical or algebraic.

1. $\frac{8(7) + 2}{12}$ **2.** $\frac{2x}{4 - 8}$ **3.** $a^2 + b^2$

In 4–9, evaluate.

4. $12 - 2 \cdot 4$ **5.** $5^2 + 2^2$ **6.** $(5 + 2)^2$

7. $3(10 - 6)^3 + 15$ **8.** $5 - \frac{4}{8}$ **9.** $15 + \frac{9}{3} - 6$

In 10–13, evaluate the expression for the given values of the variables.

10. $5m^2$ when $m = 3$ **11.** $(5m)^2$ when $m = 3$

12. $(1 + r)^3$ when $r = 3$ **13.** $\frac{a + 2b}{5}$ when $a = 11.6$ and $b = 9.2$

14. **a.** Write a BASIC expression to evaluate $0.5H(A + H)$ when $H = 32$ and $A = 0.7$.
b. Evaluate the expression.

15. What would you input in order to use a computer to evaluate $\left(\frac{7x + y}{x - 6y}\right)^4$?

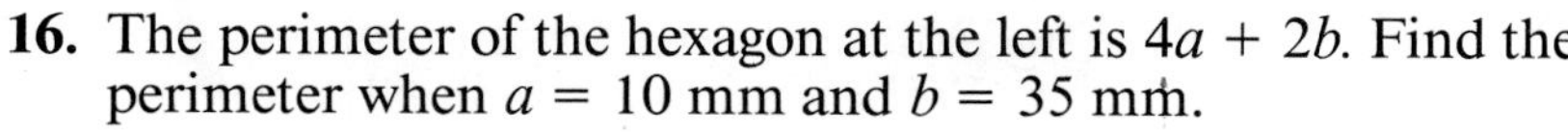

Applying the Mathematics

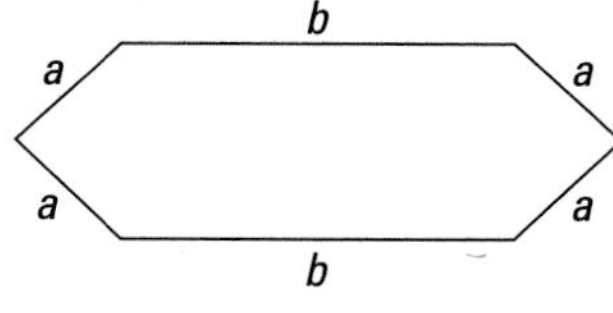

16. The perimeter of the hexagon at the left is $4a + 2b$. Find the perimeter when $a = 10$ mm and $b = 35$ mm.

17. Evaluate $3x^2 - 9x + 6 - 3(x - 1)(x - 2)$ when $x = 10$.

18. Evaluate $6 + (5 - (4 + (3 - 2)))$.

Three generations of flight. *The Wright brothers' first airplane reached a maximum speed of 30 mph during its first flight in 1903. In 1928, a Boeing 80-A airplane began passenger service between Chicago and San Francisco. It cruised at 125 mph with a full load of 14 passengers. In 1970, the first jumbo jet, the Boeing 747 was introduced. With a full capacity of 490 passengers, it cruises at speeds over 625 mph.*

In 19 and 20, use this information. If you go somewhere at an average speed g, and come back at an average speed c, then your average speed for the trip is $\frac{2gc}{g + c}$.

19. Find the average speed for the trip if you bike somewhere averaging 12 mph but then bike back averaging 10 mph.

20. Find the average speed for the trip if a plane averages 800 kilometers per hour going from A to B with a tailwind, but only 700 kilometers per hour going from B to A due to a headwind.

Review

21. Let P = the set of prime numbers and E = the set of even numbers. List all the elements of $P \cap E$. *(Lesson 1-3, Previous course)*

22. Consider the interval $-4 \leq y \leq 6$. *(Lessons 1-2, 1-3)*
a. Tell whether it is open, closed, or neither open nor closed.
b. Describe the interval as the intersection of two other sets.

23. A person makes more than \$1800 a month. Let d be the amount made.
a. What inequality describes this situation? *(Lesson 1-1)*
b. Graph the inequality. *(Lesson 1-2)*

24. Compute in your head. *(Previous course)*
a. $7 \cdot -8$ **b.** $-7 \cdot 8$ **c.** $-7 \cdot -8$

25. Molly has three and three-eighths yards of ribbon. If she uses half of it to make a bow, how much ribbon does she use? *(Previous course)*

In 26 and 27, a square is shaded. **a.** Find the area of the square, using □ as one square unit. **b.** Explain how you got your answer. *(Previous course)*

26.

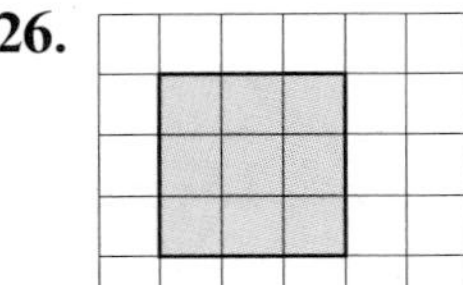

27.

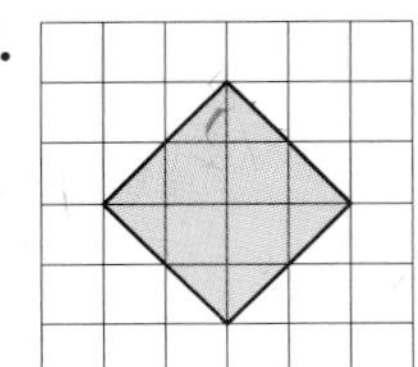

Exploration

28. **a.** Evaluate these expressions:

1	$4 - (3 - (2 - 1))$
$2 - 1$	$5 - (4 - (3 - (2 - 1)))$
$3 - (2 - 1)$	$6 - (5 - (4 - (3 - (2 - 1))))$

b. What patterns do you notice?
c. What would you predict the value of
$10 - (9 - (8 - (7 - (6 - (5 - (4 - (3 - (2 - 1))))))))$ to be? Why? Find its value. Was your prediction correct?

LESSON

1-5

Variables in Formulas

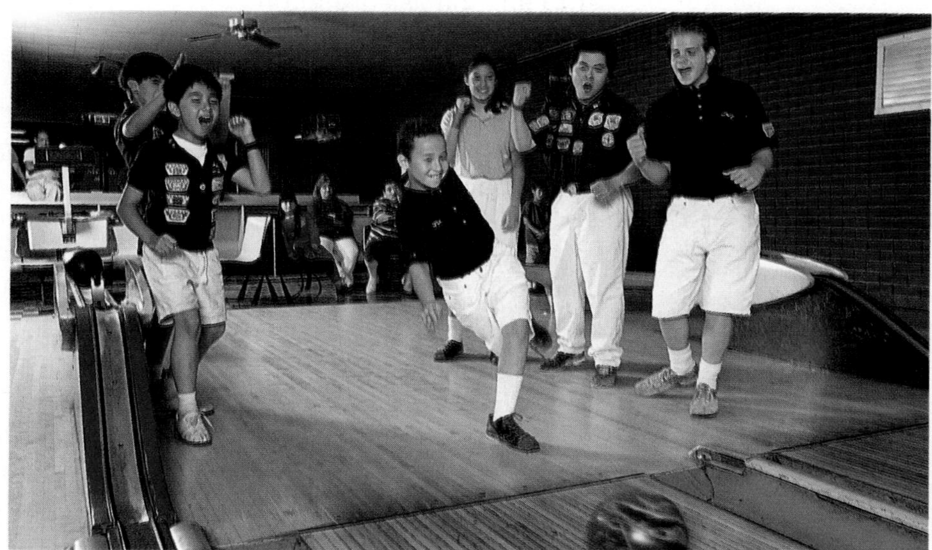

On a roll. *Bowling is one of the most popular indoor sports. About 64 million Americans enjoy this sport each year. When bowlers bowl in a league, they are often given a handicap to help balance the differences in past performances.*

A **formula** is an equation in which one variable is given in terms of other variables or numbers. Formulas are used in many real-life situations. You can evaluate any formula by substituting numbers for the variables.

A Formula Involving Two Variables

In bowling, 300 is a perfect score, and 200 is considered very good. In some leagues, bowlers whose average score is under 200 have *handicaps* added to their score. Example 1 shows a formula for finding a handicap.

Example 1

The handicap H of a bowler whose average is A is often found by using the formula $H = .8(200 - A)$. A bowler's final score for a game is the actual score plus the bowler's handicap.

a. Tony's average score is 145. What is his handicap?

b. What would Tony's final score be if he actually bowled 120?

c. What happens when an average greater than 200 is substituted for A?

Solution

a. Substitute 145 for A in the formula and follow the order of operations.

$$\begin{aligned} H &= .8(200 - 145) \\ &= .8(55) \\ &= 44 \end{aligned}$$

Tony's handicap is 44.

b. Add the handicap to the actual score. Tony's final score is

$$120 + 44 = 164.$$

c. When the average is greater than 200, then $200 - A$ is negative. Thus averages over 200 are not used in the formula. The domain of A is the set of whole numbers less than or equal to 200.

The equation $H = .8(200 - A)$ is called a *formula for H in terms of A.* We also say that *H depends on A.*

A Formula Involving Four Variables

Almost everyone has been in heavy traffic. Perhaps you have been in a traffic jam on your way to an amusement park, a sports event, or a concert. Have you ever wondered how many cars are on a given road? City planners and traffic safety engineers must be able to estimate the number of cars on roads to make decisions about road construction, speed limits, and safety improvements.

An estimate of the number of cars can be calculated using the formula below. Here N is the estimated maximum number of cars allowable on a road if the cars are traveling at safe distances from one another.

$$N = \frac{20Ld}{600 + s^2}$$

L = number of lanes of road
d = length of road (in feet)
s = average speed of the cars (in miles per hour)

This formula was derived by using principles from geometry and physics. In the formula, we say that N is given *in terms of L, d,* and *s.* We also say that *N depends on L, d,* and *s.* You can find N by substituting numbers for each of the other variables.

Example 2

Use the above formula to estimate the maximum number of cars there can safely be on a 1-mile stretch of a 2-lane highway, if the cars are traveling an average speed of 30 miles per hour.

Solution

Determine the value of each variable. Note that the length of the road is given in miles but the formula requires feet. Recall 5280 feet = 1 mile.

L = 2 lanes
d = 1 mile = 5280 feet
s = 30 miles per hour

Write the formula.

$$N = \frac{20Ld}{600 + s^2}$$

Replace the variables in the formula with their values.

$$N = \frac{20 \cdot 2 \cdot 5280}{600 + 30^2}$$

Evaluate this expression using the order of operations. A calculator can help. Here is a typical key sequence. Notice that you need parentheses to group the numbers in the denominator.

20 [×] 2 [×] 5280 [÷] [(] 600 [+] 30 [x^2] [)] [=]

N = 140.8

There can be about 140 cars on the road.

Often formulas give values so large that a computer or calculator must display the result in scientific notation. For instance, the volume of Earth can be approximated using the formula $V = \frac{4}{3}\pi r^3$, with the radius $r \approx 4000$ miles. One calculator displays the result as 2.67946 11. Another displays 2.67946 E11. This stands for $2.67946 \cdot 10^{11}$, or 267,946,000,000. (To review scientific notation, see Appendix B.)

QUESTIONS

Covering the Reading

1. The opener for Chapter 1 on page 5 shows a formula known to Egyptians over 3500 years ago. What is this formula?

In 2–4, refer to Example 1.

2. Debra's bowling average is 120. What is her handicap?
3. *Multiple choice.* Which average does not entitle a bowler to a handicap?
(a) 95 (b) 145 (c) 195 (d) 205
4. Substitute 200 for A in the handicap formula. Explain what you get.

Something worth chirping about. *In many countries, the presence of a cricket is a sign of good luck. In the Orient, crickets are caged for their songs and are kept as pets.*

In 5–7, use the formula preceding Example 2, page 28.

5. In this formula, N is given in terms of which variables?
6. About how many cars can safely be on a 2-mile stretch of a 3-lane highway if the average speed of the cars is 50 mph?
7. About how many cars can safely be on a 1.5-mile part of a 4-lane highway if the average speed of the cars is 20 mph?
8. Use $V = \frac{4}{3}\pi r^3$ to estimate the volume of the planet Jupiter, whose radius is about 142,000 km. Write your answer in scientific notation.

Applying the Mathematics

9. The circumference C of a circle with diameter d is given by the formula $C = \pi d$. To the nearest centimeter, what is the circumference of a circle with diameter 7.3 cm?

In 10–12, note that some crickets, such as the Snowy Tree Cricket, chirp at a regular rate. You can estimate the Fahrenheit temperature T by counting the number of chirps C a cricket makes in one minute, and applying the formula

$$T = \frac{1}{4}C + 37.$$

10. What is a reasonable domain for C?
11. A cricket chirps 200 times per minute. Estimate the temperature.
12. Estimate the temperature when a cricket chirps 150 times per minute.

13. a. Fill in the blanks to complete this BASIC program which computes area and perimeter of rectangles.

```
10 PRINT "AREA AND PERIMETER OF A RECTANGLE"
20 INPUT "LENGTH"; L
30 INPUT "WIDTH"; W
40 LET A = _?_
50 LET P = _?_
60 PRINT "AREA", "PERIMETER"
70 PRINT A, P
80 END
```

b. What will the computer find for the area and perimeter when L = 52.5 and W = 38?

Review

14. Evaluate $77x - (29x + 2)$ when $x = 8$. *(Lesson 1-4)*

15. Give the BASIC symbol for each operation. *(Lesson 1-4)*
a. addition **b.** division **c.** multiplication **d.** powering

Taking a break from the books. *Albert Einstein, world-renowned physicist, was fond of classical music, played the violin, and enjoyed taking his bicycle out for a spin.*

16. Write $\left(\frac{3}{y+2}\right)^{10}$ in the BASIC language. *(Lesson 1-4)*

17. The graph below is the union of the solution sets to what two inequalities? *(Lesson 1-3)*

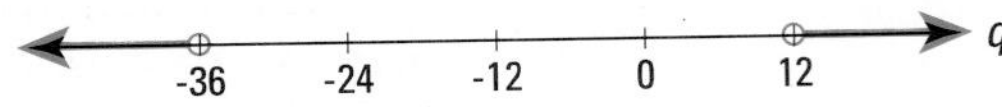

18. Graph all solutions to $4 < x \leq 10$ when
a. x is the length of a table (in feet);
b. x is the number of sticks in a package of chewing gum. *(Lesson 1-2)*

19. Give an example of each. *(Lesson 1-2)*
a. a whole number that is not positive
b. an integer that is not a whole number
c. a real number that is not an integer

20. Round π to the nearest hundred-thousandth. *(Previous course)*

21. A recipe for a cake that serves 12 people calls for $3\frac{1}{2}$ cups of flour. If you wish to halve the recipe to make the cake for 6 people, how much flour will you need? *(Previous course)*

Exploration

22. One of the world's most famous formulas was discovered by Albert Einstein in 1905. It is $E = mc^2$.
a. What do E, m, and c stand for?
b. What physical phenomenon does the formula describe?

LESSON

1-6

Square Roots and Variables

"A wonderful square root. Let us hope it can be used for the good of mankind."

What Are Square Roots?

The term *square root* comes from the geometry of squares and their sides. Pictured below are squares with sides of lengths 4 and 4.5 units.

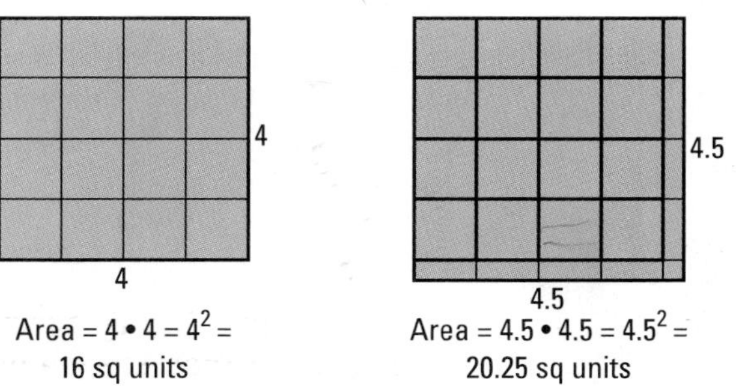

Because $16 = 4^2$, we say that 4 is a square root of 16. Similarly, 20.25 is the square of 4.5, so 4.5 is the square root of 20.25. In general, if $A = s^2$, then s is called a **square root** of A.

The symbol for square root, $\sqrt{}$, is called a radical sign. From the above, $\sqrt{16} = 4$ and $\sqrt{20.25} = 4.5$. Thus, if the area of a square is A, the length of a side is $\sqrt{A}$.

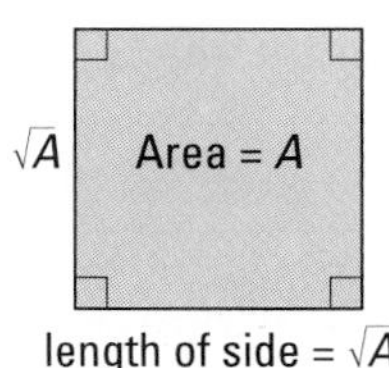

Example 1

The area of a square is 196 in^2. Give the length of a side:
a. using a radical symbol;
b. as a whole number or a decimal approximated to two decimal places.

Solution

a. Since the area is 196 in^2, *a side is* $\sqrt{196}$ *in.*
b. Use a calculator to evaluate $\sqrt{196}$.
On some calculators a key sequence is 196 [√x].
On others you may use [√] 196 [EXE].
On still others, you may need to use the 2nd function key:
[2nd] [√] 196 [ENTER].
You should see 14 on the display. *The length of a side is 14 in.*

Check

Does $14 \cdot 14 = 196$? $(14)^2 = 196$, so it checks.

Check how your calculator evaluates square roots by finding $\sqrt{196}$.

Like the fraction bar, the radical sign $\sqrt{}$ is a grouping symbol for any expression contained within it.

Example 2

Evaluate $\sqrt{12^2 + 5^2}$.

Solution

Because the radical sign is a grouping symbol, begin by doing operations under the radical sign. First square 12 and 5; then add the result.

$$\begin{aligned}\sqrt{12^2 + 5^2} &= \sqrt{144 + 25}\\ &= \sqrt{169}\\ &= 13\end{aligned}$$

Square Roots That Are Not Whole Numbers

You are familiar with squares of whole numbers.
$0^2 = 0$ $1^2 = 1$ $2^2 = 4$ $3^2 = 9$ $4^2 = 16$ and so on.

Squares of whole numbers are called **perfect squares.** Their square roots are whole numbers.
$0 = \sqrt{0}$ $1 = \sqrt{1}$ $2 = \sqrt{4}$ $3 = \sqrt{9}$ $4 = \sqrt{16}$ and so on.

You can get a rough estimate of a square root that is not a whole number by locating it between whole numbers. A good approximation can be found using a calculator.

Example 3

Estimate $\sqrt{2}$.

Solution 1

A rough estimate: *Since* $\sqrt{1} = 1$ *and* $\sqrt{4} = 2$, $\sqrt{2}$ *is between 1 and 2.*

Solution 2

A good approximation: Use a calculator.

An 8-digit display will show 1.4142136.

The actual decimal for $\sqrt{2}$ is infinite and does not repeat. An estimate of $\sqrt{2}$ is the number 1.4142136.

Check

Multiply 1.4142136 by itself. One calculator shows that

$$1.4142136 \cdot 1.4142136 \approx 2.0000001.$$

When the value of a square root is not an integer, your teacher may expect two versions: (1) the exact value and (2) a decimal approximation rounded to the nearest hundredth or thousandth. For instance, the square root of 15 is exactly $\sqrt{15}$. Rounded to the nearest hundredth, $\sqrt{15}$ is 3.87.

The check to Example 3 is approximate because 1.4142136 is used to approximate $\sqrt{2}$. But $\sqrt{2} \cdot \sqrt{2} = 2$ exactly. In general, the following property is true.

Square of the Square Root Property
For any nonnegative number n, $\sqrt{n} \cdot \sqrt{n} = n$.

You can use this property to simplify or to evaluate expressions without using a calculator.

Example 4

Without using a calculator, multiply $4\sqrt{10} \cdot \sqrt{10}$.

Solution

Think of $4\sqrt{10}$ as being $4 \cdot \sqrt{10}$.

$$4\sqrt{10} \cdot \sqrt{10} = 4 \cdot \sqrt{10} \cdot \sqrt{10}$$

By the Square of the Square Root Property, $\sqrt{10} \cdot \sqrt{10} = 10$.

So $$4\sqrt{10} \cdot \sqrt{10} = 4 \cdot 10$$
$$= 40$$

Check 1

Use your calculator. If you use the proper key sequence for $4\sqrt{10} \cdot \sqrt{10}$, you should see 40. (You should check this.)

Check 2

Estimate $4\sqrt{10} \cdot \sqrt{10}$. $\sqrt{10}$ is a little more than 3, so $4\sqrt{10}$ is little more than 12. Multiply this by $\sqrt{10}$, which is a little more than 3, and the product of 40 seems reasonable.

Squares and Square Roots in Equations

The equation $W^2 = 49$ has an obvious solution 7, because $7 \cdot 7 = 49$. However, $-7 \cdot -7 = 49$, so -7 is also a solution. Every positive number has *two* square roots, one positive and one negative. The radical sign symbolizes only the *positive* one. $\sqrt{49}$ means "the positive square root of 49," so $\sqrt{49} = 7$. The symbol for the *negative* square root of 49 is $-\sqrt{49}$. So, $-\sqrt{49} = -7$. The two square roots are opposites of each other.

When you solve equations such as $W^2 = 49$ or $97 = n^2$, you should assume that the domain of the variable contains positive and negative numbers and give all solutions, unless something from the situation limits the domain.

Example 5

Solve $97 = n^2$.

Solution

The exact solutions are $n = \sqrt{97}$ or $n = -\sqrt{97}$.
Rounded to the nearest hundredth, $n \approx 9.85$ or $n \approx -9.85$.

QUESTIONS

Covering the Reading

In 1 and 2, fill in the blanks.

1. Because $169 = 13 \cdot 13$, 13 is called a __?__ of 169.

2. $72.25 = 8.5 \cdot 8.5$; so __?__ is a square root of __?__.

3. The area of a square is 289 m^2. What is the length of a side?

4. If the area of a square is A, then the length of a side is __?__.

In 5–7, compute in your head.

5. $\sqrt{100}$ **6.** $-\sqrt{81}$ **7.** $10\sqrt{81}$

In 8–10, evaluate.

8. $\sqrt{3^2 + 4^2}$ **9.** $\sqrt{3^2 \cdot 4^2}$ **10.** $\sqrt{17^2 - 15^2}$

11. a. At the right is a table of squares and square roots for the integers from 1 to 10. The approximations are rounded to the nearest thousandth. Copy the table and extend it to include all whole-number values of n through 20.

b. Describe some patterns you observe in the table.

$\sqrt{n}$	n	n^2
1	1	1
1.414	2	4
1.732	3	9
2	4	16
2.236	5	25
2.449	6	36
2.646	7	49
2.828	8	64
3	9	81
3.162	10	100

12. a. Approximate $\sqrt{407}$ to the nearest hundredth using your calculator.
b. Check your answer.

13. Evaluate without using a calculator.
a. $(\sqrt{8})^2$ b. $\sqrt{8^2}$ c. $3\sqrt{8} \cdot \sqrt{8}$

14. Every positive number has __?__ square root(s).

15. Name the two square roots of 16.

In 16 and 17, an equation is given.
a. Describe each solution using the radical symbol.
b. Write each solution as a decimal rounded to the nearest hundredth.

16. $x^2 = 121$

17. $301 = b^2$

Can pendulums swing by themselves? *Yes. The pendulum shown here is in the Museum of Science and Industry in Chicago, Illinois. Each evening, the pendulum's swing is stopped by securing it with a string. Each morning, a museum employee burns the string to allow the pendulum to swing freely.*

Applying the Mathematics

18. The area of the square at the left is 18 square units. (Count to check this.)
a. Give the exact length of a side.
b. Estimate the length of a side to the nearest hundredth.

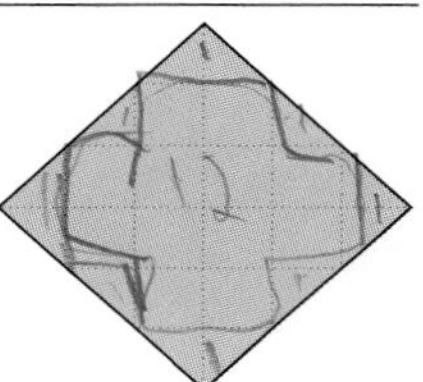

19. Tell how you can determine, without using a calculator, what two consecutive integers $\sqrt{32}$ is between.

20. Find the two solutions to $m^2 + 64 = 100$.

21. For clarity, the product of a number y and the square root of x is usually written as $y\sqrt{x}$ instead of as $\sqrt{x} \cdot y$ or $\sqrt{x}y$. ($\sqrt{x} \cdot y$ can be mistaken for $\sqrt{xy}$.) When $x = 4$ and $y = 9$, evaluate.
a. $y\sqrt{x}$ b. $\sqrt{xy}$ c. $x\sqrt{y}$

22. a. Use a calculator to evaluate $\sqrt{12.25}$.
b. Draw a square that has area 12.25 cm^2.
c. Write a sentence or two to tell how parts **a** and **b** are related.

23. If an object is dropped, it takes about $\sqrt{\frac{d}{16}}$ seconds to fall d feet. If an object is dropped from a 100-foot high building, about how long does it take the object to hit the ground?

24. A *pendulum* is an object that is suspended from a fixed point and swings freely under the force of gravity. Scientists have established this formula for the time T it takes for one complete swing in terms of the length L of the pendulum and the acceleration due to gravity g:

$$T \approx 2\pi\sqrt{\frac{L}{g}}$$

A grandfather's clock has a pendulum of length 0.85 m. Use the formula to find T to the nearest hundredth, where $g \approx 9.8$ m/sec^2.

Review

25. You invest $1000 in an account where interest is compounded yearly. The formula $A = 1000 + 1000r$ gives you the amount in the account after one year. Evaluate A when:

a. r is 3%. **b.** r is 4.5%. *(Lesson 1-5, Previous course)*

In 26–28, evaluate. *(Lesson 1-4)*

26. $5^2 \cdot 5^3$ **27.** $57 - 3 \cdot 18$ **28.** $(11 - 7)^3$

29. Copy the Venn diagram below and shade in $(A \cap B) \cup C$. *(Lesson 1-3)*

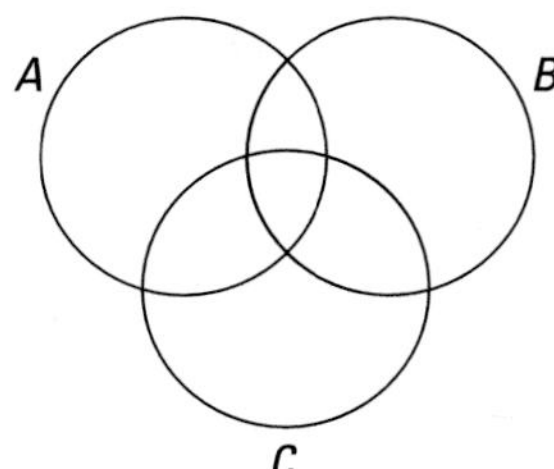

In 30 and 31, a situation and a variable are given. *(Lesson 1-2)*

a. State a reasonable domain for the variable.
b. Rewrite the information as a sentence using the given variable.
c. Graph the solution set for the sentence.

30. A house cat usually has no more than 6 kittens in a litter. Let n = the number of kittens in a typical litter.

31. A typical adult female cat weighs at least 2 kg and no more than 6 kg. Let w = the weight in kg.

32. Which of the numbers -4, 2, and 6 make the sentence $15 + n + {}^-11 > 0$ true? *(Lesson 1-1)*

33. *Skill sequence.* Compute. *(Previous course)*

a. $\frac{2}{3} + \frac{4}{3}$ **b.** $\frac{2}{3} + \frac{4}{9}$ **c.** $\frac{2}{3} + \frac{4}{13}$

Exploration

34. Try to evaluate $\sqrt{-4}$ using your calculator.

a. Write the key sequence you are using.
b. What does your calculator display?
c. Why does the calculator display what it does?

35. Augustus De Morgan, an English mathematician who lived in the 19th century, once said, "I was x years old in the year x^2." In what year was he born? (Remember that you were born in the 20th century.)

LESSON

1-7

Variables and Patterns

Celebrating *Kwanza*. *This African-American family celebrates* Kwanza, *a holiday centered around ideals such as unity and creativity. The bright colors and patterns in the family's clothes are based on traditional African dress.*

The formulas in Lesson 1-4 are sentences that describe numerical *patterns.* A **pattern** is a general idea for which there are many examples. An example of a pattern is called an **instance.** When values are substituted for the variables in a formula, an instance of the formula results. The formula $A = s^2$ tells how to calculate the area of a square with side of length s. If a square has sides 3 cm long, then $A = 3^2 = 9 \text{ cm}^2$. This is an instance of the area formula.

Patterns Involving One Variable

A pattern may be described in several ways. Here is a pattern written in words.

Example 1

Consider the following pattern: If you double a number and then triple the answer, the result is the same as six times the original number.

a. Give three instances of this pattern. Tell if the pattern is true for each instance.

b. Use a variable to describe the pattern algebraically.

Solution

a. Choose three numbers, say 20, 1.5, and -4. For each number, write the sentence in symbols. Recall that "to double" means to multiply by 2, and "to triple" means to multiply by 3.

$(2 \cdot 20) \cdot 3 = 6 \cdot 20$ True, since $40 \cdot 3 = 120$

$(2 \cdot 1.5) \cdot 3 = 6 \cdot 1.5$ True, since $3 \cdot 3 = 9$

$(2 \cdot -4) \cdot 3 = 6 \cdot -4$ True, since $-8 \cdot 3 = -24$

▶ **b.** Notice what remains the same in the left column from instance to instance and what changes. Each instance has an equal sign; multiplication by 2, followed by multiplication by 3 on the left; and multiplication by 6 on the right. The things that stay the same in the instances are also the same in the pattern. Put blanks for the numbers that change.

$$(2 \cdot \_\_) \cdot 3 = 6 \cdot \_\_$$

What goes in the blanks? Each instance has a different number, but the numbers are the same for both blanks. Use a variable for this number that varies. We use x. The pattern is

$$(2 \cdot x) \cdot 3 = 6 \cdot x.$$

When you see a pattern in a number of instances, you may be able to describe that pattern with a formula, a sentence, or an algebraic expression.

Example 2

Throughout the 1980s, the African-American population of the United States increased by about 350,000 people per year. For example,
it increased by about 350,000 people in 1 year;
it increased by about 350,000 · 2 people in 2 years; and
it increased by about 350,000 · 3 people in 3 years.

Describe this pattern, letting the variable y = the number of years.

Solution

Think of the first instance this way:

The population increased by 350,000 · 1 people in 1 year.

Now all the instances look alike. The only things that change are the number of years and the number multiplied by 350,000.
Write:

The population increased by about 350,000 · __ people in __ years.

In each instance, the numbers in the blanks are the same, so y can represent both of them. Here is the pattern.

The population increased by 350,000 · y people in y years.

The mosaic art of St. Mark's. *The Basilica of St. Mark, in Venice, Italy is richly decorated with this artistic, mosaic tile floor. Notice the many squares and repeating geometric patterns in the tile's intricate design.*

Patterns Involving Two or More Variables

You can use more than one variable to describe a pattern. In Example 2, if we let I = the population increase in y years, we could describe the pattern with the formula $I = 350{,}000y$. Example 3 shows a pattern with several variables.

A putter's paradise. *Miniature golf, a fad of the late 1920s, has staged a major comeback in the 1990s. The boom is credited to the popularity of new larger-scale links. Some modern courses are built around adventure themes with waterfalls, mountains, and boat rides.*

Example 3

A family consisting of two adults and three children under age 12 spent the following amounts on tickets while on vacation last month:

Activity	a = cost for 1 adult ($)	c = cost for 1 child ($)	T = total cost ($)
movie	6	3	$2 \cdot 6 + 3 \cdot 3$
ferryboat	10	4	$2 \cdot 10 + 3 \cdot 4$
miniature golf	4	2.50	$2 \cdot 4 + 3 \cdot 2.50$

Find a formula for T in terms of a and c.

Solution

Look at each expression in the *total cost* column. The instances are:

$$2 \cdot 6 + 3 \cdot 3$$
$$2 \cdot 10 + 3 \cdot 4$$
$$2 \cdot 4 + 3 \cdot 2.50$$

The pattern is $2 \cdot \_ + 3 \cdot \_$,

where the first blank is the cost of an adult ticket, and the second is the cost of a child's ticket. So the total cost is $2 \cdot a + 3 \cdot c$.

The formula is $T = 2a + 3c$.

Example 4

Consider these true instances. Describe the pattern using variables.

$$6 + 1.2 = 6 \cdot 1.2$$
$$5 + 1\tfrac{1}{4} = 5 \cdot 1\tfrac{1}{4}$$
$$\tfrac{4}{3} + 4 = \tfrac{4}{3} \cdot 4$$
$$2 + 2 = 2 \cdot 2$$

Solution

First, describe the pattern in your own words. You may think of something like, "When you add two numbers you get the same result as multiplying them." Because two numbers vary, you need two variables. Choose two letters, perhaps a and b. Translate the pattern from words to variables.

$$a + b = a \cdot b$$

Notice that the fourth instance of the pattern has $a = 2$ and $b = 2$. It is possible for different variables to have the same value.

The description of the pattern in Example 4 is correct for the four instances given, but the pattern is not true in general. Writing the pattern with variables makes it easier to find instances where the rule does not work. If $a = 2$, and $b = 3$, the pattern would say

$$2 + 3 = 2 \cdot 3.$$

But $5 \neq 6$. This type of instance, which shows that a pattern is not always true, is called a **counterexample.** The instance in which $a = 2$ and $b = 3$ is a counterexample to the pattern $a + b = a \cdot b$.

Example 5

Find a counterexample to the following statement: If n is an integer, then $n + 1$ is an odd integer.

Solution

Try some values for n.
If $n = 2$, then $n + 1 = 3$, which is an odd integer. The statement is true. So 2 is not a counterexample. Try another value.
If n = 3, then n + 1 = 4, which is not an odd integer. So the pattern "If n is an integer, then $n + 1$ is an odd integer" is not true when $n = 3$. This means that 3 is a counterexample.

QUESTIONS

Covering the Reading

1. The following sentence describes a pattern. "When a number is multiplied by ten and then divided by two, the result is five times the original number."
  a. Give three instances of the pattern.
  b. Write an equation to describe the pattern. (Use n to stand for the number.)

In 2–4, give two instances of the pattern.

2. $n \cdot 1 = n$ **3.** $xy = yx$ **4.** d dogs have $4d$ legs.

5. The population of Vacaville is decreasing by 250 people each month.
  a. At this rate, by how many people will the population decrease in 10 months?
  b. At this rate, by how many people will the population decrease in m months?
  c. Does your answer to part **b** work when $m = 0$?

In 6 and 7, describe the general pattern using one variable.

6.
$$(3 + 9) + 2 = 5 + 9$$
$$(3 + 4) + 2 = 5 + 4$$
$$(3 + 90) + 2 = 5 + 90$$

7.
$$15 + 2 \cdot 15 = 3 \cdot 15$$
$$\frac{1}{3} + 2 \cdot \frac{1}{3} = 3 \cdot \frac{1}{3}$$
$$47.1 + 2 \cdot 47.1 = 3 \cdot 47.1$$

8. Use two variables to write a pattern that describes the following instances.
$$6 \cdot 3 + 6 \cdot 4 = 6 \cdot (3 + 4)$$
$$6 \cdot 11 + 6 \cdot \frac{1}{3} = 6 \cdot (11 + \frac{1}{3})$$
$$6 \cdot 7 + 6 \cdot 1000 = 6 \cdot (7 + 1000)$$

9. Refer to Example 3. How much would the family pay for tickets to a circus that cost \$8 for each adult and \$5 for each child?

10. What is a *counterexample?*

11. Refer to the pattern in Example 4. Give another counterexample.

12. Give a counterexample to show that the pattern
$$(a - 3) + b = a - (3 + b)$$
does not hold for all real numbers a and b.

Vegetables grow in popularity—as well as in gardens. *Health concerns about fat and cholesterol levels in food have led many people to eat more vegetables. California is the leading state in the production of lettuce and tomatoes.*

Applying the Mathematics

13. Here are two instances of a pattern:
2 heads of lettuce and 3 tomatoes cost $2 \cdot 89¢ + 3 \cdot 24¢$.
5 heads of lettuce and 2 tomatoes cost $5 \cdot 89¢ + 2 \cdot 24¢$.
a. Describe the pattern using variables. Let L = the number of heads of lettuce, and T = the number of tomatoes.
b. Write another instance of this pattern.
c. If C = the total cost of lettuce and tomatoes, write a formula for C in terms of L and T.

In 14–16, a pattern is given.
a. Give three instances of the pattern.
b. Tell whether you think the pattern is true for all real numbers. Explain your reasoning.

14. $y + y + y = 3 \cdot y$ **15.** $10(x - y) = 10x - y$ **16.** $(a^2)a = a^3$

17. a. Describe this pattern using one variable.
$$2 \cdot 2 > 2$$
$$3 \cdot 3 > 3$$
$$4 \cdot 4 > 4$$
$$100 \cdot 100 > 100$$
b. Find another integer that is an instance of the pattern.
c. Find an integer that is a counterexample to the pattern.
d. Find a non-integer that is a counterexample to the pattern.

18. A pizza is cut into pieces by making each cut go through the center.

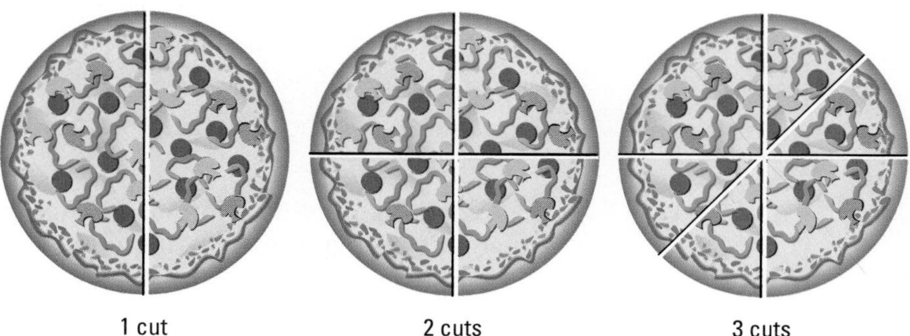

a. How many pieces are made with 4 cuts?
b. Find a formula for the number of pieces p of pizza you get from c cuts.
c. What does your formula predict for the number of pieces from 12 cuts?

19. Below are four designs made with pennies.

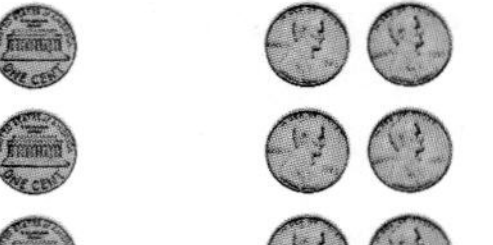  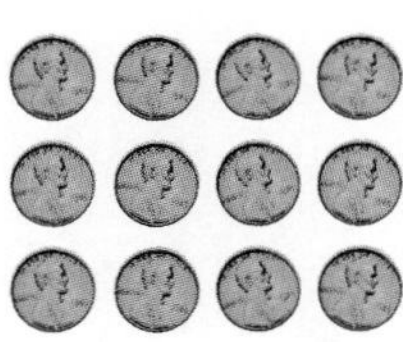

a. Draw what you think the next design should be.
b. Describe in words the patterns you see in these five designs.
c. If the designs continue to follow the same pattern, how many pennies will be needed to make the 20th design?

Review

In 20–22, evaluate without a calculator. *(Lesson 1-6)*

20. $\sqrt{49}$ **21.** $\sqrt{400}$ **22.** $\sqrt{401} \cdot \sqrt{401}$

23. A bag of grass seed can cover an area of 3000 square feet. *(Lesson 1-6)*
a. What is the side length of the largest square plot it can seed?
b. Give the dimensions of a non-square plot it can seed.

24. Suppose each small square in the figure below has area 1 square unit. *(Lesson 1-6)*
a. Find the area of *DFGH*.
b. What is the length of one side?

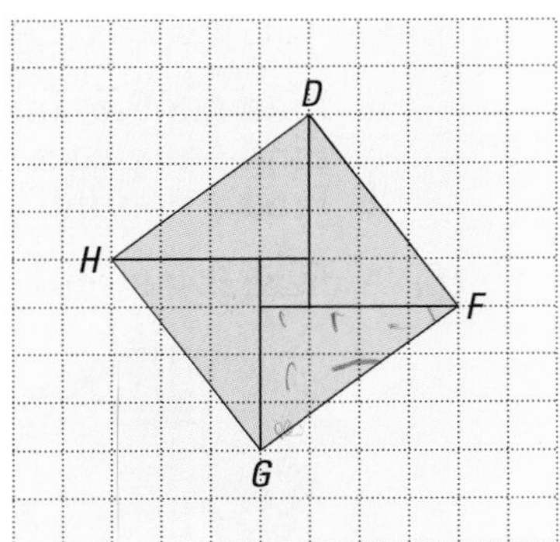

In 25 and 26, use this information. According to some medical charts, an adult's normal weight w (in pounds) can be estimated by the formula $w = \frac{11}{2}h - 220$ when his or her height h (in inches) is known. *(Lesson 1-5)*

25. Estimate the normal weight of a person who is 6 feet tall.

26. **a.** According to this formula, what is the normal weight of a person who is 40 inches tall?
b. Explain your answer to part **a** by relating it to a reasonable domain of the variable h.

27. Let $d = 3$. Find a value of n so that the following sentences are true. *(Lessons 1-1, 1-5)*

a. $\frac{n}{d} > 1$ **b.** $\frac{n}{d} < 1$ **c.** $\frac{n}{d} = 1$

28. Suppose $T = \{10, 12, 14, 16, 18, 20\}$, $V = \{12, 15, 18\}$ and $W = \{5, 10, 15, 20\}$. *(Lesson 1-3)*

a. Does $T \cup (V \cap W) = (T \cup V) \cap W$?

b. Explain how you got your answer.

29. A machinist is making metal rods for lamps. The metal rods are to be 1.25 inches in diameter with an allowable error of at most 0.005 inch. (The quantity 0.005 inch is called the *tolerance.*) Give an interval for the possible diameters d of the rods. *(Lesson 1-2)*

Exploration

30. A monthly calendar contains many patterns.

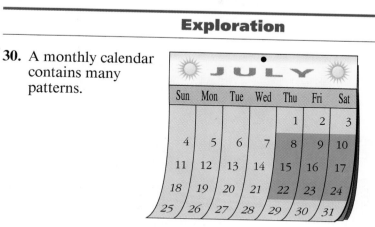

JULY

Sun	Mon	Tue	Wed	Thu	Fri	Sat
				1	2	3
4	5	6	7	8	9	10
11	12	13	14	15	16	17
18	19	20	21	22	23	24
25	26	27	28	29	30	31

a. Consider a 3×3 square such as the one drawn on the calendar. Add the nine dates. What is the relationship between the sum and the middle date? Try this again. Does it always seem to work?

b. In a 3×3 square portion of the calendar, if the middle date is expressed as N, then the date below it would be $N + 7$ because it is 7 days later. Copy the chart below, and then fill in the other blanks.

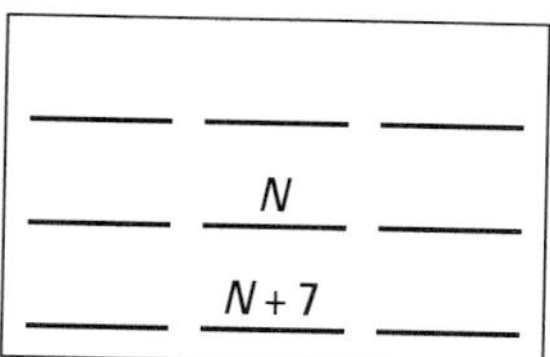

c. Show how your result from part **b** can be used to explain your conclusion in part **a.**

Introducing Lesson 1-8

Patterns with Squares and Right Triangles

IN-CLASS ACTIVITY

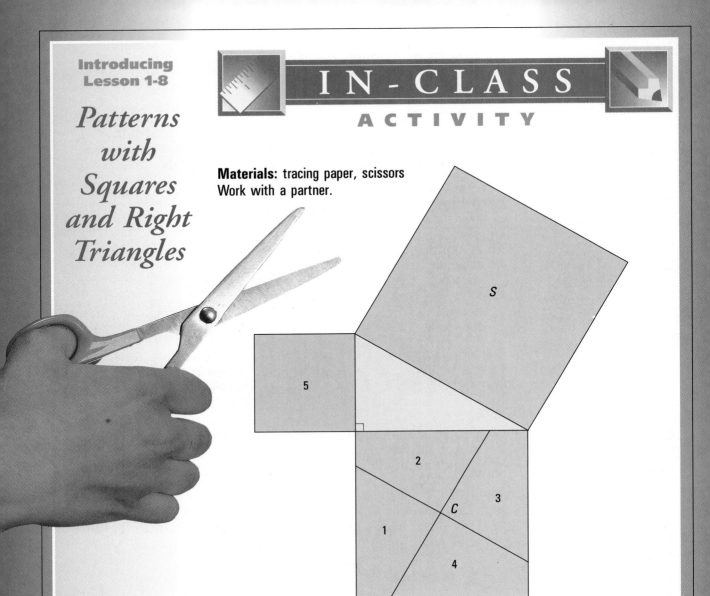

Materials: tracing paper, scissors
Work with a partner.

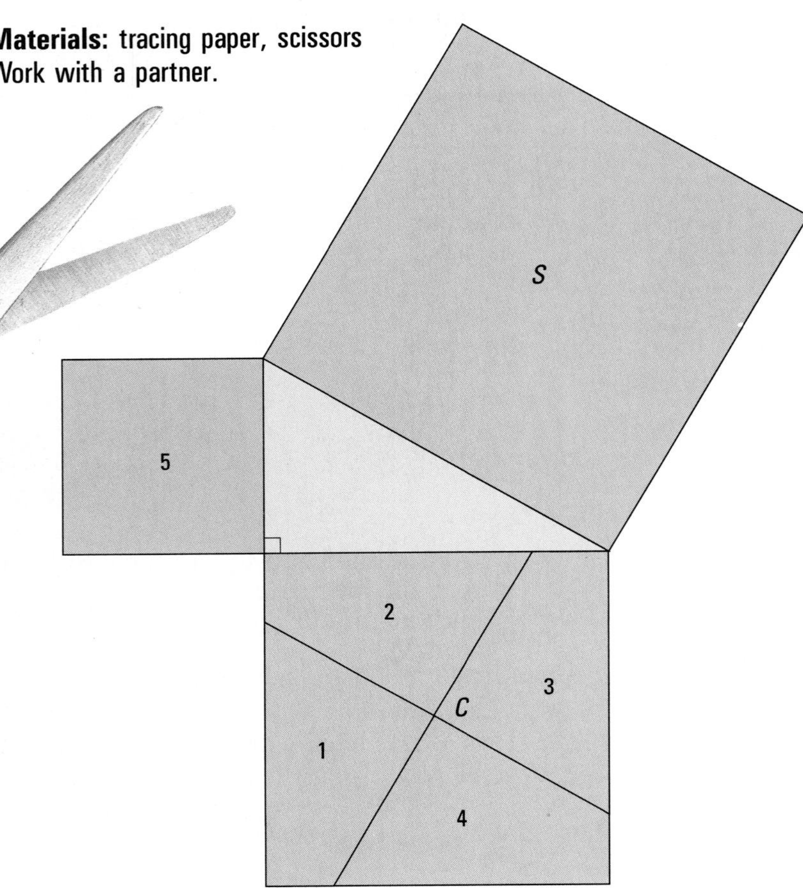

In the figure above, the blue region is enclosed by a right triangle. A square is drawn on each side of the right triangle. Point C is the center of the bottom square, and regions 1, 2, 3, and 4 are formed by lines parallel to sides of the largest square S.

1 Trace the figure. Cut out pieces 1, 2, 3, 4, and 5. With your partner, arrange the five pieces to form a square the size of S, and tape them in place.

2 Draw a right triangle of a different shape. Draw squares on the three sides as was done above. Then form regions 1–5 with the new squares. Finally, repeat step 1 for your figure.

3 ***Draw conclusions.*** This activity illustrates a very famous mathematical pattern. The pattern involves a relation between the areas of the squares drawn on the three sides of a right triangle. Describe the pattern using words or variables.

LESSON

1-8

The Pythagorean Theorem

Pythagoras the teacher. *Pythagoras was a Greek mathematician, philosopher, and teacher born around 580 B.C. The scene above, a detail from* School of Athens *by Raphael, depicts Pythagoras (forefront holding a book) surrounded by students.*

What Is the Pythagorean Theorem?

The In-class Activity on page 44 involves a very famous pattern known as the *Pythagorean Theorem.* To describe this pattern in words, we need some language about right triangles. Recall that in a right triangle, one of the angles must be 90°. In the figure below, the right angle is formed by the sides with lengths a and b. These sides are called the **legs** of the right triangle. The longest side of the triangle, the **hypotenuse,** is opposite the right angle. Here the hypotenuse has length c.

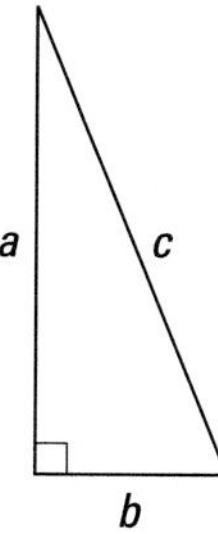

Look carefully at the figure in the In-class Activity. The purpose of the activity was for you to see that the squares built on the legs of a right triangle can be cut apart and reassembled to fill the square on the hypotenuse. You could do this because the sum of the areas of the two smaller squares equals the area of the largest square.

In some situations, this pattern can be verified by counting. Look carefully at the two instances at the top of the next page. Each shows a right triangle in the center with a square built on each side. Count to find the area of each square.

Instance 1

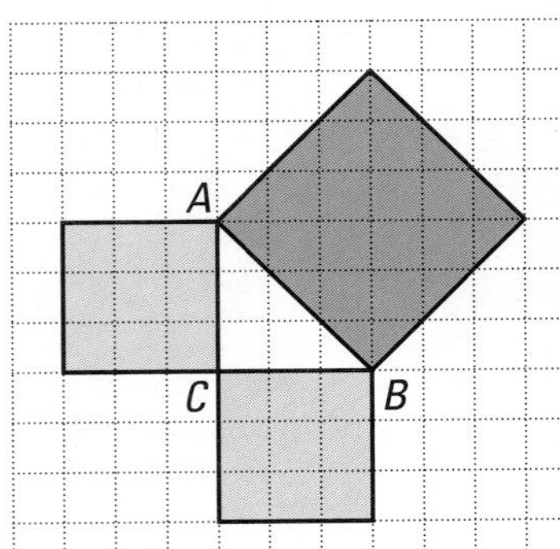

Instance 2

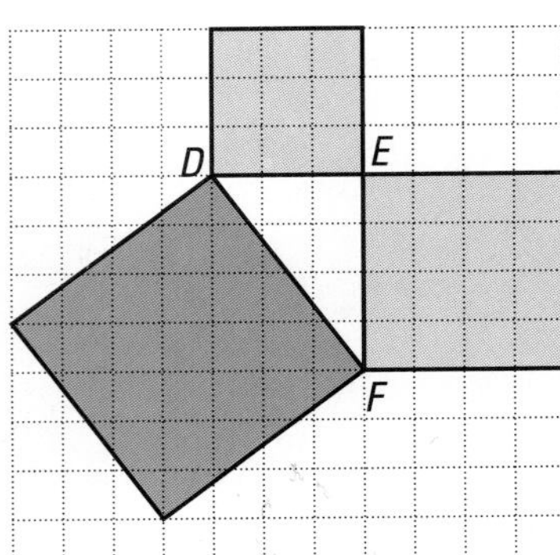

In Instance 1, you should find that the square built on leg $\overline{AC}$ has area 3^2, or 9. The square built on leg $\overline{BC}$ also has area 9. The area of the square on the hypotenuse $\overline{AB}$ of $\triangle ABC$ is 18. Notice that $9 + 9 = 18$.

In Instance 2, the area of the square on leg $\overline{DE}$ is $3^2 = 9$. The area of the square on leg $\overline{EF}$ is 4^2 or 16. The area of the square on the hypotenuse $\overline{DF}$ is 25. Notice that $9 + 16 = 25$.

If the legs of a right triangle have lengths a and b, and the hypotenuse has length c, then the areas of the squares built on the sides are a^2, b^2, and c^2.

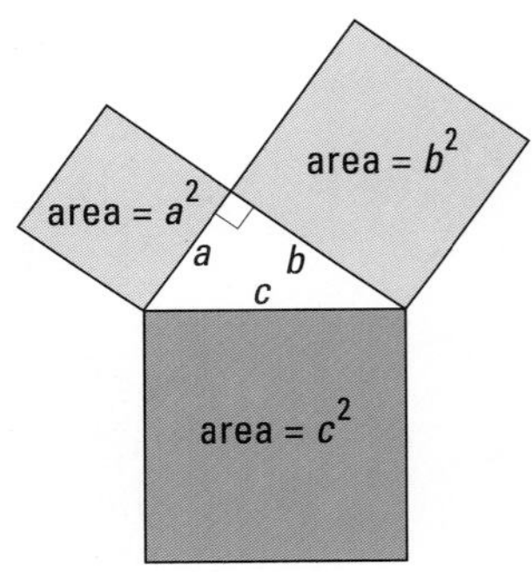

The pattern can be written as follows.

Pythagorean Theorem
In a right triangle with legs of lengths *a* and *b* and hypotenuse of length *c*,

$$a^2 + b^2 = c^2.$$

Theorems are important properties that have been proved to be true. The Pythagorean Theorem is named after the Greek mathematician Pythagoras, who lived about 2,500 years ago. Pythagoras seems to have been the first to show that this pattern is true for every right triangle. It was applied even earlier by the Babylonians, and perhaps more than a thousand years before them by the architects of the pyramids in Egypt and the builders of Stonehenge in England. The Chinese also discovered this property independently. It is one of the most famous theorems in mathematics.

Using the Pythagorean Theorem to Find the Hypotenuse

The Pythagorean Theorem allows you to find the length of one side of a right triangle if you know the lengths of the other two sides. The following examples show how to find the length of the hypotenuse. In a later chapter you will learn to use the Pythagorean Theorem to find the length of a leg.

Example 1

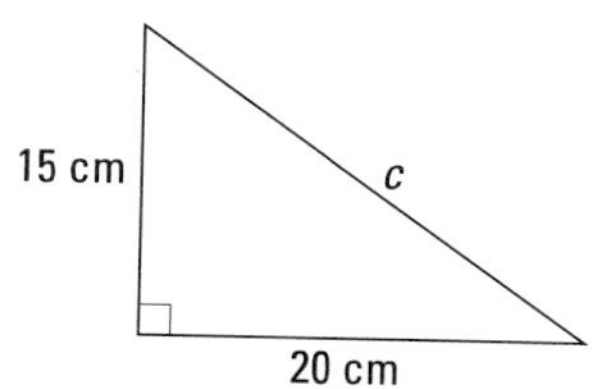

What is the length of the hypotenuse of the right triangle drawn on the left?

Solution

Substitute the lengths of the legs of the triangle for a and b in the Pythagorean Theorem. The hypotenuse is c.

In a right triangle, $a^2 + b^2 = c^2$.

Substituting,

$$15^2 + 20^2 = c^2$$
$$225 + 400 = c^2$$
$$625 = c^2.$$

This equation has two solutions, the positive and negative square roots of 625.

$$c = \sqrt{625} \text{ or } c = -\sqrt{625}$$

These square roots are integers.

$$c = 25 \text{ or } c = -25$$

However, the length of the hypotenuse cannot be negative, so use only the positive solution. The hypotenuse is 25 cm.

Example 2

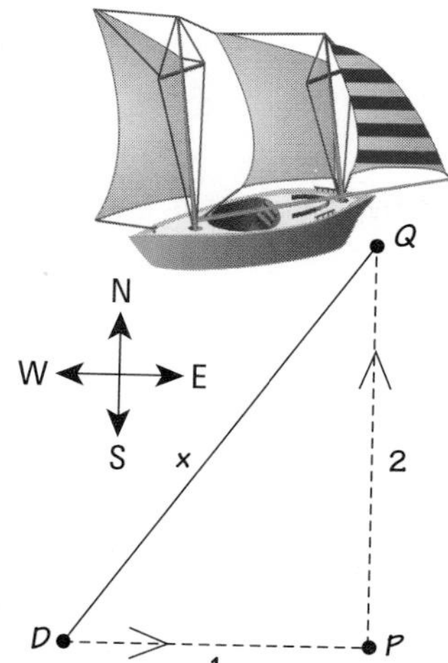

A sailboat leaves its dock and travels 1 mi due east. Then it turns and sails 2 mi due north. At this point, how far is it from the dock?

Solution

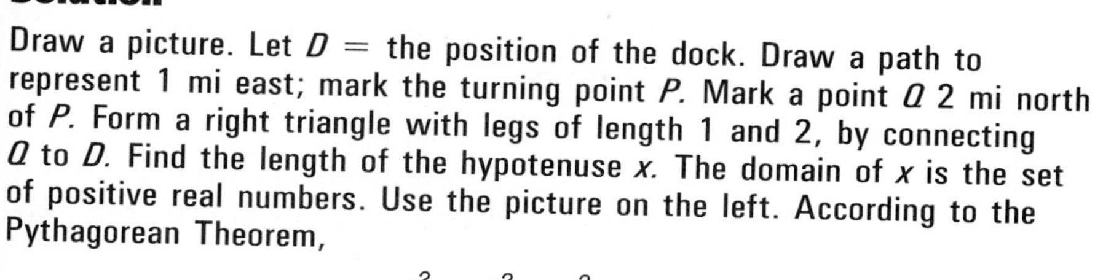

Draw a picture. Let D = the position of the dock. Draw a path to represent 1 mi east; mark the turning point P. Mark a point Q 2 mi north of P. Form a right triangle with legs of length 1 and 2, by connecting Q to D. Find the length of the hypotenuse x. The domain of x is the set of positive real numbers. Use the picture on the left. According to the Pythagorean Theorem,

$$1^2 + 2^2 = x^2$$
$$1 + 4 = x^2$$
$$5 = x^2$$
$$x = \sqrt{5} \text{ or } x = -\sqrt{5}$$

We can discard $-\sqrt{5}$ because x must be positive.

The exact distance between the boat and the dock is $\sqrt{5}$ mi. This is approximately 2.2 mi.

The Pythagorean Theorem is helpful when a right triangle can be identified in figures other than triangles.

Example 3

The figure below is a rectangle with sides of 2.5 cm and 6 cm. Find x, the length of the diagonal.

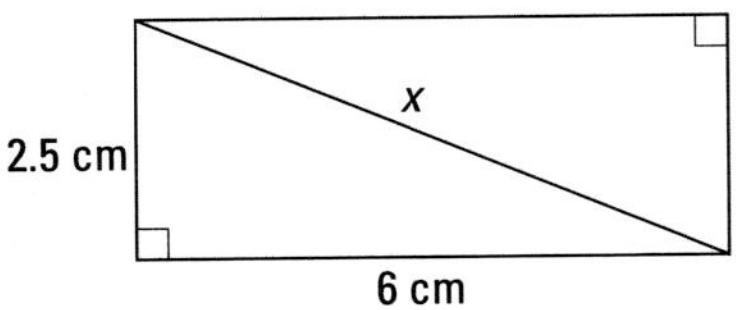

Solution

The diagonal forms two right triangles. Each has legs of length 2.5 cm and 6 cm. The diagonal of the rectangle is the hypotenuse of the right triangles. Use the Pythagorean Theorem to find x. Remember that x must be positive.

$$2.5^2 + 6^2 = x^2$$
$$6.25 + 36 = x^2$$
$$42.25 = x^2$$

So $$x = \sqrt{42.25} = 6.5.$$

The diagonal is 6.5 cm long.

QUESTIONS

Covering the Reading

1. What is the longest side of a right triangle called?

2. Examine the right triangle shown at the right.
   a. Which sides are the legs?
   b. Which side is the hypotenuse?

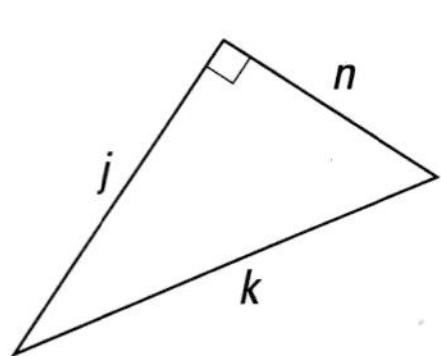

3. $\triangle PQR$ is a right triangle with hypotenuse $\overline{PR}$. Describe the relation among the areas of squares I, II, and III.

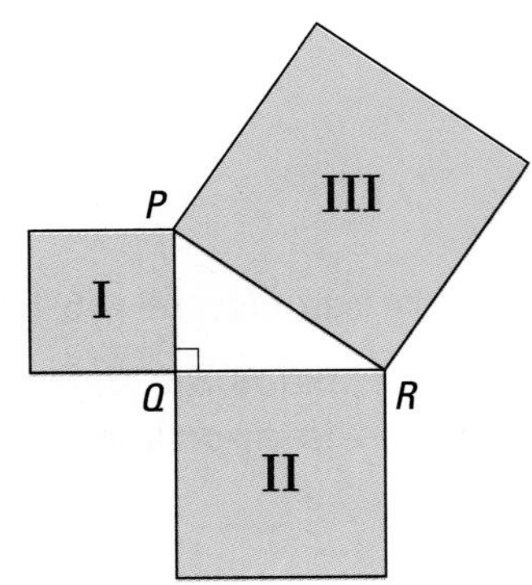

4. State the Pythagorean Theorem.

5. $\triangle ABC$ is a right triangle with $\angle B$ the right angle.
   a. What is the area of the square with side $\overline{AC}$?
   b. What is the length of $\overline{AC}$?

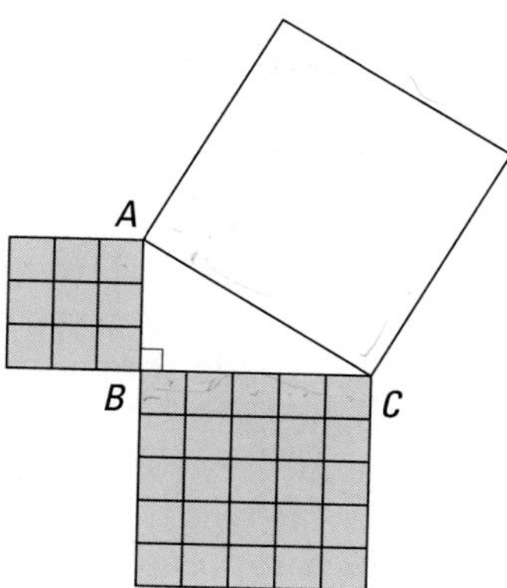

In 6–8, find the length of the hypotenuse. If the answer is not a whole number, find both its exact value and an approximation rounded to the nearest hundredth.

6.

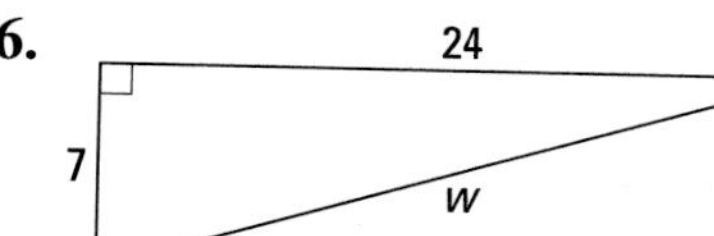

7.

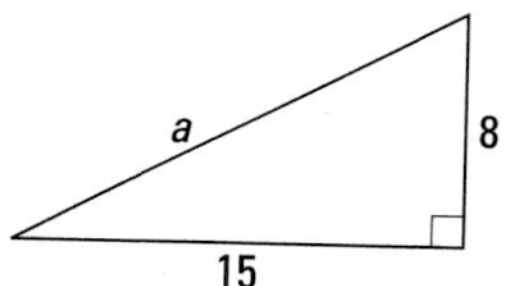

8.

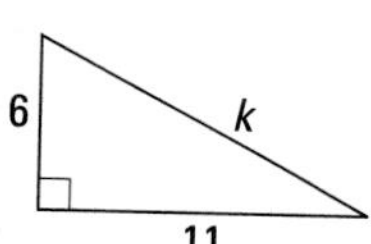

9. A small plane travels 10 miles due west from an airport, then turns and flies due south for 15 miles. At this point how far is the plane from the airport?

In 10 and 11, refer to rectangle $WXYZ$ below.

10. *True or false.* $\triangle WXY$ is a right triangle. Justify your answer.

11. If a rectangular garden has width 9 m and length 15 m, what is the length of the diagonal?

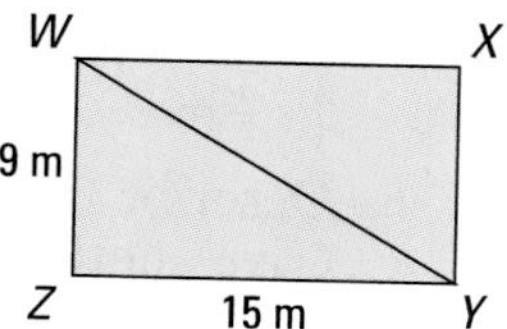

Applying the Mathematics

12. A carpenter wants to make a garden gate with a cross brace for support. If the gate is to be 36 in. wide and 48 in. high, how long must the cross brace be? (Give your answer to the nearest inch.)

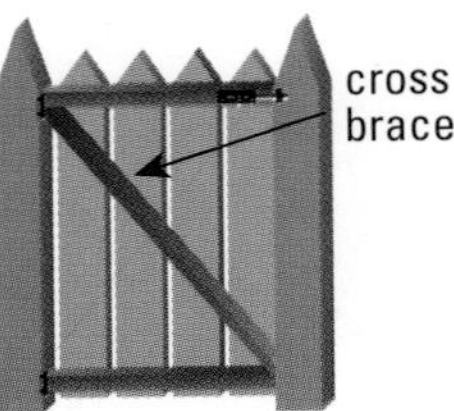

13. Some pedestrians want to get from point A to point B. The two roads shown meet at right angles.

a. If they follow the roads, how far will the pedestrians walk?

b. Suppose that, instead of walking along the roads, they took the shortcut from A to B. Use the Pythagorean Theorem to find the length of the shortcut, rounded to the nearest tenth of a km.

c. How much distance would they save by taking the shortcut?

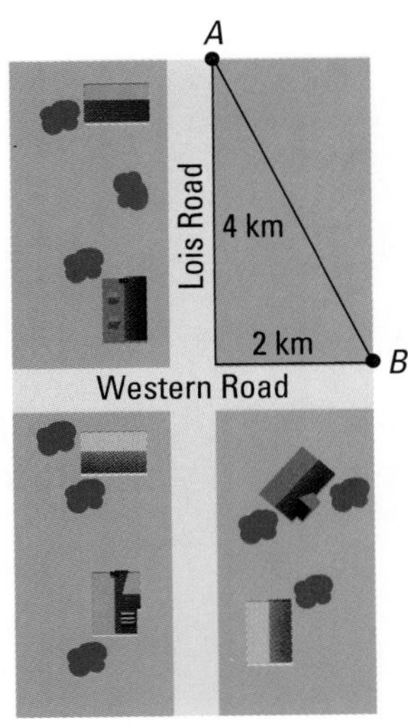

14. The area of a square is 64 cm^2.

a. Find the length of a side.

b. Use the Pythagorean Theorem to find the length of the diagonal.

15. A right triangle has legs of lengths 3 cm and $\sqrt{3}$ cm.

a. How long is the hypotenuse?

b. Draw a right triangle with the given dimensions and measure its hypotenuse.

c. How close are your answers to parts **a** and **b?**

Review

16. a. Evaluate the expressions below.

$\frac{2}{3} \cdot \frac{3}{2} = ?$ $\quad \frac{4}{5} \cdot \frac{5}{4} = ?$ $\quad \frac{9}{10} \cdot \frac{10}{9} = ?$ $\quad \frac{-3}{5} \cdot \frac{5}{-3} = ?$

b. Describe the pattern using the variables a and b.

c. Give another instance of your pattern. *(Lesson 1-7)*

In 17 and 18, fill in the blank with =, <, or >. *(Lesson 1-6)*

17. $\sqrt{25} + \sqrt{4}$ _?_ $\sqrt{29}$

18. $(\sqrt{887})^2$ _?_ 887

19. In 1638, Galileo claimed that an object propelled into the air will reach a height that can be determined by a formula. If h is the height in meters, and the object is propelled from a height of 1 meter with a vertical velocity of 30 meters a second, then

$$h = -4.9t^2 + 30t + 1,$$

where t is the number of seconds that the object is in the air. Find the height of a batted ball 2 seconds after it is hit. *(Lesson 1-5)*

20. The domain for g is {-2, 0, 7}. For each element of the domain, evaluate $10g^2$. *(Lessons 1-2, 1-4)*

In 21–23, determine which is larger. *(Lesson 1-1)*

21. $3\frac{1}{2}$ or 3.45

22. $\frac{9}{4}$ or $\frac{9}{5}$

23. $\frac{4}{5}$ or $\frac{5}{6}$

24. Order from largest to smallest. *(Previous course or Appendix B)*
$6.5 \cdot 10^{14}$, $7.2 \cdot 10^{15}$, $9.4 \cdot 10^{13}$

25. Often 15% of a restaurant bill is left for a tip. If a bill is $29.88, then what would be a 15% tip? (Try to answer this question in your head by rounding $29.88 to the nearest dollar.) *(Previous course)*

Exploration

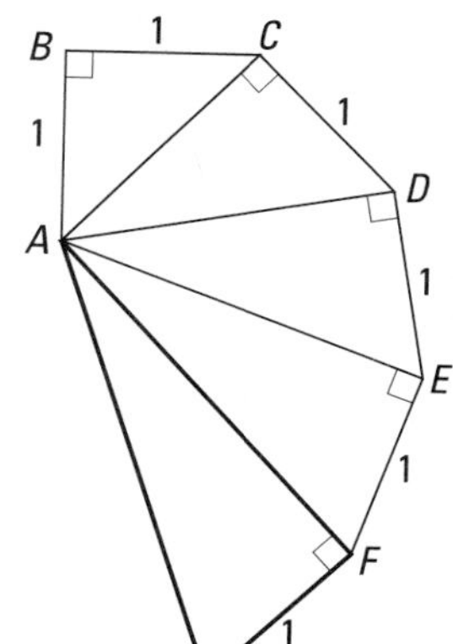

26. Six of the segments in the figure on the left have length one unit.

a. Find the length of each of $\overline{AC}$, $\overline{AD}$, $\overline{AE}$, $\overline{AF}$, and $\overline{AG}$.

b. Copy the drawing and add to it to make a segment whose length is $\sqrt{8}$.

27. Use a sheet of graph paper to draw a square that has an area of 13 square units. (Hint: Find two perfect squares whose sum is 13.)

28. Fill in the blank with =, <, >, ≤, or ≥ to make a true sentence for all nonnegative real numbers.

$$\sqrt{x^2 + y^2} \ \underline{\ ?\ }\ x + y.$$

Is your sentence always true? How does this sentence relate to the Pythagorean Theorem and right triangles with legs of lengths x and y?

Introducing Lesson 1-9

Variables and Tables

IN-CLASS ACTIVITY

Materials: about 30 square tiles per group
Work in small groups.

In this chapter you have used many sentences and expressions. You may wonder where these sentences and expressions came from. In many cases, the answer is simple: Formulas, sentences, and expressions describe patterns that people have found. In this activity, you will make designs with square tiles and look for numerical patterns in the designs.

1 Arrange the tiles in the four designs shown below.

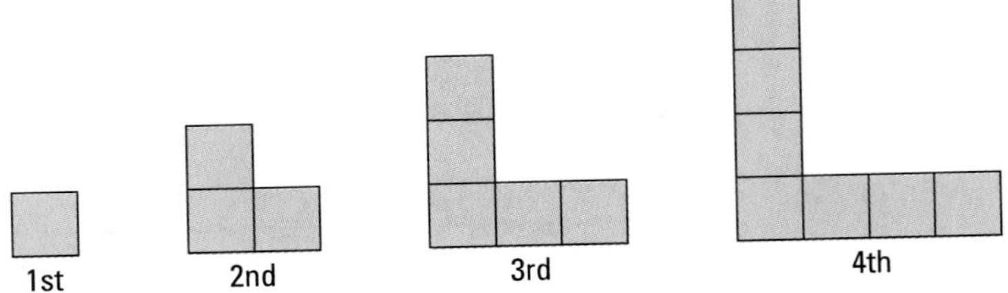

Use tiles to make the 5th and 6th designs. Draw these designs.

2 Suppose the length of one side of a tile is 1. Then the perimeter of the first design is 4, and the perimeter of the second is 8. What is the perimeter of the third design?

perimeter = 4

perimeter = 8

3 Let n be the number of the design and p be its perimeter. For each design find p, then copy and complete the table at the right.

n	p
1	4
2	8
3	
4	
5	
6	

4 **a.** If you continued to make L-shaped designs like those in step 1, what would be the perimeter of the 10th design?

b. Explain how you got your answer.

5 ***Draw conclusions.*** Write a sentence or a formula that describes the relation between p and n.

LESSON

1-9

Variables from Tables

Art patterns in architecture. *Many homes in San Francisco, California, are two-story wooden buildings that share at least one wall with the house next door. Example 1 deals with patterns from rows of houses.*

When looking for patterns with variables, it sometimes helps to organize information in a table. For instance, suppose that some students are making designs from congruent triangular blocks. One student makes the following designs.

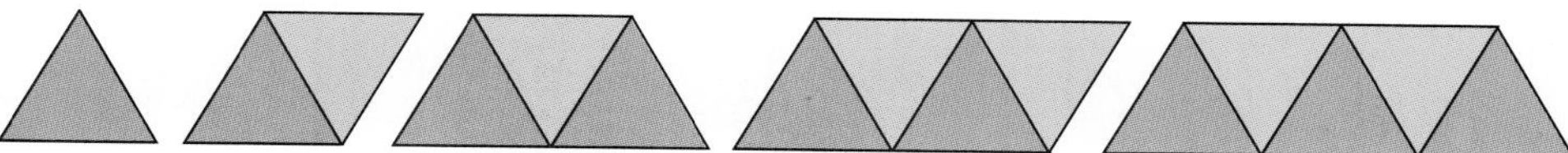

Each block design after the first is outlined by a trapezoid. How does the perimeter of the trapezoid change with the number of triangular blocks used?

Assume each side of the triangular block has length 1 unit. Add to determine the perimeter of each trapezoid.

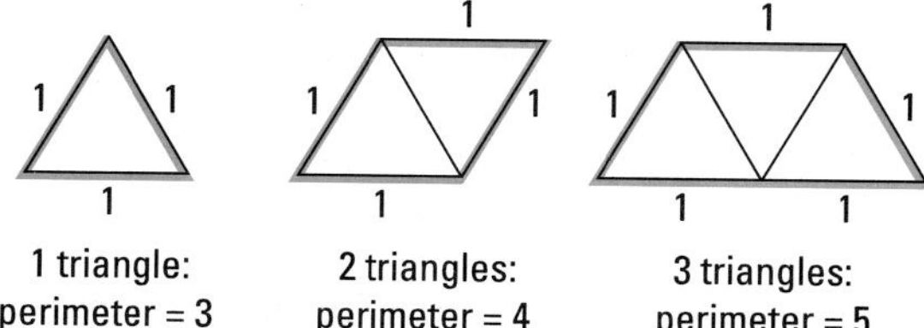

Record the data in a table, as shown at the top of the next page. There T = the number of triangles, and P = the perimeter of the trapezoid.

T	P
1	3
2	4
3	5
4	6
⋮	⋮
T	$T + 2$

In this situation, to find a formula relating the perimeter to the number of triangles, you need to see a relation between the numbers in *each row*. Ask, "What single rule relates 1 and 3? 2 and 4? 3 and 5? 4 and 6?"

In each case, the second number is 2 more than the first, so $P = T + 2$.

This formula is true for all whole numbers T, when $T \geq 1$. For instance, if $T = 1000$, then $P = 1000 + 2 = 1002$. When there are 1000 blocks placed in this design, the figure formed will have perimeter 1002.

Tables may be written horizontally or vertically. In Example 1, the numbers are written horizontally in a table.

Example 1

These designs show houses made from triangular and rectangular blocks.

a. How many blocks are needed to construct a row of 10 such houses?

b. Let h = the number of houses, and b = the number of blocks. Find a formula for b in terms of h.

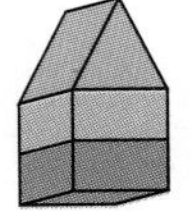 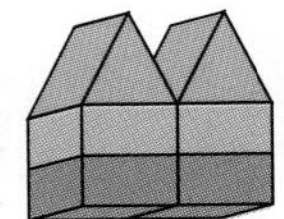  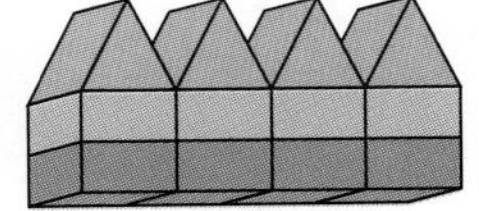

Solution

a. Count the number of blocks needed to make 1, 2, 3, and 4 houses as shown above. Organize the data in a table.

h	1	2	3	4	• • •	10
b	3	6	9	12		?

Look for a pattern relating the values of h and b. Here the values of one variable are all in a row. So you must look down the columns to find pairs of variables. Notice that each value of b is 3 times the value of h. For 10 houses $3 \cdot 10 = 30$ blocks are needed.

b. $b = 3h$

Check

a. Draw the design with 10 row houses. There are 30 blocks needed.
b. Check that the numbers in the table are instances of the formula. For example, is $h = 3$ and $b = 9$ an instance of $b = 3h$? Yes, $9 = 3 \cdot 3$.

You should be able to recognize when two variables are related by addition, subtraction, multiplication, division, powers, or square roots.

Example 2

Multiple choice. Which formula describes the numbers in the table?

x	2	3	4	5
y	4	9	16	25

(a) $y = x + 2$ (b) $y = 2x$ (c) $y = \sqrt{x}$ (d) $y = x^2$

Solution

Substitute numbers from the table for x and y. If each pair of x and y values gives a true sentence, then the formula describes the pattern. If you find a counterexample to the formula, the formula does not describe the numbers in the table. Formula (a) holds when $x = 2$ and $y = 4$ $(4 = 2 + 2)$, but it does not hold for any other values. For instance, $9 \neq 3 + 2$. Similarly, formula (b) does not hold when $x = 3$ and $y = 9$ $(9 \neq 2 \cdot 3)$. Formula (c) does not hold for any pairs of values in the table. For instance, $25 \neq \sqrt{5}$. Only formula (d) is true for all four values in the table.

$$4 = 2^2, \quad 9 = 3^2, \quad 16 = 4^2, \quad 25 = 5^2$$

So $y = x^2$.

Four generations. *Celebrating Thanksgiving are a boy, his two parents, his four grandparents, and one of his great-grandmothers.*

Example 3

You have two biological parents. Each of your parents had two parents, and each of them had two parents. The table below looks back four generations. If you go back n generations, how many biological ancestors are in that generation?

Generation name	Number of generations back	Number of ancestors
parents	1	2
grandparents	2	4
great-grandparents	3	8
great-great-grandparents	4	16

Solution

Look for patterns in the table. Notice that each number of ancestors is double the number in the previous generation. Each is also a power of 2. So rewrite the given numbers of ancestors as powers of 2.

Generation name	Number of generations back	Number of ancestors
parents	1	2^1
grandparents	2	2^2
great-grandparents	3	2^3
great-great-grandparents	4	2^4

Now read across the rows. Notice that in each row the generation number is the same as the exponent for the number of ancestors.

$n \qquad 2^n$

So, the pattern of the first four rows is that n generations back, a person has 2^n ancestors.

Let A be the number of ancestors n generations back. Then $A = 2^n$.

QUESTIONS

Covering the Reading

1. Refer to the opening example about the designs made from triangular blocks.
  a. Use the formula to find the perimeter of the polygon formed by placing 8 triangular blocks in a line.
  b. Draw a figure to justify your answer.

In 2 and 3, refer to Example 1.

2. How many blocks are needed to construct a row of 25 such houses?

3. *True or false.* The formula $h = \frac{b}{3}$ describes the data in the table.

In 4–6, *multiple choice.* Which formula describes the numbers in the table?

(a) $y = x + 5$ (b) $y = 5x$ (c) $y = 5^x$ (d) $y = x^5$

4.

x	1	2	3	4
y	5	10	15	20

5.

x	1	2	3	4
y	5	25	125	625

6.

x	1	2	3	4
y	6	7	8	9

In 7–10, refer to Example 3.

7. How many biological grandparents does every person have?

8. How many biological great-great-grandparents does a person have?

9. In the formula $A = 2^n$ for the number of biological ancestors, what do A and n stand for?

10. Explain why the formula $A = 2^n$ might not give the correct number of a person's ancestors 20 generations back.

Applying the Mathematics

11. A small manufacturer makes windows with two rows of panes of glass. Below are four examples of their designs.

a. Complete the table below.

w = width of window	1	2	3	4	5
p = number of panes					

b. Describe some patterns you find in the table in part **a.**
c. Find a formula for p in terms of w.

12. Below are the first three instances of a pattern made with triangular and square blocks.

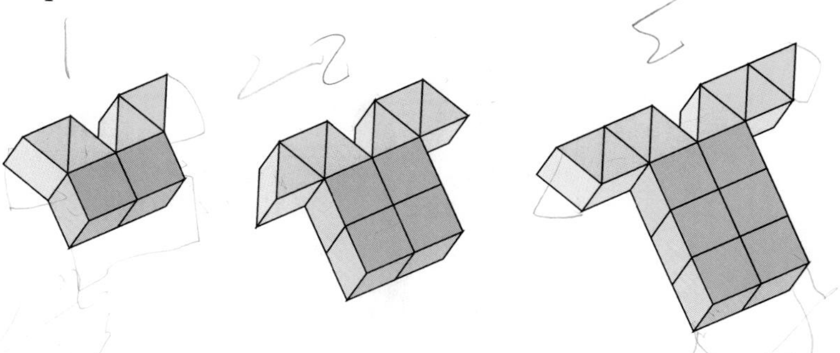

a. Draw the fourth instance.
b. How many blocks will be needed to make the 7th instance of this pattern? How many will be triangular? How many will be square? Explain how you got your answers.

13. Take a sheet of newspaper. Call this a thickness of 1.

a. Fold it in half and record the thickness of the folded paper.

b. Fold it in half again (so you now have two folds) and record the thickness of the folded paper.

c. Continue folding the paper in half and complete the table below.

n = number of folds	1	2	3	4
t = thickness of folded paper				

d. Find a formula for t in terms of n.

e. How thick would the folded paper be if you could do 9 folds?

14. Refer to the table below. Find a formula relating x and y.

x	y
1	3
2	9
3	27
4	81

15. *Multiple choice.* In the Chinese game of *wei ch'i* (way key), a player's pieces are captured and removed if they are surrounded by enemy pieces with no path to an empty point on the board. The drawing below shows situations in which black can capture the white pieces by placing a black piece on the marked point.

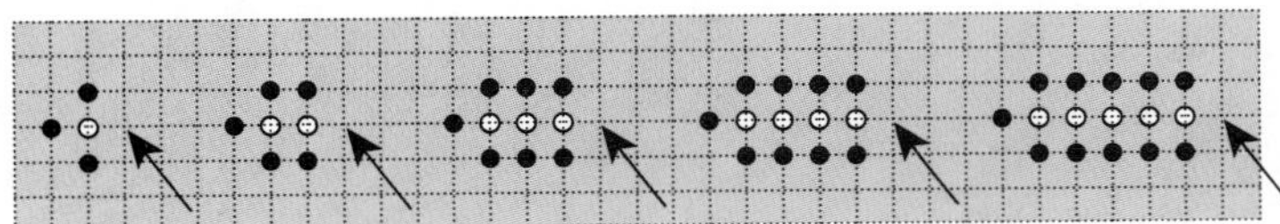

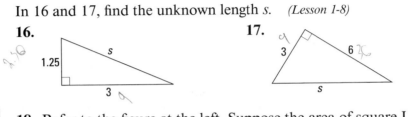

The number of black pieces required to capture each row of white pieces is shown in the table below. Which formula gives the number of black pieces, b, needed to capture a row of w white pieces?

(a) $b = 2w + 1$
(b) $b = 2w + 2$
(c) $b = w + 3$
(d) $b = 3w$

w	b
1	4
2	6
3	8
4	10
5	12

Review

In 16 and 17, find the unknown length s. *(Lesson 1-8)*

16.

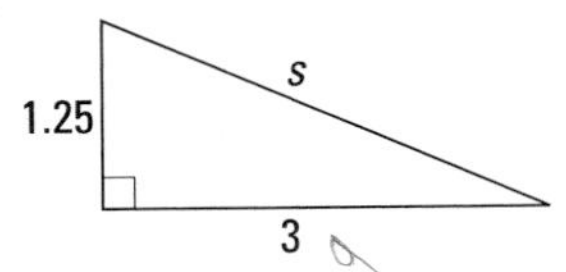

17.

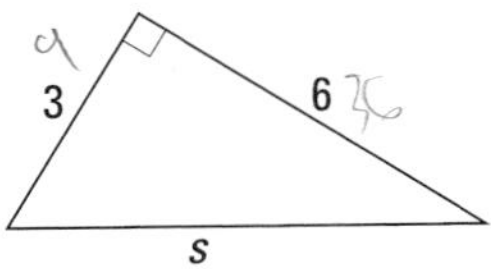

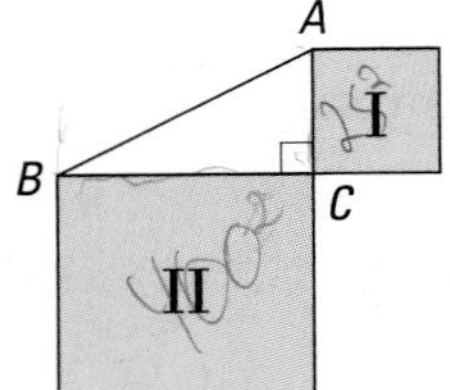

18. Refer to the figure at the left. Suppose the area of square I is 5 cm^2, and the area of square II is 20 cm^2. Find the length of each side. *(Lesson 1-8)*

a. $\overline{AC}$ **b.** $\overline{BC}$ **c.** $\overline{AB}$

Out of this world. *The first U.S. astronaut to walk in space was Edward H. White II on June 3, 1965. White was firmly anchored to the* Gemini 4 *spacecraft by a 26-foot cord.*

19. A museum charges a \$5.00 entry fee for adults and a \$1.50 fee for children.

a. How much would it cost for 6 adults and 5 children?

b. How much would it cost for a adults and c children? *(Lesson 1-7)*

20. A formula for the sum S of two fractions $\frac{a}{b}$ and $\frac{c}{d}$ is $S = \frac{ad + bc}{bd}$. *(Lesson 1-5, Previous course)*

a. What values would a, b, c, and d be when finding the sum of $\frac{5}{8}$ and $\frac{3}{7}$?

b. Use the formula to compute the sum of $\frac{5}{8}$ and $\frac{3}{7}$.

c. Use the formula to compute the sum of $-\frac{2}{3}$ and $\frac{1}{2}$.

21. Suppose a person weighs w kg at sea level. Scientists have determined that the person's weight W at height h kilometers above the Earth is given by the formula $W = w\left(\frac{6400}{6400 + h}\right)^2$. An astronaut weighs 70 kg at sea level. How much does the astronaut weigh when traveling in space 16,000 km above Earth? *(Lesson 1-5)*

22. *Multiple choice.* The graph pictures the solutions to which inequality? *(Lesson 1-2)*

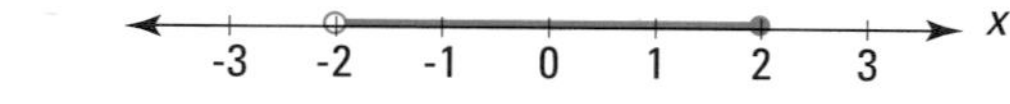

(a) $-2 \le x \le 2$ (b) $-2 < x < 2$ (c) $-2 \le x < 2$ (d) $-2 < x \le 2$

23. a. What is the smallest whole number solution to $p > -7$?

b. What is the smallest integer solution to $p > -7$?

c. Graph the solution set for $p > -7$, using the set of real numbers as domain.

d. Is there a smallest real number that solves $p > -7$? Why or why not? *(Lessons 1-1, 1-2)*

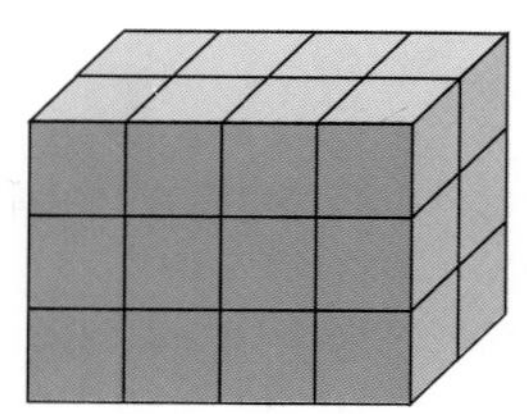

24. One-inch cubes are stacked to form a rectangular solid as shown at left. *(Previous course)*

a. How many cubes are used?

b. What is the volume of the rectangular solid?

Exploration

In 25 and 26, refer to Question 13.

25. Suppose you started with a very large sheet of paper, and could fold it 100 times.

a. How many thicknesses of paper would you have?

b. Would this be thicker than your local telephone book? Justify your answer.

c. If you answered "yes" to part **b,** about how many phone books would you need to stack to equal the thickness of the paper in part **a?**

26. How many times can you fold a newspaper page? Explain how you arrived at your answer.

PROJECTS 1
CHAPTER ONE

A project presents an opportunity for you to extend your knowledge of a topic related to the material in this chapter. You should allow more time for a project than you do for typical homework questions.

1 Interview with Pythagoras

Use a library to do some research on Pythagoras. Conduct an interview in which you, the interviewer, ask Pythagoras questions about his work. Write the interview as if it were to be printed in a magazine. Include responses you think Pythagoras might have given. You may want to perform your interview (with a partner) for your class.

2 Order of Operations

Write a letter to a student who has not yet taken algebra explaining the order of operations. Use examples to illustrate your explanation. You may want to give your letter to a younger student to read before you hand it in. If he or she has difficulty understanding your explanation, you may want to revise it before you give it to your teacher.

3 Estimating Square Roots

Al-Karkhi (also known as al-Karaji) was an Arabian mathematician who lived during the early 11th century. He approximated the square root of positive integers by using a process equivalent to the formula

$$\sqrt{a} = w + \frac{a - w^2}{2w + 1}.$$

The variable w stands for the whole number portion of the square root. Investigate al-Karkhi's method by making a table with different values for a from 1 to 50. Calculate $\sqrt{a}$ to the nearest thousandth using his method. Compare the results to what you obtain by finding $\sqrt{a}$ with a calculator. What conclusions can you make concerning al-Karkhi's method?

4 Formulas

Find ten formulas you have used that are not mentioned in this chapter. Make a poster or write a report to explain what the variables in the formulas represent and why you might want to use the formulas.

5 Squares Surrounding Triangles

On a piece of grid paper, draw ten squares of different sizes. (Note: They don't *all* have to be different from one another.) Then carefully cut them out.

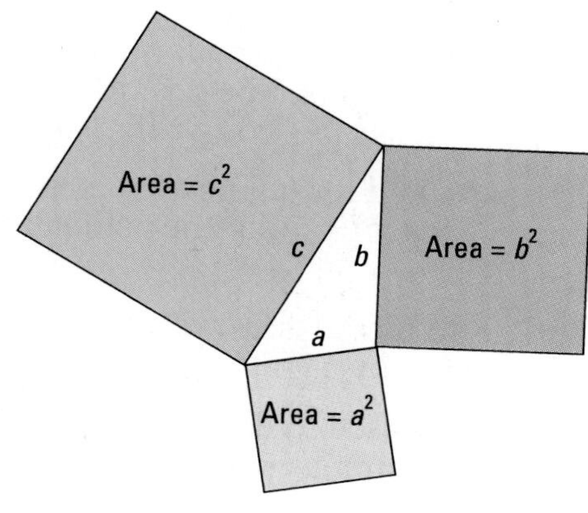

a. **(i)** Lay three of the squares on the table to form a triangle, as shown above. Refer to the longest side as c and the shorter sides as a and b.

(ii) Add the areas of the two smaller squares, and write down the sum. Is the sum $(a^2 + b^2)$ equal to the area of the largest square (c^2)? If not, is the sum larger or smaller?

(iii) Use a protractor to measure the angle across from the longest side. Does it have the largest measure? Is it a right angle? If not, is it acute or obtuse?

b. Repeat this procedure for at least ten different triangles. Record your information in a table, and look for patterns in the data.

c. Write a brief report about what you have learned about triangles, their largest angles, and the lengths of their sides.

6 Figurate Numbers

Some numbers are called *figurate numbers* because they can be easily represented geometrically. Pictured below are the first four triangular numbers, the first four square numbers, and the first four pentagonal numbers.

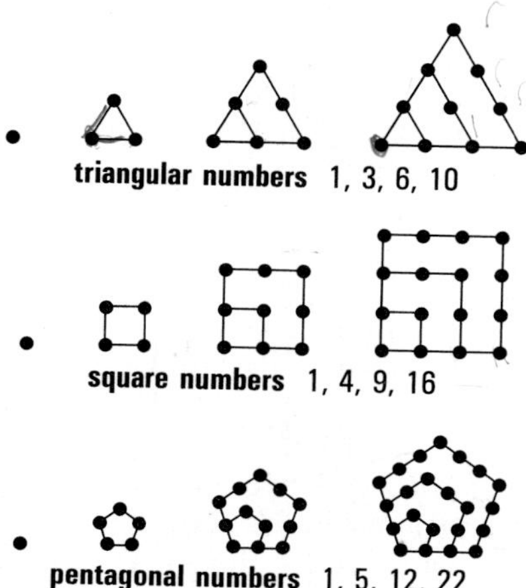

a. Draw a picture of the 5th triangular, 5th square, and 5th pentagonal numbers.

b. Based on the patterns in number of dots on the side of each figure, find the 10th triangular number, the 10th square number, and the 10th pentagonal number.

c. Make a poster or write a report about figurate numbers for your classroom.

SUMMARY

A variable is a letter or other symbol that can be replaced by (or represent) any element from a set, called its domain. Some common domains are the sets of whole numbers, integers, real numbers, and positive real numbers.

Two important operations with sets are union, denoted by the symbol $\cup$, and intersection, denoted by $\cap$. The union of two sets consists of all elements in the first set or the second set (or both). The intersection of two sets consists of those elements common to both sets.

Four uses of variables are described in this chapter. Variables are the language for translating some English expressions and sentences involving numbers into algebraic expressions and sentences. In algebraic expressions, the rules for order of operations are followed: work in parentheses or other grouping symbols first, then do powers, then do multiplications or divisions from left to right, and then additions or subtractions from left to right. Scientific calculators and computer languages usually follow the same rules.

Variables may represent numbers or quantities in formulas like $A = lw$. Some formulas involve square roots of numbers. The positive square root of a positive number r is written $\sqrt{r}$.

Variables may stand for unknowns, as in the open sentence $x + 5 = 70$ or the inequality $3y - 2 > 8$.

The solutions to inequalities with one variable are often pictured by a graph on a number line.

Variables may be used to describe patterns. Among the most famous patterns is the Pythagorean Theorem: in a right triangle with legs a and b and hypotenuse c, $a^2 + b^2 = c^2$. General patterns can sometimes be found from studying specific instances or arranging information in tables.

VOCABULARY

Below are the most important terms and phrases for this chapter. You should be able to give a general description and a specific example of each.

Lesson 1-1
variable
sentence
equation, inequality
$=, \neq, <, \leq, \approx, >, \geq$
open sentence, solution

Lesson 1-2
set, element, { . . . }, equal sets
whole numbers, integers, real numbers
domain, solution set
interval, endpoints
open interval, closed interval
discrete

Lesson 1-3
intersection, $\cap$
union, $\cup$
Venn diagram
empty set, null set, ø, { }

Lesson 1-4
numerical expression
algebraic expression
evaluating an expression
order of operations

Lesson 1-5
formula
in terms of, depends on

Lesson 1-6
square root, $\sqrt{}$, radical sign
perfect squares
Square of the Square Root Property

Lesson 1-7
pattern, instance
counterexample

Lesson 1-8
leg, hypotenuse
Pythagorean Theorem

 CHAPTER ONE

PROGRESS SELF-TEST

Take this test as you would take a test in class. You will need a ruler and calculator. Then check your work with the solutions in the Selected Answers section in the back of the book.

In 1–5, evaluate each expression.

1. $2(a + 3b)$, when $a = 3$ and $b = 5$

2. $5 \cdot 6^n$, when $n = 4$

3. $\frac{p + t^2}{p - t}$, when $p = 5$ and $t = 2$

4. $(\sqrt{50})^2$

5. 10 * y ^ 2 + 5, when $y = 3$

6. Find the value of $3\sqrt{42}$ rounded to the nearest tenth.

7. Let $M = \{2, 4, 6, 8, 10, 12, 14, 16\}$ and $N = \{3, 6, 9, 12, 15\}$. List the members of each set.

a. $M \cup N$ **b.** $M \cap N$

8. A road has a 25 mph speed limit. A person is driving at S mph and is speeding. Express the possible values of S with an inequality.

9. *Multiple choice.* t is the number of towels at a swimming pool. Which is the most appropriate domain for t?
(a) set of whole numbers
(b) set of integers
(c) set of real numbers

10. Which of the numbers 2, 5, and 8 make the open sentence $4y + 7 = 2y + 23$ true?

11. Give three values for x that make $3x < 24$ true.

12. The formula $C = 23(n - 1) + 29$ gives the cost of first-class postage in 1993. In the formula, C is the cost in cents and n is the weight of the mail rounded up to the nearest ounce. What does it cost to mail a letter weighing 3.2 ounces?

13. The area of a circle is given by the formula $A = \pi r^2$. To the nearest square meter, what is the area of a circle with radius 3 meters? (Use $\pi \approx 3.14159$.)

14. *True or false.* $\sqrt{100} + \sqrt{36} = \sqrt{136}$. Explain your reasoning.

15. Let C = the set of original states in the U.S. How many elements does C have?

16. One ticket costs \$3.50. Four tickets cost \$3.50 · 4. Ten tickets cost \$3.50 · 10. Describe the pattern using one variable.

17. Write three instances of the following pattern.

$$\frac{a}{5} - \frac{b}{5} = \frac{a - b}{5}$$

18. Find a formula relating x and y in the table below.

x	y
1	8
2	16
3	24
4	32

CHAPTER ONE

PROGRESS SELF-TEST

19. On a number line, graph the solution set for $n < 8$, where the domain is the set of whole numbers.

20. a. Write an inequality to describe the graph below.

b. Describe a real world situation that fits this graph.

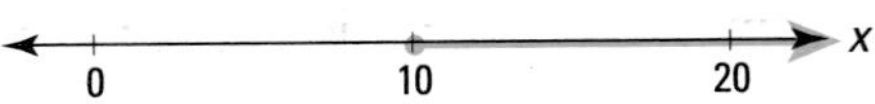

21. Graph $x > 5$ or $x \leq -3$ on a number line using the set of real numbers as the domain.

22. Sir Gawain plans to use a ladder to get to the top of the wall of a castle. The wall is 45 ft tall and is protected by a moat that is 10 ft wide. How long must the ladder be? Round to the nearest tenth.

23. An acre is 4840 yd^2, so a 40-acre field contains 193,600 yd^2. If a 40-acre field is in the shape of a square, how long is each side?

24. The "size" of a television screen is described by the length of its diagonal. What is the size of the TV screen below?

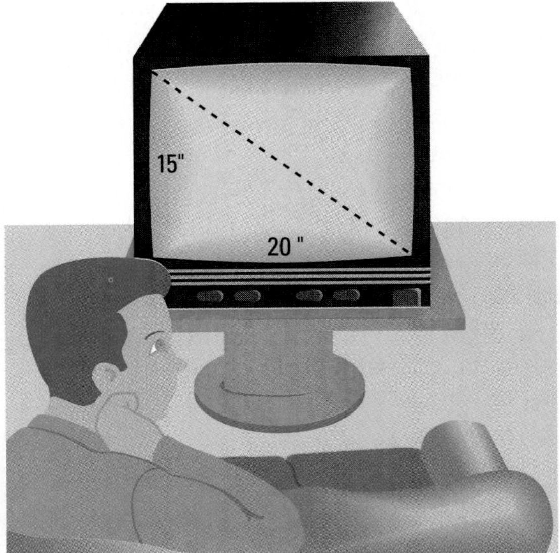

After taking and correcting the Self-Test, you may want to make a list of the problems you got wrong. Then write down what you need to study most. If you can, try to explain your most frequent or common mistakes. Use what you write to help you study and review the chapter.

CHAPTER REVIEW

Questions on SPUR Objectives

SPUR stands for **S**kills, **P**roperties, **U**ses, and **R**epresentations. The Chapter Review questions are grouped according to the SPUR Objectives for this chapter.

SKILLS DEAL WITH THE PROCEDURES USED TO GET ANSWERS.

Objective A: *Find solutions to open sentences using trial and error.* *(Lesson 1-1)*

1. Which of the numbers 3, 4, or 5 is a solution to $2x + 13 = 3x + 9$?

2. Using {1, 4, 7} as a domain, give the solution set of $7x - 13 < 2x$.

3. Find three values for x so that $-5 \leq x \leq -3$.

4. Give three values for y so that $2y \geq 150$.

In 5 and 6, find all solutions.

5. $x^2 = 9$ **6.** $x^2 = 90$

Objective B: *Find unions and intersections.* *(Lesson 1-3)*

7. Let $A = \{11, 15, 19, 23, 25\}$, $B = \{10, 15, 20, 25, 30\}$.

a. Find $A \cap B$. **b.** Find $A \cup B$.

8. Suppose $C = \{2, 8, 9\}$, $D = \{4, 8, 12\}$, and $E = \{6, 8, 9\}$.

a. Find $(C \cup D) \cap E$. **b.** Find $C \cup (D \cap E)$.

9. Let W = the set of whole numbers and $X = \{-1, 0, 1, 2\}$. Describe each set.

a. $W \cup X$ **b.** $W \cap X$

10. Let O = the set of odd whole numbers, and P = the set of prime numbers. List the five smallest numbers in each set.

a. $O \cap P$ **b.** $O \cup P$

Objective C: *Evaluate numerical and algebraic expressions.* *(Lessons 1-4, 1-6)*

In 11–13, evaluate the expression.

11. a. $3 - \frac{2}{5} + 6$ **b.** $(3 - 2)(5 + 6)$

12. $-35 + 5 \cdot 2$

13. (3 + 4 * 5) ^ 2

In 14–18, evaluate the expression for the given values of the variables.

14. $-2p$ when $p = 3.5$

15. $4x^2$ when $x = 12$

16. $4(p - q)$ when $p = 13.8$ and $q = 5.4$

17. $5(M + N)$ when $M = \frac{2}{5}$ and $N = \frac{1}{5}$

18. $\left(\frac{n}{4}\right)^3$ when $n = 36$

Objective D: *Evaluate square roots with and without a calculator.* *(Lesson 1-6)*

19. $\sqrt{81}$

20. $-\sqrt{49}$

21. $3\sqrt{100}$

22. $\sqrt{36} + \sqrt{64}$

23. *True or false.* $\sqrt{25} + \sqrt{4} = \sqrt{29}$. Explain your answer.

24. Fill in the blank with one of >, =, or <. $\sqrt{144} \cdot \sqrt{9}$ __?__ 36. Explain how you got your answer.

25. $\sqrt{20}$ is between which two consecutive whole numbers?

26. Find two consecutive whole numbers a and b such that $a < \sqrt{8} < b$.

In 27 and 28, approximate the value to the nearest thousandth.

27. $\sqrt{199}$ **28.** $\sqrt{200^2 + 300^2}$

PROPERTIES DEAL WITH THE PRINCIPLES BEHIND THE MATHEMATICS.

Objective E: *Read and interpret set language and notation.* *(Lessons 1-2, 1-3)*

29. If S = the set of states in the United States, how many members does S have?

30. Let B = the solution set to $-2 < n < 4$, where n is an integer. List the elements of B.

31. a. What does the symbol ø represent?

b. Give an example of a set equal to ø.

32. *True or false.* The number -3 is an element of the set of whole numbers.

Objective F: *Use the Square of a Square Root Property.* *(Lesson 1-6)*

In 33–35, evaluate without using a calculator.

33. $\sqrt{7}\cdot\sqrt{7}$ **34.** $8(\sqrt{15}\cdot\sqrt{15})$ **35.** $(\sqrt{39})^2$

36. Find the area of a square with sides of length $\sqrt{6}$.

Objective G: *Give instances or counterexamples of patterns.* *(Lessons 1-7, 1-9)*

In 37 and 38, a pattern is given. Write three instances of the pattern.

37. $a + a = 2a$ **38.** $8(x + y) = 8x + 8y$

39. A hardware store finds that the number of hammers h and the number of wrenches w that it sells are related by the formula $w = 4.5h$. Give two instances of this pattern.

40. Hal simplified the following few fractions. He used the wrong method, but got the right answers!

$$\frac{16}{64} = \frac{1\not{6}}{\not{6}4} = \frac{1}{4} \qquad \frac{22}{22} = \frac{2\not{2}}{\not{2}2} = \frac{2}{2} = \frac{1}{1}$$

$$\frac{19}{95} = \frac{1\not{9}}{\not{9}5} = \frac{1}{5} \qquad \frac{49}{98} = \frac{4\not{9}}{\not{9}8} = \frac{4}{8} = \frac{1}{2}$$

Give a counterexample to show that Hal's method will not always work.

Objective H: *Use variables to describe patterns in instances or tables.* *(Lessons 1-7, 1-9)*

In 41 and 42, three instances of a pattern are given. Use one or two variables to describe the pattern.

41. One sheep has 1 · 4 legs, 80 sheep have 80 · 4 legs, and six sheep have 6 · 4 legs.

42. Two shirts and three pair of jeans cost 2 · \$19 + 3 · \$27.
Four shirts and one pair of jeans cost 4 · \$19 + 1 · \$27.
Two shirts and five pair of jeans cost 2 · \$19 + 5 · \$27.

43. Since 1960 the population of Lincoln, Nebraska, has been increasing by about 2100 people per year. Assume that this rate continues indefinitely.

a. How much will the population increase in a decade?

b. How much will the population increase in y years?

In 44 and 45, find a formula relating x and y.

44.

x	y
1	11
2	22
3	33
4	44

45.

x	y
1	4
2	16
3	64
4	256

46. The diagram below shows frames for photographs.

a. Complete the table.

b. Find a formula for t in terms of n.

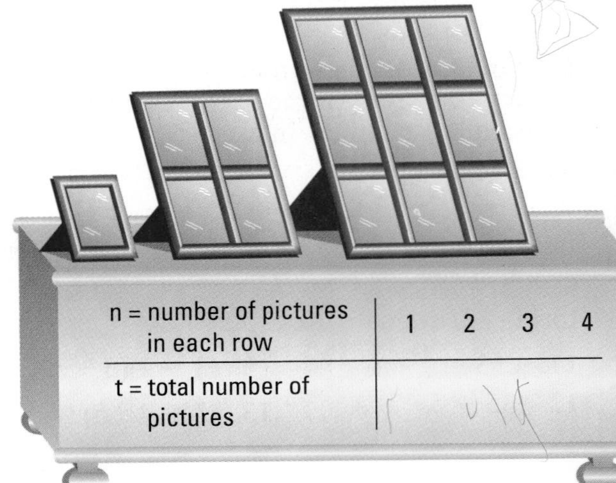

n = number of pictures in each row	1	2	3	4
t = total number of pictures				

CHAPTER ONE

USES DEAL WITH APPLICATIONS OF MATHEMATICS IN REAL SITUATIONS.

Objective I: *In real situations, choose a reasonable domain for a variable.* (Lesson 1-2)

In 47–50, *multiple choice.* Choose a domain for the variable from these sets.

(a) set of whole numbers
(b) set of integers
(c) set of positive real numbers
(d) set of real numbers

47. d, the distance that a jogger runs in a day

48. the weight w of a molecule

49. P, the altitude of a point on the surface of Earth

50. roast beef sandwiches for n people

Objective J: *Evaluate formulas in real situations.* (Lesson 1-5)

51. The percent discount p on an item is given by the formula, $p = 100\left(1 - \frac{n}{g}\right)$, where g is the original price and n is the new price. Find the percent discount on a pair of jeans whose price is reduced from \$20 to \$15.

52. If a can of orange juice is h centimeters high and has a bottom radius of r centimeters, then the volume equals $\pi r^2 h$ cubic centimeters. What is the volume of a can 12 cm high with a radius of 6 cm? (Use $\pi \approx 3.14159$.)

In 53 and 54, use this information. The cost c of carpeting a room is given by $c = p\left(\frac{\ell w}{9}\right)$, where p is the price of the carpeting per square yard and ℓ and w are the length and width of the room in feet.

53. Find the cost of carpeting a 12′ by 15′ room with carpeting that sells for \$19.95 per square yard.

54. At \$8.99 per square yard, what is the cost of carpeting an 8′ by 30′ deck?

Objective K: *Apply the Pythagorean Theorem to solve problems in real situations.* (Lesson 1-8)

55. Ben uses a guy wire to support a young tree. He attaches it to a point 6 ft up the tree trunk and stretches the wire to a stake 8 ft from the tree trunk. How long must the wire be?

56. Megan is flying a kite. The kite is 20 m from her along the ground and 10 m above her. How long is the string between Megan and the kite?

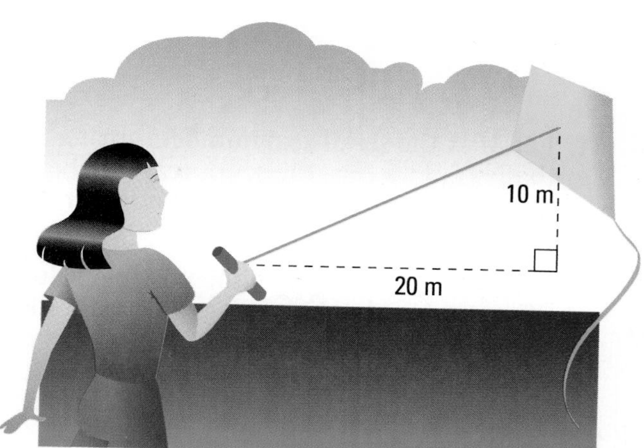

57. A builder designing a house needs to consider the cost of the roof. How much longer is a rafter in Plan *B* than one in Plan *A?*

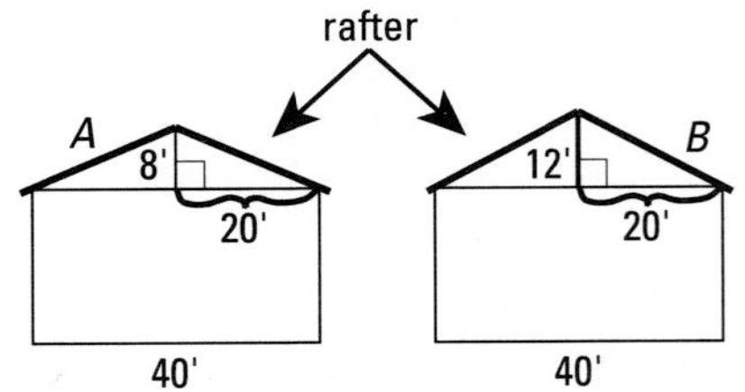

58. A rectangular field is 300 feet long and 100 feet wide.

a. To the nearest foot, how far will you walk if you cut across the field diagonally?

b. How much shorter is this than walking along the edges of the field?

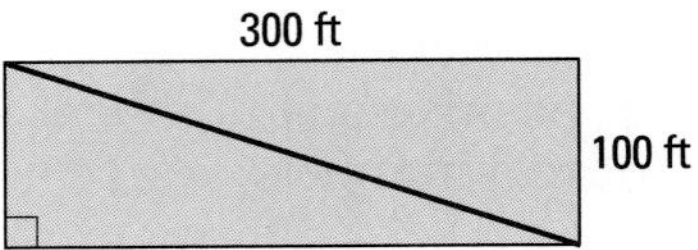

REPRESENTATIONS DEAL WITH PICTURES, GRAPHS, OR OBJECTS THAT ILLUSTRATE CONCEPTS.

Objective L: *Draw and interpret graphs of solution sets to inequalities.* *(Lessons 1-2, 1-3)*

59. Kim bought stamps for less than $25. On a number line, graph what she might have spent.

60. Consider the sign pictured below.

a. Express the interval of the legal speeds as an inequality using s to represent speed.

b. Graph all legal speeds.

61. Let M = the set of real numbers greater than -10, and N = the set of real numbers less than 6. Graph the set on a number line.

a. $M \cap N$ **b.** $M \cup N$

62. If $x \geq 9$ or $x \leq 4$, graph the possible values of x on a number line using as domain the set of real numbers.

63. Graph the solution set to $y \geq 19$, if y has the given domain.

a. set of real numbers **b.** set of integers

In 64 and 65, *multiple choice.* Choose from the three graphs below.

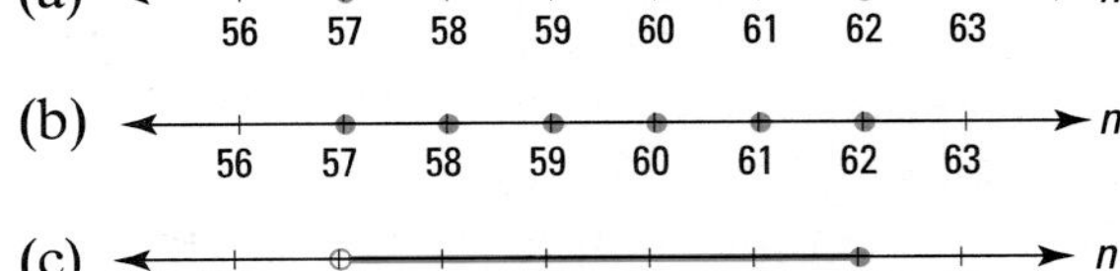

64. Which could be a graph of the solutions to $57 < n \leq 62$ if n is a real number?

65. Which represents the statement "There are from 57 to 62 students with green eyes in the school"?

In 66 and 67, a graph is given.

a. Write an inequality to describe the graph.

b. Describe a real-world situation that fits this graph.

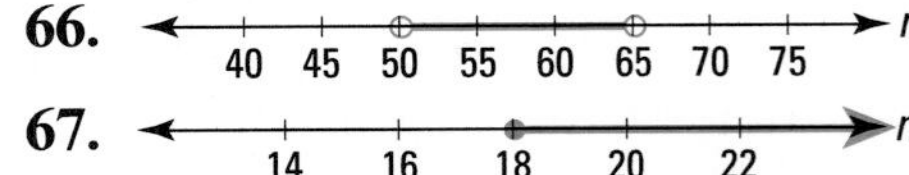

REFRESHER

Chapter 2, which discusses multiplication in algebra, assumes that you have mastered certain objectives in your previous mathematics courses. Use these questions to check your mastery.

A. Multiply any positive numbers or quantities.

1. $4.7 \cdot 3.21$
2. $0.04 \cdot 312$
3. $666 \cdot 0.00001$
4. $.17 \cdot .02$
5. $\frac{2}{3} \cdot 30$
6. $\frac{5}{2} \cdot 11$
7. $\frac{2}{9} \cdot \frac{3}{4}$
8. $\frac{1}{4} \cdot \frac{1}{3} \cdot \frac{1}{2}$
9. $1\frac{1}{4} \cdot 2\frac{1}{8}$
10. $30\% \cdot 120$
11. $3\% \cdot \$6000$
12. $5.25\% \cdot 1500$

B. Apply multiplication in rate-factor situations.

13. A school cafeteria sells 150 cartons of milk per day. How many cartons does it sell in a 180-day school year?

14. A sales clerk makes $6.50 an hour and works 37.5 hours a week. How much does the person earn per week?

15. The manufacturer claims that a new car gets 33.5 miles per gallon for highway driving. The gas tank holds 12 gallons. How far can the car travel on the highway on a full tank of gas?

16. If a plane travels at 625 miles per hour for $2\frac{1}{2}$ hours, how far does it travel?

C. Multiply positive and negative integers.

17. $3 \cdot -2$
18. $11 \cdot -11$
19. $-6 \cdot 4$
20. $-5 \cdot -5$
21. $0 \cdot -1$
22. $-14 \cdot 130$
23. $-60 \cdot -59$
24. $-3 \cdot -2 \cdot -1 \cdot -1$

D. Solve equations of the form $ax = b$, when a and b are positive integers.

25. $3x = 12$
26. $5y = 110$
27. $10z = 5$
28. $1 = 9w$
29. $6 = 50a$
30. $b \cdot 21 = 14$
31. $8c = 4$
32. $7 = 2d$

E. Determine the area of a rectangle given its dimensions.

33. length 15″, width 12″
34. length 4.5 cm, width 3.2 cm
35. length 2 m, width 4 m
36. length 100 ft, width 82 ft

F. Determine the volume of a rectangular solid given its dimensions.

37. length 8 cm, width 4 cm, height 3 cm
38. length 12 in., width 9 in., height $1\frac{1}{2}$ in.
39. length $\frac{1}{2}$ ft, width $\frac{1}{2}$ ft, height $\frac{1}{2}$ ft
40. length 60 m, width 50 m, height 10 m

CHAPTER 2

MULTIPLICATION IN ALGEBRA

Our galaxy, the Milky Way, looks somewhat like the Andromeda Galaxy pictured here. Ever since it was discovered that the other planets were bodies like Earth, rotating around the sun, people have wondered: Is there life on some other planet in our galaxy?

One formula for computing the number of planets N currently supporting intelligent life in our galaxy involves the product of seven numbers.

$$N = T \cdot P \cdot E \cdot L \cdot I \cdot C \cdot A$$

Of course you need to know what the letters in the formula represent. Here are their meanings and an estimated value for each from *Space,* a novel by James Michener.

T = *Total* number of stars in our galaxy = 400,000,000,000

P = fraction of stars with a *Planetary* system = $\frac{1}{4}$

E = fraction of planetary systems with a planet that has an *Ecology* able to sustain life = $\frac{1}{2}$

L = fraction of planets able to sustain life on which *Life* actually developed = $\frac{9}{10}$

I = fraction of planets with *Intelligent* life = $\frac{1}{10}$

C = fraction of planets with intelligent life that could *Communicate* outwardly = $\frac{1}{3}$

A = fraction of planets with communicating life which is *Alive* now = $\frac{1}{100{,}000{,}000}$

This yields a value of 15 for N, meaning that there may be 15 planets in our galaxy with intelligent life. However, no one knows the values of the variables in the formula exactly, and people disagree on them. Changing them can change the value of N by a great deal. For instance, if $P = \frac{1}{10}$ rather than $\frac{1}{4}$, then $N = 6$ instead of 15. This is only one of many situations involving multiplication you will see in this chapter.

LESSON

2-1

Areas, Arrays, and Volumes

An early skyscraper. *The Pyramid of Kukulan is located in Chichen Itza, Mexico. Built by the Mayas before 800 A.D., the pyramid rises to a height of 24.1 m. What dimensions are needed to find the pyramid's base area and volume?*

Area of Rectangles and Rectangular Arrays

The operation of multiplication can be pictured using area. The area of any rectangle is the product of its two dimensions. The rectangles below are actual size.

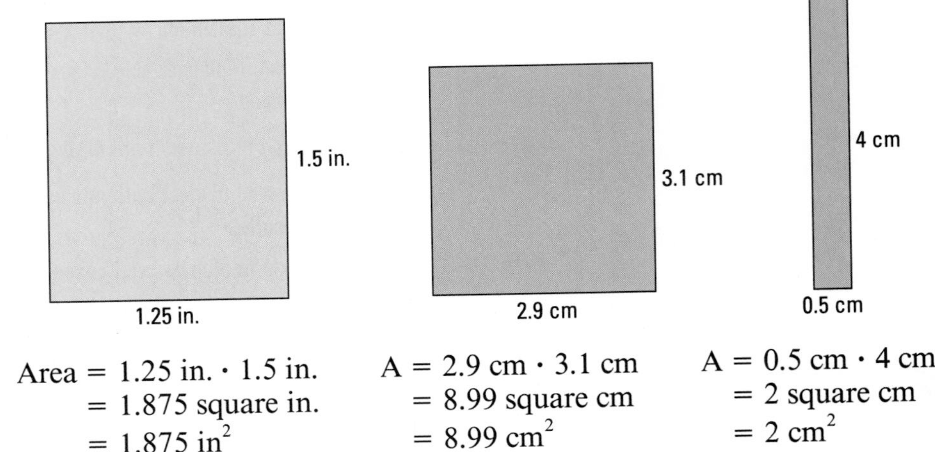

Area = 1.25 in. · 1.5 in.
= 1.875 square in.
= 1.875 in^2

A = 2.9 cm · 3.1 cm
= 8.99 square cm
= 8.99 cm^2

A = 0.5 cm · 4 cm
= 2 square cm
= 2 cm^2

The areas are in *square units* because the units are multiplied, as well as the numbers.

A **model** for an operation is a general pattern that includes many of the uses of the operation. The examples above are instances of the *Area Model for Multiplication.*

Area Model for Multiplication
The area A of a rectangle with length ℓ and width w is ℓw.

Rectangle I below left has been rotated to give rectangle II. The two rectangles have the same dimensions. So they must have the same area. Thus the area formula could be written $A = w\ell$ as well as $A = \ell w$.

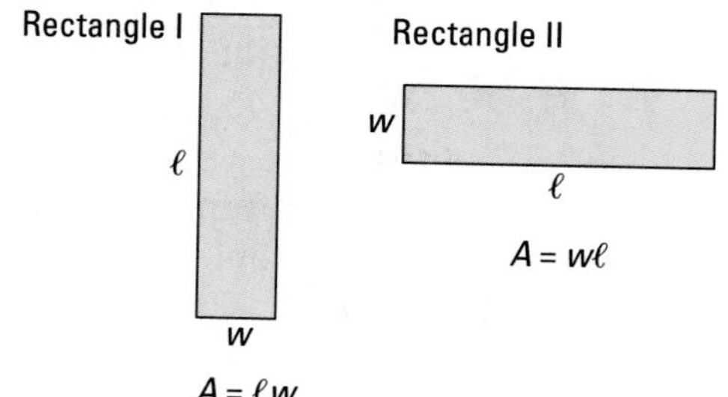

In this way, the area model pictures a general property named by François Servois (Fron swah Sayr vwah) in 1814. He used the French word *commutatif,* which means "switchable." The English name is "commutative."

Commutative Property of Multiplication
For any real numbers a and b, $ab = ba$.

In algebra and geometry, areas of many shapes can be found by working with rectangles.

Example 1

A driveway to a house has the shape shown at the right. All angles are right angles. Find the area of the driveway.

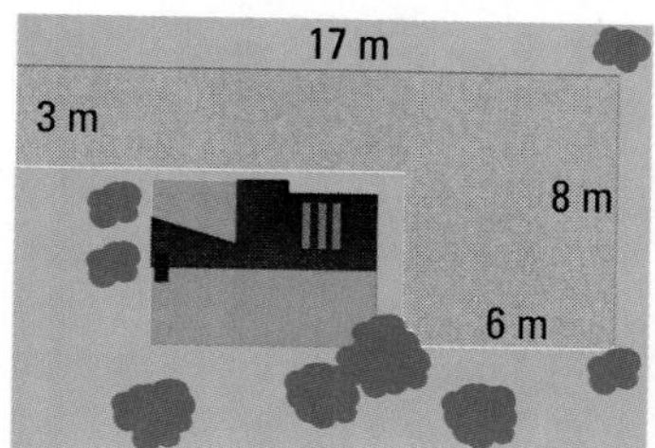

Solution

There is no simple formula for an irregular shape like this. But the shape can be split into two rectangles. One way to split the figure is shown below. This splits the right side into lengths of 3 m and 5 m. Call the smaller areas A_1 and A_2. To find their areas, multiply the dimensions of the rectangles.

$A_1 = 3\text{ m} \cdot 17\text{ m} = 51\text{ m}^2$
$A_2 = 5\text{ m} \cdot 6\text{ m} = 30\text{ m}^2$

The total area is the sum of the areas of these parts.

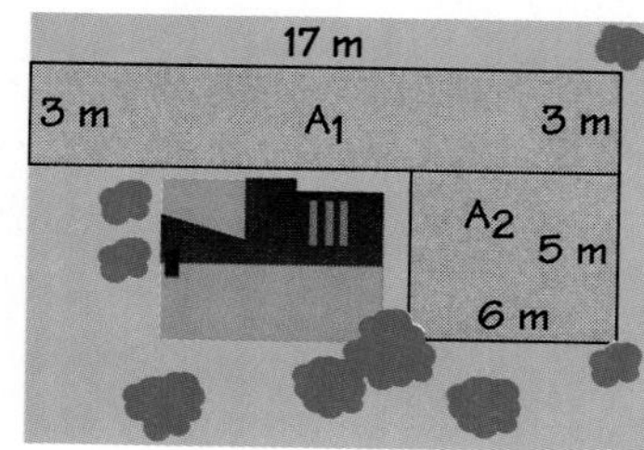

$$\begin{aligned}\text{Area of driveway} &= A_1 + A_2\\ &= 51\text{ m}^2 + 30\text{ m}^2\\ &= 81\text{ m}^2\end{aligned}$$

A set of objects that can be counted is a **discrete set.** The discrete form of a rectangle is a **rectangular array.** The stars at the left form a rectangular array with 5 rows and 9 columns, or a 5-by-9 array. The numbers 5 and 9 are the *dimensions* of the array. The total number of stars is 45, the product of the dimensions.

The general pattern is the discrete form of the Area Model for Multiplication.

> **Area Model for Multiplication (discrete form)**
> The number of elements in a rectangular array with r rows and c columns is rc.

Volume of Rectangular Solids

The area model can be extended to three-dimensional figures. The **volume** of a **rectangular solid** is the product of its three dimensions.

$$\text{Volume} = \text{length} \cdot \text{width} \cdot \text{height}$$
$$V = \ell wh$$

Example 2

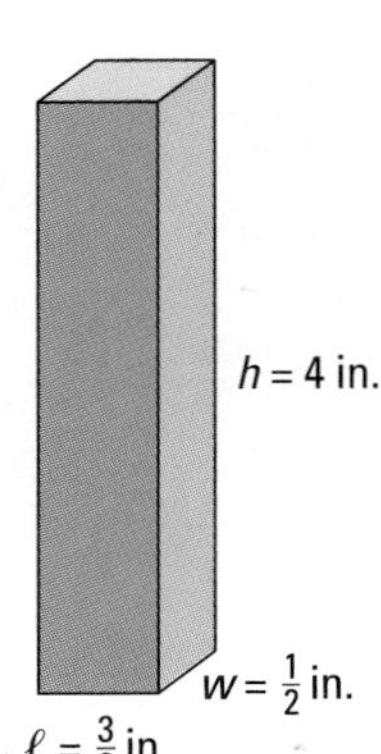

Find the volume of the rectangular solid pictured at the left.

Solution

Substitute the given dimensions for ℓ, w, and h in the volume formula.

$$V = \frac{3}{8} \text{ in.} \cdot \frac{1}{2} \text{ in.} \cdot 4 \text{ in.}$$

Recall that the product of fractions is the product of the numerators divided by the product of the denominators. So $\frac{3}{8} \cdot \frac{1}{2} = \frac{3}{16}$, and

$$\begin{aligned} V &= \frac{3}{16} \text{ in}^2 \cdot 4 \text{ in.} \\ &= \frac{12}{16} \text{ in}^3 \\ &= \frac{3}{4} \text{ in}^3 \end{aligned}$$

(Volume is measured in cubic units; in^3 means cubic inches.)

Check

Rewrite the dimensions as decimals.

$$\begin{aligned} \frac{3}{8} \cdot \frac{1}{2} \cdot 4 &= 0.375 \cdot 0.5 \cdot 4 \\ &= 0.1875 \cdot 4 \\ &= 0.75 \end{aligned}$$

Because $\frac{3}{4} = 0.75$, the answer checks.

The area and volume models for multiplication date back to the time of the ancient Greeks. Pythagoras himself thought of multiplication this way and used area in the proof of his famous theorem. That is how the names "x squared" and "x cubed" became associated with x^2 and x^3.

In Example 2, we did not have to follow order of operations and multiply from left to right. You could first multiply $\frac{1}{2}$ in. by 4 in. This gives 2 in^2, and 2 $\text{in}^2 \cdot \frac{3}{8}$ in. $= \frac{3}{4} \text{in}^3$. This illustrates that

$$(\ell \cdot w) \cdot h = \ell \cdot (w \cdot h).$$

Doing left multiplication first = Doing right multiplication first.

The general property, true for all numbers, is called the *Associative Property of Multiplication.* It was given its name in 1835 by the Irish mathematician Sir William Rowan Hamilton.

Associative Property of Multiplication
For any real numbers a, b, and c, $(ab)c = a(bc)$.

Enormous rectangular solids. *The 110-story, 1,350-foot-high World Trade Center, located along the Hudson River in New York City, is the world's largest office complex.*

Both the commutative and associative properties involve changing order. The commutative property says you can change the order of the *numbers* being multiplied. The associative property says you can change the order of the *multiplications* by *regrouping* the numbers. Together these properties let you multiply as many numbers as you please in any order. For instance, in the formula $N = TPELICA$ on page 71, you could multiply the 7 variables in any order, such as $PLICATE$.

In the volume formula $V = \ell wh$, since ℓw is the area of the base, the formula may be expressed as $V = Bh$, where B is the area of the base.

In the next example, the associative property is used to rearrange factors in a multiplication. This allows us to simplify the multiplication of expressions involving variables.

Example 3

What is the volume of a box in which the height is $4xy$ and the area of the base is $5x^2$?

Solution

Volume = Bh

$V = (5x^2)4xy$	
$= 4(5x^2)xy$	Commutative Property of Multiplication
$= (4 \cdot 5)(x^2 \cdot x)y$	Associative Property of Multiplication
$= 20x^3y$	$x^2 \cdot x = x^3$

The volume of the box is $20x^3y$.

Check

Pick values for x and y. See if the formula works for those values. We pick $x = 2$ and $y = 3$. Then $B = 5x^2 = 20$ and $h = 4xy = 24$, so the volume is 480. Now evaluate the answer $20x^3y$ with the same values of x and y.

$$20x^3y = 20 \cdot 2^3 \cdot 3 = 20 \cdot 8 \cdot 3 = 480, \text{ which checks.}$$

When units are not given, as in Example 3, you can assume they are the same—all inches, all centimeters, or all something else. Otherwise, make sure units are the same before multiplying.

Sharper images in space. *The Hubble Space Telescope, launched by NASA in 1990, is a reflecting telescope built as an orbiting observatory. The Hubble telescope produces images about 10 times as sharp and observes objects 50 times as faint as any telescope on Earth. One goal of the telescope is to detect planets around stars other than our sun.*

QUESTIONS

Covering the Reading

1. A formula for computing the number of planets in our galaxy was discussed by a Congressional committee in 1975. Some astronomers from Green Bank, West Virginia, gave for their *low* estimate the following values of the variables: $T = 100{,}000{,}000{,}000$, $P = 0.4$, $E = 1$, $L = 1$, $I = 1$, $C = 0.1$, $A = \frac{1}{100{,}000{,}000}$. With these values, how many planets in our galaxy support intelligent life?

2. State the Area Model for Multiplication.

3. *Multiple choice.* If the length and width of a rectangle are measured in inches, in what unit is the area measured?
(a) inches (b) square inches (c) cubic inches

In 4 and 5, find the area of a rectangle with the given dimensions. Include units in your answer.

4. length 7.2 cm and width 4.3 cm

5. length 8 in. and width y in.

6. Explain how to find the number of dots in the following rectangular array without counting them all.

7. In the figure at the right all angles are right angles. Find its area, and show how you found it.

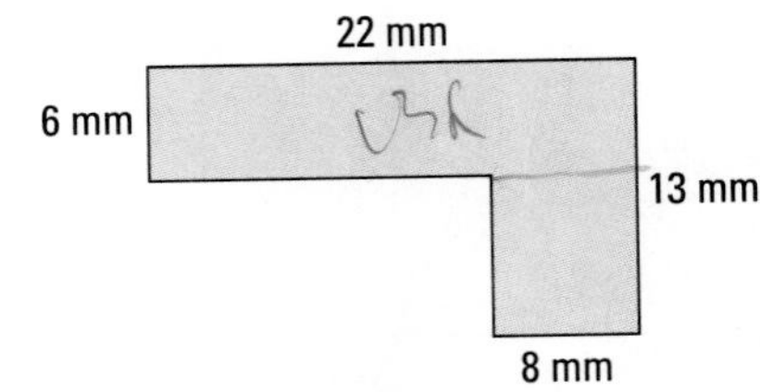

In 8–10, find the volume of the rectangular solid described. Be sure to include units where appropriate.

8. The dimensions are $\frac{1}{4}$ ft by $\frac{2}{3}$ ft by $\frac{1}{8}$ ft.

9. Its length is 3, width is $7x$, and height is y.

10. The area of its base is $9p^2$ and its height is $12p$.

11. Which property guarantees that the two multiplication problems at right have the same answer?

$$\begin{array}{r} 38.2 \\ \times\ 65 \\ \hline \end{array} \qquad \begin{array}{r} 65 \\ \times\ 38.2 \\ \hline \end{array}$$

12. Describe the differences and similarities between the Commutative and Associative Properties of Multiplication.

Applying the Mathematics

13. In each situation, tell whether "followed by" is a commutative operation.

a. Putting on your socks followed by putting on your shoes

b. Putting cream in your coffee followed by putting sugar in your coffee

c. Writing on the blackboard followed by erasing the blackboard

d. Make up an example of your own. Tell whether it is commutative.

In 14 and 15, use the associative and commutative properties to do the multiplication in your head.

14. $25 \cdot x \cdot 4 \cdot 341$

15. $(2 \cdot 3x)(8x \cdot 5)$

16. The O'Learys' garden is a rectangle 45 ft by 60 ft.

a. What is the area of their garden?

b. If they put a fence around the garden, how long will the fence be?

17. Refer to the rectangle at the right.

a. Write an expression for its area.

b. Write an expression for its perimeter.

18. The largest rectangle below is made of six identical smaller ones, each with width a and length b.

a. What is the width of the largest rectangle?

b. What is the length of the largest rectangle?

c. Express the area of the largest rectangle as width · length.

d. Simplify your answer to part **c.**

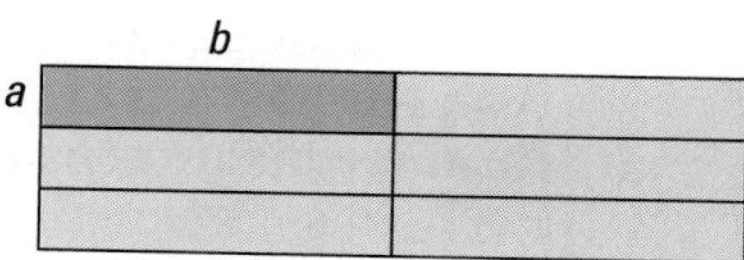

19. In the figure below, all angles are right angles. The unit is meters. Find the area.

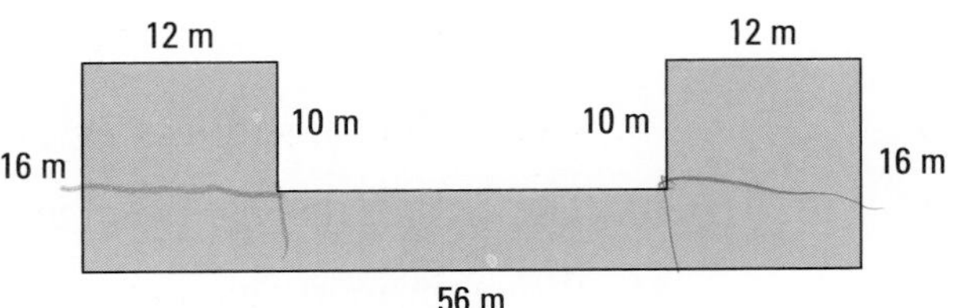

20. What is the volume of a rectangular solid in which the width is x cm, the length is $2x$ cm, and the height is $6x$ cm?

21. How many cubic inches are in a box 5 inches wide, 1 foot long, and 3 inches high?

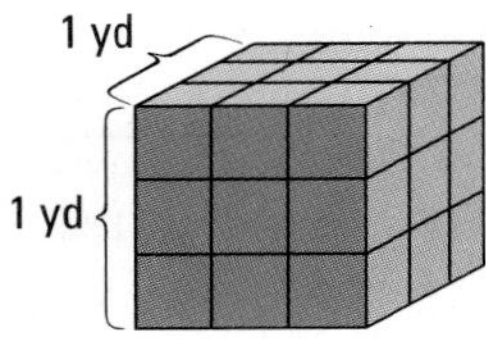

22. One cubic yard is the volume of a cube with each edge of length 1 yard. How many cubic feet equal one cubic yard?

23. In your own words, explain the difference between area and volume. Use an example or problem from this lesson to help illustrate your idea, or make up your own.

Review

24. Carl runs more than three miles every morning.
  a. Write an inequality to describe how many miles he runs.
  b. Graph on a number line. *(Lesson 1-2)*

In 25 and 26, let L = the length of a segment. Write an expression for each of the following. *(Previous course)*

25. twice that length

26. one third that length

27. Rewrite the fraction $\frac{36}{120}$ in lowest terms. *(Previous course)*

28. There are 392 students in a school. Of these, $\frac{3}{8}$ play on an athletic team. How many students play on a team? *(Previous course)*

Exploration

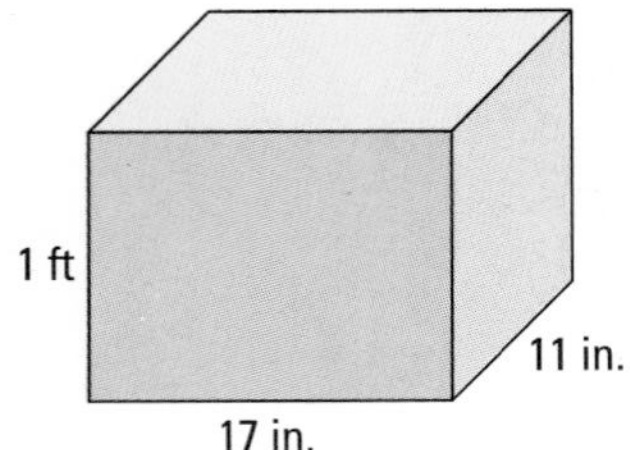

29. **a.** Measure the dimensions of this textbook. At most, how many texts could you fit inside the box at the left? Explain how you got your answer.
  b. Give the dimensions of a box that could be used to pack 24 books. Explain how you would pack the books in the box.

LESSON

2-2

Special Numbers in Multiplication

One singular sensation! *The musical,* A Chorus Line, *is the longest-running play in Broadway history. It ran for about 15 years for a total of 6,137 performances! The play's most popular song, "One," is a special number about a special number.*

The numbers 1 and 0 have special roles in multiplication.

Some Properties of the Number 1

Multiplying a number by the number 1 keeps the identity of that number. So 1 is called the **multiplicative identity.**

> **Multiplicative Identity Property of 1**
> For any real number a, $a \cdot 1 = 1 \cdot a = a$.

If the length of a rectangle is x units and the width is 1 unit, the area is $x \cdot 1 = x$ square units.

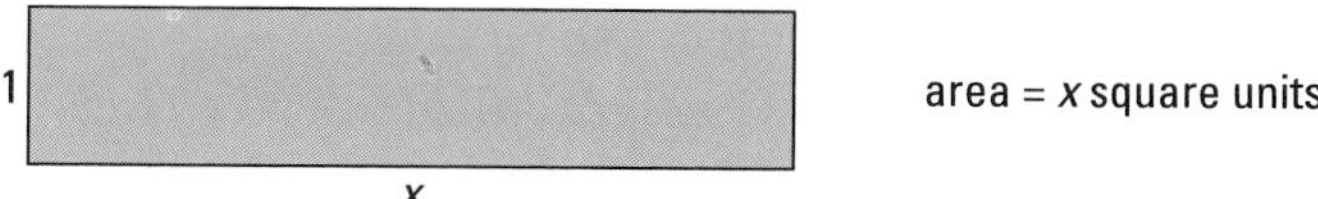

Because the number 1 is so important in multiplication, numbers whose product is 1 are also important. Such numbers are called *reciprocals* or *multiplicative inverses.* The **reciprocal** of a is the number that gives a product of 1 when multiplied by a. For instance, $4 \cdot \frac{1}{4} = 1$. So 4 and $\frac{1}{4}$ are reciprocals. Every number but 0 has a reciprocal.

> **Property of Reciprocals**
> Suppose $a \neq 0$. The reciprocal of a is $\frac{1}{a}$. That is, $a \cdot \frac{1}{a} = \frac{1}{a} \cdot a = 1$

In general, if the length of a rectangle is x units and its area is 1 square unit, its width is $\frac{1}{x}$ unit.

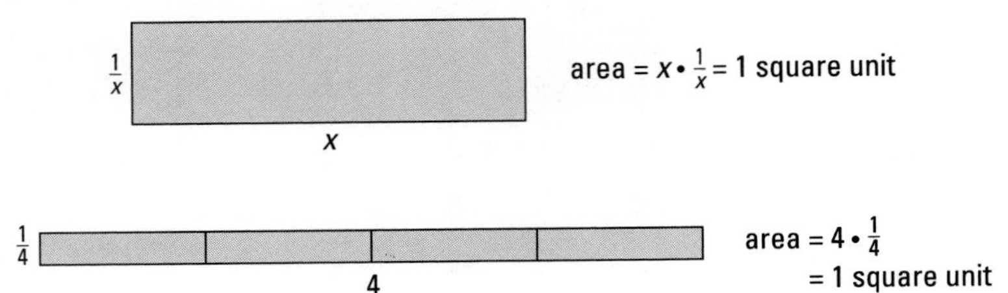

The names of many metric measurement units are based on reciprocals. For instance, the prefix *centi-* means $\frac{1}{100}$. A centimeter is $\frac{1}{100}$ of a meter. So, 100 centimeters equals $100 \cdot \frac{1}{100}$ meter, or 1 meter.

Calculating Reciprocals

Because the reciprocal of a is $\frac{1}{a}$, you can calculate the reciprocal of a number by dividing 1 by the number.

Every scientific and graphics calculator has a reciprocal key, usually [1/x] or [x^{-1}]. To find the reciprocal of the number n, key in: n [1/x]. On some calculators you may need to press [ENTER] to see the result. The answer will be displayed as a decimal.

Example 1

Give the reciprocal of 1.25.

Solution

The reciprocal of 1.25 is $\frac{1}{1.25}$. To find the decimal for $\frac{1}{1.25}$, you can divide 1 by 1.25. A calculator shows that $1 \div 1.25 = 0.8$. Or you can use the reciprocal key. Key in

1.25 [1/x] or 1.25 [x^{-1}] [ENTER].

You will see [0.8].

The reciprocal of 1.25 is 0.8.

You know that $0.8 = \frac{8}{10} = \frac{4}{5}$. So, $\frac{8}{10}$ and $\frac{4}{5}$ are other names for the reciprocal of 1.25. You also know that $1.25 = 1\frac{1}{4} = \frac{5}{4}$. So, the reciprocal of $\frac{5}{4}$ is $\frac{4}{5}$. This is an instance of the following property.

Reciprocal of a Fraction Property

Suppose $a \neq 0$ and $b \neq 0$. The reciprocal of $\frac{a}{b}$ is $\frac{b}{a}$.

Example 2

Find the reciprocal of $\frac{8}{3}$.

Solution 1

Use the Reciprocal of a Fraction Property. The reciprocal of $\frac{8}{3}$ is $\frac{3}{8}$.

Solution 2

Use the Property of Reciprocals stated earlier. Divide 1 by $\frac{8}{3}$. On a calculator you must use parentheses. Key in

1 [÷] [(] 8 [÷] 3 [)] [=].

You should see 0.375 on the display. The reciprocal of $\frac{8}{3}$ is 0.375. You may want to check the above on your calculator.

Properties of the Number 0

The Property of Reciprocals, $a \cdot \frac{1}{a} = 1$, requires that $a \neq 0$ because 0 does not have a reciprocal. If $\frac{1}{0}$ existed, then it would have to be true that $0 \cdot \frac{1}{0} = 1$. But in fact, you know that 0 times any real number is 0. This property also has a name.

Multiplication Property of Zero
For any real number a, $a \cdot 0 = 0 \cdot a = 0$.

The Multiplication Property of Zero can save work in evaluating some otherwise complicated algebraic expressions.

Example 3

Evaluate $(w + 4.7)(2.6 - w)(w + 7.1)$ when $w = -4.7$.

Solution

Substituting gives

$$(-4.7 + 4.7)(2.6 - -4.7)(-4.7 + 7.1).$$

Because $-4.7 + 4.7 = 0$, the product is 0. There is no need to do the computation in the second or third set of parentheses.

QUESTIONS

Covering the Reading

1. The title of this lesson is "Special Numbers in Multiplication." Name the two special numbers that are discussed.

2. What number is the multiplicative identity?

3. Complete the following:
The numbers 18 and $\frac{1}{18}$ are reciprocals because their product is __?__.

4. What is another name for *reciprocal?*

5. Explain in your own words why 0 does not have a reciprocal.

In 6–11, give the reciprocal and check your answer.

6. 10
7. $\frac{1}{9}$
8. $\frac{6}{7}$
9. $\frac{13}{12}$
10. 2.5
11. y

In 12 and 13, find the area of each rectangle.

12.

p by 1

13. 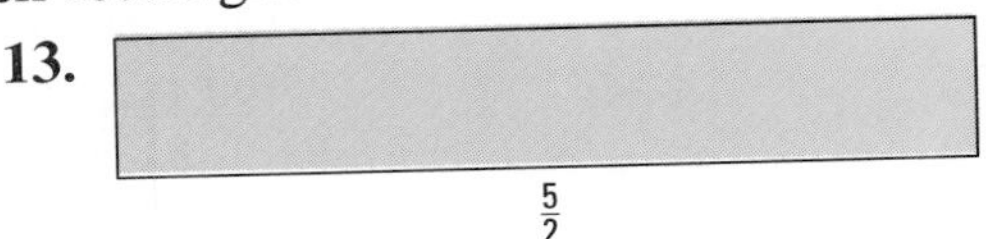

$\frac{5}{2}$ by $\frac{2}{5}$

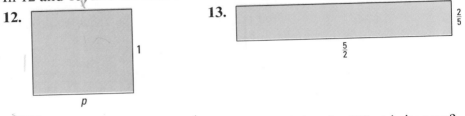

14. a. A rectangle has length $\frac{1}{3}$ unit and width 3 units. What is its area?
b. Draw a picture to justify your answer.

15. Explain how the word *millimeter* is related to the idea of reciprocal.

16. Evaluate $(x + 1)(x + 2)(x + 3)(x + 4)$ when **a.** $x = -3$, **b.** $x = 3$.

What's your best time in the mile run? *In 1994, the men's world record for the mile run on an indoor track was 3:49.78. The men's world record for the mile run on an outdoor track in 1994 was 3:44.39.*

Applying the Mathematics

In 17 and 18, *multiple choice.*

17. Which does *not* equal 1?
(a) $\frac{0.8}{0.8}$
(b) $0.8 \cdot \frac{5}{4}$
(c) $4.1 - 4.1$
(d) $0.9 + 0.1$

18. For which value of n is the expression $(n - 2)(n - 1)(n + 1)(n + 2)$ *not* equal to 0?
(a) 1
(b) 0
(c) -1
(d) -2

In 19–22, there is a pair of numbers. **a.** Tell whether the numbers are or are not reciprocals. **b.** Justify your answer.

19. 200 and 0.005
20. $\frac{1}{4}$ and 0.25
21. 1.5 and $\frac{2}{3}$
22. $\frac{3}{5}$ and $-\frac{3}{5}$

23. If an indoor track has length $\frac{1}{4}$ mile, you must run around it 4 times to run a mile. How many times must you run around a track of length $\frac{2}{5}$ mile in order to run a mile?

24. Suppose $D = (w + 2)(w - 3)(w + 6)$. For what three values of w will D have a value of 0?

25. Compute in your head: $9875 \cdot (1676 - 1675 - 1)$

In 26 and 27, give the reciprocal. Assume that x, p, and $q \neq 0$.

26. $\frac{x}{4}$

27. $\frac{p}{q}$

28. To attract people to a grocery store, the manager decided to sell milk at the same price he paid for it, thus making a 0¢ profit for each gallon sold. How much profit is made when b gallons of milk are sold?

Review

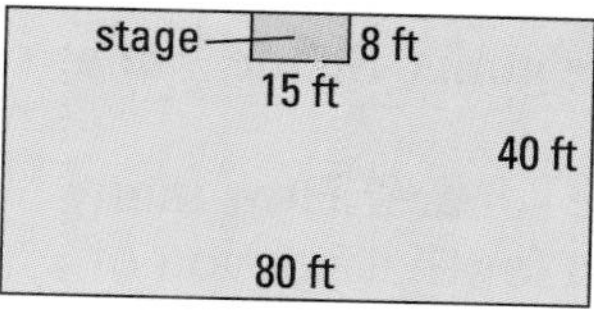

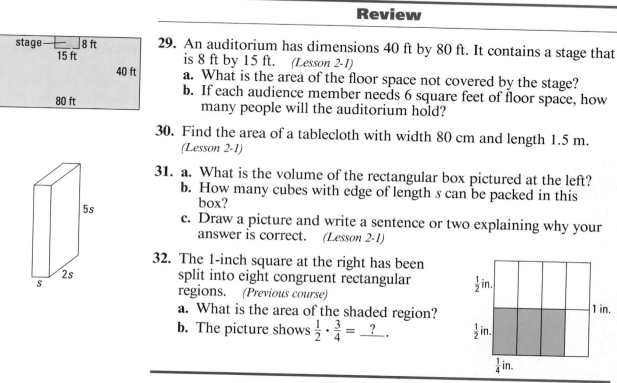

29. An auditorium has dimensions 40 ft by 80 ft. It contains a stage that is 8 ft by 15 ft. *(Lesson 2-1)*
 a. What is the area of the floor space not covered by the stage?
 b. If each audience member needs 6 square feet of floor space, how many people will the auditorium hold?

30. Find the area of a tablecloth with width 80 cm and length 1.5 m. *(Lesson 2-1)*

31. a. What is the volume of the rectangular box pictured at the left?
 b. How many cubes with edge of length s can be packed in this box?
 c. Draw a picture and write a sentence or two explaining why your answer is correct. *(Lesson 2-1)*

32. The 1-inch square at the right has been split into eight congruent rectangular regions. *(Previous course)*
 a. What is the area of the shaded region?
 b. The picture shows $\frac{1}{2} \cdot \frac{3}{4} = \underline{\ ?\ }$.

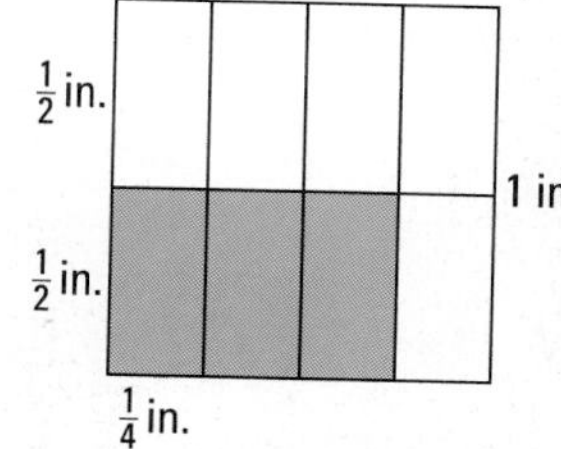

Exploration

33. The expression *numero uno* is Spanish for "number one." Translate *one* into three other languages.

34. Find 3 positive numbers a, b, and c such that $a^b \cdot c^a = abca$, where $abca$ stands for a 4-digit number.

Introducing Lesson 2-3

Multiplying Algebraic Fractions

IN-CLASS

ACTIVITY

You will need paper, colored pens or pencils, and a ruler to do this activity. Work with a partner or in a small group.

1 **a.** Draw a square or use a square piece of paper. Assume each side of the square is 1 unit. Divide the square vertically into fourths as shown. Use one color to shade $\frac{3}{4}$ of it.

b. Divide the same square horizontally into eighths. Shade $\frac{1}{8}$ in a second color.

c. How many little rectangles are there in the square? How many are shaded in both colors? Write a fraction for that portion of the square that is shaded in both colors.

d. Multiply $\frac{1}{8} \cdot \frac{3}{4}$. Discuss among yourselves how the work you did in steps **a** and **b** pictures the product $\frac{1}{8} \cdot \frac{3}{4}$.

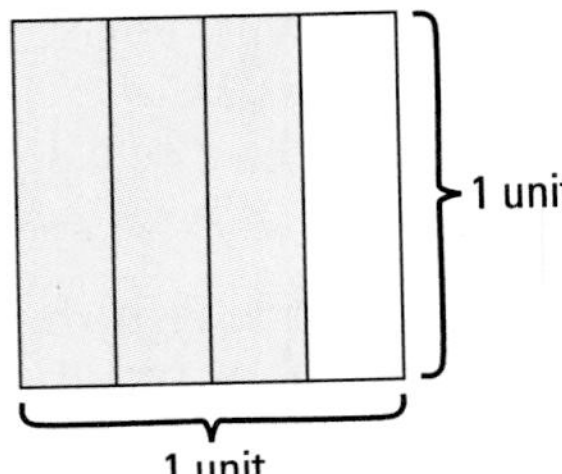

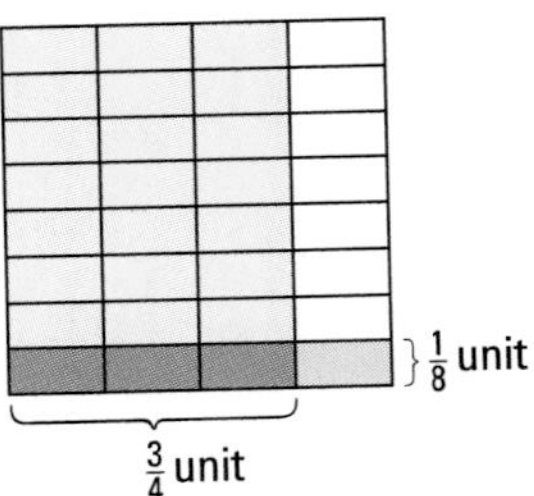

2 Use a square to picture each product. Then compute the product.

a. $\frac{3}{8} \cdot \frac{3}{4}$ **b.** $\frac{2}{3} \cdot \frac{5}{6}$

3 ***Draw conclusions.*** Discuss in your group a rule for multiplying fractions. Write a few sentences about how the models you made in Questions 1–3 illustrate this rule.

4 In Question 1, suppose the length of each side of the square is *s* units. What product does the double shaded region represent?

LESSON

2-3

Multiplying Algebraic Fractions

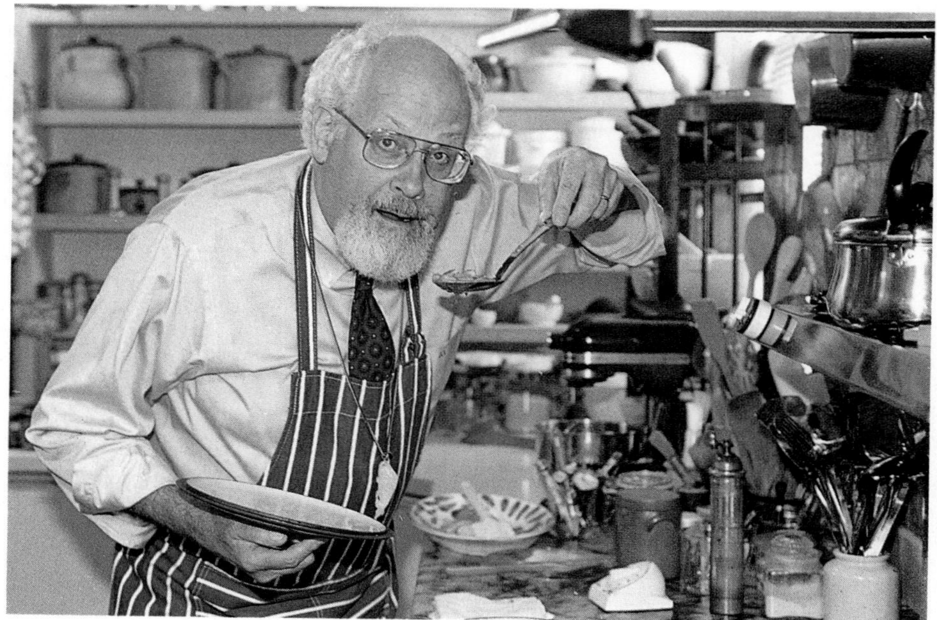

A half of this, a third of that . . . *Jeff Smith is known to millions for his TV program, "The Frugal Gourmet." Operations with fractions are often necessary when using recipes.*

Multiplying Fractions

Every fraction between 0 and 1 can be considered to be part of some whole. In the previous activity, you saw three instances of how multiplying two such fractions is related to area.

Here is another example. To picture the product of the fractions $\frac{2}{7}$ and $\frac{4}{5}$, first draw a unit square. Then find $\frac{2}{7}$ of one side and $\frac{4}{5}$ of an adjacent side. The rectangle that has dimensions $\frac{2}{7}$ and $\frac{4}{5}$ is shaded. Notice that its area is $\frac{8}{35}$ of the total. This is because there are $2 \cdot 4$ shaded small rectangles out of the total of $7 \cdot 5$, or 35. Since the total area is 1 square unit, the area of the shaded region is $\frac{8}{35}$ square unit.

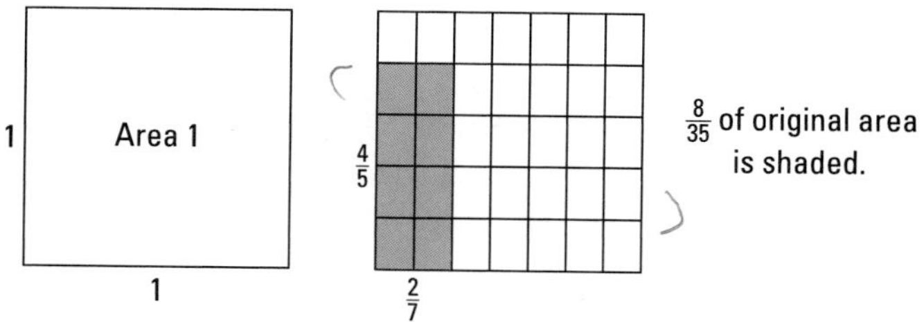

The pattern can be generalized. To multiply $\frac{a}{b}$ by $\frac{c}{d}$, you could shade ac rectangles out of bd small rectangles in the unit area. So the area would be $\frac{ac}{bd}$. This pictures the common rule for multiplying fractions.

Multiplying Fractions Property
For all real numbers a, b, c, and d, with b and d not zero,

$$\frac{a}{b} \cdot \frac{c}{d} = \frac{ac}{bd}.$$

Whenever variables represent real numbers you may apply this property to **algebraic fractions,** that is, to fractions that have variables in the numerator or denominator.

Example 1

Multiply $\frac{b}{3} \cdot \frac{h}{2}$.

Solution

Use the Multiplying Fractions Property.

$\frac{b}{3} \cdot \frac{h}{2} = \frac{bh}{6}$

Check 1

Substitute values for b and h, say $b = 6$ and $h = 10$.

Does $\frac{6}{3} \cdot \frac{10}{2} = 6 \cdot \frac{10}{6}$? Yes, it does, because of the Multiplying Fractions Property.

Check 2

Use an area model. First draw a rectangle with base b and height h. Then find $\frac{b}{3}$ or $\frac{1}{3}b$ of one side and $\frac{h}{2}$ or $\frac{1}{2}h$ of the other side.

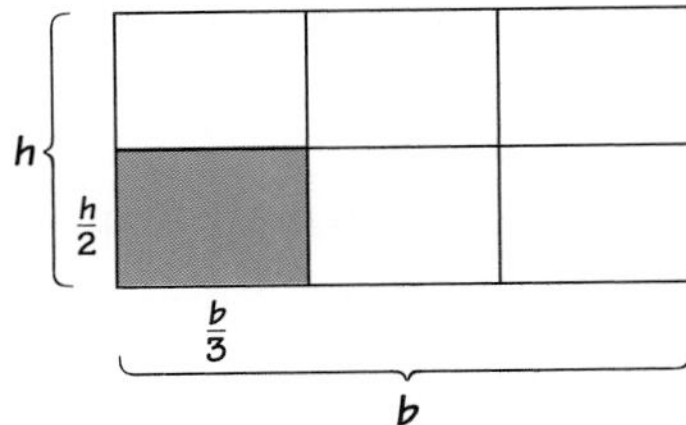

Notice that the area of the shaded rectangle is $\frac{1}{6}$ the area of the original rectangle, or $\frac{bh}{6}$.

Notice that in Example 1, $\frac{bh}{6}$ is another way of writing $\frac{1}{6}bh$. Example 2 shows how the Multiplying Fractions Property can be used to explain why many different expressions are equal to each other.

Example 2

Show that each of the following expressions equals $\frac{2h}{3}$.

a. $\frac{2}{3}h$ **b.** $\frac{1}{3} \cdot 2h$ **c.** $2 \cdot \frac{h}{3}$

Solution

Notice how each line below uses the Multiplying Fractions Property and the property that $x = \frac{x}{1}$.

a. $\frac{2}{3}h = \frac{2}{3} \cdot \frac{h}{1} = \frac{2h}{3}$

b. $\frac{1}{3} \cdot 2h = \frac{1}{3} \cdot \frac{2h}{1} = \frac{2h}{3}$

c. $2 \cdot \frac{h}{3} = \frac{2}{1} \cdot \frac{h}{3} = \frac{2h}{3}$

We have shown that $\frac{2h}{3}$, $\frac{2}{3}h$, $\frac{1}{3} \cdot 2h$, and $2 \cdot \frac{h}{3}$ are all equal to each other.

Simplifying Fractions

You should be familiar with simplifying numerical fractions. For instance, $\frac{30}{50} = \frac{3}{5}$. This is an instance of the *Equal Fractions Property.*

Equal Fractions Property
If $b \neq 0$ and $k \neq 0$, then

$$\frac{ak}{bk} = \frac{a}{b}.$$

The Equal Fractions Property holds for all fractions $\frac{a}{b}$ as long as the denominator is not 0. If there are common factors in the numerator and denominator of an algebraic fraction, you can use the Equal Fractions Property to simplify it.

Example 3

Simplify $\frac{36cd}{3d}$.

Solution 1

Look for common factors in the numerator and denominator. Notice that $3d$ is a factor of each. The solution below applies the Associative Property of Multiplication in the numerator and the Multiplicative Identity Property of 1 in the denominator.

$$\frac{36cd}{3d} = \frac{12c \cdot 3d}{1 \cdot 3d}$$

$$= \frac{12c}{1} \quad \text{Equal Fractions Property}$$

$$= 12c$$

Solution 2

Experts often skip steps. They sometimes strike out the common factors with slashes.

$$\frac{\overset{12}{\cancel{36}}c\overset{1}{\cancel{d}}}{\underset{1}{\cancel{3}}\underset{1}{\cancel{d}}} = 12c$$

In Example 4, both the Multiplying Fractions Property and the Equal Fractions Property are used with algebraic fractions.

Example 4

Assume $a \neq 0$ and $m \neq 0$. Multiply $\frac{5m}{12a}$ by $\frac{3a}{10m^2}$, and simplify the result.

Solution 1

Here we show all the steps.

$$\begin{aligned}
\frac{5m}{12a} \cdot \frac{3a}{10m^2} &= \frac{5m \cdot 3a}{12a \cdot 10m^2} \\
&= \frac{15ma}{120am^2} \\
&= \frac{1 \cdot 15 \cdot m \cdot a}{8 \cdot 15 \cdot m \cdot m \cdot a} \\
&= \frac{1}{8} \cdot \frac{15}{15} \cdot \frac{m}{m} \cdot \frac{1}{m} \cdot \frac{a}{a} \\
&= \frac{1}{8} \cdot \frac{1}{m} \\
&= \frac{1}{8m}
\end{aligned}$$

Solution 2

This is what others might write. They would look for common factors in the numerator and denominator.

$$\frac{5m}{12a} \cdot \frac{3a}{10m^2} = \frac{\overset{1}{\cancel{5}} \cdot \overset{1}{\cancel{m}} \cdot \overset{1}{\cancel{3}} \cdot \overset{1}{\cancel{a}}}{\underset{4}{\cancel{12}} \cdot \underset{1}{\cancel{a}} \cdot \underset{2}{\cancel{10}} \cdot \underset{1}{\cancel{m}} \cdot m} = \frac{1}{8}m$$

Ask your teacher how much detail he or she wants you to provide when multiplying or simplifying algebraic fractions.

Caution: The Equal Fractions Property is a property related to multiplication. It does not work when the same terms are *added* to the numerator and denominator.

QUESTIONS

Covering the Reading

1. State the Multiplying Fractions Property.

2. a. Draw a picture to represent the product $\frac{2}{3} \cdot \frac{4}{5}$.
   b. How does your picture illustrate the Multiplying Fractions Property?

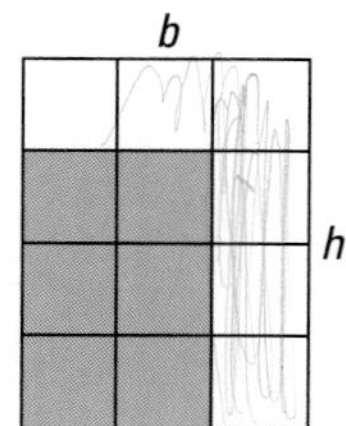

3. The rectangle at the left has base b and height h.
   a. If all the small rectangles have the same dimensions, what is the area of the shaded region?
   b. What product of algebraic fractions is represented by the shaded area?

In 4 and 5, multiply the fractions.

4. $\frac{a}{7} \cdot \frac{b}{2}$

5. $\frac{x}{3} \cdot \frac{y}{z}$

6. *True or false.* $\frac{1}{5}n = \frac{n}{5}$.

7. Show that $\frac{4}{9}x$ is equal to $\frac{4x}{9}$.

8. *Multiple choice.* Which does not equal the others?
(a) $\frac{7t}{12}$ (b) $\frac{7}{12}t$ (c) $7t \cdot \frac{1}{12}$ (d) $\frac{7}{t} \cdot 12$

In 9–12, use the Equal Fractions Property to simplify each fraction.

9. $\frac{800}{1900}$

10. $\frac{20y}{5y}$

11. $\frac{3mn}{9mt}$

12. $\frac{24gr}{18gr^2}$

In 13–16, multiply and simplify the result.

13. $\frac{3m}{n} \cdot \frac{7m}{9}$

14. $\frac{6a}{b} \cdot \frac{b}{6a}$

15. $\frac{24c}{5d} \cdot \frac{20d}{21}$

16. $\frac{50}{9x} \cdot \frac{18x^2}{25y}$

Applying the Mathematics

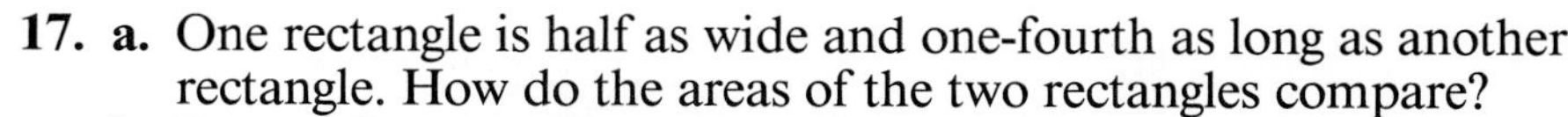

17. **a.** One rectangle is half as wide and one-fourth as long as another rectangle. How do the areas of the two rectangles compare?
b. Draw a figure to illustrate your answer.

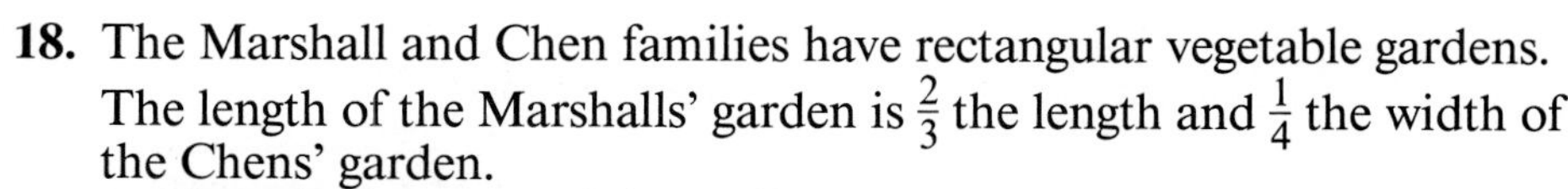

18. The Marshall and Chen families have rectangular vegetable gardens. The length of the Marshalls' garden is $\frac{2}{3}$ the length and $\frac{1}{4}$ the width of the Chens' garden.
a. How do the areas of the gardens compare?
b. Check your answer by using a specific length and width for the Chens' garden.

"Community gardening" has become popular in big cities where backyards are not available for planting. Several families may share the work and the harvest of a city-owned plot.

19. *Skill sequence.* Compute in your head.
a. $\frac{5}{3} \cdot 3$ **b.** $\frac{9}{x} \cdot x$ **c.** $\frac{a}{b} \cdot b$ **d.** $n^2 \cdot \frac{a}{n^2}$

In 20 and 21, *multiple choice.* Find the fraction that is *not* equal to the other three.

20. (a) $\frac{9a}{11a}$ (b) $\frac{99}{121}$ (c) $\frac{90}{100}$ (d) $\frac{450}{550}$

21. (a) $\frac{100}{260}$ (b) $\frac{35t}{91t}$ (c) $\frac{38}{100}$ (d) $\frac{500x^2}{1300x^2}$

In 22–25, multiply and simplify where possible.

22. $\frac{a}{b} \cdot \frac{c}{d} \cdot \frac{e}{f}$

23. $\frac{a}{b} \cdot \frac{b}{c} \cdot \frac{c}{a}$

24. $\frac{22a^3}{7b} \cdot \frac{21b}{11a^2}$

25. $\frac{9x^2}{10y^2} \cdot \frac{14y^3}{3x}$

Acquiring aquariums. *When setting up a new aquarium, you should not overstock. The total length of all the fish in inches should be no greater than the number of gallons of water the tank will hold.*

26. a. Find the volume of the aquarium at the left.
b. Check your answer by letting $b = 12$.
c. Think of a cube with sides of length b. How many of these aquariums would fit into the cube? How can you tell?

27. Find two *algebraic* fractions that when multiplied yield $\frac{12x^2}{5y^3}$.

Review

28. *Skill sequence.* Write the reciprocal of each number. *(Lesson 2-2)*
a. 4 **b.** $\frac{1}{9}$ **c.** $\frac{4}{9}$

In 29 and 30, compute in your head using the Associative and Commutative Properties of Multiplication. *(Lesson 2-1)*

29. $2 \cdot 7 \cdot 4 \cdot 5$

30. $2.5 \cdot 4 \cdot 2 \cdot 9$

31. A single-story house is to be built on a lot 75 feet wide by 100 feet deep. The shorter side of the lot faces the street. The house must be set back from the street at least 25 feet. It must be 20 feet from the back lot line and 10 feet from each side lot line. What is the maximum square footage (area) the house can have? *(Lesson 2-1)*

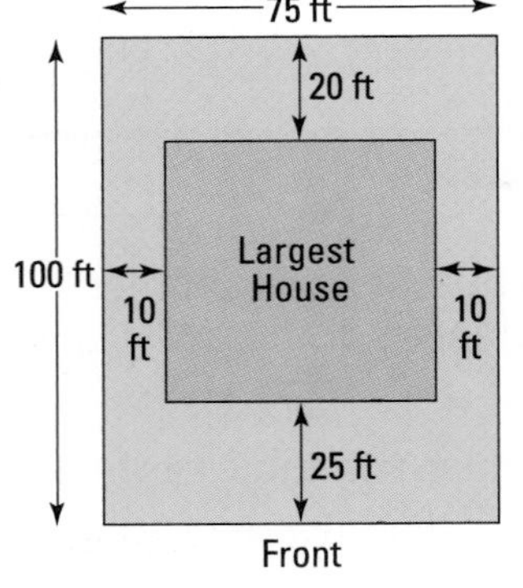

32. *True or false.* *(Lesson 1-1)*
a. $4x = 18$ if $x = 4.5$
b. $-9y = 42$ if $y = -\frac{14}{3}$
c. $\frac{4}{5}z = -96$ if $z = 120$
d. $\frac{10}{3} = \frac{-15}{2}w$ if $w = \frac{-4}{9}$

Exploration

33. a. Calculate the following products.

$\frac{1}{2} \cdot \frac{2}{3}$

$\frac{1}{2} \cdot \frac{2}{3} \cdot \frac{3}{4}$

$\frac{1}{2} \cdot \frac{2}{3} \cdot \frac{3}{4} \cdot \frac{4}{5}$

b. Write a sentence or two describing the patterns you observe.
c. Predict the following products.

$\frac{1}{2} \cdot \frac{2}{3} \cdot \frac{3}{4} \cdot \ldots \cdot \frac{1996}{1997}$

$\frac{1}{2} \cdot \frac{2}{3} \cdot \frac{3}{4} \cdot \ldots \cdot \frac{n}{n+1}$

LESSON

2-4

Multiplying Rates

Racers, start your engines. *The Indianapolis 500 is one of the most famous races in the U.S., attracting about 300,000 people annually. Drivers reach speeds of more than 220 mph (354 km/h) as they race the 2.5-mile oval track.*

Rates appear often in everyday life. Some common rates are "55 miles per hour" or "$4.25 an hour."

Fraction forms often appear in situations involving rates. Units for rates can be expressed using a slash "/" or a horizontal bar "—". The slash and the bar are read "per" or "for each."

rate	with a slash	with a bar
99 cents per pound	99 cents/lb	$99 \frac{\text{cents}}{\text{lb}}$
88 kilometers per hour	88 km/hr	$88 \frac{\text{km}}{\text{hr}}$
W words per minute	*W* words/min	$W \frac{\text{words}}{\text{min}}$

Some rates are expressed with abbreviations. For instance, "miles per hour" is abbreviated "mph," and "miles per gallon" is abbreviated "mpg."

How to Multiply Rates

Rates can be multiplied by other quantities. The units are multiplied as if they were numerical fractions.

Example 1

At the Indianapolis Motor Speedway, the race track is 2.5 miles long. How many miles would a race car go if it went 200 times around the track?

Solution

The path around a race track is called a *lap*. The rate described is $2.5 \frac{\text{miles}}{\text{lap}}$. Multiply this rate by the number of laps. The lap units "cancel."

$$200 \text{ laps} \cdot 2.5 \frac{\text{miles}}{\text{lap}} = 500 \text{ miles}$$

The car would travel 500 miles.

This is one instance of the *Rate Factor Model for Multiplication.*

Rate Factor Model for Multiplication
When a rate r is multiplied by another quantity x, the product is rx. So the unit of rx is the product of the units for r and x.

Here is another instance. Suppose that, when resting, your heart rate is 70 beats per minute. In 5 minutes there will be

$$70\ \frac{\text{beats}}{\cancel{\text{min}}} \cdot 5\ \cancel{\text{min}} = 350 \text{ beats}.$$

In x minutes, there will be

$$70\ \frac{\text{beats}}{\cancel{\text{min}}} \cdot x\ \cancel{\text{min}} = 70x \text{ beats}.$$

Rates are used in many formulas. One of the most important formulas is $d = rt$, which gives the distance d traveled by an object moving at a constant rate r during a time t. (r is often called the speed.)

Example 2

If an airplane averages 540 mph for $2\frac{1}{4}$ hours, about how far does it travel?

Solution

You are given $r = 540$ mph and $t = 2\frac{1}{4}$ hr. Substitute into the formula

$$d = rt.$$
$$d = 540 \text{ mph} \cdot 2\tfrac{1}{4} \text{ hr}$$

Rewrite the abbreviation mph to see the units more clearly, and convert $2\frac{1}{4}$ to an improper fraction.

$$d = 540\ \frac{\text{miles}}{\cancel{\text{hr}}} \cdot \frac{9}{4}\ \cancel{\text{hr}}$$
$$= 1215 \text{ miles}$$

The plane travels about 1200 miles.

Converting Measurement Units

Rates can be used to convert units in measurements. Since 1 hour = 60 minutes, dividing one unit by the other equals 1.

$$\frac{1 \text{ hour}}{60 \text{ min}} = \frac{60 \text{ min}}{1 \text{ hour}} = 1$$

Multiplying a quantity by one of these rates does not change its value.

Example 3

If an airplane travels at 540 mph, how far does it travel per second?

Solution

Convert miles per hour to miles per second. Multiply by a rate to change hours to minutes, and by another rate to change minutes to seconds.

$$540\ \frac{\text{miles}}{\cancel{\text{hr}}} \cdot \frac{1\ \cancel{\text{hr}}}{60\ \cancel{\text{min}}} \cdot \frac{1\ \cancel{\text{min}}}{60 \text{ sec}} = \frac{3 \text{ miles}}{20 \text{ sec}} = \frac{3}{20} \text{ mile per sec}$$

Reciprocal Rates

Sometimes a rate does not help simplify the computation, but its reciprocal would. Example 3 used $\frac{1 \text{ hr}}{60 \text{ min}}$ rather than $\frac{60 \text{ min}}{1 \text{ hr}}$. These are reciprocal rates. In reciprocal rates both the numbers and units are reciprocals. For instance, $2.5 \frac{\text{miles}}{\text{lap}}$ and $\frac{1}{2.5} \frac{\text{lap}}{\text{mile}}$ are *reciprocal rates.* **Reciprocal rates** describe the same situation from different points of view. For instance, if 200 words are read in a minute, then it takes $\frac{1}{200}$ of a minute to read one word. The reciprocal rate of $200 \frac{\text{words}}{\text{min}}$ is $\frac{1}{200} \frac{\text{min}}{\text{word}}$.

Example 4 shows that you must often decide which of two reciprocal rates to use. Choose the one that cancels correctly, resulting in an answer with the appropriate units.

Marching for the March of Dimes. *President Franklin Delano Roosevelt started the March of Dimes to combat polio in 1938. The "Dimes" in the name refers to the dimes sent to the White House during the agency's first fund-raiser.*

Example 4

People sometimes go for a walk after a big meal to "burn it off." To lose 1 pound, a person must burn 3500 calories. An average-sized person burns about 300 calories per hour by walking. If a person walks for 2 hours, how much weight will he or she lose?

Solution

The unit of the answer should be a weight, in this case pounds. The two rates $3500 \frac{\text{calories}}{\text{pound}}$ and $300 \frac{\text{calories}}{\text{hour}}$ or their reciprocals might be used. To eliminate the calorie unit we must multiply by a rate having calorie in the denominator. We also must have the pound unit left after the multiplication, so the rate used must be $\frac{\text{pound}}{\text{calories}}$. So we use the reciprocal of $3500 \frac{\text{calories}}{\text{pound}}$.

$$2 \cancel{\text{hours}} \cdot \frac{1}{3500} \frac{\text{pound}}{\cancel{\text{calories}}} \cdot 300 \frac{\cancel{\text{calories}}}{\cancel{\text{hour}}} = \frac{2 \cdot 300}{3500} \text{ pound}$$
$$\approx .17 \text{ pound}$$

If a person walks for 2 hours, he or she will lose about $\frac{1}{6}$ pound.

QUESTIONS

Covering the Reading

In 1 and 2 a rate is given. **a.** Copy the sentence and underline the rate. **b.** Write the rate unit using a slash. **c.** Write the rate unit using a fraction bar.

1. A secretary types 70 words per minute.

2. There are exactly 2.54 centimeters per inch.

3. Write the rate "6 dollars an hour" using the following symbols.
a. a slash **b.** a horizontal bar

Faster than a speeding bullet! *The world's fastest electric train is the French TGV (train à grande vitesse, or high-speed train). The TGV gets its power from an overhead wire system.*

4. While Tonisha exercises, her heart rate is 150 beats per minute. How many times will her heart beat while exercising for the following times?

a. 10 minutes **b.** m minutes

5. In the annual car race at LeMans, France, teams of drivers take turns driving for 24 hours. On the 1992 winning team, two drivers were British and one was French. They had an average speed of 123.89 mph. How many miles did the team drive during the race?

6. A small plane flies 380 miles per hour. How far will it fly in $2\frac{1}{2}$ hours?

7. The high-speed French TGV train can go 300 km/h. Convert this rate to kilometers per second.

8. *Multiple choice.* What is the reciprocal rate of $40 \frac{\text{miles}}{\text{gallon}}$?

(a) $40 \frac{\text{gallons}}{\text{mile}}$ (b) $\frac{1}{40} \frac{\text{gallon}}{\text{mile}}$

(c) $\frac{1}{40} \frac{\text{mile}}{\text{gallon}}$ (d) 40 mpg

9. Refer to Example 4. There are 16 ounces per pound. If an average person walks for 2 hours, how many ounces will he or she lose?

Applying the Mathematics

10. In the decade from 1980 to 1990, the population of West Virginia was decreasing by an average of 14,300 people per year. What was the total population loss during those 10 years?

11. There are 24 bottles per case. Each bottle contains 12 ounces of liquid.

a. How many ounces are in 10 cases?

b. How many ounces are in c cases?

In 12 and 13, use the fact that 1 inch = 2.54 cm.

12. Elise is 60 inches tall. What is her height in centimeters?

13. In his home country, Carlos wears an 80-cm belt. His friend Gloria wants to send him a belt from the United States where belts are made in 2-inch increments (28, 30, 32, and so on). What size belt should Gloria send? Explain your reasoning.

14. In 1994, Al Unser, Jr. won the Indianapolis 500 with an average speed of 160.872 mph. At this average speed, how long would it take him to go once around the $2\frac{1}{2}$-mile track?

15. Marty can wash k dishes per minute. His sister Sue is twice as fast.

a. How many dishes can Sue wash per minute?

b. How many minutes does Sue spend per dish?

16. If shrimp costs \$6/lb, and you get 30 shrimp/lb, how many shrimp can you buy for a dollar? Hint: Use a reciprocal rate.

Climb every mountain. *Most major mountains, except those in remote areas, have been climbed. The tallest peak in North America, Mt. McKinley, was first climbed in 1913. Mt. Everest, the world's highest mountain, was first scaled in 1953.*

17. Suppose an alpine climber begins 2400 meters from the summit of a mountain. If she climbs 48 meters per hour, and she climbs 10 hours each day, how many days will it take her to reach the summit?

In 18 and 19, write a rate multiplication problem whose answer is the given quantity.

18. 25 miles

19. $3 \frac{\text{minutes}}{\text{page}}$

20. In 1992, the birth rate in the U.S. was 16.2 babies per 1,000 people. That year the population was about 252,000,000. Use this product:

$$16.2 \frac{\text{babies}}{1{,}000 \text{ people}} \cdot 252{,}000{,}000 \text{ people}$$

to determine how many babies were born in 1992.

Review

21. *Multiple choice.* Which is *not* equal to the others? *(Lesson 2-3)*
(a) $\frac{42}{4}x$ (b) $\frac{42x}{4}$ (c) $\frac{7}{4} \cdot 6x$ (d) $\frac{42}{4x}$

22. Multiply and simplify $\frac{8b}{7c} \cdot \frac{21a}{2x} \cdot 5c$. *(Lesson 2-3)*

23. Find two algebraic fractions that have a product of $\frac{8x}{15y}$. *(Lesson 2-3)*

In 24 and 25, an expression is given. **a.** Simplify. **b.** Name the multiplication property used. *(Lessons 2-2, 2-3)*

24. $6x \cdot \frac{1}{6x}$

25. $\frac{17t^2}{x} \cdot \frac{x^4}{901t^3} \cdot 0$

26. Compute in your head $(9998 + 1) \frac{1}{9999}$. *(Lesson 2-2)*

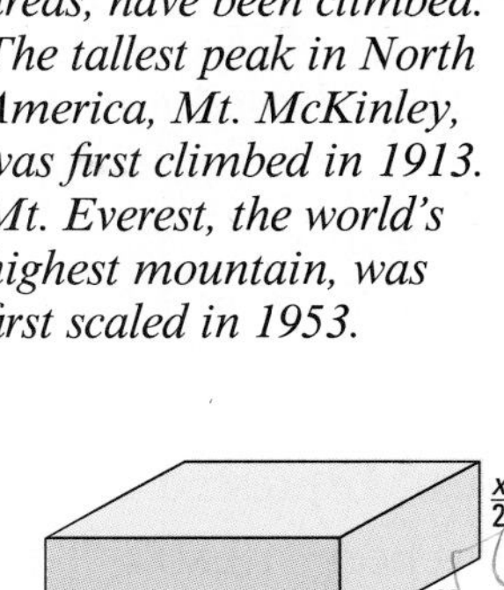

27. a. Find the volume of the box shown at the left.
b. Give the dimensions of two other boxes that have the same volume. *(Lessons 2-1, 2-3)*

Exploration

28. For what rate does the abbreviation stand, and where is it used?
a. rpm **b.** psi

29. Suppose that on January 1 in the year 2000, you begin counting from 1 at the rate of 1 number per second. If this counting continues around the clock, during which month and year would you reach the following numbers? Explain how you got your answer.
a. 1 million **b.** 1 billion **c.** 1 trillion

LESSON

2-5

Products and Powers with Negative Numbers

A clear-cut problem. *This photo shows the clear-cutting of a part of the rain forest of Belize, Central America. With 50 million acres of rain forests worldwide being destroyed yearly, many species of animals and plants may become extinct.*

Multiplication with Negative Numbers

The rate factor model can be applied to both positive and negative rates. For example, recently in Brazil, an average of about 13,800 square kilometers of rain forest has been destroyed each year. Expressed as a rate, the change in the area of Brazil's rain forest has been $-13{,}800 \frac{\text{km}^2}{\text{year}}$. If this rate continues unchanged for 20 years, multiplying gives the total amount destroyed.

$$20 \text{ years} \cdot -13{,}800 \frac{\text{km}^2}{\text{year}} = -276{,}000 \text{ km}^2$$

The final answer is negative, which means that 276,000 square kilometers (about 107,000 square miles) of forest would be *lost* during those twenty years. This example shows that *the product of a positive and a negative number is negative.*

To multiply two negatives, again consider the decrease in Brazil's forested land. One year ago, there were 13,800 square kilometers more forest than there are now. Going back in time is represented by a negative number. Expressed as a product using rates:

1 year ago at a loss of 13,800 km^2 per year = 13,800 km^2 more than now.

$$\text{So, } -1 \text{ year} \cdot -13{,}800 \frac{\text{km}^2}{\text{year}} = 13{,}800 \text{ km}^2.$$

The positive answer indicates there was *more* forest last year than now.

The numbers -13,800 and 13,800 are **opposites.** The last instance illustrates that multiplication by -1 changes a number to its opposite.

Multiplication Property of -1
For any real number a, $a \cdot -1 = -1 \cdot a = -a$.

What happens when the opposite of one number is multiplied by the opposite of another number?

Since $-x = -1x$, and $-y = -1y$, then

$-x \cdot -y = -1x \cdot -1y$	
$= (-1 \cdot -1)xy$	Commutative Property of Multiplication
$= 1xy$	Multiplication Property of -1
$= xy.$	

The work above shows that the product of the opposites of two numbers is the same as the product of the two original numbers. When x and y are positive, $-x$ and $-y$ are negative. Thus, *the product of two negative numbers is positive.*

Rules for Multiplying Positive and Negative Numbers
If two numbers have the same sign, their product is positive.
If two numbers have different signs, their product is negative.

The rules for multiplying positive and negative numbers also apply to numerical and algebraic expressions.

Example 1

Multiply and simplify where possible.

a. $-4x \cdot -3y$

b. $-15n \cdot \frac{2}{n}$

c. $-\frac{3}{7} \cdot -\frac{7}{3}$

Solution

a. $-4x \cdot -3y = 4 \cdot 3 \cdot x \cdot y = 12xy$

b. $-15\cancel{n} \cdot \frac{2}{\cancel{n}} = -30$

c. $-\frac{3}{7} \cdot -\frac{7}{3} = \frac{3}{7} \cdot \frac{7}{3} = 1$

In part **c**, because $-\frac{3}{7} \cdot -\frac{7}{3} = 1$, the reciprocal of $-\frac{3}{7}$ is $-\frac{7}{3}$. In general, if neither a nor b is zero, the reciprocal of $-\frac{a}{b}$ is $-\frac{b}{a}$.

Example 2

Evaluate $y = ax^2 + bx + c$ when $a = -2$, $b = 3$, $c = 4$, and $x = -5$.

Solution

Substitute the values for the appropriate variables in the formula.

$$y = -2(-5)^2 + 3(-5) + 4$$

Follow the order of operations given in Lesson 1-4. Square first; then multiply; then add.

$$\begin{aligned} y &= -2(25) + -15 + 4 \\ &= -50 + -15 + 4 \\ &= -61 \end{aligned}$$

The integer powers of negative numbers follow a simple pattern. Examine these powers of -2.

$$\begin{aligned} (-2)^1 &= -2 \\ (-2)^2 &= -2 \cdot -2 = 4 \\ (-2)^3 &= -2 \cdot -2 \cdot -2 = -8 \\ (-2)^4 &= -2 \cdot -2 \cdot -2 \cdot -2 = 16 \\ (-2)^5 &= -2 \cdot -2 \cdot -2 \cdot -2 \cdot -2 = -32 \end{aligned}$$

Notice the pattern. Odd powers of -2 are negative. Even powers of -2 are positive. There is a more general property.

Properties of Multiplication of Positive and Negative Numbers

1. The product of an odd number of negative numbers is negative.
2. The product of an even number of negative numbers is positive.

Example 3

Without computing, determine whether the number is positive, negative, or zero.

a. $(-3)^6$ **b.** $5 \cdot 3 \cdot 1 \cdot -1 \cdot -3 \cdot -5$ **c.** $3 \cdot 2 \cdot 1 \cdot 0 \cdot -1 \cdot -2 \cdot -3$

Solution

a. This is an even power of a negative number, so $(-3)^6$ is positive.

b. Three of the numbers multiplied are negative, so the product is negative.

c. Zero is one of the factors, so the product is 0.

Recall that parentheses are important when evaluating expressions. This is particularly true when negative numbers and powers are involved. For instance,

$$(-3)^6 = -3 \cdot -3 \cdot -3 \cdot -3 \cdot -3 \cdot -3 = 729$$

In contrast,

-3^6 means the opposite of 3^6, or $-1 \cdot 3^6$.

$$-3^6 = -1 \cdot 3^6 = -1 \cdot 3 \cdot 3 \cdot 3 \cdot 3 \cdot 3 \cdot 3 = -729$$

Graphics calculators usually allow you to calculate powers of negative numbers. Other scientific calculators often do not. Here are key sequences that work on some calculators.

$(-3)^6$:	[( ] [(–)] 3 [ )] [^] 6 [ENTER]	display	(-3)^6 729
	[( ] [(–)] 3 [ )] [x^y] 6 [EXE]	display	(-3) x^y 6 729
-3^6:	[(–)] 3 [^] 6 [ENTER]	display	(-3)^6 -729
	[(–)] 3 [x^y] 6 [EXE]	display	(-3)x^y 6 -729
	3 [x^y] 6 [=] [±]	display	-729

Check your calculator to see whether it can evaluate powers with negative numbers.

QUESTIONS

Covering the Reading

1. a. Copy and complete this table. Assume the area of Brazil's rain forest is decreasing by the amount mentioned in the reading.

Years from present	Change in Brazil's rain forest compared to now
20	
5	
1	
-1	
-5	

b. What is meant by "-5 years from now"?

2. In recent years, a farm's topsoil has eroded at a rate of 0.3 inch per year.

a. What was the total change after 4 years?
b. What rate factor multiplication is needed to answer the question in part **a?**
c. How much deeper was the topsoil 5 years ago?
d. What rate factor multiplication is needed to answer the question in part **c?**

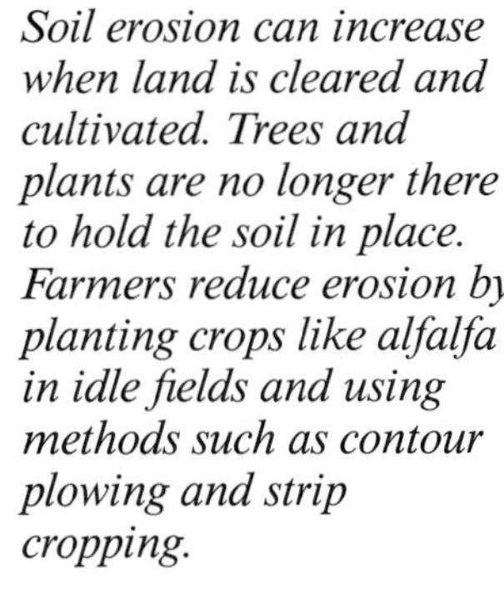

Soil erosion can increase when land is cleared and cultivated. Trees and plants are no longer there to hold the soil in place. Farmers reduce erosion by planting crops like alfalfa in idle fields and using methods such as contour plowing and strip cropping.

In 3–6, compute in your head.

3. a. $6 \cdot -3$ **b.** $6x \cdot -3y$

4. a. $4 \cdot -9$ **b.** $4p \cdot -9s$

5. a. $-7 \cdot -5$ **b.** $-7t \cdot -5t$

6. a. $-\frac{4}{3} \cdot -\frac{3}{4}$ **b.** $\frac{-4}{m} \cdot \frac{m}{-4}$

7. *Skill sequence.* State the reciprocal of each number.

a. $\frac{1}{9}$ **b.** $-\frac{1}{9}$

c. $-\frac{11}{9}$ **d.** $-\frac{11k}{9}$

8. Evaluate $y = -3x^2 + 6x + 1$ when $x = -2$.

In 9–11, tell whether the product of the numbers is positive or negative.

9. two positive numbers and two negative numbers

10. one hundred negative numbers

11. three negative numbers and one positive number

In 12 and 13, evaluate without a calculator.

12. $-1 \cdot 2 \cdot -3 \cdot 4$ **13.** $(-1)^5$

14. a. Evaluate each expression.

i. $(-3)^6$ **ii.** $(-3)^7$ **iii.** $(-3)^8$ **iv.** $(-3)^9$

b. Which powers of -3 are positive?

c. Which powers of -3 are negative?

15. *True or false.* Justify your answer.

a. $(-5)^3 = -5^3$ **b.** $(-5)^4 = -5^4$

Applying the Mathematics

In 16–19, multiply and simplify.

16. $(-4a)(-a)$ **17.** $(-3x)^2$

18. $(-10n)^3$ **19.** $(2a)^2 \cdot (-2a)^3$

20. Evaluate and simplify.

a. $-\frac{1}{2} \cdot -\frac{2}{3}$

b. $-\frac{1}{2} \cdot -\frac{2}{3} \cdot -\frac{3}{4}$

c. $-\frac{1}{2} \cdot -\frac{2}{3} \cdot -\frac{3}{4} \cdot -\frac{4}{5}$

d. $-\frac{1}{2} \cdot -\frac{2}{3} \cdot -\frac{3}{4} \cdot \ldots \cdot -\frac{9}{10}$

21. *Skill sequence.* Tell whether the number is positive, negative, or zero.

a. $(-5)^{10}$ **b.** $(-1)(-5)^{10}$ **c.** $(-1)^{10}(-5)^{10}$

d. $(5)^{10}(-5)^{10}$ **e.** $(5 + -5)^{10}$ **f.** $(-1)^{10}(-5)$

22. The number $(-3)^{500}$ cannot be evaluated on most calculators because it is too large for the memory. Describe how you can determine whether $(-3)^{500}$ is positive or negative.

Bear facts. *Brown bears, known also as grizzly bears, are found mainly in Alaska and Canada. Grizzlies may grow to a height of 8 feet and weigh close to 400 pounds. They get angry quickly, but usually don't attack unless threatened.*

Review

In 23–25, tell whether the quantity is a rate. *(Lesson 2-4)*

23. 520 miles **24.** $35 \frac{\text{miles}}{\text{gallon}}$ **25.** 55 mph

26. An apartment rents for \$600 per month. Find the rent for the given time. *(Lesson 2-4)*
a. two years **b.** y years

27. The maximum speed a grizzly bear can run is 50 km/hr. How far could a grizzly bear run in 10 seconds? *(Lesson 2-4)*

28. A contractor is planning to build a driveway. The driveway can be thought of as a rectangular solid 10 ft wide, 36 ft long, and 6 in. thick.
a. Convert these dimensions to yards.
b. How many cubic yards of concrete should be ordered? (Only whole numbers of yards may be ordered.)
c. If the concrete costs \$125 per cubic yard, find the total cost of the concrete *(Lessons 2-1, 2-4)*

29. Multiply and simplify $\frac{3}{7}a \cdot \left(\frac{3}{7}a \cdot \frac{7}{a}\right)$. *(Lesson 2-3)*

30. A right triangle has legs of length 2 and $\sqrt{2}$. What is the length of the hypotenuse? *(Lesson 1-8)*

Exploration

31. The command "about-face" in the military signals a soldier to rotate 180°. Two commands of "about-face" result in the soldier facing forward again.

Number of About-faces	Facing
1	Reverse
2	Forward
3	Reverse
4	Forward
⋮	⋮

How does this relate to $(-1)^n$?

LESSON

2-6

Solving $ax = b$

Pictured here is a balance scale.

Four small boxes, each of unknown weight w ounces, are on the left side of the scale. They balance the 8 one-ounce weights on the right side. This situation can be described by the equation

$$4w = 8.$$

A way to find the unknown weight is to take $\frac{1}{4}$ of each side.

This pictures the result of multiplying both sides of the equation by $\frac{1}{4}$.

$$\frac{1}{4}(4w) = \frac{1}{4} \cdot 8$$

The result can be seen in the picture. Each box weighs 2 ounces. In the language of equations,

$$w = 2.$$

In general, multiplying both sides of an equation by any nonzero number will not affect the solutions. This property is called the *Multiplication Property of Equality.*

Multiplication Property of Equality
For all real numbers a, b, and c, if $a = b$, then $ca = cb$.

This property is important in solving equations. Examples 1 and 2 show how this property and other properties you have studied are used.

Solving Equations Using the Multiplication Property of Equality

Example 1

An auditorium can seat 40 people in each row. How many full rows will be needed if 600 people are expected to attend a lecture?

Solution

Draw a picture.

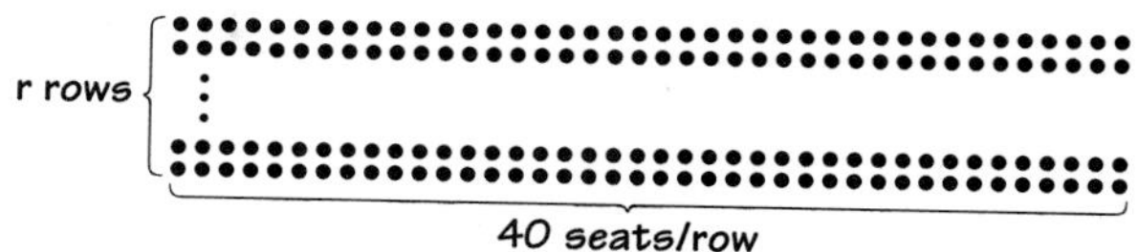

The total number of seats is $40r$. You need to solve the equation

$$40r = 600.$$

To solve this equation, multiply both sides by $\frac{1}{40}$, the reciprocal of 40.

$\frac{1}{40} \cdot 40r = \frac{1}{40} \cdot 600$	Multiplication Property of Equality
$\left(\frac{1}{40} \cdot 40\right)r = \frac{600}{40}$	Associative Property of Multiplication
$1 \cdot r = \frac{600}{40}$	Property of Reciprocals
$r = 15$	Multiplicative Identity Property of 1

So 15 rows will be needed.

Check

Are there 600 seats in 15 rows? Yes. $15 \cdot 40 = 600$.

In Example 1, both sides of the equation were multiplied by $\frac{1}{40}$ because $\frac{1}{40}$ is the reciprocal of 40. A similar thing happens in Example 2.

Example 2

Solve $\frac{7}{2}w = 4$. Check the solution.

Solution

$\frac{7}{2}w = 4$	
$\frac{2}{7} \cdot \frac{7}{2}w = \frac{2}{7} \cdot 4$	Multiplication Property of Equality
$\left(\frac{2}{7} \cdot \frac{7}{2}\right)w = \frac{2}{7} \cdot 4$	Associative Property of Multiplication
$1 \cdot w = \frac{8}{7}$	Property of Reciprocals
$w = \frac{8}{7}$	Multiplicative Identity Property of 1

Check

Substitute $w = \frac{8}{7}$ into the equation. Does $\frac{\overset{1}{\cancel{7}}}{\cancel{2}} \cdot \frac{\overset{4}{\cancel{8}}}{\cancel{7}} = 4$? Yes, $4 = 4$.

To solve $ax = b$ for x (when a is not zero), multiply both sides of the equation by the reciprocal of a.

The number a in the expression ax is the *coefficient* of the variable x. In Example 3, the coefficient of y is negative. We show the solution to the equation as you might write it, without naming the properties used.

Example 3

Solve $-6y = 117$. Check the solution.

Solution

The reciprocal of the coefficient -6 is $-\frac{1}{6}$. So, multiply both sides of the equation by $-\frac{1}{6}$.

$$-6y = 117$$
$$-\frac{1}{6} \cdot -6y = -\frac{1}{6} \cdot 117$$
$$y = -\frac{117}{6}$$
$$y = -19\frac{1}{2}$$

Check

Substitute -19.5 for y in the original equation and see if it works.

Does $-6(-19.5) = 117$? Yes, so it checks.

Since -19.5 makes the equation true, -19.5 is the solution.

Cincinnati. *Pictured is the Roebling Bridge over the Ohio River with Cincinnati in the background. Cincinnati is the world's leading manufacturer of soap and is the main producer of machine tools in the U.S.*

Equations and Formulas

You are familiar with the formula $d = rt$, which states that if you are moving at a constant speed, the distance d you travel is equal to your rate r multiplied by the time t you travel.

Example 4

How long will it take to drive from Detroit to Cincinnati, a distance of about 270 miles, if you travel at 55 mph?

Solution 1

Use the formula $d = rt$. Let t be the length of time, in hours.

Write: $d = r \cdot t$

Think: $270 \text{ miles} = 55 \frac{\text{miles}}{\text{hour}} \cdot t \text{ hours}$

Write: $270 = 55t$

Now solve the equation: $\frac{1}{55} \cdot 270 = \frac{1}{55} \cdot 55t$

$$\frac{270}{55} = t$$
$$4.9 \approx t$$

It will take about 5 hours.

Solution 2

Use a reciprocal rate. $270 \text{ mi} \cdot \frac{1}{55} \frac{\text{hr}}{\text{mi}} \approx 4.9 \text{ hours}$

Check

If you travel at 55 mph for 4.9 hours, will you travel about 270 miles? Yes, because

$$55 \frac{\text{mi}}{\text{hr}} \cdot 4.9 \text{ hr} = 269.5 \text{ mi} \approx 270 \text{ mi}.$$

Example 5

In March of 1861, the Pony Express made its fastest mail delivery. That month, the riders covered the 1,966 miles from St. Joseph, Missouri, to Sacramento, California, at a rate of about 255 miles per day. How long did it take the riders to make the delivery?

WANTED
OUNG SKINNY WIRY FELLOWS
ot over eighteen. Must be expert
riders willing to risk death daily.
Orphans preferred. WAGES $25 per
week. Apply, *Central Overland Express, Alta Bldg., Montgomery St.*

Solution

Use the same strategy as in Example 4.

$$d = rt$$
$$1966 = 255t$$

Now multiply each side by the reciprocal of 255.

$$\frac{1}{255} \cdot 1966 = \frac{1}{255} \cdot 255t$$

It took the riders about 7.7 days.

If you have to do many calculations like those in Examples 4 and 5, it may be easier to rearrange the formula $d = rt$ so that the variable t is alone on one side.

Example 6

Solve $d = rt$ for t.

Solution

This problem is similar to solving $270 = 55t$ in Example 4. There the coefficient of t was 55; both sides were multiplied by $\frac{1}{55}$. In $d = rt$, the coefficient of t is r, so in order to isolate t we multiply by $\frac{1}{r}$. Write:

$$d = rt.$$
$$\frac{1}{r} \cdot d = \frac{1}{r} \cdot rt$$
$$\frac{d}{r} = t$$

Notice the pattern. To solve $270 = 55t$ in Example 4, we multiplied each side by $\frac{1}{55}$, the reciprocal of 55. To solve $1966 = 255t$ in Example 5, we multiplied by $\frac{1}{255}$. In both cases, the time of travel is calculated by dividing the distance by the rate of travel. This is the same calculation described in general by the formula $t = \frac{d}{r}$.

When you solve equations, we strongly recommend that you arrange your work so that the equal signs of each line are directly below each other (as the examples show). This arrangement helps avoid confusion.

QUESTIONS

Covering the Reading

1. Each box on the left side of the balance scale has the same weight.
   a. Use an equation to describe the situation pictured below.
   b. How much does each box weigh?

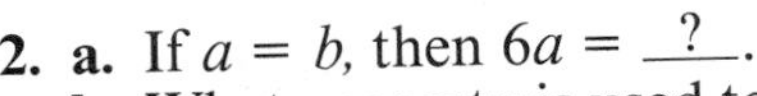

2. a. If $a = b$, then $6a = \underline{\ ?\ }$.
   b. What property is used to answer part **a?**

3. In your own words, explain why multiplying each side of the equation by the reciprocal "works."

4. An auditorium has rows with 28 seats in each row. It is desired to rope off 500 seats for a lecture. How many rows need to be roped off?

An art deco auditorium. *Art deco was a decorative style of the 1920s and 1930s that utilized vivid colors and geometric patterns. This style is visible in the decor of the Pickwick Theatre pictured above.*

In 5–10, an equation is given. **a.** What is the coefficient of the variable? **b.** By what number can you multiply both sides to solve the equation? **c.** Solve the equation.

5. $5n = 61$
6. $-32x = 416$
7. $-12 = \frac{1}{4}p$
8. $\frac{3}{32}A = \frac{3}{4}$
9. $-210 = -4.2y$
10. $36.3 = -16.5r$

11. To solve $ax = b$ for x, multiply both sides of the equation by $\underline{\ ?\ }$.

12. Julie thinks $\frac{1}{4}$ is the solution to the equation $\frac{1}{3} \cdot m = \frac{4}{3}$. Is she correct? Why or why not?

13. Refer to Example 4. If you average 60 mph traveling from Detroit to Cincinnati instead of 55 mph, about how much time would you save?

14. Refer to Example 5. On its slowest runs, the Pony Express mail delivery took 10 full days. Then how many miles per day did the riders travel?

15. Solve $d = rt$ for r.

Applying the Mathematics

Bike to better health. *The President's Council on Physical Fitness recommends that everyone exercise daily for 30 minutes to increase endurance and strength.*

16. Solve each equation.

a. $\frac{5x}{3} = 60$ **b.** $60 = \frac{-5y}{3}$

17. **a.** 1 inch = 2.54 cm, so 12 inches = __?__.

b. Explain how the Multiplication Property of Equality can be applied to get the answer to part **a.**

18. Consider the equation $1.5(8x) = 300$.

a. Simplify the left side of the equation.

b. Solve.

c. Check your solution.

19. Solve and check $6.5 = 5(10x)$.

20. Bicycling at a moderate speed burns about 660 calories per hour for an average-sized person. How many hours would a person need to bicycle to burn 3500 calories (about 1 pound of fat)?

21. The volume of a box needs to be 500 cubic centimeters. If the base of the box has dimensions 12.5 cm and 5 cm, how high must the box be?

22. Recall that the circumference C of a circle is given by the formula $C = \pi d$, where d is the diameter.

a. Find d to the nearest hundredth, when $C = 39$ cm.

b. Solve the formula $C = \pi d$ for d.

23. The formula $F = ma$ (force = mass · acceleration) is used in physics. Solve this formula for a.

24. According to the Museum of Natural History, a cheetah has a top speed of about 70 mph, while a greyhound has a top speed of about 39 mph. If these speeds could be maintained for one minute, how much farther would the cheetah travel?

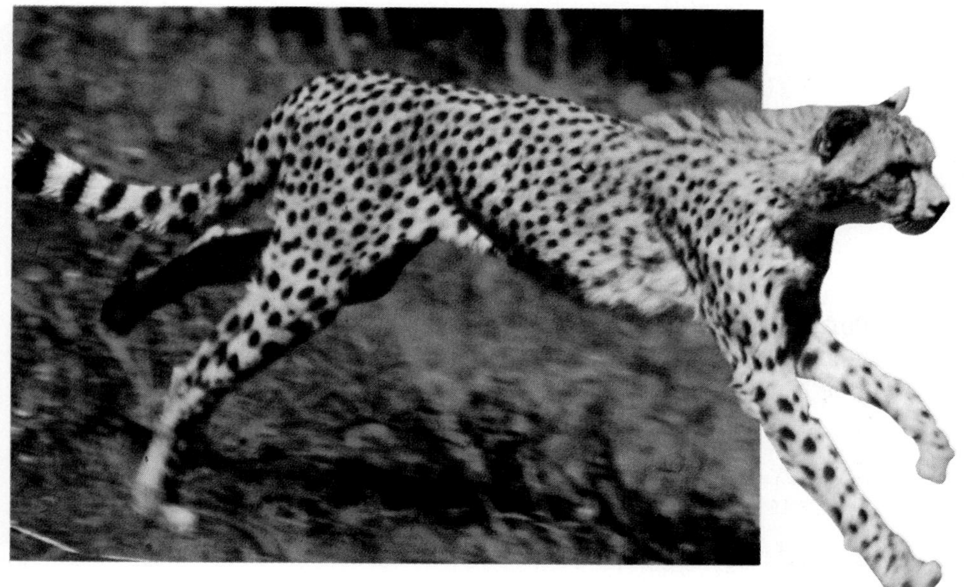

Fast cat. *The cheetah, a large cat found mainly in Africa, is the world's fastest animal when running short distances.*

Review

25. Suppose $y = x^3$. Tell whether y is positive, negative, or zero:
a. when $x > 0$. **b.** when $x < 0$. **c.** when $x = 0$. *(Lesson 2-5)*

26. *Skill sequence.* Evaluate. *(Lesson 2-5)*
a. 2^8 **b.** -2^8 **c.** $(-2)^8$
d. -2^9 **e.** $(-2)^9$

27. Use two rates to change 4.5 yards to centimeters. *(Lesson 2-4)*

28. If Irma dribbles a basketball twice a second and moves 4.5 ft per second, how many dribbles will she make moving 60 ft downcourt? *(Lesson 2-4)*

29. A magician performs at birthday parties. If the magician charges D dollars per party, how much will be earned from P parties? *(Lesson 2-4)*

30. Simplify $-x \cdot \frac{2}{15x} \cdot \frac{5}{7x}$. *(Lesson 2-3)*

31. The sketch at the right is a floor plan or map of a house. All angles are right angles.
a. Find the area of the floor plan.
b. Check your work by finding the area in another way.
c. If the house has two stories, what is the total area of the living space?
d. Another house with the same area has only one story, and has a square floor plan. What are the dimensions of its floor plan?
(Lessons 1-6, 2-1)

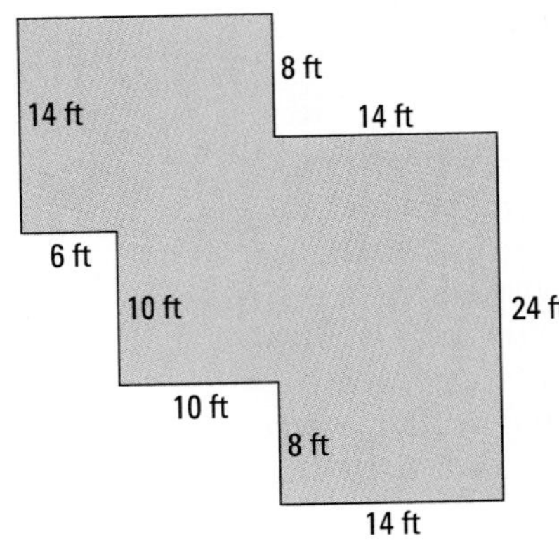

32. Tell whether the number is or is not a solution to $3x \le 20$. *(Lesson 1-2)*
a. 6 **b.** 6.5 **c.** 7

Exploration

33. A number of well-known formulas involve only multiplication. Three of them are listed below. Tell what the variables represent in each formula.
a. $A = \frac{1}{2}bh$ **b.** $C = 2\pi r$ **c.** $I = prt$

LESSON

2-7

Special Numbers in Equations

Zero in on this. *Some stores have sales that offer 0% interest if you pay the balance due on your purchase before a set period of time. Of course, you should always read the fine print. Equations involving zero also require careful attention.*

In the last lesson, you learned that all equations of the form $ax = b$ where $a \neq 0$ have exactly one solution. This solution is found by multiplying each side of the equation by $\frac{1}{a}$. Because the numbers 0 and -1 sometimes cause students trouble, in this lesson we examine what happens when these numbers are used as coefficients in equations.

Equations with 0

Example 1

Solve $0x = 4$.

Solution

By the Multiplication Property of 0, for any value of x, $0x = 0$. So $0x$ cannot equal 4. There is no solution.
Write: There is no solution.

Example 2

Solve $0x = 0$.

Solution

This is the Multiplication Property of 0. It is true no matter what value is assigned to x.
Write: All real numbers are solutions.

Notice that in Examples 1 and 2, you cannot multiply both sides by the reciprocal of 0. Why?—because 0 has no reciprocal! Be careful when 0 is the coefficient of the unknown. Zero is the only number that causes such problems.

Consider the equation $ax = b$, where $a \neq 0$, but $b = 0$. Then there is exactly one solution. This is because the coefficient a has a reciprocal.

Example 3

Solve $13x = 0$.

Solution 1

$$13x = 0$$

Multiply both sides by $\frac{1}{13}$, the reciprocal of 13.

$$\frac{1}{13} \cdot 13x = \frac{1}{13} \cdot 0$$

$$x = 0$$

Solution 2

Do it in your head. The only number that 13 can be multiplied by to get 0 is 0 itself. So $x = 0$. Write: The solution is 0.

To avoid writing sentences as explanations of unusual solutions like these, solution sets can be written. For Examples 1, 2, and 3, the solution sets are as follows.

Equation	Sentence	Solution set
$0x = 4$	There is no solution.	{ } or ø
$0x = 0$	All real numbers are solutions.	set of real numbers
$13x = 0$	The solution is 0.	{0}

Notice that {0} and { } are different. The set { }, or ø, has no elements and indicates that no number works in $0x = 4$. The set {0} contains the element 0 and indicates that 0 works in $13x = 0$.

Equations with -1

When -1 appears as a coefficient, some people find it helpful to use the Multiplication Property of -1 to write $-x$ as $-1 \cdot x$.

Example 4

Solve $-x = 3.824$.

Solution 1

Rewrite $-x$ as $-1 \cdot x$. So,

$$-1 \cdot x = 3.824.$$

Multiply both sides by -1, the reciprocal of -1.

$$-1 \cdot -1 \cdot x = -1 \cdot 3.824$$

$$x = -3.824$$

Solution 2

Translate the equation into words and find the solution in your head. The opposite of what number is 3.824? Answer: -3.824.

The following computer program will solve equations of the form $ax = b$ when the user enters values for a and b. The program finds the solution by computing $\frac{b}{a}$ except when $a = 0$. If a computer is instructed to divide by zero, the program will stop running and an error message will appear on the screen. To avoid this situation, an IF-THEN command instructs the computer to make a decision. When the sentence between the words IF and THEN is true, the computer executes the instruction following the THEN. When the sentence is false, the instruction following the THEN is completely ignored by the computer which proceeds to the next line. Lines 60, 70, and 80 use IF-THEN statements to deal with the different situations that arise when a or b is equal to zero. Notice that the BASIC symbol for "≠" is <>. Therefore, *A* <> 0 means that $a \neq 0$.

```
10 PRINT "SOLVE AX = B"
20 PRINT "ENTER A"
30 INPUT A
40 PRINT "ENTER B"
50 INPUT B
60 IF A <> 0 THEN PRINT "SOLUTION IS"; B/A
70 IF A = 0 AND B = 0 THEN PRINT "ALL REAL NUMBERS ARE SOLUTIONS"
80 IF A = 0 AND B <> 0 THEN PRINT "NO SOLUTION"
90 END
```

What will happen if this program is used to solve the equations in this lesson? Let's begin with the last example. To solve $-x = 3.824$, you enter -1 (the coefficient of x) for A and 3.824 for B. At line 60 the computer checks to see if A differs from zero. Since $-1 \neq 0$ (A is not zero), it will then print SOLUTION IS -3.824 (the result of 3.824/-1). The "then" parts of lines 70 and 80 are only carried out if A = 0, so the computer ignores these parts. The program would act much the same to solve $13t = 0$, the equation in Example 3.

In Example 2 however, 0 will be entered for both A and B. At line 60, the computer checks to see if A is different from 0. Since it is not, the rest of line 60 is skipped. At line 70, since both A = 0 and B = 0 are true, the computer prints ALL REAL NUMBERS ARE SOLUTIONS.

In solving $0x = 4$ from Example 1, A is 0, but B is 4. So line 80 is the one that fits this equation. The computer will print NO SOLUTION.

QUESTIONS

Covering the Reading

1. Why can't both sides of $3 = 0x$ be multiplied by the reciprocal of 0?

In 2–4, describe the solutions **a.** with a sentence; **b.** with the solution set.

2. $7y = 0$ **3.** $0 \cdot w = 14$ **4.** $0 = a \cdot 0$

5. *Multiple choice.* Which set is the same as ø?
(a) {ø} (b) {0} (c) 0 (d) { }

6. What is the reciprocal of -1?

In 7–9, solve.

7. $-1 \cdot x = 40$ 8. $-y = -3$ 9. $-z = 0$

In 10–12, suppose you want to solve the given equation using the computer program at the end of the lesson.
a. Give the input for A and B.
b. Tell which of lines 60, 70, and 80 will cause the computer to print.
c. Write the output of the program.

10. $0x = 1.8$ 11. $24 = -x$ 12. $0x = 0$

Applying the Mathematics

13. Refer to the formula $N = T \cdot P \cdot E \cdot L \cdot I \cdot C \cdot A$ on page 71. Some people believe that, other than Earth, the value of $L = 0$. What effect does this have on the value of N?

14. Consider the following impossible situation: The car was parked for t hours. During this time it traveled 70 miles.
a. Write an equation of the form $d = rt$ to describe this situation.
b. Explain why your equation does not have a solution.

15. Example 3 concerns the equation $13x = 0$. Describe a real-world situation that can be modeled by this equation.

16. Solve the equation.
a. $-(-x) = 18.5$ **b.** $-(-(-x)) = 18.5$

17. Solve and check: $15 = (6 - 7)x$.

18. Write an equation not given in this lesson with more than one solution.

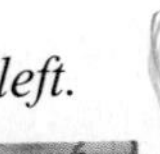

A lot of baloney. *The butcher is holding bologna in his right hand and other sausages in his left.*

Review

19. Solve and check. *(Lesson 2-6)*
a. $-4p = 12$ **b.** $12p = -4$

In 20–23, solve. *(Lesson 2-6)*

20. $\frac{7}{9}q = 140$ 21. $\frac{3n}{5} = 2$

22. $-20 = -0.04m$ 23. $\frac{1}{3} = 4x$

24. A crate contains 12 cases. Each case holds 24 boxes. Each box holds 60 packages of batteries. Each package holds 2 batteries. How many crates will it take to ship 100,000 batteries? *(Lessons 2-4, 2-6)*

25. Suppose eight ounces of sausage cost c cents. If x = cost per ounce, write a multiplication equation relating 8, c, and x. *(Lesson 2-6)*

In 26 and 27, solve an equation of the form $ax = b$ to answer the question. *(Lesson 2-6)*

26. Three fourths of the company's employees are covered by dental insurance. If 165 employees are covered, how many employees does the company have?

27. The manager of a manufacturing company knows that the company's workers can produce 340 parts per hour. A customer has ordered 5,100 parts. How many hours will it take to fill the order?

Sports of the Mayas. *Mayan ball courts can be found throughout Mexico. This one is near Oaxaca.*

28. a. Evaluate $-1 \cdot -1 \cdot -1 \cdot -1 \cdot -1 \cdot -1 \cdot -1 \cdot -1$ **b.** Evaluate $(-1)^{25}$.
c. Give a general rule for answering questions about powers of -1. *(Lesson 2-5)*

29. From about 200 B.C. to 900 A.D., the Mayas of Central America and Mexico played a ball game called *pok-ta-pok.* Playing fields for pok-ta-pok varied in size and shape. One playing field had the shape shown below. All angles are right angles. What is its area? *(Lesson 2-1)*

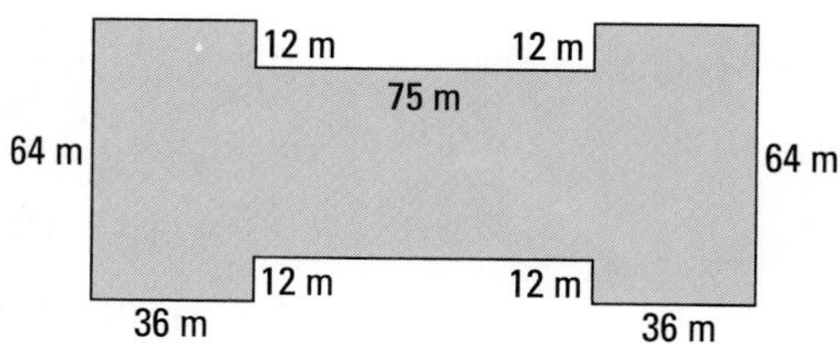

30. Graph the solution set to $x \geq 7$. Use the set of real numbers as the domain. *(Lesson 1-2)*

31. Consider the inequality $-2x < 10$. Tell whether the number is or is not a solution to the sentence. *(Lesson 1-1)*
a. 4 **b.** -4 **c.** 5 **d.** -5

Exploration

32. Consider the pattern at the right.
a. What will be written in row n?
b. What is the solution to the equation in row n?
c. What will be written in row 100?
d. What is the solution to the equation in row 100?
e. As n gets larger, to what equation are the equations getting closer and closer? What is happening to the solutions?

row 1	$x = 10$
row 2	$\frac{1}{2}x = 10$
row 3	$\frac{1}{3}x = 10$
row 4	$\frac{1}{4}x = 10$
row 100	$\underline{?}\ x = \underline{?}$
row n	$\underline{?}\ x = \underline{?}$

LESSON

2-8

Solving $ax < b$

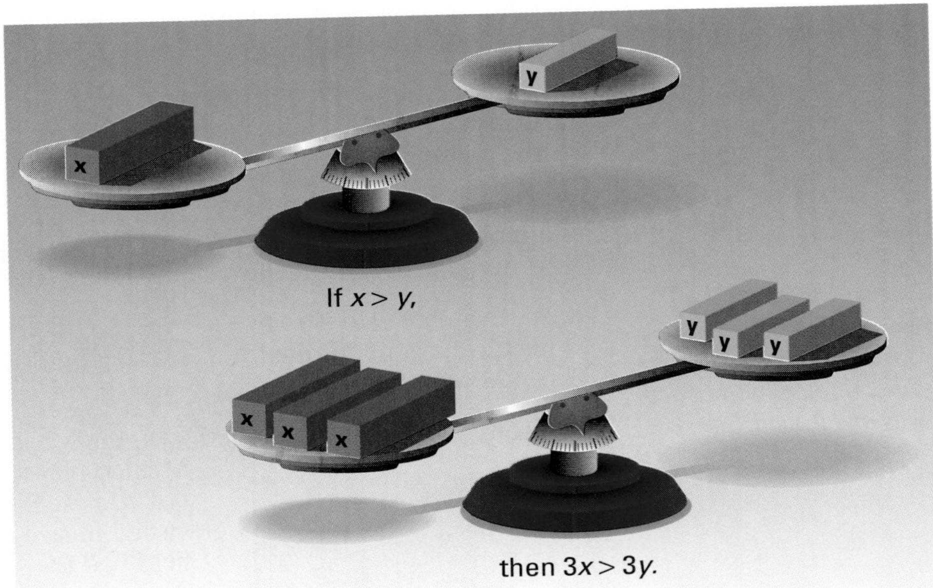

The Multiplication Property of Inequality

Here are some numbers in increasing order. Because the numbers are in order, you can put the inequality sign < between any two of them.

$$-10 < -6 < 5 < 30 < 30.32 < 870$$

Now multiply these numbers by some fixed *positive* number, say 10. Here are the products.

$$-100 \quad -60 \quad 50 \quad 300 \quad 303.2 \quad 8700$$

The order stays the same. You could still put a < sign between any two of the numbers. This illustrates that if $x < y$, then $10x < 10y$. In general, multiplication by a *positive* number maintains the order of a pair or list of numbers.

> **Multiplication Property of Inequality (Part 1)**
> **If $x < y$ and a is positive, then $ax < ay$.**

The signs >, ≤, and ≥ between numbers or expressions also indicate order. The Multiplication Property of Inequality works with any of those signs. For instance, if $x > y$, then $3x > 3y$. This is pictured on the scales above.

Solving Inequalities with Positive Coefficients

To solve an inequality of the form $ax < b$ where a is positive, use the Multiplication Property of Inequality to multiply each side by $\frac{1}{a}$.

Example 1

a. Solve $4x \leq 20$.
b. List three elements of the solution set.
c. Graph the solution set.

Solution

a. Multiply both sides by $\frac{1}{4}$. Since $\frac{1}{4}$ is positive, the Multiplication Property of Inequality tells you that

$$\frac{1}{4} \cdot 4x \leq \frac{1}{4} \cdot 20.$$
$$x \leq 5$$

b. Pick any three real numbers less than or equal to 5. *Some elements of the solution set are -10, 3, and 5.*

c. The graph pictures all real numbers less than or equal to 5.

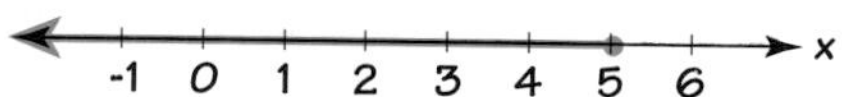

Check

Since inequalities often have infinitely many solutions, you cannot check the answer by substituting a single number. You must do two things.

Step 1: Check the boundary point by substituting it into the original inequality. It should make both sides of the inequality *equal.* The boundary point in $x \leq 5$ is 5.

Does $4 \cdot 5 = 20$? Yes.

Step 2: Check whether the sense of the inequality is correct. Pick some number that works in $x < 5$. This number should also work in the original inequality. *Check the number 3. Is $4 \cdot 3 < 20$? Yes, $12 < 20$.* Since both steps worked, $x \leq 5$ is the solution to $4x \leq 20$.

Example 2

The length of a rectangle is 50 cm. Its area is greater than 175 cm^2. Find the width of the rectangle.

Solution

50 cm

w cm

It may help to draw a picture. You know that $A = \ell w$, and $\ell = 50$ cm. So the area of the rectangle is $50w$ cm^2.

Thus, $$50w > 175.$$
$$\frac{1}{50} \cdot 50w > \frac{1}{50} \cdot 175 \quad \text{Multiply both sides by } \tfrac{1}{50}.$$
$$w > 3.5$$

The width is greater than 3.5 cm.

Check

Step 1: *Does $50 \cdot 3.5 = 175$? Yes.*

Step 2: Pick some value that works in $w > 3.5$. We pick 10. *Is $50 \cdot 10 > 175$? Yes.*

Solving Inequalities with Negative Coefficients

Here are the numbers from the beginning of this lesson.

$$-10 < -6 < 5 < 30 < 30.32 < 870$$

Multiplying these numbers by -10 yields the products below.

$$100 \quad 60 \quad -50 \quad -300 \quad -303.2 \quad -8700$$

Notice that the numbers in the first row are in *increasing* order, while the numbers in the second row are in *decreasing* order. The order has been reversed. If you multiply both sides of an inequality by a *negative* number, you must *change the direction of the inequality.*

For instance, $-6 < 5$, but $-10 \cdot -6 > -10 \cdot 5$. This idea can be generalized:

Multiplication Property of Inequality (Part 2)
If $x < y$ and a is negative, then $ax > ay$.

As in Part 1 of the Multiplication Property of Inequality, Part 2 of the property also holds for the signs $>$, $\leq$, and $\geq$. To solve an inequality of the form $ax < b$ where a is negative, multiply each side by $\frac{1}{a}$ and change the direction of the inequality sign.

Example 3

Solve $126 \leq -7x$ and check.

Solution

Multiply both sides by $-\frac{1}{7}$, the reciprocal of -7. Since $-\frac{1}{7}$ is a negative number, Part 2 of the Multiplication Property of Inequality tells you to *change the inequality sign* from $\leq$ to $\geq$.

$$-\frac{1}{7} \cdot 126 \geq -\frac{1}{7} \cdot -7x$$

Now simplify.

$$-\frac{126}{7} \geq x$$

$$-18 \geq x$$

Check

Step 1: Try -18. Does $-7 \cdot -18 = 126$? Yes.

Step 2: Try a number that works in $x \leq -18$. We use -20.
Is $-7 \cdot -20 \geq 126$? Yes, $140 \geq 126$.

Changing from $<$ to $>$, or from $\leq$ to $\geq$, or vice-versa, is called **changing the sense** of the inequality. This is the same as changing the direction of the inequality sign. The only time you have to change the sense of an inequality is when you are multiplying both sides by a negative number. Otherwise, solving $ax < b$ is similar to solving $ax = b$.

You can see why the two-step check of an inequality is important. The first step checks the boundary point in the solution. The second step checks the sense of the inequality.

Questions

Covering the Reading

In 1–3, consider the inequality $20 < 30$. What inequality results if you multiply both sides of the inequality by the given number?

1. 6 **2.** $\frac{2}{5}$ **3.** -4

4. Consider the inequality $8x \le 42$.
- **a.** Solve the inequality.
- **b.** List three elements of the solution set.
- **c.** Graph the solution set.

5. Consider the inequality $-9x < -18$.
- **a.** Tell whether or not the number is a solution.
  i. 2 **ii.** -2 **iii.** 3 **iv.** -3 **v.** -1 **vi.** 0
- **b.** Graph all solutions.

6. In what way does Part 2 of the Multiplication Property of Inequality differ from Part 1?

In 7 and 8, change the sense of each inequality.

7. $<$ **8.** $\ge$

In 9–14, solve and check each sentence.

9. $5x \ge 10$ **10.** $-3y < 300$ **11.** $-4A < -124$

12. $13 > 2z$ **13.** $-2 \le 5a$ **14.** $0.09 > -9c$

15. The length of a rectangle is 20 cm. Its area is less than 154 cm^2. Find the width of the rectangle.

A wedding scene in Seoul. *Western clothing styles are popular in Korea. However, for special occasions, Koreans wear traditional, colorful clothing made of satin or cotton.*

Applying the Mathematics

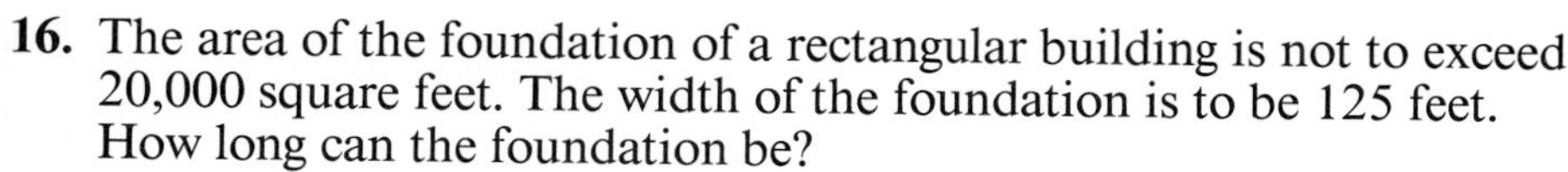

16. The area of the foundation of a rectangular building is not to exceed 20,000 square feet. The width of the foundation is to be 125 feet. How long can the foundation be?

17. An auditorium has more than 1500 seats. There are 48 seats in each row. How many rows does the auditorium have?

18. Parents of the bride have budgeted \$2500 for the dinner after the wedding. Each person's dinner will cost \$27.50. At most how many people can attend the dinner?

In 19–22, solve the inequality.

19. $-m < 8$ **20.** $-2 \ge -n$

21. $\frac{1}{4}x \ge 96$ **22.** $\frac{2}{3}p \le \frac{1}{4}$

23. Three-fourths of a number is less than two hundred four. What are the possible values of the number?

24. Use the clues to find x.
Clue 1: x is an integer.
Clue 2: $2x < 10$
Clue 3: $-3x < -9$

A note on the harpsichord. *The harpsichord first appeared in the 1300s. It has been played as a solo instrument, with orchestras, and in chamber-music ensembles. Famous composers of the 1700s, such as Bach and Mozart, played ornately decorated harpsichords similar to this one.*

Review

25. *Skill sequence.* Solve the equation. *(Lesson 2-7)*
a. $1n = 4$ **b.** $0n = 3$ **c.** $-1n = 2$ **d.** $-2n = 1$

In 26 and 27, solve. *(Lesson 2-6)*

26. $30\pi = \pi d$

27. $200 = -4(5x)$

28. A trill is a musical term for alternating very quickly between two notes a step or half-step apart. In Johann Sebastian Bach's *Two-Part Invention #4,* the harpsichordist or pianist is required to trill a note with the left hand for five measures. If a trill has four notes per beat, and there are three beats in each measure, how many notes are played? *(Lesson 2-4)*

29. Multiply and simplify $\frac{1}{6}x^3 \cdot \frac{2}{5x}$. *(Lesson 2-3)*

30. A box is made by folding the pattern at right along the blue lines and taping the edges. What is the volume of the box? *(Lesson 2-1)*

31. The formula $A = \frac{1}{2}bh$ for the area of a triangle is true for a right triangle because the area of a right triangle with legs b and h is half the area of a rectangle with sides b and h. Write a formula for the area of a triangle whose area is half the area of a square with sides of length s. *(Lesson 1-5)*

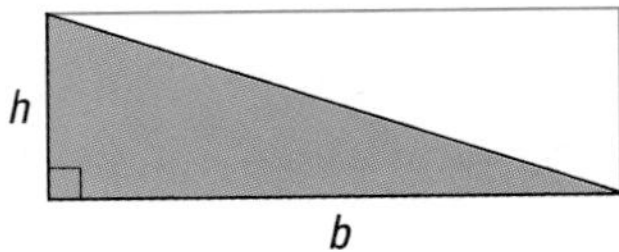

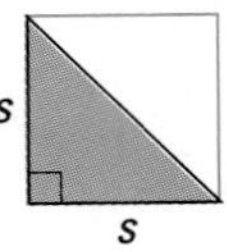

32. Write using exponents: $3 \cdot 3 \cdot x \cdot x \cdot x \cdot y \cdot y \cdot y \cdot y \cdot y$. *(Previous course)*

Exploration

33. Find a number $x < 0$ such that $0.05 < x^2 < 0.06$. (A calculator may help.)

LESSON

2-9

The Multiplication Counting Principle

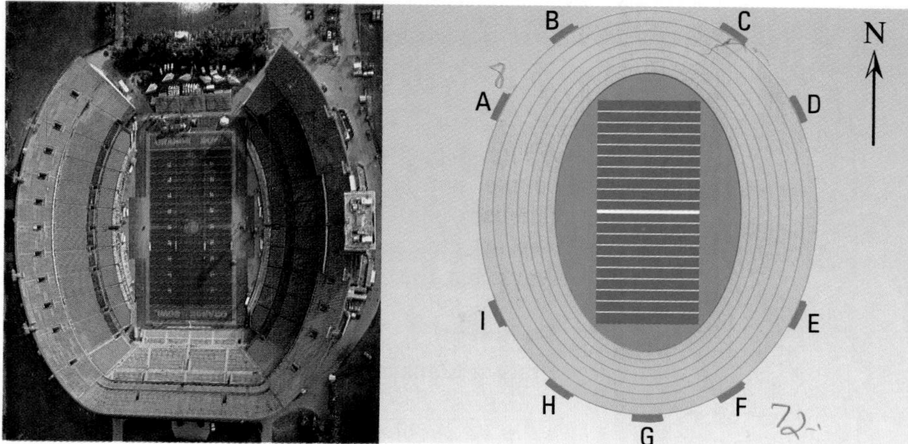

In mathematics, a procedure is said to be *elegant* if it is both clever and simple. For instance, it may surprise you that multiplication can be cleverly used to solve many types of counting problems. If you can organize the items being counted into rectangular arrays, the area model of multiplication allows you to multiply to get the counts.

Example 1

Suppose a stadium has 9 gates as in the drawing above. Gates *A, B, C,* and *D* are on the north side. Gates *E, F, G, H,* and *I* are on the south side. In how many ways can you enter the stadium through a north gate and leave through a south gate?

Solution

Create a rectangular array in which each entry is an ordered pair. The first letter represents a gate you enter; the second stands for an exit gate.

		Exit Gate (South)				
		E	F	G	H	I
	A	(A,E)	(A,F)	(A,G)	(A,H)	(A,I)
Entry	B	(B,E)	(B,F)	(B,G)	(B,H)	(B,I)
Gate	C	(C,E)	(C,F)	(C,G)	(C,H)	(C,I)
(North)	D	(D,E)	(D,F)	(D,G)	(D,H)	(D,I)

Notice that the array has 4 rows and 5 columns, so there are $4 \cdot 5 = 20$ pairs in the table. There are 20 ways of entering through a north gate and leaving through a south gate.

This elegant use of multiplication occurs often enough that we give it a special name, the *Multiplication Counting Principle.*

Multiplication Counting Principle
If one choice can be made in m ways and a second choice can be made in n ways, then there are mn ways of making the first choice followed by the second choice.

The Multiplication Counting Principle can be extended to situations where more than two choices must be made.

Example 2

A high-school student wants to take a foreign-language class, a music course, and an art course. The language classes available are French, Spanish, and German. The music classes available are chorus and band. The art classes available are drawing and painting. In how many different ways can the student choose the three classes?

Solution

Draw a blank for each decision to be made.

________ · ________ · ________
ways to choose language · ways to choose music · ways to choose art

Now fill in the blanks with the number of ways each subject can be chosen. There are 3 choices in foreign language, 2 choices in music, and 2 choices in art. Use the Multiplication Counting Principle.

3 · 2 · 2
ways to choose language · ways to choose music · ways to choose art

There are $3 \cdot 2 \cdot 2 = 12$ choices.

In Example 2, the Multiplication Counting Principle quickly told *how many* ways a student can choose his courses, but it did not tell *what* the choices are. One way to see all of them is to make an organized list.

French-chorus-drawing	Spanish-chorus-drawing	German-chorus-drawing
French-chorus-painting	Spanish-chorus-painting	German-chorus-painting
French-band-drawing	Spanish-band-drawing	German-band-drawing
French-band-painting	Spanish-band-painting	German-band-painting

A second way is to use a *tree diagram,* which requires less writing.

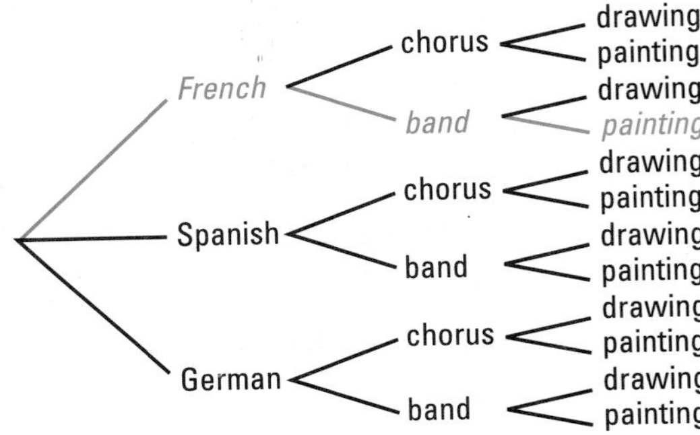

Each choice can be found by following a path from the left to the right in the diagram. One possible choice is shown in blue: *French–band–painting.*

You can count 12 paths. That is, twelve different choices are possible.

Example 3

Mr. Lorio is giving his algebra class a quiz with five questions. Since Angie has not done her homework, she has to guess. The quiz has two multiple-choice questions with choices A, B, C, and D, and three true-false questions.

a. How many possible ways are there for Angie to answer all five questions?

b. What is the probability that Angie will get all the questions correct?

Solution

a. There are 4 choices for each multiple-choice question and 2 choices for each true-false question. Use the Multiplication Counting Principle.

4	•	4	•	2	•	2	•	2
choices for question #1		choices for question #2		choices for question #3		choices for question #4		choices for question #5

There are $4 \cdot 4 \cdot 2 \cdot 2 \cdot 2 = 128$ different ways of answering the five questions.

b. Only one of the 128 possible outcomes has the correct answer for each question. With random guessing, Angie has only 1 chance out of 128 of getting all the questions correct. The probability that she will get all the answers correct is $\frac{1}{128}$.

When the number of choices is not precisely known, variables may appear in counting problems.

Example 4

Ms. Alvarez has written a chapter test. It has three multiple-choice questions each with m possible answers, two multiple-choice questions each with n possible answers, and 5 true-false questions. How many ways are there to answer the questions?

Solution

Make a blank for each of the 10 questions. Fill each blank with the number of possible answers for that question.

Question Number	1	2	3	4	5	6	7	8	9	10
Choices	m	m	m	n	n	2	2	2	2	2

Apply the Multiplication Counting Principle to get

$$m \cdot m \cdot m \cdot n \cdot n \cdot 2 \cdot 2 \cdot 2 \cdot 2 \cdot 2 \text{ sets of answers.}$$

With exponents, this product can be expressed as $m^3n^22^5$ or $32m^3n^2$ sets of answers.

QUESTIONS

Covering the Reading

In 1–3, refer to Example 1.

1. In how many ways can a person enter through a north gate and leave through gate G?

2. Using an array, list the ways to enter the stadium through a north gate and leave through a south gate if gates D, G, and H are closed.

3. **a.** Suppose in Example 1 that a person could enter through any gate and leave through any gate. In how many ways can this be done?
   b. Suppose in Example 1 that a person could enter through any gate and leave through any *other* gate. In how many ways can this be done?

4. State the Multiplication Counting Principle.

5. In mathematics, when is a procedure said to be *elegant?*

6. Suppose the school in Example 2 offered Russian as a fourth language choice.
   a. Now how many ways could a student choose a schedule?
   b. Draw a tree diagram showing all the possible choices of schedules.
   c. Noah is taking Russian, chorus, and painting. Underline his schedule.

7. Each of 20 questions on a quiz can be answered *true* or *false.* Suppose a student guesses randomly.
   a. How many ways are there of answering the test?
   b. What are the chances of getting all of the answers correct?

8. Suppose Ms. McCullagh gives a quiz that has two questions with x choices and three true-false questions. Give the number of ways to answer the items on the test with an expression:
   a. not using exponents; **b.** using exponents.

9. If a test has five multiple-choice questions each with q possible answers and five true-false questions, how many different ways would there be to answer the test?

WBEZ. *This woman is the news director of Chicago's public radio station WBEZ. Public radio stations feature in-depth coverage of national and local issues, talk shows, and interviews.*

Applying the Mathematics

10. Radio station call letters, such as WNEW, must start with W or K.
   a. How many choices are there for the first letter?
   b. How many choices are there for the second letter?
   c. How many different 4-letter station names are possible?

11. Vince is buying a new suit. He must decide if he wants wide or narrow lapels. He must also choose among four colors—blue, gray, tan, and brown. Draw a tree diagram showing all the possible choices he has.

12. At the Fulton High School cafeteria, students had a choice of chicken or fish. The vegetable choices were carrots or beans. There were three dessert choices: an apple, pudding, or yogurt.

a. Organize the possible meals consisting of one main dish, one vegetable, and one dessert using either a tree diagram or a list.

b. How many different such meals are possible?

13. Aram, Brad, Carl, and Dave are candidates for Winter Carnival King. Janice, Kara, and Leshawn are the nominees for Winter Carnival Queen.

a. Using initials, write the names of all the possible "Royal Couples."

b. If the king and queen are chosen at random, what is the probability that the Royal Couple will be Aram and Kara?

14. Telephone area codes consist of 3 digits. Prior to 1989, they fit the following rule: The first digit must be chosen from 2 through 9, the second digit must be 0 or 1, and the third digit cannot be 0. In 1989, this policy changed so that the first and third digits could be any digit 0 through 9. (The second digit still must be 0 or 1.)

a. How many area codes were possible before 1989?

b. How many area codes were possible after 1989?

c. How many area code possibilities were added by the policy change?

15. Write a Multiplication Counting Principle problem whose answer is $2 \cdot 3 \cdot 5$.

16. The Cayuga Indians played a game called Dish using a wooden bowl and six pits from peaches. The pits were blackened on one side and uncolored on the other. When pits were tossed, they landed on the black and uncolored sides with about the same frequency. A player scored five points if the six pits were tossed and six black or six uncolored sides landed up.

a. Use the Multiplication Counting Principle to determine how many possible ways the six pits could land.

b. What is the probability that a player would score 5 points with one toss?

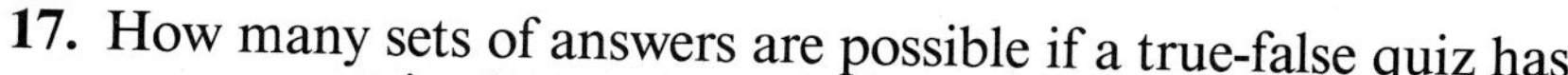

17. How many sets of answers are possible if a true-false quiz has

a. one question? **b.** two questions?

c. three questions? **d.** ten questions?

e. n questions?

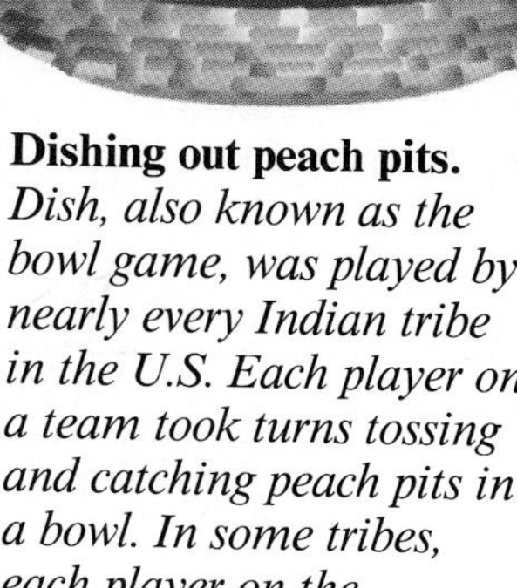

Dishing out peach pits. *Dish, also known as the bowl game, was played by nearly every Indian tribe in the U.S. Each player on a team took turns tossing and catching peach pits in a bowl. In some tribes, each player on the winning side would be given a pony by members of the losing side.*

Review

In 18–21, solve. *(Lessons 2-6, 2-8)*

18. $4j < 13$

19. $-20k = \frac{4}{5}$

20. $\frac{1}{3} > -6m$

21. $-96 = -0.08n$

22. Last year a family paid $12,750, more than one third of their earned income, in income taxes. *(Lesson 2-8)*

a. Let I = the family's earned income last year. Write a sentence that describes the situation above.

b. Solve the sentence.

23. *Skill sequence.* Solve the sentence. *(Lessons 2-7, 2-8)*

a. $0x = 5$ **b.** $5x = 0$

c. $5x < 0$ **d.** $-5x < 0$

24. a. Multiply: $\frac{2}{5}x \cdot 20$. **b.** Solve: $\frac{2}{5}x = 20$.

c. Explain in your own words how the questions in parts **a** and **b** are different. *(Lessons 2-3, 2-6)*

25. Solve $\frac{-x}{6} = 17$. *(Lesson 2-6)*

26. If a farmer harvests 40 bales of hay per acre, how many acres will produce a harvest of 224 bales? *(Lessons 2-4, 2-6)*

27. All angles in this figure are right angles. Write an expression for the area of the figure. *(Lesson 2-1)*

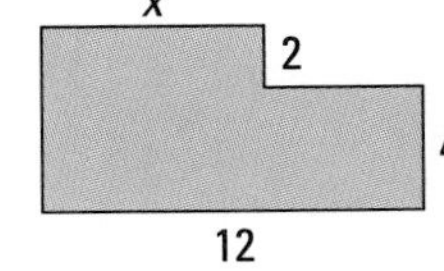

28. Evaluate $6(3xy)$ when $x = -\frac{7}{3}$ and $y = -\frac{32}{5}$. Write your answer as a fraction. *(Lessons 1-4, 2-3)*

29. Evaluate when $d = -\frac{1}{2}$. *(Lessons 1-4, 2-5)*

a. d^3 **b.** $3d + 4$ **c.** $-8d^4$

30. At Harwood High, $\frac{3}{8}$ of the students take French, and $\frac{1}{4}$ of the French students are in Ms. Walker's French class. What fraction of Harwood High students are in Ms. Walker's French class? *(Lesson 2-3)*

Exploration

In 31 and 32, use the fact that an *acronym* is a name made from first letters of words or parts of words.

31. Here are some famous acronyms. Tell what the letters stand for.

a. NASA **b.** UNICEF **c.** CIA

d. IBM **e.** ICBM **f.** AFL-CIO

g. NFL **h.** NATO **i.** UNESCO

32. Is it true that over a half million 4-letter acronyms are possible in English? Explain your answer using the Multiplication Counting Principle.

LESSON

2-10 Factorials and Permutations

Mount Rushmore. *The construction of Mount Rushmore began in 1927 and took over 14 years to complete. The four busts were cut out of the granite cliff with dynamite and drills.*

The Factorial Symbol

A special case of the Multiplication Counting Principle occurs when a list of things is to be ranked or ordered.

On Mount Rushmore in South Dakota, the sculptor Gutzon Borglum carved busts of four presidents of the United States. From left to right they are George Washington, Thomas Jefferson, Theodore Roosevelt, and Abraham Lincoln. Some students were asked to rank these men in order of greatness. Here are three possible rankings.

1st place	2nd place	3rd place	4th place
Washington	Lincoln	Roosevelt	Jefferson
Jefferson	Lincoln	Washington	Roosevelt
Lincoln	Washington	Jefferson	Roosevelt

How many rankings are possible? This question can be answered using the Multiplication Counting Principle. There are 4 people who could be ranked first. After choosing someone for first place, there are only 3 people left who could be second. Then, after 1st and 2nd places have been chosen, there are only 2 people left who could be third, and the remaining person will be last.

4	·	3	·	2	·	1
ways to choose 1st place		ways to choose 2nd place		ways to choose 3rd place		ways to choose 4th place

The answer is $4 \cdot 3 \cdot 2 \cdot 1 = 24$.

A shortcut way to write $4 \cdot 3 \cdot 2 \cdot 1$ is 4!. This is read "four *factorial.*"

Definition: When n is a positive integer, the symbol $n!$ (read **n factorial**) means the product of the integers from n to 1.

Example 1

Evaluate 5!.

Solution

$$5! = 5 \cdot 4 \cdot 3 \cdot 2 \cdot 1 = 120$$

Presidential facelift. *These workers are restoring the sculpture to its original beauty.*

Scientific calculators usually have a factorial key [x!]. To evaluate 5! on such calculators, key in 5 [x!]. On some calculators you may have to use a second function key, [2nd] or [INV]. Then key in 5 [INV] [x!]. On graphics calculators there is often no factorial key visible. However, these machines usually have a [MATH] key. When you press the [MATH] key you will see a menu that includes a way to calculate factorials. Check your calculator now by using it to evaluate 5!. You should get 120.

Permutations

An ordered arrangement of letters, names, or objects is called a **permutation.** In the discussion of Mount Rushmore, we calculated that there are 4! permutations of four names. In general, when you are making arrangements of all n items in a set, the following theorem applies.

Permutation Theorem
There are $n!$ possible permutations of n different items when each item is used exactly once.

Example 2

A baseball manager is setting a batting order for the 9 starting players. How many batting orders are possible?

Solution

The batting order is an arrangement of the starting players. Each player is used only once so the Permutation Theorem applies.
There are 9 starting players to be ordered. So, there are 9! possible batting orders. With a calculator, evaluate 9!.

$$9! = 362{,}880$$

Businesses are often interested in a type of problem that involves permutations. One type of permutation problem is called a "traveling salesman problem." It involves finding the shortest path between locations. For instance, if a salesman needs to visit 12 cities there are 12!, or 479,001,600 different orders in which he could make his stops. If there are 532 cities, there are 532! possible routes.

Problems of this type need not involve just salespeople. Deciding in what order to connect telephone lines is another example of a "traveling salesman problem." In 1986, a computer was used to find the shortest route connecting the 532 cities with central office switching systems of local telephone companies. The solution is shown below.

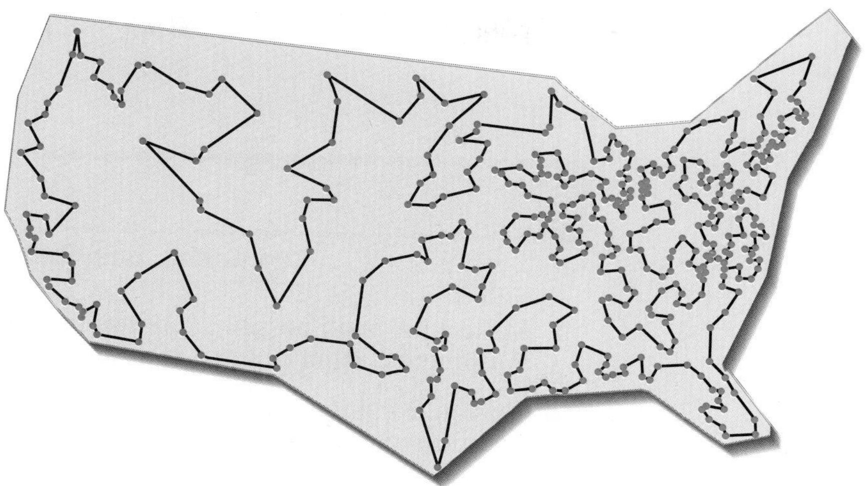

Shortcuts for Evaluating Expressions with Factorials

Problems involving $n!$ often become so difficult that even with computers, cleverness is needed to simplify expressions. Some calculations with factorials can be done more quickly with pencil and paper than with a calculator. This is often the case when factorials are divided.

Example 3

Evaluate $\frac{12!}{10!}$.

Solution 1

Express the factorials as products and use the Equal Fractions Property to simplify.

$$\begin{aligned}\frac{12!}{10!} &= \frac{12 \cdot 11 \cdot \cancel{10} \cdot \cancel{9} \cdot \cancel{8} \cdot \cancel{7} \cdot \cancel{6} \cdot \cancel{5} \cdot \cancel{4} \cdot \cancel{3} \cdot \cancel{2} \cdot \cancel{1}}{\cancel{10} \cdot \cancel{9} \cdot \cancel{8} \cdot \cancel{7} \cdot \cancel{6} \cdot \cancel{5} \cdot \cancel{4} \cdot \cancel{3} \cdot \cancel{2} \cdot \cancel{1}} \\ &= 12 \cdot 11 \\ &= 132\end{aligned}$$

Solution 2

Write out only as many of the factors of the factorials as necessary.

$$\begin{aligned}\frac{12!}{10!} &= \frac{12 \cdot 11 \cdot \cancel{10!}}{\cancel{10!}} \\ &= 12 \cdot 11 \\ &= 132\end{aligned}$$

Check

Use a calculator to evaluate the factorials. Key in 12 [x!] [÷] 10 [x!] [=]

$$\frac{12!}{10!} \approx \frac{4.79 \cdot 10^8}{3{,}628{,}000} \approx 132$$

(Notice that 12! is so large that many calculators must express it in scientific notation.)

QUESTIONS

Covering the Reading

1. Make a list of all possible rankings in a poll with the three presidents Washington, Jefferson, and Lincoln.

2. Gutzon Borglum did not have to put the presidents in the order from left to right as he did. How many other orders are possible?

3. What is a short way to write $6 \cdot 5 \cdot 4 \cdot 3 \cdot 2 \cdot 1$?

4. What does the symbol $n!$ mean?

5. Evaluate $n!$ when n equals each of the following.

a. 1 **b.** 2 **c.** 3
d. 4 **e.** 5 **f.** 6

In 6 and 7, evaluate with a calculator.

6. 15! **7.** 30!

8. What is a permutation?

9. Five singers show up to audition for a job. Explain why there are 5! possible orders in which they can be asked to sing.

10. In how many ways can a saleswoman arrange a trip to visit 18 cities?

11. How many permutations of n different objects are possible when each object is used exactly once?

12. **a.** Explain how to evaluate $\frac{11!}{9!}$ without using a calculator.

b. Check by using a calculator.

In 13 and 14, evaluate.

13. $\frac{6!}{3!}$ **14.** $\frac{25!}{24!}$

Many art museums employ docents who serve as tour guides and art lecturers for small groups. This group is at the Mint Museum of Art in Charlotte, North Carolina.

Applying the Mathematics

15. The curator of a museum must arrange twenty pictures in a line on a wall. In how many ways can the pictures be arranged?

16. Suppose eight horses are in a race.
a. In how many ways can first and second places be won?
b. In how many different orders can all eight horses finish?

17. a. *True or false.* $8! = 8 \cdot 7!$
b. *True or false.* $11! = 11 \cdot 10!$
c. Use one variable to write a generalization of the pattern in parts **a** and **b**.

18. a. Evaluate $\frac{n!}{(n-1)!}$ when $n = 10$.
b. Give another instance of $\frac{n!}{(n-1)!}$.
c. Generalize parts **a** and **b**.

19. Because of their limited memories, most calculators cannot evaluate 100! or 100^{100}. Explain how to determine which of 100! or 100^{100} is larger.

Review

20. Suppose you have just won a new car. You have your choice of body style (2-door, 4-door, or station wagon), transmission (automatic or standard), and color (white, black, red, silver, or green). *(Lesson 2-9)*
a. How many different ways can you make your choices?
b. If another color choice of yellow is given to you, how many more choices do you have?

21. A multiple-choice test has five questions. The first three questions have four choices each. Questions 4 and 5 have five choices each. If Mary guesses on all the questions, what are the chances that Mary will get them all correct? *(Lesson 2-9)*

22. *Skill sequence.* Solve each sentence. *(Lessons 2-6, 2-8)*
a. $-2a = 7$ **b.** $7a = -2$
c. $7a \geq -2$ **d.** $-7a \geq 2$

In 23 and 24, a sentence is given. **a.** Make up a question about a real situation that can be answered by solving the equation or inequality. **b.** Solve the sentence and answer your question. *(Lesson 2-6)*

23. $10x = 723$ **24.** $130 > 2.5x$

In 25 and 26, solve. *(Lessons 2-6, 2-8)*

25. $\frac{3}{8}y < \frac{5}{4}$ **26.** $(2 - 3)t \leq 8$

27. Which property of multiplication is used to conclude from $\frac{2}{3}x = 18$, that $\frac{3}{2}\left(\frac{2}{3}x\right) = \frac{3}{2} \cdot 18$? *(Lesson 2-6)*

28. Lamar earns \$7.28 per hour. Last week he earned \$101.92. How many hours did he work? *(Lessons 2-4, 2-6)*

29. There are about 16,000 grains of sand per cubic inch, and 1728 cubic inches per cubic foot. About how many grains of sand are in a 25-cubic foot sandbox? *(Lesson 2-4)*

30. Write the reciprocal of $\frac{50n^3}{233m^5}$. *(Lesson 2-2)*

31. Give an instance of the Commutative Property of Multiplication. *(Lesson 2-1)*

32. One rectangular solid is half as long, $\frac{2}{3}$ as wide, and $\frac{3}{4}$ as high as another. How do the volumes of the two solids compare? *(Lessons 2-1, 2-3)*

33. Consider the statement $a + b > a$. *(Lesson 1-7)*
a. Give an instance of the statement.
b. Give a counterexample to the statement.
c. For what values of b is the statement true?

Exploration

34. What is the largest value of n for which your calculator can calculate or estimate $n!$?

35. What is the smallest value of n for which $n!$ is divisible by 100?

36. Key in 2.5 [x!] on a calculator. Explain what happens.

37. Some puzzles involve permutations of letters.
a. Unscramble these letters to spell four mathematical terms.

L U Q A E　　　E S H P R E
I R I D A　　　R I N G I O

b. Unscramble the last letters of the words you found to spell another mathematical term.

A project presents an opportunity for you to extend your knowledge of a topic related to the material of this chapter. You should allow more time for a project than you do for typical homework questions.

1 Extraterrestrial Life

Interview a number of people about their views of life on other planets in our galaxy. What values do they give for $T, P, E, L, I, C,$ and A on page 71? Calculate N for their values. Does the calculated value agree with their views?

2 Properties of Operations

Multiplication and addition are both commutative and associative.

a. Are subtraction and division commutative and associative? How do you know?

b. Consider the operation # in which, for all x and y, $x \# y = xy + 2$. For example,

$$1 \# 2 = 4$$
$$2 \# 2 = 6$$
$$3 \# 1 = 5$$
$$5 \# 2 = 12.$$

c. Is # a commutative operation? Is # an associative operation? Explain your reasoning. Design your own operation. Give a rule and several examples. Is your operation associative? Is it commutative?

3 Making a Box

How large a box can you make by cutting out square corners from a rectangular piece of paper?

a. Begin with an $8\frac{1}{2}''$ by $11''$ sheet of paper. Cut a $1''$ square from each corner. Fold and tape to make an open box.

b. What is the volume of the box you have made?

c. By cutting different-sized corners, you can create boxes of different volumes. Experiment to see how large a volume you can create using this method. Organize your data in a table and write a brief report describing your methods and results.

4 The King's Gold Coins

A king has eight gold coins. One of the coins is counterfeit and weighs less than the others. The king has only a balance scale. How can the king find the counterfeit coin using the *least* number of weighings?

(continued)

5 Rates

As you have seen in this chapter, rates have many uses. Choose a situation that interests you, such as playing a favorite sport, taking a trip, and so on. Make up a short story about that situation. Use at least ten different rate factor multiplications.

7 Telephone Numbers

How many different telephone numbers would be possible for a given area code if there were no restrictions? In all areas, however, there are restrictions. For example, in most places, no 7-digit telephone number can begin with 911. Find out what restrictions exist for your area code. Considering the restrictions, how many telephone numbers are possible for your area code?

6 Window Washer

How long would it take one window washer to wash all the windows in the Sears Tower or in some other famous skyscraper? What information do you need and how are you going to get it? If each of the windows is to be washed at least twice a year, how many window washers need to work at once to keep the windows clean? Explain your reasoning, estimates, and answers.

SUMMARY

Multiplication has many uses. The product xy may stand for any one of the following.

Area Model: xy is the area of a rectangle with length x and width y.

Area Model (discrete version): xy is the number of elements in a rectangular array with x rows and y columns.

Rate Factor Model: xy is the result of multiplying a rate x by a quantity y.

Multiplication Counting Principle: xy is the number of ways of making a first choice followed by a second choice, if the first choice can be made in x ways and the second can be made in y ways.

The product of three numbers xyz may be the volume of a box with dimensions, x, y, and z. The product of an odd number of negative numbers is negative; the product of an even number of negative numbers is positive. The product of the integers from 1 to n, written $n!$, is the number of ways of arranging n objects. The commutative and associative properties allow you to rearrange multiplication expressions.

The numbers 0, 1, and -1 are special in multiplication. Multiplying any number by zero gives the same result: 0. For this reason, equations of the form $0x = b$ are either true for all real numbers or for none of them. Multiplying any number by 1 yields that number. A conversion factor is the number 1 written using different units, so multiplying by it does not change the value of a quantity. Multiplying any number by -1 changes it to its opposite.

Because of the many uses of multiplication, equations of the form $ax = b$ and inequalities of the form $ax < b$ are quite common. When $a \neq 0$, such sentences can be solved by multiplying both sides by the number $\frac{1}{a}$, the reciprocal of a. Remember to change the sense of the inequality if a is negative.

VOCABULARY

Below are the most important terms and phrases for this chapter. You should be able to give a general description and a specific example of each.

Lesson 2-1
model
Area Model for Multiplication
Commutative Property of Multiplication
discrete set
rectangular array, dimension
rectangular solid, box
volume
Associative Property of Multiplication

Lesson 2-2
Multiplicative Identity Property of 1
reciprocals, multiplicative inverses
Property of Reciprocals
Reciprocal of a Fraction Property
Multiplication Property of Zero

Lesson 2-3
Multiplying Fractions Property
algebraic fractions
Equal Fractions Property

Lesson 2-4
Rate Factor Model for Multiplication
reciprocal rates

Lesson 2-5
opposites
Multiplication Property of -1
Rules for Multiplying Positive and Negative Numbers
Properties of Multiplication of Positive and Negative Numbers

Lesson 2-6
Multiplication Property of Equality

Lesson 2-7
IF-THEN

Lesson 2-8
Multiplication Property of Inequality
direction of an inequality sign
sense of an inequality

Lesson 2-9
Multiplication Counting Principle
tree diagram

Lesson 2-10
n factorial, $n!$
permutation
Permutation Theorem

PROGRESS SELF-TEST

Take this test as you would take a test in class. You will need graph paper. Then check your work with the solutions in the Selected Answers section in the back of the book.

1. Evaluate $\frac{22!}{20!}$.
2. *True or false.* $(-5)^{10} = -5^{10}$. Explain your reasoning.

In 3–5, multiply and simplify the result.

3. $\frac{20x}{3y} \cdot \frac{5}{4x}$
4. $\frac{4}{x^2} \cdot \frac{11}{2x}$
5. $-5a \cdot \frac{a}{5}$
6. Give an instance of the Commutative Property of Multiplication.

In 7–10, solve.

7. $50x = 10$
8. $\frac{1}{4}k = -24$
9. $15 \leq 3m$
10. $-y \leq -2$
11. **a.** Solve $-2n < 18$. **b.** Graph the result on a number line.
12. Solve $-48 = -\frac{4}{3}n$, and show how to check your answer.

In 13 and 14, give the reciprocal.

13. $-\frac{3}{n}$
14. 3.2
15. Write using symbols: The product of a number and the opposite of its reciprocal is -1.
16. How many centimeters are there in 8 inches? Use 1 in. = 2.54 cm.
17. A town is planning a new tax on cars owned by its citizens. The town planners want to charge $15 per car. In the U.S. there is an average of 0.57 car per person. If the population of the town is 24,000, about how much tax should the planners expect to collect?
18. What is the volume of a box with dimensions $4n$, $8n$, and $1.5n$?
19. How long will it take to drive from Chicago to St. Louis, a distance of about 300 miles, if you travel at 55 mph?

The Gateway Arch in St. Louis, Missouri

 CHAPTER TWO

PROGRESS SELF-TEST

20. A cement contractor has been hired to pave this courtyard. It is in the shape of a square that is 80 ft on a side. In the center is a fountain that will not be paved. The fountain is also square, with sides 15 ft long. How many square feet will be paved?

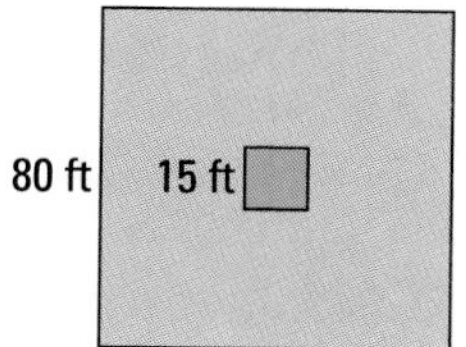

In 21 and 22, a number is given.

a. Tell if the number is positive or negative.

b. Explain how you know.

21. (-528,500)(-3,000,000)(4,150,000)

22. $(-287)^{51}$

23. Refer to the shaded rectangle below.

a. What are its length and width?

b. What is its area?

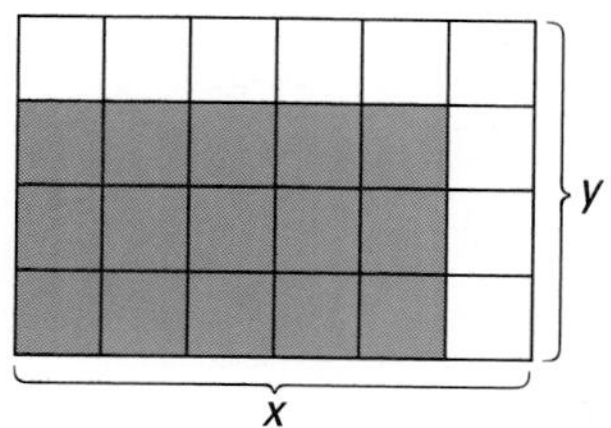

24. a. Give an example of an equation that has no real solution.

b. Explain why your equation has no real solution.

25. Mr. and Mrs. Williams bought a load of 600 used bricks to build a garden wall. They measured and found that they would need 40 bricks in each row.

a. Let r = the number of rows in the wall. Write an equation to describe this situation.

b. How many rows high can they make their wall?

26. At the Central Park Day Camp, a child must choose a sport (7 choices), a craft activity (5 choices), and a nature activity (3 choices). How many different programs can be designed?

27. In how many ways could you arrange five musicians in a row for a photo?

28. How many different sets of answers are possible on a 25-question true-false test?

After taking and correcting the Self-Test, you may want to make a list of the problems you got wrong. Then write down what you need to study most. If you can, try to explain your most frequent or common mistakes. Use what you write to help you study and review the chapter.

CHAPTER REVIEW

Questions on SPUR Objectives

SPUR stands for **S**kills, **P**roperties, **U**ses, and **R**epresentations. The Chapter Review questions are grouped according to the SPUR Objectives for this chapter.

SKILLS DEAL WITH THE PROCEDURES USED TO GET ANSWERS

Objective A: *Multiply and simplify algebraic fractions.* *(Lesson 2-3)*

In 1–6, multiply the fractions. Simplify if possible.

1. $\frac{9x}{10} \cdot \frac{3}{4x}$ **2.** $\frac{3}{5} \cdot \frac{n}{2}$

3. $\frac{7x}{2} \cdot \frac{2}{7y}$ **4.** $\frac{ax}{3} \cdot \frac{6x}{a}$

5. $\frac{3n}{4} \cdot 4$ **6.** $y \cdot \frac{5y}{3z} \cdot z^2$

In 7 and 8, simplify the fraction.

7. $\frac{13pqr}{39pq}$ **8.** $-\frac{48xy^2}{32x^3y^2}$

Objective B: *Multiply positive and negative numbers.* *(Lesson 2-5)*

In 9–12, evaluate.

9. $-6 \cdot 15 \cdot -3$ **10.** $-24 \cdot -\frac{1}{2} \cdot -2$

11. $(-2)^3$ **12.** $(-5)^4$

In 13 and 14, tell if the expression is positive or negative.

13. $-3(6.2)(-872)(-6)$ **14.** $(-395)^{10}$

15. Evaluate.

a. $(-6)^2$ **b.** -6^2 **c.** $(-6)^3$ **d.** -6^3

16. *True or false.* $(-9)^6 = -9^6$. Explain your reasoning.

17. If $a = -8$, is $-3a$ positive or negative?

18. Describe the property that is used to determine if the product of two numbers is positive or negative.

Objective C: *Solve and check equations of the form $ax = b$.* *(Lessons 2-6, 2-7)*

In 19–24, solve and check.

19. $2.4\ m = 360$ **20.** $-\frac{1}{2}k = -10$

21. $-2 = 0.4h$ **22.** $12 = 36m$

23. $\frac{5}{3}n = -45$ **24.** $4(5x) = 0$

25. Solve $d = cg$ for g.

26. Solve $ky = 2z$ for y.

In 27 and 28, give an example of an equation of the form $ax = b$ that has:

27. no solution.

28. all real numbers as solutions.

Objective D: *Solve and check inequalities of the form $ax < b$.* *(Lesson 2-8)*

In 29–32, solve and check.

29. $8m \le 16$ **30.** $-250 < 5y$

31. $-6u > 12$ **32.** $-x \ge -1$

In 33 and 34, a sentence is given. **a.** Solve. **b.** Graph the solution set on a number line.

33. $-\frac{1}{2}g \ge 5$ **34.** $3.6h < 720$

Objective E: *Evaluate expressions containing a factorial symbol.* *(Lesson 2-10)*

In 35 and 36, evaluate without using a calculator.

35. $4! + 3!$ **36.** $\frac{16!}{14!}$

In 37 and 38, use a calculator to evaluate.

37. $15!$ **38.** $24!$

In 39 and 40, **a.** true or false? **b.** Justify your answer.

39. $\frac{10!}{10} = 1!$ **40.** $18! = 18 \cdot 17!$

41. Simplify $\frac{102!}{99!}$.

PROPERTIES DEAL WITH THE PRINCIPLES BEHIND THE MATHEMATICS.

Objective F: *Identify and apply the following properties of multiplication.* *(Lessons 2-1, 2-2, 2-5, 2-6, 2-7, 2-8)*

Commutative Property of Multiplication
Associative Property of Multiplication
Multiplicative Identity Property of 1
Property of Reciprocals
Multiplication Property of Zero
Multiplication Property of Equality
Multiplication Property of Inequality
Multiplication Property of -1

42. Give an instance of the Associative Property of Multiplication.

43. **a.** Simplify in your head: $4 \cdot x \cdot 25 \cdot 22$

b. What properties can aid you in the simplification?

44. $3 \cdot a = a \cdot 3$ is an instance of what property?

In 45–47, write the reciprocal of the given number.

45. -2 **46.** 0.6 **47.** $\frac{3}{4}x$

48. Write in symbols: The product of a number and its reciprocal is the multiplicative identity.

49. Of what property is this an instance? If $m = n$, then $12m = 12n$.

50. If $-12x < 4$, what inequality results from multiplying both sides by $-\frac{1}{12}$?

51. Multiplication by -1 changes a number to its __?__.

USES DEAL WITH APPLICATIONS OF MATHEMATICS IN REAL SITUATIONS.

Objective G: *Apply the Area Model for Multiplication in real situations.* *(Lessons 2-1, 2-8)*

52. Consider the sketch of the 9′ × 12′ area rug below. What is the area of the shaded part?

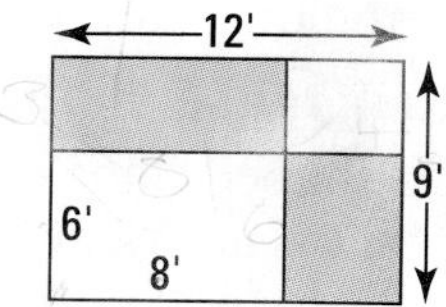

53. The Cohens have a rectangular flower garden with length ℓ and width w. The Banerjils' vegetable garden is half as long and two thirds as wide.

a. What are the dimensions of the Banerjils' garden?

b. How do the areas of the two gardens compare?

54. A box of grass seed will cover about 5000 square feet. The median strip on a boulevard is 60 ft wide. How long a strip can be seeded with one box?

55. The volume of a storage bin needs to be 100,000 cubic feet. The base of the bin is a rectangle with dimensions 40 feet and 80 feet. What should the height be?

56. What is the volume of a box that is 12 cm long, 15 cm high, and 8 cm wide?

57. How many cubes of sugar with sides of length s can fit in a rectangular solid with dimensions $10s$ by $12s$ by $6s$? Explain how you get your answer.

Objective H: *Apply the Rate Factor Model for Multiplication to real situations.* *(Lessons 2-4, 2-6, 2-8)*

58. Express the rent for k months on an apartment that rents for $450 per month.

59. At 30 miles per gallon and $1.00 per gallon of gas, what is the cost per mile?

60. There are 43,560 sq ft/acre. How many square feet are there in 24 acres?

61. A hairdresser charges 15 dollars per cut. How much will he earn in 5 hours if he does 3 cuts per hour?

62. On the average, B books fit on 1 foot of shelf space. If one bookcase has 24 feet of shelf space, how many books can fit on C such bookcases?

63. How long does it take to drive from Baton Rouge to Tallahassee, a distance of 446 miles, if a person can average 50 mph?

64. Daniel budgeted \$550 for accommodations. The hotel costs \$45 a day. At most how many days can he stay at the hotel?

65. a. What is the reciprocal rate of the rate in italics below?
A ballet dancer makes *2 revolutions per second* while pirouetting.

b. Explain the meaning of the result in part **a.**

66. Use the rates below to find the price per box.
$$24\ \frac{\text{boxes}}{\text{carton}} \text{ and } 75.30\ \frac{\text{dollars}}{\text{carton}}$$

Objective I: *Apply the Multiplication Counting Principle and the Permutation Theorem.* *(Lessons 2-9, 2-10)*

ΑΒΓΔΕΖΗΘΙΚΛΜΝΞΟΠΡΣΤΥΦΧΨΩ

67. The Greek alphabet has the 24 letters shown above. How many 3-letter monograms are possible if you may use a letter more than once? (3-letter monograms often are used to name fraternities and sororities.)

68. All 10 questions on a quiz are multiple choice, each with 5 possible choices. How many different sets of answers are possible on the test?

69. A parade is going to contain 18 different groups (bands, dancers, and so on). In how many different orders can they march?

70. A class of 30 students is lined up to go into the school after recess. In how many ways can they be arranged?

71. A restaurant serves enchiladas. The customer has a choice of f fillings and s sauces. How many different types of enchiladas are possible?

72. A catering service is setting up a buffet table with p platters of food. In how many ways can the platters be arranged in a line?

REPRESENTATIONS DEAL WITH PICTURES, GRAPHS, OR OBJECTS THAT ILLUSTRATE CONCEPTS.

Objective J: *Use rectangles, rectangular solids, or rectangular arrays to picture multiplication.* *(Lessons 2-1, 2-3)*

73. Below are two rectangles with $A_1 = A_2$. What property of multiplication is illustrated?

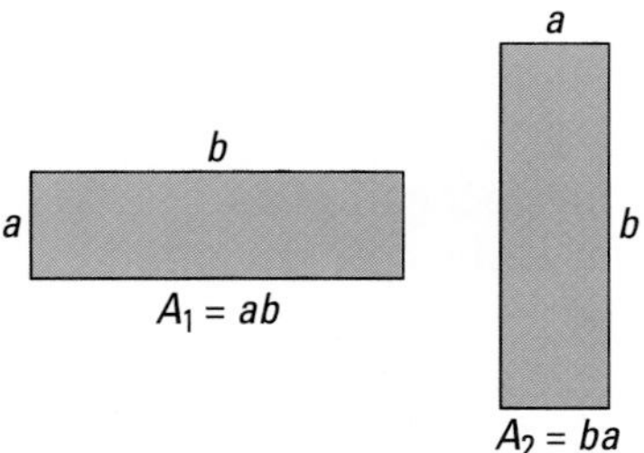

74. The square below has side of length 1. What multiplication of fractions does the drawing represent?

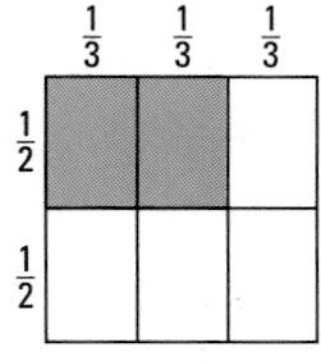

75. Draw a square with side of length s.

a. Shade or color the diagram to show $\frac{s}{2} \cdot \frac{3s}{4}$.

b. What is the result of the multiplication?

76. All angles in the L-shaped floor plan of a house illustrated below are right angles. What is the area of the base of the house?

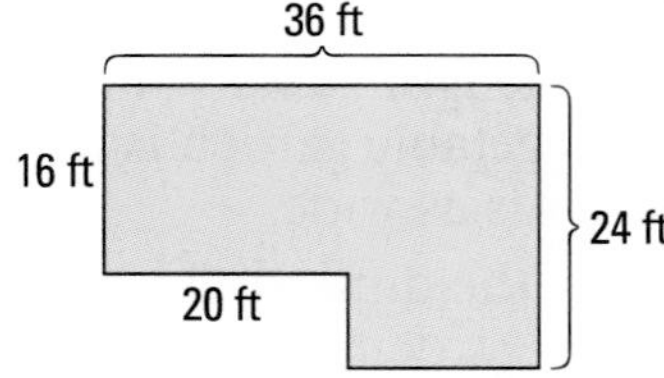

77. Find the volume of the rectangular solid pictured below.

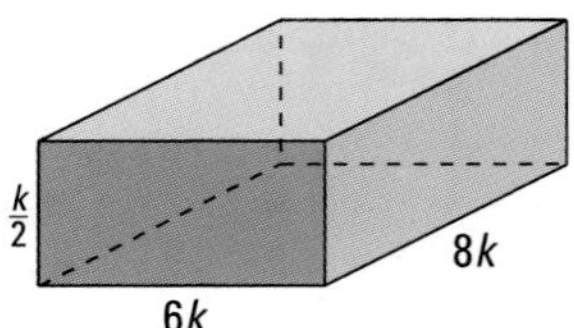

78. What is the volume of a box with dimensions 0.3 meter, 0.45 meter, and 4 meters?

79. A spreadsheet has rows numbered 1 to 100 and columns labeled A through Z. How many cells are in the spreadsheet?

80. How many dots are in the array at right?

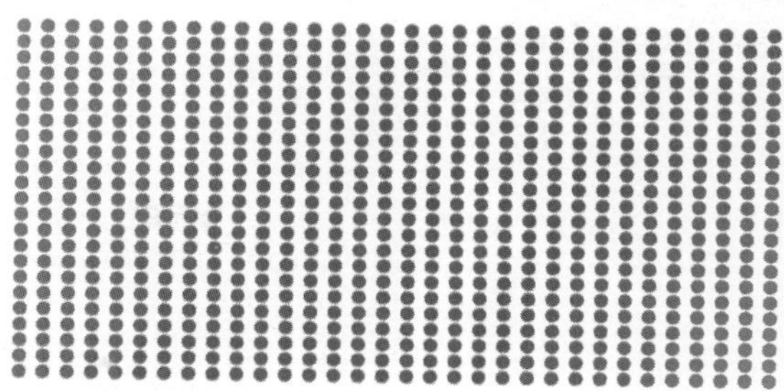

REFRESHER

Chapter 3, which discusses addition in algebra, assumes that you have mastered certain objectives in your previous mathematics courses. Use these questions to check your mastery.

A. Add positive numbers or quantities. In 1–14, find the sum.

1. 3.5 + 4.3 **2.** 122.4 + 11 + .16

3. 3.024 + 7.9999 **4.** $1\frac{1}{2} + 2\frac{1}{4}$

5. $\frac{2}{3} + 8\frac{1}{3}$ **6.** $\frac{2}{5} + \frac{1}{6} + \frac{3}{7}$

7. 6% + 12% **8.** 20% + 11.2%

9. 11 cm + .03 cm **10.** 0.4 km + 1.9 km

11. 2′3″ + 9″ **12.** 6′ + 11″ + 4″

13. 30 oz + 8 lb

14. 4 lb 13 oz + 2 lb 12 oz

B. Add positive and negative integers. In 15–20, find the sum.

15. 30 + -6 **16.** -11 + -4

17. -1 + -1 + 3 **18.** -99 + 112

19. -2 + 4 + -6 **20.** 8+-8+-8+-8

In 21 and 22, what addition problem is pictured on the number line?

21.

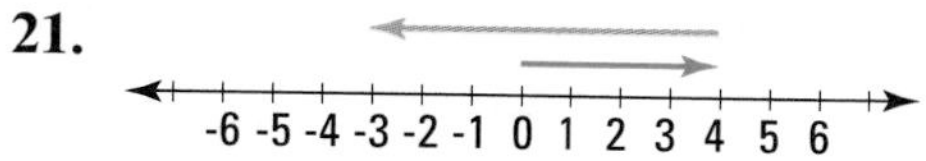

22.

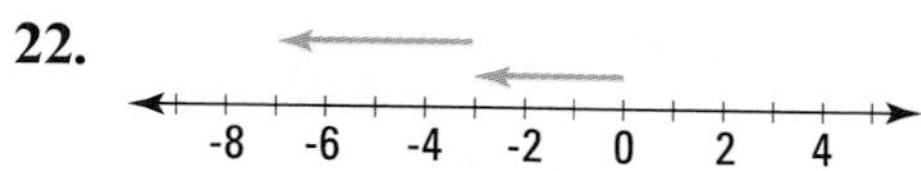

C. Graph ordered pairs on the coordinate plane. In 23–31, draw a set of axes as shown below. Plot and label each point.

23. $A = (4, 3)$ **24.** $B = (5, -2)$

25. $C = (-2, 4)$ **26.** $D = (-3, -1)$

27. $E = (0, 4)$ **28.** $F = (0, 2)$

29. $G = (-3, 0)$ **30.** $H = (3, 0)$

31. $I = (0, 0)$

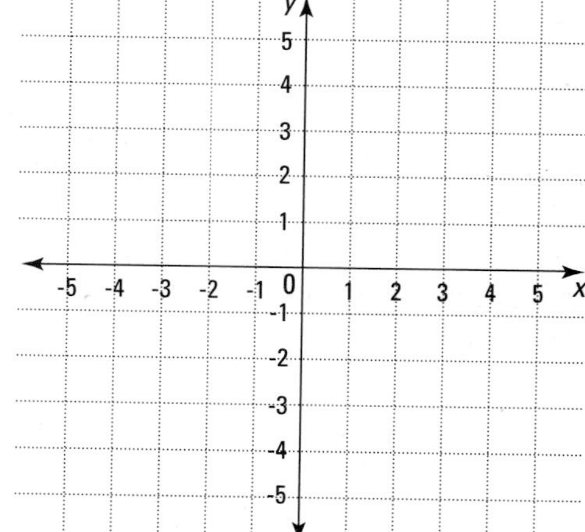

D. Solve equations of the form $x + a = b$, where a and b are positive integers. In 32–37, solve the equation.

32. $x + 3 = 11$

33. $9 + z = 40$

34. $665 + w = 1072$

35. $7 = m + 2$

36. $2000 = n + 1461$

37. $472 = 173 + s$

38. The sum of a number and 75 is 2000. What is the number?

CHAPTER 3

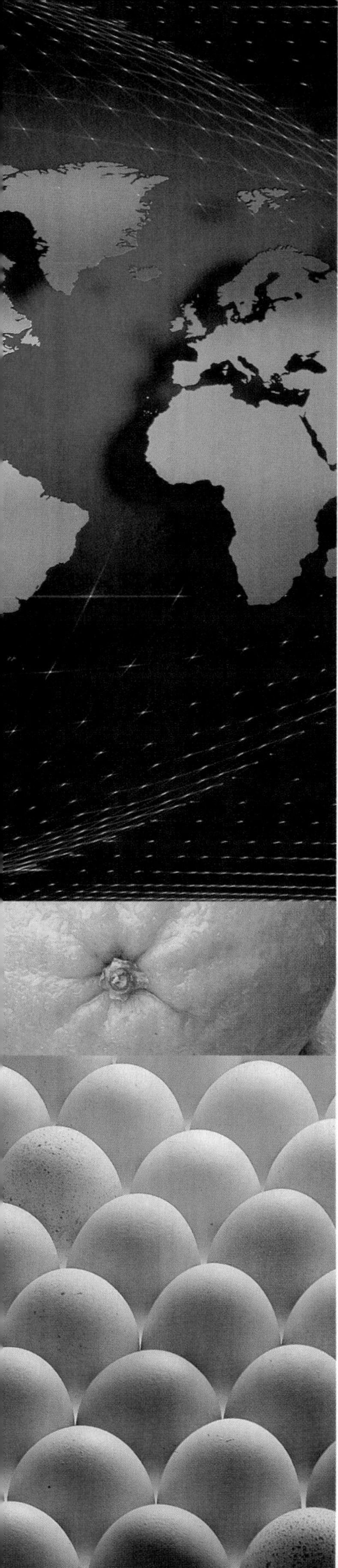

ADDITION IN ALGEBRA

There is an old saying, "You are what you eat." It is a shortened version of, "Tell me what you eat, and I will tell you what you are," written by the Frenchman Anthelme Brillat-Savarin in 1825, in a book entitled *The Physiology of Taste.*

Savarin was speaking about the tendency of different types of people to eat different foods. Today we know that foods affect not only personality but also health. For this reason, many people keep track of various ingredients in the foods they eat.

Suppose for breakfast today you ate the foods listed in the table below. The data are taken from the U.S. government document *Home and Garden Bulletin No. 72* and from information supplied by the American Heart Association.

Food	Protein	Energy	Vitamin A	Fat	Cholesterol
serving	*grams*	*calories*	*I.U.*	*grams*	*milligrams*
Orange juice (1 cup)	2	120	540	0.6	0
Corn flakes (1 cup)	2	95	1180	0.1	0
Eggs (1)	6	95	310	7.1	248
White toast (1 slice)	2	70	0	0.9	0
Whole milk (1 cup)	8	150	310	8	33

To determine the total number of calories you consumed or the amount of cholesterol in your breakfast, multiply the entries in each row by the number of portions you ate.

Suppose you had j cups of orange juice, c cups of corn flakes, e eggs, b pieces of toast, and m cups of milk. Then you would have consumed approximately

$$120j + 95c + 95e + 70b + 150m \text{ calories.}$$

This pattern involves multiplying variables by real numbers and adding the products. It is an example of a *linear expression.* Related ideas are studied in the next few chapters.

LESSON

3-1

Models and Properties of Addition

Window shopping. *One South Wacker Drive is both the name and address of this office building. The skyscraper was designed with a step-like facade of reflective glass, allowing scenic views of downtown Chicago.*

The Putting-Together Model for Addition

Morgan Windows, Inc. (a fictitious company) has two plants, one in New York and one in New Orleans. The *bar graph* with separate bars for the two plants on the left below shows the profit made by each plant each year from 1990 to 1994. What patterns do you notice in this display?

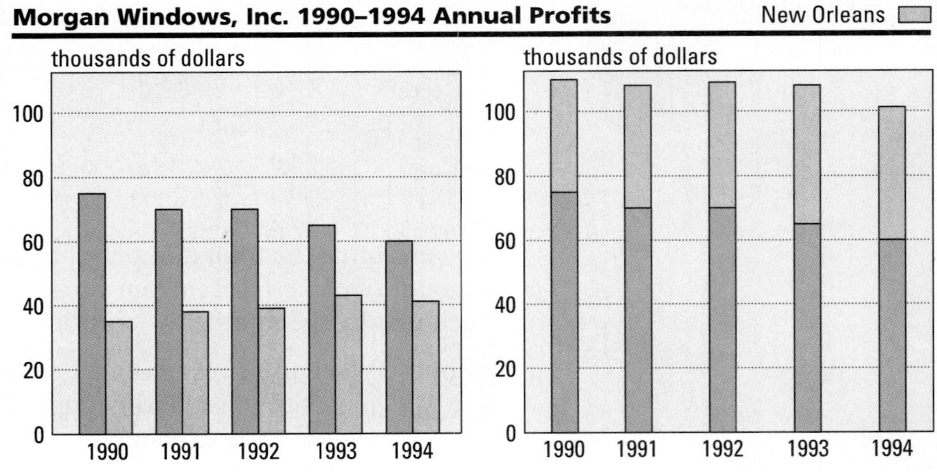

Refer to the graph on the left. Note how the bars representing profits in the New York plant get shorter as you read from left to right. This means that the profits in New York went down over the five-year period. In contrast, the profits for the New Orleans plant increased until 1993 and then decreased in 1994.

Are total profits for the two plants increasing or decreasing? To determine this, the data in the first bar graph can be rearranged into the *stacked bar graph* at the right. Each bar for the profit made in New Orleans is placed above the bar representing the profit made in New York for the same year. You can see that profits for the company were nearly constant until 1994.

To make this stacked bar graph, the two profits for each year shown in the left graph are added. For instance, in 1994, the New York plant had a profit of $60,000 and the New Orleans plant had a profit of $41,000. In total, they had a profit of $101,000. So the bar for 1994 is drawn to the 101 mark.

The stacked bar graph illustrates an important model for addition.

Putting-Together Model for Addition
If a quantity x is put together with a quantity y *with the same units,* and there is no overlap, then the result is the quantity $x + y$.

On the road again. *The cost of most business trips is covered by the company for whom the employee works.*

Example 1

Ms. Kumar is given a daily allowance of $200 by her company for her travel expenses on a business trip. She spends $109.95 for her hotel room. She estimates that she will spend about $20.00 for gas and $7.50 for parking. Let E be the amount still available for other expenses. Write an equation to show how E is related to the other quantities.

Solution

The different expenses do not overlap, so the Putting-Together Model applies. The total allowance is $200.00. So

$$109.95 + 20.00 + 7.50 + E = 200.$$

This can be rewritten as

$$137.45 + E = 200.$$

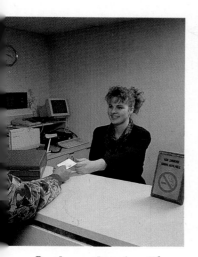

The Slide Model for Addition

A second important model for addition is the *Slide Model.* In the Slide Model for Addition, positive numbers are shifts, or slides, in one direction. Negative numbers are slides in the opposite direction. The + sign means "followed by." The sum indicates the net result of the two slides.

Slide Model for Addition
If a slide x is followed by a slide y, the result is the slide $x + y$.

Example 2

A football team lost 4 yards on the first play and gained 6 yards on the second play. What is the net result of these two plays?

Solution

Think of -4 as a slide to the left. Think of 6 as a slide to the right. The net result is a slide to the right of 2 units. This means a net gain of 2 yards.

$$-4 + 6 = 2$$

Check

The two plays can be represented on a number line where 0 is the "line of scrimmage," or starting point, for the first play.

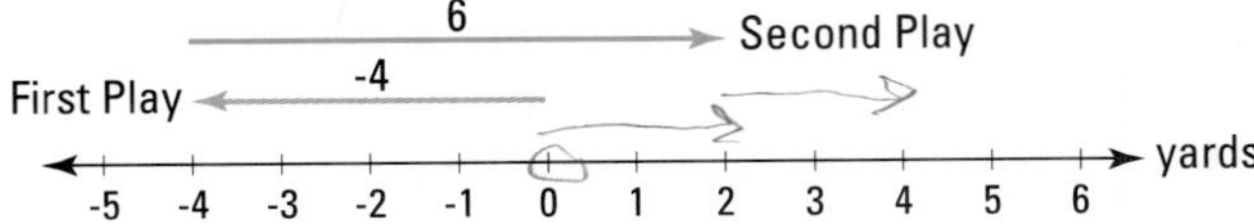

The first arrow, starting at 0 and ending at -4, indicates a loss of 4 yards. The next arrow, from -4 to 2, represents a gain of 6 yards. Notice that the result is a net gain of 2 yards.

The Slide Model can also describe the end result of changes from any arbitrary starting value.

Example 3

The temperature at 11 A.M. is $T°$. If it rises 3° during the next hour, what will be the temperature at noon?

Solution

This, too, is a slide. You can think of the mercury sliding up or down on a thermometer. At noon the temperature will be $T° + 3°$.

The Commutative and Associative Properties of Addition

Addition, like multiplication, has many properties. The examples for the Putting-Together and Slide Models can illustrate these properties.

When two quantities are put together, either quantity may come first. If the stacked bar graph at the start of the lesson had the bars representing profit in New York above the bars representing profit in New Orleans, the total heights would still be the same. For 1994, the total profit is

$$\$41{,}000 + \$60{,}000 = \$60{,}000 + \$41{,}000.$$

In Example 3, you could have written $3° + T°$ or $T° + 3°$. These are instances of a general pattern, the *Commutative Property of Addition.*

Commutative Property of Addition
For any real numbers a and b, $a + b = b + a$.

You can also regroup numbers being added without affecting the sum. For instance, the sum

$$(67 + 98) + 2$$

might be easier to calculate mentally if it is regrouped.

$$(67 + 98) + 2 = 67 + (98 + 2) = 67 + 100 = 167$$

This is an instance of the *Associative Property of Addition.*

Associative Property of Addition
For any real numbers a, b, and c, $(a + b) + c = a + (b + c)$.

As in multiplication, the Commutative and Associative Properties of Addition are used to evaluate or simplify expressions.

Example 4

Simplify $(-8 + y) + -4$.

Solution

First change the order, and then regroup.

$$\begin{aligned}(-8 + y) + -4 &= (y + -8) + -4 && \text{Commutative Property of Addition}\\ &= y + (-8 + -4) && \text{Associative Property of Addition}\\ &= y + -12\end{aligned}$$

Experts often skip steps. They apply the Commutative and Associative Properties mentally. They simply write

$$(-8 + y) + -4 = y + -12, \text{ or } (-8 + y) + -4 = -12 + y.$$

Check

For any particular value of y, $(-8 + y) + -4$ must give the same value as $y + -12$. We substitute 7 for y, and then follow the order of operations.

Original: $(-8 + y) + -4 = (-8 + 7) + -4 = -1 + -4 = -5$
Answer: $y + -12 = 7 + -12 = -5$

Since both expressions have a value of -5, they are equal.

QUESTIONS

Covering the Reading

1. Refer to page 141. A person had 1 cup of orange juice, 1 cup of corn flakes, 2 slices of white bread, and $1\frac{1}{2}$ cups of milk for breakfast. How many calories did the person consume?

2. State the Putting-Together Model for Addition.

New Orleans. *Pictured is part of the French Quarter in New Orleans, Louisiana. Notice the distinctive geometric patterns on the iron grillwork.*

In 3–5, refer to the graphs of the profits for Morgan Windows, Inc.

3. In which year did both the New York and New Orleans plants show a decrease in profits?

4. Approximately what were the combined profits of both plants in 1990?

5. *True or false.* The total profits for Morgan Windows for the five years 1990–1994 can be found by adding the profits for each year in any order.

6. A businessman spent \$72.50 for a hotel room, \$23.75 for food, and \$20 for a cab ride. Write an equation relating these amounts and the amount E still available for other expenses from his \$150 daily allowance.

7. On the first play, a football team gained 7 yards. On the second play, the team lost 3 yards.
a. Represent this situation on a number line.
b. What was the net result of the two plays?

8. If the temperature is $T°$ and then it goes up 5°, what is the new temperature?

In 9 and 10, simplify the expression.

9. $28 + (k + 30)$

10. $(p + -139) + 639$

In 11 and 12, *multiple choice.* Which addition property is illustrated?

(a) commutative only
(b) both commutative and associative
(c) associative only
(d) neither commutative nor associative

11. $2L + 2W = 2W + 2L$

12. $(2x + 3) + 4 = 2x + (3 + 4)$

Applying the Mathematics

In 13 and 14, make up your own example to illustrate the property.

13. Commutative Property of Addition

14. Associative Property of Addition

15. Write an addition expression suggested by this situation. The temperature goes up $x°$, then falls 3°, and then rises 5°.

16. The symbol below at the left means $R = P + Q$. Fill in the empty light green squares in the large figure.

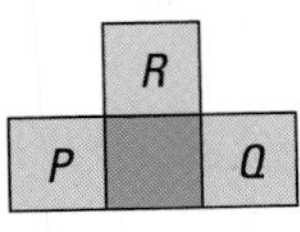

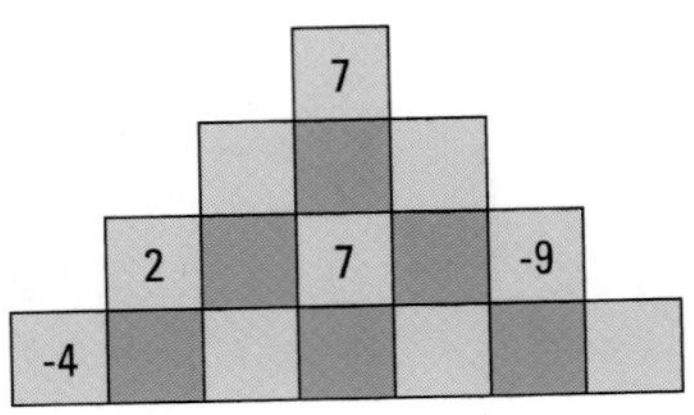

17. If Andy's age is now A, write an expression for his age at each time.
a. 3 years from now
b. 4 years ago

18. Use the bar graphs below.

Years of School Completed by Gender: 1991

(for persons 25 years old and over)

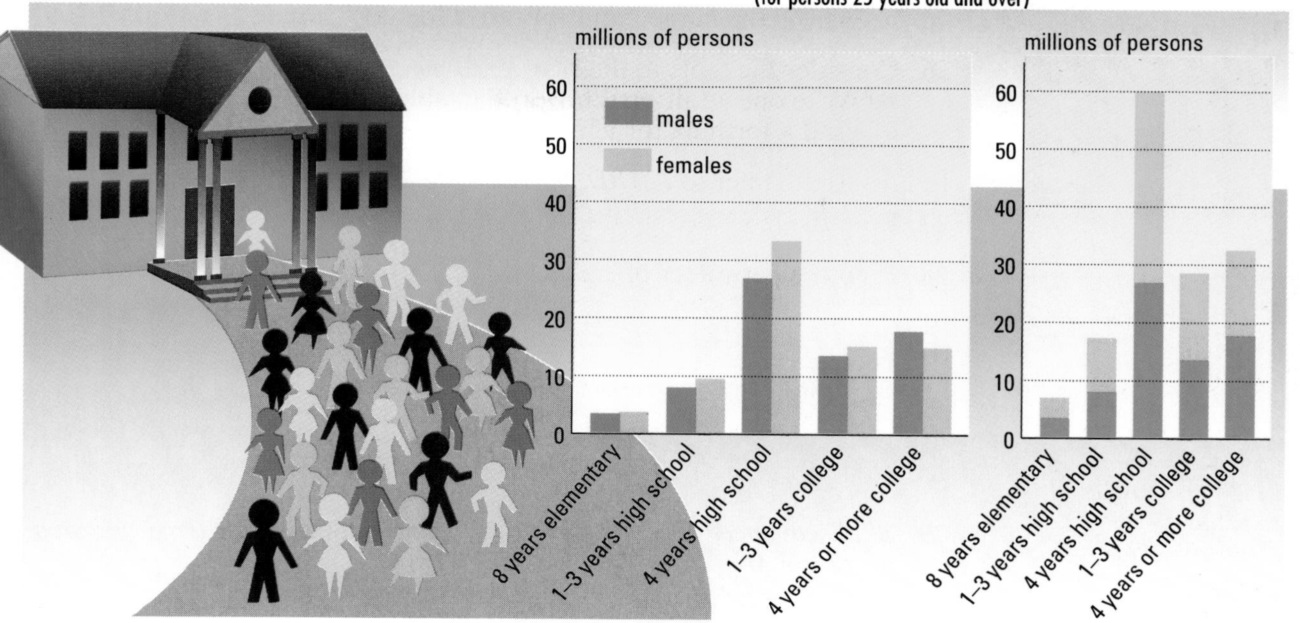

a. How many males have completed at least one year of college?
b. How many people have completed at least one year of college?
c. Write a sentence or two describing some other conclusion you can make from these data.

19. When Rita got on a scale it registered 50 kg. Then she held her baby in her arms, and the scale went up by x kg. It then read 54 kg. Write an equation using addition relating x, 50, and 54.

20. **a.** Write an equation to express the area of the largest rectangle in terms of the areas of the smaller rectangles.
b. Find x. Use trial and error if necessary.
c. Find the dimensions of each of the smaller rectangles.

6 in. | 24 in^2 | x in^2 | 18 in^2 | 9 in.

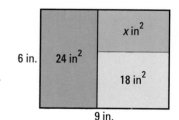

21. Write an inequality relating the three numbers mentioned in the following situation: The bill for lunch came to \$10.50. Milo had M dollars. His sister Nancy had \$4.25. Together they did not have enough to pay the bill.

In 22 and 23, simplify.

22. $(x + 4) + (5 + y)$

23. $(a + {}^{-}2) + (b + 7)$

24. Use the associative and commutative properties to add mentally:
$$49.95 + 59.28 + 0.05 + 0.72.$$

Review

25. a. What number is the multiplicative identity?
b. Why is the name "multiplicative identity" used? *(Lesson 2-2)*

26. Consider the table at the right. *(Lesson 1-9)*
a. Write one additional row in the table.
b. Find a formula for y in terms of x.

x	y
1	1
2	4
3	9
4	16

In 27 and 28, refer to $\triangle ABD$. *(Lesson 1-8)*

27. Find the length of x.

28. Find the perimeter of $\triangle BCD$.

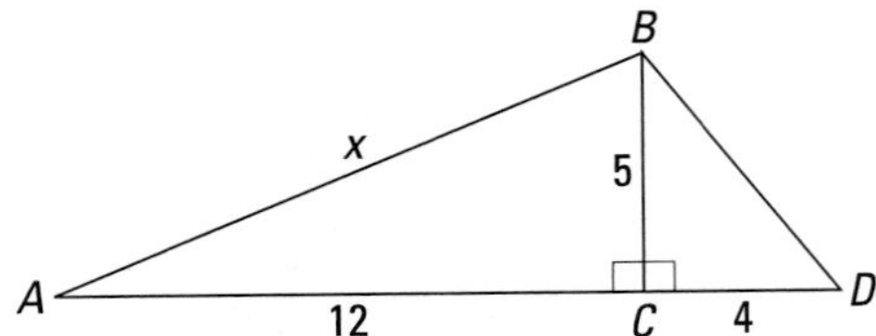

29. *Skill sequence.* Simplify without using a calculator. *(Previous course)*
a. 12.6 + 0 **b.** 12.6 + -12.6 **c.** -(-12.6 + 12.6)

30. *Skill sequence.* *(Previous course)*
a. What is 16% of 24?
b. What is 16% of 2400?
c. Of the 24 million people aged 12–18 in the U.S., 16% like jazz. How many people is this?

Exploration

31. Seafood is sometimes cleaned with a saline solution. If the solution is made from $\frac{1}{4}$ cup salt and $\frac{3}{4}$ cup water, why won't there be a full cup of solution?

LESSON

3-2

More Properties of Addition

Football for everyone. *Shown with their mascot are some members of a girl's high school football team formed for a fun, exhibition game. Addition can be used to find the total yards gained or lost by a team.*

Identity and Opposite Properties

Suppose you deposit \$480 in a bank. If you make no withdrawal or deposit, you will have \$480 + 0 or \$480. Adding 0 to a number keeps the *identity* of that number. So 0 is called the **additive identity.**

Additive Identity Property
For any real number a, $a + 0 = 0 + a = a$.

The role of the number 0 in addition is similar to the role of the number 1 in multiplication. Recall that 1 is the multiplicative identity because

$$a \cdot 1 = 1 \cdot a = a.$$

Recall also that every number except zero has a multiplicative inverse or reciprocal. When $a \neq 0$, its multiplicative inverse is $\frac{1}{a}$. Every number, including 0, has an *additive inverse.*

Additive inverses can be illustrated by one slide followed by another of the same length, but in the opposite direction. For instance, suppose a football team were to first gain 7 yards and then lose 7 yards. After these two plays the team would be back at the original line of scrimmage. This situation shows that $7 + -7 = 0$. Here is a picture using slides.

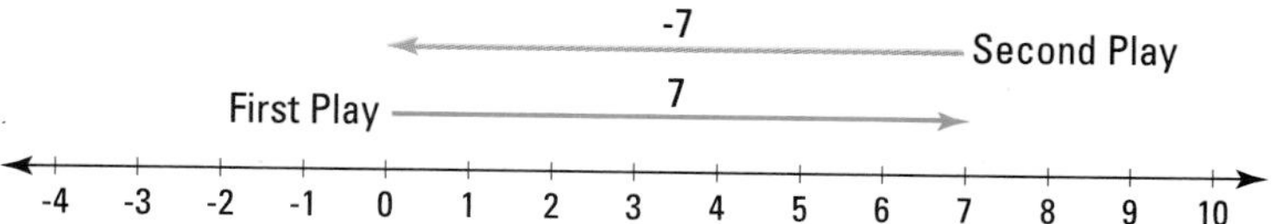

The numbers 7 and -7 are called *opposites,* or *additive inverses,* of each other. In general, the **opposite** of any real number a is written $-a$.

Property of Opposites
For any real number a, $a + -a = -a + a = 0$.

Remember that when you multiply two reciprocals, the product is 1, the multiplicative identity. Similarly, when you add two opposites, the sum is 0, the additive identity.

Example 1

Evaluate the following expressions.

a. $5 \cdot 10^{32} + -(5 \cdot 10^{32})$ **b.** $\left(\frac{11}{5} + -\frac{3}{5}\right) + \frac{3}{5}$

Solution

a. The numbers are opposites. So, $5 \cdot 10^{32} + -(5 \cdot 10^{32}) = 0$.

b. Here we write all the steps and the properties that justify them.

$\left(\frac{11}{5} + -\frac{3}{5}\right) + \frac{3}{5} = \frac{11}{5} + \left(-\frac{3}{5} + \frac{3}{5}\right)$	Associative Property of Addition
$= \frac{11}{5} + 0$	Property of Opposites
$= \frac{11}{5}$	Additive Identity

Experts often skip steps or do them in their head. They can often see immediately that $\left(\frac{11}{5} + -\frac{3}{5}\right) + \frac{3}{5} = \frac{11}{5}$.

Every number has only one opposite. For instance, 10 is the only opposite of -10. But notice that the opposite of -10 can also be written as -(-10), read "the opposite of the opposite of 10." So, -(-10) must equal 10.

In general, the numbers a and $-a$ are additive inverses, as are $-a$ and $-(-a)$. A number has only one additive inverse, so $-(-a) = a$. We call this the *Opposite of Opposites Property,* or the *Op-op Property,* for short.

Opposite of Opposites (Op-op) Property
For any real number a, $-(-a) = a$.

Caution: The expression $-a$ does not always represent a negative number. If $a = 60$, then $-a$ is its opposite; so $-a = -60$. But if $a = -60$, then $-a$ is the opposite of -60, so $-a = 60$. *When a is negative, -a is positive.*

The Addition Property of Equality

Suppose Tina's age is T years and Robert's age is R years. If Tina and Robert are the same age, then $T = R$.

Eight years from now, Tina's age will be $T + 8$ years and Robert's age will be $R + 8$ years. They will still be the same age, so

$$T + 8 = R + 8.$$

Similarly, Tina's age three years ago was $T + -3$ and Robert's age was $R + -3$. Since they would have been the same age then as well,

$$T + -3 = R + -3.$$

The equations relating Tina's and Robert's ages are instances of the general property known as the Addition Property of Equality.

Addition Property of Equality
For all real numbers a, b, and c, if $a = b$, then $a + c = b + c$.

The Addition Property of Equality is useful for solving equations. This property indicates that you can add any number c to both sides of the equation without changing its solutions. It is very much like the Multiplication Property of Equality that you studied in Lesson 2-6. Notice how the Addition Property of Equality is used in the first step of Solution 1 in Example 2.

Example 2

Solve $x + -26 = 83$, and check the answer.

Solution 1

We want the left side to simplify to just x. So we add 26 to both sides since 26 is the opposite of -26. Beginners put in all the steps.

$(x + -26) + 26 = 83 + 26$	Addition Property of Equality
$x + (-26 + 26) = 109$	Associative Property of Addition
$x + 0 = 109$	Property of Opposites
$x = 109$	Additive Identity Property

Solution 2

Experts do some work mentally and may write out fewer steps.

$x + -26 = 83$	
$x = 83 + 26$	Addition Property of Equality
$x = 109$	

Check

Substitute 109 for x in the original equation. Does $109 + -26 = 83$? Yes. So 109 is the solution.

All the steps were shown in Solution 1 to Example 2 to illustrate the properties that justify this process. Like the expert, you may not always need to include all steps in solving an equation. Directions in the problem and your teacher's instructions will guide you in choosing what steps to include.

The key to solving equations is knowing what should be done to both sides. For equations of the $x + a = b$ type, there is only one step to remember.

To solve an equation of the form $x + a = b$, add $-a$ to both sides and simplify.

The CD boom. *In 1983, compact-disc players and recordings were introduced in the U.S. By 1986, Americans had purchased more CD players than turntables, making records almost obsolete.*

Example 3

Emily is saving money for a \$179.95 compact-disc (CD) player. She has \$75.50 so far. How much more money does Emily need?

Solution

Use an equation. Let x = the amount of money Emily still needs.

$$75.50 + x = 179.95$$

Add the opposite of 75.50 to each side of the equation.

$$x = 179.95 + -75.50$$

Simplify. $x = 104.45$

She needs another \$104.45 to buy the CD player.

Check

Does $75.50 + 104.45 = 179.95$? Yes, so the solution checks.

QUESTIONS

Covering the Reading

1. a. $0 + -10 = \underline{\ ?\ }$.
b. Why is zero called the additive identity?
c. What number is the multiplicative identity?

2. What is another name for an additive inverse?

In 3–6, simplify.

3. $-\frac{9}{4} + \frac{9}{4}$

4. $(x + -93.2) + 93.2$

5. $-(-7)$

6. $-41 + -(-41)$

7. a. What is the additive inverse of $-x$?
b. Give a value of n for which $-n$ is positive.

In 8 and 9, an instance of what property is given?

8. $\sqrt{2} + -\sqrt{2} = 0$

9. $-y = 0 + -y$

In 10 and 11, suppose your age is A, a friend's age is B, and $A = B$. What does each sentence mean?

10. $A + 4 = B + 4$

11. $A + -5 = B + -5$

12. State the Addition Property of Equality.

13. You wish to solve $m + 42 = 87$.
a. What number should be added to each side?
b. Solve and check.

In 14 and 15, solve and check.

14. $-12 + y = -241$

15. $z + 14 = 60$

In 16 and 17, consider these steps used to solve $-173 + x = 209$.

Step 1: $-173 + x = 209$
Step 2: $173 + -173 + x = 173 + 209$
Step 3: $0 + x = 382$
Step 4: $x = 382$

16. What property was used to get from step 1 to step 2?

17. What property was used to get from step 3 to step 4?

18. Hank has \$35.00. A graphics calculator costs \$72.95. Let n equal the amount of money he needs.
  a. What addition equation can be solved to find out how much more money he needs?
  b. Solve this equation.

Applying the Mathematics

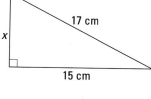

19. A right triangle has hypotenuse of length 17 cm and one leg of length 15 cm. Find the length of the other leg.

20. Describe a real situation not involving football that leads to $-9 + 9 = 0$.

21. Display the number 3.14 on your calculator. Press the [+/−] key twice.
  a. What number appears in the display?
  b. What property has been checked?

22. If $p = -8$ and $q = -10$, then $-(p + q) =$ __?__.

In 23 and 24, solve and check.

23. $15.2 = f + 2.15$

24. $C + -8 + -5 = -15$

In 25 and 26, a question is asked. **a.** Write an equation whose solution answers the question. **b.** Solve your equation. **c.** Check your answer.

25. At noon Monday the temperature was 17°F. At midnight Tuesday the temperature was -12°F. Let C be the change in temperature. By how much did the temperature change?

26. A trail begins at an altitude of 562 meters and ends at 1321 meters. If a hiker climbs from the beginning to the end of the trail, what is the net change in altitude? Let A be the change in altitude.

The king of rock 'n roll. *Shown is a commemorative U.S. stamp honoring Elvis Presley. Elvis recorded over 170 hit singles from 1955–1977.*

Review

27. Mary collects and trades old records. She now has 40 Elvis Presley records. If she sells n of them and buys twice as many as she sells, how many Elvis records will she have? *(Lesson 3-1)*

28. a. Write a simplified expression for the area of the figure shown.
b. What model for addition is being applied? *(Lesson 3-1)*

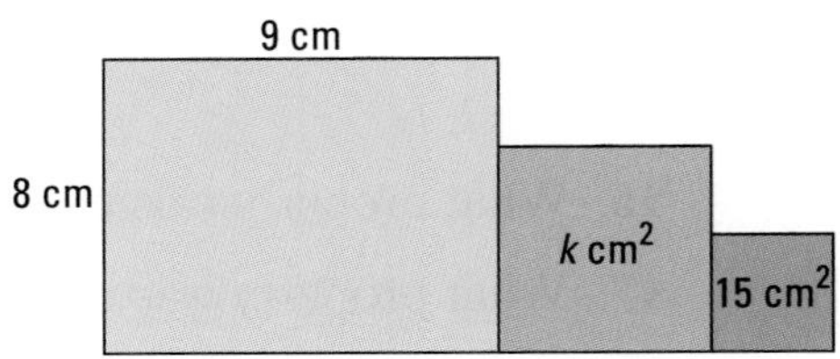

29. *Multiple choice.* In which expression would the answer be the same if the parentheses were deleted? *(Lessons 1-4, 2-5)*

(a) $(8 + 3) \cdot -5$
(b) $7 - (2 \cdot 2)$
(c) $(2 + 6) \cdot 8 + 5$
(d) $1 + 6 \div (2 \cdot 3)$

30. Freezing temperatures are those at or below 32° Fahrenheit.
a. Graph all possibilities for freezing temperatures on a vertical number line. *(Lesson 1-2)*
b. Write an inequality representing all freezing temperatures in the Fahrenheit scale using x as the variable. *(Lesson 1-1)*

31. What is a "horizontal line"? *(Previous course)*

32. What term describes the pairs of lines below? *(Previous course)*

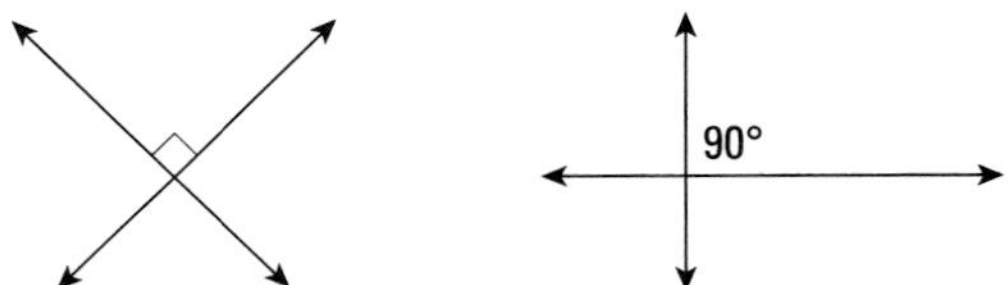

33. Give the coordinates of each point on this graph. *(Previous course)*

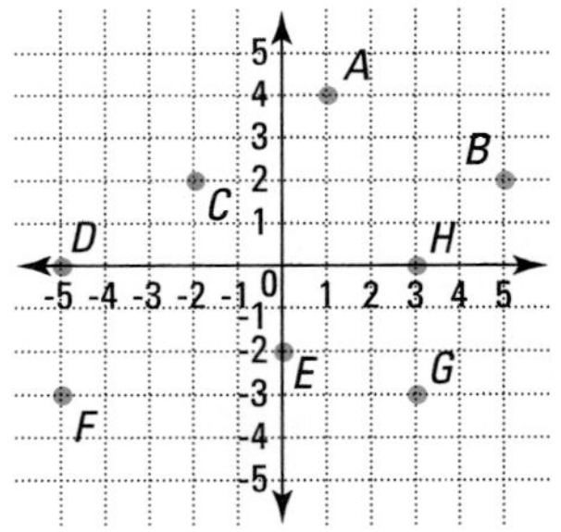

Exploration

34. Negative numbers and their opposites appear in many situations. What real situation might each number represent? What would the opposite represent?

a. $-5\frac{5}{8}$ in stock market values
b. -9 in rocket launches
c. -3 in golf
d. -399 billion in the Federal budget

LESSON

3-3

The Coordinate Plane

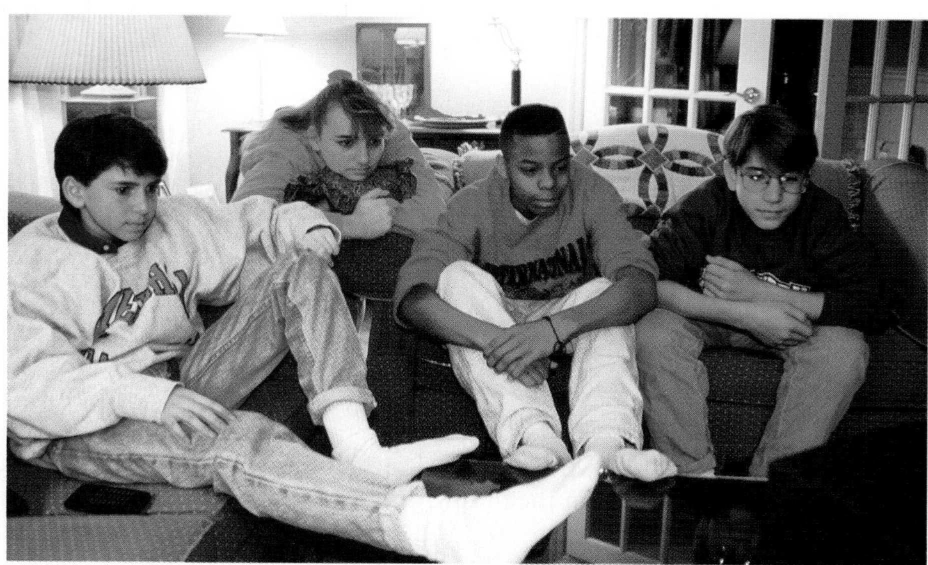

Watch your watching. *Several studies have linked watching a lot of TV with poor academic grades. One study found that 22% of 13-year-old students in the U.S. watch 5 or more hours of TV daily.*

A **plane** is a flat surface that stretches forever in all directions. Graphs on a plane can display a great deal of information and show trends in a small space as illustrated by the following example.

Mrs. Hernandez read the following headline: "TV Linked to Drop in Homework." So she asked her class to keep track of the time spent at home studying and the time spent watching TV. The table below shows what her students reported the next day.

Time Spent on TV and Homework (minutes)

Student	TV	Homework	Student	TV	Homework
Alex	60	30	Jim	120	75
Beth	0	60	Kerry	30	45
Carol	120	30	Lawanda	120	45
David	75	90	Meg	150	60
Evan	210	0	Nancy	180	15
Frank	150	30	Paula	90	75
Gary	0	90	Quincy	60	45
Harper	90	60	Ria	60	120
Irene	120	0			

To explore the relation between the time students spent watching television and the time spent studying, Mrs. Hernandez used a two-dimensional *coordinate graph.* A two-dimensional coordinate graph is needed for this data since each response involves two numbers. She drew two perpendicular number lines called **axes.** She used the horizontal axis for the time spent watching TV and the vertical axis for the time spent on homework. The axes intersect at the point labeled 0 on each number line. This point is called the **origin.**

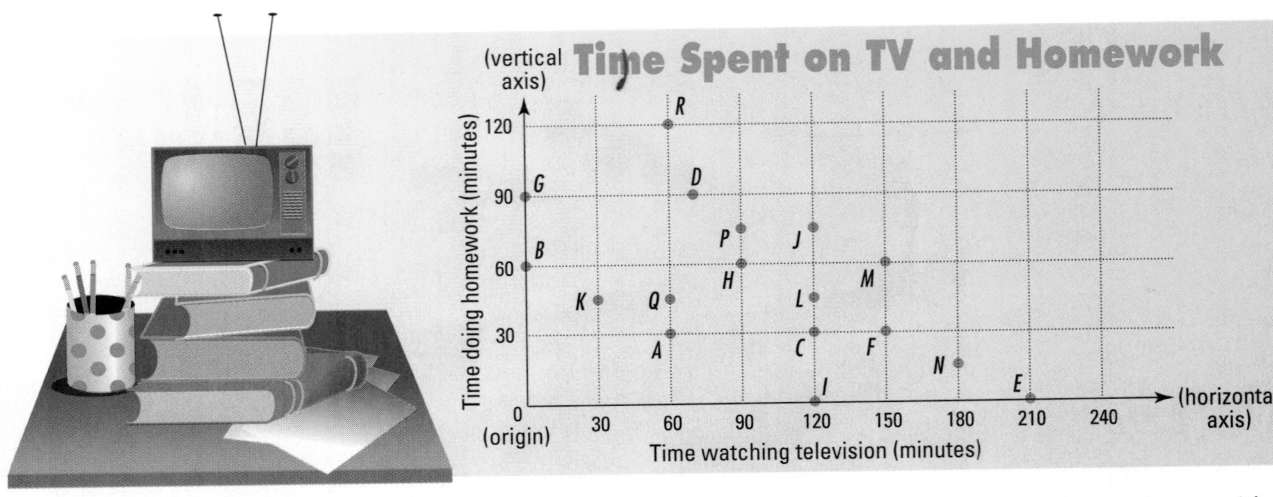

Mrs. Hernandez then located a point for each student's data, plotted it on the graph, and coded it with the student's initial. For instance, Paula spent 90 minutes watching TV and 75 minutes on homework; so Paula's data point is 90 units to the right and 75 units up from the origin. This point can be expressed as the ordered pair (90, 75) and is labeled as point P on the graph. Notice that David's data point is (75, 90) and is labeled as point D on the graph.

A graph like this, in which individual points are plotted, is called a **scatterplot.** Every point in the plane of the graph can be identified with a pair of numbers called its **coordinates.** Such a plane is a **coordinate plane.**

Example 1

Refer to the scatterplot above.

a. How many student responses are shown on the scatterplot?
b. How many students reported doing homework for exactly 90 minutes?
c. How many students watched at least 120 minutes of television?
d. Who is represented by the ordered pair (60, 120)?

Solution

a. There are 17 student responses because there are 17 data points on the scatterplot.

b. Homework time is shown by the distance above the horizontal axis. Two students, Gary and David, did exactly 90 minutes of homework. They are shown by the two data points on the horizontal line for 90.

c. "At least 120" means 120 or more. Television time is shown by distance to the right of the origin. Points showing 120 minutes of TV time are in a vertical line in the middle of the graph. Look for points on this line or to the right of it. There are 8 such points, *J, L, C, I, M, F, N,* and *E*. So 8 students watched at least 120 minutes of TV.

d. The point 60 right and 120 up is named R. It represents Ria.

From the pattern of dots, Mrs. Hernandez's class decided the headline was generally true. As students watched more TV, they did less homework.

Recall that coordinates of points may be positive, negative, or zero. When the first coordinate of a point is negative, the point lies to the left of the vertical axis. When the second coordinate of a point is negative, the point lies below the horizontal axis.

U.S. exports. *The United States is one of the largest car producers in the world, but it exports only a small portion of its production. Pictured above is a shipment of U.S. cars being loaded for export.*

Example 2

The table below shows net *exports* of the United States in billions of dollars from 1975 through 1992. The *net exports* for a given year represents the difference of value between goods and services exported *to* other countries and goods and services imported *from* other countries. A positive net-export value represents money flowing into the United States. A negative value indicates money leaving the USA.

Year	1975	1976	1977	1978	1979	1980	1981	1982	1983
Amount	9	-8	-29	-31	-28	-24	-27	-32	-58

Year	1984	1985	1986	1987	1988	1989	1990	1991	1992
Amount	-108	-132	-153	-152	-119	-110	-102	-65	-84

a. Graph the information in this table.
b. In which year was the value of net exports the highest? the lowest?
c. Give a general description of changes in net exports between 1975 and 1980.
d. Give a general description of changes in net exports after 1980.

Solution

a. Time is usually graphed on the horizontal axis. Begin at 1975, where the table starts, and mark the horizontal axis off in years. The vertical axis will show the net-export amount. Look at the table to see the size of the numbers that must fit on the graph. The largest is 9 and the smallest is -153, so the vertical axis needs to be marked off in equal intervals that span this range. Units of 25 work well. Now plot a point for each ordered pair in the table. A graph is drawn below.

Net Exports by Year

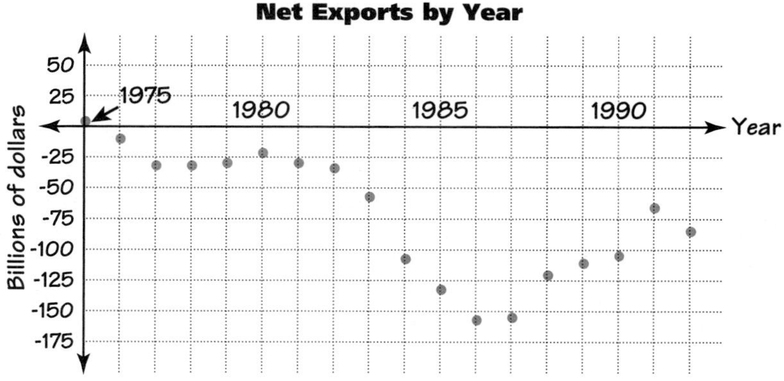

b. Look at the points that are graphed and pick out the highest and lowest ones. Net exports were highest in 1975 and lowest in 1986. You could also find this information in the table of values.

c. Find 1980 on the horizontal axis and look to the left. The points are fairly close to the horizontal axis. You might write: Between 1975 and 1978, net exports declined. Between 1978 and 1980 they rose slightly.

d. Look to the right of 1980. The points drop sharply until 1986 when they start to rise. From 1981 to 1986 the value of net exports was negative and decreased each year. From 1987 to 1991 the value of net exports was still negative, but increasing. In 1992, the value decreased again.

The coordinate plane is a natural way to represent locations on a small part of Earth. As with most maps, east is usually at the right and north at the top. They become the positive directions of the horizontal and vertical axes. Their opposites, west and south, respectively, represent the negative directions.

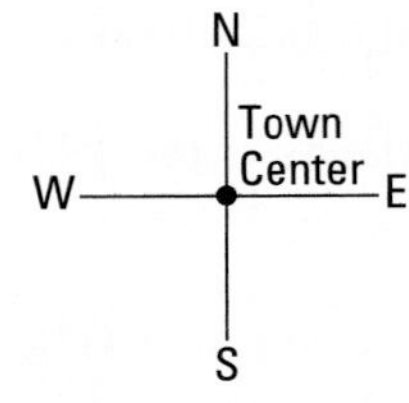

Example 3

A city or town often has a central point from which street addresses are numbered north, south, east, and west. Julie lives 0.5 mile west and 2.1 miles north of the Town Center. Frank lives 1.3 miles west and 3 miles south of the Town Center. Sally lives 2 miles due east of the Town Center. Al lives 3.4 miles east and 0.7 mile north of the Town Center. Graph these locations on a coordinate plane with the Town Center as its origin.

Solution

The first coordinate is the distance east or west. The second coordinate is the distance north or south. The points are graphed.

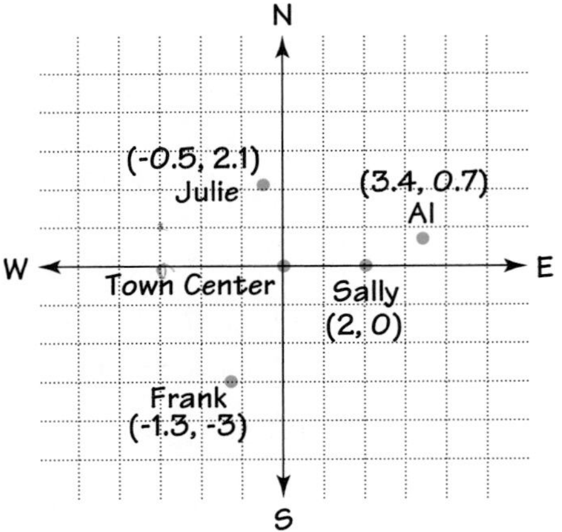

Notice that the points on the graphs in Examples 1 to 3 are not connected. Such a graph is called *discrete.* The graph in Example 4 is **continuous;** all points are connected.

Example 4

Here is a graph indicating the speed Harold Hooper traveled as he drove from home to work. Write a story that explains the graph.

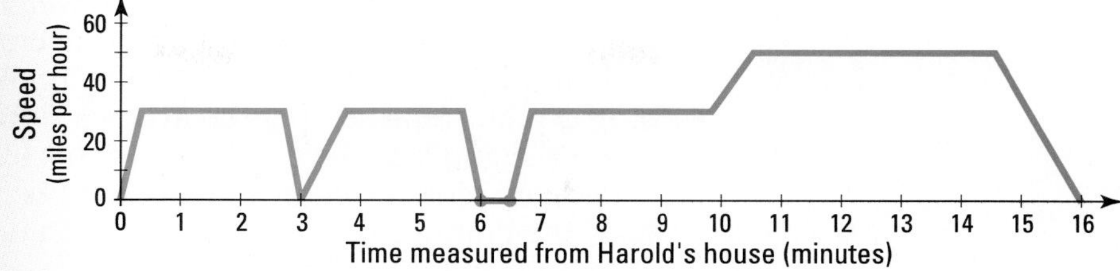

Solution

Here is one possible story: Harold began on a road driving at the 30-mph speed limit. After 3 minutes, he stopped for a stop sign. He stopped at a stoplight for about $\frac{1}{2}$ minute beginning at the six-minute mark. Harold then resumed driving at 30 mph for about $3\frac{1}{2}$ minutes. Then the road widened and the speed limit increased to 50 mph. He drove at 50 mph for about 5 minutes, until he neared work and slowed down.

QUESTIONS

Covering the Reading

1. What are the number lines used in a coordinate graph called?

In 2–4, refer to Example 1.

2. Which ordered pair describes Alex's responses?

3. **a.** According to the graph, how many students did not watch TV?
b. How much time did each of these students spend on homework?

4. What is the trend in this graph?

In 5–7, refer to Example 2.

5. Why are the numbers on the vertical axis given in billions of dollars?

6. When were the net exports the greatest?

7. How much less were the net exports in 1985 than in 1975?

In 8 and 9, refer to Example 3.

8. A bowling alley is 3 miles west and 1 mile north of the Town Center.
a. What are the coordinates of the location of the bowling alley?
b. Which of the four people lives closest to the bowling alley?

9. Bill lives at (1.4, -2.9) on the graph. Describe his location from the Town Center using the words north, south, east, and west.

10. In Example 4, how long did it take Harold to drive to work?

Applying the Mathematics

11. A 1990 study examined the way American and Japanese high school students spend their time. It found that the average American student spends 122 minutes per day watching TV and 46 minutes doing homework. Japanese students average 152 minutes per day watching TV and 163 minutes doing homework.

a. If a point representing the average American student were plotted on the graph at the beginning of this lesson, which of the points already plotted on the graph would be closest to it?

b. If the point for a typical Japanese student were plotted, which point would be closest to it?

c. Would you judge Mrs. Hernandez's class to be more typical of American students or of Japanese students? Explain your answer.

12. a. Make a coordinate graph showing the following set of points:
$$\{(2, 3), (4, 5), (6, 7), (8, 9), (10, 11)\}.$$

b. If the pattern continues, what is the missing coordinate in (100, __?__)?

c. If the pattern continues, what is the missing coordinate in (m, __?__)?

A chilling experience. *The wind chill index gives an estimated calm-air temperature based upon a combination of wind speed and actual temperature. The faster the wind blows, the faster a person loses body heat. The feeling of cold increases as the wind speed increases.*

13. a. The table at the right shows the wind chill index for various temperatures when there is a 10 mph wind. Make a graph of this data. Show the actual temperature on the horizontal axis and the wind chill on the vertical axis.

b. Describe some pattern you observe in the scatterplot.

actual temp °F	wind chill
30°	16
20°	3
10°	-9
0°	-22
-10°	-34
-20°	-46
-30°	-58

In 14 and 15, *multiple choice.* Which of the following situations is represented by the graph?

(a) the distance traveled in h hours at 50 mph

(b) the distance traveled in h hours at 75 mph

(c) the distance you are from home if you started 150 miles from home and traveled toward home at 50 mph

14. **15.**

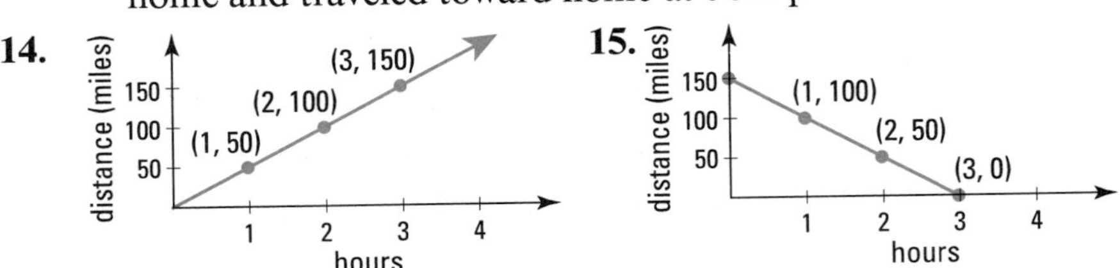

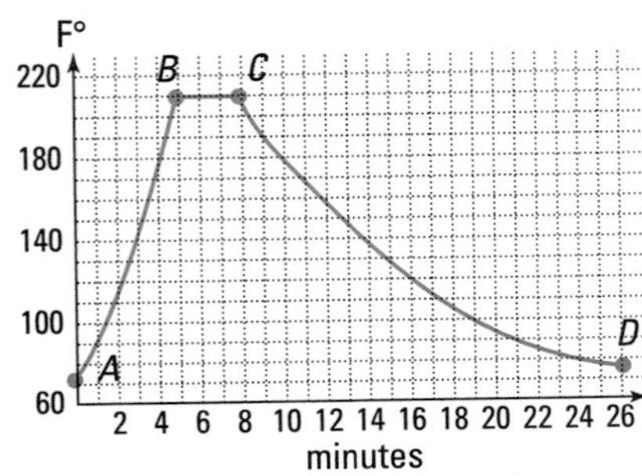

16. The graph at the left shows the temperature of water in a pot on a stove. Write a story that explains the graph. Be sure to mention times and temperatures of important points on the graph.

In 17–20, use the graph below. It shows how long the average American had to work to earn enough money, before taxes, to purchase the goods and services listed.

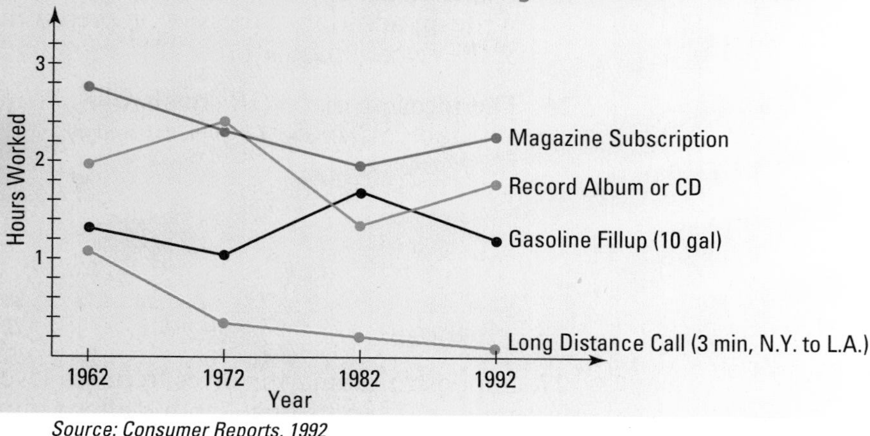

Source: Consumer Reports, 1992

17. On the average, how many hours did a worker have to work in 1982 to earn enough to fill up a car with gasoline?

18. Between 1972 and 1982, which item showed the greatest drop in hours needed? Justify your answer.

19. *Multiple choice.* In hours of work required, the cost of the items in 1992 as compared with 1962 is
(a) always more.
(b) always less.
(c) sometimes more, sometimes less.

20. a. Describe the trend in the cost of a long-distance call between 1962 and 1972.
b. Make a prediction about the cost of a long-distance call in 2002, and explain how you arrived at your prediction.

Review

21. Suppose $h_1 + h_2 = 0$ and $h_1 = 17.3$.
a. What is the value of h_2?
b. What property is illustrated? *(Lesson 3-2)*

22. a. Solve the equation $x + 3.5 = 10.8$.
b. Make up a question that can be answered by solving the equation in part **a.** *(Lesson 3-2)*

23. *Skill sequence.* Solve the equation. *(Lessons 2-6, 3-2)*
a. $n + 10 = 19$
b. $n + {}^-10 = 19$
c. ${}^-10 = n + 19$
d. ${}^-10 = 19n$

24. The value of Wurthmore stock went down $2\frac{3}{8}$ points on July 30. On July 31 and August 1 it went up $\frac{3}{8}$ of a point each day. Find the net change in Wurthmore stock over this three-day period. *(Lesson 3-1)*

25. The river dropped two feet, rose p feet, and then dropped q feet. Write an addition expression, representing the change in the height of the river. *(Lesson 3-1)*

26. The measure of $\angle PQR$ equals 140°. Write an equation that relates 140 and y. *(Lesson 3-1, previous course)*

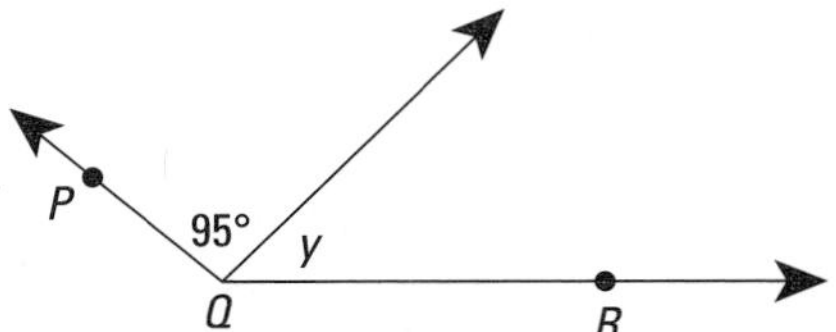

27. Suppose a submarine dives from sea level at a rate of 25 meters per minute. Then its final depth d after t minutes is given by

$$d = -25t.$$

How many minutes does it take to reach a depth of -225 meters? *(Lesson 2-6)*

28. To estimate the number N of bricks needed in a wall some bricklayers use the formula $N = 7LH$, where L and H are the length and height of the wall in feet. About how many bricks would a bricklayer need for a wall 8.5 feet high and 24.5 feet long? *(Lesson 1-5)*

29. Evaluate mentally given $t = -4$. *(Lesson 1-4)*
a. $t + -9$ **b.** $-3t$ **c.** $-2t^2$

Exploration

30. **a.** Survey at least 10 of your friends. Ask them how much time they spent watching TV and doing homework yesterday.
b. Plot your results on a scatterplot.
c. Do your data agree with the newspaper headline in this lesson?

LESSON

3-4

Two-Dimensional Slides

The invasion of video games. *Soon after the 1979 introduction of a video game called Space Invaders, video games became the toy industry's hottest items.*

You probably have played video games in which a figure moves across a video screen. Programmers of games move the figures by first imagining them on a coordinate plane. Diagram 1 below shows how you might see a screen. Diagram 2 shows how a programmer might see the screen.

1. How you see the screen

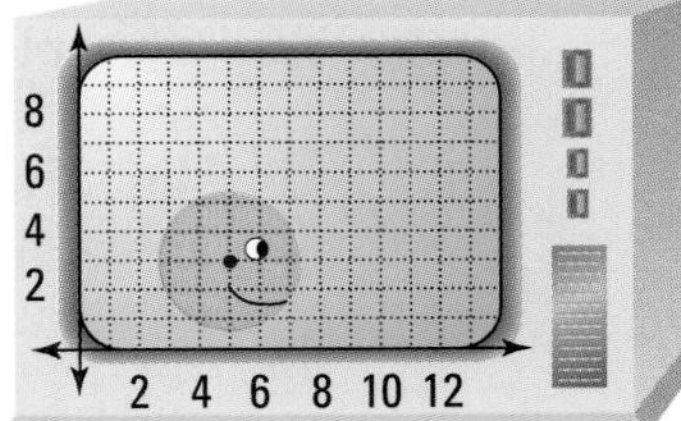

2. How the programmer sees the screen

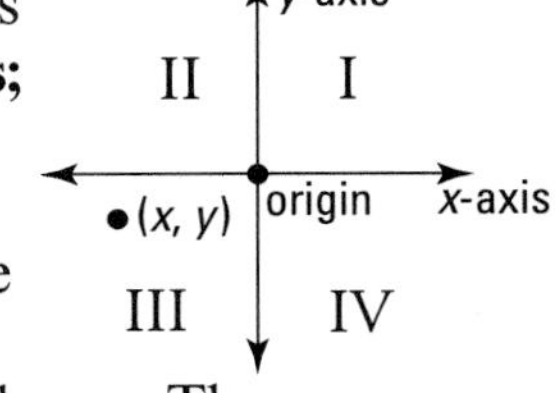

Some of the customary language for describing graphs is shown at the right. The horizontal axis is the ***x*-axis;** the vertical axis is the ***y*-axis.** A general point is often labeled (x, y). The first coordinate of any point is called the ***x*-coordinate;** the second coordinate is the ***y*-coordinate.** The axes separate the coordinate plane into four **quadrants** identified by I, II, III, and IV as shown. The diagrams above show only part of the first quadrant of a coordinate plane.

The signs of the coordinates of points in the quadrants are as follows:

quadrant	*x*-coordinate	*y*-coordinate
I	positive	positive
II	negative	positive
III	negative	negative
IV	positive	negative

One basic movement of a figure is a **two-dimensional slide,** or **translation.** The movement from the original position, or **preimage,** to the final position, or **image,** is shown in Diagram 3. The arrow shows the path a center point of the figure takes. The programmer could describe this movement as a horizontal slide followed by a vertical slide, as shown in Diagram 4.

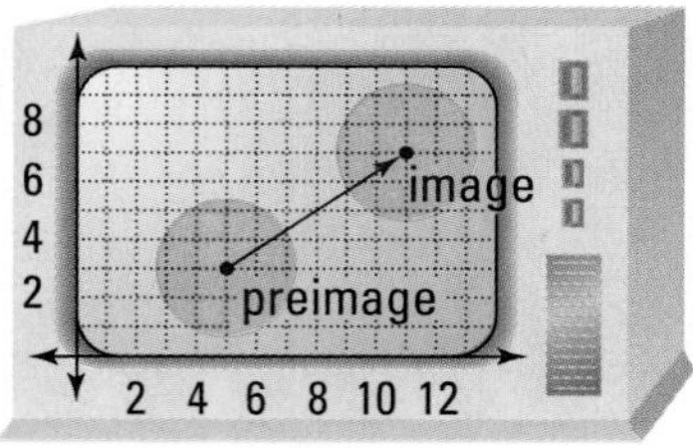

3. A slide of the figure

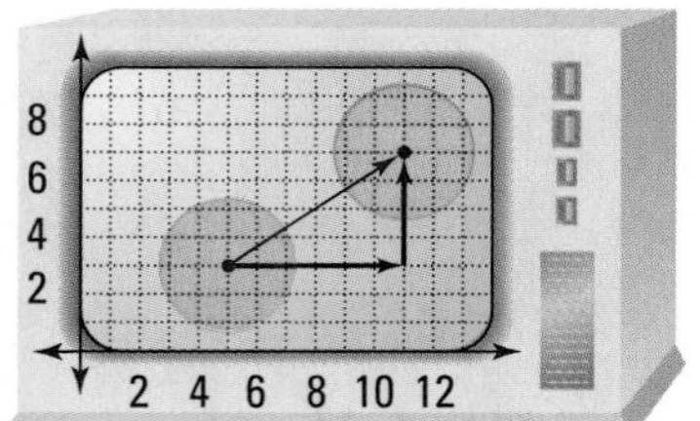

4. One way of describing the slide

In Diagrams 3 and 4, the slide is 6 units to the right and 4 units up. To slide a point 6 units to the right, you must add 6 to the first coordinate of the point. To slide 4 units up, add 4 to the second coordinate. For instance, the figure on the video screen is based on a circle. The center of the figure was originally (5, 3). The new center is (5 + 6, 3 + 4), which is (11, 7). The pattern is as follows:

> If a *preimage point* is (x, y), then the *image point* after a slide 6 units to the right and 4 units up is $(x + 6, y + 4)$.

Adding a negative number to the x-coordinate results in a slide left, and adding a negative number to the y-coordinate results in a slide down.

Example 1 illustrates the translation of a figure with more than one point. The more points the figure has, the easier it may be to see the slide. Here we adopt the common practice of placing a prime after the letters naming the points of the image. If a preimage is point P, then its image is labeled P', read "P prime."

Example 1

a. Plot $\triangle ABC$, where $A = (-2, 4)$, $B = (-1, 7)$, and $C = (2, 3)$.
b. Draw the image of $\triangle ABC$ after a slide of 1 unit to the right and 6 units down.

Solution

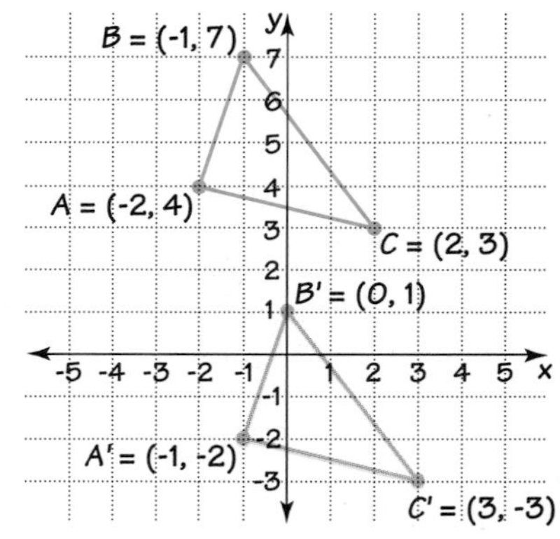

a. $\triangle ABC$ is drawn at the right.
b. To find the vertices of the image after the slide, add 1 to each x-coordinate and -6 to each y-coordinate:
$A' = (-2 + 1, 4 + -6) = (-1, -2)$
$B' = (-1 + 1, 7 + -6) = (0, 1)$
$C' = (2 + 1, 3 + -6) = (3, -3)$
Graph these points, connect them, and call the image $\triangle A'B'C'$.

Check

$\triangle ABC$ looks as if it has been slid 1 unit to the right and 6 units down to get $\triangle A'B'C'$. (That's why this use of addition is called a two-dimensional slide.)

The rule for the slide in Example 1 can be written algebraically as "the image of (x, y) is $(x + 1, y + -6)$." When you have a rule for a slide, you can use it to compute the images of complex figures. Since a slide doesn't change the size or shape of a figure, you can compute the images of a few special points, like vertices, and then use a ruler or other tools to complete the image figure.

A cartoonist at work. *In 1990, Matt Groening's animated cartoon,* The Simpsons, *became the first prime-time animated TV series in many years.* The Flintstones *was the first prime-time cartoon series, debuting in 1960.*

Activity

A graphics artist is working on a cartoon involving a rabbit. Figure *PQRST* shows a small part of the design. Using the rule that the image of (x, y) is $(x + -5, y + 2)$, draw the image of this figure. Label the image $P'Q'R'S'T'$.

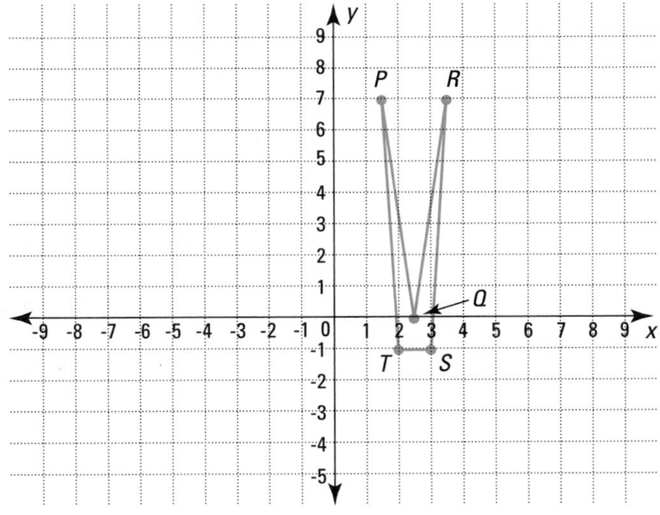

You have done the activity correctly if you slid the rabbit's ears 5 units to the left and 2 units up.

Example 2

If the point (x, y) is slid 8 units up, what is its image?

Solution

There is no move left or right, so 0 is added to the first coordinate. To move 8 units up, add 8 to the second coordinate.

The image of $(x, y) = (x + 0, y + 8) = (x, y + 8)$.

Check

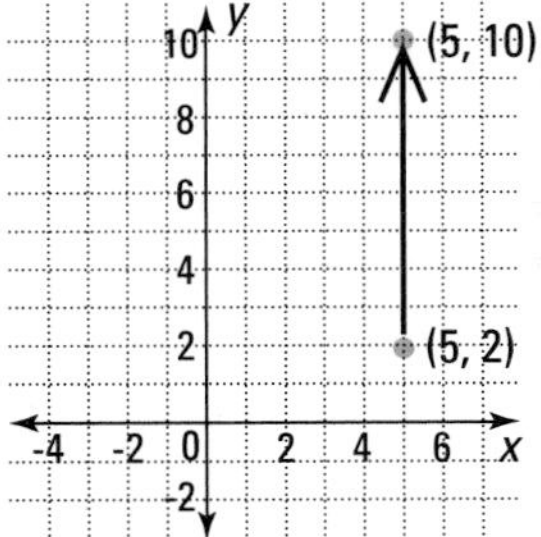

Substitute values for x and y and graph. If $x = 5$ and $y = 2$, the preimage (x, y) is $(5, 2)$. The image is $(5, 2 + 8) = (5, 10)$. Is $(5, 10)$ eight units above $(5, 2)$? The graph at the left shows this, so it checks.

QUESTIONS

Covering the Reading

1. a. The first coordinate of a point is also called its ? -coordinate.
   b. The second coordinate is also called its ? -coordinate.

2. If the x-coordinate of a point is negative and the y-coordinate is positive, in which quadrant is the point?

3. Which two quadrants are located to the left of the y-axis?

4. a. When you slide a figure, what is the resulting figure called?
   b. What is the original figure called?

5. Find the image of (-2, -1) after a slide 0.5 units to the left and 6 units up.

6. Find the image of (0, 0) after a slide 45 units up.

7. Draw a coordinate plane and graph the point P = (3, 5) and its image after a slide of 2 units to the right and 2 units down.

8. A preimage is (-3, 1.5).
   a. Graph the preimage and its image after a slide 0.5 unit to the left and 4 units up.
   b. In what quadrant is the image?

9. a. Copy the graph of $\triangle PQR$ with P = (-5, -2), Q = (-7, -5), and R = (-3, -8).
   b. On the same axes, graph the image of this figure after a slide of 3 units to the right and 4 units up.

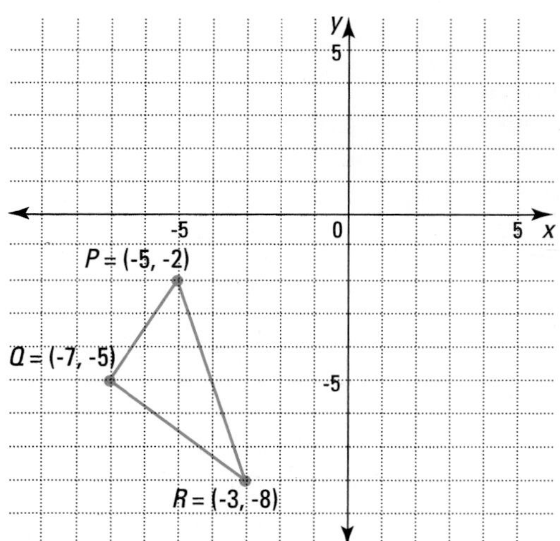

10. a. Draw $\triangle ABC$ with A = (-2, 2), B = (1, 2), and C = (-2, 3).
    b. Suppose the image of each point (x, y) on $\triangle ABC$ is $(x + 7, y + -3)$. Draw the image $\triangle A'B'C'$.

In 11 and 12, let point $P = (x, y)$. **a.** Write the image of P after the slide. **b.** Check your answer to part **a** by picking values for x and y and graphing.

11. 3 units right, 7 units down

12. 3 units left, 1 unit down

Applying the Mathematics

13. Copy the figure drawn below. Slide the figure using the rule that the image of (x, y) is $(x + 5, y + -3)$.
a. What are the coordinates of Q'?
b. Graph the image of the figure.

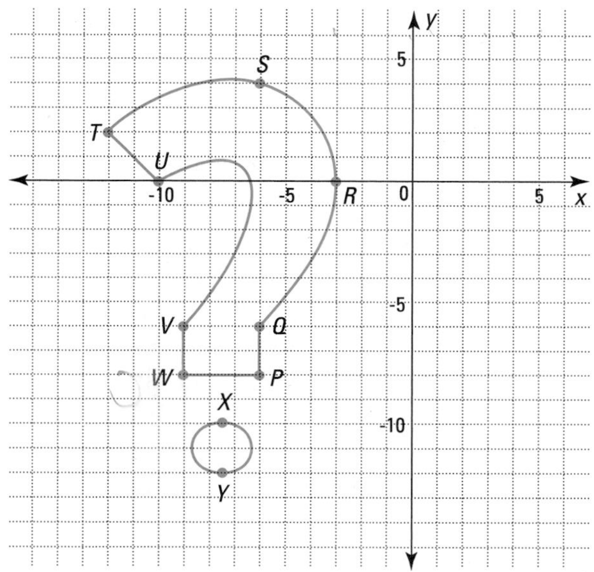

Incan pottery. *Shown is a sample of the excellent-quality pottery created by the Incas. The Incan empire prospered during the 15th century in what today is Peru and other South American countries.*

14. a. What is the image of the point (x, y) after a slide 9 units to the left?
b. Explain how your answer to part **a** is related to the Additive Identity Property.

15. A point is (7, 2) and its image is (15, -4). Describe the slide: _?_ units to the (left or right) and _?_ units (up or down).

16. In the Incan empire, artists often painted simple strip patterns on pottery. Refer to the Incan strip pattern shown. Take each repetition of the pattern to be 1 unit wide. Pick one part of the design as a preimage. Give a rule for a slide that causes the image to look exactly like the preimage.

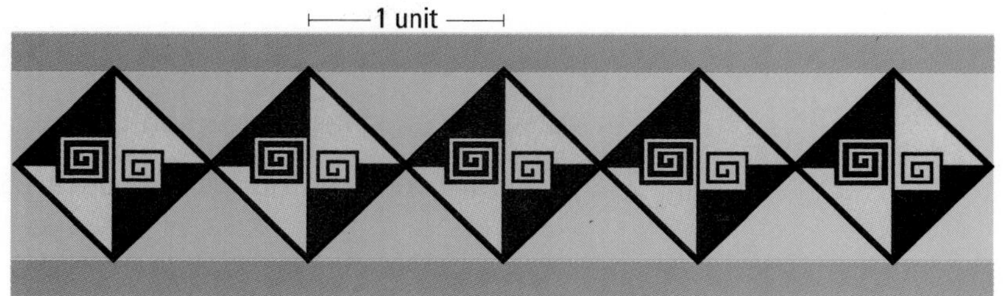

17. Examine the two-dimensional slide below.

a. Under this slide, the image of any point is __?__ units right and __?__ units above the preimage.

b. Under this slide, the image of (x, y) is $(x +$ __?__, $y +$ __?__$)$.

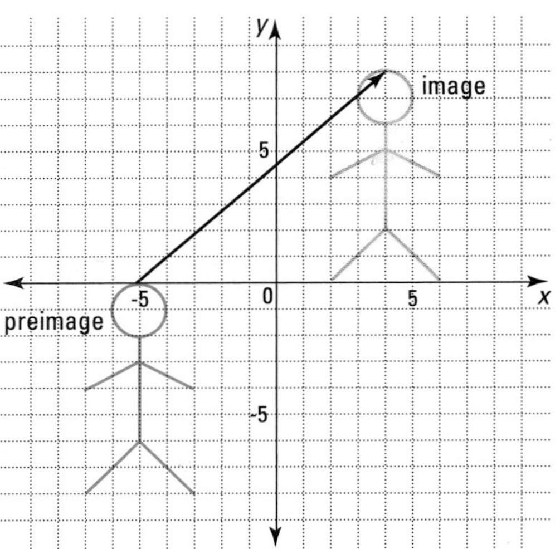

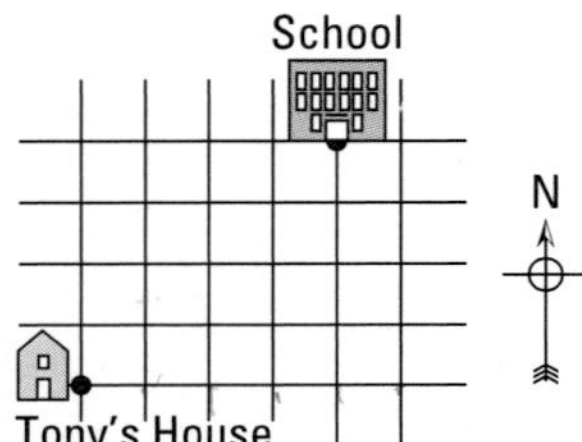

18. Use the diagram at the left. One route Tony can take to get to school is by going 2 blocks east, 4 blocks north, and another 2 blocks east. Name three other routes Tony can take to get from his house to school.

19. After a slide 3 units right and 9 units up, an image is $(7, -1)$. What are the coordinates of its preimage?

Review

In 20 and 21, a person begins standing 24 ft from a sensor that measures distances to objects. The graph shows how the distance between the person and the sensor changed. *(Lesson 3-3)*

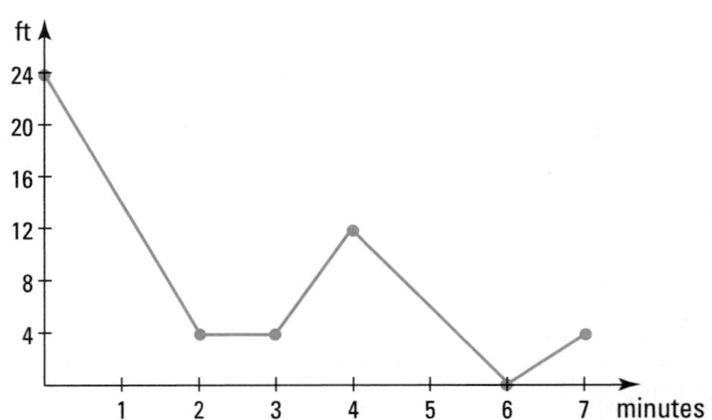

20. At what time was the person moving closer to the sensor?

21. At what time was the person touching the sensor?

22. A can of juice sells for \$.59 at the grocery store. On a coordinate graph make *cost* the unit on the vertical axis and *number of cans* the unit on the horizontal axis. Plot a graph showing the cost of 1, 2, 3, 4, 5, and 6 cans. *(Lesson 3-3)*

In 23 and 24, simplify. *(Lessons 3-1, 3-2)*

23. $(x + 14) + -14$

24. $(-7 + 5x) + 7$

25. The Ceramco Co. showed a loss of \$5 million in 1983 and a profit of \$3.2 million in 1984. What was the net change during this time? *(Lesson 3-1)*

26. **a.** Solve $2.75x = 8.25$.
b. What is done to solve an equation of the form $ax = b$?
c. Solve $2.75 + x = 8.25$.
d. What is done to solve an equation of the form $a + x = b$? *(Lessons 2-6, 3-2)*

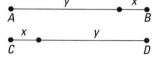

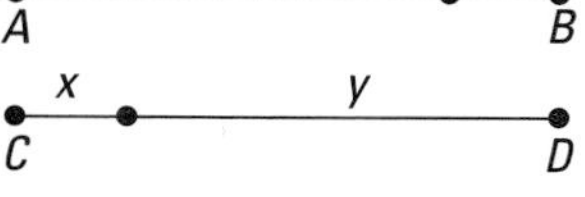

27. **a.** Which property tells you segments $\overline{AB}$ and $\overline{CD}$ at the left have the same length?
b. Express that length in terms of x and y. *(Lesson 3-1)*

28. Consider $\frac{2}{3}x = 24$,
a. Solve this equation.
b. Make up a problem which can be solved with this equation. *(Lesson 2-6)*

29. An air conditioning unit with a high energy efficient ratio (EER) gives more cooling with less electricity. To find the EER of a unit, divide the BTU (British Thermal Unit) number by the number of watts. The higher the EER, the more efficient the air conditioner.

$$\text{EER} = \frac{\text{BTU}}{\text{watts}}$$

a. Find the EER to the nearest tenth for an air conditioner having BTU = 12,600 and watts = 1315.
b. Find the EER to the nearest tenth for an air conditioner having BTU = 5,000 and watts = 850.
c. Which air conditioner, in part **a** or **b,** is more efficient? Justify your answer. *(Lesson 1-5)*

30. Is $x = -6$ a solution to $4x + 15 = -7$? Explain how you know. *(Lesson 1-1)*

At the center of things. *Pictured is Falling Springs Mill in Mark Twain National Forest. The forest covers an area of 608,719 acres.*

Exploration

31. In 1990, the U.S. center of population was near Steelville, Missouri, in Mark Twain National Forest. According to American Demographics, the center of population has been moving daily about 58 feet west and 29 feet south.
a. At this rate, how far will it have moved by the year 2020?
b. Where will it be? (You will need to consult a map.)

LESSON

3-5

Solving $ax + b = c$

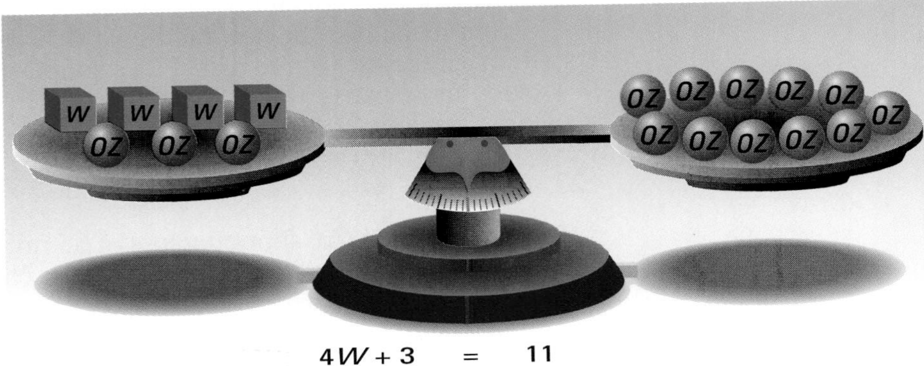

$4W + 3 \quad = \quad 11$

So far you have solved equations that involve either multiplication or addition *but not both.* Now you are ready to solve equations that involve multiplication *and* addition. One such equation is $4W + 3 = 11$.

This equation is pictured above with a balance scale. On the left side of the scale are 4 boxes, each of unknown weight W, and 3 one-ounce weights. They balance with the 11 ounces on the right.

You can find the weight W of one box in two steps. Each step keeps the scale balanced.

Step 1: Remove 3 oz from each side of the scale.

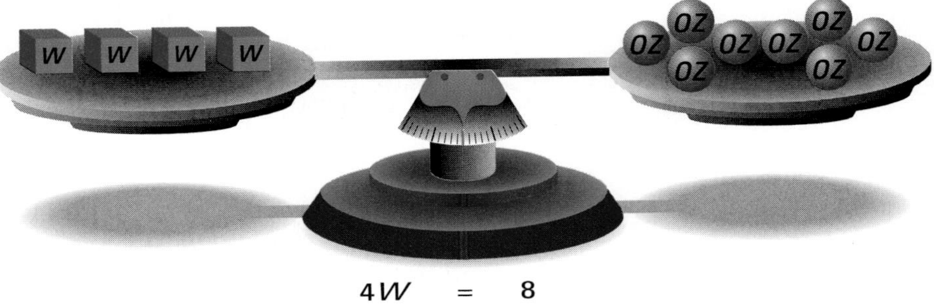

$4W \quad = \quad 8$

Step 2: Leave $\frac{1}{4}$ of the contents on each side.

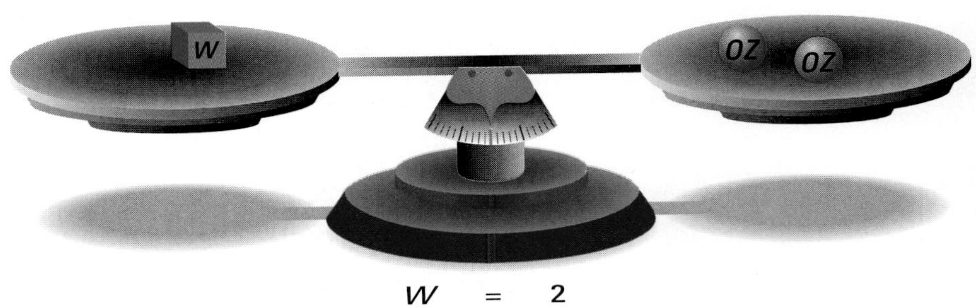

$W \quad = \quad 2$

Example 1 shows the same steps without the balance scale.

Example 1

Solve $4W + 3 = 11$.

Solution

$$4W + 3 = 11$$
$$4W + 3 + (-3) = 11 + (-3)$$ Addition Property of Equality (Add -3 to each side.)
$$4W = 8$$
$$\frac{1}{4}(4W) = \frac{1}{4}(8)$$ Multiplication Property of Equality (Multiply both sides by $\frac{1}{4}$.)
$$W = 2$$

Check

Substitute 2 for W in the original equation. Does $4 \cdot 2 + 3 = 11$? Yes.

Any equation of the form $ax + b = c$ can be solved in two steps. First add the opposite of b to both sides. Then multiply both sides by the reciprocal of a.

Example 2

Solve $\frac{2}{3}x + 19 = 7$, and check your answer.

Solution

This equation is of the form $ax + b = c$, with $a = \frac{2}{3}$, $b = 19$, and $c = 7$.

$$\frac{2}{3}x + 19 = 7$$
$$\frac{2}{3}x + 19 + -19 = 7 + -19$$ Add -19 to each side.
$$\frac{2}{3}x = -12$$ Simplify.
$$\frac{3}{2}\left(\frac{2}{3}x\right) = \frac{3}{2} \cdot -12$$ Multiply each side by the reciprocal of $\frac{2}{3}$.
$$x = -18$$ Simplify.

Check

Substitute $x = -18$ in the original equation. Do you get a true sentence?

Does $\frac{2}{3}(-18) + 19 = 7$?

$$\frac{2}{\cancel{3}_1} \cdot \frac{\cancel{-18}^{-6}}{1} + 19 = 7?$$

$$-12 + 19 = 7?$$

Yes, the sentence is true. So -18 checks.

Often equations are complicated but can be simplified into ones that you can solve. Simplifying each side is an important step in equation solving.

Leaping lemurs. *Most lemurs in the wild live in trees. The ring-tailed lemur, however, usually dwells on the ground. This ring-tailed lemur is in the Royal Zoological Gardens in Melbourne, Australia.*

Example 3

When Val works overtime at the zoo on Saturday, she earns \$9.80 per hour. She is also paid \$8.00 for meals and \$3.00 for transportation. Last Saturday she earned \$77.15. How many hours did she work?

Solution

Let h = the number of hours Val worked. In h hours she earned $9.80h$ dollars. So

$$9.80h + 8.00 + 3.00 = 77.15.$$

Solve for h.

$$9.80h + 11 = 77.15$$
$$9.80h + 11 + -11 = 77.15 + -11$$
$$9.80h = 66.15$$
$$\frac{1}{9.80} \cdot 9.80h = \frac{1}{9.80} \cdot 66.15$$
$$h = 6.75$$

Val worked $6\frac{3}{4}$ hours.

Check

If she worked 6.75 hours at \$9.80 per hour, she earned 6.75 · 9.80 dollars. That comes to \$66.15. Now add \$8 for meals and \$3 for transportation. The total is \$77.15, as it needs to be.

Even if the coefficient of x in $ax + b = c$ is negative, the equation is solved in the same way. If the equation is written with the variable on the right side, such as $c = ax + b$, the solution is still obtained by adding the opposite of b, and multiplying by the reciprocal of a.

Example 4

Solve $-2 = -8x + -6$.

Solution

First add the opposite of -6 to both sides. The opposite is 6.

$$-2 = -8x + -6$$
$$6 + -2 = -8x + -6 + 6$$
$$4 = -8x$$

Now multiply each side by $-\frac{1}{8}$, which is the reciprocal of -8.

$$-\frac{1}{8}(4) = -\frac{1}{8} \cdot -8x$$
$$-\frac{1}{2} = x$$

Check

Substitute $x = -\frac{1}{2}$ into the equation.

Does $-2 = -8\left(-\frac{1}{2}\right) + -6$?

$-2 = 4 + -6$?

Yes, since $-2 = -2$ the equation checks.

QUESTIONS

Covering the Reading

1. The boxes are of unknown equal weight W.
   a. What equation is pictured by this balance scale?

   b. What two steps can be done with the weights on the scale to find the weight of a single box?
   c. How much does each box weigh?

2. a. When solving $4n + 8 = 60$, first add _?_ to both sides. Then _?_ both sides by _?_.
   b. Solve and check $4n + 8 = 60$.

3. The steps used to solve $55v + 61 = 556$ are shown here.

   Given: $55v + 61 = 556$
   Step 1: $55v = 495$
   Step 2: $v = 9$

   a. What was done to arrive at Step 1?
   b. What was done to arrive at Step 2?

In 4–7, solve and check.

4. $8x + 15 = 47$
5. $7y + 11 = 74$
6. $-2z + 32 = 288$
7. $2 = 9x + -3$

8. a. What should be the first step in solving $3.5 + 2x + 5.6 = 10$?
   b. Solve and check this equation.

9. Refer to Example 3. If Val's pay two Saturdays ago was \$89.40, how many hours did she work that day?

In 10–13, solve and check.

10. $\frac{3}{4}x + 12 = 27$
11. $5 = -4x + 15$
12. $-8n + -18 = 88$
13. $16 = \frac{2}{3}a + 20$

New England beauty. *Pictured is a view of Starke, New Hampshire. In the fall, tourists visit to see the changing colors of the leaves.*

Applying the Mathematics

In 14–16, a situation is given. **a.** Write an equation to describe the situation. **b.** Solve the equation and answer the question.

14. Luisa lives in New Hampshire, where there is no sales tax. She bought three hamburgers with a \$5 bill and received \$0.53 change. What was the price of one hamburger?

15. On June 2, Carlos's savings account showed a balance of \$4347.59. During the next month he deposited a total of \$752.85 and withdrew \$550.00. His July 1 bank statement reported that Carlos had \$4574.14 including interest. How much interest had he earned during June?

16. The Kuderskis are saving for their children's education. They have 150 shares of a single stock and \$1500 cash. If the total value of their savings is currently \$3637.50, what is the value of one share of the stock?

In 17 and 18, solve and check.

17. $3\frac{1}{4} + x = 10\frac{1}{2}$

18. $\frac{2}{3}t + \frac{1}{3} = 7$

Review

In 19 and 20, graph $A = (-3, -4)$ and $B = (-9, -4)$, and connect the points with a line segment.

19. If the image of (x, y) is $(x + 3, y + 6)$, draw the image of $\overline{AB}$. *(Lesson 3-4)*

20. What rule would describe a slide of $\overline{AB}$ 5 units down? *(Lesson 3-4)*

21. Triangle $D'E'F'$ is a slide image of triangle DEF. $D = (0, 0)$, $E = (1, 4)$, $F = (3, 6)$, and $F' = (5, 2)$. *(Lesson 3-4)*

a. Describe the slide choosing the appropriate directions: __?__ units (left or right) and __?__ units (up or down).

b. What are the coordinates of D' and E'?

22. a. Evaluate $-(-39) + -(-(-39))$.

b. Explain how you got your answer. *(Lesson 3-2)*

23. The numbers in the table below represent the cost c (in dollars) of n note pads. *(Lessons 1-9, 2-4)*

n	1	2	3	4
c	.50	1.00	1.50	2.00

a. How much would you expect to pay for 6 note pads?

b. Write a formula for the cost of n note pads.

24. The rectangle at right is made from a square with sides x units long and two rectangles with sides of lengths 1 and x. Find the area of the rectangle. *(Lesson 2-1)*

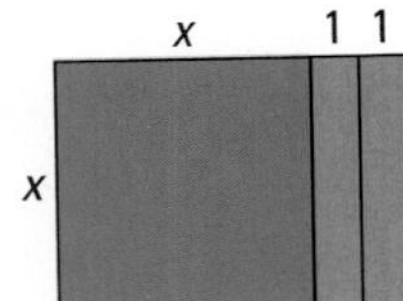

25. *Skill sequence.* Write as a decimal. *(Previous course)*
   a. 20%
   b. 2%
   c. 102%
   d. 120%

In 26–29, a consumer research organization evaluated the flavor and texture of popular strawberry ice creams. The rating scale had a maximum possible score of 6 (flavor and texture very good) and a minimum possible of -6 (very bad). The data are presented in the table and scatterplot. *(Lesson 3-3)*

Cost and Ratings for Popular Strawberry Ice Cream Brands

Brand	Cost per Serving	Rating
Berry N'ice	57	3
Perfect Parfait	55	5
Delicious	52	5
Merry Berry	49	-1
Sundae Special	46	3
Gourmet	43	2
Fabulous Flavors	26	1
Bon Appetit	22	2
Betty's Best	19	3
Select	18	-1
P. Good	18	1
Creamy Creations	17	-2
Mix-in Magic	17	-4
I. Scream	16	3
Ambrosia	16	-1
Tasty Treat	12	-2
Nuts and Berries	12	-1
Sweet Swirl	12	-3

26. Which point (*A, B, C, D,* or *E*) represents Gourmet?

27. Which point represents Select?

28. Which ice cream seems to be a poor value; that is, it costs a lot but has a low rating?

29. Is there a tendency for ice cream which costs more to taste better? Justify your answer.

30. a. Make up an equation of the form $x + a = b$, where a and b are positive, and the solution is negative.
    b. Solve your equation. *(Lesson 3-2)*

Exploration

31. Consider equations of the form $ax + b = c$, where $a \neq 0$.
    a. Write a program for a calculator or computer that accepts values of a, b, and c as input, and gives the value of x as output.
    b. Run your program with different values of a, b, and c leading to both positive and negative solutions.

Introducing Lesson 3-6

Using Algebra Tiles to Add Expressions

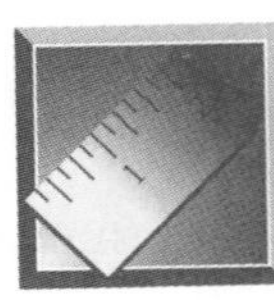

IN-CLASS ACTIVITY

Work on this activity in small groups. Each group will need about eight tiles of each size.

Many algebraic expressions can be illustrated by using *algebra tiles.* Algebra tiles are rectangular tiles whose areas represent 1, x, and x^2.

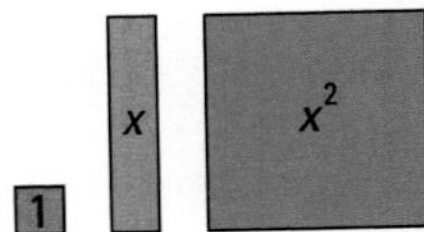

You can arrange tiles to represent an expression. The expressions $2x^2 + 3x$ and $2x + 3$ are shown below.

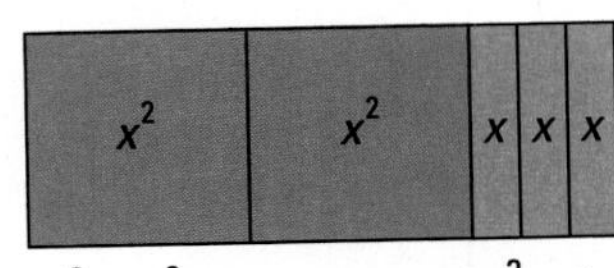

$x^2 + x^2 + x + x + x = 2x^2 + 3x$

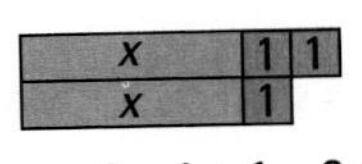

$x + x + 1 + 1 + 1 = 2x + 3$

In 1 and 2, use tiles to represent the expression. Draw pictures of your tiles.

1. a. $4x^2$
b. $4x^2 + 2x$
c. $4x^2 + 2x + 2$

2. a. $4x$
b. $3x^2 + 4x$
c. $3x^2 + 4x + 1$

In 3 and 4, what expression is shown?

3.

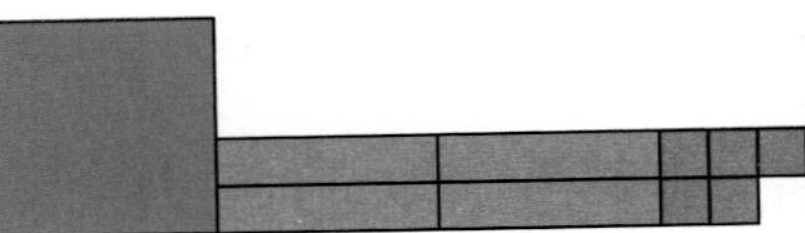

4.

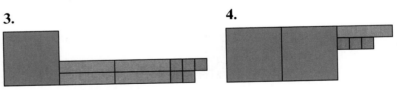

5. To add algebraic expressions, arrange tiles representing each addend and count like tiles. Use tiles to illustrate this sum. Then copy and complete the equation.

$$(x^2 + 3x) + (2x^2 + x + 2) = \underline{\ ?\ }\,x^2 + \underline{\ ?\ }\,x + \underline{\ ?\ }$$

In 6, use tiles to illustrate the sum, and give the sum.

6. a. $(2x^2 + 4x + 1) + (x^2 + 4)$
b. $(x^2 + 3) + (2x + 2)$
c. $(x^2 + 3x) + (x^2 + x)$
d. $(4 + 2x + x^2) + (3x^2 + x + 1)$

LESSON

3-6

The Distributive Property and Adding Like Terms

The latest addition. *As this library obtained new editions, it found it needed a new addition. The total area of the library, after its completed addition, can be found by using the Distributive Property.*

Suppose two rectangles, each having width c, are placed end to end. What is the total area? You can calculate the area two ways. First, you can add the areas of the two smaller rectangles. The total area is $ac + bc$.

Area = $ac + bc$

c Area = ac Area = bc

a b

Or, you can multiply the width c by the total length $a + b$. The total area is $(a + b)c$.

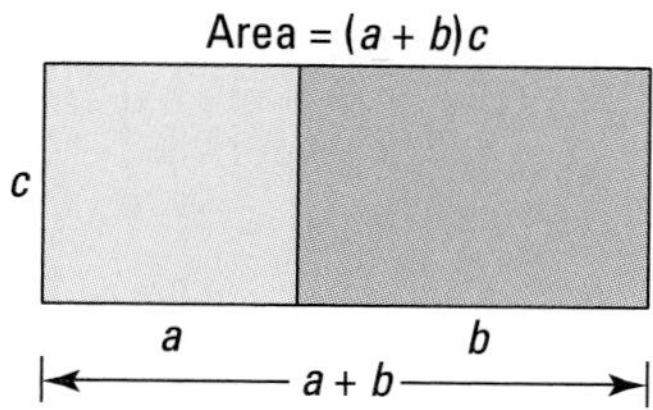

The area is the same, regardless of how you calculate it:
$ac + bc = (a + b)c$. This is an instance of the *Distributive Property.*

Distributive Property: Adding or Subtracting Like Terms Forms
For any real numbers a, b, and c,

$$ac + bc = (a + b)c \text{ and}$$
$$ac - bc = (a - b)c.$$

Activity

Take a rectangular piece of paper of length 11 in. and any width x in. Cut off a piece of length 4 in. What is the area of the remaining piece?

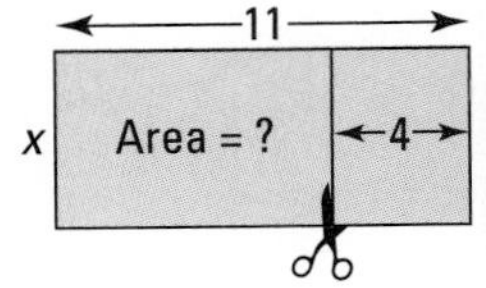

Adding Like Terms.

A **term** is either a single number or a variable, or a product of numbers and variables. The expressions $7s$ and $-2s$ are called **terms,** and the 7 and -2 are the **coefficients** of s. When the variables in the terms are the same, they are called **like terms,** and the addition can be performed by adding the coefficients. This simplification is called **adding like terms.** Adding like terms is an instance of the Distributive Property.

Example 1

Simplify $7s + -2s$.

Solution 1

Use the Distributive Property to add the like terms.

$7s + -2s = (7 + -2)s = 5s$

Recall the Multiplicative Identity Property of 1. For any real number n, $n = 1n$. When this property is combined with the Distributive Property, many expressions can be simplified. For instance, $k + 8k = 1k + 8k = 9k$.

Example 2

Simplify $(3s + d + 6) + (2d + s)$.

Solution 1

Use the Commutative and Associative Properties to group like terms.

$$\begin{aligned}(3s + d + 6) + (2d + s) &= (3s + s) + (d + 2d) + 6 && s = 1 \cdot s\\ &= (3s + 1s) + (1d + 2d) + 6 && d = 1 \cdot d\\ &= 4s + 3d + 6 && \text{Add like terms.}\end{aligned}$$

So $(3s + d + 6) + (2d + s) = 4s + 3d + 6$.

Solution 2

Think of segments of lengths s, d, and 6. Use the Putting-Together Model of Addition.

If $s\ s\ s\ d\ 6$ is put together with $d\ d\ s$, the result is $s\ s\ s\ d\ 6\ d\ d\ s$.

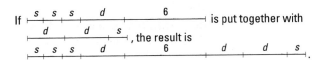

This segment has length $4s + 3d + 6$.
So $(3s + d + 6) + (2d + s) = 4s + 3d + 6$.

Adding Like Terms in Equations

The Distributive Property can be used in solving equations.

Example 3

Solve $-3x + 5x = -12$.

Solution

Apply the Distributive Property to add like terms.

$$(-3 + 5)x = -12$$
$$2x = -12$$
$$\frac{1}{2} \cdot 2x = \frac{1}{2} \cdot -12 \qquad \text{Multiply both sides by } \tfrac{1}{2}.$$
$$x = -6$$

Check

Substitute -6 for x in the original equation.

Does $-3 \cdot -6 + 5 \cdot -6 = -12$?
$18 + -30 = -12$?
$-12 = -12$? Yes, -6 checks.

Estate Sale. *Sometimes a home is included in a person's estate. This two-story frame house would sell for about $150,000 in some sections of the U.S.*

Example 4

A $150,000 estate is to be split among three children, a grandchild, and a charity. Each child gets the same amount, while the grandchild gets half as much. If the charity receives $10,000, how much will each child receive?

Solution

Write an equation. Let c represent each child's portion. Then $\frac{1}{2}c$ is the grandchild's portion.

$$c + c + c + \frac{1}{2}c + 10{,}000 = 150{,}000$$
$$1c + 1c + 1c + \frac{1}{2}c + 10{,}000 = 150{,}000 \qquad \text{Use the Multiplicative Identity Property of 1.}$$
$$3\frac{1}{2}c + 10{,}000 = 150{,}000 \qquad \text{Use the Distributive Property to add like terms.}$$
$$3\frac{1}{2}c = 150{,}000 + -10{,}000 \qquad \text{Add } -10{,}000 \text{ to each side.}$$
$$3\frac{1}{2}c = 140{,}000$$
$$c = 40{,}000 \qquad \text{Multiply each side by the reciprocal of } 3\tfrac{1}{2}, \text{ or by } \tfrac{2}{7}.$$

Each child should receive $40,000.

Check

The grandchild receives half as much as each child does, and so gets $20,000. Does $40{,}000 + 40{,}000 + 40{,}000 + 20{,}000 + 10{,}000 = 150{,}000$? Yes.

Note: The arithmetic in Example 4 could have been made easier by thinking of all amounts "in thousands." Then the equation $3\frac{1}{2}c + 10 = 150$ would describe the situation. It would still be solved using the same steps as in the example.

QUESTIONS

Covering the Reading

1. State the Distributive Property.

2. What answer did you get for the Activity in the lesson?

3. What instance of the Distributive Property is pictured below?

21 8

n

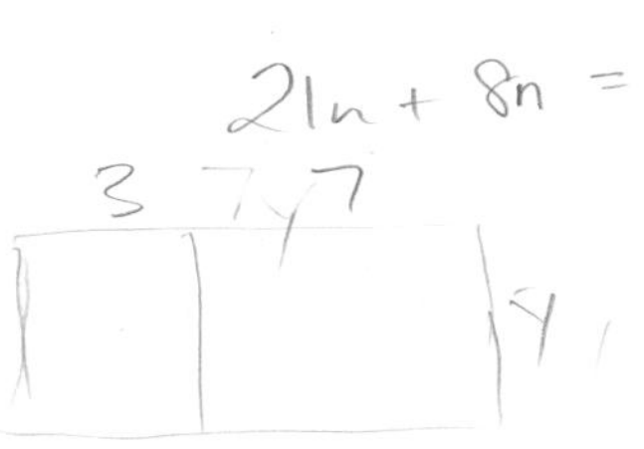

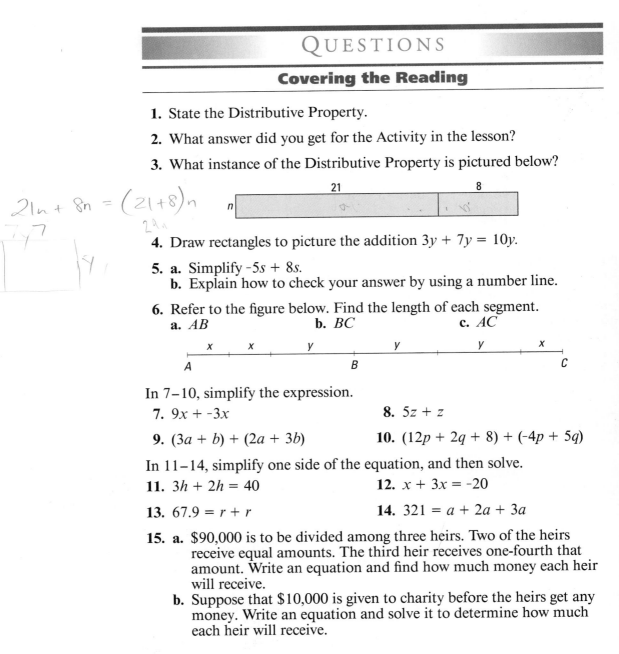

4. Draw rectangles to picture the addition $3y + 7y = 10y$.

5. a. Simplify $-5s + 8s$.
   b. Explain how to check your answer by using a number line.

6. Refer to the figure below. Find the length of each segment.
   a. AB b. BC c. AC

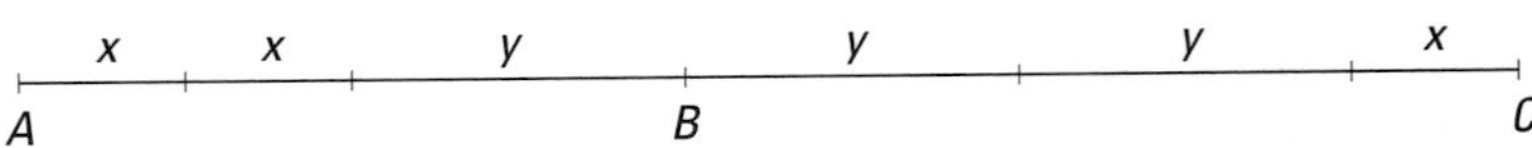

In 7–10, simplify the expression.

7. $9x + -3x$

8. $5z + z$

9. $(3a + b) + (2a + 3b)$

10. $(12p + 2q + 8) + (-4p + 5q)$

In 11–14, simplify one side of the equation, and then solve.

11. $3h + 2h = 40$

12. $x + 3x = -20$

13. $67.9 = r + r$

14. $321 = a + 2a + 3a$

15. a. \$90,000 is to be divided among three heirs. Two of the heirs receive equal amounts. The third heir receives one-fourth that amount. Write an equation and find how much money each heir will receive.
    b. Suppose that \$10,000 is given to charity before the heirs get any money. Write an equation and solve it to determine how much each heir will receive.

Universal hobby. *Stamp collecting has been called "the hobby of kings and the king of hobbies." Collectors value stamps for different reasons. Some stamps are valuable because they are rare.*

Applying the Mathematics

16. Ramon and Ramona collect stamps. Ramon has s stamps, and Ramona has 5 times as many stamps as Ramon. How many stamps do they have together?

17. Some taxicab companies allow their drivers to keep $\frac{3}{10}$ of all fares collected. The rest goes to the company. If a driver receives F dollars from fares, write an expression for the company's share.

In 18 and 19, simplify.

18. $(x + -6) + (2x + 4) + (3x + -3)$

19. $5f + (4f - 6) + (6 + -9f)$

20. In a boxing match, the money divided between the fighters is called the purse. If the loser receives one-fifth of what the winner will get and the total purse is \$500,000, how much will each boxer receive?

21. Around 800 A.D., the Islamic law of inheritance stated that when a woman died her husband received one-fourth of the estate and the rest was the children's share with a son receiving twice as much as a daughter. How many camels would a daughter inherit if the estate had 60 camels and her father and two brothers were still living?

22. Solve and check $-n + 2n + -5n + 7 = -9$.

Review

23. The scale below is balanced. Each box has the same weight. The other weights are 1 ounce each.

a. What equation is pictured?

b. How much does each box weigh? *(Lesson 3-5)*

24. Solve and check $93 = \frac{2}{3}x + 17$. *(Lesson 3-5)*

25. Explain how to solve any equation of the form $ax + b = c$, in which $a \neq 0$. Make up your own equation(s) to illustrate your explanation. *(Lesson 3-5)*

26. On a coordinate graph, plot and connect in order points P, Q, R, and S. $P = (-4, -3)$, $Q = (6, -3)$, $R = (6, 4)$, $S = (-4, 4)$.
   a. What geometric figure is $PQRS$?
   b. What is the area of $PQRS$?
   c. Draw $P'Q'R'S'$, the image of $PQRS$ under the slide in which the image of (x, y) is $(x + -3, y + 1)$.
   d. How is the area of $P'Q'R'S'$ related to the area of $PQRS$? *(Lessons 2-1, 3-4)*

In 27 and 28, a number is given. **a.** State its opposite. **b.** State its reciprocal. *(Lessons 2-2, 3-2)*

27. 17

28. $-\frac{39}{2}$

29. The elevation of Death Valley, California, is listed as -282 feet. What does the minus sign mean in this situation? *(Lesson 3-1)*

Sizzling temperatures. *The highest temperature ever recorded in the U.S. was in Death Valley. The temperature soared to 134°F (57°C) on July 10, 1913. The valley's warm winter and geological attractions have made it a popular winter resort area.*

30. *Skill sequence.* Add. *(Previous course)*
   a. $\frac{3}{10} + \frac{1}{10}$
   b. $\frac{3}{10} + \frac{1}{5}$
   c. $\frac{3}{10} + \frac{2}{15}$

31. Suppose a laser printer prints 8 pages per minute. How long will it take to print 2400 documents with 3 pages per document? *(Lesson 2-4)*

32. The average mass of air molecules is $30 \cdot 1.66 \cdot 10^{-24}$ grams. Write this number in scientific notation. *(Previous course)*

Exploration

33. Some people overgeneralize the Distributive Property. They think that because $6x + 2x = 8x$, all of the following should be true.
   (a) $6x \cdot 2x = 12x$
   (b) $\frac{6x}{2x} = 3x$
   (c) $6^x + 2^x = 8^x$
   (d) $6^x \cdot 2^x = 12^x$
   (e) $6\sqrt{x} + 2\sqrt{x} = 8\sqrt{x}$

   Only two of the above statements are true for all positive values of x. Which two are true? Give a counterexample for each of the others to show that they are false.

LESSON

3-7

The Distributive Property and Removing Parentheses

Removing parentheses. *The display shows that the three expressions have the same result. Some people prefer the third expression because it has no parentheses.*

You have used the Distributive Property to add like terms. Another form of the Distributive Property results when properties of addition and multiplication are applied.

$ac + bc = (a + b)c$	Add like terms.
$(a + b)c = ac + bc$	Switch sides of the equation.
$c(a + b) = ca + cb$	Use the Commutative Property of Multiplication.

We call this the **Removing Parentheses** form of the Distributive Property.

The Distributive Property: Removing Parentheses
For all real numbers a, b, and c,
$$c(a + b) = ca + cb \text{ and}$$
$$c(a - b) = ca - cb.$$

This form of the Distributive Property can be used to multiply a sum by a single term. It shows that c times the quantity $a + b$ is the same as the sum of c times each of its terms. The c is "distributed" over each term.

Example 1

Multiply the following.

a. $2(2x + 3)$

b. $x(2x + 3)$

Solution

a. Each term of $2x + 3$ is multiplied by 2.

$$2(2x + 3) = 2 \cdot 2x + 2 \cdot 3$$
$$= 4x + 6$$

b. Each term of $2x + 3$ is multiplied by x.

$$x(2x + 3) = x \cdot 2x + x \cdot 3$$
$$= 2x^2 + 3x$$

Check

Use algebra tiles.

a. Consider a rectangle with length $2x + 3$ and width 2.

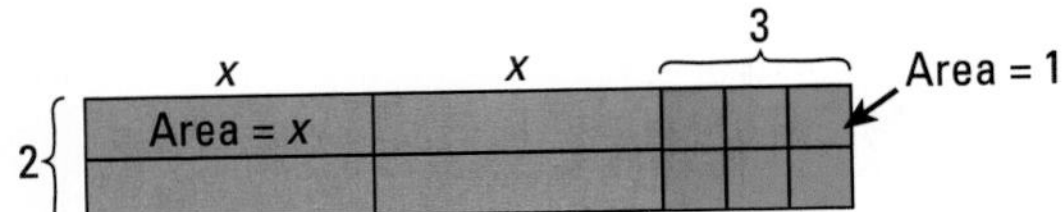

Its area is $2(2x + 3)$. This equals the sum of the area of the four rectangles and six squares. So the area equals $4x + 6$.

b. Draw a rectangle with length $2x + 3$ and width x.

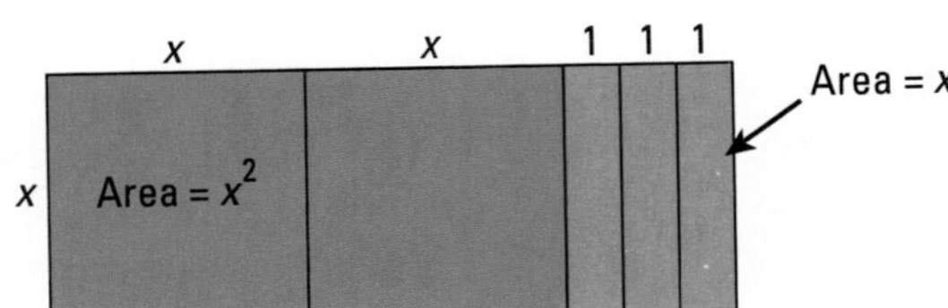

Its area is $x(2x + 3)$. You can also think of the area as being made up of the 5 individual parts: two squares of area x^2, and 3 rectangles, each with area $1 \cdot x = x$. So the area equals $2x^2 + 3x$.

How much would one CD cost?

The Distributive Property and Mental Arithmetic

The Distributive Property can be used to help you perform some calculations mentally.

Example 2

Calculate mentally how much 5 tapes cost if they sell for \$8.97 each.

Solution

Think of \$8.97 as \$9.00 – 3¢.

So 5 · \$8.97 = 5(\$9 – 3¢)
= 5 · \$9 – 5 · 3¢   Do all this in your head.
= \$45 – 15¢
= \$44.85

Check

Use pencil and paper or calculator to show that $5(8.97) = 44.85$.

Working overtime. *Many workers, including certain airline employees, earn overtime pay when they work more than a certain number of hours per day or per week.*

Example 3

Monica's hourly wage is $12.00. If she receives time and a half for overtime, what is her overtime hourly wage?

Solution

Time and a half means $1\frac{1}{2}$ times the regular hourly wage. Since $1\frac{1}{2} = 1 + \frac{1}{2}$, use the Distributive Property to calculate the wage mentally.

$$\begin{aligned} 12 \cdot 1\tfrac{1}{2} &= 12 \cdot (1 + \tfrac{1}{2}) \\ &= 12 \cdot 1 + 12 \cdot \tfrac{1}{2} \\ &= 12 + 6 \\ &= 18 \end{aligned}$$

Monica receives $18 an hour for overtime.

Check

$18 is halfway between her hourly wage ($12) and twice that amount ($24).

The Distributive Property and Solving Equations

Examples 4 and 5 show how the Distributive Property is used to remove parentheses and to add like terms in the same problem.

Example 4

Solve $-5(x + 2) + 3x = 8$.

Solution 1

$-5 \cdot x + -5 \cdot 2 + 3x = 8$	Distributive Property: Removing Parentheses
$-5x + -10 + 3x = 8$	
$-5x + 3x + -10 = 8$	Commutative Property of Addition
$-2x + -10 = 8$	Distributive Property: Adding Like Terms
$-2x + -10 + 10 = 8 + 10$	Addition Property of Equality
$-2x = 18$	Property of Opposites
$x = -9$	Multiplication Property of Equality

Solution 2

Experts do some work mentally and may write down only a few steps.

$$\begin{aligned} -5x - 10 + 3x &= 8 \\ -2x - 10 &= 8 \\ -2x &= 18 \\ x &= -9 \end{aligned}$$

Check

Substitute -9 for x in the original sentence. Follow the order of operations. *Does* $-5(-9 + 2) + 3 \cdot -9 = 8$?

$-5 \cdot -7 + 3 \cdot -9 = 8$?

$35 + -27 = 8$? *Yes.*

Example 5

Suppose the cost of a phone call is 40¢ for the first 3 minutes and 13¢ for each additional minute. An equation that gives the cost C in terms of the whole number n of minutes you talk is $C = .40 + .13(n - 3)$. (Here $n \geq 3$, and n is a whole number.) How long can you talk for $5.00?

Solution

The cost is to equal $5. So substitute $C = 5.00$ and solve.

$$.40 + .13 \cdot (n - 3) = 5.00$$

Apply the Distributive Property to remove parentheses.

$$.40 + .13n - .13 \cdot 3 = 5.00$$
$$.40 + .13n - .39 = 5.00$$
$$.13n + .01 = 5.00$$
$$.13n = 4.99$$
$$n \approx 38.385$$

You can talk for 38 minutes for $5.00.

QUESTIONS

Covering the Reading

1. State the Removing-Parentheses form of the Distributive Property.

In 2 and 3, multiply. Check your answer by drawing rectangles.

2. $5x(x + 2)$ **3.** $5(x + 2)$

In 4–6, use the Removing-Parentheses form of the Distributive Property to eliminate parentheses.

4. $4(n + 6)$ **5.** $12(k - 5)$ **6.** $10b(b + c)$

7. Show how the Distributive Property can help you mentally compute the price of 5 CDs if each CD costs $9.96.

8. Mentally compute the total cost of four gallons of milk at $2.07 each.

In Questions 9 and 10, mentally compute the overtime hourly wage (at time and a half for overtime) if the normal hourly wage is the given amount.

9. $14.00 **10.** $6.50 **11.** $9.99

In 12–17, solve and check.

12. $2(x + 3.1) = 9.8$ **13.** $6(m + -1) = 10$

14. $7(u + -3) = 0$ **15.** $9 = 2(2x + 2) + 2$

16. $2 + 3(v + 4) = 5$ **17.** $-5(t + 2) + 3t = 8$

In 18 and 19, refer to Example 5.

18. Why was the answer not stated as 38.39 minutes?

19. At these prices how long could you talk for $2.00?

20. Suppose the cost of a phone call is 49¢ for the first 3 minutes and 16¢ for each additional minute.
a. What will it cost to talk for an hour?
b. How long can you talk for less than $6.00?

Applying the Mathematics

21. Mentally compute 6 times 999,999.

22. For each hour of television, there is an average of $8\frac{1}{2}$ minutes of commercials. If you watch 6 hours of television in a week, compute mentally how many minutes of commercials you will see.

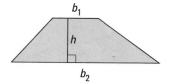

23. The area A of a trapezoid with parallel bases b_1 and b_2 and height h is given by the formula $A = \frac{1}{2}h(b_1 + b_2)$. A trapezoid has base $b_1 = 5$ cm, height 6 cm, and area 60 cm^2.
a. Substitute these values into the area formula above.
b. Solve to find b_2, the length of the other base.

In 24 and 25, the daily charge at a car rental company is $19.95 with the first 100 miles free. The charge for each mile after that is $.25.

24. How much would you pay to rent a car for a day if you drove the indicated distances?
a. 76 mi **b.** 100 mi **c.** 150 mi

25. A formula for the cost C of renting a car from this company and driving m miles, where $m \geq 100$, is $C = 19.95 + .25(m - 100)$. A sales representative was charged $25.70 to rent a car. How many miles did the sales representative drive that day?

Talking turkey. *Your call to a distant city could be to Istanbul, Turkey. Istanbul is unique in that it is in two continents—Asia and Europe. The waterway shown, the Bosporus, divides the city into its Asian and European parts.*

Review

In 26–28, add and simplify. *(Lesson 3-6)*

26. $(x^2 + 3x + 1) + (2x^2 + x + 8)$

27. $(3a + 2b + c) + (-a + 4b + -3c)$ **28.** $n + .04n + .15n$

29. a. Solve and check $5m + 2m < 84$. *(Lessons 2-8, 3-6)*
b. Is -5 a solution of $5m + 2m < 84$? How can you tell? *(Lesson 1-1)*

In 30 and 31, multiply in your head. *(Lessons 1-4, 2-5)*

30. $(-1)(-2)(-3)(4)$ **31.** $(\sqrt{3} \cdot \sqrt{3})(\sqrt{4} \cdot \sqrt{4})$

Exploration

32. Investigate phone rates to a distant city you'd like to call. How do the rates compare to those given in Example 5?

LESSON

3-8

Writing Linear Expressions

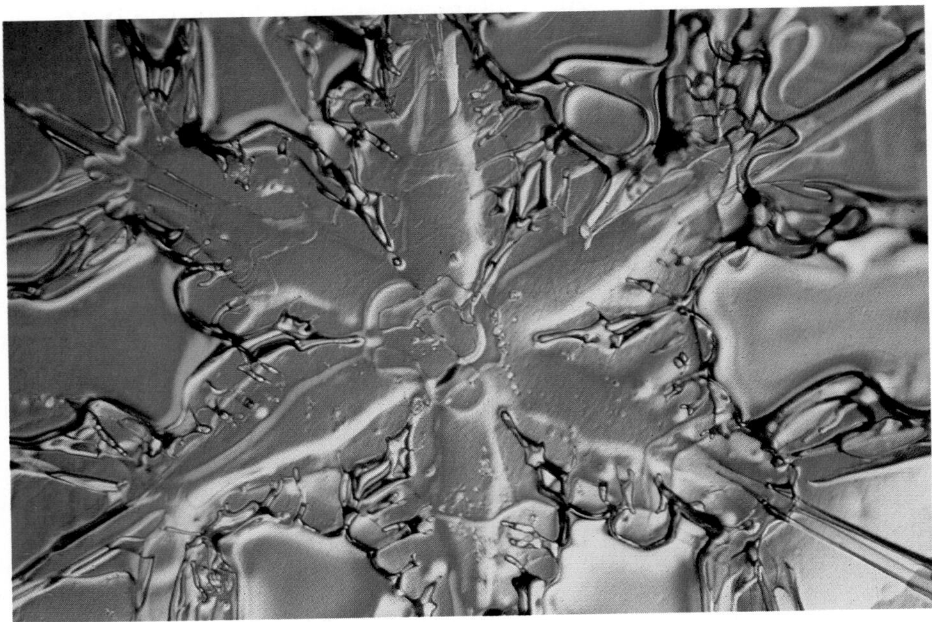

Facts about flakes. *Snowflakes are made up of masses of tiny ice crystals. Although they differ in shape, all ice crystals have six sides. The crystals collide and adhere to each other to produce snowflakes.*

Patterns of the form $ax + b$ or equations of the form $ax + b = c$ often arise from situations of repeated addition. For instance, Suzie made a *sequence* of designs, which she called snowflake designs, using circular chips. She began each design by placing one chip and surrounding it with six other chips. Here are the first three designs. What patterns do you see as Suzie continues to make designs?

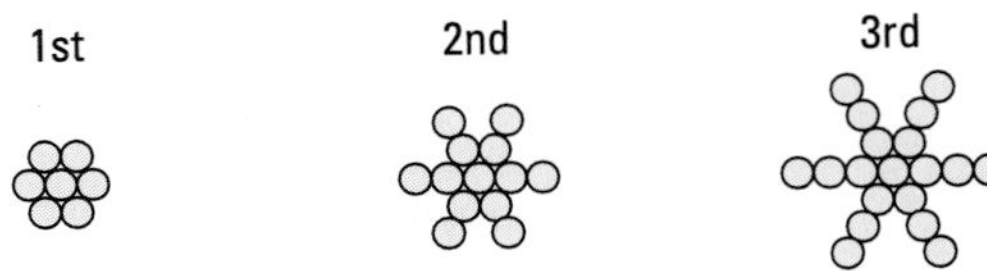

Suzie noticed the following pattern. Each snowflake design can be thought of as having six "spokes." Here is the sketch she drew to illustrate the spokes for a friend.

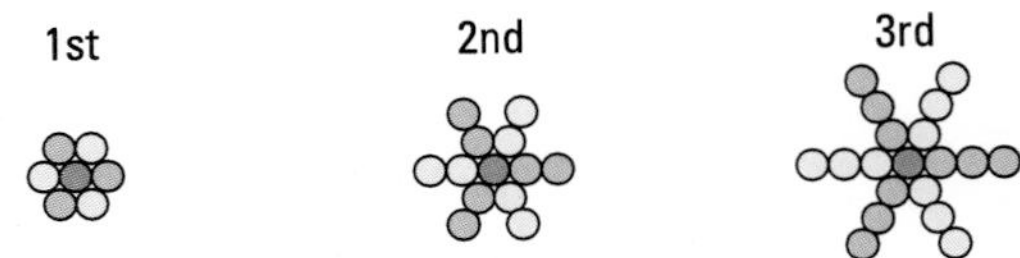

She then noticed in each case that the number of chips in each spoke equals the design's place in the sequence. For instance, the 2nd design has 2 chips per spoke, and the 3rd design has 3 chips per spoke.

Example 1

Suppose that Suzie makes a 4th snowflake design by adding 6 more chips to the 3rd snowflake. How many chips will she need in all?

Solution 1

Draw or make the 4th snowflake.

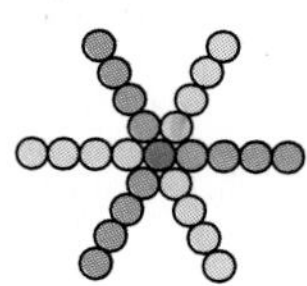

Count the number of chips needed. There are 25 in the fourth snowflake.

Solution 2

Generalize Suzie's observation. There will be one chip in the center, and 6 spokes with 4 chips each. So there will be $1 + 6 \cdot 4$ or 25 chips.

Each solution can be used to determine how many chips Suzie will need to make such a snowflake of any size. However, the first solution strategy might be tedious for a very large snowflake.

Example 2

a. How many chips will Suzie need to make the 10th such snowflake design?
b. If Suzie makes the dth design, how many chips will she need?

Solution

a. Generalize the pattern used in Solution 2 of Example 1.

Design Number	Number of chips/spoke	Total number of chips
1	1	$1 + 6 \cdot 1 = 7$
2	2	$1 + 6 \cdot 2 = 13$
3	3	$1 + 6 \cdot 3 = 19$
4	4	$1 + 6 \cdot 4 = 25$
⋮	⋮	⋮

In the 10th design there will be 1 center chip and 6 spokes with 10 chips each. So there will be $1 + 6 \cdot 10$ or 61 chips.

b. There will be 1 chip in the center and 6 spokes with d chips each. So there will be $1 + 6d$ chips. If you let $c =$ the number of chips, then you can write $c = 1 + 6d$.

Once you have found an algebraic expression or formula for the number of chips, you can use it to work backward.

Going around in spirals. *Marina City Towers provided the first high-rise spiral parking in the U.S. Built in Chicago in 1962, the towers include 19 levels of parking that can accommodate 503 cars.*

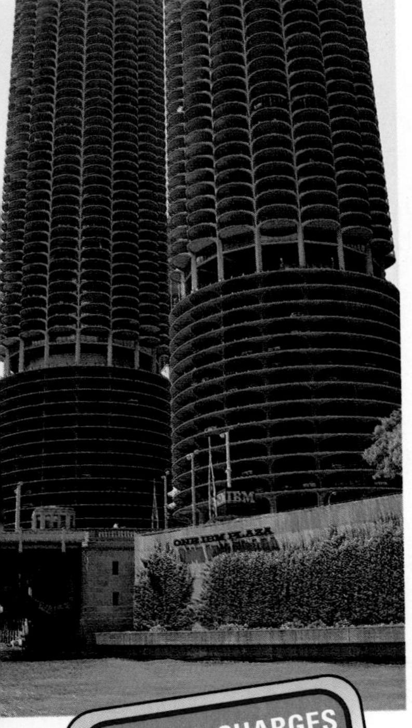

PARKING CHARGES
up to 1 HOUR $2.75
2 HOURS 4.00
3 HOURS 5.25
4 HOURS 6.50
5 HOURS 7.75
Each add'l hr $1.25

Example 3

Suzie has made a big snowflake design with 97 chips. What design number in the sequence is it?

Solution

Use the formula in part **b** of Example 2. You are given that $c = 97$. Substitute into the formula and solve for d.

$$\begin{aligned} c &= 1 + 6d \\ 97 &= 1 + 6d \\ 96 &= 6d \\ 16 &= d \end{aligned}$$

97 chips are in the 16th snowflake pattern.

The expression $1 + 6d$ is a *linear expression* in one variable. In **linear expressions,** all variables are to the first power. If there are two or more variables, they are either added or subtracted. Numbers may have any position. The equations $c = 1 + 6d$ and $97 = 1 + 6d$ are called *linear equations* because each side is either an arithmetic expression or a linear expression.

Real situations in which there is some initial amount followed by a constant amount repeatedly added or subtracted can often be modeled by linear expressions.

Example 4

As shown in the sign at the left, a parking garage charges \$2.75 for parking up to 1 hour and \$1.25 for each additional hour or fraction thereof. If the car is left for h hours, what is the parking charge?

Solution 1

Make a table and look for a pattern. Notice that each entry in the sign is \$1.25 more than the previous one. This suggests that it may be helpful to rewrite each of the costs after the first using multiples of \$1.25.

Hours	Charges	Cost Pattern
1	2.75	2.75
2	4.00	2.75 + 1.25 · 1
3	5.25	2.75 + 1.25 · 2
4	6.50	2.75 + 1.25 · 3
5	7.75	2.75 + 1.25 · 4

Notice how the hours and cost pattern are related. The number on the far right is one less than the number of hours. This suggests the following: The cost c for parking h hours is

$$c = 2.75 + 1.25(h - 1).$$

This could also be written using the Distributive Property as

$$\begin{aligned} c &= 2.75 + 1.25h - 1.25 \\ &= 1.25h + 1.50. \end{aligned}$$

Solution 2

Translate the given information directly into an equation. Let h = the number of hours parked and c = the cost for h hours. Then h − 1 = the number of hours over 1. Each of these hours costs $1.25. Therefore, the cost of parking h − 1 hours is 1.25(h − 1). The total cost is c = 2.75 + 1.25(h − 1).

The examples in this lesson involve patterns in which a constant amount is repeatedly added. Computers are very useful when jobs require speed or a lot of repetition. The following computer program will make a list of the first 200 terms of the sequence in Examples 1 to 3. The program uses a FOR/NEXT loop. The FOR statement tells the computer the number of times to execute the loop. The first time through the loop, the variable D is 1. Each time through, the NEXT statement increases D by one, and sends the computer back to the FOR statement.

```
10 PRINT "NUMBER OF CHIPS IN SNOWFLAKE DESIGN"
20 FOR D = 1 TO 200      The loop will be executed 200 times.
30 LET C = 1 + 6 * D     The formula from Example 2 is evaluated.
40 PRINT D, C            Print the value of the term.
50 NEXT D                Go back to line 20 with a new value for D.
60 END
```

When the program is run, the computer will print 200 terms. The first three terms printed are:

```
NUMBER OF CHIPS IN SNOWFLAKE DESIGN
1       7
2       13
3       19
.       .
.       .
.       .
```

The last three terms are:

```
198     1189
199     1195
200     1201
```

QUESTIONS

Covering the Reading

In 1–4, consider the snowflake designs described in the first part of the lesson.

1. a. How many chips will Suzie need to make the 5th design?
b. Draw the 5th design.

2. Refer to the formula $c = 1 + 6d$.
a. What does the variable d represent?
b. What does c represent?

3. How many chips does Suzie need for the 13th design?

4. Suzie made a snowflake design with 103 chips. Which design in the sequence was it?

In 5–8, use the parking lot formula $c = 1.50 + 1.25h$.

5. a. What does the h in the formula represent?
   b. What does the 1.25 in the formula represent?
   c. What would it cost to park 24 hours in this garage?

6. a. Does the formula work for parking one hour?
   b. Does it work for $2\frac{1}{2}$ hours?
   c. What is the domain for h in the formula?

7. Your aunt parked in this garage for 10 hours. How much did she pay for parking?

8. Your uncle was charged $17.75 for parking in this garage. How many hours did he park?

9. Refer to the computer program in this lesson. Suppose line 20 reads `FOR D = 1 TO 999`.
   a. How many times would the loop be executed?
   b. What would be the first term printed?
   c. What would be the last term printed?
   d. Find the value of `C` in line 30 when `D` = 86.

Applying the Mathematics

10. *Multiple choice.* Which of these graphs pictures the parking rate data from Example 4?

(a) 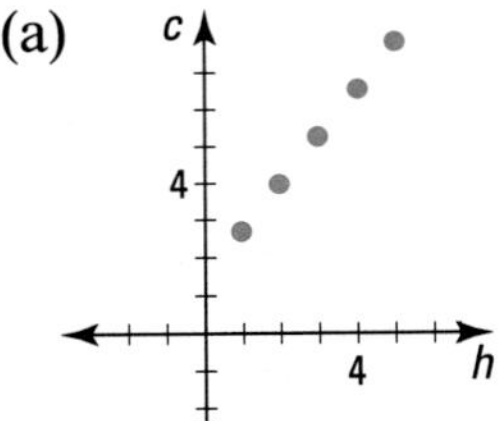

(b) 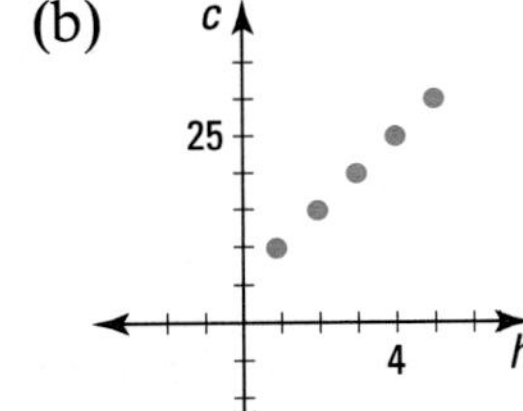

(c) 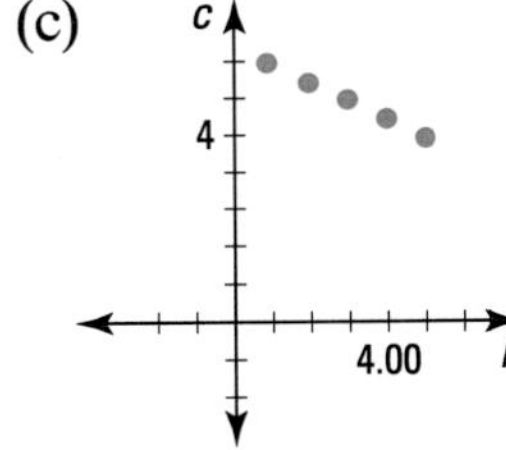

In 11–13, refer to the sequence of designs made from toothpicks shown below.

Let n = the number of triangles, and t = the number of toothpicks used.

11. Copy and complete the table below.

n	1	2	3	4
t	?	?	?	?

12. a. If the toothpick designs are continued, how many toothpicks will be needed to form 7 triangles?
b. Justify your answer by making or drawing such a design.

13. a. How many toothpicks are needed to make a design with n triangles?
b. How many triangles are formed by a design with 23 toothpicks?

14. The Huichol Indians of Mexico often stretch strands of yarn onto bamboo sticks to form squares having patterns like those shown below. Each new set of dots represents the vertices of a different colored square.

Mexican Art. *These colorful, woven yarn designs are known as "Eyes of God" or* Ojos de Dios. *Mexicans sometimes use them as decorations during their Green Squash Festival.*

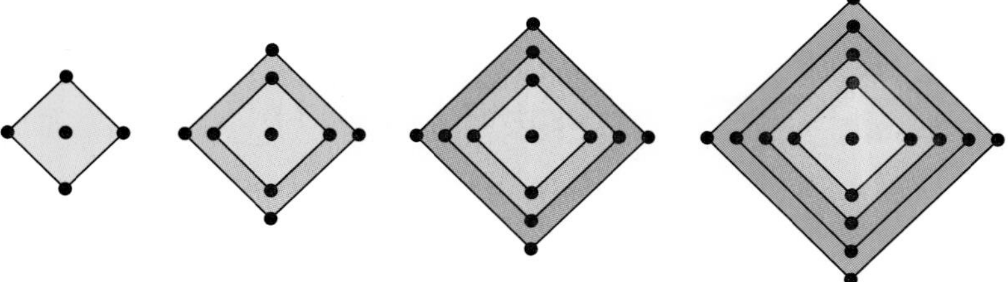

a. How many dots are needed for a pattern with 10 colors?
b. Let c = the number of colors and d = the number of dots. Find a formula for d in terms of c. Explain how you got this formula.

15. A cellular phone company charges $65.95 for up to 200 minutes of phone calls per month, and 21¢ per minute thereafter.
a. Make a table showing the cost of using this phone for 100, 200, 300, 400, and 500 minutes during the month.
b. Write a formula for the cost c in terms of the number m of minutes the phone is used.
c. To what values of m does your formula apply?

In 16 and 17, suppose a school uses 7500 gallons of gas per day for buses.

16. a. Make a table showing gas consumption for 1, 2, 3, 4, 5, and 6 days.
b. Find a formula for gas consumed in n days.
c. Calculate the number of gallons of gas consumed in 20 days.
d. If 7.5 million gallons of gas were used, how many days passed?

17. Write a computer program that will print a list of the gallons of gas consumed for 1 through 20 days.

Review

In 18 and 19, explain how to use the Distributive Property to calculate the cost mentally. *(Lesson 3-7)*

18. the cost of four T-shirts at $9.95 each

19. the cost of 25 notebooks at $1.99 each

20. Find the area. *(Lessons 2-1, 3-7)*

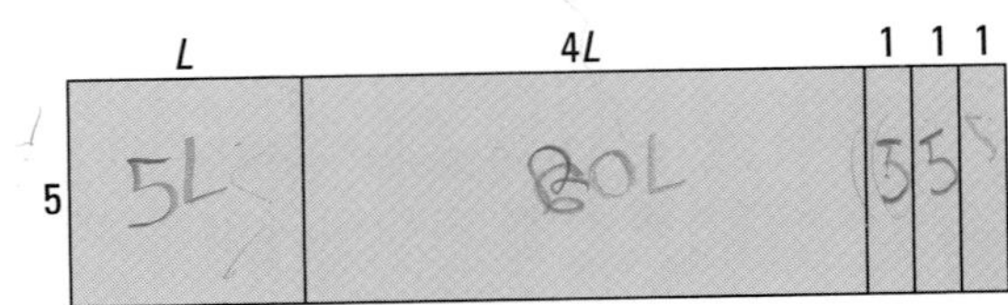

In 21–24, solve. *(Lessons 3-5, 3-6, 3-7)*

21. $3x + -5x + 12 + -15 = -4$

22. $t + 0.1t = 1.8$

23. $\frac{5}{2}x + 1 = 26$

24. $-3 = -6n + 1 + 2(n + 1)$

25. In which quadrant is the x-coordinate negative and the y-coordinate positive? *(Lesson 3-3)*

26. Refer to the graph below. Douglas, Arizona, is typical of a monsoon area in which a wind system causes yearly rain to be concentrated in a few months.
   a. During which months is the average rainfall in Douglas less than 1 inch? *(Lesson 3-3)*
   b. Which months have the highest average rainfall?
   c. Give the average rainfall for April in Douglas.
   d. Between which two consecutive months is there the greatest change in rainfall?

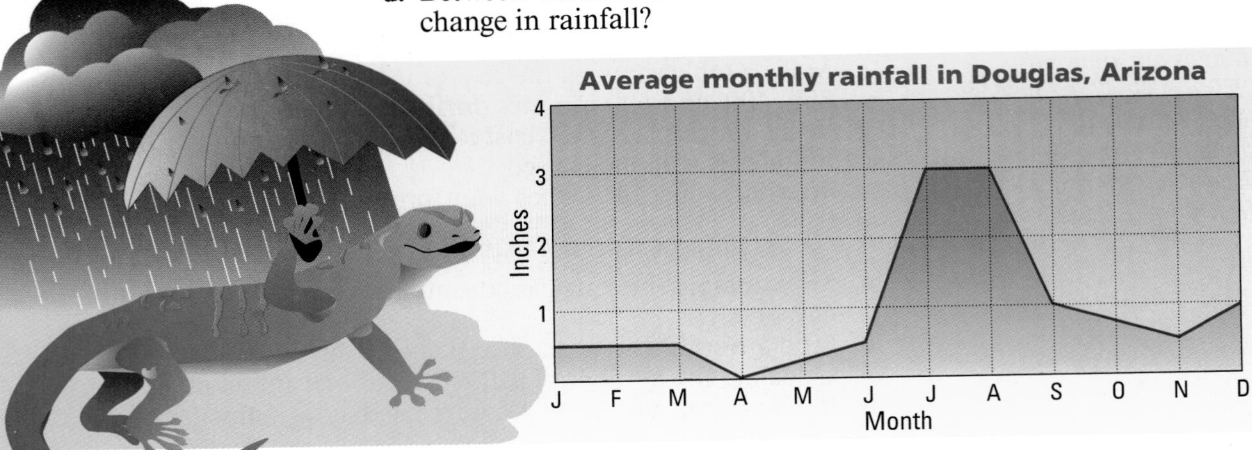

27. a. Solve: $3z < 231$. b. Graph the solution set on a number line. *(Lessons 2-8, 1-2)*

28. Roberto averaged 53 mph for $2\frac{1}{2}$ hours. About how many miles did he travel? *(Lesson 2-4)*

Exploration

29. Make up or find a situation in which a pattern leads to a linear expression. Describe your situation to a friend. Ask your friend to write a formula to describe the pattern.

LESSON

3-9 Adding Algebraic Fractions

A note on fractions. *Composers need to add fractions. In this piece, there are 3 counts per measure. The quarter notes get 1 count each, the eighth notes get one-half count each, and the sixteenth notes get one-fourth count each.*

The slide model for addition can help you picture addition of fractions. For instance, the slide model confirms that $\frac{1}{4} + \frac{3}{4} = 1$.

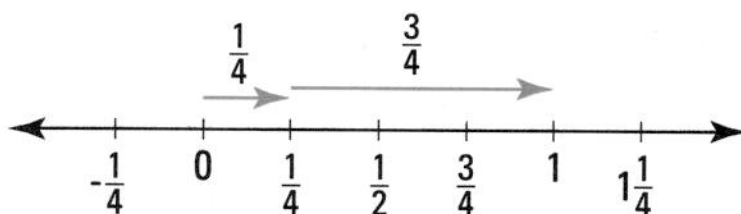

To add fractions with the same denominator, add the numerators and keep the same denominator, called the **common denominator.** This is the Adding Fractions form of the Distributive Property.

Distributive Property: Adding Fractions
For all real numbers a, b, and c, with $c \neq 0$, $\frac{a}{c} + \frac{b}{c} = \frac{a+b}{c}$.

Notice the similarity with adding like terms. The statement

$$a \cdot \frac{1}{c} + b \cdot \frac{1}{c} = (a + b) \cdot \frac{1}{c}$$

is the same as $ax + bx = (a + b)x$, when $\frac{1}{c}$ is substituted for x.

Example 1

Simplify $\frac{x}{3} + \frac{2y}{3}$.

Solution

By the Adding-Fractions form of the Distributive Property,

$$\frac{x}{3} + \frac{2y}{3} = \frac{x + 2y}{3}.$$

Because x and y are unlike terms, $\frac{x + 2y}{3}$ cannot be simplified further.

Check

Substitute values for x and y. We let $x = 5$ and $y = 6$.
Then in the original expression, $\frac{x}{3} + \frac{2y}{3} = \frac{5}{3} + 2 \cdot \frac{6}{3} = \frac{5}{3} + \frac{12}{3} = \frac{17}{3}$.
In the answer, $\frac{x + 2y}{3} = \frac{5 + 2 \cdot 6}{3} = \frac{17}{3}$ also.

Example 2

Simplify $\frac{-9 + 3b}{b} + \frac{9}{b}$.

Solution

Since the denominators are the same, the Adding-Fractions form of the Distributive Property can be applied.

$$\frac{-9 + 3b}{b} + \frac{9}{b} = \frac{-9 + 3b + 9}{b}$$

$$= \frac{3b}{b} \quad \text{Adding like terms; Additive Identity Property}$$

$$= 3 \quad \text{Equal Fractions Property}$$

The expression equals 3, regardless of the value of b, provided $b \neq 0$.

Check

Pick a value for b. We substitute 2 for b in the original expression and follow the order of operations
$\frac{-9 + 3 \cdot 2}{2} + \frac{9}{2} = \frac{-9 + 6}{2} + \frac{9}{2} = \frac{-3}{2} + \frac{9}{2} = \frac{6}{2} = 3$. It checks.

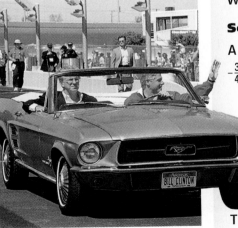

A classic car. *In 1994, President Clinton joined thousands of other people for the 30th birthday of the Ford Mustang car. President Clinton drove his 1967 Mustang at the Charlotte Motor Speedway, where the celebration took place.*

The procedure for adding fractions is more complicated when the denominators are different. The next example uses negative numbers in fraction form.

Example 3

On Monday, Ford Motor Co. stock fell $\frac{3}{4}$ of a point; on Tuesday it rose $1\frac{1}{8}$. What was the net change?

Solution

A loss of $\frac{3}{4}$ is a change of $-\frac{3}{4}$. The answer can be found by computing $-\frac{3}{4} + 1\frac{1}{8}$. A common denominator of 8 can be used.

$$-\frac{3}{4} + 1\frac{1}{8} = -\frac{6}{8} + 1\frac{1}{8}$$

$$= \frac{-6}{8} + \frac{9}{8}$$

$$= \frac{-6 + 9}{8}$$

$$= \frac{3}{8}$$

The stock rose $\frac{3}{8}$ of a point for a net change of $\frac{3}{8}$.

Example 4

Write $\frac{3x}{4} + \frac{x}{3}$ as a single fraction.

Solution 1

Use a common multiple of 3 and 4. The least common multiple is 12. Rewrite each fraction with denominator equal to 12.

$$\begin{aligned}\frac{3x}{4} + \frac{x}{3} &= \frac{3x}{4} \cdot \frac{3}{3} + \frac{x}{3} \cdot \frac{4}{4}\\ &= \frac{9x}{12} + \frac{4x}{12}\\ &= \frac{13x}{12}\end{aligned}$$

Solution 2

Show the use of the Distributive Property.

$$\begin{aligned}\frac{3x}{4} + \frac{x}{3} &= \frac{3}{4} \cdot x + \frac{1}{3} \cdot x\\ &= \left(\frac{3}{4} + \frac{1}{3}\right)x\\ &= \left(\frac{9}{12} + \frac{4}{12}\right)x\\ &= \frac{13}{12}x\end{aligned}$$

Check

Substitute a number for x. We use $x = 2$.
Then $\frac{3x}{4} + \frac{x}{3} = \frac{6}{4} + \frac{2}{3} = 1.5 + .\overline{6} = 2.1\overline{6}$ and $\frac{13x}{12} = \frac{26}{12} = 2.1\overline{6}$.
It checks.

An engineering feat. *Engineers build reservoirs by constructing a dam across a valley or by digging a basin in a level tract of land. Man-made reservoirs can be used to supply water, to generate power, or to provide a place for recreation.*

QUESTIONS

Covering the Reading

1. To add two fractions by adding the numerators, what must be true of the denominators?

2. One day the water level in a reservoir rose $\frac{3}{10}$ inch. The next day it fell $\frac{2}{5}$ inch. What was the total change in water level in the two days?

In 3–8, perform the additions.

3. $\frac{2}{5} + -\frac{1}{5} + 3\frac{4}{5}$

4. $-\frac{11}{3} + -\frac{2}{3} + -\frac{8}{3}$

5. $\frac{2a}{3} + \frac{28a}{3}$

6. $\frac{6y + 11}{2y} + -\frac{11}{2y}$

7. $\frac{x}{5} + \frac{x}{5} + \frac{2x}{5}$

8. $\frac{a}{6} + \frac{2a}{6}$

9. Consider the sum of $\frac{x}{5}$ and $\frac{3x}{4}$.
 a. What common denominator will help you add?
 b. Add and simplify.

In 10–17, simplify each sum.

10. $\frac{a}{6} + \frac{a}{3}$

11. $\frac{2d}{7} + \frac{3d}{4}$

12. $C + \frac{C}{3} + \frac{C}{3}$

13. $B + -\frac{B}{2}$

The timeless tug-of-war. *Tug-of-war games were common as far back as the Middle Ages. Knights and squires engaged in this contest while training for combat.*

Applying the Mathematics

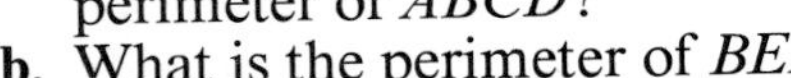

14. $\frac{k}{2} + \frac{2k}{3} + \frac{k}{4}$

15. $\frac{5}{x} + \frac{3}{x} - \frac{9}{x}$

16. $\frac{2n}{3} + 4 \cdot \frac{n}{3} + 5\left(\frac{1}{3}n\right)$

17. $-\frac{2}{x} + \frac{3}{2x} - \frac{1}{x}$

18. Find values of a, b, c, and d so that $\frac{a}{b} + \frac{c}{d} = \frac{a+c}{b+d}$ is false.

19. Rectangle $ABCD$ has length L and width W. Rectangle $BEFG$ has half the length and half the width of $ABCD$.

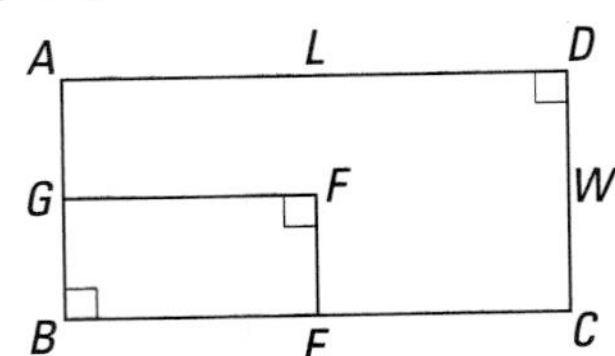

a. In terms of L and W, what is the perimeter of $ABCD$?

b. What is the perimeter of $BEFG$?

20. In a tug-of-war game, team A pulled the rope $3\frac{1}{2}$ ft toward its side, and then team B pulled it $1\frac{2}{3}$ ft toward its side. After another minute, team B pulled the rope $4\frac{5}{6}$ ft more in its direction. Describe where the middle of the rope is now.

Review

21. A photo shop has the following charges for making prints from slides.

number of prints	cost
1	\$0.75
2	\$1.25
3	\$1.75
4	\$2.25
5	\$2.75

a. Describe some patterns you notice in the data.

b. Find the cost of making 8 prints of one slide. Describe how you found your answer.

c. A proud grandmother paid \$7.75 for some copies of a slide of her new grandson. How many copies did she order? *(Lesson 3-8)*

22. Refer to the figure at the right.

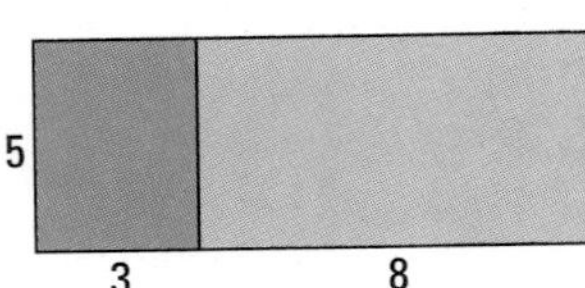

a. How many rectangles are pictured?

b. Find the area of the largest rectangle in two different ways.

c. What property is illustrated here? *(Lessons 3-6, 3-7)*

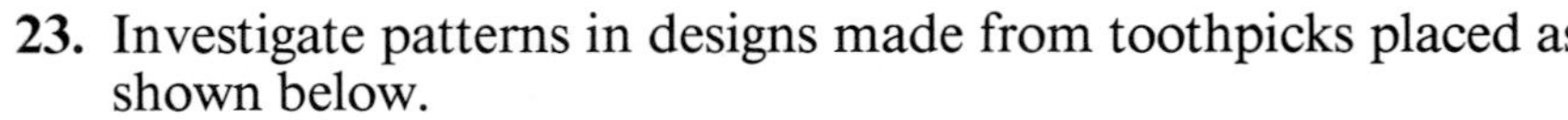

23. Investigate patterns in designs made from toothpicks placed as shown below.

Let n = the number of squares and t = the number of toothpicks used.

a. Make a table showing values for n and t.
b. Describe patterns you find in the numbers in your table.
c. How many toothpicks would you need to make a design with 60 squares?
d. Suppose $t = 301$. Find n, and write a question about squares and toothpicks that has the answer you just found. *(Lesson 3-8)*

In 24 and 25, solve and check. *(Lessons 3-6, 3-7)*

24. $x + 2x + 3x + 4x = 5$

25. $8 = 2(n + 3) + 4(5n + 6)$

26. *Multiple choice.* Which of the following *must* be negative? *(Lesson 3-2)*
(a) $-x$ (b) $-(-3)$ (c) $-\left(-\left(-\frac{1}{3}\right)\right)$ (d) $-(-a)$

A box-office hit. *Released in 1983,* Return of the Jedi *was the third of the* Star Wars *spectacles. It was also the final chapter of the Luke Skywalker trilogy. Pictured above are Princess Leia, the robot C-3PO, and Chewbacca.*

In 27 and 28, write an expression to describe the total change in each situation. *(Lesson 3-1)*

27. Marie climbed u meters up the hill and then d meters back down.

28. Stanley earned $7e$ dollars and then paid back $5e$ dollars to one friend while collecting c dollars from another.

29. Without a factorial key [x!], how can you calculate 11! on a calculator? *(Lesson 2-10)*

30. **a.** Solve $-2n > 10$. **b.** Graph the solution set. *(Lesson 2-8)*

31. Jamie earns $14.40 per hour as a critical-care nurse. How many hours must she work in a week in order to earn more than $400? *(Lessons 2-4, 2-8)*

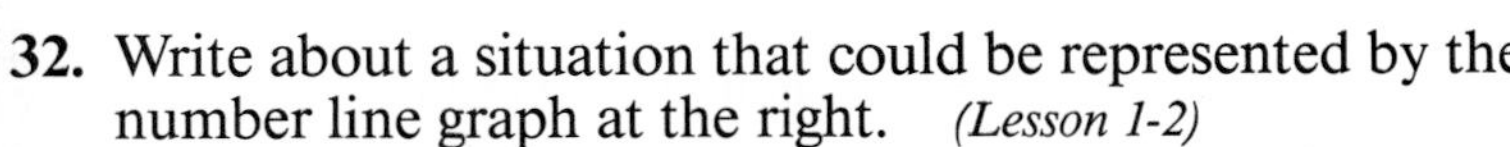

32. Write about a situation that could be represented by the number line graph at the right. *(Lesson 1-2)*

33. In the nine-year period from January 1, 1984, through January 1, 1993, the movie *The Return of the Jedi* grossed approximately $169,000,000. What was the average gross per year? *(Previous course)*

Exploration

34. **a.** *Skill sequence.* Compute each sum.
(i) $\frac{1}{2} + \frac{1}{4} + \frac{1}{8}$ (ii) $\frac{1}{2} + \frac{1}{4} + \frac{1}{8} + \frac{1}{16}$
(iii) $\frac{1}{2} + \frac{1}{4} + \frac{1}{8} + \frac{1}{16} + \frac{1}{32}$
b. If you could do an infinite addition problem, what sum would you predict for $\frac{1}{2} + \frac{1}{4} + \frac{1}{8} + \frac{1}{16} + \frac{1}{32} + \frac{1}{64} + \ldots$?

LESSON

3-10

Solving $ax + b < c$

I will always be younger than you. *These photos were taken several years apart. The boy is younger than his sister. In algebra, as in real life, if one quantity is less than another, it will always be less if the same amount is added to both.*

If you are x years old and an *older* friend's age is y, then

$$x < y.$$

Five years from now you will still be younger than your friend. In other words,

$$x + 5 < y + 5.$$

In general, J years from now you will still be younger than your friend. This can be written

$$x + J < y + J.$$

These examples illustrate the *Addition Property of Inequality.*

Addition Property of Inequality
For all real numbers a, b, and c,

$$\text{if } a < b,$$
$$\text{then } a + c < b + c.$$

The Addition Property of Inequality can be pictured with a balance scale. Suppose a and b represent the weights of two packages and $a < b$.

If the same weight c is added to each side of the scale, then $a + c < b + c$.

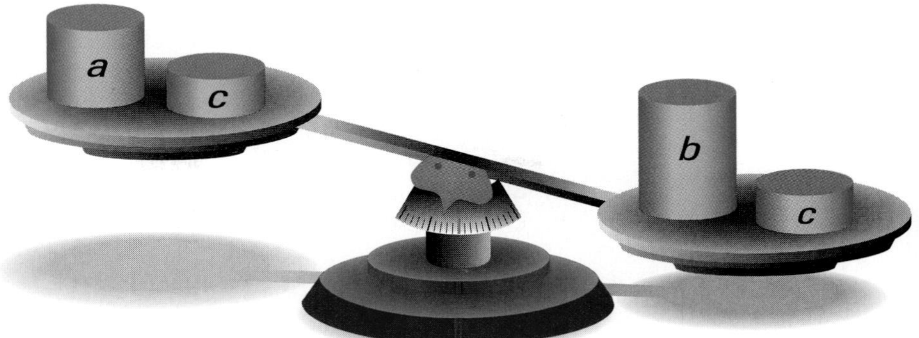

So, if $a < b$, then $a + c < b + c$. The same idea works for $>$, $\leq$, and $\geq$.

Thus, sentences with $=$, $<$, $>$, $\leq$, or $\geq$ can all be solved in the same way.

You may add the same number to both sides of an equation or inequality without affecting the solutions.

Where do lemons grow? *Four-fifths of the nation's lemons are harvested in California. About half of the lemon crop is processed into juice and oil. The other half is sold as fresh fruit.*

Solving Inequalities with Positive Coefficients

Example 1

A crate weighs 6 kg when empty. A lemon weighs about 0.2 kg. For shipping, the crate and lemons must weigh at least 50 kg. How many lemons should be put in the crate?

Solution

Let n be the number of lemons. Then the weight of n lemons is $0.2n$. The weight of the crate with n lemons is $0.2n + 6$, so the question can be answered by solving the inequality

$$0.2n + 6 \geq 50.$$

This inequality is of the form $ax + b \geq c$ and is solved the same way you solve $ax + b = c$.

$0.2n + 6 + -6 \geq 50 + -6$ First, add -6 to both sides and simplify.

$0.2n \geq 44$

$\frac{0.2n}{0.2} \geq \frac{44}{0.2}$ Multiply both sides by $\frac{1}{0.2}$, and simplify.

$n \geq 220$

At least 220 lemons should be put in the crate.

Check

Part 1: Does $0.2(220) + 6 = 50$? $44 + 6 = 50$? Yes.

Part 2: Pick some value that works for $n > 220$. We choose 250.
Is $0.2(250) + 6 > 50$? $50 + 6 > 50$? Yes, so $n \geq 220$ checks.

Solving Inequalities with Negative Coefficients

Recall that when you multiply each side of an inequality by a negative number, you must reverse the direction of the inequality. Otherwise, these sentences are solved just as in Example 1.

Example 2

a. Solve $-4n + 9 \leq 1$.
b. Graph the solution set.

Solution

a. Add the opposite of 9 to each side of the inequality.

$$-4n + 9 \leq 1$$
$$-4n + 9 + -9 \leq 1 + -9$$
$$-4n \leq -8$$

Multiply each side by the reciprocal of -4. Remember to switch the sign of the inequality.

$$-\frac{1}{4}(-4n) \geq -\frac{1}{4}(-8)$$
$$n \geq 2$$

b. A graph of the solution set is below.

-4 -3 -2 -1 0 1 2 3 4 n

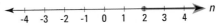

Check

Again, you should check two numbers: the boundary point 2 and a number larger than 2. Substitute $n = 2$ in the original inequality. You should get equality.
For $n = 2$, $-4(2) + 9 = -8 + 9 = 1$. It checks.
Try $n = 3$. $-4(3) + 9 = -12 + 9 = -3$. It checks because $-3 \leq 1$.

QUESTIONS

Covering the Reading

1. Use the symbol $>$ to state the Addition Property of Inequality.

2. **a.** What inequality is suggested below?
   b. What is the solution to the inequality?

3. Refer to Example 1. If a loaded crate of lemons can weigh no more than 200 kg, at most how many lemons can be put in the crate?

4. An empty crate weighs 10 kg. A grapefruit weighs about 0.5 kg. How many grapefruit can be packed in the crate and still keep the total weight under 50 kg?

In 5–8, solve, graph, and check.

5. $3x + 4 < 19$

6. $6 \leq 4b + 10$

7. $5 \leq -3n + 2$

8. $101 + 102x > 103$

Applying the Mathematics

9. Three consecutive integers can be expressed as n, $n + 1$, and $n + 2$. Write an inequality and use it to find the three smallest consecutive integers whose sum is greater than 79.

In 10–13, solve.

10. $3(x + 4) < 12$

11. $-.02y + \frac{1}{2} \geq 0.48$

12. $15 \geq 12 + \frac{1}{3}y$

13. $\frac{-5x}{6} + 30 < 120$

14. Find a negative number which is a solution to $-4x + 7(x + -2) > -18$.

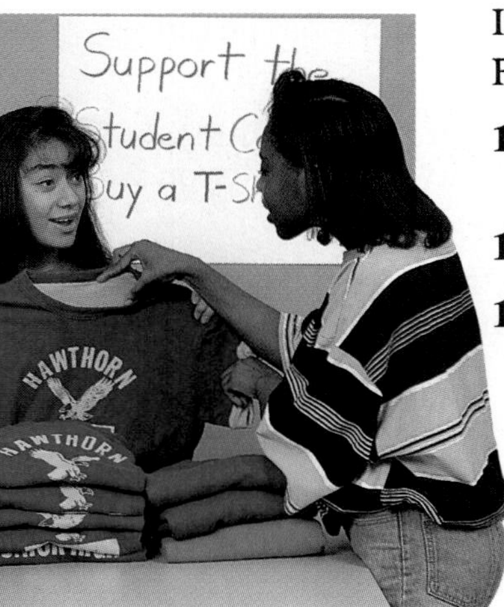

In 15 and 16, use the formula $C = \frac{5}{9}(F - 32)$ relating Celsius and Fahrenheit temperatures.

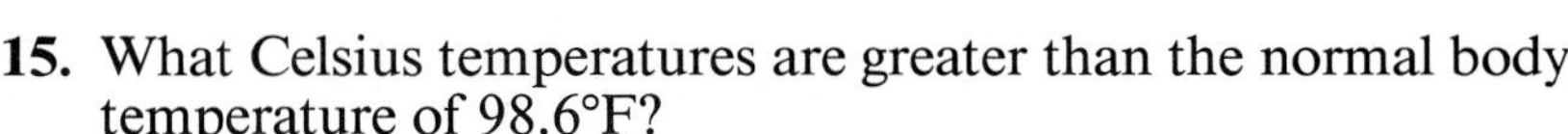

15. What Celsius temperatures are greater than the normal body temperature of 98.6°F?

16. What Celsius temperatures are below 68°F?

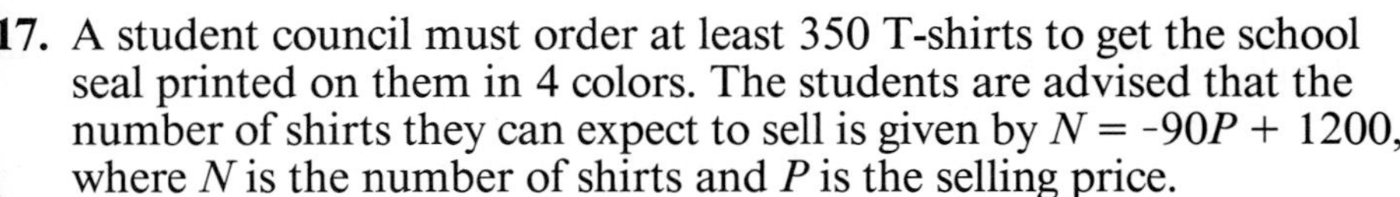

17. A student council must order at least 350 T-shirts to get the school seal printed on them in 4 colors. The students are advised that the number of shirts they can expect to sell is given by $N = -90P + 1200$, where N is the number of shirts and P is the selling price.
   a. What price should they charge if they estimate 350 shirts will be sold?
   b. What price might they charge if the council is willing to sell as few as 300 shirts?

18. a. Solve for x in the sentence: $ax + b < c$ when $a > 0$.
   b. How does the result from part **a** change if $a < 0$?

Review

In 19–21, add and simplify. *(Lesson 3-9)*

19. $\frac{3a}{5} + \frac{7a}{5}$ **20.** $\frac{3}{x} + \frac{-2}{x} + \frac{4}{x}$ **21.** $\frac{x}{3} + \frac{-2x}{5}$

x	y
1	10
2	14
3	18
4	22
5	26

22. Consider the table at the left. *(Lesson 3-8)*
a. Describe the pattern.
b. Find y when $x = 8$.
c. Find x when $y = 96$.

23. The three L-shaped figures begin a pattern. How many cubes are needed to make each figure? *(Lesson 3-8)*
a. the 1st **b.** the 4th **c.** the nth

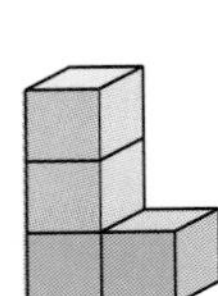
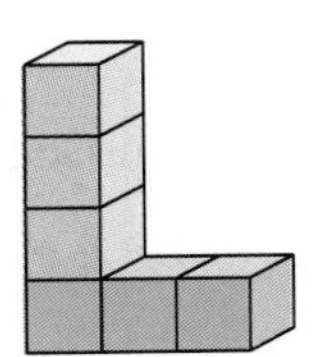

Pictured above is a banner from Indiana University in Bloomington. Their athletic teams are known as the Indiana Hoosiers.

In 24 and 25, solve and check. *(Lessons 3-5, 3-6, 3-7)*

24. $(4x + 3) + 2x = -9$ **25.** $6(4y + -1) + -2y = 82$

26. Darrell was asked to simplify $-3(x + 2y + -6)$. His answer was $-3x + -6y + -6$. Write a note to Darrell telling him what he did wrong. *(Lesson 3-7)*

27. If you work 40 hours a week for \$5.25 an hour and h additional hours earning time and a half for overtime, how long must you work to earn \$250 in one week? *(Lessons 3-5, 3-6, 3-7)*

28. The Booster Club ordered a bolt of cloth 100 yd long. They will use 2 yds for a table cloth, and the rest for banners showing the school mascot. If each banner takes $\frac{2}{3}$ yd of cloth, how many banners can they make? *(Lesson 3-5)*

29. a. Which number is larger, $75!$ or 75^{75}?
b. Explain your reason for your answer to part **a.** *(Lesson 2-10)*

30. Use the clues to find x. *(Lessons 1-1, 1-4, 2-8)*
Clue 1: x is greater than the average of 5, 11, and 2.
Clue 2: $-3x > -33$.
Clue 3: $x \neq \sqrt{64}$.
Clue 4: x is an integer.
Clue 5: x is not an odd number.

Exploration

31. Make up five inequalities of the form $ax + b < c$ whose solutions are $x < 24$.

PROJECTS 3 CHAPTER THREE

A project presents an opportunity for you to extend your knowledge of a topic related to the material of this chapter. You should allow more time for a project than you do for typical homework questions.

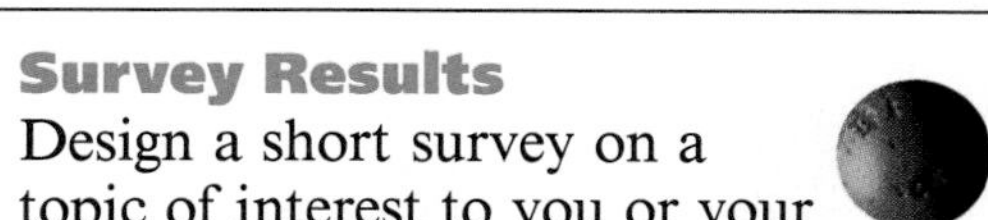

1 Survey Results

Design a short survey on a topic of interest to you or your class, such as favorite lunch food, type of pet, time spent on homework, or favorite radio station. Design a short survey of the question you chose. Ask 20 students in your class or grade and 20 students in a different class or grade (or 20 adults) to respond to your survey. Design a stacked bar graph, like that in Lesson 3-1 that compares the responses of your class to the other class. Write a paragraph to explain your graph and the results of your survey.

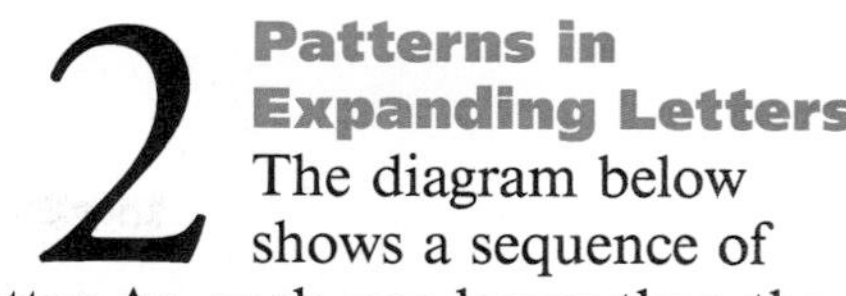

2 Patterns in Expanding Letters

The diagram below shows a sequence of letter As, each one larger than the previous one. Find a formula for the number of squares that make up the *n*th letter A. Choose several letters of the alphabet (perhaps your initials). For each one, design a way to make the letter out of squares. Make a sequence of enlarged versions of that letter, and then find a formula for the number of squares that make up the *n*th term in the sequence.

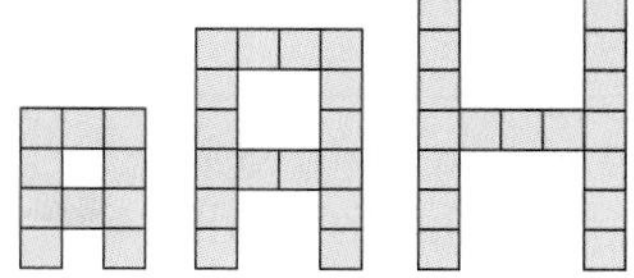

3 Graphs of Everyday Quantities

In Lesson 3-3, a graph shows changes in Harold Hooper's speed while driving. Choose several other quantities that vary during the course of a person's day. For each one, write a paragraph describing how the quantity changes and draw a graph. Some suggestions for topics are:

- The outside air temperature in town during the day
- The amount of gasoline in the tank of a car during the day

4 Staircase Patterns

Consider the following situation: a staircase is made from Cuisenaire® rods of a single color. Each time a rod is added to the staircase, it is offset by the space of a white (one unit) rod. For instance, shown below is a staircase made from three purple rods (each is 4 units in length).

a. Pick any color rod other than the unit cube. Record its volume and surface area.

b. Build staircases of different sizes with that color rod. Each time you build a staircase, record the number of rods used, the volume of the staircase, and the surface area of the staircase.

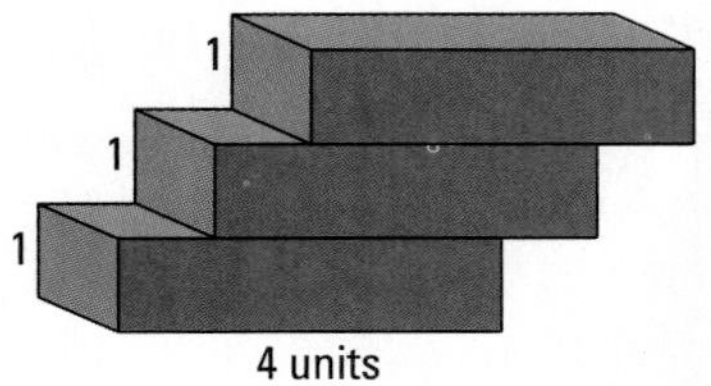

c. What patterns do you find in your data? What do you predict will be the volume and surface area of the staircase that has 10 rods? 25 rods? n rods?

d. How do the patterns change if you build a staircase from rods of a different length and different color?

5 Slides in Activities

In activities such as marching band, football, basketball, dance ensembles, and show choirs, participants need to move in a manner very similar to a slide. Use a coordinate plane and graph paper to design a move in a performance or an offensive play in a sport. Using a drawing and a written explanation, show how each person or group must move.

6 BASIC Solution of Inequalities

Write a BASIC program that will solve inequalities of the form $ax + b > c$. Your program should ask the user for the values of a, b, and c, and then print an appropriate solution. You will probably need to use an IF-THEN statement in your program. This statement allows the computer to make a choice, depending on what values of a, b, and c are given. You may also need to determine how to use the INPUT statement and the PRINT statement on your computer. Test your program to make sure it works for all values of a, b, and c.

SUMMARY

In algebra as in arithmetic, addition is a basic operation. The most frequent applications of addition occur in situations which can be represented by putting together or a slide. Putting together occurs when quantities that do not overlap are combined. A slide occurs when you start with a quantity and go higher or lower by a given amount. Slides can help picture addition of integers, fractions, and like terms.

The properties of addition can be verified through their uses. For example, putting together quantities in a different order yields the same sum, so addition is commutative.

Graphing provides a picture that can help clarify solutions to a problem or trends in data. If two quantities are being considered, a coordinate graph in a plane can show both values. A two-dimensional slide can be represented as a combination of a horizontal and vertical slide on a coordinate graph.

Sentences of the forms $ax + b = c$ or $ax + b < c$ (in which a $\neq 0$) combine multiplication and addition. To solve this type of sentence, first add $-b$ to both sides, and then multiply both sides by $\frac{1}{a}$. Sometimes it is necessary to simplify expressions on either side of the inequality first.

When addition and multiplication occur in the same expression, the Distributive Property may provide a link between the operations. It tells how to change between a form where the multiplication is done first, $ac + bc$, and one where the addition is done first, $(a + b)c$. Forms of the Distributive Property are used to add fractions, to add like terms, and to remove parentheses from expressions.

VOCABULARY

Below are the most important terms and phrases for this chapter. You should be able to give a general description and specific examples of each.

Lesson 3-1
bar graph, stacked bar graph
Putting-Together Model for Addition
Slide model for Addition
Commutative Property of Addition
Associative Property of Addition

Lesson 3-2
additive identity
Additive Identity Property
opposite, additive inverse
Property of Opposites
Opposite of Opposites (Op-op) Property
Addition Property of Equality

Lesson 3-3
plane
coordinate graph, axes, origin
scatterplot
coordinates, coordinate plane
continuous

Lesson 3-4
x-axis, y-axis
x-coordinate, y-coordinate
quadrant
two-dimensional slide, translation
preimage, image

Lesson 3-6
Distributive Property: Like Terms
term, coefficient
like terms, adding like terms

Lesson 3-7
Distributive Property: Removing Parentheses

Lesson 3-8
sequence
linear expression
linear equation
FOR-NEXT loop

Lesson 3-9
common denominator
Distributive Property: Adding Fractions

Lesson 3-10
Addition Property of Inequality

PROGRESS SELF-TEST

Take this test as you would take a test in class. Then check your work with the solutions in the Selected Answers section in the back of the book.

In 1–7, simplify.

1. $m + 3m$
2. $\frac{5}{2}(4v + 100 + w)$
3. $-9k + 3(k + 3)$
4. $(x + 5 + x) + (-8 + -x)$
5. $-(-(-p))$
6. $\frac{2}{n} + \frac{5}{n} + \frac{-3}{n}$
7. $\frac{3x}{2} + \frac{5x}{3}$

In 8–11, solve.

8. $8r + 14 = 74$
9. $-4q + 3 + 9q = -12$
10. $3(x + 2) + 100 = 54$
11. $85 = x + 2(3x + 4)$
12. Solve and graph the solutions to $30v + -18 > 15$.

In 13 and 14, name the property that is illustrated.

13. If $y + 11 = 3$, then $y + 15 = 7$.
14. $8x + -15x = -7x$
15. Explain why -100 is or is not an element of the solution set of $15 \leq x + 87$.
16. You have $137.25 in your savings account, and you add $2.50 each week. Disregarding interest, how much will be in the account after w weeks?
17. Irving bought 6 pairs of socks at $2.99 per pair. Show how Irving could use the Distributive Property to calculate mentally the total cost of the socks.
18. If $F = \frac{9}{5}C + 32$, find the Celsius equivalent of a Fahrenheit temperature of 50°.
19. Juana and Jill had a lemonade stand. Juana worked twice as long as Jill, so Juana received twice as much of the profits as Jill. If the total profits were $58.50, how much should Jill receive?
20. **a.** Write two expressions to describe the total area of the figure below.
    b. Explain how your expressions represent the Distributive Property.

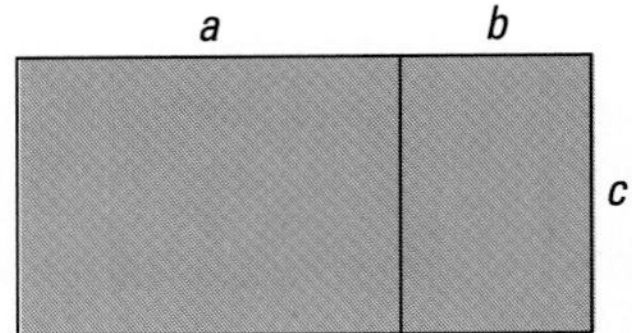

21. Find the image of (5, -2) after a slide of 4 units to the left and 5 units up.
22. $\triangle A'B'C'$ is the image of $\triangle ABC$ under a slide. Find the coordinates of B'.

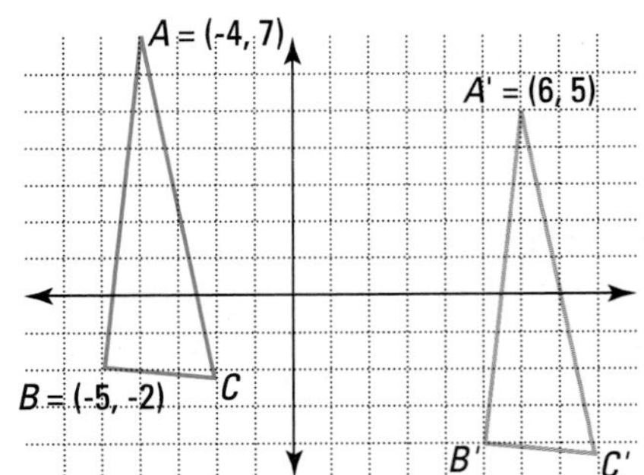

 CHAPTER THREE

PROGRESS SELF-TEST

23. The graph below shows the tornado death rate per 100,000 population since 1917.

a. What conclusion can you draw from the graph?

b. Give some possible reasons for your conclusion.

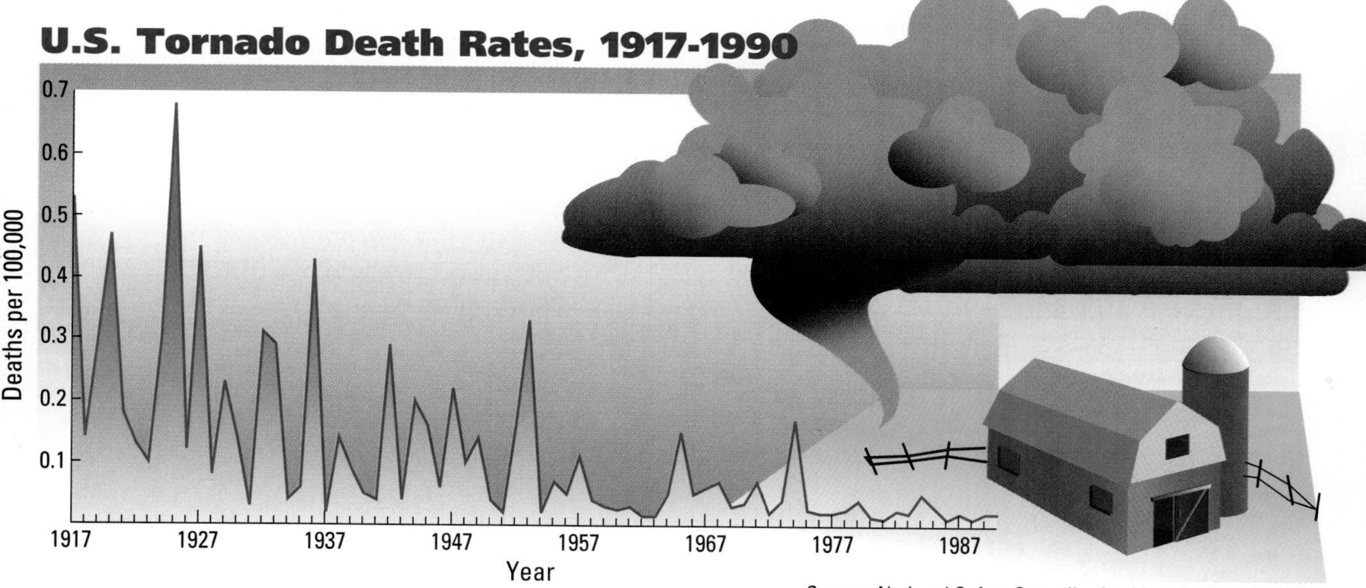

Source: National Safety Council – Accident Facts 1992 Edition

24. The following chart shows total public secondary-school enrollment in the United States. Draw a scatterplot of the data. Plot years on the horizontal axis.

Year	Number of Students
1955	8,521,000
1965	15,504,000
1975	19,151,000
1985	15,219,000
1995 (estimate)	16,431,000

After taking and correcting the Self-Test, you may want to make a list of the problems you got wrong. Then write down what you need to study most. If you can, try to explain your most frequent or common mistakes. Use what you write to help you study and review the chapter.

CHAPTER REVIEW

Questions on SPUR Objectives

SPUR stands for **S**kills, **P**roperties, **U**ses, and **R**epresentations. The Chapter Review questions are grouped according to the SPUR Objectives for this chapter.

SKILLS DEAL WITH THE PROCEDURES USED TO GET ANSWERS.

Objective A: *Use the Distributive Property and the properties of addition to simplify expressions.* *(Lessons 3-1, 3-2, 3-6, 3-7)*

In 1–6, simplify.

1. $8x + -3x + 10x$ **2.** $-3m + 4m + -m$

3. $c + \frac{1}{2}c$ **4.** $-\frac{2}{3}(6 + -9v + 4v)$

5. $11(3x + 2) + 4(5x + 6)$

6. $3w + 4 + 5(4w + 6)$

Objective B: *Solve and check equations of the form $x + a = b$ and $ax + b = c$.* *(Lessons 3-2, 3-5, 3-6, 3-7)*

In 7–18, solve and check.

7. $2.5 = t + 3.1$ **8.** $x + -11 = 12$

9. $(3 + n) + -11 = -5 + 4$

10. $\left(2\frac{1}{2} + 1\frac{1}{4}\right) + -\frac{3}{4} = (a + 6) + -2$

11. $4n + 3 = 15$ **12.** $-470 + 2n = 1100$

13. $\frac{2}{3}x + 14 = 15$ **14.** $5m + -3m + 6 = 12$

15. $17r + 12 + 9r = 1312$

16. $5(x + 3) = 95$ **17.** $2x + 3(1 + x) = 18$

18. $16 = \frac{3}{4}b + 22$

Objective C: *Add algebraic fractions.* *(Lesson 3-9)*

In 19–24, simplify each sum.

19. $\frac{x}{3} + \frac{y}{3}$ **20.** $\frac{30}{a} + \frac{10}{a} + \frac{20}{a}$

21. $\frac{2}{3}x + \frac{1}{3}x$ **22.** $\frac{x}{3} + \frac{x}{4}$

23. $\frac{x}{5} + -\frac{3x}{2}$ **24.** $\frac{2x}{5} + \frac{3y}{5} + \frac{-3x}{5}$

Objective D: *Solve and check inequalities of the form $ax + b < c$.* *(Lesson 3-10)*

In 25–30, solve and check.

25. $2x + 11 < 201$ **26.** $\frac{3}{4}t + 21 > 12$

27. $-2 + (5 + x) > 4$ **28.** $-28 \le 17 + (y + 5)$

29. $4 < -16g + 7g + 5$

30. $p + 2p + 3p + 4p \le 85$

PROPERTIES DEAL WITH THE PRINCIPLES BEHIND THE MATHEMATICS.

Objective E: *Identify and apply properties of addition or the Distributive Property.* *(Lessons 3-1, 3-2, 3-6, 3-7, 3-9)*

In 31–36, what property has been applied?

31. $2(L + W) = 2(W + L)$

32. $(28 + -16) + -23 = 28 + (-16 + -23)$

33. $-(-31) = 31$ **34.** $\frac{x}{31} + \frac{y}{31} = \frac{x + y}{31}$

35. $8x + -13x = -5x$

36. If $t + 18 < -3$, then $t + 18 + -18 < -3 + -18$.

37. Hillary adds -14 to both sides of $x + -7 = 14$. What sentence results?

38. If x is a negative integer, what is $-x$?

Objective F: *Use the Distributive Property to perform calculations in your head.* *(Lesson 3-7)*

In 39–42, explain how the Distributive Property can be used to do the calculations mentally.

39. $7 \cdot \$3.04$ **40.** $101 \cdot 35$ **41.** $3 \cdot 95$

42. the cost of 9 shirts if each one costs $19.99

USES DEAL WITH APPLICATIONS OF MATHEMATICS IN REAL SITUATIONS.

Objective G: *Apply models for addition to write linear expressions or to solve sentences of the form $x + a = b$, $ax + b = c$, or $ax + b < c$.* (Lessons 3-1, 3-2, 3-5, 3-6, 3-10)

43. If the temperature is -11°C, by how much must it increase to become 13°C?

44. The temperature was T_1 degrees. It changed by C degrees. Now it is more than T_2 degrees. Give a sentence relating T_1, C, and T_2.

45. If Wisconsin produced w billion pounds of milk and California produced c billion pounds, how much milk did the two states produce together?

46. Mark has $5.40 and would like to buy a pair of jeans for $26. He earns d dollars babysitting and $7.50 for mowing the lawn, but still does not have enough money. What sentence relates $5.40, $26, $7.50, and d?

47. Eli needs $5 more for a concert ticket. How much must he earn to go to the concert and have at least $4 for bus fare and food?

48. Katy earns $7.80 per hour at the zoo. She also receives a $25.00 meal allowance, $15.00 for transportation, and $7.50 for dry cleaning. Last week she was paid a total of $320.50. How many hours did she work?

49. A $67,500 estate is to be split among four children and a grandchild. Each child gets the same amount and the grandchild gets half that amount. How much will each receive?

Objective H: *Write expressions and solve problems involving linear patterns with two variables.* (Lesson 3-8)

50. Anna opened a savings account with $45. She plans to deposit $6 each week.

a. Disregarding interest, how much will be in the account after w weeks?

b. When will she have saved $195?

51. Refer to the table.

a. What is the next row in the table?

b. Find a formula that relates x and y.

x	y
1	10
2	13
3	16
4	19
5	22

52. In 1993 the cost of mailing a letter was 29¢ for one ounce or less and 23¢ for each additional ounce, rounded up. The table shows different mailing costs.

Weight (oz)	Charges	Cost Pattern
1	.29	.29 + .23 · 0
2	.29 + .23	.29 + .23 · 1
3	.29 + .23 + .23	.29 + .23 · 2
4	.29 + .23 + .23 + .23	.29 + .23 · 3

a. Write the next row of the table.

b. Write an expression for the cost of mailing a letter that weighs n ounces.

53. A business with a 900 exchange charges callers $2.95 for a 3-minute call and $.75 for each additional minute. Find the cost of calling for each of the times.

a. 4 minutes **b.** 10 minutes **c.** n minutes

REPRESENTATIONS DEAL WITH PICTURES, GRAPHS, OR OBJECTS THAT ILLUSTRATE CONCEPTS.

Objective I: *Draw and interpret two-dimensional graphs.* (Lesson 3-3)

54. The table shown, from *Places Rated Almanac of 1989,* lists some of the metropolitan areas expecting new jobs between 1989 and 1995.

a. What do the negative signs mean?

b. Draw a graph using number of blue-collar jobs created as the horizontal axis. Plot and label the points given in the table.

Jobs to be Created 1989–95

Metro area	Blue Collar	White Collar
Akron, OH	-5,540	13,800
Bellingham, WA	180	3,320
Casper, WY	-1540	-120
Danbury, CT	1110	13,490
Elkhart-Goshen, IN	5,540	2,880
Fort Smith, AR	4,190	4,650

CHAPTER THREE

In 55 and 56, the graph below shows the height of a boy's head from the ground as he rides in a Ferris wheel.

55. Where is the boy (top, bottom, or halfway up) after 40 seconds on the ride?

56. After everyone is on, how many times does the Ferris wheel go around before it begins to let people off?

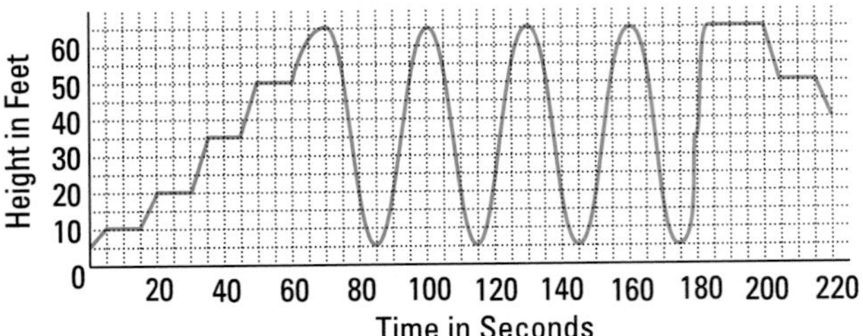

In 57–59, use the graph below.

57. Which state had more people in 1930, Texas or Ohio?

58. In which decade were the populations of Ohio and Texas the same?

59. In which decade did Ohio have its greatest increase in population?

Population of Texas and Ohio in various years

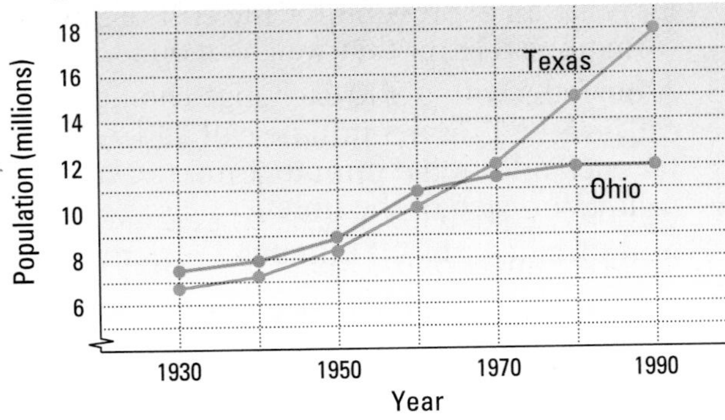

60. Draw a graph to illustrate these data (year, the U.S. population per square mile): (1800, 6.1), (1850, 7.9), (1900, 50.7), (1980, 62.6). Label the axes.

Objective J: *Draw and interpret two-dimensional slides on a coordinate graph.* *(Lesson 3-4)*

61. Find the image of (2, -4) after a slide of 40 units to the left and 60 units up.

62. Find the image of (x, y) after a slide of 4 units to the right and 10 units down.

63. The image of the point (4, 9) after a slide is (3, 17). Describe the slide with a rule or words.

64. If $R = (-3, 2)$, use the rule $(x, y) \rightarrow (x + 3, y)$ to find R'.

65. After a slide, the image of C is $C' = (6, 4)$. Graph the image of $\triangle ABC$ shown at the right by finding the image of each vertex.

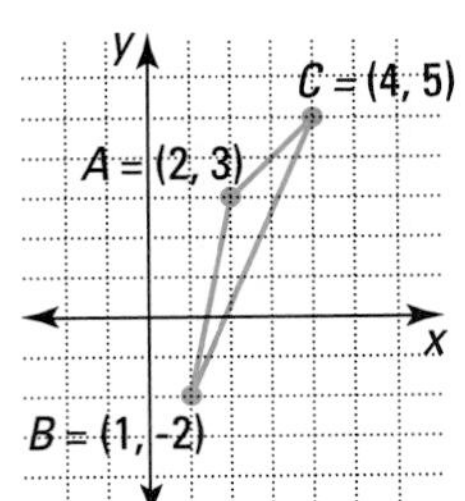

Objective K: *Use balance scales or area models to represent expressions or sentences.* *(Lessons 3-5, 3-6, 3-7, 3-10)*

66. Use the picture of a balance below. All the cylinders weigh the same.

a. Write an equation describing the situation with b representing the weight of one cylinder.

b. What does one cylinder weigh?

67. Use the picture below.

a. Write an equation to describe this situation with W representing the weight of one box.

b. What is the weight of one box?

68. Draw a picture to illustrate the inequality $3w + 4 < 11$.

In 69–70, write two different expressions to describe the total area of each large rectangle.

69.

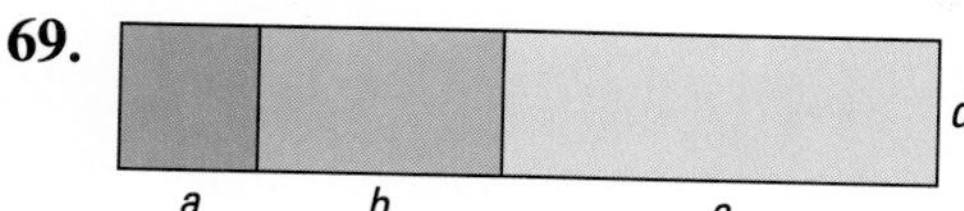

70.

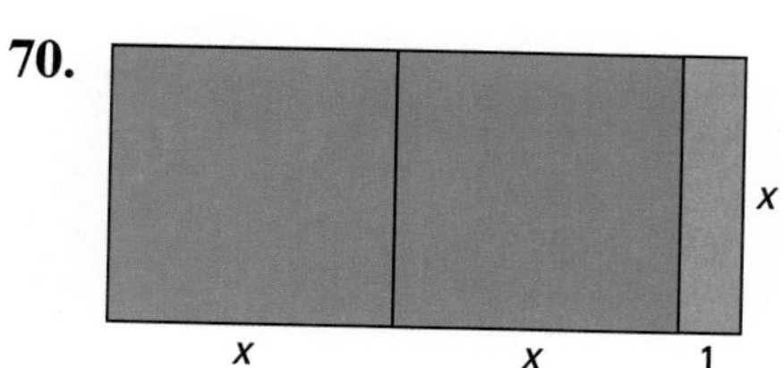

Objective L: *On a number line graph solutions to inequalities of the form $ax + b < c$.* *(Lesson 3-10)*

In 71–74, graph all solutions.

71. $12 + y \leq 48$

72. $\frac{-3}{5} > z + 20$

73. $-2x + 4 < 17$

74. $5(2 + 3x) + 6 \geq 106$

REFRESHER

Chapter 4, which discusses subtraction in algebra, assumes that you have mastered certain objectives in your previous studies of mathematics. Use these questions to check your mastery.

A. Subtract positive numbers or quantities. In 1–6, find the difference.

1. $8\frac{2}{3} - 3\frac{1}{3}$

2. $13.96 - 4.89$

3. $12.5 - 6.85$

4. $18\frac{1}{2} - 10\frac{3}{4}$

5. $100\% - 4\%$

6. $100\% - 8.5\%$

B. Subtract positive and negative integers. In 7–12, find the difference.

7. $40 - 200$

8. $76 - 79$

9. $-2 - 6$

10. $-12 - -11$

11. $111 - -88$

12. $-2 - -3$

C. Solve simple equations of the form $x - a = b$ when a and b are integers. In 13–16, solve the equation.

13. $x - 40 = 11$

14. $878 = y - 31$

15. $w - 64 = 49$

16. $-100 = z - 402$

D. Measure angles.

17. *Multiple choice.* Without using a protractor, tell whether the measure of angle V below is:
(a) between 0° and 45°,
(b) between 45° and 90°,
(c) greater than 90°.

18. *Multiple choice.* Without using a protractor, tell whether the measure of angle W below is:
(a) between 0° and 45°,
(b) between 45° and 90°,
(c) greater than 90°.

19. Measure $\angle V$ to the nearest degree.

20. Measure $\angle W$ to the nearest degree.

21. Draw an angle whose measure is 110°.

22. Draw an angle whose measure is 11°.

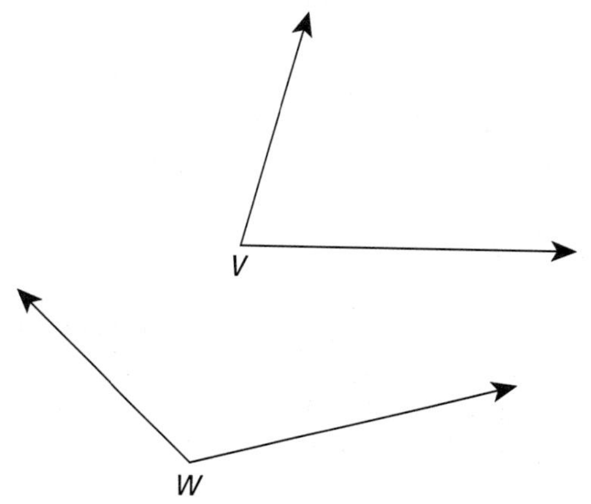

CHAPTER

SUBTRACTION IN ALGEBRA

Many situations lead to linear sentences involving subtraction. For instance, answers to the questions below can be found by solving the equation

$$5000 - 30x = 1000.$$

1. You want to go to Kenya to see lions and elephants in their native habitats. The trip will cost $5000. Your parents tell you that if you can save all but $1000, they will provide that amount. You think that you can get a job and save $30 a week. How many weeks will it take for you to save the necessary amount?

2. The area of rectangle $ABCD$ is 5000 square units, and the area of rectangle $AEFD$ is 1000 square units. What is the length of $\overline{FC}$?

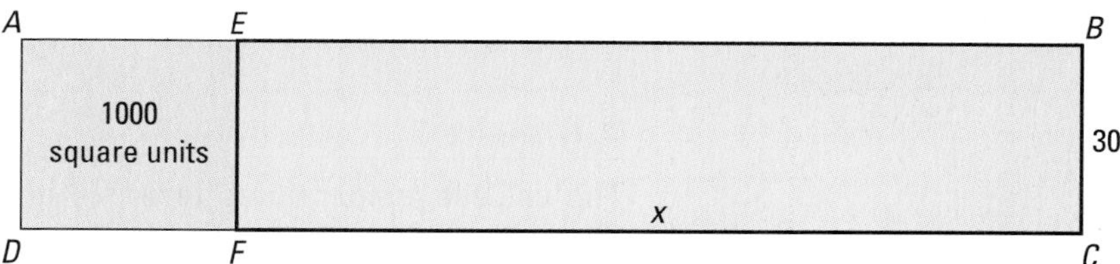

In this chapter, you will learn that you can use variations on techniques that you learned in Chapter 3 to solve the equation $5000 - 30x = 1000$. You will also learn to make and use graphs, such as the one at the right, to solve equations.

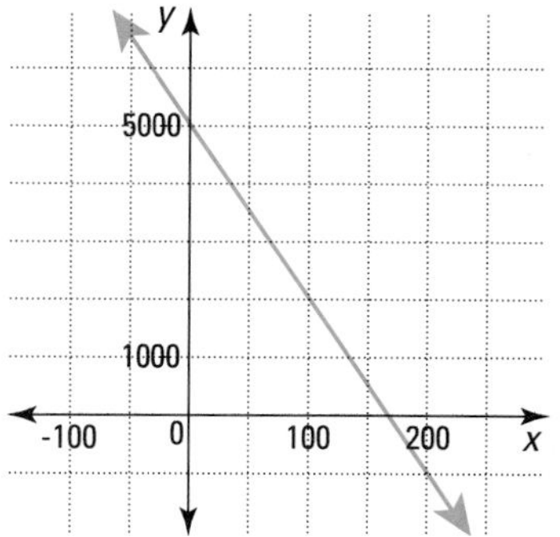

LESSON

4-1

Subtraction of Real Numbers

Sub-zero subtraction. *Plants like the mosses and lichen pictured can survive in the frigid climate of Antarctica. Subtraction of negative numbers is required to find Antarctica's daily temperature range.*

Defining Subtraction in Terms of Addition

Many situations can be thought of either as addition or as subtraction. Consider the following situation.

Example 1

The coldest temperatures recorded in the world have been in Antarctica. Suppose a temperature of -50°C is recorded there. If the temperature then falls 30°, what is the new temperature?

Solution 1

Begin with -50. Subtract 30.

$$-50 - 30 = -80$$

The new temperature is -80°C.

Solution 2

Begin with -50. Add -30.

$$-50 + -30 = -80$$

The new temperature is -80°C.

Notice that, in this instance, subtracting a number is the same as adding its opposite. This pattern is true for all real numbers. It is known as the *Algebraic Definition of Subtraction.*

Algebraic Definition of Subtraction
For all real numbers a and b,

$$a - b = a + -b.$$

Evaluating Expressions Using the Algebraic Definition of Subtraction

You should be able to subtract any two real numbers using this definition.

Example 2

Evaluate $x - y$ when $x = 7.31$ and $y = -5.62$.

Solution

Substitute the values for x and y.

$$x - y = 7.31 - -5.62$$

The opposite of -5.62 is 5.62, so by the definition of subtraction,

$$x - y = 7.31 + 5.62$$
$$= 12.93.$$

Check

Use a calculator.

7.31 [−] 5.62 [±] [=] or 7.31 [−] [(−)] 5.62 [ENTER]

You should see 12.93 displayed.

Caution: Subtraction is *not* associative. For example, in $3 - 9 - 1$ you will get one answer if you do $3 - 9$ first and another answer if you do $9 - 1$ first. You must follow the order of operations and subtract from left to right. (Do $3 - 9$ first.) However, since *addition* is associative, you can gain flexibility by changing the subtractions to adding the opposites.

$$3 - 9 - 1 = 3 + -9 + -1.$$

In $3 + -9 + -1$, either addition can be done first, and the answer is -7.

Simplifying Expressions Using the Definition of Subtraction

You should also be able to use the definition of subtraction and other properties of real numbers to simplify expressions.

Example 3

Simplify $-10 - (-y)$.

Solution

$-10 - (-y) = -10 + y$	Algebraic Definition of Subtraction
$= y + -10$	Commutative Property of Addition
$= y - 10$	Algebraic Definition of Subtraction

Check

The original and final values of the expression should be the same for any value of y. We let $y = 25$. Then $-10 - (-y) = -10 - (-25) = -10 + 25 = 15$. Also, $y - 10 = 25 - 10 = 15$. It checks.

The answer $-10 + y$ is also correct in Example 3, but the form $y - 10$ uses fewer symbols. So, many people think $y - 10$ is easier to read and check than is $-10 + y$.

Example 4

Simplify $5x + 3y - 2 - x$.

Solution

First, think of x as $1x$.

$5x + 3y - 2 - 1x = 5x + 3y - 2 + -1x$	Algebraic Definition of Subtraction
$= 5x + -1x + 3y - 2$	Commutative and Associative Properties of Addition
$= 4x + 3y - 2$	Combining Like Terms form of Distributive Property

QUESTIONS

Covering the Reading

1. State the Algebraic Definition of Subtraction.

In 2–5, **a.** apply the Algebraic Definition of Subtraction to rewrite the subtraction as an addition; **b.** evaluate the expression.

2. $-2 - 7$ **3.** $28 - -63$ **4.** $\frac{3}{5} - -\frac{7}{10}$ **5.** $-3.5 - 0.9$

6. The temperature is 12°F. It falls 15°.
a. Write an addition expression and find the new temperature.
b. Write a subtraction expression and find the new temperature.

7. Write the key sequence to do $-73 - -91$ on your calculator.
a. Use the [−] key. **b.** Use the [+] key.

8. a. *True or false.* $(3 - 9) - 1 = 3 - (9 - 1)$
b. What property is or is not verified in part **a**?

In 9 and 10, calculate.

9. $20 - 4 - 3$ **10.** $-7 - 30 - 20$

In 11–14, simplify.

11. $x - (-d)$ **12.** $-y - (-5)$

13. $10p - 2q + 4 + 8q$ **14.** $-2a - 3a + 4b - b$

Applying the Mathematics

In 15–17, evaluate the expression using the given value(s).

15. $3 - x^2$, when $x = 5$ **16.** $-x - y$, when $x = -12$ and $y = 2$

17. $a - y - b$, when $a = -1$, $b = 2$, and $y = -3$

Take stock in this. *Shown is the trading floor of the New York Stock Exchange. Stockbrokers act as agents for the public in buying and selling shares of stock of various companies.*

18. During a six-week period, the value of a stock in dollars per share changed as follows.

Week 1	Week 2	Week 3	Week 4	Week 5	Week 6
down $4\frac{7}{8}$	up $1\frac{3}{4}$	down $1\frac{3}{4}$	up $1\frac{1}{4}$	up $2\frac{1}{8}$	down $\frac{5}{8}$

Find the net change of the stock over the six weeks.

19. Mr. Whittaker's doctor advised him to exercise more. The changes in Mr. Whittaker's weight were:

First Week	Second Week	Third Week	Fourth Week
lost 4 lb	lost 3 lb	lost 3 lb	gained 5 lb

a. Write an expression for the net change using addition.
b. Write an expression for the net change using subtraction.
c. What was the net change for the four weeks?

20. Let t be Toni's age. Let f be Fred's age. Suppose $t - f = 35$.
a. Find a value for t and a value for f such that $t - f = 35$.
b. Who is older, Toni or Fred?
c. Find the value of $f - t$.

21. Find all possible values for x using the clues.
Choices for x: -7, -4, -3, -2, 1, 3, 4, 7

Clue 1: $x > 0 - 2$
Clue 2: $x < 4 - -1$
Clue 3: $x \neq 2 - -1$
Clue 4: $-x \neq -4$

22. In parts **a–d** evaluate $p - q$ and $q - p$ when:
a. $p = 5$ and $q = -1$;
b. $p = 1$ and $q = 3$;
c. $p = -2$ and $q = 0$;
d. $p = -3$ and $q = -6$.
e. Based upon parts **a** to **d,** does subtraction seem to be commutative? Why or why not?
f. Describe the relationship between $p - q$ and $q - p$.

Review

23. Simplify $n + .2(n + 15)$. *(Lessons 3-6, 3-7)*

In 24 and 25, solve. *(Lessons 3-2, 3-5)*

24. $-3 = x + (-7)$

25. $2.7 + 7y = 3.4$

26. How many four-digit numerals do not contain zeros and have no repeated digits? *(Lesson 2-9)*

27. Find 30% of 74.95. *(Previous course)*

28. Tell which sentence best describes each graph. *(Lesson 3-3)*
   a. The number of people in a restaurant from 6 A.M. to 6 P.M.
   b. The number of people in a school from 6 A.M. to 6 P.M.
   c. The number of people in a hospital from 6 A.M. to 6 P.M.

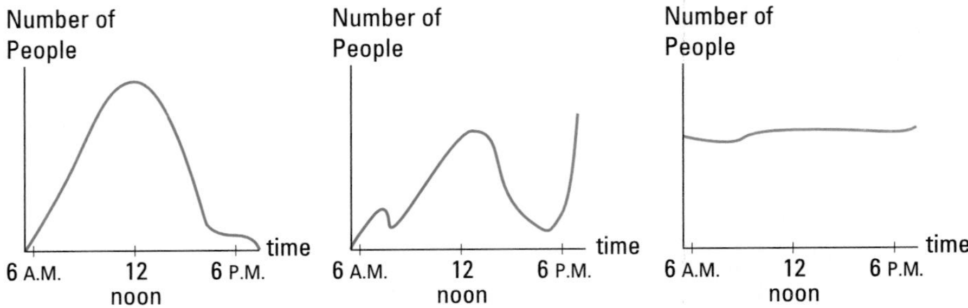

Exploration

29. Roman numerals were the most common way of writing numbers in Western Europe from about 100 B.C. to 1700 A.D. Even in this century, they have often been used to give dates on buildings and to name the hours on timepieces. Here are the values of the individual letters in a Roman numeral.

I	V	X	L	C	D	M
1	5	10	50	100	500	1000

When a letter appears to the left of a letter with a higher value, the value of the first letter is subtracted. So IV = 5 − 1 = 4, and CM = 1000 − 100 = 900. Match each year with its corresponding Roman numeral.

   a. the year Benjamin Banneker, one of the first African-American mathematicians, was born
   b. the year Christopher Columbus first sailed for the New World
   c. the year Jeannette Rankin, the first woman to serve in the U.S. Congress, was elected to the House of Representatives
   d. the year the character "Mickey Mouse" first appeared in a cartoon

(i) MDCCXXXI
(ii) MCDXCII
(iii) MCMXXVIII
(iv) MCMXVII

A man of many talents. *Benjamin Banneker was an astronomer, farmer, and mathematician.*

Mickey Mouse's debut. *Mickey Mouse debuted in Walt Disney's cartoon,* Steamboat Willie. *It was the first cartoon to use synchronized sound.*

LESSON

4-2

Models for Subtraction

Strong storms. *Beaches along a coastline are often eroded by hurricanes. Subtraction can be used to find the amount of beach area left after the storm passes.*

The Take-Away Model for Subtraction

An island had an area of 27.8 square miles. During a hurricane, 1.6 square miles of beach were washed away. The area of the island left was 27.8 − 1.6, or 26.2, square miles. This situation illustrates an important model for subtraction.

Take-Away Model for Subtraction
If a quantity y is taken away from an original quantity x, the quantity left is $x - y$.

The take-away model leads to algebraic expressions involving subtraction.

Example 1

Suppose a giant submarine sandwich has length L inches. You eat 6 inches of it. How much is left?

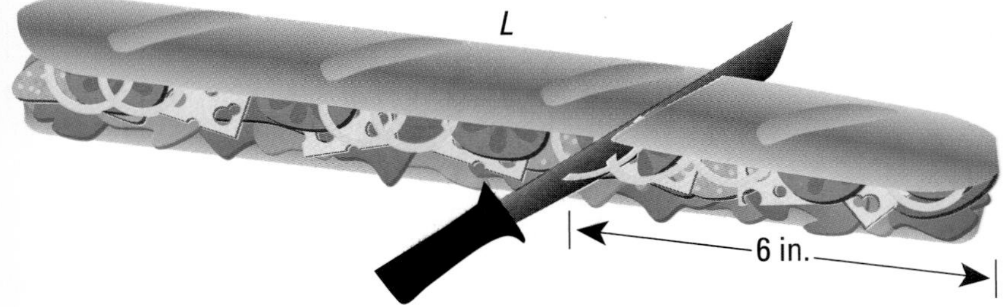

Solution

Take 6 inches away from L inches. There are L − 6 inches left.

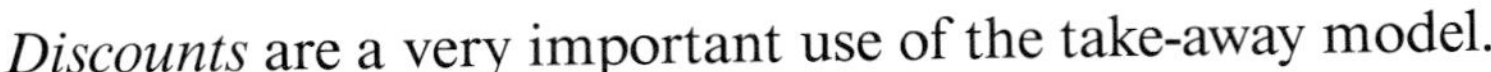

Discounts are a very important use of the take-away model.

Example 2

Josie bought a new bicycle. The *list price* was \$225. The store gave a 30% discount. What was the *sale price?*

Solution 1

Take away the discount from the list price.

$$\begin{aligned}\text{Sale price} &= \text{list price} - \text{discount}\\ &= 225 - (30\% \text{ of } 225)\\ &= 225 - .30(225)\\ &= 225 - 67.50\\ &= 157.50\end{aligned}$$

The sale price was \$157.50.

Solution 2

Compute the percent left after the discount. If an item is on sale for 30% off, you pay 100% − 30%, or 70%, of the original price.

$$70\% \text{ of } 225 = .70(225) = 157.50$$

The sale price was \$157.50.

If an item is discounted x%, you pay $(100 - x)$% of the original price.

Example 3

A microwave oven is on sale for 15% off the regular price. If the regular price is P, what is the sale price?

Solution

The discount is 15% of the regular price or $.15P$.

$$\begin{aligned}\text{sale price} &= \text{regular price} - \text{discount}\\ &= P - .15P\\ &= 1P - .15P && \text{Multiplicative Identity Property of 1}\\ &= .85P && \text{Subtract like terms.}\end{aligned}$$

The sale price is $.85P$ or .85 of the regular price.

If the original price of an item is increased by a given percent, a *markup* results. Taxes and profits often involve markups.

Example 4

The Alvarez family is buying a new van. The purchase price is D dollars. They must also pay a 6.5% sales tax. What is the total cost of the van?

Solution

The sales tax is 6.5% of D, or $.065D$.

$$\begin{aligned}\text{total cost} &= \text{purchase price} + \text{sales tax}\\ &= D + .065D\\ &= 1D + .065D\\ &= 1.065D\end{aligned}$$

The total cost is $1.065D$ dollars.

The ideas illustrated in Examples 2 to 4 are summarized below.

> If an item is discounted $x\%$, you pay $(100 - x)\%$ of the original or listed price. If an item is marked up or taxed $x\%$, you pay $(100 + x)\%$ of the original or listed price.

The Comparison Model for Subtraction

In Example 2, Josie paid \$157.50 for the bicycle. Another store advertised the same item for \$170. To find out how much money Josie saved, you subtract.

$$\$170 - \$157.50 = \$12.50$$

This is an instance of a second model for subtraction.

> **Comparison Model for Subtraction**
> The quantity $x - y$ tells how much the quantity x differs from the quantity y.

Example 5

The Transamerica Building in San Francisco is shaped like a pyramid. Each floor is approximately square. Let x be the side of one floor and let y be the side of a lower floor. How much greater is the area of the lower floor than the area of the upper floor?

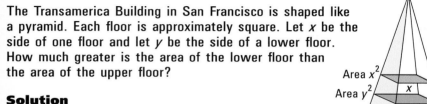

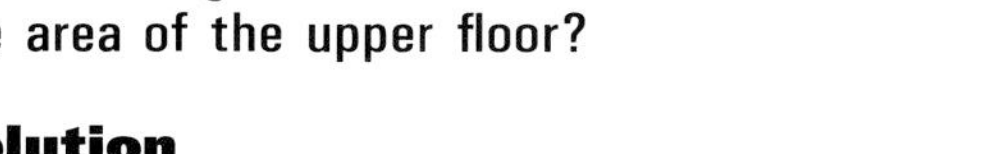

Solution

The two floors have areas x^2 and y^2. Since y^2 is the greater area, the difference in area is $y^2 - x^2$.

Area x^2
Area y^2
x
y

A statue of Lewis, Clark, and Sacagawea stands at Ft. Benton, Montana. Sacagawea, a Shoshone woman, acted as an interpreter for the explorers Lewis and Clark.

What Is the Range of a Set of Numbers?

An important application of the comparison model is the calculation of the *range* of a set of numbers. The **range** of a set is the difference obtained when the set's **minimum** (least) value is subtracted from its **maximum** (greatest) value.

Example 6

The greatest recorded difference in temperature during a single day occurred in Browning, Montana, in January of 1916. During one 24-hour period, the low temperature was -56°F and the high temperature was 44°F. What was the range of the temperatures?

Solution

Compare the numbers by subtracting.

$$\begin{aligned}\text{range} &= \text{maximum} - \text{minimum}\\ &= 44 - (-56)\\ &= 44 + 56 = 100\end{aligned}$$

The range of the temperatures was 100°F.

QUESTIONS

Covering the Reading

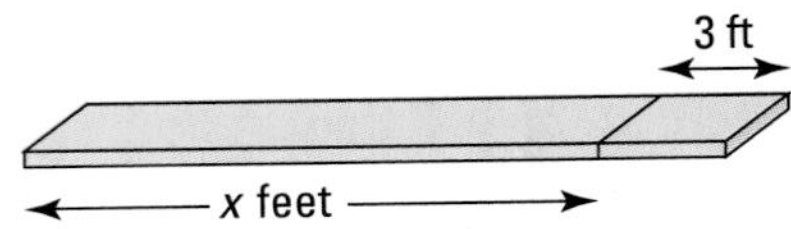

1. A carpenter has a board that is x feet long. He cuts a 3-foot piece from it. Write an expression for the length of the remaining piece.

2. Refer to the sales tag at the left.
   a. What is the amount of the discount?
   b. What is the percent of discount?
   c. What is the sale price?

3. Suppose a coat originally cost C dollars, but it is on sale now for 25% off.
   a. What is the amount (in dollars) of the discount?
   b. What is the sale price of the coat?

4. Suppose a pair of jeans is on sale for 40% off. If the regular price of the jeans is J dollars, what is the sale price of the jeans?

5. The price of an item is P dollars, and there is a 4% sales tax.
   a. Express the tax paid in terms of P.
   b. What is the total amount paid for the item?
   c. If P = \$65, what is the total amount paid?

A modern pyramid. *The Transamerica Building is one of San Francisco's landmarks. Built in 1972, this skyscraper towers 853 feet (260 m).*

6. Refer to Example 5. Suppose one side of a floor in the Transamerica Building has length 40 meters. Let b meters be the length of a side of a higher floor. Write an expression for the positive difference in the areas of the two floors.

7. An unusual temperature change was recorded in Spearfish, South Dakota, on January 22, 1943. Over a period of just two minutes, the temperature rose from -4°F to 45°F. By how much did the temperature change during this time?

8. As of the summer of 1993, the highest temperature ever recorded in the United States occurred on July 10, 1913, at Greenland Ranch, California, where the temperature reached 134°F. The lowest temperature ever recorded occurred on January 23, 1971, in Prospect Creek, Alaska, where the temperature fell to -80°F. What is the range of the temperatures in the U.S.?

Applying the Mathematics

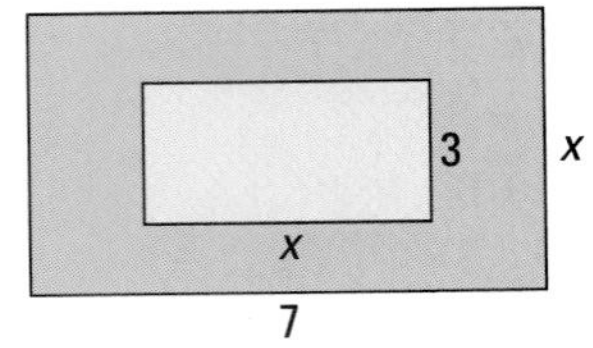

9. What is the difference in area between the two rectangles at the left?

10. A company bought a piece of property which has an area of 14,000 square feet. On it the company built a store with area S square feet and a parking lot with area 2580 square feet. Write an expression for the area left for the lawn.

In 11 and 12, Bernie's age is B, John's age is $B - 3$, and Robin's age is $B - 7$.

11. **a.** Who is older, Robin or Bernie? **b.** How much older?

12. **a.** Who is older, John or Robin? **b.** How much older?

A glimpse of NYC. *New York City is representative of America's multicultural society. The five largest ethnic groups in the city are African-American, Irish, Italian, Jewish, and Puerto Rican. These groups constitute 80% of the population.*

13. Sam's Super Saver Store is offering 20% off on all merchandise sold this week. An item with a regular price R is purchased.
a. Give the sale price of the item in terms of R.
b. Freda receives an additional 10% employee discount off the customer price. Express the cost of this item to Freda in terms of R.
c. If the state sales tax is 3%, what is the total cost of the item to Freda?

14. The thermometers below show a hospital patient's temperature at three times on the same day.
a. What was the change from 3 P.M. to 6 P.M.?
b. What was the change from 3 P.M. to 9 P.M.?

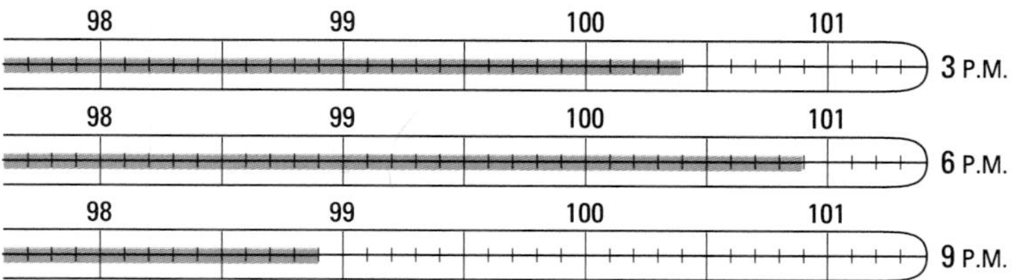

15. **a.** In New York City, the sun's rays make a $72\frac{1}{2}°$ angle with the ground at noon on the first day of summer (about June 21). On the first day of winter (about December 21), the angle is $25\frac{1}{2}°$. By how much do the angles differ?
b. If in another city the angles the sun's rays make with the ground are a and b, what are the possible values of the difference in angle measures?

16. Some students took a chapter pretest and posttest. Suppose *Change* = *Posttest* − *Pretest*. **a.** Complete the table.
b. Which student's test scores showed most improvement?

Student	Pretest	Posttest	Change
Chui, L.	57	65	8
Fields, S.	43	41	?
Ivan, J.	63	?	7
Washington, C.	?	51	-3

Review

In 17–20, simplify. *(Lesson 4-1)*

17. $-3 - (-3)$

18. $-8.7 - 16.03$

19. $-p - (-q)$

20. $-7ab + 2a - 5b - 6ab - 4a + b$

21. E is 4 units to the left of A. F is n units to the right of A. *(Lesson 4-1)*
a. What is the coordinate of E?
b. Write an expression for the coordinate of F.
c. Why do you think point F was not included on the number line below?

In 22–25, solve. *(Lessons 3-5, 3-10)*

22. $7t + 6 > 41$

23. $3x + 8 = 5$

24. $16 + x = 0$

25. $-28.3 > -x + 17.5$

26. The image of (x, y) is $(7, -1)$. It results from a slide four units to the right and two units down. Find x and y. *(Lesson 3-4)*

27. Three instances of a pattern are given below.

$$9^2 - 4^2 = (9 + 4)(9 - 4)$$
$$31^2 - 29^2 = (31 + 29)(31 - 29)$$
$$3.5^2 - 2.5^2 = (3.5 + 2.5)(3.5 - 2.5)$$

a. Do a mental check or use a calculator to verify that the sentences are true.
b. Describe the general pattern using two variables.
c. Write another instance of this pattern. Is your instance true? *(Lesson 1-7)*

28. Find the area of each triangle. *(Lesson 1-5)*

a.
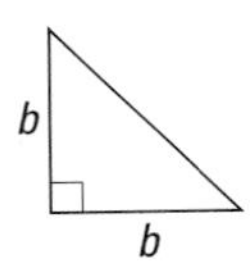

b.
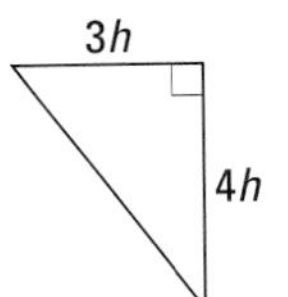

c.
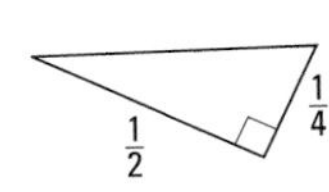

Important inkstand. *This silver inkstand was used to sign both the Declaration of Independence in 1776 and the U.S. Constitution in 1787. It is displayed in Independence Hall in Philadelphia, PA.*

Exploration

29. The number of years between two historical events can be calculated by subtracting the years in which the events occurred. You will be within a year of the number of years between them.
a. Nicholas and Nicole were born in two consecutive years. What is the least and greatest number of days between their birthdays?
b. Maria Gaetana Agnesi, an Italian mathematician after whom a special mathematical curve is named, was born in 1718 and died in 1799. How old was she when she died?
c. How many years apart are the Declaration of Independence and the end of the Civil War?
d. How many years are between the death of Archimedes in 212 B.C. and the death of the Roman emperor Nero in 68 A.D.? (Be careful! There was no year 0.)

LESSON

4-3

Solving Sentences Involving Subtraction

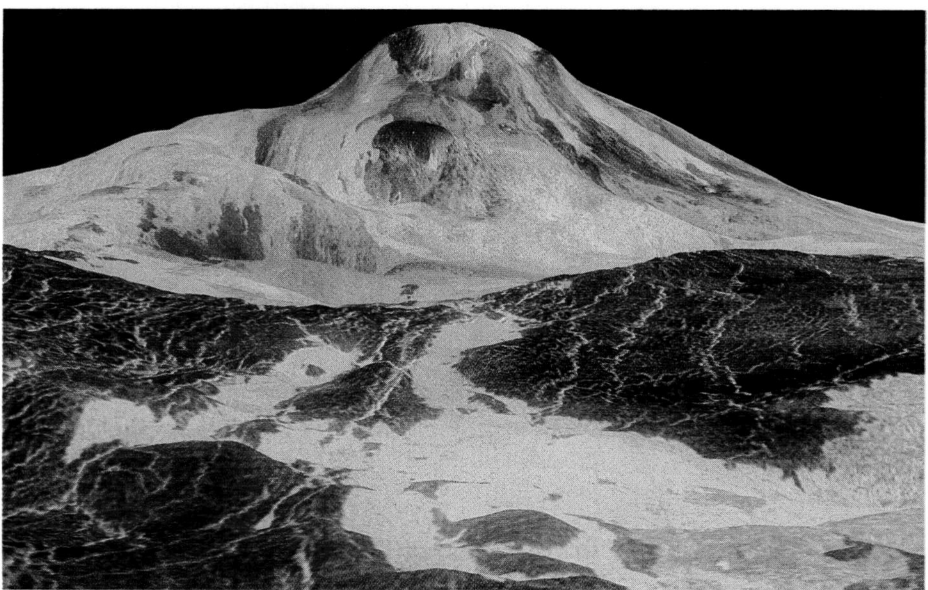

Venus Volcano. *Shown is Maat Mons, an active volcano on Venus, discovered in 1991 by the U.S. space probe* Magellan.

Solving $ax - b = c$

In Chapter 3, you solved sentences of the form $ax + b = c$ by adding the same number to both sides. Sentences of the form $ax - b = c$ are solved in much the same way.

For instance, scientists have found that the range of temperatures recorded on the planets of our solar system is about 695°C. The minimum temperature of -220°C was recorded on Pluto. The maximum was recorded on Venus. To find the maximum recorded temperature, you can use a subtraction equation.

Let T = the maximum temperature recorded. Substitute into the formula for range.

$$\begin{aligned} \text{range} &= \text{maximum} - \text{minimum} \\ 695 &= T - (-220) \\ 695 &= T + 220 \end{aligned}$$

Notice that you can change the subtraction to addition. Now solve as you would an addition equation. Add -220 to each side.

$$\begin{aligned} 695 + -220 &= T + 220 + -220 \\ 475 &= T \end{aligned}$$

The maximum recorded temperature is about 475°C.

The subtraction equation above was changed to one involving addition by using the Algebraic Definition of Subtraction. You can do this with *any* sentence involving subtraction. Then you can apply anything you already know about solving sentences.

Example 1

Solve $14x - 21 = -133$.

Solution

Change the subtraction to addition of the opposite.

$$14x + -21 = -133$$

This is now an equation of the same type you solved in the last chapter.

$14x + -21 + 21 = -133 + 21$	Add 21 to each side.
$14x = -112$	Simplify.
$x = -8$	Multiply both sides by $\frac{1}{14}$ and simplify.

Check

Substitute -8 for x in the original equation. Does $14 \cdot -8 - 21 = -133$? $14 \cdot -8 - 21 = -133$, so the solution is -8.

Solving Equations When the Unknown Is Subtracted

In Example 1, you probably noticed that changing subtraction to addition was not essential. However, in the next example you will see how this technique can be helpful.

Example 2

A metal bar was 1.2 cm too long. It was shortened and found to be 0.03 cm too short. How much was cut off?

Solution

Let the amount cut off be c. "Too short" means -0.03. Use the Take-Away Model for Subtraction. The answer is the solution to

$$1.2 - c = -0.03.$$

$1.2 + -c = -0.03$	Definition of Subtraction
$-1.2 + 1.2 + -c = -1.2 + -0.03$	Addition Property of Equality
$-c = -1.23$	
$c = 1.23$	Multiplication Property of Equality

The amount cut off was 1.23 cm.

Check

If 1.23 cm is cut off from a bar 1.2 cm too long, will it be 0.03 cm too short? Yes, it will.

Solving Inequalities When the Unknown Is Subtracted

The questions from page 215 can be answered using the same idea as used in Example 2.

Example 3

You want to go to Kenya to see lions and elephants in their native habitats. The trip will cost $5000. Your parents say that if you save all but $1000, they will provide that amount. You think you can get a job and save $30 per week. How many weeks will it take to save the necessary amount? ▶

Solution

Let x = the number of weeks you must save until the balance is \$1000 or less. Solve $5000 - 30x \le 1000$.

Change the subtraction to addition of the opposite.

$$5000 + -30x \le 1000$$

Now add -5000 to each side. This isolates the expression with the variable.

$$-5000 + 5000 + -30x \le -5000 + 1000$$

$$-30x \le -4000$$

Multiply both sides by $-\frac{1}{30}$ and simplify. This changes the sense of the inequality.

$$-\frac{1}{30} \cdot (-30x) \ge -\frac{1}{30} \cdot -4000$$

$$x \ge \frac{400}{3}$$

$$x \ge 133.\overline{3}$$

Because your savings occur only once a week, you will need to save for at least 134 weeks (a little over $2\frac{1}{2}$ years) in order for the balance to be \$1000 or less.

Caution: The most common student error in solving sentences like those in Examples 2 and 3 is ignoring the subtraction sign. You can lessen your chances of making this error by changing the subtraction to an addition *on paper* (not just in your head) and by checking the answer you get.

QUESTIONS

Covering the Reading

1. The range of temperatures recorded on Mercury one day was 573°C. The minimum temperature recorded that day was -154°C.
 a. Using the formula for range, write an equation to find Mercury's maximum recorded temperature that day.
 b. Solve to find the maximum temperature.

2. In each example in this lesson, the first step in solving the equation is the same. What is done in that step?

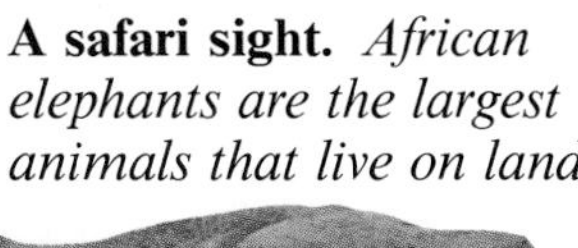

A safari sight. *African elephants are the largest animals that live on land.*

In 3–6, solve and check.

3. $12y - 9 = -3$

4. $\frac{1}{2}x - 7 = 8$

5. $5z - 3.4 = 2.9$

6. $-9A - 1 = 0$

7. A bar was 3.01 mm too long. It was shortened by c and still found to be 0.54 mm too long. Solve an equation to find c.

In 8 and 9 solve.

8. $-3 - t = -1$

9. $-10 = 400 - A$

10. Use Example 3 of this lesson to answer Question 2 on page 215.

11. *Skill sequence.* Solve.
a. $2x - 16 = 20$
b. $2x - 16 \geq 20$
c. $16 - 2x = 20$
d. $16 - 2x \geq 20$

Applying the Mathematics

12. Here is Ali's work to solve $4 - 3x = 5$. Answer his last question.

$$4 - 3x = 5$$
$$4 + -3x = 5$$
$$-4 + 4 + -3x = -4 + 5$$
$$3x = 1$$
$$x = \frac{1}{3}$$

Check. Substitute $\frac{1}{3}$ for x in the original equation.

$$\text{Does } 4 - 3 \cdot \frac{1}{3} = 5?$$
$$4 - 1 = 5?$$

No. It doesn't check.
Where did I go wrong?

Bonnie's bounty. *U.S. speedskater Bonnie Blair is shown after winning the 1,000-meter race at the 1994 Winter Olympics. Bonnie has won five Olympic gold medals, more than any woman in U.S. history.*

13. Three less than 10 times a number is 84.
a. Write an equation which represents this situation. Let n = the unknown number.
b. What is the number?

14. Hometown Bank and Trust requires a minimum balance of $1500 for free checking. If Mr. Archer can withdraw $3276 and still have free checking, how much is in his account?

15. A small town's population was 11,200 in 1990. Each year since 1990, about 60 people moved away. If this trend continues, in what year will the town's population reach 8,000? (Hint: Make a table to help determine the equation.)

16. At the 1994 Winter Olympics, Bonnie Blair won her third consecutive gold medal in the women's 500-meter speed skating event with a time of 39.25 seconds. This was 0.15 second slower than when she won for the first time, in 1988. What was her 1988 Olympic time?

In 17–20, solve.

17. $-12.2 - p = -0.56$

18. $-1 \leq -1 - y$

19. $\frac{3}{4}t - 11 > 7$

20. $-3(2n + 1) - 4 = -11$

Review

21. The Dolans celebrated their 50th (golden) wedding anniversary in year y. In what year were they married? *(Lesson 4-2)*

22. The Valases will celebrate their nth wedding anniversary in 2000. In what year were they married? *(Lesson 4-2)*

23. Susan currently earns \$20.00 per hour.
   a. Suppose she receives a 10% raise. What is her new hourly wage?
   b. After receiving the raise in part **a,** her employer's profits decrease, so her wages are cut 10%. What is her hourly wage at this point?
   c. Explain why the result in part **b** is not \$20.00. *(Lesson 4-2)*

24. A stock was valued at $120\frac{1}{4}$ one day. The next day its value fell $9\frac{3}{8}$. What was its value at that point? *(Lessons 3-9, 4-2)*

In 25–27, *multiple choice.* Suppose that x is negative and y is positive. Tell whether the value of the expression is (a) always negative, (b) always zero, (c) always positive, or (d) none of these. *(Lessons 2-5, 3-2, 4-1)*

25. $x \cdot y$ **26.** $x + y$ **27.** $x - y$

In 28 and 29, simplify. *(Lessons 1-4, 4-1)*

28. $5 - (4 - (3 - (2 - 1)))$ **29.** $1 - (2 - (3 - (4 - 5)))$

30. Simplify $\frac{4a}{5} + \frac{a}{3}$. *(Lesson 3-9)*

31. Use the Distributive Property to simplify $-1(3x + -5)$. *(Lesson 3-7)*

32. *Multiple choice.* Which of the following real-world situations is represented by the graph below? Write a brief explanation to justify your answer(s). *(Lesson 3-3)*
(a) the distance traveled in h hours at 30 mph
(b) the distance traveled in h hours at 50 mph
(c) the distance you are from home if you started 90 miles from home and traveled home at 30 mph

Exploration

33. During gym class, the students formed a circle and counted off. (The first student counted "1," the second student counted "2," and so on.)
   a. Student number 7 was directly across the circle from student number 28. How many students are in the class?
   b. Suppose student number 7 was directly across from student number n. Then how many students are in this class?

LESSON 4-4

Spreadsheets

Then and now. *This house was custom-built in Chicago in 1946 for a cost of $31,500. This is equivalent to a cost of over $300,000 today.*

Construction Statement

		Total Contract	Amount Paid	Balance Due
Excavating	340 00			
Work Order #1	15 30	355 30	225 00	130 30
Foundation		1731 00	1150 00	581 00
Brickwork	4385 00			
Work Order #1	10 00	4395 00	3750 00	645 00
Ornamental Iron		230 00		230 00
Structural Iron		142 00	142 00	
Metal Windows & Damper		49 00	49 00	
Carpenter	6560 00			
Work Order #1	375 00	6935 00	700 00	6235 00
Insulation		110 00		110 00
Roofing	195 00			
Work Order #1	25 00	220 00		220 00
Heating	1220 00			
Work Order #1	245 00			
Credit Order #1	(53 00)	1412 00		1412 00

What Is a Spreadsheet?

Shown above is a copy of part of a bill sent in 1946 by a builder of a home to the family that was having the home built. It is an example of a *spreadsheet.* In many situations, very large pages or pages taped together were used for records like these. These pages were sometimes folded for storage and then spread out for examination. That is how "spreadsheets" got their name.

Notice that some numbers are added or subtracted to yield other numbers in the spreadsheet. Consequently, maintaining a spreadsheet used to be quite difficult. If one entry was changed, many other numbers would need to be changed as a result. By the mid 1970s, large companies were using computers for this purpose, but the process was still tedious.

In 1978, Dan Bricklin and Bob Frankston, graduate students at the Harvard University School of Business, recognized that desktop computers could save them a great deal of work. They prepared a computer spreadsheet program that would recompute their work as they tried new values for costs or prices. Their program, *VisiCalc,* was a major advance in computer applications because it had the power to hold formulas. Each formula indicates how a number in a particular place in the chart is computed from other numbers on the spreadsheet. When one number is changed, the computer uses the formulas to change the corresponding numbers.

Computer spreadsheets were an immediate success. Nowadays they are used to organize data, experiment with complicated computations, determine patterns, and test properties. Many people use spreadsheets to keep personal or household financial records.

Cells in Spreadsheets

A typical computer spreadsheet has *rows* identified by numbers and *columns* identified with letters. The locations or boxes formed by the intersection of rows and columns are called **cells.** A spreadsheet can have hundreds of columns and thousands of rows. How many depends on the software and on the amount of memory in the computer.

Typically the location of a cell is named by the column letter first and the row number second.

The spreadsheet below shows the number of coupon books sold by students to raise money for their school. This spreadsheet tells you that Susan sold 50 books. Her name is in cell A2, and the number 50 is in cell C2. You can think of a cell name as a variable because the number or other entry in the cell can change.

	A	B	C	D
1	Student		Books	Sales
2	Susan		50	
3	Sam		25	
4	Rashana		33	
5				
6	Total		108	

The number 108 in C6 is the result of adding 50 + 25 + 33. It is an instance of the formula C6 = C2 + C3 + C4. To enter this formula in C6 on many spreadsheets you can type the phrase = C2 + C3 + C4 in cell C6. The "=" symbol tells the computer to calculate the value of the expression and to display the result, not the formula. (Some spreadsheets allow the use of other symbols, such as a + or − sign instead of an equal sign.)

Example 1

Use the spreadsheet pictured on the previous page.

a. Which cell contains the number 33?
b. What is in cell A3?
c. What cell contains the word "Books"?
d. Suppose D6 is computed by typing = C2 + C4. What number would appear in D6?
e. What is in cell B6?
f. Suppose Rashana sells one more coupon book, so the number in C4 changes from 33 to 34. What other entry will change?

Solution

a. The number 33 is in C4.
b. The word "Sam" is in A3.
c. C1
d. 83
e. Nothing; it is empty.
f. C6 will change to 109.

Expressions and Formulas in Spreadsheets

The language of expressions and formulas in most spreadsheets is similar to that in many computer languages. The arithmetic symbols are + for addition, – for subtraction, * for multiplication, / for division, ^ for powering, and ( ) for grouping. SQRT( ) is usually used for square root. Typically, every operation must be shown by a symbol. So to multiply the value in C7 by 5 you would type = C7 * 5.

Spreadsheets can quickly give instances of properties.

Example 2

Use a spreadsheet to give examples of the Distributive Property, $a(b - c) = ab - ac$.

Solution

The examples will involve different values of a, b, and c, so make columns for these numbers. Then make a column for $a(b - c)$ and a column for $ab - ac$. The columns are named in row 1. Since = signs do not precede these expressions, the spreadsheet program does no calculations in this row.

	A	B	C	D	E
1	a	b	c	a(b – c)	ab – ac
2					
3					

▶ Now pick some values for a, b, and c. We pick 10, 15, and -3. Type these in cells A2, B2, and C2. In cell D2, type the formula = A2∗(B2 − C2). In cell E2, type the formula = A2∗B2 − A2∗C2. When these formulas are entered, the spreadsheet will look like this.

	A	B	C	D	E
1	a	b	c	a(b − c)	ab − ac
2	10	5	-3	80	80
3					

You can put any real numbers you wish in cells A3, B3, and C3. Now in cell D3, enter the formula = A3∗(B3 − C3). In cell E3, enter the formula = A3∗B3 − A3∗C3. Because the property $a(b - c) = ab - ac$ holds for all real numbers, you should see the same numbers in cells D3 and E3.

Replicating Formulas

It is tedious to type in each formula needed in Example 2, and it is easy to make a mistake. Fortunately, all spreadsheets have a feature that enables you to copy formulas from one cell to another in a way that changes the cell names in the formula to agree with the column and row position of the new cell. Consider again the formulas needed in column D of the spreadsheet of Example 2.

Cell	Formula
D2	= A2∗(B2 − C2)
D3	= A3∗(B3 − C3)
D4	= A4∗(B4 − C4)

The formulas follow a pattern as you go down the column. The formulas that are to be entered in column D are the same except for the cell names used in the formula. Each formula refers to cells in the same positions relative to the cell containing the formula. When this occurs, you can type the first formula (in cell D2), and then command the program to copy the formula to the cells below D2. The references to the cells in columns A, B, and C will automatically change as it does the copying. This way of copying is called **replication.** The formulas are replicated from the first cell to reflect the changes in cell references.

Different spreadsheets have different commands for replicating formulas. If you have access to a spreadsheet program, you should determine how you can replicate a formula in it.

Many different kinds of formulas can be used in spreadsheets.

Example 3

Alex Ploor is learning about the Pythagorean Theorem. Alex wants to see how increasing the length of one leg in a right triangle affects the length of the hypotenuse. So Alex created the following spreadsheet. Row 1 contains the names of the columns. Row 2 lists the variables assigned to the column names.

	A	B	C	D	E
1	Leg 1	Leg 2		Hypotenuse	
2	a	b	c^2	c	
3	1	1			
4	1	2			
5	1	3			
6	1	4			

a. What formula should be put in cell C3?
b. What formula should be put in cell D3?
c. What number will appear in cell D3?
d. What formulas should be put in each of cells C4 through C6 and D4 through D6 to complete the spreadsheet?
e. What will the completed spreadsheet show on the screen?

Solution

a. By the Pythagorean Theorem, $c^2 = a^2 + b^2$. Thus C3 should contain the sum of the squares of the numbers in A3 and B3. So in cell C3, Alex should type the following:

```
= A3^2 + B3^2
```

b. The number c is the positive square root of c^2. So in cell D3, Alex should type the following:

```
= SQRT(C3)
```

c. Since $1^2 + 1^2 = 2$, the number 2 will appear in C3, and 1.414213 . . . (the decimal for $\sqrt{2}$) will appear in D3.

d. The formula for cell C4 would be = A4^2 + B4^2. The formula for cell C5 would be = A5^2 + B5^2, and for C6 it would be = A6^2 + B6^2. Alex can replicate the formula in cell C3 in cells C4 to C6 to avoid all this typing. Similarly, he can replicate the formula in cell D3 to cells D4 to D6.

e. Columns A and B will stay as shown. In cells C3 through C6 and D3 through D6 the formulas given in parts *a, b,* and *d* will be evaluated. The spreadsheet will show the following:

	A	B	C	D	E
1	Leg 1	Leg 2		Hypotenuse	
2	a	b	c^2	c	
3	1	1	2	1.414	
4	1	2	5	2.236	
5	1	3	10	3.162	
6	1	4	17	4.123	

QUESTIONS

Covering the Reading

1. Refer to the spreadsheet that begins this lesson. Suppose the cost of excavating were $350 instead of $340. What other numbers would change?

2. Who designed the first electronic spreadsheet software, and when?

3. What is the term for the places in a spreadsheet into which you can type words, numbers, or formulas?

4. Suppose cell A5 in a spreadsheet contains the number 18 and B5 contains the number -3. If C5 contains the formula = A5 + B5 ^ 2, what number will appear in cell C5?

In 5 and 6, refer to Example 2.

5. If .6 is typed in cell A3, 4 in B3, and 5 in C3, what numbers will appear in cells D3 and E3?

6. **a.** What formula should be typed in cell D4?
b. What formula should be typed in cell E4?
c. Why will these formulas yield the same values?

7. What is the term for copying formulas in a way that allows the spreadsheet to change cell references?

8. Refer to Example 3. Suppose the table is extended so A7 contains the number 1 and B7 contains the number 5.
a. What formula should be typed in cell C7?
b. What will appear in C7?
c. What formula should be typed in cell D7?
d. What will appear there?

Applying the Mathematics

In 9–12, use the spreadsheet below. It shows a listing of students who took two tests. The mean is computed from their test scores.

9. What is the entry in cell B4?

10. What cell shows the number 75?

11. What does the 80 in D5 represent?

12. The mean of John's scores is computed in cell D2 by the formula = (B2 + C2) / 2.

a. Write the formula for Marcel's mean.

b. What cell contains this formula?

	A	B	C	D
1	Name	Test 1	Test 2	Mean
2	John	90	95	92.5
3	Gordon	70	80	75
4	Anne	86	86	86
5	Marcel	64	96	80

13. Write a formula for the mean of the values in cells J5, K5, J6, and K6.

14. Suppose that cell F6 contains the number x and cell F7 contains y. Write the expression in x and y represented by the formula = (F6 − F7) / (2 * F6 + F7 ^ 3).

15. Cell T5 contains the number 16, and cell T6 contains the number 25. Cell T7 shows the number 9. Could the formula in T7 be = SQRT(T5 + T6)? Why or why not?

16. A student wants to make a table of powers of numbers. He types in names of columns as shown here.

	A	B	C	D	E
1	x	square	cube	4th power	5th power
2					

He wants to be able to enter a number in cell A2 and have the program automatically calculate the values in B2, C2, D2, and E2. What formulas should he enter in those cells?

CLASSIFIED ADS

$5 for the first 25 words

50¢ for each additional word

Call 1-800-

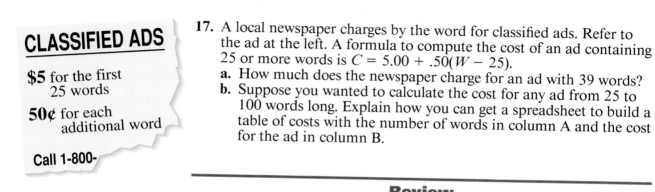

17. A local newspaper charges by the word for classified ads. Refer to the ad at the left. A formula to compute the cost of an ad containing 25 or more words is $C = 5.00 + .50(W - 25)$.

a. How much does the newspaper charge for an ad with 39 words?

b. Suppose you wanted to calculate the cost for any ad from 25 to 100 words long. Explain how you can get a spreadsheet to build a table of costs with the number of words in column A and the cost for the ad in column B.

Review

In 18–21, solve. *(Lesson 4-3)*

18. $9x - 18 = 432$

19. $9(x - 18) = 432$

20. $\frac{y}{2} - 7 = -6$

21. $-4 < 8z - 4$

22. *Multiple choice.* Below is a table produced by one of the four formulas listed. Which formula is it? *(Lesson 1-9)*

(a) $y = 3x$
(b) $y = x + 3$
(c) $y = 3x + 1$
(d) $y = 3x + 3$

x	y
0	3
1	6
2	9
3	12
4	15

23. One ski lift at the Alpine Ski Resort in East Troy, Wisconsin, is represented by the diagram. How far is it from A to B? *(Lesson 1-8)*

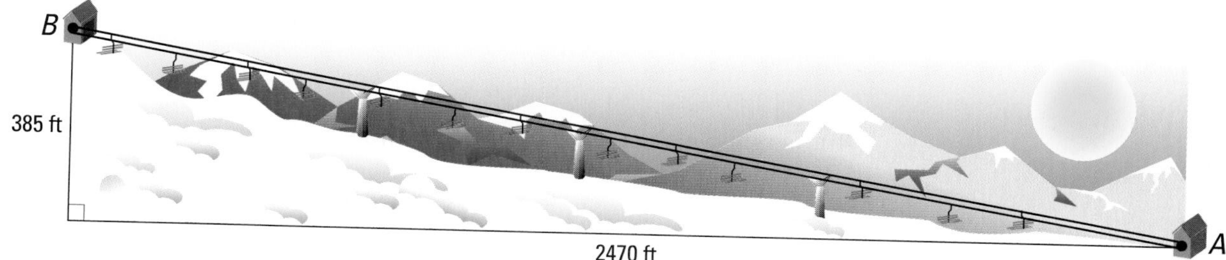

24. Without using a calculator, find two consecutive whole numbers a and b so that $a < \sqrt{53} < b$. Explain how you found your answer. *(Lesson 1-6)*

Exploration

25. Find an actual spreadsheet program and try it out. What is the name of the cell in the bottom right-hand corner of this spreadsheet? How many rows and columns does this spreadsheet have? How many cells does it have? What would be the answers to Example 2 on this spreadsheet?

LESSON

4-5

The Opposite of a Sum or Difference

Could you bank on them? *Shown is a re-creation of an 1850s town. Some banks in those days failed, and depositors often lost their savings. Today, most deposits up to $100,000 are insured by a branch of the U.S. government.*

An Example of the Opposite of a Sum

Suppose Anton has $500 in his savings account. He withdraws a dollars from his account. Deciding that this is not enough, he makes another withdrawal of b dollars. Notice that the amount of money left in his savings account can be expressed in several different ways.

Anton can subtract one amount from 500 and then subtract the other from the difference. This can be written

$$500 - a - b, \text{ or } 500 + -a + -b.$$

Anton could also add the two withdrawals and then subtract the total from 500. This is written as

$$500 - (a + b), \text{ or } 500 + -(a + b).$$

The fact that $-(a + b)$ is the same as $-a - b$ is related to the Distributive Property. To see how, start with $-(a + b)$. Recall that the Multiplication Property of -1 states that for any real number n, $-n = -1n$. So,

$-(a + b) = -1(a + b)$	Multiplication Property of -1
$= -1a + -1b$	Distributive Property
$= -a + -b$	Multiplication Property of -1
$= -a - b.$	Definition of Subtraction

Opposite of a Sum Property
For all real numbers a and b,

$$-(a + b) = -a + -b = -a - b.$$

The first part of this property, $-(a + b) = -a + -b$, says that *the opposite of a sum is the sum of the opposites of its terms.*

Example 1

Simplify $-(2k + 14)$.

Solution

Apply the Opposite of a Sum Property.

$$-(2k + 14) = -2k - 14$$

The Opposite of a Difference

Suppose the expression in parentheses involves subtraction rather than addition. How can you rewrite its opposite? Again, the Distributive Property can be used.

$-(a - b) = -(a + -b)$	Algebraic Definition of Subtraction
$= -a + -(-b)$	Opposite of a Sum
$= -a + b$	Op-op Property

Now we can rewrite the opposite of a difference.

Opposite of a Difference Property
For all real numbers a and b,

$$-(a - b) = -a + b.$$

Example 2

Simplify $-(4a - 7)$.

Solution

$$-(4a - 7) = -4a + 7$$

When a difference is subtracted from another number or expression, the Opposite of a Difference Property can be used. First use the Algebraic Definition of Subtraction to change the subtraction of the difference to the addition of its opposite.

Example 3

Simplify $(10a + 6) - (4a - 7)$.

Solution 1

Change the subtracting of $(4a - 7)$ to addition.

$$x \quad - \quad y \quad = \quad x \quad + \quad -y$$

$(10a + 6) - (4a - 7) = (10a + 6) + -(4a - 7)$	Algebraic Definition of Subtraction
$= 10a + 6 + -4a + 7$	Opposite of a Difference
$= 6a + 13$	Add like terms.

▶

Solution 2

An expert might write the following:

$$\begin{aligned}(10a + 6) - (4a - 7) &= 10a + 6 - 4a + 7 \\ &= 6a + 13\end{aligned}$$

Opposites of Expressions in Equations

In the next example, the left side of the equation may seem complicated. It shows that first $2 + 4x$ is multiplied by 3, and then the result is subtracted from $13x$. Again we begin work by using the Algebraic Definition of Subtraction.

Example 4

Solve $13x - 3(2 + 4x) = 25$.

Solution 1

Change the subtraction to an addition.

$13x - 3(2 + 4x) = 25$	
$13x + -3(2 + 4x) = 25$	Algebraic Definition of Subtraction
$13x + -6 + -12x = 25$	Distributive Property (Remove Parentheses.)
$x + -6 = 25$	Distributive Property (Add like terms.)
$x = 31$	Addition Property of Equality

Solution 2

Experts sometimes apply the Algebraic Definition of Subtraction, the Distributive Property, and the Opposite of a Sum Property in one step. They might write the following:

$$\begin{aligned}13x - 3(2 + 4x) &= 25 \\ 13x - 6 - 12x &= 25 \\ x - 6 &= 25 \\ x &= 31\end{aligned}$$

Check

Does $13 \cdot 31 - 3(2 + 4 \cdot 31) = 25$?

Does $403 - 3(126) = 25$? Yes, so it checks.

Opposites of Expressions in Fractions

To subtract fractions, you must have a common denominator. When numerators involve sums of differences, you must be careful with the signs.

Example 5

Simplify $\frac{4x}{5} - \frac{2x + 1}{10}$.

Solution

A common denominator is 10.

$$\begin{aligned}\frac{4x}{5} - \frac{2x+1}{10} &= \frac{8x}{10} - \frac{2x+1}{10} \\ &= \frac{8x - (2x+1)}{10} \\ &= \frac{8x - 2x - 1}{10} \\ &= \frac{6x - 1}{10}\end{aligned}$$

Check

Substitute a value for x in the given and the answer.

$$\frac{4 \cdot 2}{5} - \frac{2 \cdot 2 + 1}{10} = \frac{8}{5} - \frac{5}{10} = \frac{16}{10} - \frac{5}{10} = \frac{11}{10}$$

$\frac{6 \cdot 2 - 1}{10} = \frac{11}{10}$. It checks.

QUESTIONS

Covering the Reading

1. A clerk has a 20-yard bolt of cloth. She first cuts r yards of cloth from the bolt and then cuts t yards from the same bolt.
 a. Express the amount of cloth left in two different ways.
 b. Let $r = 3$ yards and $t = 7$ yards to check that the two expressions in part **a** are equal.

2. You begin the day with D dollars. You spend L dollars for lunch and $10 for a book. Write two expressions for the amount of money you have left.

In 3 and 4, *multiple choice.*

3. Which of the following equals $-(x + 4)$?
(a) $-x + 4$ (b) $x + -4$ (c) $-x + -4$

4. Which expression does not equal $-(x - y)$?
(a) $-x + y$ (b) $-1x + -1y$ (c) $-x - (-y)$

In 5–10, simplify.

5. $-(x + 15)$

6. $-(4n - 3m)$

7. $x - (x + 2)$

8. $3y - 5(y + 1)$

9. $(3k + 4) - (7k - 9)$

10. $-(5 + k) + (k - 18)$

In 11–14, solve and check.

11. $-(A - 9) = 11$

12. $2 - (x + 3) = 4$

13. $12 - (2y - 4) = 18$

14. $5x - 3(5 - 2x) = -15$

In 15 and 16, simplify.

15. $\frac{8x}{3} - \frac{3x + 2}{6}$

16. $\frac{n + 1}{2} - \frac{n - 1}{3}$

Applying the Mathematics

17. Rewrite without parentheses: $-(a + 2b - c)$.

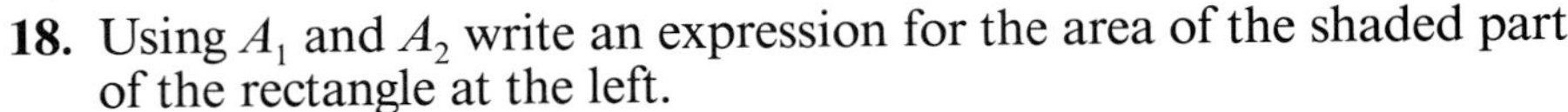

18. Using A_1 and A_2 write an expression for the area of the shaded part of the rectangle at the left.

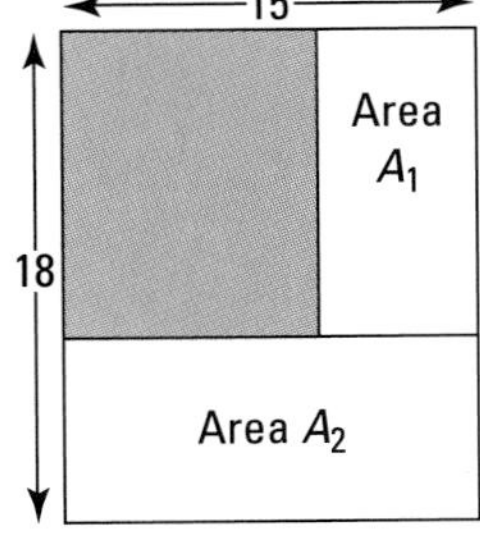

19. In baseball, the batter hits into a playing field. The foul lines form a 90° angle. Suppose that each of the four infielders can cover an angle of about $f°$. The pitcher can cover about $p°$. How much of the infield is left for the hitter to hit through?

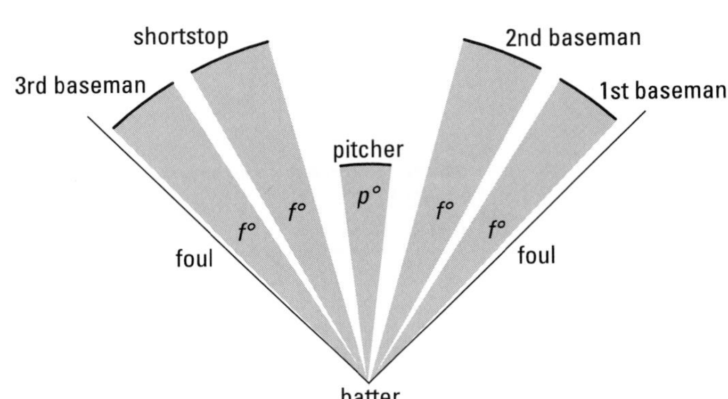

In 20 and 21, solve and check.

20. $3x - 2(x + 6.5) < 25.5$

21. $3(t + 9) - (9 + t) + 6(t + 9) = 80$

22. Create a spreadsheet with the following entries.

	A	B	C	D	E
1	Month	Income	Expenses	Profit 1	Profit 2
2		I	E	I – E	I + -E
3	January	10,312	9,080		
4	February	11,557	12,364		

a. Fill in cells D3, D4, E3, and E4 using the formulas in cells D2 and E2.
b. Compare what you get in column D to what you get in column E.
c. Explain your results for columns D and E.

Review

In 23 and 24, use these formulas relating shoe sizes S and approximate foot length L in inches. *(Lessons 1-5, 4-3)*

$$\text{for men: } S = 3L - 26$$
$$\text{for women: } S = 3L - 22$$

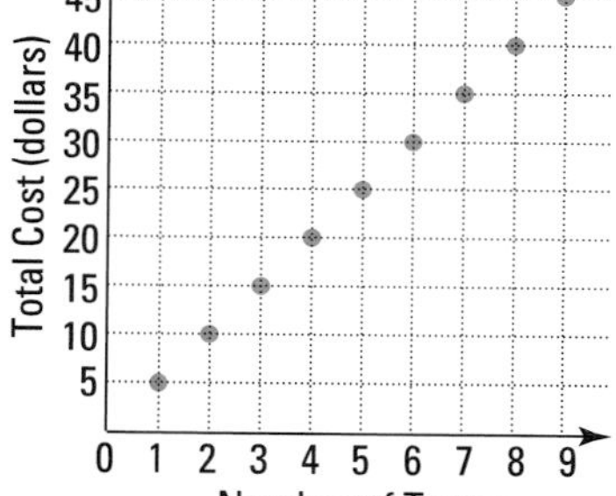

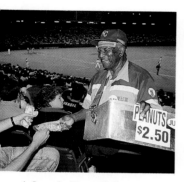

A baseball tradition. *This peanut vendor is working in Kauffman Stadium, home of the Kansas City Royals.*

23. If a man wears a size 10 shoe, about how long are his feet?

24. If a woman wears a size $6\frac{1}{2}$ shoe, about how long are her feet?

25. A person buys dinner for D dollars and has a 6% tax and a 15% tip (on the dinner only) added to the bill. Express the amount paid in terms of D. *(Lessons 3-1, 4-2)*

26. A peanut vendor at Bulldog baseball games is paid $15 a game plus $.10 for each bag sold. How many bags must be sold to earn at least $25? *(Lessons 3-8, 3-10)*

27. a. Plot the points $A = (-3, 1)$, $B = (0, 4)$, $C = (2, 6)$, and $D = (3, 7)$.
b. What pattern(s) do you notice in the graph?
c. What patterns do you see in the coordinates of the points?
(Lessons 3-3, 3-8)

Total Cost (dollars)
45 40 35 30 25 20 15 10 5
0 1 2 3 4 5 6 7 8 9
Number of Tapes

28. If the dots continue on a line in the graph at the left, how much would you pay for 20 tapes? Explain how you got your answer. *(Lesson 3-3)*

29. Make a table showing what will be printed when this program is run. *(Lesson 3-3)*

```
10 PRINT "TABLE OF (X, Y) VALUES"
20 PRINT "X VALUE", "Y VALUE"
30 FOR X = 0 TO 5
40 LET Y = 12 - X
50 PRINT X, Y
60 NEXT X
70 END
```

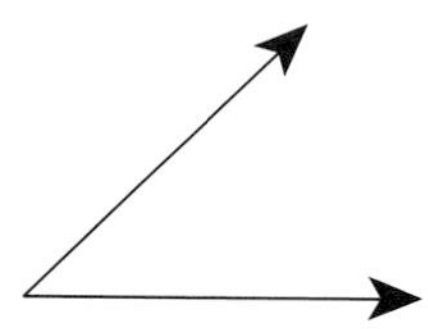

30. *Multiple choice.* What is the measure of the angle drawn at the left? *(Previous course)*
(a) 15° (b) 45° (c) 75° (d) 90°

Exploration

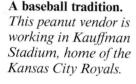

31. The difference of two numbers is subtracted from their sum. What can be said about the answer? Write a short paragraph to explain how you explored this problem.

32. A Japanese mathematics textbook for eighth-grade students has the following problem. Try to solve it.

$$\frac{x-1}{3} - \frac{2x+5}{4} > -2$$

LESSON

4-6

Graphing $x + y = k$ *and* $x - y = k$

No time to loaf. *Bakery inspectors check to ensure that goods meet production standards. If you want to buy five loaves of bread from the bakery—wheat or rye—you can show all of the possibilities by graphing.*

Graphing Constant-Sum Situations

Tim was asked to buy a dozen muffins. He was told to buy only raisin or bran muffins.

At the left below is a table of all the ways he can purchase a dozen muffins. Notice that there are 13 ordered pairs. For example he may buy 0 raisin and 12 bran, or 1 raisin and 11 bran.

The patterns in the table can be described with two variables. If Tim buys R raisin muffins and B bran muffins, then Tim's choices can be described by the equation $R + B = 12$. You could solve for B and write $B = 12 - R$. You could also solve for R and write $R = 12 - B$.

Because the solutions to the equation $R + B = 12$ are ordered pairs, they can be graphed. Below at the right is a graph of Tim's possible choices for a dozen muffins.

R Raisin	B Bran	(R, B) Ordered Pairs
0	12	(0, 12)
1	11	(1, 11)
2	10	(2, 10)
3	9	(3, 9)
4	8	(4, 8)
5	7	(5, 7)
6	6	(6, 6)
7	5	(7, 5)
8	4	(8, 4)
9	3	(9, 3)
10	2	(10, 2)
11	1	(11, 1)
12	0	(12, 0)

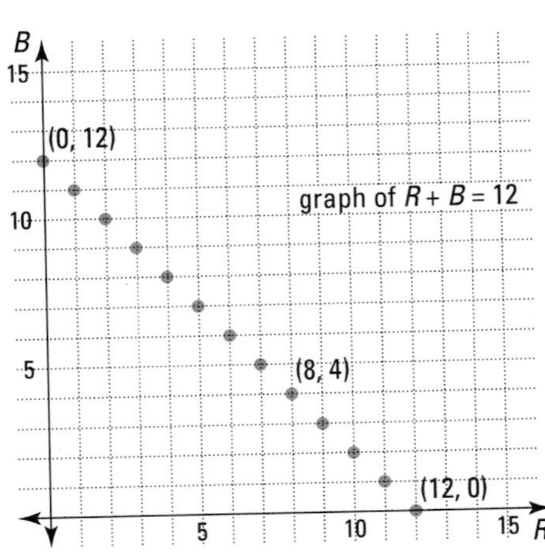

The points are all on the same line. There are 13 points on the graph because there are 13 combinations of muffins Tim can buy. He cannot buy a fraction of a muffin and he certainly can't have a negative number of muffins.

Suppose you wanted to graph *all the pairs* of numbers x and y whose sum is 12. Then you are graphing $x + y = 12$, where x and y can be any real numbers. The graph includes (15, -3), (4.5, 7.5), and infinitely many other pairs whose sum is 12. The points still all lie in a straight line. The graph of $x + y = 12$ is continuous and is the line through the points (0, 12) and (12, 0).

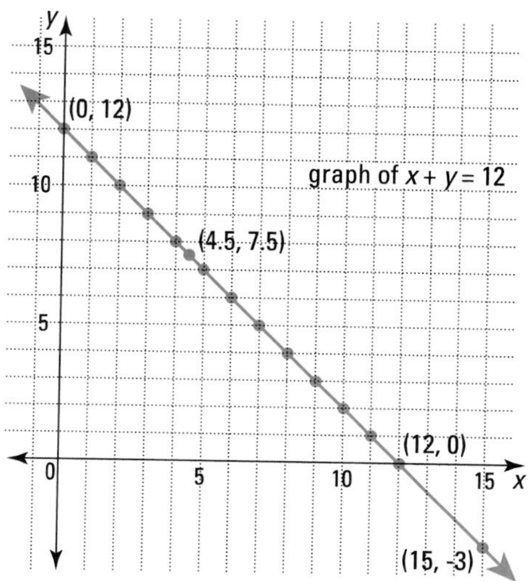

The situations leading to sentences such as $R + B = 12$ or $x + y = 12$ are called *constant-sum* situations. The pairs of numbers change, but the sum is always 12. The graph of any constant-sum situation will look much like the graph above and on page 246.

Graphing Constant-Difference Situations

There are situations in which the difference between two expressions is a constant number. The graph of a *constant-difference* situation is also a line.

Example

Graph all ordered pairs of temperatures which would produce a daily range of 20°.

Solution

Let x equal the maximum daily temperature, and let y equal the minimum daily temperature. Then the range equals $x - y$.
So $x - y = 20$.

Prepare a table with various values of x and y. You can find some ordered pairs in your head. Think of two numbers whose difference is 20, say 50 and 30. So $x = 50$ and $y = 30$ is one solution. You can find other solutions by choosing a value of x, substituting it into $x - y = 20$, and solving for y. For example, if $x = -30$, then $-30 - y = 20$, and $y = -50$. The ordered pair is (-30, -50). The table below shows six possible ordered pairs.

x	y	(x, y)
50	30	(50, 30)
40	20	(40, 20)
30	10	(30, 10)
10	-10	(10, -10)
-30	-50	(-30, -50)
-5	-25	(-5, -25)

Now plot the ordered pairs as shown in the graph below. Since x and y can be any real number, the graph should include points whose coordinates are fractions and negative numbers. Connect the points to form a line. This line has equation $x - y = 20$.

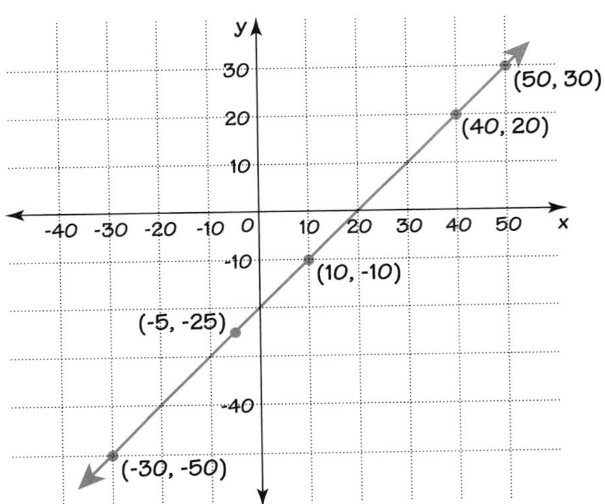

Check

Try a value of x that is not listed in the table. Let $x = 15°$. The corresponding y value on the line appears to be $-5°$. Do these values satisfy the equation?

$x - y = 15 - -5 = 20$. So, the solution checks.

When the scales on the axes are the same, the lines in constant-sum and constant-difference situations form a 45° angle with the y-axis. Later in this chapter you will study equations for other lines.

QUESTIONS

Covering the Reading

In 1–3, consider the muffin example in this lesson.

1. If Tim buys 10 raisin muffins, how many bran muffins must he buy?

2. *True or false.* The pair (0, 12) is a solution to $R + B = 12$.

3. a. How many solutions does the equation $R + B = 12$ have?
b. How is each solution represented on the graph?

4. Describe the graph of all pairs of real numbers whose sum is 12.

5. The sum of two numbers is 7.
a. Using x and y, write an equation that describes this relationship.
b. Copy and complete this table to show ordered pairs that satisfy your equation.

x	y	(x, y)
4		
3		
2		
1		
0		
−1		

c. Graph all ordered pairs of real numbers whose sum is 7.

The triple jump. *This athlete is competing in the triple jump. After a hop and a step, the athlete jumps and lands on both feet in a sandpit. The total distance covered is then measured.*

6. Suppose in the Example of this lesson that x has the values as indicated in the table below.

x	y	(x, y)
4		
5		
6		
7.5		

a. Find the corresponding values of y and (x, y).
b. Are these points on the line graphed in the Example?

7. Xandra and Yvonne each have earned medals in track-and-field events. Yvonne has 5 more medals than Xandra. Let x = the number of medals Xandra has, and y = the number of medals Yvonne has.
a. Which equation, $y - x = 5$ or $x + y = 5$, describes this situation?
b. Copy and complete the chart to show some possible numbers of medals for the girls.
c. Graph the ordered pairs from the chart.
d. In this situation does it make sense to connect the points on the graph? Why or why not?

x	y	(x, y)
1	6	
2		
3		
4		
5	10	
6		

Red-eyes. *This tropical fish is known as the red-spotted cardinal. It can be found in the warm waters of the Indian Ocean.*

Applying the Mathematics

8. In the muffin example, suppose Tim buys twice as many bran muffins as raisin muffins. Which ordered pair fits this situation?

9. Sal and Al each have tropical fish. Together they have eight fish.
  a. Make a table listing all possible number pairs of fish Sal and Al may have.
  b. If Sal has two more fish than Al, how many fish does Al have?

10. a. Copy and complete the chart below to show ordered pairs (x, y) in which the y-coordinate is the opposite of the x-coordinate.

x	y	(x, y)
-2	2	(-2, 2)
-1		
0		
1		
2		

  b. On a coordinate plane, graph all ordered pairs of real numbers for which $y = -x$.
  c. Write an equation which shows that this is a constant-sum situation.

In 11 and 12, identify all listed equations that describe the points graphed.

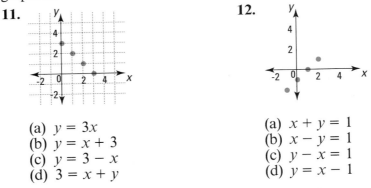

11. 

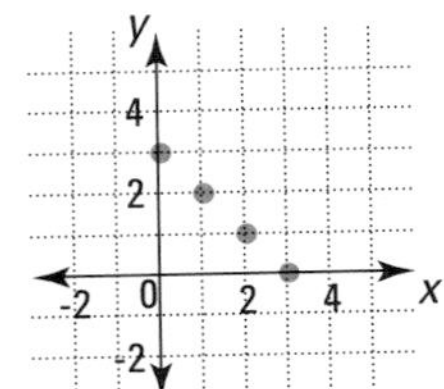

(a) $y = 3x$
(b) $y = x + 3$
(c) $y = 3 - x$
(d) $3 = x + y$

12.

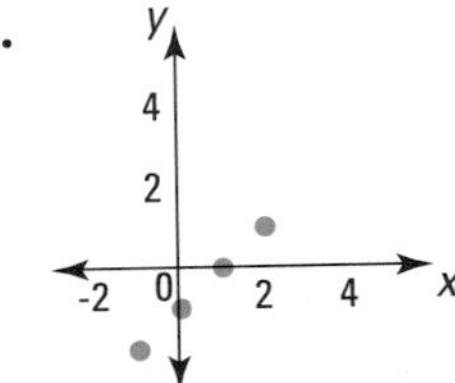

(a) $x + y = 1$
(b) $x - y = 1$
(c) $y - x = 1$
(d) $y = x - 1$

13. Calvin is giving 10 cassettes to Jorge and Maria.
  a. Make a table showing the possible ways to divide the cassettes.
  b. Graph the possibilities with the number of Jorge's cassettes on the horizontal axis and the number of Maria's cassettes on the vertical axis.

14. Make up a situation that leads to an equation of the form $x + y = 20$ in which x and y may be fractions.

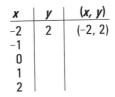

Review

In 15 and 16, simplify. *(Lesson 4-5)*

15. $12a - 3(5a + 4)$

16. $\frac{11n + 1}{5} - \frac{3n + 4}{5}$

17. Distances after various times when traveling at 55 mph are shown on the spreadsheet at the right. *(Lessons 1-5, 2-4, 4-4)*

a. Complete the table.

b. What formula could have been used to get the value in cell B6?

	A	B
1	Time (hrs)	Distance (mi)
2	1.0	
3	1.5	
4	2.0	110
5	2.5	137.5
6	3.0	165
7	3.5	

In 18 and 19, solve. *(Lessons 4-3, 4-5)*

18. $16 - 5x = 21$

19. $y - (7 - 4y) = 11$

20. The table below gives the maximum and minimum temperatures on four continents. *(Lesson 4-2)*

continent	maximum temp.	minimum temp.
Africa	58°C; El Azizia, Libya	-24°C; Ifrane, Morocco
Australia	53°C; Cloncurry, Queensland	-22°C; Charlottes Pass, N.S.W.
North America	57°C; Death Valley, CA, U.S.A.	-63°C; Snag, Yukon, Canada
South America	49°C; Rivadavia, Argentina	-33°C; Sarmiento, Argentina

Boumalne, Morocco, about 250 km from Ifrane, Morocco

a. Which of these continents has the smallest range of temperatures?

b. Which of these continents has the largest range of temperatures?

In 21 and 22, use the following information. A factory packs mixed 5-pound packages of peanuts and cashews. If c pounds of cashews are in a package, the price p of the package can be found using the formula

$$p = 2.39c + 1.69(5 - c).$$

21. If a package contains 1.5 pounds of cashews, find its price. *(Lesson 1-5)*

22. If a package costs $9.95, how many pounds of cashews does it contain? *(Lesson 3-7)*

23. *Skill sequence.* Evaluate in your head when $x = 10$. *(Lesson 1-4)*

a. x^3 **b.** $2x^3$ **c.** $\left(\frac{1}{2}x\right)^3$

24. At 3:00, what is the measure of the angle between the hands of a clock? *(Previous course)*

25. The top of a portable computer swings up 120° from the keyboard. What is the angle b between the top and the back of the case? *(Previous course)*

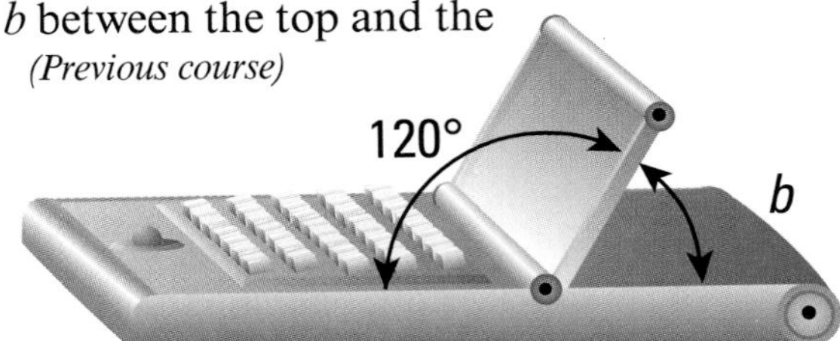

Exploration

26. a. The year 1995 roughly corresponds to the year 5755 in the Jewish calendar. The Norman conquest of England occurred in 1066. What year is 1066 in the Jewish calendar?

b. If you graph ordered pairs (J, G), where J is the year in the Jewish calendar and G the year in the Gregorian calendar (the official one in the United States), the graph is a line. Does the line slant up or down as you go to the right?

c. Look in an almanac or other reference book. What other calendars are there?

27. This graph for the number of Democrats and Republicans in the U.S. Senate is almost perfectly symmetric; that is, the graphs are nearly reflection images of each other. Write an explanation for what causes this.

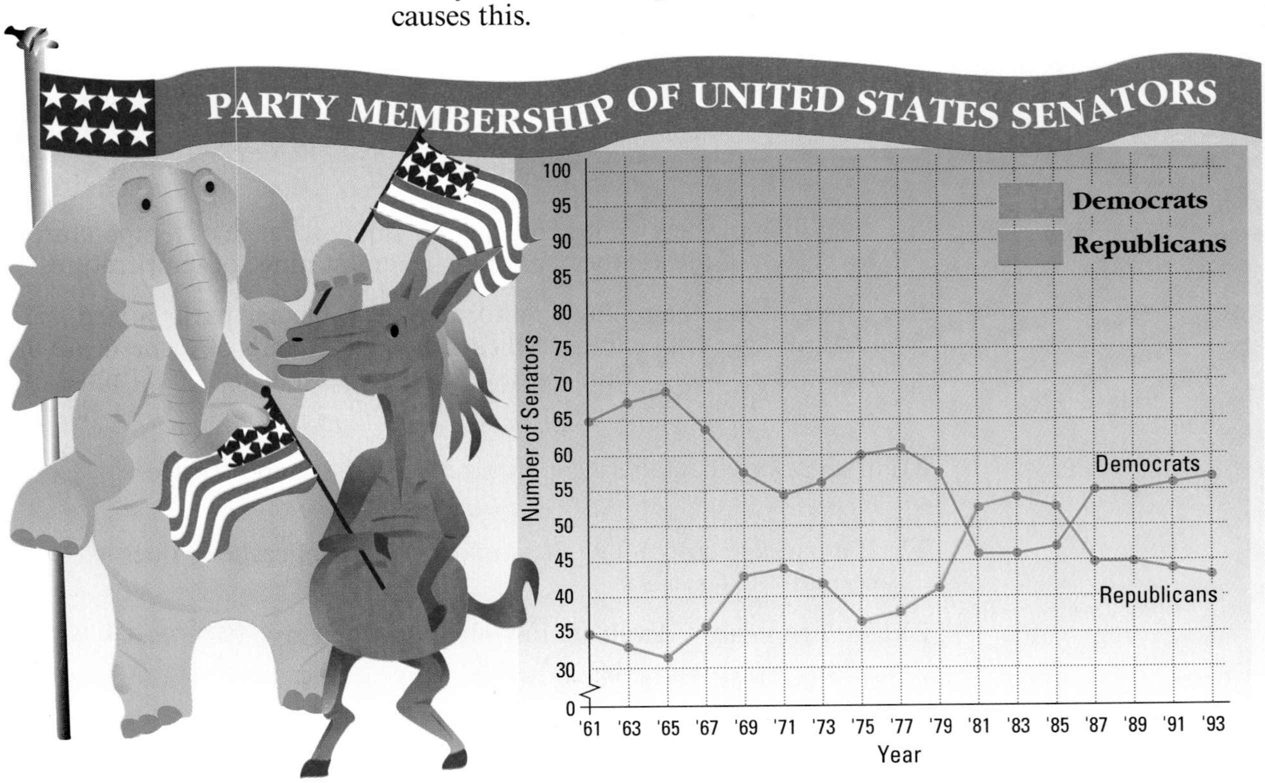

PARTY MEMBERSHIP OF UNITED STATES SENATORS

Democrats
Republicans

Number of Senators: 0, 30, 35, 40, 45, 50, 55, 60, 65, 70, 75, 80, 85, 90, 95, 100

Year: '61 '63 '65 '67 '69 '71 '73 '75 '77 '79 '81 '83 '85 '87 '89 '91 '93

LESSON

4-7

Sums and Differences in Geometry

Airport angles. *The sum of the measures x and y of the angles in the above runway is 180°.*

Two Angles Whose Sum Is 180°

Constant sums occur in many places in geometry. In each picture below, the pendulum makes two angles with the crossbar of the clock. As the pendulum swings, the measures x and y of the angles between the pendulum and the horizontal frame vary. But the sum of the measures of the angles is always 180°. That is, $x + y = 180$.

Often it is useful to solve the equation for y.

$$\begin{aligned} x + y &= 180 \\ -x + x + y &= 180 + -x \\ 0 + y &= 180 + -x \\ y &= 180 - x \end{aligned}$$

So if x is known, y can be found by subtracting x from 180.

Example 1

Using the angles shown on page 253, find *y* if *x* is 128°.

Solution 1

$$\begin{aligned} y &= 180 - x \\ &= 180 - 128 \\ &= 52 \end{aligned}$$

Solution 2

$$\begin{aligned} x + y &= 180 \\ 128 + y &= 180 \quad \text{Substitute 128 for } x. \\ y &= 52 \quad \text{Solve for } y. \end{aligned}$$

Two angles whose sum is 180° are quite common, so they are given a name: **supplementary angles** or simply **supplements.**

Below is a table of measures of some supplementary angles. If the pairs of possible measures of supplementary angles are graphed, they lie on the part of the line $x + y = 180$ that is in the first quadrant. Below on the right is a graph of these pairs. The points from the table are identified. Because the measure of an angle is always positive, neither (0, 180) nor (180, 0) is on the graph.

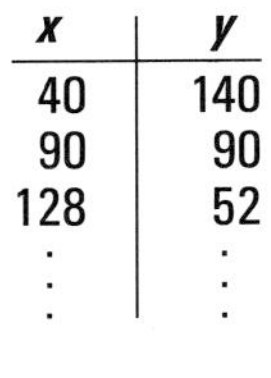

x	*y*
40	140
90	90
128	52
⋮	⋮

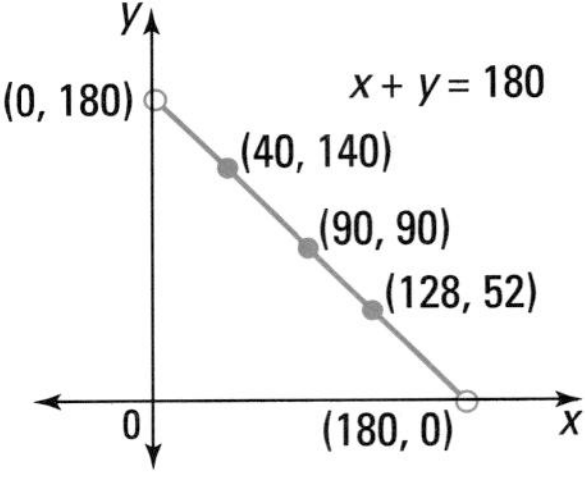

The Sum of Angle Measures in a Triangle

Another constant-sum situation involves triangles.

Activity

a. Measure all the angles in each triangle below, and find the sum of the angle measures.
b. Draw another triangle of a different size and shape. Measure its angles and find the sum of its angle measures.
c. What patterns do you notice?

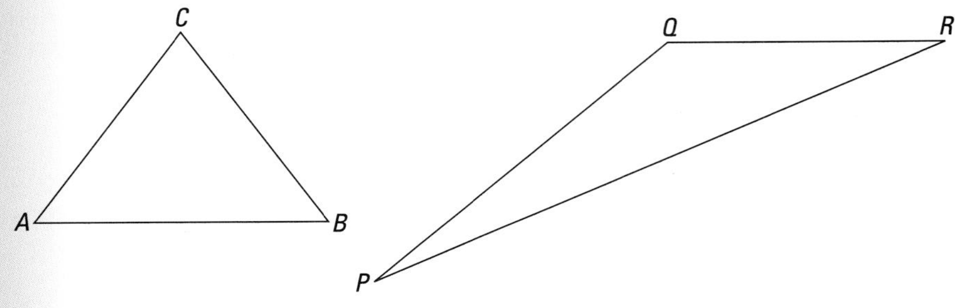

In a triangle, regardless of its shape or size, the sum of the measures of the three angles is 180. This gives a constant sum with three variables.

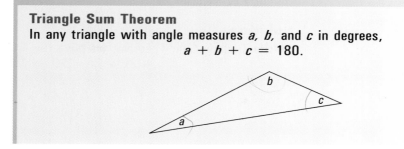

Triangle Sum Theorem
In any triangle with angle measures a, b, and c in degrees,

$$a + b + c = 180.$$

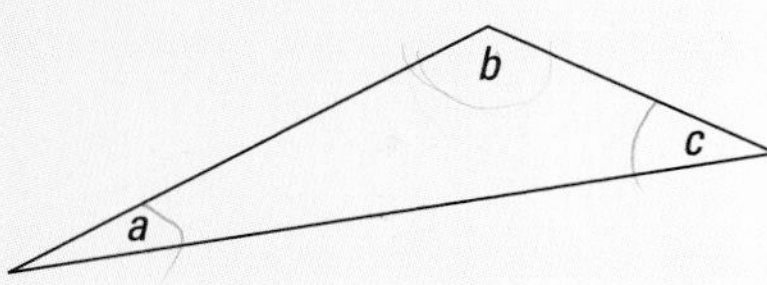

Using this theorem, if you know the measures of two angles of a triangle, you can find the measure of the third angle.

Example 2

Find the measure of the third angle in a triangle whose other angles measure 51° and 19°.

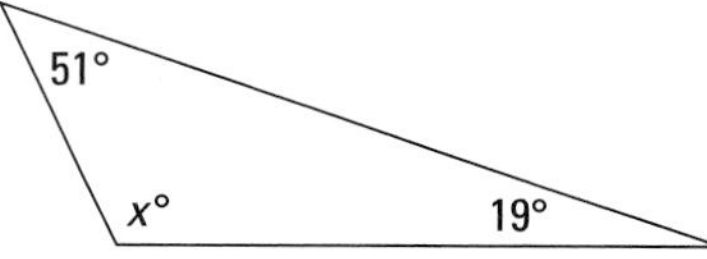

Solution

Use the Triangle Sum Theorem.

$$\begin{aligned} 51 + 19 + x &= 180 \\ 70 + x &= 180 \\ x &= 110 \end{aligned}$$

Avoiding radar. *This U.S. fighter plane is known as the* Stealth. *It is designed with small, flat surfaces (including triangles and quadrilaterals) that make it difficult for radar to detect. It has delta wings that are triangular-shaped to enable it to fly at speeds greater than Mach 1 (about 742 mph).*

Because the numbers in Example 2 are easy to add, some people solve it without paper or pencil. In Example 3, the relation between the angles is more complicated.

To save space, the following symbols are often used.

△	triangle
∠	angle
m∠	measure of angle

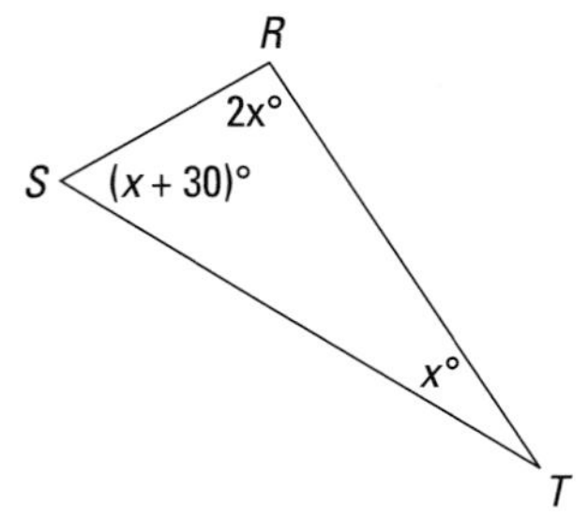

Example 3

In $\triangle RST$ at the left, the angles have measures as shown. Find the measure of each angle.

Solution

By the Triangle Sum Theorem, we have the following:

$$\begin{aligned} x + 2x + (x + 30) &= 180 \\ 4x + 30 &= 180 \\ 4x &= 150 \\ \tfrac{1}{4}(4x) &= \tfrac{1}{4}(150) \\ x &= 37.5 \end{aligned}$$

So, $m\angle T = 37.5°$,
$m\angle R = 2x° = 2(37.5°) = 75°$,
$m\angle S = (x + 30)° = (37.5 + 30)° = 67.5°$.

Two Angles Whose Sum Is 90°

Recall that in a right triangle one of the angles measures 90°. In drawings, a 90° angle is often marked with the symbol ⌐.

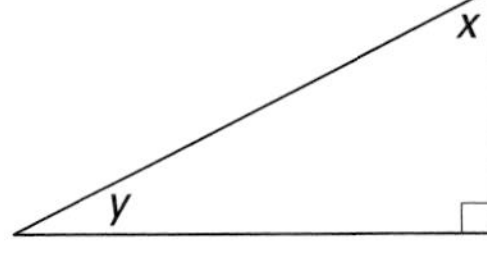

Example 4

Write a formula for y in terms of x in the triangle at the left.

Solution

Triangle Sum Theorem	$y + x + 90 = 180$
Solve for y and simplify.	$y = 180 + -90 + -x$
	$y = 90 - x$

Two angles are **complementary angles,** or **complements,** if the sum of their measures is 90°. In Example 4, the angles with measures y and x are complements.

QUESTIONS

Covering the Reading

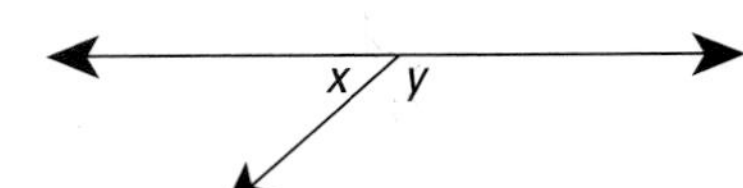

In 1–3, use the drawing at the right.

1. Write an equation that relates y and x.

2. If $x = 42°$, what is y?

3. Find x if y is 137.5°.

4. If $m\angle F = 58°$ and $m\angle G = 132°$, are $\angle F$ and $\angle G$ supplementary? Explain your reasoning.

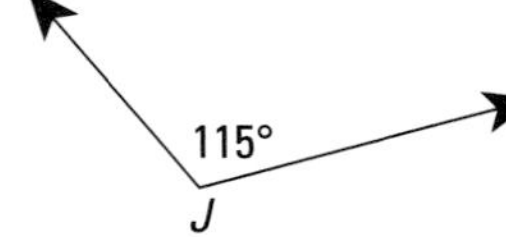

5. Find the measure of a supplement of $\angle J$, shown at the left.

6. a. What are the measures of each angle in the triangles in the Activity?
b. Describe how your results are related to the Triangle Sum Theorem.

7. Find the measure of the third angle in a triangle whose other angles measure 114° and 46°.

8. Two angles of a triangle have measures of 37° and 53°.
   a. Find the measure of the third angle.
   b. What kind of triangle is this triangle?

In 9–11, find the value of the variable.

9.

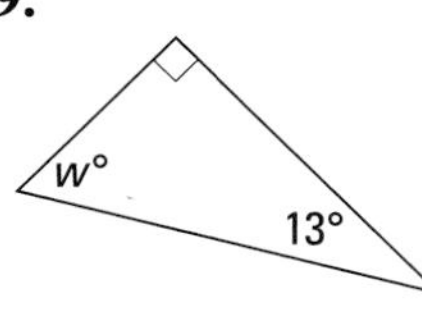

10.

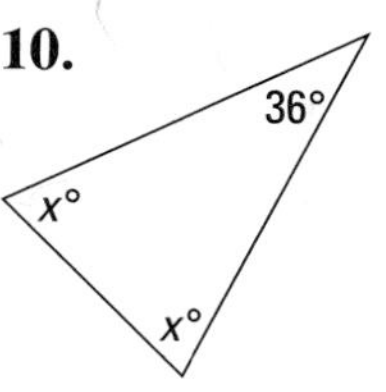

11.

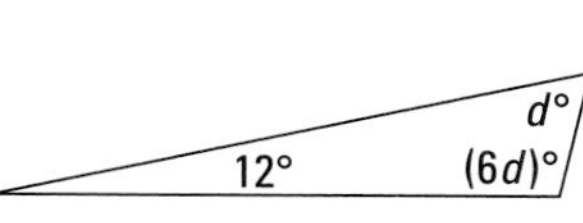

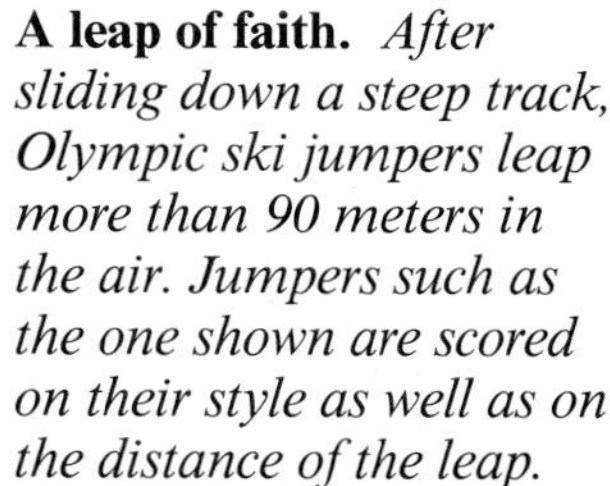

A leap of faith. *After sliding down a steep track, Olympic ski jumpers leap more than 90 meters in the air. Jumpers such as the one shown are scored on their style as well as on the distance of the leap.*

12. *True or false.* 65° and 25° are measures of complementary angles.

13. If $m\angle Q$ is 29°, what is the measure of a complement?

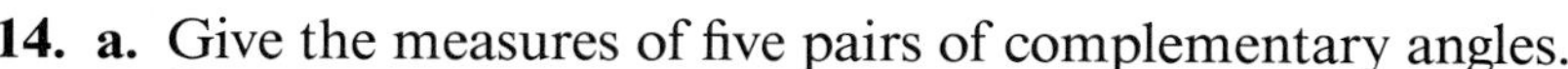

Applying the Mathematics

14. **a.** Give the measures of five pairs of complementary angles.
    b. Graph your pairs on the coordinate plane.
    c. Show the possible measures of all pairs of complementary angles on your graph.
    d. Find an equation to describe the graph in part **c.**

15. Angles A and B are complementary. If $m\angle A = x$, write an expression to represent the measure of $\angle B$.

16. **a.** Use a protractor to draw a 75° angle.
    b. Draw a complement of the angle in part **a.**
    c. Draw a supplement of the angle in part **a.**
    d. How much greater is the measure of the supplement than the measure of the complement?

17. To achieve a long jump, a ski jumper needs to lean forward quite a bit, as shown in the photo. The angle formed by the jumper's legs and the front part of the skis measures 15°. Describe and find the measure of another angle in this situation.

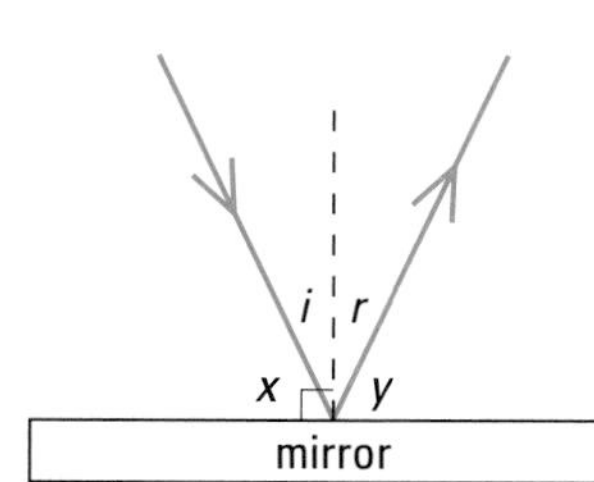

18. When a light ray strikes a smooth surface, the light ray is reflected so that the measure of the angle of incidence (i) equals the measure of the angle of reflection (r). Refer to the diagram at the left. If $i = 32°$, find the angle measures x and y.

19. Two angles in a triangle have equal measure; the third measure is 12° less than either of the others. Find the measure of each angle.

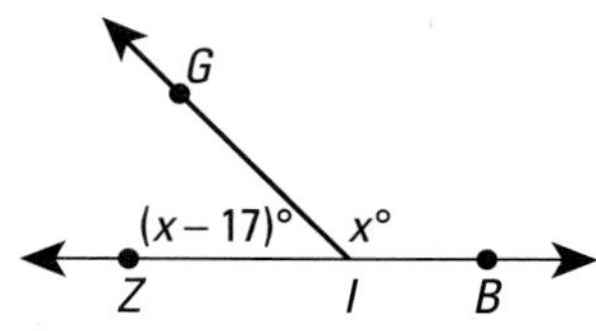

20. **a.** What is the measure of $\angle BIG$ at the left?
    b. What is the measure of $\angle ZIG$ at the left?

21. Refer to the triangle at the right. Write a formula to determine $m\angle C$ in terms of $m\angle A$ and $m\angle B$.

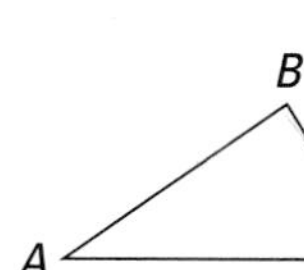

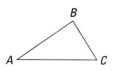

Review

22. Gordon puts \$25 in the bank each week. He puts some of the money in his savings account and some in his checking account.

a. Make a table showing 5 possibilities of his savings and checking deposits.

b. Write this situation as a constant-sum equation. Let x = amount in savings, and y = amount in checking.

c. Graph the possibilities. *(Lesson 4-6)*

23. Graph all solutions to $x - y = -5$. *(Lesson 4-6)*

In 24 and 25, solve. *(Lessons 3-10, 4-5)*

24. $50 + x > 30$

25. $12n + 5 - (2n + 20) = -20$

26. *Multiple choice.* Which is not equal to $w + -k$? *(Lesson 4-1)*
(a) $w - k$ (b) $k - w$ (c) $-k + w$

In 27 and 28, simplify. *(Lessons 3-9, 4-1)*

27. $\frac{1}{2}x - \frac{1}{3}x - \frac{3}{4}x$

28. $\frac{y}{2} - \frac{y}{3} - \frac{3y}{4}$

29. Suppose the cost to repair a refrigerator is \$50 for the service call plus \$20 for each half hour of labor.

a. Make a table showing total charges for repairs which take $\frac{1}{2}$, 1, $1\frac{1}{2}$, 2, $2\frac{1}{2}$, and 3 hours.

b. Write a formula which gives the total cost c in terms of the number n of half-hour periods. *(Lesson 3-8)*

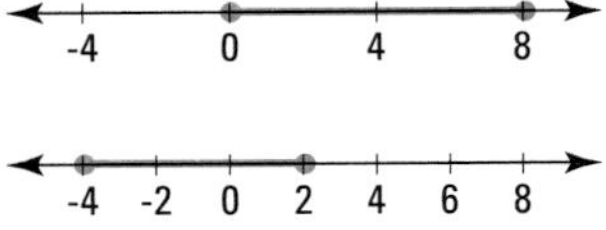

In 30 and 31, consider the two segments pictured at the left. *(Lesson 1-3)*

30. Draw their union, and describe it with an inequality.

31. Draw their intersection and describe it with an inequality.

Exploration

32. Below is a map of Centerville. John lives at the corner marked J. His friend Mary lives by the corner marked M. John and Mary each walk to the theater. The sum of the distances they walk is 6 blocks. Copy the diagram and show all possible locations of the theater.

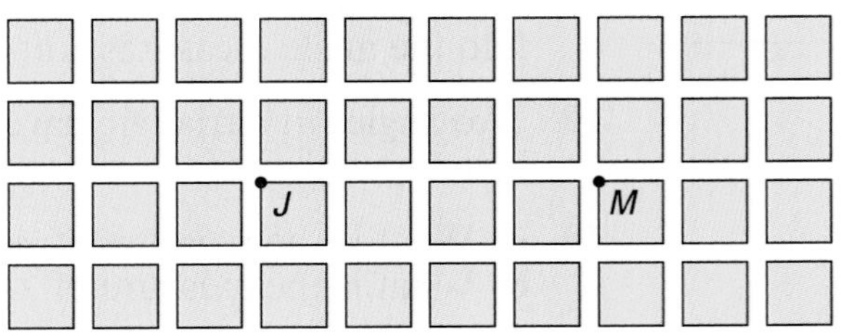

Introducing Lesson 4-8

The Triangle Inequality

IN-CLASS ACTIVITY

Work in small groups. Cut straws into the following lengths:

1 in., 2 in., 2 in., 3 in., 3 in., 4 in., 5 in., 6 in.

Then make a table like the one shown below.

1 Place the three straws listed in each of parts **a–h** on your desk. With the straws touching only at the ends, try to make a triangle. In the second column of the table below, indicate whether you could make a triangle.

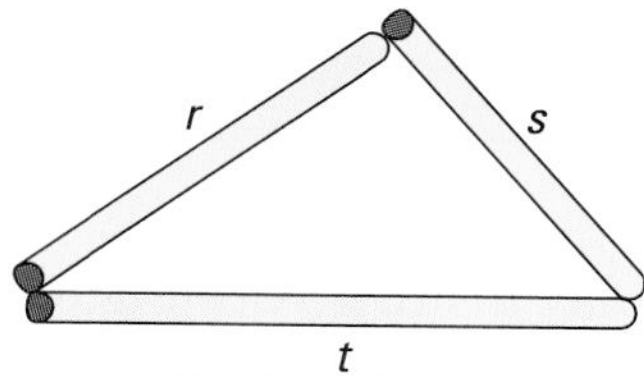

This is a triangle.

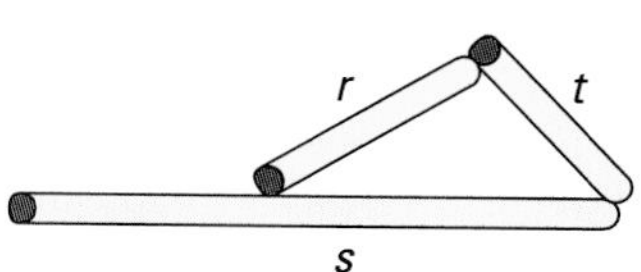

This is not a triangle.

	Length of Straws r	s	t	Triangle? (Yes or No)	Compare, Use <, >, or =. $r + s$ _?_ t	$s + t$ _?_ r	$r + t$ _?_ s
a.	1 in.	2 in.	2 in.				
b.	3 in.	3 in.	6 in.				
c.	3 in.	6 in.	1 in.				
d.	2 in.	3 in.	4 in.				
e.	2 in.	2 in.	3 in.				
f.	5 in.	2 in.	3 in.				
g.	3 in.	4 in.	5 in.				
h.	1 in.	2 in.	5 in.				

Now complete the last three columns of the table in which the sum of the lengths of two sides is compared to the length of the third side. Discuss the results with your group.

2 ***Draw conclusions.*** Complete the following statement using one of the following:

less than, greater than, or *the same as.*

The sum of the lengths of two sides of any triangle must be _?_ the length of the third side.

LESSON

4-8 The Triangle Inequality

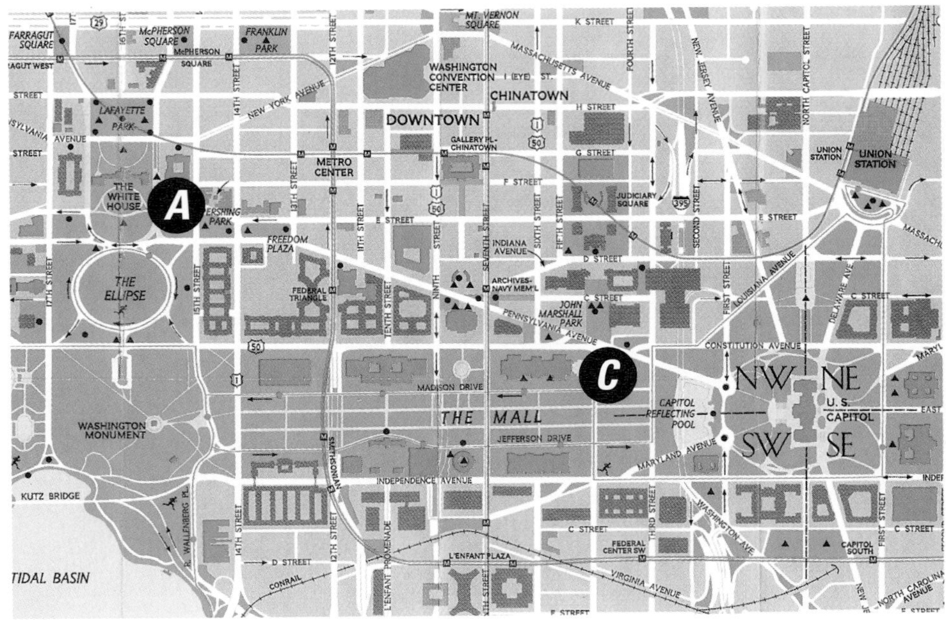

Some students are visiting Washington, D.C. They have just toured the White House, and are standing where point A is located on the map. Their next tour is at the U.S. Capitol building. They agree to split into small groups and to meet in an hour at the Mall on the corner of Pennsylvania Avenue and 4th Street, point C on the map. Notice that the students have their choice of many paths from A to C.

What Is the Triangle Inequality?

A fundamental property of distance is that the shortest path from A to C is along the line segment connecting points A and C. We denote that segment as $\overline{AC}$ and its length as the number AC. On the map $\overline{AC}$ corresponds to the direct route from A to C along Pennsylvania Avenue.

Suppose each group decides to stop for lunch at some point B. Now compare the distance AC with the length of a path from point A to a selected point B and then to point C, where B is any point other than A or C. There are two possibilities, as shown.

(1) B is on $\overline{AC}$.

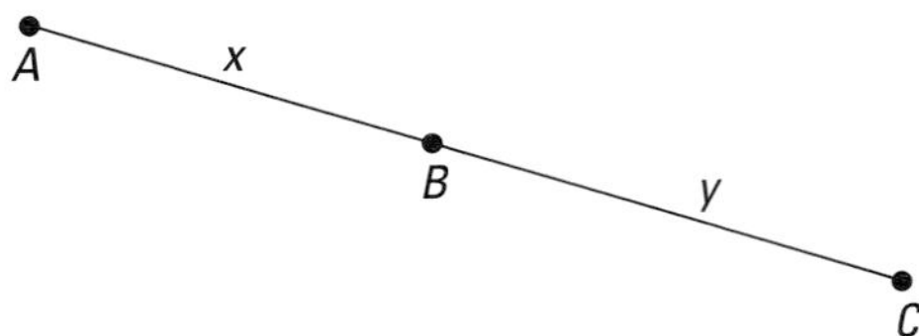

Then $AB + BC = AC$

$x + y = AC$

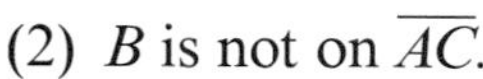

(2) B is not on $\overline{AC}$.

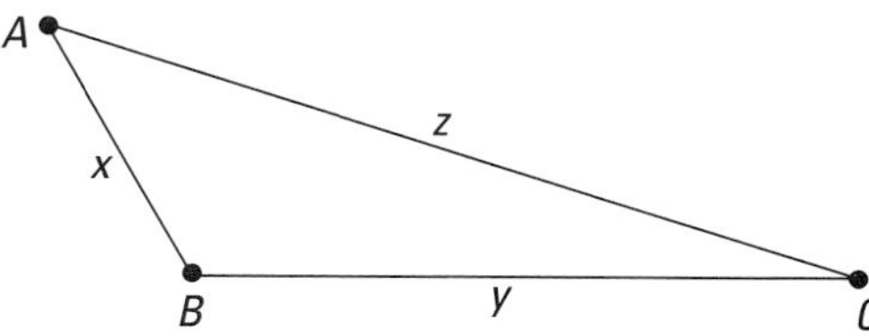

Then $AB + BC > AC$

$$x + y > z$$

That is, if you stop for lunch at any point B that is not between A and C on Pennsylvania Avenue, you will be going out of your way.

These ideas are summarized in the following general properties, known together as the *Triangle Inequality.*

Pennsylvania Avenue. *Shown is a view of the Capitol Building as seen from the White House along Pennsylvania Avenue.*

Triangle Inequality
Part 1: If A, B, and C are any three points, then $AB + BC \geq AC$.

Part 2: If A, B, and C are vertices of a triangle, then $AB + BC > AC$.

Part 2 of the Triangle Inequality can be stated as follows: the sum of the lengths of two sides of any triangle is greater than the length of the third side. This is the relation you found in the Activity on page 259.

In $\triangle ABC$ as drawn above with sides of length x, y, and z, the following three inequalities are true:

$$x + y > z \qquad \text{and} \qquad x + z > y \qquad \text{and} \qquad y + z > x.$$

How Is the Triangle Inequality Used?

A surprising application of the Triangle Inequality is that you can calculate the possible lengths of the third side of a triangle if you know the lengths of the other two sides.

Example 1

Suppose two sides of a triangle have lengths 42 and 30. What are the possible lengths of the third side?

Solution 1

Draw a picture, naming the sides x, y, and z.

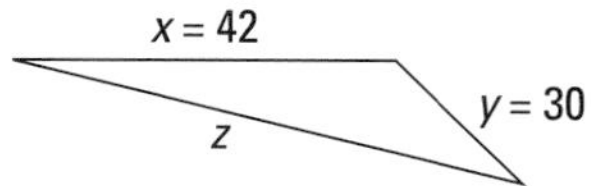

By the Triangle Inequality, the following are true:

$$x + y > z \qquad \text{and} \qquad x + z > y \qquad \text{and} \qquad y + z > x.$$

Now substitute 42 for x and 30 for y, and solve the three inequalities.

$42 + 30 > z$ and $42 + z > 30$ and $30 + z > 42$

$z < 72$ and $z > -12$ and $z > 12$

The solution to the problem is the intersection of these three inequalities.

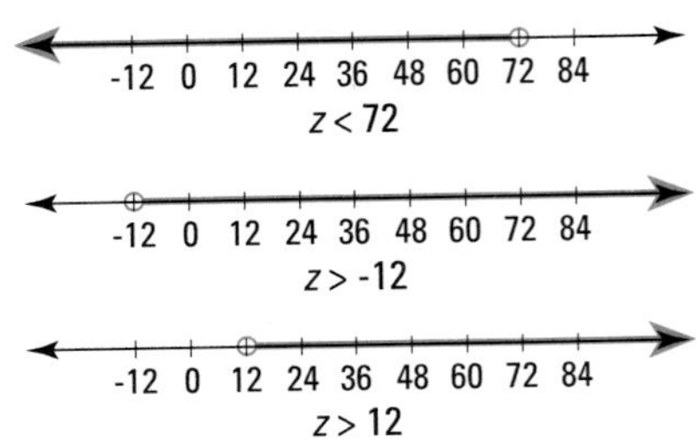

This indicates that z is between 12 and 72. The third side can have lengths between 12 and 72.

Solution 2

Imagine that the side of length 42 is a wall and the side of length 30 is a door. The segment that connects their ends along the floor is the third side of a triangle. Imagine how that length changes as the door swings.

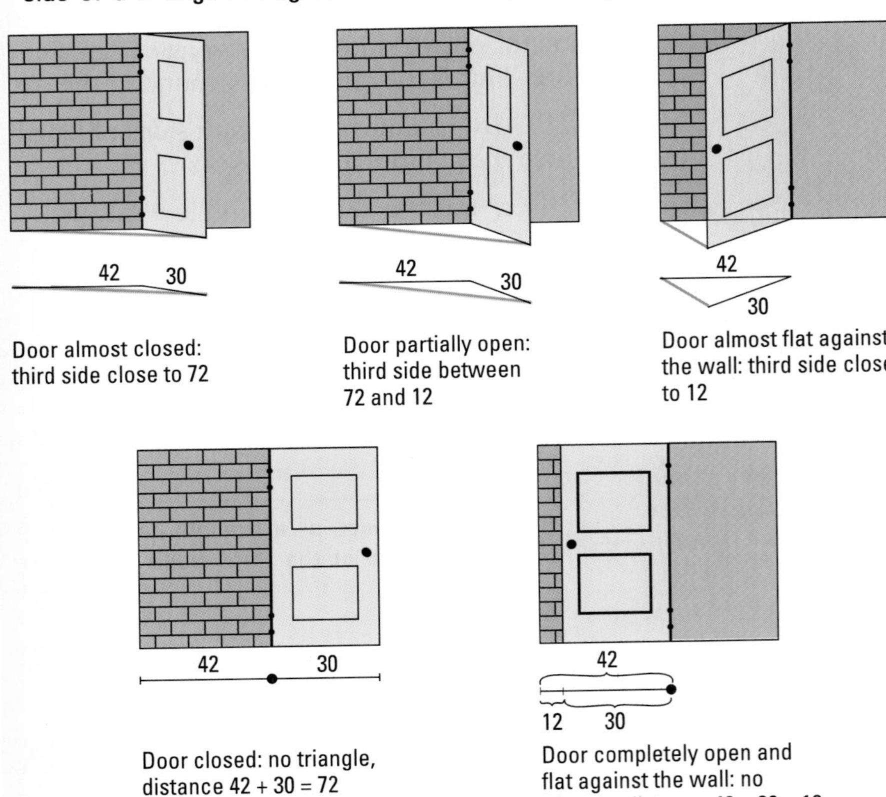

Door almost closed: third side close to 72

Door partially open: third side between 72 and 12

Door almost flat against the wall: third side close to 12

Door closed: no triangle, distance 42 + 30 = 72

Door completely open and flat against the wall: no triangle, distance 42 – 30 = 12

So when a triangle is formed, the length of the third side is between 12 and 72.

Notice in Example 1 that $42 + 30 = 72$ and $42 - 30 = 12$. That is, the possible lengths of the third side of the triangle are between the *sum* and the *difference* of the two given sides. We can use algebra to prove that this is *always* the case. Consider a triangle with two sides x and y. We want to describe all possible lengths z of the third side.

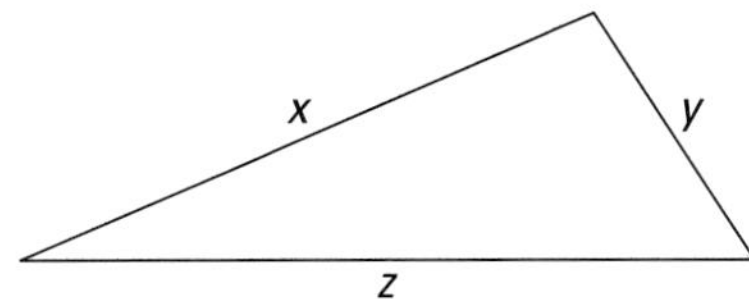

(For convenience, we assume that if x and y differ, x is the longer side.) The Triangle Inequality tells us that the sum of two sides is greater than the third. So

$$y + z > x \quad \text{and} \quad x + y > z \quad \text{and} \quad z + x > y.$$

Now solve these inequalities for z.

$$z > x - y \quad \text{and} \quad z < x + y \quad \text{and} \quad z > y - x$$

The third inequality, $z > y - x$, yields no new information because if x is the longest side, $y - x$ is negative. We summarize the relation as follows.

Third Side Property
If x and y are the lengths of two sides of a triangle, and $x > y$, then the length z of the third side must satisfy the inequality

$$x - y < z < x + y.$$

Example 2

Two sides of a triangle are 4 cm and 6 cm. How long is the third side?

Solution

Use the Third Side Property. Let $x = 6$ and $y = 4$. (Remember that x is the longer side.) The third side has a length z that satisfies

$$6 - 4 < z < 6 + 4.$$

That is, $2 < z < 10$.

The third side can be any length between 2 cm and 10 cm.

The Third Side Property enables you to obtain information involving unknown distances. In some situations you must deal with three points, and you do not know if they form a triangle or if they all lie on a line.

Example 3

May lives one mile from school and 0.4 mile from Larry. How far does Larry live from school?

Solution

Let d be the distance from Larry's house to school. Think of the extreme cases first. If the houses and the school lie on a line, the distance from Larry's house to school can be as large as $1.0 + 0.4 = 1.4$ miles or as small as $1.0 - 0.4 = 0.6$ mile.

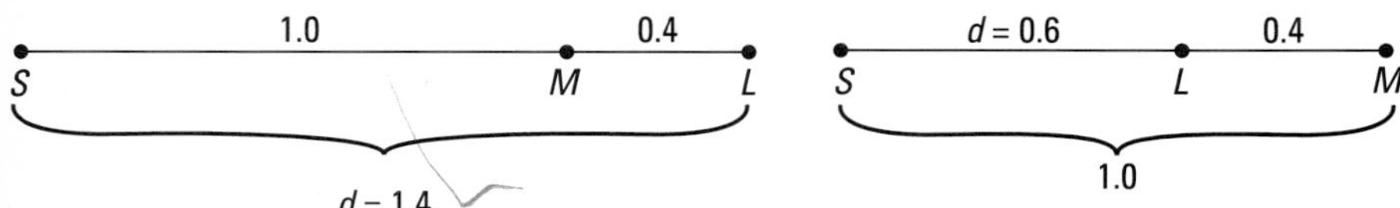

If the houses and school do not lie on a line, they form a triangle, and by the Third Side Property, $0.6 < d < 1.4$.

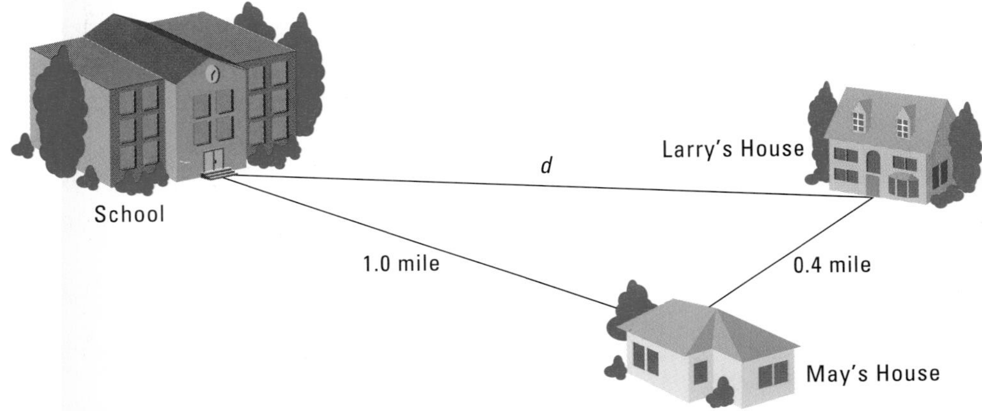

So $0.6 \leq d \leq 1.4$. Larry lives from 0.6 to 1.4 miles from school.

QUESTIONS

Covering the Reading

1. In the drawing below, A is on $\overline{BC}$. How long is $\overline{BC}$?

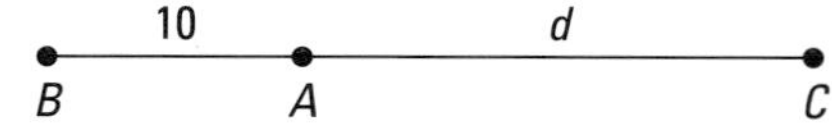

2. If M is between P and Q on a line, $PM = 5$, and $PQ = 7$, what is MQ?

3. State the Triangle Inequality.

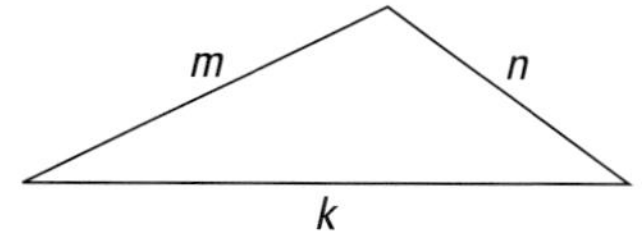

4. Refer to the triangle at the left. Copy and complete.
   a. $k + n >$ ___?___
   b. $n + m >$ ___?___
   c. ___?___ + ___?___ $> n$

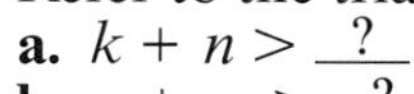

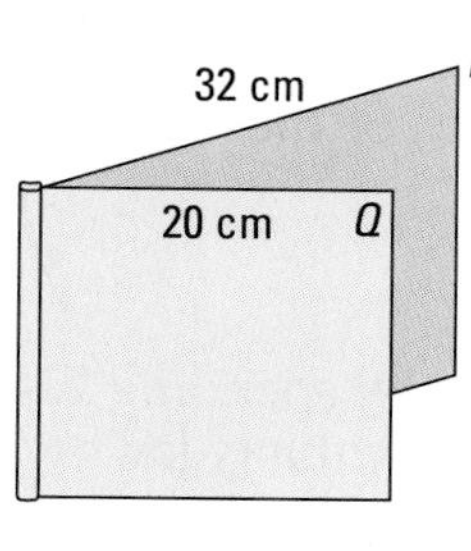

5. Two metal plates are joined by a hinge as in the drawing at the left.
   a. PQ can be no shorter than _?_ cm.
   b. PQ can be no longer than _?_ cm.

6. Refer to the triangle at the right.
   a. x must be less than _?_.
   b. x must be greater than _?_.
   c. _?_ $< x <$ _?_.

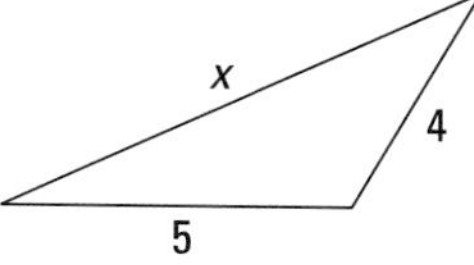

7. If two sides of a triangle are 6.5 cm and 8.2 cm, what inequality must the length of the third side satisfy?

8. Suppose Larry lives 1.3 km from school and 0.8 km from May. The distance May lives from school must be greater than or equal to _?_ but less than or equal to _?_.

Applying the Mathematics

9. In an *isosceles triangle,* at least two sides are the same length. If the congruent sides are each 3 in. long, how long can the third side be?

10. If two sides of a triangle have lengths $100x^2$ and $75x^2$, the third side can have any length between _?_ and _?_.

11. Why is there no triangle with sides of lengths 1 cm, 2 cm, and 4 cm?

A, B, C; $8x + 1$; 223; $13x + 12$

12. a. Refer to the triangle at the left. Use the Triangle Inequality to write a sentence using x that compares the sum of AB and BC with AC.
    b. Solve the inequality for x.

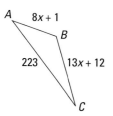

13. Sirius, the brightest star in the nighttime sky, is 8.7 light-years from Earth. Procyon, a bright star near Sirius, is 11.3 light-years from Earth. Let m be the distance between Sirius and Procyon. Knowing only this information, what are the smallest and largest values that m can have?

14. Betty can walk to school in 25 minutes. She can walk to her boyfriend's house in 10 minutes. Let t be the length of time for her boyfriend to walk to school. If they walk at the same rate, what are the largest and smallest possible values of t?

15. A road atlas states that the driving mileage from Boston to Los Angeles is 3028 miles. Do you think that the flying distance is greater or less than the driving distance? Explain why.

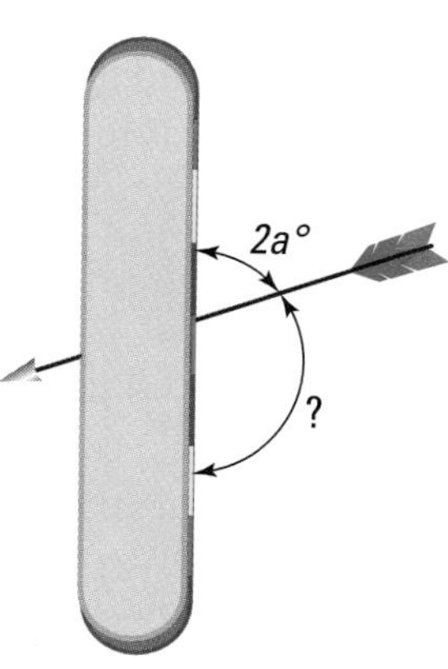

Review

16. An arrow enters a target at an angle of $2a°$. What is the measure of the supplementary angle as shown at the left? *(Lesson 4-7)*

17. One angle in a triangle is twice the measure of the smallest angle. The third angle is four times the measure of the smallest angle. What are the measures of the three angles? *(Lesson 4-7)*

18. In a right triangle, one of the acute angles measures $5b°$. The other measures $4b°$. What are the numerical measures of these two angles? *(Lesson 4-7)*

19. Plot the line $x + y = 7$. *(Lesson 4-6)*

20. Suppose two girls start a lawn-mowing business by spending $200 for a mower and then earn $15 per lawn cut. How many lawns do they need to mow if they wish to earn a profit of at least $350? *(Lesson 3-10)*

21. Simplify $4x + -3(u + 2x + v^2) + 2x$. *(Lessons 3-6, 3-7)*

22. *Skill sequence.* Simplify. *(Lessons 2-1, 3-6)*
   a. $p + p + p$
   b. $p \cdot p \cdot p$

23. Find the image of the point $P = (3, 6)$ after a slide of 4 units to the right and 6 units up. *(Lesson 3-4)*

24. The graph below gives the life expectancy at birth for four groups of Americans with years of birth from 1920 to 1990. *(Lesson 3-3)*

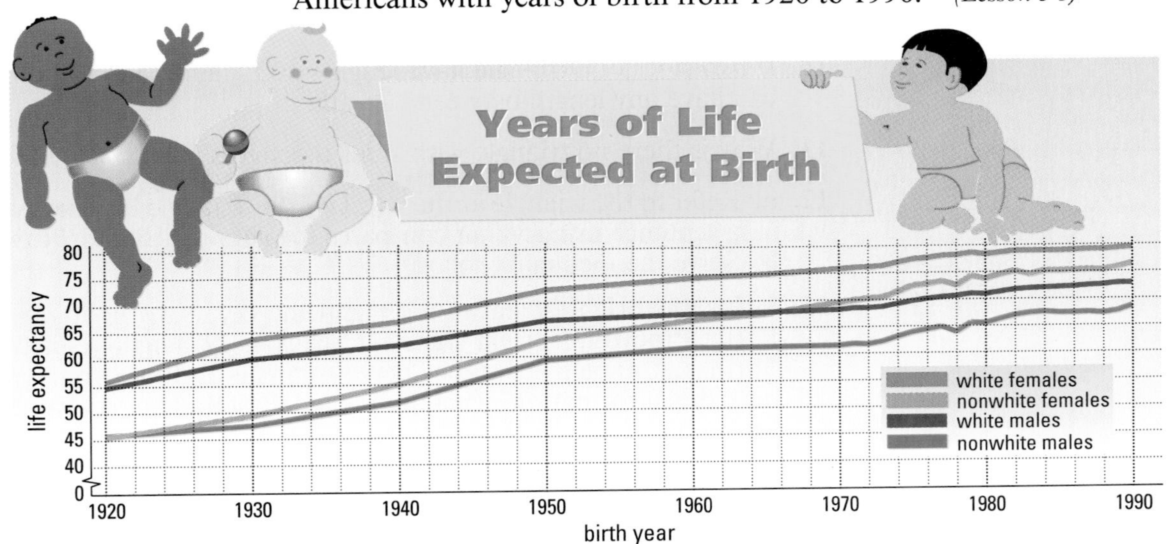

Source: *The World Almanac and Book of Facts, 1994*

   a. Which group has had the greatest increase in life expectancy since 1920?
   b. In your year of birth, how long could a newborn be expected to live?
   c. Identify three or more conclusions that you can make from studying the graph.

Exploration

25. The perimeter of a triangle is 15. The sides all have different integer lengths. How many different combinations of side lengths are possible? (Hint: The answer may be fewer than you think. Use trial and error and the Triangle Inequality.)

LESSON

4-9 Graphing Linear Patterns

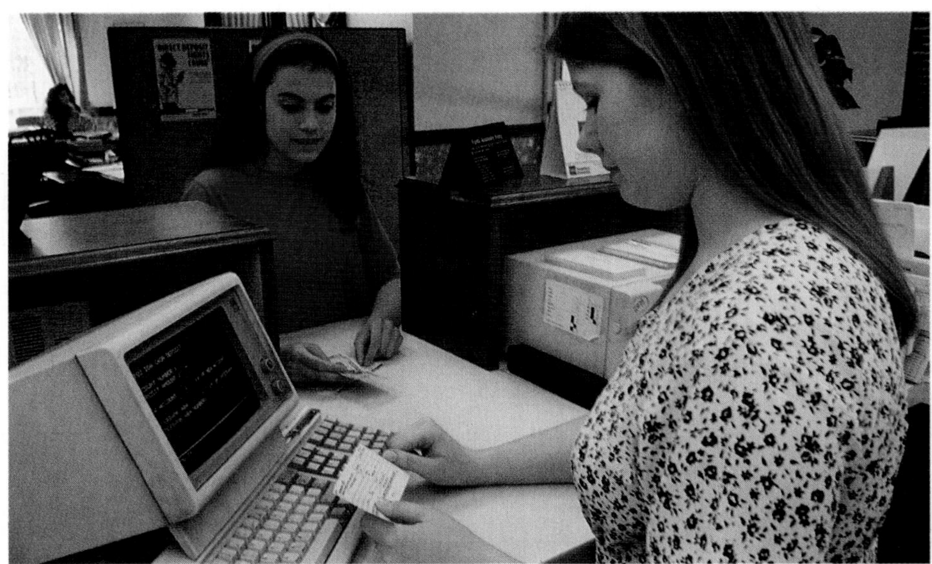

Profitable practice. *Shown is Beth making her weekly deposit of $5 into a savings account. By saving money at a constant rate and by using a table or making a graph, Beth can accurately predict her bank balance at any given time.*

Graphing a Constant-Increase Pattern

Suppose Beth begins with $10 in the bank and adds $5 to her account each week. After w weeks she will have $10 + 5w$ dollars. Let t be the total amount in her account at the end of w weeks. Beth's bank balance can be described in three different ways.

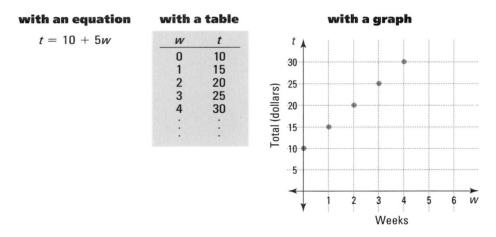

with an equation

$t = 10 + 5w$

with a table

w	t
0	10
1	15
2	20
3	25
4	30
⋮	⋮

with a graph

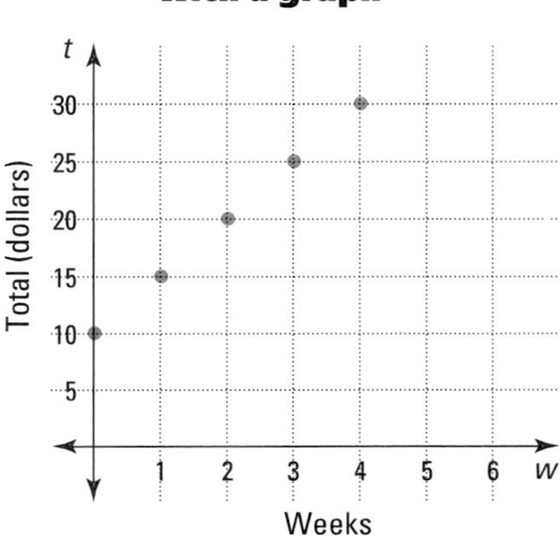

The table lists the ordered pairs (0, 10), (1, 15), (2, 20), (3, 25), and (4, 30). All these pairs make the equation $t = 10 + 5w$ true. The equation $t = 10 + 5w$ is called a **linear equation** because all the points of its graph lie on the same line. Because Beth puts money into her account at specific whole-number intervals, the domain of w is the set of whole numbers. It does not make sense to connect the points, because numbers such as $2\frac{1}{2}$ are not in the domain of w.

Graphing a Constant-Decrease Pattern

Other situations lead to equations whose variables may take on any real numbers as values. Their graphs are straight lines.

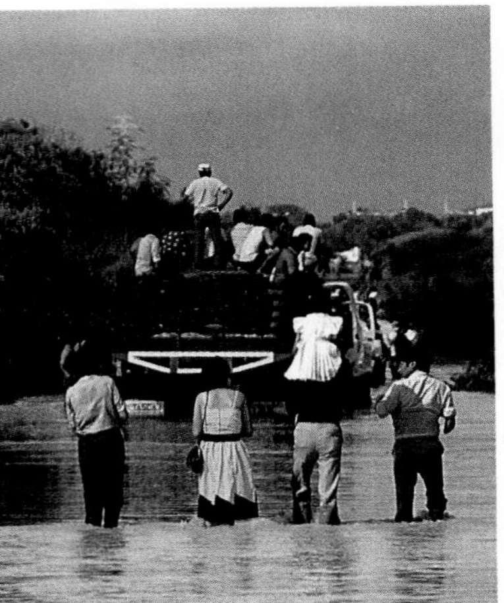

Flash flood. *Heavy rains can produce flash floods if small rivers or streams rise suddenly and overflow. Shown is a flooded road in Zulia State, Venezuela.*

Example 1

A flooded stream is now 14 inches above its normal level. The water level is dropping 2 inches per hour. Its height y in inches above normal after x hours is given by the equation $y = 14 - 2x$. Graph this relationship.

Solution

Find the height at various times and make a table. Below is a table for 0, 1, 2, 3, and 4 hours.

hour (x)	height (y = 14 − 2x)	ordered pair (x, y)
0	14 − 2 · 0 = 14	(0, 14)
1	14 − 2 · 1 = 12	(1, 12)
2	14 − 2 · 2 = 10	(2, 10)
3	14 − 2 · 3 = 8	(3, 8)
4	14 − 2 · 4 = 6	(4, 6)

Plot the ordered pairs in the table and look for patterns. You should see that the five points lie on the same line. Now think about the domain of x. Time in hours can be any nonnegative real number, such as $1\frac{3}{4}$ or $3\frac{1}{2}$. This means that other points lie between the ones you have already plotted. So, draw the line through them for $x \geq 0$.

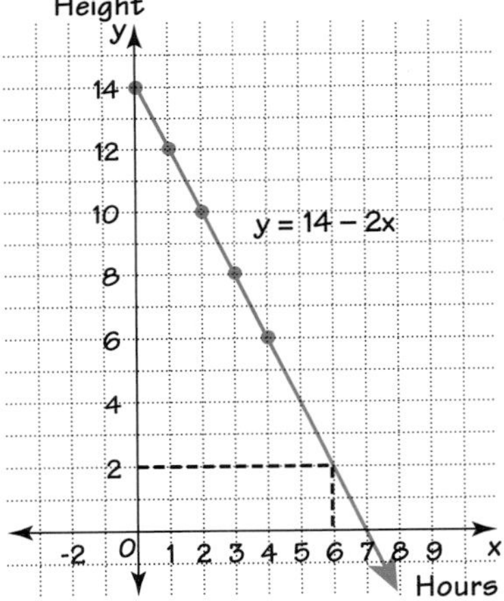

Once the graph of an equation has been drawn, you can use it to answer questions about its points.

Example 2

In Example 1, how many hours will it take for the water level to fall to 2 inches above normal?

Solution 1

Look at the graph on page 268. The height of the water above normal level is given by y. Find the point for 2 inches on the y-axis, and then look across to find the point on the graph that has this height. The x-coordinate of this point is 6. This is shown by the dashed path on the graph. The water will be 2 inches above normal after 6 hours.

Solution 2

Use the equation $y = 14 - 2x$. The height "2 inches above normal" means that $y = 2$. So solve $2 = 14 - 2x$.

$2 = 14 - 2x$	
$-12 = -2x$	Add -14 to each side.
$6 = x$	Multiply both sides by $-\frac{1}{2}$.

Graphing a Linear Equation

If you are not told the domain of a variable in a linear pattern, you should assume that any real number may be substituted for x. The graph will then be a straight line.

Example 3

Draw the graph of $y = -2x + 1$.

Solution

You can choose *any* values for x. We choose -1, 0, 1, 2, 3, and 4. Make a table of solutions. Plot the points. Because there are no restrictions on the values of x and y, they may be any real numbers. Draw a line through the points.

x	-2x + 1 = y
-1	$-2 \cdot -1 + 1 = 3$
0	$-2 \cdot 0 + 1 = 1$
1	$-2 \cdot 1 + 1 = -1$
2	$-2 \cdot 2 + 1 = -3$
3	$-2 \cdot 3 + 1 = -5$
4	$-2 \cdot 4 + 1 = -7$

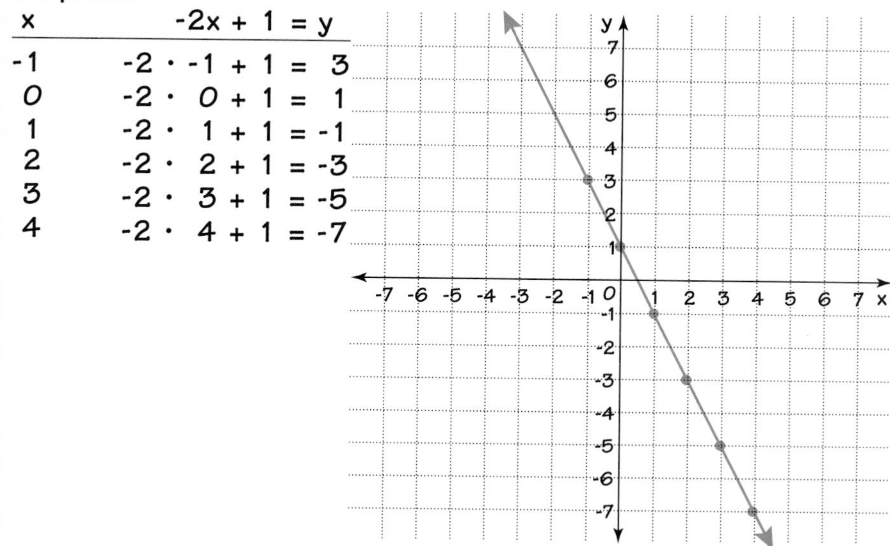

You have seen that a computer can be used to make tables. The following program will print a table of values similar to the one calculated on page 269. Lines 30–60 use a FOR/NEXT loop in which the first time through the loop, x is -1.

```
10 PRINT "TABLE OF (X,Y) VALUES"
20 PRINT "X VALUE", "Y VALUE"
30 FOR X = -1 TO 4
40 LET Y = -2 * X + 1
50 PRINT X,Y
60 NEXT X
70 END
```

When the program is run the computer will print the following.

```
TABLE OF (X, Y) VALUES
X VALUE          Y VALUE
  -1                3
   0                1
   1               -1
   2               -3
   3               -5
   4               -7
```

You can change line 30 to specify a different set of x values for your table. You can change line 40 to specify a different equation.

A spreadsheet can also be used to make a table of values for an equation. Many spreadsheet programs will automatically make a scatterplot for any identified set of ordered pairs, as well. Here is how one spreadsheet can be started.

	A	B
1	x	y
2	-1	= -2*A2+1
3	=A2+1	
4		
5		
6		

To fill in the columns, copy A3 down the A column as far as you want. Then copy B2 down the B column. The result should be the same as found with the computer program.

QUESTIONS

Covering the Reading

1. Consider Beth's savings plan described in the lesson.
   a. How much money will she have in her account after 3 weeks?
   b. How much money will she have in her account after 6 weeks?

2. Suppose Miguel begins with \$5 in an account and adds \$2 per week.
   a. Copy and complete the chart below, showing t, the total amount Miguel will have at the end of w weeks.

weeks (w)	total (t)
0	
1	
2	
3	
4	

   b. Graph the ordered pairs (w, t). That is, plot w along the horizontal axis, and t along the vertical axis.
   c. Write an equation that represents t in terms of w.
   d. What is the domain of w?

In 3–6, refer to Examples 1 and 2.

3. After how many hours will the stream be 10 inches above normal?

4. How high above normal will the stream be after 4 hours?

5. After how many hours will the stream level be back to normal? (Hint: y will be equal to zero.)

6. What do the points in Quadrant IV on the graph represent?

In 7 and 8, refer to Example 3.

7. **a.** When $x = \frac{1}{2}$, what is y?
   b. Will this point lie on the line that is graphed?

8. Find the x-coordinate of the point on the graph which has a y-coordinate of -2.

9. Refer to the computer program after Example 3.
   a. Rewrite line 30 so that ordered pairs are printed for $x = 0, 1, 2, 3, 4, 5, 6$, and 7 when the program is run.
   b. Rewrite the program to print a table of values for $y = 8x - 3$ from $x = -5$ to $x = 5$.

10. Refer to the spreadsheet in the lesson.
   a. Describe how you would modify the spreadsheet to show values for $x = 0, 2, 4, 6, 8$, and 10.
   b. What would change if you wanted a table of values for $y = 3x + 40$?

World's tallest trees. *The redwood tree stump shown below is in Myers Flat, California. You can tell how fast this tree grew by examining the width of its rings. Redwoods often grow 200 to 275 feet high and have trunks that are 8 to 12 feet in diameter.*

Applying the Mathematics

11. Oprah begins with \$30 in the bank and withdraws \$5 per week. Let b = the balance after w weeks.

a. Describe her bank balance with a chart using $w = 0, 1, 2, 3, 4$.

b. Describe her bank balance with an equation using variables b and w for balance and number of weeks respectively.

c. Make a graph of all solutions to the equation in part **b.**

d. In what week will Oprah withdraw her last dollar?

12. a. Make a table of values which satisfy $y = 4x - 2$.

b. Graph the equation.

13. A tree now has a trunk with a 10 cm radius. The radius is increasing by 0.5 cm per year. Its radius y after x years is described by $y = 10 + 0.5x$.

a. Make a table of values for this relationship.

b. What is the domain of x?

c. Draw a graph of this situation.

d. After how many years will the radius equal 20 cm?

14. Use the graph below to complete the table.

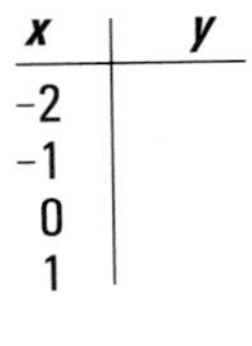

x	y
-2	
-1	
0	
1	

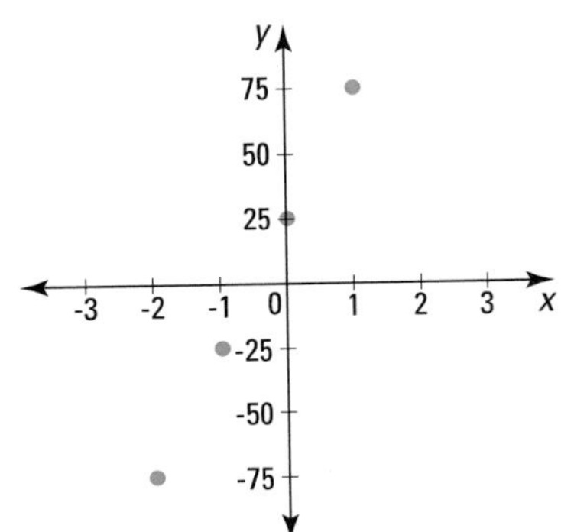

15. a. Draw the graph of $y = 3x$. Choose your own values for x.

b. On the same grid that you used in part **a,** draw the graph of $y = -3x$.

c. At what point(s) do the graphs of parts **a** and **b** intersect?

d. Describe any patterns you observe in these graphs.

16. The *St. Louis Post-Dispatch* reported that the level of the Mississippi River in St. Louis County on August 16, 1993, was 39.1 feet. It was expected to drop 0.6 feet by the next day. Let x equal the number of days since August 16 and y equal the level of the Mississippi River (in feet).

a. Suppose the river continued to drop at the same speed. Write an equation for y in terms of x.

b. Graph your equation from part **a.**

c. Use your graph from part **b** to estimate when the Mississippi River was expected to drop to the lowest "flood stage level," of 30 feet, in St. Louis County.

17. What is the minimum number of points needed to graph a straight line?

Japanese fishermen in Tokyo Bay Harbor unload the day's catch.

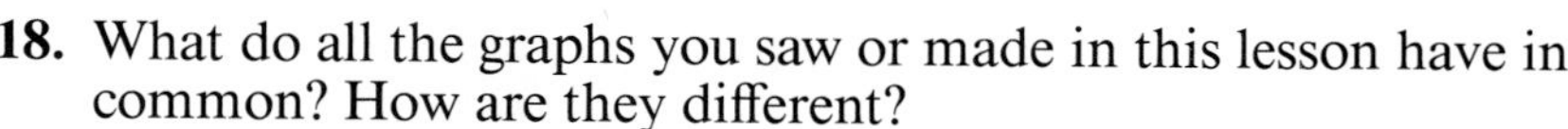

18. What do all the graphs you saw or made in this lesson have in common? How are they different?

Review

19. A triangle has sides of length 20 and 27. What are the possible lengths of the third side? *(Lesson 4-8)*

20. Osami lives 2 km from Tokyo Bay and 1.4 km from Toshiki's house. How far does Toshiki live from Tokyo Bay? *(Lesson 4-8)*

21. The angles in a triangle have measures $x - 10$, $x + 10$, and $x + 30$. What are the measures of these three angles? *(Lesson 4-7)*

22. The measure of an angle is $x°$.
a. What is the measure of its supplement?
b. What is the measure of its complement?
c. What is the difference between the measures of the supplement and complement? *(Lesson 4-7)*

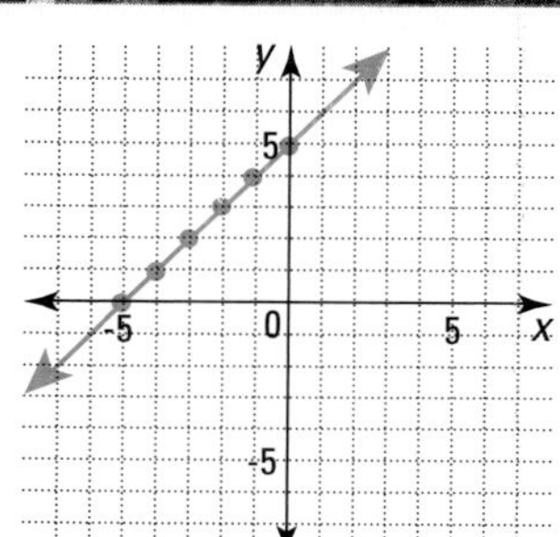

23. *Multiple choice.* The graph of which equation is pictured at the left? *(Lesson 4-6)*
(a) $x + y = 5$ (b) $x + y = -5$ (c) $x - y = 5$ (d) $x - y = -5$

24. Solve $3(w - 4) = 48$. *(Lessons 3-7, 4-3)*

25. Simplify $\frac{x}{7} - \frac{2x}{5}$. *(Lessons 3-9, 4-1)*

26. Write $\frac{12!}{3!\,4!}$ in scientific notation. *(Lesson 2-10, Appendix B)*

27. *Skill sequence.* Write as a decimal. *(Previous course)*
a. 1 divided by 4 **b.** 1 divided by .4
c. 1 divided by .04 **d.** 1 divided by .000004

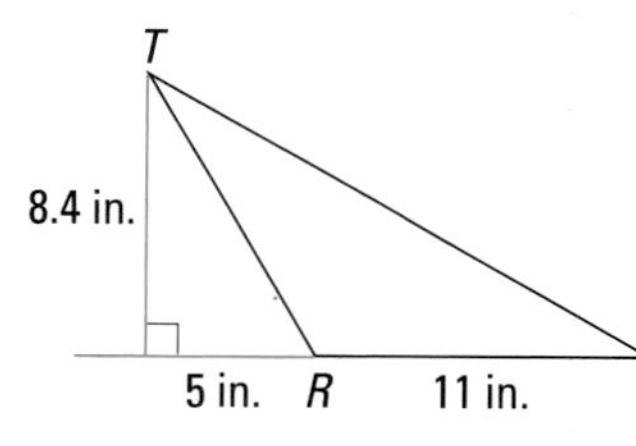

28. Find the area of $\triangle RST$ at the left. *(Lesson 1-5)*

Exploration

29. Which activity listed below could produce a graph like this? Explain what is happening during the activity over the time period shown and how this relates to the graph.

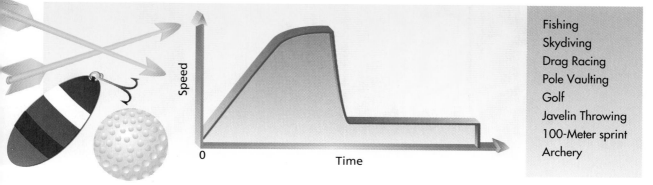

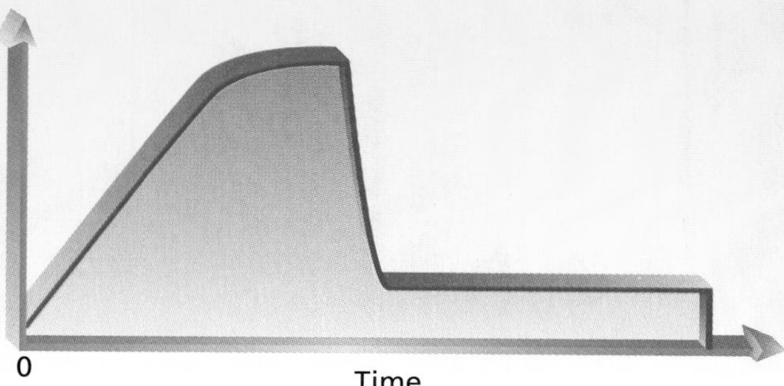

Fishing
Skydiving
Drag Racing
Pole Vaulting
Golf
Javelin Throwing
100-Meter sprint
Archery

A project presents an opportunity for you to extend your knowledge of a topic related to the material of this chapter. You should allow more time for a project than you do for typical homework questions.

1 Patterns in Fraction Products

Find each product.

a. $\left(1 - \frac{1}{2}\right)\left(1 - \frac{1}{3}\right)\left(1 - \frac{1}{4}\right)$

b. $\left(1 - \frac{1}{2}\right)\left(1 - \frac{1}{3}\right)\left(1 - \frac{1}{4}\right)\left(1 - \frac{1}{5}\right)$

c. $\left(1 - \frac{1}{2}\right)\left(1 - \frac{1}{3}\right)\left(1 - \frac{1}{4}\right)\left(1 - \frac{1}{5}\right)\left(1 - \frac{1}{6}\right)$

Write your answers as fractions. Do you see any patterns? Find this product. (There are 99 terms in the product.)

d. $\left(1 - \frac{1}{2}\right)\left(1 - \frac{1}{3}\right)\left(1 - \frac{1}{4}\right) \ldots \left(1 - \frac{1}{99}\right)\left(1 - \frac{1}{100}\right)$

Can you predict what this product will be?

e. $\left(1 - \frac{1}{2}\right)\left(1 - \frac{1}{3}\right)\left(1 - \frac{1}{4}\right) \ldots \left(1 - \frac{1}{n-1}\right)\left(1 - \frac{1}{n}\right)$

Write about the patterns you see in these products.

2 Credit-Card Balances

Mr. Spendalot bought a \$5,000 computer and charged it to his Gotta-Have-It Credit Card which had no previous balance. The Gotta-Have-It Credit Card charges 1.5% interest per month, based on the unpaid balance. So if Mr. Spendalot makes a partial payment of only \$200, there will be a finance charge added to next month's bill.

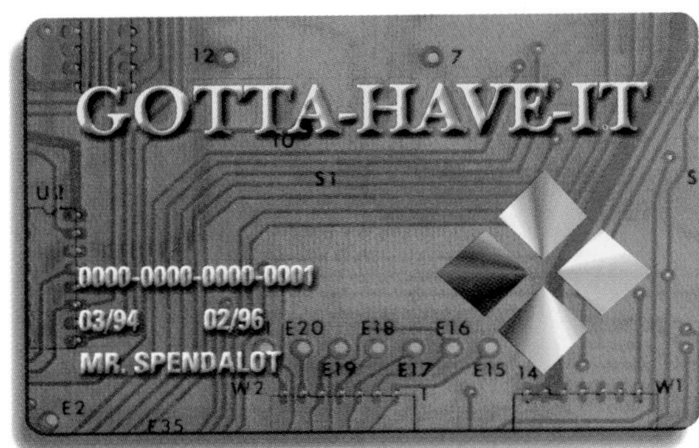

Use a spreadsheet to show the month-to-month changes in Mr. Spendalot's account. Assume he will pay \$200 a month toward his debt and will not charge any more purchases. How long will it take him to pay off the \$5,000? How much money will he end up paying for the computer? Compare the purchase price to the actual cost.

3 Price versus Sales of CDs

Choose a popular CD title, and ask 20 of your classmates whether or not they would buy the CD if the price were \$1, \$2, \$3, . . . up to \$18. Make a graph of the price versus the number of people who would buy the CD. Explain the graph. If you knew that the CD cost the store \$4, what price would you recommend that the store charge for the CD, and why?

4 Situations Leading to $300 - 5x > 40$

On page 215 you read about two situations leading to the equation $5000 - 30x = 1000$. Write four different questions that could be answered by the linear sentence $300 - 5x > 40$. Make sure that one of the questions leads to meaningful negative values for x.

5 Triangles with Integer Sides

Consider triangles with sides of integer lengths and with no two sides having the same length. Explain why there are only three such triangles with all sides of length 5 or less. Then find all such triangles with all sides of length 10 or less.

6 Angle Sums in Polygons

The sum of the measures of the three angles of a triangle always equals 180°. The sum of the measures of the angles of a quadrilateral equals 360°. Consider polygons with more than four sides. Do you think that the sum of the measures of the angles of a pentagon is always a particular number? Draw several different pentagons and measure all five angles with a protractor. From one vertex, draw segments that connect with each of the other vertices. What is formed? Use what you know about triangles to explain the sum of the angle measures in a pentagon. Repeat this experiment with some hexagons (six sides) and heptagons (seven sides). Do all polygons with the same number of sides have angles whose measures yield a constant sum? Why or why not?

SUMMARY

You can think of this chapter as having three major topics. The first topic includes the algebraic definition and uses of subtraction and solving subtraction sentences. The second topic is graphs of lines. Properties of geometric figures that relate to addition and subtraction are the third topic.

The Algebraic Definition of Subtraction is in terms of addition: $a - b = a + -b$. This definition enables you to solve subtraction sentences of the form $ax - b = c$ or $ax - b < c$ by converting them to addition. Such sentences arise from the uses of subtraction. The two major models for subtraction are take-away and comparison. Discounts apply the take-away model. Finding the range of a data set is an application of the comparison model.

When an equation has two variables, its solutions may be graphed by making a table of values and plotting the resulting points on a coordinate plane.

Several important geometry properties involve sums and differences. The Triangle Inequality leads to the fact that if two sides of a triangle have lengths x and y, where $y > x$, the length of the third side is between $y - x$ and $y + x$. Two angles whose measures add to 180° are supplementary; two angles whose measures add to 90° are complementary. The three angle measures of a triangle must total 180°.

Any of the properties in this and previous chapters can be illustrated on a spreadsheet. Formulas relating the cells of a spreadsheet make spreadsheets powerful tools for storing and obtaining information.

VOCABULARY

Below are the most important terms and phrases for this chapter. You should be able to give a general description and a specific example of each.

Lesson 4-1
Algebraic Definition of Subtraction

Lesson 4-2
Take-Away Model for Subtraction
Comparison Model for Subtraction
range of a set, maximum, minimum

Lesson 4-4
spreadsheet
row, column, cell

Lesson 4-5
Opposite of a Sum Property
Opposite of a Difference Property

Lesson 4-6
constant sum
constant difference

Lesson 4-7
supplementary angles, supplements
Triangle Sum Theorem
⌐ symbol for 90° angle
complementary angles, complements

Lesson 4-8
Triangle Inequality
Third Side Property

Lesson 4-9
linear equation

PROGRESS SELF-TEST

Take this test as you would take a test in class. You will need graph paper. Then check your work with the solutions in the Selected Answers section in the back of the book.

1. According to the Algebraic Definition of Subtraction, adding -7 to a number is the same as doing what else?
2. Yesterday it was 53°. Today it is warmer, and the temperature is D degrees. How much warmer is it today than it was yesterday?

In 3–7, simplify.

3. $9p - 7q - (-14z)$
4. $n - 16 - 2n - (-12)$
5. $-8x - (2x - x)$
6. $-(2b - 6)$
7. $\frac{1}{2}m - \frac{7m}{2}$
8. A set of skis costing S dollars is on sale for 20% off. Write an expression for the sale price of the skis.

In 9–12, solve.

9. $5n - 6 = 54$
10. $8 < 6 - 3p$
11. $\frac{3}{4} - \frac{1}{4}m = 12$
12. $201 = 15f - 2(3 + 6f)$
13. If $C = \frac{5}{9}(F - 32)$, find the Fahrenheit equivalent of 50°C.
14. Each minute a computer printer prints 6 sheets of paper. Suppose the printer starts with 1100 sheets of paper and prints continuously. For how long will there be more than 350 sheets left? Explain how you got your solution.
15. The 9 members of the U.S. Supreme Court each vote "yes" or "no" on an issue. Graph all possible ordered pairs representing the number of possible yes votes and no votes.
16. a. Make a table of values of x and y that satisfy the equation $y = 2x - 3$, using $x = -1, 0, 1, 2,$ and 3.
    b. Graph all solutions to $y = 2x - 3$.
17. The distance from Los Angeles to El Paso is 786 miles and from El Paso to San Antonio is 582 miles. What is the greatest possible distance from Los Angeles to San Antonio?
18. An angle has measure 18°.
    a. What is the measure of its supplement?
    b. What is the measure of its complement?
19. Graph the set of ordered pairs that represent the possible measures of all pairs of complementary angles.
20. Find the measure of $\angle L$. Show your work.

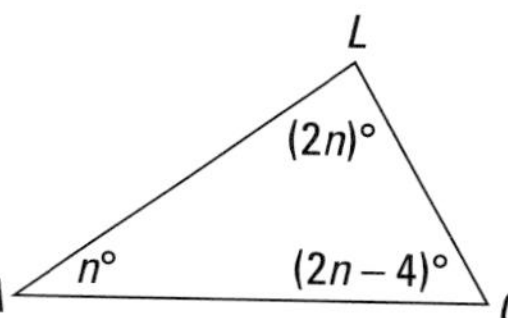

21. Find the possible values of p. Explain how you got your answer.

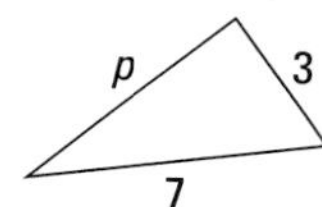

22. Use the spreadsheet below, which is similar to that used in many businesses. It has columns for base pay, federal tax, state tax, and take-home pay.
    a. What is in cell B2?
    b. Which cell contains the name Wojak, D.?
    c. The formula in cell C3 is = .3565 · B3. What number will be printed in cell C3?
    d. An employee's take-home pay is found by subtracting federal and state taxes from the base pay. Write the formula that should be used in cell E2.

	A	B	C	D	E
1	NAME	BASE PAY	FED TAX	STATE TAX	TAKE-HOME
2	Chavez, M.	428.75	152.85	38.21	
3	Lee, S.	401.18		35.76	
4	Wojak, D.	410.34	146.29		

CHAPTER REVIEW

Questions on SPUR Objectives

SPUR stands for **S**kills, **P**roperties, **U**ses, and **R**epresentations. The Chapter Review questions are grouped according to the SPUR Objectives for this chapter.

SKILLS DEAL WITH THE PROCEDURES USED TO GET ANSWERS.

Objective A: *Simplify expressions involving subtraction.* *(Lesson 4-1)*

In 1–8, simplify.

1. $3x - 4x + 5x$ **2.** $-\frac{2}{3} - \frac{4}{5}$

3. $\frac{3a}{2} - \frac{9a}{2}$ **4.** $\frac{1}{3}x - \frac{5x}{3}$

5. $c - \frac{c}{3} - 2c - 2$ **6.** $3x + y - 4x - 7y$

7. $z^3 - 7 + 8 - 4z^3$ **8.** $3(x - 6) - 5x$

Objective B: *Solve and check linear equations involving subtraction.* *(Lesson 4-3)*

9. $x - 47 = -2$ **10.** $2.5 = t - 3.34$

11. $\frac{3}{2} + y - \frac{1}{4} = \frac{3}{4}$ **12.** $8 = \frac{3}{4}a - 10$

13. $4n - 3 = 17$ **14.** $470 - 2n = 1100$

15. $m - 3m = 10$ **16.** $46n - 71n - 6 = 144$

17. $0 = 4a - 6$ **18.** $18(2x - 4) - 6 = -168$

Objective C: *Solve and check linear inequalities involving subtraction.* *(Lesson 4-3)*

19. $2x - 11 < 201$ **20.** $-8y + 4 \le 12$

21. $32 - y > 45$

22. $0.9(90n - 14) - 3n \ge 455.4$

Objective D: *Use the Opposite of a Sum or Difference Property to simplify expressions and solve sentences.* *(Lesson 4-5)*

In 23–28, simplify.

23. $-(4a + 7)$ **24.** $-(3f - 4g + 6)$

25. $1 - (z - 1)$ **26.** $3x - (2x - 9)$

27. $2(a - 3) - 5(a + 2)$ **28.** $-3(n + 6) - 6(n - 3)$

In 29–34, solve.

29. $-(p - 6) = 14$ **30.** $-(r + 3) > 9$

31. $5 - 2(x - 3) < -9$ **32.** $1\frac{1}{2} - \left(\frac{3}{4} - y\right) = 7$

33. $(5x - 8) - (3x + 1) = 36$

34. $75 = 4e - 5(3 + 2e)$

PROPERTIES DEAL WITH THE PRINCIPLES BEHIND THE MATHEMATICS.

Objective E: *Apply the Algebraic Definition of Subtraction.* *(Lesson 4-1)*

In 35 and 36, rewrite each subtraction as an addition.

35. $x - y + z$ **36.** $-8 - v = 42$

37. *True or false.* The sum of $m - k$ and $k - m$ is zero.

38. *Multiple choice.* Which does not equal the others?

(a) $a - b$ (b) $a + -b$

(c) $-b + a$ (d) $b + -a$

Objective F: *Use the definitions of supplements and complements, and the Triangle Sum Theorem.* *(Lesson 4-7)*

39. If $m\angle Q = 17°$, find the measure of

a. its complement. **b.** its supplement.

40. $\angle R$ and $\angle S$ are complements. If $m\angle R = x°$ and $m\angle S = z°$, write an equation relating x and z.

41. *True or false.* The measure of the supplement of any angle is always greater than the measure of its complement. Justify your answer.

42. Find the measure of each angle in the figure below. Explain how you found your answers.

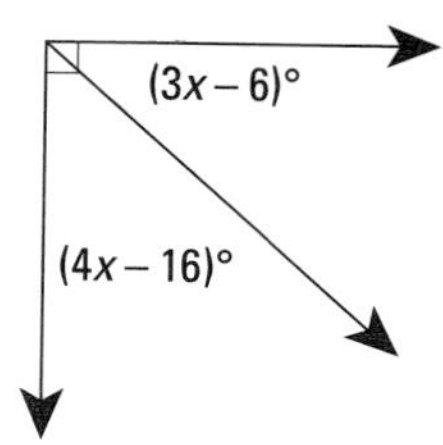

43. Two angles of a triangle measure 75° and 32°. What is the measure of the third angle?

44. In a triangle, the largest angle has measure 10 times the smallest; the other angle has measure 10° more than the smallest. Find the measure of each.

45. A triangle has a 40° angle. Two of the angles of the triangle have equal measures. Find all possible measures of the other two angles.

46. Find the measure of each angle in $\triangle MAC$.

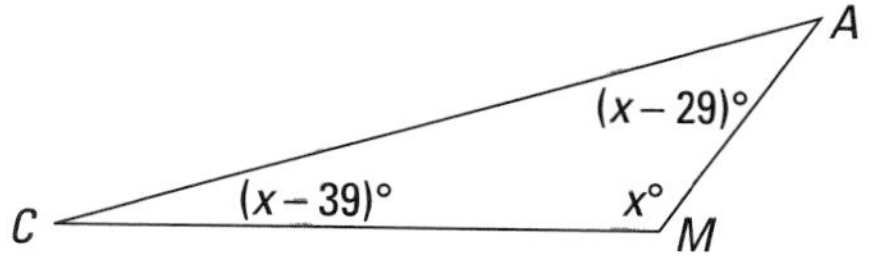

Objective G: *Use the Triangle Inequality to determine possible lengths of sides of triangles.* (Lesson 4-8)

In 47 and 48, tell whether the three numbers can be the lengths of sides of a triangle. If they cannot be, justify your answer.

47. 16, 3, 5

48. 16, 8, 10

In 49 and 50, use the Triangle Inequality to write the three inequalities which must be satisfied by lengths of sides in the triangle.

49.

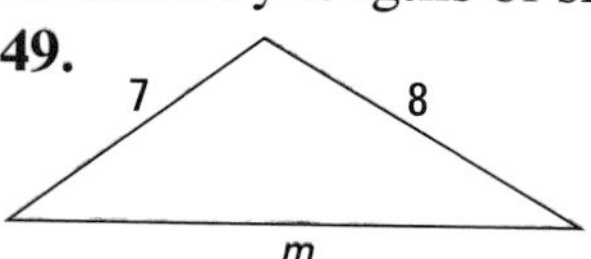

50.

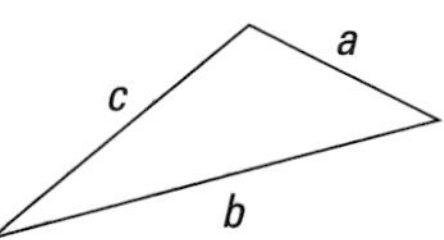

In 51 and 52, find the possible values for y.

51.

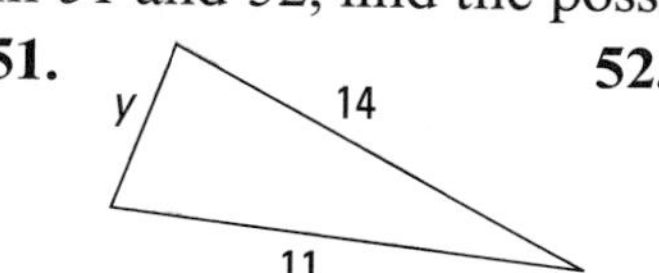

52.

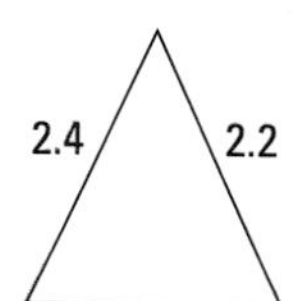

USES DEAL WITH APPLICATIONS OF MATHEMATICS IN REAL SITUATIONS.

Objective H: *Use models for subtraction to write expressions and sentences involving subtraction.* (Lessons 4-2, 4-3)

53. Last week Carla earned E dollars, saved S dollars, and spent P dollars. Relate E, S, and P in a subtraction sentence.

54. An elevator won't run if it holds more than L kilograms. A person weighing 80 kg gets on a crowded elevator and an "overload" light goes on. How much did the other passengers weigh?

55. The total floor area of a three-story house is advertised as 3500 sq ft. If the first floor's area is F sq ft and the third floor's area is 1000 sq ft, what is the area of the second floor?

56. After spending $40, Mort has less than $3 left. If he started with S dollars, write an inequality to describe the possible values of S.

57. Donna is 5 years older than Eileen. If Donna's age is D, how old is Eileen?

58. A plane 30,000 feet above sea level is radioing a submarine 1500 feet below sea level. What is the difference in their altitudes?

59. A video game is regularly priced at V dollars. Save-a-Buck is selling the game for 30% off.

a. Write an expression for the amount of the discount you get at Save-a-Buck.

b. Write an expression for the sale price of the game.

60. A family went out for dinner. The total cost of the food was F dollars. The restaurant automatically adds 15% as the standard tip for service.

a. Write an expression for the cost of the food and tip.

b. The family has a coupon worth $5 off the cost of the food. How much does the dinner, including tip, actually cost them?

CHAPTER FOUR

Objective I: *Solve problems using linear sentences involving subtraction.* *(Lesson 4-3)*

61. Liz saves \$15 per week. She has \$750 in the bank and needs \$1500 in the bank before she can afford to go on vacation. For how many weeks must she save before she has more than the required amount? Explain how you got the answer.

62. If $F = \frac{9}{5}C + 32$, find the Celsius equivalent of 100°F.

63. Radio station WARM has a trivia contest. The first day that a question is asked, the prize for a correct answer is \$200. If no one wins the money, the prize increases by \$92 per day until a correct answer is received. If the program director budgets \$1400 for the contest, for how many days will the potential prize remain within the budget?

64. Each minute, a computer printer prints 8 sheets of paper. Suppose the printer starts with 1500 sheets of paper and prints continuously.

a. Write an expression that represents the number of sheets left after m minutes.

b. After how many minutes will 200 sheets be left?

Objective J: *Apply the Triangle Inequality in real situations.* *(Lesson 4-8)*

65. It is 346 miles from El Paso to Phoenix and 887 miles from Dallas to Phoenix. Based on this information, what is the greatest possible distance from Dallas to El Paso?

66. Malinda lives 20 minutes by train from Roger and 30 minutes by train from Charles. By train, how long would it take her to get from Roger's place to Charles's place? (Assume all trains go at the same rate.)

REPRESENTATIONS DEAL WITH PICTURES, GRAPHS, OR OBJECTS THAT ILLUSTRATE CONCEPTS

Objective K: *Use a spreadsheet to show patterns and make tables from formulas.* *(Lesson 4-4)*

67. Use the spreadsheet below.

a. The formula in cell B1 is = A1^2 + A1. This was replicated down the column. What is the formula for B6?

b. Find the value that belongs in cell B7.

c. Describe what would happen if you changed the value in cell A6 to -20.

	A	B
1	3	12
2	2	
3	1	2
4	0	
5	-1	0
6	-2	2
7	-3	

68. This spreadsheet shows the evaluation of the formula $d = 180n - 360$.

a. What formula gives the values in cell B2?

b. What formula gives the values in cell B6?

c. Complete the table.

	A	B
1	*n*	*d*
2	2	0
3	3	180
4	4	
5	5	
6	6	
7	7	

CHAPTER FOUR

69. The spreadsheet below calculates the total ticket income for a movie theater on a series of days. Adult tickets cost \$6.00 each. Children's tickets cost \$4.00 each.

a. Complete the totals in column D.

b. Write a formula for computing the value of cell D4 from cells B4 and C4.

c. What day recorded the most income?

	A	B	C	D
1	Day	Adult	Children	Total
2	August 26	100	20	680
3	August 27	110	20	
4	August 28	91	10	
5	August 29	145	62	
6	August 30	170	40	
7	August 31	60	40	

70. The spreadsheet below gives the scoring for a team of basketball players. Column B gives the number of free throws they completed (each worth one point). Column C gives the number of 2-point field goals for each player. Column D gives the number of 3-pointers scored by each player. Column E gives the total number of points scored by the player in the game. Ken scored 16 points, since $5 \cdot 1 + 4 \cdot 2 + 1 \cdot 3 = 16$.

a. Complete column E in the score sheet.

b. What cell contains the number 8?

c. What formula is in cell E5?

d. Write a formula for cell C8 that would total the number of 2-point field goals.

	A	B	C	D	E
1	Player	Free Throws	Field Goals	3-pointers	POINTS
2	Jose	2	6	1	
3	Djin	1	4	0	
4	Ken	5	4	1	16
5	Arunas	1	8	2	
6	Bill	2	2	0	
7	Mike	0	1	1	
8					

Objective L: *Graph equations of the forms $x \pm y = k$ or $y = ax \pm b$ by making a table of values.* (Lessons 4-6, 4-9)

71. Xavier is four years older than his sister Yvonne. Graph all possible ordered pairs that represent their ages. Let x represent Xavier's age and y represent Yvonne's age.

72. The sum of two numbers is 0. Graph all possible pairs of numbers.

73. Mary and Peter have a total of 5 pets. Graph all possible ways the pets may be divided between them.

In 74–77, graph all ordered pairs (x, y) that satisfy the equation.

74. $x - y = 3$

75. $y = \frac{1}{2}x + 10$

76. $y = 7 - 2x$

77. $x + y = 100$

CHAPTER

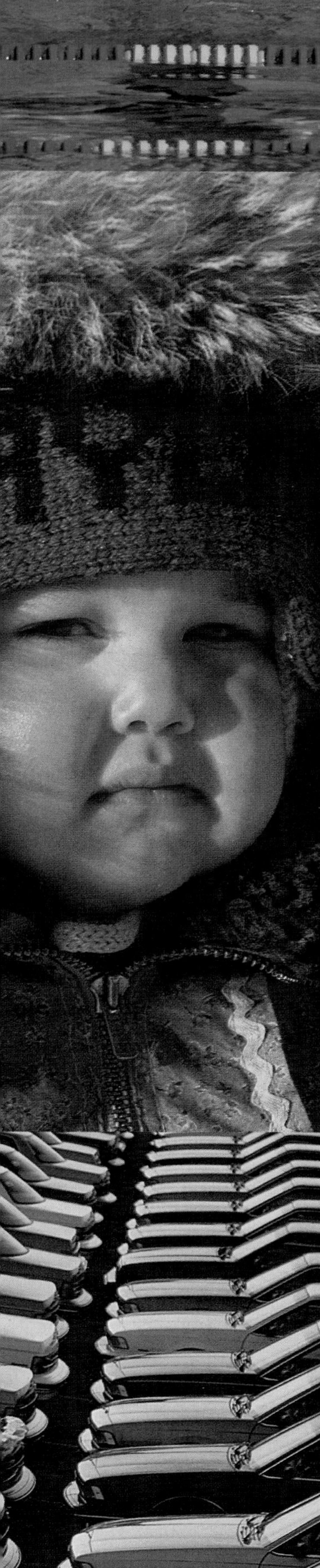

LINEAR SENTENCES

A light-bulb manufacturer produces two kinds of bulbs—regular bulbs and new energy-efficient (EE) bulbs. The EE bulbs use less electricity, but they cost more. Does the money you save in electricity make up for the higher initial cost of the bulb?

This question and other related questions can be answered by using tables or spreadsheets, graphs, and algebraic sentences, as shown below.

1. In a table:

	A	B	C
1	Hours	EE	Regular
2	0	\$0.92	\$0.70
3	100	\$1.34	\$1.18
4	200	\$1.76	\$1.66
5	300	\$2.18	\$2.14
6	400	\$2.60	\$2.62
7	500	\$3.02	\$3.10
8	600	\$3.44	\$3.58
9	700	\$3.86	\$4.06
10	800	\$4.28	\$4.54
11	900	\$4.70	\$5.02
12	1000	\$5.12	\$5.50

2. With a graph:

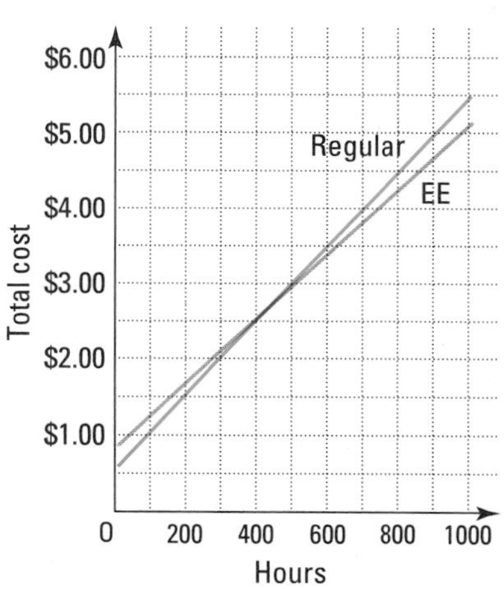

3. By algebraic sentences:

	Total cost with EE		Total cost with regular
When is the regular bulb cheaper?	$.92 + .0042h$	$>$	$.70 + .0048h$
When is the EE bulb cheaper?	$.92 + .0042h$	$<$	$.70 + .0048h$
When are the costs the same?	$.92 + .0042h$	$=$	$.70 + .0048h$

In this chapter, you will study all three of these ways to solve linear sentences and learn how they are related to each other.

LESSON

5-1 Horizontal and Vertical Lines

This music's a gas. *In 1973, lines at gas stations were so long that in some places entertainers performed to keep customers from becoming angry.*

Equations for Horizontal Lines

Before 1974, there was no national speed limit in the United States. Each state set its own limit, and some states had none. But in 1973, oil became scarce and its price jumped. Long lines appeared at many gas stations because gasoline (a by-product of oil) was in short supply. To reduce gasoline usage, the U.S. Congress passed a law that set a national highway speed limit of 55 $\frac{\text{miles}}{\text{hr}}$, beginning in 1974. Even though the oil crisis passed, a benefit of this law was that at reduced speeds there were fewer highway deaths. Congress increased the national highway speed limit to 65 $\frac{\text{miles}}{\text{hr}}$ on July 1, 1987.

The preceding paragraph has described in *writing,* or *prose,* what have been the national speed limits since 1974. Here are three other ways.

In a table:

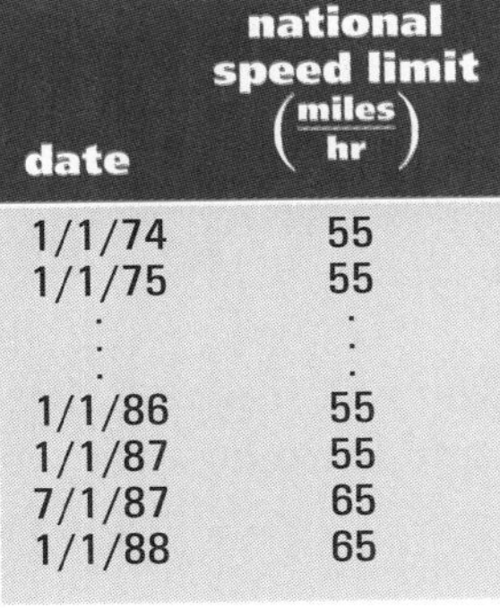

date	national speed limit $\left(\frac{\text{miles}}{\text{hr}}\right)$
1/1/74	55
1/1/75	55
⋮	⋮
1/1/86	55
1/1/87	55
7/1/87	65
1/1/88	65

With a graph:

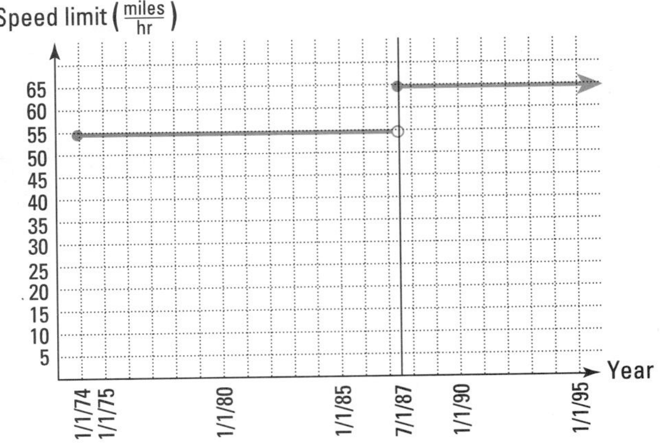

By algebraic sentences:

Let x = the date and y = the national speed limit $\left(\text{in } \frac{\text{miles}}{\text{hr}}\right)$.
When $1974 \le x < 1987.5$, $y = 55$.
When $1987.5 \le x$, $y = 65$.

Each description has its advantages. The description by algebraic sentences is brief and precise.

The graph shows that all the points from 1/1/74 to 7/1/87, not including 7/1/87, have the same *y-coordinate.* For all of these points, the y-coordinate equals 55. Therefore, an equation for the line is $y = 55$. Beginning halfway through 1987, the value of y is 65, so these points lie on a second horizontal line, with equation $y = 65$. Every *horizontal line* has an equation of this type.

Police occasionally set up radar to inform motorists of their speeds.

Every **horizontal line** has an equation of the form $y = k$, where k is a fixed real number.

Example 1

Give an equation describing all points on line m graphed below.

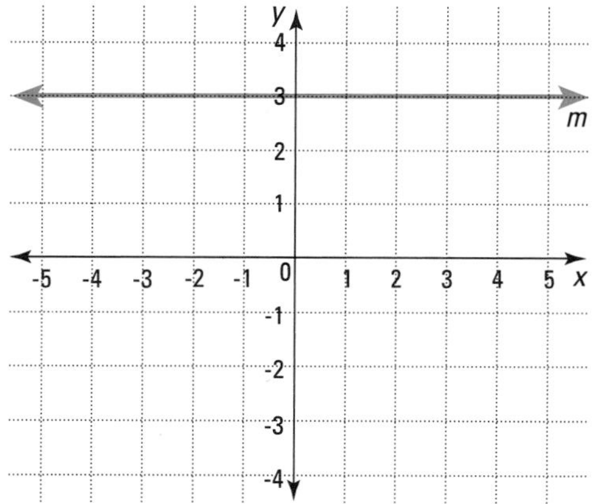

Solution

The points are on a horizontal line that crosses the y-axis at 3.
An equation for line m is y = 3.

Check

Two of the points on line m have coordinates (0, 3) and (5, 3). These numbers satisfy the equation $y = 3$.

Equations for Vertical Lines

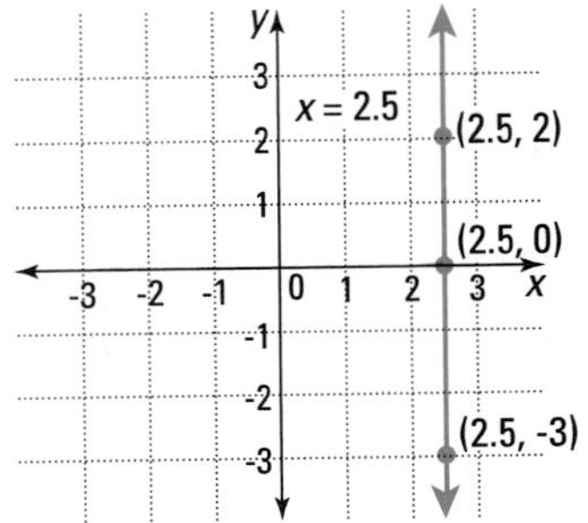

A *vertical line* is drawn at the right. Notice that each point on the line has the same *x-coordinate,* 2.5. Thus an equation for the line is $x = 2.5$. This means x is fixed at 2.5, but y can be any number.

Every **vertical line** has an equation of the form $x = h$, where h is a fixed real number.

An equation with only one variable, such as $x = -5$ or $y = 8$, can be graphed on a number line (in which case its graph is a point), or on a coordinate plane (in which case its graph is a line). The directions or the context of the problem will usually tell you which type of graph to make.

Example 2

a. Graph $x = -5$ on a number line.
b. Graph $x = -5$ on a coordinate plane.

Solution

a. Draw a number line. Mark the point with coordinate equal to -5.

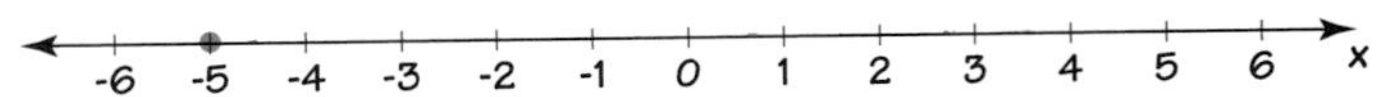

The graph of $x = -5$ on a number line is the single point with coordinate -5.

b. Draw x- and y-axes. Plot points whose x-coordinate is -5. Some points are (-5, 2), (-5, 0), and (-5, -3). Draw the line through these points.

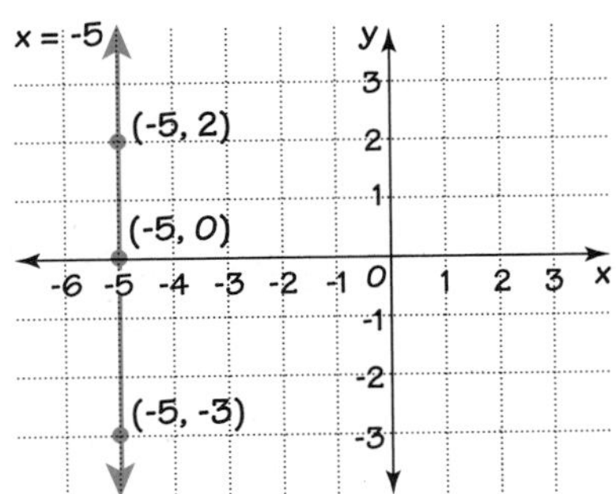

The graph of $x = -5$ in a coordinate plane is the vertical line that crosses the x-axis at -5.

Using Horizontal and Vertical Lines

Horizontal and vertical lines can be used to solve equations or inequalities.

Example 3

At the beginning of the year, Mac had $580 in his savings account. As long as he has at least $300 in the account, he does not have to pay a service fee. For how long can he withdraw $20 per week without paying a service fee?

Solution 1

Let y equal Mac's bank balance, and x equal the number of weeks after the start of school. Then, $y = 580 - 20x$.
This line is graphed using a table of (x, y) values. Also graphed is the horizontal line $y = 300$ to represent the minimum balance that he must keep to avoid service charges.

x	y
0	580
5	480
10	380
15	280

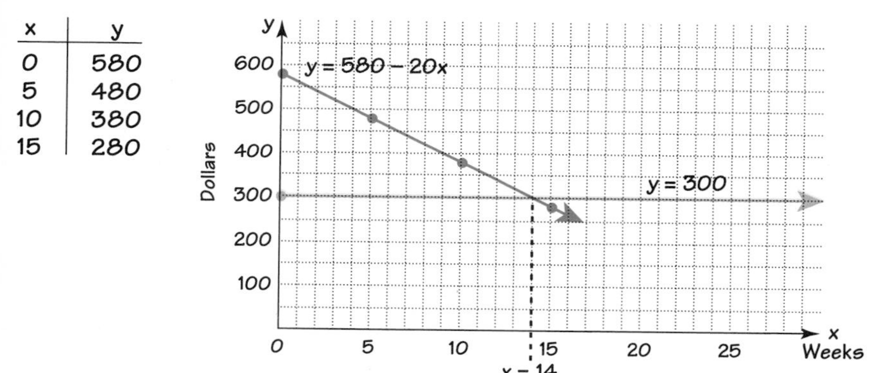

As long as Mac's balance is at or above the line $y = 300$, he does not have to pay a service fee. The lines appear to cross at about $x = 14$. For $x \le 14$ weeks, Mac's balance is high enough that he does not have to pay a service charge.

Solution 2

Translate "Mac's balance is at least $300" into an inequality.
Let x = the number of weeks after the start of school.

Solve $580 - 20x \ge 300$.

$580 - 20x - 580 \ge 300 - 580$ Subtract 580 from each side.

$-20x \ge -280$

$-\frac{1}{20} \cdot -20x \le -\frac{1}{20} \cdot -280$ Multiply each side by $-\frac{1}{20}$, and change the sense of the inequality.

$x \le 14$

For the first 14 weeks Mac does not pay a service fee.

In Solution 2, we began solving $580 - 20x \ge 300$ by subtracting 580 from each side. This is equivalent to adding -580 to each side. We use both techniques in this book. You or your teacher may prefer one method over the other.

QUESTIONS

Covering the Reading

In 1 and 2, what was the national speed limit at the indicated time?

1. before 1/1/74

2. on July 1, 1987

In 3 and 4, let y = the national speed limit. Describe y during the indicated time period.

3. between 1975 and 1977

4. between 1989 and 1991

5. All points on a horizontal line have the same __?__ -coordinate.

6. All points on a vertical line have the same __?__ -coordinate.

In 7 and 8, write an equation for each graph.

7.

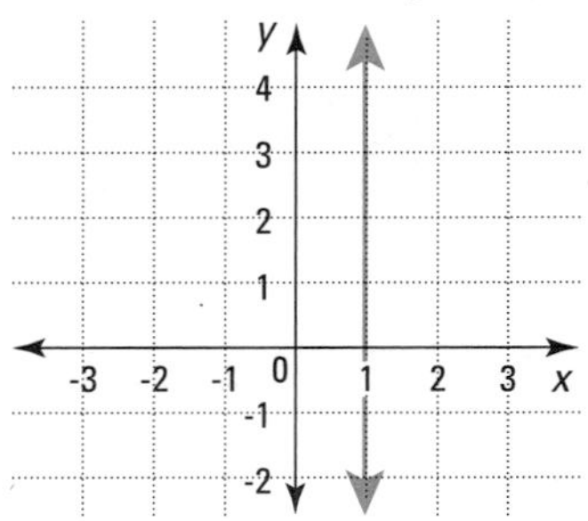

8.

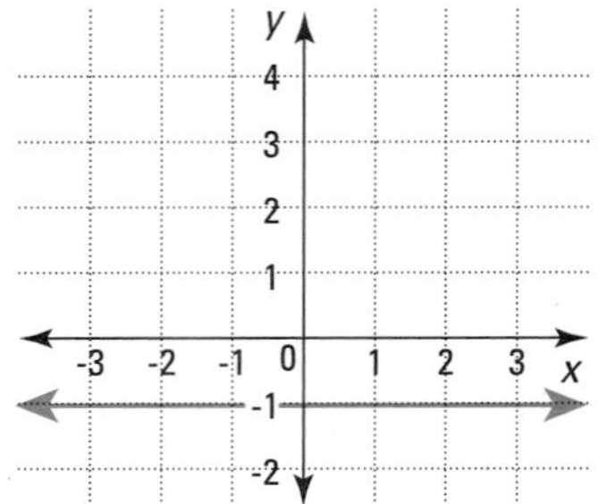

In 9–12, an equation is given.

a. Graph the points on a number line which satisfy the equation.

b. Graph the points in the coordinate plane which satisfy the equation.

9. $x = 4$

10. $y = 4$

11. $y = -\frac{1}{2}$

12. $x = -15$

In 13 and 14, use Example 3.

13. Suppose the bank raised the minimum balance required for avoiding a service fee from \$300 to \$400.

a. Use the graph to estimate the number of weeks that Mac would not have to pay the service fee.

b. Use an inequality to find the number of weeks Mac would not have to pay the service fee.

14. Suppose Mac started the school year with \$680. Draw a coordinate graph to show the amount of money that he will have each week after making a \$20 withdrawal.

Applying the Mathematics

15. In recent years, the cost of mailing a first-class letter weighing 1 ounce or less has been as follows:

from 2/17/85 to 4/2/88	22¢
from 4/3/88 to 2/2/91	25¢
from 2/3/91 on	29¢.

a. Graph the relationship between date and cost.
b. With algebraic sentences, describe the graph that you drew in part **a.**

In 16 and 17, write an equation for the line containing the points given.

16. (-9, 12), (4, 12), (0.3, 12) **17.** (-6, -3), (-6, 0), (-6, 200)

18. Graph the lines $y = -4$ and $x = 15$. Find the coordinates of the point of intersection.

19. a. Write an equation of the horizontal line through (7, -13).
b. Write an equation of the vertical line through (7, -13).

20. Horizontal lines are parallel to the _?_ axis and perpendicular to the _?_ axis.

21. Ron's Refrigerator Repair Service charges $25 for travel time plus $35 per hour to repair coolers. Sasha's Steak House has a cooler that needs repairs. Sasha is willing to spend no more than $250 on repairs.
a. Write an equation to relate the repair cost y to the time spent x.
b. Draw a graph of the equation that you wrote in part **a.**
c. Use the same coordinate axes as part **b.** Draw the line $y = 250$ to represent the money that Sasha is willing to spend on repairs.
d. Use your graph to determine the maximum whole number of hours that Ron could work and still keep Sasha's bill under $250.
e. Check your answer to part **d** by solving an inequality.

Review

22. The geometric board shown at right is used in the game called Cows and Leopards. The game is popular in southern Asia. Find the angle measures w, x, y, and z. *(Lesson 4-7)*

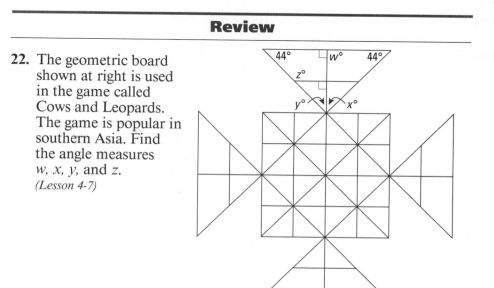

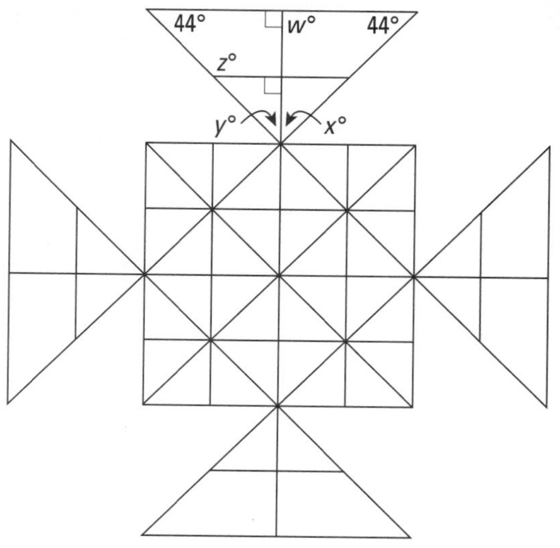

23. The regular price of a compact-disc player is d dollars. What is the sale price if a discount of 20% is offered? *(Lesson 4-2)*

24. A glacier has already moved 25 inches and is moving at a rate of 2 inches per week. Let t be the total number of inches the glacier has moved after w additional weeks. Write a formula for t in terms of w. *(Lesson 3-8)*

25. Show two ways to find the total area of the rectangles below. *(Lesson 3-6)*

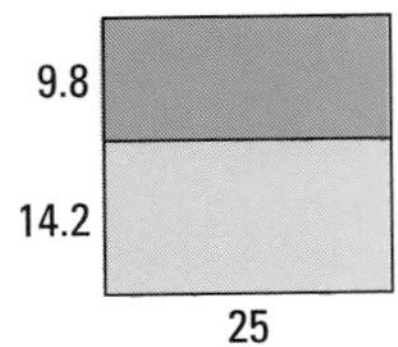

In 26 and 27, simplify. *(Lessons 2-2, 2-3, 2-5)*

26. $(-a \cdot b) \cdot -a$

27. $-\frac{3}{2} \cdot \left(-\frac{2}{3}a\right)$

28. Shown at the left is part of an ad that appeared in the June 19, 1992, edition of *USA Today*.

"Al and Scott are in separate automobiles traveling in the same direction at 232 mph. If Al reaches his destination 0.043 seconds before Scott, how far ahead of Scott does Al reach his destination?"
TARGET

"Al and Scott" refer to Al Unser, Jr., and Scott Goodyear, who finished first and second in the 1992 Indianapolis 500 motor race. The margin of victory was the closest in the history of the race, with only 0.043 second separating Al and Scott. In a follow-up article in *USA Today,* on June 25, 1992, it was reported that the answer given in the original ad, 6 feet, was incorrect. Over 200 readers of the newspaper had called or written the company to report the error. To the nearest tenth of a foot, how far ahead of Scott did Al reach his destination? *(Lesson 2-4)*

Split-second service. *Pit stops are important in racing events such as the Indianapolis 500. Pit crews are expected to refuel a car and change the tires during the race, usually within about 20 seconds.*

29. a. In your own words, state the Multiplicative Identity Property of 1.
b. Give an instance of this property. *(Lesson 2-2)*

Exploration

30. What is the maximum speed limit on highways in your state?

LESSON

5-2

Using Tables to Compare Linear Expressions

No free parking. *In many cities, parking spaces are scarce and costly. Motorists are wise to compare the rates of different parking facilities.*

Tables are often useful for comparing values from two or more situations.

Example 1

The table below gives the charges for parking a car at two different garages.

a. When does it cost less to park at Gil's Garage?

b. When does it cost less to park at Patsy's Parking?

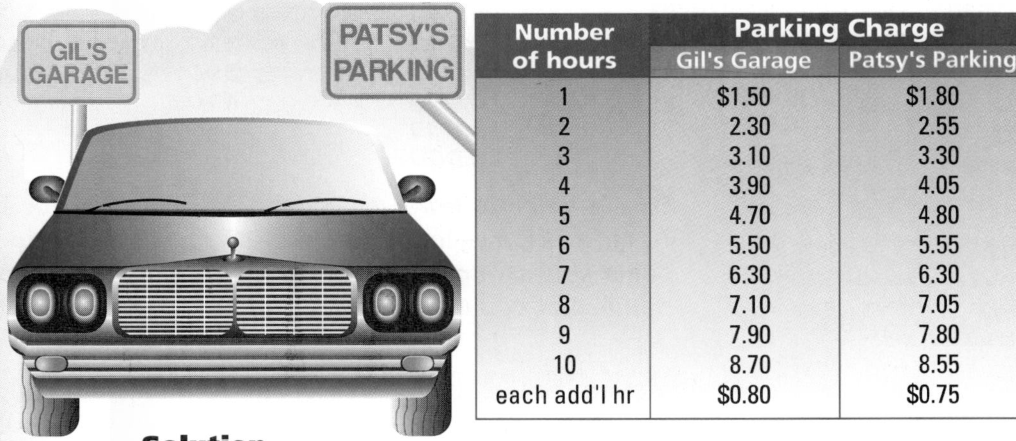

Number of hours	Parking Charge	
	Gil's Garage	Patsy's Parking
1	\$1.50	\$1.80
2	2.30	2.55
3	3.10	3.30
4	3.90	4.05
5	4.70	4.80
6	5.50	5.55
7	6.30	6.30
8	7.10	7.05
9	7.90	7.80
10	8.70	8.55
each add'l hr	\$0.80	\$0.75

Solution

Read down the columns for charges at Gil's and Patsy's.

Let h = the number of hours parked.

a. When $h \le 6$ it costs less to park at Gil's.

b. When $h \ge 8$ it costs less to park at Patsy's.

If a table is not given, it is often helpful to make one.

Is "Best" the better buy? *Many businesses rent photocopy machines rather than buy them. Office managers need to determine which rental company offers the best rate.*

Example 2

You are the manager of an office. Your company needs another copy machine. After contacting several office supply firms, you have obtained the following rental rates.

Acme Copiers offers a copier for $250 per month and an additional charge of 1 cent ($.01) per copy.

Best Printers offers the same machine for $70 per month and a per-copy charge of 3 cents ($.03).

Records show that your office made as few as 482 copies during a holiday month and as many as 17,386 copies during inventory month. Make a table showing the costs for 0, 2000, 4000, 6000, . . . , 20,000 copies per month. Use the table to draw conclusions about which company charges less for making copies.

Solution

Let n = the number of copies made per month.
Cost from Acme Copiers: 250 + 0.01n
Cost from Best Printers: 70 + 0.03n

Substitute the values 0, 2000, 4000, 6000, . . . , 20,000 for n into each of the expressions and record the results in a table.

Number of Copies n	Acme's Charges 250 + 0.01n	Best's Charges 70 + 0.03n
0	250	70
2,000	270	130
4,000	290	190
6,000	310	250
8,000	330	310
10,000	350	370
12,000	370	430
14,000	390	490
16,000	410	550
18,000	430	610
20,000	450	670

Best's price is lower than Acme's for 8,000 or fewer copies. Acme's price is lower for 10,000 or more copies. From this table you cannot be sure which company charges less if the number of copies made per month is between 8,000 and 10,000.

To determine the "break-even" point for the two rental companies, you can evaluate the expressions $250 + 0.01n$ and $70 + 0.03n$ for values of n between 8,000 and 10,000. This is easily done with a computer program that prints tables, or with a spreadsheet.

You can set up the spreadsheet with column A representing the number of copies. Start at 8000. Continue down the column in increments of 200. In cell B3 put the Acme charge formula = 250+.01*A3. In cell C3 put the Best charge formula =70+.03*A3. Replicate the formulas in B3 and C3 down to B13 and C13, respectively. In the spreadsheet shown, you can see that 9000 copies (row 8) produces the same charge for both companies. If you make fewer than 9000 copies, Best is cheaper. If you make more than 9000 copies, Acme offers the better deal.

Copy cats. *Today's copy centers offer a variety of services such as typesetting résumés, creating colorful brochures, and printing business forms.*

	A	B	C
1	# OF COPIES	ACME COST	BEST COST
2	N	250+.01*N	70+.03*N
3	8000	330	310
4	8200	332	316
5	8400	334	322
6	8600	336	328
7	8800	338	334
8	9000	340	340
9	9200	342	346
10	9400	344	352
11	9600	346	358
12	9800	348	364
13	10000	350	370

To answer the question "When is renting from Acme cheaper or the same as renting from Best?" in algebraic terms, solve the inequality:

$$250 + 0.01n \leq 70 + 0.03n.$$

As shown in the spreadsheet above, there are many solutions to this inequality. The spreadsheet shows the solutions 9000, 9200, . . . , 10,000 copies. In the next three lessons you will learn other methods for solving linear sentences like this one.

QUESTIONS

Covering the Reading

1. The table below gives the charges for copies of photos at two different shops.

Number of copies	Charges at Phil's Photos	Charges at Peggy's Prints
1	.75	.80
2	1.25	1.40
3	2.00	2.00
4	2.75	2.60
5	3.50	3.20

   a. For how many copies does it cost less at Phil's?
   b. For how many copies does it cost less at Peggy's?

In 2 and 3, refer to Example 2.

2. What would it cost to rent a copier from Best Printers for a month and use it to print 4,292 copies?

3. If your office averaged about 4,500 copies per month, which company should you select? Why?

In 4–6, refer to the spreadsheet on page 293.

4. Write the formula for cell B10.

5. Write the formula for cell C11.

6. a. Use the spreadsheet to tell which company provides the better deal for 9100 copies a month.
   b. Compute the price each company charges for 9100 copies per month.

7. Suppose that Best Printers wishes to get more customers by reducing their charge per copy to 2 cents while keeping the basic monthly charge of $70. In response, Acme drops their charge per month to $200 plus 1¢ per copy.
   a. Copy and complete the table below.
   b. When will the charges be the same for both companies?

Number of Copies	Cost with Acme	Cost with Best
10,000		
11,000		
12,000		
13,000		
14,000		
15,000		

Applying the Mathematics

8. Suppose price changes in Question 7 expired at the end of the month, and the original prices were reinstated. At that point, Best Printers realized that they were not competitive in offices that copied in high volume. To be more competitive with these customers, Best advertised "Maximum Rental Charge: \$600 per month." How many copies could be copied for \$600 at Acme?

9. Kim starts with \$20 and is *saving* at a rate of \$6 per week. Jenny starts with \$150 and is *spending* at a rate of \$4 per week.

a. Write an expression for the amount Kim has after w weeks.

b. Write an expression for the amount Jenny has after w weeks.

c. Make a table. Use it to determine when Kim and Jenny would have the same amount.

10. Rufus is offered two sales positions. With Company A he would earn \$800 per month plus 5% commission on sales. With Company B he would earn \$600 per month plus 6% commission on sales.

a. If Rufus expects sales of about \$10,000, which company would pay him more?

b. Complete this table and determine how much he must sell to be paid more at Company B than at Company A.

Sales	Earnings at Company A	Earnings at Company B
12,000	1400	1320
14,000	1500	1440
16,000	1600	1560
18,000	1700	1680
20,000		
22,000		
24,000		
26,000	2100	2160
28,000	2200	2280
30,000	2300	2400

Review

11. Give the coordinates of three points on the line $y = -6$. *(Lesson 5-1)*

12. Write an equation for the line k graphed at the left. *(Lesson 5-1)*

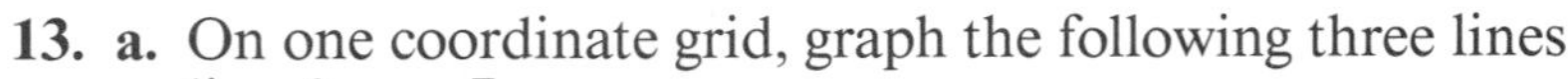

13. **a.** On one coordinate grid, graph the following three lines.
line ℓ: $y = 7$
line m: $x = 2$
line n: $y = 2x - 3$
b. At what point do lines ℓ and m intersect?
c. At what point do lines ℓ and n intersect?
d. Find the area of the triangle formed by the three lines.
(Lessons 4-9, 5-1, Previous course)

14. By air, it is 1061 miles from Miami to St. Louis and 1724 miles from St. Louis to Seattle. Based on this information and the Triangle Inequality, what is the longest possible air distance from Miami to Seattle? *(Lesson 4-8)*

15. **a.** On the same coordinate plane, graph all solutions to $x + y = 10$ and $x - y = 8$.
b. At what point do the graphs intersect? *(Lesson 4-6)*

In 16 and 17, solve. *(Lessons 3-10, 4-3, 4-5)*

16. $1 = 9 - (c - 2)$

17. $42(m - 4) + 210 < 252$

18. A farm worker picking grapes earns \$5.40 per hour plus \$.28 per box. If the worker picks grapes 8 hours, how many boxes would he or she have to pick to earn at least \$50.00 for the day? *(Lessons 3-5, 3-10)*

19. The equation below is incorrect. Correct it by changing the right side of the equation. Explain the error that was made. *(Lesson 3-2)*

$$x + y - 3(z + w) = x + y - 3z + w$$

20. After school Alonzo needs to practice the piano, visit a friend in the hospital, and do homework for algebra, biology, and Spanish.
a. In how many different orders can he do these things?
b. If he visits his friend first, in how many orders can he do the other things? *(Lesson 2-10)*

In 21 and 22, simplify. *(Lessons 1-4, 1-6)*

21. $(\sqrt{37})^2$

22. $(8c)^2$

23. In 1991, the national debt was about 3.46 trillion dollars. Write 3.46 trillion in scientific notation. *(Previous course, Appendix)*

In 24 and 25, estimate mentally to the nearest whole number. *(Previous course)*

24. $12 \cdot 9.95$

25. $\frac{5}{6} + \frac{7}{8} + \frac{13}{12}$

Exploration

26. Computer programs in BASIC can be used to create tables like the one that is asked for in Question 9. Suppose Kim is saving \$9 per week. Write and run a BASIC program to find out when Kim and Jenny will have the same amount of money.

LESSON

5-3

Solving $ax + b = cx + d$

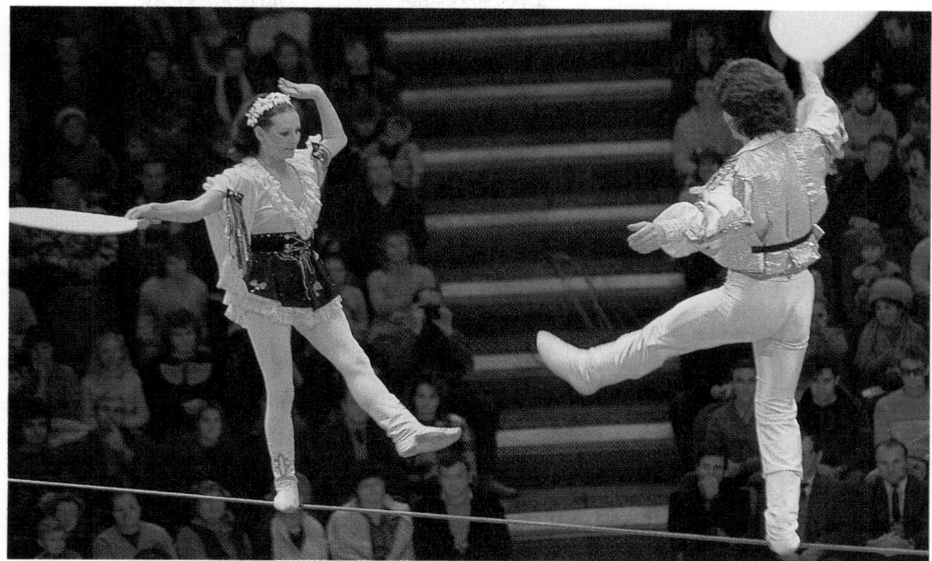

A delicate balance. *In balancing equations, as in balancing on a high-wire, you need to keep equal amounts on both sides of the middle.*

Representing $ax + b = cx + d$

This balance scale is similar to the one pictured in Lesson 3-5. Each ball represents 1 ounce, and all boxes have equal weight w. If the two sides of the scale are balanced, what is the weight of one box? The answer can be found algebraically. If w is the weight of one box, then the situation is described by the equation

$$5w + 1 = 2w + 13.$$

The difference between this equation and ones that you have solved before is that the variable w is on each side of the equation.

To solve this equation pictorially, remove 2 boxes from each pan.

$$3w + 1 = 13$$

This equation is of the form $ax + b = c$, so remove 1 oz from each pan.

$3w = 12$

Then leave one-third of the contents of each pan.

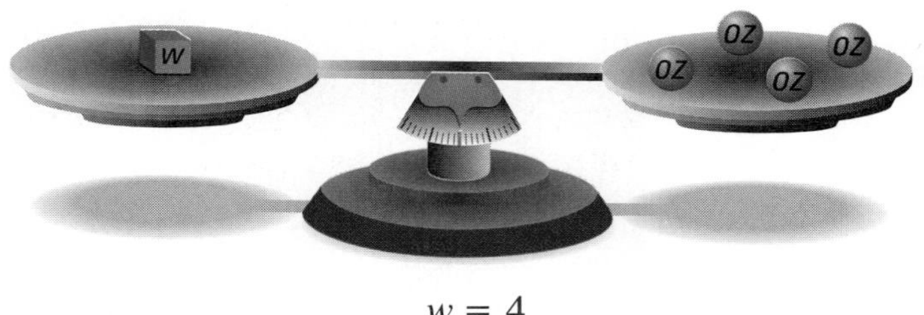

$w = 4$

One box weighs 4 ounces.

Solving $ax + b = cx + d$ Algebraically

Example 1 shows this process algebraically. There are three major steps.

Example 1

Solve $5w + 1 = 2w + 13$.

Solution 1

$-2w + 5w + 1 = -2w + 2w + 13$	Addition Property of Equality, (Add $-2w$ to each side.)
$3w + 1 = 13$	
$3w + 1 + -1 = 13 + -1$	Addition Property of Equality,
$3w = 12$	(Add -1 to each side.)
$\frac{1}{3}(3w) = \frac{1}{3}(12)$	Multiplication Property of Equality,
$w = 4$	(Multiply each side by $\frac{1}{3}$.)

Solution 2

Some people do the additions and multiplications in their heads.

$$5w + 1 = 2w + 13$$
$$3w + 1 = 13$$
$$3w = 12$$
$$w = 4$$

Check

Substitute 4 for w in the original equation.
Does $5 \cdot 4 + 1 = 2 \cdot 4 + 13$?
Does $20 + 1 = 8 + 13$? Yes.

The equation in Example 1 has the unknown variable on each side. It is of the form

$$ax + b = cx + d$$

and is called the **general linear equation.** To solve equations of this type, you can add either $-cx$ or $-ax$ to both sides. Or, subtract either cx or ax from both sides. This step removes the variable from one side and leaves an equation of the kind you have solved in previous lessons.

Applying $ax + b = cx + d$

Let us return to the copier example of Lesson 5-2. There we compared costs of Acme Copiers and Best Printers. Acme's charge for n copies was given by $C = 250 + 0.01n$ and Best's was $C = 70 + 0.03n$. By solving an equation, we can determine precisely when the charges are equal.

Example 2

Solve $250 + 0.01n = 70 + 0.03n$. Check your answer.

Solution

Either $0.01n$ or $0.03n$ can be subtracted from both sides. We choose to subtract $0.01n$.

$$250 + 0.01n - 0.01n = 70 + 0.03n - 0.01n$$

$$250 = 70 + 0.02n \quad \text{Combine like terms.}$$

$$250 - 70 = 70 + 0.02n - 70 \quad \text{Subtract 70 from each side.}$$

$$180 = 0.02n$$

$$\frac{1}{0.02} \cdot 180 = 0.02n \cdot \frac{1}{0.02} \quad \text{Multiply both sides by } \frac{1}{0.02} \text{ (or 50).}$$

$$9000 = n$$

Check

This is the same solution we found in Lesson 5-2 by using a table. Substitute 9000 for n wherever n appears in the original equation.

Does $250 + 0.01 \cdot 9000 = 70 + 0.03 \cdot 9000$?

$$250 + 90 = 70 + 270$$

$$340 = 340 \quad \text{Yes, it checks.}$$

So when 9000 copies are made in a month, the charges of Acme and Best are equal.

Solving Linear Equations with Subtraction

In Example 2, subtracting $0.03n$ from both sides would have introduced a negative coefficient. Still, the solution would have been the same, and it would have taken the same number of steps. Sometimes the choice of what term to add to both sides of an equation can lead to a shorter solution, as shown in Example 3.

Example 3

Solve $-7y - 55 = 4y$.

Solution 1

Add $7y$ to both sides. The equation can now be solved in two additional steps.

$$7y + -7y - 55 = 7y + 4y$$

$$-55 = 11y \quad \text{Add like terms.}$$

$$-5 = y \quad \text{Multiply by } \tfrac{1}{11}.$$

Solution 2

Add $-4y$ to both sides. The solution takes three additional steps.

$$-4y + -7y - 55 = 4y + -4y$$

$$-11y - 55 = 0 \quad \text{Add like terms.}$$

$$-11y = 55 \quad \text{Add 55 to each side.}$$

$$y = -5 \quad \text{Multiply each side by } -\tfrac{1}{11}.$$

Check

Does $-7(-5) - 55 = 4(-5)$?

$35 - 55 = -20$? Yes, it checks.

In general, solving equations of the form $ax = bx + c$, or $bx + c = ax$, by adding $-bx$ to both sides of the equation requires fewer steps than adding $-ax$ to both sides.

Example 4 shows another real application of solving a linear equation.

El Paso, Texas

Example 4

The 1990 population of El Paso, Texas, was 515,000. In the 1980s, its population increased at an average rate of 9000 people per year. In 1990, the population of Milwaukee, Wisconsin, was 628,000. In the 1980s, its population decreased at the rate of 800 people a year. If these rates continue, in what year will the populations of the cities be the same?

Solution

Find expressions for the populations in years after 1990. In El Paso, n years after 1990 the population will be $515{,}000 + 9000n$. In Milwaukee, n years after 1990 the population will be $628{,}000 - 800n$. The populations will be the same when the two expressions are equal. Solve this equation for n.

$$515{,}000 + 9000n = 628{,}000 - 800n$$

Add $800n$ to each side.

$$515{,}000 + 9000n + 800n = 628{,}000 - 800n + 800n$$

$$515{,}000 + 9800n = 628{,}000$$

Add $-515{,}000$ to both sides.

$$-515{,}000 + 515{,}000 + 9800n = -515{,}000 + 628{,}000$$

$$9800n = 113{,}000$$

$$n \approx 11.53 \quad \text{Multiply by } \tfrac{1}{9800}.$$

At these rates, about 11.53 years after 1990, that is, in 2001 or 2002, the populations would be the same.

Check

In 11.53 years after 1990, the population of El Paso will be 515,000 + 11.53 · 9000, or 618,770. The population of Milwaukee will be 628,000 − 11.53 · 800, or 618,776. These figures are close enough, given the accuracy of the information.

Because there are many situations that lead to linear equations, solving linear equations is considered by many people to be the most important skill in beginning algebra.

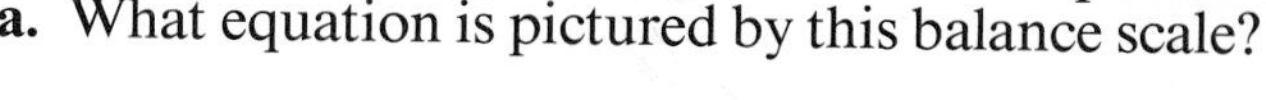

QUESTIONS

Covering the Reading

1. The boxes are of equal weight. Each ball represents 1 ounce.
   a. What equation is pictured by this balance scale?

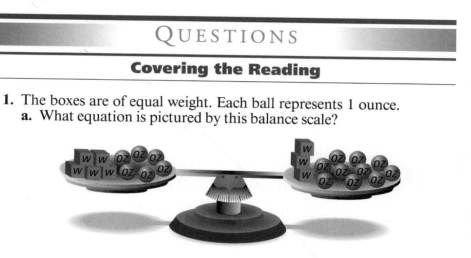

   b. Describe the steps you could use to find the weight of a box using a scale.
   c. What is the weight of one box?

2. a. To solve $8x + 7 = 2x + 9$, what could you add to both sides so x is on only one side of the equation?
   b. Solve the equation in part **a.**

3. Solve the equation of Example 2 by subtracting $0.03n$ from both sides.

4. In solving Example 3, what advantage does adding $7y$ to both sides of the equation have over adding $-4y$?

In 5–12, solve and check.

5. $2p + 38 = 5p + 5$
6. $4n = -2n + 3$
7. $12x + 1 = 3x - 8$
8. $43 - 8w = 25 + w$
9. $7y = 5y - 3$
10. $2 - z = 3 - 4z$
11. $12 + 0.6f = 1.2f$
12. $2.85p - 3.95 = 9.7p + 9.75$

13. Alaska's population, which had been increasing at a rate of 15,000 people a year, reached 550,000 in 1990. Delaware's population, which had been increasing at a rate of 7,000 people a year reached 666,000 in 1990. If the rates of increase do not change, when will the populations be equal?

14. Refer to Example 4. The 1990 population of Jacksonville, Florida, was 673,000. It had been increasing at a rate of 13,200 people each year. Assuming the rates had not changed, when *were* the populations of El Paso and Jacksonville the same?

15. An equation is solved below. Fill in the blanks to explain the steps of the solution.

$x - 1 = -2x - 3$	Add **a.** to each side.
$3x - 1 = -3$	**b.** to each side.
$3x = -2$	**c.** each side by **d.**
$x = -\frac{2}{3}$	

Applying the Mathematics

Swimming for the gold. *In 1988, U.S. Olympic swimmer Janet Evans won 3 gold medals and set an Olympic record in the 800-meter freestyle with a time of 8:20.20 minutes.*

In 16–19, solve.

16. $1.5c + 17 = 0.8c - 32$

17. $3d + 4d + 5 = 6d + 7d + 8$

18. $3(x - 4) = 4(x - 3)$

19. $7(3y - 6) = 14y$

20. Five more than twice a number is three more than four times the number. What is the number?

21. In 1992, the women's Olympic winning time for the 100-meter freestyle in swimming was 54.64 seconds. The winning time had been decreasing at an average rate of 0.33 second per year. The men's winning time was 49.36 seconds and had been decreasing by an average of 0.18 second a year. Assume that these rates continue.
   a. What will the women's 100-meter winning time be x years after 1992?
   b. What will the men's 100-meter winning time be x years after 1992?
   c. After how many years will the winning times be the same?

22. Refer to the figures below. The perimeter of the triangle is equal to the perimeter of the square. Find the length of a side of the square.

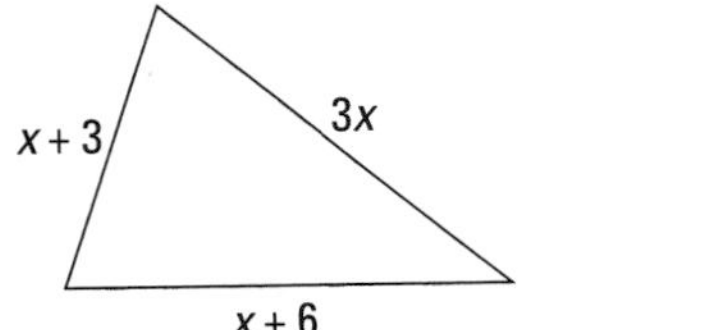

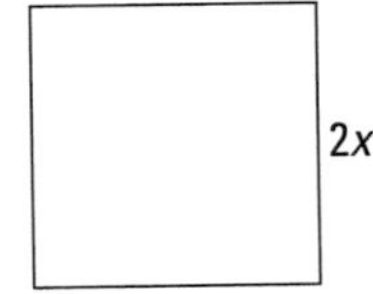

Review

23. Lake Mills is 24 miles by road north of Columbus. Lamont bicycles north from Lake Mills at a rate of 9 mph. Chris leaves Columbus at the same time, also traveling north but at a rate of 13 mph. *(Lesson 5-2)*
   a. Make a table showing how far each bicyclist is from Columbus after hours 0 through 6.
   b. How long does it take Chris to catch up to Lamont?

24. Write an equation for the line containing the points $(4, n)$, $(0, n)$, and $(-2, n)$. *(Lesson 5-1)*

25. Graph $y = 4x - 6$. *(Lesson 4-8)*

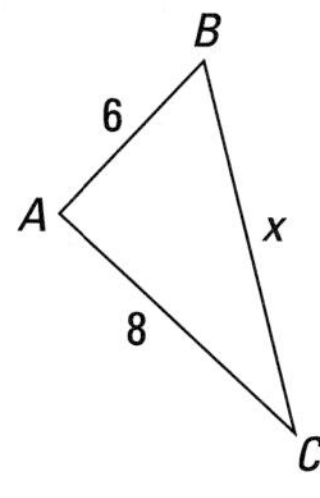

26. Refer to the triangle at the left.
   a. What are the possible values of x?
   b. If $\angle A$ is a right angle, what is the value of x?
   c. Redraw the picture making $\angle B$ a right angle and find x. *(Lessons 1-8, 4-7)*

27. Rewrite $3(x + 2)$ using the stated property.
   a. Commutative Property of Addition
   b. Commutative Property of Multiplication
   c. Distributive Property *(Lessons 2-1, 3-1, 3-6)*

28. The opposite of 4 is added to the reciprocal of 4. What is the result? *(Lessons 2-2, 3-2)*

29. Tell whether the number is a solution to $3x \leq 2x + 6$. *(Lesson 1-1)*
   a. 6 b. 10 c. $\sqrt{31}$

Exploration

30. This puzzle is from the book *Cyclopedia of Puzzles,* written by Sam Loyd in 1914. If a bottle and a glass balance with a pitcher, a bottle balances with a glass and a plate, and two pitchers balance with three plates, how many glasses will balance with a bottle?

LESSON

5-4

Using Graphs to Compare Linear Expressions

The Tortoise and the Hare. *This detail from a picture of Aesop's famous fable was drawn by Arthur Rackham in 1912. See Example 2.*

In Lesson 5-2, situations leading to linear expressions were described with tables. In Lesson 5-3, equations were used. Here, we use graphs.

Consider again the costs of renting copiers from Acme Copiers and Best Printers. Recall that Acme charges \$250 per month plus \$.01 per copy, and Best charges \$70 per month plus \$.03 per copy. The table below is from Lesson 5-2. The graph is of the cost equations $y = 250 + 0.01x$ for Acme and $y = 70 + 0.03x$ for Best.

Number of Copies x	Acme's Charges $y = 250 + 0.01x$	Best's Charges $y = 70 + 0.03x$
0	250	70
2,000	270	130
4,000	290	190
6,000	310	250
8,000	330	310
10,000	350	370
12,000	370	430
14,000	390	490

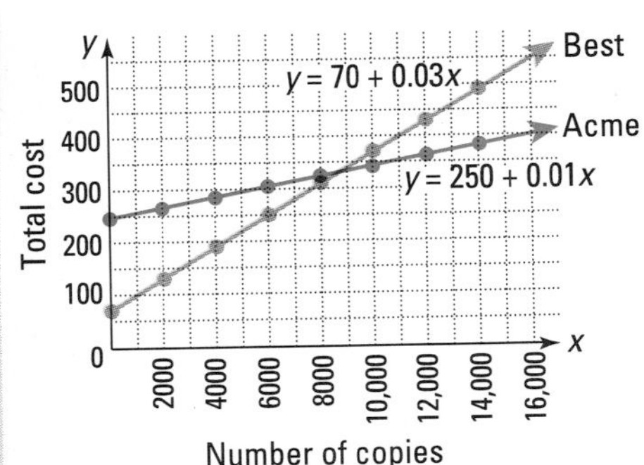

Notice that the graph of each equation is a line. The line for the cost of renting a copier from Best starts lower on the graph. Look on the y-axis for the cost when 0 copies are made; Best's price is \$70. The basic fee for an Acme copier is higher; when $x = 0$, the price is \$250.

Now look to the right of the y-axis. Notice that as the number of copies goes up, the price charged by each plan goes up. The cost increases faster for a Best copier, because its "per-copy" charge is greater.

Example 1

Use the graph to estimate the answer to each question.

a. For how many copies per month are the costs of Acme and Best equal?
b. For how many copies per month is Best more expensive than Acme?
c. For how many copies per month is Best less expensive than Acme?

Solution

a. Find the point where the two lines intersect. The x-coordinate of this point is the "break-even" value. The two lines cross at about $x \approx 9000$. So the costs are equal when $x \approx 9000$.
b. Best is more expensive than Acme when Best's line is higher. That is about when $x > 9000$.
c. Best is less expensive than Acme when Best's line is lower. That is about when $x < 9000$.

Check

The equation $250 + 0.01x = 70 + 0.03x$ tells when the costs are equal. This was solved as Example 2 in the last lesson. The solution is $x = 9000$.

Graphs, tables, and algebraic sentences each have advantages in describing information. Graphs can picture a great deal of information, and they are very useful for comparing values. Graphs may be time-consuming to make, but automatic graphers draw them quickly. Tables can include important specific values. Long tables might be time-consuming to make, but computer programs such as spreadsheets can make them easily. Solving sentences is often preferred because it is the most efficient tool, and the result is a precise answer.

For the next example, recall Aesop's fable about the tortoise and the hare. Their race can be pictured with a graph.

Example 2

The tortoise and the hare decide to have a race. To be generous, the hare decides to give the tortoise a 100-ft head start. The hare runs at 5 ft/sec, and the tortoise's speed is 0.1 ft/sec.

a. Write equations to show the distance d each animal is from the starting point after t seconds.
b. Graph the equations over the interval $0 \le t \le 30$ seconds.
c. Use the graph to tell when the hare reaches the tortoise.
d. Use an equation to tell when the hare reaches the tortoise.

Solution

a. Let t = the number of seconds an animal runs, and let d = the number of feet the animal is from the starting point. The hare's distance at t sec is given by $d = 5t$. Due to the head start, when the race begins the tortoise is already 100 ft from the starting point. After t seconds, the tortoise has gone $100 + 0.1t$ ft. The tortoise's distance is given by $d = 100 + 0.1t$.

▶

b. To graph each equation over the interval $0 \le t \le 30$, make a table with $t = 0$ and $t = 30$. You should also choose a third point (here, we use $t = 15$) to check that the other two points and the line you draw between them are correct. Plot the points and draw the two lines.

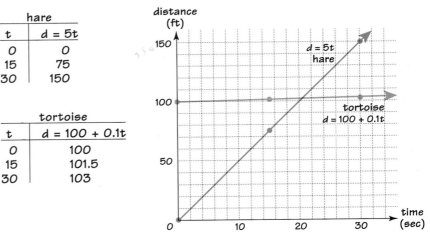

hare

t	d = 5t
0	0
15	75
30	150

tortoise

t	d = 100 + 0.1t
0	100
15	101.5
30	103

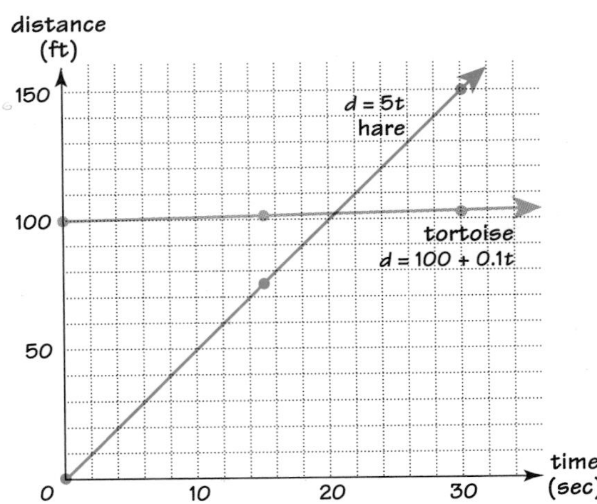

c. Look to see where the lines cross. This happens at a point whose first coordinate is between 20 and 22 seconds. The hare passes the tortoise after about 21 seconds.

d. When the hare reaches the tortoise, their distances will be equal.

$$\text{tortoise's distance} = \text{hare's distance}$$
$$100 + 0.1t = 5t$$

Now solve the equation.

$$100 + 0.1t - 0.1t = 5t - 0.1t$$
$$100 = 4.9t$$
$$\frac{1}{4.9} \cdot 100 = \frac{1}{4.9} \cdot 4.9t$$
$$20.4 \approx t$$

The hare reaches the tortoise after about 20.4 seconds. This result verifies the answer found in part **c** and is more accurate.

QUESTIONS

Covering the Reading

In 1–5, use Example 2.

1. Find the distance each animal is from the starting point after 12 seconds.

2. Find the distance each animal is from the starting point after 60 seconds.

3. When the hare reaches the tortoise, about how far are they from the starting point?

4. Suppose the race course is 110 ft long.
   a. Who will win the race? **b.** How long will it take?

5. The hare had another race, one with a rat that ran 2 ft/sec. The rat had a 50-foot head start.
   a. Write an equation to describe the rat's distance d after t seconds.
   b. Draw a graph showing the hare's and rat's distances over the interval $0 \le t \le 60$ seconds.
   c. Use your graph to find out when the hare reaches the rat.

In 6 and 7, use the graph at the right. It shows the graph from the beginning of this lesson with another line added to show the prices for renting from Carlson's Copy Machines. Carlson's charges a \$10-per-month basic fee and 5¢ a copy. So for Carlson's, $y = 10 + 0.05x$.

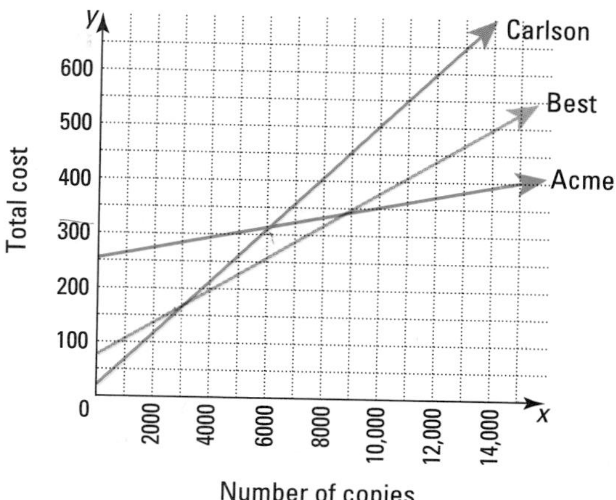

6. **a.** For about how many copies per month is Carlson's price the same as Best's price?
   b. About how much will it cost for the number of copies in part **a?**
   c. Solve an equation to verify your answer to part **a.**

7. **a.** For how many copies do Carlson's and Acme charge the same price?
   b. About how much will it cost for the number of copies in part **a?**
   c. Solve an equation to verify your answer to part **a.**

Applying the Mathematics

8. Julio needs to rent a car for a week. For comparable cars, American Rental charges \$70 per week plus 10¢ per mile, and Coast-to-Coast Rent-a-Car charges \$50 per week plus 13¢ per mile.
   a. Write equations to describe the cost of the rental car at each agency. Let x = the number of miles driven, and y = the cost in dollars.
   b. How much would it cost to rent a car from American for 1 week if you drove 600 miles?
   c. Draw a graph showing the cost of renting from each company for trips up to 1000 miles during the week.
   d. According to your graph, for what mileage will the two rental agencies cost the same amount?
   e. Use an equation to tell when the two rental agencies will cost the same amount.
   f. Julio expects to drive between 700 and 1000 miles during the week. What advice would you give him?

9. Refer to the data at the start of the lesson. Acme Copiers decides to change prices. Their copy machines will now cost \$200 plus $1\frac{1}{2}$ cents per copy. Best Printers will keep their prices the same.

a. Use a graph to determine what number of copies will cost the same for both companies.

b. Use an equation to find out when the costs will be equal.

10. Drucilla and Phil traveled from Omaha, Nebraska, to Boise, Idaho, a distance of about 1250 miles. Drucilla drove, while Phil left later and flew. The graph shows the distance y each had traveled x hours after Drucilla began her trip.

a. Who arrived in Boise first? How can you tell?

b. How long did Phil's trip take?

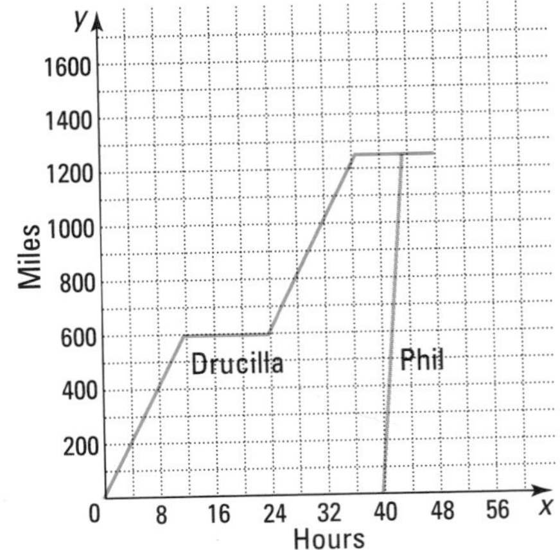

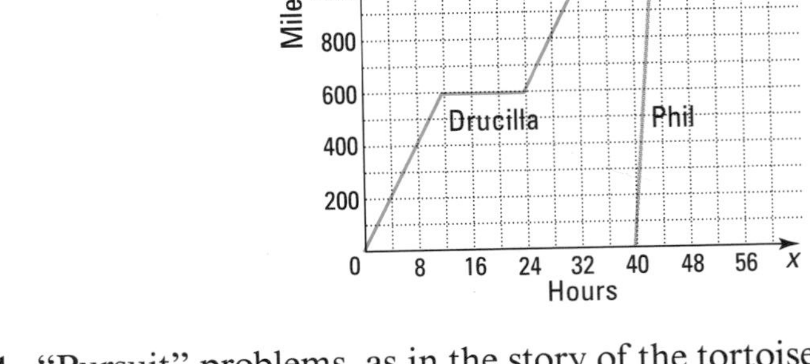

11. "Pursuit" problems, as in the story of the tortoise and the hare in Example 2, have been around for a long time. The following problem appeared in an ancient Chinese textbook around the first century A.D.

> A hare runs 50 pu (paces) ahead of a dog. The dog pursues the hare for 125 pu but the hare is still 30 pu ahead. How many pu will the dog have traveled altogether when he overtakes the hare?

Solve the problem, and explain how you solved it.

This is a modern Chinese algebra book. The title (written vertically far right) is Junior High School Algebra I.

Review

In 12–15, solve. *(Lesson 5-3)*

12. $3p - 5 = 2p + 12$

13. $-7r + 54 = 2r - 36$

14. $6(x + 3) + 4(3x - 75) = 258$

15. $6m = 2m + 2$

16. GE sells a type of light bulb called Energy Choice™ that uses less electricity than its regular Soft White bulbs. However, the initial cost of the bulb is more than that of the regular Soft White brand. The table below shows the cost of using a 60-watt bulb of each type in one part of the country. *(Lesson 5-2)*

A bright idea. *This is a replica of the first light bulb invented by Thomas Edison in 1879. Edison's first bulb burned brightly for only two days.*

hours	Energy Choice	Soft White
0	$0.75	$0.59
100	$1.17	$1.07
200	$1.59	$1.55
300	$2.01	$2.03
400	$2.43	$2.51
500	$2.85	$2.99
600	$3.27	$3.47
700	$3.69	$3.95
800	$4.11	$4.43
900	$4.53	$4.91
1000	$4.95	$5.39

a. How much does a new Soft White 60-watt bulb cost?
b. How much more expensive is an Energy Choice bulb?
c. After approximately how many hours of use are the total costs of the two bulbs the same?
d. If a bulb lasts for 1000 hours of use (the average life according to GE), how much money would you save by buying an Energy Choice bulb?

17. **a.** On one coordinate grid graph $y = -3$ and $x = 4$.
b. Give the coordinates of the point the two lines have in common. *(Lesson 5-1)*

18. *Multiple choice.* Which of the following does *not* equal $\frac{x}{6}$? *(Lesson 2-3)*
(a) $\frac{1}{6}x$ (b) $\frac{2}{12}x$ (c) $\frac{3x}{18}$ (d) $\frac{4x}{10}$

19. What will be printed when this BASIC program is run? *(Lessons 1-4, 1-6)*

```
10 PRINT "SQUARE ROOTS AND SQUARES"
20 FOR A = 1 TO 6
30 PRINT SQR(A), A, A^2
40 NEXT A
50 END
```

Exploration

20. Find examples of two similar situations that lead to linear expressions, as in Example 1 of this lesson. You may want to look in newspapers or magazines or visit a store. Draw a graph to compare the values of these expressions. What conclusions can you make based on your graph?

LESSON

5-5

Using an Automatic Grapher

Graph it your way. *This student is displaying the same graph on both the computer screen and the graphics calculator. Both tools allow you to display multiple graphs at one time and find points of intersection using a zoom feature.*

What Is an Automatic Grapher?

Graphs of equations are so helpful in solving problems that there are now several makes of calculators that will automatically display part of a graph. Also, there are programs for every personal computer that will display graphs. Because computer screens are larger than calculator screens, they can show a greater part of a graph, but graphing calculators are less expensive and easier to carry around.

Graphing calculators and computer graphing programs work in much the same way. As such we call them **automatic graphers** and do not distinguish between them. Of course, no grapher is completely automatic. For each you must learn particular keys to press. Here we discuss some of the general features of automatic graphers.

Instructions

The part of the coordinate grid that is shown is called a **window.** The screen below displays a window in which

$$-2 \le x \le 12$$
$$\text{and } -3 \le y \le 7.$$

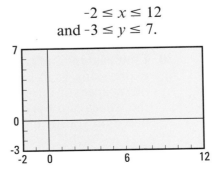

Most graphers have a **default window,** set by the manufacturer. In this lesson we use $-15 \le x \le 15$ and $-10 \le y \le 10$ as the default window. If your graph does not appear on the default window, you will need to determine a new window to use.

You usually enter the instructions for the window first, and then enter the equation to be graphed. On almost all graphers, the equation to be graphed must be a formula for y in terms of x.

$y = 4x - 5$ can be handled.
$x + y = 180$ cannot be handled.

On calculator screens, the intervals for x and y are often left unmarked.

Graphing a Linear Equation on an Automatic Grapher

Example 1

Draw the graph of $y = 4x - 5$ using an automatic grapher's default window.

Solution

Notice that y is given in terms of x, so the equation may be entered directly. Enter the equation. Here is a key sequence for one calculator.

[y =] 4 [×] [X, T, θ] [−] 5 [ENTER]

Press [GRAPH] to display the graph.

Here is the graph as it appears on our default screen.

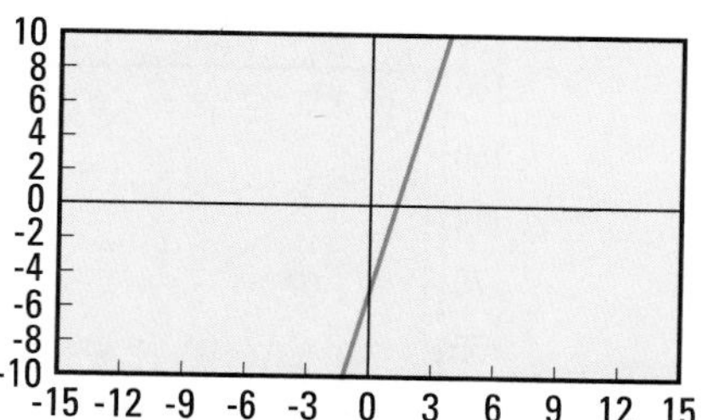

You need to clear each equation before you graph the next one unless you wish both graphs to appear.

Activity 1

Draw the graph of $y = 2x - 5$ on your default window. Copy it on paper. Describe one thing your graph has in common with the graph in Example 1.

Example 2

Draw a graph of $x + y = 180$.

Solution

To graph this equation on most automatic graphers you must first solve for y. That is, you must enter $y = -x + 180$ or $y = 180 - x$. Enter one of these equations on your grapher, and graph it on the default window. You should not see any of the graph—all the points on the graph are off the screen. So on most graphers, you must change the window. You need to pick the x values for each end of the window. These are sometimes called x-min (for minimum x value) and x-max (for maximum x value). Some graphers automatically pick the y values after you have chosen the x values so all the key features of the graph can be seen. On other graphers, you must also enter y-min and y-max. If your grapher shows **tick marks** (marks on the x- and y-axes), you may change the intervals for the tick marks too.

Below are graphs of $y = 180 - x$ on four different windows.

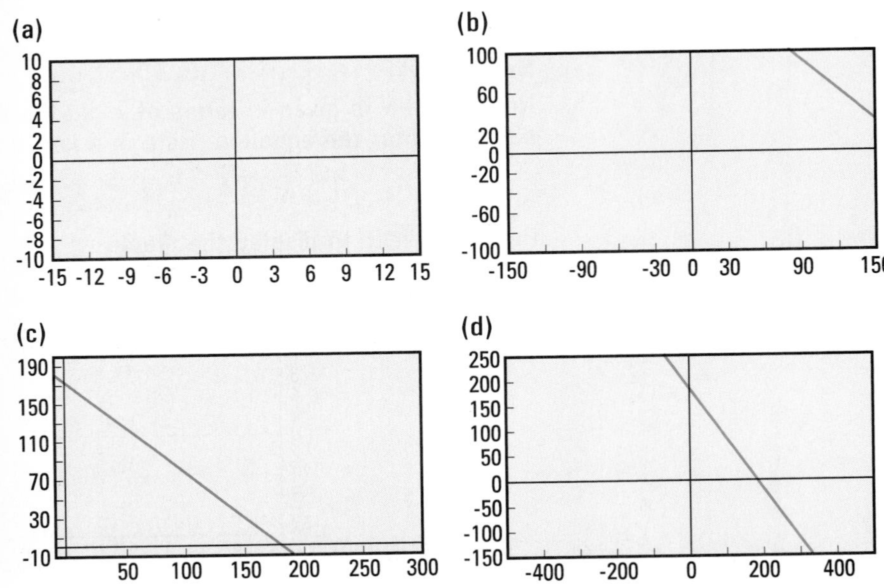

Notice how changing the window can change the appearance of the graph. In (d) the graph appears steep; in (b) and (c) it appears less steep. On (b) you see points in only the first quadrant; on others you see points in quadrants I, II, and IV. The mathematical properties of the graph, however, do *not* change when you change the window. For instance, each graph contains the points (0, 180) and (60, 120) even though they may not appear on the screen.

Most automatic graphers have a **trace** option. This allows you to move a **cursor,** often an arrow or blinking pixel, along the graph. As the cursor moves, it tells you one or both coordinates of the point it is on.

Activity 2

Clear the equation from Activity 1. Draw a graph of $x + y = 180$ on a window different from those in Example 2. Trace along the line on the screen. Sketch the graph on paper. You should see that, as the cursor moves to the right along the graph of $y = 180 - x$, the x-coordinate increases, and the y-coordinate decreases. You should also verify that the sum of the x- and y-coordinates is always 180. Write the coordinates of three points that are on your graph.

Graphing Two Equations at a Time

Automatic graphers can graph more than one equation at a time. They also can easily graph complicated equations. This enables them to help you solve problems like those in the previous lesson.

Example 3

The cost per month y of renting a copier and making x copies from Acme Copiers is given by $y = 250 + 0.01x$. The cost from Best Printers is given by $y = 70 + 0.03x$. For what number of copies is Acme cheaper?

Solution

With an automatic grapher, first enter the equation $y = 250 + 0.01x$ to be graphed. (You may need to enter the multiplication symbol between 0.01 and x.) Then enter the second equation $y = 70 + 0.03x$. The window $0 \le x \le 16{,}000$, $0 \le y \le 550$ will give a graph like the one on page 304.

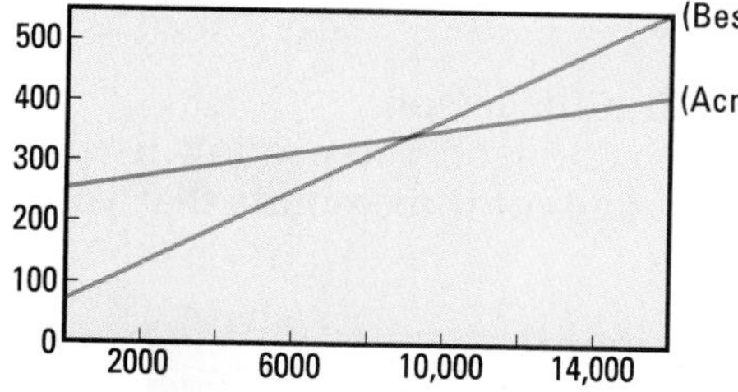

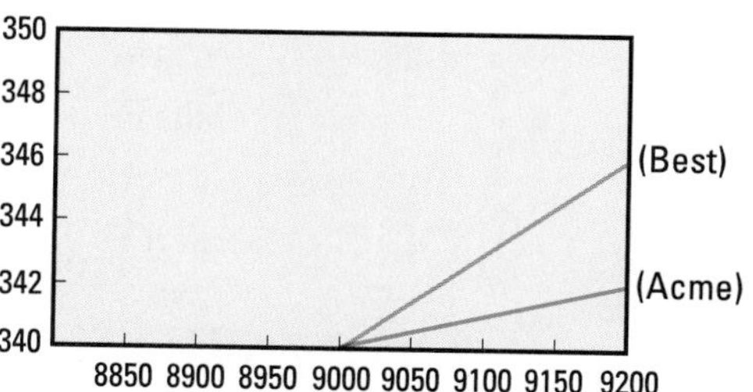

Now use the trace option to move the tracer near the point of intersection of the two lines. Our grapher indicates a point of intersection near (9095, 343). To be more accurate, either **zoom in** on the point of intersection or create a new window. We created a new window, $8800 \le x \le 9200$ and $340 \le y \le 350$. It is shown at the right above. This window indicates that the intersection is very near (9000, 340). Check these values in the equations to confirm that this is the point of intersection. Thus, the line for Acme is below the line for Best when $x > 9000$. This means that Acme is cheaper when the number of copies is greater than 9000.

Many automatic graphers will print a **hard copy**, which is a paper copy of the graph from the screen. When copying or printing graphs, be sure to identify the limits of the window, as we have done.

QUESTIONS

Covering the Reading

1. On an automatic grapher,
   a. what is a window?
   b. What is a default window?

2. What are the dimensions of the default window on your automatic grapher?

3. Use two inequalities to describe the window shown below.

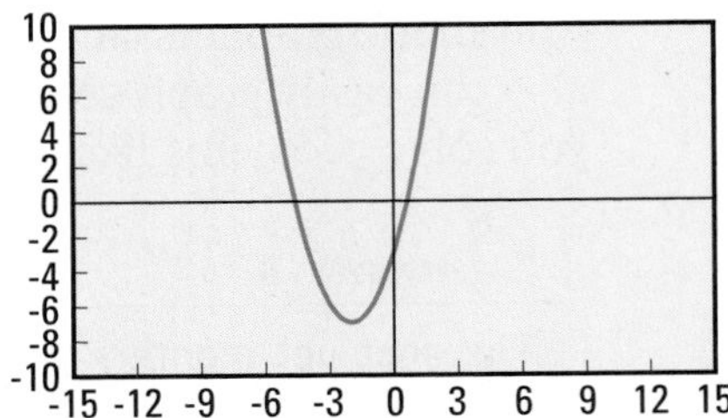

In 4–6, tell whether the equation is in a form with which it can be entered into your grapher.

4. $y = 4 + 3x^2$
5. $y = 2.7x$
6. $x - y = 7$

7. Write your answer to Activity 1 in the lesson.

In 8 and 9, use an automatic grapher to graph the equation on your default window.

8. $y = -2x + 1$
9. $x - y = -5$

10. Refer to Activity 2 in the lesson.
    a. Sketch the graph and label the window used.
    b. Give the coordinates of three points that your cursor shows are on this graph.

11. a. Graph the solutions to $x + y = 90$ on three different windows.
    b. *True or false.* The point (5, 85) is on each graph you drew in part **a.**
    c. Trace along the graph from left to right with a cursor. Describe what happens to the x- and y-coordinates.

12. Refer to Example 3.
    a. Sketch the two graphs on an appropriate default window.
    b. What point on the grid is closest to the point where the two lines intersect?

Applying the Mathematics

It's a tossup. *Some pizza chefs toss the dough to stretch it to the correct thickness. Other chefs say flipping the dough isn't essential for good pizza.*

13. Set the window of your grapher to $-10 \le x \le 10$, $-15 \le y \le 15$.
a. Graph $y = x$, $y = 2x$, $y = 3x$, and $y = 4x$ on the same screen.
b. Describe the graphs. How are they similar? How are they different?
c. Predict what the graph of $y = 5x$ will look like. Check by graphing.

14. Battaglia's Pizza Parlor charges $10 for a large cheese pizza plus $0.90 for each additional topping. Tonelli's charges $12 for the same pizza but only $0.40 for each additional topping. Graph the equations $y = 10 + .90x$ and $y = 12 + .40x$ to find the number of toppings for which Tonelli's is cheaper than Battaglia's.

15. a. Use the window $-15 \le x \le 15$, $-10 \le y \le 10$ to graph $y = 2x$ and $y = x^2$. Copy the graphs.
b. Use the graphs to explain why $x + x$ is or is not equal to x^2.

16. a. Graph $y = \sqrt{x}$ when $0 \le x \le 100$ and $0 \le y \le 10$.
b. Use your cursor to find the coordinates of five points on the graph.

Review

17. Solve $7(d + 5) = 11(2d - 20)$. *(Lessons 3-7, 5-3)*

18. Answer Question 14 by making a table. *(Lesson 5-2)*

19. *Multiple choice.* Which equation describes the ordered pairs? *(Lesson 4-6)*
a. $y = 5x$
b. $y = x + 5$
c. $y = x - 5$
d. $y = \frac{x}{5}$

x	y
0	-5
1	-4
2	-3
3	-2

20. Simplify $(17a + 12) - (a - 4)$. *(Lesson 4-5)*

21. Solve $(4x - 1) - 3(x + 2) = 2$. *(Lesson 4-5)*

22. *Skill sequence.* Solve. *(Lesson 4-3)*
a. $2W - 7 < 1$
b. $7 - 2x < 1$
c. $2y - 7 < -1$
d. $7 - 2z < -1$

In 23 and 24, a cook has x slices of pastrami for sandwiches. The cook needs s slices for a small sandwich and m slices for a medium sandwich. Write an expression for the amount of pastrami left in each situation. *(Lessons 3-8, 4-2)*

23. a. The cook takes meat first for one small sandwich, then for two medium ones.

b. The cook takes enough at one time for one small and two medium sandwiches. (Use parentheses.)

24. a. The cook takes enough for 3 smalls and a medium, then later takes enough for 1 small and 2 mediums.

b. Simplify your answer to part **a.**

25. *Multiple choice.* Suppose $j < n$. What is true about $j - n$?

(a) It is always positive.
(b) It is always negative.
(c) It can be either positive or negative. *(Lesson 4-1)*

26. A tortoise is walking at a rate of $3\ \frac{\text{feet}}{\text{minute}}$. Assume this rate continues.

a. How long will it take the tortoise to travel 20 feet?

b. How long will it take the tortoise to travel f feet? *(Lesson 2-4)*

27. *Skill sequence.* Solve in your head. *(Lesson 1-4)*

a. $x + \sqrt{9} = \sqrt{25}$

b. $\sqrt{y} + \sqrt{9} = \sqrt{25}$

c. $\sqrt{z} + \sqrt{z} = \sqrt{64}$

Exploration

28. Find four equations whose graphs are lines that form a square, as shown below.

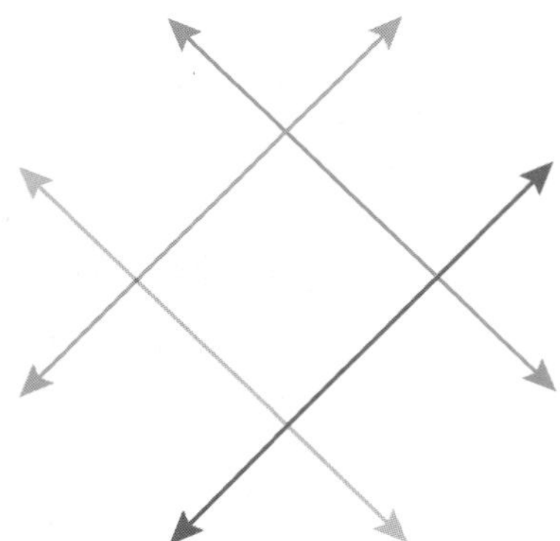

29. Determine whether or not your automatic grapher has the following features. If so, describe what the feature does and how to access it.

a. zoom in

b. zoom out

c. scatterplot of individual points

LESSON

5-6

Solving $ax + b < cx + d$

Arbor Day. *These students are observing Arbor Day by planting trees on the school grounds. Arbor Day is observed at different times throughout the U.S.*

The Addition and Multiplication Properties of Inequality from Chapters 2 and 3 can be used to solve linear inequalities of the form $ax + b < cx + d$.

Consider the following situation. When they are young, many species of trees grow at rates that are nearly constant. Some years ago the Lee family planted two trees, a beech tree that was 8 feet tall and a maple tree that was only 3 feet tall. Maple trees grow at a rate of about $1\ \frac{\text{ft}}{\text{year}}$. Beech trees grow more slowly, at about $\frac{1}{2}\ \frac{\text{ft}}{\text{year}}$. So t years after planting, the height of the beech tree is $8 + 0.5t$ feet and the height of the maple tree is $3 + t$ feet.

Example 1

Mr. Lee looked at an old photograph of the two trees. In it, the beech tree was taller than the maple tree. He wondered when the photo was taken.

Solution 1

The beech was taller than the maple, so the heights satisfy the inequality

$$8 + 0.5t > 3 + t.$$

Solve this inequality as you would an equation with a variable on both sides.

$8 + 0.5t + -0.5t > 3 + t + -0.5t$	Add $-0.5t$ to both sides.
$8 > 3 + 0.5t$	Add like terms.
$8 + -3 > 3 + -3 + 0.5t$	Add -3 to both sides.
$5 > 0.5t$	Add like terms
$\frac{1}{0.5} \cdot 5 > \frac{1}{0.5} \cdot 0.5t$	Multiply both sides by $\frac{1}{0.5}$.
$10 > t$	

So $t < 10$. The photo was taken less than 10 years after the Lees planted the trees. ▶

Solution 2

Graph the two equations giving the heights of the two trees after t years, $h = 8 + 0.5t$ and $h = 3 + t$. Then compare the graphs.

t	height h in ft	
number of years	maple	beech
0	3	8
2	5	9
4	7	10
6	9	11
8	11	12
10	13	13
12	15	14

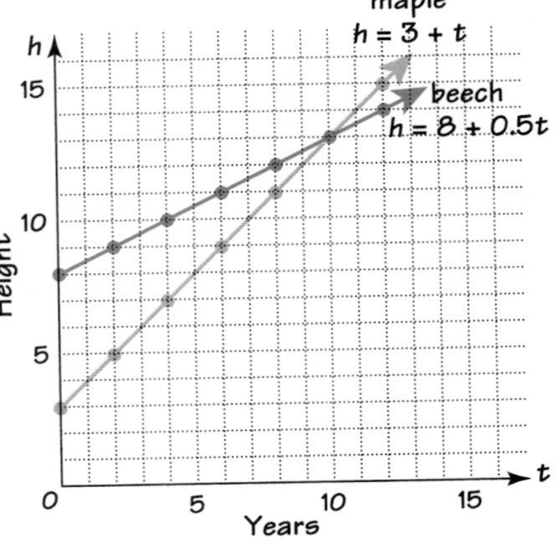

Trees and their leaves. *Beech trees have thin papery leaves that turn gold-colored in the fall. Maple leaves are known for their broad, flat shape with three to seven lobes that resemble fingers. Their dark-green leaves turn red, orange, or yellow in autumn.*

The lines cross at the point where $t = 10$. So ten years after they were planted, the trees were the same height. We are looking for the times when the beech was taller than the maple. So look for values of t where the beech's line is *above* that for the maple. These times lie to the left of the intersection point (10, 13). This is where $t < 10$. The photo was taken less than 10 years after the trees were planted.

The Addition and Multiplication Properties of Inequality can be used to solve any inequality of the form $ax + b < cx + d$, where $a \neq 0$ and $c \neq 0$. In Example 2, two algebraic solutions are given. Solution 2 involves multiplying the inequality by a negative number. Recall that multiplying an inequality by a negative number reverses the sense of the inequality.

Example 2

a. Solve $6 - 9x \leq 5x - 1$.

b. Check your answer.

Solution 1

a. It is reasonable to add either $9x$ or $-5x$ to each side of the sentence.

$6 - 9x + 9x \leq 5x - 1 + 9x$	Begin by adding $9x$ to each side.
$6 \leq 14x - 1$	Add like terms.
$7 \leq 14x$	Add 1 to both sides.
$\frac{1}{14} \cdot 7 \leq 14x \cdot \frac{1}{14}$	Multiply both sides by $\frac{1}{14}$.
$\frac{1}{2} \leq x$	Simplify.

Thus $x \geq \frac{1}{2}$.

Solution 2

a. $6 - 9x + -5x \le 5x - 1 + -5x$ Add $-5x$ to each side.

$6 - 14x \le -1$ Add like terms.

$6 - 14x + -6 \le -1 + -6$ Add -6 to both sides.

$-14x \le -7$ Add like terms.

$-\frac{1}{14}(-14x) \ge -7\left(-\frac{1}{14}\right)$ Multiply each side by $-\frac{1}{14}$, so reverse the inequality sign.

$x \ge \frac{1}{2}$ Simplify.

b. Recall that checking an inequality requires two steps.

Step 1: Check that $x = \frac{1}{2}$ makes each side of the original sentence true as an equality.

Does $6 - 9 \cdot \frac{1}{2} = 5 \cdot \frac{1}{2} - 1$?

$6 - 4\frac{1}{2} = 2\frac{1}{2} - 1$?

Yes, each side equals $1\frac{1}{2}$.

Step 2: Try a number that satisfies $x > \frac{1}{2}$. Does this number satisfy the original inequality? Try $x = 1$.

Is $6 - 9 \cdot 1 \le 5 \cdot 1 - 1$?

$-3 \le 4$? Yes.

Example 3

Three times a number is less than two times the same number. Find the number.

Solution

It may seem that there is no such number. But work it out to see. Let n be such a number. Then n must be a solution to

$$3n < 2n.$$

Solve this as you would any other linear inequality. Add $-2n$ to each side.

$$3n + -2n < 2n + -2n$$

$$n < 0$$

So n can be any negative number.

Check

Step 1: If $n = 0$, $3 \cdot 0 = 2 \cdot 0$, and $0 = 0$.

Step 2: Pick a value of n that is less than 0. We let $n = -5$.
Is $3 \cdot -5 < 2 \cdot -5$? $-15 < -10$? Yes, it checks.

So if n is any negative number, 3 times n will be less than 2 times n.

QUESTIONS

Covering the Reading

In 1–4, refer to Example 1.

1. Which tree was taller when they were planted?

2. Which tree was taller 7 years after they were planted?

3. Which tree will be taller 20 years after they were planted?

4. **a.** Solve $8 + 0.5t < 3 + t$.
b. What does the answer indicate?

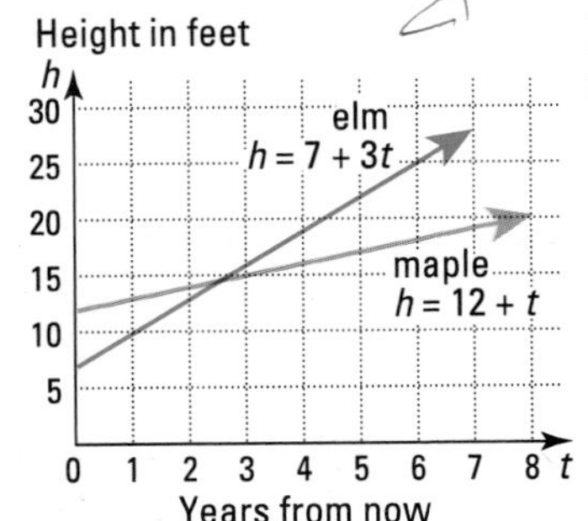

In 5 and 6, use the graph at the left showing the growth habits of two trees.

5. Which tree will be taller 4 years from now? How much taller will it be?

6. When is the maple tree taller than the elm?

7. **a.** Solve $4k + 3 > 9k + 18$ by first adding $-4k$ to each side.
b. Solve $4k + 3 > 9k + 18$ by first adding $-9k$ to each side.
c. Should you get the same answer for parts **a** and **b**?
d. Describe how the steps in the solutions to parts **a** and **b** are different.

In 8–11, **a.** solve, and **b.** check.

8. $5n + 7 \geq 2n + 19$

9. $-48 + 10a \leq -8 + 20a$

10. $4x + 12 < -2x - 6$

11. $12 - 3y > 27 + 7y$

12. Five times a number is less than three times the same number. Find such a number.

Applying the Mathematics

In 13 and 14, solve.

13. $5 - 3(x + 2) < 10x$

14. $-3t \geq -t + 3 + 9t$

15. The Smith family is planning to advertise their used car for sale. The *Gazette* charges \$2.00 plus 8¢ per word. The *Herald* charges \$1.50 plus 10¢ per word. For what length are the *Gazette's* ad rates cheaper? Justify your answer.

16. Sending a package by Fast Fellows shipping service costs \$3.50 plus 25 cents per ounce. Speedy Service charges \$4.75 plus 10 cents per ounce.
a. If your package weighs 1 lb 4 oz, which service is cheaper? Justify your answer.
b. What advice would you give someone who needs to use one of these services?

Review

17. **a.** Graph $y = 10 + x$ and $y = 2 + 3x$ on the same set of axes.
b. According to the graph, for what value(s) of x is $10 + x = 2 + 3x$? *(Lesson 5-4)*

In 18–21, solve. *(Lesson 5-3)*

18. $4x + 12 = -2(x + 3)$

19. $60t - 1 = 48t$

20. $109 - m = 18m - 5$

21. $3n - n + 5 = 4n - n + 20$

22. **a.** Use an automatic grapher to find the intersection of the lines $y = 8$ and $y = 4 + 0.75x$.
b. Solve $8 = 4 + 0.75x$. *(Lessons 3-5, 5-1, 5-5)*

23. Simplify $\frac{t}{4} + \frac{t}{3}$. *(Lesson 3-9)*

24. The distance from Los Angeles to New Orleans, along Interstate 10 as shown, is 1946 miles. At an average speed of $60\frac{\text{miles}}{\text{hour}}$, estimate the driving time from San Antonio to Houston. *(Lessons 2-6, 3-1)*

The River Walk (Paseo del Rio) *is a popular dining and shopping area that stretches along the San Antonio River in San Antonio, Texas.*

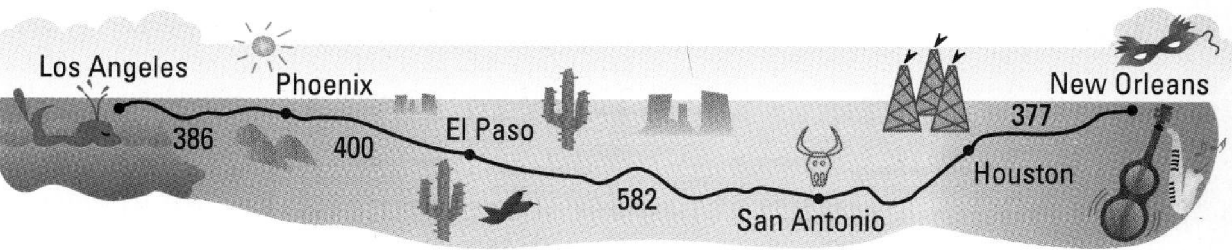

25. Are 12! and 2 · 6! equal? Explain your answer. *(Lesson 2-10)*

26. *Multiple choice.* Which of the following does not equal $\frac{4x}{3}$? *(Lesson 2-3)*

(a) $\frac{4}{3}x$ (b) $4\left(\frac{x}{3}\right)$ (c) $4\left(\frac{1}{3}x\right)$ (d) $4\left(\frac{1}{3x}\right)$

27. *Skill sequence.* Multiply. *(Lesson 2-3)*

a. $30 \cdot \frac{1}{6}$ **b.** $30 \cdot \frac{5}{6}$ **c.** $30 \cdot \frac{x}{6}$ **d.** $30 \cdot \frac{5}{6}x$

28. Four instances of a general pattern are given below. *(Lesson 1-7)*

$$-(7 - 4) = 4 - 7$$
$$-\left(\frac{8}{3} - \frac{5}{6}\right) = \frac{5}{6} - \frac{8}{3}$$
$$-(1.2 - 2.1) = 2.1 - 1.2$$
$$-(v - 9) = 9 - v$$

a. Write the pattern using the variables a and b.
b. Is the pattern true for all real number values of a and b? Justify your answer.

Exploration

29. The square of a number is less than the product of one less than the number and two greater than the number.
a. Find one such number. **b.** Find all such numbers.

LESSON

5-7 Equivalent Formulas

A drive-in volcano. *Kilauea is an active volcano in Hawaii. A nearby road allows visitors to view the lava flow. The stick this tourist is holding ignites as it touches the hot lava that is about 2000°F (about 1100°C).*

Two different temperature scales are in common use throughout the world. The scale used wherever people use the metric system is the **Celsius scale.** That scale is also called the **centigrade scale** because of the 100-degree interval between the freezing point of water, 0°C (read "0 degrees Celsius" or "0 degrees centigrade"), and the boiling point 100°C. (The Latin word for 100 is "centum.") The other scale is the **Fahrenheit scale,** which is now used only in the United States and a few other countries. The freezing and boiling points of water are 32°F (read "32 degrees Fahrenheit") and 212°F, respectively.

A person who visits the United States from most other countries may wish to translate Fahrenheit temperatures into the Celsius temperatures more familiar to himself or herself. One formula for C in terms of F is

$$C = \frac{5}{9}(F - 32).$$

But if you visit another country, you may wish to convert in the other direction, from Celsius to Fahrenheit. This can be done by solving the above formula for F.

Example 1

Solve $C = \frac{5}{9}(F - 32)$ for F.

Solution

Start with the given equation.

$$C = \frac{5}{9}(F - 32)$$

One way to solve for F is to "undo" the multiplication by $\frac{5}{9}$.

$$\frac{9}{5}C = \frac{9}{5} \cdot \frac{5}{9}(F - 32) \quad \text{Multiply each side by } \frac{9}{5}.$$

$$\frac{9}{5}C = F - 32$$

$$\frac{9}{5}C + 32 = F \quad \text{Add 32 to each side.}$$

The desired formula is $F = \frac{9}{5}C + 32$.

Although the formulas $C = \frac{5}{9}(F - 32)$ and $F = \frac{9}{5}C + 32$ look different, they are **equivalent formulas** because every pair of values of F and C that works in one formula also works in the other. One important use of equivalent formulas arises when using most automatic graphers, where formulas to be entered must give y in terms of x.

Example 2

Use an automatic grapher to graph $5x - 2y = 100$.

Solution

It is often necessary to solve for y. Do this as if you were solving any other linear equation for y.

$$5x - 2y = 100$$

$$-5x + 5x - 2y = -5x + 100 \quad \text{Add } -5x \text{ to both sides.}$$

$$-2y = -5x + 100 \quad \text{Add like terms.}$$

$$\left(-\frac{1}{2}\right)(-2y) = -\frac{1}{2}(-5x + 100) \quad \text{Multiply both sides by } -\frac{1}{2}.$$

$$y = -\frac{1}{2}(-5x + 100) \quad \text{Use the Property of Reciprocals.}$$

$$y = 2.5x - 50 \quad \text{Use the Distributive Property.}$$

Now enter the equation $y = 2.5x - 50$ into the grapher, and instruct it to graph. The graph is a line. A window of $-20 \le x \le 30$ and $-60 \le y \le 60$ is shown here.

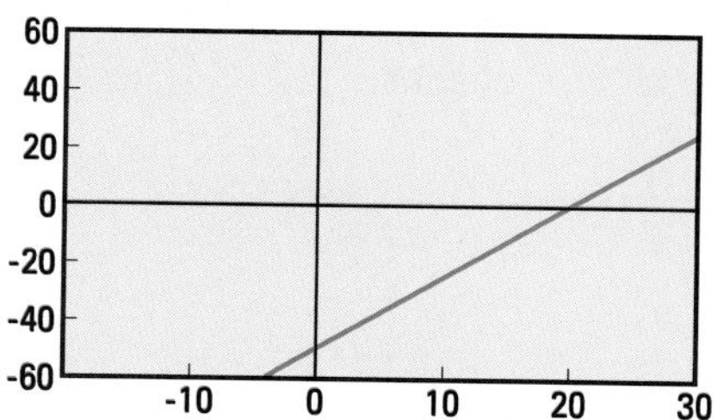

Check 1

Use the trace feature to read the coordinates of some points on the line. Check that these satisfy the original equation. For instance, our trace showed the point with $x \approx 10.5$, $y \approx -23.7$ on the graph.

Does $5(10.5) - 2(-23.7) = 100$?
Is $99.9 \approx 100$? Yes. It checks.
The point $(10.5, -23.7)$ is very close to the graph of $5x - 2y = 100$.

Check 2

Compute the coordinates of a point on the line. For instance,
when $x = 0$,

$$\begin{aligned} 5(0) - 2y &= 100. \\ -2y &= 100 \\ y &= -50 \end{aligned}$$

The trace on our grapher shows that the point $(0, -50)$ is on the line.

Examples 1 and 2 each have two variables. Some important formulas have more than two variables. These can also be rewritten in equivalent forms.

Example 3

The formula $A = p + pr$ gives the amount A (in dollars) after one year in a bank account which started with p dollars and which has an annual yield of r. Solve this formula for r. (This gives you the rate if you know what you started with and what you ended up with.)

Solution

To solve the formula $A = p + pr$ for r, you must isolate the variable r.

$$\begin{aligned} A &= p + pr \\ A - p &= pr \end{aligned}$$

Now multiply each side by $\frac{1}{p}$.

$$\begin{aligned} \frac{1}{p}(A - p) &= \frac{1}{p} \cdot pr \\ \frac{A - p}{p} &= r \end{aligned}$$

So an equivalent formula is $r = \frac{A - p}{p}$.

Check

Pick values for all variables that work in the original formula. They should work in the equivalent formula. For instance, if $p = \$100$ and $r = 5\%$, then in the original equation $A = 100 + 100 \cdot 0.05 = 105$. Now substitute these values in $r = \frac{A - p}{p}$. Does $5\% = \frac{105 - 100}{100}$? Yes, it checks.

QUESTIONS

Covering the Reading

Anchorage, Alaska. *With a population of about 226,000, Anchorage is Alaska's largest city.*

1. As of 1994, the highest recorded temperature in both Alaska and Hawaii was 100°F. Change this record temperature to degrees Celsius.

2. Change 200°C to degrees Fahrenheit.

In 3 and 4, solve for y.

3. $5x + y = 6$

4. $3x - 6y = 12$

5. Name a use for solving equations for y such as those in Questions 3 and 4.

6. What formula can be used to find the amount of money after one year in a bank account?

7. Below the formula in Example 3 is solved for p. Fill in the blanks to explain the steps used.

$A = p + p \cdot r$	
$A = p \cdot 1 + p \cdot r$	Use the __a.__ Property of 1.
$A = p(1 + r)$	Use the __b.__ Property.
$\frac{1}{1+r} \cdot A = p(1 + r) \cdot \frac{1}{1+r}$	Multiply each side by __c.__
$\frac{A}{1+r} = p$	Simplify.

In 8 and 9, solve the formula for the indicated variable.

8. $t = \frac{d}{r}$; d

9. $P = 2(L + W)$; L

10. $F = m \cdot a$; a

11. $A = \frac{1}{2}(b_1 + b_2)h$; h

Applying the Mathematics

12. The formula $C = K - 273$ converts temperatures from the Kelvin scale to the Celsius scale. Solve this formula for K.

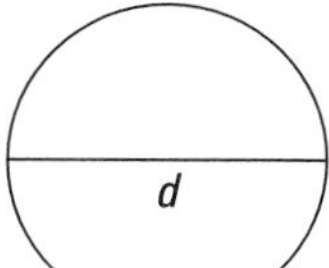

13. A formula for the circumference of a circle is $C = \pi d$.
   a. Solve this formula for π.
   b. How could you use the formula to find a value of π?
   c. Use your answer to part **b** to estimate π from the measurements of some circular object you have.

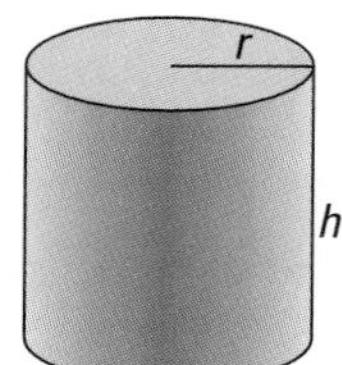

14. The formula $V = 2\pi r^2 + 2\pi rh$ gives the total surface area of a cylindrical solid with radius r and height h. Solve this formula for h.

15. Many calculators will convert angle measures from degrees D to grads G, using the formula $G = \frac{10}{9}D$.
   a. Solve the formula for D.
   b. Convert 100 grads to degrees.

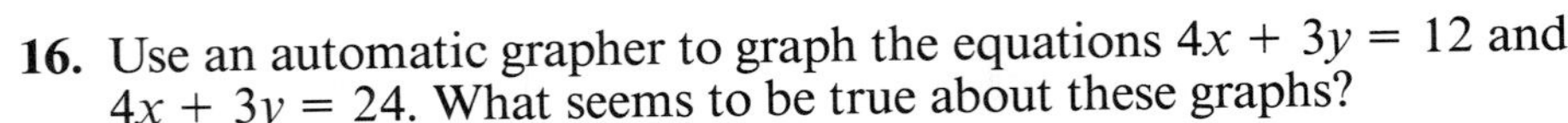

16. Use an automatic grapher to graph the equations $4x + 3y = 12$ and $4x + 3y = 24$. What seems to be true about these graphs?

17. Some banks have "time and temperature" displays that often alternately display temperature in degrees Fahrenheit and Celsius.
a. What temperature gives the same reading on both the Fahrenheit and Celsius scales?
b. Explain how you obtained your answer to part **a.**

Review

In 18 and 19, solve. *(Lesson 5-6)*

18. $-3z - 4 > 2z - 24$

19. $7(x - 2) + 5(2 - x) \geq 4(3x - 4)$

20. **a.** Make up a question about a real situation that can be answered by solving the equation $150 + 2x = 100 + 5x$.
b. Solve the equation.
c. Answer the question in part **a.** *(Lesson 5-3)*

21. Graph $x = 1$ in the coordinate plane. *(Lesson 5-1)*

22. Give an equation for the line. *(Lesson 5-1)*
a. the x-axis
b. the y-axis

23. In 1992 the population of Lagos, Nigeria, was 8,487,000 and growing at a rate of about 395,000 people per year. In 1990 the population of metropolitan London, England was 9,170,000 and declining at a rate of about 55,000 per year. If these rates continue, in what year will the population of Lagos first exceed the population of London? Explain how you got your answer. *(Lessons 5-2, 5-4, 5-6)*

Rush hour in Lagos. *With a population of 8,487,000, Lagos is the largest city in Nigeria.*

24. *Skill sequence.* Simplify. *(Lesson 3-9)*
a. $\frac{2}{3} + \frac{1}{8}$
b. $\frac{2}{3}a + \frac{1}{8}a$
c. $\frac{2a}{3} + \frac{a}{8}$

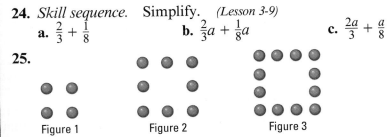

25.

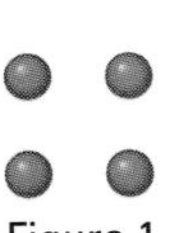

Figure 1

Figure 2

Figure 3

These figures are made from dots. Suppose the pattern continues.
a. How many dots would make up Figure 10?
b. How many dots would make up Figure n? *(Lessons 1-7, 3-8)*

26. Mentally compute 6 times 999,999. *(Lesson 3-7)*

Exploration

27. Find an example of a formula different from those in this lesson. Explain what the variables represent. Solve the formula for one of its other variables.

LESSON

5-8

Advanced Solving Technique I: Multiplying Through

Multiplying through in baseball. *In 1993, Andres Galarraga led the National League with a batting average of .370. Batting averages are read as if the reader had multiplied through by 1000. So .370 is read, "three seventy."*

Clearing Fractions

With the techniques you have learned you can solve any linear equation. However, when you want to solve an equation with several fractions, for instance,

$$\frac{1}{4}t = 21 - \frac{1}{3}t,$$

you may want to *clear the fractions* before you do anything else. A technique called *multiplying through* can then help. The idea is to multiply both sides of the equation by a common multiple of the denominators. The result is an equation in which all the coefficients are integers.

Example 1

Solve $\frac{1}{4}t = 21 - \frac{1}{3}t$.

Solution 1

Multiply each side of the equation by a number that will produce an equation that has only *integer* coefficients. What number should be used? Use 12, the same number that would be the least common denominator if the fractions were added.

$12 \cdot \frac{1}{4}t = 12(21 - \frac{1}{3}t)$ Multiply both sides by 12.

$3t = 12 \cdot 21 - 12 \cdot \frac{1}{3}t$ Apply the Distributive Property.

$3t = 252 - 4t$ Simplify.

Notice—now there are no fractions!

$7t = 252$ Add $4t$ to both sides.

$t = 36$ Multiply both sides by $\frac{1}{7}$.

Solution 2

Use the techniques of Lesson 5-3.

$$\frac{1}{4}t = 21 - \frac{1}{3}t$$

$\frac{1}{4}t + \frac{1}{3}t = 21$ Add $\frac{1}{3}t$ to each side.

$\frac{7}{12}t = 21$ Add like terms. $\left(\frac{1}{4}t + \frac{1}{3}t = \frac{7}{12}t\right)$

$\frac{12}{7} \cdot \frac{7}{12}t = \frac{12}{7} \cdot 21$ Multiply both sides by $\frac{12}{7}$.

$t = 36$ Simplify.

A vacation in Acapulco. *This is the atrium inside a large hotel in Acapulco, Mexico.*

In Solution 1, we say that the fractions were cleared by multiplying each side of the equation by 12.

> To clear fractions in a linear sentence:
> 1. Choose a common multiple of each of the denominators in the sentence.
> 2. Multiply each side of the sentence by that number.

Example 2

The DeVries family took a vacation. They spent $\frac{1}{3}$ of the vacation cost on transportation, $\frac{2}{5}$ on hotels, and $\frac{1}{5}$ on food. They spent $350 on other things. What was the total amount they spent for their trip?

Solution

Let t = total dollar amount spent on vacation. Notice that

$$\text{transportation} + \text{hotel} + \text{food} + \text{other} = \text{total}$$

$$\frac{1}{3}t + \frac{2}{5}t + \frac{1}{5}t + 350 = t$$

Multiply each side of the equation by a common multiple of 3 and 5. We use 15.

$$15\left(\frac{1}{3}t + \frac{2}{5}t + \frac{1}{5}t + 350\right) = 15 \cdot t$$

Apply the Distributive Property.

$$15 \cdot \frac{t}{3} + 15 \cdot \frac{2t}{5} + 15 \cdot \frac{t}{5} + 15 \cdot 350 = 15 \cdot t$$

$$5t + 6t + 3t + 5250 = 15t$$

$$14t + 5250 = 15t$$

$$5250 = t$$

The family spent $5250 for the vacation.

Check

If the DeVries spent $5250 in all, then they spent:

$\frac{1}{3} \cdot \$5250 = \$1750$ for transportation;

$\frac{2}{5} \cdot \$5250 = \$2100$ for hotels;

$\frac{1}{5} \cdot \$5250 = \$1050$ for food; and had $350 left over.

Does $1750 + 2100 + 1050 + 350 = 5250$? Yes, it checks.

The techniques described in this lesson for solving equations with fractions may also be used to solve inequalities with fractions.

Example 3

Solve $\frac{n}{4} + \frac{3n}{10} + 7 < 3n$.

Solution

The two denominators in the sentence are 4 and 10. Their least common multiple is 20. So multiply each side of the inequality by 20.

$$20\left(\frac{n}{4} + \frac{3n}{10} + 7\right) < 20 \cdot 3n$$

$$20 \cdot \frac{n}{4} + 20 \cdot \frac{3n}{10} + 20 \cdot 7 < 60n$$

$$5n + 6n + 140 < 60n$$

$$11n + 140 < 60n$$

$$140 < 49n$$

$$\frac{140}{49} < n$$

$$\text{or } n > \frac{20}{7}$$

Clearing Decimals

Like fractions, decimals can be cleared from an equation to give a simpler equation with integer coefficients. A decimal can be thought of as a fraction whose denominator is given by 10 to the power that is the number of decimal places. For instance, the "denominator" of 0.8 is 10, because $0.8 = \frac{8}{10}$. Similarly, the "denominators" of 38.15 and -2.001 are 100 and 1000, respectively.

Example 4

Solve $0.92m + 2 = m - 0.4$.

Solution

The equation involves two decimal fractions, 0.92 and 0.4. Their "hidden denominators" are 100 and 10. Since 100 is divisible by both 100 and 10, multiply each side of the equation by 100.

$$100(0.92m + 2) = 100(m - 0.4)$$

$$100 \cdot 0.92m + 100 \cdot 2 = 100 \cdot m - 100 \cdot 0.4$$

$$92m + 200 = 100m - 40$$

Now that the coefficients are integers, the equation can be solved easily without a calculator.

$$240 = 8m$$

$$30 = m$$

Clearing Large Numbers

Sometimes the numbers that appear in a sentence are quite large. Then it is wise to consider multiplying through by a fraction or a decimal to get an equivalent equation with smaller coefficients. Compare the solution to Example 5 with the solution to Example 4 of Lesson 5-3, where this technique was not used.

Summer fun. *The "City of Festivals" parade, which takes place each summer in Milwaukee, Wisconsin, features multiethnic costumes and food.*

Example 5

The 1990 population of El Paso, Texas, was 515,000. In the 1980s its population had been increasing at an average rate of 9000 people per year. In 1990, the population of Milwaukee, Wisconsin, was 628,000. In the 1980s, its population decreased at the rate of 800 people a year. If these rates continue, in what year will the population of El Paso and Milwaukee be the same?

Solution

Let n be the number of years after 1990. The populations are the same when

$$515{,}000 + 9000n = 628{,}000 - 800n.$$

Each number is a multiple of 100, so multiply both sides by $\frac{1}{100}$.

$$\frac{1}{100}(515{,}000 + 9000n) = \frac{1}{100}(628{,}000 - 800n)$$

Again apply the Distributive Property. We do the arithmetic mentally.

$$5150 + 90n = 6280 - 8n$$

You might now notice that each number is divisible by 2. This suggests multiplying both sides by $\frac{1}{2}$. But that does not reduce the amount of work. The rest of the solution is left up to you.

Multiplying through is simply a special use of the Multiplication Properties of Equality and Inequality. So this technique can be applied to solving all sentences, not just the linear sentences of this lesson.

QUESTIONS

Covering the Reading

1. Suppose $\frac{1}{3}x + 5 = \frac{5}{12}x$.
   a. Multiply each side of the equation by 12.
   b. Solve the resulting equation.
   c. Check your answer.

2. Consider the equation $\frac{a}{4} = 20 - \frac{a}{6}$.
   a. Multiply each side by 12 and solve the resulting equation.
   b. Multiply each side by 24 and solve the resulting equation.
   c. What conclusions can you make from your work in parts **a** and **b?**

3. Check the solution to Example 1.

4. On what property does the "multiplying through" technique rely?

5. The first $\frac{1}{5}$ of Toni's drive to work is on I-90. The next leg, about $\frac{1}{2}$ of the trip, is on Elm Street. The last leg takes 18 minutes. How long does the whole commute take?
a. Write an equation to describe this situation.
b. Solve by first clearing the fractions.

In 6 and 7, a sentence and a number are given. **a.** Write the sentence that results if both sides of the given sentence are multiplied by the given number. **b.** Solve the sentence.

6. $\frac{8a}{15} - 2 = \frac{a}{5}$; multiply by 15.

7. $\frac{x}{2} + \frac{x}{6} + 10 < \frac{5}{9}x$; multiply by 18.

8. Finish the solution to Example 5.

In 9–14, solve and check.

9. $\frac{5}{6}x + \frac{1}{2} = \frac{2}{3}$

10. $\frac{a}{5} - 1 = \frac{a}{30}$

11. $0.03y - 1.5 = 0.09y - 0.48$

12. $6000n + 9000 = 11000 - 2000n$

13. $\frac{3x}{5} - \frac{x}{10} < 5$

14. $\frac{n}{2} - 1 \geq \frac{4}{5} + \frac{3n}{10}$

Due to the wealth from its oil exports, Saudi Arabia has become a leading economic power in the Middle East. Shown is an oil-pumping station located in the eastern part of the country.

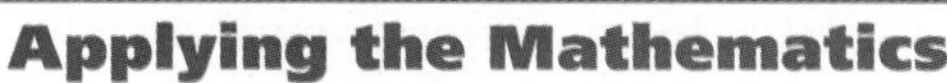

Applying the Mathematics

15. Mr. Bigbear owns $\frac{3}{8}$ of the stock in Amalgamated Industries and Mrs. Bigbear owns $\frac{1}{4}$ of it. This means that they receive, respectively, $\frac{3}{8}$ and $\frac{1}{4}$ of the dividends paid to stockholders.
a. Last year, the Bigbears together earned \$25,400 from this stock. What was the total amount of dividends paid to stockholders?
b. How much did stockholders other than Mr. and Mrs. Bigbear receive in dividends?

16. Saudi Arabia is reported to have about $\frac{1}{4}$ of the world's crude-oil reserves, while the rest of the Middle East has about 40%, or $\frac{2}{5}$, of the world's reserves. Altogether the Middle East's crude-oil reserves amount to 660 billion barrels. How many barrels are estimated to be in the world's total crude-oil reserves?

17. In solving $x - 2000 = 4000x$, a student first multiplies both sides by $\frac{1}{1000}$. Is this a good idea? Why or why not?

In 18–21, solve the sentence.

18. $\frac{n}{3} - 5 = \frac{n}{4} - 2n$

19. $\frac{1}{4}(x + 6) + \frac{1}{8}x = \frac{1}{2}(x + 1)$

20. $\frac{a + 3}{3} = \frac{2a - 3}{6} + \frac{a + 2}{4}$

21. $\frac{y - 2}{6} - \frac{1}{15} < \frac{2y + 1}{10}$

Review

In 22 and 23, solve for v. *(Lesson 5-6)*

22. $x = \frac{2uv}{g}$

23. $d = \frac{1}{2}gt - vt$

24. **a.** Solve for x: $3(x - 9) < 9(3 - x)$.
b. Graph the solutions on a number line. *(Lessons 1-2, 2-8, 5-6)*

In 25 and 26, the values of two cars, A and B, are compared. *(Lesson 5-5)*

Value (dollars)
15,000
10,000
5,000
0
1 2 3 4 5 6 7
Years after purchase
B
A

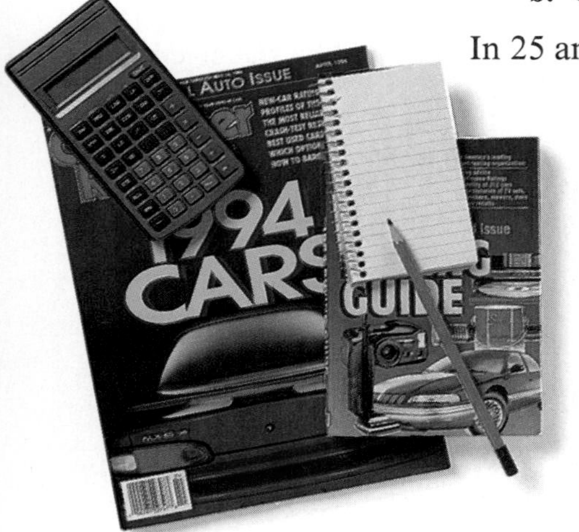

Auto advice. *Before buying a new or used car, consult with a car-buying guide. These magazines rate features such as safety, repair frequency, and resale value.*

25. **a.** Which car is depreciating in value faster?
b. About how much does the value of that car change each year?

26. **a.** After about how many years are the cars the same in value?
b. If you are trading the car in after 4 years, for which car will you get more money, and about how much more will you get?

27. Town A has a population of 25,000 and is growing at a rate of 1200 people per year. Town B has 35,500 people and is declining at a rate of 300 people per year.
a. Make a table showing the population of each town for 0, 1, 2, 3, . . . , 8 years from now.
b. In how many years will the populations of the towns be equal? *(Lessons 3-8, 5-2, 5-3)*

28. Simplify $\sqrt{9} \cdot x + 3! \cdot x - 50\% \cdot x$. *(Lessons 1-6, 2-10, 3-2)*

29. Solve for p: $p^2 = 10000$. *(Lesson 1-1)*

Exploration

30. Recall that Diophantus was the first known person to use letters to stand for unknown numbers. Very little is known about the life of Diophantus. Over 800 years ago someone wrote this problem in his honor:

> Diophantus passed one sixth of his life in childhood, one twelfth in youth, and one seventh more as a bachelor. Five years after his marriage his son was born, who died four years before his father, at half his father's final age. How long did Diophantus live?

LESSON

5-9

Advanced Solving Technique II: Chunking

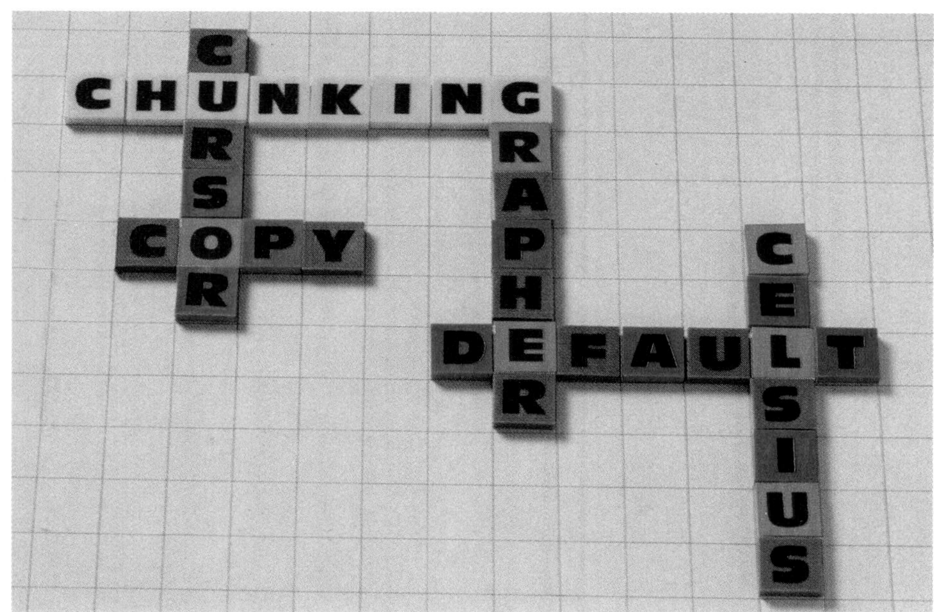

Word puzzles. *In solving word puzzles, letter tiles are arranged to form words. Forming words from letters can be thought of as a type of "chunking."*

What Is Chunking?

When an expression involves repeated occurrences of the same expression, you can often simplify it in more than one way.

Example 1

Simplify $8(x + 4) - 3(x + 4)$.

Solution 1

Use the Distributive Property. Then add like terms.

$$\begin{aligned} 8(x + 4) - 3(x + 4) &= (8x + 32) + -3(x + 4) \\ &= 8x + 32 - 3x - 12 \\ &= 5x + 20 \end{aligned}$$

Solution 2

The expression has the form $8n - 3n$ where $n = x + 4$. Just as you can subtract the like terms $8n - 3n$ to get $5n$, you can simplify the given expression in one step.

$$\begin{aligned} 8(x + 4) - 3(x + 4) &= 5(x + 4) \\ &= 5x + 20 \end{aligned}$$

The idea used in Example 1 is called *chunking.* **Chunking** is a term psychologists use to describe the process of grouping some small bits of information into a single piece of information. For instance, when reading the word "store," you don't think "s, t, o, r, e." You group the five letters into one word. In algebra, chunking can be done by viewing an entire algebraic expression as one variable. Example 2 shows how chunking can be used to simplify fractions.

Example 2

Simplify $\frac{4}{3x + 1} + \frac{3}{3x + 1}$.

Solution

Think of $(3x + 1)$ as a single chunk c. Then the addition looks like $\frac{4}{c} + \frac{3}{c}$ and its sum is $\frac{7}{c}$.

$$\frac{4}{3x + 1} + \frac{3}{3x + 1} = \frac{7}{3x + 1}$$

Check

Substitute a number for x. We use $x = 5$.

Does $\frac{4}{3 \cdot 5 + 1} + \frac{3}{3 \cdot 5 + 1} = \frac{7}{3 \cdot 5 + 1}$?

Does $\frac{4}{16} + \frac{3}{16} = \frac{7}{16}$? Yes, it checks.

In Example 2, since $3x + 1$ is in the denominator it cannot be 0. Your teacher may want you to include $3x + 1 \neq 0$ in your solution.

Using Chunking in Linear Equations

Example 3

Use chunking to solve $4(3x + 5) + 8(3x + 5) + 1 = 6(3x + 5) + 13$.

Solution

Notice that $3x + 5$ occurs in three places in the equation. Think of $3x + 5$ as one variable.

$12(3x + 5) + 1 = 6(3x + 5) + 13$	Add like chunks.
$6(3x + 5) + 1 = 13$	Add -6 chunks to each side.
$6(3x + 5) = 12$	Add -1 to each side.
$3x + 5 = 2$	Multiply each side by $\frac{1}{6}$.
$3x = -3$	Simplify.
$x = -1$	Multiply each side by $\frac{1}{3}$.

If the value of an algebraic expression is known, it can be substituted as a chunk into other expressions.

Example 4

If $3y = 8.5$, find $6y + 5$.

Solution

This can be done without solving for y. Think of $3y$ as a chunk. Since you know the value of $3y$, you can double it to get $6y$. Then add 5.

$$3y = 8.5$$
$$6y = 2 \cdot 8.5 = 17$$
$$6y + 5 = 17 + 5 = 22$$

Check

If $3y = 8.5$, then $y = \frac{1}{3}(8.5) \approx 2.83$. So $6y + 5 \approx 6 \cdot 2.83 + 5 = 21.98$.

Using Chunking to Solve Other Equations

You know that the solutions to $x^2 = 81$ are $x = 9$ and $x = -9$. Using this fact, you can solve any equation of the form $(\text{chunk})^2 = 81$. Example 5 shows how.

Example 5

Solve $(k + 6)^2 = 81$.

Solution

The two numbers which give 81 when squared are 9 and -9. So the expression $k + 6$ must equal 9 or -9.

$$(k + 6)^2 = 81$$

$k + 6 = 9$	or	$k + 6 = -9$
$k = 3$	or	$k = -15$

There are two solutions, 3 and -15.

Check

Each solution must be checked.

Check 3. Does $(3 + 6)^2 = 81$?
$9^2 = 81$? Yes, it checks.

Check -15. Does $(-15 + 6)^2 = 81$?
$(-9)^2 = 81$? Yes, it checks.

The technique used in Example 5 generalizes so you can solve many other equations of the form: $(\text{linear expression})^2 = \text{a constant}$.

QUESTIONS

Covering the Reading

1. What is chunking?

In 2 and 3, use chunking to simplify the expression.

2. $8(12t - 7) - 3(12t - 7)$

3. $12(10 - a) - 4(10 - a)$

4. a. When adding $\frac{11}{2y - 6} + \frac{4}{2y - 6}$, what should you think of as a chunk?

b. Perform the addition.

c. What value can $2y - 6$ *not* have?

5. **a.** Simplify $\frac{x}{x + 8} - \frac{5}{x + 8}$.
   b. What value(s) can x *not* have?

6. **a.** Solve the equation $10(2a + 3) - 7(2a + 3) = 39$ using chunking.
   b. Solve the equation in part **a** another way.
   c. Which method do you prefer?

7. If $3x = 8.5$, find $6x - 1$.

8. If $18a = 12$, find $9a + 7$.

9. Follow Example 5 to solve these equations.
   a. $(x + 1)^2 = 81$
   b. $(4x + 1)^2 = 81$

10. **a.** If $(m - 11)^2 = 64$, what two values can $m - 11$ have?
    b. Find two solutions to $(m - 11)^2 = 64$.

11. Find all solutions to $(p + 3)^2 = 225$.

Applying the Mathematics

In 12–17, simplify.

12. $3\sqrt{5} + 6\sqrt{5} - 2\sqrt{5}$

13. $5\sqrt{a} - 7\sqrt{a}$

14. $7(x^2 - 9) - 4(x^2 - 9)$

15. $3(x^2 - 2y) + 6(x^2 - 2y) - 4(x^2 - 2y)$

16. $\frac{8(x + 7)}{5a} \cdot \frac{3a}{2(x + 7)}$

17. $\frac{x + 10}{x + 5} + \frac{2x + 5}{x + 5}$

18. Approximate the solutions of $(d + 11)^2 = 57$ to two decimal places.

19. If $5y = 7$, find the value of $(5y)^2 + 3$.

20. If $a + 7 = 91$, what is the value of $2a + 14$?

21. If $18y - 12t = 25$, find $9y - 6t$.

22. If $5x + 4y = 32$ and $2y = 1$, give the value of each expression.
    a. $5x$
    b. x

In 23 and 24, find all real-number solutions.

23. $(3p + 5)^2 = 625$

24. $(x^2)^2 = 256$.

Review

25. State a property that can be used to transform $\frac{a + 3}{4} = \frac{1}{2}$ to $a + 3 = 2$. *(Lesson 5-8)*

In 26 and 27, solve. *(Lesson 5-8)*

26. $\frac{3w}{4} - 2 \leq \frac{1}{2}$

27. $\frac{x}{2} + \frac{x}{3} - \frac{1}{4} = \frac{1}{6}$

28. The Koenigs bought a farm. They did not know the capacity of the heating-oil storage tank. At one point, the tank was $\frac{1}{8}$ full. After a delivery of 450 gallons, the tank's gauge showed that it was $\frac{7}{8}$ full. How many gallons of oil does the tank hold? *(Lesson 5-8)*

29. Bertrand spends half his monthly income on housing and food, and budgets the other half as follows: $\frac{1}{4}$ of the half on clothes, $\frac{1}{3}$ on entertainment, $\frac{1}{4}$ on transportation, and the remaining $40 in savings. What are his monthly earnings? *(Lesson 5-8)*

30. Newton's laws of motion state that the velocity v of an object moving in a circular path is related to its acceleration a and the radius of the path r by the formula $v^2 = ar$. Solve this formula for r. *(Lesson 5-7)*

31. **a.** Solve for y: $2x + 5y = 10$.
    b. Graph the line. *(Lessons 4-9, 5-7)*

32. **a.** Make up a question about a real situation that can be answered by solving the equation $200 - 3x = 240 - 5x$.
    b. Solve the equation. *(Lessons 3-8, 5-3)*

33. Give an instance of the Commutative Property of Addition. *(Lesson 3-1)*

34. A catalog company that sells travel bags sews customers' 3-letter monograms onto the bags. However, some bags are returned. As a promotion, the company advertises that if you could find your monogram on a bag among those they have on hand, the bag is free. How many different 3-letter monograms are possible? *(Lesson 2-9)*

35. Refer to the figures below. The triangle and rectangle have equal perimeters. What is the length of a side of the triangle? *(Lesson 5-3)*

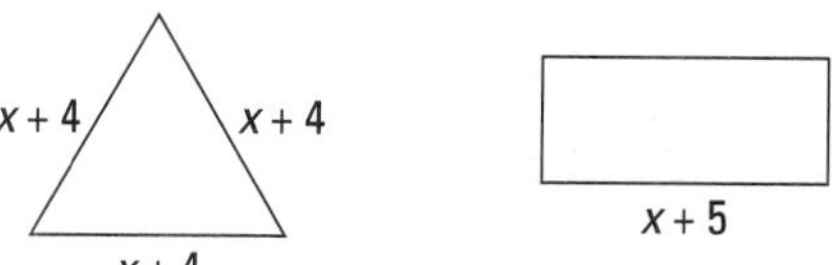

Exploration

36. Often people use chunking as a help in memorization. Here are the first twenty decimal places of π.

$$3.14159265358979323846$$

Try memorizing these 20 digits by memorizing the following four chunks of five digits each.

14159 26535 89793 23846

PROJECTS 5 CHAPTER FIVE

A project presents an opportunity for you to extend your knowledge of a topic related to the material of this chapter. You should allow more time for a project than you do for typical homework questions.

1 Chunking

Explain how the following equation can be solved by repeated applications of chunking:

$$\frac{\frac{x+3}{4}+5}{2}+6=7.$$

Then make up a few similar equations and solve them.

2 Tree growth

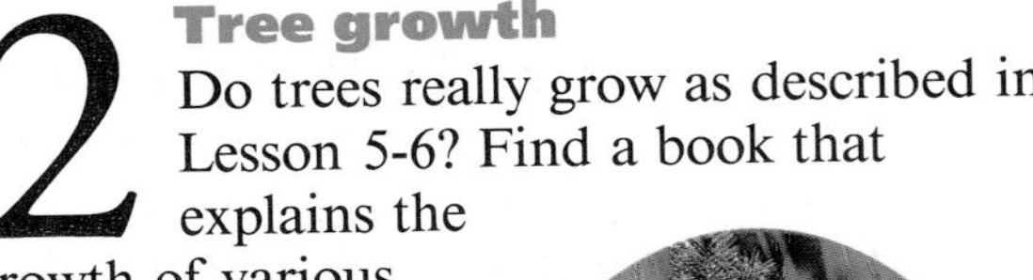

Do trees really grow as described in Lesson 5-6? Find a book that explains the growth of various species of trees. For at least three species, graph the height of a typical tree as it grows to its mature size.

3 Lifetime Cost of an Appliance

Visit a store where appliances are sold, or check your phone book for such a store. New major appliances, like refrigerators, are tested for their expected energy consumption per year. That information is available to consumers to aid them in their decisions. Pick a range of sizes of refrigerators (for instance 16 to 21 cu ft), and find the price and expected yearly energy cost for at least four different models. For each model develop an equation that describes its total cost over its life span. Graph these equations. Is the least expensive model always the most economical? Which model is the best value over 10 years?

4 Equivalent formulas

Below are formulas for the areas of various common figures studied in geometry. For each formula, draw a figure and indicate on the figure what each variable represents. Solve every formula for each of the other variables in it.

a. triangle $A = \frac{1}{2}hb$

b. parallelogram $A = bh$

c. kite $A = \frac{1}{2}d_1d_2$

d. trapezoid $A = \frac{1}{2}h(b_1 + b_2)$

e. ellipse $A = \pi ab$

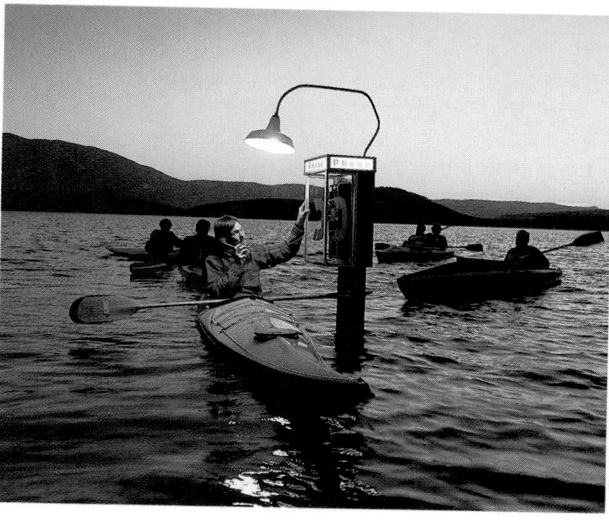

5 Light Bulbs

On the opening page of this chapter, a table of prices of two kinds of light bulbs is given. Visit a store and find the costs of these or other similar bulbs in your area. Also find the cost of electricity, perhaps by calling your local electric company. Organize the information you collect in a table and a graph, and based upon your information, answer the questions.

a. How much does a regular light bulb cost?

b. How much does an energy-efficient bulb cost?

c. After approximately how many hours of use will the total costs (purchase price of bulb plus cost of electricity) of the two bulbs be the same?

d. If the average life of a bulb is 1000 hours, how much money would you save by buying and using an energy-efficient bulb?

6 Long-distance Telephone Rates

Find out the rates for telephone calls from where you live to a city or town that has a different area code. Compare the rates from two long-distance companies for calls of various lengths. You may have to consider different times of the day and week. Do not make the calls to determine their costs!

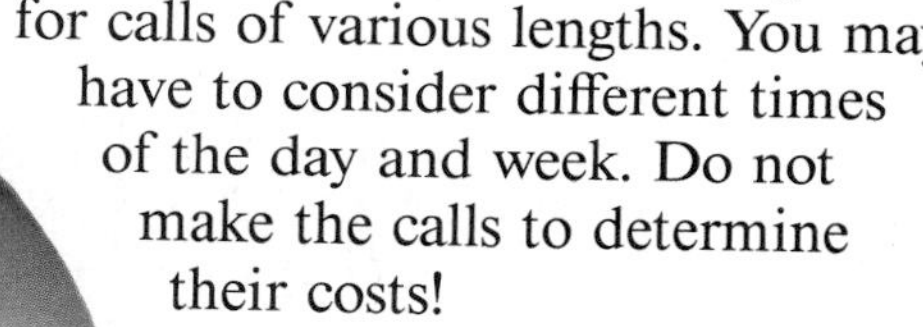

SUMMARY

The chapter begins by examining two special forms of equations: $x = h$ and $y = k$. If an equation of the form $x = h$ is graphed in the coordinate plane, its graph is a vertical line. If $y = k$ is graphed in the coordinate plane, its graph is a horizontal line.

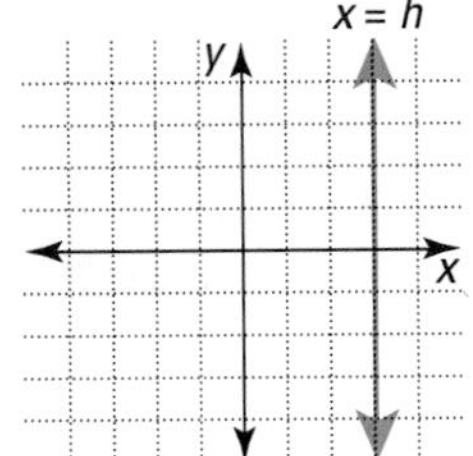

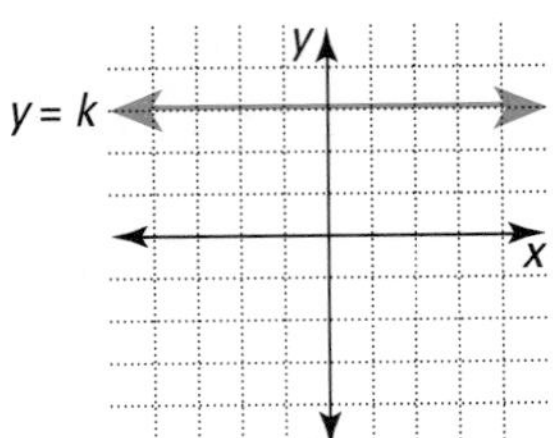

The next five lessons of the chapter deal with ways to compare two quantities described by linear expressions such as $ax + b$ and $cx + d$. One way to compare them is to make a table that shows the value of each expression for different values of x. This can be done with paper and pencil or with a spreadsheet. Questions can then be answered by examining the numbers that appear in the chart.

A second method of comparison is to write an equation to describe each pattern, in the form $y = ax + b$ and $y = cx + d$. Graph the two equations, and examine where the two graphs cross or where one is below the other. Graphing can be done with paper and pencil or with an automatic grapher.

The third technique discussed is to solve an equation or inequality. To find when the two quantities are equal, solve $ax + b = cx + d$. To find when the first is less than the second, solve $ax + b < cx + d$. Solving sentences has the advantage of yielding exact solutions.

The last lessons discuss special techniques involved in solving sentences. Many formulas or equations that contain more than one variable can be "solved" for any of the variables in them. The process is similar to that for equations with just one variable. To solve a sentence containing fractions, find the least common multiple of the denominators, and then multiply each term by it. If numbers in a sentence are all large, then multiply both sides by a small fraction.

Chunking is a problem-solving technique by which an expression is considered as a single number. Many complicated expressions and equations can be handled by recognizing their similarities with simpler patterns.

VOCABULARY

Below are the most important terms and phrases for this chapter. You should be able to give a general description and a specific example of each.

Lesson 5-1
horizontal line, $y = k$
vertical line, $x = h$

Lesson 5-3
general linear equation

Lesson 5-5
automatic grapher
window, default window
tick marks
trace, cursor
hard copy

Lesson 5-7
Fahrenheit, Celsius, Centigrade, Kelvin temperature scales
equivalent formulas

Lesson 5-8
clearing fractions
multiplying through

Lesson 5-9
chunking

 CHAPTER FIVE

PROGRESS SELF-TEST

Take this test as you would take a test in class. Use graph paper, a ruler, and a calculator. Then check your work with the solutions in the Selected Answers section in the back of the book.

1. a. To solve the equation $5y - 9 = 12 - 3y$ an effective first step is to add __?__ to each side of the equation.

b. What property is being applied in part **a**?

2. For the equation $\frac{3}{5}x - \frac{2}{3} = \frac{1}{5}x + 7$, by what number can each side of the equation be multiplied to clear the fractions?

In 3–9, solve. Show your work.

3. $4x - 3 = 3x + 14$

4. $3.9z - 56.9 = 6.1 - 4.7z$

5. $5n \geq 2n + 12$

6. $5(10 - y) = 6(y + 1)$

7. $-5a + 6 < -11a + 24$

8. $\frac{1}{2}m - \frac{3}{4} = \frac{2}{3}$

9. $5000 - 4000v = 11000v + 680{,}000$

10. If $4y = 2.6$, find the value of $20y + 3$.

In 11 and 12, simplify.

11. $\frac{4}{t+7} + \frac{5}{t+7}$

12. $8(x^2 - 5) + 3(x^2 - 5)$

13. Find all solutions to $(n + 3)^2 = 49$.

14. Solve the sentence $3x + 5y = 15$ for y.

15. Solve the formula $C = np$ for p.

16. Graph the points in the coordinate plane satisfying the equation $x = 3$.

17. Suppose the perimeter of the triangle is equal to the perimeter of the square. Then find the lengths of the sides of the triangle and the length of a side of the square.

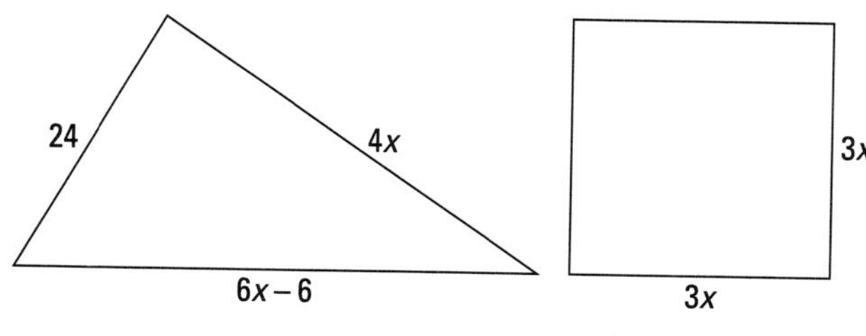

18. Two children in a family want to see who can save more money. The younger child has saved \$1200 and is saving \$50 per month. The older child has saved \$1500 and is saving \$25 per month. After how many months will the younger child have more money saved than the older child?

a. Make a graph to represent this situation. Explain how the graph can be used to answer the question.

b. Answer the question by solving an inequality.

19. Jolisa is considering two different sales positions. Sun Fashions pays a total salary of \$400 per month plus 12% of her sales. Her second option is to work at Today's Outerwear, where she would receive \$750 per month plus 10% of her sales.

a. Complete the table.

Monthly Sales	Sun Fashions Total Salary	Today's Outerwear Total Salary
\$ 0	\$ 400	
\$5000		
\$10000		\$1750
\$15000		
\$20000		
\$25000	\$3400	

b. Use the table to find amounts of sales for which Sun Fashions will pay Jolisa a greater total salary.

c. Jolisa thinks she can sell \$12,000 of merchandise each month. Explain which job you think she should take and why.

PROGRESS SELF-TEST

20. This year, the Green family bought a new home and a new car. They anticipate that their $80,000 home will increase in value by $3500 per year, but their $14,000 car will decrease in value by $1800 per year.

a. Copy and complete the spreadsheet below to show the values of their home and auto over the next six years.

b. Give a formula for the value of their home t years from now.

	A	B	C
1	years from now	House value	Car value
2	0	80,000	14,000
3	1		
4	2		
5	3		
6	4		
7	5		
8	6		

21. The graph below shows the cost of having film developed at two different stores.

a. For how many photos is it cheaper to go to Picture Perfect?

b. For how many photos is it more expensive to go to Picture Perfect?

c. For how many photos do the two stores charge the same price?

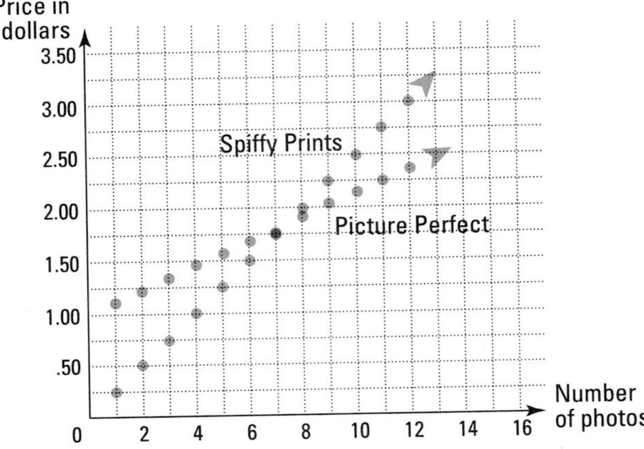

22. Use an automatic grapher. The graph of $y = 0.5x + 7$ passes through three quadrants.

a. Find a window that shows the graph in all three quadrants. Sketch the graph on that window and label the window used.

b. Use the trace key to estimate the x-coordinate when $y = 0$.

c. Describe what happens to the y-coordinate when the x-coordinate increases.

CHAPTER REVIEW

Questions on SPUR Objectives

SPUR stands for **S**kills, **P**roperties, **U**ses, and **R**epresentations. The Chapter Review questions are grouped according to the SPUR Objectives for this chapter.

SKILLS DEAL WITH THE PROCEDURES USED TO GET ANSWERS.

Objective A: *Solve linear equations of the form $ax + b = cx + d$.* *(Lessons 5-3, 5-8)*

In 1–12, solve.

1. $A + 5 = 9A - 11$ **2.** $-12k + 144 = 3k - 45$

3. $2a - 6 = -2a$ **4.** $3n = 2n + 4 + 5n$

5. $4x - 11.7 = -0.3x$ **6.** $10e = 4e - 5(3 + 2e)$

7. $3(x + 10) = -2(2x - 1)$ **8.** $5(4 - y) = 2(y + 10)$

9. $\frac{n}{2} + \frac{3}{2} = 3$ **10.** $\frac{1}{4}a - 1 = -2a$

11. $\frac{x}{3} + \frac{x}{5} + 21 = x$ **12.** $193.4x + 193.4 = 1934$

Objective B: *Solve linear inequalities of the form $ax + b < cx + d$.* *(Lessons 5-6, 5-8)*

In 13–16, solve.

13. $11h + 71 \geq 13h - 219$

14. $4x - 1 < 2x + 1$

15. $5(5 + z) < 3(2 + 2z)$ **16.** $-3x + 7 \geq -9x + 25$

In 17 and 18, **a.** solve; and **b.** graph all solutions on a number line.

17. $9(x + 3) + 2 \leq 4 - x$ **18.** $4x + 3 < 6x$

In 19–22, solve.

19. $\frac{3}{10}n + \frac{3}{5} < -\frac{1}{2}n - \frac{11}{5}$ **20.** $\frac{x}{2} + \frac{5}{3} > \frac{2x}{3} - 2$

21. $5000x - 9000 > 3500x - 15,000$

22. $0.12x + 4 \leq 0.2x - 0.32$

Objective C: *Use chunking to simplify or evaluate expressions or to solve equations.* *(Lesson 5-9)*

23. If $7y = 21.2$, find $21y + 4$.

24. If $11x + 5 = 29.5$, find $22x + 10$.

In 25–28, simplify the expression.

25. $13(x - 7) - 10(x - 7)$

26. $99(12a + 19) - 98(12a + 19)$

27. $\frac{-8(x + y)}{z} + \frac{9(x + y)}{z}$ **28.** $\frac{4(n + 3)}{n - 7} \cdot \frac{n - 7}{n + 3}$

In 29–32, find all solutions.

29. $(m + 2)^2 = 64$ **30.** $(z - 4)^2 = 144$

31. $(2x)^2 = 400$ **32.** $(3x + 7)^2 = 676$

Objective D: *Find equivalent forms of formulas and equations.* *(Lesson 5-7)*

In 33–36, solve for the stated variable.

33. $A = \frac{1}{2}bh$ for b **34.** $V = \ell wh$ for h

35. $P = 2(\ell + w)$ for w **36.** $S = 2\pi r^2 + 2\pi rh$ for h

37. Solve $y = \frac{x}{z}$ for x.

38. Solve $T = a + (n - 1)d$ for n.

In 39 and 40, solve for y.

39. $10x + 8y = 40$ **40.** $6y - 5x = 12$

PROPERTIES DEAL WITH THE PRINCIPLES BEHIND THE MATHEMATICS.

Objective E: *Apply and recognize properties associated with linear sentences.* *(Lessons 5-3, 5-6, 5-7, 5-8)*

41. Consider the equation $6x + 3 = 8x + 5$.

a. Solve the equation by first adding $-6x$ to each side.

b. Solve by first adding $-8x$ to each side.

c. Compare your answers to parts **a** and **b.**

CHAPTER FIVE

42. Consider the inequality $a + 2 < 3a + 4$.

a. Solve it by first adding $-a$ to each side.

b. Solve it by first adding $-3a$ to each side.

c. How are your answers to parts **a** and **b** related?

In 43 and 44, a sentence is solved. Write the steps in the solution replacing the blanks with a number, operation, or property.

43.

$\frac{3}{4}x-5 = \frac{1}{8}x+10$	
$16\left(\frac{3}{4}x-5\right) = 16\left(\frac{1}{8}x+10\right)$	_a_ each side by _b_.
$16 \cdot \frac{3}{4}x - 16 \cdot 5 = 16 \cdot \frac{1}{8}x + 16 \cdot 10$	Apply the _c_ property.
$12x - 80 = 2x + 160$	Simplify.
$10x - 80 = 160$	_d_ to each side.
$10x = 240$	_e_ to each side.
$x = 24$	

44.

$3y + 8 \geq -5y - 2$	
$8y + 8 \geq -2$	Add _a_ to each side.
$8y \geq -10$	_b_ to each side.
$y \geq -\frac{10}{8}$	_c_ each side by _d_.

In 45 and 46, multiply each side by a number to result in an equation in which all numbers are integers. (You need not solve the equation.)

45. $\frac{1}{4} - 2x = \frac{5}{6}x + 9$ **46.** $3.6y = 0.15 - 0.04y$

In 47 and 48, multiply each side by a number to result in an equation in which all numbers are smaller integers. (You need not solve the equation.)

47. $4800t - 120{,}000 = 3600t$

48. $35w + 21 = 105(w + 2)$

USES DEAL WITH APPLICATIONS OF MATHEMATICS IN REAL SITUATIONS.

Objective F: *Use linear equations and inequalities of the form $ax + b = cx + d$ or $ax + b < cx + d$ to solve real-world problems.* *(Lessons 5-3, 5-6, 5-8)*

49. Kate has \$1500 in an account and adds \$45 each month. Melissa has \$2000 and adds \$20 a month.

a. How much will each girl have in her account after n months?

b. After how many months will they have the same amount of money in their accounts?

50. Len has \$25 and is saving at the rate of \$9 a week. Basil has \$100 and is spending \$5 a week.

a. Let x = the number of weeks that have passed. Write a linear inequality which can be used to find out when Len will have more money than Basil.

b. Solve the sentence in part **a.**

51. Willow trees grow quickly, about 3 feet per year, while American elms grow about 1.5 feet per year. If a landscaping company plants an 8-foot willow tree and a 12-foot elm tree, after how many years will the willow tree be taller than the elm?

52. For what number of miles driven will the cost of renting a car for \$100 per week and \$.15 per mile equal the cost of renting a car with fees of \$150 and \$.12 per mile?

53. The new Newton Bakery can produce 550 loaves of bread per day. The old bakery has produced 109,000 loaves over the years and still makes 200 loaves per day. When will the new bakery have made more loaves of bread than the old one?

54. Angie is racing her younger sister, so she gives her sister a 50-m head start. If Angie can run at 5 $\frac{\text{m}}{\text{sec}}$ and her sister runs at 3 $\frac{\text{m}}{\text{sec}}$, how long will it take Angie to catch up with her sister?

55. An automobile gas tank was about $\frac{1}{8}$ full. 10.4 gallons of gas were needed to fill the tank. What is the capacity of the tank, rounded to the nearest gallon?

56. A couple budgets their after-tax income as follows: 25% for housing; 15% for food; 12% for savings; and 10% each for transportation, health care, clothing, and education. If next year the couple wants to have at least \$3000 left for entertainment and other expenses, what must their after-tax income be?

Objective G: *Use tables or spreadsheets to solve real-world problems involving linear situations. (Lesson 5-2)*

57. Two mail-order CD clubs offer discount prices. The first one has a $15 membership fee and charges $9 per CD. The second club charges $11 to join, plus $9.50 for each CD.

a. Copy and complete the chart below.

Number of CDs	Charges First club	Charges Second club
2		
4		
6		
8		
10		

b. How many CDs must you buy for the two clubs' charges to be equal?

c. When is the first club's price better?

d. When is the second club's price better?

58. Alicia is considering two different sales positions. Appliance World would pay a total salary of $1000 per month plus a commission of 4% of sales. Better Kitchens would pay a total salary of $600 plus a commission of 6% of sales.

a. Complete the table.

Sales	Appliance World Total Salary	Better Kitchens Total Salary
$ 0	$1,000	$600
5,000		
10,000		
15,000		
20,000		
25,000		$2,100
30,000	$2,200	

b. For what amounts of sales will Better Kitchens pay a greater total salary?

59. A farmer plans to increase his crop production over the next 10 years by adding acreage and improving farming techniques. This year he harvested 10,200 bushels of corn. He feels he can increase this by 100 bushels each year. Similarly, this year he harvested 6750 bushels of soybeans and plans to increase the output by 500 bushels per year.

a. Complete a spreadsheet like the one below to show the farmer's production over the next ten years. (The abbreviation for bushels is bu.)

	A	B	C
1	yrs from now	bu. corn	bu. soybeans
2	0	10200	6750
3	1		
4	2		
5	3		
6	4		
7	5		
8	6		
9	7		
10	8		
11	9		
12	10		

b. What formulas are used in cells B5 and C5?

c. In what year will the number of bushels of soybeans produced exceed the number of bushels of corn?

d. Assume that the farmer can sell his corn for $2.60 per bushel and his soybeans for $7.70 per bushel. Add a column showing the dollar value of the farmer's corn crop for each of the years and a column for the value of the soybean crop.

e. Write a short paragraph describing the values of the two crops over this 10-year period.

REPRESENTATIONS DEAL WITH PICTURES, GRAPHS, OR OBJECTS THAT ILLUSTRATE CONCEPTS.

Objective H: *Graph horizontal and vertical lines.* (*Lesson 5-1*)

In 60 and 61, graph the points in the coordinate plane satisfying each equation.

60. $x = 4$ **61.** $y = 1$

In 62 and 63, *true* or *false*.

62. The graph of all points in the plane satisfying $x = -0.4$ is a vertical line.

63. The graph of all points in the plane satisfying $y = 73$ is a horizontal line.

64. Write a sentence describing the line in the graph below.

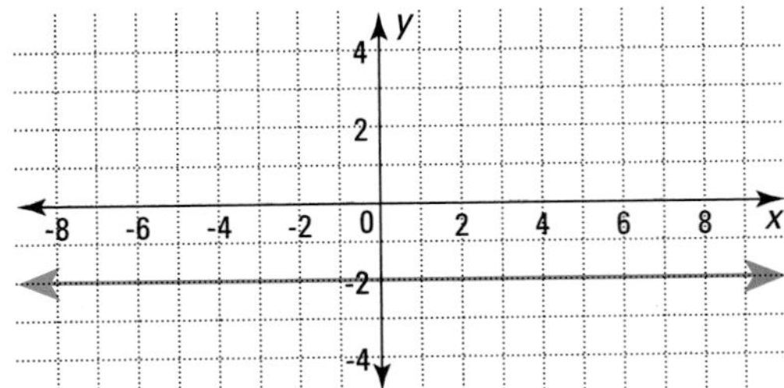

65. Write an equation for the line containing the points (5, 11), (5, 4), and (5, -7).

Objective I: *Use graphs to solve problems involving linear expressions.* (*Lessons 5-1, 5-4*)

66. A plane is cruising at 35,000 feet when it begins its descent into the airport at 1500 feet per minute.

a. Write an equation that relates the plane's height h and the time m since starting to descend.

b. Graph your equation from part **a.**

c. Use the graph to determine how many minutes it will take to land.

67. Picture Perfect will develop a roll of film for 22¢ a photo with no developing charge. You Oughta Be in Pictures charges 14¢ a photo plus a \$3.20 developing charge. Make a graph to determine what number of photos would cost the same at both developers.

Objective J: *Given an equation, be able to use an automatic grapher to draw and interpret a graph.* (*Lesson 5-5*)

68. a. On an automatic grapher, graph the line $18x + 6y = 90$ using the window $-5 \le x \le 5$ and $-10 \le y \le 20$.

b. For what values of x is $y > 0$?

69. On an automatic grapher draw the graphs of $y = -0.7x - 16.8$ and $y = -3.2x + 4.4$. Use the trace feature to determine what values of x make the following sentences true.

a. $-0.7x - 16.8 = -3.2x + 4.4$

b. $-0.7x - 16.8 < -3.2x + 4.4$

c. $-0.7x - 16.8 > -3.2x + 4.4$

70. What is a *window* on an automatic grapher?

71. Use two inequalities to describe the window pictured below.

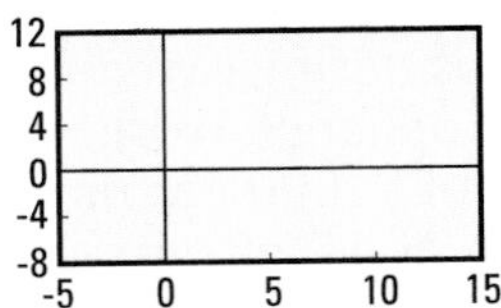

72. What does the trace command on an automatic grapher do?

In 73–75, use an automatic grapher.

73. Graph $y = x + 4$ on the window $-5 \le x \le 10$, $-10 \le y \le 15$. Trace to find the value of x for which $y = 0$.

74. a. Graph $y = x - 30$ on two different windows.

b. Describe what happened when you changed your window.

75. Graph $y = 3^x$ on the window $0 \le x \le 5$, $0 \le y \le 300$. Then use the trace key to solve the equation $3^x = 81$.

REFRESHER

Chapter 6, which discusses division in algebra, assumes that you have mastered certain objectives in your previous mathematics work. Use these questions to check your mastery. Do not use a calculator when it is easy to do the question without it.

A. Identify the **a.** divisor; **b.** dividend; and **c.** quotient in a division situation.

1. $21 \div 3 = 7$ **2.** $0.2 = 20 \div 100$

3. $\frac{56}{7} = 8$ **4.** $9\overline{)11.7}$ with quotient 1.3

B. Divide positive numbers.

5. a. $40 \div 100$ **b.** $100 \div 40$

6. a. Divide 30 by $\frac{1}{2}$. **b.** Divide $\frac{1}{2}$ by 30.

7. a. What is $\frac{1}{4}$ divided by $\frac{1}{5}$?

b. What is $\frac{1}{5}$ divided by $\frac{1}{4}$?

In 8–19, divide without using a calculator.

8. $7.2 \div 3$ **9.** $80 \div 0.05$

10. $\frac{0.06}{0.3}$ **11.** $\frac{6.8}{34}$

12. $\frac{12}{7} \div \frac{2}{7}$ **13.** $\frac{3}{5} \div \frac{3}{4}$

14. $\frac{2}{9} \div \frac{1}{3}$ **15.** $\frac{3}{2} \div \frac{4}{5}$

16. 6 ft ÷ 2 **17.** 10 m ÷ 4

18. 100 kg ÷ 7 **19.** 6 lb ÷ 25

C. Divide positive and negative integers.

In 20–25, divide without using a calculator.

20. $-8 \div -4$ **21.** $-40 \div 5$

22. $60 \div -120$ **23.** $2 \div -80$

24. $\frac{-3}{-6}$ **25.** $\frac{400}{-4}$

D. Convert any simple fraction to a decimal and percent.

In 26–28, give the decimal and percent equivalents.

26. $\frac{3}{4}$ **27.** $\frac{1}{40}$ **28.** $\frac{73}{100}$

In 29–31, write as a decimal. Round to the nearest hundredth.

29. $\frac{1}{7}$ **30.** $\frac{20}{3}$ **31.** $\frac{110}{17}$

In 32–34, write as a percent. Round to the nearest whole percent.

32. $\frac{11}{5}$ **33.** $\frac{27}{100}$ **34.** $\frac{8}{9}$

E. Convert a percent to **a.** a decimal and **b.** a fraction.

35. 30% **36.** 1% **37.** 300%

38. 2.46% **39.** .03% **40.** $\frac{1}{4}$%

F. Find a percent of a number or quantity.

41. 32% of 750

42. 94% of 72 questions

43. 7.3% of 40,296 voters

44. 100% of 12,000 square miles

45. 0% of 60

46. 150% of $10,000

G. Find the mean of a set of numbers.

47. 14, 9, 47, 17

48. 3, 3.1, -6, 0, 14, 5.5

H. Find the area of a circle.

49. What is the area of the circle shown?

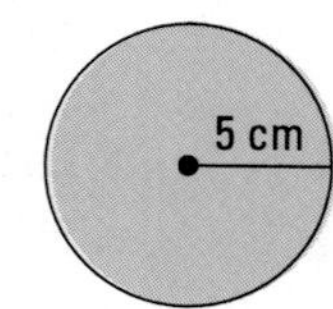

50. Find the area of a circle with diameter 8 inches.

CHAPTER

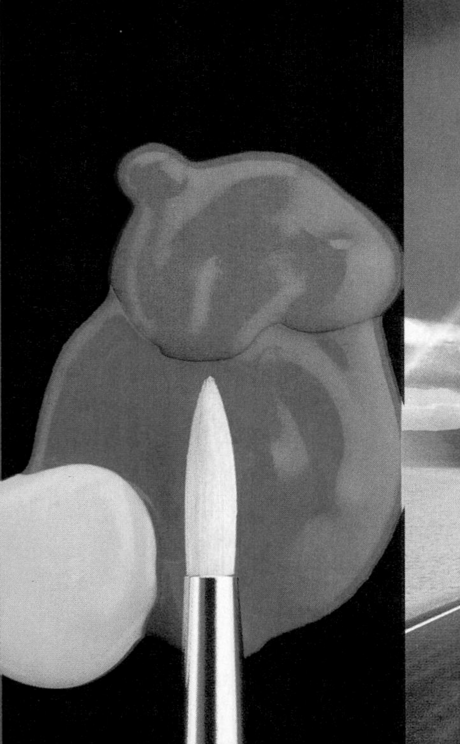

Division in Algebra

When manufacturers cut sheets of metal or wood or paper, they try to minimize the waste and cost of leftover material. Mathematicians are also concerned about minimizing waste. Consider this problem.

Can the 6-by-7 rectangle on the left below be split into little regions of the T shape shown with no squares left over?

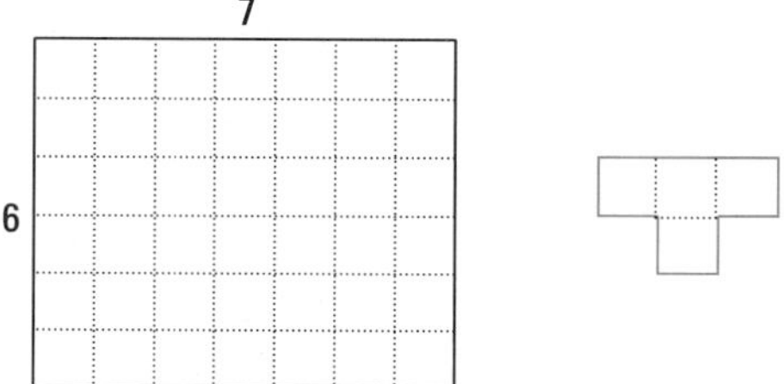

You could try to solve this problem by using trial and error. Shown below is a start that leads to a situation which does *not* work. But maybe the start was wrong.

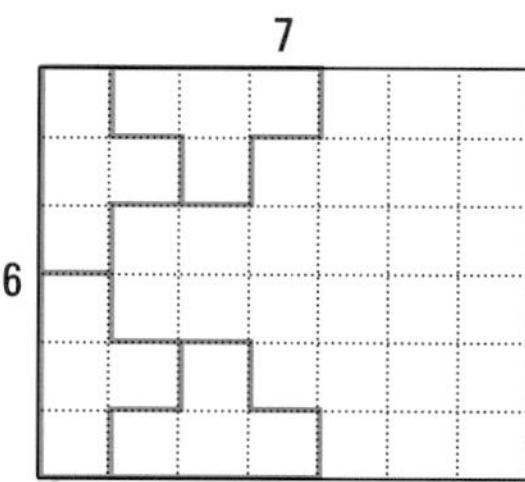

There is another way to examine the problem. The area of the rectangle is 42 square units. The area of the T-shaped region is 4 square units. If the rectangle can be cut into these T-shapes, there will be $\frac{42}{4}$, or $10\frac{1}{2}$ T-shaped regions. Since the number of regions must be a whole number, no attempt at splitting the rectangle will work.

When you first studied division you answered questions about splitting objects into equal parts with no "leftovers." In this chapter, you will extend this idea of division to other uses, including applications to rates, ratios, percents, probability, and similarity.

LESSON

6-1 The Algebraic Definition of Division

Appealing fruit. *Shown is an orange grove in Australia. New South Wales and South Australia produce most of the country's oranges. Although Australia is not a leading producer of oranges, citrus fruit is grown there.*

The Relation Between Division and Multiplication

Abe divided a quart of orange juice equally among his five children. How many ounces of orange juice did each child receive?

This question can be answered by either division or multiplication. Either way, you must change 1 quart to 32 ounces. You can view Abe's problem as dividing 32 ounces among 5 children,

$$32 \div 5 = 6.4,$$

or as giving each child $\frac{1}{5}$ of the orange juice,

$$32 \cdot \frac{1}{5} = 6.4.$$

Both ways the answer is the same, 6.4 ounces. Dividing by 5 is the same as multiplying by $\frac{1}{5}$. In general, dividing by b is the same as multiplying by its reciprocal $\frac{1}{b}$.

Algebraic Definition of Division
For any real numbers a and b, $b \neq 0$,

$$a \div b = a \cdot \frac{1}{b}.$$

Recall that $a \div b$ can also be written as $\frac{a}{b}$ or a/b. So if $b \neq 0$,

$$a \div b = \frac{a}{b} = a/b = a \cdot \frac{1}{b}.$$

The Algebraic Definition of Division allows any division situation to be converted to multiplication. For instance, consider the division of fractions $\frac{a}{b} \div \frac{c}{d}$. According to the Algebraic Definition of Division, dividing by $\frac{c}{d}$ is the same as multiplying by its reciprocal $\frac{d}{c}$.

$$\frac{a}{b} \div \frac{c}{d} = \frac{a}{b} \cdot \frac{d}{c}.$$

For example, $\frac{5}{2} \div \frac{9}{7} = \frac{5}{2} \cdot \frac{7}{9} = \frac{35}{18}$. This is sometimes called the "invert and multiply" rule. Algebraic fractions are divided in the same way.

Dividing Algebraic Fractions

Example 1

Simplify $\frac{x}{5} \div \frac{3}{4}$.

Solution

Dividing by $\frac{3}{4}$ is the same as multiplying by $\frac{4}{3}$.

$$\frac{x}{5} \div \frac{3}{4} = \frac{x}{5} \cdot \frac{4}{3} = \frac{4x}{15}$$

Check

Substitute some value for x. Use this number to evaluate the original expression and your answer. Suppose $x = 2$. Does $\frac{2}{5} \div \frac{3}{4} = \frac{4 \cdot 2}{15}$? To determine this, change each fraction to a decimal. Does $0.4 \div 0.75 = \frac{8}{15}$? Yes, each side equals $0.5\overline{3}$.

Dividing Complex Fractions

Recall that a fraction bar also indicates division. In the next example, there are three fractions. One is the numerator and the second is the denominator of a third "bigger" fraction. Fractions of the form $\frac{\frac{a}{b}}{\frac{c}{d}}$ are called **complex fractions.** To simplify a complex fraction, first rewrite the complex fraction as an expression using division.

Example 2

Simplify $\frac{\frac{7\pi}{3}}{-\frac{\pi}{21}}$ and check your answer.

Solution

Rewrite the fraction using division. Then use the Algebraic Definition of Division to rewrite using multiplication.

$$\frac{7\pi}{3} \div -\frac{\pi}{21} = \frac{7\pi}{3} \cdot -\frac{21}{\pi}$$

Now, use the Multiplying Fractions Property.

$$= \frac{7\cancel{\pi}^{1}}{\cancel{3}_{1}} \cdot -\frac{\cancel{21}^{7}}{\cancel{\pi}_{1}}$$

$$= -49$$

For the check, use a calculator. Notice how we are careful to use parentheses to group the numerator and denominator separately.

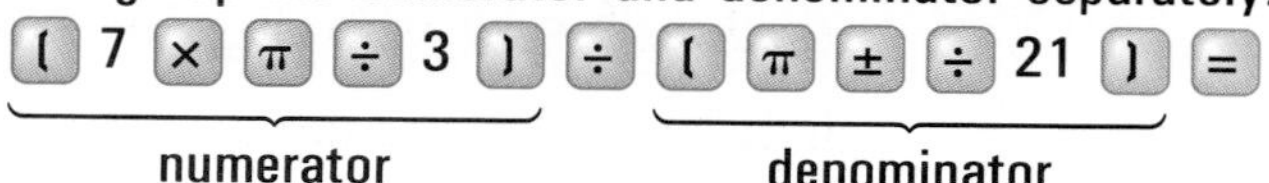

In Example 2, a positive number is divided by a negative number. The divisor is replaced by its reciprocal when converting the division into a multiplication. Since a number and its reciprocal have the same sign, the signs of answers to division problems follow the same rules as those for multiplication. This is why the answer to Example 2 is negative.

If two numbers have the same + or − sign, their quotient is positive. If two numbers have different signs, their quotient is negative.

Notice that in Example 2, the negative sign is written in front of the fraction, $-\frac{\pi}{21}$. The rules of division tell us that the negative sign could also have been in the numerator or in the denominator without altering the value. This means that

$$-\frac{\pi}{21} = \frac{-\pi}{21} = \frac{\pi}{-21}.$$

In general, for all a and b, $b \neq 0$,

$$-\frac{a}{b} = \frac{-a}{b} = \frac{a}{-b}.$$

Solving Equations Using Division

The Algebraic Definition of Division leads to another method for solving equations of the form $ax = b$. In the first solution of Example 3, both sides of the equation are multiplied by $-\frac{1}{31}$. This is the same as dividing by -31, as shown in the second solution.

Example 3

Solve $-31m = 527$.

Solution 1

$$-31m = 527$$

$$-\frac{1}{31} \cdot -31m = -\frac{1}{31} \cdot 527 \quad \text{Multiply both sides by } -\frac{1}{31}.$$

$$m = -\frac{527}{31}$$

$$m = -17$$

Solution 2

$$-31m = 527$$

$$\frac{-31m}{-31} = \frac{527}{-31} \quad \text{Divide both sides by } -31.$$

$$m = -17$$

Check

Substitute -17 for m. Does $-31(-17) = 527$? Yes. It checks.

Both methods of Example 3 are acceptable.

QUESTIONS

Covering the Reading

1. Suppose Abe had divided a quart of orange juice equally among his five children *and himself.* Show two ways to determine how many ounces of orange juice each person received.

2. State the Algebraic Definition of Division.

In 3 and 4, fill in the blanks.

3. a. $\frac{m}{n} = m \div \underline{\ ?\ }$ **b.** $\frac{m}{n} = m \cdot \underline{\ ?\ }$

4. a. $\frac{\frac{p}{q}}{\frac{r}{s}} = \frac{p}{q} \div \underline{\ ?\ }$ **b.** $\frac{p}{q} \div \frac{r}{s} = \frac{p}{q} \cdot \underline{\ ?\ }$

In 5–10, simplify.

5. $\frac{\frac{3}{4}}{-\frac{4}{5}}$

6. $\frac{\frac{x}{4}}{-\frac{4}{y}}$

7. $\frac{5}{6} \div \frac{n}{10}$

8. $-\frac{12\pi}{5} \div \frac{\pi}{4}$

9. $1\frac{2}{3} \div 3\frac{1}{3}$

10. $\frac{\frac{3\pi}{5}}{\frac{6}{\pi}}$

11. Suppose you are asked to solve $-3j = -48$ in one major step.

a. By what could you multiply both sides?

b. By what could you divide both sides?

The United States produces about 18% of the world's annual orange harvest, or about 15.5 billion pounds of oranges.

12. *Multiple choice.* Which of these expressions is *not* equivalent to the others?

(a) $\frac{-7}{2}$ (b) $-\frac{7}{2}$ (c) $\frac{-7}{-2}$ (d) $\frac{7}{-2}$

13. Round to the nearest thousandth.

a. $\frac{-3}{11}$ **b.** $\frac{3}{-11}$ **c.** $-\frac{3}{11}$

14. Generalize the idea of Questions 12 and 13 using two variables.

In 15–20, solve.

15. $-143 = 13x$

16. $-1.5q = -75$

17. $2.5k = -0.7$

18. $-1900 = -0.2n$

19. $\frac{2}{3}t = \frac{5}{6}$

20. $3x = -\frac{5}{3}$

Scents for cents. *The cost of one ounce of perfume can range from $150 to over $300. The cost of one ounce of cologne can range from $25 to over $40.*

Applying the Mathematics

21. Le Parfum Company produces perfume in 200-ounce batches and bottles it in quarter-ounce bottles. Suppose you want to know how many bottles will be filled by a batch.
  a. Write a division problem that will tell you.
  b. Find the answer.

22. A gallon of milk was divided equally among x people. How many ounces did each person receive?

23. Half of a pizza was divided equally among 3 people. How much of the original pizza did each person receive?

In 24 and 25, simplify.

24. $b \div \frac{1}{b}$

25. $\frac{xy}{21} \div \frac{x}{4y}$

26. **a.** Evaluate $x \div y$ and $y \div x$ for each of the following.
  i. $x = 12$ and $y = 2$
  ii. $x = 20$ and $y = -5$
  iii. $x = \frac{2}{3}$ and $y = \frac{4}{5}$
  b. Does your answer to part **a** indicate that division is commutative? Explain.
  c. Describe how $x \div y$ and $y \div x$ are related in general.

27. In 1577, Guillaume Gosselin published a book titled *De Arte Magna.* It contained some of the first work on positive and negative numbers. It was written in Latin. Here are some rules Gosselin wrote. Translate these rules into English.
  a. P in P diviso quotus est P.
  b. M in M diviso quotus est P.
  c. M in P diviso quotus est M.
  d. P in M diviso quotus est M.

Review

28. A $100 investment grows at 7% interest annually for 5 years. The interest is added to the investment each year. The value of the investment at the end of each year is shown in the table below. *(Lesson 4-2)*

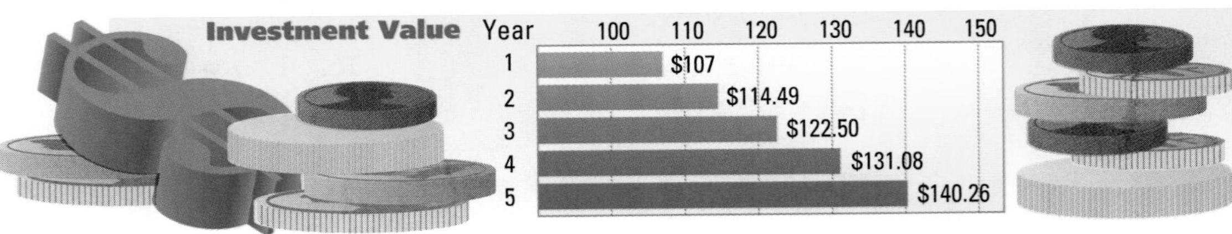

  a. How much did the investment grow in value between years 2 and 3?
  b. In what year is the growth in the investment the largest?

29. Use the picture of a balance scale below. The cylinders are equal in weight and the balls are one-kilogram weights. *(Lesson 3-5)*

a. Write an equation describing the situation with B representing the weight of one cylinder.
b. What is the weight of one cylinder?

30. Solve $V + 0.06V + 100 = 14{,}289.16$. *(Lessons 3-5, 3-6)*

31. If $100 < n! < 200$, find n. *(Lesson 2-10)*

Longest lake, too. *These fishermen prepare their net for fishing on Lake Tanganyika. With a length of 680 km, it is the longest freshwater lake in the world.*

32. Let y = the depth of a point in Lake Tanganyika in Africa, the second deepest lake in the world.
a. Give a reasonable domain for y.
b. It is known that the deepest point in Lake Tanganyika is 1470 meters below the surface of the lake. What inequality does y satisfy?
c. Graph the solution set to part **b.** *(Lesson 1-2)*

33. A circle has a radius of 1.2 m. Find its area to the nearest tenth of a square meter. *(Previous course)*

Exploration

34. Congruent figures are figures with the same size and shape. Split this region into 6 congruent pieces.

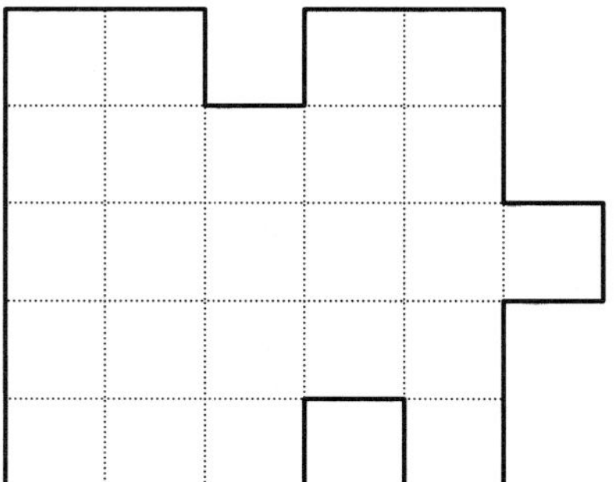

LESSON

6-2

Rates

Southwest is Key West. *Shown is part of the Old Town section of Key West, a popular resort town about 100 miles southwest of the Florida mainland.*

Situations involving splitting up can lead to rates. If 100 tickets are split up evenly among 4 people, then there are

$$\frac{100 \text{ tickets}}{4 \text{ people}} = \frac{25 \text{ tickets}}{\text{person}}.$$

The result is a *rate,* like the ones you multiplied in Lesson 2-4.

The situation above is an instance of the *Rate Model for Division.*

Rate Model for Division
If a and b are quantities with different units, then $\frac{a}{b}$ is the amount of quantity a per quantity b.

The rate model says that if you are not given a rate, you can calculate one using division.

Using Division to Calculate Rates

Example 1

Ivan and Katya drove from Key West to Tampa, Florida, a trip of 400 miles. The trip took 8 hours. What was their average rate?

Solution 1

Divide the distance in miles by the time in hours.

$$\frac{400 \text{ miles}}{8 \text{ hours}}$$

Separate the measurement units from the numerical parts.

$$\frac{400 \text{ miles}}{8 \text{ hours}} = \frac{400}{8} \frac{\text{miles}}{\text{hours}}$$

$$= 50 \text{ miles per hour}$$

Their average rate was 50 miles per hour.

Solution 2

You could also divide the time by the distance.

$$\frac{8 \text{ hours}}{400 \text{ miles}} = \frac{8}{400} \frac{\text{hours}}{\text{miles}}$$
$$= \frac{1}{50} \text{ hour per mile}$$

This means that, on the average, it took them $\frac{1}{50}$ of an hour to travel each mile.

Is the marker correct?
Yes. Hawaii is farther south than Key West, but Hawaii is not part of the continental *U.S.A.*

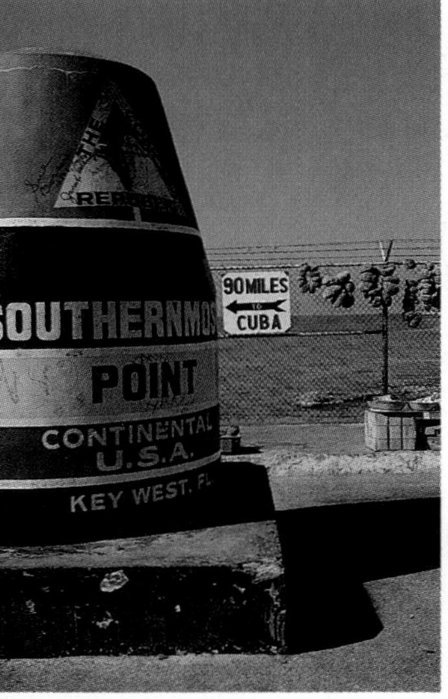

The first solution gives the rate in *miles per hour.* You are familiar with the meaning of this rate from speed limit signs or from the speedometer in a car. The second solution gives the rate in *hours per mile,* or how long it takes to travel one mile. Notice that $\frac{1}{50}$ of an hour is $\frac{1}{50} \cdot 60$ minutes, or 1.2 minutes. In other words, it takes a little over a minute to go one mile. Either rate is correct. The one to use depends on the situation in which you use it.

Averages can also be thought of as rates.

Example 2

In 1991, total medical costs in the United States were about $728,600,000,000. That year the population was about 252,700,000. What was the average cost per person for medical care?

Solution

This situation should prompt you to find dollars per person. To get this rate, divide the total spent by the number of people.

$$\frac{728{,}600{,}000{,}000 \text{ dollars}}{252{,}700{,}000 \text{ people}} = \frac{728{,}600{,}000{,}000}{252{,}700{,}000} \frac{\text{dollars}}{\text{people}}$$
$$\approx 2883 \text{ dollars/person}$$

In 1991 the average cost per person for medical care was about $2883.

In dividing by a rate, the arithmetic of units follows the Algebraic Definition of Division.

Example 3

Alice reads 12 $\frac{\text{pages}}{\text{day}}$. At this rate how long will it take her to read a book that is 190 pages long?

Solution

Think of splitting up the 190 pages into chunks of 12 $\frac{\text{pages}}{\text{day}}$.

$$\frac{190 \text{ pages}}{12 \frac{\text{pages}}{\text{day}}} = 190 \text{ pages} \cdot \frac{1}{12} \frac{\text{day}}{\text{page}} = \frac{190}{12} \text{ days} = 15.8\overline{3} \text{ days}$$

It will take her about 16 days.

Rates and Negative Numbers

Rates can be negative quantities.

Example 4

If the temperature goes down 12 degrees in 5 hours, what is the rate of temperature change in degrees per hour?

Solution

To find the rate, divide the number of degrees changed by the number of hours.

$$\text{rate of temperature change} = \frac{\text{drop of 12 degrees}}{\text{5 hours}}$$
$$= \text{drop of 2.4 degrees per hour}$$

This can be written as

$$\frac{\text{-12 degrees}}{\text{5 hours}} = \text{-2.4 degrees per hour.}$$

The temperature drops at a rate of 2.4 degrees per hour.

Division by Zero and Rates

Consider a rate such as $\frac{\text{0 meters}}{\text{10 seconds}}$, which has 0 in the numerator. This means that you did not travel at all in 10 seconds. So your rate is 0 m/sec. Thus, $\frac{0}{10} = 0$. In contrast, consider a rate such as $\frac{\text{10 meters}}{\text{0 seconds}}$, which has 0 in the denominator. It would mean you are traveling 10 meters in 0 seconds. To be true, you would have to be in two places at the same time! Since that is impossible, this model illustrates that it is meaningless to think of $\frac{10}{0}$. *Division by zero is impossible.* Thus, the denominator of a fraction cannot be zero.

When variables appear in expressions involving division, you must make sure that you don't attempt to divide by zero.

Example 5

What value(s) can k *not* have in $\frac{k-7}{k-5}$?

Solution

Solve $k - 5 = 0$ to find out when the denominator is 0.

If $k - 5 = 0$, then $k = 5$. Thus k cannot be 5.

Check

If $k = 5$, then $\frac{k-7}{k-5} = \frac{5-7}{5-5} = \frac{-2}{0}$, which is impossible.

If a calculator or computer is instructed to divide by zero, it won't do it! Instead it will display an error message. When we tried to evaluate the expression $\frac{5-7}{5-5}$ with a calculator, we keyed in

(5 − 7) ÷ (5 − 5) = .

One calculator showed E 0.

Another showed Error 02 Math.

Caution: Note in Example 5 that it is possible to have a numerator equal to 0. When $k = 7$, the value of $\frac{k-7}{k-5}$ is $\frac{7-7}{7-5} = \frac{0}{2} = 0$.

QUESTIONS

Covering the Reading

1. Suppose Carmen and Ricardo traveled 300 miles and it took them 8 hours.
   a. What was their average speed in miles per hour?
   b. On the average, how long did it take them to travel a mile?

2. State the Rate Model for Division.

3. Give a rate suggested by each situation.
   a. A family drove 24 miles in $\frac{2}{3}$ hour.
   b. A family drove m miles in $\frac{2}{3}$ hour.
   c. A family drove 24 miles in h hours.
   d. A family drove m miles in h hours.

In 4–6, calculate a rate suggested by each situation.

4. In 9 days Julian earned $495.

5. You buy c cans of natural lemonade for $2.10.

6. You use 7.8 gallons of gasoline in traveling 270 miles.

Oldest U.S. college. *Founded in 1636, Harvard University is located in Cambridge, MA. These Harvard sculling teams are practicing on the nearby Charles River.*

7. In the 1992–93 school year, Harvard University had 6,799 undergraduate students and 931 faculty members. What was the average number of undergraduate students per faculty member?

In 8–10, calculate the rate of temperature change for each of the following situations.

8. The temperature drops 11 degrees in 5 hours.

9. The temperature rose 6 degrees in 12 hours.

10. The temperature stayed the same for 4 hours.

11. What happens on your calculator when you divide by 0?

In 12–14, what value can the variable *not* have in each expression?

12. $\frac{18}{k - 4}$ **13.** $\frac{x - 6}{x + 1}$ **14.** $\frac{w + 1}{3}$

Applying the Mathematics

Traffic tangles. *Pedestrians and pedicabs in the cities of Bangladesh often create traffic jams. With the eighth largest population, Bangladesh is one of the world's most densely populated countries.*

In 15 and 16, divide by the rate to obtain a new piece of information.

15. The family needed \$120 a week for food and had only \$300 on hand.

16. Willie can type 35 words a minute. He needs to type a 500-word essay.

17. In one store a 20-ounce can of pineapple cost \$1.78 and a 6-ounce can of the same kind of pineapple cost \$.78. The unit cost is calculated as the rate *cost per ounce.*
a. Calculate the unit cost of the 20-ounce can.
b. Calculate the unit cost of the 6-ounce can.
c. Based on the unit cost, which is the better buy?

18. a. For each place, find the number of people per square mile. (This is the **population density.**)
Bangladesh: 1993 population 122,255,000 (estimated); area 55,813 sq mi
Greenland: 1993 population 57,000 (estimated); area 840,000 sq mi
b. The population density of Bangladesh is how many times the population density of Greenland?

19. A very fast runner can run a half mile in 2 minutes. Express the average rate in each of these units.
a. miles per minute
b. miles per hour

20. *Multiple choice.* In t minutes, a copy machine made n copies. At this rate, how many copies per hour can be made?
(a) $\frac{60t}{n}$ (b) $\frac{60n}{t}$ (c) $\frac{n}{60t}$ (d) $\frac{t}{60n}$

21. In baseball, batting averages are computed by dividing the number of hits by the official number of at bats and rounding to three decimal places. In the 1993 regular season, Chicago White Sox first baseman Frank Thomas had 174 hits in 549 at bats, and Toronto Blue Jays first baseman John Olerud had 200 hits in 551 at bats. Which player had the better batting average?

22. Give the domain of the variable in each expression.
a. $\frac{5}{2n + 3}$ **b.** $\frac{5n + 1}{7}$ **c.** $\frac{5n + 1}{7(2n + 3)}$

23. Suppose a spreadsheet program is opened, `100` is typed in cell A1, and `0` is typed in cell B1.
a. If `= A1/B1` is typed in cell C1, what is printed in cell C1, and why?
b. If `= B1/A1` is typed in cell D1, what is printed in cell D1, and why?

Review

In 24–27, simplify. *(Lesson 6-1)*

24. $\frac{\frac{5}{2}}{3}$ **25.** $\frac{-3}{4} \div \frac{-3}{2}$ **26.** $\frac{x}{2y} \div \frac{11y}{3}$ **27.** $6x \div \frac{x}{2}$

28. A cook has x pounds of ground beef. If it takes $\frac{1}{4}$ lb of ground beef to make one hamburger, how many hamburgers can the cook make? *(Lesson 6-1)*

29. Let a be the measure of an angle.
a. Write an expression for the supplement of the angle.
b. Can the supplement of the angle ever equal the complement of the angle? Explain.
c. Can the supplement of an angle ever be 4 times the complement of the angle? Explain. *(Lessons 4-7, 5-3)*

30. Use the equation $5x - 4y = 40$. *(Lessons 4-3, 4-6)*
a. If $x = 10$, find the value of y.
b. What point on the graph of $5x - 4y = 40$ have you found in part **a**?

31. *Skill sequence.* Solve. *(Lessons 2-8, 3-10, 5-6)*
a. $-x > 8$ **b.** $-2x > 8$
c. $-3x + 4 > 8$ **d.** $-5x + 6 > 8 - 7x$

Exploration

32. Refer to the BASIC program below or use an equivalent calculator program.

```
10 PRINT "EVALUATE 7/(K – 5)"
20 PRINT "ENTER VALUE OF K"
30 INPUT K
40 PRINT "VALUE OF 7/(K – 5)"
50 IF K = 5 THEN PRINT "DIVISION BY ZERO IS IMPOSSIBLE"
60 IF K <> 5 THEN PRINT 7/(K – 5)
70 END
```

a. Run the program and input the values 8, 7.5, 7, 6.5, 6, 5.5, and 5. Record your results in a table.
b. Input the values 4.5, 4, 3.5, 3, 2.5, and 2, and record the results.
c. Describe some patterns you find in the data you collect.

33. Use an almanac or some other source to find the national debt. Find the U.S. population. Then calculate the average debt per capita. (The phrase *per capita* means "per person.")

34. See Question 18. Find the population density of the community or place where you live.

LESSON

6-3 Ratios

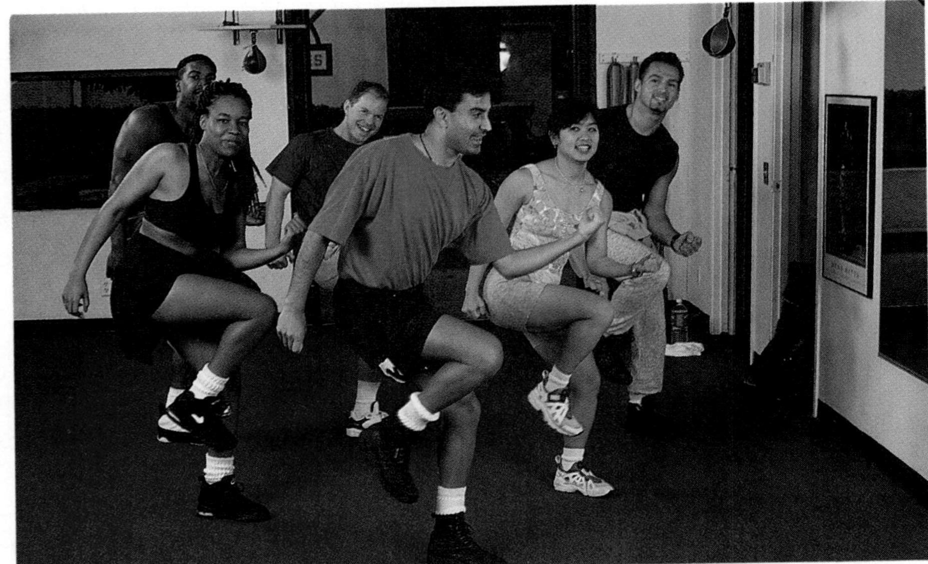

Steps in the right direction. *Many people join aerobics classes to increase endurance and energy and to help maintain proper weight. Regular aerobic exercise may also help lessen the risk of heart disease.*

What Is a Ratio?

Being overweight increases the probability that a person will suffer from heart disease. The chart below shows a way to test whether an adult has an increased risk. Dividing the waist measure by the hip measure results in a *ratio* which can be used to compare them.

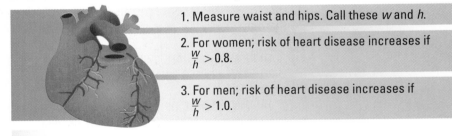

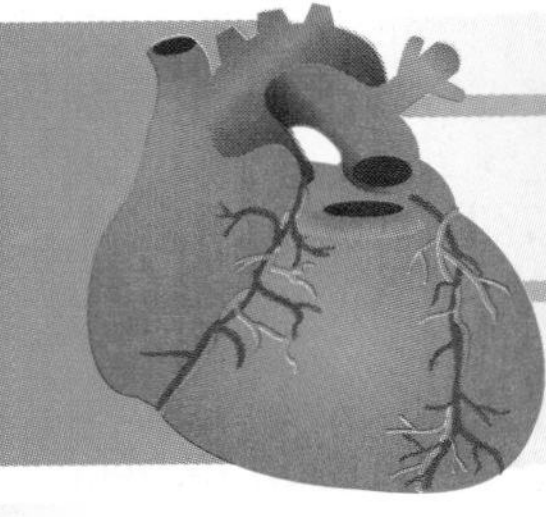

1. Measure waist and hips. Call these w and h.
2. For women; risk of heart disease increases if $\frac{w}{h} > 0.8$.
3. For men; risk of heart disease increases if $\frac{w}{h} > 1.0$.

Example 1

Ms. Mott's waist measure is 26 in. Her hip measure is 35 in. According to the chart above, does she run an increased risk of heart disease?

Solution

$$\frac{\text{waist measure}}{\text{hip measure}} = \frac{26''}{35''} \approx 0.74.$$

Since $0.74 < 0.8$, her risk (according to this test) is not increased.

The direction of the comparison is important. If Ms. Mott compared her hip measure to her waist measure, the result would be $\frac{35''}{26''} \approx 1.35$, which is greater than 0.8. She would be using the wrong number and would misinterpret the test.

As you know, subtraction provides one way to compare quantities. Division is another way.

Ratio Model for Division
Let a and b be quantities with the same units.
Then the **ratio** $\frac{a}{b}$ compares a to b.

Similarly, the ratio $\frac{b}{a}$ compares b to a.

Notice the difference between a rate and a ratio. In the rate a per b, the units for a and b are different. In a ratio, the units are the same.

$\frac{40 \text{ km}}{2 \text{ hr}} = 20 \frac{\text{km}}{\text{hr}}$ is a rate. $\frac{40 \text{ km}}{2 \text{ km}} = 20$ is a ratio.

Example 2

It takes Mr. Garcia $\frac{3}{4}$ hour to go to work and it takes Ms. Wang 10 minutes to go to work. Find a ratio comparing Ms. Wang's time to Mr. Garcia's time.

Solution

The units of measure for 10 minutes and $\frac{3}{4}$ hour are not the same. So we change hours to minutes. Since Ms. Wang's time is to be compared to Mr. Garcia's, her time is the numerator.

$$\frac{\text{Ms. Wang's time}}{\text{Mr. Garcia's time}} = \frac{10 \text{ minutes}}{\frac{3}{4} \text{ hour}} = \frac{10 \text{ minutes}}{\frac{3}{4} \cdot 60 \text{ minutes}} = \frac{10 \text{ minutes}}{45 \text{ minutes}} = \frac{2}{9}$$

This means that, on the average, Ms. Wang travels 2 minutes for every 9 minutes that Mr. Garcia travels. It takes Ms. Wang $\frac{2}{9}$ of the time it takes Mr. Garcia to go to work.

Ratios and Percents

Ratios often are written as percents. In Example 2, $\frac{2}{9} = .\overline{22} \approx 22\%$, so you could say that it takes Ms. Wang about 22% of the time it takes Mr. Garcia to go to work. A percent can always be interpreted as a ratio, in this case about $\frac{22}{100}$.

In problems involving discounts, the **percent of discount** is the ratio of the discount to the original price. The **percent of tax** is the ratio of the tax to the selling price.

The first U.S. jeans. *In the 1850s according to legend, a miner asked Levi Strauss, a canvas peddler, for a pair of durable canvas pants. Before long, Strauss was making pants from a tough, cotton fabric now known as denim. In 1879, Levi's denim jeans with rivet-reinforced pockets sold for \$1.46 a pair.*

Example 3

A pair of jeans originally cost \$29.95. They were reduced to \$23.95. The tax on the reduced price is \$1.20.

a. What is the percent of discount?

b. What is the percent of tax?

Solution

a. $\text{Percent of discount} = \frac{\text{amount of discount}}{\text{original price}} = \frac{29.95 - 23.95}{29.95}$

$= \frac{6.00}{29.95}$

$\approx 0.2003 \approx 20\%$

b. Calculate the ratio between the amount of tax and the reduced price.

$\text{Percent of tax} = \frac{1.20}{23.95} \approx 0.0501 \approx 5\%$

Using Ratios to Set Up Equations

Suppose in Example 2 you knew only that Ms. Wang travels 2 minutes for every 9 minutes that Mr. Garcia travels. Then, whatever times W and G they travel, the ratio of W to G equals $\frac{2}{9}$.

$$\frac{W}{G} = \frac{2 \text{ minutes}}{9 \text{ minutes}} = \frac{2}{9}$$

But you also know that for all nonzero values of x, $\frac{2x}{9x} = \frac{2}{9}$.

So there is a value of x with $W = 2x$ and $G = 9x$. This idea is very useful in many problems, because directions for mixing foods or chemicals are frequently given as ratios. You can use equations to determine actual quantities from the ratios.

Example 4

Instructions for preparing a lemonade drink call for 3 parts of water for each part of lemonade concentrate. How much of each ingredient is needed to make 10 gallons of lemonade?

Solution

Because the water and concentrate are in the ratio of 3 to 1, let $3x$ be the number of gallons of water and $1x$ be the number of gallons of concentrate. The total needed is 10 gallons, so

$$3x + 1x = 10.$$
$$4x = 10$$
$$x = \frac{10}{4} = 2.5$$

So $3x = 3 \cdot 2.5 = 7.5$.

You need 2.5 gallons of lemonade concentrate and 7.5 gallons of water.

Check

$\frac{7.5 \text{ gal}}{2.5 \text{ gal}} = \frac{3}{1}$ and 7.5 gal + 2.5 gal = 10 gal. It checks.

QUESTIONS

Covering the Reading

In 1–3, refer to the method for testing heart disease risk.

1. Does a woman run an increased risk of heart disease if her waist and hip measurements are 32 in. and 37 in., respectively?

2. If w is a man's waist measure and h is his hip measure, write a sentence describing when a man's risk of heart disease increases.

3. Does a man run an increased risk of heart disease if his waist is 34 in. and his hips are 36 in.?

4. Let x and y be two quantities with the same units. Write two ratios comparing x and y.

5. What is the difference between a rate and a ratio?

6. *Multiple choice.* Which is not a ratio?
(a) $\frac{\text{14 seconds}}{\text{23 seconds}}$ (b) $\frac{\text{150 miles}}{\text{3 hours}}$ (c) $\frac{\text{27 cookies}}{\text{13 cookies}}$

7. Suppose it takes Ms. Lopez 25 minutes to complete a particular job and it takes Mr. Sampson half an hour to complete the same job. Write a ratio which:
a. compares Ms. Lopez's time to Mr. Sampson's time;
b. compares Mr. Sampson's time to Ms. Lopez's time.

8. An item is on sale for \$15. It originally cost \$21.
a. What is the discount (in dollars)?
b. What is the percent of discount?

9. You pay 64¢ tax on a \$16.00 purchase.
a. Write a ratio of tax to purchase price.
b. What is the percent of tax?

10. Tell whether the fraction equals the ratio of 4 to 3.
a. $\frac{4}{3}$ **b.** $\frac{3}{4}$ **c.** $\frac{8}{6}$
d. $\frac{6}{5}$ **e.** $\frac{4x}{3x}$ **f.** $\frac{3x}{4x}$

11. In Example 4, determine the amounts of lemonade concentrate and water you would use to make 2 quarts of lemonade drink.

Applying the Mathematics

12. Banner High School has won 36 of its last 40 football games.
a. Write a ratio of games won to games played.
b. What percent of these 40 games has it won?
c. Winning percentages are often written as three-place decimals. Write the answer from part **a** as a three-place decimal.

13. According to a teacher group, the average teacher salary for 1991–92 was $34,413. In 1870 it was $189. The 1991–92 salary is about how many times the 1870 salary?

In a class by itself. The Country School, *a painting by Winslow Homer, shows a one-room school that was common in rural U.S. during the 1800s. Most had one teacher who taught all grade levels.*

14. Frieda owns x sweaters while her twin brother Fred owns y sweaters.
    a. Fred owns _?_ times as many sweaters as Frieda does.
    b. Is the answer to part **a** a rate or a ratio?

15. The ratio of x to y is the _?_ of the ratio of y to x.

16. The ratio of adults to children at a concert is expected to be 4 to 1. If 200 people attend, how many are expected to be children?

17. If two quantities are in the ratio a to b and the first is ax, then what is the second?

18. A paint mixture calls for 7 parts of linseed oil, 5 parts of solvent, and one part of pigment. How much of each ingredient is needed to make 50 gallons of paint?

19. The circles below have radii 2 and 3.
    a. Find the ratio of the diameter of Circle I to the diameter of Circle II.
    b. Give the ratio of the area of Circle I to the area of Circle II, in lowest terms.
    c. Write the ratio in part **b** as a decimal.

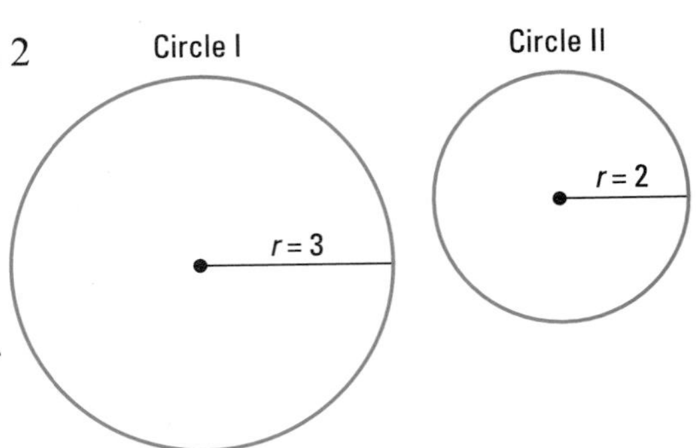

Review

20. In the five years from 1987 to 1992, the U.S. federal debt increased by about $1,500,000,000,000.
    a. Write this amount in words.
    b. Estimate the amount of increase per year. *(Lesson 6-2, previous course)*

Martin Lopez-Zubero

21. In the 200-meter backstroke competition at the 1992 Olympics, Martin Lopez-Zubero of Spain set an Olympic record by completing the race in 118.47 seconds. What was his average rate in meters per second? *(Lesson 6-2)*

22. For what values of x is $\frac{5 - 2x}{5 + 2x}$ undefined? *(Lesson 6-2)*

23. Simplify $\frac{3x}{4} \div \frac{x}{2}$. *(Lesson 6-1)*

24. *Multiple choice.* Which of the following is *not* equal to $-\frac{a}{b}$? *(Lesson 6-1)*

(a) $\frac{a}{-b}$ (b) $\frac{-a}{b}$ (c) $\frac{-a}{-b}$ (d) $-\frac{-a}{-b}$

25. If $5a = 36$, find the value of $\sqrt{5a} + 7\sqrt{5a}$. *(Lessons 3-6, 5-9)*

26. *Skill sequence.* Solve for x. *(Lessons 2-6, 3-5)*
a. $5x = 19$
b. $0.05x = 19$
c. $x + 0.05x = 19$

27. A student has earned the following scores on algebra tests this term: 75, 90, 86, 78, 80. What is the student's test average? *(Previous course)*

Exploration

28. Consider this right triangle with an angle of about 37°. The three sides can form six different ratios.

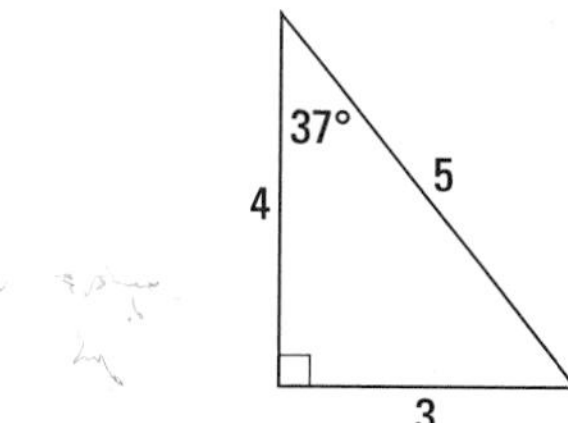

a. Write the values of all six of these ratios.
b. Some of these ratios have special names. One of these is called the *sine.* For the triangle above it is written sin 37°. Find the value of sin 37° on your calculator using the key sequence 37 [sin] or [sin] 37 (depending on your calculator), and determine which of the six ratios in part **a** is the sine.
c. Another special ratio is called *cosine.* Find the value of cos 37° on your calculator and determine which of the six ratios in part **a** is the cosine.
d. A third special ratio is called the *tangent.* Find the value of tan 37° on your calculator and determine which of the six ratios in part **a** is the tangent.

Introducing
Lesson 6-4

Relative Frequencies and Probabilities

IN-CLASS ACTIVITY

Work in groups of 2 or 3 students.
Materials: Two dice per group

1 When two dice are tossed and the numbers on top are added, the sum can be any whole number from 2 to 12. Which sum do you think is most likely to occur? Why did you pick this number?

2 Toss the dice 50 times. Each time, record the result in the column labeled frequency in a table like the one below. The table shows the sums of the first 10 tosses of one group.

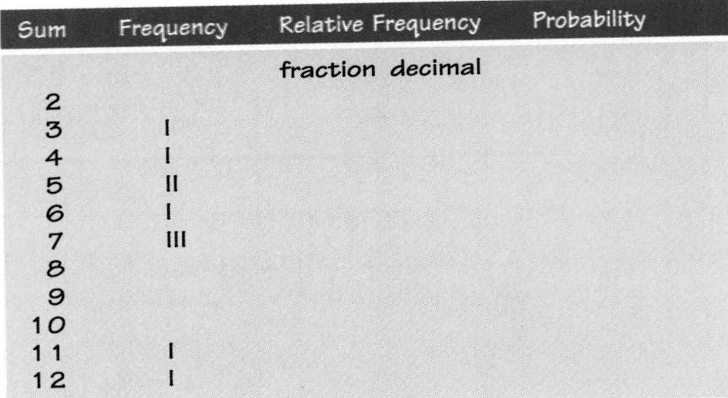

Sum	Frequency	Relative Frequency		Probability
		fraction	decimal	
2				
3	I			
4	I			
5	II			
6	I			
7	III			
8				
9				
10				
11	I			
12	I			

3 After you have tossed the dice 50 times, calculate the relative frequency of getting each sum. For instance, the relative frequency of getting a 2 is the ratio between the number of 2s you got and 50. If you got three 2s in 50 tosses of the dice, the relative frequency of getting a sum of 2 is $\frac{3}{50} = .06$. Write each relative frequency as both a fraction and a decimal.

4 ***Draw a conclusion.*** Based on your table, how, if at all, would you revise your prediction in part 1?

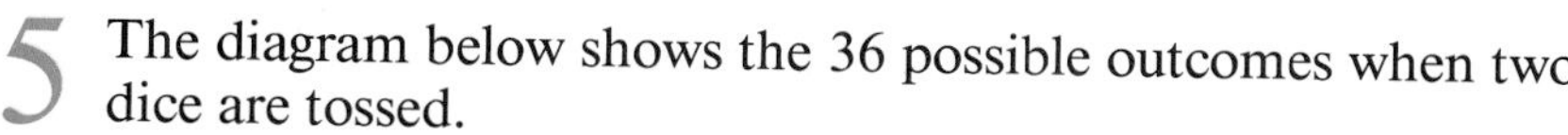

5 The diagram below shows the 36 possible outcomes when two dice are tossed.

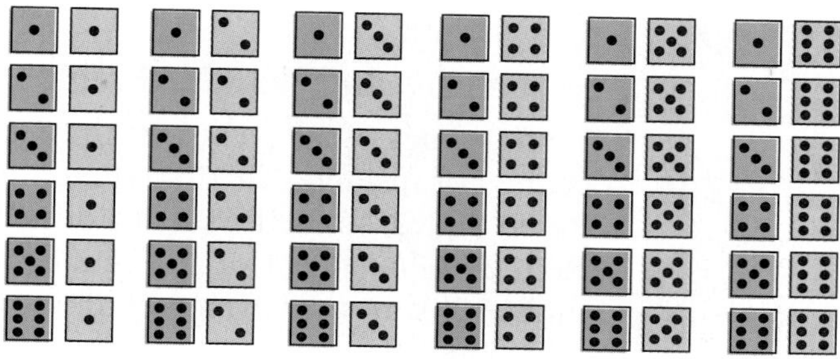

Notice that there is only 1 way to get a sum of 2. We say that the *probability* of getting a sum of 2 is $\frac{1}{36} \approx .028$. In contrast, there are 3 ways of getting a sum of 10 (, ,), so the probability of getting a sum of 10 is $\frac{3}{36} \approx .083$.

a. Calculate the probability of getting each of the other sums from 2 to 12, and record these numbers in the last column of the table.

b. How close are your group's relative frequencies to the probabilities?

6 **a.** Combine the results of all the groups in your class. Calculate the relative frequency for each sum from 2 to 12 for the entire class. Make a new column labeled class relative frequency and record the results from the whole class.

b. Which set of relative frequencies—those from your small group, or those from the whole class—are closer to the probabilities?

LESSON

6-4

Relative Frequency and Probability

Frequencies and Relative Frequency

The **frequency** of an event is simply the number of times the event occurs. For instance, in 1990 there were 2,129,000 boys born in the United States, so the frequency of male births that year was 2,129,000. The *relative frequency* of an event compares the frequency with the total number of times the event could have occurred. In 1990, there were 4,158,000 children born in the United States. The relative frequency of boys to births is found by dividing the number of boys born by the total number of births.

$$\text{relative frequency} = \frac{\text{number of boys born}}{\text{total number of births}} = \frac{2{,}129{,}000}{4{,}158{,}000} \approx 0.512 = 51.2\%$$

Relative Frequency of an Event

Suppose a particular event has occurred with a frequency of F times in a total of T opportunities for it to happen. Then the relative frequency of the event is $\frac{F}{T}$.

Like other ratios, relative frequencies may be written as fractions, decimals, or percents.

Example 1

The Illinois Department of Public Health reported that for the years 1980 to 1990, 3840 skunks were examined for rabies. Of that number, 1446 actually had rabies. From 1980 to 1990 what was the relative frequency of skunks with rabies among those tested?

Solution

Here the event is "having rabies." You are asked to compare the skunks with rabies to the entire group of skunks.

$$\frac{F}{T} = \frac{\text{1446 skunks with rabies}}{\text{3840 skunks in all}} = \frac{1446}{3840} \approx 0.377 = 37.7\%$$

The relative frequency of skunks with rabies was about 38%.

In Example 1, the relative frequency could have been as low as $\frac{0}{3840}$ or 0, if no skunks had rabies. It could have been as high as $\frac{3840}{3840}$, or 1, if all the skunks tested had rabies. In the table you made for the dice activity, all the relative frequencies you calculated should also be between 0 and 1.

Probability

You have already calculated probabilities. Now we examine this idea in more detail. The *probability* of an event measures how likely it is that the event happens. Like relative frequency, a probability is a number between 0 and 1. Sometimes a relative frequency is used to estimate a probability. For instance, for many years the relative frequency of male births has been about 0.51, so we say that the probability of a newborn being a boy is about 0.51.

At other times, the probability of an event is determined by some theoretical assumptions. When a coin is tossed, we think that heads and tails are equally likely. Then the probability of getting a head is $\frac{1}{2}$ and the probability of getting a tail is $\frac{1}{2}$.

A possible result of an experiment is an **outcome.** For instance, suppose slips of paper with the integers from 1 to 25 are put into a hat. If one slip is drawn, the set S of outcomes is $\{1, 2, 3, \ldots, 25\}$, and the number of elements in set S, written $N(S)$, equals 25. An **event** is a set of outcomes chosen from S. The event "picking a number greater than 18" is the set $\{19, 20, 21, 22, 23, 24, 25\}$. If all 25 outcomes are assumed to be equally likely, then the probability of this event is $\frac{7}{25}$. This is the ratio of the number of outcomes in the event to the total number of possible outcomes.

Whirling probabilities. *This NASA photo of Hurricane Elena was taken from the space shuttle* Discovery *in 1985. The probability of a hurricane reaching a specific area is partially based on past records of such storms reaching that area.*

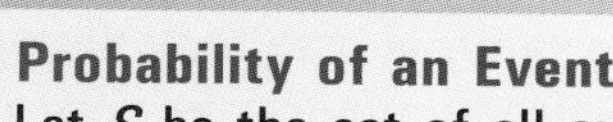

Probability of an Event
Let S be the set of all outcomes of an experiment and let E be an event. If the outcomes are equally likely, then the **probability of the event E, $P(E)$,** is $\frac{N(E)}{N(S)}$.

Example 2

Hurricanes happen often enough in some states that public officials must have evacuation plans ready. If hurricanes are equally likely to occur on any day of the week, what is the probability that one occurs on Saturday or Sunday, times when there are the fewest people in schools and offices?

Solution

There are seven possible outcomes, which are the days of the week. Thus here $N(S) = 7$. The event that we are interested in is $E = \{\text{Saturday, Sunday}\}$. So $N(E) = 2$. If a hurricane occurs, the probability that it hits on Saturday or Sunday = P(Saturday or Sunday) = $\frac{N(E)}{N(S)} = \frac{2}{7}$.

When were playing cards invented? *Playing cards may have originated in China around 800 A.D. The four suits originated in France in the 1500s.*

When all outcomes in an experiment have the same probability of occurring, then they are said to occur **at random** or **randomly.**

Example 3

A standard deck of playing cards has 52 cards. There are four suits: clubs ♣, diamonds ♦, hearts ♥, and spades ♠. Each suit has 13 cards: ace, 2, 3, 4, 5, 6, 7, 8, 9, 10, jack, queen, king. Suppose you shuffle the cards well and pull one card at random. Find:

a. P(ace of hearts)

b. P(queen)

Solution

a. There are 52 possible outcomes, the number of cards in the deck. So $N(S) = 52$. Only one outcome is in the event: the ace of hearts. So $N(E) = 1$.

$$P(\text{ace of hearts}) = \frac{N(E)}{N(S)} = \frac{1}{52}$$

b. The event "getting a queen" has 4 outcomes: Q♣, Q♦, Q♥, and Q♠.

$$P(\text{getting a queen}) = \frac{4}{52} = \frac{1}{13}$$

Probability and Complementary Events

If the probability of a hurricane occurring on a week*end* is $\frac{2}{7}$, then the probability of a hurricane occurring on a week*day* is $\frac{5}{7}$. The events "occurring on a weekend" and "occurring on a weekday" are called *complements* of each other. Two events are **complements** if their intersection is the empty set and their union is the set of all possible outcomes.

Forecasting the future. *Shown is TV meteorologist Spencer Christian. Meteorologists use computers to help analyze weather data gathered from satellites and observation stations.*

The sum of the probability of an event and the probability of its complement is 1. The same is true for relative frequencies. Thus the probability or relative frequency of the complement of an event is found by subtracting the probability or relative frequency of the original event from 1.

$$P(E) + P(\text{complement of } E) = 1$$
$$P(\text{complement of } E) = 1 - P(E)$$

Example 4

A weather forecaster reports that the probability of rain tomorrow is 70%. What is the probability that it does not rain?

Solution

The events "rain tomorrow" and "no rain tomorrow" are complements. So the probability of no rain is 100% − 70% = 30%.

The table below summarizes some of the important similarities and differences between relative frequencies and probabilities.

Relative frequency	Probability
1. calculated from an experiment	1. deduced from assumptions (like randomness) or assumed to be close to some relative frequency
2. the ratio of the number of times an event has occurred to the number of times it could occur	2. if outcomes are equally likely, the ratio of the number of outcomes in an event to the total number of possible outcomes
3. 0 means that an event did not occur. 1 means the event occurred every time it could.	3. 0 means that an event is impossible. 1 means that an event is sure to happen.
4. The more often an event occurred relative to the number of times it could occur, the closer its relative frequency is to 1.	4. The more likely an event is, the closer its probability is to 1.
5. If the relative frequency of an event is r, then the relative frequency of its complement is $1-r$.	5. If the probability of an event is p, then the probability of its complement is $1-p$.

QUESTIONS

Covering the Reading

1. a. How many girls were born in 1990?
b. What was the relative frequency of female births in 1990?

2. The Illinois Department of Health reported that for the years 1971 to 1991, of 474 horses tested for rabies only 22 actually had the disease.
a. What percent of horses tested had rabies?
b. What percent did not have rabies?

3. What is the meaning of a relative frequency of 0?

4. What is the meaning of a relative frequency of 1?

5. There were 93 million households in the U.S. with televisions in 1992. Of these homes, 67 million also had a VCR.
a. Find the relative frequency of households with a television that also have a VCR.
b. Find the relative frequency of households with a television that have no VCR.

6. *Multiple choice.* A student flipped a coin 100 times and counted 47 heads. Which phrase best describes the ratio $\frac{47}{100}$?
(a) the probability of a toss coming up heads
(b) the relative frequency of a toss coming up heads

7. Suppose a multiple-choice question has 4 choices, A, B, C, and D, and you guess randomly.
   a. What is the probability that you will get the question correct?
   b. If choice C is the correct answer, identify the event E, $N(E)$, the set of all outcomes S, $N(S)$, and $P(E)$ for the situation in part **a.**
   c. What is the probability that you will miss the question?
   d. What are the events in parts **a** and **c** called?

8. Suppose that slips containing the numbers from 1 to 50 are put in a hat. A number x is drawn. Determine each probability.
   a. $P(x > 32)$ b. $P(x < 32)$ c. $P(x = 32)$

In 9–11, a card is drawn randomly from a standard deck of cards. Find:

9. P(selecting the 7 of clubs)

10. P(selecting a king)

11. P(selecting a diamond)

12. What is the meaning when a probability equals 0?

13. What is the meaning when a probability equals 1?

14. If p is the probability of an event, and q is the probability of its complement, what is the value of $p + q$?

Applying the Mathematics

15. A letter is picked randomly from the English alphabet. Calculate P(A, E, I, O, or U), the probability of picking a vowel.

16. A student mails an entry to a magazine sweepstakes. The student says, "The probability of winning is $\frac{1}{2}$, since either I will win or I won't." What is wrong with this argument?

17. Suppose X, Y, and Z are events, $P(X) = \frac{3}{4}$, $P(Y) = \frac{1}{2}$, and $P(Z) = \frac{2}{3}$.
   a. If you could find actual relative frequencies of these events, which event would you expect to have the largest relative frequency? Explain.
   b. Event Y has the smallest probability. Must it have the smallest relative frequency? Explain.

Mileage until worn out	Number of tires
10,000–14,999	1
15,000–19,999	3
20,000–24,999	6
25,000–29,999	15
30,000–34,999	14
35,000–39,999	7
40,000 or more	4

In 18 and 19, consider the following. A tire company tested 50 tires to see how long they last under typical road conditions. The results they obtained are given in the table at the left.

18. What is the relative frequency that a tire lasted less than 25,000 miles? (Write as a fraction.)

19. What is the relative frequency that a tire lasted at least 10,000 miles?

In 20 and 21, *multiple choice.*

20. An event occurred c times out of t possible occurrences. The relative frequency of the event was 30%. Which is true?

(a) $\frac{c}{t} = 0.3$ (b) $\frac{t}{c} = 0.3$ (c) $ct = 0.3$ (d) $t - c = 0.3$

21. Of the people surveyed, $\frac{4}{9}$ thought the American League team would win the World Series. If $36n$ people were surveyed, how many thought the American League team would win?

(a) $\frac{4}{9}n$ (b) $4n$ (c) $9n$ (d) $16n$

Review

22. a. Sneakers were originally $79.95 and now are on sale for $59.95. Give the percent of discount.

b. Sneakers were originally F dollars and now are on sale for S dollars. Give the percent of discount.

c. If in part **a** you must pay $3.30 tax, what is the tax rate? *(Lesson 6-3)*

23. An orange punch is made by mixing two parts orange juice with three parts ginger ale. Six gallons of punch are needed. How many quarts of orange juice and how many quarts of ginger ale will it take? *(Lesson 6-3)*

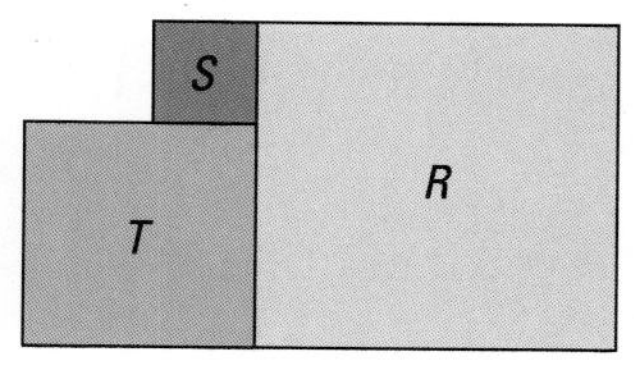

24. What is the domain of t in the expression $t + \frac{t-1}{t}$? *(Lesson 6-2)*

25. Refer to the figure at the left. If the area of square S is 9 and the *perimeter* of square T is 20, find the area of square R. *(Lesson 1-5)*

26. *Skill sequence.* Suppose a and b are not zero. Simplify the expression. *(Lessons 2-3, 3-9, 4-1, 6-1)*

a. $\frac{2a}{5} + \frac{a}{b}$ **b.** $\frac{2a}{5} - \frac{a}{b}$ **c.** $\frac{2a}{5} \cdot \frac{a}{b}$ **d.** $\frac{2a}{5} \div \frac{a}{b}$

In 27 and 28, solve. *(Lessons 2-6, 4-3)*

27. $\frac{5}{8}m = \frac{10}{3}$

28. $P - 0.06P - 14 = 98.8$

29. The list price of a bicycle is b dollars. Find the selling price according to the following conditions.

a. You get a 25% discount, and there is no tax.

b. You pay 4% sales tax and get no discount.

c. You get a 25% discount and pay a 4% sales tax. *(Lesson 4-2)*

Paper airplanes can take many forms.

Exploration

30. a. Make a paper airplane. After flying it 5 times, guess the probability that when it is flown, it will land right side up.

b. Fly the plane a large number of times and calculate the relative frequency that it landed right side up.

c. Having done the experiment, decide whether or not you should change the probability you guessed. Explain your decision.

LESSON

6-5

Solving Percent Problems Using Equations

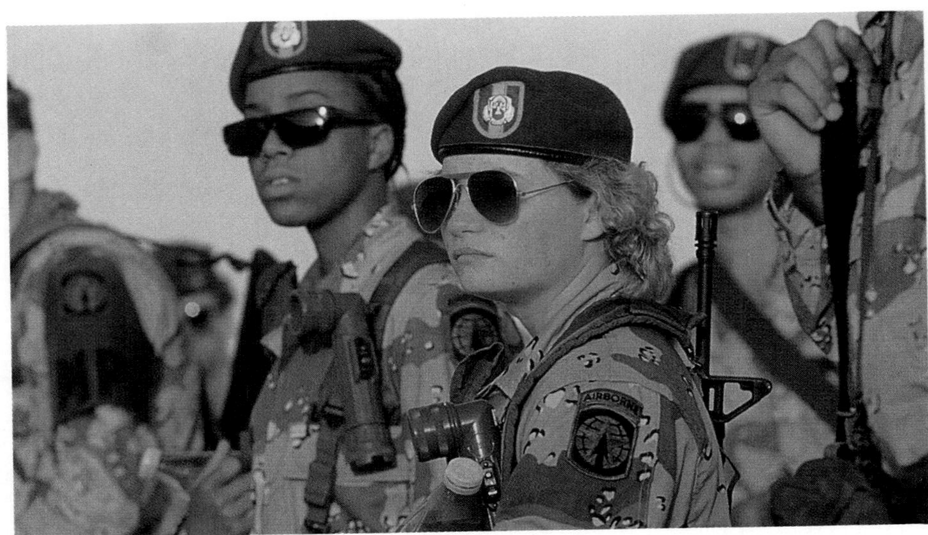

Women in uniform. *In 1991, 35,000 U.S. women participated in the Persian Gulf War. They represented 6% of the U.S. troops who served there. See Example 3.*

The word **percent** (often written as the two words **per cent**) comes from the Latin words "per centum," meaning "per 100." So 7% literally means 7 per 100, or the ratio $\frac{7}{100}$, or 0.07. The symbol % for percent is only about 100 years old.

Recall that to find a percent p of a given quantity q you can multiply q by the decimal or fraction form of p.

To find 50% of 120,000 calculate $0.5 \cdot 120{,}000 = 60{,}000$.
To find $5\frac{1}{4}$% of \$3000 calculate $0.0525 \cdot 3000 = \$157.50$.

This gives a straightforward method for solving many percent problems. Just translate the words into an equation of the form $pq = r$.

Example 1

a. 112% of 650 is what number?
b. 7% of what number is 31.5?
c. What percent of 3.5 is 0.84?

Solution

a. 112% of 650 is what number?

$$\begin{array}{ccccc} \downarrow & \downarrow & \downarrow & \downarrow & \downarrow \\ p & \cdot & q & = & r \end{array}$$

$1.12 \cdot 650 = r$ Change 112% to 1.12.

$728 = r$

112% of 650 is 728.

Check

$\frac{728}{650} = 1.12 = 112\%$.

Solution

b. 7% of what number is 31.5?

$$\begin{aligned} p \cdot q &= r \\ 0.07 \cdot q &= 31.5 && \text{7\% is 0.07.} \\ \frac{0.07q}{0.07} &= \frac{31.5}{0.07} && \text{Divide both sides by 0.07.} \\ q &= 450 \end{aligned}$$

31.5 is 7% of 450.

Check

7% of 450 = 0.07 · 450 = 31.5

Solution

c. What % of 3.5 is 0.84?

$$\begin{aligned} p \cdot q &= r \\ p \cdot 3.5 &= 0.84 \\ \frac{p \cdot 3.5}{3.5} &= \frac{0.84}{3.5} && \text{Divide both sides by 3.5.} \\ p &= 0.24 && \text{This is the solution to the equation.} \\ p &= 24\% && \text{Rewrite the solution as a percent.} \end{aligned}$$

Check

24% · 3.5 = 0.24 · 3.5 = 0.84

In 1960, about 32% of married women were in the labor force. By 1980, that figure had jumped to about 50%.

Example 2

It was reported that in 1991 about 59% of the 52.5 million married couples in the United States had two incomes. Approximately how many couples had two incomes in 1991?

Solution

Let c be the number of millions of couples with two incomes.

$$0.59 \cdot 52.5 = c$$
$$30.975 = c$$

In 1991, approximately 31 million couples had two incomes.

Example 3

In 1992, there were 207,828 women serving in the Armed Forces of the United States, accounting for about 11.5% of total military personnel. In all, how many persons were serving in the Armed Forces in 1992?

Solution

Let x be the number of persons in the Armed Forces.

$$11.5\% \cdot x = 207{,}828$$
$$0.115 \cdot x = 207{,}828$$
$$x = \frac{207{,}828}{0.115}$$
$$x = 1{,}807{,}200$$

About 1,807,200 persons served in the Armed Forces in 1992.

Many problems related to business involve percents. If the problems involve discounts (or markups), you often must use subtraction (or addition) as well as multiplication and division to solve them.

Example 4

The total cost of a video game was \$51.89. This included the 6% sales tax. What was the price of the game without the tax?

Solution

Let V be the price of the video game.

$$\text{total cost} = \text{price} + 6\% \text{ sales tax}$$
$$51.89 = V + .06V$$
Remember, $V = 1V$
$$51.89 = (1 + .06)V$$
$$51.89 = 1.06V$$
$$\frac{51.89}{1.06} = \frac{1.06V}{1.06}$$
$$48.95 \approx V$$

The price without sales tax was \$48.95.

Check

$48.95 + 0.06(48.95) = 51.887 \approx 51.89.$

Percent problems are so common that many calculators have a [%] key. If you press 5 [%] on such a calculator you will see 0.05, which equals 5%. On many calculators when the [%] key is used after an addition or subtraction sign, it will calculate discounts or markups without having to enter the original amount again.

Activity

If your calculator has a [%] key, enter the following key sequence.

48.95 [+] 6 [%] [=]

What does your calculator display? If it shows 51.887, it is interpreting the "[+] 6 [%]" to mean "plus 6% of the previous amount"; the calculator evaluates $48.95 + 6\% \cdot 48.95$. If it displays 49.01, your calculator has evaluated $48.95 + 0.06$.

QUESTIONS

Covering the Reading

1. 123% of 780 is what number?

2. 40% of what number is 440?

3. What percent of 4.7 is 0.94?

4. What number is 62% of 980?

5. **a.** In 1991, how many married couples were there in the United States?
   b. Of these, how many did not have two incomes?

6. **a.** In 1992, what percent of U.S. military personnel were men?
   b. How many men were in the U.S. Armed Forces in 1992?

7. About 67,000 or 13.1% of U.S. Navy personnel are officers.
   a. Write an equation that you can use to determine how many Navy personnel there are in all.
   b. Solve the equation.

8. In Example 4, if the price with sales tax was \$42.39, find the price without sales tax.

9. The total cost of a camera including an 8.5% sales tax is \$59. How much tax was paid in this purchase?

10. The Richardsons bought a new car. The total amount they paid was \$14,064.75, including the 5% sales tax. What was the price before the sales tax was added?

Goat awareness. *These goats are grazing on the roof of a Swedish restaurant in Wisconsin. In Sweden, many homes in small villages are designed so that goats may graze on the roofs.*

Applying the Mathematics

11. On a mathematics test there were 8 *A*s, 12 *B*s, 10 *C*s, 2 *D*s and 0 *F*s. What percent of the students earned *A*s?

12. Clearwater High School expects a 14% increase in enrollment next year. There are 1850 students enrolled this year.
    a. How many students will the school gain?
    b. What is the expected enrollment next year?

13. In the decade of the 1980s the U.S. Congress passed many bills designating commemorative days, weeks, months, and so on. These included National Prune Day, Tap Dance Day, and Dairy Goat Awareness Week. This type of law amounted to about 250 bills per session or 38% of all laws passed. What was the total number of laws passed per session?

In 14 and 15, use this information. Sucrose, or common table sugar, is composed of carbon, hydrogen, and oxygen. Suppose an experiment calls for 68.4 grams of sucrose.

14. If 4.2% of the weight of sucrose is carbon, how much carbon is in the 68.4 grams? (Round your answer to the nearest tenth of a gram.)

15. If 35.2 of the 68.4 grams is oxygen, what percent of the weight of sucrose is oxygen? (Round your answer to the nearest whole number percent.)

16. Pat bought some new clothes at a "30% off" sale. She spent \$73.50. What was the price of the clothes before the discount?

17. A TV originally cost \$320. It is on sale for \$208. What is the percent of discount?

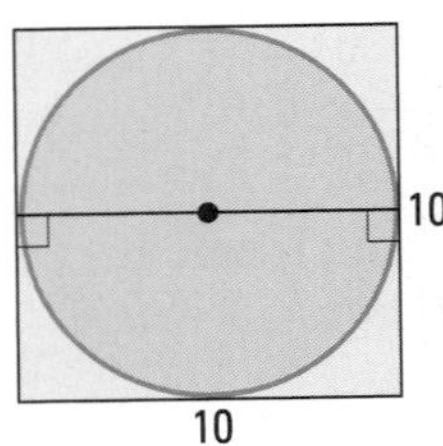

18. Refer to the figure at the left. The area of the circle is what percent of the area of the square?

Review

19. In Seattle, Washington, it rained or snowed an average of 156 days per year in the 30 years from 1961 through 1990. What was the relative frequency of a day without rain or snow? *(Lesson 6-4)*

20. A fair die is tossed once.
  a. Which event is more likely, "the number showing is divisible by three" or "the number showing is even"?
  b. Explain your answer to part **a.** *(Lesson 6-4)*

21. Stanley took $1\frac{1}{4}$ hours to do his homework; Jenile took 35 minutes.
  a. What is the ratio of Jenile's homework time to Stanley's?
  b. What is the ratio of Stanley's homework time to Jenile's? *(Lesson 6-3)*

22. The Olsons' farm has x acres. The Kramers' farm has y acres. Use a ratio to compare the number of acres on the Olsons' farm to the number of acres on the Kramers' farm. *(Lesson 6-3)*

23. The Nguyen family earned $19,600 last year on their 80-acre farm.
  a. What is the income per acre?
  b. Is the income per acre a ratio or a rate? *(Lessons 6-2, 6-3)*

24. When the expression $\frac{4}{b-7}$ is written, what is assumed about the value of b? *(Lesson 6-2)*

25. Solve $(m - 6)^2 = 100$. *(Lesson 5-9)*

26. **a.** Solve for y: $-4x + 3y = 10$.
  b. Graph the equation on the coordinate plane. *(Lesson 5-7)*

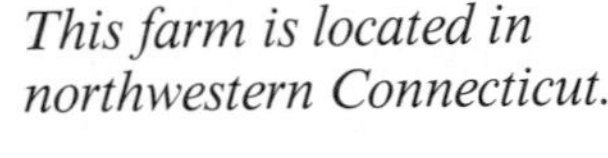
This farm is located in northwestern Connecticut.

27. Two thirds of a number is 87. Find the number. *(Lesson 2-6)*

28. What fraction of a complete turn is a turn of 45°? *(Previous course)*

29. Convert 0.325823224 to the nearest percent. *(Previous course)*

Exploration

30. **a.** Pick a number and increase it by 30%. Decrease this result by 30%. Did you end up with the number you started with?
  b. Repeat part **a** three times using a different number to start with each time.
  c. Explain how the final result is related to the original number.

LESSON

6-6 Probability Without Counting

On target. *Archery is enjoyed both as an amateur and as a professional sport. Through practice, the probability of hitting the target should increase.*

When a situation has equally likely outcomes, the probability of an event is the ratio of the number of outcomes in the event to the total number of outcomes. But sometimes the number of outcomes is infinite. In such cases, if you can set up a geometric model, you can still use division to find probabilities.

Example 1

Suppose a dart is thrown at a 20-inch square board containing a circle of radius 3 inches. Assuming that the dart hits the board, and that it is equally likely to land on any point on the board, what is the probability that the dart lands in the circle?

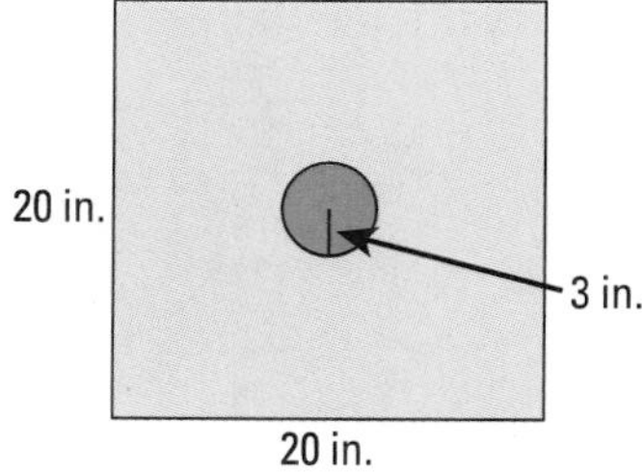

Solution

Recall that the area of a circle with radius r is πr^2. Compare the area of the circle to the area of the square.

$$\begin{aligned} P(\text{dart lands in circle}) &= \frac{\text{area of circle}}{\text{area of square}} \\ &= \frac{(\pi \cdot 3^2)\text{ sq in}}{(20 \cdot 20)\text{ sq in}} \\ &= \frac{9\pi}{400} \\ &\approx 0.071 \text{ or about } 7\% \end{aligned}$$

So, the probability that the dart lands in the circle is about 7%.

Example 1 illustrates the *Probability Formula for Geometric Regions.*

Probability Formula for Geometric Regions
Suppose points are selected at random in a region, and part of that region's points represent an event *E*. The probability *P*(*E*) of the event is given by

$$P(E) = \frac{\text{measure of region in the event}}{\text{measure of entire region}}.$$

In Example 1, the probability is the ratio of two areas. In Example 2, the probability is the ratio of two lengths.

Example 2

Points *A, B, C, D,* and *E* below represent exits on an interstate highway.

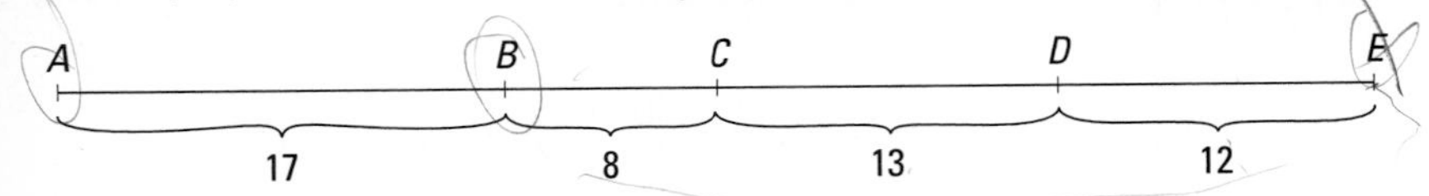

The distances between the exits are measured in miles. If accidents occur at random along the highway between points *A* and *E,* what is the probability that an accident occurs between *B* and *C*?

Solution

Use the Putting-Together Model for Addition to find the length of *AE.*
AE = 17 + 8 + 13 + 12 = 50.

$$P(\text{the accident is in } \overline{BC}) = \frac{BC}{AE} = \frac{8 \text{ miles}}{50 \text{ miles}} = 0.16 \text{ or } 16\%$$

The probability that an accident occurs between B and C is about 16%.

Traffic-safety engineers might compare the probabilities in Example 2 with the actual relative frequency of accidents. If the relative frequency along one stretch of the highway is too large, then that part of the highway might be a candidate for repair or new safety features.

Probabilities can also be determined by finding ratios of angle measures.

Example 3

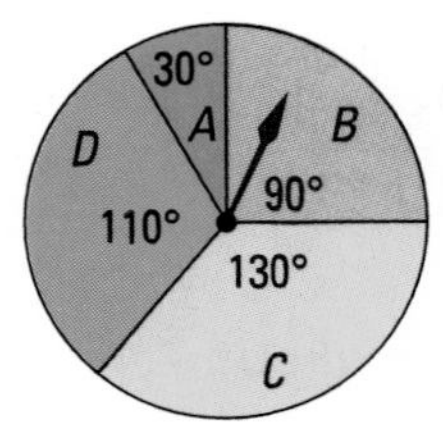

At the left is a picture of a spinner like that used in many games. Suppose the spinner is equally likely to point in any direction. What is the probability that it stops in region *A*?

Solution

The sum of the measures of all the angles around the center of the circle is 360°. P(spinner stops in A) = $\frac{30°}{360°} = \frac{1}{12}$.

Sometimes to compute a probability you must add or subtract first.

Example 4

A target consists of four evenly spaced concentric circles as shown at the left. The smallest circle (called the "bull's-eye") has radius equal to 2 inches. If a point is selected at random from inside the target, what is the probability that it lies in the blue region?

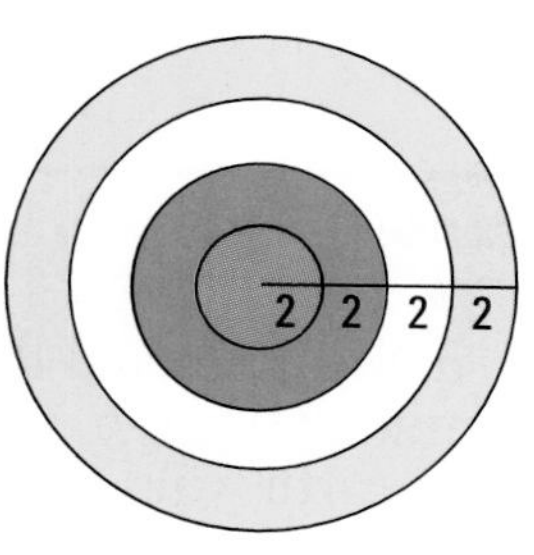

Solution

Probability that a point lies in blue region $= \frac{\text{area of blue region}}{\text{area of largest circle}}$.

The area of the blue region equals the difference in the areas of the circles with radii 4 and 2.

$$\text{Area of blue region} = \pi(4)^2 - \pi(2)^2 = 16\pi - 4\pi = 12\pi$$

Area of largest circle $= \pi(8)^2 = 64\pi$

Thus the probability of choosing a point in the blue region is $\frac{12\pi}{64\pi}$, or $\frac{3}{16}$.

QUESTIONS

Covering the Reading

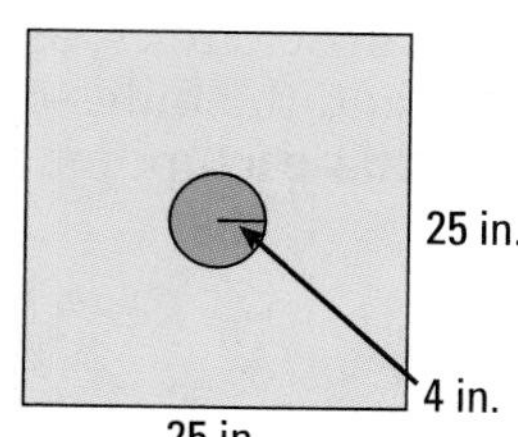

1. Consider the square archery target board at the left.
   a. What is the area of the bull's-eye?
   b. What is the area of the entire target board?
   c. What is the probability that an arrow hitting the board at random will land in the bull's-eye?
   d. What is the probability that the arrow will land on the target outside the circle?

In 2 and 3, refer to Example 2. What is the probability that an accident occurring between exits A and E occurs between the given exits?

2. B and E
3. C and E

In 4–6, refer to Example 3.

4. What is the probability that the spinner will land in region B?
5. What is the probability that the spinner will land in region C?
6. What is the probability that the spinner will not land in region C?
7. In general, what is the probability of an event involving a geometric region?

In 8 and 9, refer to Example 4. Suppose a point on the circular target is chosen at random.

8. What is the probability that it lies inside the bull's-eye?

9. What is the probability that it lies in the outermost ring?

Applying the Mathematics

10. The land area of the earth is about 57,510,000 square miles and the water surface area is about 139,440,000 square miles. Give the probability that a meteor hitting the surface of the earth will:
a. fall on land. **b.** fall on water.

11. In a rectangular yard of dimensions q by p is a rectangular garden of dimensions b by a. If a newspaper is thrown randomly into the yard, what is the probability that it lands on a point in the garden?

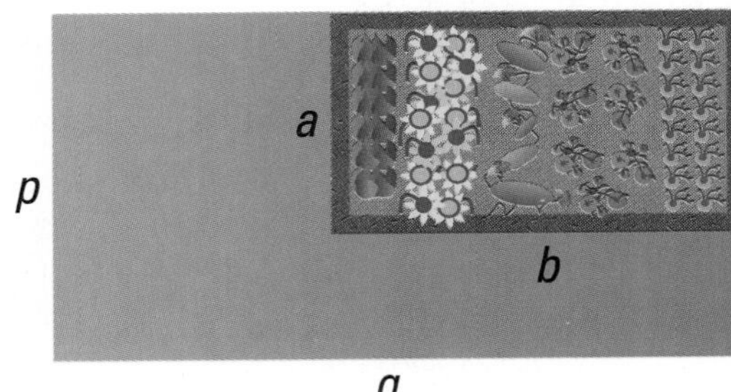

12. An electric clock was stopped by a power failure. What is the probability that the minute hand stopped between the following two numerals on the face of the clock?
a. 12 and 3 **b.** 1 and 5 **c.** 11 and 1

13. Suppose a 40-cm square dart board has two concentric circles, as shown below. The larger has a radius of 17.5 cm, and the bull's-eye has radius 7.5 cm. Suppose a dart is thrown and that each point on the board is equally likely to be hit.

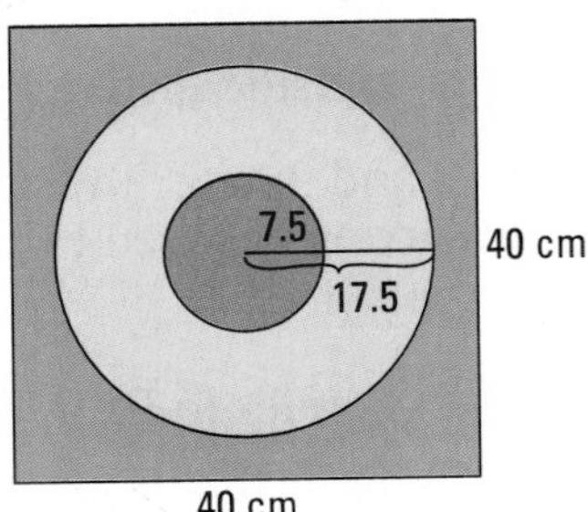

a. What is the probability that the dart lands inside the larger circle?
b. What is the probability that the dart lands inside the larger circle but outside the bull's-eye?
c. What is the probability that the dart lands outside both circles?

14. The basketball court used in the NBA (National Basketball Association) has a wider free-throw lane than that used in high schools and colleges. The dimensions of the NBA lane are 16′ by 19′. For high schools they are 12′ by 19′. Assuming that rebounds are equally likely to land anywhere in the NBA lane, what is the probability that a rebound which lands in the NBA lane would also land in the high school lane?

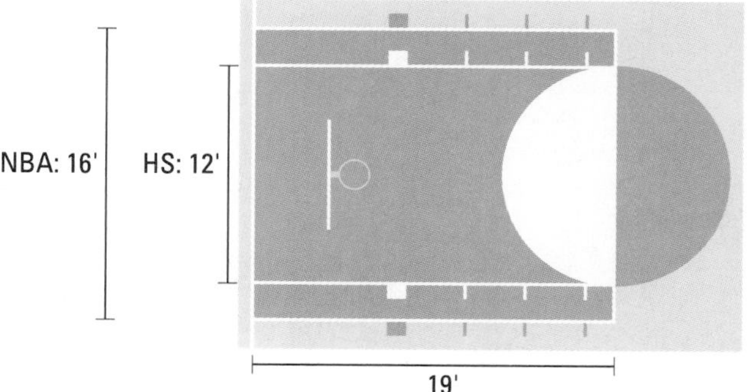

15. The table at the right gives the membership in a high school club. Design a spinner that can be used to select a representative group from the club.

Grade	Enrollment
9	5
10	15
11	17
12	23

The rebounding Rocket. *Nigerian-born Hakeem Olajuwon of the Houston Rockets did not start playing basketball until age 15. He was named NBA's Most Valuable Player in 1994.*

Review

16. Compute in your head. *(Lesson 6-5)*
    a. What is 25% of 60?
    b. 50% of what number is 13?
    c. 8 is what percent of 24?

17. A television station has scheduled n hours of news, c hours of comedy, d hours of drama, s hours of sports, and x hours of other programs during the week.
    a. What is the ratio of hours of comedy to hours of sports?
    b. What is the ratio of hours of drama to total number of hours of programs during the week?
    c. If you turn on this station at some randomly chosen time, and a show is on the air, what is the probability that it is a news program? *(Lessons 6-4, 6-3)*

18. A 10-foot-long board is cut so that the two pieces formed have lengths in a ratio of $\frac{5}{3}$. How long is each piece? *(Lesson 6-3)*

A marathon winner. *In 1994, Jean Driscoll of Champaign, Illinois, won her fifth straight women's wheelchair title in the Boston Marathon with a record time of 1:34.21. This is an average rate of about 7.46 meters per second!*

19. In July of 1994, a new world record in the 100-meter dash was set by American Leroy Burrell with a time of 9.85 seconds. The world record in the 200-meter dash, set by Italian Pietro Mennea in 1979, was 19.72 seconds.

a. Find the average number of meters per second in each record run.

b. By this measure, which runner is faster? Explain your answer. *(Lesson 6-2)*

In 20–22, perform the operation and simplify. *(Lessons 3-9, 6-1)*

20. $1\frac{5}{9} \div 2\frac{1}{7}$ **21.** $\frac{2x}{s} \div \frac{x}{10}$ **22.** $\frac{q}{2} + \frac{q}{3}$

23. Solve $\frac{x}{2} - 4 = \frac{3x}{4}$. *(Lesson 5-8)*

24. Solve for y: $x + 10y = 15$. *(Lesson 5-7)*

25. Graph $y = 4$ and $x = -2$ in the coordinate plane. *(Lesson 5-1)*

26. The volume of a box is to be more than 1700 cm^3. The base has dimensions 8 cm by 15 cm. What are the possible heights of the box? *(Lessons 2-1, 2-8)*

27. In the U.S. Army a *squad* is usually 10 enlisted soldiers. A *platoon* is 4 squads. A *company* is 4 platoons. A *battalion* is 4 companies. A *brigade* is 3 battalions. A *division* is 3 brigades. A *corps* is 2 divisions. A *field army* is 2 corps. How many enlisted personnel are there in a field army? *(Lesson 2-4)*

Exploration

28. In 1760, the French mathematician Buffon performed an experiment and discovered an amazing property. He drew four parallel lines as close to exactly ℓ units apart as possible, where ℓ was the length of a needle. He dropped the needle onto the lines and counted how often it touched a line and how often it did not. Buffon discovered that if the needle is dropped randomly onto the lines, then the probability the needle touches a line is $\frac{2}{\pi}$.

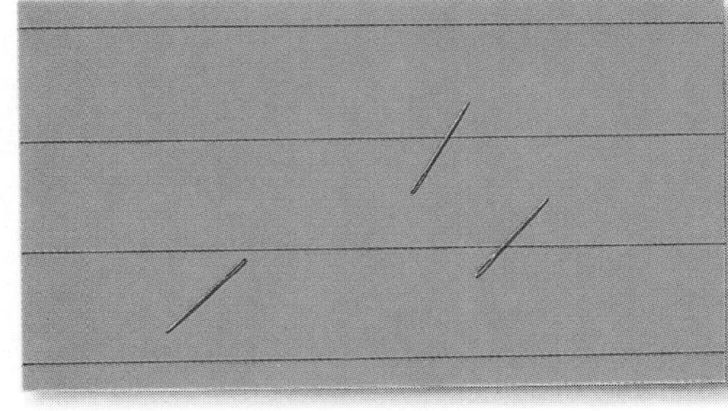

That is, $\frac{\text{number of times needle touches line}}{\text{number of times needle is dropped}} = \frac{2}{\pi}$.
Try Buffon's experiment, drawing the lines and then dropping the needle at least 100 times. How close do you get to $\frac{2}{\pi}$?

LESSON

6-7

Size Changes

Honey of a size change. *This scene from a Disney film,* Honey, I Shrunk the Kids, *shows what happened to the kids when their father accidentally shrunk them with a gadget that had a size change factor of about 1%.*

Size Changes and Multiplication

Some photocopy machines can reduce or enlarge a preimage to a given percentage of its original size. The percent of reduction or enlargement is called a *size change factor.*

For instance, the figure below was reduced to 75% of its original size to form image *A*. It was reduced to 64% of its original size to form image *B*. It was enlarged to 120% of its original size to form image *C*. Notice that the lengths and widths of the original figure are multiplied by each size change factor.

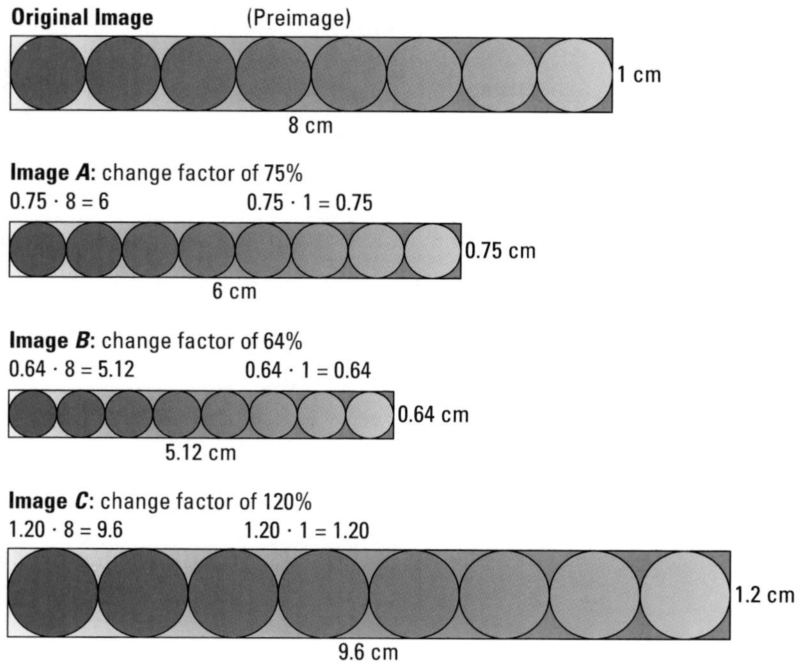

These are instances of the *Size Change Model for Multiplication.*

Size Change Model for Multiplication
Let k be a nonzero number without a unit. Then kx is the result of applying a size change of magnitude k to the quantity x.

If you know the size of an image, by working backwards you can find the size of the preimage.

Example 1

After a size change of 75% has been applied, an image is 10 cm long. What is the length of the original figure?

Solution

An equation describing this situation is 75% · original length = 10 cm. Let L be the original length. Change 75% to 0.75. The equation becomes:

$$0.75L = 10$$

$$\frac{0.75L}{0.75} = \frac{10}{0.75} \quad \text{Divide both sides by 0.75.}$$

$$L = 13.\overline{3}$$

The original length was $13\frac{1}{3}$ cm.

Check

75% of 13.3 is just about 10. So the solution checks.

Types of Size Changes

When the size change factor k is greater than 1 or less than -1, the size change is called an **expansion.** Since 120% = 1.2, image C on page 387 illustrates an expansion. An expansion always produces an image larger than the preimage figure. If k is between -1 and 1, the size change is a **contraction.** Since 75% = 0.75 and 64% = 0.64, images A and B picture contractions. A contraction always produces an image smaller than the preimage. If the size change factor is 1 or -1, the image is the same size as the preimage.

Expansions and contractions can be done easily when figures are graphed in a coordinate plane. In geometric situations, the size change factor is called the **magnitude of the size change.** To find the image of a figure in the coordinate plane under a size change, just multiply the coordinates of each point on the figure by the magnitude of the size change.

Example 2

A quadrilateral in a coordinate plane has vertices $A = (3, 0)$, $B = (0, 3)$, $C = (6, 6)$, and $D = (6, 3)$.

a. Draw its image under a size change of magnitude $\frac{1}{3}$.

b. Is the image an expansion, a contraction, or neither?

▶

Solution

a. Multiply the coordinates of each vertex of $ABCD$ by $\frac{1}{3}$ to find the new image points A', B', C', and D'.

Preimage Point	Image point
$A = (3, 0)$	$A' = \left(\frac{1}{3} \cdot 3, \frac{1}{3} \cdot 0\right) = (1, 0)$
$B = (0, 3)$	$B' = \left(\frac{1}{3} \cdot 0, \frac{1}{3} \cdot 3\right) = (0, 1)$
$C = (6, 6)$	$C' = \left(\frac{1}{3} \cdot 6, \frac{1}{3} \cdot 6\right) = (2, 2)$
$D = (6, 3)$	$D' = \left(\frac{1}{3} \cdot 6, \frac{1}{3} \cdot 3\right) = (2, 1)$

Draw $A'B'C'D'$. Below we show both the preimage and image.

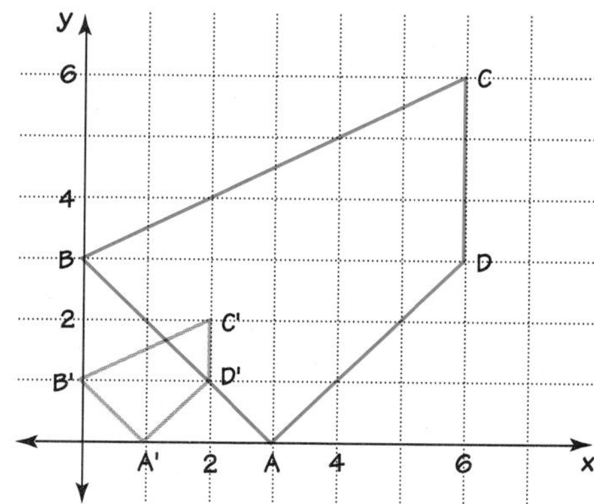

b. The size change factor $k = \frac{1}{3}$ is less than 1 but greater than -1, so the size change is a contraction. You can check this by looking at the graph.

In Example 2, notice that under a size change of magnitude $\frac{1}{3}$, the image of the point (x, y) is $\left(\frac{1}{3}x, \frac{1}{3}y\right)$. The general pattern is quite simple.

> Multiplying the coordinates of all points of a figure by a nonzero number performs a size change of magnitude k. The image of (x, y) under a size change of magnitude k is (kx, ky).

In Example 3 a size change with a negative size change factor is shown. That is, the coordinates of the points are multiplied by a negative number.

Example 3

A quadrilateral has vertices $E = (-3, 0)$, $F = (-3, -4)$, $G = (-4, -2)$, and $H = (-5, -2)$.

a. Find its image under a size change of magnitude -2.

b. Describe the size and location of the image in relation to the original figure.

Solution

a. Multiply the coordinates of each vertex of *EFGH* by -2. Label the new figure *E'F'G'H'*.

Preimage	Image
E = (-3, 0)	E′ = (6, 0)
F = (-3, -4)	F′ = (6, 8)
G = (-4, -2)	G′ = (8, 4)
H = (-5, -2)	H′ = (10, 4)

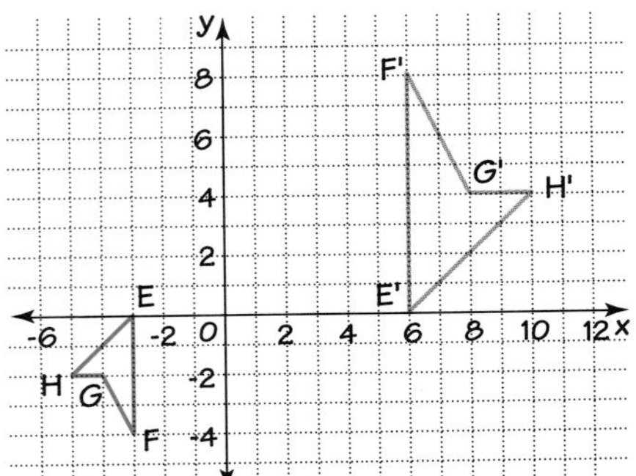

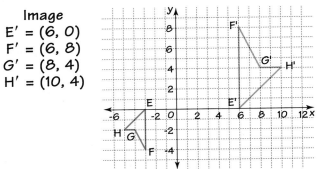

b. The size change is an expansion. Each side of E′F′G′H′ is twice as long as the corresponding side of EFGH. The image seems to be rotated a half turn from the position of the original figure.

In general, if the magnitude is negative, the original figure is rotated 180° to obtain the image. Turn the book upside down to see that the image in Example 3 looks like the original, only larger.

Size Changes Not Involving Figures

Size changes can involve quantities other than lengths or geometric figures.

The tools of the trade. *This tool and die maker is using a metal lathe to produce a die. A die is a precision tool used to shape or cut metals.*

Example 4

A tool and die maker receives time and a half for overtime. In February of 1993, the average overtime wage for this occupation in St. Louis, Missouri, was \$28.28 per hour. What was a tool and die maker's normal hourly wage?

Solution

Time and a half means that the overtime wage is $1\frac{1}{2}$ times the normal wage.

Let W be the normal hourly wage.

$$1\frac{1}{2} \cdot W = \$28.28$$

$$1.5W = 28.28$$

$$W = \frac{28.28}{1.5} \quad \text{Divide both sides by 1.5.}$$

$$W = 18.85\overline{3}$$

The normal hourly wage is about \$18.85.

Check

Half of \$18.85 is about \$9.43, and that added to \$18.85 equals \$28.28.

QUESTIONS

Covering the Reading

1. In the photocopy machine discussed at the beginning of this lesson, what are the magnitudes of the size change factors mentioned?

In 2–4, tell if a size change of the given magnitude produces an image larger than, smaller than, or the same size as the original figure.

2. 1.5 **3.** 100% **4.** $\frac{1}{2}$

5. A picture with dimensions 10 inches by 15 inches is reduced on a photocopy machine by a factor of 64%. What are the dimensions of the image?

6. The photo at the left is 5.5 cm tall. It is to be enlarged on a photocopy machine by a factor of 120%. How tall will the image be?

7. After a size change of 120%, the image of a figure is 18 cm long.
a. Write an equation using L for the length of the original figure.
b. What is the length of the original figure?

8. What is the image of (3, -9) under a size change of magnitude -7?

9. What is the image of (x, y) under a size change of magnitude 8?

10. **a.** Graph the quadrilateral $ABCD$ in Example 2.
b. Graph its image under a size change of magnitude 2.
c. Graph its image under a size change of magnitude -2.
d. Are the size change images in parts **b** and **c** expansions, contractions, or neither?

11. **a.** Draw $\triangle ABC$ with $A = (-8, 2)$, $B = (-4, -2)$, and $C = (-4, -10)$.
b. Draw $\triangle A'B'C'$, the image of $\triangle ABC$ under a size change of magnitude $\frac{1}{4}$.
c. Draw $\triangle A^*B^*C^*$, the image of $\triangle ABC$ under a size change of magnitude $-\frac{1}{4}$.
d. Describe in your own words the relationships among the three triangles.

12. The magnitude k of a size change is given. Tell whether the size change is an expansion, a contraction, or neither.
a. $-1 < k < 1$ **b.** $k > 1$ **c.** $k < -1$ **d.** $k = -1$

13. $J'K'L'M'N'$ is the image of pentagon $JKLMN$ under a size change with rule that the image of (x, y) is $\left(\frac{3}{2}x, \frac{3}{2}y\right)$.
a. Find the coordinates of J', K', L', M' and N'.
b. Graph pentagon $J'K'L'M'N'$.

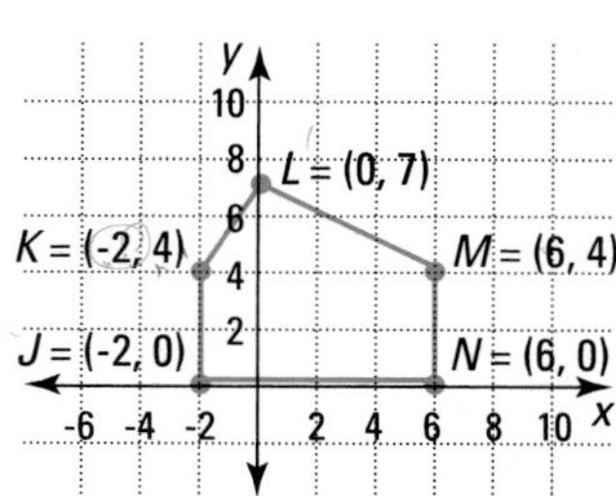

14. Maintenance electricians were averaging $30.98 per overtime hour in San Jose, California, in August of 1992. If an electrician receives time and a half for overtime, what was the normal hourly wage?

Applying the Mathematics

15. Lorenzo is one year old and weighs 10 kg. This is $3\frac{1}{2}$ times his birth weight. How much did he weigh at birth?

16. A human hair is 0.1 mm thick. Suppose that under a certain microscope it appears 15 mm thick. By how many times has it been magnified?

17. Under a size change of magnitude 6, the image P' of a point P is (9, -42). What are the coordinates of P?

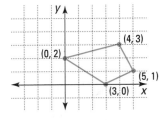

18. **a.** Describe the graph of the image of the quadrilateral at the left under a size change of magnitude 1.
b. What property of multiplication does this size change represent?

19. Refer to the four figures at the start of this lesson.
a. Find the area of each rectangle.
b. *True or false.* If a figure undergoes a size change of magnitude k, the area of the image is k times the area of the preimage. Justify your answer.

20. **a.** Draw the rectangle with vertices $A = (1, -1)$, $B = (3, -1)$, $C = (3, 2)$, and $D = (1, 2)$.
b. Draw its image $A'B'C'D'$ under a size change of magnitude 3.
c. Find the ratio of the perimeter of the larger figure to the perimeter of the smaller.
d. Find the ratio of the area of the larger figure to the area of the smaller.

Review

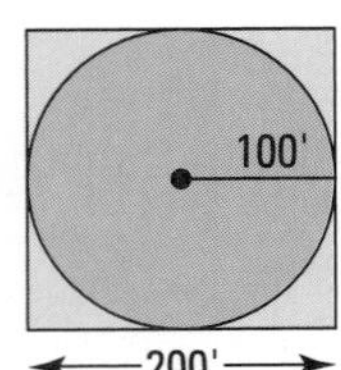

21. A farmer is thinking of installing an irrigation system on his farm. The system consists of a 100-ft-long pipe that rotates around a fixed point. The farm's crops are planted in rows in a 200-ft by 200-ft square as shown at the left.
a. What percent of the crops planted in the square would be irrigated by this system?
b. If a wind-blown seed falls randomly into the square, what is the probability that it will fall in the irrigated region? *(Lessons 6-5, 6-6)*

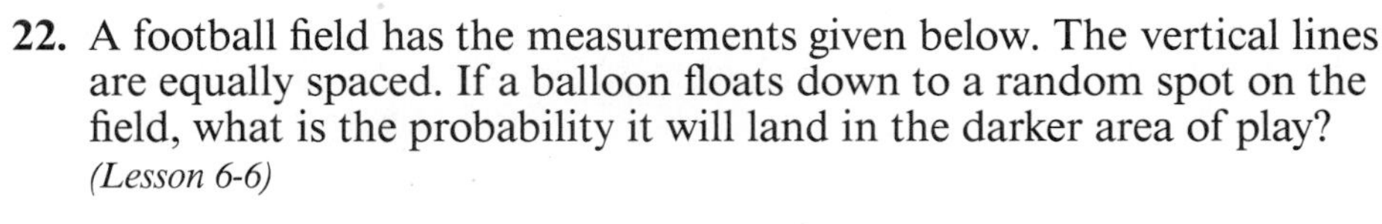

22. A football field has the measurements given below. The vertical lines are equally spaced. If a balloon floats down to a random spot on the field, what is the probability it will land in the darker area of play? *(Lesson 6-6)*

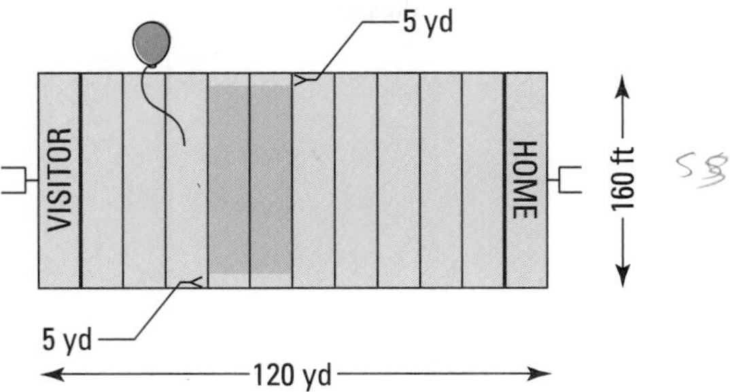

23. The Environmental Protection Agency calculated that in 1990, of 195.7 million tons of solid waste generated in the U.S., 73.3 million tons consisted of paper. What percent of the total waste is paper? *(Lesson 6-5)*

24. A scarf normally sells for $23.95. It is on sale for 30% off.
a. What percent of the price does the customer pay?
b. Write an equation to find the sale price.
c. How much does the scarf cost on sale? *(Lesson 6-5)*

25. About one in 86 pregnancies in the United States results in twins.
a. What is the relative frequency of having twins?
b. What is an estimate of the probability that a pregnancy results in twins? *(Lessons 6-3, 6-4)*

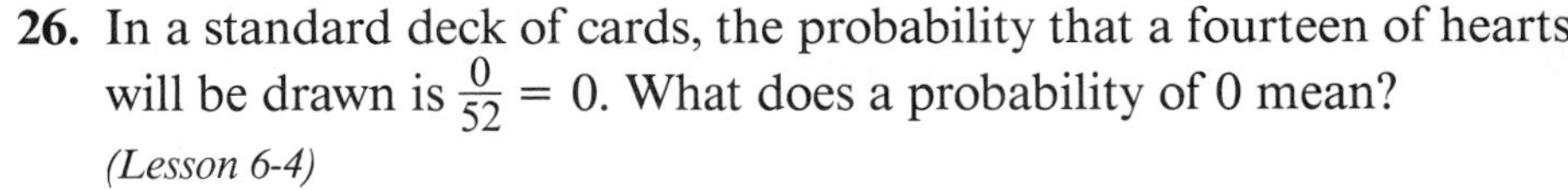

26. In a standard deck of cards, the probability that a fourteen of hearts will be drawn is $\frac{0}{52} = 0$. What does a probability of 0 mean? *(Lesson 6-4)*

Twins Tia and Tamera Mowry star in the TV comedy series, Sister, Sister.

27. Give an example of an event with a probability of 1. *(Lesson 6-4)*

28. *Skill sequence.* Simplify. *(Lessons 2-3, 3-9, 4-1, 6-1)*
a. $\frac{x}{2} + \frac{x}{3}$ **b.** $\frac{x}{2} - \frac{x}{3}$ **c.** $\frac{x}{2} \cdot \frac{x}{3}$ **d.** $\frac{x}{2} \div \frac{x}{3}$

Exploration

29. a. Using grid paper, draw a rectangle with dimensions 24 units by 36 units. Draw a rectangle whose sides are $\frac{1}{4}$ the lengths of sides of the first rectangle. How many of the small rectangles can fit in the large rectangle?
b. Repeat part **a,** except make the sides of the small rectangle $\frac{1}{6}$ of the sides of the first rectangle.
c. Draw a rectangle with dimensions 15 by 36. Draw a rectangle whose sides are $\frac{1}{3}$ the lengths of the sides of the first rectangle. How many of the small rectangles can fit in the large rectangle?
d. What pattern(s) do you see from parts **a, b,** and **c?** Test your pattern(s) with another rectangle and another fraction.

LESSON

6-8

Proportions

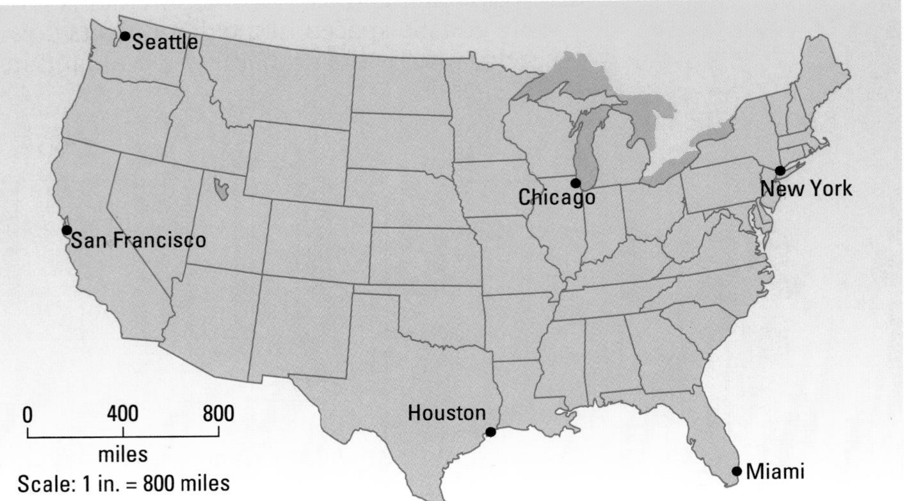

Map scales. *A map scale is a ratio of the map distance to the actual distance. From a map, a proportion can be used to find actual distance, as in Example 2.*

A **proportion** is a statement that two fractions are equal. Thus any equation of the form

$$\frac{a}{b} = \frac{c}{d}$$

is a proportion. This equation is sometimes written $a{:}b = c{:}d$, and is read "a is to b as c is to d." Because of the order of reading the proportion, the numbers b and c are called the **means** of the proportion; the numbers a and d are called the **extremes.** For example, in the proportion $\frac{30}{x} = \frac{6}{7}$, the means are x and 6. The extremes are 30 and 7. The equation $\frac{x}{2} + \frac{5}{4} = \frac{1}{2}$ is not a proportion because the left side of the equation is not a single fraction.

The Means-Extremes Property

Suppose two fractions are equal. What is true about the means and extremes of the proportion? Below we show a specific case on the left and the general case on the right.

$$\frac{9}{12} = \frac{30}{40} \qquad\qquad \frac{a}{b} = \frac{c}{d}$$

To clear the fractions, multiply both sides by the product of the denominators: $12 \cdot 40 = 480$, and $b \cdot d = bd$.

$$480 \cdot \frac{9}{12} = 480 \cdot \frac{30}{40} \qquad\qquad bd \cdot \frac{a}{b} = bd \cdot \frac{c}{d}$$

Simplify the equation.

$$\overset{40}{\cancel{480}} \cdot \frac{9}{\underset{1}{\cancel{12}}} = \overset{12}{\cancel{480}} \cdot \frac{30}{\underset{1}{\cancel{40}}} \qquad\qquad \overset{1}{\cancel{b}}d \cdot \frac{a}{\underset{1}{\cancel{b}}} = b\overset{1}{\cancel{d}} \cdot \frac{c}{\underset{1}{\cancel{d}}}$$

$$40 \cdot 9 = 12 \cdot 30 \qquad\qquad ad = bc$$

The general pattern is that the product of the means equals the product of the extremes.

Means-Extremes Property
For all real numbers *a, b, c,* and *d* (with *b* and *d* not zero),
if $\frac{a}{b} = \frac{c}{d}$, then $ad = bc$.

Many questions can be answered by writing a proportion from two equal ratios or two equal rates, and then using the Means-Extremes Property to solve the proportion.

Using the Means-Extremes Property to Solve Proportions

Example 1

A motorist traveled 283.5 miles on 9.0 gallons of gas. With the same driving conditions, how far could the car go on 14 gallons of gas?

Solution

Let x be the number of miles traveled on 14 gallons. Since conditions are the same, the rate $\frac{283.5 \text{ mi}}{9 \text{ gal}}$ should equal the rate $\frac{x \text{ mi}}{14 \text{ gal}}$.

$$\frac{283.5 \text{ miles}}{9 \text{ gallons}} = \frac{x \text{ miles}}{14 \text{ gallons}}$$

$$\frac{283.5}{9} = \frac{x}{14}$$

Use the Means-Extremes property to solve for x.

$$283.5 \cdot 14 = 9x$$

$$3969 = 9x$$

$$441 = x$$

The car could travel about 441 miles on 14 gallons.

Example 1 was set up by equating two *rates.* Proportions can also be set up using equal *ratios.* In Example 2 below, Solution 1 uses rates. Solution 2 uses ratios.

Example 2

In the map on page 394, 1 inch represents 800 miles. If New York and Miami are $1\frac{5}{8}$ inches apart on the map, how many miles apart are they?

Solution 1

Let x be the actual distance between the cities. Set up a proportion using equal rates in inches per mile.

$$\frac{1 \text{ inch}}{800 \text{ miles}} = \frac{1\frac{5}{8} \text{ inches}}{x \text{ miles}}$$

Use the Means-Extremes Property.

$$1 \cdot x = 800 \cdot 1\frac{5}{8}$$

$$x = 1300$$

The cities are about 1300 miles apart.

Solution 2

Let x be the actual distance between the cities. Set up a proportion using equal ratios. One ratio compares inches; the other compares miles.

$$\frac{1 \text{ inch}}{1\frac{5}{8} \text{ inches}} = \frac{800 \text{ miles}}{x \text{ miles}}$$

$$1 \cdot x = 800 \cdot 1\frac{5}{8}$$

$$x = 1300$$

The cities are about 1300 miles apart.

Often several correct proportions can be used to solve a problem. Here are two other proportions which could have been used to find the distance from New York to Miami.

$$\frac{800 \text{ miles}}{1 \text{ inch}} = \frac{x \text{ miles}}{1\frac{5}{8} \text{ inches}} \text{ or } \frac{1\frac{5}{8} \text{ inches}}{1 \text{ inch}} = \frac{x \text{ miles}}{800 \text{ miles}}$$

Each gives the same final answer.

Activity

Measure the distance on the map between Chicago and San Francisco. About how far apart are the cities in miles?

If the numerators or denominators in a proportion are expressions with variables, you can treat them as chunks. You can solve the resulting equations using techniques you have seen in earlier chapters.

Example 3

Solve $\frac{3g + 4}{5} = \frac{4g - 8}{4}$.

Solution 1

Use the Means-Extremes Property.

$4(3g + 4) = 5(4g - 8)$	
$12g + 16 = 20g - 40$	Apply the Distributive Property.
$16 = 8g - 40$	Add -12g to both sides.
$56 = 8g$	Add 40 to each side.
$7 = g$	Divide each side by 8.

Solution 2

Clear the fractions by multiplying each side of the equation by the least common multiple of the denominators.

$$\overset{4}{\cancel{20}} \cdot \frac{3g + 4}{\underset{1}{\cancel{5}}} = \overset{5}{\cancel{20}} \cdot \frac{4g - 8}{\underset{1}{\cancel{4}}}$$

$$4(3g + 4) = 5(4g - 8)$$

Now apply the Distributive Property and find the value of g as in Solution 1.

Check

Substitute. Does $\frac{3 \cdot 7 + 4}{5} = \frac{4 \cdot 7 - 8}{4}$?

$\frac{25}{5} = \frac{20}{4}$?

$5 = 5$? Yes.

Some proportions have more than one solution.

Example 4

Solve $\frac{p}{2} = \frac{32}{p}$.

Solution

Apply the Means-Extremes Property.

$p^2 = 64$

There are two solutions.

$p = 8$ or $p = -8$

Check

Check each solution by substitution.

Is $\frac{8}{2} = \frac{32}{8}$? Yes, $4 = 4$.

Is $\frac{-8}{2} = \frac{32}{-8}$? Yes, $-4 = -4$.

QUESTIONS

Covering the Reading

1. What is a *proportion?*

2. *Multiple choice.* Which of the following is a proportion?
   (a) $2 + \frac{x}{3} = \frac{1}{5}$ (b) $\frac{a}{b} = \frac{b}{x}$ (c) $\frac{a}{b} \cdot \frac{c}{d}$

3. According to the Means-Extremes Property, if $\frac{x}{y} = \frac{z}{w}$, then __?__.

In 4 and 5, a proportion is given. **a.** Use the Means-Extremes Property to solve the proportion. **b.** Check your work.

4. $\frac{x}{12} = \frac{3}{18}$

5. $\frac{-15}{12} = \frac{x}{-20}$

6. A motorist keeps records of his car's gas mileage. The last time he filled the tank, he had gone 216 miles on 13.8 gallons of gas. At this rate, how far can the car go on a full tank of 21 gallons?

7. In parts **a–c**, write the equation that results from using the Means-Extremes Property.
   a. $\frac{3}{5} = \frac{n}{7}$ **b.** $\frac{3}{7} = \frac{n}{5}$ **c.** $\frac{7}{5} = \frac{n}{3}$
   d. Which of the proportions in parts **a–c** has a different solution from that of the other two?

In 8 and 9, refer to the map on page 394.

8. On the map the distance between Seattle and Houston is $2\frac{7}{8}$ in.
   a. Write a proportion that can be used to estimate the distance between the two cities.
   b. Use your proportion in part **a** to estimate the distance between Seattle and Houston.
   c. Write another proportion that can be used to estimate the distance between Seattle and Houston.
   d. Solve the proportion in part **c**.

9. About how far apart in miles are Chicago and San Francisco?

In 10–15, solve using the Means-Extremes Property.

10. $\frac{2}{a} = \frac{-14}{15}$ **11.** $\frac{15}{2a} = \frac{3}{10}$ **12.** $\frac{g+3}{4} = \frac{g-2}{2}$

13. $\frac{3x}{4} = \frac{3x+1}{6}$ **14.** $\frac{x}{4} = \frac{9}{x}$ **15.** $\frac{98}{x} = \frac{x}{8}$

Applying the Mathematics

Snow in Sapporo. *Long winters and heavy snowfalls make Sapporo, Japan, an ideal spot for winter sports. Shown is a child enjoying the snow slide during the annual Winter Festival in Sapporo.*

16. A basketball team scores 17 points in the first 6 minutes of play. At this rate, how many points will the team score in a 32-minute game?

17. One of the heaviest snowfalls in recent history occurred in Bessans, France, on April 5–6, 1969, when about 173 cm of snow fell in 19 hours. At this rate, how many centimeters of snow would fall in 24 hours?

18. The fastest scheduled passenger train in the United States in 1992 traveled from Wilmington, Delaware, to Baltimore, Maryland, a distance of 68.4 miles, in 42 minutes. At this rate, what distance could this train cover in an hour?

19. Helen of Troy was described as having "a face that could launch a thousand ships." If a face could launch a thousand ships, what would it take to launch five ships?

20. **a.** Multiply $\frac{10}{16} \cdot \frac{15}{x}$. **b.** Solve $\frac{10}{16} = \frac{15}{x}$.

21. Consider the equation $\frac{2}{d} = \frac{d}{11}$. Express the solution(s)
    a. exactly; **b.** as decimals rounded to the nearest hundredth.

22. For each gear setting on a bicycle, there is a ratio of the number of turns of the pedals to the number of turns of the rear wheel.

Gear	Number of pedal turns	Number of rear-wheel turns
1st	9	14
2nd	4	7
3rd	1	2
4th	3	7
5th	5	14

While in 2nd gear, Bonnie turned the pedals 15 times. How many times did the rear wheel turn?

In 23–25, use the Means-Extremes Property to determine whether the given fractions are equal.

23. $\frac{1}{3}$, $\frac{33}{100}$

24. $\frac{4.5}{-5}$, $\frac{-153}{170}$

25. $\frac{388{,}162}{171{,}958}$, $\frac{430{,}262}{190{,}603}$

26. Solve $\frac{13x - 78}{x} = \frac{0}{7953}$.

27. Is there a fraction equal to $\frac{3}{4}$ whose numerator is five more than its denominator? If there is, find the fraction. If there is not, explain why such a number cannot exist.

Review

28. **a.** Plot $A = (-3, 6)$ and $B = (6, 9)$, and draw $\overline{AB}$.
b. Find the image of $\overline{AB}$ under a size change of magnitude $-\frac{5}{3}$.
c. Is the size change an expansion or a contraction? *(Lesson 6-7)*

29. **a.** Copy the figure below. Draw the image of $\triangle QRS$ using the rule that the image of (x, y) is $(3x, 3y)$. Call the image $Q'R'S'$.
b. What is QR?
c. What is $Q'R'$?
d. Describe how the image and preimage are related.
(Lessons 1-7, 6-7)

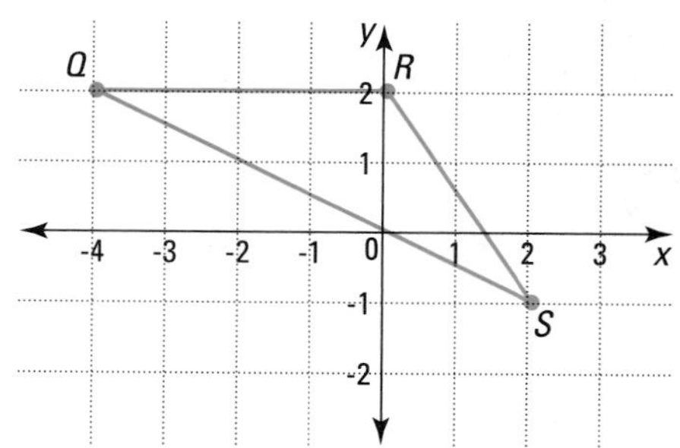

30. What is the probability of a randomly picked point inside rectangle $ABCD$ at the right being in the shaded region? *(Lesson 6-6)*

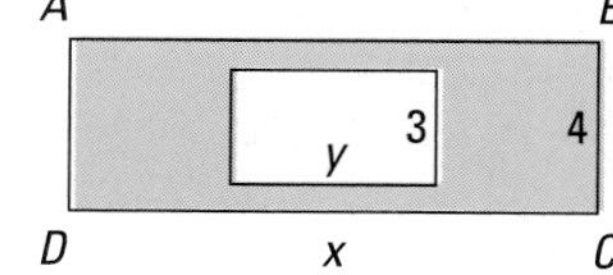

In 31 and 32, use the fact that a *karat* is a measure of fineness used for gold and other precious materials. Pure gold is 24 karats. Gold of 18-karat fineness is 18 parts pure gold and 6 parts other metals giving 24 parts altogether. *(Lessons 6-3, 6-5)*

31. A bracelet is 18-karat gold. What percent gold is this?

32. A necklace weighing 5 ounces is 10-karat gold. How many ounces of pure gold is in the necklace?

33. What value(s) can m *not* have in the expression $\frac{m - 7}{m + 5}$? *(Lesson 6-2)*

Golden treasures. *In 1989, an archeological dig in Nimrud, Iraq, uncovered this gold jewelry that dates back to the Assyrians. The Assyrian empire began in the 800s* B.C. *and declined after 612* B.C.

Exploration

34. Below are some data from the 1990 census about the states whose names begin with "new."

State	Population	Number of Motor Vehicles Registered
New Hampshire	1,109,252	945,743
New Jersey	7,730,188	5,652,382
New Mexico	1,515,069	1,301,261
New York	17,990,455	10,196,153

a. Rank these states from greatest to lowest in the number of motor vehicles registered per 100 people.

b. What reasons do you think account for the differences in motor vehicle registration rates you found in part **a?** *(Lesson 6-2)*

35. a. The statement "driver is to car as pilot is to airplane" is called an *analogy*. This analogy corresponds to the proportion $\frac{a}{b} = \frac{c}{d}$, where a = driver, b = car, c = pilot, and d = airplane. Find the missing word in each of the following analogies.

i. Soup is to bowl as water is to __?__.
ii. Inch is to centimeter as __?__ is to kilogram.
iii. __?__ is to Earth as Earth is to sun.
iv. Cow is to __?__ as hen is to chick.
v. Hoop is to __?__ as net is to soccer.
vi. Author is to novel as __?__ is to symphony.
vii. __?__ is to Maryland as Sacramento is to California.
viii. Motorist is to car as __?__ is to bicycle.

b. Make up two analogies similar to those in part **a.**

Introducing Lesson 6-9

Similar Figures

IN-CLASS ACTIVITY

Work in a small group.
Materials: graph paper, ruler, protractor

1 **a.** Draw a set of axes about in the middle of the graph paper. Draw $\triangle ABC$ with $A = (0, -2)$, $B = (4, 2)$, and $C = (7, 0)$.

b. Draw the image of $\triangle ABC$ under a size change of magnitude 2.

c. Measure the sides and angles of each triangle and record your results in a table like the one below.

$\triangle ABC$						$\triangle A'B'C'$					
AB	BC	AC	$m\angle A$	$m\angle B$	$m\angle C$	$A'B'$	$B'C'$	$A'C'$	$m\angle A'$	$m\angle B'$	$m\angle C'$

d. Calculate the ratios $\frac{A'B'}{AB}$, $\frac{B'C'}{BC}$, and $\frac{A'C'}{AC}$. What conclusions can you draw?

e. What is true about the angle measures in $\triangle ABC$ and $\triangle A'B'C'$?

f. Pick a scale factor k other than 1 or 2. For instance, use $k = \frac{1}{2}$ or $k = 3$. (Each person should use a different scale factor.) Draw $\triangle A''B''C''$, the image of $\triangle ABC$ under the size change with your scale factor k.

g. Calculate the ratios $\frac{A''B''}{AB}$, $\frac{B''C''}{BC}$, and $\frac{A''C''}{AC}$ for your figures. How are these ratios related?

h. Measure $\angle A''$, $\angle B''$, and $\angle C''$. How are these angle measures related to the measures of angles A, B, and C?

i. ***Draw conclusions.*** Compare your results with those of the others in your group. Write several sentences to summarize what you found.

2 **a.** Use the other side of the graph paper. Draw a set of axes in the middle of the paper. Draw quadrilateral $WXYZ$ with $W = (-3, 2)$, $X = (-3, -3)$, $Y = (9, -4)$, and $Z = (2, 7)$.

b. Pick a scale factor k other than 1. (Each person in the group should pick a different value.) Describe how you think the image of $WXYZ$ will be related to $WXYZ$ under a size change of magnitude k.

c. Draw the image of $WXYZ$ under the size change of magnitude k. Label it $W'X'Y'Z'$.

d. Measure the sides and angles of each quadrilateral. Calculate ratios of lengths of corresponding sides.

e. Discuss your work in parts **a** to **d** with the other members of your group. Write several sentences summarizing what you found.

LESSON

6-9

Similar Figures

Similar stacking dolls. *Russian stacking dolls, painted to look like peasants, are hollowed and fit inside one another. Woodcarvers make these dolls, called* matreshka, *in sets of 4 or more.*

What Are Similar Figures?

The image of a figure under a size change has the same shape as the preimage. Such figures are called **similar figures.** All the triangles you drew in Question 1 of the In-class Activity are similar to each other. All the quadrilaterals in Question 2 are similar.

Two fundamental properties of similar figures are given below.

Fundamental Properties of Similar Figures
If two figures are similar, then
(1) corresponding angles have the same measure.
(2) ratios of lengths of corresponding sides are equal.

You saw several instances of these properties in the In-class Activity.

Not only are the ratios of corresponding sides equal, but they equal the size change factor. This ratio is called a **ratio of similitude.** In Question **1d** of the In-class Activity you should have found that

$$\frac{A'B'}{AB} = \frac{B'C'}{BC} = \frac{A'C'}{AC} = 2.$$

If $\triangle ABC$ were considered to be the image of $\triangle A'B'C'$, then the reciprocals of these ratios would be taken. The ratio of similitude would then be $\frac{1}{2}$.

Finding Lengths in Similar Figures

When two figures are similar, a true proportion can be written using corresponding sides. If three of the four lengths in the proportion are known, the fourth can be found by solving an equation.

Example 1

The two quadrilaterals below are similar, with corresponding sides parallel. Find *x*, the length of *CD*.

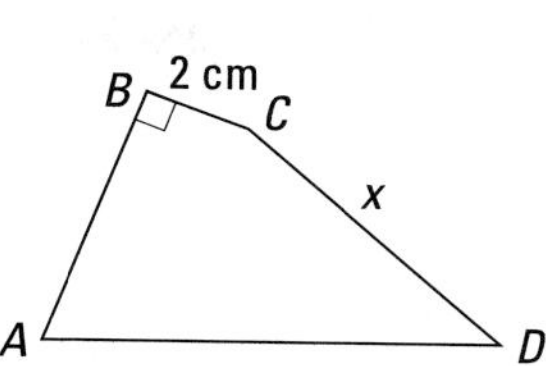

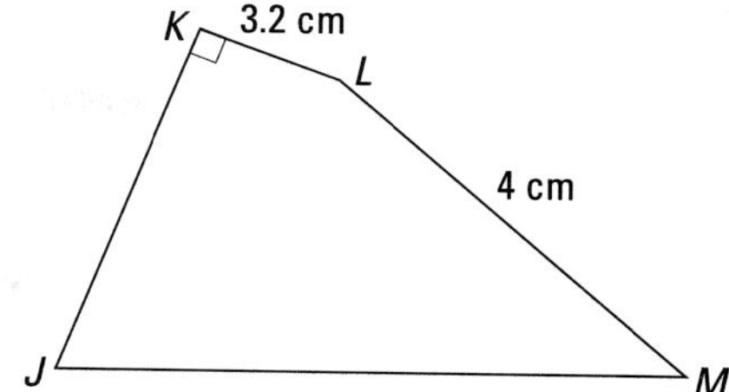

Solution

The side corresponding to $\overline{CD}$ is $\overline{LM}$. Now find a pair of corresponding sides whose lengths are known. These are $\overline{BC}$ and $\overline{KL}$. Since the figures are similar, the ratios of lengths of these corresponding sides are equal.

$$\frac{CD}{LM} = \frac{BC}{KL}$$

Substitute the known lengths.

$$\frac{x}{4} = \frac{2}{3.2}$$

Use the Means-Extremes Property to solve for *x*.

$$3.2 \cdot x = 2 \cdot 4$$
$$3.2x = 8$$
$$x = \frac{8}{3.2} = 2.5$$

Check

Find the ratio of similitude first using *BC* and *KL*, then using *CD* and *LM*. $\frac{BC}{KL} = \frac{2}{3.2} = \frac{20}{32} = \frac{5}{8}$ and $\frac{CD}{LM} = \frac{2.5}{4} = \frac{25}{40} = \frac{5}{8}$. They are equal.

Using Similar Figures to Find Lengths Without Measuring

Similar figures have many uses. For instance, you can use similar triangles to find the height of an object you cannot measure easily.

Suppose you want to find the height *h* of the flagpole in front of your school. Here is how you can do it. Hold a yardstick parallel to the flagpole and measure the length of its shadow. Then measure the length of the shadow of the flagpole. The following picture illustrates one possible set of measurements.

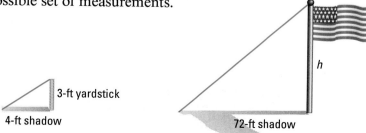

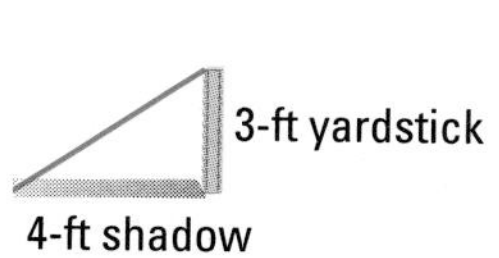

h

72-ft shadow

Example 2

Use the measurements on page 403 to find the height h of the flagpole.

Solution

Two similar right triangles are formed. Now, use ratios of corresponding sides.

$$\frac{3}{h} = \frac{4}{72}$$
$$4h = 72 \cdot 3$$
$$4h = 216$$
$$h = 54.$$

The flagpole is 54 feet tall.

QUESTIONS

Covering the Reading

1. In similar figures, what is true about lengths of corresponding line segments?

In 2–5, the two triangles below are similar. Corresponding sides are parallel.

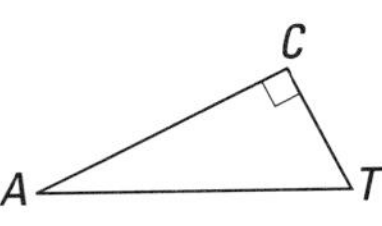

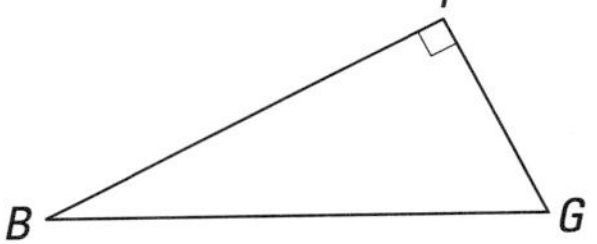

2. Which side of $\triangle BIG$ corresponds to the given side of $\triangle ACT$?
a. $\overline{AT}$ **b.** $\overline{CT}$ **c.** $\overline{AC}$

3. Find two ratios equal to $\frac{AC}{BI}$.

4. Suppose $AC = 15$, $CT = 8$, $AT = 17$, and $BI = 30$. Find
a. IG. **b.** BG.

5. Suppose $\triangle BIG$ is the image of $\triangle ACT$ under a size change of magnitude 3 and $BG = 12$. What other lengths can be found?

6. The rectangles below are similar.

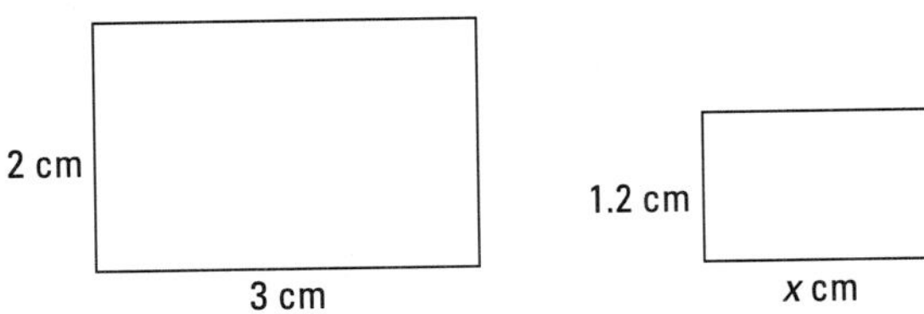

a. Write a proportion that could be used to find x.
b. Solve the proportion you wrote in part **a.**
c. What are the two possible ratios of similitude?

7. $\triangle ABC$ is similar to $\triangle A'B'C'$.

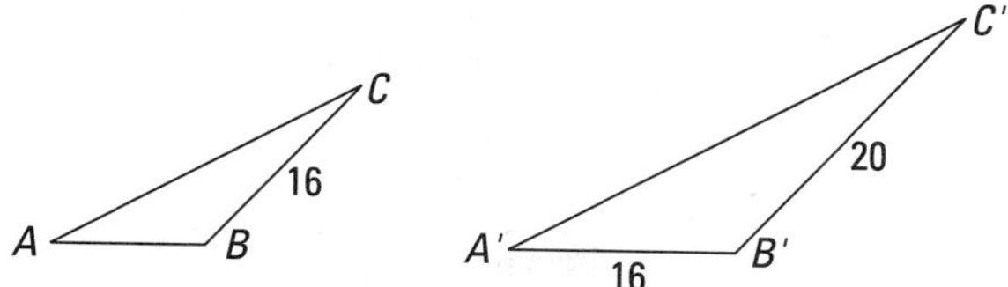

a. Write a proportion that could be used to find the length of $\overline{AB}$.
b. Find AB.

8. A tree casts a shadow that is 4.25 m long. A meter stick casts a shadow that is 0.7 m long.
a. Copy this diagram and label the given lengths.

b. Write a proportion that can be used to find the height of the tree.
c. How tall is the tree?

Applying the Mathematics

9. The two quadrilaterals below are similar. Corresponding sides are parallel.

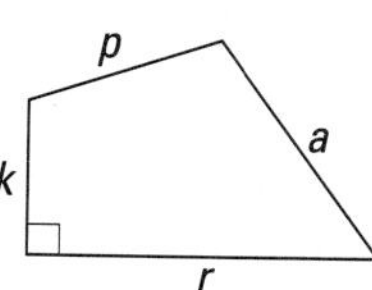

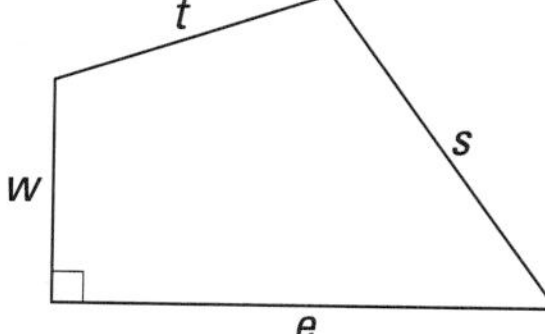

a. Write another ratio which must equal $\frac{s}{a}$.
b. Write three ratios which must equal $\frac{p}{t}$.

10. The two rectangles at the right are similar. Find the length and width of the larger rectangle.

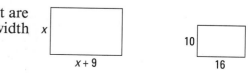

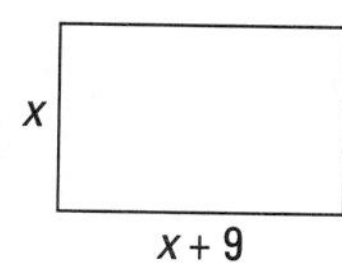

11. At a certain time on a sunny day Jim, who is 6 feet tall, casts a shadow that is 10 feet long. A nearby building, which is t feet tall, casts a shadow that is 25 feet long.

a. Draw a diagram of this situation. Write in the lengths.

b. Write a proportion that describes this situation.

c. How tall is the building?

12. Copy the second drawing and draw the complete pentagon $PQRST$ which is similar to $ABCDE$, given that $\overline{PT}$ corresponds to $\overline{AE}$.

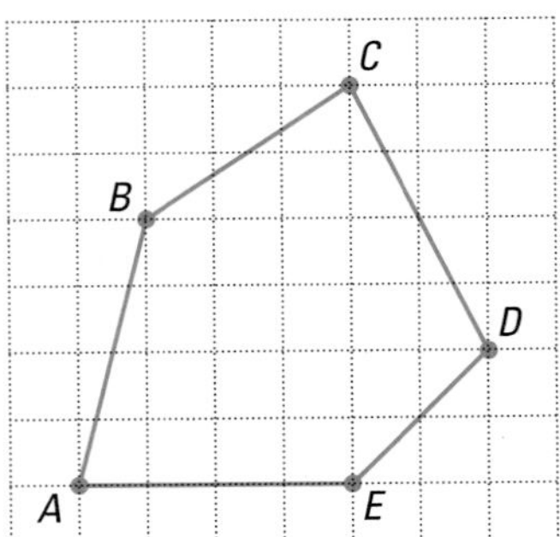

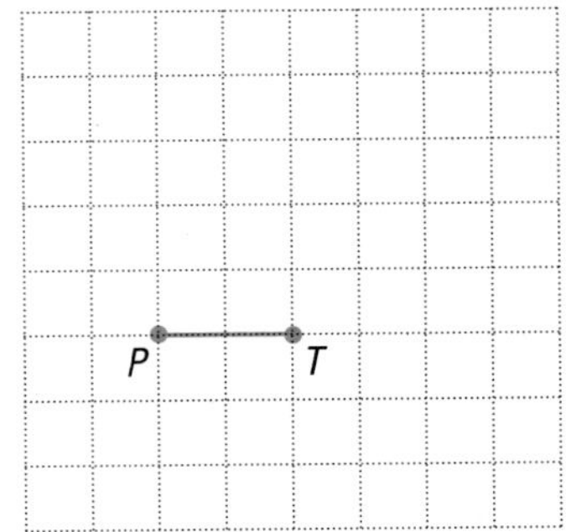

13. For this question, you need to use a ruler and properties of similar figures. A scale drawing of a house is shown below. The actual width (across the front) of the house is 9 m.

a. Write a ratio comparing the width of the house in the drawing to the actual width of the house.

b. Write a proportion you could use to find the actual distance from the ground to the peak of the roof.

c. Solve the proportion in part **b.**

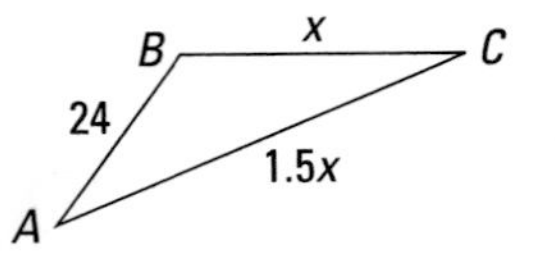

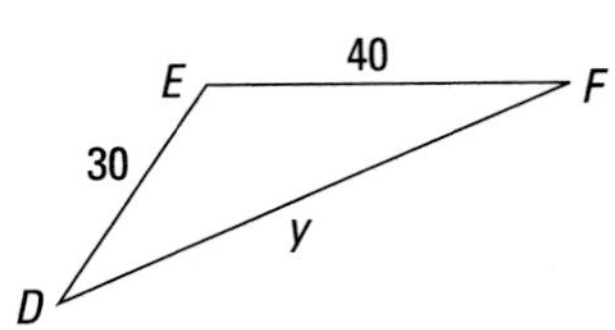

14. Triangles ABC and DEF at the left are similar. Corresponding sides are parallel.

a. *Multiple choice.* Which proportion could you use to find x?

(a) $\frac{24}{30} = \frac{1.5x}{y}$ (b) $\frac{x}{y} = \frac{1.5x}{40}$ (c) $\frac{24}{30} = \frac{x}{40}$

b. Solve the equation you choose for x.

c. Use your solution to part **a** to find y.

d. Based on your answers, what is the ratio of the *perimeter* of $\triangle ABC$ to the *perimeter* of $\triangle DEF$?

Review

In 15 and 16, solve. *(Lesson 6-8)*

15. $\frac{48}{2x} = \frac{9}{5}$ **16.** $\frac{1}{x-1} = \frac{3}{x-2}$

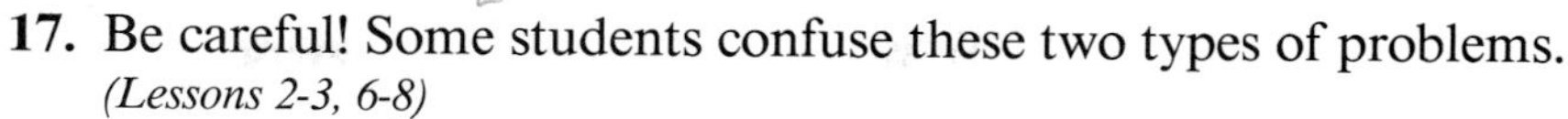

17. Be careful! Some students confuse these two types of problems. *(Lessons 2-3, 6-8)*

a. Multiply $\frac{9}{n} \cdot \frac{n}{25}$. **b.** Solve $\frac{9}{n} = \frac{n}{25}$.

18. a. Draw $\triangle ABC$ with $A = (0, 0)$, $B = (1, 0)$, and $C = (4, 5)$.
b. Draw the image of $\triangle ABC$ under a size change of magnitude -2. *(Lesson 6-7)*

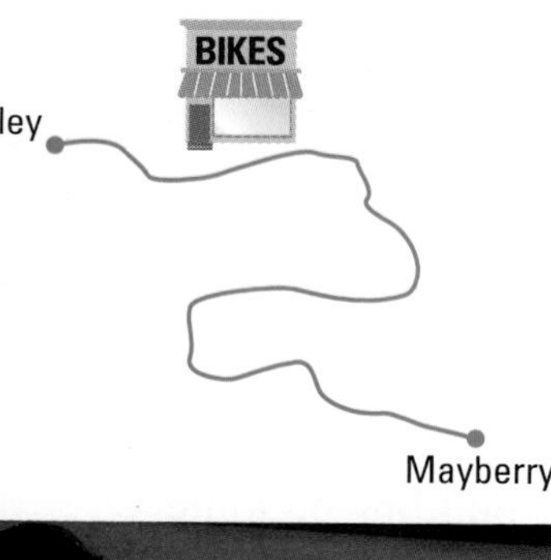

19. Pixley is 38 kilometers by road from Mayberry. A bicycle shop is somewhere between 5 and 10 km from Pixley. If a bicyclist has a flat tire at some point between the two towns, what is the probability of its occurring within $\frac{1}{2}$ km of the shop? *(Lesson 6-6)*

20. The dinner bill at some restaurants includes the tip. If a \$38.81 bill included a 15% tip, what was the amount of the original bill before the tip was added? *(Lesson 6-5)*

21. On sale, a sweater costs \$17. It originally cost \$21. What is the percent of discount? *(Lesson 6-5)*

22. A letter is chosen randomly from the English alphabet. What is the probability that it is a letter in the word MATH? *(Lesson 6-4)*

23. How many times as large as $\frac{1}{2}x$ is $3x$? *(Lesson 6-3)*

24. Simplify $\frac{3\pi}{y^2} \div \frac{\pi}{5}$. *(Lesson 6-1)*

Time to retire.
Bicycle mechanics perform a variety of repair jobs, such as fixing flat tires, replacing worn or broken parts, and reconditioning bikes.

In 25 and 26, use the expression $\frac{c-d}{a-b}$. *(Lessons 1-5, 6-1)*

25. Evaluate the expression when $a = 3$, $b = 4$, $c = 5$, and $d = 2$.

26. If $a = 7$, and $d = 8$, for what values is the expression not defined?

27. *Skill sequence.* Solve. *(Lessons 2-6, 3-5, 4-3)*

a. $p + 8 = 23$ **b.** $8p = 23$ **c.** $8 - p = 23$ **d.** $\frac{1}{8}p = 23$

28. Light travels at the incredibly rapid rate of about 186,000 miles per second. About how many miles does light travel in one hour? *(Lesson 2-4)*

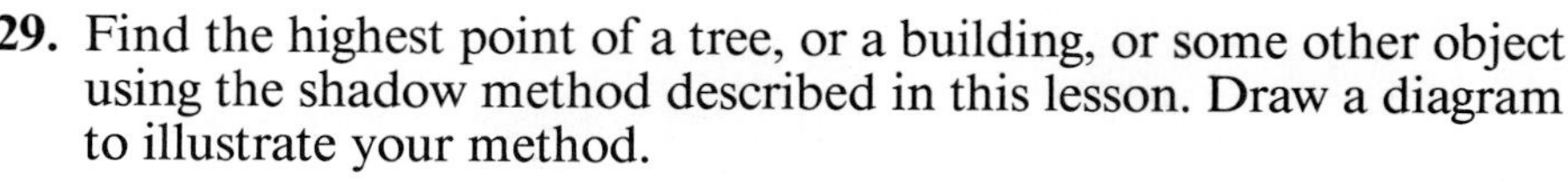

Exploration

29. Find the highest point of a tree, or a building, or some other object, using the shadow method described in this lesson. Draw a diagram to illustrate your method.

PROJECTS 6 CHAPTER SIX

A project presents an opportunity for you to extend your knowledge of a topic related to the material of this chapter. You should allow more time for a project than you do for typical homework questions.

1 Population Densities

A *population density* is a rate, defined as the number of people living in a region divided by the area of that region.

a. Find the population and area of your town or city, your state, the United States, and the world. Calculate the population densities in people per square mile. On the basis of population density, rank these four places from least crowded to most crowded.

b. Which states in the United States have the highest population densities? Which have the lowest?

c. Name some countries that have much higher population densities than that of the United States; name some countries that have much lower population densities.

2 Indirect Measurement

In Lesson 6-9, we described how similar triangles can be used to find the height of an object indirectly, that is, without measuring the object's height. There are many other ways to measure heights indirectly. Find at least four ways to determine the height of your school building or some other tall object in your neighborhood. Describe the methods you used, and discuss the advantages and disadvantages of using each.

3 Area, Perimeter, and Size Changes

a. Draw an interesting figure on a coordinate grid. Find its perimeter and area.

b. Draw its image under size changes of magnitude 2, 3, 4, and $\frac{1}{2}$. Find the perimeter and area of each image.

c. Describe the patterns you find in your data. How are the perimeter and the area of the image related to the perimeter and the area of its preimage?

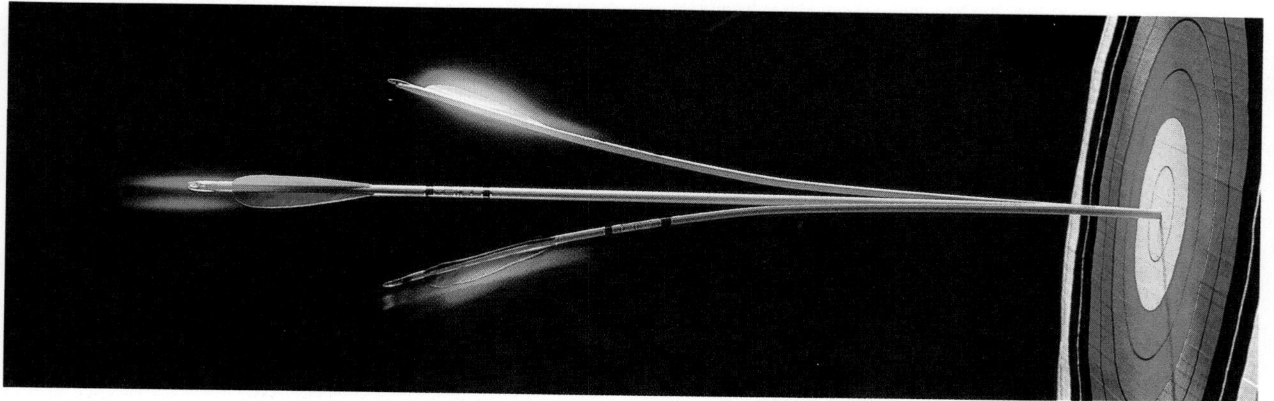

4 Archery

A sport that often uses targets with concentric circles like those used in Lesson 6-6 is archery. Find out the sizes of the targets, the sizes of the different regions of the targets, and how scores are calculated in this sport. What is the relationship between the sizes of the regions and the points for a "hit" in each region? Make up some mathematics questions based on archery and probability.

5 Lotteries

Many states have lotteries to raise money for education, public works, or other needs. If your state has a lottery, find out how it works. Include information about how winners are determined, how prize

money is calculated, and how the chances of winning are determined. How are the concepts of relative frequency and probability applied in your state's lottery?

6 Density and Floating

In science the *density* of an object is determined by dividing the mass (weight in grams) of the object by its volume (in cubic centimeters). For instance, the density of cool tap water is one gram per cubic centimeter.

a. Find some blocks of different types of wood, such as pine, oak, maple, and cedar. Calculate the density of each block.
b. Fill a tub with enough tap water to see if the blocks of wood will float. Which blocks of wood float?
c. Write up your methods and conclusions. How does density appear to be related to an object's ability to float?

SUMMARY

Division is closely related to multiplication. The definition of division states that to divide by a number is the same as to multiply by its reciprocal. This definition is applied directly to divide fractions. Because zero has no reciprocal, division by zero is impossible.

Rates and ratios are models for division. A rate compares quantities with different units; ratios compare quantities with the same units. A statement that two fractions are equal is called a proportion. The Means-Extremes Property can be used to find missing values in a proportion. One important place proportions appear is in similar figures.

Percent, relative frequency, probability in geometry, and size changes are applications involving rates and ratios. It is possible to translate most percent problems to equations of the form $ab = c$. Solving the equation then gives an answer to the problem. Since it is impossible to count the infinite number of points in a geometric region, a ratio of lengths or areas is used to compute probabilities in geometric situations. Size changes yield similar figures. If the magnitude is k and $k > 1$ or $k < -1$, then the size change is an expansion. If $-1 < k < 1$, the size change is a contraction.

VOCABULARY

Below are the most important terms and phrases for this chapter. You should be able to give a general description and a specific example of each.

Lesson 6-1
Algebraic Definition of Division
complex fraction

Lesson 6-2
rate
Rate Model for Division
population density

Lesson 6-3
ratio
Ratio Model for Division
percent of discount
percent of tax

Lesson 6-4
frequency
relative frequency
probability
outcome, event
at random, randomly
complement

Lesson 6-5
percent

Lesson 6-6
Probability Formula for Geometric Regions

Lesson 6-7
size change factor
Size Change Model for Multiplication
expansion, contraction
magnitude of a size change

Lesson 6-8
proportion
means, extremes
Means-Extremes Property

Lesson 6-9
similar figures
Fundamental Properties of Similar Figures
ratio of similitude

CHAPTER SIX

PROGRESS SELF-TEST

Take this test as you would take a test in class. You will need a calculator and graph paper. Then check your work with the solutions in the Selected Answers section in the back of the book.

In 1–3, simplify.

1. $15 \div -\frac{3}{2}$ **2.** $\frac{x}{9} \div \frac{2}{3}$ **3.** $\frac{\frac{2b}{3}}{\frac{b}{3}}$

In 4–6, solve. Show your work.

4. $\frac{y}{11} = \frac{2}{23}$ **5.** $\frac{b}{49} = \frac{25}{b}$ **6.** $\frac{4g - 3}{26} = \frac{g}{8}$

7. If 14% of a number is 60, what is the number?

8. $\frac{1}{2}$ is what percent of $\frac{4}{5}$?

9. When rolling a single die, what is the probability that at least a 5 will result?

10. What value can v not have in the expression $\frac{10v}{v + 1}$? Why can it not have this value?

11. In the proportion $\frac{2}{3} = \frac{x}{10}$, which numbers are the means?

12. Horatio spent 36 minutes on his English homework. Mary Ellen spent a half hour on her English homework.

a. Horatio's time is what percent of Mary Ellen's time?

b. Horatio studied __?__ percent longer than Mary Ellen.

13. A kennel has c cats and d dogs, but no other animals. What is the ratio of the number of dogs to the number of animals?

14. The two triangles below are similar with corresponding sides parallel.

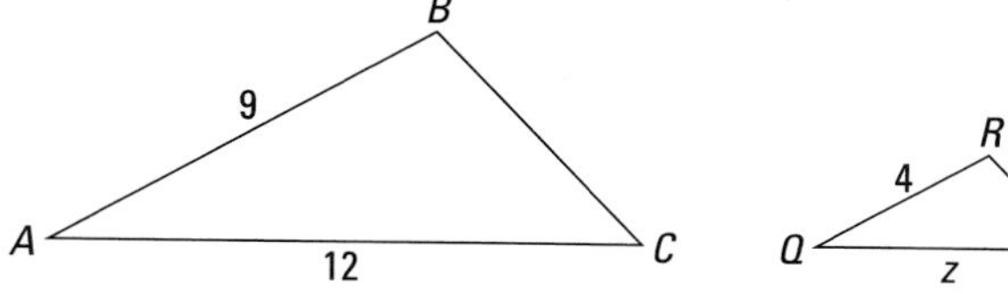

a. What are the two possible ratios of similitude in this situation?

b. Find z.

15. A television cost $236 after a 20% discount. What was the original price?

16. Which is faster, reading p pages in $7y$ minutes or p pages in $8y$ minutes? Justify your answer.

17. In a survey of 400 people over the age of 70, 36 were found to have Alzheimer's disease. What is the relative frequency of Alzheimer's disease in this group of people? Explain how you found your answer. Write your answer as a percent.

18. If the electricity goes out and a clock stops, what is the probability that the second hand stops between 2 and 3?

19. If a car travels 280 miles on 12 gallons of gas, about how far (to the nearest mile) can the car travel on 14 gallons of gas?

20. a. Graph the triangle with vertices $A = (2, 3)$, $B = (-1, -3)$, and $C = (3, 1)$.

b. Graph the image of $\triangle ABC$ under a size change of -2.

21. a. What is the image of (-6, 4) under the size change with the rule that the image of (x, y) is $\left(\frac{2}{3}x, \frac{2}{3}y\right)$?

b. Is the size change an expansion, a contraction, or neither?

22. The rectangles below are similar. Find the length and width of the smaller rectangle. Show your work.

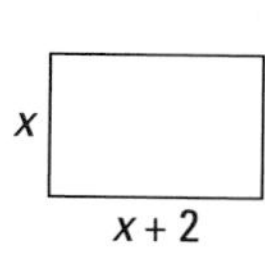

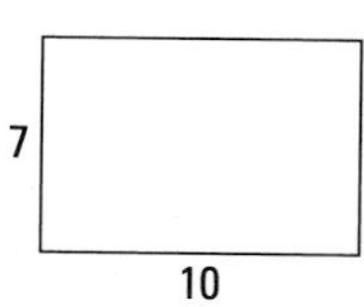

23. In a certain video game, if villains appear in the shaded area of the screen you will not be able to destroy them. Suppose the villains are equally likely to be in any part of the screen. What is the probability that you will not be able to destroy the villains?

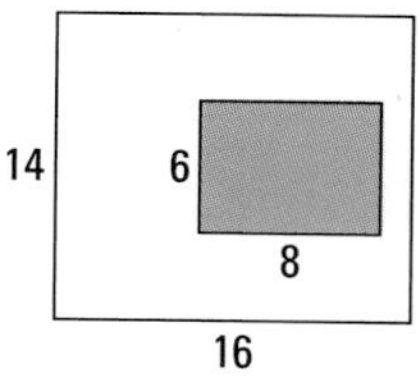

 CHAPTER SIX

CHAPTER REVIEW

Questions on SPUR Objectives

SPUR stands for **S**kills, **P**roperties, **U**ses, and **R**epresentations. The Chapter Review questions are grouped according to the SPUR Objectives for this chapter.

SKILLS DEAL WITH PROCEDURES USED TO GET ANSWERS.

Objective A: *Divide real numbers and algebraic fractions.* (Lesson 6-1)

In 1–8, simplify.

1. $-25 \div \frac{1}{5}$ **2.** $\frac{a}{14} \div \frac{7}{2}$ **3.** $\frac{2\pi}{9} \div \frac{\pi}{6}$ **4.** $\frac{x}{z} \div \frac{x}{y}$

5. $\dfrac{\frac{3x}{2}}{-\frac{15x}{16}}$ **6.** $\dfrac{\frac{a}{b}}{\frac{c}{d}}$ **7.** $\dfrac{60}{\frac{1}{4}}$ **8.** $\dfrac{x}{-\frac{x}{4}}$

In 9 and 10, what value(s) can the variable not have?

9. $\frac{x + 1}{x + 4}$ **10.** $\frac{15x}{x - 0.2}$

Objective B: *Solve percent problems.* (Lesson 6-5)

11. What is 75% of 32?

12. What is 20% of 18?

13. 10 is what percent of 5?

14. 1.2 is what percent of 0.8?

15. 85% of what number is 170?

16. 30% of a number is $\frac{3}{4}$. What is the number?

Objective C: *Solve proportions.* (Lesson 6-8)

In 17–22, solve.

17. $\frac{x}{130} = \frac{6}{5}$ **18.** $\frac{6}{25} = -\frac{10}{m}$ **19.** $\frac{4}{x} = \frac{x}{225}$

20. $\frac{y}{7} = \frac{10}{y}$ **21.** $\frac{2m - 1}{21} = \frac{m - 5}{24}$

22. $\frac{5 + r}{3} = \frac{r + 2}{2}$

PROPERTIES DEAL WITH THE PRINCIPLES BEHIND THE MATHEMATICS.

Objective D: *Use the language of proportions and the Means-Extremes Property.* (Lesson 6-8)

23. In the proportion $\frac{5}{8} = \frac{15}{24}$,

a. which numbers are the means?

b. which numbers are the extremes?

24. If $\frac{m}{n} = \frac{p}{q}$, then by the Means-Extremes Property __?__ = __?__.

25. If $\frac{2}{3} = \frac{x}{5}$, what does $\frac{5}{x}$ equal?

26. If $\frac{a}{b} = \frac{c}{d}$, then $\frac{c}{a} =$ __?__.

USES DEAL WITH APPLICATIONS OF MATHEMATICS IN REAL SITUATIONS.

Objective E: *Use the Rate Model for Division.* (Lesson 6-2)

27. Assume a 22-mile bike trip took 2 hours.

a. What was the rate in miles per hour?

b. What was the rate in hours per mile?

28. A train travels from Newark to Trenton, a distance of 48.1 miles, in 30 minutes. At what average speed, in miles per hour, does the train travel?

29. Marlene worked $3\frac{1}{2}$ hours and earned $21. How long does it take her to earn a dollar?

In 30–32, make up a question based on the given information and calculate a rate to answer the question.

30. The Johnsons drove 30 miles in $\frac{3}{4}$ hours.

31. In d days Tony spent \$400.

32. 4 weeks ago the puppy weighed 3 kilograms less than it does today.

33. In one store a 46-ounce can of tomato juice costs \$1.77, and a 6-ounce can costs 28 cents.

a. Calculate the cost per ounce for each size.

b. Based on the cost per ounce, which is the better buy?

34. Which is faster, reading w words in m minutes or $6w$ words in $2m$ minutes? How can you tell?

Objective F: *Use ratios to compare two quantities.* *(Lesson 6-3)*

35. In an algebra class of 27 students, there are 10 girls. The number of boys is how many times the number of girls?

36. Life expectancy increased from an estimated 18 years in 3000 B.C. to about 76 years in 1990. Compare the life expectancy in 1990 to life expectancy in 3000 B.C.

a. using a ratio. **b.** using a percent.

37. An item selling for \$36 is reduced by \$6. What is the percent of discount?

38. The profit on an item selling for \$20 is \$8. What percent of the item's selling price is profit?

39. Carl has x CDs and Carla has y CDs. Express the ratio of the number of Carl's CDs to the total number of CDs.

40. What is the ratio of the total number of heads to the total number of feet among m cows and n chickens?

Objective G: *Calculate relative frequencies or probabilities in situations with a finite number of equally likely outcomes.* *(Lesson 6-4)*

41. A number is selected randomly from the integers {-5, -4, -3, . . . , 5}. What is the probability that the number is less than 4?

42. A fair die is thrown once. Find the probability of getting an odd number.

In 43 and 44, one card is drawn at random from a standard deck of playing cards.

43. What is the probability of drawing a club?

44. What is the probability of drawing a jack or a king?

45. If the probability of winning a prize in a lottery is $\frac{1}{1{,}000{,}000}$, what is the probability of not winning?

46. Event A has a probability of 0.3. Event B has a probability of $\frac{4}{9}$. Event C has a probability of 33%. Which event is

a. most likely to happen?

b. least likely to happen?

47. In one community, 116 of 200 people surveyed had at least one pet. What is the relative frequency of pet owners in this survey?

48. If the relative frequency of adults who smoke in one town is 14.2%, what is the relative frequency of adults who don't smoke?

In 49 and 50, consider the following. A company with forty salespeople keeps records of how many sales calls are made each week. The results from one week are given in the table.

Number of sales calls	Number of salespersons
10–14	1
15–19	2
20–24	5
25–29	11
30–34	12
35–39	6
40 or more	3

49. What is the relative frequency of a salesperson making fewer than 25 sales calls? (Write as a fraction.)

50. What is the relative frequency of a salesperson making 10 or more sales calls?

Objective H: *Solve percent problems in real situations.* *(Lessons 6-3, 6-5, 6-7)*

51. A sofa is on sale for \$450. It originally cost \$562.52. What percent of the original price is the sale price?

 CHAPTER SIX

52. A $15.99 tape is on sale for $11.99. To the nearest percent, what is the percent of discount?

53. In August of 1992, *Popular Mechanics* reported that the magnesium valve covers of a BMW automobile weighed 3.7 pounds each and were 37% lighter than the prior year's aluminum covers. How much did the prior year's aluminum valve covers weigh?

54. In 1992, 48% of all accidental deaths occurred in motor vehicle accidents. If 86,000 people died accidentally that year, how many were killed in motor-vehicle accidents?

55. After a 30% discount and a 5% tax, you paid $7.30 for a shirt. What was the price of the shirt before the discount?

56. Model trains of HO gauge are $\frac{1}{87}$ actual size (no fooling!). If a model locomotive is 30 cm long, how long is the real locomotive?

57. In 1991, accounting clerks in Boston were earning an average of $437 per week, an increase of 5.4% from 1990. What was their average salary in 1990?

58. Between September 1991 and August 1992 Super Nintendo© dropped in price by 55.3% to $89. What was the September 1991 price?

Objective I: *Solve problems involving proportions in real situations.* (*Lessons 6-8, 6-9*)

59. Anne was saving for a class trip. For every $35 that Anne earned her mother added an extra $15. If Anne earned $245, how much would her mother add?

60. If $\frac{3}{4}$ cup of sugar equals 12 tablespoons of sugar, how many tablespoons are there in 3 cups of sugar?

61. In the first seven days of November, one family made 45 calls. At this rate, about how many calls will they make during the month?

62. In 1993 for every dollar, you could get 1.60 deutschemarks (the currency in Germany). If a crystal vase cost 120 deutschemarks in 1993, what would its cost be in dollars?

In 63 and 64, *multiple choice.*

63. Suppose s sweaters cost d dollars. At this rate, how many sweaters can be bought for $75?
(a) $\frac{s}{75d}$ (b) $\frac{75s}{d}$ (c) $\frac{75d}{s}$ (d) $\frac{d}{75s}$

64. People planning a party estimate they will need one 2-liter bottle of soda pop for 5 people. How many liters of soda pop will they need for n people?
(a) $\frac{n}{5}$ (b) $\frac{5}{n}$ (c) $\frac{2n}{5}$ (d) $\frac{2}{5n}$

REPRESENTATIONS DEAL WITH PICTURES, GRAPHS, OR OBJECTS THAT ILLUSTRATE CONCEPTS.

Objective J: *Find probabilities involving geometric regions.* (*Lesson 6-6*)

65. A 3-cm square inside a 4-cm square is shown at the right. If a point is selected at random from the figure, what is the probability that it lies in the shaded region?

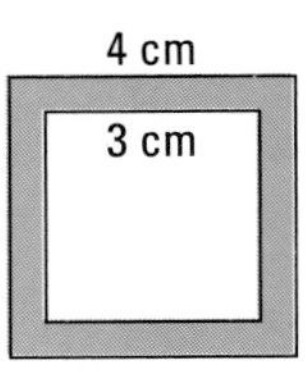

66. The map below shows the roads from town A to town B. Accidents occur randomly on these roads. What is the probability that an accident on these roads occurs on the 8 km stretch of road drawn in blue?

5 km
B
2 km
8 km
1 km
A

67. A target consists of a set of 4 concentric circles with radii of 3″, 6″, 9″, and 12″. The largest circle is inscribed in a square. A sharpshooter shoots blindly so that all points inside the square are equally likely to be hit. A bullet hits somewhere inside the square.

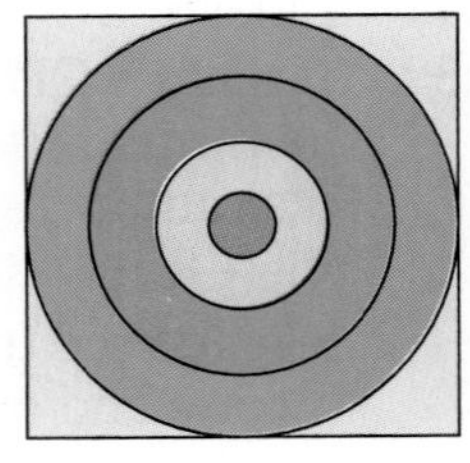

a. What is the probability that it hits the bull's eye?

b. What is the probability that it hits within one of the two outer rings (but not within the innermost rings)?

68. At the right is a picture of a spinner. Suppose all positions of the spinner are equally likely. What is the probability that the spinner lands in region A or B?

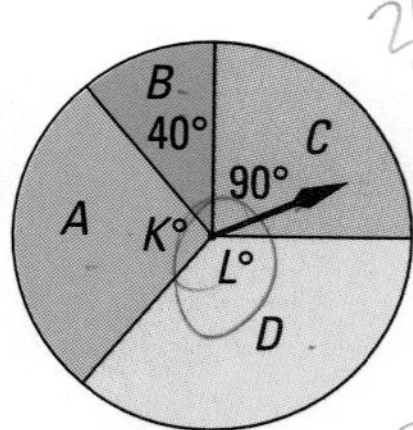

69. In a storm, the electricity sometimes goes out and clocks stop. What is the probability that a clock's second hand will stop between the 5 and the 7?

Objective K: *Apply the Size Change Model for Multiplication.* *(Lesson 6-7)*

70. A drawing of height 6 cm undergoes an expansion of 120%. What is the height of the image?

71. Give the image of (2, 4) under a size change of magnitude 3.

72. Give the image of (-8, -12) under a size change of magnitude $-\frac{1}{4}$.

73. a. Draw $\overline{AB}$, where $A = (2, 4)$ and $B = (-8, -2)$.

b. Draw the image of $\overline{AB}$ under the size change in which the image of (x, y) is $\left(\frac{3}{4}x, \frac{3}{4}y\right)$.

c. Is this size change an expansion or a contraction?

In 74 and 75, copy quadrilateral *ABCD*. Then draw its image under the size change with the given magnitude.

74. $\frac{1}{2}$ **75.** -3

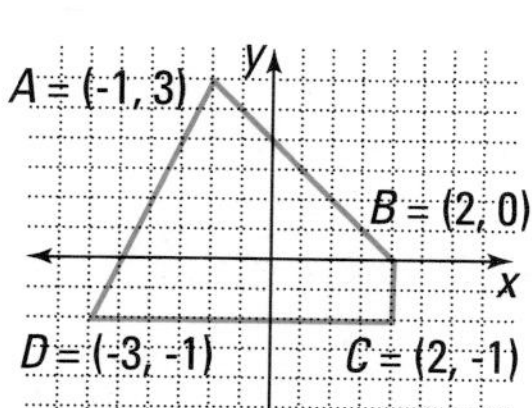

76. If a size change has magnitude -1, what is its effect on the preimage? Justify your answer.

Objective L: *Find lengths and ratios of similitude in similar figures.* *(Lesson 6-9)*

In 77 and 78, refer to the similar rectangles below. Corresponding sides are parallel.

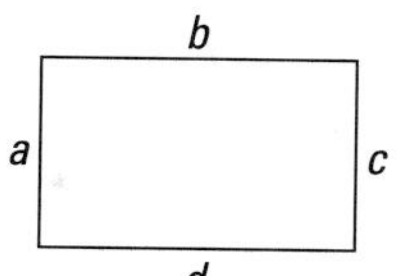

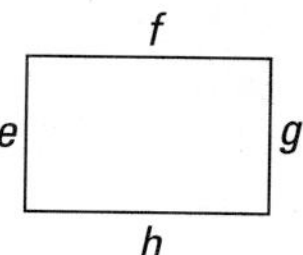

77. *Multiple choice.* Which proportion is incorrect?

(a) $\frac{d}{h} = \frac{c}{g}$ (b) $\frac{e}{a} = \frac{f}{b}$ (c) $\frac{g}{c} = \frac{a}{e}$ (d) $\frac{h}{d} = \frac{f}{b}$

78. If $a = 12$, $b = 15$, and $e = 10$, what is f?

79. One rectangle has dimensions 6 cm by 10 cm; another rectangle has dimensions 5 cm by 9 cm. Are the rectangles similar? Explain your reasoning.

80. Refer to the similar triangles below. Corresponding sides are drawn in the same color.

a. Give the two possible ratios of similitude.

b. Find the length of $\overline{AB}$.

c. Find the length of $\overline{DF}$.

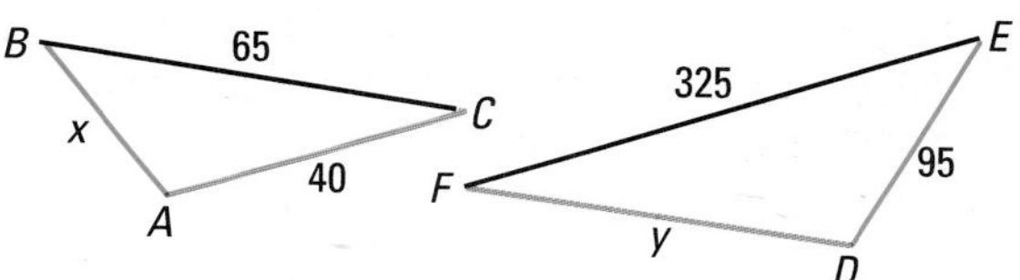

81. The quadrilaterals below are similar. Corresponding sides are parallel.

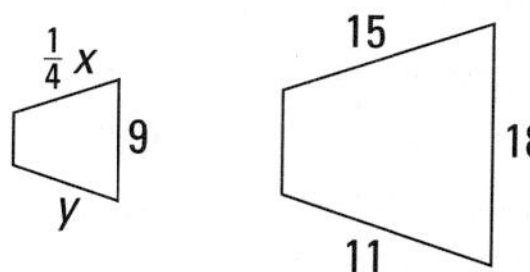

a. Solve for y. **b.** Solve for x.

In 82 and 83 **a.** draw a sketch of the situation; **b.** show how a proportion can be used to solve the problem.

82. A man 6 feet tall casts a shadow 15 feet long. At the same time, the shadow of a tree is 140 feet long. How tall is the tree?

83. A tree casts a shadow that is 9 feet long. A yardstick casts a shadow n feet long. How tall is the tree?

CHAPTER

SLOPES AND LINES

Below is a graph of the population of Manhattan Island (part of New York City) every ten years from 1790 to 1990. Coordinates of some of the points are shown.

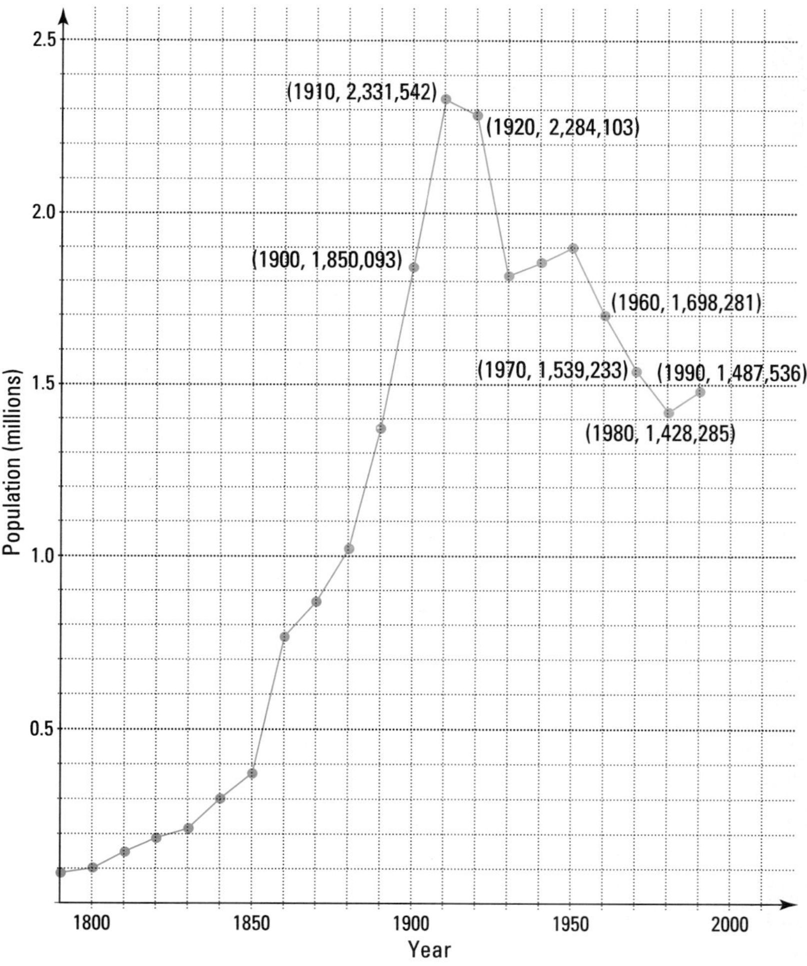

The slopes of the lines connecting the points indicate how fast the population increased or decreased. In this chapter, you will study many examples of lines and slopes.

LESSON

7-1

Rate of Change

Different rates. *The fastest rate of growth for girls usually begins between the ages of 9 and 11. Boys usually begin their growth spurt about 2 years later than girls.*

What Is a Rate of Change?

At age 9 Karen was 4′3″ (4 feet 3 inches) tall. At age 11 she was 4′9″ tall. How fast did she grow from age 9 to age 11? To answer this question, we calculate the *rate of change* of Karen's height, that is, how much she grew per year.

The rate of change in her height per year from age 9 to age 11 is

$$\frac{\text{change in height}}{\text{change in age}} = \frac{4'9'' - 4'3''}{(11 - 9)\text{ years}} = \frac{6\text{ inches}}{2\text{ years}} = 3\,\frac{\text{inches}}{\text{year}}.$$

At age 14 Karen was 5′4″ tall. How fast did she grow from age 11 to age 14? Use the same method. The rate of change in her height per year from age 11 to age 14 is

$$\frac{\text{change in height}}{\text{change in age}} = \frac{5'4'' - 4'9''}{(14 - 11)\text{ years}} = \frac{64'' - 57''}{3\text{ years}} = \frac{7\text{ inches}}{3\text{ years}} = 2.\overline{3}\,\frac{\text{inches}}{\text{year}}.$$

Notice that Karen grew at a faster rate from age 9 to age 11 than from age 11 to 14.

At the right, the data points are plotted and connected. The rate of change measures how fast the segment goes up as you read from left to right along the x-axis.

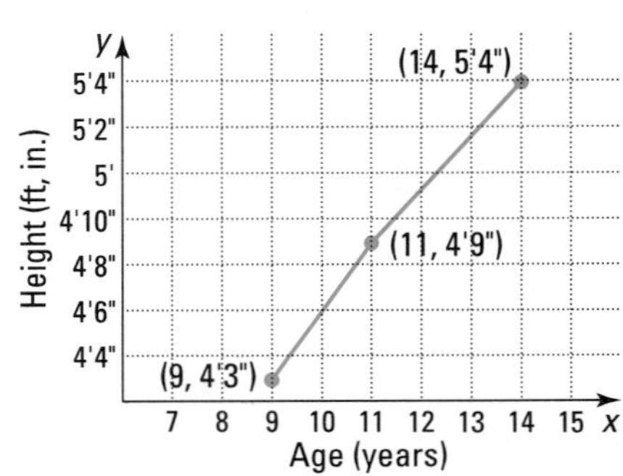

Since the rate of change of Karen's height is greater from age 9 to age 11 than from age 11 to age 14, the segment connecting (9, 4′3″) to (11, 4′9″) is steeper than the one connecting (11, 4′9″) and (14, 5′4″).

As illustrated in the following example, rates of change may be positive or negative.

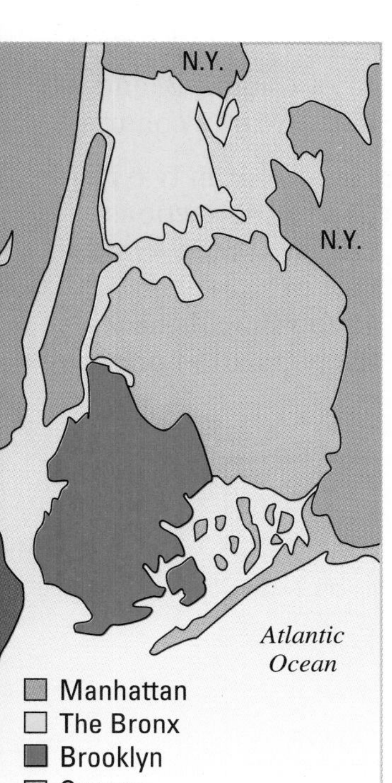

New York City. *New York City consists of five boroughs: The Bronx, Brooklyn, Manhattan, Queens, and Staten Island.*

Example 1

The graph on page 417 shows the population of Manhattan. Find the rate of change of the population of Manhattan (in people per year) during the given time, and tell how that rate is pictured on the graph

a. between 1900 and 1910. **b.** between 1910 and 1920.

Solution

To find the rate of change in people per year, calculate $\frac{\text{change in population}}{\text{change in years}}$.

a. Between 1900 and 1910, the rate of change is

$$\frac{2{,}331{,}542 - 1{,}850{,}093}{1910 - 1900} = \frac{481{,}449 \text{ people}}{10 \text{ years}}$$
$$= 48{,}144.9 \text{ people per year.}$$

Between 1900 and 1910, the population increased. So the rate of change is positive. As you read from left to right, the graph slants upward.

b. Between 1910 and 1920 the rate of change is

$$\frac{2{,}284{,}103 - 2{,}331{,}542}{1920 - 1910} = \frac{-47{,}439 \text{ people}}{10 \text{ years}}$$
$$= -4{,}743.9 \text{ people per year.}$$

Between 1910 and 1920, the population decreased. So the rate of change is negative. Between those dates, as you read from left to right, the graph slants downward.

When you read graphs, read from left to right just as you read prose. A positive rate of change indicates that the graph slants upward. A negative rate of change indicates that the graph slants downward.

There is a general formula for rate of change, in terms of coordinates. In Example 1, the year is the *x*-coordinate on the graph on page 417 and the population is the *y*-coordinate. The rate of change between two points is calculated by dividing the difference in the *y*-coordinates by the difference in the *x*-coordinates. We use the subscripts $_1$ and $_2$ to identify the coordinates of the two points. x_1 is read "*x* one" or "*x* sub one" or "the first *x*."

Definition

The **rate of change** between points (x_1, y_1) and (x_2, y_2) is $\frac{y_2 - y_1}{x_2 - x_1}$.

Because every rate of change comes from a division, the unit of a rate of change is a rate unit. For instance, in Example 1, since a number of people is divided by a number of years, the unit of the rate of change is

$$\frac{\text{number of people}}{\text{year}}.$$

Rates of Change in Tables

Spreadsheets and other table generators can be used to calculate rates of change. For instance, the spreadsheet below shows the years from 1900 to 1990 in column A, the population of Manhattan in column B, and the rate of change of population for the decade ending that year in column C.

Each rate of change is found using years and populations from two rows of the spreadsheet. Each formula in column C does two subtractions, one to find the change in population and one to find the change in years. Then the population change is divided by the change in years. For instance, the formula for C4 is = (B4−B3)/(A4−A3). You should check this. Notice that cell C3 cannot be filled because the population previous to 1900 has not been entered.

	A	B	C
1			Rate of Change
2	Year	Population	for Previous Decade
3	1900	1850093	
4	1910	2331542	48144.9
5	1920	2284103	-4743.9
6	1930	1876412	
7	1940	1889924	
8	1950	1960101	
9	1960	1698281	
10	1970	1539233	
11	1980	1428285	
12	1990	1487536	

A deer problem. *In some places, the deer population has grown too large for its natural habitat. So deer often search outside their area for food.*

Constant Rates of Change

Sometimes quantities do not change over a certain interval.

Example 2

The graph at the right shows the number of deer in a park between 1975 and 1990.

a. During what time period did the deer population not change?

b. Find the rate of change of population during this time.

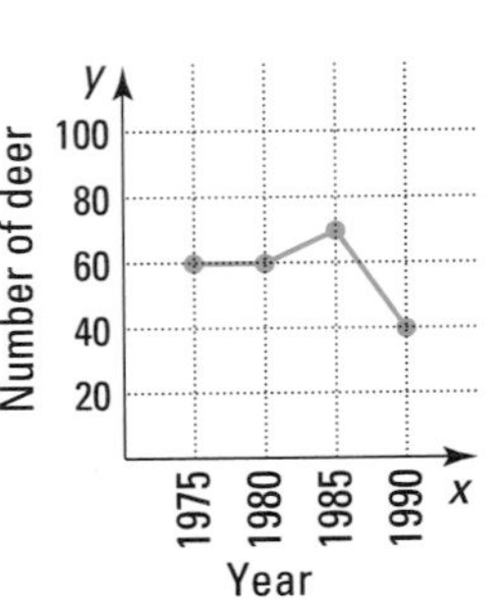

Solution

a. When the number of deer has been constant, there is zero change in the y-coordinate. This means that the line segment is horizontal. The deer population did not change between 1975 and 1980.

b. Use the coordinates of the endpoints, (1975, 60) and (1980, 60), of the horizontal segment.

$$\text{rate of change} = \frac{60 - 60}{1980 - 1975} = \frac{0}{5} = 0\ \frac{\text{deer}}{\text{year}}$$

In general, a rate of change of 0 corresponds to a horizontal segment on a graph.

The following table summarizes the relationship between rates of change and their graphs.

Situation	Rate of Change	Slant (from left to right)	Sketch of Graph
increase	positive	upward	
no change	zero	horizontal	
decrease	negative	downward	

QUESTIONS

Covering the Reading

1. The rate of change in height per year is the change in __?__ divided by the change in __?__.

In 2–4, refer to the situation at the beginning of the lesson about Karen's height. Suppose Karen is 5′5″ tall at age 18 and 5′5″ tall at age 19.

2. a. Copy the graph of her height on page 418. Extend the axes and plot her height at ages 18 and 19.
   b. Which is steeper, the segment connecting (9, 4′3″) to (11, 4′9″) or the segment connecting (11, 4′9″) to (14, 5′4″)?

3. a. What is the rate of change of Karen's height from age 14 to age 18?
   b. Was the rate of change of her height greater from age 9 to age 11 or from age 14 to age 18?

4. What is the rate of change in her height from age 18 to age 19?

In 5–10, use the graph of the population of Manhattan on page 417.

5. If the rate of change in population in a given time period is positive, has the population increased or decreased?

6. Identify all decades in which the rate of change of Manhattan's population was negative.

7. Between 1950 and 1960, did the population increase or decrease?

8. Find the rate of change in population between 1980 and 1990.

9. Find the rate of change in population from 1900 to 1990.

10. During which 10-year period—1850 to 1860 or 1860 to 1870—did the population change more rapidly? Justify your answer.

11. In terms of coordinates, the rate of change between two points is the __?__ of the y-coordinates divided by the difference of the __?__.

12. **a.** Copy the spreadsheet below and fill in the information for the deer populations graphed in Example 2.

	A	B	C
1	Year	Deer	Rate of change in Previous 5-Yr. Period
2			
3			
4			
5			

b. What is the formula in cell C4?
c. What is the value of C4?
d. Which cell cannot be filled in?

13. Describe the graph of the segment connecting two points when the rate of change between them is **a.** positive; **b.** negative; **c.** zero.

Scavenger hunt. *This Rüppell's griffon vulture lives in the dry bush and desert areas of Kenya.*

Applying the Mathematics

In 14–17, use this graph showing the altitude of a vulture over a period of time.

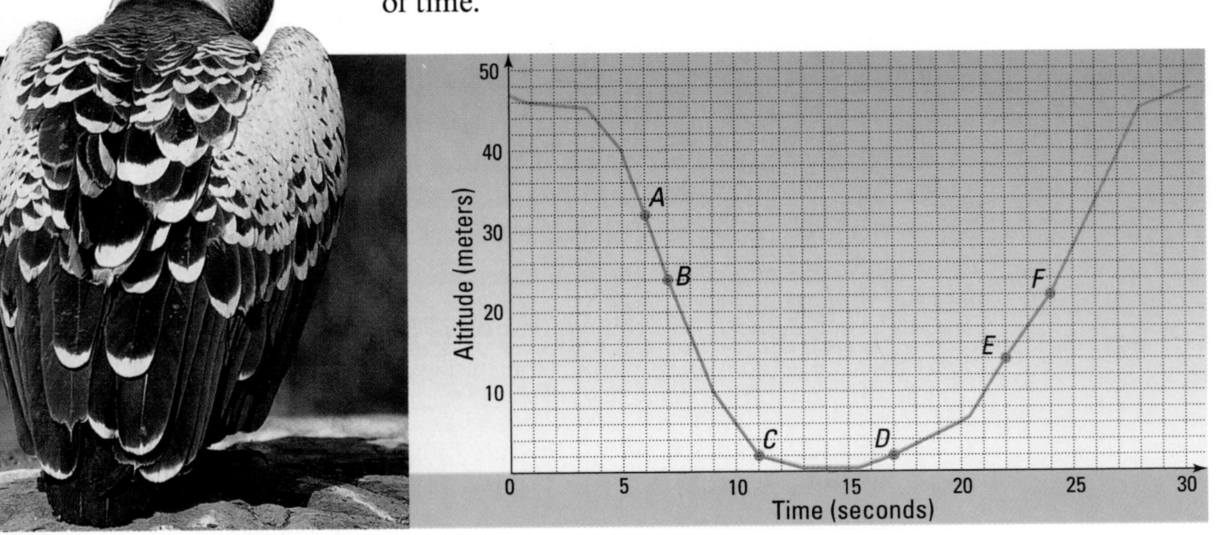

In 14–17, refer to the graph on page 422.

14. **a.** Give the coordinates of point A.
b. Give the coordinates of point B.
c. What is the rate of change of altitude (in meters per second) between points A and B?

15. What is the rate of change of altitude between points C and D?

16. After about how many seconds does the altitude of the vulture begin to increase?

Miniature trees. *Bonsai is the art of growing miniature trees. Growers keep bonsai trees small by pruning the roots and branches, by pinching off new growth, and by wiring the branches. Bonsai, which means tray-planted, originated in China and Japan in the 11th century.*

17. Is the rate of change between E and F positive, negative, or zero?

18. Below are heights (in inches) for a boy from age 9 to age 15.

Age	9	11	13	15
Height	51″	58″	63″	65″

a. Accurately graph these data and connect the points.
b. In which two-year period did the boy grow the fastest? How can you tell?
c. Calculate his rate of growth in that two-year period.

19. Older people tend to lose height. Tim reached his full height at age 20, when he was 74″ tall. He stayed at that height for 35 years, and then started losing height. At age 65 his height was 73″. What was the rate of change of his height from age 55 to age 65?

20. When the bonsai tree was bought, it was 12 years old and h cm tall. When the tree was 15 years old, it was H cm tall.
a. Write an expression for the rate of change in the height of the tree from ages 12 to 15.
b. If the bonsai did not change during these years, what must be true about the rate of change in its height?

21. In the graph at the right, find the rate of change in meters per minute.

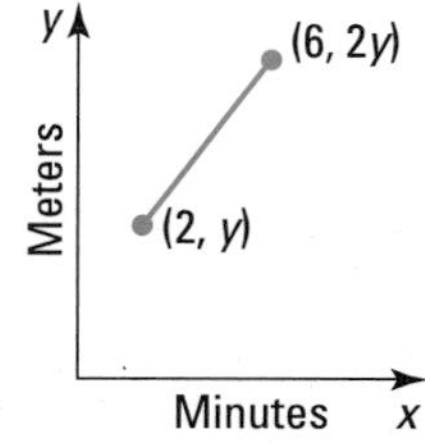

22. The graph below shows the population of Hong Kong at seven different times. Which line segment—$\overline{AB}$, $\overline{BC}$, $\overline{CD}$, $\overline{DE}$, $\overline{EF}$, or $\overline{FG}$—matches the following description?
a. The population decreased.
b. The rate of change is 0.
c. The rate of change was greatest.

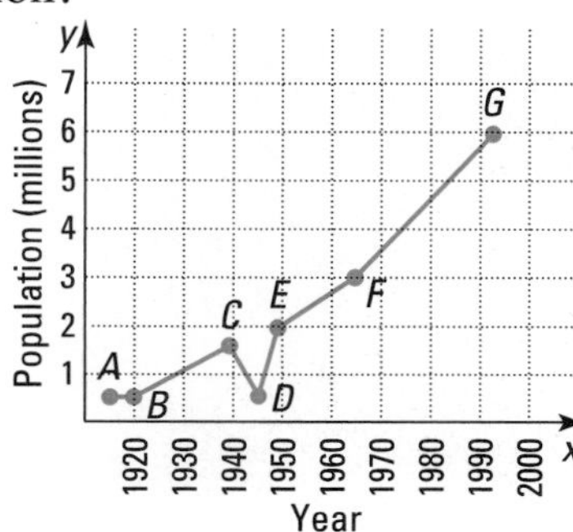

Review

23. *Skill sequence.* *(Lessons 3-9, 5-9, 6-8)*

a. Simplify $\frac{2}{a} - \frac{1}{a}$.

b. Simplify $\frac{2}{3 + 4x} - \frac{1}{3 + 4x}$.

c. Solve $\frac{2}{3 + 4x} - \frac{1}{3 + 4x} = \frac{4}{x}$.

24. Simplify $-\frac{\frac{2}{5}}{4}$. *(Lesson 6-1)*

25. *Skill sequence.* Solve for y. *(Lesson 5-7)*

a. $33 - 4y = 12$ **b.** $3x - 4y = 12$ **c.** $ax - 4y = 12$

26. **a.** Graph $y = 3x + 2$ and $y = -3x + 2$ on the same axes.

b. Where do the two lines intersect? *(Lesson 4-9)*

27. Suppose a stamp collection now contains 10,000 stamps. If it grows at 1000 stamps per year, how many stamps will there be in x years? *(Lesson 3-8)*

28. *True or false.* The rate "3 pizzas for 4 people" is equal to the rate "$\frac{3}{4}$ pizza for 1 person." Explain your answer. *(Lesson 6-2)*

In 29 and 30, evaluate $\frac{y_2 - y_1}{x_2 - x_1}$ at the given values. *(Lesson 1-5)*

29. $y_2 = 5$, $y_1 = 6$, $x_2 = -2$, and $x_1 = -4$

30. $y_2 = 8$, $y_1 = 4$, $x_2 = 2$, and $x_1 = \frac{1}{3}$

Exploration

31. **a.** Find the population of the town or city where you live (or near where you live) in 1940, 1950, 1960, 1970, 1980, and 1990. (You may want to call a local historical society or public library.)

b. In which 10-year period did the population change the most? Was the rate of change positive or negative?

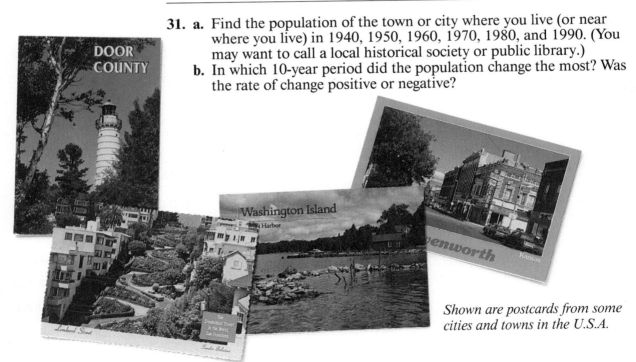

Shown are postcards from some cities and towns in the U.S.A.

LESSON

7-2

The Slope of a Line

Scuba-diving pressure. *As scuba divers descend, pressure increases at a constant rate. For example, in salt water, pressure increases by about 15 pounds per square inch for every 33 feet of descent.*

Consider the following situation. An ant, 12 feet high on a flagpole, walks down the flagpole at a rate of 8 inches (which is $\frac{2}{3}$ foot) per minute.

This is a *constant-decrease* situation. Each minute the height of the ant decreases by $\frac{2}{3}$ foot. You can see the constant decrease by graphing the height of the ant after 0, 1, 2, 3, 4, . . . minutes of walking. Below are the ordered pairs (time, height) charting the ant's progress.

Time (min)	Height (ft)
0	12
1	$11\frac{1}{3}$
2	$10\frac{2}{3}$
3	10
4	$9\frac{1}{3}$
5	$8\frac{2}{3}$
6	8

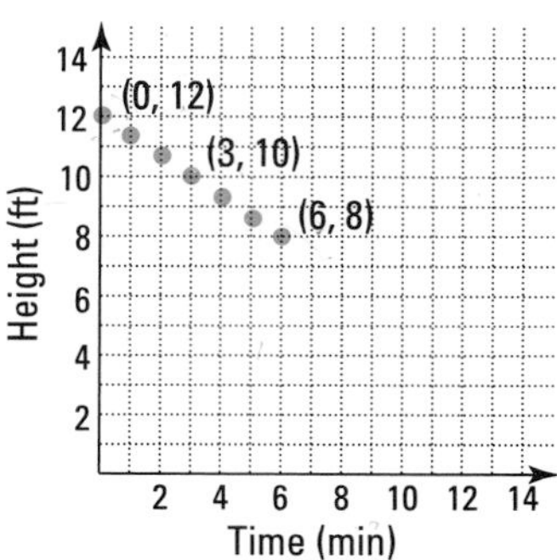

Notice that the rate of change between the points (0, 12) and (3, 10) is

$$\frac{\text{change in height}}{\text{change in time}} = \frac{10 \text{ feet} - 12 \text{ feet}}{3 \text{ minutes} - 0 \text{ minutes}} = \frac{-2 \text{ feet}}{3 \text{ minutes}}.$$

The rate of change is $-\frac{2}{3}$ foot per minute, the same as the ant's rate.

Similarly, after 1 minute the ant is $11\frac{1}{3}$ feet high. After 5 minutes the ant is $8\frac{2}{3}$ feet high. The rate of change of height in $\frac{\text{ft}}{\text{min}}$ between these points is

$$\frac{\text{change in height}}{\text{change in time}} = \frac{8\frac{2}{3} - 11\frac{1}{3}}{5 - 1} = \frac{-2\frac{2}{3}}{4} = \frac{-\frac{8}{3}}{4} = \frac{-8}{3} \cdot \frac{1}{4} = \frac{-2}{3}.$$

As long as the ant has a constant rate of walking, the rate of change in its height will always equal that constant rate.

Notice that all the points on the graph of the ant's height lie on the same line. In *any* situation in which there is a constant rate of change between points, the points lie on the same line.

Contrast this situation to the Manhattan population situation graphed on page 417. There, no three consecutive points lie on the same line. The rates of change of Manhattan's population are different for each decade.

This leads to an important property of lines. On a line the rate of change between any two points is always the same. The constant rate of change is called the *slope* of that line.

Definition
The slope of the line through (x_1, y_1) and (x_2, y_2) is

$$\frac{y_2 - y_1}{x_2 - x_1}.$$

Example 1

Find the slope of line ℓ at the right.

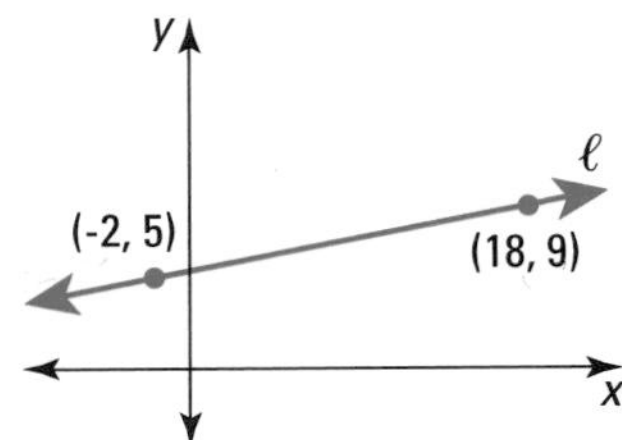

Solution 1

Note that the coordinates of two points are given. Let $(x_1, y_1) = (-2, 5)$ and $(x_2, y_2) = (18, 9)$. Apply the formula for slope.

$$\text{slope} = \frac{9 - 5}{18 - -2} = \frac{4}{20} = \frac{1}{5}$$

Solution 2

Let $(x_1, y_1) = (18, 9)$ and $(x_2, y_2) = (-2, 5)$. Apply the formula for slope.

$$\text{slope} = \frac{5 - 9}{-2 - 18} = \frac{-4}{-20} = \frac{1}{5}$$

Check

As you look from left to right, the graph slants upward. This checks that the slope is positive.

Notice that it does not matter which point is chosen as (x_1, y_1) and which one is (x_2, y_2). The resulting slope is the same.

When lines are drawn on a coordinate grid where each tick mark represents one unit, you can find the slope by counting or by measuring.

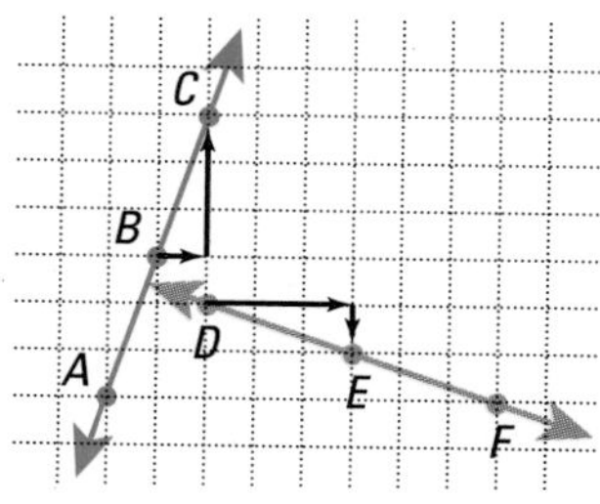

Example 2

In the coordinate grid at the left, the side of each square is one unit. Find the slope of the line.

a. $\overleftrightarrow{AC}$ b. $\overleftrightarrow{DF}$

Solution

a. Pick two points on $\overleftrightarrow{AC}$. We pick points B and C. The change in the x-coordinate as we read from left to right (B to C) is 1 unit. The change in the y-coordinate is 3 units. The slope of $\overleftrightarrow{AC} = \frac{3}{1} = 3$.

b. We pick points D and E. The change in the x-coordinate as we read from left to right (D to E) is 3 units. The change in the y-coordinate is -1 unit. The slope of $\overleftrightarrow{DF} = \frac{-1}{3}$.

Lines with negative slope, such as $\overleftrightarrow{DF}$ in Example 2, go downward as you read from left to right. Lines with positive slope, such as $\overleftrightarrow{AB}$ in Example 2, go upward as you read from left to right.

If the rate of change or slope is the same for a set of points, then all the points lie on the same line. If the rate of change is different for different parts of a graph, the graph is *not* a line.

Example 3

a. Show that (0, 3), (4, 1), and (-8, 7) lie on the same line.
b. Give the slope of that line.

Solution

a. Pick pairs of points and calculate the rate of change between them. The rate of change between (0, 3) and (4, 1) is

$$\frac{1 - 3}{4 - 0} = \frac{-2}{4} = -\frac{1}{2}.$$

The rate of change between (-8, 7) and (4, 1) is

$$\frac{1 - 7}{4 - -8} = \frac{-6}{12} = -\frac{1}{2}.$$

The rate of change between (-8, 7) and (0, 3) is

$$\frac{3 - 7}{0 - -8} = \frac{-4}{8} = -\frac{1}{2}.$$

Since the rate of change between any pair of the given points is $-\frac{1}{2}$, the points lie on the same line.

b. The slope of the line is the constant rate of change, $-\frac{1}{2}$.

Check

Graph the points. They do lie on the same line. The line goes down as you read from left to right, so a negative slope is correct. Counting the change in the units from (0, 3) to (4, 1), we see that the slope is $\frac{-2}{4}$ or $\frac{-1}{2}$.

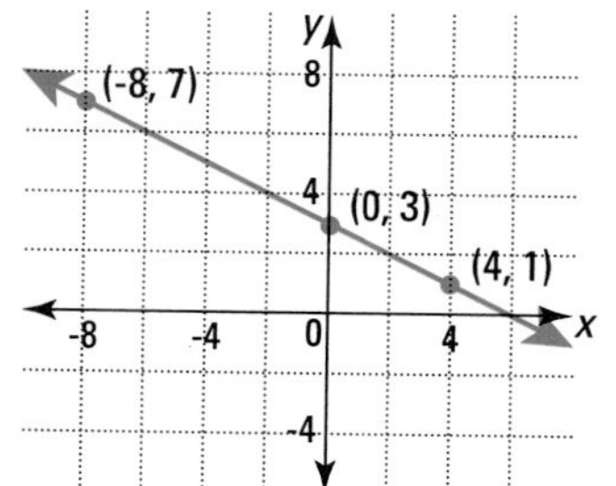

Given an equation for a line, it is easy to find the slope of the line. Just find two points on it and calculate the rate of change between them.

Example 4

Find the slope of the line with equation $3x + 4y = 6$.

Solution

First find two points that satisfy the equation. Pick values for x or y, then substitute into $3x + 4y = 6$. To make the calculations easier, let $x = 0$ for the first point and then $y = 0$ for the second point.

Let $x = 0$:

$$\begin{aligned} 3 \cdot 0 + 4y &= 6 \\ 0 + 4y &= 6 \\ y &= \frac{6}{4} = \frac{3}{2} \end{aligned}$$

The point $(0, \frac{3}{2})$ is on the line.

Let $y = 0$:

$$\begin{aligned} 3x + 4 \cdot 0 &= 6 \\ 3x + 0 &= 6 \\ x &= 2 \end{aligned}$$

The point $(2, 0)$ is on the line.

Substitute $\left(0, \frac{3}{2}\right)$ and $(2, 0)$ into the slope formula.

$$\text{Slope} = \frac{\frac{3}{2} - 0}{0 - 2} = \frac{\frac{3}{2}}{-2} = \frac{3}{2} \cdot -\frac{1}{2} = -\frac{3}{4}$$

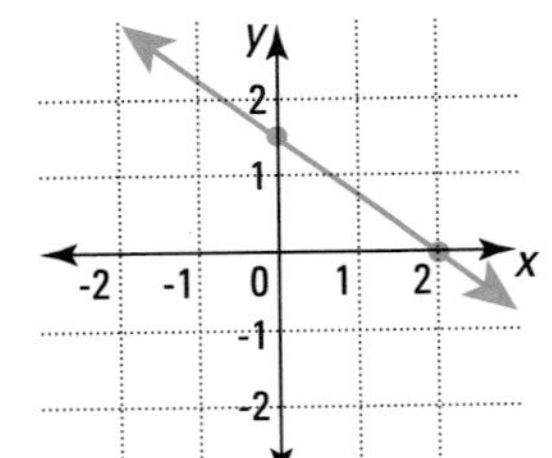

Check 1

Graph the line. Does its slope seem to be $-\frac{3}{4}$? Yes, the graph slants downward, so the slope must be negative.

Check 2

Find another ordered pair that satisfies $3x + 4y = 6$. We substitute $x = 1$ in the equation.

$$\begin{aligned} 3(1) + 4y &= 6 \\ 4y &= 3 \\ y &= \frac{3}{4} \end{aligned}$$

So $(1, \frac{3}{4})$ is on the line. Now calculate the slope determined by it and one of the points in the solution, say $(2, 0)$. This gives $\frac{\frac{3}{4} - 0}{1 - 2}$, which equals $-\frac{3}{4}$. This checks.

QUESTIONS

Covering the Reading

1. In a constant-increase or constant-decrease situation, all points lie on the same ___?___.

2. What is the constant rate of change between any two points on a line called?

3. An equation for the height y of the ant in this lesson after x minutes is $y = -\frac{2}{3}x + 12$.
   a. Find two points on this line not graphed in this lesson.
   b. Find the rate of change between those points.

4. An ant starts 5 feet from the base of a flagpole and climbs $\frac{1}{3}$ foot up the pole each minute.
   a. Graph the ant's progress for the first 6 minutes using ordered pairs (time, height).
   b. Find the rate of change between any two points on the graph.

In 5 and 6, calculate the slope of the line through the given pair of points.

5. (0, 1) and (2, 7)

6. (4, 1) and (-2, 4)

7. a. Calculate the slope of the line through (1, 2) and (6, 11).
   b. Calculate the slope of the line through (6, 11) and (-10, -16).
   c. Do the points (1, 2), (6, 11), and (-10, -16) lie on the same line? How can you tell?

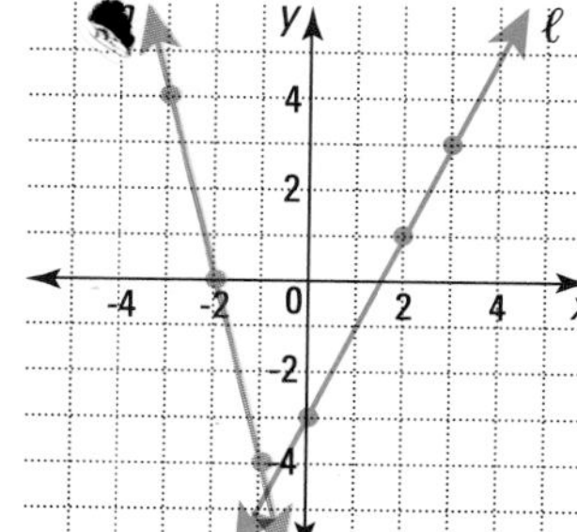

In 8 and 9, refer to the graphs at the left. Find the slope of the line.

8. line ℓ

9. line m

In 10 and 11, an equation is given. **a.** Find two points on the line. **b.** Find the slope of the line. **c.** Check your work by graphing the line.

10. $5x - 2y = 10$

11. $x + y = 6$

Applying the Mathematics

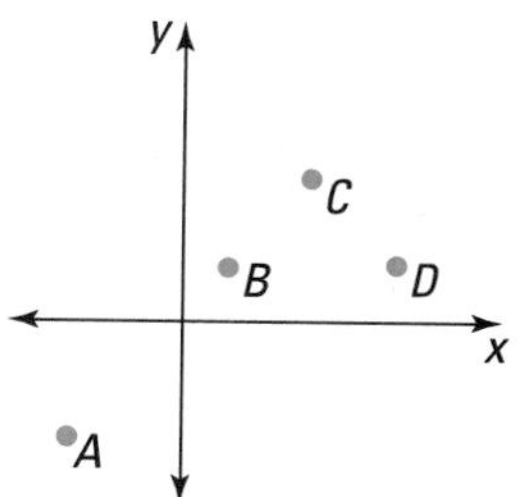

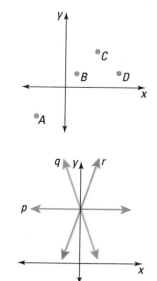

In 12–14, use the figure at the left.

12. The slope determined by points A and B seems to equal the slope determined by B and what other point?

13. The slope determined by __?__ and __?__ is negative.

14. The slope determined by __?__ and __?__ is zero.

15. Consider lines p, q, and r graphed at the left.
   a. Which line has slope $\frac{5}{2}$?
   b. Which line has slope $-\frac{5}{2}$?
   c. Which line has slope zero?

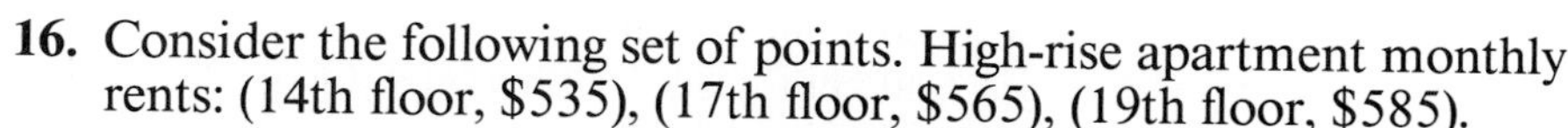

16. Consider the following set of points. High-rise apartment monthly rents: (14th floor, \$535), (17th floor, \$565), (19th floor, \$585).
   a. Calculate the slope of the line through these points.
   b. What does the slope represent in this situation?

17. The points (2, 5) and (3, y) are on the same line. Find y when the slope of the line is -2.

In 18 and 19, coordinates of points are given in columns A and B of a spreadsheet.

a. Fill in the rate-of-change column, leaving C2 blank.

b. Do the points lie on a line? How can you tell?

18.

	A	B	C
1	x	y	rate of change
2	-5	19	
3	0	9	
4	7	-5	
5	10	-11	

19.

	A	B	C
1	x	y	rate of change
2	0	1	
3	4	6	
4	8	7	
5	12	11	

20. Refer to the graph below at the left.

a. Find the slope of the horizontal line $y = 2$.

b. Find the slope of the horizontal line $y = -3$.

c. From parts **a** and **b,** what do you think can be said about the slope of all horizontal lines?

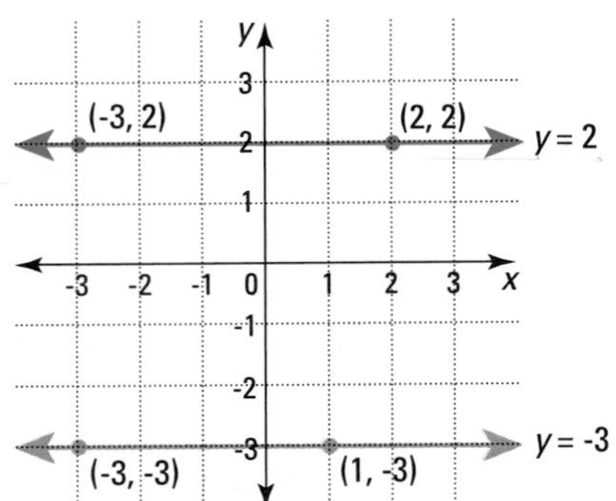

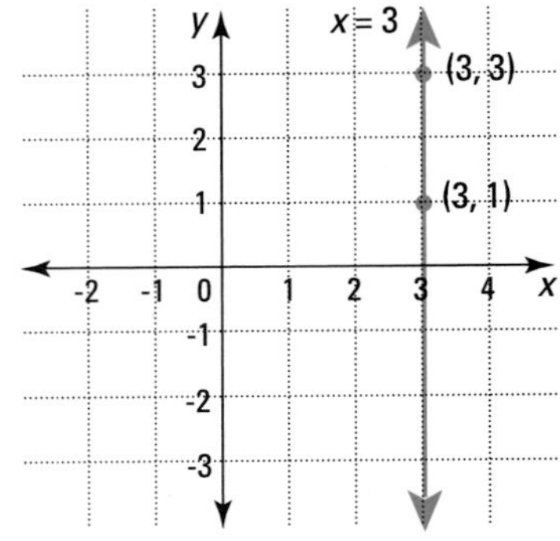

21. Consider the vertical line $x = 3$ graphed above at the right.

a. What happens when you try to find its slope?

b. From your answer to part **a,** do you think that vertical lines have slope?

22. The program below determines the slope of a line given two points.

```
10 PRINT "DETERMINE SLOPE FROM TWO POINTS"
20 PRINT "GIVE COORDINATES OF FIRST POINT"
30 INPUT X1, Y1
40 PRINT "GIVE COORDINATES OF SECOND POINT"
50 INPUT X2, Y2
60 LET M = (Y2 − Y1)/(X2 − X1)
70 PRINT "THE SLOPE IS "; M
80 END
```

What would you input at lines 30 and 50 to answer Question 5?

Review

23. Below is a graph showing the cost (in cents) of sending a one-ounce first-class letter. The graph is shown in 5-year intervals beginning in 1960. *(Lesson 7-1)*

a. During which 5-year period did the cost of postage increase the most?

b. During which 5-year periods was the increase in postage the same?

c. Calculate the average increase per year from 1965 to 1975.

d. Give a reason why the points on the graph are not connected.

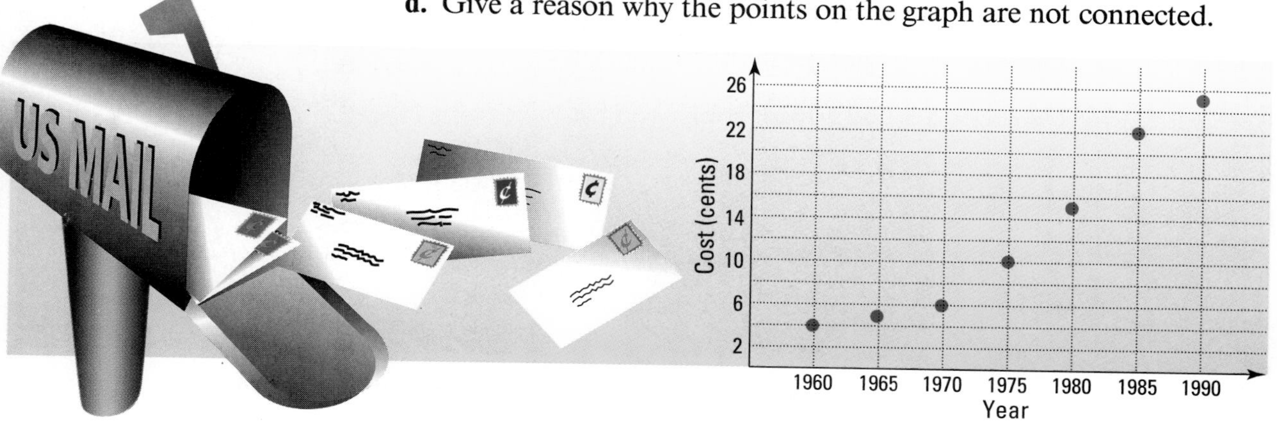

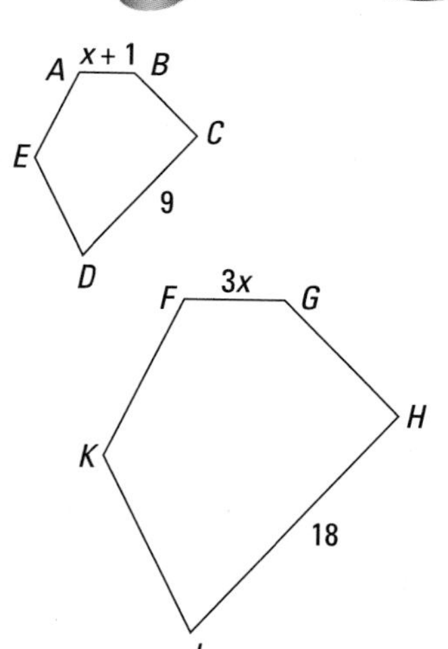

24. The two figures at the left are similar. Corresponding sides are parallel. Find AB and FG. *(Lessons 6-8, 6-9)*

In 25 and 26, graph the line. *(Lessons 4-9, 5-1)*

25. $y = 5x$

26. $y = 5$

27. Is the point (7, 6) on the line $y = 2x - 5$? Explain your reasoning. *(Lessons 4-6, 4-9)*

In 28 and 29, write an expression for the amount of money the person has after x weeks. *(Lesson 3-8)*

28. Eddie is given \$100 and spends \$4 a week.

29. Gretchen owes \$350 on a stereo and is paying it off at \$5 a week.

Exploration

30. Find a record of your height at some time over a year ago. Compare it with your height now. How fast has your height been changing from then until now?

LESSON

7-3

Properties of Slope

Land movers. *Bulldozers are machines used in grading and excavating land for construction projects such as highways. In road construction, the slant or angle of the roadway is carefully calibrated to meet certain specifications.*

A test ramp for bulldozers goes down 0.6 foot for each foot it goes across, as illustrated below at the left.

The part of the ramp with the triangle is enlarged and graphed below at the right. Two points are shown on the graph. The rate of change between these points is the slope of the line.

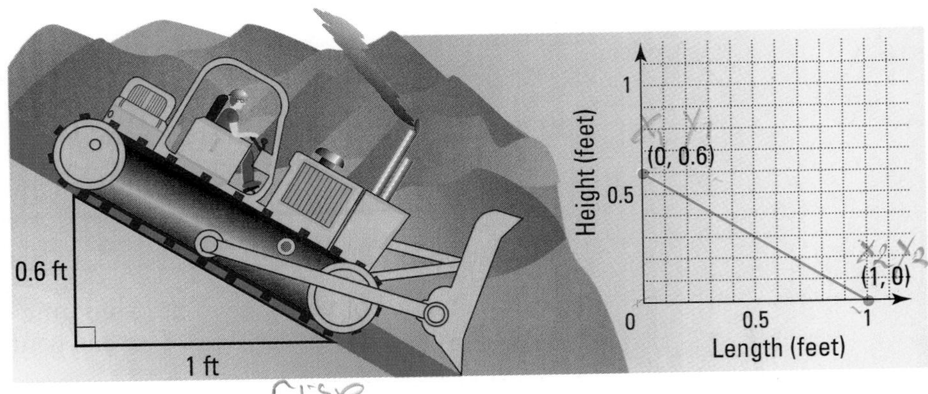

$$\text{slope} = \frac{\text{vertical change}}{\text{horizontal change}} = \frac{\text{change in height}}{\text{change in length}} = \frac{0 - 0.6}{1 - 0} = -0.6$$

This verifies an important property of slopes of lines that is given at the top of page 433.

The slope of a line is the amount of change in the height of the line for every change of one unit to the right.

For instance, if a line has slope 3, as you move one unit to the right, the line goes up three units.

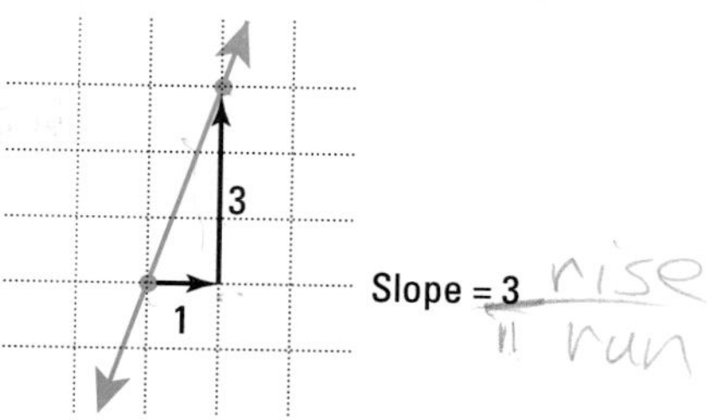

If the slope is $-\frac{3}{2}$, for every change of one unit to the right, the line goes down $\frac{3}{2}$ units.

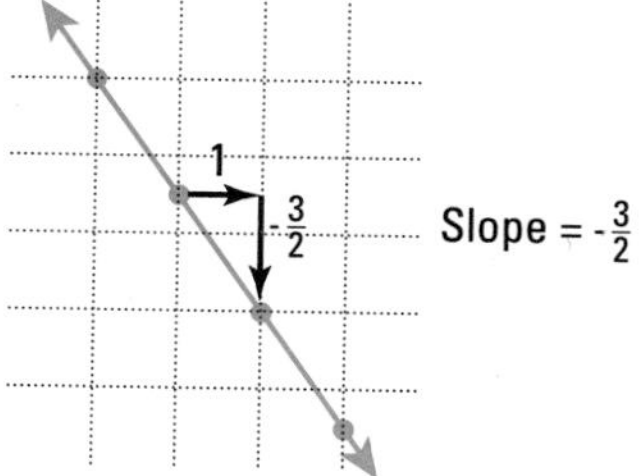

Graphing Lines by Using Properties of Slope

Example 1

Graph the line which passes through (3, −1) and has a slope of −2.

Solution

Plot the point (3, −1). Since the line has slope −2, start at (3, −1), and then move one unit to the right and two units down. Plot that point and draw the line.

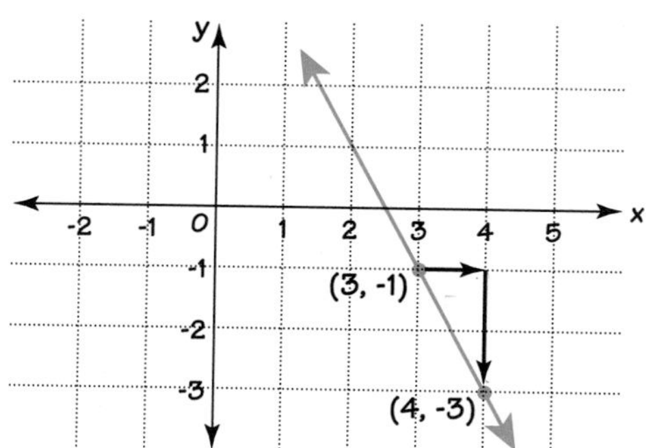

To graph lines having slopes that are given as fractions, it may be helpful to split unit intervals.

Example 2

a. Graph the line through (-2, 1) with slope $\frac{1}{4}$.

b. Name another point on the line with integer coordinates.

Solution 1

Draw the axes with the unit intervals split into fourths. Plot (-2, 1), then move right 1 unit and up $\frac{1}{4}$ unit. Plot the resulting point, $\left(-1, 1\frac{1}{4}\right)$, and draw the line through the two points.

a.

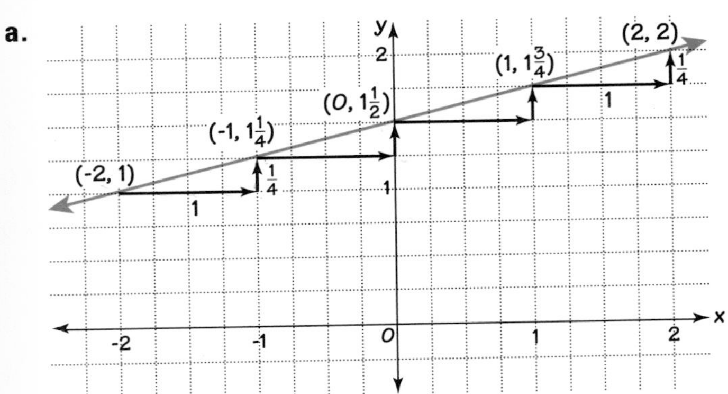

b. One point with integer coordinates (-2, 1) is given. Continue plotting points by moving right 1 unit and up $\frac{1}{4}$ unit, until you reach another point with integral coordinates. As shown in the graph, the point (2, 2) is also on the line.

Solution 2

a. Plot (-2, 1) as in Solution 1. However, instead of going across 1 and up $\frac{1}{4}$, go across $4 \cdot 1$ and up $4 \cdot \frac{1}{4}$. That is, go across 4 and up 1 to the point (2, 2). The graph is shown below.

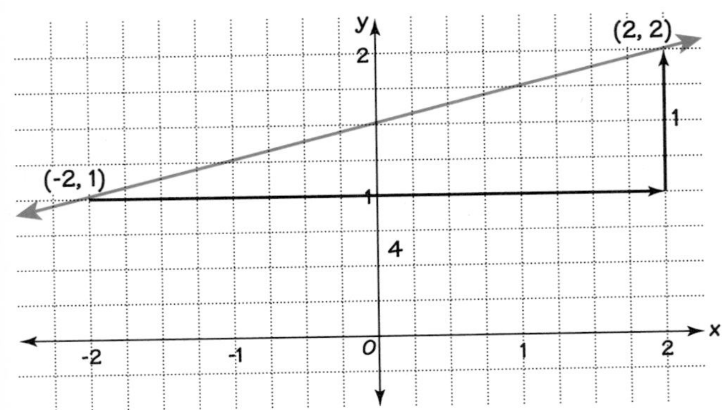

b. The solution to part a shows that (2, 2) is also on the line.

Sky-high cable cars. *The upward path of a cable car illustrates a situation of constant increase. This cable car takes tourists up to Masada, an ancient fortress in Israel atop a huge rock that is almost 430 m tall. The Dead Sea is seen in the background.*

The idea in Examples 1 and 2 helps in drawing graphs of situations of constant increase.

Example 3

Beginning in 1991, postage rates for first-class mail were \$.29 for the first ounce and \$.23 for each additional ounce up to 10 ounces. Graph the relation between the weight of a letter and its mailing cost for whole-number weights.

Solution

The starting point is (1, 0.29). Because the rate goes up \$.23 for each ounce, the slope is \$.23/oz. The points are on a line because the situation is a constant-increase situation.

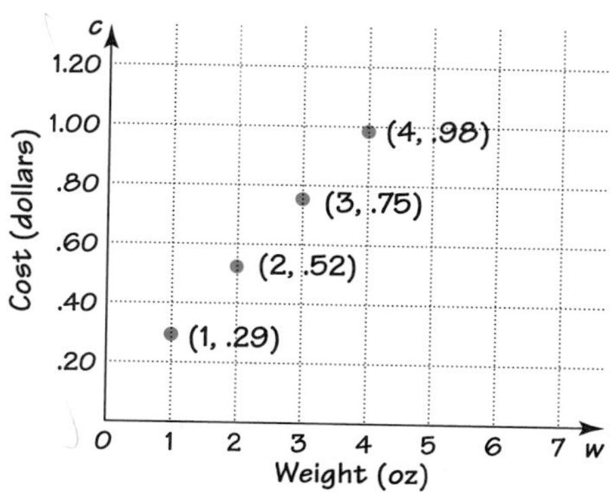

An equation for the cost C to mail a first-class letter weighing w whole ounces is $C = 0.29 + 0.23(w - 1)$. This is an equation for the line that contains the points graphed in Example 3. Notice that the given numbers 0.29 and 0.23 both appear in this equation. In the next lesson, you will learn how the slope can be easily determined from the equation.

Zero Slope and Undefined Slope

Lines with slope 0 may be drawn using the method of Examples 2 and 3.

Example 4

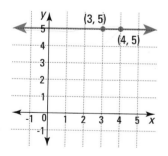

Draw a line through the point (3, 5) with slope 0.

Solution

Plot the point (3, 5). From this point move one unit to the right, and 0 units up. Draw the line through these two points as shown at the left.

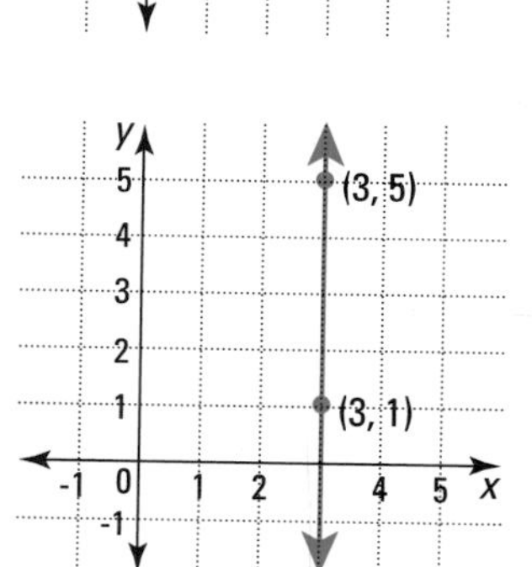

There is one kind of line for which the methods of Examples 2 and 3 does not work. You cannot go to the right on a vertical line. Furthermore, if you try to calculate the slope from two points on a vertical line, the denominator will be zero. For instance, to find the slope of the vertical line through (3, 5) and (3, 1), you would be calculating $\frac{5-1}{3-3} = \frac{4}{0}$, which is not defined. Thus the slope of a vertical line is *not defined.*

The slope of every horizontal line is 0.
The slope of every vertical line is undefined.

QUESTIONS

Covering the Reading

In 1 and 2, a test ramp for bulldozers goes down 0.5 unit for each unit across.

1. What is its slope?

2. Draw a picture of the test ramp.

3. Slope is the amount of change in the __?__ of a graph for every change of one unit to the right.

4. If the slope of a line is 8, then you move __?__ as you move one unit to the right.

In 5 and 6, give the slope of the line.

5.

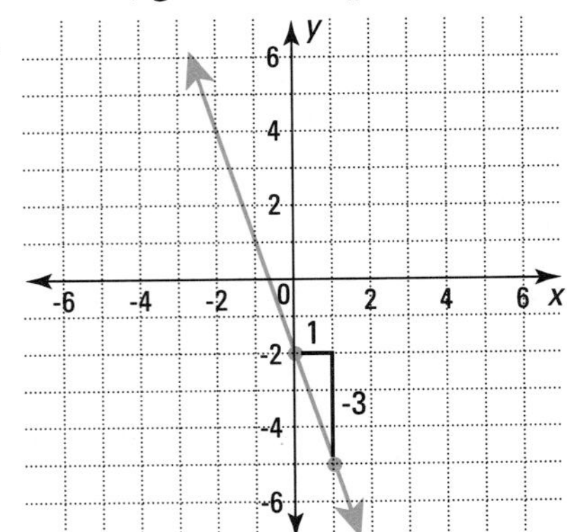

6.

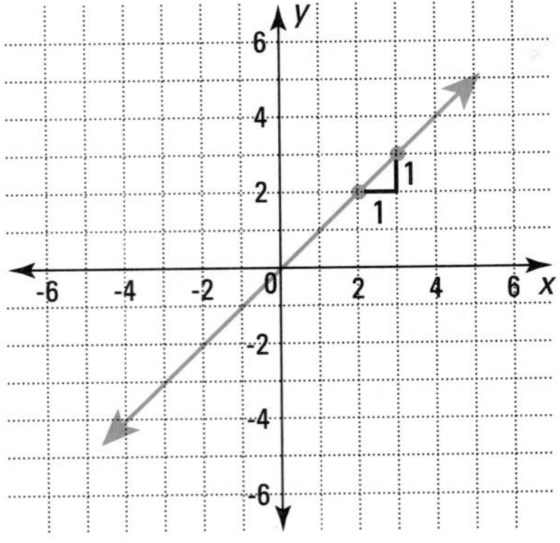

In 7–10, a point and slope are given. **a.** Graph the line with the given slope that goes through the given point. **b.** Find one other point on this line.

7. point (0, 2), slope 3

8. point (-5, 6), slope -4

9. point (-1, 2), slope $-\frac{2}{5}$

10. point (-2, 1), slope 0

11. A line has slope $\frac{1}{3}$ and passes through the point (-4, -2). Name one other point with integer coordinates on this line.

12. In 1991, what was the cost to mail a letter weighing 9 oz?

13. In 1987, postal rates were 22¢ for the first ounce and 17¢ for each additional ounce up to 10 ounces. Graph at least five points showing the relation between weight and cost for whole-number weights.

14. Explain why the slope of a vertical line is not defined. Include a specific example not in the text.

15. Graph the line through (-2, 1) with an undefined slope.

Applying the Mathematics

In 16–18, the slope of a line is given. Fill in the cells of the spreadsheets so that the coordinates are points on the line.

16. slope = 7 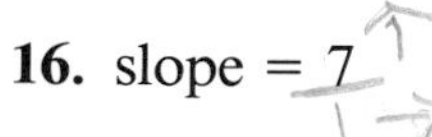

	A	B
1	x	y
2	-5	2
3	-4	
4	-3	
5	-2	

17. slope = -8

	A	B
1	x	y
2	6	10
3	7	
4	8	
5	9	

18. slope = $\frac{5}{4}$

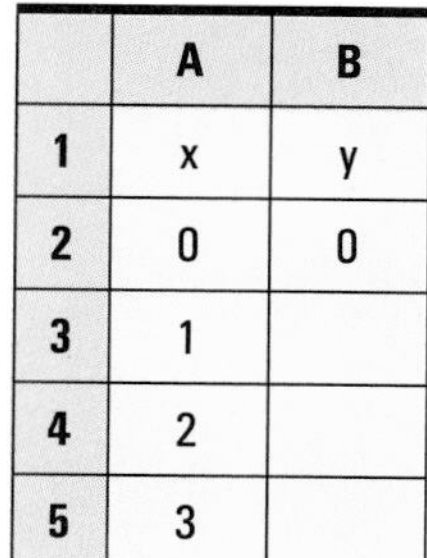

	A	B
1	x	y
2	0	0
3	1	
4	2	
5	3	

19. The steepest street in the world is Baldwin Street in Dunedin, New Zealand, where the maximum *gradient* is 1 in 1.266. This means that for every horizontal change of 1.266 units, the road goes 1 unit up. Assume that $\overleftrightarrow{UP}$ represents Baldwin Street. What is the slope of $\overleftrightarrow{UP}$?

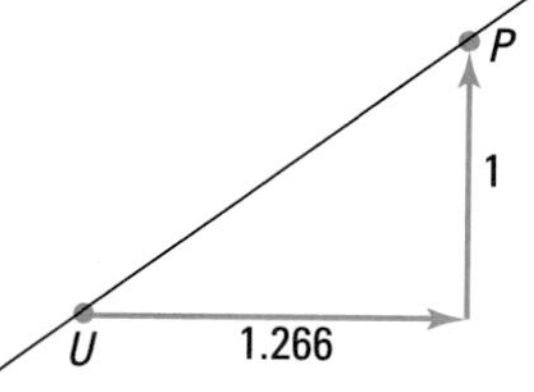

Hills of Dunedin. *In addition to having the world's steepest street, Dunedin is also known for its cathedrals, chocolate factories, and scenic railway station.*

20. Kareem collects comic books. His budget allows him to buy 3 comic books a week. This year, after 10 weeks he has 66 comic books in his collection. Draw a graph to represent the growth of Kareem's collection over the weeks.

21. **a.** Draw a set of coordinate axes and plot the point (0, 0).
b. Draw the lines through (0, 0) with the following slopes:
line *p*: slope 2; line *q*: slope -2;
line *r*: slope $\frac{1}{2}$; line *s*: slope $-\frac{1}{2}$.
c. Describe how the slopes of the lines are related to the appearance of their graphs.

Review

In 22 and 23, calculate the slope determined by the two points. *(Lesson 7-2)*

22. (5, 3) and (8, -2)

23. (-6, -9) and (-13, 5)

24. Find the slope of the line $3x - y = 15$. *(Lesson 7-2)*

25. A rental truck costs $39 for a day plus $.25 a mile. After *x* miles, the total cost will be *y* dollars.
a. Write an equation relating *x* and *y*.
b. Graph the line.
c. Find the rate of change between any two points on the graph. *(Lessons 3-8, 4-9, 7-2)*

In 26–28, refer to the table and graph below. *(Lesson 7-1)*

Average Home Runs per Game

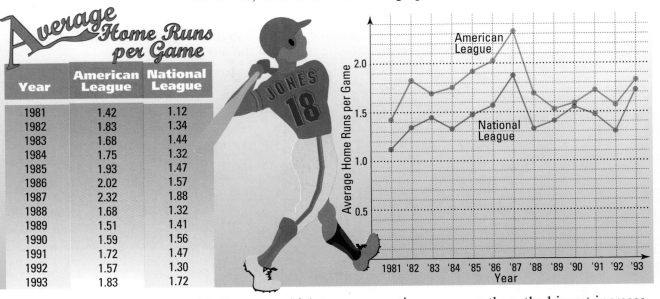

Year	American League	National League
1981	1.42	1.12
1982	1.83	1.34
1983	1.68	1.44
1984	1.75	1.32
1985	1.93	1.47
1986	2.02	1.57
1987	2.32	1.88
1988	1.68	1.32
1989	1.51	1.41
1990	1.59	1.56
1991	1.72	1.47
1992	1.57	1.30
1993	1.83	1.72

26. Between which two consecutive years was there the biggest increase in average home runs per game for the
a. American League? **b.** National League?

27. Between which two consecutive years was there the biggest decrease in average home runs per game for the
a. American League? **b.** National League?

28. What is the rate of change from 1981 to 1990 in average home runs per game for the
a. American League? **b.** National League?

29. If $4a = 15$, find $12a - 5$. *(Lesson 5-8)*

30. If the triangle below has perimeter 71, what is z? *(Lessons 3-5, 4-6)*

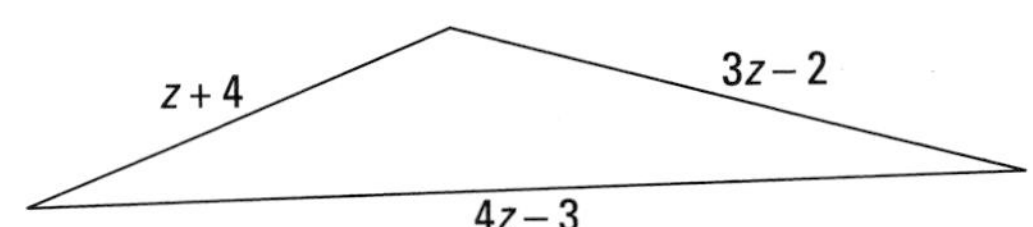

31. Calculate in your head. *(Lesson 3-7)*
a. 15% of \$8.00 **b.** 200% of x dollars

32. Find b if $y = mx + b$, $y = -4$, $m = 3$, and $x = 2$. *(Lessons 1-5, 3-7)*

Exploration

33. Here is a famous puzzle. Beware of the trick. Henri Escargot, a snail, is $14\frac{1}{2}$ feet deep in a well. Every day he climbs 3 feet up the walls of the well. At night the walls are damper and he slips down 2 feet. How many days will it take him to climb out of the well? Explain how you arrived at your answer.

LESSON

7-4

Slope-Intercept Equations for Lines

Saving for the future. *Many people, such as this painter, save a fixed amount of their weekly earnings in a savings account. When a bank balance grows at a constant rate, the savings can be represented by a graph of a linear equation.*

Shawn has \$300 saved for a used car and saves an additional \$25 a week. After 3 weeks, Shawn will have $300 + 3 \cdot 25$ dollars, or \$375. After x weeks, Shawn will have y dollars, where $y = 300 + 25x$.

The line $y = 300 + 25x$ is graphed below. There are two key numbers in the equation for this line. The number 25 is the slope of the line. The number 300 indicates where the line crosses the y-axis. That number is the y-intercept of the line. In general, when a graph intersects the y-axis at the point $(0, b)$, the number b is a **y-intercept** for the graph.

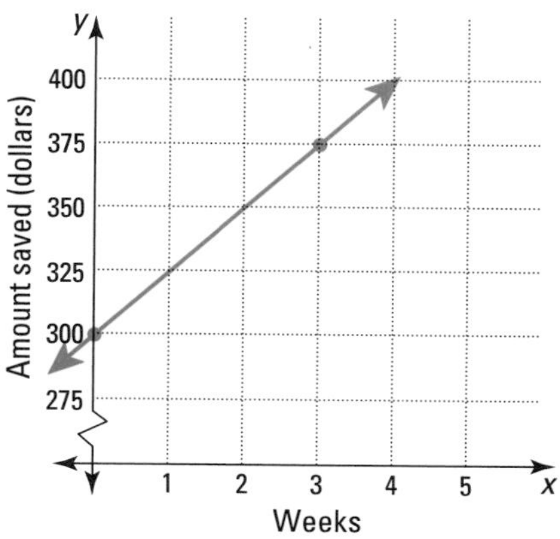

The graph of this line is completely determined by its slope, 25, and y-intercept, 300. When the equation is rewritten in the form $y = 25x + 300$, it is said to be in *slope-intercept* form.

Slope-Intercept Property
The line with equation $y = mx + b$ has slope m and y-intercept b.

The form $y = mx + b$ is called the **slope-intercept form** for the line.

Finding the Slope and y-intercept from the Equation of a Line

Example 1

Give the slope and the y-intercept of $y = \frac{1}{2}x + 4$.

Solution

The equation is already in slope-intercept form. Compare $y = \frac{1}{2}x + 4$ to $y = mx + b$. $m = \frac{1}{2}$ and $b = 4$. The slope is $\frac{1}{2}$ and the y-intercept is 4.

Check

A y-intercept of 4 means the line must contain the point (0, 4). Does (0, 4) satisfy $y = \frac{1}{2}x + 4$? Yes, because $4 = \frac{1}{2} \cdot 0 + 4$. A slope of $\frac{1}{2}$ implies that the point (2, 5) is also on the line. Does $5 = \frac{1}{2}(2) + 4$? Yes; it checks.

The advantage of slope-intercept form is that it tells you much about the line. So it is often useful to convert other equations for lines into slope-intercept form.

Example 2

Give the slope and y-intercept of $y = -8 - 3x$.

Solution

Rewrite $y = -8 - 3x$ in the form $y = mx + b$ using the definition of subtraction and the Commutative Property of Addition.

$$y = -3x - 8$$

The slope is -3 and the y-intercept is -8.

Check

Draw a graph of $y = -8 - 3x$. Verify that (0, -8) is on the graph. Note that the point (1, -11) is also on the graph. This verifies that the slope is -3.

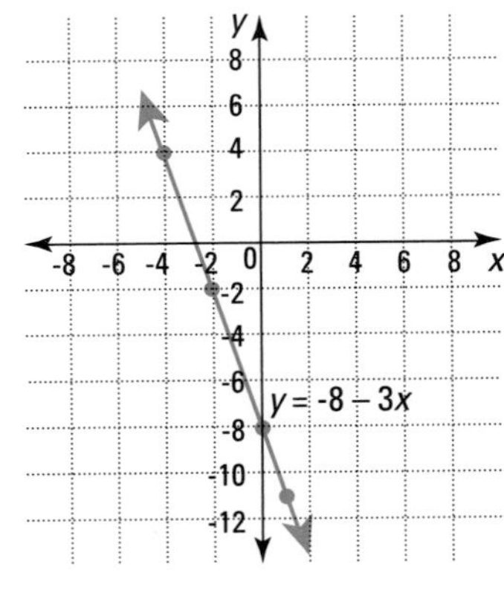

Being able to change the form of an equation is an important skill. Many automatic graphers will not accept the equation $3x + 4y = 9$ for graphing. It needs to be solved for y. On such graphers, the slope-intercept form of a linear equation is acceptable.

Example 3

Write the equation $3x + 4y = 9$ in slope-intercept form. Give the slope and y-intercept of the line.

Solution

Solve $3x + 4y = 9$ for y.

$$3x + 4y = 9$$
$$4y = -3x + 9$$
$$\frac{4y}{4} = \frac{-3x + 9}{4}$$
$$y = -\frac{3}{4}x + \frac{9}{4}$$

The slope is $-\frac{3}{4}$. The y-intercept is $\frac{9}{4}$ or $2\frac{1}{4}$.

Writing an Equation for a Line with Given Slope and y-intercept

Equations for all nonvertical lines can be written in slope-intercept form.

Example 4

a. Write an equation for the line with slope 5 and y-intercept -1.
b. Graph the line.

Solution

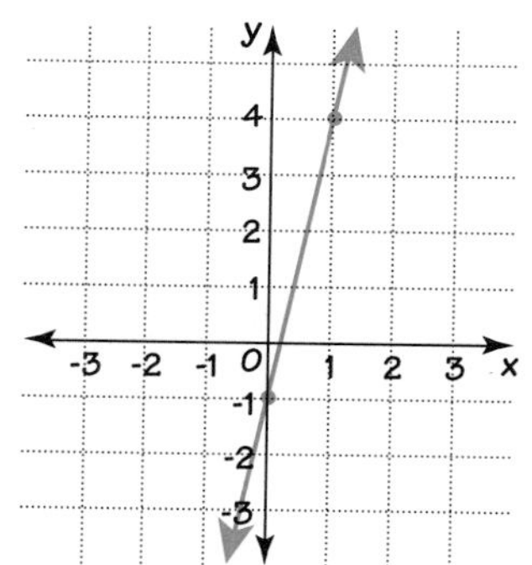

a. Use $y = mx + b$. Substitute 5 for m and -1 for b.
$y = 5x + -1$ or $y = 5x - 1$ is an equation for the line.

b. Since the y-intercept is -1, the graph contains (0, -1). Plot this point. Since the slope is 5, the graph goes up 5 units for each unit it goes to the right. Plot another point 1 unit over and 5 units up from (0, -1). That point is (1, 4).

Check

Both points (0, -1) and (1, 4) satisfy the equation $y = 5x - 1$; and the slope determined by the points is 5. This agrees with the given information.

Every constant-increase or constant-decrease situation can be described by an equation whose graph is a line. The y-intercept of that line can be interpreted as the starting amount. The slope of that line is the amount of increase or decrease per unit.

Example 5

Pam received \$100 for her birthday and spends \$4 of it a week.

a. Find an equation for the amount y she has after x weeks.

b. What are the slope and the y-intercept of the graph?

Solution

a. The equation is found by methods you have learned in previous chapters.

$$y = 100 - 4x$$

b. Rewrite in slope-intercept form.

$$y = -4x + 100.$$

The slope is -4 and the y-intercept is 100.

Recall that every vertical line has an equation of the form $x = h$ where h is a fixed number. Equations of this form clearly cannot be solved for y. Thus, equations of vertical lines cannot be written in slope-intercept form. (They cannot be graphed on many automatic graphers.) This confirms that the slope of vertical lines cannot be defined.

Rising graduation rates. *In 1950, only about 59% of teenagers in the U.S. graduated from high school with their age group. By 1993, that figure had risen to about 74%.*

QUESTIONS

Covering the Reading

1. In what form is the equation of the line $y = mx + b$?

In 2 and 3, an equation of a line is given. **a.** What is its slope? **b.** What is its y-intercept? **c.** Graph the line.

2. $y = 4x + 2$

3. $y = -\frac{1}{3}x + 6$

In 4–7, an equation of a line is given. **a.** Rewrite the equation in slope-intercept form. **b.** Give its slope. **c.** Give its y-intercept.

4. $y = 7.3 - 1.2x$

5. $x + 6y = 7$

6. $3x + 2y = 10$

7. $y = x$

In 8 and 9, write an equation of the line with the given characteristics.

8. slope -3, y-intercept 5

9. slope $\frac{2}{3}$, y-intercept -1

10. Suppose you receive \$100 for a graduation present, and you deposit it in a savings account. Then each week thereafter, you add \$5 to the account but no interest is earned.

a. Find an equation for the amount y you have after x weeks.

b. Draw a graph of the equation in part **a.**

c. What are the slope and the y-intercept of the graph in part **b?**

11. When a constant-increase situation is graphed, the y-intercept can be interpreted as the _?_.

12. Equations of _?_ lines cannot be written in slope-intercept form.

Applying the Mathematics

In 13–15, consider these three graphs. **a.** Match the situation with its graph. **b.** Give the slope of the line. **c.** Give the y-intercept.

(i)

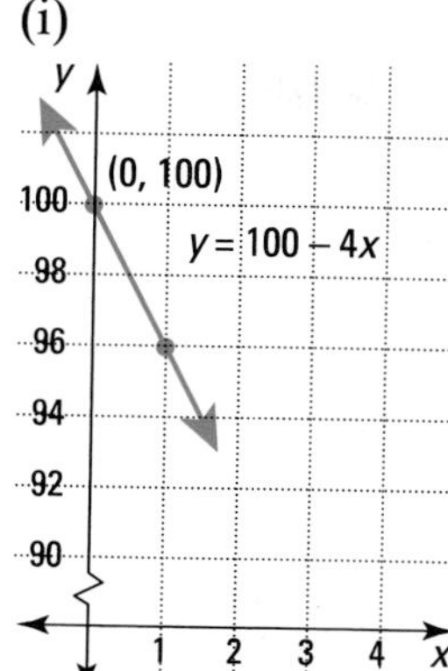

(ii)

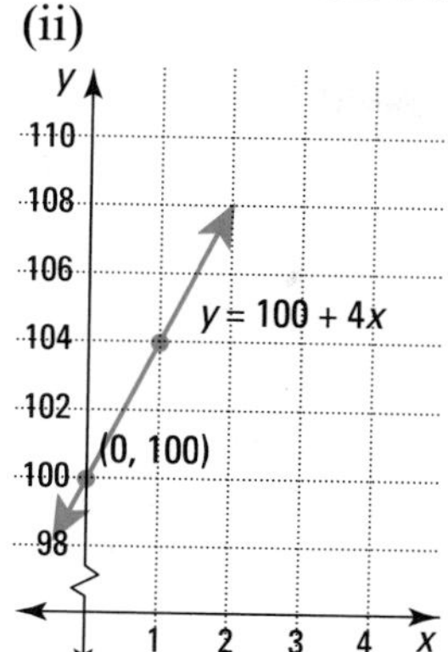

(iii)

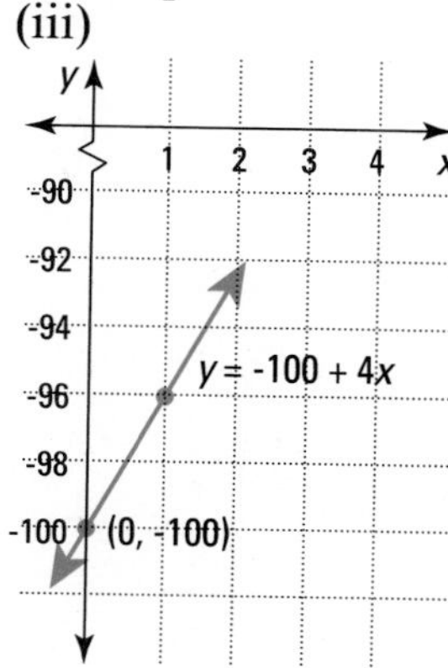

13. Shawn began with \$100 and saves \$4 a week.

14. Pam received \$100 for her birthday and spends \$4 a week.

15. The Carter family owes \$100 on luggage and is paying it off at \$4 a week.

16. Write an equation of the horizontal line with y-intercept -4.

In 17 and 18, each situation leads to an equation of the form $y = mx + b$. **a.** Give the equation. **b.** Graph the equation.

17. Begin with \$8.00. Collect \$.50 per day.

18. A boat is 9 miles from you. It travels towards you at a rate of 6 mph.

19. Match each line n, p, q, and r with its equation.

a. $y = x$ **b.** $y = x + 1$

c. $y = x + 3$ **d.** $y = x - 3$

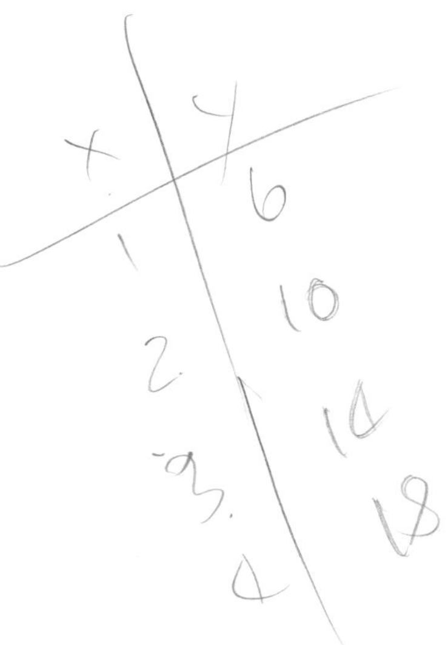

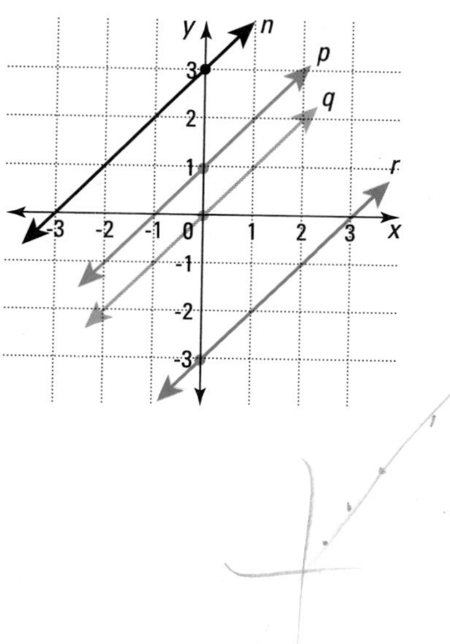

Shown is the Katoomba Scenic Railway located in the Blue Mountains of New South Wales, Australia. The railway is a series of cable cars that drops at an angle close to 52° into the Jamison Valley.

20. Match each line s, t, u, and v with its equation.

a. $y = 2x - 2$ **b.** $y = \frac{2}{3}x - 2$

c. $y = -2x - 2$ **d.** $y = -\frac{2}{3}x - 2$

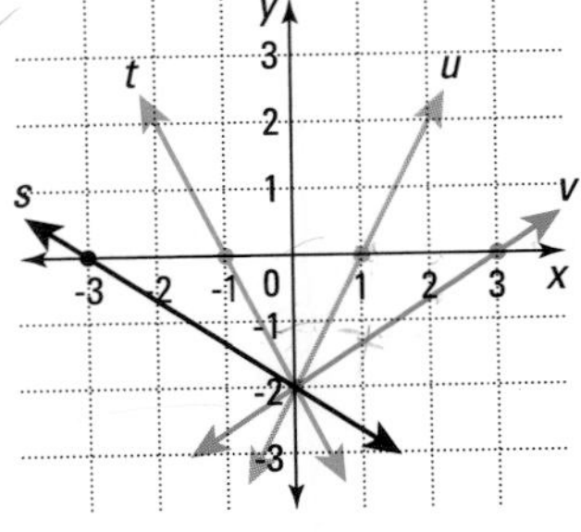

21. An automatic grapher may be helpful in this question.

a. Graph the lines with equations $y = 3x + 5$, $y = 4x + 5$, and $y = 5x + 5$.

b. Describe the graph of $y = mx + 5$.

c. Graph $y = -2x + 5$. Is your description in part **b** true when m is negative?

Review

In 22 and 23, the gradient of a railway is given. Find the slope of the line made as the train goes up the incline. *(Lesson 7-3)*

22. The Katoomba Scenic Railway in the Blue Mountains of New South Wales, Australia, is the steepest railway in the world with gradient of 1 in 0.82.

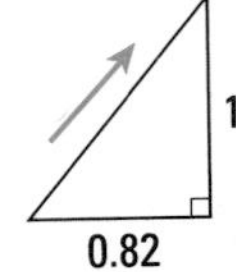

23. The steepest standard gauge railway in the world is between Chedde and Seovoz, in France, with a gradient of 1:11.

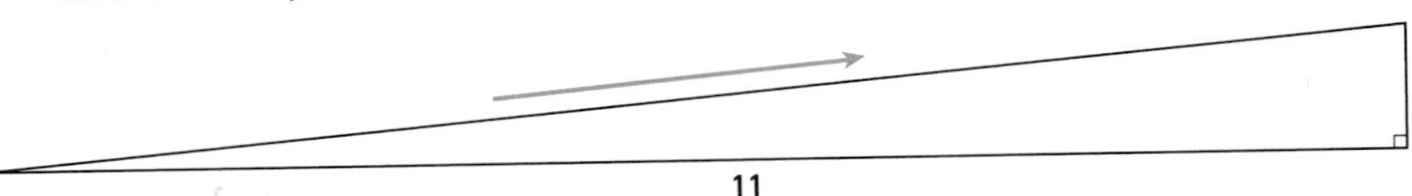

24. Which lines have no slope? Explain your reasoning. *(Lesson 7-3)*

25. Find the slope of the line through (5, -3) and (2, 7). *(Lesson 7-2)*

In 26 and 27, solve. *(Lesson 6-8)*

26. $\frac{3x + 5}{2 - 4x} = \frac{2}{3}$

27. $\frac{z + 2}{2} = 2z - 2$

28. A rectangle is 5 cm longer than twice its width. Its perimeter is 58 cm.

a. Write an expression for the length.

b. Find its width and length.

c. What is its area? *(Lessons 2-1, 3-5)*

Exploration

29. Find equations for four lines, none of them horizontal or vertical, which intersect at the vertices of a rectangle. What is true about the slopes of these lines?

LESSON

7-5

Equations for Lines with a Given Point and Slope

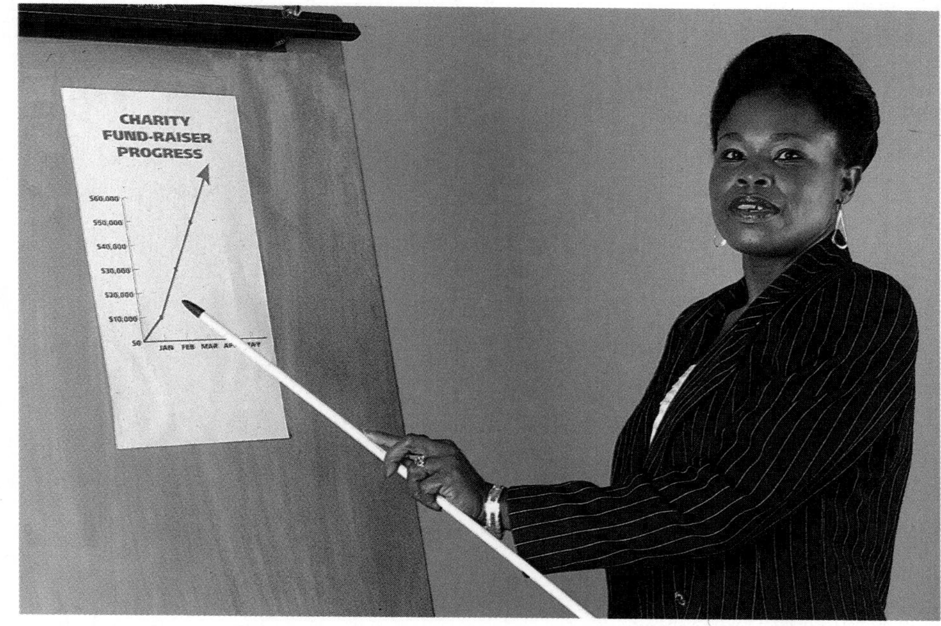

Reaching the goal. *This businesswoman is showing the progress made in the company's charity fund-raiser. The steep slope of the line indicates that large donations have increased over time.*

In the last lesson you found an equation of a line given its slope and a particular point on the line, its *y*-intercept. You can, in fact, find an equation in slope-intercept form of a line given its slope and any point on the line, not necessarily the *y*-intercept.

Example 1

Find an equation in slope-intercept form for the line through (-3, 5) with a slope of 2.

Solution

You know that $y = mx + b$ is the slope-intercept equation of a line. In this case you are given $m = 2$. All that is needed is b. Follow these three steps.

1. Substitute 2 for m and the coordinates (-3, 5) for (x, y) into $y = mx + b$. This gives
$$5 = 2 \cdot -3 + b.$$
2. Solve this equation for b.
$$5 = -6 + b$$
$$11 = b$$
3. Substitute the slope for m and the value you found for b in $y = mx + b$.
The equation is $y = 2x + 11$.

Check 1

Plot the point (-3, 5). Draw the line through it having slope 2. Notice that the graph as shown at the right has y-intercept 11.

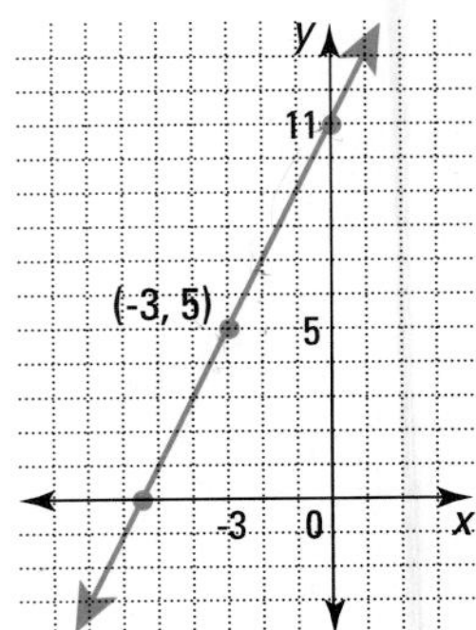

Check 2

Use an automatic grapher to graph the equation $y = 2x + 11$. Use the trace feature to check that the line appears to pass through the point (-3, 5), and that a slope of 2 seems to fit the line.

The method of Example 1 may be used to find an equation for any nonvertical line, if you know its slope and the coordinates of one point on the line.

To find an equation of a nonvertical line given one point and the slope:

Step 1: Substitute the slope for m and the coordinates of the given point for x and y in the equation $y = mx + b$.

Step 2: Solve the equation from Step 1 for b.

Step 3: Substitute the slope for m and the value you found for b in $y = mx + b$.

In particular, you can use the algorithm above to find an equation for a line whose slope and x-intercept are known. The **x-intercept** of a graph is the x-coordinate of a point where the graph intersects the x-axis.

Example 2

A line has slope -4, and its x-intercept is 6. Find an equation for the line in slope-intercept form.

Solution

Because the x-intercept is 6, we know that (6, 0) is a point on this line. Follow the steps above with $m = -4$, $x = 6$, and $y = 0$.

1. Substitute for m, x, and y. $\quad 0 = -4 \cdot 6 + b$
2. Solve for b. $\quad 24 = b$
3. Substitute for m and b. $\quad y = -4x + 24$

Check

A check is left for you to do.

This procedure is often useful for finding an equation to describe a real-life situation when you know a rate of change and information that gives coordinates for one point.

A scenic square. *Nathan Phillips Square is located in Toronto in Ontario. Ontario has the largest population of the Canadian provinces. More than 20 million tourists visit Ontario each year.*

Example 3

The population of the province of Ontario in Canada was 10,085,000 in 1991. At that time, the population was increasing at a rate of about 80,000 people per year. Assume this rate continues.

a. Find an equation relating the population y of Ontario to the year x.

b. Use the equation to predict the population of Ontario in the year 2001.

Solution

a. This is a constant-increase situation, so it can be described by a line with equation $y = mx + b$. The rate 80,000 people per year is the slope, so m = 80,000. The population of 10,085,000 in 1991 is described by the point (1991, 10,085,000). Now follow the steps listed on page 446.

1. Substitute for m, x, and y. $\quad 10{,}085{,}000 = 80{,}000 \cdot 1991 + b$
2. Solve for b. $\quad 10{,}085{,}000 = 159{,}280{,}000 + b$
   $-149{,}195{,}000 = b$
3. Substitute for m and b. $\quad y = 80{,}000x - 149{,}195{,}000.$

b. Evaluate the formula $y = 80{,}000x - 149{,}195{,}000$ when $x = 2001$.

$$y = 80{,}000(2001) - 149{,}195{,}000$$
$$= 160{,}080{,}000 - 149{,}195{,}000$$
$$= 10{,}885{,}000.$$

A prediction for the population in the year 2001 is about 10,885,000.

Check

a. When $x = 1991$, does $y = 10{,}085{,}000$?
Does $80{,}000(1991) - 149{,}195{,}000 = 10{,}085{,}000$? Yes, so it checks.

b. Does the rate of change in population between 1991 and 2001 equal 80,000 people per year?

$$\frac{10{,}885{,}000 - 10{,}085{,}000}{2001 - 1991} = \frac{800{,}000}{10} = 80{,}000.$$ Yes, it checks.

QUESTIONS

Covering the Reading

1. Describe the steps for finding an equation for a line, given the slope and one point on the line.

2. Check the answer to Example 2.

In 3–6, find an equation for the line given its slope and one point.

3. point (2, 3); slope 4

4. point (-10, 3); slope -2

5. point (-6, 0); slope $\frac{1}{3}$

6. point (4, $-\frac{1}{2}$); slope 0

7. Find an equation for the line with slope -4 and x-intercept 7.

8. Refer to Example 3. If the rate of population increase stays constant, what will be the population of Ontario in the year 2005?

Applying the Mathematics

9. The population of Anchorage, Alaska, was 226,300 in 1990. Suppose the population is increasing at a rate of 5,200 people per year, and the rate stays constant indefinitely.
   a. Find an equation relating the population of Anchorage to the year.
   b. Predict the population of Anchorage in the year 2002.
   c. At this rate of growth, in what year will the population of Anchorage first reach 500,000?

Are koalas bears? *No. Koalas are marsupials. These Australian mammals spend the first seven months of their life in their mother's pouch and the next six months riding on her back.*

10. A newborn koala (age 0 months) is 2 cm long. Until maturity, it grows at an average rate of 1.5 cm per month. Koalas mature at about 4 years.
    a. How long are mature koalas?
    b. The given information represents the slope m and the (x, y) coordinates of a point. Give the values of m, x, and y.
    c. Find an equation estimating the Koala's length y at age x months.
    d. Estimate the koala's length at age 6 months.

11. a. Write an equation for the line graphed at the right.
    b. At what point does the line cross the y-axis?
    c. What are the coordinates of point P?
    d. Check your answer to part **a** by showing that the coordinates of P make the equation in part **a** true.

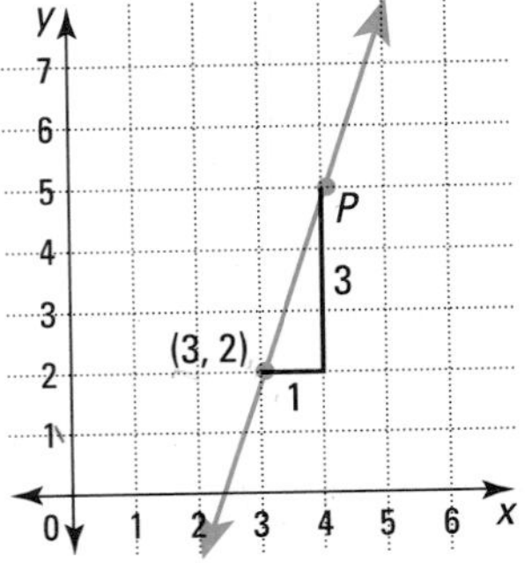

12. The slopes of two lines are reciprocals. The equation of one of the lines is $y = 2x + 1$.
    a. Find the slope of the second line.
    b. Find an equation for the second line if it passes through the point (2, 3).

In 13 and 14, some information is given. **a.** Write the slope and an ordered pair described by the information. **b.** Write an equation relating x and y.

13. Marty is spending money at the average rate of \$3 a day. After 14 days he has \$68 left. Let y be the amount left after x days.

14. Diane knows a phone call to a friend costs 25¢ for the first 3 minutes and 10¢ for each additional minute. Let y be the cost of a call that lasts x minutes.

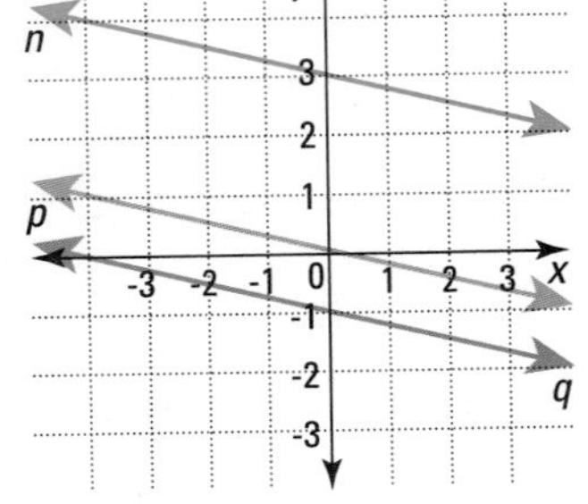

Review

15. Match each of lines n, p, and q at the left with its equation. *(Lesson 7-4)*
    a. $y = -\frac{1}{4}x$ b. $y = -\frac{1}{4}x + 3$ c. $y = -1 - \frac{1}{4}x$

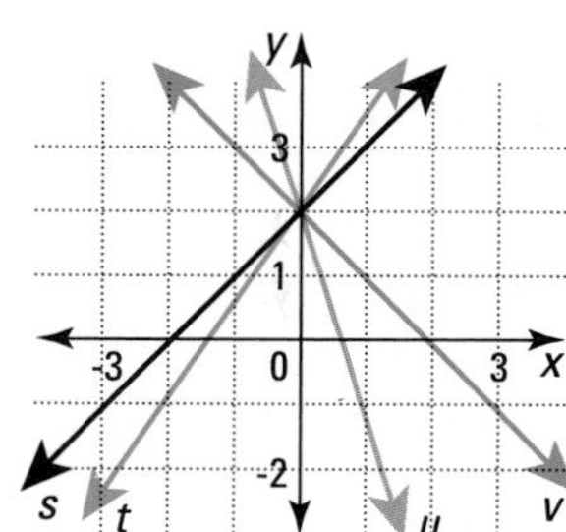

16. Match each of lines s, t, u, and v with its equation. *(Lesson 7-4)*

a. $y = x + 2$
b. $y = -x + 2$
c. $y = 2 - 3x$
d. $y = \frac{3}{2}x + 2$

17. Graph the line with equation $y = -x$. *(Lesson 7-4)*

18. Describe a real-world situation that could fit the equation $y = 15x + 45$. *(Lesson 7-4)*

19. Do the points (1, 3), (-3, -5), and (3, 6) lie on the same line? Justify your answer. *(Lesson 7-2)*

20. The following two points give information about an overseas telephone call: (5 minutes, \$5.91), (10 minutes, \$10.86). Calculate the slope and describe what it stands for. *(Lesson 7-2)*

21. Which section, or sections, of this graph shows:
a. the fastest increase? **b.** the slowest decrease? *(Lesson 7-1)*

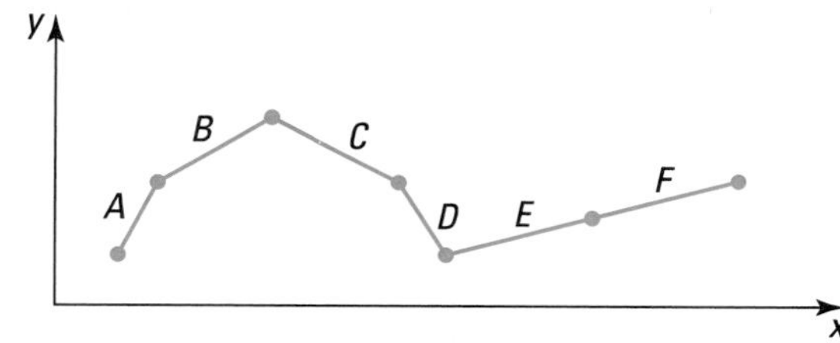

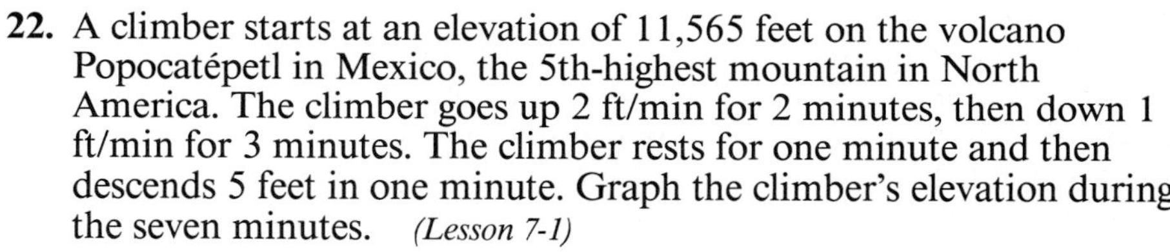

22. A climber starts at an elevation of 11,565 feet on the volcano Popocatépetl in Mexico, the 5th-highest mountain in North America. The climber goes up 2 ft/min for 2 minutes, then down 1 ft/min for 3 minutes. The climber rests for one minute and then descends 5 feet in one minute. Graph the climber's elevation during the seven minutes. *(Lesson 7-1)*

23. A survey showed that 7 out of 10 adults drink coffee in the morning. If 650 adults live in an apartment complex, how many of them would you expect to drink coffee in the morning? *(Lesson 6-8)*

24. *Skill sequence.* Simplify. *(Lesson 6-1)*

a. $\frac{15}{2} \div \frac{15}{3}$ **b.** $\frac{a}{2} \div \frac{a}{3}$ **c.** $\frac{a}{b} \div \frac{a}{c}$

25. Solve for n. *(Lessons 1-6, 2-10)*

a. $n^2 = 24$ **b.** $\sqrt{n} = 24$ **c.** $n! = 24$

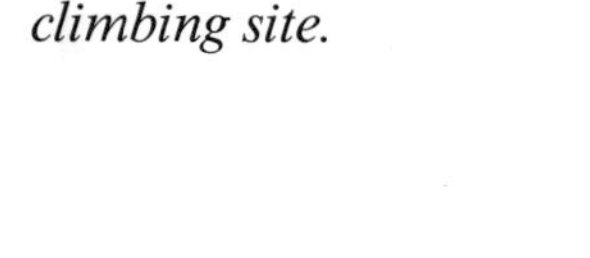

Mexican mountain. *Popocatépetl, a volcanic mountain in Cholula, Mexico, is 5,452 m high. It is a popular mountain-climbing site.*

Exploration

26. Examine Example 3.

a. Find the population of the state where you live at the last census.
b. Estimate or find out how fast your state is growing per year.
c. Using your estimate, find a linear equation relating the population y to the year x.
d. Use this equation to estimate the population of your state when you will be 50 years old.

LESSON

7-6

Equations for Lines Through Two Points

Cellular phone costs. *The average cost of a cellular phone call is greater than the cost of a regular telephone call. One reason cellular costs are higher is that cellular telephones use a combination of radio, telephone, and computer technology.*

A phone book might tell you that a call to a friend costs 25¢ for the first 3 minutes and 10¢ each additional minute. This is like being given the point (3, 25) and the slope 10. If you know an equation of the line through this point with this slope, then you have a formula which can help you determine the cost of any phone call to that friend. In Lesson 7-5 you learned how to find such an equation.

Sometimes in a constant-increase situation, the given information does not include the slope, but includes two points. For instance, you might be told that a 5-minute call overseas costs \$5.91 and a 10-minute call costs \$10.86. This means that you have two data points, (5, 5.91) and (10, 10.86). To obtain an equation for the line through these points, first calculate the slope. Then work as before.

Before finding this equation for the cost of a phone call, we work through an example with simpler numbers.

Example 1

Find an equation for the line through (5, -1) and (-3, 3).

Solution

1. First find the slope m.

$$m = \frac{3 - -1}{-3 - 5} = \frac{4}{-8} = -\frac{1}{2}$$

Now you have the slope and two points. This is more information than is needed. Pick one of these points. We pick (5, -1). Now follow the steps given in Lesson 7-5 to find an equation for the line through (5, -1) and with slope $-\frac{1}{2}$.

2. Substitute $-\frac{1}{2}$ and the coordinates of (5, -1) into $y = mx + b$.

$$-1 = -\frac{1}{2}(5) + b$$

3. Solve for b.

$$-1 = -\frac{5}{2} + b$$

$$\frac{3}{2} = b$$

4. Substitute the values of m and b into the equation.

$$y = -\frac{1}{2}x + \frac{3}{2}$$

Check 1

Substitute the coordinates of the point not used to see if they work.

Point (-3, 3): Does $3 = -\frac{1}{2}(-3) + \frac{3}{2}$?

Does $3 = \frac{3}{2} + \frac{3}{2}$? Yes.

Check 2

Use an automatic grapher to draw a graph of $y = -.5x + 1.5$. Use the trace feature to verify that (5, -1) and (-3, 3) are on the graph.

Notice that the procedure in Example 1 involves just one more step than the procedure of Lesson 7-5.

To find an equation for a nonvertical line given two points on it:

Step 1: Find the slope determined by the two points.

Step 2: Substitute the slope for m and the coordinates of one of the points (x, y) in the equation $y = mx + b$.

Step 3: Solve for b.

Step 4: Substitute the values you found for m and b in $y = mx + b$.

This method of finding the equation of a line was developed by René Descartes in the early 1600s. We now apply it to a problem about the cost of a phone call.

Example 2

Suppose a 5-minute overseas call costs \$5.91 and a 10-minute call costs \$10.86.

a. What is the cost y of a call of x minutes duration? (Assume that this is a constant-increase situation and x is a positive integer.)

b. How long could you talk for \$20.00?

Solution

a. You need an equation for the line through (5, 5.91) and (10, 10.86). Find the slope.

$$m = \frac{10.86 - 5.91}{10 - 5} = \frac{4.95}{5} = 0.99$$

Substitute 0.99 and the coordinates of one of the points into $y = mx + b$.

We pick (5, 5.91).

$$5.91 = 0.99(5) + b$$

Solve for b.

$$5.91 = 4.95 + b$$

$$b = 0.96$$

Substitute the values for m and b into $y = mx + b$.

$$y = 0.99x + 0.96$$

The equation $y = 0.99x + 0.96$ tells you that a call costs \$.96 to make plus \$.99 for each minute you talk.

b. Substitute $y = 20$ into the equation in part **a,** and solve for x.

$$0.99x + 0.96 = 20.00$$
$$0.99x = 19.04$$
$$x \approx 19.23$$

You can talk for about 19 minutes for \$20.00.

Check

a. Does this equation give the correct cost for a 10-minute call? Substitute 10 for x in the equation $y = 0.99x + 0.96$. Then $y = 0.99(10) + 0.96 = 9.9 + 0.96 = 10.86$, as it should.

b. The cost of a 19-minute call is $0.99(19) + 0.96 = 19.77$. The cost of a 20-minute call is $0.99(20) + 0.96 = 20.76$. It checks.

Relationships that can be described by an equation of a line are called *linear relationships.* They occur in many places. Here is how a relationship mentioned in Chapter 1 was found.

Example 3

Biologists have found that the number of chirps some crickets make per minute is related to the temperature. The relationship is very close to being linear. When crickets chirp 124 times a minute, it is about 68°F. When they chirp 172 times a minute, it is about 80°F. Below is a graph of this information.

a. Find an equation for the line through the two points.

b. About how warm is it if you hear 100 chirps in a minute?

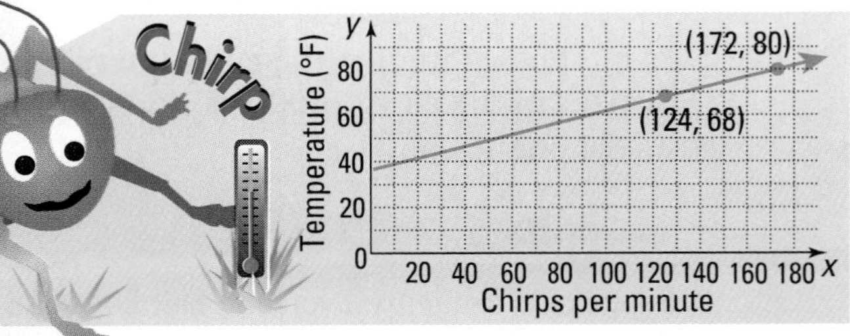

Solution

a. First, find the slope.

$$m = \frac{80 - 68}{172 - 124} = \frac{12}{48} = \frac{1}{4}$$

Substitute $\frac{1}{4}$ and the coordinates of (124, 68) into $y = mx + b$.

$$68 = \tfrac{1}{4}(124) + b$$

Solve for b.

$$68 = 31 + b$$
$$37 = b$$

Substitute for m and b. An equation is $y = \frac{1}{4}x + 37$.

b. Substitute 100 for x in the equation $y = \frac{1}{4}x + 37$.

$$y = \frac{1}{4}(100) + 37$$
$$y = 25 + 37$$
$$y = 62$$

It is about 62°F when you hear 100 chirps in a minute.

Check

a. Substitute the coordinates of the point (172, 80) which were not used in finding the equation.

Does $80 = \frac{1}{4}(172) + 37$?

Does $80 = 43 + 37$? Yes.

b. Look at the graph. Is the point (100, 62) on the line? Yes, it checks.

The equation in Example 3 enables the temperature to be estimated for any number of chirps. By solving for x in terms of y, you could get a formula for the number of chirps to expect at a given temperature. Formulas like these seldom work for values far from the given data points. Crickets tend not to chirp at all below 50°F, yet the formula $y = \frac{1}{4}x + 37$ predicts about 50 chirps a minute at 50°F.

Notice that in Example 2 you were told to let x equal the number of minutes and y equal the cost. In Example 3, the graph told you to use x for the number of chirps and y for the temperature. If you must decide which variable is x, and which is y, use a question about rate of change to suggest which makes more sense. For instance, in a situation about distances and miles, ask yourself, "Does it make more sense to think about miles per hour or hours per mile in this situation?" If you need miles per hour, let y = the number of miles and x = the number of hours. If you need hours per mile, let y = the number of hours and x = the number of miles. If you can solve the problem using either rate, then x and y can represent either quantity.

QUESTIONS

Covering the Reading

In 1–4, find an equation for the line through the two given points. Check your answer.

1. (1, 0), (4, 15)

2. (1, 9), (7, 3)

3. (6, -3), (-8, -10)

4. (0, 11), (13, 0)

5. Who developed the method used in this lesson for finding the equation of a line?

6. In Example 2, what would an 8-minute call cost?

7. If a 5-minute overseas call to Bonn costs \$4.50 and a 10-minute call costs \$8.50, find a formula relating time and cost.

In 8–10, refer to Example 3.

8. The number of times a cricket chirps in a minute and the temperature is very close to what kind of relationship?

9. By substituting in the equation $y = \frac{1}{4}x + 37$, about how many chirps per minute would you expect if the temperature is 70°F?

10. Suppose the cricket chirps 90 times per minute.
a. Use the graph to estimate the temperature.
b. Use the equation to estimate the temperature.

Applying the Mathematics

11. Find an equation for the line with x-intercept 6 and y-intercept 5.

12. a. Show that $A = (-4, 7)$, $B = (1, 5)$, and $C = (16, -1)$ lie on the same line.
b. Find an equation for that line.

13. The graph below shows the linear relationship between Fahrenheit and Celsius temperatures. The freezing point of water is 32°F and 0°C. The boiling point of water is 212°F and 100°C.

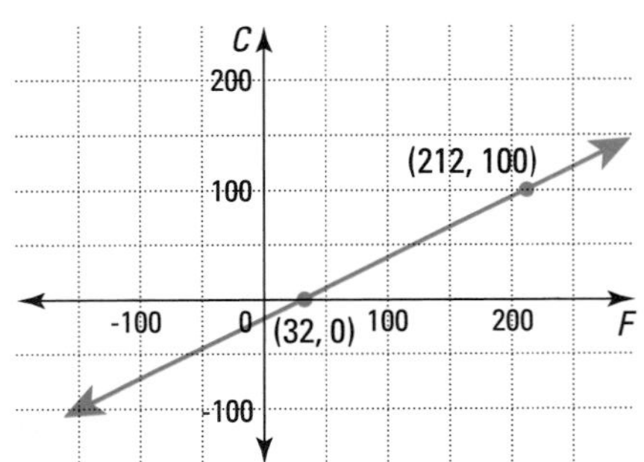

a. Find an equation that relates Celsius and Fahrenheit temperatures. (Hint: $C = mF + b$.)
b. When it is 150°F, what is the temperature in degrees Celsius?
c. When it is 150°C, what is the temperature in degrees Fahrenheit?

14. The total length y and the tail length x of females of the snake species *Lampropeltis polyzona* have close to a linear relationship. When $x = 60$ mm, $y = 455$ mm. When $x = 140$ mm, $y = 1050$ mm.
a. Find the slope of this linear relationship. (Approximate your answer to the nearest tenth.)
b. Find an equation for y in terms of x using (60, 455) and your slope from part **a.**
c. Check your equation using (140, 1050).
d. If a female of this species has a tail of length 100 mm, how long is the snake?

It's all relative. *Shown is a milk snake, a member of the* Lampropeltis *family.*

15. The sum of the measures of the angles in a triangle is 180°. In a quadrilateral, the sum is 360°. Find an equation relating *n*, the number of sides in a polygon, and the sum of its angles, *S*.

Review

16. A cab company charges a base rate plus $1.20 per mile. A 12-mile cab ride costs $15.90.

a. Write an equation relating the number of miles driven to the cost of the cab ride.

b. Make up a question about this cab company. Use your equation from part **a** to answer your question. *(Lesson 7-5)*

17. What is an equation for the line with slope 7 which passes through the point (7, 7)? *(Lesson 7-5)*

18. Graph $y = 4x - 3$ by using its slope and y-intercept. *(Lesson 7-4)*

19. The points $(a, 5)$ and $(-2, -4)$ lie on a line with slope $\frac{3}{4}$. Find a. *(Lesson 7-2)*

20. A road leads from the town of Salida. There are signs every 5 miles that tell the elevation. *(Lesson 7-1)*

Miles from Salida	Elevation (feet)
0	1744
5	1749
10	1749
15	1759
20	1757

a. Between which two signs is the rate of change of elevation negative?

b. Calculate the rate of change of elevation for the entire distance.

21. *Skill sequence.* Solve. *(Lesson 5-9)*

a. $x^2 = 64$ **b.** $(y - 7)^2 = 64$ **c.** $(3y - 7)^2 = 64$

22. *Skill sequence.* Simplify. *(Lesson 2-3)*

a. $\frac{12}{144}$ **b.** $\frac{3}{3^2}$ **c.** $\frac{d}{d^2}$

Exploration

23. In many places, a taxicab ride costs a fixed number of dollars plus a constant charge per mile.

a. Find a rate for taxi rides in your community or in a nearby city.

b. Graph your findings on coordinate axes like the ones at the right.

c. Find an equation relating distance traveled and cost.

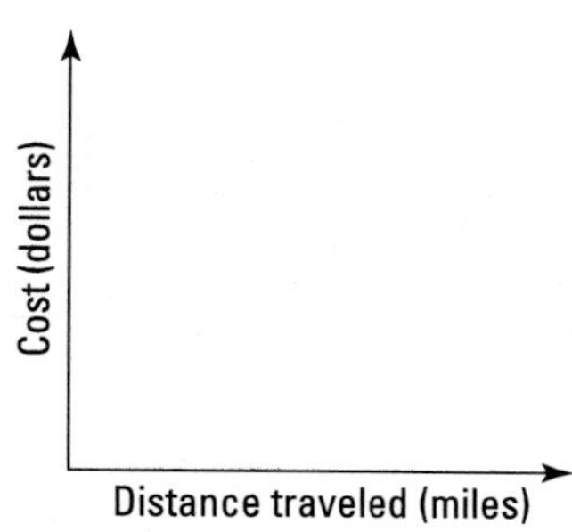

Introducing Lesson 7-7

Latitude and Temperature

IN-CLASS ACTIVITY

Materials: Graph paper
Work in small groups.

When a situation involves a relation between two variables, a graph of the relation can be drawn. If the points all lie on a line, an equation for the line can be found. When the points do not lie on a line, but are approximately on a line, an equation can be found that closely describes the relation. In this activity, you will explore how to do this.

1 The *latitude* of a place on Earth tells how far the place is from the equator. Latitudes in the Northern Hemisphere range from 0° at the equator to 90° at the North Pole. In the table below are the latitudes and mean high temperatures in April for selected cities in the Northern Hemisphere. (The mean high temperature is the mean of all the daily high temperatures for the month.) Although in all of these cities temperature is measured in degrees Celsius, we have converted the temperatures to Fahrenheit for you.

Latitude and Temperature in Selected Cities

City	North Latitude	April Mean High Temperature (°F)
Lagos, Nigeria	6	89
San Juan, Puerto Rico	18	84
Calcutta, India	23	97
Cairo, Egypt	30	83
Tokyo, Japan	35	63
Rome, Italy	42	68
Quebec City, Canada	47	46
London, England	52	56
Copenhagen, Denmark	56	50
Moscow, Russia	56	47

North Pole: latitude 90° north
Minneapolis, MN: latitude 45° north
Equator: 0° latitude

Graph the data points in the table. Let the latitude be the x-value and the temperature be the y-value in degrees. Label each point with the city it represents. We have plotted the first three entries from the table.

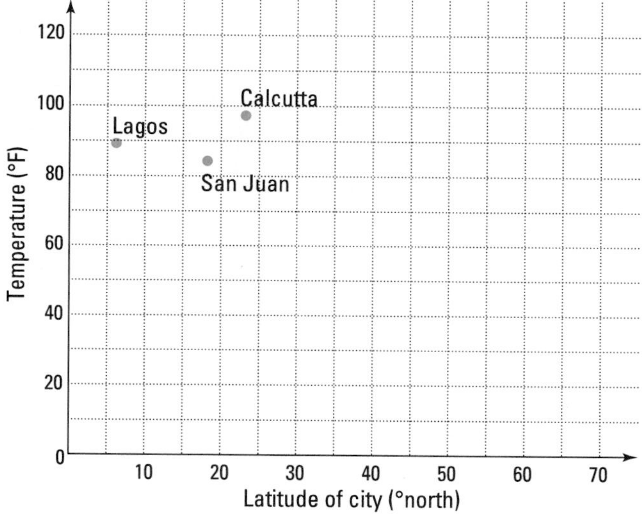

2 What pattern(s) do you notice in this scatterplot? In particular, as the latitude increases, what generally happens to the temperature in April?

3 Based on the data above, what do you expect the mean high temperature to be in April in cities at the following north latitudes?

a. Singapore at 1° north latitude
b. a city at 19° north latitude, like Mexico City or Bombay, India
c. Madrid, Spain, at 40° north latitude
d. Helsinki, Finland, at 60° north latitude

Explain how you arrived at your answers.

LESSON

7-7 Fitting a Line to Data

Denmark's charm. *Shown is Copenhagen's Nyhavn Canal with its ships, old buildings, and cafes. Copenhagen is the capital of Denmark. Because the country is almost entirely surrounded by water, the climate is mild and damp.*

If the data points of a set do not all lie on one line, but are close to being linear, you can often use an equation for a line to describe trends in the data. For instance, below is a graph of the latitude and temperature data you used in the In-class Activity on page 456.

Fitting a Line to Data

Even though no line passes through all the data points, you can find a line which describes the trend of higher latitude, lower temperature. The simplest way is to take a ruler and draw a line that seems "close" to all the points. This is called "fitting a line by eye," and one such line is graphed at the right. Notice that the line we drew does not pass through any of the original data points.

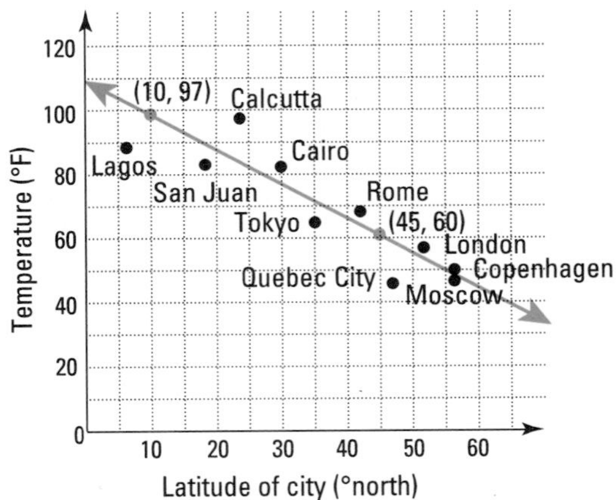

Once a line has been fitted to the data, you can find an equation for the line, and then use the equation to predict temperatures for different latitudes.

Example 1

Find an equation for the line drawn to fit the latitude and temperature data points on page 457.

Solution

Notice that the line contains the points (10, 97) and (45, 60). Use these points to find the slope of the line.

$$\text{slope} = \frac{60 - 97}{45 - 10} = \frac{-37}{35} \approx -1.06$$

Now substitute this slope and the coordinates of one of the points into $y = mx + b$ and solve. We use (10, 97).

$$97 = -1.06 \cdot 10 + b$$
$$97 = -10.6 + b$$
$$107.6 = b$$

An equation for the line is $y = -1.06x + 107.6$.

The negative slope -1.06 means as you move 1° north in latitude, the April mean high temperature is lower by about 1°F. The y-intercept 107.6 means that the high temperature at the equator (0° latitude) should be about 108°F. Notice that this agrees with the graph drawn on the previous page. The line crosses the vertical axis at about 108°F.

Activity 1

Use the scatterplot of latitude and temperature data you made in the In-class Activity on page 457.

a. Draw another line that seems to fit the data.
b. Give the coordinates of two points on your line.
c. Find an equation for your line.

Famous Spanish writer. *Miguel de Cervantes (1547–1616) ranks as one of the world's greatest writers. Statues of two of his most famous characters, Don Quixote and Sancho Panza, stand in Madrid, Spain.*

Using an Equation for a Fitted Line to Make Predictions

Using the equation $y = -1.06x + 107.6$ found in Example 1, you can estimate the April mean high temperature for any city in the Northern Hemisphere.

Example 2

Use the equation for the fitted line to predict the April mean high temperature for Madrid, Spain, which is at 40° north latitude.

Solution

Substitute 40° for x in the equation.

$$y = -1.06 \cdot x + 107.6$$
$$y = -1.06 \cdot 40 + 107.6$$
$$y = 65.2$$

You can predict that Madrid would have an April mean high temperature of about 65°F.

The mean high temperature in April is actually 64°, so the predicted temperature is quite close.

Located in the National Palace of Mexico City are the Diego Rivera murals. This one shows Hidalgo, Morelos, and Juárez—men who made important political contributions.

Activity 2

Use the equation you found in Activity 1 to predict the mean high temperature in April for Madrid. By how much does your prediction differ from the actual value?

Sometimes the fitted line does not predict temperature accurately. For instance, for a city at a latitude of 19° north, the line predicts a temperature of 87°. Both Bombay, India, and Mexico City are at this latitude. For Bombay the predicted temperature is accurate. But in Mexico City, the actual April mean temperature is 78°. The prediction is too high because Mexico City is at an altitude of about one mile, and temperatures at high altitudes are lower than those at sea level.

Being able to fit a line to data allows you to obtain a general formula from a few cases. This is such an important skill that some computer software and graphing calculators will find the line that *best fits* data. If you have access to such technology, you might want to enter these latitude and temperature data.

QUESTIONS

Covering the Reading

1. What does the latitude of a place on Earth signify?

2. What is the latitude of the equator?

3. Which city is farther north, Calcutta or Cairo?

4. What does it mean to "fit a line by eye" to a scatterplot?

5. Once a line is fitted, what is a good first step toward getting an equation for the line?

6. a. Use the graph of the fitted line in this lesson to predict the mean high temperature in April in a city at 25° north latitude.
b. Use the equation in Example 1 to predict this temperature.
c. Use the equation you found in Activity 1 to predict this temperature.
d. What is true about your answers to parts **a, b,** and **c?**

7. Refer to Activity 2. By how much does your prediction differ from the actual mean high temperature in Madrid?

Shown is an outdoor marketplace in Calcutta, India. The city lies along the east bank of the Hooghly river and serves as India's chief port for trade with Southeast Asia.

8. a. Acapulco, Mexico, is at 17° north latitude. Use the equation in Example 1 to estimate its average April high temperature.
b. The actual mean high temperature in April for Acapulco is 87°F. Give a reason why the answer in part **a** is closer to the actual value for Acapulco than the prediction was for Mexico City.

9. a. What is the latitude of the North Pole?
b. What is the predicted mean high temperature in April at the North Pole?

Applying the Mathematics

10. If a city in the Northern Hemisphere has a mean high temperature in April of about 80°, at what latitude would you expect it to be? Explain how you got your answer.

11. Below are the latitudes and April mean high temperatures for two cities in the Southern Hemisphere.

Rio de Janeiro, Brazil (23° south, 69°F)
Cape Town, South Africa (34° south, 58°F)

Could you predict temperatures for cities south of the equator by using negative values of x in the equation in Example 1? Explain your reasoning.

In 12–15, tell whether fitting a line to the data points would be appropriate.

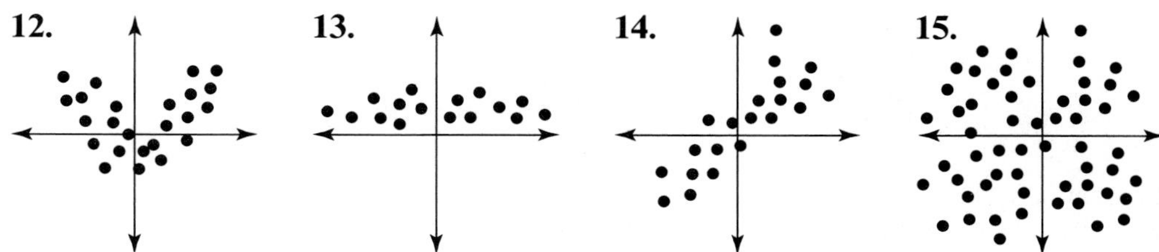

16. Use the following data.

City	North Latitude	January Mean Low Temperature (°F)
Lagos, Nigeria	6	74
San Juan, Puerto Rico	18	67
Calcutta, India	23	55
Cairo, Egypt	30	47
Tokyo, Japan	35	29
Rome, Italy	42	39
Quebec City, Canada	47	2
London, England	52	35
Copenhagen, Denmark	56	29
Moscow, Russia	56	9

a. Carefully draw a scatterplot showing a point for each city.
b. Fit a line to the data by eye and draw the line with a ruler.
c. Estimate the coordinates of two points on the line you drew.
d. Find an equation for the line through the points in part **c.**
e. Complete the following sentence: "As you go one degree north, the January mean low temperature tends to _?_"
f. What does the equation predict for a January mean low temperature at the equator?
g. Predict the January mean low temperature for the North Pole.
h. Use your equation to predict the January mean low temperature for Acapulco, which is at 17° north latitude. (Note: the actual mean low is 70°F.)

Ringling Brothers, 1907; Circus of Seville, 1993.

Review

17. Find an equation for the line with y-intercept equal to 7 and x-intercept equal to 4. *(Lesson 7-6)*

18. a. Find an equation for the line through (3, 2) with a slope of $\frac{3}{5}$.
b. Give the coordinates of one other point on this line. *(Lesson 7-5)*

19. a. Give the slope and y-intercept of the line $3x + 5y = 2$.
b. Graph the line. *(Lesson 7-4)*

In 20 and 21, give **a.** the slope; and **b.** the y-intercept of the line. *(Lesson 7-4)*

20. $y = -\frac{x}{2}$

21. $y = -2$

22. a. Draw a line which has no y-intercept.
b. What is the slope of your line? *(Lesson 7-4)*

23. Seattle scored 17 points in the first nine minutes of a game. At that rate about how many points would they score in a 48-minute game? *(Lesson 6-8)*

24. Solve for y. *(Lesson 5-7)*
a. $8y = 2 + y$
b. $By = C + y$

25. If a is the cost of an adult's ticket to the circus and c is the cost of a child's ticket, what is the cost of tickets for 2 adults and 4 children? *(Lessons 2-4, 3-1)*

26. Graph the solution set to $-3x \geq 15$ using the given domain. *(Lessons 1-2, 2-8)*
a. set of integers
b. set of real numbers

Exploration

27. a. Find the latitude of your school, to the nearest degree.
b. Predict the April mean high temperature in °F for your latitude using the graph in Example 1.
c. Check your prediction in part **b** against some other source of the April mean high temperature. (Newspapers or TV weather records are possible sources.)

28. What is meant by *longitude?*

LESSON

7-8

Equations for All Lines

In this chapter, lines have been used to describe situations involving constant increase or decrease. We have also fitted lines to data. The slope-intercept form $y = mx + b$ arises naturally from these applications. All lines except vertical lines can have equations in slope-intercept form.

Some situations, such as the one below, naturally lead to equations of lines in a different form.

The Ramirez family bought 4 sandwiches and 3 salads. They spent \$24.00. If x is the cost of a sandwich and y the cost of a salad, then

$$4x + 3y = 24.$$

The pairs of values of x and y that make this equation true are the possible costs of the items. For instance, because

$$4(4.50) + 3 \cdot 2 = 18 + 6 = 24$$

each sandwich could have cost \$4.50 and each salad \$2. This yields the point (4.5, 2). Other possible costs of sandwiches and salads can be found by rewriting this equation in slope-intercept form.

$$4x + 3y = 24$$
$$3y = -4x + 24$$
$$y = -\frac{4}{3}x + 8$$

The graph of this equation is a line with slope $-\frac{4}{3}$ and y-intercept 8. Since the graph shows the possible costs of the sandwiches and salads, we use only the first quadrant where both x and y are positive numbers.

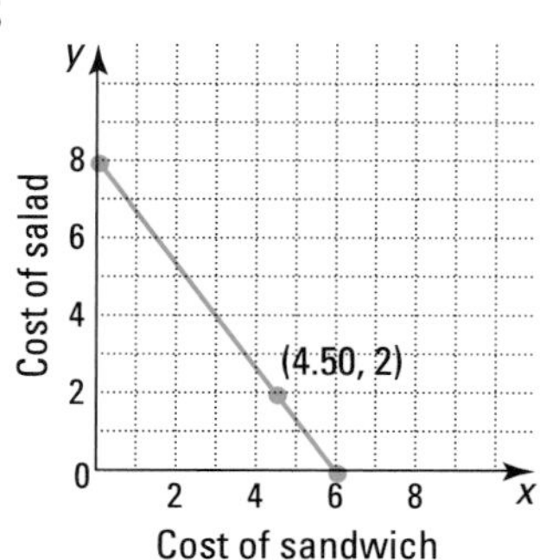

The Standard Form of an Equation for a Line

The equation $4x + 3y = 24$ has the form

$$Ax + By = C,$$

where A, B, and C are constants. The variables x and y are on one side of the equation. The constant term C is on the other. The equation $Ax + By = C$, where A, B, and C are constants, is the **standard form** for an equation of a line.

To graph a line whose equation is in standard form, you do not need to rewrite the equation in slope-intercept form. Instead, you can find the intercepts and draw the line that contains the intercepts.

Example 1

Graph $4x + 3y = 24$.

Solution

Find the x- and y-intercepts.

Let $x = 0$. $\quad 4 \cdot 0 + 3y = 24$

$3y = 24$

$y = 8$

The y-intercept is 8, so one point on the line is (0, 8).

Let $y = 0$. $\quad 4x + 3 \cdot 0 = 24$

$4x = 24$

$x = 6$

The x-intercept is 6, so (6, 0) is also on the line.

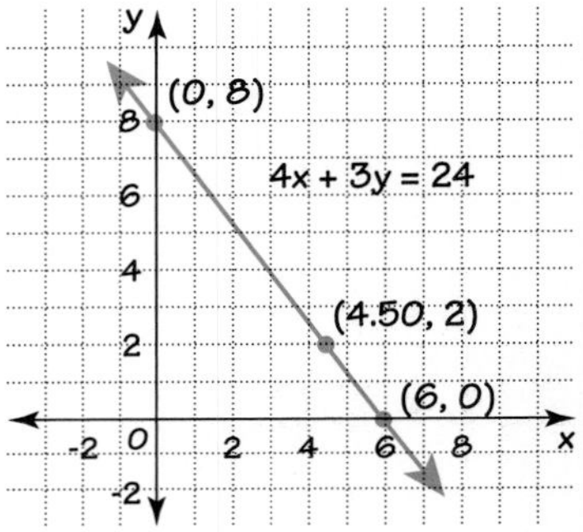

Now plot (6, 0) and (0, 8) and draw the line through them, as shown.

Check

Find a third point satisfying the equation. Earlier we noted that (4.50, 2) satisfies the equation. Is this point on the graph? Yes, it is.

You should be able to recognize real situations that lead to equations of lines.

Example 2

Roast beef sells for \$6 a pound. Shrimp costs \$12 a pound. Andy has \$96 to buy beef and shrimp for a party. Write an equation in standard form to describe the different possible combinations Andy could buy.

Solution

Let x = pounds of roast beef bought. So $6x$ = cost of roast beef.
Let y = pounds of shrimp bought. So $12y$ = cost of shrimp.

Cost of roast beef + cost of shrimp = total cost

$$6x + 12y = 96$$

Check

To check the equation, suppose Andy buys 6 pounds of roast beef. He will have spent $6 \cdot \$6 = \$36$. He will have $\$96 - \$36 = \$60$ left to spend on shrimp. At \$12 per pound he can buy $\frac{60}{12} = 5$ pounds of shrimp. Does (6, 5) work in $6x + 12y = 96$? Yes, $6 \cdot 6 + 12 \cdot 5 = 96$.

Notice that the equation $6x + 12y = 96$ can be simplified by dividing each term by 6.

$$\frac{6x}{6} + \frac{12y}{6} = \frac{96}{6}$$
$$x + 2y = 16$$

Notice that this equation is also in standard form, so it also is an answer to Example 2.

Rewriting Equations in Slope-Intercept and Standard Form

The lines in Examples 1 and 2 are **oblique,** meaning they are neither horizontal nor vertical. Recall from Lesson 5-1, if a line is vertical, then its equation has the form $x = h$. If a line is horizontal, it has an equation $y = k$. Notice that these are already in standard form. For example,

$x = 3$ is equivalent to $x + 0y = 3$, where $A = 1$, $B = 0$, $C = 3$.
$y = -2$ is equivalent to $0x + y = -2$, where $A = 0$, $B = 1$, $C = -2$.

Thus *every line* has an equation in the standard form $Ax + By = C$.

You now have seen the two most common forms of equations of lines.

Slope-Intercept Form: $y = mx + b$
Standard Form: $Ax + By = C$

You should be able to quickly change an equation from one form into the other. In standard form, the equation is usually written with A, B, and C as integers.

Example 3

Rewrite $y = \frac{3}{5}x + \frac{8}{5}$ in standard form with integer coefficients. Find the values of A, B, and C.

Solution

First, multiply each side of the equation by 5 to clear the fractions.

$$5y = 5 \cdot \frac{3}{5}x + 5 \cdot \frac{8}{5}$$
$$5y = 3x + 8$$

Now add $-3x$ to both sides so the x and y terms are both on the left. Write the x term first.

$$-3x + 5y = 8$$

This is in standard form with $A = -3$, $B = 5$, and $C = 8$.

Some people prefer that the coefficient A be positive in standard form. To rewrite the equation in Example 3 with A positive, you could multiply each side by -1.

$$-1(-3x + 5y) = -1(8)$$
$$3x - 5y = -8$$

Every line has many equations in standard form. The equation $3x - 5y = -8$ is usually considered simplest because 3, -5, and -8 have no common factors.

Example 4

Rewrite $y = .25x$ in standard form with integer values of A, B, and C.

Solution

Multiply each side by 100 to clear the decimal.

$$y = .25x$$
$$100y = 25x$$

Now, add $-25x$ to each side to get both variable terms on the left. Write the x-term first.

$$-25x + 100y = 0$$

Here $A = -25$, $B = 100$, and $C = 0$.

QUESTIONS

Covering the Reading

In 1 and 2, refer to the situation about the Ramirez family buying food.

1. **a.** If each sandwich cost $3.75, how much did each salad cost?
b. Give the coordinates of the point on the graph in Example 1 corresponding to your answer to part **a.**

2. Give another pair of possible costs for the sandwich and salad.

3. What is the form $Ax + By = C$ called?

In 4 and 5, the equation is in the form $Ax + By = C$. Give the values of A, B, and C.

4. $4x + 2y = 5$

5. $x - 8y = 2$

In 6 and 7, an equation for a line is given.
a. Find the x- and y-intercepts.
b. Graph.

6. $3x + 5y = 30$

7. $2x - 3y = 12$

8. What lines do not have equations in slope-intercept form?

9. What lines do not have equations in standard form?

In 10 and 11, refer to Example 2.

10. Find three different combinations of pounds of roast beef and shrimp Andy could buy.

11. The store at which Andy usually shops is having a sale. Roast beef costs \$4 a pound and shrimp costs \$10 a pound.
a. Write an equation in standard form to describe the different possible combinations of roast beef and shrimp he can buy for \$96.
b. What is the greatest amount of roast beef he can buy?
c. What is the greatest amount of shrimp he can buy?
d. Graph the solutions to the equation in part **a.**

In 12–14, an equation in slope-intercept form is given.
a. Rewrite the equation in standard form with integer coefficients.
b. Give the values of A, B, and C.

12. $y = \frac{2}{3}x + 12$ **13.** $y = 4x$ **14.** $y = -8x - 3$

Applying the Mathematics

15. A 100-point test has x questions worth 2 points apiece and y questions worth 4 points apiece.
a. Write an equation in standard form that describes all possible numbers of questions that might be on the test.
b. Give three solutions to the equation in part **a.**

16. Louise has \$36 in five-dollar bills and singles. How many of each kind of bill does she have?
a. Write an equation to describe this situation.
b. Give three solutions.
c. Graph all possible solutions. (The graph will be discrete.)

17. Suppose $Ax + By = C$, with $B \neq 0$.
a. Solve this equation for y.
b. Identify the slope and the y-intercept of this line.

18. **a.** What are the x- and y-intercepts of the line graphed at the right?
b. Write an equation in standard form for the line.

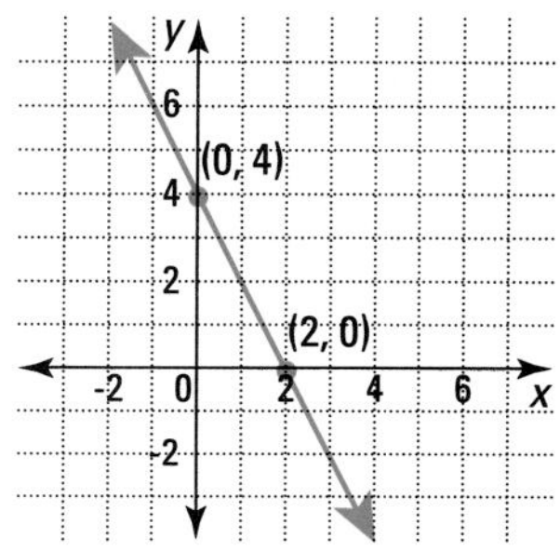

Review

19. The length of the winning toss in the men's discus throw has increased since the first Olympics in 1896. The scatterplot below shows the length of the winning tosses in meters in each Olympic year through 1992. *(Lesson 7-7)*

a. Trace the graph and fit a line to the data.

b. Find an equation for the line in part **a.**

c. Use your equation to predict the length of the winning toss in the year 2000.

d. Why might the prediction for 2000 be incorrect?

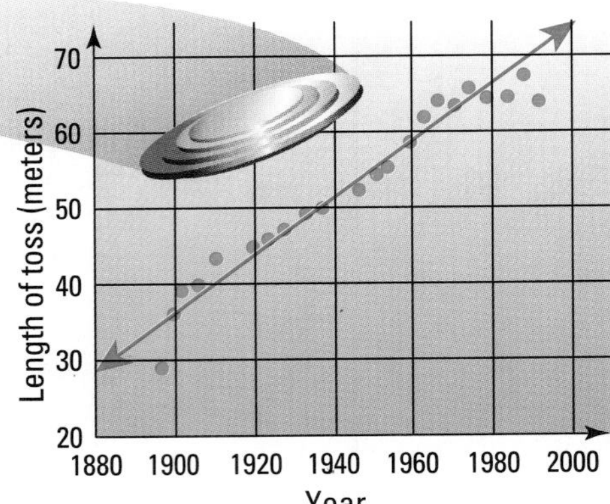

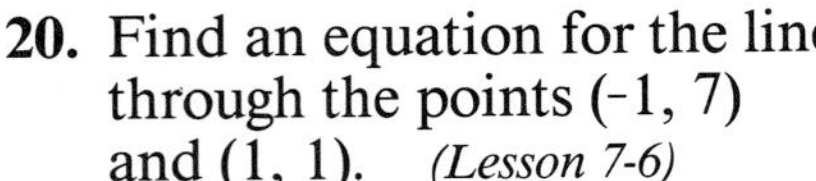

20. Find an equation for the line through the points (-1, 7) and (1, 1). *(Lesson 7-6)*

21. A robot is moving along a floor which has a coordinate grid. The robot starts at the point (8, -2) and moves along a line with slope 2. Does the robot pass through the following points? Justify your answer.

a. (10, 0) **b.** (18, 18) *(Lessons 7-2, 7-5)*

22. An airplane is flying at an altitude of 30,000 ft. The pilot is instructed to drop to an altitude of 25,000 ft in the next mile relative to the ground. Assume the pilot flies in a straight line.

a. Sketch the path of the plane.

b. What is the rate of change of altitude over horizontal distance in the path? *(Lesson 7-1)*

23. Graph on a coordinate plane: $y = 17$. *(Lesson 5-1)*

In 24 and 25, tell whether (0, 0) is a solution to the sentence. *(Lesson 3-10)*

24. $x + y < -4$ **25.** $6x - y > -6$

26. Write an inequality to describe this graph. *(Lesson 1-1)*

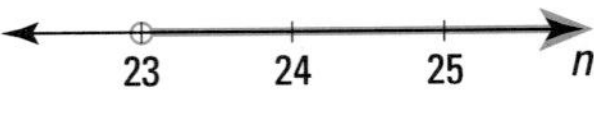

Exploration

27. An automatic grapher may be useful in this question.

a. Graph the lines with equations:

$$3x + 2y = 6$$
$$3x + 2y = 12$$
$$3x + 2y = 18.$$

b. What happens to the graph of $3x + 2y = C$ as C gets larger?

c. Try values of C that are negative. What can you say about the graphs of $3x + 2y = C$ then?

LESSON

7-9 Graphing Linear Inequalities

Surveying the land. *These two surveyors are using a transit to determine the boundary line between two areas of land in a subdivision. Boundary lines are critical in graphs of linear inequalities.*

You have graphed solutions to inequalities since the very first lesson of Chapter 1. Recall how to graph $n < 3$ on a number line. First find the point where $n = 3$.

-2 -1 0 1 2 3 4 5 n

Next, decide which part of the line contains the solutions to the inequality, that is, which part to shade. The sentence $n < 3$ states that we want values smaller than 3. Shade the points which are to the left of 3.

-2 -1 0 1 2 3 4 5 n

Recall that the 3 is marked with an open circle because 3 doesn't actually make the sentence $n < 3$ true, but it is important. It is the *boundary point* between the values that satisfy $n < 3$ and those that don't. It separates the line into two parts.

These ideas can be extended to graphs of inequalities in two dimensions. In this case, the boundary is a line instead of a point.

Inequalities Involving Horizontal or Vertical Lines

Example 1

Graph $y < 3$ on a coordinate plane.

Solution

Graph the line $y = 3$. This horizontal line is the boundary of the solution set. The line is dashed to show that the points having a y-coordinate of 3 are not included in the solution set. The solution set consists of all points having a y-coordinate less than 3. This is the region below the boundary line.

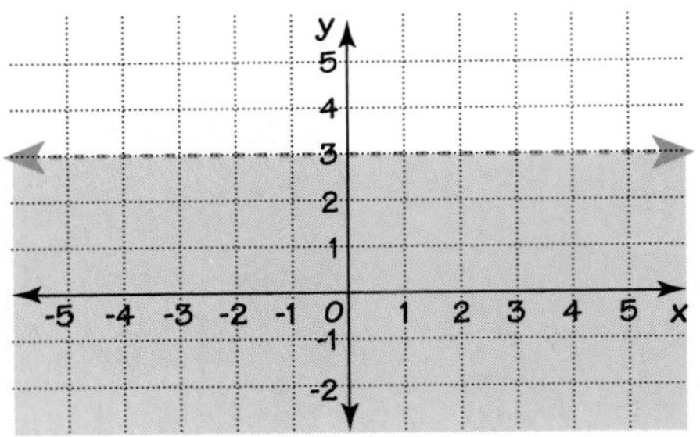

Check

Pick a point in the shaded region. We choose (0, 0). Do the coordinates of this point satisfy the inequality? Yes. $y = 0$ and $0 < 3$.

The regions on either side of a line in a plane are called **half-planes.** The boundary line is the edge of the half-plane. In Example 1, the line $y = 3$ is the edge of the half-plane $y < 3$. If Example 1 had asked you to graph $y \le 3$, then the edge $y = 3$ would be included and shown as a solid line.

Example 2

Give a sentence describing all points in each region.

a. 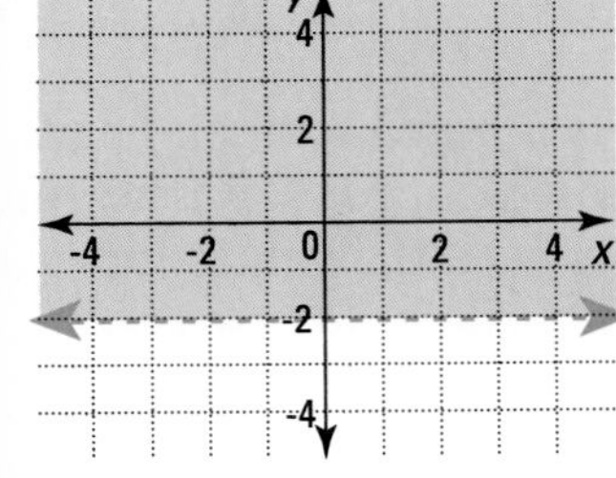

b. 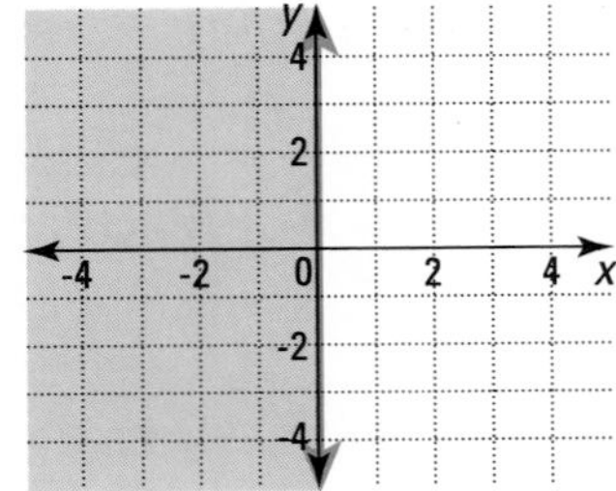

Solution

a. Every point in the half-plane has a y-coordinate greater than -2. The dotted line shows that points with a y-coordinate equal to -2 are not included. So a sentence describing this half-plane is $y > -2$.

b. Every point to the left of the y-axis has a negative x-coordinate. The solid line indicates that all points with x-coordinate equal to 0 are also to be included. So a sentence describing the region is $x \le 0$. This region is the union of a half-plane and its edge.

Inequalities Involving Oblique Lines

Every oblique line has an equation of the form $y = mx + b$. It is the boundary of two half-planes, described by $y < mx + b$ and $y > mx + b$.

Example 3

Draw the graph of $y \geq -3x + 2$.

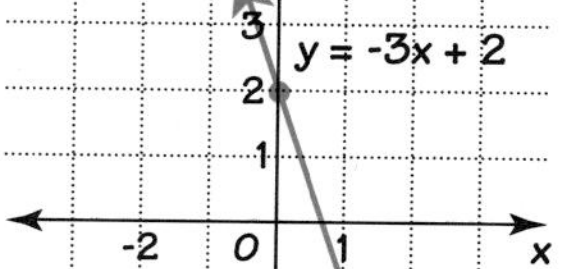

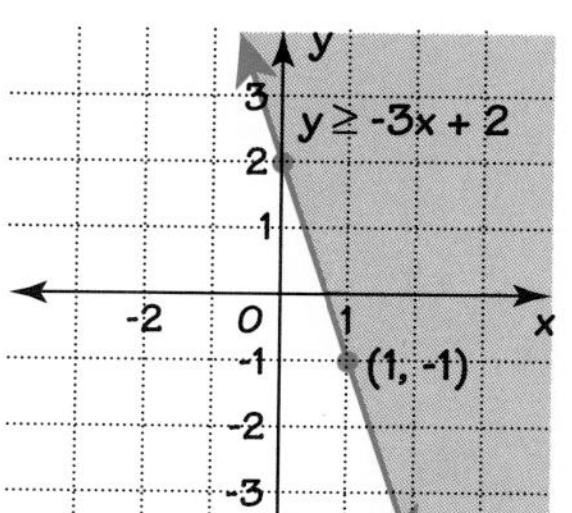

Solution

Start by graphing the boundary line $y = -3x + 2$ using techniques you have learned.

The $\geq$ sign in the inequality $y \geq -3x + 2$ indicates that points are desired whose y-coordinate is greater than or equal to the y values which satisfy $y = -3x + 2$. Since y values get larger as one goes higher, shade the entire region above the line.

Check

Try the point (0, 3), which is in the shaded region. Does it satisfy the inequality? Is $3 > -3 \cdot 0 + 2$? Yes, 3 is greater than 2. So the correct side of the line has been shaded.

When an inequality is written in slope-intercept form, it is easy to tell which half-plane of the boundary line to shade. For $y < mx + b$ shade below; for $y > mx + b$ shade above. But when an inequality is in the standard form $Ax + By > C$, you cannot use the inequality sign to determine which side of the line to shade. A more direct method, the testing of a point, is usually used. The point (0, 0) is usually chosen if it is not on the boundary line. If (0, 0) is a solution to the inequality, then the half-plane that contains it is shaded. If (0, 0) does not satisfy the inequality, shade the other side of the boundary line.

Example 4

Graph $3x - 4y > 12$.

Solution

First graph the boundary line $3x - 4y = 12$. This line is dashed because it is not part of the solution. To determine which side of the line is the solution set, substitute (0, 0) into the original equation. Is $3 \cdot 0 - 4 \cdot 0 > 12$? No. Since (0, 0) is in the upper half-plane and is *not* a solution, shade the lower half-plane. The graph of $3x - 4y > 12$ is shown at the right.

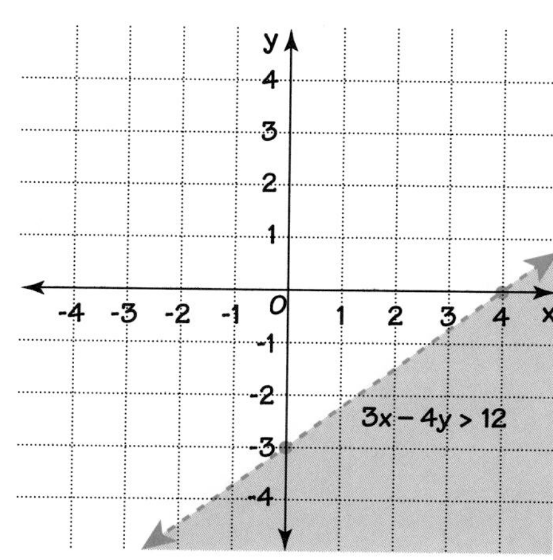

Sentences equivalent to $Ax + By < C$ or $Ax + By \leq C$ are called **linear inequalities.** The preceding examples show that there are two steps to graphing linear inequalities.

Step 1: Graph a dashed or solid line as the solution to the corresponding linear equation.

Step 2: Shade the half-plane that makes the inequality true. (You may have to test a point. If possible, use (0, 0).)

Sometimes the graph of all solutions to a linear inequality is not an entire half-plane.

Watching your money grow. *Some "piggy banks" are made of transparent plastic. That makes it easy to see how much money you have.*

Example 5

Suppose you have less than \$5.00 in nickels and dimes. Find an inequality and sketch a graph to describe how many of each coin you might have.

Solution

Let n = the number of nickels you have.
Let d = the number of dimes you have.
Since each nickel is worth .05 and each dime is worth .10,

$$0.05n + 0.10d < 5.00.$$

Multiply both sides by 100 to clear the decimals.

$$5n + 10d < 500$$

This inequality is graphed at the left.

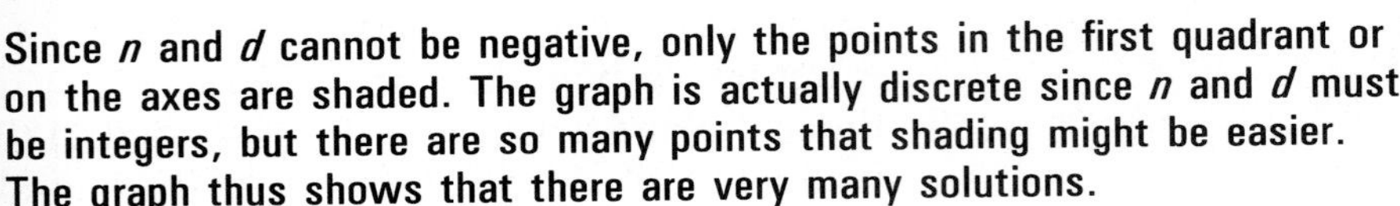

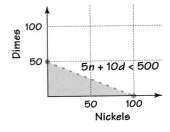

Since n and d cannot be negative, only the points in the first quadrant or on the axes are shaded. The graph is actually discrete since n and d must be integers, but there are so many points that shading might be easier. The graph thus shows that there are very many solutions.

Check

The point (0, 0) is in the shaded region. If you have 0 nickels and 0 dimes, do you have less than \$5.00? Yes.

QUESTIONS

Covering the Reading

In 1 and 2, a sentence is given.
a. Graph the set of points satisfying the sentence on a number line.
b. Graph the points satisfying the sentence on a coordinate plane.

1. $y \geq 2$

2. $x < -3$

In 3 and 4, write an inequality describing each graph.

3. 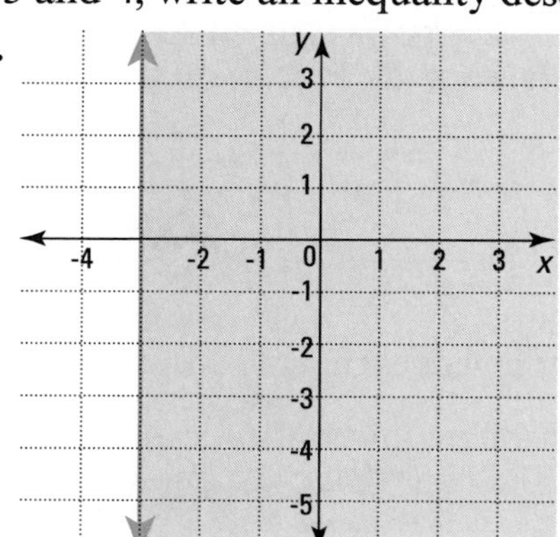

4. 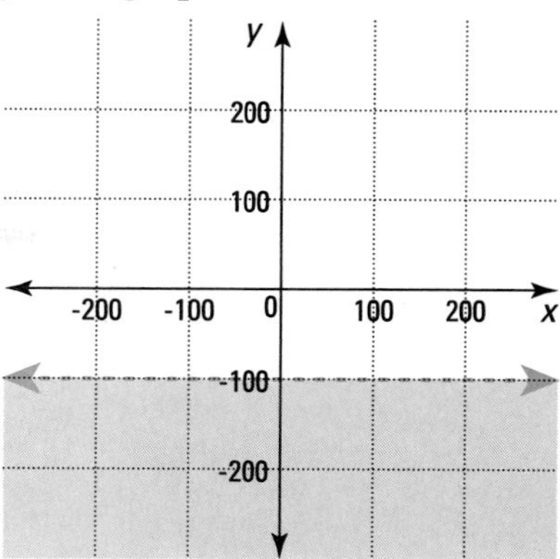

In 5 and 6, an inequality is given. Tell whether the boundary line should be drawn solid or dashed. (You do not need to draw the graph.)

5. $y > -4x - 7$

6. $4x + y < -7$

7. a. What point is usually chosen to test which half-plane is the solution set?

b. When would you not choose that point?

In 8–11, graph all points (x, y) that satisfy the inequality.

8. $x + y > 4$

9. $y \geq -3x - 2$

10. $5x - y > 3$

11. $5x - y \leq 3$

12. Suppose a person has less than \$4.00 all in quarters or dimes. Let x = the number of quarters and y = the number of dimes the person has.

a. Write an inequality to describe this situation.

b. Graph the possible numbers of quarters and dimes the person might have.

Stanley Cup champs. *Shown is Mark Messier of the New York Rangers holding the Stanley Cup trophy after his team won the championship in 1994. This was the Rangers' first championship in 54 years.*

Applying the Mathematics

13. "It will take at least 20 points to make the playoffs," the hockey team coach told the players. "We get 2 points for a win and 1 for a tie." Let W be the number of wins and T the number of ties.

a. Write an inequality to describe the values of W and T that will enable the team to make the playoffs.

b. Graph these values.

14. Suppose CDs cost \$10 and tapes cost \$8. Miguel has \$40 to spend. Let x = the number of CDs, and y = the number of tapes Miguel buys.

a. Write an inequality to describe the number of CDs and tapes Miguel can buy.

b. Graph the solution set to the inequality.

c. Which points on your graph represent ways Miguel can buy CDs and still have money left?

15. Suppose m and n are positive integers.
  a. How many points (m, n) satisfy $m + n < 5$?
  b. Graph them all. Plot m on the x-axis and n on the y-axis.

Review

16. a. Rewrite $4x - 28 = 3y$ in standard form.
  b. Give the values of A, B, and C. *(Lesson 7-8)*

17. The table below gives latitude and the average annual snowfall of some cities in the U.S. that have an average of at least 1″ of snow per year.

City	Latitude (°North)	Average Annual Snowfall (in.)
New York, NY	41	29
Chicago, IL	42	40
Philadelphia, PA	40	20
Boston, MA	42	42
Louisville, KY	38	17
Charlotte, NC	35	6
Dallas, TX	33	3

  a. In general, how is the average annual snowfall of a city related to its latitude?
  b. Find an equation of a line that seems to fit these data.
  c. Use the equation to predict the average annual snowfall of a U.S. city at 36° north latitude.
  d. Nashville, Tennessee, and Las Vegas, Nevada, are both at 36° north latitude. Nashville has about 11″ of snow per year and Las Vegas only 1″. What factors other than latitude affect the amount of snowfall a city receives? *(Lesson 7-7)*

In 18 and 19, consider the rectangular field pictured below.

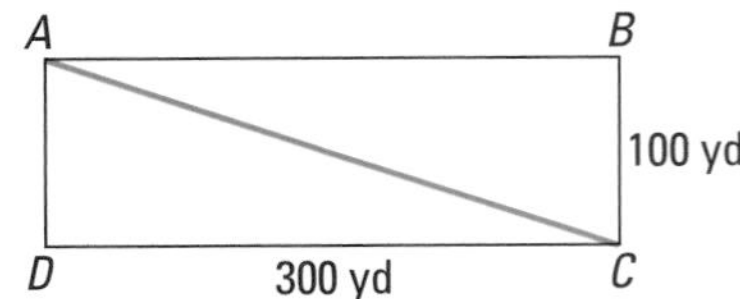

18. How much shorter is the distance from A to C if you walk diagonally across the field instead of along its sides? *(Lesson 1-8)*

19. If D is the origin of a coordinate system with y-axis on $\overleftrightarrow{AD}$ and the x-axis on $\overleftrightarrow{DC}$, what is the slope of $\overleftrightarrow{AC}$? *(Lesson 7-2)*

In 20 and 21, use the graph below.

20. Of the 10,886,000 males aged 25–29 in 1990, how many were single? *(Lesson 5-4)*

21. **a.** What was the average yearly increase in the percentage of females aged 20–24 staying single from 1970 to 1980?
b. Answer part **a** for the decade from 1980 to 1990.
c. Is the rate of increase constant from 1970 to 1990? *(Lesson 7-1)*

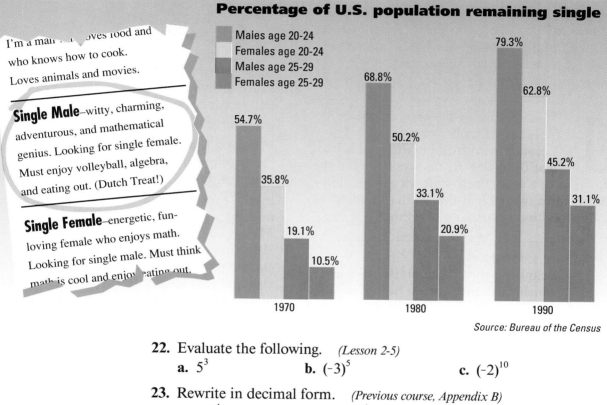

I'm a man ... loves food and who knows how to cook. Loves animals and movies.

Single Male–witty, charming, adventurous, and mathematical genius. Looking for single female. Must enjoy volleyball, algebra, and eating out. (Dutch Treat!)

Single Female–energetic, fun-loving female who enjoys math. Looking for single male. Must think math is cool and enjoy eating out.

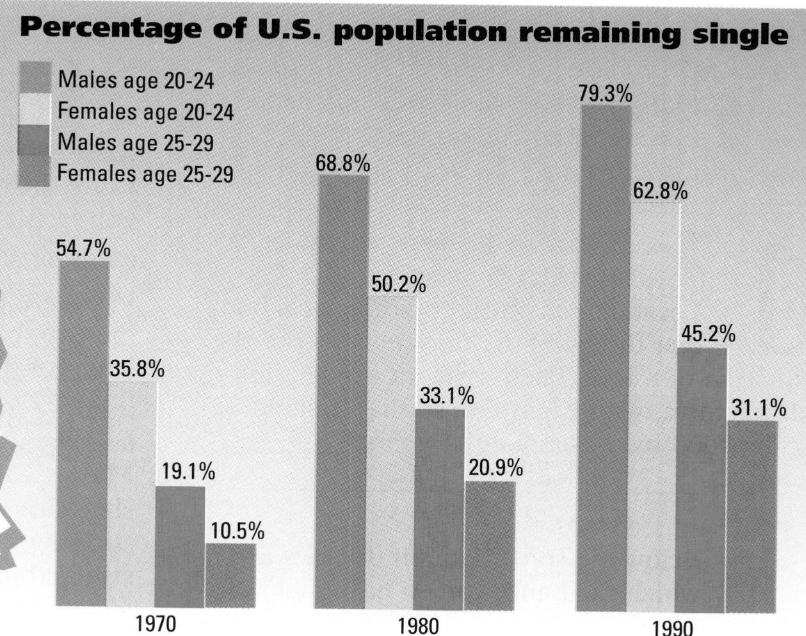

22. Evaluate the following. *(Lesson 2-5)*

a. 5^3 **b.** $(-3)^5$ **c.** $(-2)^{10}$

23. Rewrite in decimal form. *(Previous course, Appendix B)*

a. 10^{-1} **b.** $3 \cdot 10^{-2}$ **c.** $2 \cdot 5 \cdot 10^{-6}$

Exploration

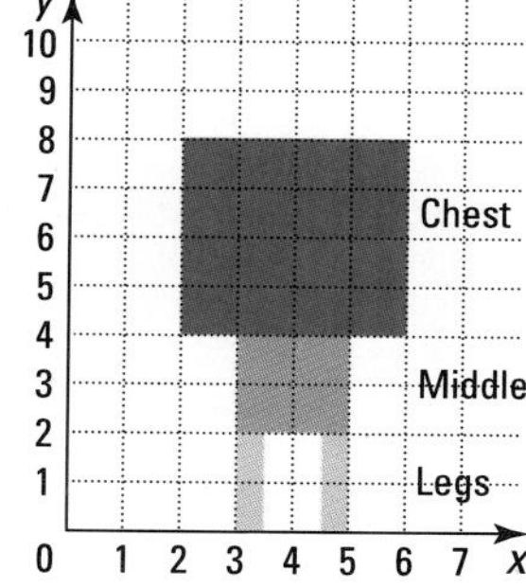

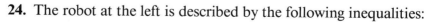

24. The robot at the left is described by the following inequalities:

Chest:	$2 \le x \le 6$	and	$4 \le y \le 8$.
Middle:	$3 \le x \le 5$	and	$2 \le y \le 4$.
Legs:	$3 \le x \le 3.5$	and	$0 \le y \le 2$.
	$4.5 \le x \le 5$	and	$0 \le y \le 2$.

Copy the diagram and put a head on the robot. Describe the head with inequalities.

A project presents an opportunity for you to extend your knowledge of a topic related to the material of this chapter. You should allow more time for a project than you do for typical homework questions.

1 The Intercept Form of an Equation of a Line

Locate an advanced mathematics book that discusses the *intercept form* of the equation of a line. Then write an explanation of this form, with examples, so that it could be understood by a classmate of yours.

2 Marriage Age

Below are some data on the ages at which men and women have first married in the past 40 years.

Median Age (in years) at First Marriage

Year	Men	Women
1950	22.8	20.3
1960	22.8	20.3
1970	23.2	20.8
1980	24.7	22.0
1990	26.1	23.9

a. Plot all the data on one graph. Describe some trends in your graph.

b. Use an automatic grapher to find the lines that best fit the data. Make some predictions based on these.

c. Find some earlier data (for the years 1900, 1910, . . . , 1940) on age of first marriage. Do the earlier data suggest any ways you might modify the predictions you made?

3 Foot Length and Shoe Size

Find ten males and ten females who agree to have you measure their feet.

a. For each person measure each foot (without shoes) to the nearest $\frac{1}{8}$ of an inch, and record the person's shoe size. If the two feet are different lengths, take the larger of the two measurements. For instance, one woman's feet are $9\frac{5}{8}''$ long, and she wears a size 7 shoe.

b. Make two graphs, one showing how men's shoe size varies with foot length, the other showing the relation between women's shoe size and foot length. In each graph, plot foot length on the x-axis.

c. Are the graphs approximately linear? If so, find equations for them. Compare them to the equations given in Lesson 4-4.

4 Describing Figures with Coordinates

In the Exploration for Lesson 7-9, you were given a drawing of a robot in a coordinate plane. Design your own figure in the coordinate plane. Write a set of clear instructions using equations and inequalities to graph that figure. Ask a friend to use your instructions to reproduce the figure you designed.

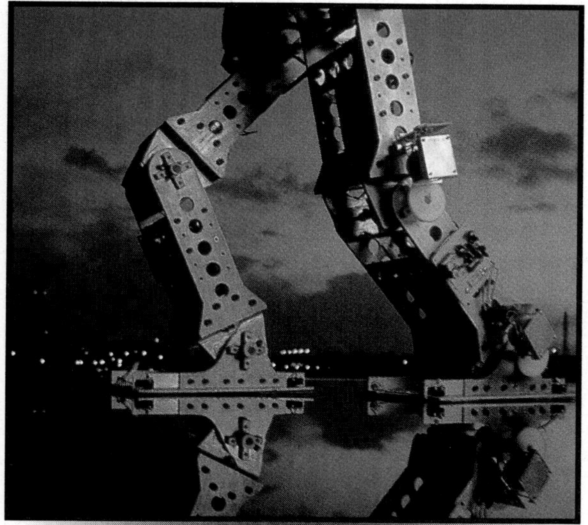

5 Populations of Cities

The population of Manhattan grew at a great pace from 1790 to 1910. Since that time, it has tended to decline. In 1990, its population was only about $\frac{2}{3}$ what it was in 1910. Has this period of great growth followed by a period of sizeable decline happened in other large cities in the U.S.? Examine the census data for the largest cities to see whether Manhattan is unusual, or whether this is a common pattern. Write a summary of what you find.

6 Paper Towels: Price vs. Absorbency

Are more expensive paper towels more absorbent than less expensive ones? Get samples of about six different kinds of paper towels, record their prices, and calculate the price paid per towel. Perform the following experiment to measure absorbency. Fold one piece of towel in half vertically, and then in half again horizontally.

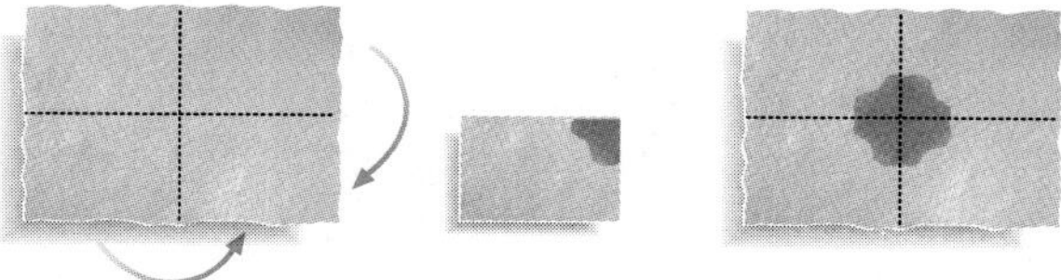

Fill an eyedropper with a fixed amount of water and drop it on the corner with the folds. Open the towel and record and measure the diameter of the circular area that is wet. Repeat for each type of towel you have, making sure you use the same amount of water each time.

a. Plot your data with unit price on the horizontal axis and diameter on the vertical axis. Is the relation linear? If so, find a line to describe your data.

b. Plot your data with unit price on the horizontal axis and the area of the wet region on the vertical axis. If the relation is linear, describe it with a line.

c. What advice would you give someone shopping for paper towels?

SUMMARY

The rate of change between two points (x_1, y_1) and (x_2, y_2) is $\frac{y_2 - y_1}{x_2 - x_1}$. When points all lie on the same line, the rate of change between them is constant and is called the slope of the line. The slope tells how much the line rises or falls for every move of one unit to the right. When the slope is positive, the line goes up to the right. When the slope is negative, the line falls to the right. When the slope is 0, the line is horizontal. The slope of vertical lines is not defined.

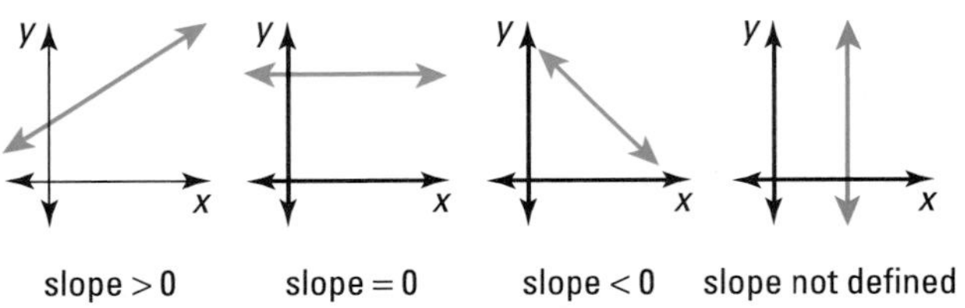

Constant-increase or constant-decrease situations lead naturally to linear equations of the form $y = mx + b$. The graph of the set of points (x, y) satisfying this equation is a line with slope m and y-intercept b.

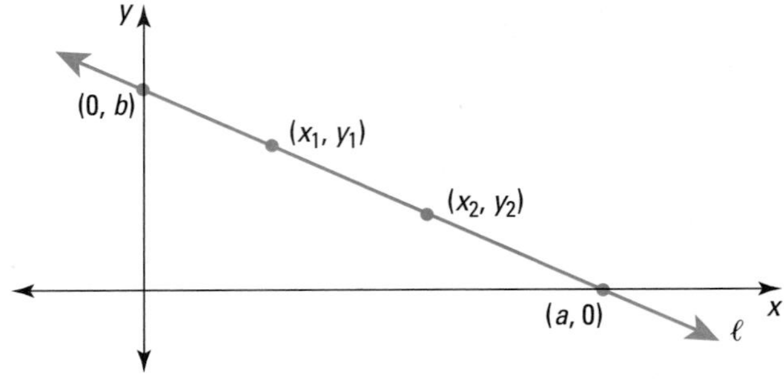

Line ℓ has slope $m = \frac{y_2 - y_1}{x_2 - x_1}$, y-intercept b, and x-intercept a.

Other situations lead to linear equations in the standard form $Ax + By = C$. When the = sign in equations of either form is replaced by $<$ or $>$, the graph of the resulting linear inequality is a half-plane, the set of points on one side of a line.

A line is determined by any point on it and its slope, and its equation can be found from this information. Likewise, an equation can be found for the line containing two given points. If more than two points are given, then there may be more than one line determined. You can then draw a line that comes close to all the points, and use these points to determine an equation for the line.

VOCABULARY

Below are the most important terms and phrases for this chapter. You should be able to give a definition or general description and a specific example of each.

Lesson 7-1
rate of change

Lesson 7-2
constant decrease
slope

Lesson 7-4
slope-intercept form for an equation of a line
Slope-Intercept Property
y-intercept

Lesson 7-5
x-intercept

Lesson 7-7
latitude
fitting a line to data

Lesson 7-8
standard form for an equation of a line
oblique line

Lesson 7-9
boundary point
half-plane
linear inequality

PROGRESS SELF-TEST

In 1–3, refer to the line graphed at the right.

1. Give its y-intercept.
2. Give its x-intercept.
3. Calculate its slope.

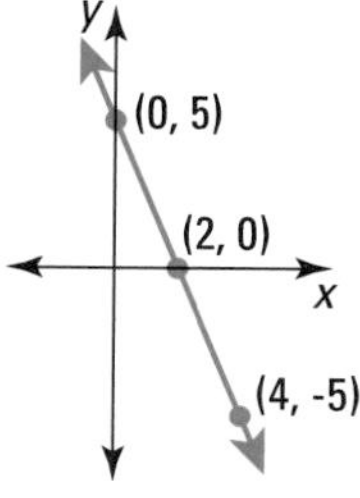

4. Do the points (4, -5), (-2, 1) and (20, -20) all lie on the same line? Explain your answer.

In 5 and 6, find the slope and y-intercept.

5. $y = 8 - 4x$
6. $5x + 2y = 1$
7. Find an equation of the line with slope $\frac{3}{4}$ and y-intercept 13.
8. Rewrite the equation $y = 5x - 2$ in standard form $Ax + By = C$ and give the values of A, B, and C.
9. What is the slope of every horizontal line?
10. If a line's slope is $\frac{3}{5}$, how will the y-coordinate change as you go one unit to the right?

In 11 and 12, use the following data of total U.S. Army personnel during and after World War II.

Year	Total Personnel
1942	3,074,184
1943	6,993,102
1944	7,992,868
1945	8,266,373
1946	1,889,690

11. Between which two consecutive years was there the greatest increase in U.S. Army personnel?
12. What is the rate of change in Army personnel per year from 1942 to 1946?
13. A basketball team scored 67 points, from x baskets worth 2 points each and y free throws worth 1 point each. Write an equation that describes all possible values of x and y.
14. a. Graph the line with slope -2 containing the point (-5, 6).
    b. Find an equation for this line.
15. At age 12 Patrick weighed 43 kg; at 14 he weighed 50 kg. Find a linear equation relating Patrick's weight y to his age x.

In 16–18, graph.

16. $y = 5x - 4$
17. $-3x + 2y = 12$
18. $y \leq x + 1$
19. *Multiple choice.* Which of the four lines below left could be the graph of $y = -x + 3$?

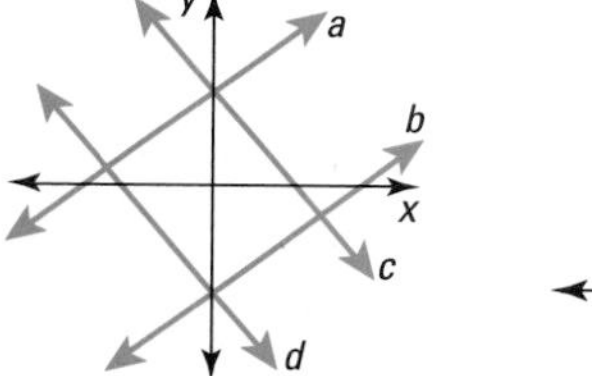

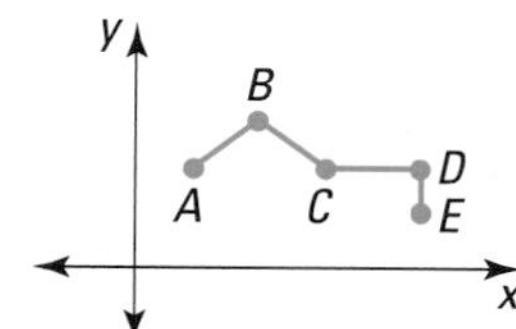

20. Using the graph above at the right, name the segment that
    a. has a positive slope;
    b. does not have a slope.
21. The scatterplot below illustrates the wingspan and length of 2-engine and 3-engine jet planes.
    a. Estimate the coordinates of two points on a line that would fit the data.
    b. Find the slope of your line.
    c. Write an equation for your line.
    d. Use the equation to estimate the length of a jet with a wingspan of 100 feet. Show your work.

Length and wingspan of 2- and 3-engine jets

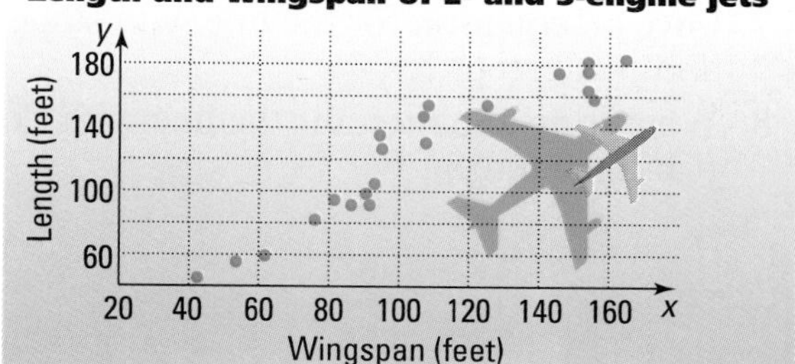

CHAPTER REVIEW

Questions on SPUR Objectives

SPUR stands for **S**kills, **P**roperties, **U**ses, and **R**epresentations. The Chapter Review questions are grouped according to the SPUR Objectives for this chapter.

SKILLS DEAL WITH THE PROCEDURES USED TO GET ANSWERS.

Objective A: *Find the slope of the line through two given points.* *(Lesson 7-2)*

1. Calculate the slope of the line containing (2, 4) and (6, 2).
2. Calculate the slope of the line through (1, 5) and (-2, -2).
3. Find the slope of line ℓ at the left below.

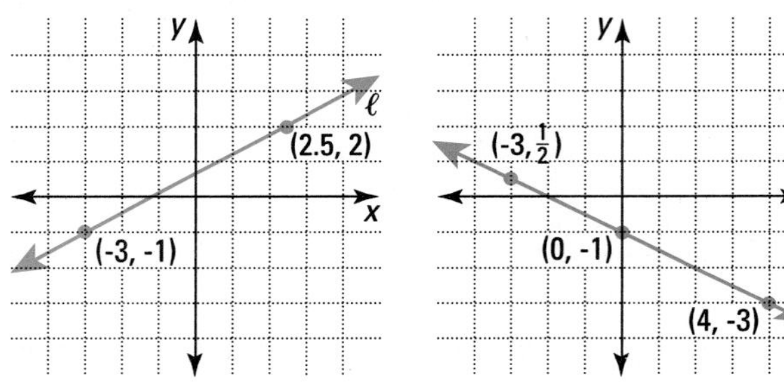

4. Use two different pairs of points to confirm that the slope of line m above is $-\frac{1}{2}$.

In 5 and 6, the points (4, 10) and (8, y) are on the same line. Find y when the slope of the line is the number given.

5. -2
6. $\frac{5}{4}$

Objective B: *Find an equation for a line given two points on it, or its slope and one point on it.* *(Lessons 7-4, 7-5, 7-6)*

7. Give an equation for the line with slope 4 and y-intercept 3.
8. What is an equation for the line with slope p and y-intercept q?

In 9–11, find an equation for the line through the given point with slope m.

9. (-4, 1), $m = -2$
10. (6, 10), $m = 0$
11. $\left(3, \frac{1}{4}\right)$, $m = 30$
12. What is an equation for the line through (3, -1) with undefined slope?

On 13–16, find an equation for the line through the two given points.

13. (5, -2), (-7, -8)
14. (0.5, 6), (0, 4)
15. (6, 9), (6, 0)
16. (-5, 2), (3, 2)

Objective C: *Write an equation for a line in standard form or slope-intercept form, and from either form find its slope and y-intercept.* *(Lessons 7-4, 7-8)*

In 17 and 18, write in the form $Ax + By = C$. Then give the values of A, B, and C.

17. $x - 22 = 5y$
18. $y = \frac{2}{5}x + 7$

In 19 and 20, rewrite the equation in slope-intercept form.

19. $2x + y = 4$
20. $x + 3y = 11$

In 21–24, find the slope and the y-intercept of the line.

21. $y = 7x - 3$
22. $4x + 5y = 1$
23. $y = -x$
24. $48x - 3y = 30$

PROPERTIES DEAL WITH THE PRINCIPLES BEHIND THE MATHEMATICS.

Objective D: *Use the definition and properties of slope.* *(Lessons 7-2, 7-3)*

25. Find the slope of the line through (*a, b*) and (*c, d*).
26. The slope determined by two points is the change in the _?_ coordinates divided by the _?_ in the *x*-coordinates.
27. Slope is the amount of change in the _?_ of the graph for every change of one unit to the _?_.

In 28 and 29, refer to the graph at the right.

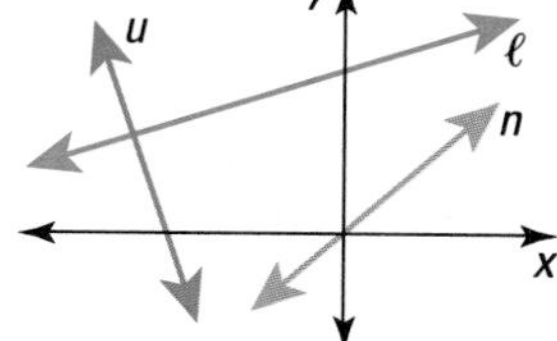

28. Which line or lines have negative slope?
29. Which line or lines have positive slope?
30. What is the slope of any horizontal line?
31. What is true about the slope of a vertical line?
32. Explain how you can use slope to determine whether three points all lie on the same line.

USES DEAL WITH APPLICATIONS OF MATHEMATICS IN REAL SITUATIONS.

Objective E: *Calculate rates of change from real data.* *(Lessons 7-1, 7-3)*

33. Assume an ascending jet plane climbs 0.46 km for each kilometer it travels away from its starting point.
    a. Draw a picture of this situation.
    b. What is the slope of the ascent?
34. The picture at the right represents part of a ski slope. What is the slope of this part?

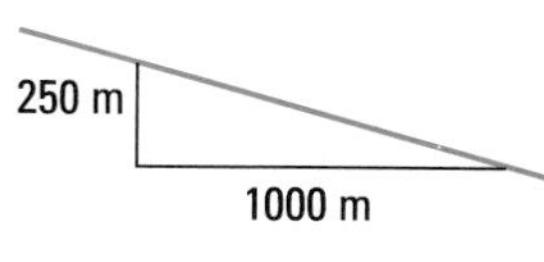

In 35–37, use the average height (in cm) for girls between birth and age 14 given below.

Age (yr)	Height (cm)
birth	50.8
2	83.8
4	99.0
6	111.7
8	127.0
10	137.1
12	147.3
14	157.5

35. Find the rate of change of height from age 10 to 12.
36. Find the rate of change of height between age 2 and age 14.
37. a. According to these data, in which two-year period do girls gain height fastest?
    b. What is this rate of change?

In 38–40, use the chart below of temperatures recorded at Capitol City Airport in Lansing, Michigan, starting at 8 P.M. on January 15, 1994.

	January 15 P.M.					January 16 A.M.											P.M.
time	8	9	10	11	12	1	2	3	4	5	6	7	8	9	10	11	12
temp. (°F)	-3	-8	-9	-6	-3	-3	-3	-4	-4	-9	-11	-11	-11	-8	-8	-5	-5

38. Find the rate of change in temperature per hour from midnight to 8 A.M.
39. What was the rate of change of temperature per hour over the period of time reported?
40. a. During which two-hour period did the temperature decrease the most?
    b. In that period, what was the average rate of change of the temperature per hour?

Objective F: *Use equations for lines to describe real situations.* *(Lessons 7-4, 7-5, 7-6, 7-8)*

In 41 and 42, each situation can be represented by a straight line. Give the slope and the *y*-intercept of the line describing this situation.

41. Julie rents a truck. She pays an initial fee of \$15 and then \$0.25 per mile. Let *y* be the cost of driving *x* miles.
42. Nick is given \$50 to spend on a vacation. He decides to spend \$5 a day. Let *y* be the amount Nick has left after *x* days.

CHAPTER SEVEN

In 43 and 44, each situation leads to an equation of the form $y = mx + b$. Find that equation.

43. A student takes a test and gets a score of 50. He gets a chance to take the test again. He estimates that every hour of studying will increase his score by 3 points. Let x be the number of hours studied and y be his score.

44. A plane loses altitude at the rate of 5 meters per second. It begins at an altitude of 8500 meters. Let y be its altitude after x seconds.

45. Julio plans a diet to gain 0.2 kg a day. After 2 weeks he weighs 40 kg. Write an equation relating Julio's weight w to the number of days d on his diet.

46. Each month about 50 new people come to live in a town. After 3 months the town has 25,500 people. Write an equation relating the number of months m to the number of people p in the town.

47. Robert babysat for \$2.50 an hour and mowed lawns for \$5 an hour. He earned a total of \$25. Write an equation that describes the possible babysitting hours B and lawn mowing hours L he could have spent at these jobs.

48. The games of the 21st Modern Olympiad were in 1976. The games of the 20th Olympiad were 4 years earlier. Let y be the year of the nth summer Olympic games. Give a linear equation which relates n and y.

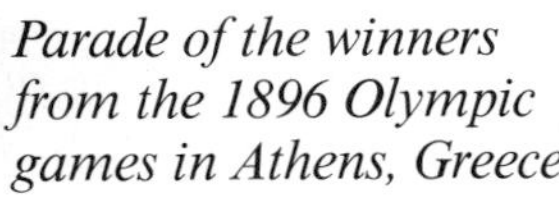

Parade of the winners from the 1896 Olympic games in Athens, Greece

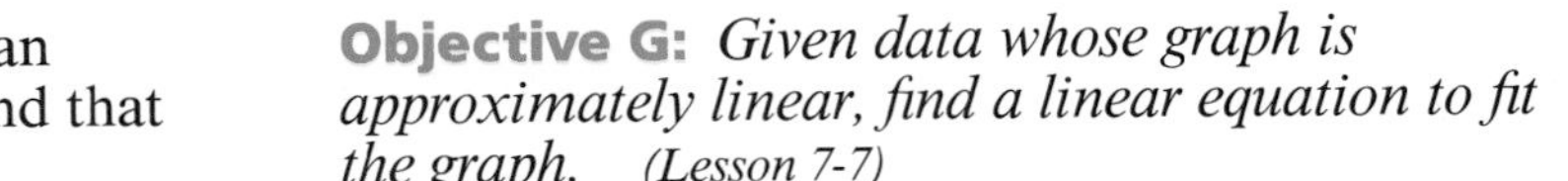

Objective G: *Given data whose graph is approximately linear, find a linear equation to fit the graph.* (*Lesson 7-7*)

49. Olympic swimmers tend to get faster over time. The scatterplot below shows the times for the winners of the Women's 400-meter freestyle for the years 1924 to 1992.

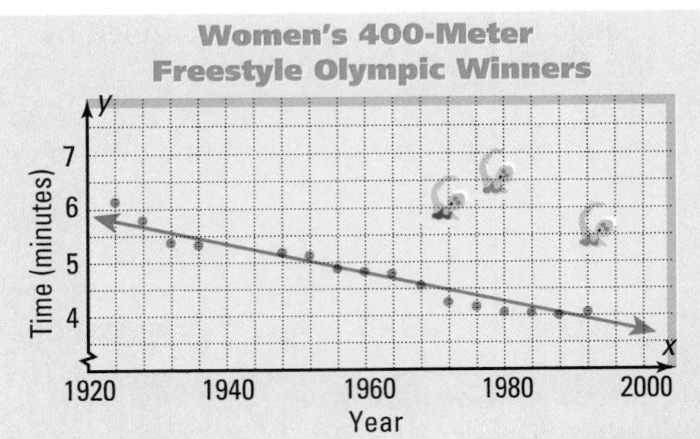

Year	Winner	Time	Decimal Time
1924	**Martha Norelius,** U.S.	6:02.2	6.04
1928	**Martha Norelius,** U.S.	5:42.8	5.71
1932	**Helene Madison,** U.S.	5:28.5	5.48
1936	**H. Mastenbroek,** Netherlands	5:26.4	5.44
1948	**Anne Curtis,** U.S.	5:17.8	5.30
1952	**Valerie Gyenge,** Hungary	5:12.1	5.20
1956	**Lorraine Crapp,** Australia	4:54.6	4.91
1960	**Chris von Saltza,** U.S.	4:40.6	4.68
1964	**Virginia Duenkel,** U.S.	4:43.3	4.72
1968	**Debbie Meyer,** U.S.	4:31.8	4.53
1972	**Shane Gould,** Australia	4:19.44	4.32
1976	**Petra Thümer,** E. Germany	4:09.89	4.17
1980	**Ines Diers,** E. Germany	4:08.76	4.15
1984	**Tiffany Cohen,** U.S.	4:07.10	4.12
1988	**Janet Evans,** U.S.	4:03.85	4.06
1992	**Dagmar Hase,** Germany	4:07.18	4.12

The times have been converted to decimal parts of a minute: 6:02.2 is graphed as 6.04 minutes. Use the converted times to complete the instructions below.

a. Draw a line to fit the data.

b. Find the slope of the fitted line.

c. Explain what the slope tells you about the trend in these data.

d. Find an equation for the line.

e. Predict the winning time in 1996.

CHAPTER SEVEN

50. The amount of gold mined in the world increased throughout the 1980s, as shown in the table below.

Year	World Gold Production (millions troy ounces)
1980	39.2
1983	45.2
1986	51.5
1989	65.3
1992	72.3

(Source: Bureau of Mines, U.S. Department of the Interior)

a. Graph the data and a line to fit it.

b. Find an equation for the line.

c. Use the equation to predict the amount of gold produced in the world in 1995.

Shown are two men who are "gold-washing" at a mine in Brazil.

REPRESENTATIONS DEAL WITH PICTURES, GRAPHS, OR OBJECTS THAT ILLUSTRATE CONCEPTS.

Objective H: *Graph a straight line given its equation, or given a point and the slope.* *(Lessons 7-3, 7-4, 7-8)*

In 51–54, graph the line with the given equation.

51. $y = -2x + 4$ **52.** $y = \frac{1}{2}x - 3$

53. $8x + 5y = 400$ **54.** $x - 3y = 11$

In 55–58, graph the line satisfying the given condition.

55. passes through (0, 4) with a slope of 4

56. passes through (-2, 4) with a slope of $-\frac{3}{4}$

57. slope 0 and y-intercept 3

58. slope 2 and y-intercept -4

Objective I: *Graph linear inequalities.* *(Lesson 7-9)*

59. What are the regions on either side of a line in a plane called?

60. If you have only x nickels and y quarters and a total of less than \$2.00, graph all possible values of x and y.

In 61–68, graph.

61. $x \geq 5$ **62.** $y \leq 4$

63. $y < -3$ **64.** $x > 0$

65. $y \geq x + 1$ **66.** $y < -3x + 2$

67. $3x + 2y > 5$ **68.** $x - 8y \leq 0$

CHAPTER 8

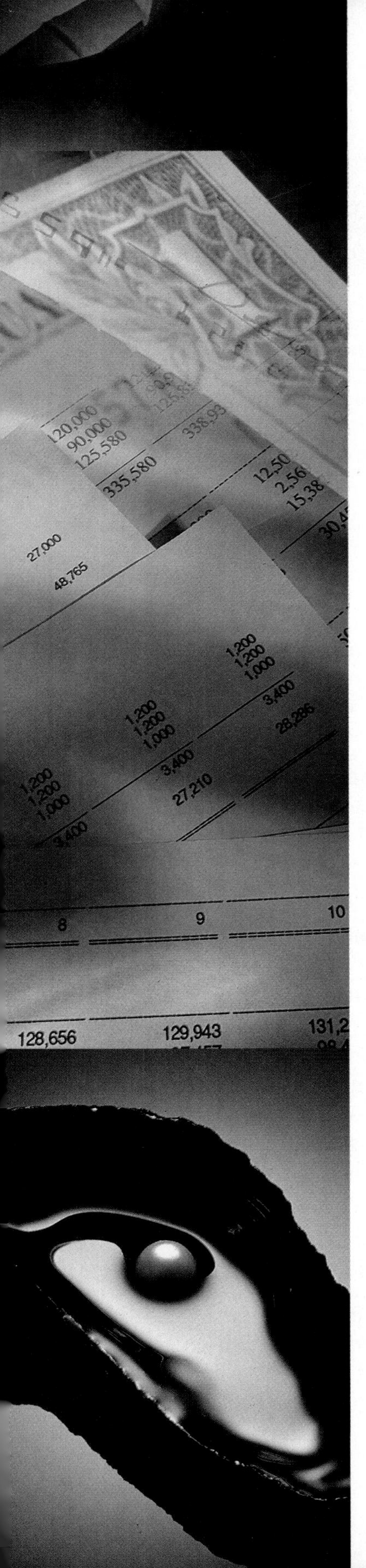

EXPONENTS AND POWERS

A city of 100,000 people is planning for the future. The planners want to know how many schools the town will need during the next 50 years. They consider three possibilities.

(1) The population will stay the same.
(2) The population will increase by 2,000 people per year. (Increases by a constant amount.)
(3) The population will grow by 2% a year. (Increases at a constant growth rate.)

Here is a graph of what would happen under the three possibilities. P is the population n years from now.

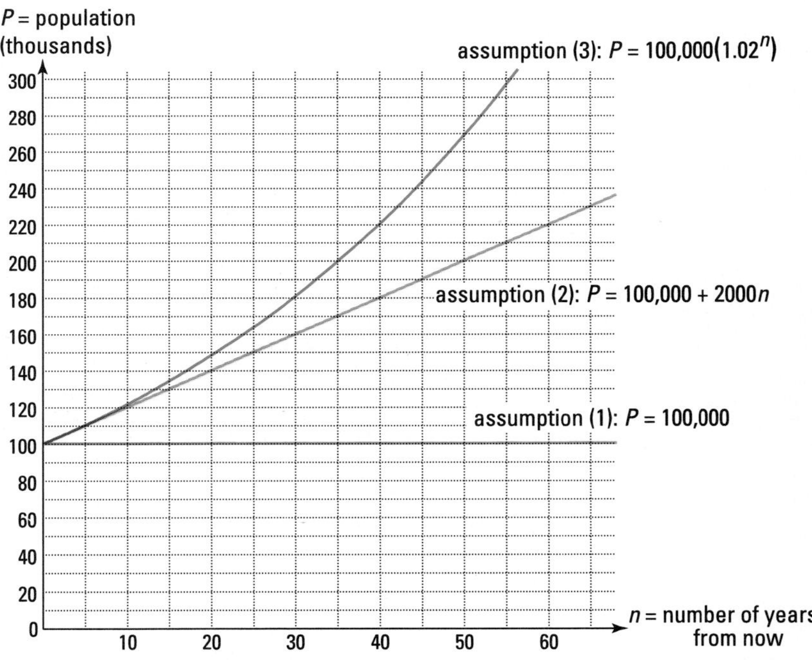

Possibility (3) is often considered the most reasonable. Under this assumption, $P = 100{,}000(1.02)^n$. Because the variable n is an exponent, this equation is said to represent *exponential growth.* Exponential growth is among the many applications of exponents and powers you will study in this chapter.

LESSON

8-1

Compound Interest

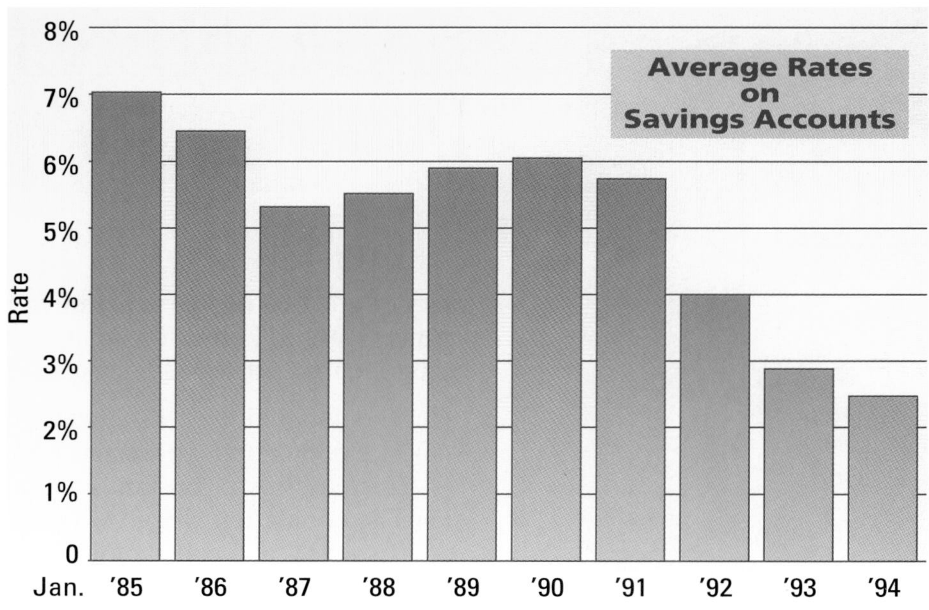

Changing bank interest rates. *Interest rates used by U.S. banks are based on many economic factors. Between 1985 and 1994, the trend was toward lower interest rates on savings accounts.*

Powers and Repeated Multiplication

A number having the form x^n is called a **power.** Certain powers can be changed to decimals by repeated multiplication. For instance, $3^4 = 3 \cdot 3 \cdot 3 \cdot 3 = 81$ and $10^7 = 10 \cdot 10 \cdot 10 \cdot 10 \cdot 10 \cdot 10 \cdot 10 = 10{,}000{,}000$. These are instances of the following model.

Repeated Multiplication Model for Powering
When n is a positive integer, $x^n = \underbrace{x \cdot x \cdot \ldots \cdot x}_{n \text{ factors}}$.

The number x^n is called the ***n*th power** of x and is read "x to the nth power" or just "x to the n." In the expression x^n, x is the **base** and n is the **exponent.** Thus 3^4 is read "3 to the 4th power," or "3 to the 4th"; 3 is the base and 4 is the exponent. In the expression $100{,}000(1.02)^n$ found on the previous page, 1.02 is the base and n is the exponent. The number 100,000 is the **coefficient** of the power 1.02^n.

How Is Interest Calculated?

An important application of exponents and powers involves savings accounts. Banks, savings and loan associations, and credit unions will pay you to give them your money to keep for you. The amount you give them is called the **principal.** The amount they pay you is called **interest.**

Interest is always a percent of the principal. The percent the money earns per year is called the **annual yield.**

Example 1

Suppose you deposit P dollars in a savings account upon which the bank pays an annual yield of 3%. If you make no other deposits or withdrawals, how much money will be in the account at the end of a year?

Solution 1

$$\begin{aligned} \text{Total} &= \text{principal} + 3\% \text{ of principal} \\ &= P + 0.03P \\ &= (1 + 0.03)P \\ &= 1.03P \end{aligned}$$

Solution 2

$$\begin{aligned} \text{Total} &= 100\% \text{ of principal} + 3\% \text{ of principal} \\ &= 103\% \text{ of principal} \\ &= 1.03P \end{aligned}$$

So, if you deposited \$1000 in a savings account with an annual yield of 3%, a bank would pay you 0.03(\$1000), or \$30, interest. You would have 1.03(\$1000), or \$1030, after one year.

Compound Interest, and How It Is Calculated

Savings accounts pay **compound interest,** which means that the interest earns interest.

Example 2

Suppose you deposit \$100 in a savings account upon which the bank pays an annual yield of 3%. If the account is left alone, how much money will be in it at the end of four years?

Solution

Refer to Example 1. Each year the amount in the bank is multiplied by $1 + 0.03$, or 1.03.

End of first year:

$$100(1.03) = 100(1.03)^1 = 103.00$$

End of second year:

$$100(1.03)(1.03) = 100(1.03)^2 = 106.09$$

End of third year:

$$100(1.03)(1.03)(1.03) = 100(1.03)^3 \approx 109.2727 \approx 109.27$$

End of fourth year:

$$100(1.03)(1.03)(1.03)(1.03) = 100(1.03)^4 \approx 112.5509 \approx 112.55$$

At the end of four years there will be \$112.55 in the account.

Examine the pattern in the solution to Example 2. At the end of n years there will be

$$100(1.03)^n$$

dollars in the account. The general formula for compound interest uses this expression, but variables replace the principal and annual yield.

Compound Interest Formula
If a principal P earns an annual yield of i, then after n years there will be a total T, where $T = P(1 + i)^n$.

The compound interest formula is read "T equals P times the quantity 1 plus i, that quantity to the nth power."

Example 3

$1500 is deposited in a savings account. What will be the total amount of money in the account after 10 years at an annual yield of 6%?

Solution

Here $P = \$1500$, $i = 6\%$, and $n = 10$.
Substitute the values in the Compound Interest Formula. Use $6\% = 0.06$.

$$T = P(1 + i)^n$$
$$= 1500(1 + 0.06)^{10}$$

Use a calculator key sequence such as one of the following:

1500 [×] 1.06 [y^x] 10 [=],
or 1500 [×] 1.06 [^] 10 [ENTER].

Displayed will be 2686.2715, which is approximately $2686.27. After 10 years, the account will contain $2686.27.

Home improvements. *Many banks offer home equity loans to customers who own homes. People often use the money from these loans to make home improvements, such as adding a deck.*

In Example 3, because the $1500 will increase to $2686.27, the saver will earn $1186.27. Since 6% of $1500 is $90, if the interest is taken out each year for 10 years, the saver will earn $900. The compounding will increase the interest by $286.27.

Ten years may seem like a long time, but it is not an unusual amount of time for money to be in retirement accounts or in accounts parents set up to save for their children's college expenses.

Why Do You Receive Interest on Savings?

Banks and other savings institutions pay you interest because they want money to lend to other people. Banks earn money by charging a higher rate of interest on the money they lend than the rate they pay customers who deposit money. Thus, if the bank loans your $1000 (perhaps to someone buying a car or a house) at 12%, the bank receives 0.12($1000), or $120, from that person. If the bank pays you 3% interest, or $30, the bank earns $120 − $30, or $90 on your money.

QUESTIONS

Covering the Reading

1. **a.** Calculate 6^3 without a calculator.
   b. Calculate 6^3 with a calculator (show your key sequence).

2. Refer to assumptions (1), (2), and (3) on page 485. What is the predicted population in 50 years under each assumption?

3. Consider the expression $50x^{10}$. Name each.
   a. base
   b. power
   c. exponent
   d. coefficient

4. Match each term with its description.
   a. money you deposit
   b. interest paid on interest
   c. yearly percentage paid

   (i) annual yield
   (ii) compound interest
   (iii) principal

In 5 and 6, write an expression for the amount in the bank after one year if P dollars are in an account with an annual yield as given.

5. 6%

6. 2.75%

7. Consider the situation of Example 2.
   a. How much money will you have in your savings account at the end of 5 years?
   b. How much interest will you have earned?

8. Follow Example 2 to explain why $100(1.054)^3$ is the total value of \$100.00 invested for 3 years at an annual yield of 5.4%.

9. **a.** Write the Compound Interest Formula.
   b. What does T represent?
   c. What does P stand for?
   d. What is i?
   e. What does n represent?

10. Write a calculator key sequence for your calculator to evaluate $573(1.063)^{24}$.

11. Suppose you deposit \$150 in a new savings account paying an annual yield of 4%. If the account is left alone, how much money will be in the account at the end of 5 years?

12. A bank advertises an annual yield of 3.81% on a 6-year CD (certificate of deposit). If the CD's original amount was \$2,000, how much will it be worth after 6 years?

13. How much interest will be earned in 7 years on a principal of \$500 at an annual yield of 5.6%?

Applying the Mathematics

14. Susan invests \$100 at an annual yield of 5%. Jake invests \$100 at an annual yield of 10%. They leave the money in the bank for 2 years.
a. How much interest does each person earn?
b. Does Jake earn exactly twice the interest that Susan does?

15. Which results in more interest, (a) an amount invested for 5 years at an annual yield of 6%, or (b) the same amount invested for 3 years at an annual yield of 10%? Justify your answer.

In 16 and 17, use the spreadsheet below. It was produced to show the interest and total owed on a loan account of \$1000 with a monthly interest rate of 1.5%, assuming none was paid back.

	A	B	C
1	Month	Total Owed	Monthly Interest
2	1	1015.00	15.00
3	2	1030.23	15.23
4	3	1045.68	
5	4	1061.36	15.69
6	5	1077.28	15.92
7	6	1093.44	16.16
8	7		16.40
9	8	1126.49	16.65
10	9	1143.39	16.90
11	10	1160.54	17.15
12	11	1177.95	17.41
13	12	1195.62	17.67

16. a. What formula should be typed into cell B8?
b. What value should be in B8?
c. What is the *yearly* interest rate?

17. a. What formula should be typed into cell C4?
b. What value should be in C4?

18. Use your calculator. If a principal of \$1000 is saved at an annual yield of 8% and the interest is kept in the account, in how many years will it double in value?

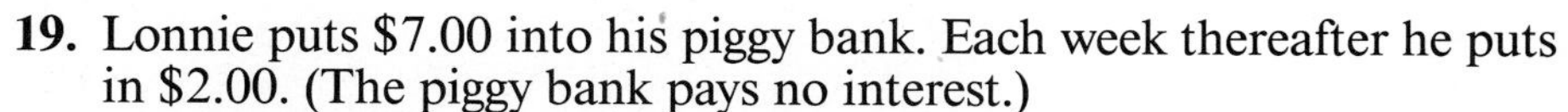

Review

19. Lonnie puts $7.00 into his piggy bank. Each week thereafter he puts in $2.00. (The piggy bank pays no interest.)

a. Write an equation showing the total amount of dollars T after W weeks.

b. Graph the equation. *(Lesson 7-2)*

In 20 and 21, find the probability of the event. *(Lesson 6-4)*

20. getting a prime number in one toss of a fair die

21. getting a face card (jack, queen, or king) when one card is pulled at random from a standard deck

22. Solve $(1 + x)^2 = 1.1664$. *(Lesson 5-9)*

23. *Skill sequence.* Simplify. *(Lessons 2-1, 3-9)*

a. $12(3n)$ **b.** $12(3n - 7)$ **c.** $-12(3n - 7)$ **d.** $n(3n - 7)$

24. *Multiple choice.* Which formula describes the numbers in this table? *(Lesson 1-9)*

x	1	2	3	4	5
y	2	4	8	16	32

(a) $y = 2x$ (b) $y = x + (x + 1)$ (c) $y = x^2$ (d) $y = 2^x$

25. Evaluate $-4x^5$ when $x = \frac{1}{2}$. *(Lesson 1-3)*

26. The volume V of a sphere with diameter d is given by the formula $V = \frac{\pi}{6}d^3$. Find, to the nearest cubic millimeter, the volume of a ball bearing with a diameter of 8 mm. *(Lesson 1-3)*

27. Find t if $2^t = 64$. *(Previous course)*

Exploration

28. Find out the yield for a savings account in a bank or other savings institution near where you live. (Often these yields are in newspaper ads.)

29. Explore the meaning of 0 as an exponent on your calculator.

a. Enter 25 [y^x] 0 [=] on your calculator. (You may have to use [x^y] or [^] instead of [y^x].) What is displayed?

b. Enter 0 [y^x] 0 [=] on your calculator. What is displayed?

c. Enter [(] 1 [+/−] [)] [y^x] 0 [=] on your calculator. What is displayed?

d. Try other numbers and generalize what happens.

LESSON

8-2

Exponential Growth

A vegetarian diet. *Wild rabbits, such as these from Australia, feed on grass, weeds, bushes, and trees. Crops are sometimes damaged because rabbits eat vegetable sprouts and the bark of fruit trees.*

Powering and Population Growth

An important application of powers is in population growth situations. As an example, consider rabbit populations, which can grow quickly. In 1859 in Australia, 22 rabbits were imported from Europe as a new source of food. Rabbits are not native to Australia, but conditions there were ideal for rabbits, and they flourished. Soon, there were so many rabbits that they damaged grazing land. By 1889, the government was offering a reward for a way to control the rabbit population.

Example 1

Twenty-five rabbits are introduced to an area. Assume that the rabbit population doubles every six months. How many rabbits will there be after 5 years?

Solution

Since the population doubles twice each year, in 5 years it will double 10 times. The number of rabbits will be

$$25 \underbrace{\cdot 2 \cdot 2 \cdot 2 \cdot 2 \cdot 2 \cdot 2 \cdot 2 \cdot 2 \cdot 2 \cdot 2}_{\text{10 factors}}.$$

To evaluate this expression on a calculator rewrite it as

$$25 \cdot 2^{10}.$$

Use the y^x or ^ key. After 5 years there will be 25,600 rabbits.

What Is Exponential Growth?

The rabbit population in Example 1 is said to grow exponentially. In **exponential growth,** the original amount is repeatedly *multiplied* by a positive number called the *growth factor.*

> **Growth Model for Powering**
> If a quantity is multiplied by a positive number g (the **growth factor**) in each of x time periods, then after the x periods, the quantity will be multiplied by g^x.

In Example 1, the population doubles (is multiplied by 2) every six months, so $g = 2$. There are 10 time periods (think 5 years = 10 half-years), so $x = 10$. The original number of rabbits, 25, is multiplied by g^x, or 2^{10}.

Compound interest is another example of the growth model. Suppose an annual yield is 5%. Then the growth factor $g = 1.05$. If the money is kept in for 7 years, then $x = 7$. The original principal is multiplied by g^x, which is 1.05^7.

What Happens if the Exponent Is Zero?

In the growth model, x can be any real number. Consider the statement of the growth model when $x = 0$. It reads:

> If a quantity is multiplied by a positive number g in each of 0 time periods, then after the 0 time periods, the quantity will be multiplied by g^0.

In 0 time periods, no time can elapse. Thus the quantity remains the same. It can remain the same only if it is multiplied by 1. This means that $g^0 = 1$, regardless of the value of the growth factor g. This property applies also when g is a negative number.

> **Zero Exponent Property**
> If g is any nonzero real number, then $g^0 = 1$.

In words, the zero power of any nonzero number equals 1. For example, $4^0 = 1$, $(-2)^0 = 1$, and $\left(\frac{5}{7}\right)^0 = 1$. The zero power of 0, which would be written 0^0, is undefined.

What Does a Graph of Exponential Growth Look Like?

An equation of the form $y = b \cdot g^x$, where g is a number greater than 1, can describe exponential growth. Graphs of such equations are not lines. They are *exponential growth curves.*

Example 2

Graph the equation $y = 3 \cdot 2^x$, when x is 0, 1, 2, 3, 4.

Solution

Substitute $x = 0, 1, 2, 3,$ and 4 into the formula $y = 3 \cdot 2^x$. Below we show the computation and the results listed as (x, y) pairs.

$$3 \cdot 2^0 = 3 \cdot 1 = 3$$
$$3 \cdot 2^1 = 3 \cdot 2 = 6$$
$$3 \cdot 2^2 = 3 \cdot 4 = 12$$
$$3 \cdot 2^3 = 3 \cdot 8 = 24$$
$$3 \cdot 2^4 = 3 \cdot 16 = 48$$

x	$y = 3 \cdot 2^x$
0	3
1	6
2	12
3	24
4	48

Plot these points. The graph is shown below.

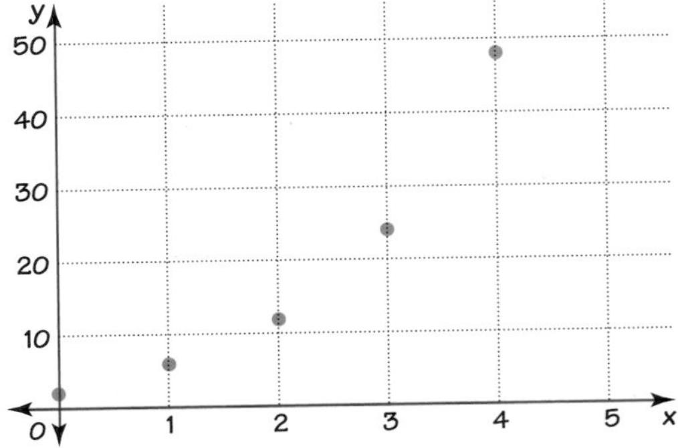

Notice that the graph in Example 2 does not have a constant rate of change. When something grows exponentially, its rate of change is continually increasing.

An Example of Exponential Population Growth

Other than money calculated using compound interest, few things in the real world grow exactly exponentially. Over the short term, however, exponential growth approximates the changes that have been observed in many different populations. For instance, consider the population of California from 1930 to 1990.

Census	Population
1930	5,677,251
1940	6,907,387
1950	10,586,223
1960	15,717,204
1970	19,971,069
1980	23,667,764
1990	29,760,021

To see how close California's population growth is to exponential growth, compare the two graphs below. On the left is a graph of California's population. On the right we have plotted points on the graph of $y = 5.68(1.32)^x$. We picked 1.32 because it is an "average" growth factor for California for the six decades. Except for the differences in the scale on the horizontal axis, they look quite a bit alike.

Streets of San Francisco. *Shown is a trolley car on one of San Francisco's busy streets. Like other parts of California, San Francisco experienced rapid growth from 1930 to 1945. In 1993, its population was 723,959.*

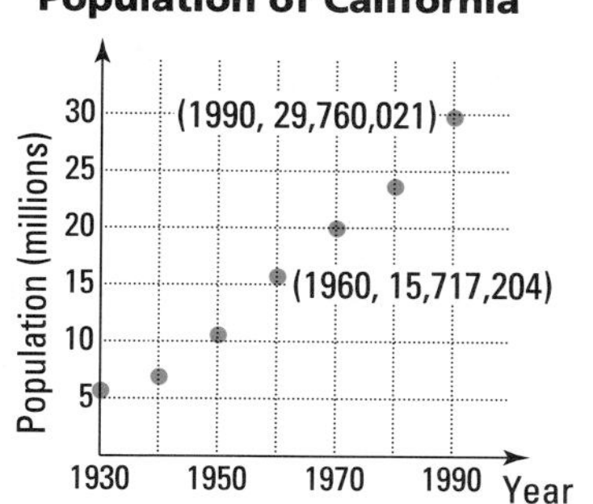

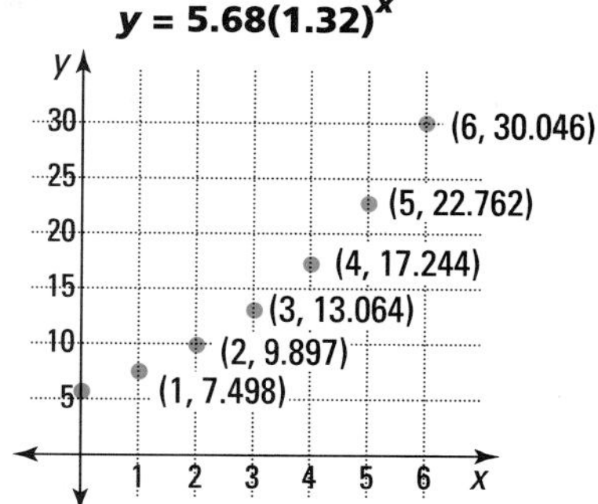

QUESTIONS

Covering the Reading

In 1–3, refer to Example 1.

1. **a.** After 7 years, how many times will the rabbit population have doubled?
b. How many rabbits will there be?

2. How many rabbits will there be after 10 years?

3. If the rabbit population triples in 6 months, rather than doubles, how many rabbits would there be after 5 years?

4. Assume the rabbit population in Australia doubled every *year* from 1859 to 1889. How many rabbits were in Australia in 1889?

5. State the Growth Model for Powering.

6. Suppose money is put into a savings account with an annual yield of 3.5%. It is left in the account for 2 years.
a. What is the growth factor?
b. What is the unit period?
c. By how much will the money be multiplied in the 2-year period?

7. *True or false.* An amount is multiplied by 10 in each of 7 time periods. Then after the 7 periods, the original amount will be multiplied by 70.

8. State the Zero Exponent Property.

9. Calculate 17^0.

10. Explain how $g^0 = 1$ fits the Growth Model for Powering.

11. **a.** Graph $y = 1.5 \cdot 2^x$ for $x = 0, 1, 2, 3$, and 4.
   b. What is the curve that contains these points called?

In 12 and 13, refer to the population data for California.

12. What has been an "average" growth factor for California for the 6 decades from 1930 to 1990?

13. Suppose California's population grows by 20% in the decade from 1990 to 2000.
   a. What would be the growth factor for that decade?
   b. What would the population be in the year 2000?

Applying the Mathematics

14. **a.** Graph $y = 1.5^x$ for $x = 0, 1, 2, 3, 4$, and 5.
   b. Calculate the rate of change on the graph from $x = 0$ to $x = 1$.
   c. Calculate the rate of change on the graph from $x = 4$ to $x = 5$.
   d. What do the answers to parts **b** and **c** tell you about this graph?

15. The following chart describes the exponential growth of a colony of bacteria. You can see that this strain of bacteria grows very fast. In only one hour it grows from 2,000 to 54,000 bacteria.

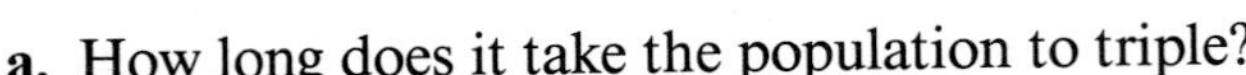

Time Intervals from Now	Time (min)	Number of Bacteria
0	0	2000
1	20	$6000 = 2000 \cdot 3^1$
2	40	$18{,}000 = 2000 \cdot 3^2$
3	60	$54{,}000 = 2000 \cdot 3^3$

   a. How long does it take the population to triple?
   b. How many times will the population triple in two hours?
   c. How many bacteria will be in the colony after two hours?
   d. How many bacteria will be in the colony after four hours?

The marketplace. *This is an outdoor market in Jamaica, an island in the West Indies. Sugar cane is Jamaica's most important crop. Other farm products include bananas, cacao, coconuts, coffee, and citrus fruits.*

16. Jamaica's population of 2,466,000 in 1990 was expected to grow exponentially by 1.1% each year for the rest of the twentieth century.
   a. With this growth, what will the population be in 1995?
   b. With this growth, what will the population be in 2010?

17. On September 30, 1992, the U.S. national debt was about 4.065 trillion dollars and was growing at a rate of about 12.5% per year. At this rate, estimate the national debt on September 30, 1994.

In 18–21, simplify.

18. $(4y)^0$ when $y = \frac{1}{2}$

19. $7^0 \cdot 7^1 \cdot 7^2$

20. $6(x + y)^0$ when $x = 3$ and $y = -8$

21. $\frac{2}{3}\left(\frac{1}{2}\right)^0 + \frac{1}{2}\left(\frac{2}{3}\right)^2$

Review

22. \$2200 is deposited in a savings account. *(Lesson 8-1)*
  a. What will be the total amount of money in the account after 6 years at an annual yield of 6%?
  b. How much interest will have been earned in those 6 years?

23. Suppose a letter from the alphabet is chosen randomly. What is the probability that it is a vowel from the first half of the alphabet? *(Lesson 6-4)*

24. *Skill sequence.* Solve. *(Lesson 5-9)*
  a. $y^2 = 144$
  b. $(4y)^2 = 144$
  c. $(4y - 20)^2 = 144$
  d. $y^2 + 80 = 144$

In 25 and 26, simplify. *(Lessons 3-9, 4-6)*

25. $6(n + 8) + 4(2n - 1)$

26. $13 - (2 - x)$

In 27 and 28, evaluate the expression. *(Lesson 1-3)*

27. $4s^9$ when $s = \frac{1}{2}$

28. $t^2 \cdot t^3$ when $t = 11$

Exploration

29. This exploration helps to explain why 0^0 is undefined.
  a. Use your calculator to give values of x^0 for $x = 1, 0.1, 0.01, 0.001$, and so on. What does this suggest for the value of 0^0?
  b. Use your calculator to give values of 0^x for $x = 1, 0.1, 0.01, 0.001$, and so on. What does this suggest for the value of 0^0?
  c. What does your calculator display when you try to evaluate 0^0? Why do you think it gives that display?

30. There is an old story about a man who did a favor for a king. The king wished to reward the man and asked how he could do so. The man asked for a chessboard with one kernel of wheat on the first square of the chessboard, two kernels on the second square, four on the third square, eight on the fourth square, and so on for the entire sixty-four squares of the board. Find how many grains of wheat would be on the last square. (The answer may amaze you. It is about 250 times the total present yearly wheat production of the world.)

LESSON

8-3

Comparing Constant Increase and Exponential Growth

Uncle Scrooge. *Walt Disney's Uncle Scrooge generously doled out money to Donald Duck's nephews, Huey, Dewey, and Louie. If they had invested this money wisely, their savings might have grown exponentially.*

What Is the Difference Between Constant Increase and Exponential Growth?

In constant-increase situations, a number is repeatedly *added.* In exponential-growth situations, a number is repeatedly *multiplied.* If the growth factor g is greater than 1, exponential growth always overtakes constant increase. This is illustrated in the following example.

Example 1

Suppose you have \$10. Your rich uncle agrees each day either to (1) increase what you had the previous day by \$50, or (2) increase what you had the previous day by 50%. Which is the better option?

Solution

Make a table to compare the two options during the first week. The exponential growth factor is $1 + 50\% = 1 + 0.5 = 1.5$.

	option (1): add 50	option (2): multiply by 1.5
1st day	$10 + 50 \cdot 1 = \$60.00$	$10 \cdot 1.5^1 = \$15.00$
2nd day	$10 + 50 \cdot 2 = \$110.00$	$10 \cdot 1.5^2 = \$22.50$
3rd day	$10 + 50 \cdot 3 = \$160.00$	$10 \cdot 1.5^3 = \$33.75$
4th day	$10 + 50 \cdot 4 = \$210.00$	$10 \cdot 1.5^4 \approx \$50.63$
5th day	$10 + 50 \cdot 5 = \$260.00$	$10 \cdot 1.5^5 \approx \$75.94$
6th day	$10 + 50 \cdot 6 = \$310.00$	$10 \cdot 1.5^6 \approx \$113.91$
7th day	$10 + 50 \cdot 7 = \$360.00$	$10 \cdot 1.5^7 \approx \$170.86$
nth day	$10 + 50n$	$10 \cdot 1.5^n$

In all the first 7 days, option (1) is better. But examine the 14th day. Substitute 14 for n in the bottom row.

14th day $10 + 50 \cdot 14 = \$710$ $\quad$ $10 \cdot 1.5^{14} = \$2919.29$

In the long run, option (2), increasing by 50% each day, is the better choice.

To see what has happened in Example 1, examine the spreadsheet below, which lists the amount received on each of the first 14 days.

	A	B	C
1	day	constant increase	exponential growth
2	x	10 + 50x	10 * 1.5^x
3	0	10	10.00
4	1	60	15.00
5	2	110	22.50
6	3	160	33.75
7	4	210	50.63
8	5	260	75.94
9	6	310	113.91
10	7	360	170.86
11	8	410	256.29
12	9	460	384.44
13	10	510	576.65
14	11	560	864.98
15	12	610	1297.46
16	13	660	1946.20
17	14	710	2919.29

How Do the Graphs of Constant Increase and Exponential Growth Compare?

Another way to compare the two options is to graph the two equations

$y = 10 + 50x$ and $y = 10 \cdot 1.5^x$.

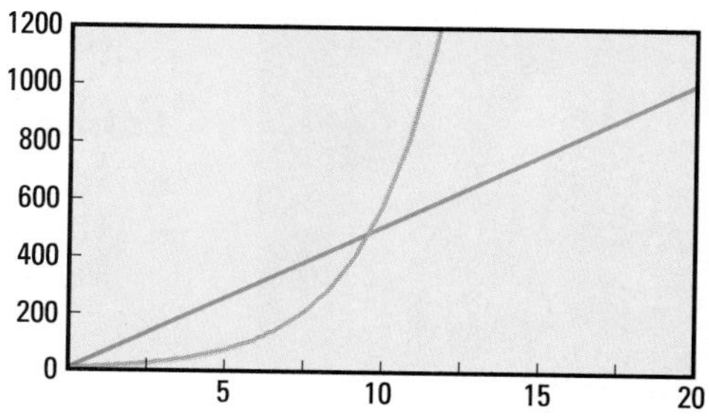

The graphs at the right were drawn with an automatic grapher. Each graph has y-intercept of 10. As you learned in Chapter 7, a line has a constant rate of change. But the graph of $y = 10 \cdot 1.5^x$ is not a line. It is a curve that gets steeper and steeper as you move to the right. Notice that at first the exponential curve is below the line. But toward the middle of the window, it intersects the line and passes above it. On later days, the graph of the curve rises farther and farther above the line. The longer your uncle gives you money, the better option 2 is when compared to option 1.

A Summary of Constant Increase and Exponential Growth

Here is how constant increase and exponential growth compare, in general.

Constant Increase	Exponential Growth
1. Begin with an amount b.	1. Begin with an amount b.
2. *Add* m (the slope) in each of x time periods.	2. *Multiply* by g (the growth factor) in each of x time periods.
3. After the x time periods, the amount will be $b + mx$.	3. After the x time periods, the amount will be $b \cdot g^x$.

The difference can be seen in the graph of each type.

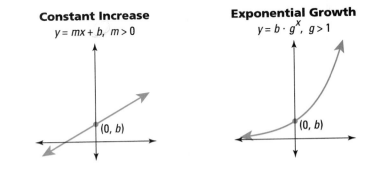

Constant Increase

$y = mx + b,\ m > 0$

(0, b)

Exponential Growth

$y = b \cdot g^x,\ g > 1$

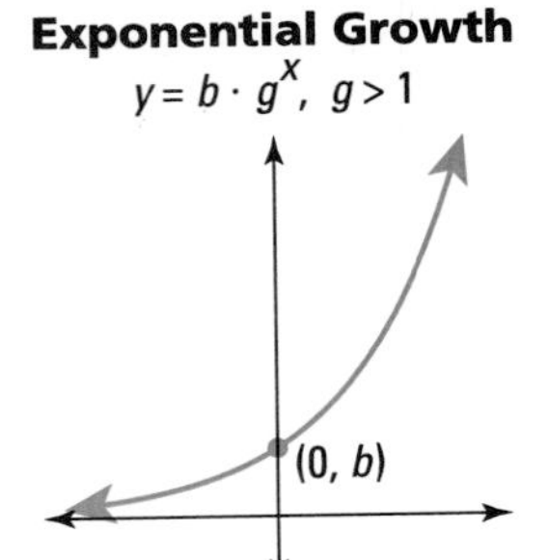

Mind over mattress. *Comedian Jack Benny played a character who stored his money in his home. However, in this scene from his TV show, even he realized it is wiser to save at a bank.*

Example 2

Suppose you have saved \$100 and are able to earn an annual yield of 6%.

a. Graph your savings if you leave your \$100 in the bank, but take the interest out and put it in a piggy bank each year. (This is *not* compound interest.)

b. On the same set of axes, graph your savings if you leave the interest in the bank account. Round values to the nearest dollar.

c. After 30 years, how much more will you have if you let your interest compound in the bank account?

Solution

a. Each year 6% of the original \$100 (or \$6) is earned in interest. If y is the total savings you have after x years, then $y = 100 + 6x$. Make a table.

number of years	0	5	10	15	20	25	30
value	\$100	\$130	\$160	\$190	\$220	\$250	\$280

b. Make a table for the amount saved at 6% compound interest. Use the formula $y = 100(1.06)^x$.

number of years	0	5	10	15	20	25	30
value	\$100	\$134	\$179	\$240	\$321	\$429	\$574

Plot the points and connect them for each graph. In the graph below, the piggy bank graph (blue) is one of constant increase. The compound interest graph (orange) is of exponential growth.

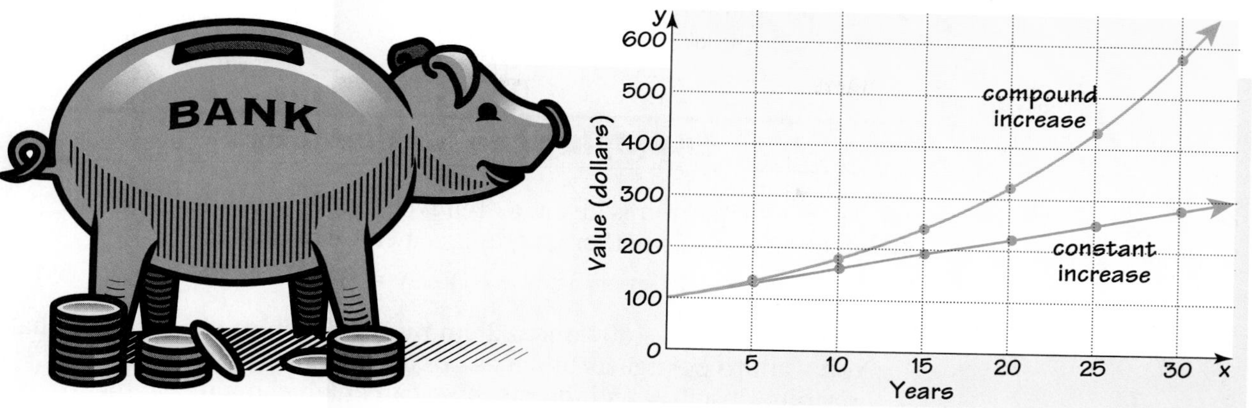

c. You can use either the table or the graph. From the table you see that after 30 years the first plan yields $280, while the second plan yields $574. *You will have $574 – $280 = $294 more with compound interest.*

QUESTIONS

Covering the Reading

1. What is the difference between a constant-increase situation and an exponential-growth situation?

2. In Example 2, as x increases, which increases more rapidly, $y = 100 + 6x$ or $y = 100(1.06)^x$?

In 3–5, use Example 1 or the spreadsheet that follows it.

3. After one month (30 days), how much money would you have under option (1)? under option (2)?

4. Which is the first day the amount received under exponential growth is greater than the amount received under constant increase?

5. Give the numbers that go in row 18 of the spreadsheet.

6. How does the graph of $y = 10 + 5x$ differ from the graph of $y = 10 \cdot 1.5^x$?

7. Give the general formula for the indicated type of situation.
a. constant increase **b.** exponential growth

In 8 and 9, refer to Example 2.

8. How much more will the amount be in 15 years at compound interest than with the piggy bank?

9. Use the graph to estimate when the investment will be worth $500 with compound interest.

Applying the Mathematics

In 10–13, an equation is given. **a.** Tell whether its graph is linear or exponential. **b.** Tell whether its graph is a curve or a line.

10. $y = 3x - 2$ **11.** $y = 3x$ **12.** $y = 3^x$ **13.** $y = 3(1.05)^x$

14. Suppose you have 60¢ and a 2 cm by 3 cm picture of a building that you want to enlarge as much as possible. Copies on a photocopy machine cost 10¢ and the machine can enlarge any image put through it to 120% of its original size. How large a picture of the building can you obtain?

Teen trends. *In 1960, 10.2 million teenagers were enrolled in U.S. high schools. By 1991, 13.1 million teenagers attended high school—about a 28% increase in student enrollment.*

15. A school board is making long-range budget plans. This year Central High went from 2400 to 2520 students. The number of students may be increasing at a constant rate of 120 students per year or exponentially by 5% per year.

a. Copy and complete the spreadsheet below. Show the future enrollments in the two possible situations.

	A	B	C
1	years from now	constant increase	exponential
2	0	2520	2520
3	1		
4	2		
5	3		
6	4		
7	5		

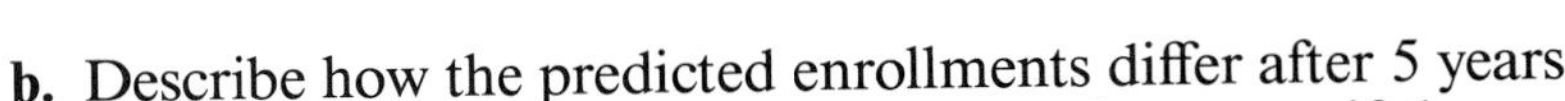

b. Describe how the predicted enrollments differ after 5 years.

c. Find the projected enrollment 15 years from now if the number of students increases at a constant rate.

d. Find the projected enrollment 15 years from now if the number of students grows exponentially.

In 16–19, *multiple choice.* Each graph is drawn on the window $0 \le x \le 10$, $0 \le y \le 4000$. Match the graph with its equation.

(a) $y = 200 + 400x$ (b) $y = 200 + 40x$
(c) $y = 200 \cdot 1.4^x$ (d) $y = 200 \cdot 1.24^x$

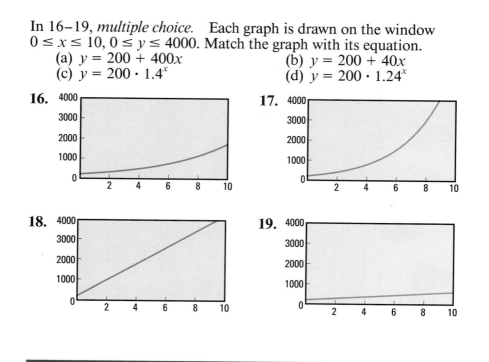

16. 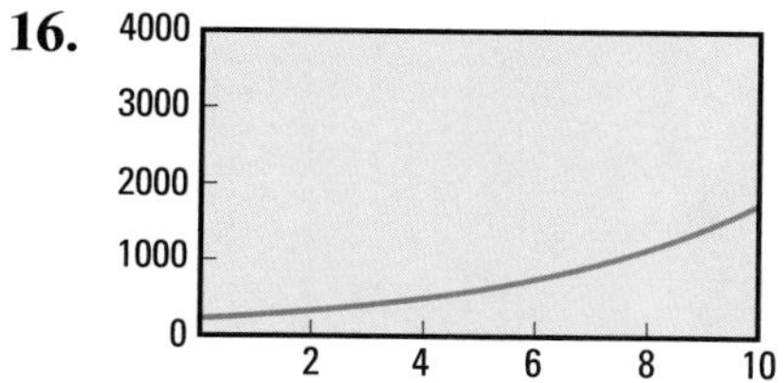

17. 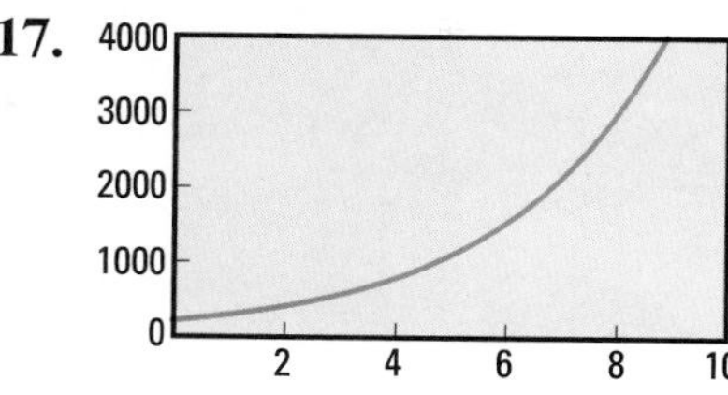

18. 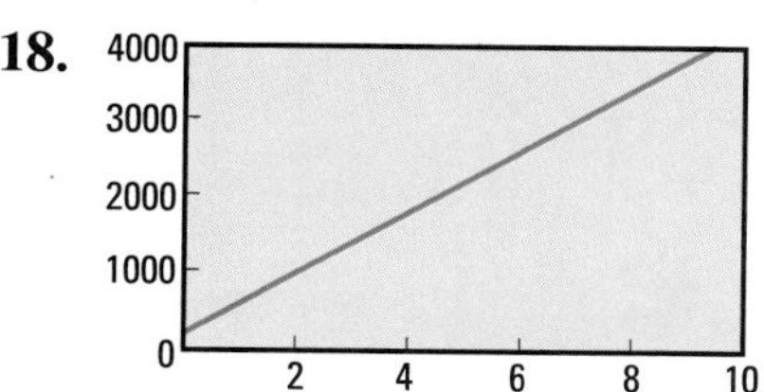

19. 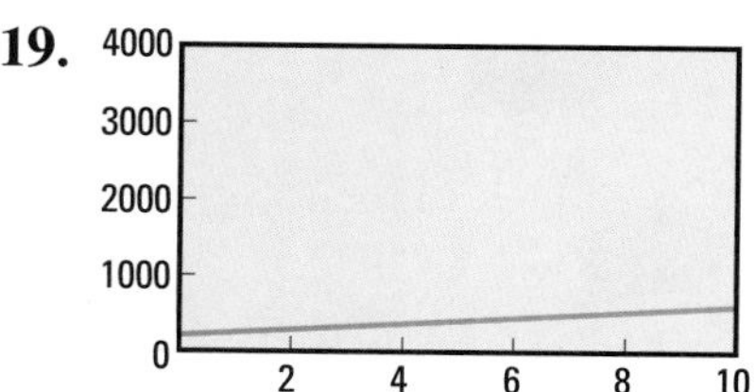

Review

20. a. Evaluate $-4x^n$ when $x = 1.2$ and $n = 3$;
b. Evaluate $4x^n$ when $x = -1.2$ and $n = 0$. *(Lessons 1-4, 8-2)*

21. a. Copy and complete the computer program below to find the total T in the bank when \$100 is invested at an annual yield of 6% throughout the next 20 years.

```
10 PRINT "$100 AT 6 PER CENT"
20 PRINT "YEARS", "TOTAL"
30 FOR YEAR = 0 TO 20
40 LET TOTAL = _?_
50 PRINT YEAR, TOTAL
60 NEXT YEAR
```

b. Write the last line that will be printed when this program is run.
c. How would you modify line 30 to print the table for YEAR = 1, 2, 3, . . . , 100?
d. After the change in part **c** is made, what will be the last line printed when the program is run? *(Lesson 8-1)*

22. According to the Repeated Multiplication Model of Powering, $x^5 =$ _?_. *(Lesson 8-1)*

23. The graph below shows the average diameter of rocks in a stream at half-mile intervals from the stream's source.

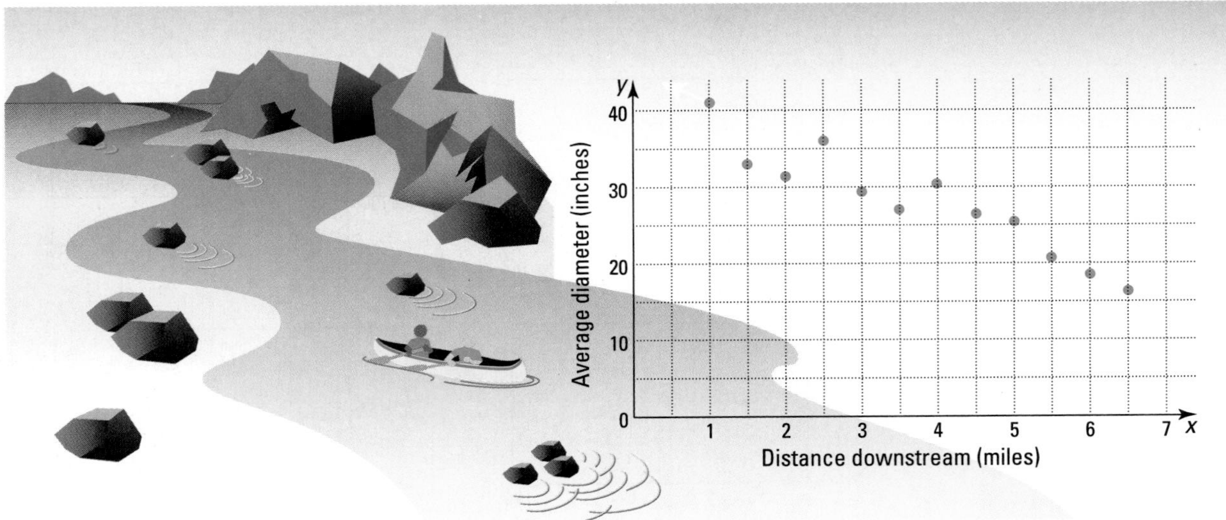

a. Fit a line to the data.
b. Find the coordinates of two points on your line in part **a.**
c. Determine an equation for your line using the points in part **b.**
d. Predict the average diameter of rocks that are 7.5 miles from the stream's source. *(Lesson 7-7)*

24. Solve $1.3d + 5.2 = 9.4d - 9.0$. *(Lessons 5-3, 5-6)*

In 25 and 26, simplify. *(Lesson 3-9)*

25. $4a^2(a + 3) + 2a(a^2 - 1)$

26. $b(b + 3) - b(b + 1)$

27. *Skill sequence.* Add and simplify. *(Lesson 3-9)*

a. $\frac{4}{x} + \frac{5}{x}$
b. $\frac{4}{y} + \frac{5}{2y}$
c. $4 + \frac{5}{2z}$

28. Recall that if an item is discounted x%, you pay $(100 - x)$% of the original price. Calculate in your head the amount you pay for a jacket originally costing \$200 and discounted the indicated amount.

a. 10%
b. 20%
c. 33% *(Lesson 3-3)*

29. Evaluate $(2x)^2(3y)^3$ when $x = 4$ and $y = -2$. *(Lesson 1-4)*

Exploration

30. Use your calculator to evaluate $y = 100(1.06)^x$ when $x = -1$ and $x = -2$. What do the answers mean for the situation of Example 2?

LESSON

8-4

Exponential Decay

French-speaking regions. *This is a street scene in Quebec, where French is the official language spoken. French is also the official language of Belgium, France, Haiti, Switzerland, parts of Africa, and the French West Indies.*

An Example of Exponential Decay

A student crams for a Friday French test, learning 100 vocabulary words Thursday night. Each day the student expects to forget 20% of the words known the day before. If the test is delayed from Friday to Monday, what will happen if the student does not review?

To answer this question, referring to a table is convenient. Since 20% of the words are forgotten each day, 80% are remembered.

Day	Day Number	Words Known
Thursday	0	100
Friday	1	$100(.80) = 80$
Saturday	2	$100(.80)(.80) = 100(.80)^2 = 64$
Sunday	3	$100(.80)(.80)(.80) = 100(.80)^3 = 51.2 \approx 51$
Monday	4	$100(.80)(.80)(.80)(.80) = 100(.80)^4 = 40.96 \approx 41$

The pattern is like that of the growth model or compound interest. After d days, this student will know about $100(0.80)^d$ words. Because the growth factor 0.80 is less than 1, the number of words remembered decreases. This type of situation is called **exponential decay.**

What Are the Characteristics of a Graph of Exponential Decay?

A table of values and a graph of this situation is shown at the top of the next page. As with exponential growth, the points lie on a curve rather than a straight line.

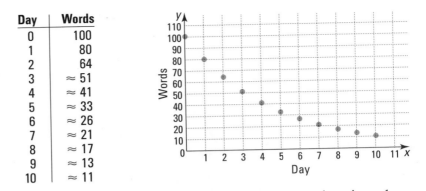

Day	Words
0	100
1	80
2	64
3	≈ 51
4	≈ 41
5	≈ 33
6	≈ 26
7	≈ 21
8	≈ 17
9	≈ 13
10	≈ 11

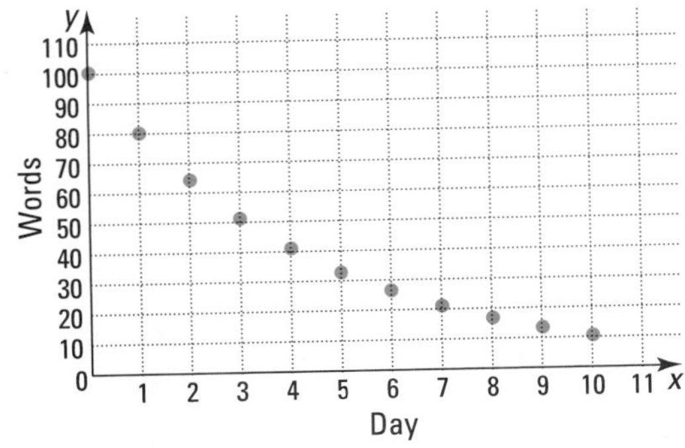

Exponential-decay situations, like constant-decrease situations, have graphs that go downward as you go to the right. However, in a constant decrease graph the slope is constant. Here the rate of change is not constant, as the chart shows. From day 0 to day 1, the rate of change is $\frac{80 - 100}{1 - 0}$, or -20 $\frac{\text{words}}{\text{day}}$. The student forgot 20 words. But from day 9 to day 10, the rate of change is $\frac{11 - 13}{10 - 9} = -2$. The student forgot only 2 words that day.

Exponential Decay with Populations

Exponential decay can occur with populations if the growth factor is less than 1.

Population: zero. *This is an abandoned ghost town in Wyoming that was once inhabited by early settlers and fur traders.*

Example

A town with population 67,000 is losing 5% of its population each year. At this rate, how many people will be left in the town after 10 years?

Solution

If 5% of the population is leaving, 95% is staying. Thus, every year, the population is multiplied by 0.95.

$$\text{population} = 67{,}000 \cdot (0.95)^{10}$$
$$\approx 40{,}115$$

After ten years, the population will be about 40,115.

Determining Growth or Decay from the Value of the Growth Factor

Exponential growth and exponential decay are both described by an equation of the same form,

$$y = b \cdot g^x.$$

The value of g determines whether the equation describes growth or decay. For exponential growth, $g > 1$. For instance, compound interest at a 3% rate means $g = 1.03$. Doubling means $g = 2$. When there is exponential decay, $0 < g < 1$. For instance, in the vocabulary situation on page 505, the growth factor is 0.8. In the population decline example, the growth factor is 0.95.

Between exponential growth and exponential decay is the situation where $g = 1$. When $g = 1$, then $y = b \cdot g^x = b \cdot 1^x = b$.

So y remains equal to b, its initial value, regardless of the value of x. So if 1 is the growth factor, then over time there is neither growth nor decay. The original value remains constant.

Exponential Growth

$y = b \cdot g^x,\ g > 1$

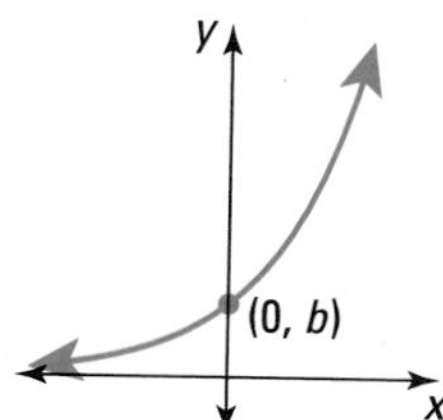

$y = b \cdot g^x,\ g = 1$

$(y = b)$

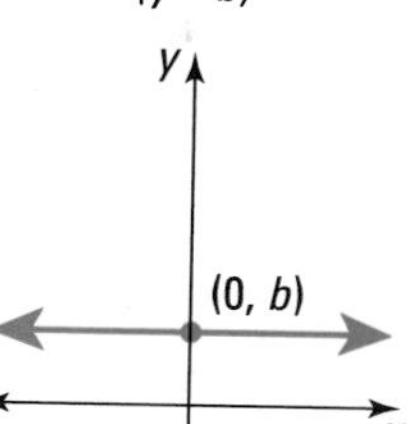

Exponential Decay

$y = b \cdot g^x,\ 0 < g < 1$

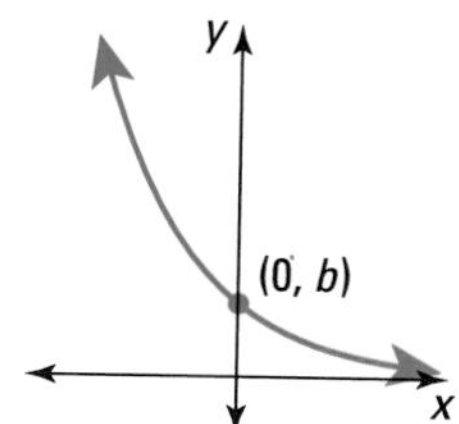

QUESTIONS

Covering the Reading

In 1–4, refer to the French test situation of this lesson. Assume the student does not review.

1. If the test is delayed a week until the next Friday, about how many words will the student know?

2. About how many words did the student forget the first 3 days?

3. About how many words did the student forget the second 3 days?

4. Evaluate $100x^5$ when $x = 0.80$. What does this answer mean?

5. State the general form of an equation for exponential decay.

6. What is a difference between the shape of the graphs of constant decrease and exponential decay?

7. *Multiple choice.* Which formula could describe an exponential decay situation?

(a) $T = 5(3)^n$ (b) $T = \frac{1}{5}(3)^n$ (c) $T = 5\left(\frac{1}{3}\right)^n$ (d) $T = 5 + \frac{1}{3}n$

8. What possible values can the growth factor have in an exponential decay equation?

In 9 and 10, refer to the Example.

9. What is the population of the town after 1 year?

10. What is the population of the town after n years?

11. If the population of a town declines by 6.3% each year, by what number would you multiply to find the population each year?

Applying the Mathematics

12. Suppose a school has 2,500 students and the number of students is decreasing by 3% each year.
   a. By what number would you multiply to find the number of students after each year?
   b. If this rate continues, write an expression for the number of students after n years.
   c. If this rate continues, how many students will the school have 10 years from now?

13. *True or false.* $\frac{1}{2}$ can be the growth factor in an exponential decay situation.

14. When one plate of tinted glass allows 60% of the light through, the amount of light y passing through x panes of glass is described by the formula $y = (0.6)^x$. Plot 5 points on the graph of this equation.

15. You fold a sheet of paper in half, then fold it again in half, and so on. Let f = the number of folds; let A = the area of what is folded; and let t = the total thickness of the sheets.
   a. As f increases, does A grow or decay exponentially? What is the growth factor?
   b. As f increases, does t grow or decay exponentially? What is the growth factor?

16. Consider the equation $y = \left(\frac{1}{2}\right)^x$.
   a. Find the value of y as both a fraction and a decimal when x takes on each value in this domain: {0, 1, 2, 3, 10, 20}.
   b. Draw the graph using the points from part **a.**
   c. Does this graph ever touch the x-axis? Justify your answer.

17. The following program finds the result when an amount grows or decays exponentially. The BASIC expression for x^n is X ^ N or X ↑ N.

```
10 PRINT "EXPONENTIAL GROWTH"
20 PRINT "WHAT IS THE AMOUNT AT BEGINNING?"
30 INPUT P
40 PRINT "WHAT ARE YOU MULTIPLYING BY?"
50 INPUT X
60 PRINT "HOW MANY TIME PERIODS?"
70 INPUT N
80 LET T = P * X ^ N
90 PRINT "TOTAL AMOUNT IS"
100 PRINT T
110 END
```

   a. What will the total be if 8, 3, and 6 are entered for P, X, and N?
   b. What will the total be if 8, 0.33, and 6 are entered for P, X, and N?
   c. What should be entered to find the population to which a city of 100,000 will grow in 10 years if the growth rate is 2% a year?

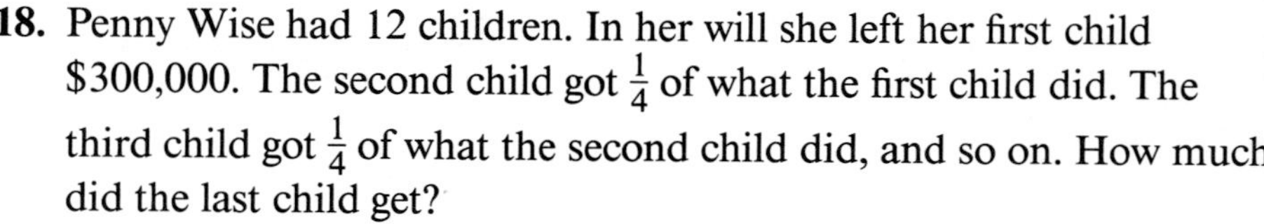

18. Penny Wise had 12 children. In her will she left her first child \$300,000. The second child got $\frac{1}{4}$ of what the first child did. The third child got $\frac{1}{4}$ of what the second child did, and so on. How much did the last child get?

Tropical fruit. *This pineapple is from a plantation in Honduras, a small Central American country.*

Review

In 19 and 20, consider the equations $k = 30 + 1.05n$ and $k = 30(1.05)^n$. *(Lesson 8-3)*

19. In which equation does k increase more rapidly as n increases?

20. Which equation could represent the value of \$30 invested at 5% compound interest for n years?

21. In the 1980s, the growth rate of the population of Honduras was about 2.7% per year (a high rate). The 1993 population of Honduras was estimated at 5,170,000. Assuming the 1980s growth rate holds throughout the 1990s, what will the population be (to the nearest thousand) in the year 2000? *(Lesson 8-2)*

22. Robert buys six guppies. Every month his guppy population doubles. Assume the population continues to grow at this rate.
a. How many guppies will there be after 4 months?
b. How many will there be after a year? *(Lesson 8-2)*

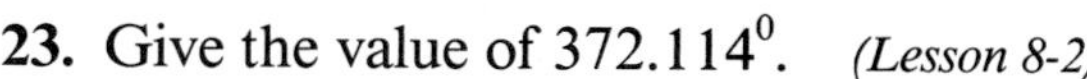

23. Give the value of 372.114^0. *(Lesson 8-2)*

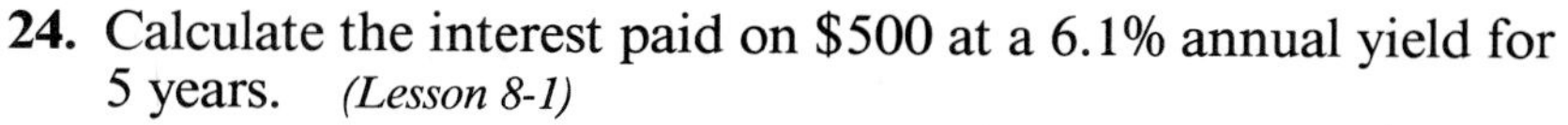

24. Calculate the interest paid on \$500 at a 6.1% annual yield for 5 years. *(Lesson 8-1)*

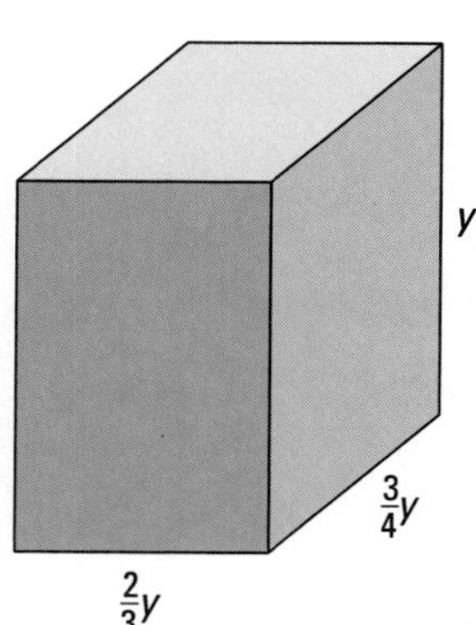

25. Refer to the box pictured at the left.
a. Write an expression for its volume.
b. Give the dimensions of a box with the same volume but different shape. *(Lessons 2-1, 2-3)*

26. Write as the product of prime numbers. *(Previous course)*
a. 14 **b.** 40 **c.** 81

Exploration

27. Exponential decay gets its name from the natural decay of elements, called *radioactive decay.* Radioactive decay is used to determine the approximate age of archaeological objects that were once alive. This can be done because all living things contain radioactive carbon 14, which has a half-life of 5600 years. When an organism dies, the amount of carbon 14 in it begins to diminish as the carbon 14 decays.
a. What is meant by the *half-life* of an element?
b. A bone is found to have $\frac{1}{16}$ of the carbon 14 that it had when the animal it came from was alive. How old is the bone?

LESSON

8-5

Products and Powers of Powers

Multiplying Powers with the Same Bases

There is no general way to simplify the sum of two powers, even when the powers have the same base. For instance, $2^5 + 2^3 = 32 + 8 = 40$, and 40 is not an integer power of 2. But some products of powers can be simplified using the repeated multiplication model of x^n. Notice the patterns in the products below.

$$2^4 \cdot 2^3 = \underbrace{(2 \cdot 2 \cdot 2 \cdot 2)}_{\text{4 factors}} \cdot \underbrace{(2 \cdot 2 \cdot 2)}_{\text{3 factors}} = \underbrace{2 \cdot 2 \cdot 2 \cdot 2 \cdot 2 \cdot 2 \cdot 2}_{\text{7 factors}} = 2^7$$

$$10^2 \cdot 10^3 = \underbrace{(10 \cdot 10)}_{\text{2 factors}} \cdot \underbrace{(10 \cdot 10 \cdot 10)}_{\text{3 factors}} = \underbrace{10 \cdot 10 \cdot 10 \cdot 10 \cdot 10}_{\text{5 factors}} = 10^5$$

$$x^2 \cdot x^5 = \underbrace{(x \cdot x)}_{\text{2 factors}} \cdot \underbrace{(x \cdot x \cdot x \cdot x \cdot x)}_{\text{5 factors}} = \underbrace{x \cdot x \cdot x \cdot x \cdot x \cdot x \cdot x}_{\text{7 factors}} = x^7$$

In each case, when we multiplied two powers with the same base the product was also a power of that base. The exponent of the product is the sum of the exponents of the factors.

$$2^4 \cdot 2^3 = 2^{4+3} = 2^7$$
$$10^2 \cdot 10^3 = 10^{2+3} = 10^5$$
$$x^2 \cdot x^5 = x^{2+5} = x^7$$

These examples are instances of the *Product of Powers Property.*

Product of Powers Property
For all m and n, and all nonzero b, $b^m \cdot b^n = b^{m+n}$.

Example 1

Multiply $x^7 \cdot x^5$.

Solution

Use the Product of Powers Property.

$$x^7 \cdot x^5 = x^{7+5} = x^{12}$$

Check

Test a special case. We let $x = 4$.

Does $(4)^7 \cdot (4)^5 = (4)^{12}$? Check with a calculator.

The key sequences 4 [y^x] 7 [×] 4 [y^x] 5 [=] and 4 [y^x] 12 [=] both give 16,777,216. (On some calculators, you need to use [^] rather than [y^x] to calculate a power.) It checks.

Recall that b^n is the growth factor in n time periods if there is growth by a factor b in each of the periods. Example 2 shows that $2^5 \cdot 2^3 = 2^8$ using this model.

Example 2

Suppose a colony of bacteria doubles in number every hour. Then, if there were 2000 bacteria in the colony at the start, after h hours there will be T bacteria, where

$$T = 2000 \cdot 2^h.$$

How many bacteria will there be after the 5th hour? How many bacteria will there be 3 hours after the 5th hour?

Solution 1

There will be $2000 \cdot 2^5$ bacteria at the end of the 5th hour. Three hours later the bacteria will have doubled three more times. So there will be $(2000 \cdot 2^5) \cdot 2^3$ bacteria. This equals $2000 \cdot (2^5 \cdot 2^3)$ bacteria.

Solution 2

Three hours after the 5th hour is the 8th hour. From the given formula, there will be $2000 \cdot 2^8$ bacteria.
Since the solutions are equal, $2^5 \cdot 2^3$ must equal 2^8.

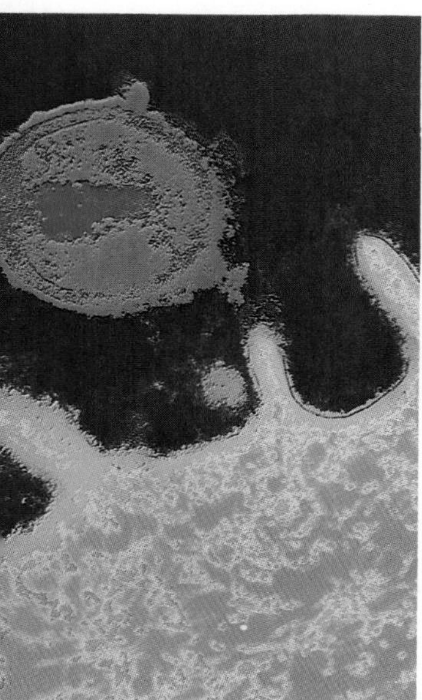

Helpful or harmful? *Certain kinds of bacteria, such as those that aid in digestion, are beneficial to humans. Other kinds of bacteria are harmful, such as this one that causes diphtheria, an infection of the upper respiratory system.*

Multiplying Powers When the Bases Are Not the Same

The Product of Powers Property tells how to simplify the product of two powers with the same base. A product with different bases, such as $a^3 \cdot b^4$, usually cannot be simplified.

Example 3

Simplify $r^4 \cdot s^3 \cdot r^5 \cdot s^8$.

Solution

Use the properties of multiplication to group factors with the same base.

$$\begin{aligned} r^4 \cdot s^3 \cdot r^5 \cdot s^8 &= r^4 \cdot r^5 \cdot s^3 \cdot s^8 \\ &= r^{4+5} \cdot s^{3+8} \quad \text{Apply the Product of Powers Property.} \\ &= r^9 \cdot s^{11} \end{aligned}$$

$r^9 \cdot s^{11}$ cannot be simplified further because the bases are different.

Check

Look at the special case when $r = 2$, $s = 3$.
Does $2^4 \cdot 3^3 \cdot 2^5 \cdot 3^8 = 2^9 \cdot 3^{11}$?
Does $16 \cdot 27 \cdot 32 \cdot 6561 = 512 \cdot 177{,}147$?
Yes, they both equal 90,699,264.

What Happens if We Take a Power of Powers?

When powers of powers are calculated, there are also some interesting patterns. Notice how we use chunking in the following examples.

Example 4

Solve $(5^2)^4 = 5^n$.

Solution

Think of 5^2 as a chunk that is raised to the 4th power.
$(5^2)^4 = 5^2 \cdot 5^2 \cdot 5^2 \cdot 5^2 = 5^{2+2+2+2} = 5^8$. So $n = 8$.

Check

Use a calculator, following order of operations. $(5^2)^4 = (25)^4 = 390{,}625$ and $5^8 = 390{,}625$. It checks.

The idea of Example 4 works with variables. For instance,

$$\begin{aligned}(x^3)^4 &= x^3 \cdot x^3 \cdot x^3 \cdot x^3\\ &= x^{3+3+3+3}\\ &= x^{12}\end{aligned}$$

The general pattern is called the *Power of a Power Property.*

Power of a Power Property
For all m and n, and all nonzero b,
$$(b^m)^n = b^{mn}.$$

In both properties in this lesson b cannot be 0, because 0^0 is not defined. However, m and n may be 0. For instance, $(5^0)^3 = 5^{0 \cdot 3} = 5^0 = 1$. This checks, because $(5^0)^3 = 1^3 = 1$.

Some expressions involve both powers and multiplication.

Example 5

Simplify $2x(x^5)^3$.

Solution 1

Recall that $x = x^1$, and rewrite $(x^5)^3$ as repeated multiplication.

$2x(x^5)^3 = 2 \cdot x^1 \cdot x^5 \cdot x^5 \cdot x^5$	Repeated Multiplication Model of Powering
$= 2 \cdot x^{16}$	Product of Powers Property

Solution 2

Use the Power of a Power Property to simplify $(x^5)^3$.

$2x(x^5)^3 = 2 \cdot x^1 \cdot x^{5 \cdot 3} = 2 \cdot x^1 \cdot x^{15} = 2x^{16}$

QUESTIONS

Covering the Reading

1. a. Explain the meaning of b^6 using the Repeated Multiplication Model of Powering.
b. Explain the meaning of b^6 using the Growth Model of Powering.

In 2 and 3, write the product as a power.

2. $3^2 \cdot 3^4$ **3.** $10^5 \cdot 10^2$

4. a. Simplify $a^2 \cdot a^3$.
b. Check your answer by letting $a = -2$.

5. State the Product of Powers Property.

6. Refer to Example 2. Give answers as powers.
a. How many bacteria will there be after 11 hours?
b. How many bacteria will there be 6 hours after that?
c. Part **b** verifies that x^{17} equals what product?

7. State the Power of a Power Property.

In 8 and 9, write the result as a single power. Then evaluate the power.

8. $(2^3)^4$ **9.** $(7^2)^3$

10. Find x when $(10^3)^5 = 10^x$.

In 11–16, simplify.

11. $x^5 \cdot x^{50}$ **12.** $(k^{10})^3$ **13.** $(n^2)^6$

14. $a^3 \cdot a^5 \cdot b^0 \cdot a^2 \cdot b^9$ **15.** $b^3(a^3b^5)$ **16.** $4n^3(n^{10})^2$

A family's keepsake. *Many families maintain records of their ancestors and living relatives by recording the information in a family tree.*

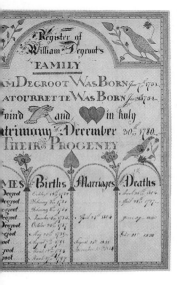

Applying the Mathematics

In 17 and 18, solve.

17. $2^5 \cdot 2^a = 2^{12}$ **18.** $(3^2)^x = 3^8$

19. In Example 1 you saw that $x^2 \cdot x^5 = x^7$. Find three other pairs of expressions whose product is x^7.

20. Write your answers as powers.
a. If you trace your family tree back through ten generations of natural parents, at most how many ancestors could you have in that generation?
b. How many times more ancestors could you have five more generations back?
c. At most how many ancestors do you have 15 generations back?
d. How are parts **a, b,** and **c** related to one of the properties in this lesson?

21. Suppose a population P of bacteria triples each day.
  a. Write an expression for the number of bacteria after 5 days.
  b. How many days after the fifth day will the bacteria population be $P \cdot 3^{17}$?

In 22–27, simplify.

22. $(x^2)^3 - (x^3)^2$ **23.** $2(x^3 \cdot x^4)$ **24.** $3m^4 \cdot 5m^2$

25. $(x^3 \cdot x^4)^2$ **26.** $a^2(a^3 + 4a^4)$ **27.** $y(y^7 - y^2)$

28. *True or false.* The tenth power of x^2 equals the square of x^{10}. Explain your answer.

Review

29. Graph $y = 5\left(\frac{1}{2}\right)^x$ for integer values of x from 0 to 5. *(Lesson 8-4)*

30. A city's current population is 2,500,000. Write an expression for the population y years from now under each assumption.
  a. The population is growing 4.5% per year.
  b. The population is decreasing 3% per year.
  c. The population is decreasing by 2,500 people each year.
  (Lessons 8-2, 8-4)

31. Give a situation which is represented by $T = 5000(1.035)^n$. *(Lesson 8-1)*

32. Rewrite $7y - 2x - 7 = 19 - 9x$
  a. in standard form. *(Lesson 7-8)*
  b. in slope-intercept form. *(Lesson 7-4)*

33. For a fund-raiser, a club mixes 50 pounds of cashews costing \$3.49/lb and 20 pounds of pecans costing \$6.99/lb. To break even on their costs, how much should they charge for a pound of the cashew-pecan mixtures? *(Lesson 6-2)*

34. Tell how many ways you could answer a 15-item true-false test by guessing. *(Lesson 2-9)*

35. **a.** Calculate $(-1)^n$ for $n = 1, 2, 3, 4, 5, 6, 7$, and 8.
  b. What is $(-1)^{100}$? *(Lesson 2-5)*

In 36 and 37, write in scientific notation. *(Appendix B)*

36. 4,000,000,000 **37.** 0.00036

Exploration

38. There are metric prefixes for many powers of 10^3. For instance, 10^3 meters is one kilometer. Give the metric prefix for each power.

a. 10^{-3} **b.** 10^6 **c.** 10^{-6} **d.** 10^9
e. 10^{-9} **f.** 10^{12} **g.** 10^{-12} **h.** 10^{15}
i. 10^{-15} **j.** 10^{18} **k.** 10^{-18}

LESSON

8-6 Negative Exponents

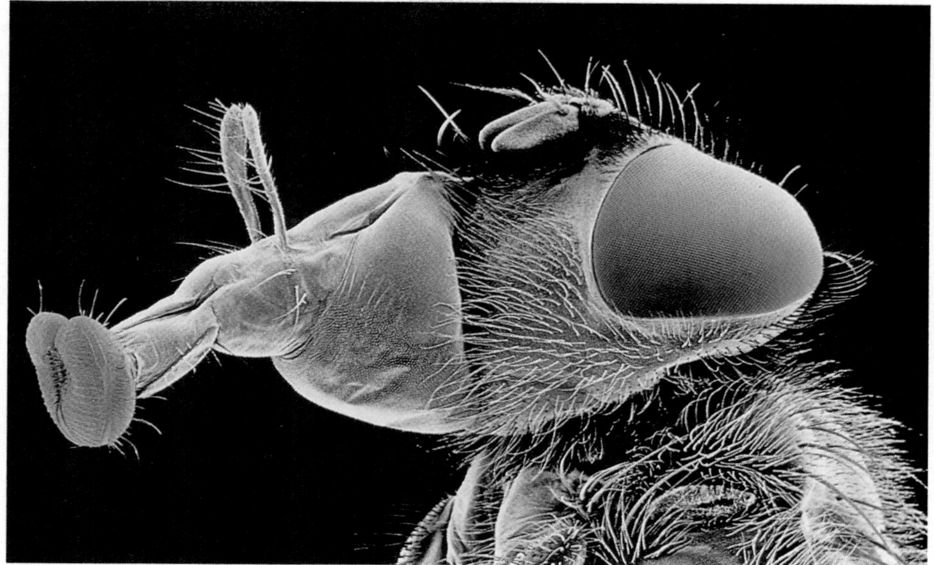

A closer look. *This is the head of a fly as seen under an electron microscope. Some electron microscopes can view objects that are less than .0001, or 10^{-4} meter in size.*

What Is the Value of a Power When Its Exponent Is Negative?

You have used the base 10 with a negative exponent to represent small numbers in scientific notation. For instance,

$$10^{-1} = 0.1 = \frac{1}{10}, \quad 10^{-2} = 0.01 = \frac{1}{10^2}, \quad 10^{-3} = 0.001 = \frac{1}{10^3}, \quad \text{and so on.}$$

Now we consider other powers with negative exponents. That is, we want to know the meaning of b^n when n is negative.

Consider this pattern of the powers of 2.

$$\begin{aligned} 2^4 &= 16 \\ 2^3 &= 8 \\ 2^2 &= 4 \\ 2^1 &= 2 \\ 2^0 &= 1 \end{aligned}$$

Each exponent is one less than the one above it. The value of each power is half that of the number above. Continuing the pattern suggests that the following are true.

$$\begin{aligned} 2^{-1} &= \frac{1}{2} \\ 2^{-2} &= \frac{1}{4} = \frac{1}{2^2} \\ 2^{-3} &= \frac{1}{8} = \frac{1}{2^3} \\ 2^{-4} &= \frac{1}{16} = \frac{1}{2^4} \end{aligned}$$

A general description of the pattern is simple: $2^{-n} = \frac{1}{2^n}$. That is, 2^{-n} is the reciprocal of 2^n.

We call the general property the *Negative Exponent Property.*

Negative Exponent Property
For any nonzero b and all n, $b^{-n} = \frac{1}{b^n}$, the reciprocal of b^n.

Example 1

a. Write 4^{-3} as a simple fraction without a negative exponent.
b. Write $\frac{1}{32}$ as a negative power of an integer.
c. Rewrite $q^5 \cdot t^{-3}$ without negative exponents.

Solution

a. Use the Negative Exponent Property.

$$4^{-3} = \frac{1}{4^3} = \frac{1}{64}$$

b. First write the denominator as a power. You must recognize 32 as a power of 2.

$$\frac{1}{32} = \frac{1}{2^5} = 2^{-5}$$

c. Substitute $\frac{1}{t^3}$ for t^{-3}.

$$q^5 \cdot t^{-3} = q^5 \cdot \frac{1}{t^3} = \frac{q^5}{t^3}$$

Notice that even though the exponent in 4^{-3} is negative, the number 4^{-3} is still positive. All negative integer powers of positive numbers are positive.

Because the Product of Powers Property applies to all exponents, it applies to negative exponents.

$$\begin{aligned} b^n \cdot b^{-n} &= b^{n+-n} && \text{Product of Powers Property} \\ &= b^0 \\ &= 1 && \text{Zero Exponent Property} \end{aligned}$$

Also,

$$\begin{aligned} b^n \cdot b^{-n} &= b^n \cdot \frac{1}{b^n} \\ &= \frac{b^n}{b^n} \\ &= 1 \end{aligned}$$

In this way, the Product of Powers Property verifies that b^{-n} must be the reciprocal of b^n. In particular, $b^{-1} = \frac{1}{b}$. That is, the -1 power (read "negative one" or "negative 1st" power) of a number is its reciprocal.

Cool choices.
Choosing ice cream flavors is usually easier than choosing answers on a multiple-choice test.

Example 2

Suppose each question on a 10-item multiple choice test has four options. Write the probability of guessing all the correct answers with a negative exponent.

Solution

The Multiplication Counting Principle gives

$$4 \cdot 4 \cdot 4 \cdot 4 \cdot 4 \cdot 4 \cdot 4 \cdot 4 \cdot 4 \cdot 4 = 4^{10}$$

different ways of doing the test. Only one of these ways is correct.

$$P(\text{perfect paper}) = \frac{1}{4^{10}}$$

Use the Negative Exponent Property. $\frac{1}{4^{10}} = 4^{-10}$, so $P = 4^{-10}$.

Evaluating Powers with Negative Exponents

You can evaluate negative exponents on your calculator. Remember to use the [+/–] or [(–)] key, not the subtraction key. Here are sample key sequences and displays for evaluating the answers to Example 2. For 4^{-10}: Key in 4 [y^x] 10 [+/–] [=]. Display: [0.000000954] or [9.5367 -7]

All the properties of powers you have learned are usable with negative exponents. They can help translate an expression with a negative power into one with only positive powers.

Example 3

Rewrite $(x^3)^{-2}$ without negative exponents.

Solution 1

Use the Power of a Power Property, which says to multiply the exponents.

$$\begin{aligned} (x^3)^{-2} &= x^{3 \cdot -2} && \text{Power of a Power Property} \\ &= x^{-6} \\ &= \frac{1}{x^6} && \text{Negative Exponent Property} \end{aligned}$$

Solution 2

Use the Negative Exponent Property first instead of last. Treat x^3 as a chunk. Find the reciprocal of the square of this chunk.

$$\begin{aligned} (x^3)^{-2} &= \frac{1}{(x^3)^2} && \text{Negative Exponent Property} \\ &= \frac{1}{x^6} && \text{Power of a Power Property} \end{aligned}$$

An Application with Negative Exponents

In the Growth Model for Powering, negative exponents stand for unit periods going back in time.

Example 4

Recall the compound interest formula

$$T = P(1 + i)^n.$$

Three years ago, Mr. Cabot put money in a CD at an annual yield of 7%. If the CD is worth $3675 now, what was it worth then?

Solution

Here $P = 3675$, $i = 0.07$, and $n = -3$ (for three years ago). So $T = 3675(1.07)^{-3}$. The calculator key sequence

3675 [×] 1.07 [y^x] 3 [±] [=]

gives 2999.8947.

So Mr. Cabot probably started with $3000. The 11¢ difference is probably due to rounding.

QUESTIONS

Covering the Reading

1. a. Complete the last three equations in this pattern. Then write the next equation in the pattern.

$$4^3 = 64$$
$$4^2 = 16$$
$$4^1 = 4$$
$$4^0 = 1$$
$$4^{-1} = \underline{?}$$
$$4^{-2} = \underline{?}$$
$$4^{-3} = \underline{?}$$

b. *True or false.* When x is negative, 4^x is negative.

2. *Multiple choice.* When $x \neq 0$, x^{-n} equals which of the following?

(a) $-x^n$ (b) $(-x)^n$ (c) $\frac{1}{x^{-n}}$ (d) $\frac{1}{x^n}$

3. a. Write 10^{-9} as a decimal.

b. Write 10^{-9} as a simple fraction.

4. b^{-1} is the __?__ of b.

In 5–7, write as a simple fraction.

5. 5^{-2} **6.** 3^{-6} **7.** $\left(\frac{1}{2}\right)^{-1}$

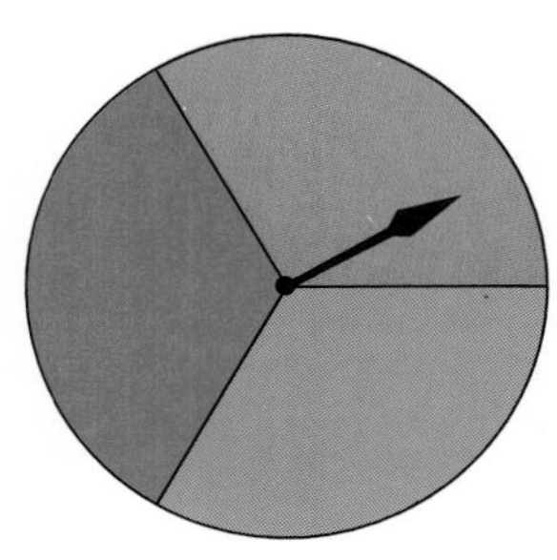

In 8–10, write without negative exponents.

8. x^5y^{-2} **9.** $3a^{-2}b^{-4}$ **10.** 2^{-n}

11. Suppose the three regions on the spinner at the left are equally likely. What is the probability that 5 spins give five reds?

12. *True or false.* If x is positive, then x^{-4} is positive.

In 13 and 14, simplify.

13. $5^3 \cdot 5^{-3}$ **14.** $c^j \cdot c^{-j}$

In 15 and 16, simplify and write without negative exponents.

15. $(x^4)^{-2}$ **16.** $(a^{-3})^4$

17. Refer to Example 4. Find the value of Mr. Cabot's CD one year ago.

18. Theresa has \$1236.47 in a savings account that has had an annual yield of 5.25% since she opened the account. Assuming no withdrawals or deposits were made, how much was in the account 8 years ago?

Applying the Mathematics

19. a. Graph $y = 2^x$ when the domain of x is {-4, -3, -2, -1, 0, 1, 2, 3}.
b. Describe what happens to the graph as x becomes smaller and smaller.

20. If the reciprocal of $(1.06)^{11}$ is $(1.06)^n$, what is n?

In 21 and 22, solve and check.

21. $3^{10} \cdot 3^{-12} = 3^z$ **22.** $5^6 \cdot 5^y = 5^{-9}$

In 23 and 24, simplify.

23. $t^{-2} \cdot t^{-4}$ **24.** $(x^5y^3) \cdot (x^{-3}y^{-5})$

25. The human population P (in billions) of the earth x years from 1985 can be estimated by the formula $P = 5 \cdot (1.017)^x$. Use this formula to estimate the earth's population in the given year.
a. 1994 **b.** 1980

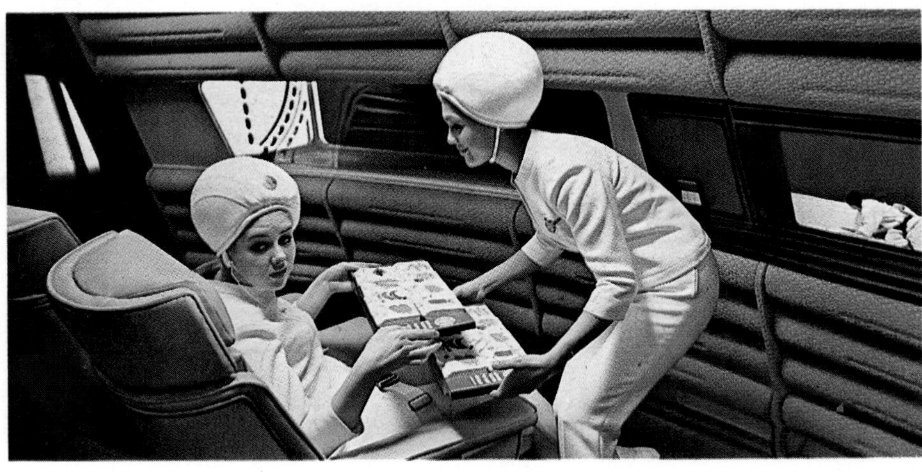

Predicting the future. *Shown is a scene from the 1968 film,* 2001: A Space Odyssey. *In this science-fiction classic, spaceships rendezvous with space stations, and crew members play chess with a soft-voiced computer, Hal.*

26. Evaluate each expression.

a. $3 \cdot 10^4 + 5 \cdot 10^2 + 6 \cdot 10^1 + 2 \cdot 10^0 + 4 \cdot 10^{-1} + 7 \cdot 10^{-3}$

b. $9 \cdot 10^3 + 8 \cdot 10^2 + 7 \cdot 10^1 + 6 \cdot 10^0 + 5 \cdot 10^{-1} + 4 \cdot 10^{-2} + 9 \cdot 10^{-4}$

Review

In 27–29, simplify. *(Lesson 8-5)*

27. $4x \cdot 3x^2$

28. $a^2 \cdot a^5 \cdot a$

29. $3c^3 \cdot 4c^4 + 5c^2 \cdot 2c^5$

30. $y^m \cdot y^n \cdot y^p$

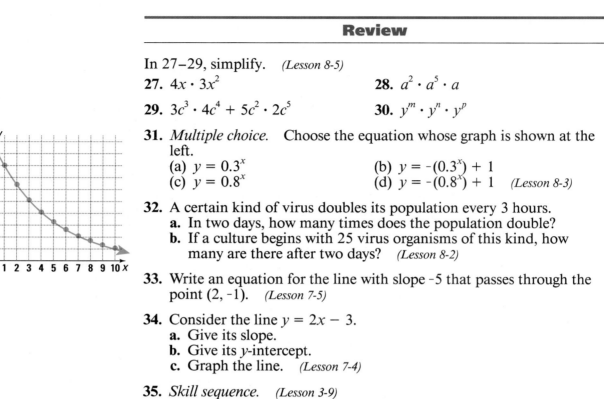

31. *Multiple choice.* Choose the equation whose graph is shown at the left.

(a) $y = 0.3^x$

(b) $y = -(0.3^x) + 1$

(c) $y = 0.8^x$

(d) $y = -(0.8^x) + 1$ *(Lesson 8-3)*

32. A certain kind of virus doubles its population every 3 hours.

a. In two days, how many times does the population double?

b. If a culture begins with 25 virus organisms of this kind, how many are there after two days? *(Lesson 8-2)*

33. Write an equation for the line with slope -5 that passes through the point (2, -1). *(Lesson 7-5)*

34. Consider the line $y = 2x - 3$.

a. Give its slope.

b. Give its y-intercept.

c. Graph the line. *(Lesson 7-4)*

35. *Skill sequence.* *(Lesson 3-9)*

a. Simplify $4(7a - 2)$.

b. Simplify $4(7a - 2) - 3(5a + 1)$.

c. Solve $4(7a - 2) - 3(5a + 1) = 15$.

36. Simplify $\frac{3x + 6x}{12x}$. *Lessons 2-3, 3-2)*

Exploration

37. a. You know that $\left(\frac{1}{2}\right)^2 = \frac{1}{2} \cdot \frac{1}{2} = \frac{1}{4}$. Find other positive and negative integer powers of the number $\frac{1}{2}$.

b. How do the powers of $\frac{1}{2}$ compare with the powers of 2?

c. Generalize to other pairs of reciprocal bases.

LESSON

8-7

Quotients of Powers

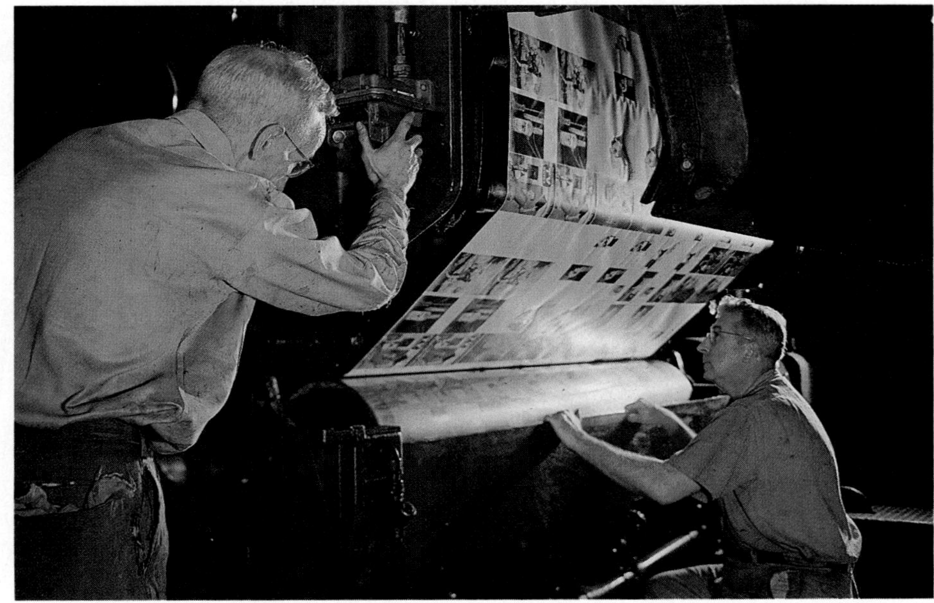

How books are made. *For many books, 32 pages are printed on each side of large sheets of paper. A machine folds each sheet several times. How many folds per sheet are needed to put the pages in order?*

Dividing Powers with the Same Base

Here is part of a list of the integer powers of 2, those from 2^{-5} through 2^{11}.

$$\ldots, \frac{1}{32}, \frac{1}{16}, \frac{1}{8}, \frac{1}{4}, \frac{1}{2}, 1, 2, 4, 8, 16, 32, 64, 128, 256, 512, 1024, 2048, \ldots$$

Multiply any two of these numbers and you will find that the product is on the list. For instance,

$$\frac{1}{8} \cdot 256 = 32.$$

When you write each number in this equation as a power of 2, you can see an instance of the Product of Powers Property.

$$2^{-3} \cdot 2^8 = 2^5.$$

It may surprise you that if you divide any two numbers on the list, their quotient is on the list as well. For instance, consider 512 and 16.

Dividing larger by smaller

$$\frac{512}{16} = 32$$

As powers: $\frac{2^9}{2^4} = 2^5$

Dividing smaller by larger

$$\frac{16}{512} = \frac{1}{32}$$

As powers: $\frac{2^4}{2^9} = \frac{1}{2^5} = 2^{-5}$

These examples illustrate the Quotient of Powers Property.

Quotient of Powers Property
For all m and n, and all nonzero b,

$$\frac{b^m}{b^n} = b^{m-n}.$$

When simplifying $\frac{b^m}{b^n}$, if the larger power is in the numerator, the result is a positive power of b. If the larger power is in the denominator, then the result is a negative power of b.

Example 1

Evaluate the following.

a. $\frac{6^{10}}{6^7}$ **b.** $\frac{6^7}{6^{10}}$

Solution 1

Use the Quotient of Powers Property.

a. $\frac{6^{10}}{6^7} = 6^{10-7}$
$= 6^3$
$= 216$

b. $\frac{6^7}{6^{10}} = 6^{7-10}$
$= 6^{-3}$
$= \frac{1}{6^3}$
$= \frac{1}{216}$

Solution 2

Use the Repeated Multiplication Model for Powering.

a. $\frac{6^{10}}{6^7} = \frac{6 \cdot 6 \cdot 6 \cdot 6 \cdot 6 \cdot 6 \cdot 6 \cdot 6 \cdot 6 \cdot 6}{6 \cdot 6 \cdot 6 \cdot 6 \cdot 6 \cdot 6 \cdot 6} = 6^3 = 216$

b. $\frac{6^7}{6^{10}} = \frac{6 \cdot 6 \cdot 6 \cdot 6 \cdot 6 \cdot 6 \cdot 6}{6 \cdot 6 \cdot 6 \cdot 6 \cdot 6 \cdot 6 \cdot 6 \cdot 6 \cdot 6 \cdot 6} = \frac{1}{6^3} = \frac{1}{216}$

Check

Use a calculator.

a. One possible key sequence is

6 [y^x] 10 [÷] 6 [y^x] 7 [=].

The display shows [216].

b. A possible key sequence is

6 [y^x] 7 [÷] 6 [y^x] 10 [=].

The display shows [0.0046296].

1 [÷] 216 [=] also gives [0.0046296].

It checks.

Using the Quotient of Powers Property

The Quotient of Powers Property may be used to divide algebraic expressions. For instance, $\frac{y^{12}}{y^5} = y^{12-5} = y^7$.

Now consider the fraction $\frac{b^m}{b^m}$. By the Quotient of Powers Property, $\frac{b^m}{b^m} = b^{m-m} = b^0$. But we know $\frac{b^m}{b^m} = 1$. This is another way of showing why $b^0 = 1$.

A note on notes. *These \$1 and \$3 bills were issued by the Continental Congress to help finance the Revolutionary War (1775–1783). So many of these notes were printed that they became almost worthless. The \$2 bill was first issued in 1862.*

The Quotient of Powers Property can be applied to divide numbers written in scientific notation.

Example 2

In 1993, there were approximately 5.3 billion one-dollar bills in circulation and about 256 million people in the United States. How many dollar bills was this per person?

Solution

Since dollars per person is a rate unit, the answer is found by division.

$$\frac{\text{number of dollar bills}}{\text{number of persons}} = \frac{5.3 \text{ billion}}{256 \text{ million}}$$

$= \frac{5.3 \cdot 10^9}{2.56 \cdot 10^8}$ Translate into scientific notation.

$= \frac{5.3}{2.56} \cdot \frac{10^9}{10^8}$ Multiplying Fractions Property

$\approx 2.07 \cdot 10^1$ Quotient of Powers Property

$\approx 21 \frac{\text{dollar bills}}{\text{person}}$

Check

Change the numbers to decimal notation and simplify the fraction.

$$\frac{5{,}300{,}000{,}000}{256{,}000{,}000} = \frac{5300}{256} \approx 21$$

Dividing Powers When the Bases Are Not the Same

To use the Quotient of Powers Property, the bases must be the same. For instance, $\frac{a^9}{b^4}$ cannot be simplified. To divide two algebraic expressions that involve different bases, group powers of the same base together and use the Quotient of Powers Property to simplify each fraction.

Example 3

Simplify $\frac{7a^3b^2c^6}{28a^2b^5c}$. Write the result as a fraction.

Solution 1

$$\frac{7a^3b^2c^6}{28a^2b^5c} = \frac{7}{28} \cdot \frac{a^3}{a^2} \cdot \frac{b^2}{b^5} \cdot \frac{c^6}{c}$$
$$= \frac{1}{4} \cdot a^{3-2} \cdot b^{2-5} \cdot c^{6-1} \qquad \text{(Remember that } c = c^1.\text{)}$$
$$= \frac{1}{4} \cdot a^1 \cdot b^{-3} \cdot c^5$$
$$= \frac{1}{4} \cdot a \cdot \frac{1}{b^3} \cdot c^5$$
$$= \frac{ac^5}{4b^3}$$

Solution 2

$$\frac{7a^3b^2c^6}{28a^2b^5c} = \frac{7 \cdot a \cdot a \cdot a \cdot b \cdot b \cdot c \cdot c \cdot c \cdot c \cdot c \cdot c}{7 \cdot 4 \cdot a \cdot a \cdot b \cdot b \cdot b \cdot b \cdot b \cdot c}$$
$$= \frac{a \cdot c \cdot c \cdot c \cdot c \cdot c}{4 \cdot b \cdot b \cdot b}$$
$$= \frac{ac^5}{4b^3}$$

Experts do all of this work in one step.

QUESTIONS

Covering the Reading

1. Rewrite the multiplication problem $64 \cdot 256 = 16{,}384$ using powers of 2.

2. Rewrite the division problem $1024 \div 16 = 64$ using powers of 2.

3. State the Quotient of Powers Property.

In 4 and 5, a fraction is given. **a.** Write the quotient as a power of 3. **b.** Check your work.

4. $\frac{3^8}{3^2}$

5. $\frac{3^2}{3^8}$

In 6–11, use the Quotient of Powers Property to simplify the fraction.

6. $\frac{x^{12}}{x^6}$

7. $\frac{y^5}{y^5}$

8. $\frac{19.2^4}{19.2^6}$

9. $\frac{9.5 \cdot 10^{12}}{1.9 \cdot 10^4}$

10. $\frac{3w^2z^6}{42w^2z^3}$

11. $\frac{4abc^{10}}{28a^2b^5c}$

12. Why can't $\frac{r^{10}}{s^7}$ be simplified?

13. In 1993, there were approximately 1.11 billion five-dollar bills in circulation and 256 million people in the United States. Convert these numbers to scientific notation and find the number of five-dollar bills per person.

Applying the Mathematics

14. If $m = n$, then $\frac{b^m}{b^n} = \underline{\ ?\ }$.

Playing with plastics. *Plastic products are often lighter, longer lasting, and less expensive to make. They can be made in any color and can be recycled and shaped into a variety of objects, such as this playground equipment.*

15. About $6 \cdot 10^9$ pounds of plastic were used in 1993 in the United States to produce trash bags, film, and other low density plastic goods. About how many pounds of plastic were used per person?

In 16–19, rewrite to eliminate the fraction.

16. $\frac{3^{-8}}{3^{-2}}$

17. $\frac{3^{-2}}{3^{-8}}$

18. $\frac{x^{-2}}{x^{-8}}$

19. $\frac{x^{-2n}}{x^{-3n}}$

In 20–23, rewrite the expression so that it has no fraction. You may need negative exponents.

20. $\frac{(7m)^2}{(7m)^3}$

21. $\frac{(x + 3)^6}{(x + 3)^6}$

22. $\frac{3x^2}{y^3} \cdot \frac{y^5}{x^9}$

23. $\frac{3p^5 + 2p^5}{p^4}$

24. Write an algebraic fraction that can be simplified to $8x^5$ using the Quotient of Powers Property.

25. In 1992, Alaska had a population of about $5.87 \cdot 10^5$ and an area of about $1.48 \cdot 10^6$ square kilometers. Find Alaska's population per square kilometer.

Review

26. **a.** Write 2^{-6} as a fraction. **b.** Write 2^{-6} as a decimal. *(Lesson 8-6)*

In 27–29, simplify. *(Lessons 8-5, 8-6)*

27. $4^x \cdot 4^y$

28. $-2(y^4)^3$

29. $-2(y^4)^3 \cdot y^4$

30. **a.** Find the probability of throwing 5 sixes on five consecutive throws of a fair die.
    b. Write your answer using negative exponents. *(Lessons 2-9, 8-6)*

31. Simplify $x^5 \cdot x^5 \cdot x^5$. Check by letting $x = 3$. *(Lesson 8-5)*

32. In 1989, Milo invested \$3000 for 10 years at an annual yield of 10%. In 1994, Sylvia invested \$6000 for 5 years at 5%. By the end of 1999, who would have more money? Justify your answer. *(Lesson 8-1)*

Going up? *This glass elevator gives people a panoramic view of the interior of a hotel in Atlanta, Georgia. Elisha G. Otis invented the first elevator with an automatic safety device in 1854.*

33. After x seconds, an elevator is on floor y, where $y = 54 - 2x$.
  a. Give the slope and y-intercept of $y = 54 - 2x$, and describe what they mean in this situation.
  b. Graph the line. *(Lesson 7-4)*

34. In your head, find the ratio of the areas of two squares, the first with a side of length 2 and the second with a side of length 3. *(Lesson 6-3)*

35. A box with dimensions 40 cm by 60 cm by 80 cm will hold how many times as much as one with dimensions 4 cm by 6 cm by 8 cm? *(Lesson 6-6)*

36. **a.** Find the perimeter of the rectangle at the right. *(Lesson 3-1)*
  b. Write a simplified expression for the area of this rectangle. *(Lessons 3-3, 3-9)*
  c. If the area is 312, find the value of x. *(Lesson 3-7)*

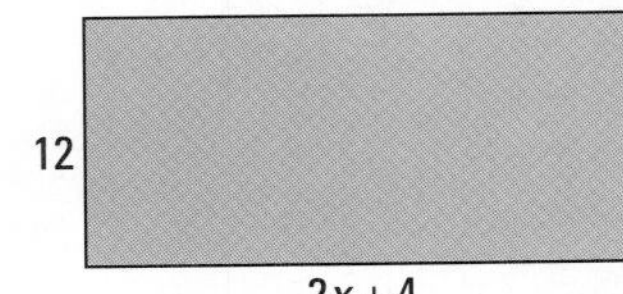

Exploration

37. The average 14-year-old has a volume of about 3 cubic feet.
  a. If you took all the students in your school, would their volume be more or less than the volume of one classroom that is 10 feet high, 30 feet long, and 30 feet wide?
  b. Assume the population of the world to be 5.5 billion people. Assume the average volume of a person to be 4 cubic feet. Is the volume of all the people more or less than the volume of a cubic mile? How much more or less? (There are 5280^3 cubic feet in a cubic mile.)

LESSON

8-8

Powers of Products and Quotients

Almost a sphere. *The distances from the center of Earth to points on Earth range from 3950 mi to 3963 mi. These are close enough that Earth is often considered a sphere. See Example 2.*

The Power of a Product

$(3x)^4$ is an example of a power of a product. It can be rewritten using repeated multiplication.

$$\begin{aligned}(3x)^4 &= (3x)\cdot(3x)\cdot(3x)\cdot(3x) \\ &= 3\cdot3\cdot3\cdot3\cdot x\cdot x\cdot x\cdot x && \text{associative and commutative properties} \\ &= 3^4\cdot x^4 && \text{Repeated Multiplication Model for Powering} \\ &= 81x^4\end{aligned}$$

You can check this answer. Consider the special case when $x = 2$. Then $(3x)^4 = 6^4 = 1296$ and $81x^4 = 81 \cdot 2^4 = 81 \cdot 16 = 1296$.

In general, any positive integer power of a product can be rewritten using repeated multiplication.

$$\begin{aligned}(ab)^n &= \underbrace{(ab)\cdot(ab)\cdot\ldots\cdot(ab)}_{n\text{ factors}} \\ &= \underbrace{a\cdot a\cdot\ldots\cdot a}_{n\text{ factors}}\cdot\underbrace{b\cdot b\cdot\ldots\cdot b}_{n\text{ factors}} \\ &= a^n\cdot b^n\end{aligned}$$

This results holds when n is any number, positive, negative, or zero.

Power of a Product Property
For all nonzero a and b, and for all n, $(ab)^n = a^n \cdot b^n$.

In the specific case that began the lesson, $a = 3$, $b = x$, and $n = 4$. So $(3x)^4 = 3^4x^4$, as we found then.

Uses of the Power of a Product

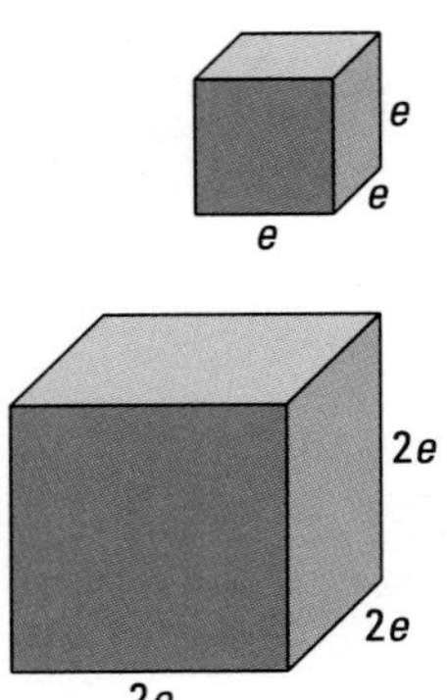

Example 1

Suppose one cube has edges twice the length of another. The volume of the larger cube is how many times the volume of the smaller?

Solution

The volume of a cube is the cube of its edge. The volume of the small cube is e^3.

$$\begin{aligned}\text{Volume of larger cube} &= (2e)^3 \\ &= 2^3e^3 \\ &= 8e^3 \\ &= 8 \cdot \text{volume of the smaller cube}\end{aligned}$$

So the larger cube has eight times the volume of the smaller.

The Power of a Product Property can be used to take powers of numbers written in scientific notation.

Ptolemy's ancient map. *Ptolemy was a Greek geographer and astronomer who lived about 150 A.D. His map was one of the first to list places with their latitude and longitude, and to depict Earth as spherical. Few people knew about his maps until they were published in an atlas in the late 1400s.*

Example 2

Earth is approximately the shape of a sphere with radius $6.36 \cdot 10^3$ km. The volume of a sphere of radius r is given by the formula

$$V = \frac{4}{3}\pi r^3.$$

Calculate the approximate volume of Earth.

Solution

Substitute $6.36 \cdot 10^3$ for r.

$$V = \frac{4}{3}\pi(6.36 \cdot 10^3)^3$$

Apply the Power of a Product Property to $(6.36 \cdot 10^3)^3$. Think of 10^3 as a single chunk.

$$\begin{aligned}V &= \frac{4}{3}\pi(6.36 \cdot 10^3)^3 \\ &\approx \frac{4}{3}\pi(257.26) \cdot 10^9 \\ &\approx 1077.61 \cdot 10^9 \\ &\approx 1.08 \cdot 10^{12} \text{ km}^3\text{, in scientific notation} \\ &\approx 1{,}080{,}000{,}000{,}000 \text{ km}^3\text{, written as a decimal}\end{aligned}$$

Earth's volume is about 1 trillion cubic kilometers.

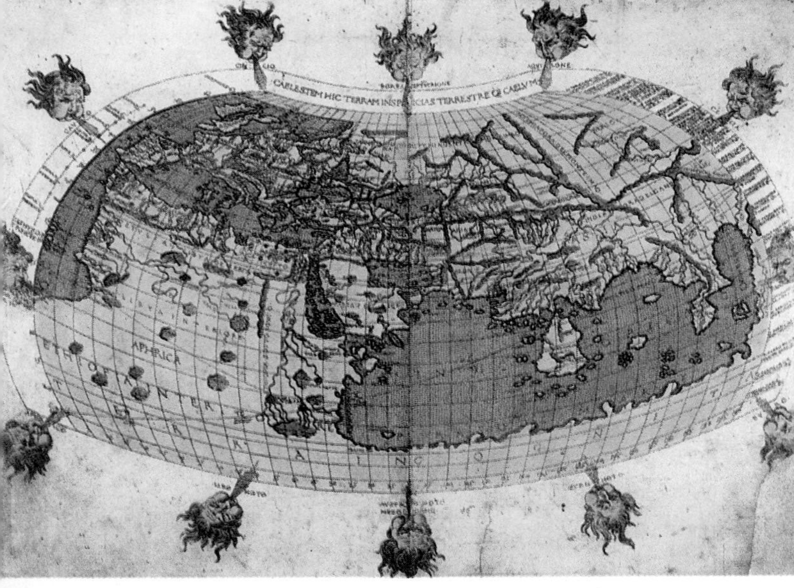

Recall that an odd power of a negative number is negative, while an even power of a negative number is positive.

Example 3

Simplify $(-xy)^6$.

Solution

First rewrite $-xy$ using $-b = -1 \cdot b$.

$$\begin{aligned}(-xy)^6 &= (-1 \cdot x \cdot y)^6 \\ &= (-1)^6 \cdot x^6 \cdot y^6 \\ &= 1 \cdot x^6 \cdot y^6 \\ &= x^6y^6\end{aligned}$$

Caution: In order of operations, powers take precedence over opposites. In $-b^2$, the power is done first, so the number $-b^2$ is negative. On the other hand, $(-b)^2$ is never negative because

$$(-b)^2 = (-b)(-b) = (-1)(b)(-1)(b) = (-1)(-1)(b)(b) = b^2.$$

So, when $b \neq 0$, $-b^2 \neq (-b)^2$. For instance, $-5^2 = -25$, while $(-5)^2 = 25$.

The Power of a Quotient

The Power of a Quotient Property is very similar to the Power of a Product Property. It enables you to easily find powers of fractions.

Power of a Quotient Property
For all nonzero a and b, and for all n,

$$\left(\frac{a}{b}\right)^n = \frac{a^n}{b^n}.$$

Example 4

Write $\left(\frac{2}{3}\right)^5$ as a simple fraction.

Solution

Use the Power of a Quotient Property.

$$\begin{aligned}\left(\frac{2}{3}\right)^5 &= \frac{2^5}{3^5} \\ &= \frac{32}{243}\end{aligned}$$

Check

Change the fractions to decimals. $\left(\frac{2}{3}\right)^5 = (0.\overline{6})^5 \approx 0.1316872\ldots$

$\frac{32}{243} \approx 0.1316872\ldots$ also.

Example 5

Rewrite $3 \cdot \left(\frac{2x}{y}\right)^4$ as a single fraction.

Solution

First rewrite the power using the Power of a Quotient Property.

$$3 \cdot \left(\frac{2x}{y}\right)^4 = 3 \cdot \frac{(2x)^4}{y^4}$$

$$= 3 \cdot \frac{2^4x^4}{y^4} \qquad \text{Power of a Product Property}$$

$$= 3 \cdot \frac{16x^4}{y^4}$$

$$= \frac{48x^4}{y^4} \qquad \text{Think of 3 as } \frac{3}{1}.$$

Check

By repeated multiplication,

$$3 \cdot \left(\frac{2x}{y}\right)^4 = 3 \cdot \frac{2x}{y} \cdot \frac{2x}{y} \cdot \frac{2x}{y} \cdot \frac{2x}{y} = \frac{48x^4}{y^4}.$$

First people on the moon. *Shown is astronaut Edwin "Buzz" Aldrin, the second person to stand on the surface of the moon, on July 20, 1969. The first was Neil Armstrong, who stepped on the moon moments earlier.*

QUESTIONS

Covering the Reading

1. **a.** Rewrite $(5x)^3$ without parentheses.
   b. Check your answer by letting $x = 2$.

2. In Example 1, suppose the length of each side of the smaller cube is 12.5 feet.
   a. Find the volume of the smaller cube.
   b. Find the volume of the larger cube.

3. The edge of one cube is k inches. The edge of a second cube is 5 times as long.
   a. Write an expression for the volume of the first cube.
   b. Write a simplified expression for the volume of the second cube.

4. Calculate $(1.3 \cdot 10^4)^5$.

5. The radius of Earth's moon is approximately $1.738 \cdot 10^3$ km. Calculate its approximate volume. Write your answer **a.** in scientific notation. **b.** as a decimal.

In 6–7, write as a simple fraction.

6. $\left(\frac{1}{2}\right)^4$

7. $\left(\frac{7}{10}\right)^3$

In 8 and 9, answer *true or false.*

8. -5^2 is negative.

9. $(-7)^2 = 7^2$

10. Simplify.

a. -3^2
b. $(-3)^2$
c. -5^3
d. $(-5)^3$
e. $(-5)^4$
f. -5^4

In 11–16, rewrite without parentheses.

11. $(ab)^3$

12. $(3x^3)^2$

13. $\left(\frac{I}{S}\right)^3$

14. $(8y)^3$

15. $(-ab)^9$

16. $\left(\frac{a}{b^5}\right)^3$

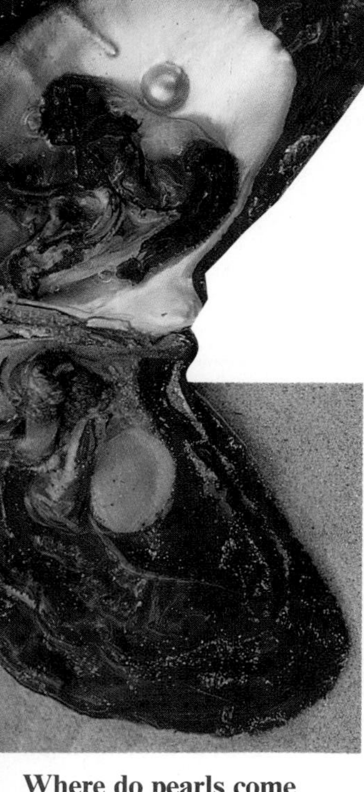

Where do pearls come from? *Pearls are found inside oyster shells. This oyster shell is from Thailand; the pearl is in the upper part of the shell.*

Applying the Mathematics

In 17–22, rewrite without parentheses and simplify.

17. $\frac{1}{2}(6x)^2$

18. $(pqr)^0$

19. $\left(\frac{u}{3}\right)^t$

20. $4L \cdot \left(\frac{5k}{L}\right)^2$

21. $\left(\frac{2}{7}z\right)^4 \cdot z$

22. $(2q)^5(3q^4)^2$

23. Suppose that about $\frac{1}{3}$ of the time, a pearl found by a pearl fisher is good enough to sell.

a. What is the probability that 5 pearls in a row will not be good enough to sell?

b. To what example of this lesson is the answer to part **a** related?

In 24–26, multiple choice. Name the property that is illustrated.

(a) Product of Powers
(b) Quotient of Powers
(c) Power of a Power
(d) Power of a Product
(e) Power of a Quotient

24. $\left(\frac{a}{2n}\right)^3 = \frac{a^3}{(2n)^3}$

25. $8^5 \cdot 8^{10} = 8^{15}$

26. $(5x^2)^3 = 5^3 \cdot (x^2)^3$

27. If $x = 3$, what is the value of $\frac{(4x)^8}{(4x)^5}$?

28. Ms. Taix incorrectly simplified $3(5x^4)^2$ as $15x^6$.

a. Find a counterexample to show that this is not true for all values of x.

b. Write out an explanation for Ms. Taix showing how to get the correct answer.

Review

29. Simplify $\frac{5n^2 - 3n^2}{10n^2}$. *(Lessons 3-3, 8-7)*

30. Other than the sun, the star nearest to us, Alpha Centauri, is about $4 \cdot 10^{13}$ km away. Earth's moon is about $3.8 \cdot 10^5$ km from us. If it took astronauts about 3 days to get to the moon in 1969, at that speed how long would it take them to get to Alpha Centauri? *(Lesson 8-7)*

In 31–33, simplify. *(Lessons 8-5, 8-6, 8-7)*

31. $\frac{k^{12}}{k^9}$ **32.** $y \cdot y^3$ **33.** $(v^{-2})^3$

34. Which is larger, $(5^4)^3$ or $5^4 \cdot 5^3$? *(Lesson 8-7)*

35. *Skill sequence.* Simplify. *(Lesson 8-5)*
a. $2(x \cdot x^4)$ **b.** $x \cdot (x^4)^2$ **c.** $(x \cdot x^4)^2$

36. a. Graph $y = 3^x$ using $x = -3, -2, -1, 0, 1,$ and 2.
b. What name is given to this curve? *(Lessons 8-2, 8-5)*

37. Calculate in your head. *(Lesson 6-2)*
a. the total cost of 6 cans of beans at \$.98 per can
b. the total cost of 4 tickets at \$15.05 per ticket
c. a 15% tip for a \$40.00 dinner bill

Exploration

38. The number 64 can be written as 8^2, or as 4^3, or as 2^6. Likewise, each of the numbers given here can be written in more than one way in the form a^n, where a and n are positive integers from 2 to 20. For each, find two pairs of values of a and n.
a. 81 **b.** 256 **c.** 32,768 **d.** 43,046,721

LESSON

8-9

Remembering Properties of Exponents and Powers

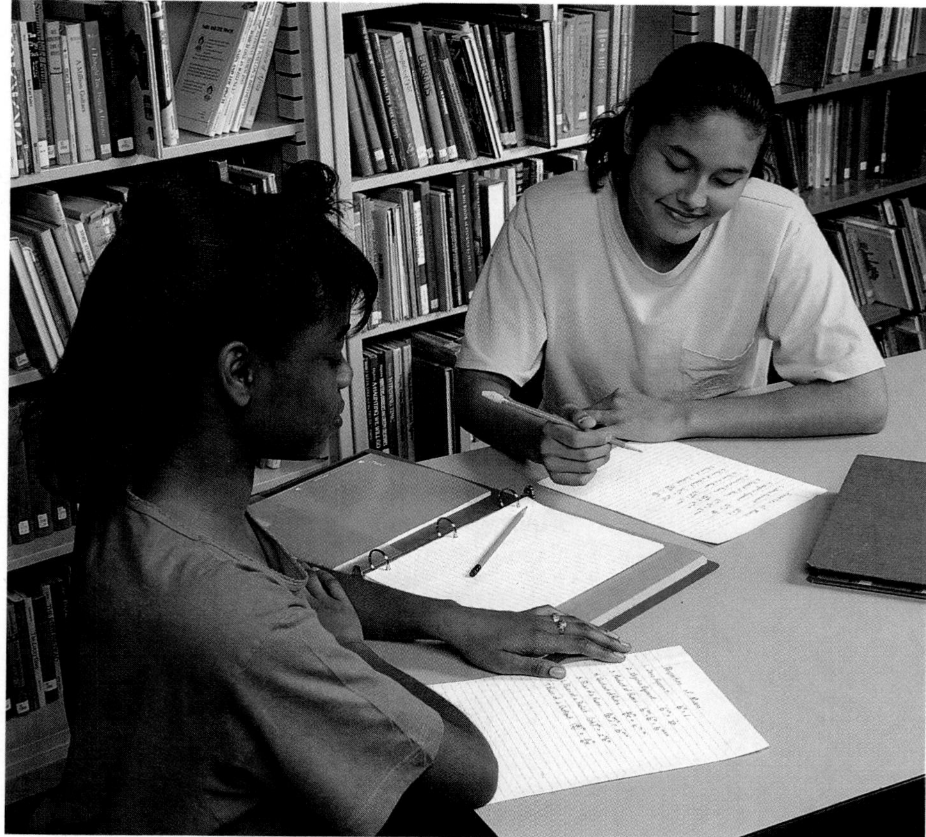

The power of studying. *These students are studying the properties of exponents and powers. By testing a special case of a property, or by showing that a pattern is not always true, these properties can easily be applied correctly.*

Seven properties of powers have been studied in this chapter. They apply to all exponents m and n and nonzero bases a and b.

Zero Exponent	$b^0 = 1$
Negative Exponent	$b^{-n} = \frac{1}{b^n}$
Product of Powers	$b^m \cdot b^n = b^{m+n}$
Quotient of Powers	$\frac{b^m}{b^n} = b^{m-n}$
Power of a Power	$(b^m)^n = b^{mn}$
Power of a Product	$(ab)^n = a^n b^n$
Power of a Quotient	$\left(\frac{a}{b}\right)^n = \frac{a^n}{b^n}$

Some people confuse these properties. Fortunately, mathematics is *consistent.* As long as you apply properties correctly, the results you get using some properties will not disagree with the results you get using other properties. We begin with what may look like a new problem: a negative power of a fraction.

Example 1

Write $\left(\frac{2}{3}\right)^{-4}$ as a simple fraction.

Solution 1

Think: The problem asks for the power of a quotient. So use that property.

$$\left(\frac{2}{3}\right)^{-4} = \frac{2^{-4}}{3^{-4}}$$

Now evaluate the numerator and denominator, using the Negative Exponent Property.

$$= \frac{\frac{1}{2^4}}{\frac{1}{3^4}} = \frac{\frac{1}{16}}{\frac{1}{81}} = \frac{1}{16} \div \frac{1}{81} = \frac{1}{16} \cdot 81 = \frac{81}{16}$$

Solution 2

Think: The problem asks for a negative exponent. So use the Negative Exponent Property first.

$$\left(\frac{2}{3}\right)^{-4} = \frac{1}{\left(\frac{2}{3}\right)^4}$$

Now use the Power of a Quotient Property.

$$= \frac{1}{\frac{2^4}{3^4}} = 1 \div \frac{2^4}{3^4} = 1 \cdot \frac{3^4}{2^4} = \frac{81}{16}$$

Solution 3

Think: $-4 = -1 \cdot 4$. Use the Power of a Power Property.

$$\left(\frac{2}{3}\right)^{-4} = \left(\left(\frac{2}{3}\right)^{-1}\right)^4$$

Now use the Negative Exponent Property to evaluate $\left(\frac{2}{3}\right)^{-1}$.

$$\left(\frac{2}{3}\right)^{-1} = \frac{1}{\frac{2}{3}}$$

$$= \frac{3}{2}$$

So $$\left(\frac{2}{3}\right)^{-4} = \left(\frac{3}{2}\right)^4.$$

Now apply the Power of a Quotient Property.

$$\left(\frac{3}{2}\right)^4 = \frac{3^4}{2^4} = \frac{81}{16}$$

Check

Use a calculator. [( ] 2 [÷] 3 [ )] [y^x] 4 [±] [=] gives 5.0625, which is $\frac{81}{16}$.

Testing a Special Case

Before the days of hand-held calculators (before the early 1970s), expressions with large exponents could not be calculated in a first-year algebra course. So it would be difficult to check some answers with arithmetic. However, with a calculator, a strategy called testing a special case is often possible.

Example 2

Norm was asked to simplify $x^8 \cdot x^6$. He wasn't sure of the answer, but knew it should be $2x^{14}$, or x^{48}, or x^{14}. Which is the correct response?

Solution 1

Use a special case. Let $x = 3$. Now calculate $3^8 \cdot 3^6$ (with a calculator) and see if it equals 3^{14} or 3^{48} or $2 \cdot 3^{14}$. A calculator shows

$$3^8 \cdot 3^6 = 4{,}782{,}969$$
$$2 \cdot 3^{14} = 9{,}565{,}938$$
$$3^{48} = 7.9766 \cdot 10^{22}$$
$$3^{14} = 4{,}782{,}969$$

The answer is 3^{14}. So $x^8 \cdot x^6 = x^{14}$.

Solution 2

Use the Repeated Multiplication Model for Powering to rewrite x^8 and x^6.

$$x^8 \cdot x^6 = (x \cdot x \cdot x \cdot x \cdot x \cdot x \cdot x \cdot x) \cdot (x \cdot x \cdot x \cdot x \cdot x \cdot x)$$

Notice that there are 14 factors of x in the product. So

$$x^8 \cdot x^6 = x^{14}.$$

Showing that a Pattern Is Not Always True

In the test of a special case, the number used should not be too special. A pattern may work for a few numbers but not all. Recall that a counterexample is a special case for which a pattern is false. To show that a pattern is not always true, it is enough to find *one* counterexample.

Example 3

Ali noticed that $2^3 = 2^2 + 2^2$ since $8 = 4 + 4$. She guessed that, in general, there is a property

$$x^3 = x^2 + x^2.$$

She tested a second case by letting $x = 0$. She found that $0^3 = 0^2 + 0^2$. She concluded that her property is always true. Is Ali right?

Solution

Try a different number. Let $x = 5$.
Does $5^3 = 5^2 + 5^2$? $125 = 25 + 25$? No.
$x = 5$ is a counterexample which shows that Ali's property is not always true.

In the questions, if you have trouble remembering a property or are not certain that you have simplified an expression correctly, try going back to the Repeated Multiplication Model for Powering or to testing a special case.

However, remember that some numbers are very special. For instance, the number 2 has properties that other numbers do not have. Squaring it gives the same result as doubling it. So beware of using 2 as a special case. Beware also of using 0 and 1.

QUESTIONS

Covering the Reading

In 1 and 2, **a.** write the number as a fraction. **b.** Check by using the Repeated Multiplication Model for Powering.

1. $\left(\frac{5}{2}\right)^{-3}$

2. $\left(\frac{3}{4}\right)^{-3}$

In 3 and 4, select the correct choice and check by testing a special case.

3. $\frac{x^6}{x^3} =$

(a) x^2 (b) x^3 (c) 2 (d) 1

4. $\left(\frac{m}{n}\right)^2 =$

(a) $\frac{2m}{n}$ (b) $\frac{m}{n}$ (c) $\frac{2m}{2n}$ (d) $\frac{m^2}{n^2}$

5. *Multiple choice.* Which of the following equals $\left(\frac{3x}{y}\right)^{-2}$? Justify your answer.

(a) $\frac{y^2}{3x^2}$ (b) $-6x^2y^2$ (c) $\frac{y^2}{9x^2}$ (d) $\frac{-6x^2}{y^2}$

6. Consider the equation $x^4 = 4x^2$. Tell if the equation is true for the special case indicated.

a. $x = 0$ **b.** $x = 2$ **c.** $x = -2$ **d.** $x = 3$

7. *True or false.* If more than two special cases of a pattern are true, then the pattern is true.

8. *True or false.* If one special case of a pattern is not true, then the general pattern is not true.

9. What is a *counterexample?*

Applying the Mathematics

10. Describe at least two different ways to simplify $\left(\frac{x^8}{x^4}\right)^{-2}$.

11. Find a counterexample to show that it is not always true that $(2x)^3 = 2x^3$.

In 12–15, an instance of a property is given. Describe the general property.

12. $(x + 1)(x + 1)^3 = (x + 1)^4$

13. $(4v)^3 = 64v^3$

14. $\left(\frac{1}{p}\right)^{10} = \frac{1}{p^{10}}$

15. $\left(\frac{4}{9}\right)^{-2} = \frac{1}{\left(\frac{4}{9}\right)^2}$

16. Consider the pattern $\frac{1}{z} + \frac{1}{y} = \frac{y + z}{yz}$.
    a. Is the pattern true when $y = 3$ and $z = 4$?
    b. Test a special case when $y = z$.
    c. Test another special case. Let $y = 5$ and $z = 2$. Convert the fractions to decimals to check.
    d. Do you think this pattern is true for all nonzero y and z?

17. Consider the pattern $a^2 + b^2 = (a + b)^2$. Test special cases to decide whether this pattern is always true. Explain how you arrived at your conclusion.

18. Use special cases to answer this question: If a price is discounted 30% and then the sale price is discounted 10%, what percent of the original price is the sale price?

Review

In 19–21, simplify. *(Lessons 8-7, 8-8)*

19. $\left(\frac{3}{5x}\right)^4 \cdot \left(\frac{2}{3}\right)^2$

20. $x^5 \cdot \left(\frac{3}{x}\right)^2$

21. $100\left(\frac{a^3}{2b}\right)^3$

22. *Skill sequence.* Evaluate $\frac{4}{3}\pi r^3$ for the given values of r. Leave your answers in terms of π. *(Lessons 8-5, 8-8)*
    a. $r = 3$
    b. $r = 3k$

23. If $\frac{6n^5}{x} = 3n$, what is x? *(Lesson 8-7)*

24. In 1993, the U.S. Department of Agriculture estimated that on farms in the United States there were 100,892 *thousand* cattle worth an average of 649 dollars each. What was the estimated total value of all these cattle? *(Lesson 8-7)*

Texas cattle. *Beef cattle, like this Texas Longhorn herd, provide about half of Texas' farm income.*

In 25 and 26, solve. *(Lessons 8-5, 8-6)*

25. $3^a \cdot 3^{10} = 3^{30}$

26. $\frac{1}{128} = 2^k$

27. Which is largest: 2^{1492}, $(2^{14})^{92}$, or $((2^{14})^9)^2$? *(Lesson 8-5)*

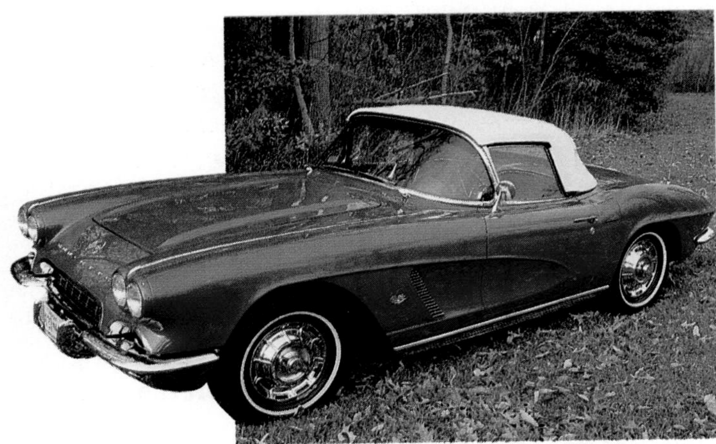

28. Suppose John bought a new car 36 years ago for $4000. It lost 10% of its value each year for the first 15 years. Then its value stayed the same for 5 years. When it was 20 years old, the car became a collector's item and its value increased 23% each year. Find the value of the car now. *(Lessons 8-2, 8-4)*

Classic sports car. *This is a 1962 Corvette convertible. Chevrolet Motor Company first began manufacturing Corvettes in 1953.*

29. a. Graph $y = 100 + 6x$ for $x = 0$ to $x = 10$.
b. Graph $y = 100(1.06)^x$ for $x = 0$ to $x = 10$.
c. Make up a question about investments that can be answered by using the graphs in parts **a** and **b.** Answer your question. *(Lesson 8-3)*

30. *Skill sequence.* $a \neq 0$. Solve for y. *(Lessons 3-8, 5-7)*
a. $2x + 3y = 4$
b. $4x + 6y = 8$
c. $6x + 9y = 12$
d. $2ax + 3ay = 4a$

31. *Multiple choice.* Choose the expression that can be simplified and simplify it. *(Lesson 3-6)*
(a) $3m^2 + 2m^2$
(b) $m^2 + m^3$
(c) $3m^2 + 2m^3$

Exploration

32. If you do not have a calculator, testing a special case can be difficult. So it helps to know the small positive integer powers of 2, 3, 4, 5, and 6. Fill in this table of values of x^n.

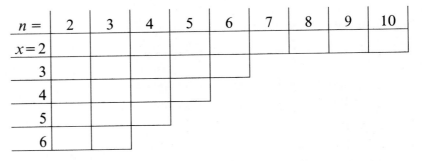

$n =$	2	3	4	5	6	7	8	9	10
$x = 2$									
3									
4									
5									
6									

How can you use the properties of powers to make it easier to remember these powers?

PROJECTS 8 CHAPTER EIGHT

A project presents an opportunity for you to extend your knowledge of a topic related to the material of this chapter. You should allow more time for a project than you do for typical homework questions.

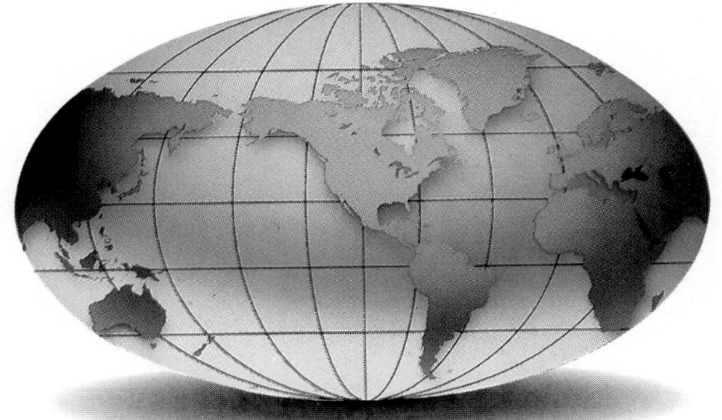

1 Population Growth over Long Periods

Not all populations grow exponentially. You learned in Chapter 7 that the population of Manhattan has not grown exponentially over the past 100 years. From an almanac or other source, examine the population of one of the following areas over the given time period. Graph the data to determine if the growth is approximately exponential. Summarize what you have found.

(a) the United States population each decade from 1790 to the present
(b) the world population since 1900
(c) the world population over the past two thousand years
(d) the population of a continent other than North America over the past 100 years

2 Applying the Properties

You have simplified some expressions like $\left(\frac{5x}{y}\right)^{-2}$ in this chapter. Devise a very complicated expression which would require you to use every property of powers in this chapter. Show how each property would be used to simplify the expression.

3 Fitting an Exponential Curve

The table below gives the speed of several microprocessors introduced since 1970. (Microprocessors function as the "brains" of computers.) The speed is described as the number of seconds it takes to scan the entire *Encyclopaedia Britannica.* Use an automatic grapher which has a "best-fitting exponential curve" option to do the following.

a. Graph the data, using the number of years after 1970 as x-values, and the time to scan as y-values.

b. Have the program determine a and b in $y = a \cdot b^x$.

Time to scan Encyclopaedia Britannica

Microprocessor	Year introduced	Speed (in seconds)
4004	1971	2250
8088	1979	400
80286	1982	45
Intel 386™	1985	13
Intel 486™	1989	4

c. According to your equation, how much time will it take a microprocessor to scan the *Encyclopaedia Britannica* in the year 2000? Do you think this prediction is a good one?

The Intel Pentium™ and Motorola PowerPC™ microprocessors are significantly faster than older processors.

4 Numbers with a Fixed Number of Factors

In this project, consider only whole numbers. Some whole numbers, called *prime numbers,* have exactly two factors, the number itself and 1. For example, 17 has only the factors 1 and 17. In this project, explore whole numbers that have the same number of factors. Do at least the following.

a. The number 49 has exactly three factors: 1, 7, and 49. Find some other numbers that have exactly three factors. Then describe *all* numbers that have exactly three factors. Explain how you determined your answer.

b. The number 8 has exactly four factors: 1, 2, 4, and 8. So does the number 10: 1, 2, 5, and 10. Find some other numbers that have exactly four factors. Then describe *all* numbers that have exactly four factors. Explain how you determine your answer.

c. Use what you have learned to find a number that has exactly 11 factors.

5 Interest Rates

There are many different kinds of interest rates, including mortgage rates, rates on car loans, rates on credit cards, and savings rates of various kinds. Use newspapers and other sources to find at least three examples of current rates of each of these kinds. Give examples of how much interest there would be on various amounts of money at each of these rates.

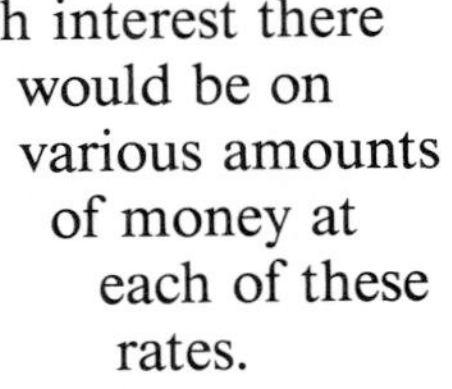

6 Cooling Water

How fast does water cool? What kind of formula describes the cooling? For this project you will need a bowl, hot water, ice cubes, a thermometer, and a watch that shows seconds.

a. As you run the tap to let the water get hot, hold the thermometer under the tap to heat it also. (Caution: Do not use a thermometer designed for body temperature. The high heat of the water may break the thermometer.) Put 1 to 2 cups of water into the bowl. Record the temperature of the water. Add 3 or 4 ice cubes. Record the temperature of the water every minute until it appears to stabilize at room temperature. Make a graph of your data. Plot time in seconds on the horizontal axis and temperature on the vertical axis. What type of equation might fit these points: linear, exponential, or neither? Justify your choice.

b. Using the same quantity of water, heated to the original temperature, add twice as many ice cubes. How does this affect the relation between time and temperature?

SUMMARY

The nth power of x is written x^n. The number n is called the exponent and x is called the base. Thus, whenever there is an exponent, there is a power. When n is a positive integer, x^n means $x \cdot x \cdot \ldots \cdot x$, where there are n factors. Because powers are related to multiplication, the basic properties of powers involve multiplication and division. For all m and n and all nonzero x and y, the following are true:

$$x^m \cdot x^n = x^{m+n} \qquad \frac{x^m}{x^n} = x^{m-n} \qquad (x^m)^n = x^{mn}$$

$$(xy)^n = x^n y^n \qquad \left(\frac{x}{y}\right)^n = \frac{x^n}{y^n}$$

The expression x^n can also be the growth model in a period of length n, when the growth factor in each unit period is x. Important applications of exponential growth and decay are population growth and compound interest. In compound interest, the growth factor is the quantity $1 + i$, where i is the annual yield. So at an annual yield of i, after n years an amount P grows to $P(1 + i)^n$.

When the growth factor is between 0 and 1, the amount gets smaller, and exponential decay occurs. Graphs of exponential growth or decay are curves.

Exponential Growth

$y = b \cdot g^x,\ g > 1$

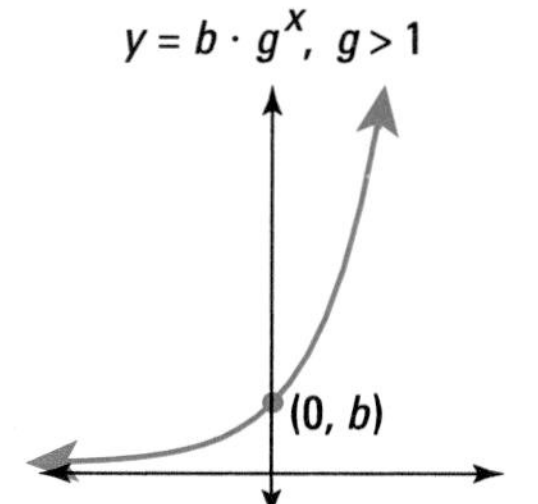

Exponential Decay

$y = b \cdot g^x,\ 0 < g < 1$

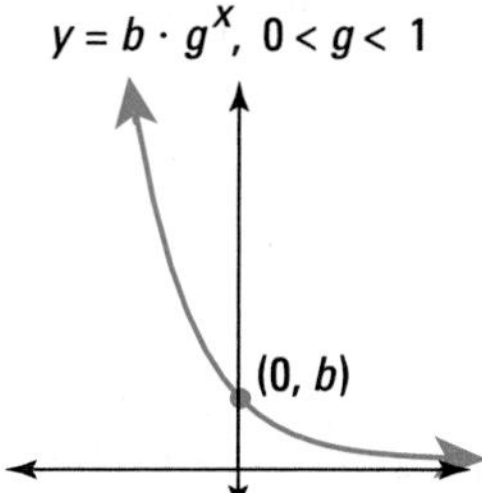

The growth model allows x^n to be interpreted when n is not a positive integer. The number x^0 is the growth factor for a period of length 0, so x^0 is the identity under multiplication. Thus $x^0 = 1$. The number x^{-n} is a growth factor going back in time, and $x^{-n} = \frac{1}{x^n}$.

VOCABULARY

Below are the most important terms and phrases for this chapter. You should be able to give a general description and a specific example of each.

Lesson 8-1
Repeated Multiplication Model for Powering
nth power, base, exponent
principal
annual yield
interest, compound interest
Compound Interest Formula

Lesson 8-2
exponential growth
growth factor
Growth Model for Powering
Zero Exponent Property

Lesson 8-4
exponential decay

Lesson 8-5
Product of Powers Property
Power of a Power Property

Lesson 8-6
Negative Exponent Property

Lesson 8-7
Quotient of Powers Property

Lesson 8-8
Power of a Product Property
Power of a Quotient Property

Lesson 8-9
testing a special case

PROGRESS SELF-TEST

Take this test as you would take a test in class. You will need graph paper and a calculator. Then check your work with the solutions in the Selected Answers section in the back of the book.

1. a. Evaluate $\frac{4^{12}}{4^6}$, and explain how you got your answer.

b. Check your answer using another method.

2. Evaluate $\frac{5 \cdot 10^{20}}{5 \cdot 10^{10}}$.

3. Write $(8)^{-5}$ as a fraction without negative exponents.

In 4–7, simplify.

4. $b^7 \cdot b^{11}$

5. $(5y^4)^3$

6. $\frac{3z^6}{12z^4}$

7. $(y^{10})^4$

8. Rewrite $\left(\frac{3}{x}\right)^2 \cdot \left(\frac{x}{3}\right)^4$ as a single fraction.

9. Simplify and rewrite $\frac{48a^3b^7}{12a^4b}$ without fractions.

10. If $q = 11$, what is the value of $6q^0$?

11. Find a counterexample to the pattern $3x^2 = (3x)^2$.

12. Name the general property that justifies $2^{10} \cdot 2^3 = 2^{13}$.

13. Felipe invests \$6500 in an account with an annual yield of 5%. Without any withdrawals or additional deposits, how much will be in the account after 5 years?

14. Darlene invests \$1900 for three years at an annual yield of 5.8%. At the end of the three years, how much interest will she earn? Show your work.

In 15 and 16, use this information. The population of a city has been growing exponentially at 3% a year. The city currently has a population of 135,000. Assume this growth rate continues.

15. What will the population be five years from now?

16. What was the population 2 years ago?

17. For each of the following equations, tell whether or not it can describe exponential growth or decay.

a. $y = \left(\frac{1}{3}\right)^x$

b. $y = 27 + 14x$

c. $y = \frac{1}{3}x$

d. $y = 27 \cdot 14^x$

18. Graph $y = 3^x$ for $x = -2, -1, 0, 1, 2$, and 3.

19. A duplicating machine enlarges a picture 30%. If that enlarger is used 3 times, how many times as large as the original picture will the final picture be?

20. Recall that the volume V of a sphere with radius r is $V = \frac{4}{3}\pi r^3$. The radius of the sun (roughly a sphere of gas) is about $6.96 \cdot 10^6$ km. Estimate the volume of the sun.

CHAPTER REVIEW

Questions on SPUR Objectives

SPUR stands for **S**kills, **P**roperties, **U**ses, and **R**epresentations. The Chapter Review questions are grouped according to the SPUR Objectives for this chapter.

SKILLS DEAL WITH THE PROCEDURES USED TO GET ANSWERS.

Objective A: *Evaluate integer powers of real numbers.* *(Lessons 8-1, 8-2, 8-6, 8-7, 8-8, 8-9)*

1. Evaluate. **a.** 3^4 **b.** -3^4 **c.** $(-3)^4$

2. Simplify $-2^5 \cdot (-2)^5$.

3. If $y = 7$, then $4y^0 = \underline{\ ?\ }$.

4. If $x = 2$, then $3x^3 - x^2 = \underline{\ ?\ }$.

In 5 and 6, simplify.

5. $(2^3)^3 \div 2^6$ **6.** $\frac{9 \cdot 10^6}{3 \cdot 10^8}$

In 7 and 8, rewrite as a fraction without an exponent.

7. 5^{-3} **8.** 2^{-5}

In 9–12, write as a simple fraction.

9. $\left(\frac{2}{7}\right)^3$ **10.** $\left(-\frac{4}{3}\right)^4$

11. $\left(\frac{1}{3}\right)^{-4}$ **12.** $10 \cdot \left(\frac{2}{5}\right)^{-3}$

Objective B: *Simplify products, quotients, and powers of powers.* *(Lessons 8-5, 8-6, 8-7)*

In 13–22, simplify.

13. $x^4 \cdot x^7$ **14.** $r^3 \cdot t^5 \cdot r^8 \cdot t^2$

15. $y^2(x^3y^{10})$ **16.** $p^4(pq^2)$

17. $\frac{n^{15}}{n^2}$ **18.** $\frac{a^{12}}{a^4} \cdot a^6$

19. $\frac{3a^4c}{3a^5}$ **20.** $\frac{15x^2y^5}{12xy^6}$

21. $(3x^5)^3 + (x^3)^5$ **22.** $(3m^4)^4 + (9m^2)^2$

23. Rewrite $\frac{4m^6}{20m^2}$ without fractions.

24. Rewrite $\frac{60w^8}{15t^2w}$ without fractions.

25. Simplify $\frac{(2 + 8)^5}{(2 + 8)^2}$.

26. Describe two different ways to evaluate $\frac{(2t - 1)^{11}}{(2t - 1)^4}$ when $t = 6$.

27. Rewrite xy^{-2} without a negative exponent.

28. Rewrite $2m^{-1}n^4p^2$ without a negative exponent.

Objective C: *Rewrite powers of products and quotients.* *(Lesson 8-8)*

In 29–40, rewrite without parentheses.

29. $\left(\frac{y}{x}\right)^{-3}$ **30.** $\left(\frac{a}{b}\right)^{-5}$

31. $(4x)^5$ **32.** $(5y)^4$

33. $\left(\frac{2}{n}\right)^5$ **34.** $\left(\frac{t^7}{2}\right)^4$

35. $(-3n)^3$ **36.** $-(2y)^3$

37. $4 \cdot \left(\frac{k}{3}\right)^3$ **38.** $45\left(\frac{t}{3}\right)^2$

39. $2(4x)^2$ **40.** $11(10k)^3$

PROPERTIES DEAL WITH THE PRINCIPLES BEHIND THE MATHEMATICS.

Objective D: *Test a special case to determine whether a pattern is true.* *(Lesson 8-9)*

41. For each case tell whether the pattern $x = x^2$ is true.

a. $x = 0$ **b.** $x = 1$

c. $x = 2$ **d.** $x = -1$

42. Consider the pattern $(x^2)^y = x^{2y}$.

a. Is the pattern true when $x = 3$ and $y = 4$?

b. Is the pattern true when $x = 5$ and $y = 2$?

c. Based on your answers to parts **a** and **b,** do you have evidence that the pattern is true, or are you sure it is not always true?

In 43 and 44, find a counterexample to the pattern.

43. $(a + b)^3 = a^3 + b^3$

44. $(x^3)^2 = x^{(3^2)}$

Objective E: *Identify properties of exponents and use them to explain operations with powers.* *(Lessons 8-2, 8-5, 8-6, 8-7, 8-8)*

Here is a list of the power properties in this chapter. For all *m* and *n,* and nonzero *a* and *b:*

Zero Exponent Property: $b^0 = 1$

Product of Powers Property: $b^m \cdot b^n = b^{m+n}$

Power of a Product Property: $(ab)^n = a^n \cdot b^n$

Negative Exponent Property: $b^{-n} = \frac{1}{b^n}$

Quotient of Powers Property: $\frac{b^m}{b^n} = b^{m-n}$

Power of a Power Property: $(b^m)^n = b^{mn}$

Power of a Quotient Property: $\left(\frac{a}{b}\right)^n = \frac{a^n}{b^n}$

In 45–52, describe the general property or properties which justify the simplification.

45. $a^7 \cdot b^7 = (ab)^7$ **46.** $a^7 \div a^2 = a^5$

47. $(4.36)^0 = 1$ **48.** $4^6 \cdot 4^9 = 4^{15}$

49. $\left(\frac{7}{g}\right)^y = \frac{7^y}{g^y}$ **50.** $6^3 \cdot 2^0 = 6^3$

51. $14^{-2} = \frac{1}{14^2}$ **52.** $\left(\frac{x}{y}\right)^{-2} = \frac{y^2}{x^2}$

53. Show two different ways to simplify $\left(\frac{x^3}{x}\right)^8$.

54. Describe two different ways to simplify $(ab^2)^4$.

USES DEAL WITH APPLICATIONS OF MATHEMATICS IN REAL SITUATIONS.

Objective F: *Calculate compound interest.* *(Lesson 8-1)*

In 55 and 56, use the advertisement at the right.

GUARANTEED
5.7% YIELD
$2,500 MINIMUM

55. Using the annual yield, how much money will there be in an account if $2,500 is deposited for 3 years?

56. Using the annual yield, calculate how much interest $3000 will earn if deposited for 4 years.

57. Susan has $1200 in a CD with an annual yield of 6%. Without any withdrawals, how much money would she have in the account after 2 years?

58. Which investment yields more money: (a) *x* dollars for 2 years at an annual yield of 10%, or (b) the same amount for 10 years at an annual yield of 2%? Explain your reasoning.

Objective G: *Solve problems involving exponential growth and decay.* *(Lessons 8-2, 8-3, 8-4, 8-6)*

59. Jennifer earns $7.25 an hour. If she gets a 5.6% raise each year, how much will she earn per hour after 4 years on the job?

60. From 1991 to 1992, the United States had an inflation rate of about 3.0% per year. Thus, an article costing $100 in 1991 cost $103.00 in 1992. Consider a book that sold for $16.95 in 1991. If the rate of inflation remains at 3.0%, how much would the book cost in 1999?

In 61 and 62, suppose that after a few hours a colony of bacteria that doubles every hour has 8000 bacteria. After n more hours there will be T bacteria where $T = 8000 \cdot 2^n$.

61. **a.** Find the value of T when $n = 4$.
b. In words describe the meaning of your answer to part **a.**

62. **a.** Find T when $n = -3$.
b. Describe the meaning of your answer to part **a.**

In 63–65, suppose that the population in a city of 1,500,000 people is decreasing exponentially at a rate of 3% per year. Let P = the population after n years.

63. Write an equation of the form $y = b \cdot g^x$ to describe the population after n years.

64. What will the population be in 6 years' time?

65. What is the population when $n = 0$? What does your answer mean?

In 66 and 67, use this information. The death rate from heart disease is decreasing each year. The formula $y = 436.4(0.983)^x$ approximates the number of deaths per 100,000 people x years after 1980.

66. Estimate the death rate per 100,000 population in 1992.

67. **a.** Find y when $x = -3$.
b. Describe the meaning of your answer to part **a.**

Objective H: *Use and simplify expressions with powers in real situations.* *(Lessons 8-6, 8-7, 8-8)*

68. A certain photographic enlarger can enlarge any picture $\frac{3}{2}$ times. By how many times will a picture be enlarged if the enlarger is used
a. twice? **b.** 5 times?

69. Water blocks out light. (At a depth of 10 meters it is not as bright as on the surface.) Suppose 1 meter of water lets in $\frac{9}{10}$ of the light. How much light will get through x meters of water?

70. In 1993 there were about $5.6 \cdot 10^9$ people on Earth. The land area is about $1.48 \cdot 10^8$ km^2. How many people are there per km^2?

71. The moon is nearly a sphere with radius of $1.08 \cdot 10^3$ miles. The volume of a sphere of radius r is $\frac{4}{3}\pi r^3$. To the nearest billion cubic miles, what is the volume of the moon?

In 72 and 73, a spinner has six equal-sized sections: two red, three blue, and one yellow. Spins are random. To win a game, you must spin yellow 3 times in a row.

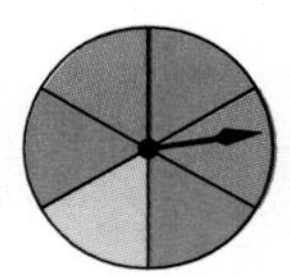

72. **a.** Find the probability of getting yellow on each of three spins.
b. Write your answer with negative exponents.

73. Find the probability of getting red on four consecutive spins.

REPRESENTATIONS DEAL WITH PICTURES, GRAPHS, OR OBJECTS THAT ILLUSTRATE CONCEPTS.

Objective I: *Graph exponential relationships.* *(Lessons 8-2, 8-3, 8-4)*

In 74–77, tell whether the graph of the equation is linear or exponential.

74. $y = 4x$ **75.** $y = \left(\frac{1}{2}\right)^x$

76. $y = 100 \cdot (3.4)^x$ **77.** $y = \frac{3}{4}x + 100$

78. Graph $y = 2^x$ for $x = -3, -2, -1, 0, 1, 2$, and 3.

79. Graph $y = .4^x$ for $x = -3, -2, -1, 0, 1, 2$, and 3.

80. Suppose an investment worth \$200 is invested in a bank at a 5% annual yield. Graph the amount after 0 to 15 years if the interest is kept in the bank.

81. When x is large, which equation's graph rises faster, $y = 56 + 0.04x$ or $y = 5 \cdot (1.04)^x$? Explain your reasoning.

82. Match each graph below with its description.
a. constant increase **b.** constant decrease
c. exponential growth **d.** exponential decay

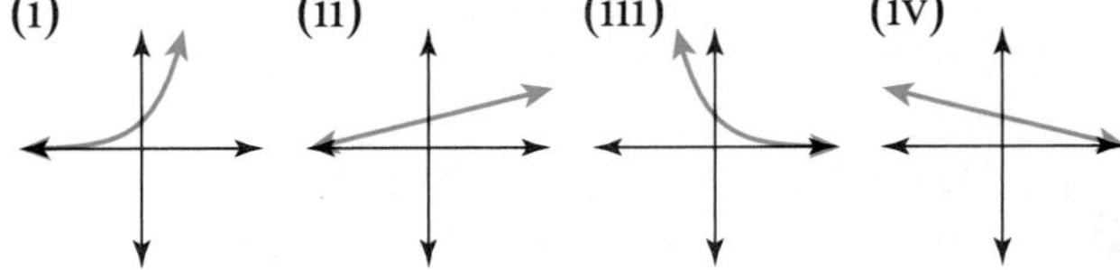

CHAPTER 9

QUADRATIC EQUATIONS AND SQUARE ROOTS

When an object is dropped from a high place, such as the roof of a building or an airplane, it does not fall at a constant speed. The longer it is in the air, the faster it falls. Furthermore, the distance d that a heavier-than-air object falls in time t does not depend on its weight.

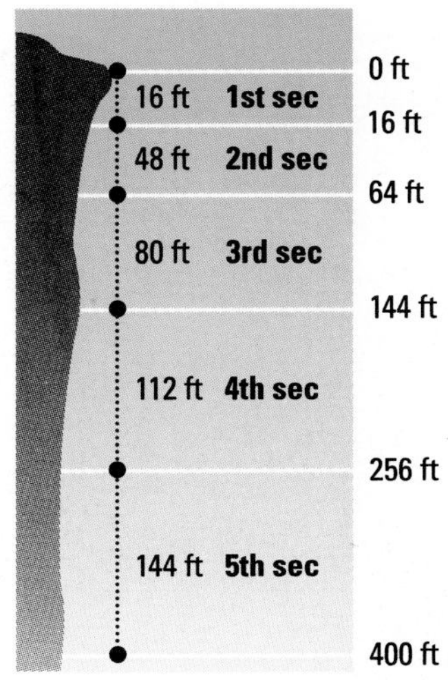

About 400 years ago, the Italian scientist Galileo described the relationship between d and t mathematically. In present-day units, if d is measured in feet and t is in seconds, then

$$d = 16t^2.$$

A table of values and a graph of this equation are shown below.

t (sec)	d (ft)
0	0
1	16
2	64
3	144
4	256
5	400

Total Distance Fallen (d)

500
400
300
200
100
0
1 2 3 4 5 6 Time in Air (t)

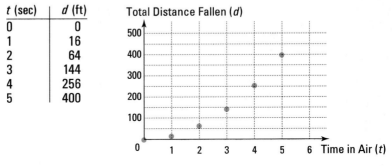

The equation $d = 16t^2$ is an example of a *quadratic equation.* The word *quadratic* comes from the Latin word for square. (Notice the t^2.) The points on the graph lie on a curve called a *parabola.* Parabolas and quadratic equations occur often, both in nature and in manufactured objects. A parabola is the shape of the path of a basketball tossed into a hoop, the stream from a water fountain, and the shape of a cross-section of a satellite dish.

In this chapter you will study quadratic equations. Solutions to quadratic equations often involve square roots, so you will also learn more about square roots.

LESSON

9-1

Graphing $y = ax^2$

Shot in slow motion. *This time-lapse photo shows the parabolic path of a basketball as it travels toward the basket.*

Graphing $y = x^2$

The simplest quadratic equation is $y = x^2$. A table of values for $y = x^2$ is given below on the left. Below on the right is a graph of the equation.

x	y
-4	16
-3	9
-2	4
-1	1
0	0
1	1
2	4
3	9
4	16

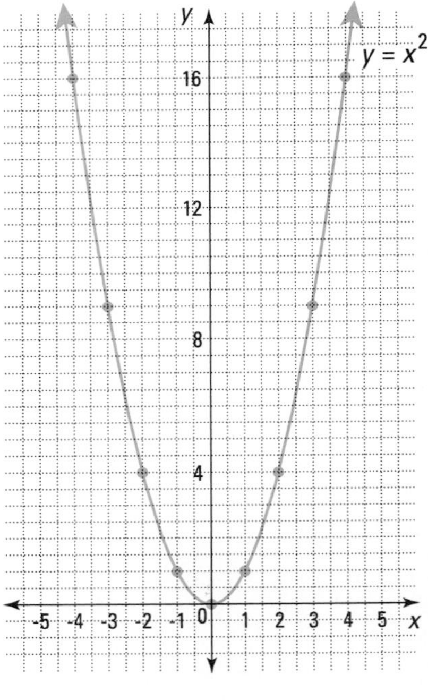

The graph of $y = x^2$ is a *parabola.* Notice that the y-axis separates this parabola into two halves. If you fold the parabola along the y-axis, the two halves coincide. For every point to the left of the y-axis, there is a matching point to the right. This is because the right half is the *reflection* or *mirror image* of the left half. For this reason we say this parabola is **symmetric** to the y-axis. The y-axis is called the **axis of symmetry** of the parabola. Because of this symmetry, every positive number is the y-coordinate for two points on the graph. For instance, 9 is the y-coordinate for both points (3, 9) and (-3, 9).

The intersection of a parabola with its axis of symmetry is called the **vertex** of the parabola. The vertex of the graph of $y = x^2$ is (0, 0).

Graphing $y = ax^2$

The equation $y = x^2$ is of the form $y = ax^2$, with $a = 1$. You should be able to draw a graph of any equation of this form.

Example 1

a. Graph $y = 2x^2$ by plotting points where x goes from -2.5 to 2.5 in intervals of 0.5.

b. Does the graph have an axis of symmetry? If so, what is an equation for the axis of symmetry?

Solution

a. Make a table of values. When evaluating $2x^2$, remember the order of operations. Square x before multiplying by 2. So when $x = -2.5$, $y = 2(-2.5)^2 = 2(6.25) = 12.5$. A table of values and graph are shown below.

x	y
-2.5	12.5
-2	8
-1.5	4.5
-1	2
-0.5	0.5
0	0
0.5	0.5
1	2
1.5	4.5
2	8
2.5	12.5

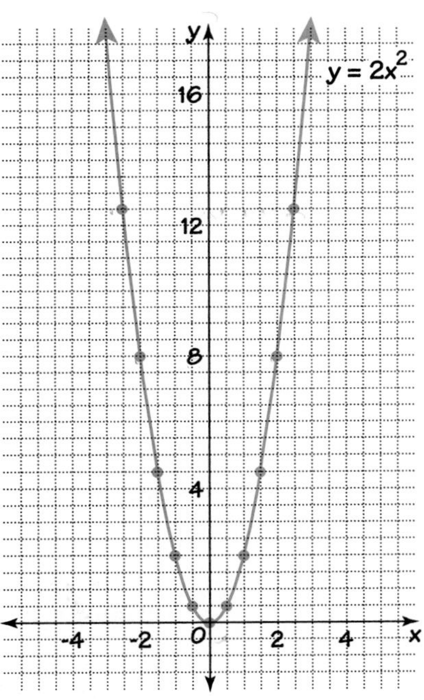

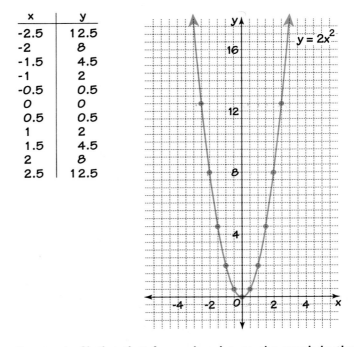

b. Examine the graph. Notice that for each point on the graph in the first quadrant, there is a corresponding point the same distance from the y-axis in the second quadrant and having the same y-coordinate. Yes, the graph has an axis of symmetry. It is the y-axis, the line with equation $x = 0$.

Equations of the form $y = ax^2$ all yield parabolas whose axis of symmetry is the y-axis. When $a > 0$ the parabola *opens up,* that is, like a cup or bowl standing upright. When $a < 0$, the graph *opens down,* like an upside-down bowl. This can be seen in Example 2.

Example 2

a. Graph $y = -\frac{1}{5}x^2$.
b. In which quadrants does the graph have points?
c. What are the coordinates of the vertex of the parabola?

Solution

a. Make a table of values. Below we use $-5 \leq x \leq 5$. Plot the points and then connect them with a smooth curve.

x	y
-5	-5
-4	$-\frac{16}{5}$
-3	$-\frac{9}{5}$
-2	$-\frac{4}{5}$
-1	$-\frac{1}{5}$
0	0
1	$-\frac{1}{5}$
2	$-\frac{4}{5}$
3	$-\frac{9}{5}$
4	$-\frac{16}{5}$
5	-5

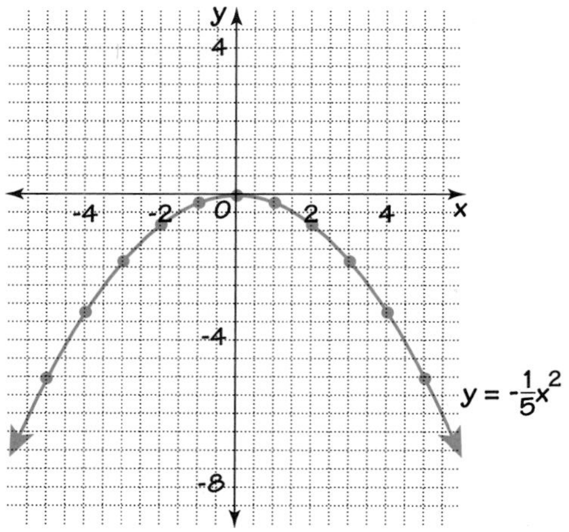

y
4
-4
-2
O
2
4
x
-4
-8
$y = -\frac{1}{5}x^2$

b. Except when $x = 0$, $-\frac{1}{5}x^2$ is negative. So except for the origin, the parabola lies below the x-axis. It has points in the 3rd and 4th quadrants.
c. The vertex is (0, 0).

Notice that if the parabola opens up, the vertex is the lowest point on the graph or the **minimum.** If the parabola opens down, the vertex is the highest point or **maximum.** Below is a summary of properties of the graph of $y = ax^2$.

> The graph of $y = ax^2$, where $a \neq 0$, has the following properties:
> **1.** It is a parabola symmetric to the y-axis.
> **2.** Its vertex is (0, 0).
> **3.** If $a > 0$, the parabola opens up.
> If $a < 0$, the parabola opens down.

Finding Points on the Graph of $y = ax^2$

In Examples 1 and 2, points were found on the graph $y = ax^2$ by first choosing values of x and then finding values of y. You should also be able to work in the reverse order. That is, if you know the y-coordinate of a point on the graph of a parabola, you should be able to find the x-coordinate or coordinates.

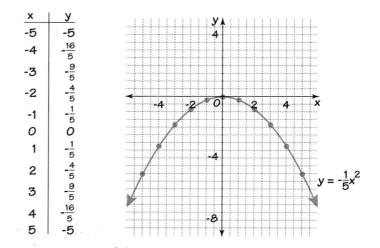

Example 3

Find the x-coordinates of the points on the graph of $y = -\frac{1}{5}x^2$ when $y = -3$.

Solution

Substitute $y = -3$ in the equation $y = -\frac{1}{5}x^2$ and solve for x.

$$-3 = -\frac{1}{5}x^2$$

$$-5(-3) = -5\left(-\frac{1}{5}x^2\right) \quad \text{Multiply each side by -5, the reciprocal of } -\frac{1}{5}.$$

So,

$$15 = x^2$$

$$x = \sqrt{15} \text{ or } x = -\sqrt{15}.$$

$$x \approx 3.87 \text{ or } x \approx -3.87.$$

Check

Draw a graph of $y = -3$ on the same axes as used in Example 2.

Notice that $y = -\frac{1}{5}x^2$ and $y = -3$ intersect at two points, labeled P and Q at the right. The x-coordinate of P is about -3.9; the x-coordinate of Q is about 3.9. It checks.

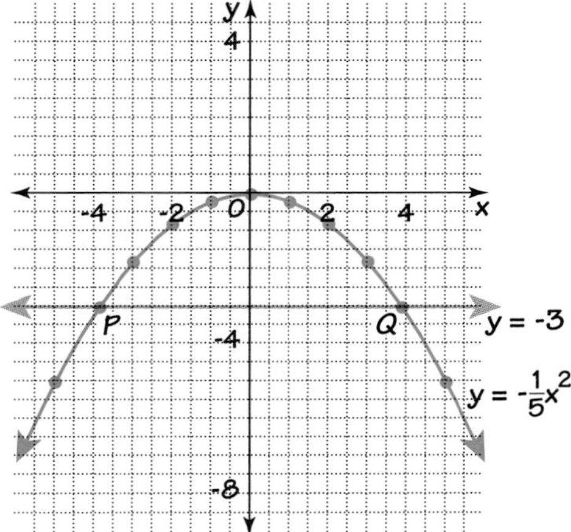

The curved path of the water from the mouth of the triceratops is part of a parabola.

QUESTIONS

Covering the Reading

1. **a.** Give an example of a quadratic equation found in this lesson.
   b. Give an example of a quadratic equation not found in this lesson.
2. What is the origin of the word *quadratic?*
3. What is the shape of the graph of every quadratic equation in this lesson?
4. Name two instances in nature or in manufactured objects where parabolas occur.

In 5 and 6, an equation is given.

a. Make a table of x- and y-values for the x-values -4, -3, -2, -1, 0, 1, 2, 3, and 4.

b. Graph the equation.

c. Describe the graph.

5. $y = \frac{1}{2}x^2$

6. $y = -\frac{1}{2}x^2$

In 7–9, consider the graph of the equation $y = ax^2$, when $a \neq 0$.

7. What is the vertex of this parabola?

8. What is an equation for its axis of symmetry?

9. a. If a is positive, the parabola opens _?_.

b. If a is negative, the parabola opens _?_.

10. Think about the equation $y = 3x^2$.

a. Without plotting any points, sketch what you think the graph will look like.

b. Make a table of values satisfying this equation. Use $x = -2, -1.5, -1, -0.5, 0, 0.5, 1, 1.5, 2$.

c. Draw a graph of this equation.

d. Tell if the vertex is a minimum or maximum.

e. For what values of x does $y = 12$?

f. Find the x-coordinates of the points where $y = 15$.

11. Refer to the parabola at the right.

a. Points A and B are symmetric to the y-axis. What are the coordinates of B?

b. Does this parabola open up or open down?

c. Does this parabola have a maximum or does it have a minimum?

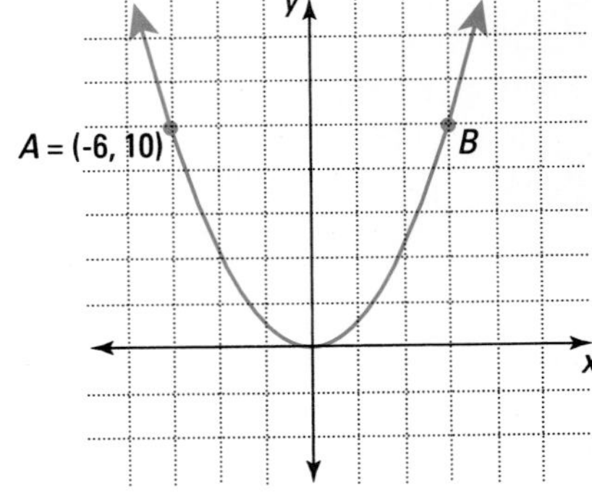

In 12 and 13, refer to the table, graph, and equation on page 547.

12. How far does an object fall in the first 3 seconds?

13. How far does an object fall in the first 2.5 seconds?

Applying the Mathematics

In 14 and 15, suppose a stone falls from a cliff.

14. How long does it take the stone to fall the first 200 feet?

15. How long does it take the stone to fall the next 200 feet?

16. Consider the formula $A = s^2$ for the area A of a square with a side of length s.
  a. Graph all possible values of s and A on a coordinate plane.
  b. Explain how the graph in part **a** is like and unlike the graph of $y = x^2$ at the beginning of this lesson.

In 17 and 18, solve.

17. $16 = \frac{1}{9}(x + 1)^2$ **18.** $\pi r^2 = 40$

Review

In 19–21, simplify. *(Lessons 8-5, 8-7, 8-8)*

19. $x^4 \cdot y \cdot x^3 \cdot y^3$ **20.** $\frac{9a^9}{6a^6}$ **21.** $\left(\frac{2a}{5}\right)^3$

22. **a.** Make up a real question that can be answered by solving $\frac{16}{21} = \frac{n}{50}$.
  b. Answer your question. *(Lesson 6-8)*

23. Julio lives 3 km from Martina and 8 km from Natasha. From this information, what can you conclude about the distance between Martina's home and Natasha's home? Justify your answer. *(Lesson 4-8)*

Enlightened science. *Scientists use the math principles of this lesson to produce the reflectors in flashlights, headlights, and searchlights.*

24. *Multiple choice.* The graph below pictures the solutions to which inequality? *(Lessons 2-7, 4-3)*

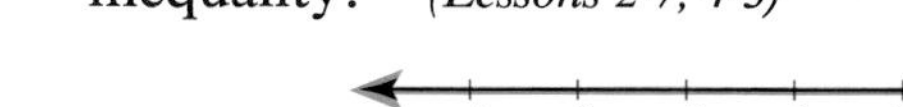

(a) $w + 6 < 5$ (b) $w - 6 < 5$
(c) $w + 5 < 6$ (d) $w - 5 < 6$

25. A house that covers 1800 square feet of ground is being built on a 7500-square-foot lot. The driveway will cover another k square feet. Write an expression for the area that is left for the lawn. *(Lesson 4-2)*

26. If $d = x + 0.05x^2$, find d when $x = 15$. *(Lesson 1-4)*

Exploration

27. Draw a set of axes on graph paper. Hold a lit flashlight at the origin so the light is centered on an axis as shown. What is the shape of the lighted area? Keep the lit end of the flashlight in the same position and tilt the flashlight to raise the other end. How does the shape of the lighted area change?

LESSON

9-2

Graphing $y = ax^2 + bx + c$

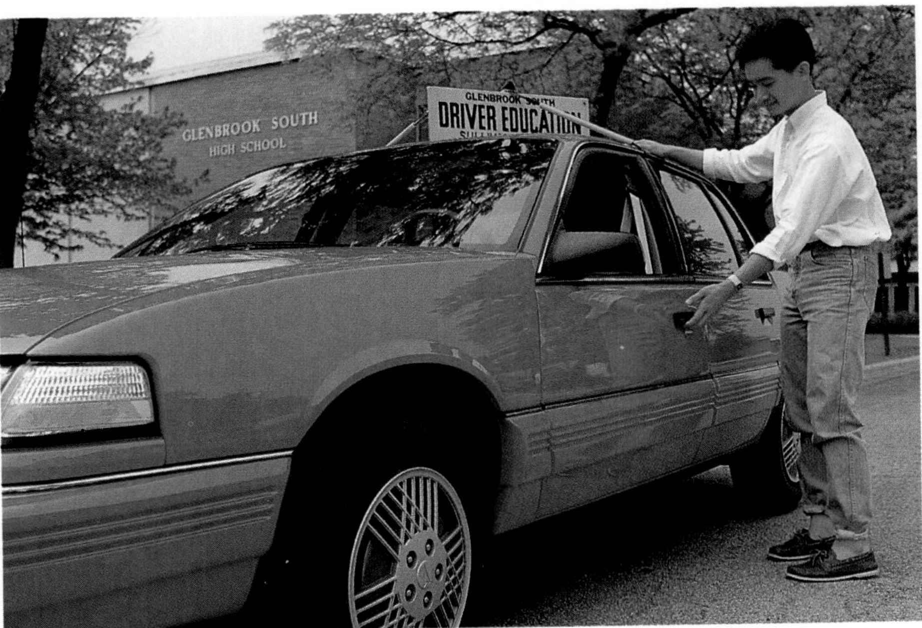

Keep a safe distance. *Driver education classes teach defensive driving techniques, such as estimating stopping distances and safe following distances at various speeds.*

How Much Distance Is Needed to Stop a Car?

When a driver decides to stop a car, it takes time to react and put the foot on the brake. Then it takes time for the car to slow down. The total distance traveled during this time is called the **stopping distance** of the car. Of course, the faster a car is traveling, the longer its stopping distance will be. A formula relating the speed x (in mph) of some cars and the stopping distance d (in feet) is

$$d = .05x^2 + x.$$

This formula is used by those who study automobile performance and safety. It is also important for determining the distance that should be maintained between a car and the car in front of it.

To find the distance needed to stop a car traveling 50 mph, you can substitute $x = 50$ in the above equation.

$$\begin{aligned} d &= .05(50)^2 + 50 \\ &= .05(2500) + 50 \\ &= 125 + 50 \\ &= 175 \end{aligned}$$

Thus, a car traveling 50 mph takes about 175 feet to come to a complete stop after the driver decides to stop.

At the top of page 555 is a table of values and a graph for the stopping distance formula. The situation makes no sense for negative values of x or d, so the graph has points in the first quadrant only. Notice that the graph is parabolic in shape.

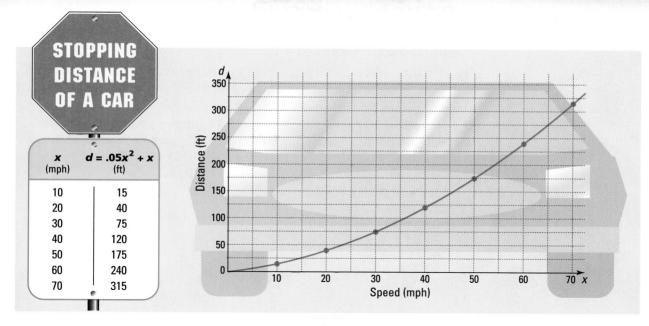

STOPPING DISTANCE OF A CAR

x (mph)	$d = .05x^2 + x$ (ft)
10	15
20	40
30	75
40	120
50	175
60	240
70	315

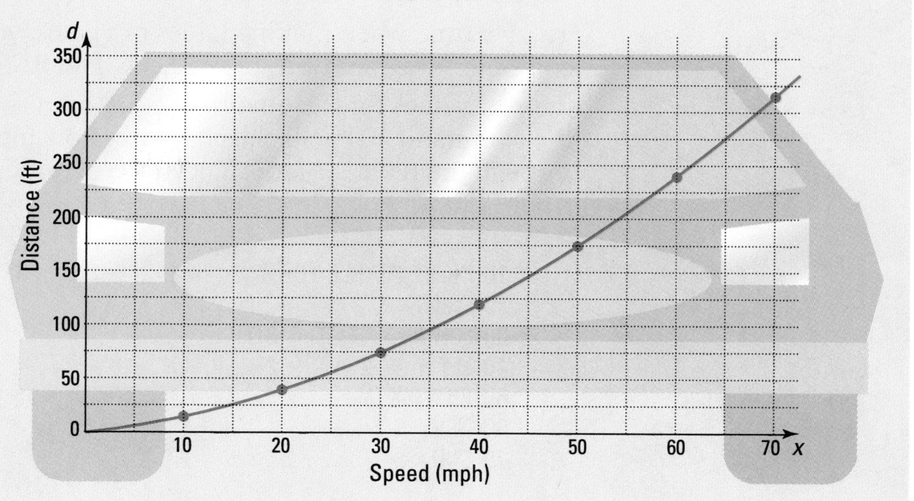

Properties of the Graph of $y = ax^2 + bx + c$

The equation $d = .05x^2 + x$ is of the form $y = ax^2 + bx + c$, with $a = .05$, $b = 1$, and $c = 0$. Many people are surprised to learn that the graph of every equation of the form $y = ax^2 + bx + c$ (with $a \neq 0$) is a parabola. The values of a, b, and c determine where the parabola is positioned in the plane, and whether it opens up or down.

The equations you studied in Lesson 9-1 are also of the form $y = ax^2 + bx + c$. As with those equations, if $a > 0$ the parabola opens up. If $a < 0$, the parabola opens down.

Example 1

a. Graph $y = -x^2 - 4$.
b. Tell whether the parabola opens up or down.
c. Identify its axis of symmetry and vertex.

Solution

a. Form a table of values and plot. Recall that $-x^2$ means $-1(x^2)$.

x	y
-2	-8
-1	-5
0	-4
1	-5
2	-8

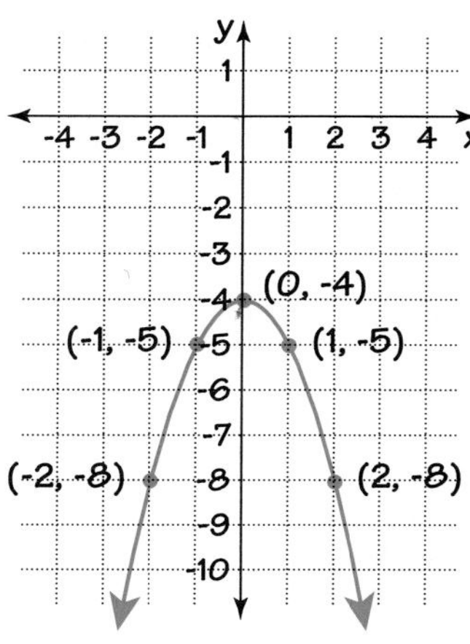

b. The parabola opens down. You can see this from the graph. You could also have predicted this from the equation, because the coefficient of x^2 is negative.

c. Examine the graph. The y-axis is the axis of symmetry. The vertex is (0, -4).

Using Tables to Graph $y = ax^2 + bx + c$

To save time when making tables of values, you can use a calculator or computer program or a spreadsheet. The BASIC program below prints a table of values for the quadratic equation $y = 2x^2 - 8x + 6$. In line 30 the command STEP tells the computer how much to add to x each time through the FOR/NEXT loop. The value of x increases in "steps" of 0.5, from -1 to 4.

```
10 PRINT "TABLE FOR Y = 2X ^ 2 − 8X + 6"
20 PRINT "X", "Y"
30 FOR X = -1 TO 4 STEP.5
40 LET Y = 2 * X ^ 2 − 8 * X + 6
50 PRINT X, Y
60 NEXT X
70 END
```

Here is what the computer prints when the program is run.

TABLE FOR Y = 2X ^ 2 − 8X + 6

X	Y
-1	16
-.5	10.5
0	6
.5	2.5
1	0
1.5	-1.5
2	-2
2.5	-1.5
3	0
3.5	2.5
4	6

You can use the symmetry of the graph of a quadratic equation to answer questions about it.

Example 2

Consider the equation $y = 2x^2 - 8x + 6$ and use the table of values produced by the BASIC program above.

a. What is the vertex of the parabola?
b. What is the axis of symmetry of the parabola?
c. Find the y-intercept of the parabola.
d. Graph the parabola.
e. What are the x-coordinates of the two points where $y = 16$?

Solution

a. Look in the table for points with the same y-values. Notice how the pairs occur on either side of (2, -2). The vertex is (2, -2).
b. The axis of symmetry is the vertical line through the vertex. The axis of symmetry is $x = 2$.
c. The y-intercept is the y-coordinate of the point where $x = 0$. From the table you can see that the y-intercept is 6.

d. Plot the points in the table and connect them with a smooth curve.

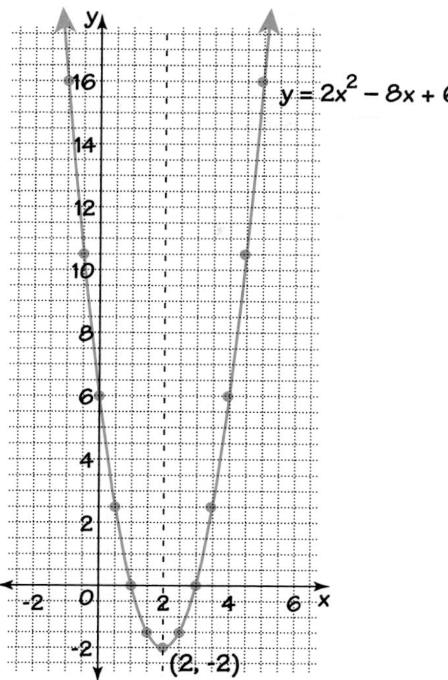

e. (-1, 16) is in the table. Because of symmetry there is another x-value for $y = 16$. Continue the pattern in the table. This results in (5, 16).

X	Y
-1	16
-.5	10.5
0	6
.5	2.5
1	0
1.5	-1.5
2	-2
2.5	-1.5
3	0
3.5	2.5
4	6
4.5	10.5
5	16

When y = 16, x = 5 or x = -1.

A driving rain. *The equation for stopping distance given in this lesson is for dry pavement. Motorists should allow more distance for stopping on wet pavement.*

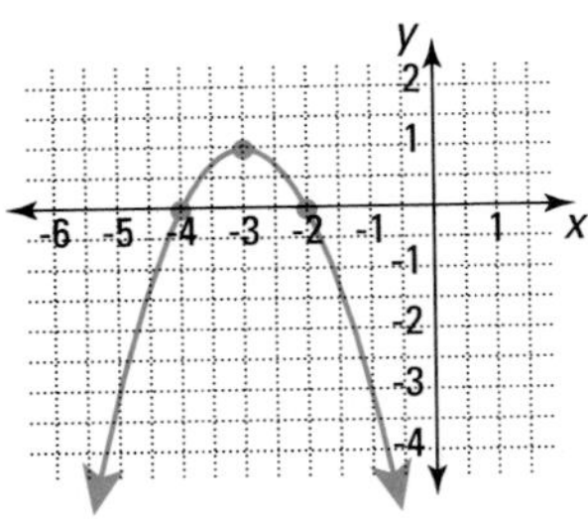

QUESTIONS

Covering the Reading

In 1–4, use the formula for automobile stopping distances in this lesson.

1. Define *stopping distance.*

2. Find the stopping distance for a car traveling 40 mph.

3. Find the stopping distance for a car traveling 55 mph.

4. *True or false.* The stopping distance for a car traveling 60 mph is double the stopping distance of a car traveling 30 mph.

5. *True or false.* The equation $d = 0.05x^2 + x$ is of the form $y = ax^2 + bx + c$.

In 6–8, an equation is given.

a. Make a table of x- and y-values for the equation when x equals -3, -2, -1, 0, 1, 2, and 3.
b. Graph the equation.
c. Identify the vertex and tell whether it is a minimum or maximum.
d. Identify the y-intercept.

6. $y = x^2 + 5$ **7.** $y = x^2 - 2x - 3$ **8.** $y = -3x^2 + 5$

9. a. What are the coordinates of the vertex of the parabola at the left?
b. Write an equation of its axis of symmetry.

In 10 and 11, consider the BASIC program below.

```
10 PRINT "TABLE FOR Y = X ^ 2 − 2X"
20 PRINT "X", "Y"
30 FOR X = -1 TO 3 STEP.5
40 LET Y = X ^2 − 2 * X
50 PRINT X, Y
60 NEXT X
70 END
```

10. What will this program print when run?

11. What is the axis of symmetry for this parabola?

12. Consider this table of values for a parabola.

x	−8	−7	−6	−5	−4	−3	−2	−1	0
y	−13	5	19	29	35	29	?	?	?

a. What is the vertex of the parabola?
b. What is the axis of symmetry of the parabola?
c. Write the y-values that are missing from the table.

13. Explain how you can tell by looking at an equation of the form $y = ax^2 + bx + c$, whether its graph will open up or down.

Applying the Mathematics

14. a. Copy the graph below on graph paper. Then use symmetry to graph more of the parabola.
b. Identify the vertex.
c. Give an equation for the axis of symmetry.
d. At what points does the parabola cross the x-axis? (These are the x-intercepts of the parabola.)

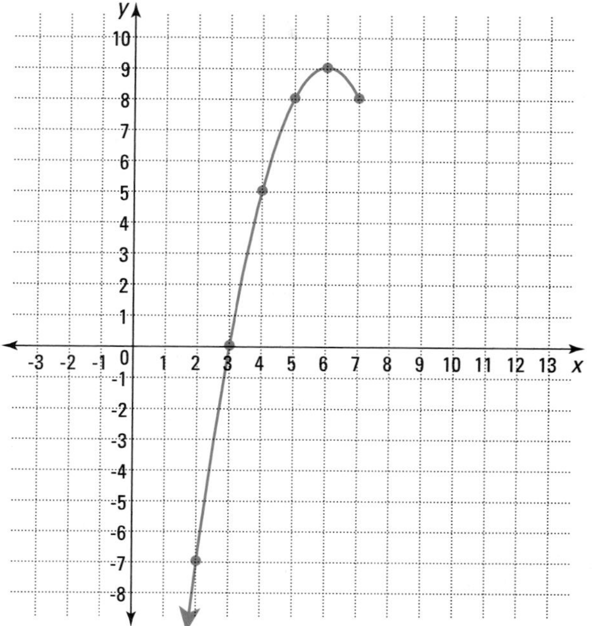

In 15–18, *multiple choice.* Match the graph with its equation.

(a) $y = x^2$ (b) $y = x^2 + 1$ (c) $y = x^2 - 1$ (d) $y = -x^2 - 1$

15.

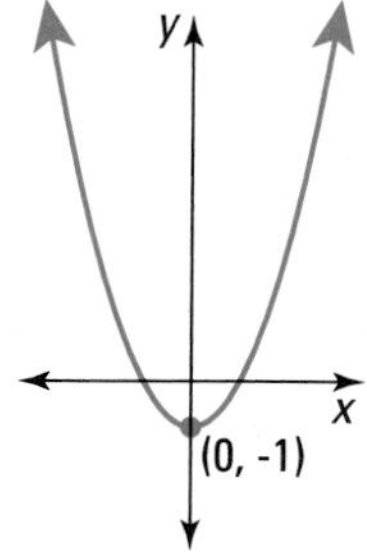

16.

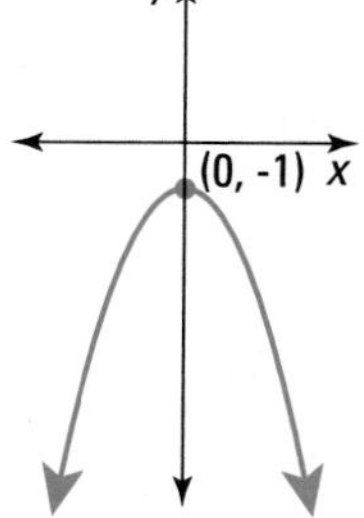

17.

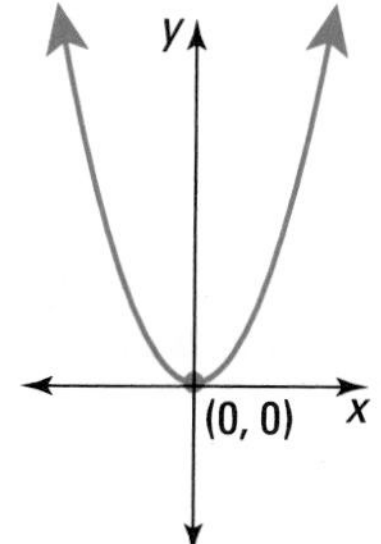

18.

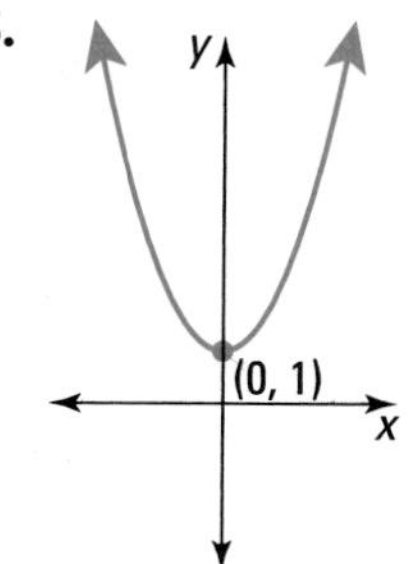

19. *Multiple choice.* Which of these could be the graph of $y = x^2 - 6x + 8$? Justify your answer.

(a)

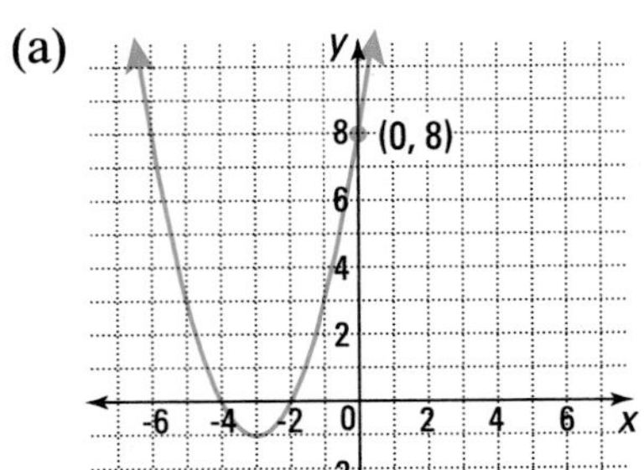

(b)

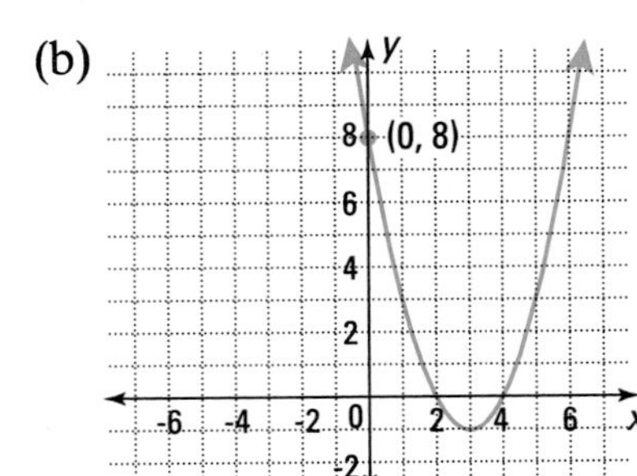

Review

20. Consider the equation $d = 16t^2$ from page 547. *(Lesson 9-1)*
 a. Find t when $d = 500$.
 b. Write a question involving distance and time that can be answered using part **a.**

21.

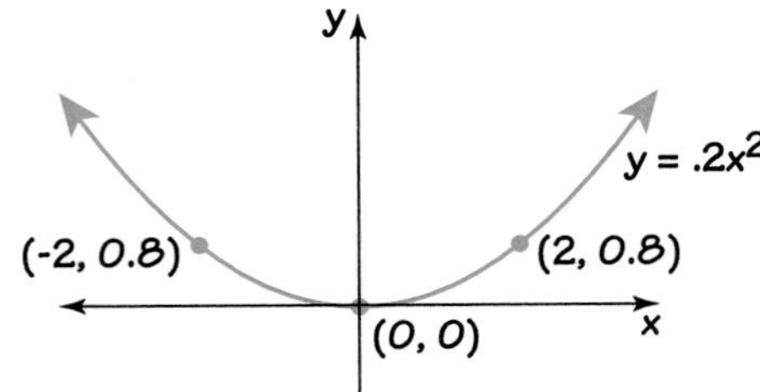

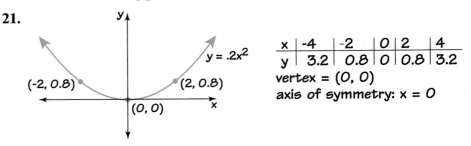

x	-4	-2	0	2	4
y	3.2	0.8	0	0.8	3.2

vertex = (0, 0)
axis of symmetry: x = 0

Above are Bill's answers to questions about $y = .2x^2$. After he wrote this, he realized that he copied the equation incorrectly. It should be $y = -.2x^2$. What does Bill need to change to correct his work? *(Lesson 9-1)*

In 22 and 23, solve. *(Lessons 6-8, 9-1)*

22. $-3n^2 = -1200$

23. $\frac{16}{x} = \frac{x}{40}$

Tale of the Manx. *The Manx cat is named after the Isle of Man in the Irish Sea where the breed originated. There are four varieties of Manx cats—rumpy, rumpy-riser, stumpy, and longie. The rumpy is the only cat without a tail.*

In 24 and 25, suppose an animal shelter has only cats, dogs, and rabbits. Three of the cats are Manx cats, which have no tails. All the dogs and rabbits have tails. Let c = the number of cats, d = number of dogs, and r = number of rabbits. *(Lessons 3-1, 4-2, 6-3)*

24. What is the ratio of the number of cats to the total number of animals?

25. What is the ratio of the number of animal heads to the number of animal tails?

Exploration

26. Use a spreadsheet to generate a table of values for one of the quadratic equations used in this lesson.

Introducing Lesson 9-3

Graphing Parabolas with an Automatic Grapher

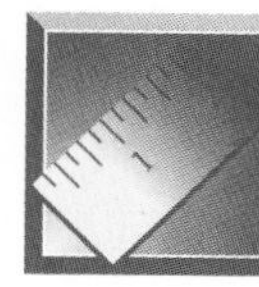

IN-CLASS ACTIVITY

Materials: Automatic Grapher

When you want to work with several graphs at a time, an automatic grapher is very helpful.

1 **a.** On the default window for your grapher, graph $y = ax^2$ for each of these values of a: 0.5, 1, 2, and 3. You should see something similar to what is shown here. Which graph is which?

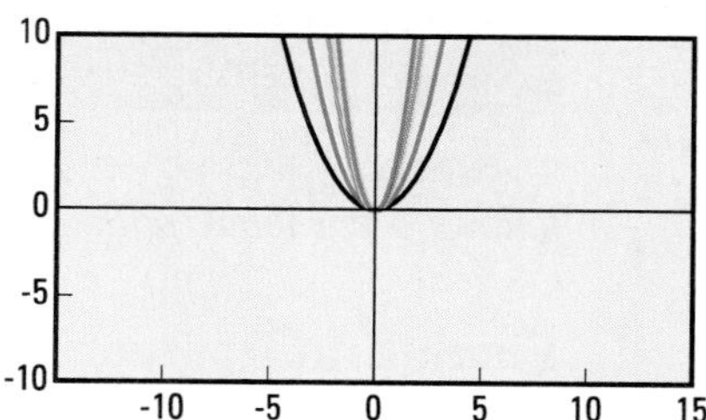

b. What happens to the graph of $y = ax^2$ as a gets larger?
c. Predict what the graph of $y = 7x^2$ will look like in relation to the graphs in part **a.** Then check your prediction by graphing.

2 Zoom in on the vertex (0, 0) of these parabolas. What happens to the way the parabolas look?

3 Clear the screen. Graph $y = x^2 - 36x + 20$. Adjust the window of your grapher so that the graph looks something like this. Describe the window.

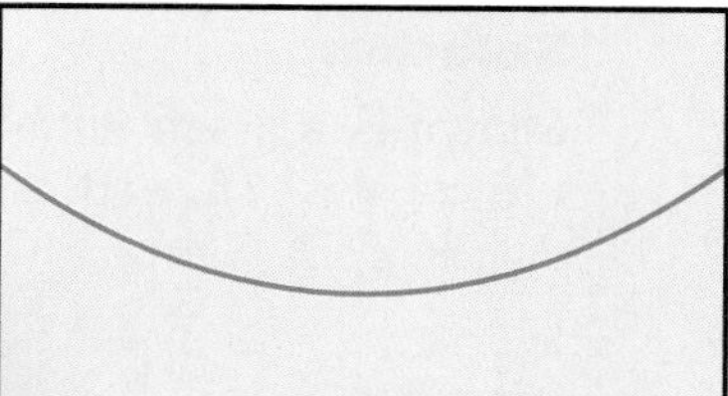

LESSON

9-3 Graphing Parabolas with an Automatic Grapher

Curves in nature. *A waterfall, such as this one at Iguazu Falls in Argentina, follows a parabolic path. A rainbow, however, is a circular arc.*

Finding an Appropriate Window

Parabolas are infinite in extent and they contain no circular arcs. This makes it quite difficult to draw an accurate parabola by hand. So we recommend that you use an automatic grapher to graph parabolas whenever you can. However, if you are not told what window to use, you may have to experiment to find values that give a good graph. You may find it convenient to try your default window first.

Example 1

Find a window that shows a graph of $y = 0.5x^2 - 20x + 100$, including its vertex, its y-intercept, and its x-intercepts.

Solution

Shown is a graph with the window $-15 \le x \le 15$, $-10 \le y \le 10$.

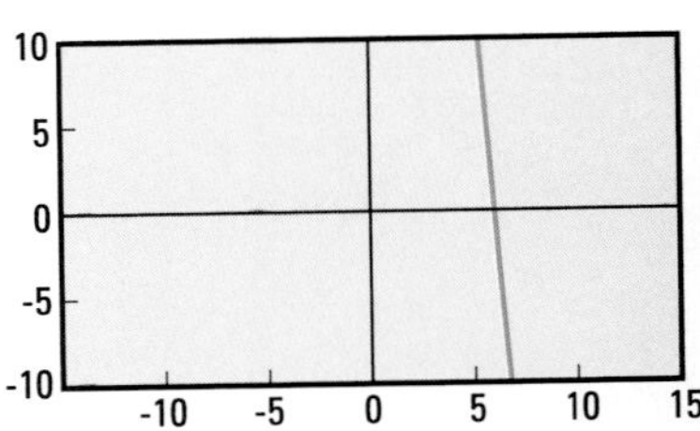

Since the equation is of the form $y = ax^2 + bx + c$, the graph will be a parabola. This window shows part of the parabola, including one x-intercept when x is a little larger than 5. However, this window does not show the other x-intercept, the y-intercept, or the vertex. The y-intercept can be calculated directly. Just substitute 0 in the equation.

$$y = 0.5(0)^2 - 20(0) + 100 = 100.$$

The vertex will be to the right of $x = 15$ and below $y = -10$. So next we graph the equation on the window $0 \leq x \leq 35$, $-50 \leq y \leq 100$. Our graph is shown at the right.

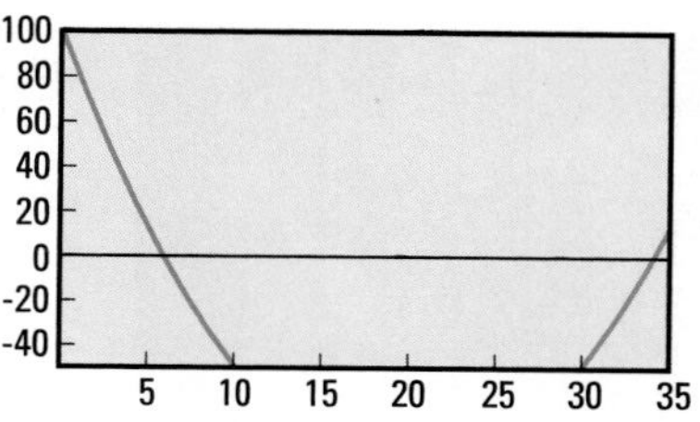

This window shows the y-intercept and the two x-intercepts, but not the vertex. It looks as if the axis of symmetry might be near $x = 20$. To check this we draw our next graph using $0 \leq x \leq 40$. To see the vertex we must use even smaller values of y. Below is a graph on the window $0 \leq x \leq 40$, $-150 \leq y \leq 100$.

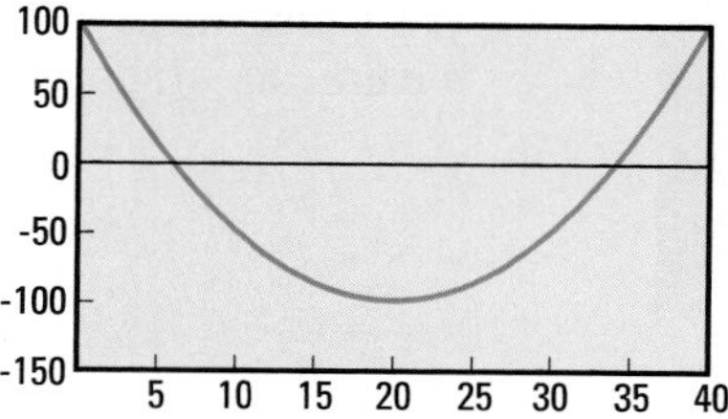

Locating Key Points on a Graph

You can use the trace feature to locate points on a parabola if it has been graphed with an automatic grapher. Key points are the parabola's vertex and the intercepts.

Example 2

Estimate the coordinates of the vertex and the x-intercepts of the graph of $y = 0.5x^2 - 20x + 100$.

Solution

You can see from the graph above that the vertex is near (20, -100), the smaller x-intercept is between 5 and 10, and the larger is between 30 and 35. Use the trace feature of your automatic grapher. A grapher shows the vertex to be about (20, -100), and the x-intercepts to be about 5.9 and 34.1.

In the parabola used for Examples 1 and 2, the average of the x-intercepts gives the x-coordinate of the vertex.

$$\frac{5.9 + 34.1}{2} = \frac{40}{2} = 20$$

This always happens. If a parabola with equation $y = ax^2 + bx + c$ has x-intercepts x_1 and x_2, then the x-coordinate of the vertex is

$$\frac{x_1 + x_2}{2}.$$

When asked to make a graph with an automatic grapher, you should either sketch the graph your grapher draws or make a printout.

An Example of a Quadratic Model

Bent out of shape. *Car insurance companies usually require written estimates from auto body shops before repair work is authorized.*

Example 3

An insurance company reports that the equation

$$y = 0.4x^2 - 36x + 1000$$

relates the age of a driver x (in years) to the driver's accident rate y (number of accidents per 50 million miles driven) when $16 \le x \le 74$.

a. Plot this equation using an automatic grapher.

b. According to this model, at what age do drivers have the fewest accidents per mile driven? How many accidents do they have?

Solution

a. We chose to use the window $0 \le x \le 100$, $-100 \le y \le 2000$ to ensure the graph would fit. A graph is shown below.

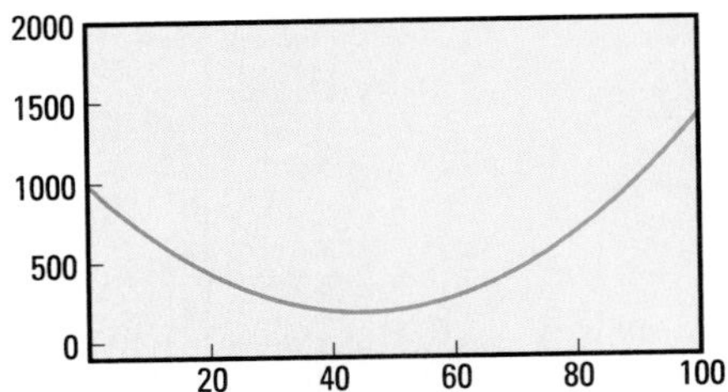

b. Use the trace feature and look for the smallest value of y. We find $y = 190.04$ when $x = 44.68$. **Drivers have the fewest accidents per mile at age 44. Then they have about 190 accidents per 50 million miles, which is about 1 accident per 260,000 miles driven.** In contrast, drivers at the age of 16 have about 1 accident per 95,000 miles driven.

If you are graphing an equation to help answer a question about a real problem as in Example 3, the portion of the graph you will want to examine depends upon the question. However, if the goal is simply to draw a graph of a parabola, you will want to draw one that, like the third graph in the solution to Example 1, shows the vertex, y-intercept, and x-intercept or intercepts, if any.

QUESTIONS

Covering the Reading

In 1 and 2, refer to the activity on page 561.

1. Describe what happens to the graphs as a increases.

2. Copy or print the graphs, and add a sketch of $y = 4x^2$ to them.

3. **a.** Graph $y = ax^2$ when $a = -\frac{1}{2}, -1, -2$, and -3.
   b. What happens to the graph as a gets smaller?

In 4 and 5, refer to Example 3.

4. At about what age do drivers have the fewest accidents per mile driven?

5. At about what ages are there 500 accidents per 50 million miles driven?

6. Graph $y = x^2 - 16$ on the following windows.
   a. $-5 \le x \le 5, -5 \le y \le 5$
   b. a window that shows the vertex, the x-intercepts, and the y-intercept of the parabola

In 7 and 8, an equation and a graph of a parabola are given. Find a different window that shows the vertex and both x-intercepts.

7. $y = x^2 - 10x + 20$

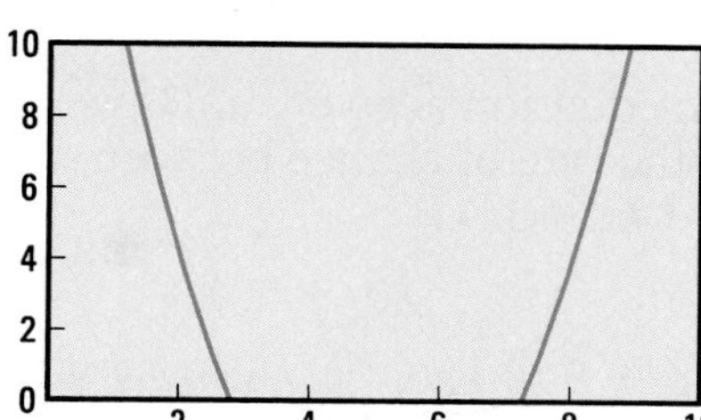

8. $y = -x^2 + 10x - 20$

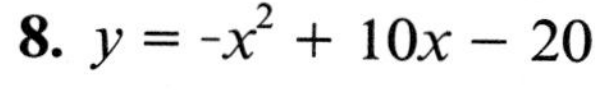

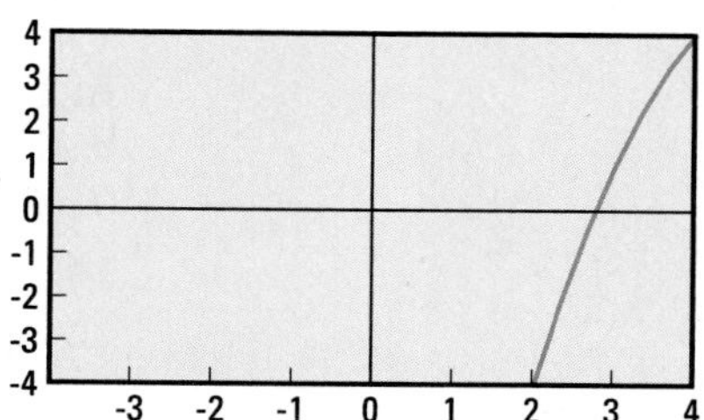

9. **a.** Graph $y = 2x^2 + 38x + 168$ on a window that shows the key points discussed in the lesson.
   b. Estimate the coordinates of the vertex of the graph.
   c. Give an equation of the axis of symmetry.
   d. Estimate the x-intercepts.

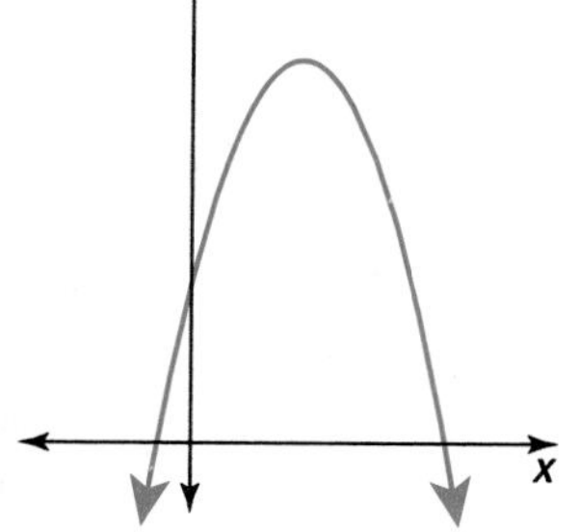

10. At the left is a sketch of the graph of $y = -x^2 + 6x + 7$. The x-intercepts are -1 and 7.
    a. What is the x-coordinate of the vertex?
    b. What is the y-coordinate of the vertex?

Applying the Mathematics

11. **a.** Graph on your default window.
    $$y = x^2 - 3$$
    $$y = x^2$$
    $$y = x^2 + 3$$
    b. Describe the graphs.
    c. Predict what the graph of $y = x^2 + 6$ will look like on the same window.
    d. Test your prediction by graphing $y = x^2 + 6$.

12. The parabola $y = x^2 - 20x + 104$ has vertex (10, 4). Find a window that will produce each of the following views of this parabola.

a.

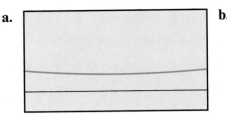

b.

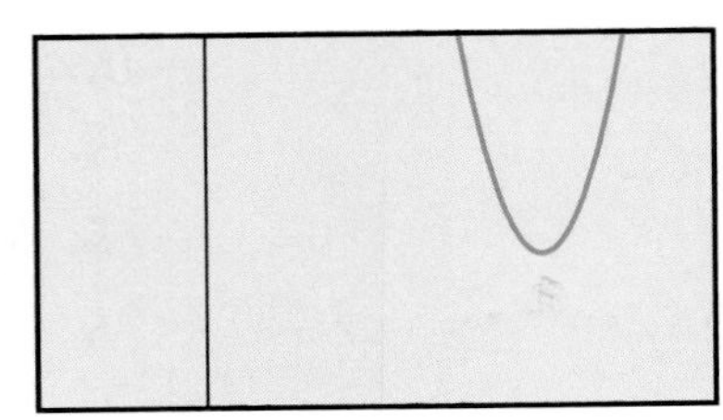

13. **a.** Graph $y = 3x + 5$ and $y = x^2$ on the same axes.
b. Estimate the two points where these graphs intersect.

Review

In 14–16, an equation is given. **a.** Draw a graph when $-1 \le x \le 4$. **b.** Calculate the rate of change between $x = 0$ and $x = 1$. *(Lessons 7-1, 7-5, 8-2, 8-3, 9-1)*

14. $y = 2x$ **15.** $y = x^2$ **16.** $y = 2^x$

17. Solve $3x^2 + 9 = 156$. *(Lessons 5-9, 9-1)*

18. Rewrite $7x^2 + 2x + 5 = x^2 - 10x$ in the form $ax^2 + bx + c = 0$. *(Lesson 5-7)*

19. In Belleville a taxi costs \$.45 plus \$1.20 for each mile. In Carrolton a taxi ride costs \$1.25 plus \$1.00 for each mile.
a. For what distance(s) traveled do rides in these cities cost the same?
b. When is it cheaper to travel in Belleville? Explain how you found your answer and why you know it is right. *(Lessons 5-2, 5-3, 5-4)*

20. A tortoise is walking at a rate of $3 \frac{\text{ft}}{\text{minute}}$. Assume this rate continues.
a. How long will it take the tortoise to travel 20 feet?
b. How long will it take the tortoise to travel f feet? *(Lesson 2-4)*

21. *Skill sequence.* If $a = -7$, $b = 6$, and $c = 15$, find the value of each expression. *(Lessons 1-4, 1-6)*
a. $-4ac$ **b.** $b^2 - 4ac$ **c.** $\sqrt{b^2 - 4ac}$

Exploration

22. **a.** Draw graphs of $y = x^2 + bx + 2$ when $b = 0$, 1, 2, and 3.
b. How does the value of b affect the graph?
c. From your answer to part **b,** predict what the graph of $y = x^2 + 100x + 2$ will look like.
d. When b is negative, where would you expect the graph of $y = x^2 + bx + 2$ to lie? Choose a negative value for b and check your prediction.

LESSON

9-4

Quadratic Equations and Projectiles

Projectile palette. *Firework flames follow parabolic paths.*

A projectile is an object that is thrown or dropped or shot, and then proceeds with no additional force of its own. Balls, bullets, and some rockets are projectiles.

Equations for Paths of Projectiles

Consider a quarterback who tosses a football to a receiver 40 yards downfield. If the ball is thrown and caught six feet above the ground and is 16 ft above the ground at the peak of the throw (the vertex), then the path of the ball can be graphed as it is below. The path is part of a parabola.

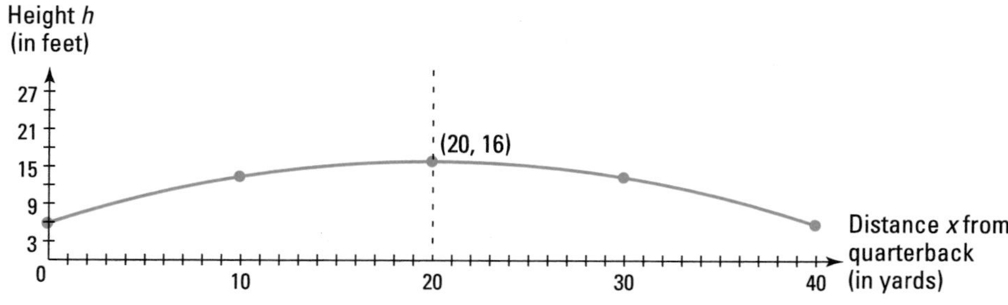

The ball reaches its maximum height halfway between the passer and the receiver, in this case after traveling 20 yards. So this parabola is symmetric to the vertical line $x = 20$. If the ball is at a height of h feet when it has traveled x yards, its path is a piece of the parabola described by the equation

$$h = -0.025x^2 + x + 6.$$

Example 1

Refer to the situation on page 567. Suppose a defender is 3 yards in front of the receiver. This means the defender is 37 yards from the quarterback. Will he be able to deflect or catch the ball?

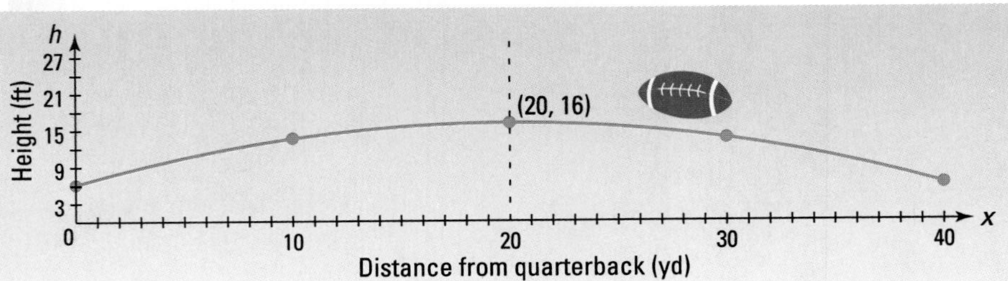

Solution

The answer depends on how high the ball is when it reaches him. The height is represented by *h* in the above equation. Substitute 37 for *x* in the equation.

$$
\begin{aligned}
h &= -0.025x^2 + x + 6 \\
&= -0.025 \cdot 37^2 + 37 + 6 \\
&= -0.025 \cdot 1369 + 37 + 6 \\
&= -34.225 + 37 + 6 \\
&= 8.775
\end{aligned}
$$

The ball will be 8.775 feet above the ground when it reaches the defender. Since $0.775 \cdot 12$ inches ≈ 9 inches, this is approximately 8 feet 9 inches. To deflect or intercept the ball, the defender would have to reach a height of 8 feet 9 inches. With a well-timed jump, this is possible for most defenders.

Equations for the Heights of Projectiles over Time

A parabola can also describe the relation between the length of time since a projectile has been thrown or shot into the air and its height above the ground.

Example 2

The equation $h = -16t^2 + 32t + 6$ gives the height *h* in feet of another ball *t* seconds after being thrown from a height of 6 feet with an initial upward velocity of 32 feet per second.

a. How high will the ball be a half second after it is thrown?

b. What is the maximum height this ball reaches?

Solution

a. Substitute 0.5 for *t* in $h = -16t^2 + 32t + 6$.

$$
\begin{aligned}
h &= -16(.5)^2 + 32(.5) + 6 \\
&= -16(.25) + 16 + 6 \\
&= -4 + 22 \\
&= 18
\end{aligned}
$$

In half a second, the ball will be 18 feet high.

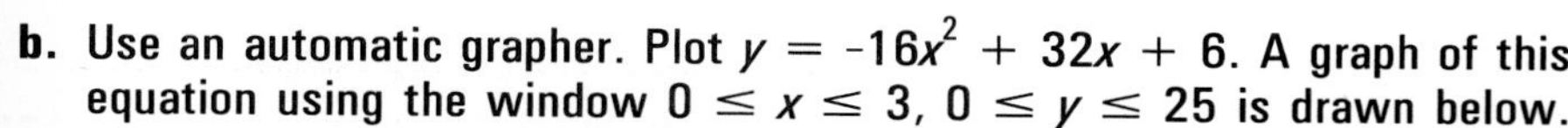

b. Use an automatic grapher. Plot $y = -16x^2 + 32x + 6$. A graph of this equation using the window $0 \leq x \leq 3$, $0 \leq y \leq 25$ is drawn below.

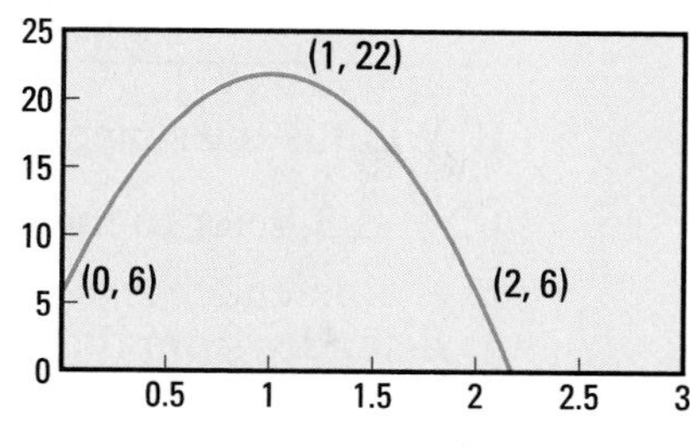

The maximum height is the largest value of h shown on the graph, the y-coordinate of the vertex. Trace along the graph; read the y-coordinate as you go. When we did this using a window $0 \leq x \leq 3$, $0 \leq y \leq 25$, our trace showed that (.989, 21.993) and (1.01, 21.998) are on the graph. So try $t = 1$. This gives $h = 22$. The maximum height reached is 22 ft.

Check

We verify the maximum height by noting that (0, 6) and (2, 6) are on the graph. This means the axis of symmetry is $t = 1$ and so the vertex is (1, 22).

Caution: The graph in Example 1 describes the actual path of an object, the height at a given *distance* from the start. The graph in Example 2 describes an object's height at a specific *time*. It is essential to read the labels of the axes to know what a graph represents.

Example 3

Refer to the situation in Example 2. Estimate when the ball is 12 feet high.

Solution 1

The values of t corresponding to $h = 12$ must be found. Draw a horizontal line at $h = 12$ feet. Read the x-coordinates at the points of intersection.

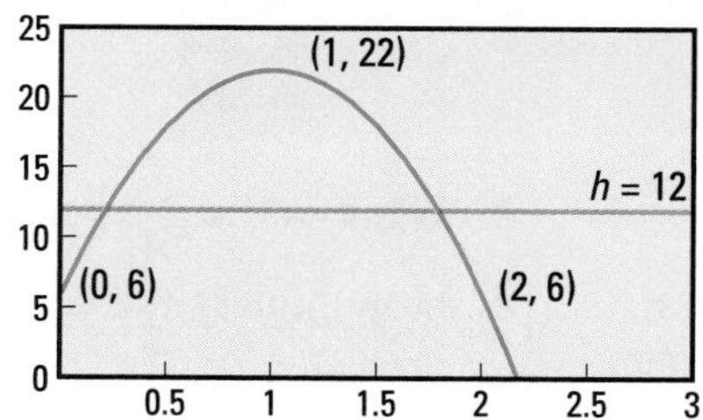

Our grapher showed that at $x \approx 0.21$ and $x \approx 1.79$, $y = 12$. The ball is 12 feet off the ground at about 0.2 second and 1.8 seconds after being thrown. Notice that both these times are 0.8 second away from 1 second, where the vertex of the parabola is located.

It is natural to want to find an exact answer to the question of Example 3. The method for finding it is discussed in Lesson 9-5.

QUESTIONS

Covering the Reading

1. What is the shape of the path of a tossed ball?

In 2 and 3, refer to the tossed football at the beginning of the lesson.

2. If a defender is 5 yards in front of the receiver, how far is the defender from the quarterback?

3. What is the height of the ball 39 yards from the quarterback? Write your answer to the nearest inch.

In 4 and 5, a quarterback throws the ball to a receiver 60 yards away. Suppose the ball's height h in feet x yards away is

$$h = -\frac{x^2}{45} + \frac{4x}{3} + 5.$$

4. At what height does the ball reach the receiver?

5. A defender is 5 yards in front of the receiver.
  a. How far is he from the quarterback?
  b. How high would he have to reach to deflect the ball?

6. When the horizontal axis shows time and the vertical axis shows the height of a tossed ball, what shape is the graph?

In 7–11, refer to the graph below. It shows h, the height in yards of a ball t seconds after it is kicked into the air.

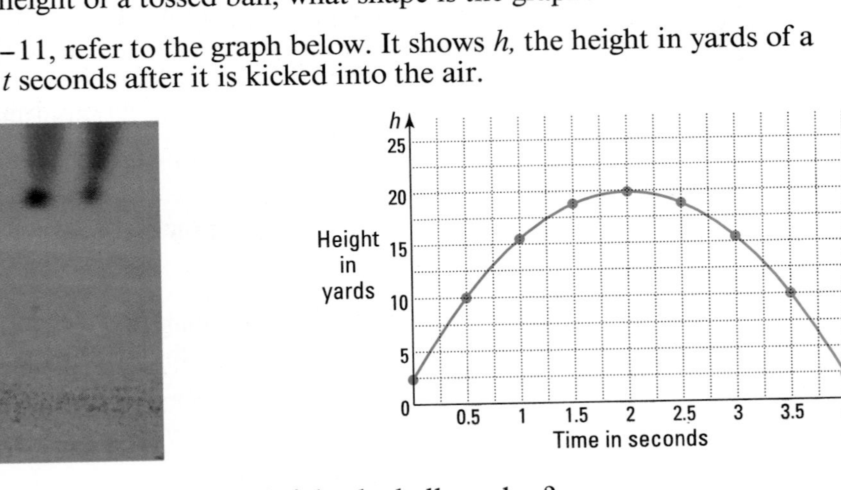

World Cup winner. *Shown is Zinho, a member of the Brazilian soccer team that won the XV World Cup in 1994. The championship was Brazil's fourth since the World Cup was started in 1930.*

7. What is the greatest height the ball reaches?

8. How high is the ball 1 second after it is kicked?

9. At what times is the ball 18 yards high?

10. How long is the ball in the air?

11. For how many seconds is the ball more than 15 yards above the ground?

Applying the Mathematics

In 12–15, a small rocket is shot from the edge of a cliff. Suppose that after t seconds, the rocket is y meters above the cliff where $y = 30t - 5t^2$. This equation is graphed below.

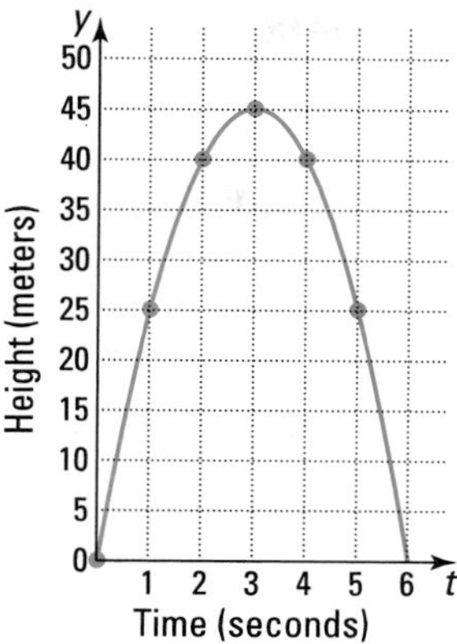

12. What is the greatest height the rocket reaches?

13. How far above the cliff edge is the rocket after 5 seconds?

14. When is the rocket 40 meters above the cliff edge?

15. How far below the edge of the cliff is the rocket after 7 seconds?

In 16–18, the path of a platform diver is described by the equation $y = -2x^2 + x + 10$, where x is the horizontal distance (in meters) of the diver from the edge of the platform, and y is the height of the diver (in meters) above the surface of the water.

16. a. Evaluate y when $x = 1$.

b. Write a sentence about the diver that describes what you found out in part **a.**

17. What is the maximum height the diver reaches?

18. In horizontal distance, how far in front of the platform will the diver hit the water?

Review

In 19 and 20 an equation is given. **a.** Make a table of values for integer values of x from -3 to 3. **b.** Graph the equation. *(Lessons 9-1, 9-2)*

19. $y = \frac{1}{2}x^2$

20. $y = 6x - x^2$

In 21 and 22, use the graph of the parabola at the right. *(Lesson 9-2)*

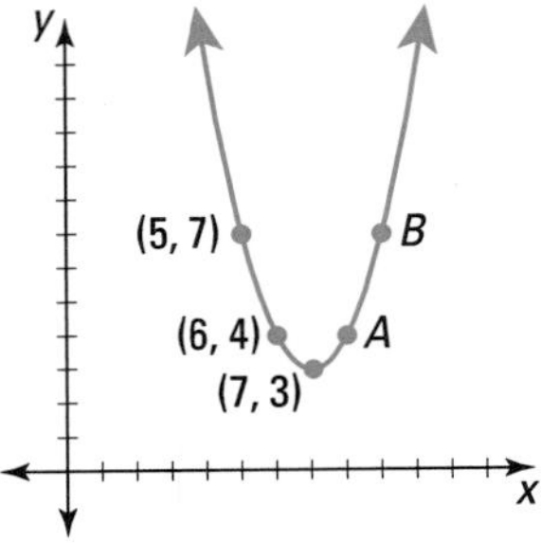

21. Use the symmetry of the parabola to find the coordinates of points A and B.

22. Write an equation of the axis of symmetry.

23. *Multiple choice.* Which of the following is *not* equal to ab^2? *(Lessons 8-5, 8-8)*
(a) $a \cdot b \cdot b$ (b) $a(b^2)$ (c) $(ab)^2$ (d) $a(b)^2$

24. *Skill sequence.* Write as a single fraction. *(Lesson 3-9)*
a. $2 + \frac{7}{11}$ **b.** $2 + \frac{x}{y}$ **c.** $a + \frac{b}{c}$

25. What is the image of (-6, 0) after a slide 3 units to the right and 5 units down? *(Lesson 3-4)*

26. *Multiple choice.* What situation can the graph at right represent? *(Lesson 3-3)*
(a) the cost c of h pencils at 10¢ each
(b) the cost c of h pencils at 2 for 25¢
(c) the cost c of h pencils at 25¢ each
(d) the cost c of h pencils at 50¢ each

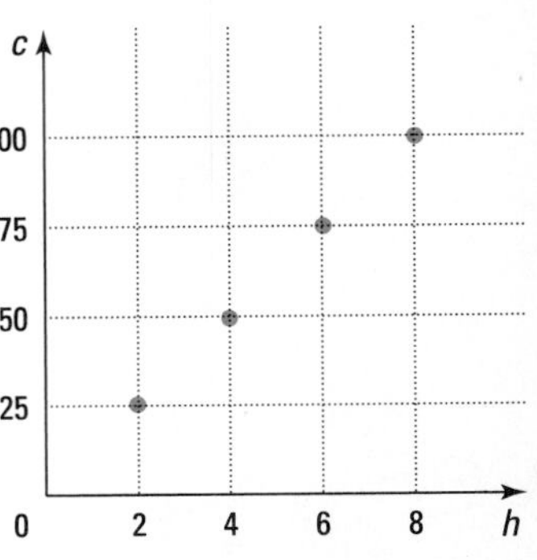

27. Find $\frac{8 + 4\sqrt{2}}{4}$ to the nearest tenth. *(Lesson 1-6)*

28. Let $a = 6$, $b = 4$, and $c = -2$. Evaluate each expression. *(Lessons 1-4, 1-6)*
a. $\sqrt{b^2 - 4ac}$ **b.** $\frac{-b + \sqrt{b^2 - 4ac}}{2a}$ **c.** $\frac{-b - \sqrt{b^2 - 4ac}}{2a}$

Exploration

29. A pitched ball in baseball does not always follow the path of a parabola. Why not?

LESSON

9-5 The Quadratic Formula

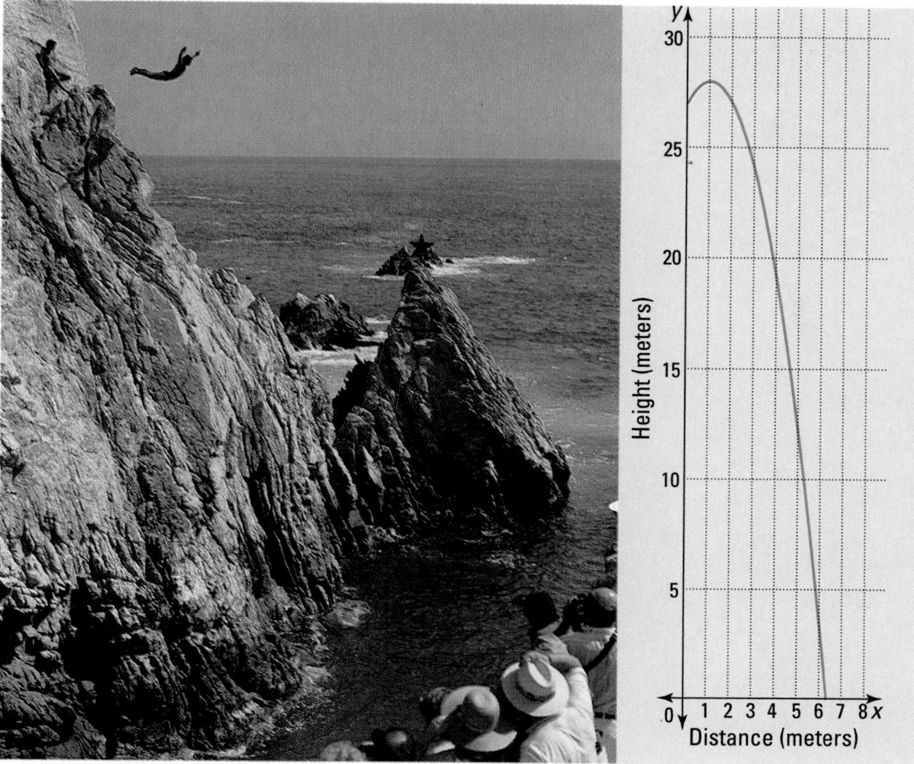

Quadratics at La Quebrada. *Divers at La Quebrada must time their dives to hit a large wave surge to ensure that the water is deep enough. The dive's path can be described with a quadratic equation.*

A Situation Leading to a Quadratic Equation

When a person dives, he or she becomes a projectile, so the person's path is part of a parabola. In Acapulco, Mexico, there is a famous place called "La Quebrada" ("the break in the rocks"). There, a diver dives from a height of approximately 27 meters to the waters below. If the diver is x meters away from the cliff and y meters above the water, then under certain conditions $y = -x^2 + 2x + 27$ describes the path of the dive.

When the diver enters the water, the height $y = 0$. So solving

$$-x^2 + 2x + 27 = 0$$

will give the number of meters from the cliff the diver enters the water. From the graph you can see that x is between 6 and 7. The equation $-x^2 + 2x + 27 = 0$ is an example of a quadratic equation. A **quadratic equation** is an equation that can be written in the form

$$ax^2 + bx + c = 0.$$

The Quadratic Formula

You can find solutions to *any* quadratic equation by using the *Quadratic Formula.* The formula tells how to calculate x using a, b, and c, the coefficients of the terms. It states that there are possibly two solutions.

If $ax^2 + bx + c = 0$, then

$$x = \frac{-b + \sqrt{b^2 - 4ac}}{2a} \quad \text{or} \quad x = \frac{-b - \sqrt{b^2 - 4ac}}{2a}$$

The two expressions differ in only one way: $\sqrt{b^2 - 4ac}$ is added to $-b$ in the numerator of the first, while it is subtracted from $-b$ in the numerator of the second.

To apply the Quadratic Formula, notice that a is the coefficient of x^2, b is the coefficient of x, and c is the constant term.

The work in calculating the two solutions is almost the same. To save writing, both solutions can be written in one expression using the symbol $\pm$, which means *plus or minus.* It shows that you should do the calculation twice, once by adding and once by subtracting.

Quadratic Formula
If $ax^2 + bx + c = 0$ and $a \neq 0$, then

$$x = \frac{-b \pm \sqrt{b^2 - 4ac}}{2a}$$

The Quadratic Formula is one of the most famous formulas in all of mathematics. *You should memorize it today.*

Caution: Many calculators have a +/– key. That key takes the opposite of a number. It does *not* perform the two operations + and – required in the Quadratic Formula.

Applying the Quadratic Formula

Example 1

Solve $-x^2 + 2x + 27 = 0$ to find the distance of the Acapulco diver from the cliff when he enters the water.

Solution

Recall that $-x^2 = -1x^2$. So rewrite the equation as

$$-1x^2 + 2x + 27 = 0.$$

Apply the Quadratic Formula with $a = -1$, $b = 2$, and $c = 27$.

$$x = \frac{-b \pm \sqrt{b^2 - 4ac}}{2a}$$

$$x = \frac{-2 \pm \sqrt{2^2 - 4 \cdot -1 \cdot 27}}{2 \cdot -1}$$

Using order of operations, work under the radical sign first.

$$= \frac{-2 \pm \sqrt{4 - (-108)}}{-2}$$

$$= \frac{-2 \pm \sqrt{112}}{-2}$$

So $x = \frac{-2 + \sqrt{112}}{-2}$ or $x = \frac{-2 - \sqrt{112}}{-2}$.

These are exact solutions. Using a calculator gives approximations. Since $\sqrt{112} \approx 10.6$,

$x \approx \frac{-2 + 10.6}{-2}$ or $x \approx \frac{-2 - 10.6}{-2}$.

So $x \approx -4.3$ or $x \approx 6.3$.

The diver cannot land a negative number of meters from the cliff, so the solution $x \approx -4.3$ does not make sense in this situation. The diver will enter the water about 6.3 meters away from the cliff.

Check

Does 6.3 work in the equation $-x^2 + 2x + 27 = 0$? Substitute 6.3 for x.

$$-(6.3)^2 + 2(6.3) + 27 = -39.69 + 12.6 + 27 = -0.09.$$

This is close enough to zero, given that 6.3 is an approximation.

When the coefficients of a quadratic equation are integers and the number under the radical sign in the Quadratic Formula equals a perfect square, its solution can be written without a radical sign.

Example 2

Solve $3x^2 - 6x - 45 = 0$.

Solution

First relate the equation to $ax^2 + bx + c = 0$ and identify a, b, and c.

$$3x^2 - 6x - 45 = 0$$

$a = 3$, $b = -6$, and $c = -45$

Now substitute these values in the Quadratic Formula.

$$x = \frac{-b \pm \sqrt{b^2 - 4ac}}{2a}$$

$$x = \frac{-(-6) \pm \sqrt{(-6)^2 - 4 \cdot 3 \cdot -45}}{2 \cdot 3}$$

$$x = \frac{6 \pm \sqrt{36 - (-540)}}{6}$$

Now simplify under the radical sign.

$$x = \frac{6 \pm \sqrt{576}}{6}$$

$$x = \frac{6 \pm 24}{6}$$

$$x = \frac{6 + 24}{6} = \frac{30}{6} = 5 \quad \text{or} \quad x = \frac{6 - 24}{6} = \frac{-18}{6} = -3$$

Check

To check, substitute each of the solutions in the original equation.

Substitute 5 for x. $3(5)^2 - 6(5) - 45 = 75 - 30 - 45 = 0$.

Substitute -3 for x. $3(-3)^2 - 6(-3) - 45 = 27 + 18 - 45 = 0$.

Both solutions check.

In Example 3, the equation has to be put into $ax^2 + bx + c = 0$ form before the formula can be applied. This form is called the **standard form for a quadratic equation.**

Example 3

Solve $m^2 - 3m = 14$. Give m to the nearest hundredth.

Solution

To apply the Quadratic Formula, you must write the equation in the form $ax^2 + bx + c = 0$. To put $m^2 - 3m = 14$ in this form add -14 to both sides.

$$m^2 - 3m - 14 = 0$$

Now the formula can be applied with $a = 1$, $b = -3$, $c = -14$.

$$m = \frac{-(-3) \pm \sqrt{(-3)^2 - 4 \cdot 1 \cdot -14}}{2 \cdot 1}$$

$$= \frac{3 \pm \sqrt{9 - (-56)}}{2}$$

$$= \frac{3 \pm \sqrt{65}}{2}$$

Thus $m = \frac{3 + \sqrt{65}}{2}$ or $m = \frac{3 - \sqrt{65}}{2}$.

$m \approx \frac{3 + 8.06}{2}$ or $m \approx \frac{3 - 8.06}{2}$.

So $m \approx 5.53$ or $m \approx -2.53$.

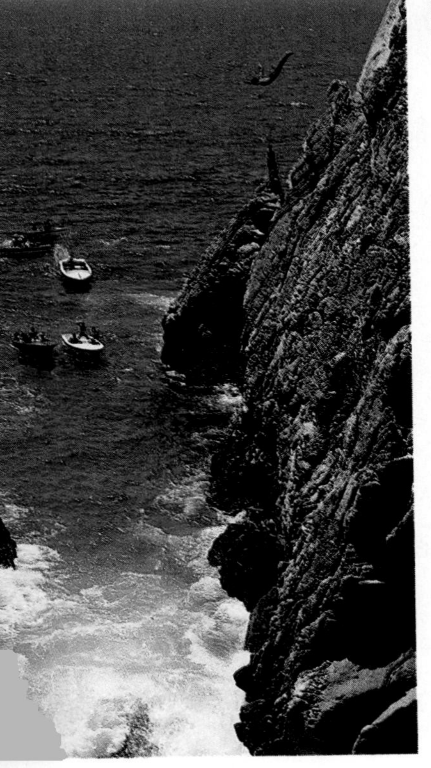

QUESTIONS

Covering the Reading

1. State the Quadratic Formula.

2. *True or false.* The Quadratic Formula can be used to solve any quadratic equation.

3. Find the two values of $\frac{-10 \pm 4}{2}$.

4. Suppose the cliff diver at the beginning of the lesson dove from a cliff 22 meters high. Then under certain conditions, solving the equation $0 = -x^2 + 2x + 22$ gives the number of meters the diver is from the cliff when entering the water.
- **a.** Give the values of a, b, and c for use in the quadratic formula.
- **b.** Apply the Quadratic Formula to give the exact solutions to the equation.
- **c.** Approximate the solutions to the nearest tenth.
- **d.** How far away from the cliff does the diver enter the water?

In 5–8, each sentence is equivalent to a quadratic equation of the form $ax^2 + bx + c = 0$. **a.** Give the values of a, b, and c. **b.** Give the exact solutions to the equation. **c.** Check the solutions.

5. $12x^2 + 7x + 1 = 0$

6. $3n^2 + n - 2 = 0$

7. $x^2 + 6x + 9 = 0$

8. $-x^2 + 4 = 0$

9. Check both decimal answers to Example 3.

In 10–13, a quadratic equation is given. **a.** Rewrite each equation in the form $ax^2 + bx + c = 0$. **b.** Solve the equation using the Quadratic Formula. Round solutions to the nearest hundredth.

10. $20m^2 - 6m = 2$

11. $3w^2 - w = 5$

12. $3 - x = 2x^2$

13. $3p^2 - 10p = 14 + 9p$

Applying the Mathematics

14. The solutions to $ax^2 + bx + c = 0$ are the x-intercepts of the graph of $y = ax^2 + bx + c$.
  a. Use the Quadratic Formula to find the x-intercepts of the graph of $y = 2x^2 + 3x - 2$.
  b. Check by using an automatic grapher.

15. When a ball on the moon is thrown upward with an initial velocity of 6 meters per second, its approximate height y after t seconds is given by $y = -0.8t^2 + 6t$.
  a. At what *two* times will it reach a height of 10 m? Give your answer to the nearest tenth of a second.
  b. Graph $y = -0.8t^2 + 6t$. Use a domain of $0 \le t \le 7.5$.
  c. Use the graph to check your answer to part **a.**

16. Some students noticed that the equation of Example 2 could be simplified by dividing both sides by 3.
  a. What is this simpler equation?
  b. Solve the simpler equation using the Quadratic Formula.

Review

In 17 and 18, a rock is tossed up from a cliff with upward velocity of 10 meters per second. The formula $h = 10t - 4.9t^2$ gives the height h of the rock *above the cliff* after t seconds.

17. How high above the cliff is the rock 1 second after it is tossed?

18. How high above the cliff is the rock 3 seconds after it is tossed? *(Lesson 9-4)*

In 19–21, use the graph below. It shows $h = -.12x^2 + 2x + 5$, the path of a basketball free throw, where h is the height of the ball when it has moved x feet forward. *(Lesson 9-3)*

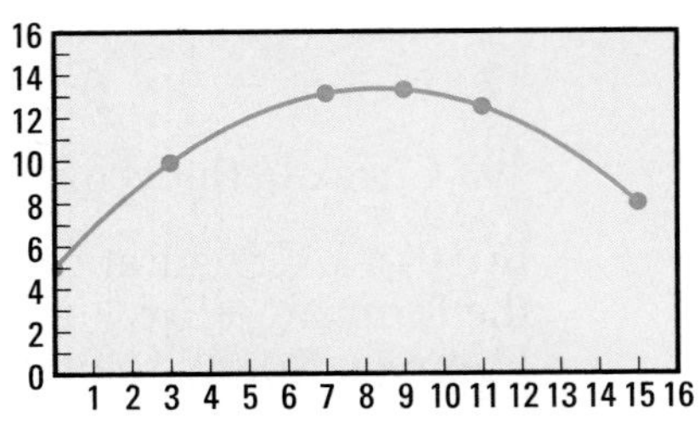

19. What is the height of the ball when it has moved 3 feet forward?

20. About how far has the ball moved forward when it was at a height of 12 feet?

21. Estimate the greatest height the ball reaches.

22. If the x-intercepts of a parabola are 3 and -1, what is the x-coordinate of the vertex? *(Lesson 9-2)*

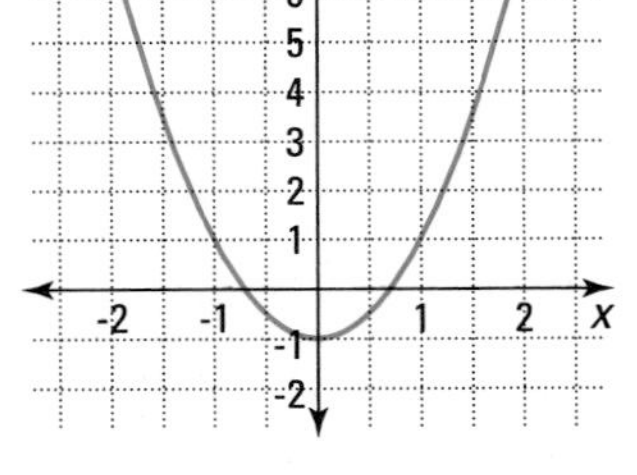

23. *Multiple choice.* Which equation is graphed at the left? *(Lesson 9-2)*
(a) $y = 2x$ (b) $y = 2x^2$
(c) $y = 2x^2 + 1$ (d) $y = 2x^2 - 1$

24. A school begins the year with 200 reams of paper. (A ream contains 500 sheets.) The teachers are using 12 reams a week. How many reams will be left after w weeks? *(Lesson 3-8)*

25. Mr. Robinson is ordering a new car. He can choose from 5 models, 12 exterior colors, and 9 interior colors. How many combinations of models, interiors, and exteriors are possible? *(Lesson 2-9)*

26. Find the volume of a box that is $15a$ inches in length, $8b$ inches in width, and $6c$ inches in height. *(Lesson 2-1)*

Exploration

27. *A team's opening batter named Nero*
Squared his number of hits, the hero!
After subtracting his score,
He took off ten and two more,
And the number resulting was zero.

How many hits did Nero have?

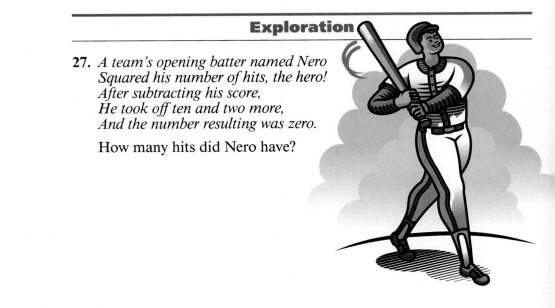

LESSON

9-6

Analyzing Solutions to Quadratic Equations

Diving dolphins. *Trained dolphins perform stunts such as leaping 20 ft to ring a bell. By graphing an equation of the path of a dive, you can find a maximum dive height.*

Using Graphs to Determine the Number of Real Solutions

In the last lesson, the path of a diver from a cliff was described by the equation

$$y = -x^2 + 2x + 27,$$

where y is the height of the diver when he or she is x meters out from the cliff. As the graph on page 573 shows, when the diver first pushes off he or she arches upward and then descends. In this lesson you will learn several ways to determine whether or not the diver reaches a particular height. For instance, will the diver ever reach a height of 27.5 meters above the water? Will the diver ever reach a height of 28 meters? Will he or she ever reach a height of 29 meters?

First let's look at how you can use graphs to answer these questions. Below on the left is a graph of $y = -x^2 + 2x + 27$ on the window $0 \le x \le 8, 0 \le y \le 40$. This is the parabola graphed on page 573.

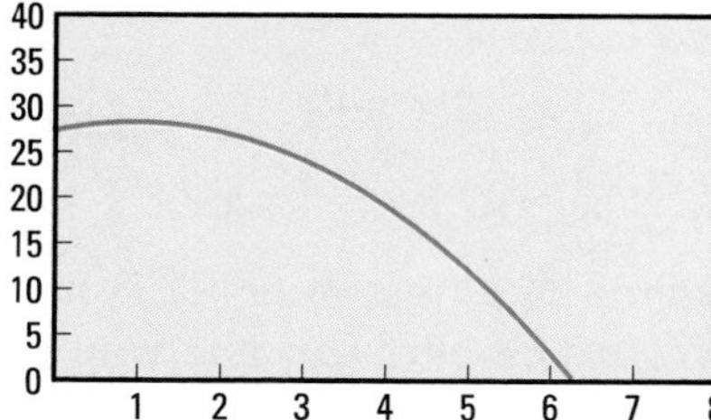

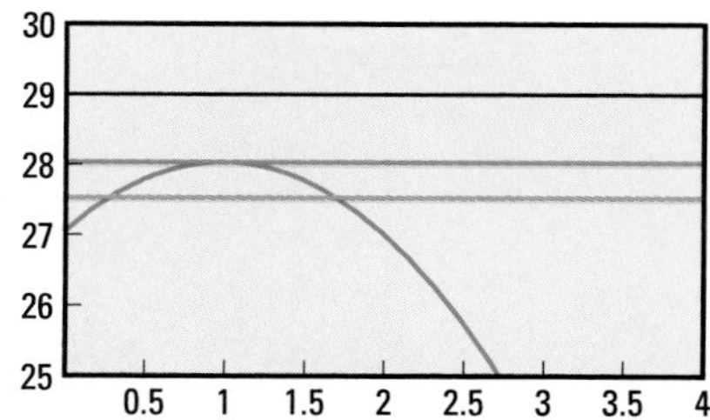

Can you tell from the graph if the height y ever reaches the values 27.5, 28, or 29?

From the graph on the left on page 579, you should be able to see that if the diver ever does get to these heights, it will be when x is between 0 and 4. So on the right we have zoomed in on the top of the parabola. This gives another graph of $y = -x^2 + 2x + 27$, this time on the window $0 \leq x \leq 4$, $25 \leq y \leq 30$. To help us see whether there are heights of 27.5, 28, and 29, we have also graphed the three horizontal lines $y = 27.5$, $y = 28$, and $y = 29$.

From the graph on the right on page 579, we can make the following conclusions:

1. The line $y = 27.5$ crosses the parabola twice. So the diver reaches a height of 27.5 meters at *two* times.
2. The line $y = 28$ appears to intersect the parabola once at the vertex of the parabola. So the diver reaches a height of 28 meters *once.*
3. The line $y = 29$ never intersects the parabola. It is completely above the parabola. So the diver *never* reaches a height of 29 meters.

Using the Quadratic Formula to Determine the Number of Real Solutions

We can answer the same questions about the diver by using the Quadratic Formula.

Example 1

Will the diver ever reach a height of 27.5 meters above the water? If so, where does this event occur?

Solution

Let $y = 27.5$ in the equation $y = -x^2 + 2x + 27$.

$$27.5 = -x^2 + 2x + 27$$

Add -27.5 to both sides to put the equation in standard form.

$$0 = -x^2 + 2x - 0.5$$

Substitute $a = -1$, $b = 2$, and $c = -0.5$ in the Quadratic Formula.

$$x = \frac{-2 \pm \sqrt{2^2 - 4 \cdot -1 \cdot -0.5}}{2 \cdot -1}$$

$$x = \frac{-2 \pm \sqrt{2}}{-2}$$

$$x \approx \frac{-2 + 1.414}{-2} \quad \text{or} \quad x \approx \frac{-2 - 1.414}{-2}$$

$$x \approx 0.3 \quad \text{or} \quad x \approx 1.7$$

The diver reaches the height of 27.5 meters twice, at about 0.3 meters from the cliff (on the way up) and about 1.7 meters from the cliff (on the way down).

Example 2

Will the diver ever reach a height of 28 meters above the water? If so, where does this occur?

Solution

Let $y = 28$ in the equation $y = -x^2 + 2x + 27$.

$$28 = -x^2 + 2x + 27$$

Write the equation in standard form by adding -28 to both sides.

$$0 = -x^2 + 2x - 1$$

Substitute $a = -1$, $b = 2$, and $c = -1$ in the Quadratic Formula.

$$x = \frac{-2 \pm \sqrt{2^2 - 4 \cdot -1 \cdot -1}}{2 \cdot -1}$$

$$x = \frac{-2 \pm \sqrt{0}}{-2}$$

$$x = \frac{-2 \pm 0}{-2}$$

Because 0 is added and subtracted, the two values for x are identical.

$$x = \frac{-2 + 0}{-2} \quad \text{or} \quad x = \frac{-2 - 0}{-2}$$

$$x = \frac{-2}{-2} \quad \text{or} \quad x = \frac{-2}{-2}$$

$$x = 1 \quad \text{or} \quad x = 1$$

Thus when $y = 28$, there is just one value of x. *The diver reaches a height of 28 meters once, when he or she is 1 meter from the cliff.* This agrees with the graph, where the height of the vertex is 28.

Example 3

Will the diver ever reach a height of 29 meters? If so, where does this occur?

Solution

Solve

$$29 = -x^2 + 2x + 27.$$

$$0 = -x^2 + 2x - 2$$

Substitute $a = -1$, $b = 2$, and $c = -2$ in the Quadratic Formula.

$$x = \frac{-2 \pm \sqrt{2^2 - 4 \cdot -1 \cdot -2}}{2 \cdot -1}$$

$$x = \frac{-2 \pm \sqrt{-4}}{-2}$$

Since no real number multiplied by itself equals -4, there is no square root of -4 in the real number system. So $29 = -x^2 + 2x + 27$ does not have a solution. *The diver never reaches a height of 29 meters.* This should not be a surprise, since we have already seen that there was no point on the parabola with a height of 29.

In general, the number of solutions to a quadratic equation is related to the number $b^2 - 4ac$ in the Quadratic Formula. This is the number under the radical sign in the expression $\frac{-b \pm \sqrt{b^2 - 4ac}}{2a}$. In Example 1, $b^2 - 4ac$ equals 2, which is positive. Adding $\sqrt{2}$ and subtracting $\sqrt{2}$ gives two different results when computing x. So there are two solutions. In Example 2, $b^2 - 4ac$ is 0. So $x = \frac{-2 + 0}{-2}$ and $x = \frac{-2 - 0}{-2}$ give the same result, 1. That quadratic equation has just one solution. There is no solution to the equation in Example 3 because $b^2 - 4ac = -4$, and $\sqrt{-4}$ is not a real number.

The Discriminant of a Quadratic Equation

Here is a summary of the examples.

height	equation	$b^2 - 4ac$	number of real solutions	number of times parabola intersects line
27.5 meters	$-x^2 + 2x + 27 = 27.5$	positive	2	$y = 27.5$, twice
28 meters	$-x^2 + 2x + 27 = 28$	zero	1	$y = 28$, once
29 meters	$-x^2 + 2x + 27 = 29$	negative	0	$y = 29$, never

Because the value of $b^2 - 4ac$ *discriminates* among the various possible numbers of real-number solutions to a quadratic equation, it is called the **discriminant** of the equation $ax^2 + bx + c = 0$. The examples above are instances of the following general property.

Discriminant Property
If $ax^2 + bx + c = 0$ and a, b, and c are real numbers with $a \neq 0$, then:
When $b^2 - 4ac > 0$, the equation has exactly two real solutions.
When $b^2 - 4ac = 0$, the equation has exactly one solution.
When $b^2 - 4ac < 0$, the equation has no real solutions.

Example 4

Without solving, determine how many real solutions the equation $8x^2 - 5x + 2 = 0$ has.

Solution

Find the value of the discriminant, $b^2 - 4ac$.
Here $a = 8$, $b = -5$, and $c = 2$.

$$\text{So, } b^2 - 4ac = (-5)^2 - 4 \cdot 8 \cdot 2 = 25 - 64 = -39.$$

By the Discriminant Property, Since $b^2 - 4ac$ is negative, there are no real solutions.

QUESTIONS

Covering the Reading

In 1–4, refer to the path of the Acapulco cliff diver described in this lesson.

1. What equation can be solved to determine how far away from the cliff the diver will be when he or she is 28 meters above the water?

2. Will the diver reach a height of 28.5 meters above the water? Justify your answer by referring to a quadratic equation.

3. Will the diver ever be 27 meters above the water? If so, where does this occur?

4. When will the diver be 10 meters above the water?

5. Tell what the discriminant is, and what use it has in solving quadratic equations.

6. Give the number of real solutions to a quadratic equation when its discriminant is **a.** positive. **b.** negative. **c.** zero.

In 7–10, a quadratic equation is given.
a. Calculate the value of the discriminant.
b. Give the number of real solutions.
c. Find all the real solutions. (If there are none, write "no real solutions.")

7. $w^2 - 16w + 64 = 0$

8. $4x^2 - 3x + 8 = 0$

9. $25y^2 = 10y - 1$

10. $13z = 5z^2 + 9$

Taking a dive. *In competition, divers are judged on their approach, take-off, grace, and entry into the water.*

Applying the Mathematics

11. Suppose an equation which describes a diver's path when diving from a platform is $d = -5t^2 + 10t + 5$, where d is the distance above the water (in meters) and t is the time from the beginning of the dive (in seconds). Round answers to the nearest tenth of a second.
a. How high is the diving platform? (Hint: let $t = 0$ seconds).
b. After how many seconds is the diver 8 meters above the water?
c. After how many seconds does the diver enter the water?

12. **a.** Solve $4x^2 + 8x = 5$.
b. Solve $4x^2 + 8x = -1$.
c. Solve $4x^2 + 8x = -10$.

13. For what value(s) of h does $x^2 + 6x + h = 0$ have exactly one solution?

A test of strength. *The shot-put is a field event in which an athlete heaves a metal ball called a shot. The weight of the shot used by women is 8 pounds 13 ounces. In 1994, the world's record for the women's shot-put event was 74 ft 3 in.*

In 14 and 15, use the graph below. It shows the height h (in feet) of a shot t seconds after a shot putter released it. An equation for this path is $y = -16t^2 + 28t + 6$.

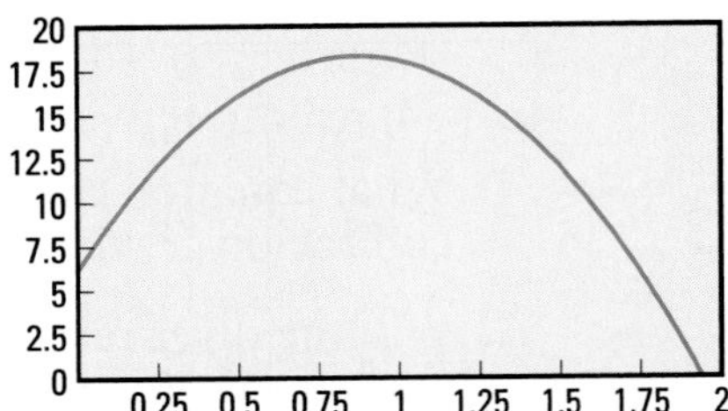

14. a. Write an equation that can be used to find if the shot is ever 10 feet above the ground.
b. Is the value of the discriminant positive, negative, or zero?
c. How many solutions does the equation have?

15. a. Write an equation that should be used to find if the shot is ever 18 feet above the ground.
b. Is the value of the discriminant positive, negative, or zero?
c. How many solutions does the equation have?

Review

16. Solve $60x^2 - 120 = 0$. *(Lessons 9-1, 9-5)*

In 17 and 18, use this information: A softball pitcher tosses a ball to her catcher 50 ft away. The height h (in feet) when it is x feet from the pitcher is given by $h = -.016x^2 + .8x + 2$. *(Lesson 9-4)*

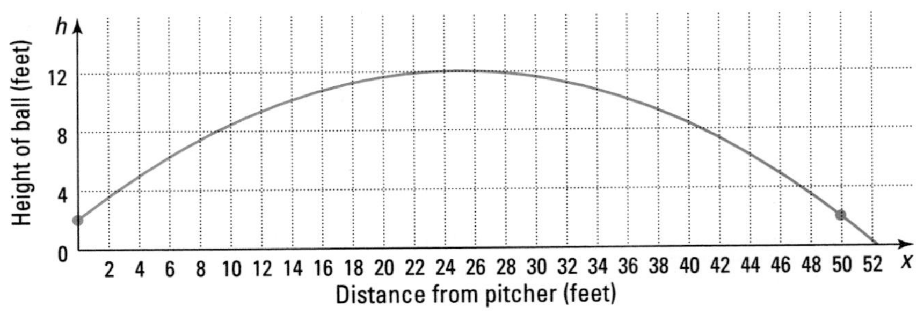

17. How high is the ball at its peak?

18. a. If the batter is 2 ft in front of the catcher, how far is she from the pitcher?
b. How high is the ball when it reaches the batter?

Parabolic pitches. *Pitchers in slow-pitch softball games must throw the ball slowly enough to make it arch on its way to the batter. Only underhand pitching is allowed.*

In 19 and 20, state whether the parabola opens up or down. *(Lessons 9-1, 9-2)*

19. $y = -\frac{1}{5}x^2 - 3x + 2$

20. $x^2 + x = y$

21. Give an equation for a parabola which has vertex at (0, 0) and opens down. *(Lesson 9-1)*

22. *Skill sequence.* Solve. *(Lesson 5-9)*

a. $\sqrt{x} = 5$ **b.** $\sqrt{y + 4} = 5$ **c.** $\sqrt{2z - 3} = 5$

23. Solve. *(Lesson 3-10)*

a. $a - \frac{1}{3} < 0$ **b.** $\frac{1}{3} - b < 0$ **c.** $-\frac{1}{3}c < 0$

24. *Skill sequence.* Add and simplify. *(Lesson 3-9)*

a. $\frac{-5 + x}{2a} + \frac{-5 - x}{2a}$

b. $\frac{-b + y}{2a} + \frac{-b - y}{2a}$

c. $\frac{-b + \sqrt{z}}{2a} + \frac{-b - \sqrt{z}}{2a}$

d. $\frac{-b + \sqrt{b^2 - 4ac}}{2a} + \frac{-b - \sqrt{b^2 - 4ac}}{2a}$

25. Calculate $\frac{15!}{8!7!}$. *(Lesson 2-10)*

26. Which numbers between 10 and 100 are perfect squares? *(Lesson 1-6)*

Exploration

27. Graph the parabola $y = 4x^2 + 8x$ with an automatic grapher or as accurately as you can by hand. Explain the answers you get to Question 12 by referring to this parabola.

LESSON

9-7

Square Roots and Products

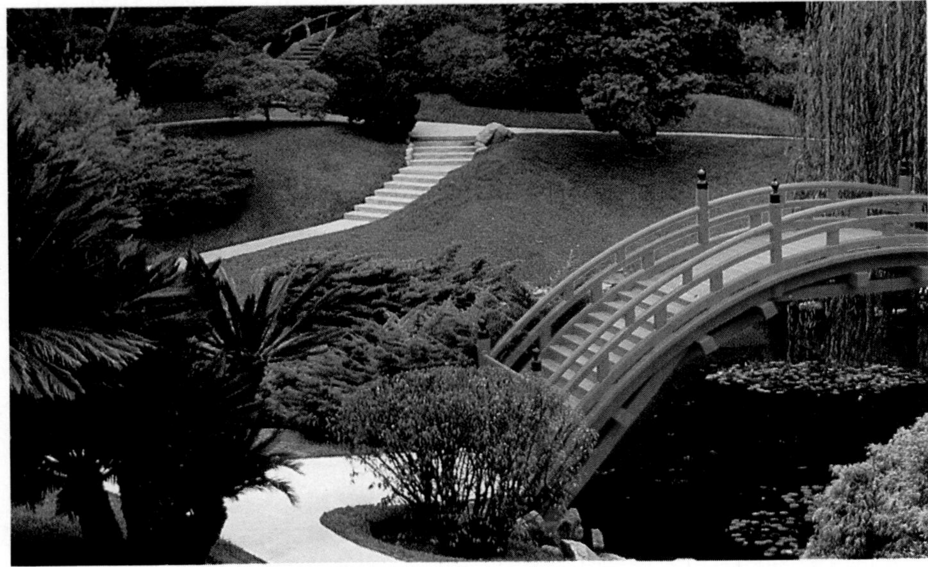

Garden roots. *Shown are three sets of stairs in the Botanical Gardens at the Huntington Library in San Marino, CA. You can find the length of the hypotenuse of each right triangle formed by these steps using square roots.*

A Problem with Two Answers?

The figure below shows a side view of some stairs planned for a garden. The contractor needs to figure out the length of $\overline{LN}$ so he can build a form to make a concrete footing for the stairs. Can you determine LN?

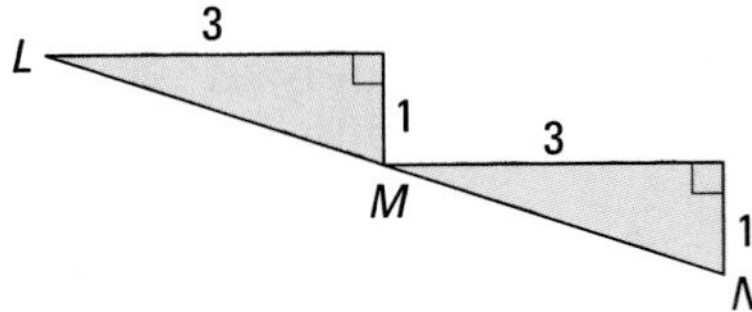

One way to solve this problem is to reason that the length of $\overline{LN}$ is twice the length of either $\overline{LM}$ or $\overline{MN}$. You can find either of these by using the Pythagorean Theorem.

$$\begin{aligned} 3^2 + 1^2 &= (LM)^2 \\ 10 &= (LM)^2 \\ \sqrt{10} &= LM \end{aligned}$$

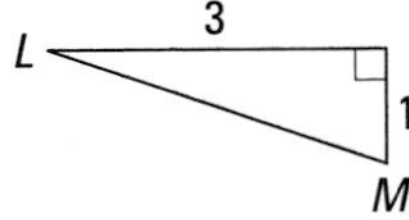

Therefore, $LN = 2(LM) = 2\sqrt{10}$.

Another technique is to imagine one large right triangle with hypotenuse $\overline{LN}$. Its legs are 6 and 2. Now use the Pythagorean Theorem on this right triangle.

$$\begin{aligned} 6^2 + 2^2 &= (LN)^2 \\ 36 + 4 &= (LN)^2 \\ 40 &= (LN)^2 \\ \sqrt{40} &= LN \end{aligned}$$

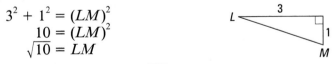

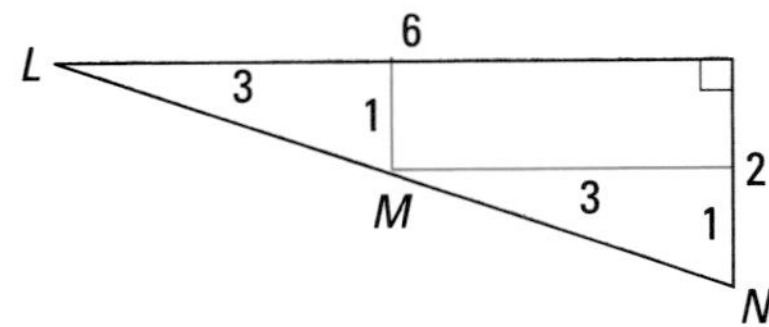

It looks as though there are two different values for *LN*: $2\sqrt{10}$ and $\sqrt{40}$. Are these two names for the same number? Example 1 shows that they are.

Example 1

Verify that $\sqrt{40} = 2\sqrt{10}$.

Solution 1

Evaluate $2\sqrt{10}$ and $\sqrt{40}$ on a calculator. On our calculator we get

$$2\sqrt{10} \approx 6.325$$
$$\sqrt{40} \approx 6.325$$

They appear to be equal.

Solution 2

Square each number. Show that the results are equal. Recall that squaring means to multiply a number by itself,

so $$(\sqrt{40})^2 = \sqrt{40} \cdot \sqrt{40} = 40.$$

Also, $$\begin{aligned}(2\sqrt{10})^2 &= 2\sqrt{10} \cdot 2\sqrt{10} \\ &= 2 \cdot 2 \cdot \sqrt{10} \cdot \sqrt{10} \\ &= 4 \cdot 10 \\ &= 40.\end{aligned}$$

So $2\sqrt{10} = \sqrt{40}$.

Multiplying Square Roots

In general, you may multiply square roots or rewrite square roots as products using the following property.

Product of Square Roots Property
For all nonnegative real numbers a and b,
$$\sqrt{a} \cdot \sqrt{b} = \sqrt{ab}.$$

In words, the square root of the product of two positive numbers is the product of their square roots.

You can multiply square roots by working from the left side of the property to the right side. For instance,

$$\sqrt{4} \cdot \sqrt{10} = \sqrt{4 \cdot 10} = \sqrt{40}.$$

Or working from the right side of the property to the left side, you can rewrite a square root as a product. For instance,

$$\begin{aligned}\sqrt{40} &= \sqrt{4 \cdot 10} \\ &= \sqrt{4} \cdot \sqrt{10} \\ &= 2\sqrt{10}.\end{aligned}$$

"Simplifying" Radicals

Some people consider $2\sqrt{10}$ to be simpler than $\sqrt{40}$ because it has a smaller integer under the radical sign. The process of rewriting a square root as the product of a whole number and a smaller square root is called **simplifying the radical.** The key to simplifying a radical is to find a perfect-square factor of the original number under the radical sign.

Example 2

Simplify $\sqrt{50}$.

Solution

A perfect-square factor of 50 is 25.

$$\begin{aligned}\sqrt{50} &= \sqrt{25 \cdot 2} \\ &= \sqrt{25} \cdot \sqrt{2} \quad \text{Product of Square Roots Property} \\ &= 5\sqrt{2}\end{aligned}$$

Check

Use a calculator. On ours, 50 [√] gives 7.0710678.
5 [×] 2 [√] [=] also gives 7.0710678.
It checks.

Is $5\sqrt{2}$ really simpler than $\sqrt{50}$? It all depends. For estimating and calculating answers to real problems, $\sqrt{50}$ is simpler. But for seeing patterns, $5\sqrt{2}$ may be easier. In the next example, the answer $8\sqrt{2}$ is related to the given information in a simple way. You would not see that without "simplifying" $\sqrt{128}$.

Example 3

Each leg of a right triangle is 8 inches long. Find the exact length of the hypotenuse.

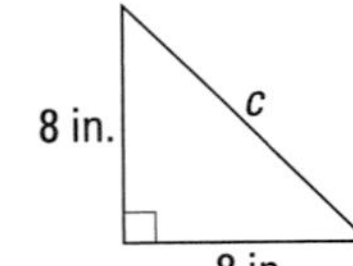

Solution

Use the Pythagorean Theorem.

$$\begin{aligned}c^2 &= 8^2 + 8^2 \\ c^2 &= 128 \\ c &= \sqrt{128}\end{aligned}$$

Now use the Product of Square Roots Property. Note that 64 is a perfect square factor of 128.

$$\begin{aligned}c &= \sqrt{64} \cdot \sqrt{2} \\ &= 8\sqrt{2}\end{aligned}$$

The exact length of the hypotenuse is $\sqrt{128}$, or $8\sqrt{2}$ inches.

Can you see how the lengths of the legs and the hypotenuse are related? You are asked to generalize the pattern in the Questions.

Applying the Product of Square Roots Property

You can combine the Product of Square Roots Property with other properties of real numbers to simplify many expressions.

Example 4

One solution to a quadratic equation is $\frac{6 + \sqrt{28}}{2}$. Simplify this expression.

Solution

A perfect-square factor of 28 is 4, so rewrite $\sqrt{28}$ as $\sqrt{4} \cdot \sqrt{7}$.

$$\frac{6 + \sqrt{28}}{2} = \frac{6 + \sqrt{4} \cdot \sqrt{7}}{2}$$
$$= \frac{6 + 2 \cdot \sqrt{7}}{2}$$

Now use the Adding Fractions form of the Distributive Property.

$$= \frac{6}{2} + \frac{2\sqrt{7}}{2}$$
$$= 3 + \sqrt{7}$$

The Product of Square Roots Property also applies to expressions containing variables.

Example 5

Assume m and n are positive. Multiply $\sqrt{7m} \cdot \sqrt{7n}$ and simplify the result.

Solution

$$\sqrt{7m} \cdot \sqrt{7n} = \sqrt{7m \cdot 7n}$$
$$= \sqrt{7^2} \cdot \sqrt{mn}$$
$$= 7\sqrt{mn}$$

Check

Does $\sqrt{7m} \cdot \sqrt{7n} = 7\sqrt{mn}$? Substitute values for m and n.
For instance, let $m = 3$ and $n = 2$.

Does $\sqrt{7 \cdot 3} \cdot \sqrt{7 \cdot 2} = 7\sqrt{3 \cdot 2}$?

Does $\sqrt{21} \cdot \sqrt{14} = 7\sqrt{6}$?

$\sqrt{21} \cdot \sqrt{14} = \sqrt{294} = \sqrt{49 \cdot 6} = 7\sqrt{6}$

It checks.

Example 6

Simplify $\sqrt{9x^2y^2}$. Assume x and y are both positive.

Solution

$$\sqrt{9x^2y^2} = \sqrt{9} \cdot \sqrt{x^2} \cdot \sqrt{y^2}$$
$$= 3 \cdot x \cdot y$$
$$= 3xy$$

▶

Check
Substitute for the variables in $\sqrt{9x^2y^2}$ and $3xy$. We let $x = 2$, $y = 4$. With these values, $\sqrt{9x^2y^2} = \sqrt{9(2)^2(4)^2} = \sqrt{9 \cdot 4 \cdot 16} = \sqrt{576} = 24$, and $3xy = 3 \cdot 2 \cdot 4 = 24$. It checks.

QUESTIONS

Covering the Reading

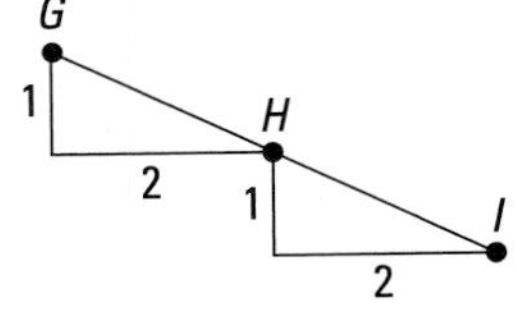

1. Refer to the drawing at the right.
   a. Find GI by first finding the length of $\overline{GH}$.
   b. Find GI by using a right triangle with $\overline{GI}$ as hypotenuse.
   c. Verify that your answers to parts **a** and **b** are equal.

2. Show that $5\sqrt{3} = \sqrt{75}$ by each method.
   a. using a calculator
   b. squaring each of $5\sqrt{3}$ and $\sqrt{75}$
   c. simplifying $\sqrt{75}$ using the Product of Square Roots Property.

3. By the Product of Square Roots Property, what must $\sqrt{p} \cdot \sqrt{q}$ equal?

4. Use a calculator to estimate the following to two decimal places.
   a. $\sqrt{7}$ **b.** $\sqrt{5}$ **c.** $\sqrt{7} \cdot \sqrt{5}$ **d.** $\sqrt{35}$

5. To use the Product of Square Roots Property to rewrite $\sqrt{ab}$, both a and b must be _?_.

In 6–8, a radical sign is given. **a.** Find a perfect-square factor of the number under the radical sign. **b.** Simplify.

6. $\sqrt{20}$
7. $\sqrt{50}$
8. $\sqrt{700}$

9. Check Example 3 by using a calculator to verify that $\sqrt{128} = 8\sqrt{2}$.

10. Each leg of a right triangle is 10 cm. Find the exact length of the hypotenuse.

In 11 and 12, simplify each expression.

11. $\frac{6 + \sqrt{18}}{2}$
12. $\frac{12 \pm \sqrt{12}}{2}$

In 13 and 14, assume a and b are both positive and simplify the result.

13. $\sqrt{6a} \cdot \sqrt{24b}$
14. $\sqrt{36a^2b^2}$

Applying the Mathematics

In 15 and 16, find the exact value of the variable. Simplify any radicals in your answers.

15.

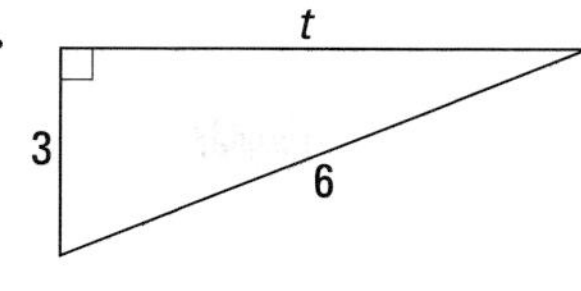

16.

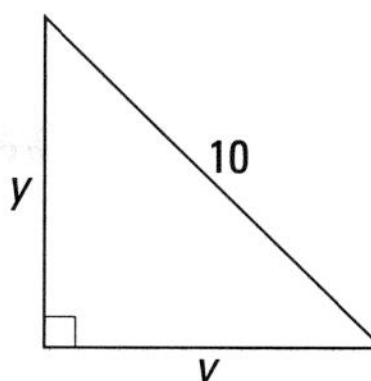

In 17–20, simplify in your head. Be able to explain what you did.

17. $\sqrt{2} \cdot \sqrt{18}$

18. $\sqrt{20} \cdot \sqrt{5}$

19. $\sqrt{5^2 \cdot 11^2}$

20. $\sqrt{100 \cdot 81 \cdot 36}$

In 21 and 22, solve and check. Give your answer in simplified radical form.

21. $(2a)^2 = 48$

22. $\frac{9}{x} = \frac{x}{6}$

23. Simplify each expression.
a. $\sqrt{75}$
b. $\sqrt{12}$
c. $\sqrt{75} + \sqrt{12}$

24. **a.** Find the area of a rectangle with base $\sqrt{18}$ and height $5\sqrt{2}$.
b. Which is longer, the base or the height? Justify your answer.

25. The area of a square is $20w^2$. Write the exact length of one side in two ways.

26. Solve $x^2 + 4x - 9 = 0$, and write the solutions in simplified radical form.

NASA vacuum chamber. *Shown is a test model satellite in vacuum Chamber "A" of the Space Environment Simulation Laboratory at the Johnson Space Center. Engineers conducted tests to ensure that the satellite's 30-foot parabolic antenna would unfold properly in a space vacuum.*

Review

27. A quadratic equation has only one solution. What can you conclude about its discriminant? *(Lesson 9-6)*

In 28–30, use this information. In a vacuum chamber on Earth, an object will drop d meters in approximately t seconds, where $d = 4.9t^2$. *(Lessons 1-6, 9-1)*

28. How far will the object drop in 3 seconds?

29. How far will an object drop in $\sqrt{10}$ seconds?

30. **a.** Write an equation that could be used to find the number of seconds it takes an object to drop 10 meters.
b. Solve the equation from part **a.**

In 31–34, determine whether the equation has 0, 1, or 2 real solutions. (You do not need to solve the equation.) *(Lessons 3-7, 9-6)*

31. $x^2 + x - 1 = 0$

32. $x^2 + x + 1 = 0$

33. $2x^2 - 12x - 18 = 0$

34. $-2x - 18 = 0$

35. Betsy has q quarters and d dimes. She has at least \$5.20. Write an inequality that describes this situation. *(Lessons 3-8, 3-10, 5-6)*

36. On the scale below, the boxes are equal in weight. The other objects are one-kilogram weights. What can you conclude about the weight of one box? Justify your answer. *(Lesson 5-6)*

Exploration

37. Generalize Example 3 and Question 10 from this lesson. That is, if the lengths of two legs of a right triangle are equal, what must be true about the hypotenuse? Justify your answer.

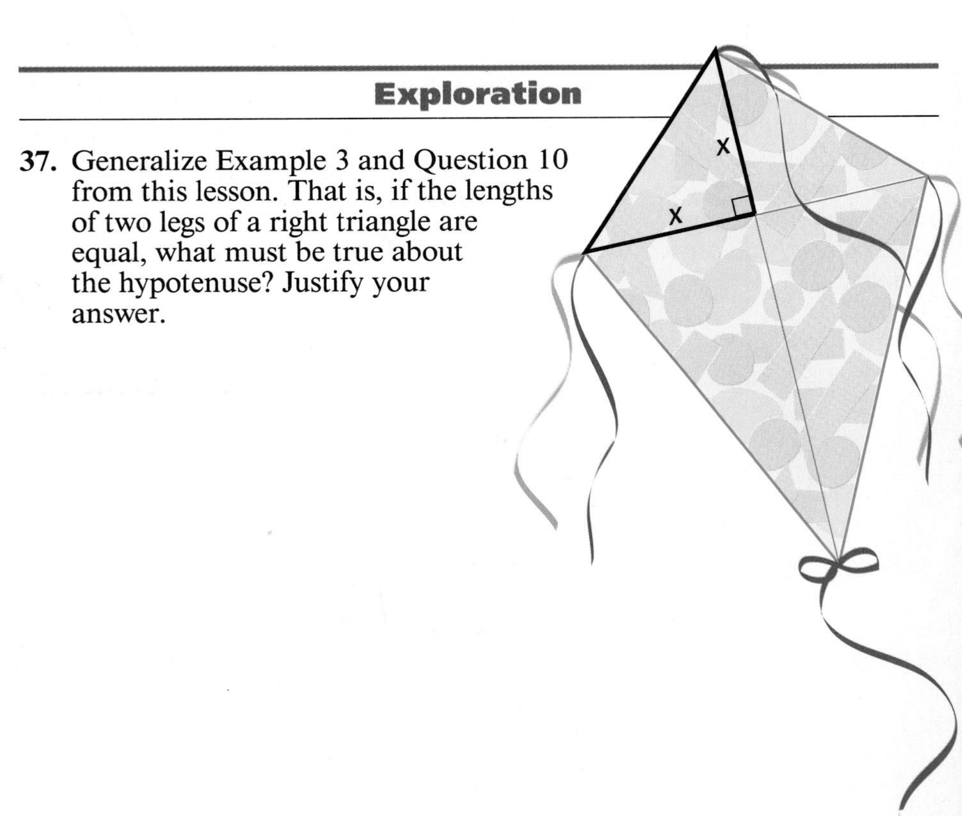

LESSON

9-8

Absolute Value, Distance, and Square Roots

Absolutely positive. *New Zealand is in the South Pacific, far from all continents but Australia. No matter in which direction you go, the distance you would travel is a positive amount.*

What Is the Absolute Value of a Number?

Recall that every real number has an opposite, and that two numbers which are opposites of each other are both the same distance from the origin. For instance, 5 and -5 are opposites; each is 5 units from the origin. The numbers $\frac{3}{2}$ and $-\frac{3}{2}$ are opposites; the distance between each and the origin is $\frac{3}{2}$.

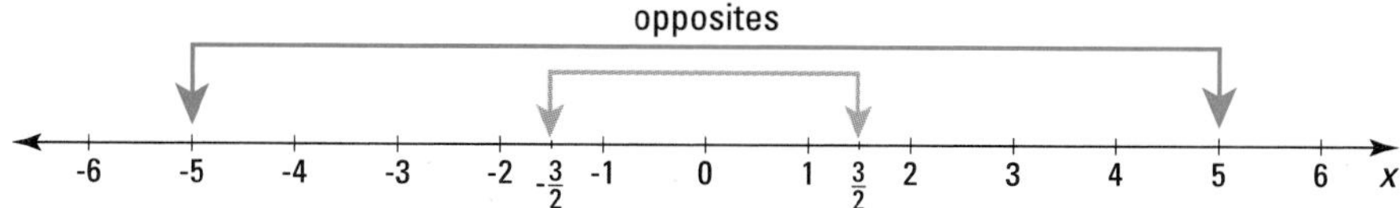

The **absolute value** of a number is the distance between its corresponding point on a number line and the origin. Because it is a distance, an absolute value is never negative. The concept of the absolute value of a number is so important in mathematics that it has its own symbol. The absolute value of x is written $|x|$. In computer languages or spreadsheets the absolute value of x is often written ABS(X).

Thus, $|5| = 5$ and $|-5| = 5$.

Also, $\left|\frac{3}{2}\right| = \frac{3}{2}$ and $\left|-\frac{3}{2}\right| = \frac{3}{2}$.

In BASIC, ABS(3.9) = ABS(-3.9) = 3.9, and ABS(0) = 0.

Example 1

Evaluate the following.

a. $|-4.5|$ **b.** $|1| - |-7|$ **c.** $|1 - 7|$

Solution

a. Think about the number line. The distance between the point -4.5 and the origin is 4.5, so

$$|-4.5| = 4.5.$$

b. Evaluate each absolute value from left to right.

$$|1| - |-7| = 1 - 7 = -6$$

c. The absolute value symbol is also a grouping symbol. So you must evaluate the expression inside it first.

$$|1 - 7| = |-6| = 6$$

Notice that when x is *negative,* $|x| = -x$. The absolute value changes the negative number to its opposite, which is a positive number.

The graph of the equation $y = |x|$ has an interesting shape. Here are a table of values and a graph.

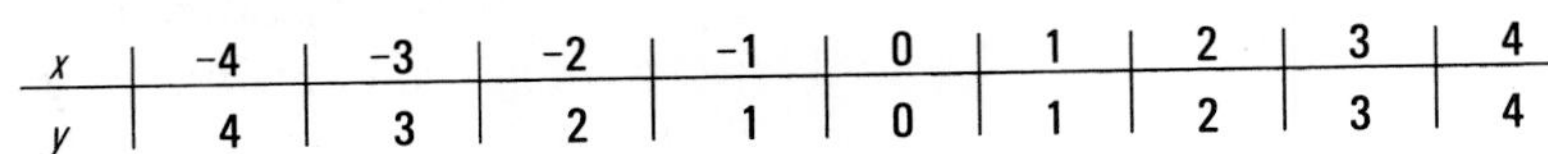

x	-4	-3	-2	-1	0	1	2	3	4
y	4	3	2	1	0	1	2	3	4

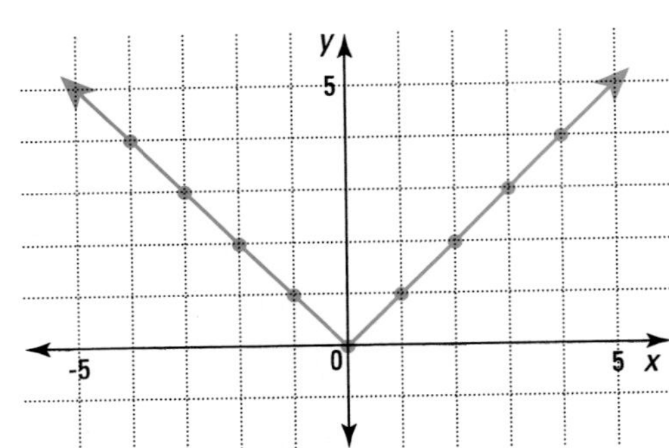

The graph has the shape of the letter V. The graph is an angle with vertex at (0, 0).

Solving Equations Involving Absolute Value

You can solve equations involving absolute value either geometrically or algebraically.

Example 2

Solve $|x| = 10$.

Solution 1

Think geometrically. What numbers on a number line are at a distance of 10 from zero? Answer: 10 and -10. So,

$$\text{if } |x| = 10, \text{ then } x = 10 \text{ or } -10.$$

The graph of the solution set consists of two points.

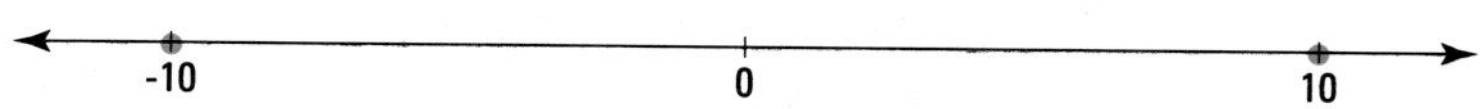

Solution 2

Think algebraically: $|-10| = 10$ and $|10| = 10$. So, if $|x| = 10$, then $x = 10$ or $x = -10$.

Absolute Value and Distance

These ideas can be extended to determine the distance between any two points on a number line. You know you can subtract to find out by how much two numbers differ. By then taking the absolute value, you can guarantee that the result is positive.

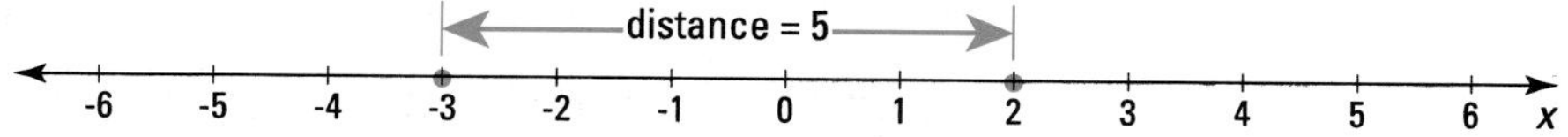

For instance, consider the distance between the points with coordinates -3 and 2. Subtracting the coordinates gives either $-3 - 2 = -5$ or $2 - (-3) = 5$. But taking the absolute value after subtracting, $|-3 - 2| = |-5| = 5$, and $|2 - -3| = |5| = 5$.

Distance Formula on a Number Line
If x_1 and x_2 are the coordinates of two points on a number line, then the **distance** between them is $|x_2 - x_1|$.

Example 3

Find the distance d between the points with coordinates -17 and 8.

Solution

Use the distance formula. $d = |8 - -17| = |8 + 17| = |25| = 25$

Check 1

Draw a number line and count.

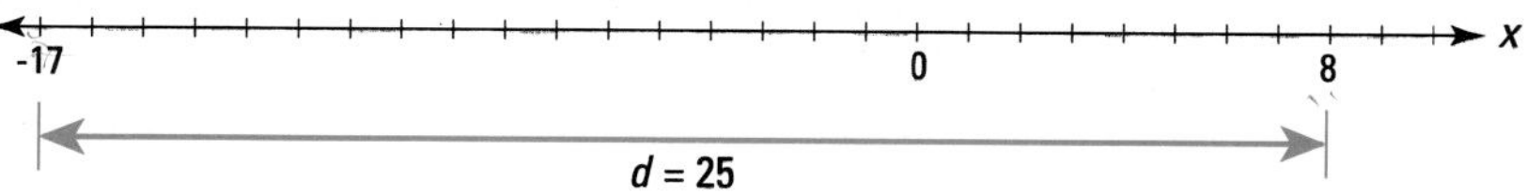

Check 2

Do the calculation in the other order. $d = |-17 - 8| = |-25| = 25$.

Example 4

Solve $|x - 4| = 6$.

Solution 1

Think geometrically: $|x - 4|$ represents the distance between the points with coordinates x and 4. This distance must be 6. Measure 6 units to either side of 4.

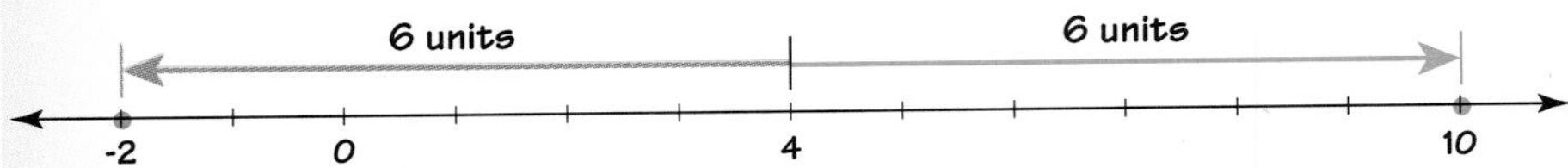

$x = -2$ or $x = 10$.

Solution 2

$|x - 4| = 6$

Think algebraically: use chunking. The only two numbers whose absolute value equals 6 are 6 itself and -6.

Either $x - 4 = 6$ or $x - 4 = -6$.

Solve each equation.

$x = 10$ or $x = -2$

Check

Substitute each solution in the equation $|x - 4| = 6$.
If $x = 10$, $|10 - 4| = |6| = 6$.
If $x = -2$, $|-2 - 4| = |-6| = 6$.
Each solution checks.

Absolute Value and Square Roots

Surprisingly, absolute values are related to square roots. Examine this table.

x	x^2	$\sqrt{x^2}$
-3	9	3
-2	4	2
-1	1	1
0	0	0
1	1	1
2	4	2
3	9	3

The general pattern is that the absolute value of a number equals the positive square root of its square.

Absolute Value–Square Root Property
For all x, $\sqrt{x^2} = |x|$.

QUESTIONS

Covering the Reading

1. On a number line, what does the absolute value of a number represent?

In 2–7, evaluate the expression.

2. $|-50|$ **3.** $|1.3|$ **4.** $|0|$

5. ABS(-16) **6.** $|8 + -4|$ **7.** $|8| + |-4|$

In 8 and 9, refer to the graph of $y = |x|$ in this lesson.

8. *True or false.* There are two points on the graph with y-coordinate 3.

9. Name a point which is on the graph of $y = |x|$ but is not pictured in this lesson.

10. a. Write a sentence that expresses the following: on a number line, the distance from the point with coordinate x and the origin is 11.
b. Solve the sentence from part **a.**

In 11–14, solve and check.

11. $|t| = 25$ **12.** $|x - 3| = 2$

13. $|x - 3| = 42$ **14.** $|300 - x| = 10$

In 15 and 16, find the distance between the two points whose coordinates are given.

15. -28 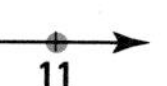 11

16. 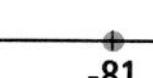 -81 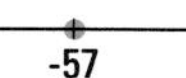 -57

17. Write an expression for the distance between the points whose coordinates are 17.5 and n.

In 18–21, evaluate.

18. $\sqrt{11^2}$ **19.** $\sqrt{n^2}$ **20.** $\sqrt{(-11)^2}$ **21.** $\sqrt{8^2}$

Applying the Mathematics

In 22–25, give the number of solutions to the equation. Then solve.

22. $|n| = 0$ **23.** $|n| = -6$ **24.** $|p| = \frac{1}{2}$ **25.** $\sqrt{q^2} = 31$

In 26–28, find two points on the number line that are 7 units from the point with the given coordinate.

26. 3 **27.** -3 **28.** a

These sensitive calipers measure the width of the metal plates made in an electric motor parts company. The width of each plate must be within specific tolerance levels.

29. Marika's age is x years and Wolfgang's age is y years. What is the difference in their ages, under the given conditions?
a. $x > y$ **b.** $y > x$
c. if you are not sure which of x or y is greater

30. Manufactured parts are allowed to vary from an accepted standard size by a specific amount called the **tolerance.** Consider a washer designed to have a diameter of 2 cm, with a tolerance of $\pm$ 0.001 cm. This means that the actual diameter may not fall outside the interval 2 ± 0.001 cm.
a. What is the smallest diameter allowed?
b. What is the largest diameter allowed?
c. Graph all allowable diameters on a number line.
d. If d is the diameter of a washer that falls within the tolerance level above, what must be true about $|d - 2|$?

Review

In 31–33, simplify. *(Lesson 9-7)*

31. $\frac{\sqrt{175}}{5}$ **32.** $\frac{\sqrt{300}}{6}$ **33.** $\frac{16 \pm \sqrt{288}}{4}$

34. *True or false.* $8\sqrt{10} = \sqrt{80}$. Explain how you got your answer. *(Lesson 9-7)*

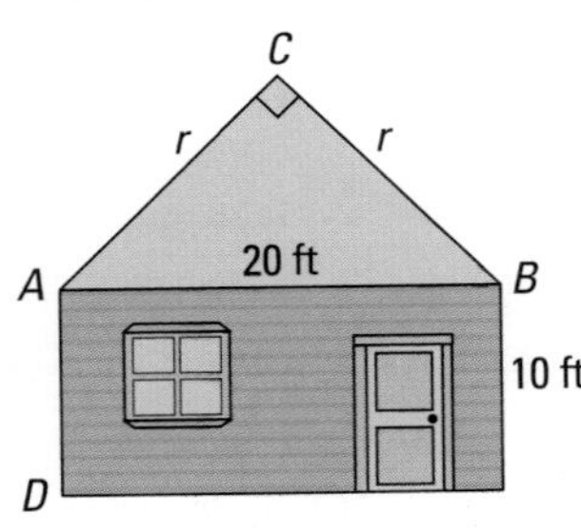

35. The figure at the left represents a side view of a building plan for a garage. The garage is to be 20 ft wide. The rafters AC and BC of the roof are to be equal in length and to meet at right angles.
a. Find the length r of each rafter as a simplified radical.
b. Round the length of a rafter to the nearest tenth of a foot.
c. Find BD. *(Lessons 1-8, 9-7)*

36. How many real solutions does the equation $17p^2 + p = 20$ have? Justify your answer. *(Lesson 9-6)*

37. Solve $2x^2 - 16x - 4 = 0$. *(Lesson 9-5)*

38. Consider these points: (0, -3), (0, 0), (0, 3).
a. Write an equation for the line containing these three points.
b. On which axis do they lie? *(Lesson 5-1)*

Exploration

39. a. Make a table of values for $y = |x - 3|$, when $-6 \le x \le 6$.
b. Graph the points you found in part **a.**
c. Describe the set of *all* points for which $y = |x - 3|$.

40. a. Use an automatic grapher to graph these three equations on the same set of axes.
$y = |x|$ $y = |x| + 4$ $y = |x| - 3$
b. Make a conjecture about the shape of the graph $y = |x| + k$, for any real number k.
c. Test your conjecture using a value of k other than 0, 4, or -3.

LESSON

9-9

Distances in the Plane

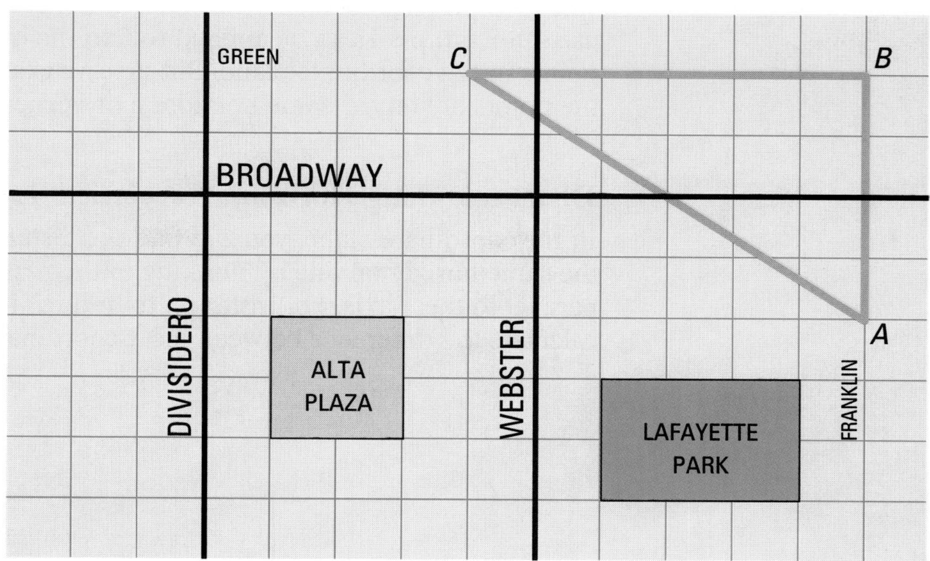

Distances Along Grids

Streets of many cities are laid out in a grid pattern like the one above. To get from one location to another by car you have to travel along streets. So the shortest distance from A to C is not the hypotenuse of triangle ABC, but the sum of the legs' lengths, $AB + BC$. But a bird or helicopter could go directly from A to C along the segment $\overline{AC}$ *"as the crow flies."*

Example 1

a. How far is it from A to C traveling by car along $\overline{AB}$ and $\overline{BC}$? (Each unit is a city block.)

b. How far is it from A to C as the crow flies?

Solution

a. Count the blocks between points A and B. There are 4 blocks. Count the blocks between B and C. There are 6 of them.

$$AB + BC = 4 + 6 = 10$$

Traveling by car, the distance between A and B is 10 blocks.

b. Use the Pythagorean Theorem on $\triangle ABC$ above.

$$\begin{aligned} AC^2 &= AB^2 + BC^2 \\ &= 4^2 + 6^2 \\ &= 16 + 36 \\ &= 52 \\ AC &= \sqrt{52} \\ &\approx 7.2 \text{ blocks} \end{aligned}$$

The distance as the crow flies from A to C is $\sqrt{52} \approx$ 7.2 blocks.

The map on page 599 looks like part of the coordinate plane. The idea used in Example 1 can be applied to find the distance between any two points in the coordinate plane. But first we examine the situations where the points are on the same horizontal or vertical line.

Distances Along Horizontal or Vertical Lines

In the coordinate plane, you can find the distance between two points on the same horizontal line by thinking of them as being on a number line parallel to the x-axis. For instance, to find RS in the figure below, calculate the difference between the x-coordinates of R and S.

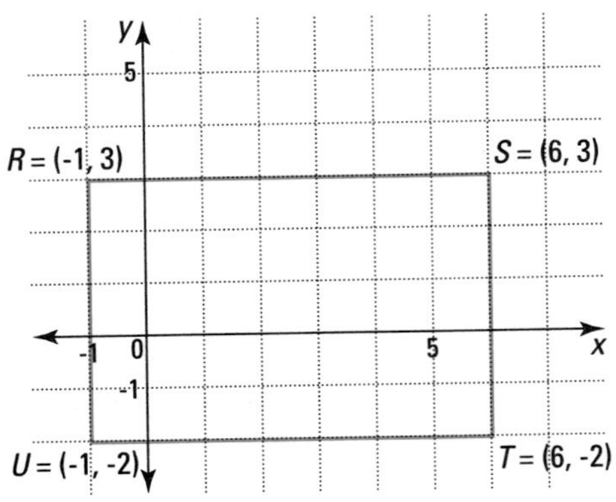

$$RS = |-1 - 6| = 7$$

Similarly, to find the distance ST, calculate the difference between the y-coordinates of S and T.

$$ST = |3 - -2| = 5$$

Distances Between Any Two Points

To find the distance between two points A and B on an oblique line, you can use the Pythagorean Theorem.

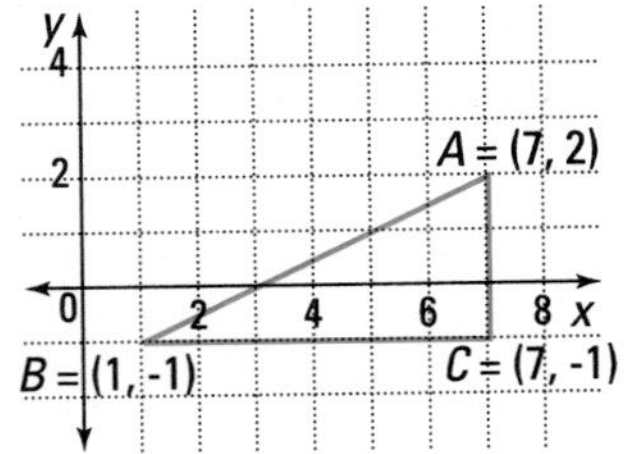

Example 2

Find AB in $\triangle ABC$ at the left.

Solution

AB is the length of the hypotenuse of a right triangle whose legs are parallel to the x- and y-axes.

$$AC = |2 - -1| = |3| = 3$$
$$BC = |1 - 7| = |-6| = 6$$

So,

$$AB^2 = AC^2 + BC^2 = 3^2 + 6^2 = 9 + 36 = 45.$$

Thus,

$$AB = \sqrt{45} = 3\sqrt{5}.$$

This method can be generalized to find the distance between any two points that lie on an oblique line. If the points are $A = (x_1, y_1)$ and $B = (x_2, y_2)$, then a right triangle can be formed with the third vertex at $C = (x_1, y_2)$.

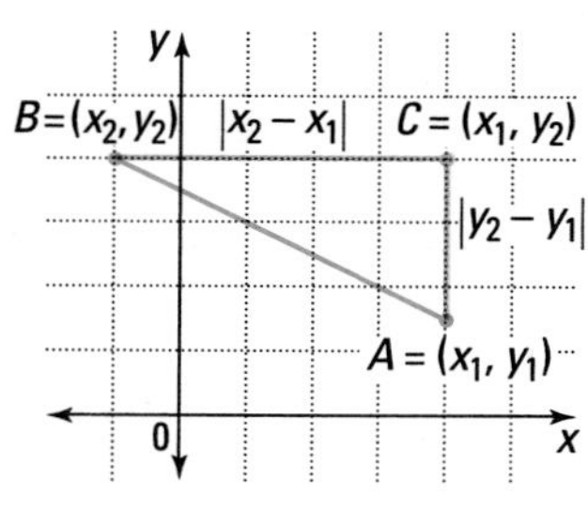

Once the third vertex has been found, the lengths of the legs are $BC = |x_2 - x_1|$ and $AC = |y_2 - y_1|$. Now use the Pythagorean Theorem.

$$AB^2 = BC^2 + AC^2$$

Substitute. $$AB^2 = |x_2 - x_1|^2 + |y_2 - y_1|^2$$

Now take the square root of each side. The result is a formula for the distance between any two points in the plane.

Pythagorean Distance Formula in the Coordinate Plane
The distance AB between the points $A = (x_1, y_1)$ and $B = (x_2, y_2)$ in the coordinate plane is $AB = \sqrt{|x_2 - x_1|^2 + |y_2 - y_1|^2}$.

Example 3

Find the distance between the points (4, 3) and (6, -2).

Solution 1

Use the Pythagorean Distance Formula. Either point may be (x_1, y_1).
Let $(x_1, y_1) = (4, 3)$ and $(x_2, y_2) = (6, -2)$.

$$\begin{aligned} \text{distance} &= \sqrt{|6 - 4|^2 + |-2 - 3|^2} \\ &= \sqrt{|2|^2 + |-5|^2} \\ &= \sqrt{2^2 + 5^2} \\ &= \sqrt{4 + 25} \\ &= \sqrt{29} \approx 5.4 \end{aligned}$$

Solution 2

Plot the points. Draw a right triangle whose hypotenuse has endpoints (4, 3) and (6, -2). At the right, the third vertex is at $C = (4, -2)$. First find the lengths of the legs.
$AC = |3 - -2| = |5| = 5$
$BC = |4 - 6| = |-2| = 2$

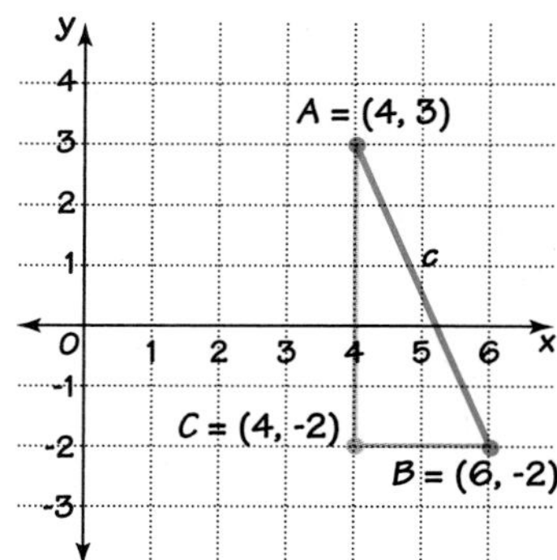

Now find AB by using the Pythagorean Theorem.

$$\begin{aligned} BC^2 + AC^2 &= AB^2 \\ 2^2 + 5^2 &= c^2 \\ 4 + 25 &= c^2 \\ 29 &= c^2 \\ \sqrt{29} &= c \end{aligned}$$

Thus, $AB = \sqrt{29} \approx 5.4$.

QUESTIONS

Covering the Reading

1. In the diagram below, each square in the grid represents a city block.

a. How many blocks does it take to go from D to E by way of F?

b. Use the Pythagorean Theorem to find the distance from D to E as the crow flies.

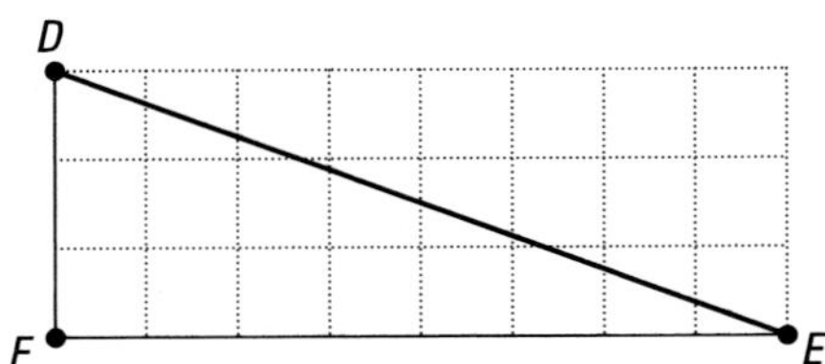

In 2 and 3, find the distance MN.

2. $M = (40, 60)$ and $N = (40, 18)$

3. $M = (-3, 10)$ and $N = (11, 10)$

In 4–6, refer to rectangle $RSTU$ on page 600. Find each length.

4. UT **5.** RU **6.** RT

7. Use the graph at the right.

a. Find the coordinates of P.

b. Find the length of $\overline{DP}$.

c. Find the length of $\overline{GP}$.

d. Use the Pythagorean Theorem to find DG.

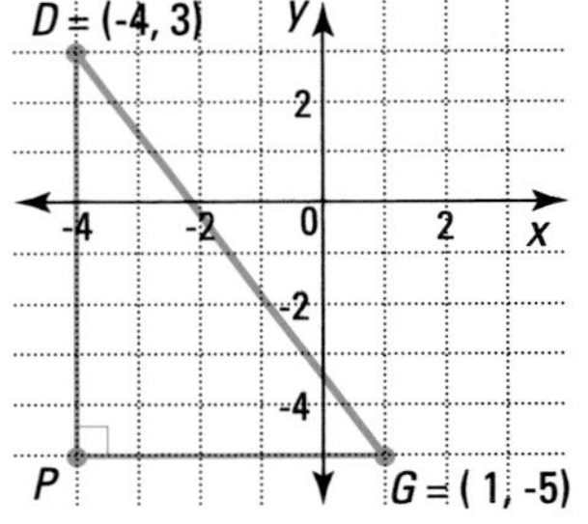

In 8 and 9, find the coordinates of point H. Then find FG.

8.

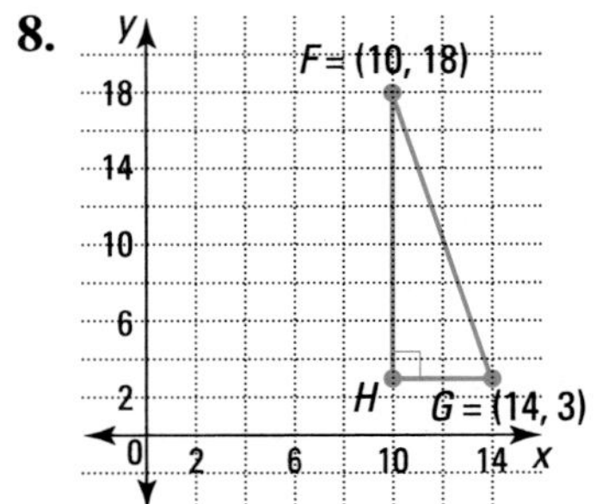

9.

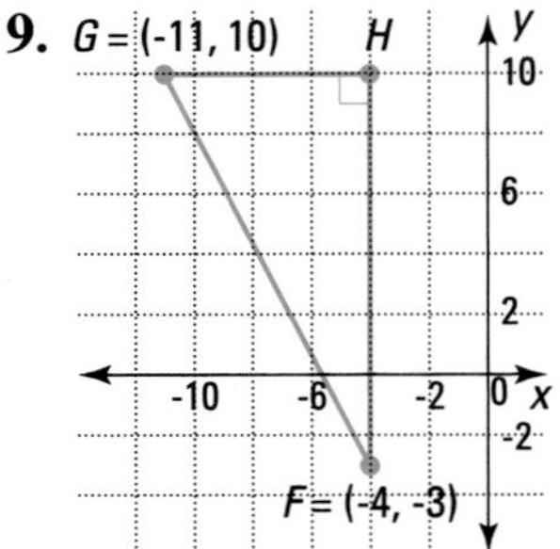

In 10–13, find the distance between the two points.

10. $A = (0, 0)$; $B = (-4, -3)$

11. $A = (4, 11)$; $B = (0, 7)$

12. $D = (5, 1)$; $C = (11, -7)$

13. $E = (-3, 5)$; $F = (-1, -8)$

Applying the Mathematics

In 14–17, the map at the right shows the location of the Singhs' house, the post office, and the school. Each small square is 1 km on a side.

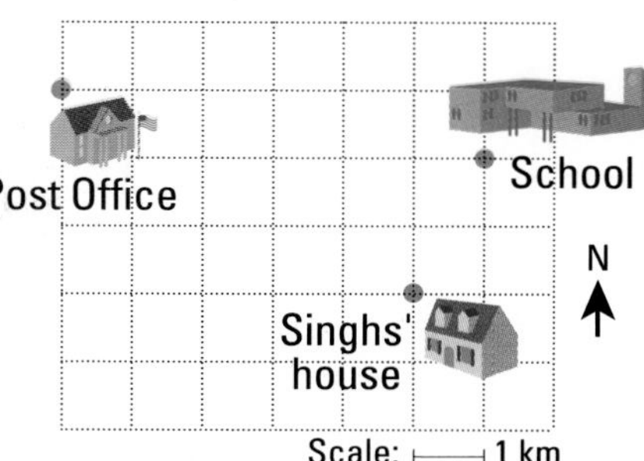

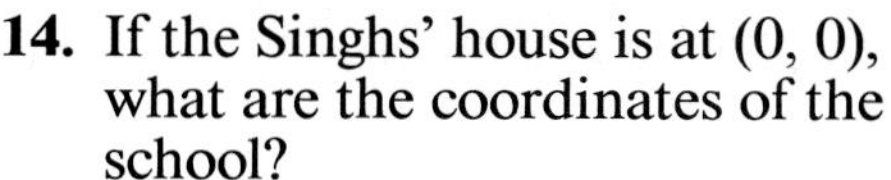

14. If the Singhs' house is at (0, 0), what are the coordinates of the school?

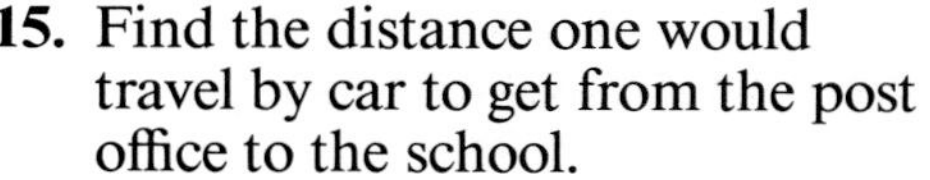

15. Find the distance one would travel by car to get from the post office to the school.

16. How far is it from the post office to school as the crow flies?

17. How far is it from the Singhs' house to the school as the crow flies?

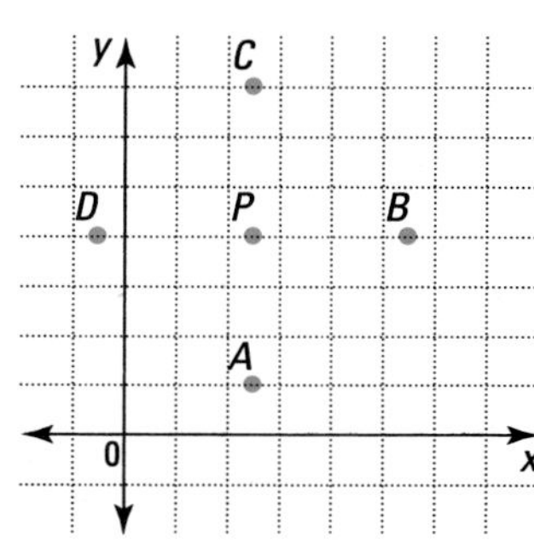

18. Refer to the diagram at the left. Point $P = (5, 8)$. Points A, B, C, and D are each 6 units from P on the horizontal or vertical line through P. Find the coordinates of these points.

In 19 and 20, write an expression for the distance between the two points.

19. $(0, 0)$ and (a, b) **20.** (a, b) and (c, d)

21. Let $J = (-5, 0)$, $K = (1, 8)$, and $L = (16, 0)$. Find the length of each side of $\triangle JKL$.

22. The distance between the points (1, 1) and $(4, y)$ is 5. Find all possible values of y.

In 23 and 24, use this computer program that asks for coordinates of points (x_1, y_1) and (x_2, y_2) in the coordinate plane, and then calculates the distance between them.

```
10 PRINT "DISTANCE IN THE PLANE"
20 PRINT "BETWEEN (X1, Y1) AND (X2, Y2)"
30 PRINT "ENTER X1, Y1"
40 INPUT X1, Y1
50 PRINT "ENTER X2, Y2"
60 INPUT X2, Y2
70 LET H = ABS(X1 - X2)
80 LET V = ABS(Y1 - Y2)
90 LET DIST = SQR(H^2 + V^2)
100 PRINT "THE DISTANCE IS"; DIST
110 END
```

23. What value for DIST will the computer print when $(x_1, y_1) = (19, -5)$ and $(x_2, y_2) = (-3, 15)$?

24. What value for DIST will the computer print when $(x_1, y_1) = (0, 0)$ and $(x_2, y_2) = (1, 1)$?

Review

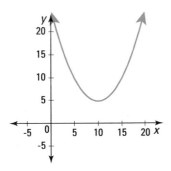

In 25 and 26, solve. *(Lesson 9-8)*

25. $|x| = 3.5$

26. $2|x + 7| = 10$

27. *Skill sequence.* Solve. *(Lessons 1-6, 9-8)*
a. $t^2 = 9$
b. $\sqrt{t} = 9$
c. $\sqrt{t^2} = 9$

28. Simplify the expression $\frac{-40 \pm \sqrt{20}}{12}$. *(Lesson 9-7)*

29. The graph at the left is of the equation $y = ax^2 + bx + c$. How many solutions does the equation $ax^2 + bx + c = 10$ have? Justify your answer. *(Lessons 9-4, 9-6)*

30. *Skill sequence.* Solve. *(Lessons 3-5, 5-3, 9-5)*
a. $3x + 6 = 2$
b. $3x + 6 = x + 2$
c. $3x + 6 = x^2 + 2$
d. $3x + 6 = x(x + 2)$

31. Find the x-intercepts of the graph of $y = x^2 + 9x - 5$ using each method.
a. by letting $y = 0$ and using the Quadratic Formula
b. by using an automatic grapher and zooming in on the intercepts
(Lessons 9-2, 9-4)

32. Copy quadrilateral $WXYZ$. Then graph the image of $WXYZ$ under a size change of magnitude $-\frac{1}{2}$. *(Lesson 6-7)*

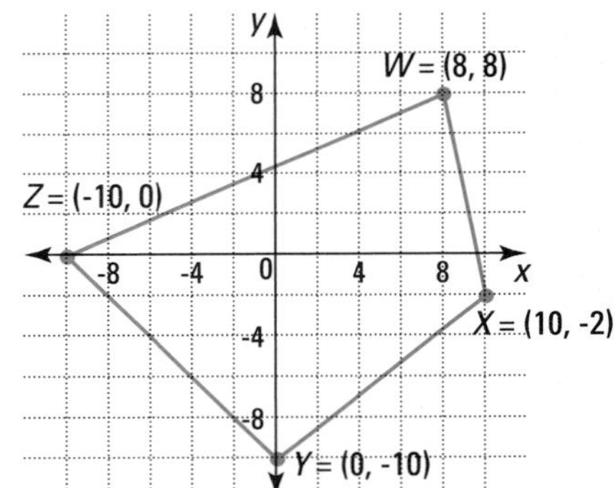

Making dough. *Croissants originated in France as a breakfast food.* Croissant *is the French word for "crescent," the shape of these rolls.*

33. In the triangle at the right, find the value of x. *(Lessons 3-5, 4-7)*

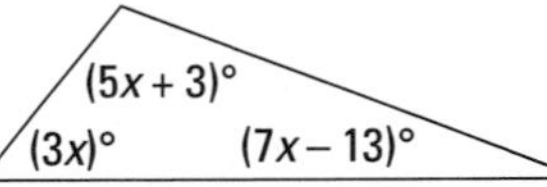

34. A bakery charges 70¢ for each croissant and 25¢ for a carry-out box. You have $5.00. At most, how many croissants can you buy if you also want them in a carry-out box? *(Lessons 3-8, 3-10)*

Exploration

35. The grid at the right pictures streets 1 block apart in a city. Graph all points 5 blocks by car from the intersection of Main and Central.

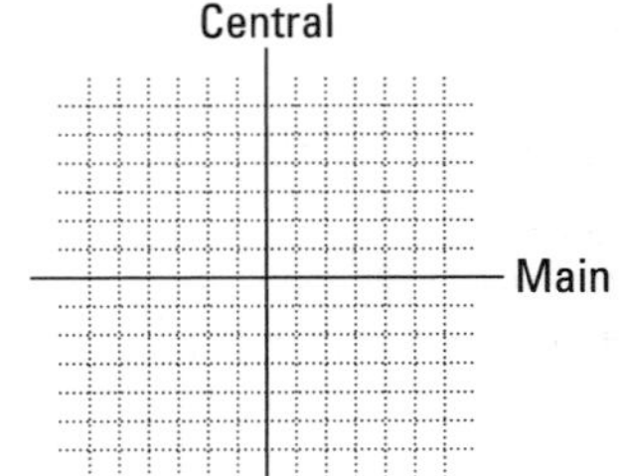

PROJECTS 9 CHAPTER NINE

A project presents an opportunity for you to extend your knowledge of a topic related to the material of this chapter. You should allow more time for a project than you do for typical homework questions.

1 Interview with Galileo

Use the library to do some research on Galileo. Write an imaginary interview with him in which you inquire about his work with falling objects. Also find out how square roots figured in his formula for *the period of a pendulum.* How did he discover the laws of the pendulum?

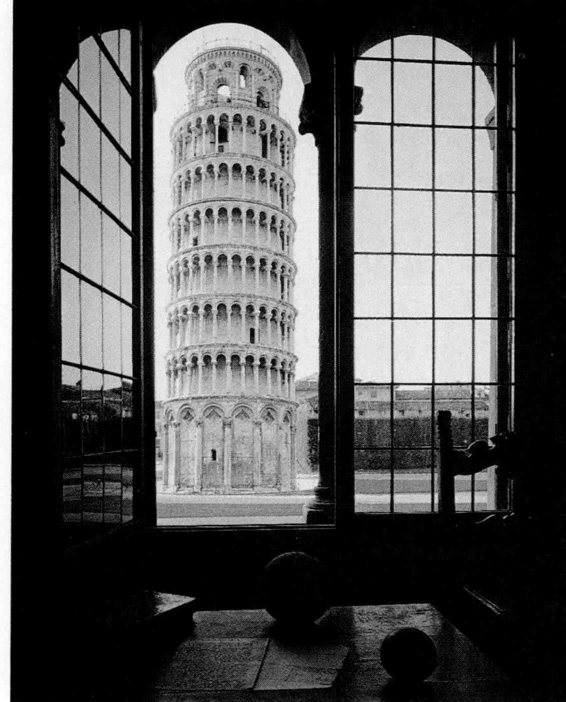

2 Quadratic Design

Make pleasing designs using parabolas (or parabolas and lines or exponential curves). For instance, McDonald's golden arch logo is a design that appears to be a parabola. Give the equations and window that you use for each design.

3 Parabolas from Folding Paper

Parabolas can be made without using equations. Follow these steps to see how to make a parabola by folding paper. You will need several sheets of unlined paper about $8\frac{1}{2}''$ by $11''$ or $9''$ by $12''$. Wax paper works very well with this activity.

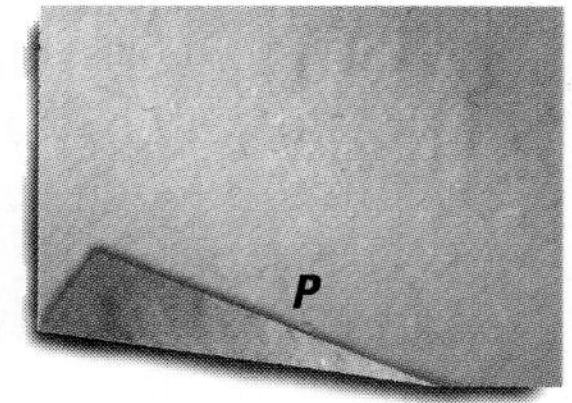

a. Mark a point P about one inch above the center of the lower edge. Fold the paper so that the lower edge touches P, make a careful crease as shown above, and then unfold the paper. Make 15 or 20 more folds each time folding in a different direction. The creases will outline a parabola. Trace the parabola.

b. Use another piece of paper the same size as you used in part **a.** This time mark P in about the center of the paper, and repeat the steps in part **a.**

c. Describe some ways in which the two parabolas are different.

4 Video of a Parabolic Path

If you have access to a camcorder, take video shots of a football player throwing a football or of a diver doing a simple dive. Hold the camera steady so that you don't move the camera during the shot. When you play back the tape, cover the TV screen with plastic wrap or other clear material. Play back the tape in slow speed, using a marker to trace on the plastic wrap the path of the ball or diver (you should follow the waist of the diver if she/he does any twists or turns). Describe the shape of the path.

5 Computer Programs for Solutions of Quadratics

Write a program for a computer or calculator to find solutions for any quadratic equation of the form $ax^2 + bx + c = 0$. Your program should allow the user to type in the three values *a, b,* and *c.* The program should first test that $a \neq 0$. Then it should evaluate the discriminant. If the discriminant is less than zero, your program should print, "There are no real number solutions." Otherwise it should print the solutions.

6 Parabolas in Manufactured Objects

Parabolas appear in mirrors, flashlights, satellite dishes, and many other manufactured objects. Consult some reference books and write a short report or make a poster explaining where and how parabolas are used.

7 Graphing Diamonds

Graph equations involving absolute values to obtain the figures shown below. Tell what equations you used. Then make up some other designs using graphs of absolute value equations.

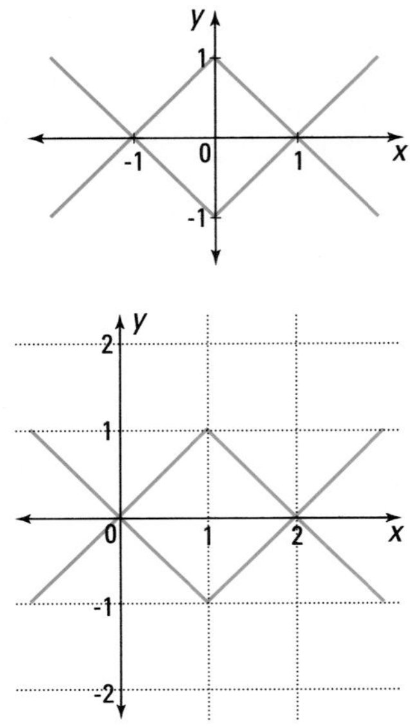

SUMMARY

The graph of $y = ax^2 + bx + c$ is a parabola. This parabola is symmetric to the vertical line through the vertex. If $a > 0$, the parabola opens up. If $a < 0$, the parabola opens down. If $b = 0$, the graph is symmetric to the y-axis.

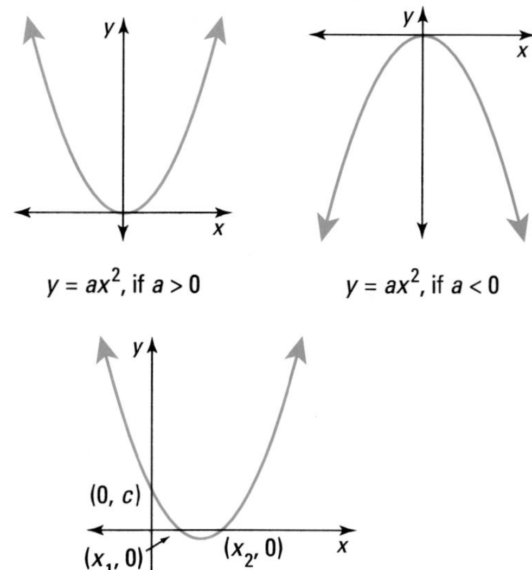

$y = ax^2$, if $a > 0$

$y = ax^2$, if $a < 0$

graph of $y = ax^2 + bx + c$ $(a > 0)$ with two x-intercepts, x_1 and x_2

To determine where this parabola crosses the horizontal line $y = k$, solve $ax^2 + bx + c = k$.

A quadratic equation (in one variable) is an equation that can be written in the form $ax^2 + bx + c = 0$, where $a \neq 0$. The solutions to the equation can be found by the Quadratic Formula,

$$x = \frac{-b \pm \sqrt{b^2 - 4ac}}{2a}.$$

The discriminant of the quadratic equation $ax^2 + bx + c = 0$ is $b^2 - 4ac$. If the discriminant is positive, there are two real solutions; if it is zero, there is one solution; if it is negative, there are no real solutions.

Solutions to quadratic equations involve square roots. Rewriting a square root so it is either a whole number or a product of a whole number and a smaller radical is called simplifying the radical. You can multiply or simplify square roots by using the Product of Square Roots Property.

If $a \geq 0$ and $b \geq 0$, then $\sqrt{a} \cdot \sqrt{b} = \sqrt{ab}$.

You can use square roots and absolute values to find distances on the number line or in the coordinate plane. There are formulas for both.

d, x_1, x_2, x

$d = |x_2 - x_1|$

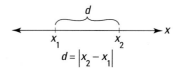

Distance Formula for a Number Line

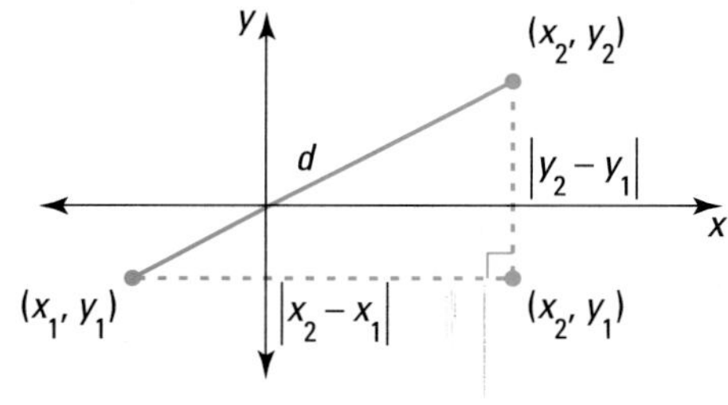

$$d = \sqrt{|x_2 - x_1|^2 + |y_2 - y_1|^2}$$

Pythagorean Distance Formula for the Coordinate Plane

Square roots and absolute value are related by the property, for all x, $\sqrt{x^2} = |x|$.

CHAPTER NINE

VOCABULARY

Below are the most important terms and phrases for this chapter. You should be able to give a general description and a specific example of each.

Lesson 9-1
parabola
axis of symmetry
symmetric
vertex, maximum, minimum
opens up
opens down

Lesson 9-2
STEP command
stopping distance

Lesson 9-4
projectile

Lesson 9-5
quadratic equation
Quadratic Formula
plus or minus, $\pm$
standard form of a quadratic equation

Lesson 9-6
discriminant
Discriminant Property

Lesson 9-7
Product of Square Roots Property
simplifying the radical

Lesson 9-8
absolute value, $|n|$, ABS(N)
Distance Formula on a Number Line
Absolute Value–Square Root Property
tolerance

Lesson 9-9
"as the crow flies"
Pythagorean Distance Formula in the Coordinate Plane

PROGRESS SELF-TEST

Take this test as you would take a test in class. Then check your work with the solutions in the Selected Answers section in the back of the book.

In 1–4, find all real solutions. Round answers to the nearest hundredth. If there is no real solution, state so.

1. $x^2 - 9x + 20 = 0$ **2.** $5y^2 - 3y = 11$

3. $8x^2 - 7x = -11$ **4.** $z^2 = 16z - 64$

5. If the discriminant of a quadratic equation is 18, how many solutions does the equation have?

6. *Multiple choice.* Which of these is the graph of $y = 2.5x^2$?

(a)

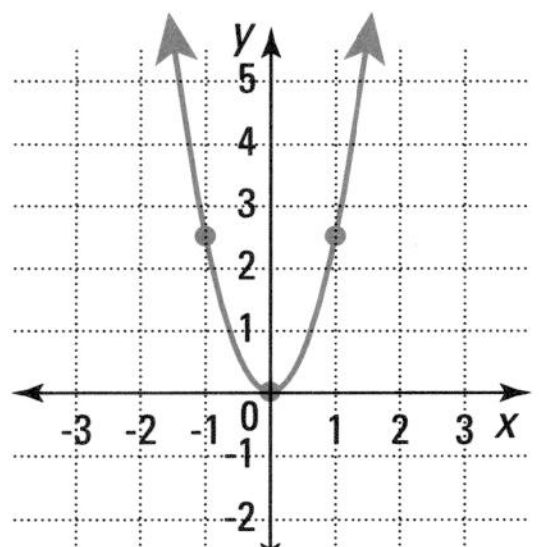

(b)

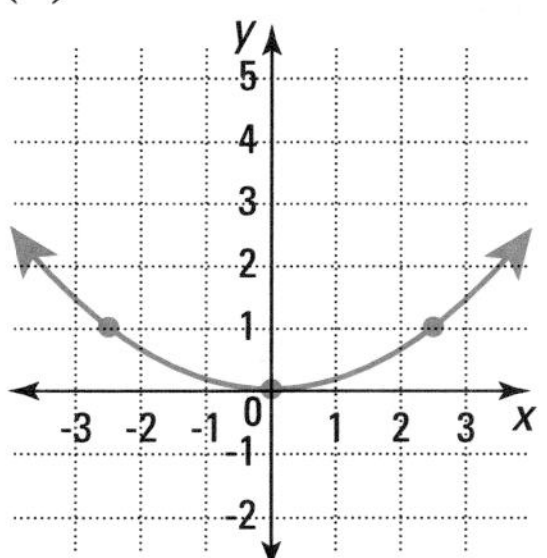

In 7 and 8, an equation is given.

a. Make a table of values of x and y for integer values of x from $x = -3$ to $x = 3$.

b. Graph the equation.

7. $y = -2x^2$ **8.** $y = x^2 - 4x + 3$

9. *True or false.* The parabola $y = 3x^2 - 7x - 35$ opens down.

In 10–12, use the graph of the parabola at the right.

10. Give the coordinates of the vertex.

11. Give the x-intercepts.

12. Give an equation for its axis of symmetry.

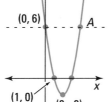

In 13–15, simplify.

13. $\sqrt{500}$ **14.** $\frac{\sqrt{75}}{5}$ **15.** $\sqrt{5x} \cdot \sqrt{45y}$

16. Given the following two facts about a parabola, sketch a possible graph.
(1) The vertex is (5, 1).
(2) When its equation is in the form $y = ax^2 + bx + c$, a is negative.

In 17–19, use the diagram at right.

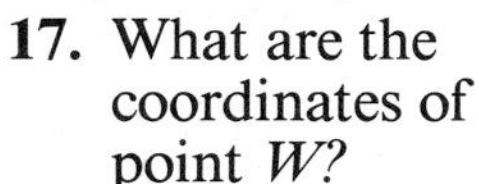

17. What are the coordinates of point W?

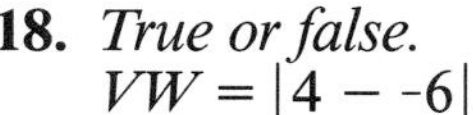

18. *True or false.* $VW = |4 - -6|$.

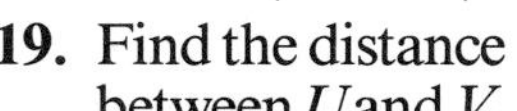

19. Find the distance between U and V.

20. Suppose $L = (3, -2)$ and $M = (x, y)$. Find LM, the distance between L and M.

In 21 and 22, a tennis ball is thrown from the top of a building. The graph shows $h = -16t^2 + 21t + 40$, giving the height h of the ball after t seconds.

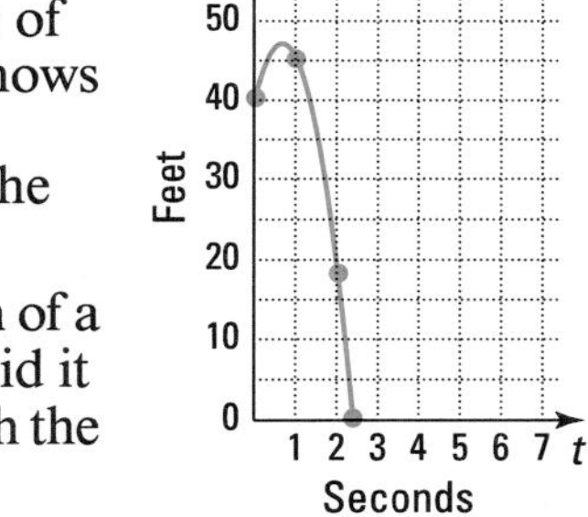

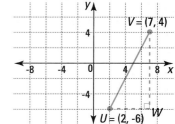

21. To the nearest tenth of a second, how long did it take the ball to reach the ground?

22. At what times was the ball 43 feet above the ground? Give your answer to the nearest tenth.

In 23 and 24, Harry tosses a ball to Fred who is 20 yards away. Melody is 2 yards in front of Fred. At x yards away from Harry, the ball is at a height h feet, where $h = -0.07x^2 + 1.4x + 5$.

23. Sketch the graph of this equation. Mark the positions of Harry, Fred, and Melody on your sketch.

24. How high is the ball when it passes above Melody? Explain how you get your answer.

In 25–27, find all solutions.

25. $\frac{1}{3}x^2 = 10$ **26.** $|x| = 57$ **27.** $|3 - n| = 0.5$

CHAPTER REVIEW

Questions on SPUR Objectives

SPUR stands for **S**kills, **P**roperties, **U**ses, and **R**epresentations. The Chapter Review questions are grouped according to the SPUR Objectives for this chapter.

SKILLS DEAL WITH THE PROCEDURES USED TO GET ANSWERS.

Objective A: *Solve quadratic equations.* *(Lessons 9-1, 9-5)*

In 1–4, solve without using the Quadratic Formula.

1. $16t^2 = 400$ **2.** $18 = \frac{1}{2}x^2$

3. $g^2 + 21 = 70$ **4.** $p^2 - 6 = 6$

In 5–8, give the exact solutions to the equation.

5. $6y^2 + 7y - 20 = 0$ **6.** $x^2 + 7x + 12 = 0$

7. $14v - 49 = v^2$ **8.** $14h - 3 = 2h^2$

In 9–14, use the Quadratic Formula to solve. Round answers to the nearest hundredth.

9. $k^2 - 7k = 2$ **10.** $2m^2 + m - 3 = 0$

11. $22a^2 + 2a + 3 = 0$ **12.** $0 = x^2 + 10x + 25$

13. $30 + 10(m^2 - 5m) = 0$

14. $16p^2 + 8p = -5$

Objective B: *Simplify square roots.* *(Lesson 9-7)*

In 15–23, simplify.

15. $\sqrt{7} \cdot \sqrt{28}$ **16.** $\sqrt{4} \cdot \sqrt{9} - \sqrt{3} \cdot \sqrt{48}$

17. $\sqrt{20^2 + 20^2}$ **18.** $\sqrt{99}$

19. $\sqrt{500}$ **20.** $\frac{\sqrt{150}}{5}$

21. $3\sqrt{72}$ **22.** $\frac{6 + \sqrt{128}}{2}$

23. $\frac{12 \pm 6\sqrt{24}}{4}$

In 24–28, simplify. Assume x and y are positive.

24. $\sqrt{24x} \cdot \sqrt{6x}$ **25.** $\sqrt{5x^2}$

26. $\sqrt{2x^2y^2}$ **27.** $\sqrt{(-17)^2}$

28. $-\sqrt{y^2}$

Objective C: *Evaluate expressions and solve equations using absolute value.* *(Lesson 9-8)*

In 29–34, evaluate the expression.

29. $|-43|$ **30.** $-|-1|$

31. $|13 - 19|$ **32.** ABS(1.6 – 1.8)

33. $|3| - |8|$ **34.** $|-20| - |15| + |-2|$

In 35–40, find all solutions.

35. $|d| = 16$ **36.** $|k| = -3$

37. $\sqrt{x^2} = 7$ **38.** $12 = \sqrt{n^2}$

39. $|r - 10| = 5$ **40.** $|300 - s| = 23$

PROPERTIES DEAL WITH THE PRINCIPLES BEHIND THE MATHEMATICS.

Objective D: *Identify and use properties of quadratic equations.* *(Lesson 9-6)*

41. Give the values of x that satisfy the equation $ax^2 + bx + c = 0$.

In 42 and 43, *true or false.*

42. Some quadratic equations have no real solutions.

43. Any quadratic equation can be solved by using the Quadratic Formula.

44. How can you use the discriminant to determine the number of real solutions a quadratic equation may have?

In 45–48, use the discriminant to determine the number of real solutions to the equation.

45. $2x^2 - 3x + 4 = 0$

46. $a^2 = 3a + 8$

47. $9d = 40 + 8d^2$

48. $n(n + 1) = -5$

USES DEAL WITH APPLICATIONS OF MATHEMATICS IN REAL SITUATIONS.

Objective E: *Use quadratic equations to solve problems about paths of projectiles.* *(Lessons 9-1, 9-4, 9-5)*

In 49 and 50, when an object is dropped near the surface of a planet or moon, the distance d (in feet) it falls in t seconds is given by the formula $d = \frac{1}{2}gt^2$, where g is the acceleration due to gravity. Near Earth, $g \approx 32$ ft/sec^2; near Earth's moon, $g \approx 5.3$ ft/sec^2.

49. A sky diver jumps from a plane at an altitude of 5000 ft. He begins his descent in "free fall," that is, without opening the parachute.

a. How far will the diver fall in 6 seconds?

b. The diver plans to open the parachute after he has fallen 2000 feet. How many seconds after jumping will this take place?

50. An astronaut on the moon drops a hammer.

a. How far will it fall in 4 seconds?

b. How long will it take the hammer to fall 100 feet?

51. One of the first astronauts who traveled to the moon hit a golf ball on the moon. Suppose that the height h in meters of a ball t seconds after it is hit is described by $h = -0.8t^2 + 10t$.

a. Sketch the graph of this equation.

b. Find the times when the ball is at a height of 20 meters.

c. Check your answer to part **b** by using the Quadratic Formula.

52. Consider again the quarterback in Lesson 9-4 who tosses a football to a receiver 40 yards down field. The ball is at height h feet, x yards down field, where $h = -0.025x^2 + x + 6$. Suppose a defender is 6 yards in front of the receiver.

a. How far is he from the quarterback?

b. Would the defender have a good chance to deflect the ball? Justify your answer.

53. Refer to the graph below of $h = -16t^2 + 64t$ which shows the height h in feet of a ball, t seconds after it is thrown from ground level at an initial upward velocity of 64 feet per second.

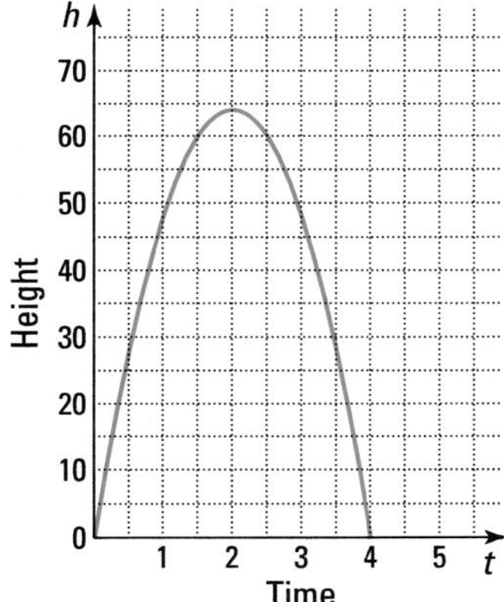

a. Give the height of the ball after 1 second.

b. Find when the ball will reach a height of 35 feet.

c. How many seconds is the ball in the air?

d. Make up a question whose answer is one of the coordinates of the vertex.

54. When a ball is thrown into the air with initial upward velocity of 20 meters per second, its approximate height y above the ground (in meters) after x seconds is given by $y = 20x - 5x^2$.

a. When will the ball hit the ground?

b. How high will the ball be at its highest point?

REPRESENTATIONS DEAL WITH PICTURES, GRAPHS, OR OBJECTS THAT ILLUSTRATE CONCEPTS.

Objective F: *Graph equations of the form $y = ax^2 + bx + c$ and interpret these graphs.* *(Lessons 9-1, 9-2, 9-3)*

55. Use the parabola below.

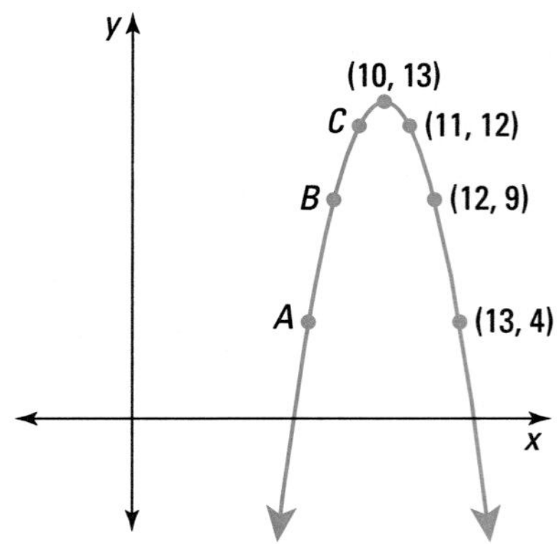

a. What are the coordinates of its vertex?

b. What is an equation for its axis of symmetry?

c. Find the coordinates of points *A*, *B*, and *C*, the reflection images of the named points over the parabola's axis of symmetry.

56. A table of values for a parabola is given below.

x	-5	-4	-3	-2	-1	0	1
y	14	9	6	5	6	?	?

a. Write an equation for its axis of symmetry.

b. What are the coordinates of its vertex?

c. Complete the table.

In 57 and 58, *true or false.*

57. Every parabola has a minimum value.

58. The parabola $y = -2x^2 + 3x + 1$ opens down.

59. What equation must you solve to find the x-intercepts of the parabola

$$y = ax^2 + bx + c?$$

60. The parabola $y = 2x^2 - 16x + 24$ has x-intercepts 6 and 2. Find the coordinates of its vertex without graphing.

In 61–64, **a.** make a table of values. **b.** Graph the equation.

61. $y = 3x^2$

62. $y = -\frac{1}{2}x^2$

63. $y = x^2 + 5x + 6$

64. $y = x^2 - 4$

65. *Multiple choice.* Which equation has the given graph?

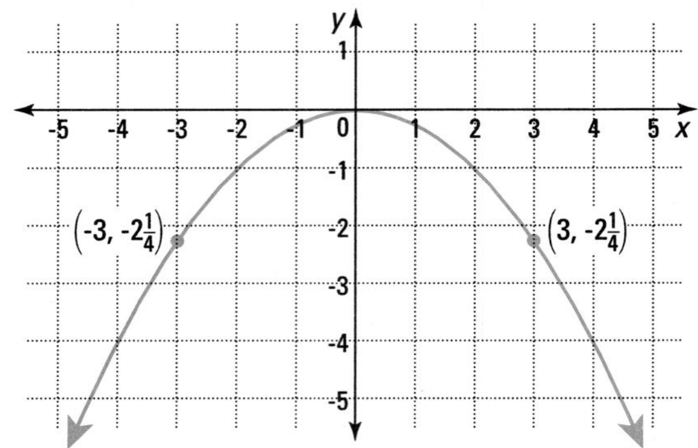

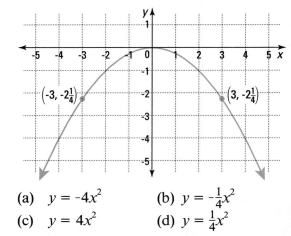

(a) $y = -4x^2$ (b) $y = -\frac{1}{4}x^2$

(c) $y = 4x^2$ (d) $y = \frac{1}{4}x^2$

66. *Multiple choice.* Which of these is the graph of $y = x^2 + 4x + 3$?

(a)

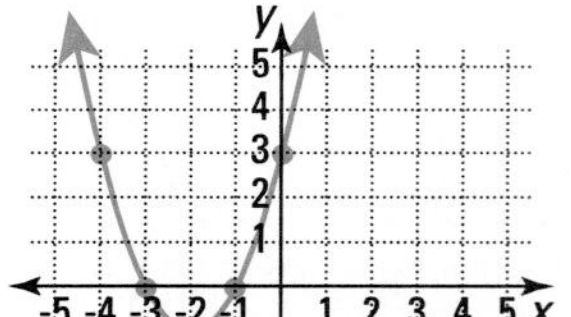

(b)

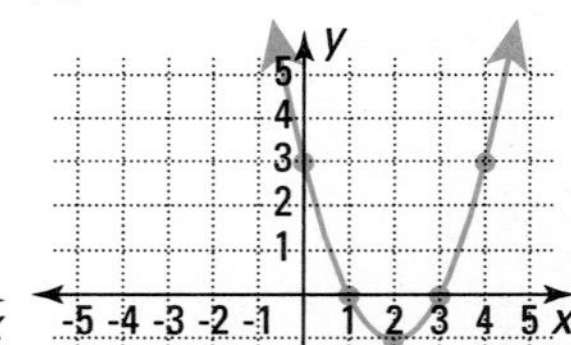

In 67 and 68, use an automatic grapher. **a.** Graph the equation. **b.** Estimate the coordinates of its vertex. **c.** Tell whether the vertex is a maximum or minimum point.

67. $y = -0.025x^2 + x + 6$, the equation of Question 52.

68. $y = 20x - 5x^2$, the equation of Question 54.

In 69 and 70, a graph of an equation is shown. Describe another window that will show the vertex and all intercepts.

69.

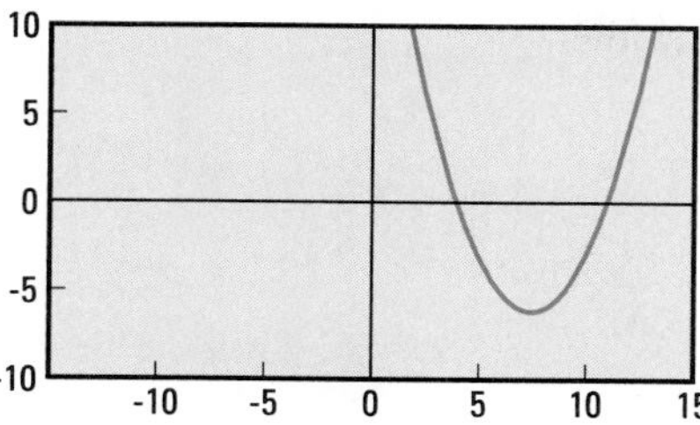

70.

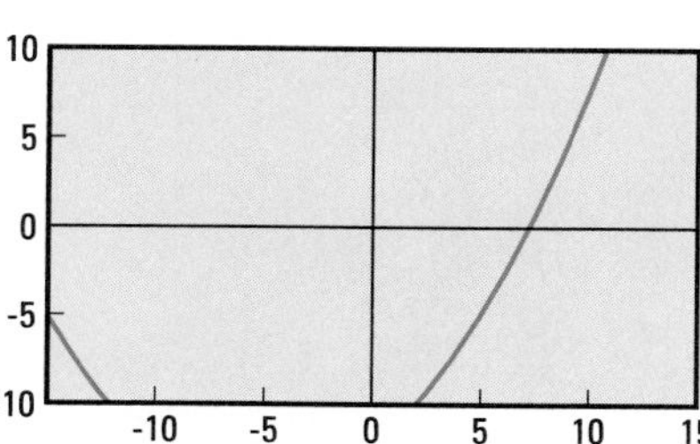

Objective G: *Calculate and represent distances on the number line or in the plane.* *(Lessons 9-8, 9-9)*

In 71 and 72, the coordinates of two points on the number line are given. Find the distance between them.

71. -47 -16

72. p and q

73. Find the coordinates of all points on a number line at a distance of 5 from the point -3.

In 74–76, use the graph below.

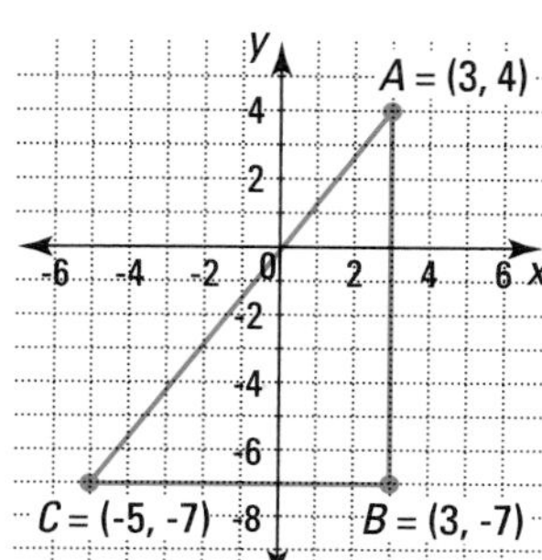

74. a. Find AB. **b.** Find AC.

75. Find the distance from A to the origin O.

76. Which is greater: BC or BO? Justify your answer.

In 77–82, find AB.

77. $A = (4, -10)$, $B = (-5, -10)$

78. $A = (3, 20)$, $B = (3, 2)$

79. $A = (50, 10)$, $B = (0, 0)$

80. $A = (14, -20)$, $B = (-2, -8)$

81. $A = (-4, -3)$, $B = (4, 12)$

82. $A = (-2, 9)$, $B = (-5, -1)$

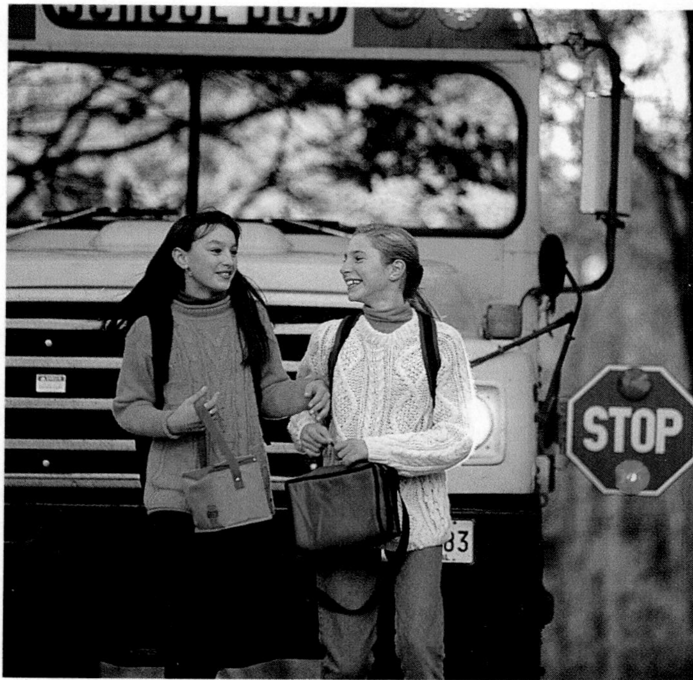

83. The school these girls attend is 3 miles east and 5 miles south of their homes. The library is 4 miles west and 2 miles north of their houses. The streets in their neighborhood are laid out in a square grid.

a. Draw a picture to represent this situation.

b. How far is it from the school to the library, as the crow flies?

84. A right triangle has vertices at (0, 4), (2, 0), and (6, 2). Find the length of its hypotenuse.

CHAPTER 10

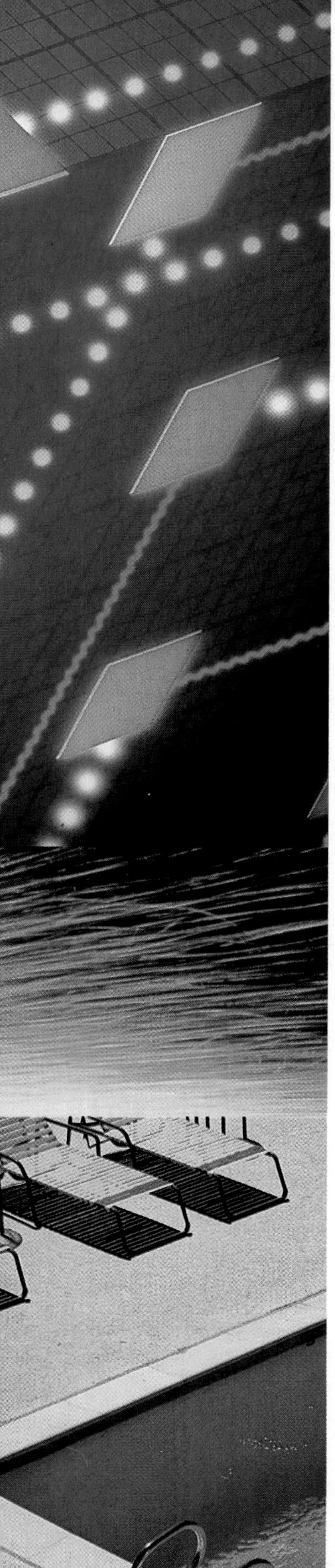

POLYNOMIALS

Expressions such as

$$6s^2,$$
$$2\ell + 2w + 4h,$$
$$\text{and } 1000x^4 + 500x^3 + 100x^2 + 200x$$

are *polynomials.*

Polynomials are found in geometry. For example, the monomial $6s^2$ represents the surface area of a cube with side s. The trinomial $2\ell + 2w + 4h$ represents the minimum length of ribbon needed to wrap a box of dimensions ℓ by w by h, as shown here.

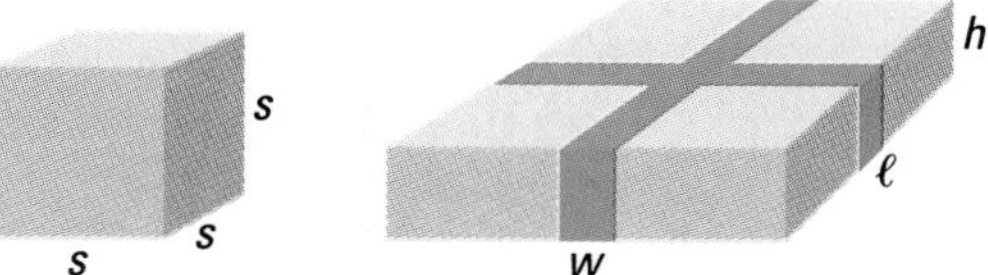

Other polynomials arise from algebra. In Chapter 8, you calculated compound interest for a single deposit. When several deposits are made, the total amount of money accumulated can be expressed as a polynomial. For instance, the polynomial $1000x^4 + 500x^3 + 100x^2 + 250x$ represents the amount of money you would have if you invested \$1000 4 years ago, added \$500 to it 3 years ago, \$100 to it 2 years ago, and \$250 to it 1 year ago, all at the same rate of $x - 1$. The quadratic expressions you studied in Chapter 9 are all polynomials.

Polynomials also form the basic structure of our base 10 arithmetic. The expanded form of a number like 1796, $1 \cdot 10^3 + 7 \cdot 10^2 + 9 \cdot 10^1 + 6$, is like the polynomial $x^3 + 7x^2 + 9x + 6$ with the base 10 substituted for x.

In this chapter you will study these and other situations that give rise to polynomials, and how to add, subtract, and multiply polynomials.

LESSON

10-1

What Are Polynomials?

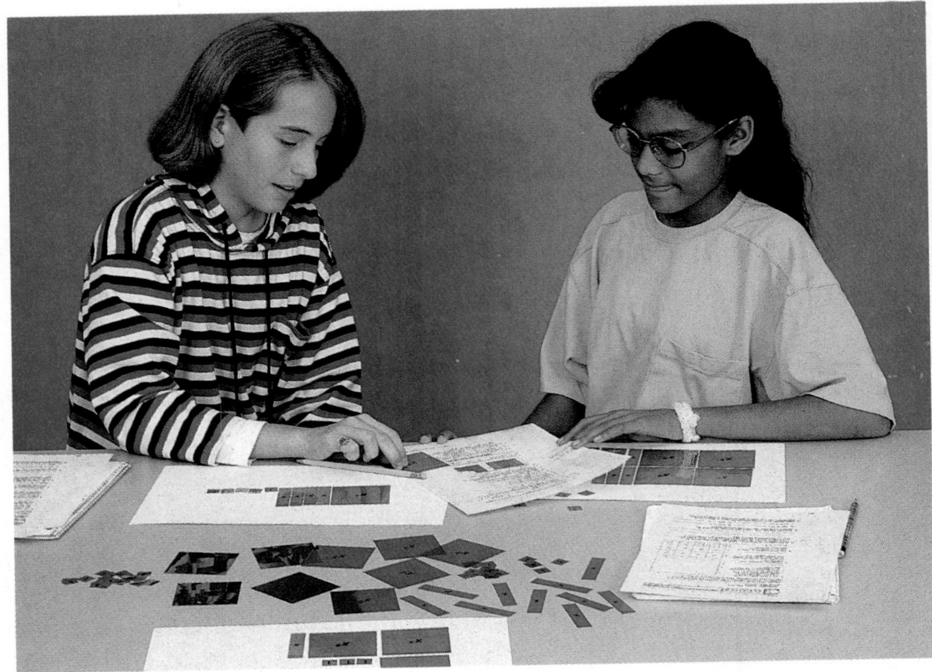

Picturing polynomials. *Polynomial expressions, such as $2x^2 + 7x + 3$, can be represented by algebra tiles. These tiles can also be used to illustrate operations with polynomials.*

You have worked with polynomials throughout your study of algebra. Now, because we are going to study polynomials in some detail, we introduce some terminology.

Monomials

Each of the expressions

$$3x, \quad 16t^2, \quad x^2y^4$$

is called a *monomial.* A **monomial** is an expression that can be written as a real number, a variable, or a product of a real number and one or more variables with nonnegative powers. The **degree of a monomial** is the sum of the exponents of the variables in the expression. For instance,

$3x$ has degree 1, (Think: $3x^1$.)
$16t^2$ has degree 2,
and x^2y^4 has degree 6.

Example 1

Tell whether the expression is a monomial or can be rewritten as a monomial. If so, identify its degree. If not, tell why not.

a. dw
b. -66
c. x^{-7}
d. $-8x^4 + 2x^4$

Solution

a. $dw = d^1w^1$. So it is a product of variables with nonnegative exponents. Yes, dw is a monomial. Its degree is $1 + 1 = 2$.

b. $-66 = -66x^0$. Zero is nonnegative. -66 is a monomial of degree 0.

c. $x^{-7} = \frac{1}{x^7}$. x^{-7} is not a monomial because it cannot be written as a product of variables with nonnegative exponents.

d. $-8x^4 + 2x^4 = (-8 + 2)x^4 = -6x^4$. Yes, $-8x^4 + 2x^4$ can be rewritten as a monomial of degree 4.

Classifying Polynomials by the Number of Terms

A **polynomial** is an expression that is either a monomial or a sum of monomials. Polynomials with two or three terms are used so often they have special names. A **binomial** is a polynomial which has two terms. A **trinomial** is a polynomial which has three terms. Here are some examples.

binomials: $x + 3$, $x^2 - y^2$, $7 - ab$

trinomials: $7x^2 - 3x + 5$, $a^2 + 2ab + b^2$, $ab + bc + cd$

The **degree of a polynomial** is the highest degree of any of its terms after the polynomial has been simplified.

Example 2

What is the degree of each polynomial?

a. $7x^2 - 3x + 5$

b. $p + q^2 + pq^2 + p^2q^3$

Solution

a. The terms have the following degrees:

$7x^2$ has degree 2;

$-3x$ has degree 1;

5 has degree 0.

$7x^2$ has degree 2, which is the highest in the polynomial. So the polynomial has degree 2.

b. Find the degree of each term by adding the exponents in that term.

$$\begin{array}{ccccccc} p & + & q^2 & + & pq^2 & + & p^2q^3 \\ \downarrow & & \downarrow & & \swarrow\downarrow & & \swarrow\downarrow \\ 1 & & 2 & & 1 + 2 = 3 & & 2 + 3 = 5 \end{array}$$

p^2q^3 has degree 5, which is the highest in the polynomial. So the polynomial has degree 5.

The polynomial

$$6x + 4x^5 + 8 + x^2$$

has degree 5. It is called a **polynomial in *x*,** because it is the sum of multiples of whole number powers of x. Its four terms could be rearranged in any order and the value of the polynomial would be the same. It is customary to write the terms in *descending order* of the exponents of its terms, with the term with the largest exponent first:

$$4x^5 + x^2 + 6x + 8.$$

This is called the **standard form of a polynomial.** The numbers 4, 1, 6, and 8 are the *coefficients* of the polynomial. Notice that 1 is the coefficient of x^2. Since this polynomial has no x^3 or x^4 term, we say that 0 is the coefficient of x^3 and x^4. Written with these coefficients, here is a polynomial equal to the preceding one:

$$4x^5 + 0x^4 + 0x^3 + 1x^2 + 6x + 8.$$

Notice what happens when 10 is substituted for x in this polynomial. The value of the polynomial is

$$400{,}168.$$

The coefficients of the polynomial have become the digits of the decimal it equals. The expanded form of 400,168 is

$$4 \cdot 10^5 + 0 \cdot 10^4 + 0 \cdot 10^3 + 1 \cdot 10^2 + 6 \cdot 10 + 8.$$

This is a polynomial in a single variable with 10 substituted for the variable. Thus the normal way of writing whole numbers in base 10 is shorthand for a polynomial.

Classifying Polynomials by Their Degree

A polynomial of degree 1, such as $3x + 5$, is called **linear.** A polynomial of degree 2, such as $20 + 24t - 16t^2$, or $5xy$, is called **quadratic.** Linear and quadratic polynomials whose coefficients are small positive whole numbers can be represented by tiles, as you have seen. The tiles below represent the polynomial $x^2 + 4x + 5$ because this polynomial is the area of the figure.

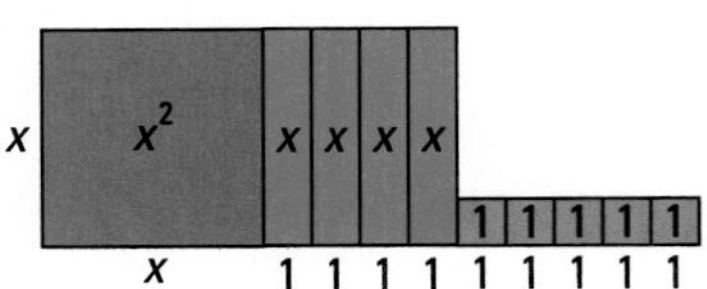

QUESTIONS

Covering the Reading

1. *Multiple choice.* Which of the following is *not* a monomial?
(a) x^4 (b) $4x$ (c) 4 (d) $x + 4$

2. Explain why $\frac{1}{y^2}$ is *not* a monomial.

In 3–7, an expression is given. **a.** Tell whether the expression is a monomial. **b.** If it is a monomial, give its degree.

3. $5t^4$ **4.** πr^2 **5.** $5t^{-4}$ **6.** $\frac{1}{2}bh$ **7.** x^3y^6

8. **a.** Is xyz a trinomial?
b. Explain your reasoning.

In 9–13, consider the following polynomials:

(a) $x^2 + 5$ (b) $x^2 + 5x + 6$
(c) $x^2 + 5xy + y^2$ (d) $x^3 + 5x^2 + 6x$

9. Which are binomials?

10. Which are trinomials?

11. Which have degree 2?

12. Which have degree 3?

13. Which are polynomials in x?

In 14 and 15, rearrange the polynomial into standard form.

14. $16d^2 - 8d^4 + d^3$ **15.** $x^9 + 2x + 4 - x^7$

16. Jacob Lawrence, noted American painter, had his first one-man exhibition at the Harlem YMCA in 1938. Write 1938 as a polynomial with 10 substituted for the variable.

The Studio, *a self-portrait painted by Jacob Lawrence in 1977.*

In 17 and 18, what polynomial is represented by the tiles?

17.

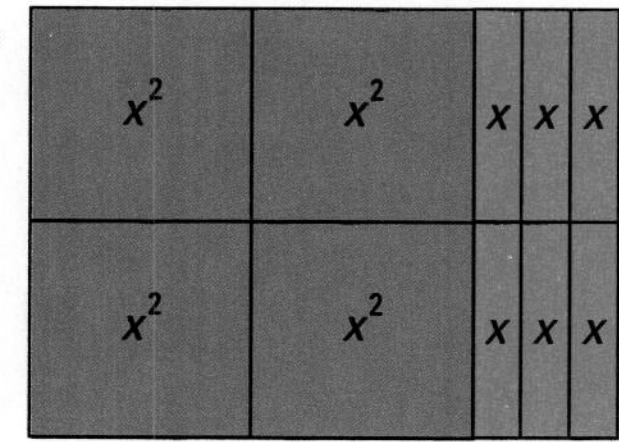

18.

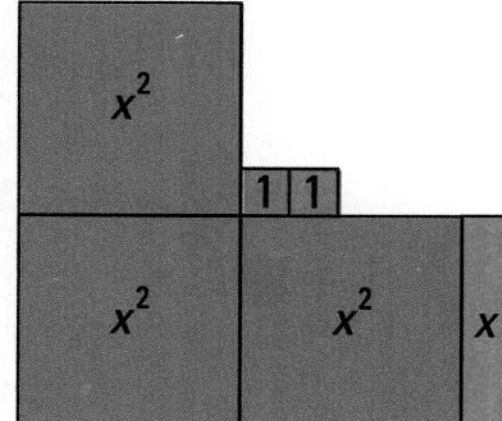

In 19 and 20, draw a representation of the polynomial using tiles.

19. $2x + 3$ **20.** $2x^2 + 3x + 1$

Applying the Mathematics

In 21–25, an expression is given. **a.** Show that the expression can be simplified into a monomial. **b.** Give the degree of the monomial.

21. $3x + x$ **22.** $3x \cdot x$ **23.** $(5n^2)(-6n^2)$

24. $2ab + 6ab$ **25.** $(2ab)^6$

26. **a.** Write a monomial with one variable whose degree is 5.
b. Write a monomial with two variables whose degree is 5.

27. Suppose you have many tiles of each of the following sizes.

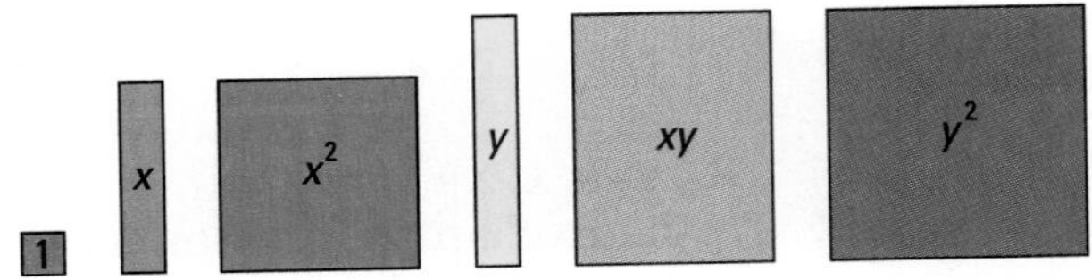

a. Draw a model for the polynomial $x^2 + 2xy + y^2$.
b. Show how the tiles in part **a** can be arranged to form a square.

28. Show how tiles representing the quantity $x^2 + 5x + 6$ can be arranged to form one large rectangle.

29. a. Write 246 and 1032 as polynomials with 10 substituted for the variable.
b. Add the polynomials from part **a.** Is your sum equal to the sum of 246 and 1032?

Review

30. If \$2000 is invested at an annual rate of 6%, compounded yearly, how much money will there be after each time period? *(Lesson 8-1)*
a. 1 year **b.** 2 years **c.** n years

In 31–32, graph the solution set **a.** on a number line; **b.** in the coordinate plane. *(Lessons 1-1, 3-10, 7-9)*

31. $x > 3$ **32.** $-6x + 9 > 3$

33. Graph in the coordinate plane: $6x + 9y > 3$. *(Lesson 7-9)*

In 34 and 35, consider a garage with a roof with a pitch of $\frac{4}{12}$. The garage is to be 30 feet wide. *(Lessons 1-7, 6-8, 7-3)*

34. What is the slope of $\overline{AC}$?

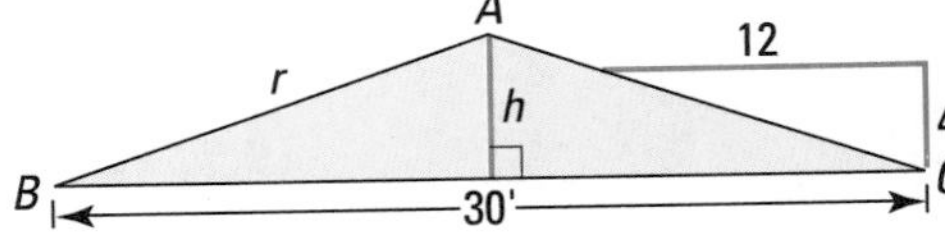

35. a. What is the height h of the roof?
b. Find the length r of one rafter.

A three-winger. *This is the* Fokker Dr-1 *triplane, flown in World War I by the Red Baron.*

Exploration

36. The words monomial, binomial, trinomial, and polynomial contain the prefixes mono-, bi-, tri-, and poly-, meaning one, two, three, and many, respectively. Find some other words, both mathematical and nonmathematical, that begin with these prefixes and give their meanings.

LESSON

10-2

Investments and Polynomials

The largest money matters most adults commonly deal with are

salary or wages,
savings,
payments on loans for cars or trips or other items, and
home mortgages or rent.

Each of these items involves paying or receiving money each month, every few months, or every year. But what is the total amount paid or received? The answer is not easy to calculate because interest may be involved. Here is an example of this kind of situation.

Example 1

Each birthday from age 12 on, Maria has been receiving $500 from her grandparents to save for college. She is putting the money in an account that pays an annual yield of 7%. How much money will she have by the time she is 16?

Solution

It helps to write down how much Maria has on each birthday. On her 12th birthday she has $500. She then receives interest on that $500 and an additional $500 on her 13th birthday. At that time she will have

$$500(1.07) + 500 = \$1035.00.$$

Each year interest is paid on all the money previously saved and each year another $500 gift is added. The totals for her 12th through 16th birthdays are given in the chart below.

Birthday		Total
12th	500	= $500
13th	$500(1.07) + 500$	= $1035.00
14th	$500(1.07)^2 + 500(1.07) + 500$	= $1607.45
15th	$500(1.07)^3 + 500(1.07)^2 + 500(1.07) + 500$	= $2219.97
16th	$500(1.07)^4 + 500(1.07)^3 + 500(1.07)^2 + 500(1.07) + 500$	= $2875.37

↑ from 12th birthday ↑ from 13th birthday ↑ from 14th birthday ↑ from 15th birthday ↑ from 16th birthday

The total of $2875.37 she has by her 16th birthday is $375.37 more than the total $2500 she received as gifts because of the interest earned.

How Investments Lead to Polynomials

When $x = 1.07$, the following polynomial gives the amount of money Maria will have (in dollars)

$$500x^4 + 500x^3 + 500x^2 + 500x + 500.$$

This polynomial in x is useful because if the interest rate changes, you only have to substitute the new value for x. We call x in this expression a *scale factor*. For instance, if Maria's account had an annual yield of 4%, the scale factor would be 104% = 1.04. After 4 years Maria would have:

$$500(1.04)^4 + 500(1.04)^3 + 500(1.04)^2 + 500(1.04) + 500 \approx \$2708.16.$$

Example 2

Cole's parents plan to give him \$50 on his 12th birthday, \$60 on his 13th, \$70 on his 14th, and \$80 on his 15th. If he invests the money in an account with a yearly scale factor x, how much money will he have on his 15th birthday?

Solution

The money from Cole's 12th birthday will earn three years worth of interest; from his 13th, two years of interest; and from his 14th, one year. The total amount of Coleman's birthday gifts will equal

$$50x^3 + 60x^2 + 70x + 80.$$

Adding Polynomials

Suppose Cole's aunt gives him \$20 on each of these 4 birthdays. If he puts this money into the same account, the amount available from the aunt's gifts would be

$$20x^3 + 20x^2 + 20x + 20.$$

To find the total amount he would have, add these two polynomials.

$$(50x^3 + 60x^2 + 70x + 80) + (20x^3 + 20x^2 + 20x + 20)$$

Simplify the sum. First, use the Associative and Commutative Properties of Addition to rearrange the polynomials so that like terms are together.

$$= (50x^3 + 20x^3) + (60x^2 + 20x^2) + (70x + 20x) + (80 + 20)$$

Then use the Distributive Property to add like terms.

$$= (50 + 20)x^3 + (60 + 20)x^2 + (70 + 20)x + (80 + 20)$$
$$= 70x^3 + 80x^2 + 90x + 100$$

Notice what the answer means in relation to Cole's birthday presents. The first year he got \$70 (\$50 from his parents and \$20 from his aunt). The \$70 has 3 years to earn interest. The \$80 from his next birthday earns interest for 2 years, and so on. Also notice that in these examples we have written the polynomials in standard form. This is common practice.

Comparing Investments

When comparing investments it is often useful to make a table.

Lion's Gate Bridge in Vancouver, Canada.

Example 3

Nellie and Joe plan to save money for a round-the-world trip when they retire 10 years from now. Nellie plans to save \$1000 per year for the first 5 years and then to stop making deposits. Joe plans to wait 5 years to begin saving but then to save \$1200 per year for 5 years. They will deposit their savings at the beginning of the year into accounts earning 6% interest compounded annually. How much will each person have after 10 years?

Solution

Make a table showing the amount of money each person will have at the end of each year. At the end of the first year, Nellie will have 1.06(1000) = \$1060. At the end of the second year she will have 106% of the sum of the previous balance and the new deposit of \$1000. In all she will have 1.06(1060 + 1000) = \$2183.60. This pattern continues. But after 5 years, she deposits no money. So her money accumulates only interest. Nellie's end-of-year balance in the spreadsheet below was computed by entering the formula = 1.06 * B3 into cell C3 and the formula = 1.06 * (C3 + B4) into cell C4. The formula in cell C4 was then replicated down column C to C12. A similar set of formulas generated Joe's end-of-year balance.

	A	B	C	D	E
1	Year	Nellie's Deposit	Nellie's Balance	Joe's Deposit	Joe's Balance
2			(end of year)		(end of year)
3	1	1000	1060.00	0	0
4	2	1000	2183.60	0	0
5	3	1000	3374.62	0	0
6	4	1000	4637.09	0	0
7	5	1000	5975.32	0	0
8	6	0	6333.84	1200	1272.00
9	7	0	6713.87	1200	2620.32
10	8	0	7116.70	1200	4049.54
11	9	0	7543.70	1200	5564.51
12	10	0	7996.32	1200	7170.38

Ten years from now Nellie will have about \$7996, and Joe will have about \$7170.

Notice that even though Nellie deposits only \$5000 and Joe deposits \$6000, compounding interest over a longer period of time will give Nellie over \$800 more than Joe. Here is what happens. After 10 years:

Nellie will have $1000x^{10} + 1000x^9 + 1000x^8 + 1000x^7 + 1000x^6$,
and Joe will have $1200x^5 + 1200x^4 + 1200x^3 + 1200x^2 + 1200x$.

When $x = 1.06$, as in Example 3, Nellie will have more than Joe.

QUESTIONS

Covering the Reading

1. Refer to Example 1. Suppose Maria is able to get an annual yield of only 5% on her investment.
  a. How much money will she have in her account by her 16th birthday?
  b. How much less is this than what she would have earned at 7% annual yield?

2. Ellery's children will give him a combined gift of \$1200 on each of his birthdays from age 61 on.
  a. If he saves the money in an account paying an annual yield of 7%, how much will he have by the time he retires at age 65?
  b. How much will he have accumulated if his investment grows by a scale factor of x each year?

In 3 and 4, refer to Example 2.

3. Consider the polynomial $70x^3 + 80x^2 + 90x + 100$. What coefficient indicates the amount Cole received on his 14th birthday from his parents and his aunt?

4. Suppose Cole also gets \$15, \$25, \$35, and \$45 from cousin Lilly on his four birthdays. He puts this money into his account also.
  a. By his 15th birthday, how much money will Cole have from just his cousin?
  b. What is the total Cole will have saved by his 15th birthday from all of his birthday presents?

In 5–7, refer to Example 3.

5. Which person—Nellie or Joe—will deposit more money in the 10 years? How much more?

6. Which person—Nellie or Joe—will have more money at the end of 10 years? How much more?

7. Suppose Nellie and Joe could earn 7% on their investments. Recalculate the balances in the table and describe the end result.

Applying the Mathematics

8. Suppose in 1989 Carrie received \$100 on her birthday. From 1990 to 1993 she received \$150 on each birthday. She put the money in a shoebox. The money is still there.
  a. How much money did Carrie have after her 1993 birthday?
  b. How much more would she have had if she had invested her money at an annual yield of 6%?

In 9–12, Huey, Dewey, and Louie are triplets. They received cash presents on their birthdays as shown in the table.

Year	Huey	Dewey	Louie
In 1992	\$100	\$150	\$50
In 1993	\$200	\$150	\$400
In 1994	\$150	\$150	nothing

Each put all his money into a bank account which paid a 6% annual yield.

9. How much money did Huey have on his 1992 birthday?

10. How much did Dewey have on his 1993 birthday?

11. How much did Louie have on his 1994 birthday?

12. In 1995, Huey received \$300 on his birthday. If he had invested all the money received from years 1992–1995 in an account with scale factor x, how much would he have had by his birthday in each of these years?
   a. 1992 **b.** 1993 **c.** 1994 **d.** 1995

In 13–16, simplify the expressions.

13. $(12y^2 + 3y - 7) + (4y^2 - 2y - 10)$

14. $(3 + 5k^2 - 2k) + (2k^2 - 3k - 10)$

15. $(6w^2 - w + 14) - (4w^2 + 3)$

16. $(x^3 - 4x + 1) - (5x^3 + 4x - 8)$

17. Solve the equation $(3x^2 + 2x + 4) + (3x^2 + 11x + 2) = 0$.

In 18 and 19, find the missing polynomial.

18. $(9x^2 + 12x - 5) + (\underline{\ ?\ }) = 13x^2 + 6$

19. $(2y^2 - y - 16) - (\underline{\ ?\ }) = -5y^2 - y + 31$

20. A wood harvester has planted trees in a forest each spring for four years, as shown in the following table.

Year	Number of trees planted
1	16,000
2	22,000
3	18,000
4	25,000

Suppose each tree when planted could provide .01 cord of wood and grows with a scale factor x each year. How many cords of wood could these trees provide in the fourth spring?

A renewable resource. *This woman is reforesting an area in the Boise National Forest in Idaho.*

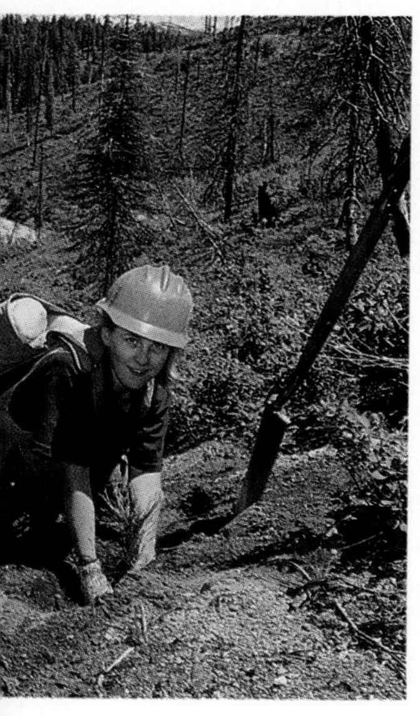

Review

In 21–23, describe a situation that might yield the polynomial. Then give the degree of the polynomial. *(Lesson 10-1)*

21. πr^2 **22.** x^3 **23.** $16t^2 + 48t$

24. a. Simplify $x(x + 5) - 10x$.
b. Solve $x(x + 5) = 10x$. *(Lessons 3-3, 3-9, 9-5)*

25. The formula $d = 0.042s^2 + 1.1s$ gives the approximate distance d in feet needed to stop a particular car traveling on dry pavement at a speed of s miles per hour. How much farther will this car travel before stopping if it is going at 65 mph instead of 55 mph? *(Lesson 9-2)*

26. Simplify $x^{-1} + -x - \frac{1}{x}$. *(Lesson 8-6)*

27. Find the slope of the line pictured at the right. *(Lesson 7-3)*

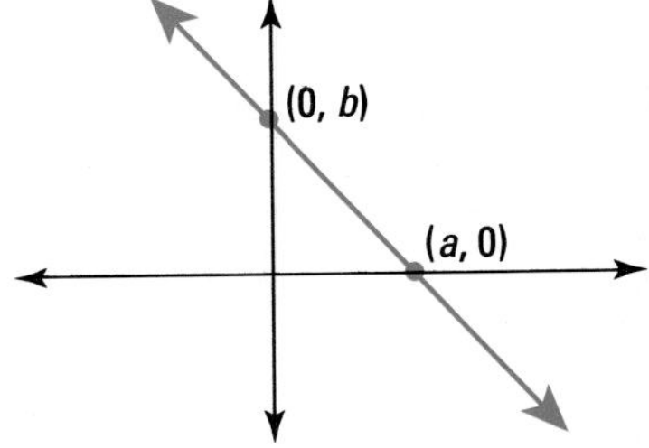

28. *Skill sequence.* Simplify. *(Lesson 3-7)*
a. $8(3y)$ **b.** $8(3y + 7)$ **c.** $8(3y + 7 - 4x)$

29. *Multiple choice.* Which is a pair of like terms? *(Lesson 3-2)*
(a) $4b$ and $4b^3$ (b) x^3 and y^3
(c) $3m^2$ and $6m^2$ (d) $50x$ and 50

30. Simplify $-5n + 2n + k + k + 10n$. *(Lesson 3-2)*

31. The figure below shows two rectangles. If the area of the shaded region is 20, what are the missing dimensions of the rectangles? *(Lessons 2-1, 2-6, 3-2)*

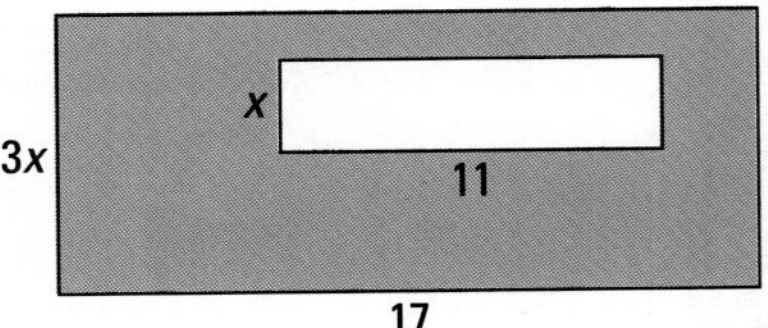

Exploration

32. Refer to Nellie and Joe in Example 3 and Questions 5–7.
a. For what interest rates will Nellie end up with more money than Joe?
b. For what interest rates will Joe end up with more than Nellie?

LESSON

10-3

Multiplying a Polynomial by a Monomial

Mall math. *To find the total area of these storefronts, you may find the area of each storefront and add the products, or multiply the height by the sum of the widths of the storefronts.*

You have already done several kinds of problems involving multiplication by a monomial. To multiply a monomial by a monomial, you can use properties of powers. For instance,

$$(2a^4b)(8a^3b^2) = 2 \cdot 8a^{(4+3)}b^{(1+2)} = 16a^7b^3.$$

To multiply a monomial by a binomial, you can use the Distributive Property, $a(b + c) = ab + ac$. For instance,

$$3x(2x + 1) = 3x \cdot 2x + 3x \cdot 1 = 6x^2 + 3x.$$

Using Area to Picture Multiplication by a Monomial

Some products of monomials and binomials can be pictured using the Area Model for Multiplication. For instance, the product $3x(2x + 1)$ is the area of a rectangle with dimensions $3x$ and $2x + 1$ as shown on the left.

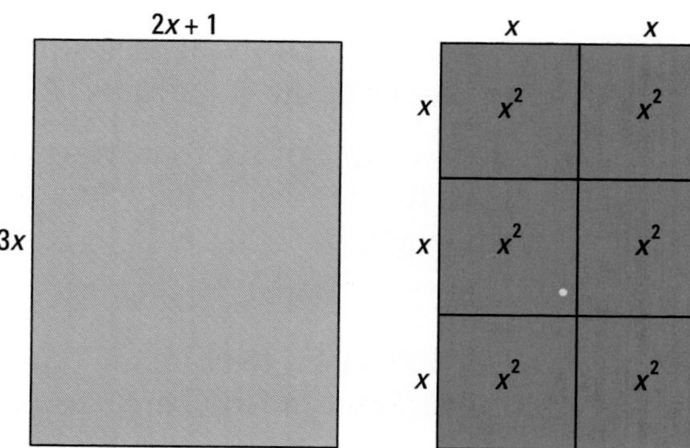

This rectangle can be split up into tiles as shown above at the right. The total area of the rectangle is $6x^2 + 3x$, which agrees with the result obtained using the Distributive Property.

Example 1

Give two equivalent expressions for the total area pictured below.

Solution

Thinking of the total area as the sum of the areas of the individual tiles, the area is $4x^2 + 8x$. Thinking of the total area as length times width, the area is $4x(x + 2)$. So $4x(x + 2) = 4x^2 + 8x$.

	x	x	x	x
x	x^2	x^2	x^2	x^2
1	x	x	x	x
1	x	x	x	x

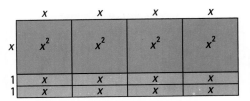

The area representation of a polynomial shows how to multiply a monomial by any other polynomial.

The pictures show a view of some storefronts at a shopping mall.

The displays in the windows are used to attract shoppers, so store owners and mall managers are interested in the areas of storefronts. Note that the height of each storefront h is a **monomial,** and the sum of the lengths of the storefronts $(L_1 + L_2 + L_3 + L_4)$ is a **polynomial.**

The total area of the four windows can be computed in two ways. One way is to consider all the windows together. They form one big rectangle with length $(L_1 + L_2 + L_3 + L_4)$ and height h. Thus, the total area equals

$$h \cdot (L_1 + L_2 + L_3 + L_4).$$

A second way is to compute the area of each storefront and add the results. Thus, the total area also equals

$$hL_1 + hL_2 + hL_3 + hL_4.$$

These areas are equal, so

$$h \cdot (L_1 + L_2 + L_3 + L_4) = hL_1 + hL_2 + hL_3 + hL_4.$$

In general, to multiply a monomial by a polynomial, simply extend the Distributive Property: multiply the monomial by each term in the polynomial and add the results. As always, you must be careful with the signs in polynomials.

Example 2

Multiply $7x(x^2 + 4.5x + 1)$.

Solution

Multiply each term in the trinomial by the monomial $7x$.

$$7x(x^2 + 4.5x + 1) = 7x \cdot x^2 + 7x \cdot 4.5x + 7x \cdot 1$$
$$= 7x^3 + 31.5x^2 + 7x$$

Check

Test a special case. We let $x = 3$.

Does $7 \cdot 3(3^2 + 4.5 \cdot 3 + 1) = 7 \cdot 3^3 + 31.5 \cdot 3^2 + 7 \cdot 3$?

$21 \cdot 23.5 = 189 + 283.5 + 21$?

Yes, $493.5 = 493.5$.

Example 3

Multiply $-4x^2(2x^3 + y - 5)$.

Solution

Distribute $-4x^2$ over each term of the polynomial.

$$-4x^2(2x^3 + y - 5) = -4x^2 \cdot 2x^3 + -4x^2 \cdot y - -4x^2 \cdot 5$$
$$= -8x^5 - 4x^2y + 20x^2$$

Multiplying a Decimal by a Power of Ten

Recall the rule for multiplying a decimal by a power of 10: To multiply by 10^n, move the decimal point n places to the right. Multiplication of a monomial by a polynomial can show why this rule works. For instance, suppose 59,078 is multiplied by 1000. We write 59,078 in expanded form which looks like a polynomial. We think of 1000 as the monomial 10^3.

$$1000 \cdot 59{,}078 = 10^3 \cdot (5 \cdot 10^4 + 9 \cdot 10^3 + 7 \cdot 10 + 8)$$

Now we use the Distributive Property.

$$= 10^3 \cdot 5 \cdot 10^4 + 10^3 \cdot 9 \cdot 10^3 + 10^3 \cdot 7 \cdot 10 + 10^3 \cdot 8$$

The products can be simplified using the Product of Powers Property and the Commutative and Associative Properties of Multiplication.

$$= 5 \cdot 10^7 + 9 \cdot 10^6 + 7 \cdot 10^4 + 8 \cdot 10^3$$

Now change the expanded form back to decimal form.

$$= 59{,}078{,}000$$

This same procedure could be repeated to explain the product of any decimal and any integer power of 10.

QUESTIONS

Covering the Reading

In 1 and 2, find the product.

1. $(3x)(4x)$

2. $(3xy^3)(4x^2y)$

In 3 and 4, **a.** find the product; **b.** draw a rectangle to represent the product.

3. $3x(x + 4)$

4. $2x(x + 5)$

In 5 and 6, a large rectangle is made up of tiles as shown.

a. Express its area as the sum of smaller areas.

b. Express its area as length · width.

c. What equality is shown?

5.

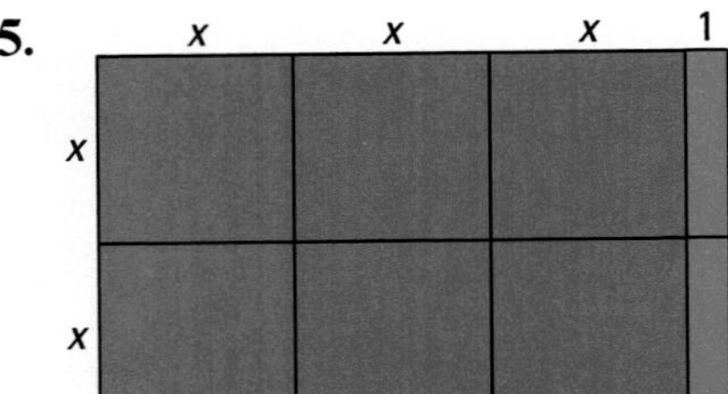

6.

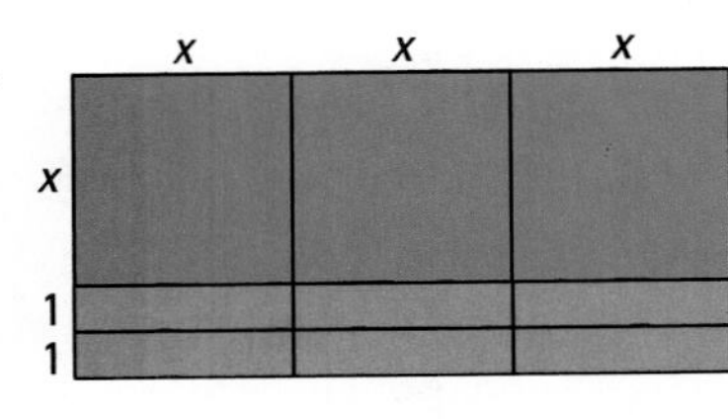

7. Suppose the height of the stores mentioned in this lesson is $2h$. Write two different expressions for the total area of the group of storefronts in the diagram.

8. Use the Distributive Property to multiply $a(b + c + d)$.

In 9–12, multiply.

9. $x(x^2 + x + 1)$

10. $5x^2(x^2 - 9x + 2)$

11. $3p(2 + p^2 + 5p^4)$

12. $-3wy(4y - 2w - 1)$

13. Use the multiplication of a monomial by a polynomial to explain why the product of 46329 and 10,000 is 463,290,000.

Applying the Mathematics

In 14 and 15, suppose the width of a rectangle is w cm. Write two equivalent expressions for the area if the rectangle's length is as described below.

14. Its length is 7 cm more than twice the width.

15. Its length is 1 cm less than five times the width.

Hopscotch. *Different versions of hopscotch are played the world over. Shown are children playing hopscotch in Soweto Township, South Africa.*

16. The formation of rectangles below is used by children in many countries for playing hopscotch. What is the total area?

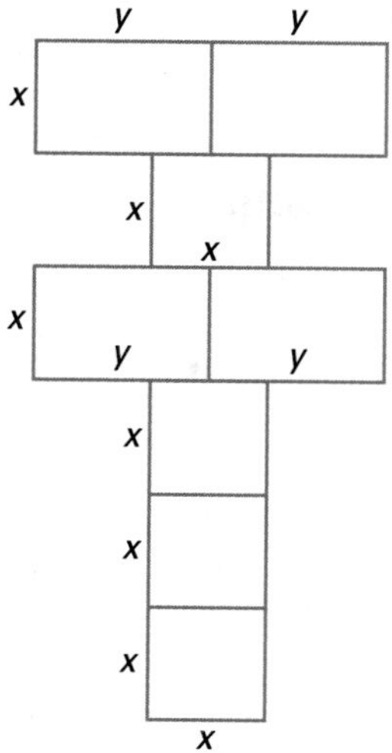

17. All angles in this figure are right angles. Find its area.

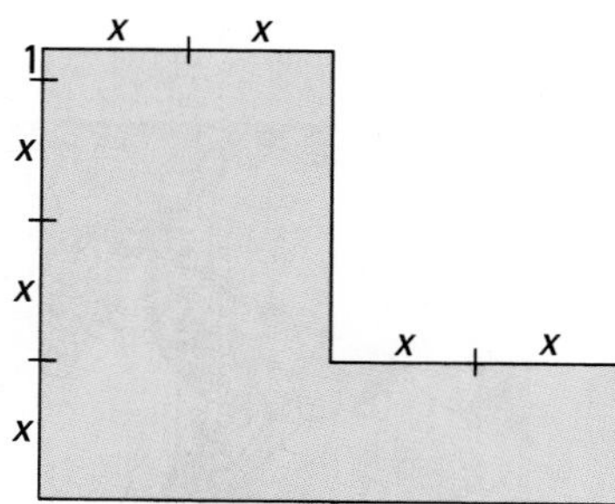

In 18–22, simplify.

18. $(5x)(2x) - (3x)(4x)$

19. $(a^2b)(2b^2c)(3ca)$

20. $2(x^2 + 3x) - 3x^2$

21. $a(y^2 - 2y) + y(a^2 + 2a)$

22. $m^3(m^2 - 3m + 2) - m^2(m^3 - 5m^2 - 6)$

23. Find the missing monomial:
$(\underline{\ ?\ }) \cdot (2n^2 + 5n - 6) = 12n^4 + 30n^3 - 36n^2$.

Review

24. After five years of birthdays, Wanda has received and saved

$$80x^4 + 60x^3 + 70x^2 + 45x + 50$$

dollars, having put the money in a savings account at a scale factor x.

a. How much did Wanda get on her most recent birthday?
b. How much did Wanda get on the first of these birthdays?
c. Give an example of a reasonable value for x in this problem.
d. If $x = 1$, how much has Wanda saved?
e. What does a value of 1 for x mean? *(Lesson 10-2)*

In 25 and 26, an expression is given. **a.** Tell whether the expression is a polynomial. **b.** If it is a polynomial, give its degree. If it is not a polynomial, tell why not. *(Lesson 10-1)*

25. $3x^2 + 2x^{-3}$

26. $a^2b + a^3b^2 + a^4b^3$

27. Write a trinomial with degree 7 and constant term 1. *(Lesson 10-1)*

In 28 and 29, simplify. *(Lessons 3-9, 8-7)*

28. $\frac{20x^2y}{2xy}$

29. $\frac{3}{2v} + \frac{4v}{2v}$

In 30 and 31, write an equation for the line with the given characteristics.

30. contains the points (-5, 8) and (-1, 16) *(Lesson 7-6)*

31. slope 3, y-intercept 5 *(Lesson 7-4)*

In 32–34, an object is described. **a.** Calculate its speed. **b.** Make a guess about what the object is. *(Lesson 6-2)*

32. It flew 200 miles in $\frac{1}{2}$ hour.

33. It slithered 14 meters in 7 seconds.

34. It took 3 days to creep 15 inches.

Hints. *Shown are some visual clues that may help you with Questions 32–34.*

Exploration

35. a. What is the rule for dividing a decimal by 1,000,000?

b. Make up an explanation like the one in this lesson to show why the rule works.

LESSON

10-4

Multiplying Polynomials

A Mondrian masterpiece. *Shown is a detail from the painting,* Composition with Red, Blue, Yellow, & Black *by the Dutch artist Piet Mondrian. The total area of the painting can be found in many ways.*

Using Area to Picture the Multiplication of Polynomials

The Area Model for Multiplication pictures how to multiply two polynomials with many terms. For instance, to multiply $a + b + c + d$ by $x + y + z$, draw a rectangle with length $a + b + c + d$ and width $x + y + z$.

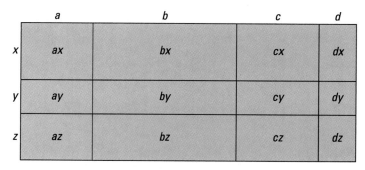

	a	b	c	d
x	ax	bx	cx	dx
y	ay	by	cy	dy
z	az	bz	cz	dz

The area of the largest rectangle equals the sum of the areas of the twelve separate smaller rectangles.

Total area $= ax + ay + az + bx + by + bz + cx + cy + cz + dx + dy + dz$

But the area of the largest rectangle also equals the product of its length and width.

Total area $= (a + b + c + d) \cdot (x + y + z)$

The Extended Distributive Property

The Distributive Property can be used to justify why the two expressions must be equal. Distribute the chunk $(x + y + z)$ over $(a + b + c + d)$ to get:

$$\begin{aligned}&(a + b + c + d) \cdot (x + y + z)\\ &= a(x + y + z) + b(x + y + z) + c(x + y + z) + d(x + y + z).\end{aligned}$$

Four more applications of the Distributive Property lead to the result found above.

$$= ax + ay + az + bx + by + bz + cx + cy + cz + dx + dy + dz$$

Because of the multiple use of the Distributive Property, we call this an instance of the Extended Distributive Property.

> **The Extended Distributive Property**
> **To multiply two sums, multiply each term in the first sum by each term in the second sum.**

If one polynomial has m terms and the second n terms, there will be mn terms in their product. This is due to the Multiplication Counting Principle. If some of these are like terms, you should simplify the product by combining like terms.

Example 1

Multiply $(5x^2 + 4x + 3)(x + 7)$.

Solution

Multiply each term in the first polynomial by each in the second. There will be six terms in the product before you simplify.

$$\begin{aligned}&(5x^2 + 4x + 3)(x + 7)\\ &= 5x^2 \cdot x + 5x^2 \cdot 7 + 4x \cdot x + 4x \cdot 7 + 3 \cdot x + 3 \cdot 7\\ &= 5x^3 + 35x^2 + 4x^2 + 28x + 3x + 21\end{aligned}$$

Now simplify by adding or subtracting like terms.

$$= 5x^3 + 39x^2 + 31x + 21$$

Check

Let $x = 2$. The check is left to you in the Questions.

After being simplified, the product of two polynomials can be a polynomial with fewer terms than one or both factors.

Example 2

Multiply $x^2 - 2x + 2$ by $x^2 + 2x + 2$.

Solution

Each term of $x^2 + 2x + 2$ must be multiplied by x^2, $-2x$, and 2. There will be nine terms before you simplify.

$$\begin{aligned} &(x^2 - 2x + 2)(x^2 + 2x + 2) \\ &= x^2(x^2 + 2x + 2) - 2x(x^2 + 2x + 2) + 2(x^2 + 2x + 2) \\ &= x^4 + 2x^3 + 2x^2 - 2x^3 - 4x^2 - 4x + 2x^2 + 4x + 4 \\ &= x^4 + 4 \quad \text{Combine like terms.} \end{aligned}$$

Check

Let $x = 10$. (Ten is a nice value to use in checks, because powers of 10 are so easily calculated.) Then $x^2 - 2x + 2 = 82$ and $x^2 + 2x + 2 = 122$. Now $82 \cdot 122 = 10{,}004$, which is the value of $x^4 + 4$ when $x = 10$.

In these long problems it is particularly important to be neat and precise. Be extra careful with problems involving negatives or several variables.

Example 3

Multiply $(3x + y - 1)(x - 5y + 8)$.

Solution

Each term of $x - 5y + 8$ must be multiplied by $3x$, y, and -1. At first there will be nine terms.

$$\begin{aligned} &(3x + y - 1)(x - 5y + 8) \\ &= 3x \cdot x + 3x \cdot -5y + 3x \cdot 8 + y \cdot x + y \cdot -5y + \\ &\quad y \cdot 8 + -1 \cdot x + -1 \cdot -5y + -1 \cdot 8 \end{aligned}$$

Watch the signs!

$$\begin{aligned} &= 3x^2 - 15xy + 24x + xy - 5y^2 + 8y - x + 5y - 8 \\ &= 3x^2 - 14xy + 23x - 5y^2 + 13y - 8 \end{aligned}$$

Check

A quick check of the coefficients can be found by letting all variables equal 1 in the last expression above.
Does $(3 + 1 - 1)(1 - 5 + 8) = 3 - 14 + 23 - 5 + 13 - 8$?
Does $3 \cdot 4 = 12$? Yes.
A better check requires using different values for both x and y.

QUESTIONS

Covering the Reading

1. **a.** What multiplication is pictured at the right?
   b. Do the multiplication.

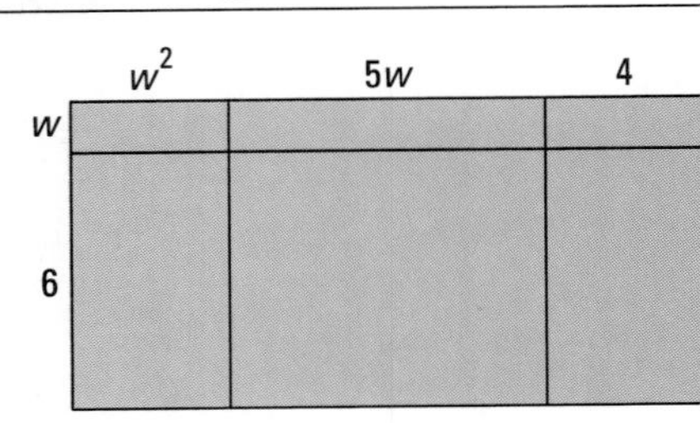

2. State the Extended Distributive Property.

3. Finish the check of Example 1.

In 4–7, multiply and simplify.

4. $(y^2 + 7y + 2)(y + 6)$
5. $(x + 1)(2x^2 + 3x - 1)$
6. $(m^2 + 10m + 3)(3m^2 - 4m - 2)$
7. $(x^2 + 4x + 8)(x^2 - 4x + 8)$

8. Check Example 3 by letting $x = 10$ and $y = 2$.

9. Find the area of the rectangle at the right, and simplify the result.

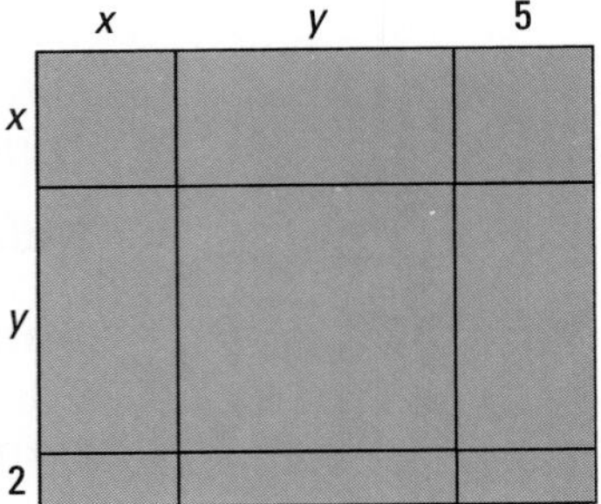

10. Multiply $(5c - 4d + 1)(c - 7d)$.

Applying the Mathematics

11. Multiply $(n - 3)(n + 4)(2n + 5)$ by first multiplying $n - 3$ by $n + 4$. Then multiply their product by $2n + 5$.

12. How much greater is the volume of a cube with sides of length $n + 1$ than the volume of a cube with sides of length n?

In 13–15, multiply and simplify.

13. $(x + 4)(x + 4) - (x - 16)^2$

14. $(a + b + c)(a + b - c) - (a + c)(a - c)$

15. $(m + 2n + 3p + 4q)(m - 2n - 3p - 4q)$

16. **a.** In Example 1, a 2nd degree polynomial is multiplied by a 1st degree polynomial. The product is a 3rd degree polynomial. Explain why this will always happen.
    b. If a polynomial of degree m is multiplied by a polynomial of degree n, what must be true about the degree of the product?

Review

17. Which investment is worth more at the end of 10 years if the annual yield is 6%? Justify your answer.
Plan A: Deposit $100 each year on January 2, beginning in 1990.
Plan B: Deposit $200 every other year on January 2, beginning in 1990. *(Lesson 10-2)*

18. Suppose x is either 7, 6, 5, 3, -2, or -8. Find x, given the clues.
Clue 1: $x > -3$.
Clue 2: x is not the degree of $a^3 + 4a^2 + 7$.
Clue 3: x is not the coefficient of b^2 in the polynomial $4b^4 + 6b^2 - 3b + 9$.
Clue 4: x is not the constant term in $2a^2 + a - 3b + c + 5$.
(Lesson 10-1)

19. a. Simplify $\sqrt{200}$.
b. Make up a question whose answer is $\sqrt{200}$. *(Lesson 9-7)*

20. *Skill sequence.* Graph on the same coordinate grid. *(Lessons 9-1, 9-2)*
a. $y = 3x^2$ **b.** $y = 3x^2 - 2$ **c.** $y = 3x^2 + 4$

In 21 and 22, simplify. *(Lesson 8-7)*

21. $\frac{14a^3b}{6ab^2}$ **22.** $\frac{-150m^5n^8}{100m^6n^3}$

23. Solve $\frac{y}{9} = 9$. *(Lesson 6-1)*

24. Subtract $3x^2 + 5$ from $2x^2 - 3x + 40$. *(Lesson 3-6)*

Exploration

25. Multiply each of the polynomials in parts a–d by $x + 1$.
a. $x - 1$
b. $x^2 - x + 1$
c. $x^3 - x^2 + x - 1$
d. $x^4 - x^3 + x^2 - x + 1$
e. Look for a pattern and use it to multiply $(x + 1)(x^8 - x^7 + x^6 - x^5 + x^4 - x^3 + x^2 - x + 1)$.
f. Predict what you think will be the product of $(x + 1)$ and $(x^{100} - x^{99} + x^{98} - x^{97} + \ldots + x^2 - x + 1)$ when simplified. Can you explain why your answer must be correct?

Introducing Lesson 10-5

Modeling Multiplication of Binomials with Algebra Tiles

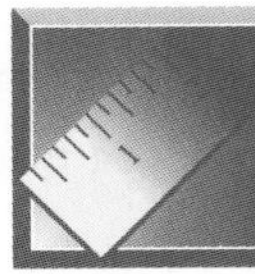

IN-CLASS ACTIVITY

Materials: Algebra tiles or ruler and graph paper

1 On a sheet of paper use tiles to form a rectangle with dimensions $2x + 1$ by $3x + 4$.

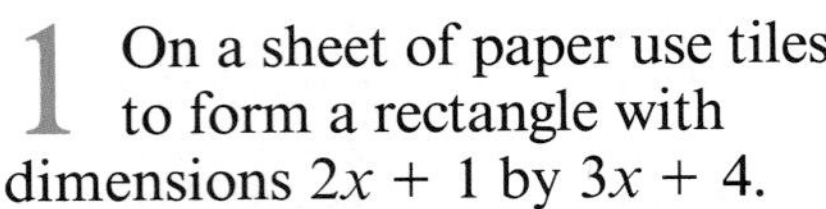

a. Give the area of the rectangle as a product of its length and width.

b. Give the area of the rectangle by writing the polynomial represented by the tiles.

c. Use your answers to parts **a** and **b** to complete the following equation:
$(2x + 1)(3x + 4) = \underline{\ ?\ }$

In 2–5, use algebra tiles to find each product.

2 $(x + 4)(x + 2)$

3 $(x + 3)(x + 2)$

4 $(2x + 1)(x + 5)$

5 $(3x + 1)(3x + 5)$

6 ***Draw conclusions.*** What patterns do you notice in the answers to Questions 1–5?

7 **a.** On a sheet of paper use tiles to form a square with a side of length $2x + 3$. Give the area of the square using the formula $A = s^2$.

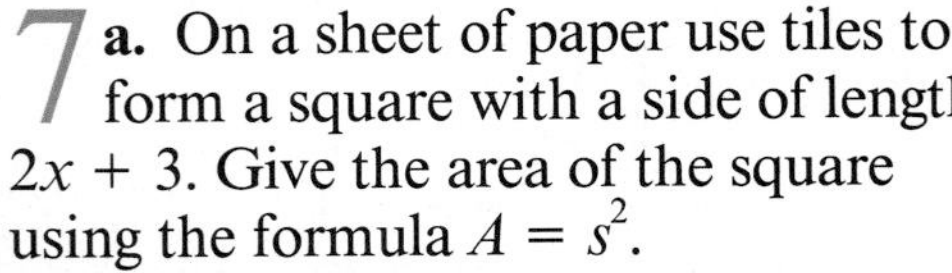

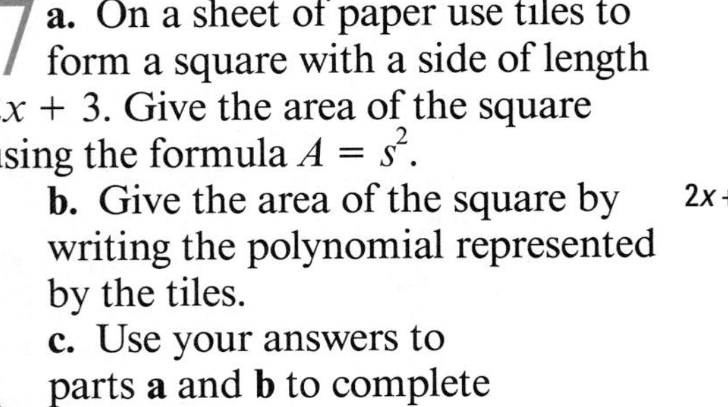

b. Give the area of the square by writing the polynomial represented by the tiles.

c. Use your answers to parts **a** and **b** to complete the following:
$(2x + 3)^2 = \underline{\ ?\ }$.

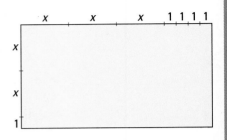

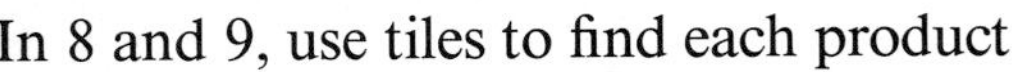

In 8 and 9, use tiles to find each product.

8 $(x + 4)^2$

9 $(3x + 1)^2$

LESSON

10-5

Multiplying Binomials

Painstaking calculation. *To find the total area, the amount of glass required for this stained-glass window, you could use a product of binomials. This kind of calculation is done in Example 3.*

Multiplying Two-Digit Numbers

Every multiplication with two two-digit numbers can be considered as the multiplication of two binomials. Consider 94 times 65. At the left is the way many people do it by hand. At the right the numbers have been rewritten as (90 + 4) and (60 + 5), and the multiplication has been done using the Extended Distributive Property.

Familiar Algorithm:

$$\begin{array}{r} 65 \\ \underline{\times\ 94} \\ 260 \\ \underline{585} \\ 6110 \end{array}$$

Extended Distributive Property:

$$\begin{aligned} &(90 + 4) \cdot (60 + 5) \\ &= 90 \cdot 60 + 90 \cdot 5 + 4 \cdot 60 + 4 \cdot 5 \\ &= 5400 + 450 + 240 + 20 \\ &= 6110 \end{aligned}$$

The Extended Distributive Property explains why the familiar algorithm works. The 260 at left is the sum of $4 \cdot 60$ and $4 \cdot 5$. The 585 is really 5850 with the 0 understood; it is the sum of $90 \cdot 5$ and $90 \cdot 60$.

Using Area to Picture the Multiplication of Two Binomials

The dimensions of the rectangle at the right are the binomials $(a + b)$ and $(c + d)$.

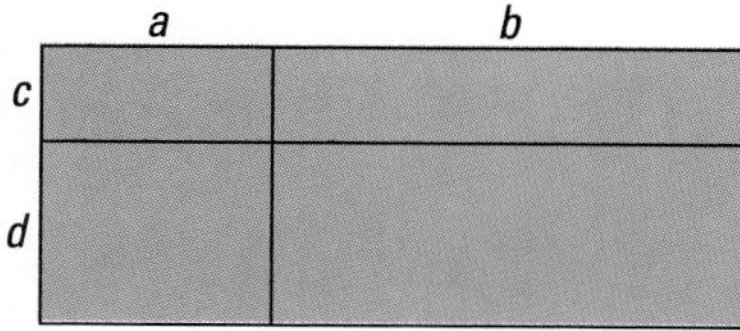

Its area is $(a + b) \cdot (c + d)$. But the area of the rectangle also must equal the sum of the areas of the four small rectangles inside it.

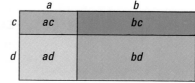

	a	b
c	ac	bc
d	ad	bd

The sum of the areas of the four small rectangles = $ac + ad + bc + bd$. So,

$$(a + b) \cdot (c + d) = ac + ad + bc + bd.$$

The FOIL Algorithm

Another way to show the pattern above is true is to think of $(c + d)$ as a chunk and to distribute it over $(a + b)$ as follows:

$$(a + b) \cdot (\mathbf{c + d}) = a(\mathbf{c + d}) + b(\mathbf{c + d}).$$

Now apply the Distributive Property twice more.

$$= ac + ad + bc + bd$$

Notice that the product of two binomials has four terms. To find them, multiply each term in the first binomial by each term in the second binomial. The face below may help you remember.

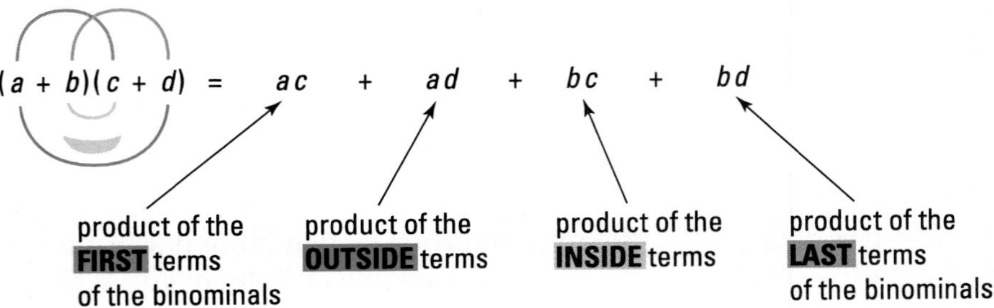

Because the product is composed of the First, Outside, Inside, and Last terms, the above pattern is sometimes called the **FOIL algorithm.** (Recall that an **algorithm** is a step-by-step recipe or procedure.)

Example 1

Multiply $(m + 4)(m + 3)$.

Solution 1

$$\begin{aligned}(m + 4)(m + 3) &= \overset{F}{m \cdot m} + \overset{O}{m \cdot 3} + \overset{I}{4 \cdot m} + \overset{L}{4 \cdot 3}\\ &= m^2 + 3m + 4m + 12\\ &= m^2 + 7m + 12\end{aligned}$$

Solution 2

Draw a rectangle with sides of $m + 4$ and $m + 3$. Find its area by dividing it into smaller rectangles. The areas of the smaller rectangles are m^2, $4m$, $3m$, and 12. So

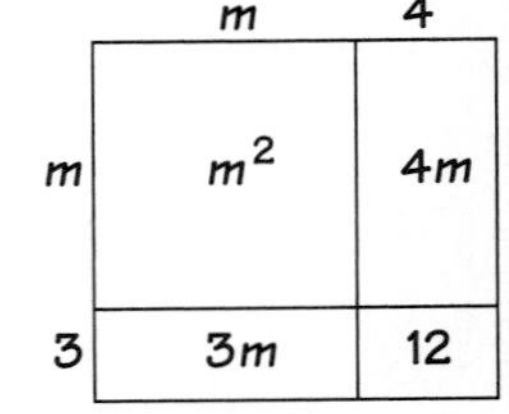

$$\begin{aligned}(m + 4)(m + 3) &= m^2 + 4m + 3m + 12\\ &= m^2 + 7m + 12.\end{aligned}$$

Check

Test a special case. Let $m = 10$. Then $m + 4 = 14$, $m + 3 = 13$, and $m^2 + 7m + 12 = 100 + 70 + 12 = 182$. It checks, because $14 \cdot 13 = 182$.

The FOIL algorithm applies to all binomials, including those with subtraction. You must, however, be careful with signs.

Example 2

Multiply $(x - 3)(x - 7)$.

Solution

Think of $x - 3$ as $x + -3$, and $x - 7$ as $x + -7$. Now use the FOIL algorithm.

$$(x + -3)(x + -7) = x^2 + -7x + -3x + 21$$

$$= x^2 + -10x + 21$$

$$\text{So } (x - 3)(x - 7) = x^2 - 10x + 21.$$

Check 1

Substitute a value for x in the original and final expression. We use $x = 5$.
$(x - 3)(x - 7) = (5 - 3)(5 - 7) = 2(-2) = -4$
$x^2 - 10x + 21 = 5^2 - 10 \cdot 5 + 21 = 25 - 50 + 21 = -4$.

It checks.

Check 2

If $(x - 3)(x - 7) = x^2 - 10x + 21$, then the graphs of $y = (x - 3)(x - 7)$ and $y = x^2 - 10x + 21$ should be identical. Below we show the output from our automatic grapher.

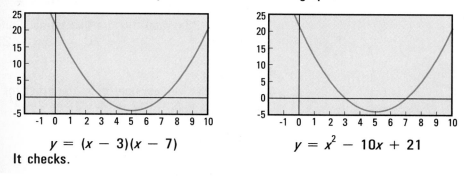

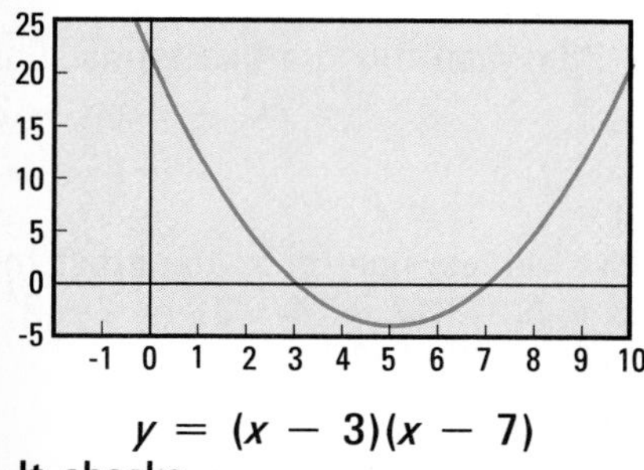

$y = (x - 3)(x - 7)$

$y = x^2 - 10x + 21$

It checks.

The FOIL algorithm applies to all real numbers *a, b, c,* and *d.* You may substitute any algebraic expression for the numbers.

Example 3

A 5″ by 7″ photo is to be surrounded on all sides by a mat of width w inches. Find an expression for the combined area of the photo and mat.

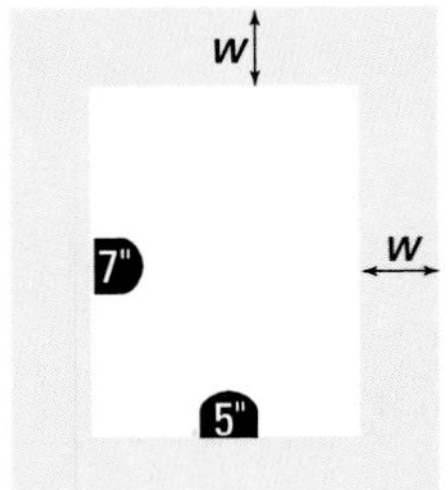

Solution

The region to be framed is a rectangle with base $5 + 2w$ and height $7 + 2w$. The area is $(5 + 2w)(7 + 2w)$. Think of $2w$ as a chunk and apply the FOIL algorithm.

$$\begin{aligned}(5 + 2w)(7 + 2w) &= 5 \cdot 7 + 5 \cdot 2w + 2w \cdot 7 + 2w \cdot 2w \\ &= 35 + 10w + 14w + 4w^2 \\ &= 35 + 24w + 4w^2\end{aligned}$$

Check

You are asked to check this result in the Questions.

Example 4 shows the FOIL algorithm in a situation where there is more than one variable in each binomial.

Example 4

Multiply $(7x + 5y)(x - 4y)$.

Solution

$$\begin{aligned}(7x + 5y)(x - 4y) &= 7x \cdot x + 7x \cdot -4y + 5y \cdot x + 5y \cdot -4y \\ &= 7x^2 - 28xy + 5yx - 20y^2\end{aligned}$$

Notice that $-28xy$ and $5yx$ are like terms. Combine them.

$$= 7x^2 - 23xy - 20y^2$$

Check

Substitute two values, one for x, the other for y. We let $x = 2$ and $y = 3$.
Does $(7 \cdot 2 + 5 \cdot 3)(2 - 4 \cdot 3) = 7 \cdot 2^2 - 23 \cdot 2 \cdot 3 - 20 \cdot 3^2$?
Does $29 \cdot -10 = 28 - 138 - 180$?

Yes, each side equals -290.

The product of two binomials generally has four terms, as in FOIL. After simplification, the products in Examples 1–4 have three terms each. Sometimes, after simplification, there are only two terms.

Example 5

Multiply $4x - 3$ by $4x + 3$.

Solution

$$\begin{aligned}(4x - 3)(4x + 3) &= 16x^2 + 12x - 12x - 9 \\ &= 16x^2 - 9\end{aligned}$$

QUESTIONS

Covering the Reading

1. Refer to the largest rectangle below.
   a. What are its dimensions? **b.** What is its area?

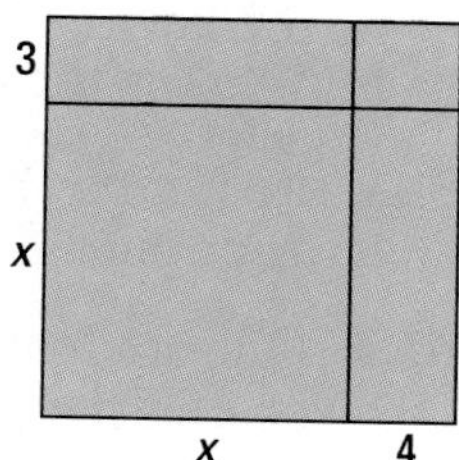

2. **a.** What multiplication is pictured below?
   b. Do the multiplication.

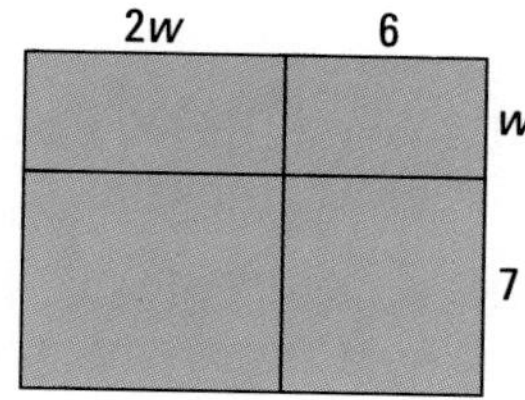

3. In the FOIL algorithm, what do the letters F, O, I, and L stand for?

In 4 and 5, **a.** multiply; **b.** draw a picture to justify your answer.

4. $(a + b)(c + d)$
5. $(n + 4)(n + 1)$

6. Refer to Example 3. Check the answer by letting $w = 2$.

In 7–12, multiply and simplify.

7. $(x - 3)(x - 4)$
8. $(a - 10)(a + 7)$
9. $(3m + 2n)(7m - 6n)$
10. $(k + 3)(9k + 8)$
11. $(a + 6)(a - 6)$
12. $(2x - 3y)(2x + 3y)$
13. A 20-ft by 40-ft pool is surrounded by a deck d ft wide.
    a. What are the outer dimensions of the deck?
    b. What is the area of the deck?

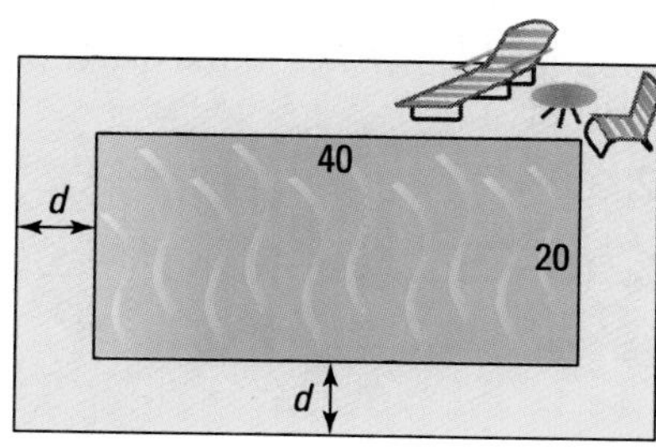

Shown are members of an organization called S.A.F.E. (St. Agatha Family Empowerment), which offers teens community-service jobs. These teens are beautifying a neighborhood garden in Chicago.

Applying the Mathematics

14. A rectangular public garden measuring x-ft by y-ft has a sidewalk surrounding the lawn. The sidewalk is 6 ft wide.

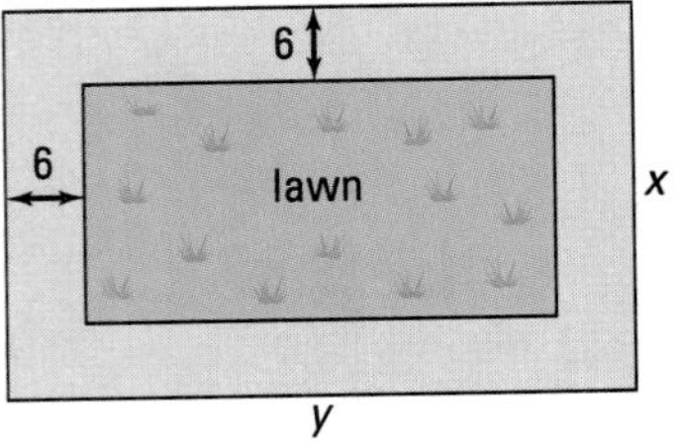

a. What are the dimensions of the lawn?
b. What is the area of the lawn?

15. Write $(3x + 4)^2$ as a polynomial. (Hint: First convert the 2nd power to multiplication.)

16. a. One student is selected from the freshman or sophomore class as co-chairperson for the school dance committee. A second student is selected from the junior or senior class as co-chairperson. How many such twosomes are possible in a school with 550 freshmen, 500 sophomores, 450 juniors, and 400 seniors?
b. Repeat part **a** if there are f freshmen, p sophomores, j juniors, and s seniors.

17. Write the area of the square at the right as a trinomial.

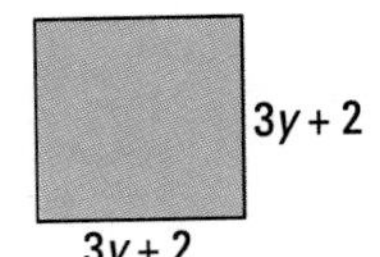

18. a. Multiply $(x^2 + 2)(x^3 - 1)$.
b. Check your answer by letting $x = 4$.

19. a. Below is a table of some solutions to the equation $y = (x + 2)(x - 3)$. Complete the table, and then plot the points.

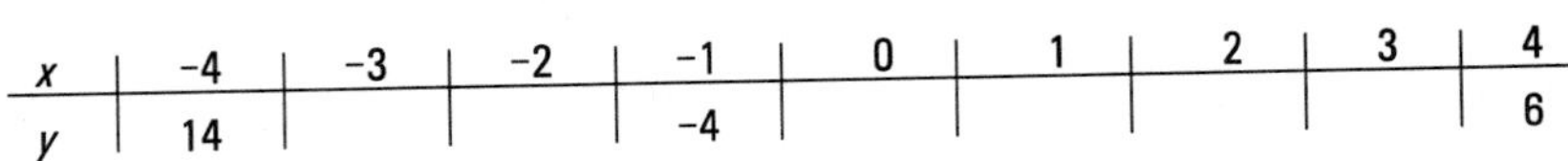

x	-4	-3	-2	-1	0	1	2	3	4
y	14			-4					6

b. Make a table of solutions to the equation $y = x^2 - x - 6$ for integer values of x from -4 to 4. Plot the points.
c. What is the relation between the tables and graphs in parts **a** and **b**? Use the content of this lesson to explain that relation.

20. a. Multiply $(\sqrt{5} - 2)(\sqrt{5} + 3)$ using the FOIL algorithm.
b. Use your calculator to evaluate the expression in part **a.**
c. Are your answers in parts **a** and **b** equal? How can you tell?

21. A rectangular box has dimensions x, $x + 1$, and $x + 2$. What is its volume?

In 22 and 23, work backwards. Fill in the blanks with the missing number.

22. $(x + 5)(x + \underline{\ ?\ }) = x^2 + \underline{\quad} x + 35$

23. $(y - \underline{\ ?\ })(y + 5) = y^2 + \underline{\quad} y - 10$

Review

24. Bill has 3 blue shirts, 2 red shirts, and 5 green shirts. He can wear them with 4 pairs of blue jeans and 2 pairs of shorts. **a.** How many outfits are possible? **b.** How is this related to the multiplication of polynomials? *(Lesson 10-4)*

25. If $3(a + b - c + d) + 2(a + b - c + d) = 5x$, solve for x. *(Lesson 10-3)*

26. Multiply $3 \cdot 10^5 + 6 \cdot 10^4 + 2 \cdot 10^3 + 10^1 + 9$ by $8 \cdot 10^3 + 9 \cdot 10^2 + 8 \cdot 10 + 1$. *(Lessons 10-1, 10-3)*

3 . . . 2 . . . 1 . . . liftoff! *The flight of a model rocket consists of several phases. Initially, the flight path is straight up, powered by a solid fuel. Next, it begins to arch as it coasts without fuel. Lastly, a parachute is released as the rocket descends to the ground.*

In 27 and 28, use the formula for the height h (in feet) of a model rocket t seconds after being fired straight up from a pad 6 feet off the ground, $h = 6 + 96t - 16t^2$.

27. a. Find the height of the rocket after 7 seconds.
b. What does this answer mean? *(Lesson 9-4)*

28. The rocket reaches its maximum height at 3 seconds. What is its maximum height? *(Lesson 9-2)*

29. Write an equation for the line which passes through the points (-3, 7) and (-9, 4). *(Lesson 7-6)*

30. Find the probability that a point selected at random from the big square at the right will be in the shaded area. *(Lesson 6-6)*

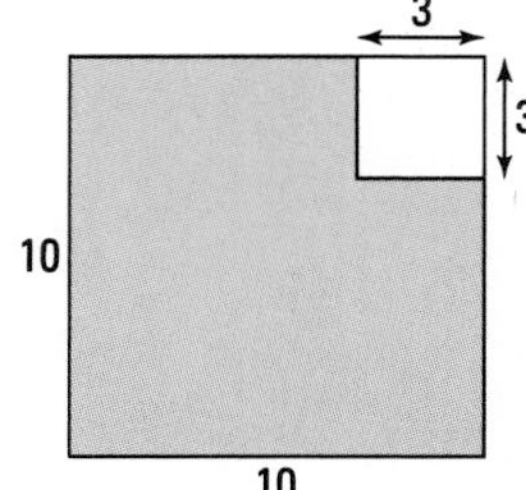

31. Solve $42(512 - x) = 0$ in your head. *(Lesson 5-8)*

32. Evaluate $(x + 1)(x - 3)(x + 4)$ in your head when $x = 3$. *(Lesson 2-2)*

Exploration

33. The largest rectangular solid below has length $(a + b)$, width $(c + d)$, and height $(e + f)$. Give at least two ways of computing the volume of the box.

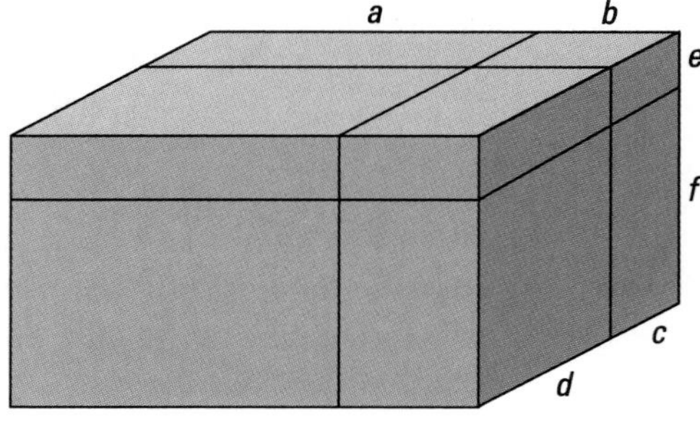

LESSON

10-6

Special Binomial Products

Some binomial products are used so frequently that they are given special names. In this lesson you will study two of them: Perfect Squares and the Difference of Two Squares.

The Square of a Sum

The expression $(a + b)^2$ (read "a plus b, the quantity squared") is the square of the binomial $(a + b)$. This product can be *expanded* by writing it as the product $(a + b)(a + b)$ and then using the Distributive Property or its special case, the FOIL algorithm.

$$(a + b)(a + b) = a^2 + ab + ba + b^2$$
$$= a^2 + 2ab + b^2 \text{ (}ab\text{ and }ba\text{ are like terms.)}$$

Geometrically, $(a + b)^2$ can be interpreted as the area of a square with side $a + b$. As the figure at the right shows, its area is $a^2 + 2ab + b^2$.

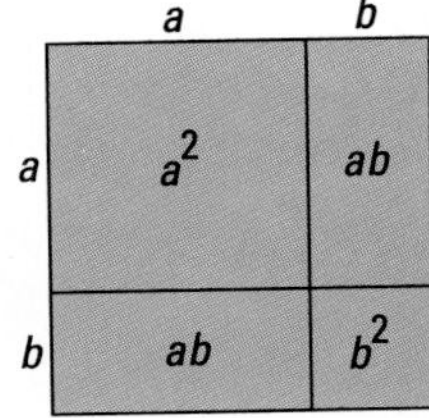

Example 1

The area of a square with side $5n + 4$ is $(5n + 4)^2$. Expand this binomial.

Solution 1

Rewrite the square as a multiplication and apply the FOIL algorithm.

$$(5n + 4)^2 = (5n + 4)(5n + 4)$$
$$= 25n^2 + 20n + 20n + 16$$
$$= 25n^2 + 40n + 16$$

Solution 2

Use the pattern for $(a + b)^2$ with $a = 5n$ and $b = 4$.

$$(a + b)^2 = a^2 + 2ab + b^2$$
$$(5n + 4)^2 = (5n)^2 + 2 \cdot 5n \cdot 4 + 4^2$$
$$= 25n^2 + 40n + 16$$

Solution 3

Draw a square with side $5n + 4$. Subdivide it into smaller rectangles and find the sum of their areas.

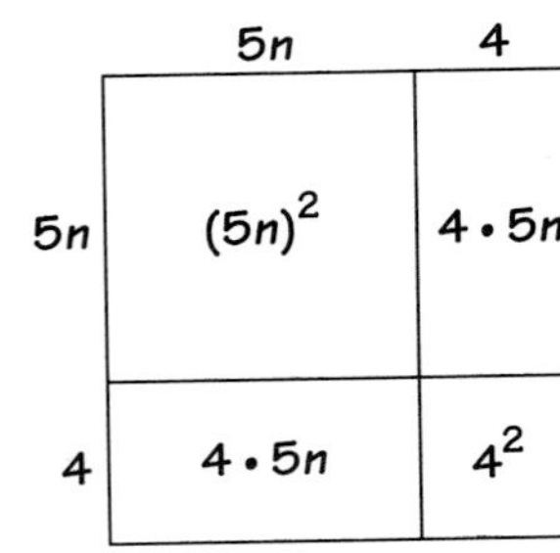

$$\text{Area} = (5n + 4)^2$$
$$= (5n)^2 + 2(4 \cdot 5n) + 4^2$$
$$= 25n^2 + 40n + 16$$

The Square of a Difference

To square the difference $(a - b)^2$, think of $a - b$ as $a + -b$.

$$\begin{aligned}(a - b)^2 &= (a - b)(a - b)\\ &= (a + -b)(a + -b)\\ &= a^2 - ab - ba + b^2\\ &= a^2 - 2ab + b^2\end{aligned}$$

Notice that after simplifying, the square of a binomial has three terms. It is a trinomial. Trinomials of the form $a^2 + 2ab + b^2$ or $a^2 - 2ab + b^2$ are called **perfect square trinomials** because each is the result of squaring a binomial.

Perfect Square Patterns
For all numbers a and b,

$$\begin{aligned}(a + b)^2 &= a^2 + 2ab + b^2,\\ (a - b)^2 &= a^2 - 2ab + b^2.\end{aligned}$$

You need to remember these patterns, and you should realize that they can be derived from the multiplication of binomials. The algebraic description is short, but many people remember these patterns in words. The square of a binomial is:

the square of its first term,
plus twice the product of its terms,
plus the square of its last term.

Example 2

Expand $(w - 3)^2$.

Solution 1

Follow the Perfect Square Pattern for $(a - b)^2$. Here $a = w$ and $b = 3$.

$$\begin{aligned}(a - b)^2 &= a^2 - 2 \cdot a \cdot b + b^2\\ (w - 3)^2 &= w^2 - 2 \cdot w \cdot 3 + 3^2\\ &= w^2 - 6w + 9\end{aligned}$$

Solution 2

Change the square to the multiplication of $w - 3$ by itself.

$$\begin{aligned}(w - 3)^2 &= (w - 3)(w - 3)\\ &= w^2 - 3w - 3w + 9\\ &= w^2 - 6w + 9\end{aligned}$$

Check

Test a special case. Let $w = 5$. Then $(w - 3)^2 = (5 - 3)^2 = 2^2 = 4$, and $w^2 - 6w + 9 = 5^2 - 6 \cdot 5 + 9 = 25 - 30 + 9 = 4$. It checks.

How Do We Know the Pythagorean Theorem Is True?

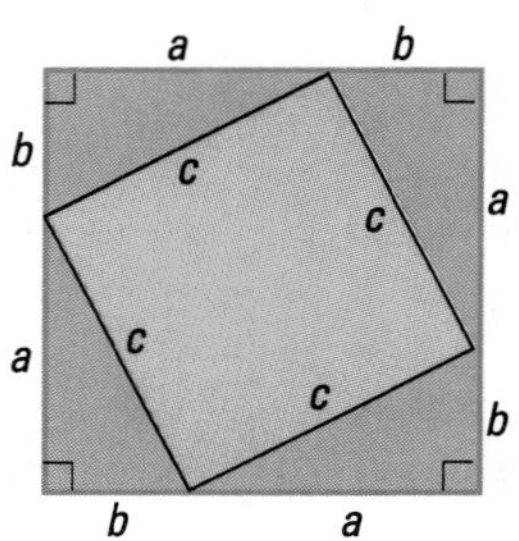

The Perfect Square Patterns help explain why the Pythagorean Theorem is true. Consider four right triangles with legs a and b and hypotenuse c placed as shown at the left. The triangles form an outer square of side $a + b$ and an inner tilted square of side c. Each triangle has area $\frac{1}{2}ab$. How is the area of the tilted square related to a and b?

The tilted square has side c. Notice that its area equals the area of the outer square with side $a + b$ minus the areas of the four right triangles.

$$c^2 = (a + b)^2 - 4\left(\frac{1}{2}ab\right)$$

Now use the Perfect Square Pattern to expand the binomial.

$$c^2 = a^2 + 2ab + b^2 - 2ab$$
$$c^2 = a^2 + b^2$$

This final equation is the Pythagorean Theorem.

The Product of the Sum and the Difference of the Same Numbers

Another interesting binomial pattern involves sums and differences of numbers. Consider two numbers a and b. Multiply their sum $a + b$ by their difference $a - b$.

$$(a + b)(a - b) = a^2 + ab - ab - b^2$$

The two middle terms are opposites, so their sum is 0.

$$= a^2 - b^2$$

In words, the product of the sum and difference of two numbers is the **difference of squares** of the two numbers.

Difference of Two Squares Pattern
For all numbers a and b,
$$(a + b)(a - b) = a^2 - b^2.$$

Example 3

Multiply $(3y + 7)(3y - 7)$.

Solution

The binomial factors are the sum and difference of the same terms. Use the Difference of Two Squares Pattern and chunking with $a = 3y$ and $b = 7$.

$$(3y + 7)(3y - 7) = (3y)^2 - 7^2$$
$$= 9y^2 - 49$$

Check

Test a special case. Let $y = 4$.

$$(3 \cdot 4 + 7)(3 \cdot 4 - 7) = 19 \cdot 5 = 95$$

Does $9 \cdot 4^2 - 49 = 95$? Yes.

The Perfect Square and Difference of Two Squares Patterns can be applied to do mental arithmetic.

Example 4

Compute in your head: **a.** 79^2; **b.** $51 \cdot 49$.

Solution

a. Use the Perfect Square Pattern. Think of 79 as 80 − 1.

$$\begin{aligned} 79^2 &= (80 - 1)^2 \\ &= 80^2 - 2 \cdot 80 \cdot 1 + 1^2 \\ &= 6400 - 160 + 1 = 6241 \end{aligned}$$

b. Think of 51 as 50 + 1, and 49 as 50 − 1. Use the Difference of Squares Pattern.

$$\begin{aligned} 51 \cdot 49 &= (50 + 1)(50 - 1) \\ &= 50^2 - 1^2 \\ &= 2500 - 1 = 2499 \end{aligned}$$

QUESTIONS

Covering the Reading

In 1 and 2, expand the binomial.

1. $(x + y)^2$ **2.** $(x - y)^2$

3. What is a *perfect square trinomial?*

4. a. What is the area of a square with side of length $n + 3$?
b. Justify your answer in part **a** by drawing a picture.
c. Check your answer by letting $n = 2$.

5. Write an expression for the area of a square with side $3x + 7$,
a. as the square of a binomial;
b. as a perfect square trinomial.

In 6–9, expand the binomial.

6. $(m - 6)^2$ **7.** $(m + 12)^2$ **8.** $(2x + 5)^2$ **9.** $(3y - 4)^2$

10. Fill in the blank: $a^2 - b^2 = (a - b) \cdot$ _?_.

In 11 and 12, multiply and simplify.

11. $(x + 13)(x - 13)$ **12.** $(3 + 8p)(3 - 8p)$

13. *Multiple choice.* Which is *not* the difference of two squares?
(a) $9 - w^2$ (b) $(x - y)^2$
(c) $x^2y^2 - 1$ (d) $121m^2 - n^2$

In 14–17, compute in your head using the methods in Example 4.

14. $(41)^2$ **15.** $(37)^2$ **16.** $69 \cdot 71$ **17.** $85 \cdot 95$

Applying the Mathematics

18. By substituting numbers for a and b, verify that $(a + b)^2 = a^2 + 2ab + b^2$.

19. By substituting numbers for a and b, verify that $(a + b)^2 \neq a^2 + b^2$.

20. Does $(x - 3)^2$ equal $(3 - x)^2$ for all values of x? Explain your reasoning.

In 21–24, write the expression as a simplified polynomial.

21. $4(9 + y)^2$

22. $(x + y)^2 - (x - y)^2$

23. $(x + \sqrt{11})(x - \sqrt{11})$

24. $(\sqrt{7} - \sqrt{2})(\sqrt{7} + \sqrt{2})$

25. A wedding cake is made from two layers, each in the shape of a rectangular solid. The bottom layer has dimensions $s + 5$ by $s + 5$ by s. The top layer has dimensions s by s by $s - 2$.

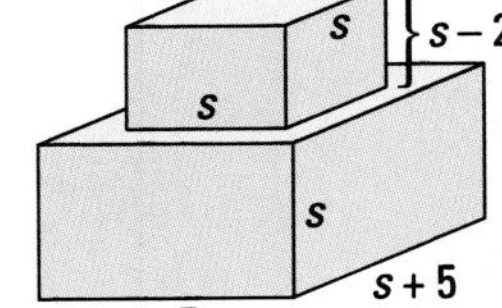

a. What is the volume of the bottom layer?
b. What is the volume of the top layer?
c. Find the volume of the cake.
d. Check your answer by letting $s = 4$.

Review

In 26 and 27, multiply and simplify. *(Lesson 10-5)*

26. $(z - 11)(z + 8)$

27. $(2c - 7d)(c + 5d)$

28. *Skill sequence.* Multiply. *(Lessons 10-3, 10-4, 10-5)*
a. $x(x + 3)$ **b.** $(x + 4)(x + 3)$ **c.** $(x + 4 - y)(x + 3)$

29. Solve. **a.** $x = 2x$ **b.** $x > 2x$ *(Lessons 6-6, 6-7)*

30. In 1992, the total sales of General Electric increased 3.3% to \$62.2 billion. What were the company's sales in 1991? *(Lessons 5-4, 6-3)*

31. Four consecutive integers (such as 7, 8, 9, and 10) can be represented by the expressions n, $n + 1$, $n + 2$, $n + 3$. The sum of four consecutive integers is 250. What are the integers? *(Lessons 3-7, 3-8)*

32. Simplify $(x^3 - 4x + 1) - (5x^3 + 4x - 8)$. *(Lesson 3-6)*

Exploration

33. a. Write three consecutive integers.
Square the second number.
Find the product of the first and the third.
b. Repeat part **a** with 3 more sets of consecutive integers.
c. What do you notice?
d. Explain why this will always happen. (Hint: Let n equal the second number.)

LESSON

10-7

The Chi-Square Statistic

Safety first! *Steelworkers follow safety guidelines to prevent accidents.*

In this lesson, we consider a statistic different from any you have yet seen. This statistic compares actual frequencies with the frequencies that would be expected by calculating probabilities. It is one of the most often used of all statistics. Here is a typical use, though the data are made up.

In a large factory, over a two-year interval, there were 180 accidents. Someone noticed that more accidents seemed to occur on Mondays. So the accidents were tabulated by the day of the week. Below is the table. Does Monday seem to be special?

Day of the week	M	Tu	W	Th	F
Actual numbers of accidents	55	38	23	22	42

The expected number of accidents is the mean number of accidents for a given day that is predicted by a probability. If accidents occurred randomly, then the expected number for each day would be the same. So the **expected number** of accidents for each day is $\frac{180}{5}$, or 36.

Day of the week	M	Tu	W	Th	F
Expected numbers of accidents	36	36	36	36	36

Recall, however, that even if events occur randomly, it is not common for all events to occur with the same frequency. When you toss a coin 10 times, you would not usually get exactly 5 heads even if the coin were fair. Similarly, if there were 35 accidents on each of 4 days and 39 on the 5th day, that would not seem to be much of a variance from the expected numbers. So we ask: Do the actual numbers deviate so much from the expected numbers that we should think the accidents are not happening randomly? Or, are the numbers so close that the differences are probably due to chance?

If we let an expected number be e and an actual observed number be a, then $|e - a|$, the absolute value of the difference between these numbers, is called the deviation of a from e. For instance, the deviation of the actual from the expected number of accidents on Monday is $|36 - 55|$, or 19.

In 1900, the English statistician Karl Pearson developed a method of determining whether the differences in two frequency distributions is greater than that expected by chance. This method uses a number called the chi-square statistic. ("Chi" is pronounced "ky" as in "sky.") The algorithm for calculating this statistic uses the squares of deviations, which is why we study it here.

Step 1: Count the number of events. Call this number n.
In the above situation, there are 5 events, one each for M, Tu, W, Th, and F.

Step 2: Let a_1, a_2, a_3, a_4, and a_5 be the actual frequencies.
In our case, $a_1 = 55$, $a_2 = 38$, $a_3 = 23$, $a_4 = 22$, and $a_5 = 42$.

Step 3: Let e_1, e_2, e_3, e_4, and e_5 be the expected frequencies.
Here $e_1 = 36$, $e_2 = 36$, $e_3 = 36$, $e_4 = 36$, and $e_5 = 36$.

Step 4: Calculate $\frac{(a_1 - e_1)^2}{e_1}$, $\frac{(a_2 - e_2)^2}{e_2}$, $\frac{(a_3 - e_3)^2}{e_3}$, and so on. Each number is the square of the deviation, divided by the expected frequency.

$$\frac{(a_1 - e_1)^2}{e_1} = \frac{(55 - 36)^2}{36} = \frac{361}{36} \qquad \frac{(a_2 - e_2)^2}{e_2} = \frac{(38 - 36)^2}{36} = \frac{4}{36}$$

$$\frac{(a_3 - e_3)^2}{e_3} = \frac{(23 - 36)^2}{36} = \frac{169}{36} \qquad \frac{(a_4 - e_4)^2}{e_4} = \frac{(22 - 36)^2}{36} = \frac{196}{36}$$

$$\frac{(a_5 - e_5)^2}{e_5} = \frac{(42 - 36)^2}{36} = \frac{36}{36}$$

Step 5: Add the n numbers found in Step 4.
This sum is the chi-square statistic.

$$\frac{361}{36} + \frac{4}{36} + \frac{169}{36} + \frac{196}{36} + \frac{36}{36} = \frac{766}{36} \approx 21.3$$

The chi-square statistic measures how different a set of actual observed scores is from a set of expected scores. The larger the differences are, the greater the chi-square statistic. But is 21.3 unusually large? You can find that out by looking in a chi-square table such as the one shown on page 653. That table gives the values for certain values of n and certain probabilities. In the table, n is the number of events. The other columns of the table correspond to probabilities of .10 (an event expected to happen $\frac{1}{10}$ of the time), .05 (or $\frac{1}{20}$ of the time), .01 (or $\frac{1}{100}$ of the time), and .001 (or $\frac{1}{1000}$ of the time). You are not expected to know how the values in the table were calculated. The mathematics needed to calculate them is beyond that normally studied before college.

How to Read This Table

The left column, titled $n - 1$, is one less than the number of events. This is because once $n - 1$ observed frequencies are known, the last frequency can be calculated by subtracting from the total of expected frequencies. For instance, once the numbers of accidents are known for Monday through Thursday, you could subtract from 180 to find the frequency for Friday. The number $n - 1$ is known as the number of **degrees of freedom** for this statistic.

The other columns give the probabilities that chi-square values as large as these will occur. For instance, examine the number 14.1, which appears in column .05, row 7. This means that, with 8 events, a chi-square value greater than 14.1 occurs with probability .05 or less.

Critical Chi-Square Values

$n - 1$	.10	.05	.01	.001
1	2.71	3.84	6.63	10.8
2	4.61	5.99	9.21	13.8
3	6.25	7.81	11.34	16.3
4	7.78	9.49	13.28	18.5
5	9.24	11.07	15.09	20.5
6	10.6	12.6	16.8	22.5
7	12.0	14.1	18.5	24.3
8	13.4	15.5	20.1	26.1
9	14.7	16.9	21.7	27.9
10	16.0	18.3	23.2	29.6
11	17.3	19.7	24.7	31.3
12	18.6	21.0	26.2	32.9
13	19.8	22.4	27.7	34.5
14	21.1	23.7	29.1	36.1
15	22.3	25.0	30.6	37.7
16	23.5	26.3	32.0	39.3
17	24.8	27.6	33.4	40.8
18	26.0	28.9	34.8	42.3
19	27.2	30.1	36.2	43.8
20	28.4	31.4	37.6	45.3
25	34.4	37.7	44.3	52.6
30	40.3	43.8	50.9	59.7
50	63.2	67.5	76.2	86.7

With the data on factory accidents, we obtained a chi-square value of 21.3 with $n = 5$ events. So we look in row $n - 1$, which is row 4. A value as large as 18.5 (the largest value in row 4) would occur with probability less than .001. Thus, a value of 21.3 is even less likely. Since .001 is a very small probability, we have evidence that the accidents are not evenly distributed among the days of the week. The factory should try to determine why there are more accidents on Monday. Perhaps it is because people come back tired from a weekend.

Suppose the frequencies of the accidents had led to a chi-square value of 8.62. Then, looking across row 4, we would see that this value is between the values 7.78 and 9.49. So 8.62 has a probability between .10 and .05.

That means that a chi-square value as high as 8.62 would occur between $\frac{1}{10}$ and $\frac{1}{20}$ of the time. Statisticians normally do not consider this probability to be low enough to think there is reason to question the expected values.

The cutoff value is usually taken as the value with probability .05 or .01. In this book, we use the probability .05. When a chi-square value as large as the one found would occur with probability less than .05, we then question the assumptions leading to the expected values. This is the case with the chi-square value for the accidents. Because a value as large as 21.3 would occur with probability less than .05, we suspect that the accidents are not occurring randomly among the days.

Example

Sixty people were asked to name the U.S. President in 1850 from the names below.

14 picked Millard Fillmore.
25 picked Abraham Lincoln.
21 picked Martin Van Buren.

Is there evidence to believe the people were just guessing?

Solution

Calculate the chi-square statistic following the steps given on page 652.

1. The number of events $n = 3$.
2. Identify the actual observed values. $a_1 = 14$; $a_2 = 25$; $a_3 = 21$.
3. Calculate the expected values. If people were just guessing, we would expect each of the three names to be picked by the same number of people. Since there were 60 people in all, each name would be picked by 20. So $e_1 = 20$; $e_2 = 20$; $e_3 = 20$.
4. Calculate $\frac{(a_1 - e_1)^2}{e_1}$, $\frac{(a_2 - e_2)^2}{e_2}$, and $\frac{(a_3 - e_3)^2}{e_3}$.

$$\frac{(a_1 - e_1)^2}{e_1} = \frac{(14 - 20)^2}{20} = \frac{36}{20} \qquad \frac{(a_2 - e_2)^2}{e_2} = \frac{(25 - 20)^2}{20} = \frac{25}{20}$$
$$\frac{(a_3 - e_3)^2}{e_3} = \frac{(21 - 20)^2}{20} = \frac{1}{20}$$

5. The sum of the numbers in step 4 is $\frac{36 + 25 + 1}{20} = \frac{62}{20} = 3.1$.

Now examine the table. When $n = 3$, $n - 1 = 2$, so look at the 2nd row. The number 3.1 is smaller than the value 5.99 that would occur with probability .05. This is not a high enough chi-square value to question the way the expected values were calculated. It is quite possible that the people were guessing.

Millard Fillmore, Abraham Lincoln, and Martin Van Buren

The chi-square statistic can be used whenever there are actual frequencies and you have some way of calculating expected frequencies. However, the chi-square statistic should not be used when there is an expected frequency that is less than 5.

QUESTIONS

Covering the Reading

1. What does the chi-square statistic measure?

2. When and by whom was the chi-square statistic developed?

In 3–6, suppose frequencies of accidents for a stretch of highway are as given for each weekday. **a.** Calculate the chi-square statistic assuming that accidents occur on random days. **b.** Is there evidence to believe that the accidents are not occurring randomly?

3.

M	T	W	T	F
22	18	20	17	23

4.

M	T	W	T	F
23	22	20	18	17

5.

M	T	W	T	F
25	15	15	30	15

6.

M	T	W	T	F
20	20	20	20	20

Forty seconds that shook L.A. *Shown is one result of the devastating earthquake that struck in the Los Angeles area on January 17, 1994.*

7. Suppose in the Example of this lesson that 30 students had picked Abraham Lincoln, 18 had picked Millard Fillmore, and 12 had picked Martin Van Buren. Would there still be evidence that students were guessing randomly?

8. For what expected frequencies should the chi-square statistic not be used?

Applying the Mathematics

9. *The World Almanac and Book of Facts 1994* lists 59 major earthquakes since 1940. Here are their frequencies by season of the year.
Autumn, 13 Winter, 13 Spring, 12 Summer, 21

Use the chi-square statistic to determine whether these figures support a view that more earthquakes occur at certain times of the year than at others.

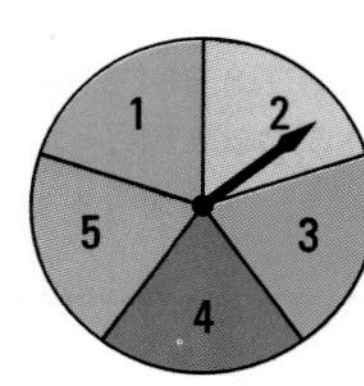

10. You build a spinner as shown at the left and spin it 40 times with the following outcomes.

Outcome	1	2	3	4	5
Frequency	10	6	4	6	14

Use the chi-square statistic to determine whether or not the spinner seems to be fair.

11. Here are the total runs scored in each of the first nine innings for the 15 Major League Baseball games played on May 1, 1994.

Inning	1	2	3	4	5	6	7	8	9	Total
Runs	24	18	17	11	8	21	17	14	8	138

Group innings 1–3, 4–6, and 7–9 together to represent the beginning, middle, and end of a game. Use the chi-square statistic to help in answering this question: Do baseball teams tend to score more runs in the beginning, middle, or end of a game, or do the runs appear to be scored equally in these sections of the game?

Review

12. **a.** Calculate 71^2 in your head by thinking of it as $(70 + 1)^2$.
b. Calculate 69^2 in your head by thinking of it as $(70 - 1)^2$.
(Lesson 10-6)

In 13 and 14, expand. *(Lessons 10-5, 10-6)*

13. **a.** $(3a - b)^2$ **b.** $(3a + b)^2$ **c.** $(3a - b)(3a + b)$

14. **a.** $(5 - y)(y - 5)$ **b.** $(4x^2 + 8xy + 4y^2)^2$

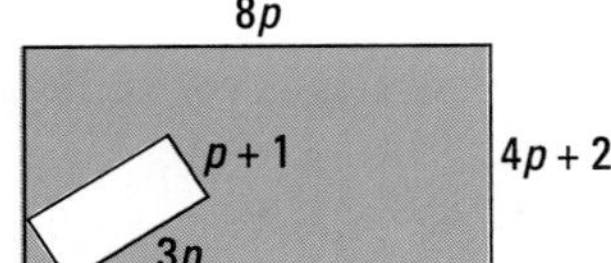

15. A rectangle with dimensions $3p$ and $p + 1$ is contained in a rectangle with dimensions $8p$ and $4p + 2$, as in the figure at the left. *(Lesson 10-3)*
a. Write an expression for the area of the big rectangle.
b. Write an expression for the area of the little rectangle.
c. Write a simplified expression for the area of the shaded region.

16. After five years of putting money into a retirement account at a scale factor x, a worker has $1000x^4 + 1100x^3 + 1200x^2 + 1400x + 1500$ dollars.
a. How much did the worker put in during the most recent year?
b. How much did the worker put in during the first year?
c. Give an example of a reasonable value for x in this problem, and evaluate the polynomial for that value of x. *(Lesson 10-2)*

In 17 and 18, simplify. *(Lessons 3-6, 3-7)*

17. $(12y^4 - 3y^3 + y) + (5y^3 - 7y^2 - 2y + 1) + (2y - 3y^2 + 6)$

18. $(x^2 - 4x + 1) - (3x^2 - 2x) - 2(7x + 4)$

19. There are 6 girls, 8 boys, 4 women, and 3 men on a community youth board. How many different leadership teams consisting of one adult and one child could be formed from these people? *(Lesson 2-9)*

Exploration

20. Repeat Question 11 with more recent data. Compare your results with those you found in Question 11.

A project presents an opportunity for you to extend your knowledge of a topic related to the material of this chapter. You should allow more time for a project than you do for typical homework questions.

1 Lifelong Savings

Sara and Sheila are twins who began to work at age 20 with identical jobs and identical salaries. At the end of each year they received identical bonuses of \$2000. In other ways the twins were not identical. Early in life Sara was conservative. Each year she invested the \$2000 bonus in a savings program earning 9% interest compounded annually. At age 30, Sara decided to have some fun and began spending her \$2000 bonus, but she let her earlier investment continue to earn interest. This continued until she was 65.

In contrast, for the first 10 years she worked, Sheila spent her \$2000 bonuses. At age 30, she began to invest her bonus every year in an account paying 9% annual interest. This continued until Sheila was 65 years old.

a. Which sister deposited more of her own money into her account? How much more?

b. Create a spreadsheet to determine how much each sister had in her account in each year.

c. How much had each sister accumulated at age 65?

d. What are the advantages of Sara's savings plan? What are the advantages of Sheila's savings plan?

e. There is a moral to the story of Sara and Sheila. What is the moral? What do you think of their story?

2 Powers of Binomials

You studied perfect square trinomials:

$$(a + b)^2 = a^2 + 2ab + b^2.$$

a. Expand $(a + b)^3$ and $(a + b)^4$.

b. What patterns do you observe in the coefficients?

c. Continue with higher powers if you can.

3 Another Proof of the Pythagorean Theorem

In the diagram below, a right triangle with legs of lengths a and b and hypotenuse of length c is replicated four times. The four triangles are placed so that the hypotenuses form a large square. Explain how to use the diagram to prove the Pythagorean Theorem.

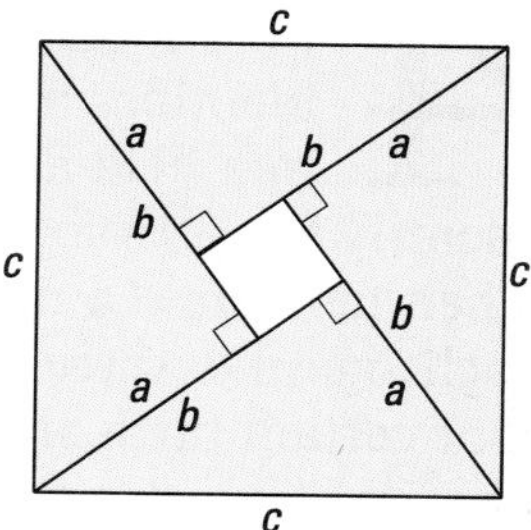

4 Testing Astrology

Find a book (like a *Who's Who*) that identifies at least 100 famous people in a field and gives their birthdates. Identify the astrological sign of each person. Then tabulate the number of people with each sign. Do these data lead you to believe that certain birth signs are more likely to produce famous people? Use a chi-square statistic assuming that a random distribution of the birthdays among the 12 astrological signs is expected. Since $n - 1 = 11$, refer to row 11 from the chi-square table on page 653.

5 Different Bases

Report on how positive integers are represented in the duodecimal system (base 12) or the hexadecimal system (base 16). Include examples on how to convert from base 10 into each system, and vice versa.

6 The Size of Products

The largest product of a 2-digit integer and a 3-digit integer is 99×999, or 98901. This equals $10^5 - 10^3 - 10^2 + 1$. The smallest product is 10×100, which equals 10^3.

a. Find a formula for the largest product of an m-digit and an n-digit integer.

b. Find a formula for the smallest product of these integers.

c. Explain why your formulas are correct.

7 Representing Positive Integers Using Powers

Do **a.** or **b.**

a. Every positive integer can be represented as a sum of different powers of 2. For instance, $100 = 2^6 + 2^5 + 2^2$. What may surprise you is that there is only one such representation. Find this power-of-2 representation for all the integers from 1 through 99. Try to explain why there is only one such representation.

b. Every positive integer can be represented as a sum or difference of different powers of 3. For instance, $100 = 3^4 + 3^3 - 3^2 + 3^0$. What may surprise you is that there is again only one such representation. Find this power-of-3 representation for all the integers from 1 through 99. Try to explain why there is only one such representation.

SUMMARY

A monomial is a product of terms. The degree of a monomial is the sum of the exponents of its variables. A polynomial is an expression that is either a monomial or a sum of monomials. The degree of a polynomial is taken to be the largest degree of its monomial terms. Linear expressions are polynomials of degree 1. Quadratic expressions are polynomials of degree 2. This chapter extends these ideas to consideration of polynomials of higher degree.

Polynomials emerge from a variety of situations. Our customary way of writing whole numbers in base 10 can be considered as a polynomial with 10 substituted for the variable. If different amounts of money are invested each year at a scale factor x, the total amount after several years is a polynomial in x. When the dimensions of a geometric figure are given as linear expressions, then areas or volumes related to the figure may be polynomials.

Addition and subtraction of polynomials are based on the Like Terms form of the Distributive Property, which you studied earlier in this book. Multiplication of polynomials is also justified by the Distributive Property. To multiply one polynomial by a second, multiply each term in the first polynomial by each term in the second, then add the products. For instance:

monomial by a polynomial:
$a(x + y + z) = ax + ay + az$

two polynomials: $(a + b + c)(x + y + z) = ax + ay + az + bx + by + bz + cx + cy + cz$

two binomials:
$(a + b)(c + d) = ac + ad + bc + bd$

perfect square patterns:
$(a + b)^2 = (a + b)(a + b) = a^2 + 2ab + b^2$
$(a - b)^2 = (a - b)(a - b) = a^2 - 2ab + b^2$

difference of two squares:
$(a + b)(a - b) = a^2 - b^2$

The square of the difference of actual and expected values in an experiment, $(a - e)^2$, appears in the calculation of the chi-square statistic. This statistic can help you to decide whether the assumptions that led to the expected values are correct.

VOCABULARY

Below are the most important terms and phrases for this chapter. You should be able to give a general description and a specific example of each.

Lesson 10-1
monomial, binomial, trinomial
degree of a monomial
polynomial
degree of a polynomial
polynomial in x
standard form of a polynomial
linear polynomial, quadratic polynomial

Lesson 10-4
Extended Distributive Property

Lesson 10-5
algorithm
FOIL algorithm

Lesson 10-6
expanding a binomial
perfect square trinomial
Perfect Square Patterns
difference of squares
Difference of Two Squares Pattern

Lesson 10-7
expected number
deviation
Chi-square statistic
degrees of freedom

CHAPTER TEN

PROGRESS SELF-TEST

Take this test as you would take a test in class. Then check your work with the solutions in the Selected Answers section in the back of the book.

In 1–3, consider the polynomial $4x^2 - 7x + 9x^2 - 12 - 11$.

1. Write this polynomial in standard form.

2. What is the degree of this polynomial?

3. Is the simplified polynomial a monomial, binomial, trinomial, or none of these?

In 4–9, perform the indicated operations and simplify.

4. Multiply $3v^2 - 9 + 2v$ by 4.

5. $-5z(z^2 - 7z + 8)$

6. $(3x - 8)(3x + 8)$

7. $(4y - 2)(3y - 16)$

8. Expand $(d - 12)^2$.

9. $(x - 3)(x^2 - 6x + 9)$

In 10–12, write as a single polynomial.

10. $(3x^2 - 10x) + (15x^3 - 7x^2 + x - 1)$

11. $8t^3 + t^2 - 7t + 1 - (5t^3 - 7t^2)$

12. $(x + y + 5)(a + b + 2)$

13. Write the area of the shaded region as a polynomial in standard form. Each polygon is a rectangle.

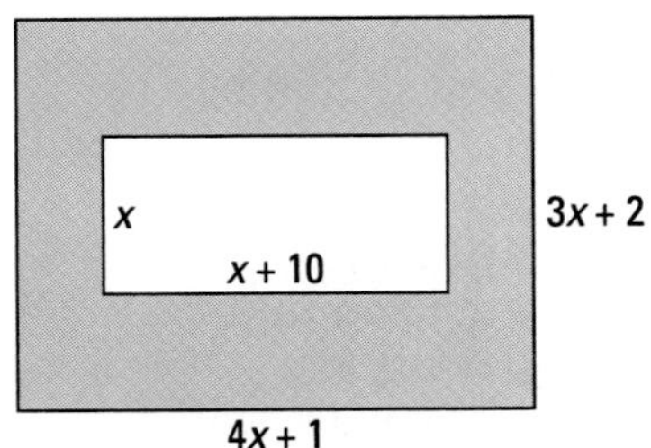

14. Represent the product $(a + b)(c + 2d)$ using areas of rectangles.

15. Show how you can compute $29 \cdot 31$ mentally.

16. Write 26,384 as a polynomial in standard form with 10 substituted for the variable x.

In 17 and 18, use this information. On his 18th birthday, Hank received \$80. He received \$60 on his 19th birthday and \$90 on his 20th birthday.

17. If he had invested all this money at a scale factor x, how much total money would he have on his 20th birthday?

18. Evaluate your answer to Question 17 when $x = 1.04$.

In 19–21, consider this situation. The Vulcan High School newspaper is supposed to give equal coverage to each of its four classes, freshmen, sophomores, juniors, and seniors. The sophomores thought that they were being shortchanged. One student totaled the numbers of lines devoted to students in each of the classes in the newspaper this year. Here is what the student found.

Class	Number of lines
Freshmen	861
Sophomores	748
Juniors	812
Seniors	939

19. Suppose the expected number of lines for each class had been the same. What would that expected number have been?

20. Calculate the chi-square statistic for this situation, assuming the expected number of lines for each class had been the same.

21. Use the Critical Chi-Square Values table on page 653. Does this support the view of the sophomores? Why or why not?

22. Write a simplified polynomial expression for the volume of the figure below.

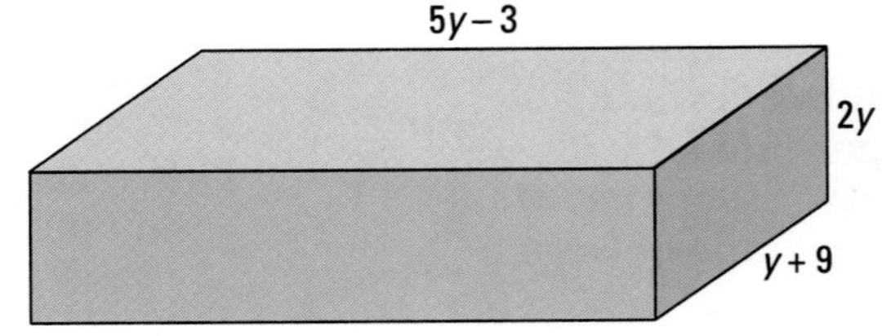

CHAPTER TEN

CHAPTER REVIEW

Questions on SPUR Objectives

SPUR stands for **S**kills, **P**roperties, **U**ses, and **R**epresentations. The Chapter Review questions are grouped according to the SPUR Objectives for this chapter.

SKILLS DEAL WITH THE PROCEDURES USED TO GET ANSWERS.

Objective A: *Add and subtract polynomials.* *(Lesson 10-2)*

In 1 and 2, **a.** simplify the expression; **b.** give the degree of the simplified expression.

1. $(5x^2 - 3x) + (2x^2 + 7x + 1)$

2. $(8m^4 - 2m^3) + (12m^3 - 6m^2 - 3m)$

3. Add $1.3x^2 + 14$, $4.7x - 1$, and $2.6x^2 - 3x + 6$.

4. Subtract $4y^5 - 2y^3 + 8y$ from $4y^5 - 6y^3 + 4y + 2$.

In 5 and 6, simplify.

5. $(k - 4) - (k^2 + 1)$ **6.** $(5p^2 - 1) - (6p^2 - p)$

Objective B: *Multiply polynomials.* *(Lesson 10-4)*

In 7–12, write as a single polynomial.

7. $(a + b + 1)(c + d + 1)$

8. $x^2(x^2 + 4) - x(x^3 - 3)$

9. $(y - 1)(y - 1)(y + 1)$

10. $(x^2 + x - 1)(x + 3)$

11. $(x + 1)(a + b + 3)$

12. $(y - 2)(y^2 + 2y + 3)$

Objective C: *Multiply a polynomial by a monomial or multiply two binomials.* *(Lessons 10-3, 10-5, 10-6)*

In 13–22, multiply and simplify, if possible.

13. $3k(k^2 + 4k - 1)$ **14.** $5xy(x + 3y^2)$

15. $2(4x^2 - x - 4) + 4(3x - 7)$

16. $(x - 3)(x + 7)$ **17.** $(y + 1)(y - 13)$

18. $(a - b)(c - d)$ **19.** $(a + 15)(a - 15)$

20. $(12b + m)(12b - m)$

21. $(4z + 1)(-z - 1)$ **22.** $(a + 3)(a^2 - 1)$

Objective D: *Expand squares of binomials.* *(Lesson 10-6)*

In 23–28, expand.

23. $(d - 1)^2$ **24.** $(2t + 3)^2$

25. $3(4x + 5)^2$ **26.** $(a_1 - e_1)^2$

27. $(x)(x + 1)^2$

28. $(m + 3n)^2 - (m - 3n)^2$

PROPERTIES DEAL WITH THE PRINCIPLES BEHIND THE MATHEMATICS.

Objective E: *Classify polynomials by their degree or number of terms.* *(Lesson 10-1)*

In 29–32, consider the following polynomials:

(a) $x^2 - 7$ (b) $x^3 - 5x^2 + 6$

(c) $x^2 + 8xy + 15y^2$ (d) $x^2 - 5x + 6$

29. Which are binomials?

30. Which are trinomials?

31. Which have degree 2?

32. Which have degree 3?

33. Give an example of a monomial of degree 4.

34. Give an example of a trinomial of degree 4.

CHAPTER TEN

Objective F: *Write whole numbers as polynomials in base 10.* *(Lesson 10-1)*

In 35 and 36, simplify.

35. $3 \cdot 10^7 + 2 \cdot 10^5 + 9 \cdot 10^2 + 1$

36. $10^4 + 2 \cdot 10^3 + 2 \cdot 10^2 + 10$

In 37 and 38, write as a polynomial in base 10.

37. 98,103

38. 4,005,600

USES DEAL WITH APPLICATIONS OF MATHEMATICS IN REAL SITUATIONS.

Objective G: *Translate investment situations into polynomials.* *(Lesson 10-2)*

39. Each birthday from age 11 on Katherine has received \$250. She puts the money in a savings account with a scale factor of x.

a. Write an expression which shows how much Katherine will have after her 15th birthday.

b. If the bank pays 8% interest a year, calculate how much Katherine will have after her 13th birthday.

40. Jose received \$25 on his 12th birthday, \$50 on his 13th birthday and \$75 on his 14th birthday, which he invested at a scale factor y. He kept his money in the same account at the same scale factor for 4 more years.

a. How much money did he have in this account at the end of that time?

b. If $y = 1.05$, how much money did he have in the account on his 15th birthday?

Objective H: *Use the chi-square statistic to determine whether or not an event is likely.* *(Lesson 10-7)*

41. In a taste test of two colas, 100 people were asked which cola they preferred. 56 preferred cola A and 44 preferred cola B.

a. If the colas are of equal taste, what are the expected numbers of preference for colas A and B?

b. Calculate the chi-square statistic for this situation, using the actual numbers and the expected numbers from part **a.**

c. Use the Critical Chi-Square Values table on page 653. Does the evidence support the fact that cola A is preferred by more people than cola B? Explain why or why not.

42. A company was open only Monday through Friday. Because it was not open Saturday or Sunday, it expected that it would get about the same amount of mail Tuesday through Friday, but three times this amount on Monday. However, some people thought there was too much mail coming on Monday. When the numbers of pieces of mail for each day for a few weeks were totaled, here were the numbers on each day.

Day of week	Mon	Tue	Wed	Thu	Fri
Pieces of mail	143	38	51	40	36

a. How many pieces of mail did the company expect each day?

b. Calculate the chi-square statistic for this situation, using the actual numbers and the expected numbers from part **a.**

c. Use the Critical Chi-Square Values table on page 653. Does the evidence support the company's expectations on how much mail to expect? Explain why or why not.

43. Some people say that the temperature of Los Angeles is "the same the year round."

Month	J	F	M	A	M	J	J	A	S	O	N	D
Actual Mean	57	58	60	61	65	69	74	75	72	68	63	58

a. What mean temperature would be expected each month if the temperature stayed the same all year?

b. Calculate the chi-square statistic for this situation, using the means given above and the expected number from part **a.**

c. Use the Critical Chi-Square Values table on page 653. Use $n = 12$. Does the chi-square statistic support the claim that the temperature is the same all year?

REPRESENTATIONS DEAL WITH PICTURES, GRAPHS, OR OBJECTS THAT ILLUSTRATE CONCEPTS.

Objective I: *Represent areas and volumes of figures with polynomials.*
(Lessons 10-1, 10-3, 10-4, 10-5, 10-6)

In 44 and 45, a rectangle is given.

a. Write the area of the figure as a polynomial.

b. Write the area as a product of polynomials.

44.

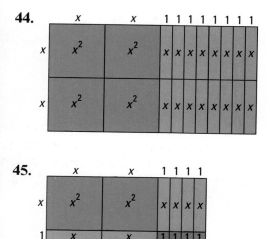

45.

46. Represent $(a + b)(c + d)$ using areas of rectangles.

47. a. Write the area of rectangle $ABCD$ below as the sum of 4 terms.

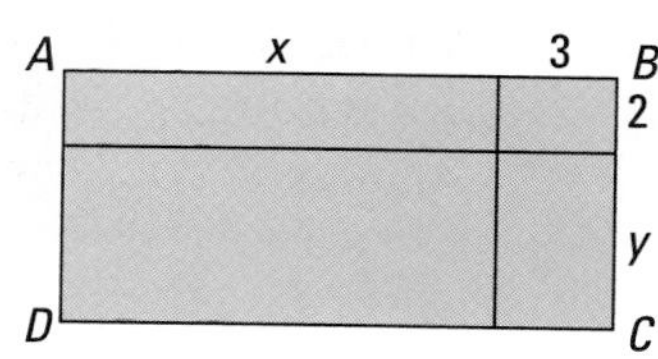

b. Write the area of $ABCD$ as the product of 2 binomials.

c. Are the answers to parts **a** and **b** equal?

In 48–51, express the area of the shaded region as a polynomial.

48.

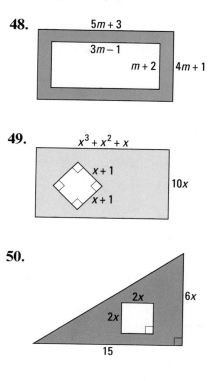

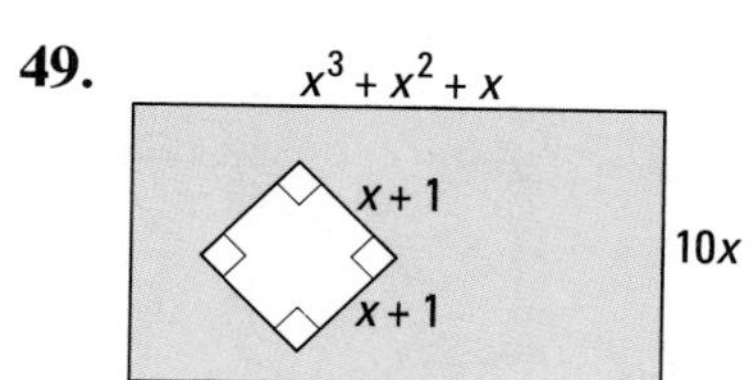

49.

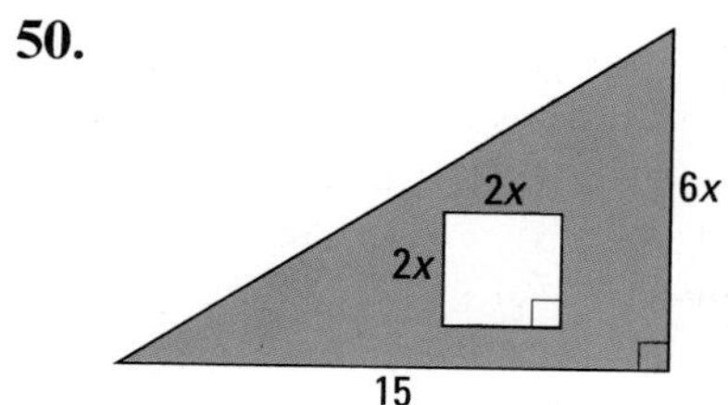

50.

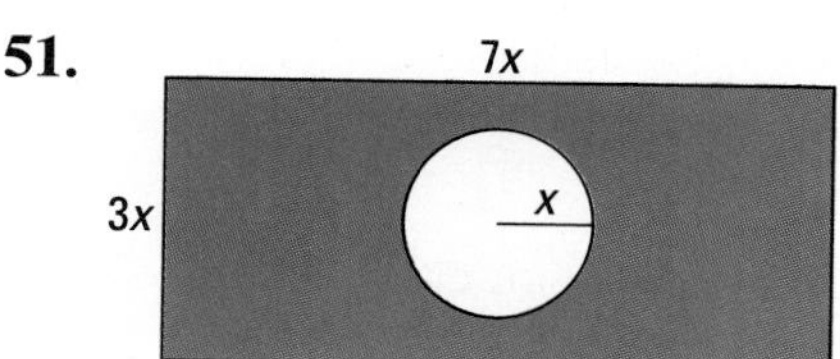

51.

52. Write a polynomial for the volume of the figure below.

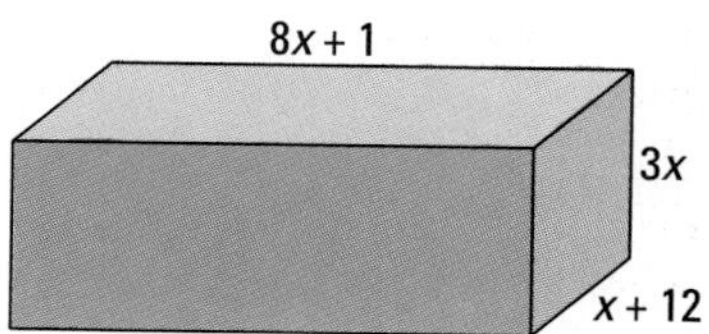

53. A box has dimensions x, $x + 1$, and $x - 1$. Write its volume as a polynomial.

CHAPTER 11

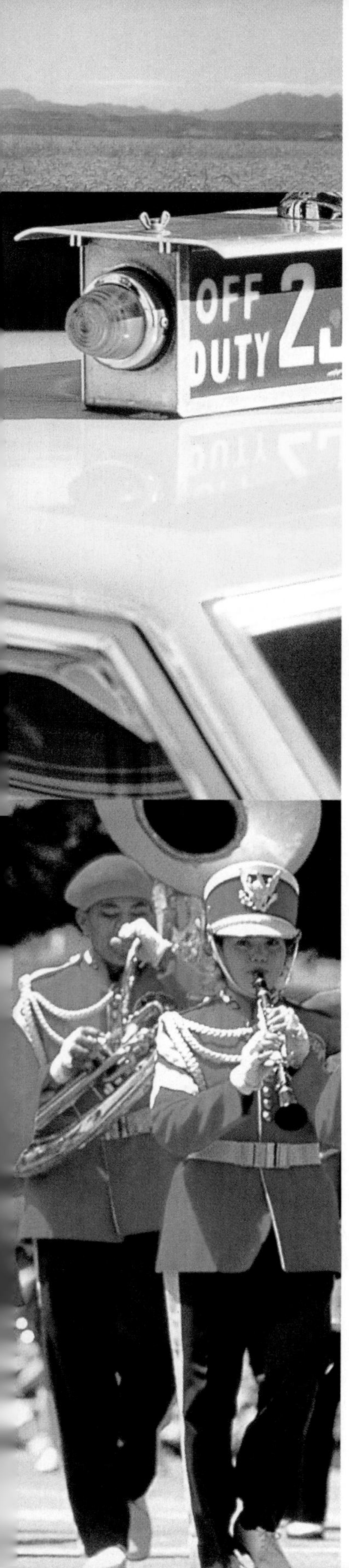

LINEAR SYSTEMS

The table and graph below give the men's and women's winning times in the Olympic 100-meter freestyle swimming race.

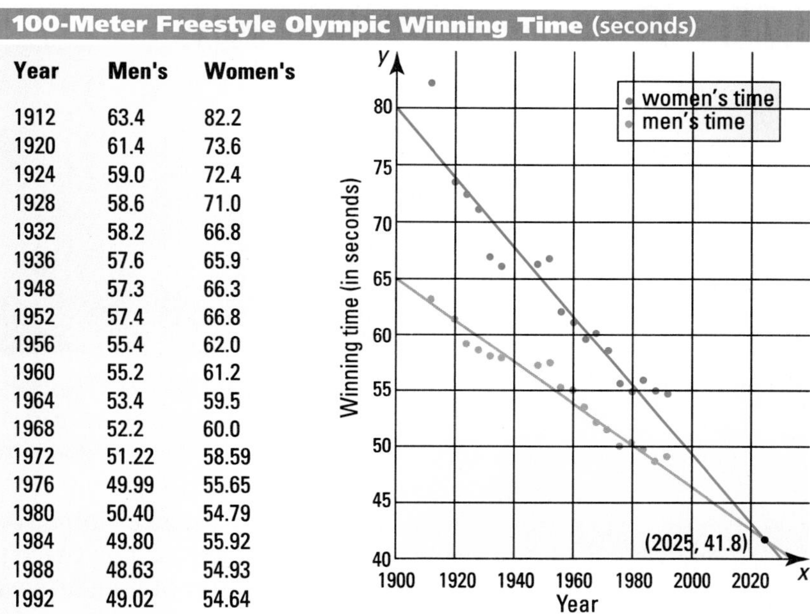

100-Meter Freestyle Olympic Winning Time (seconds)

Year	Men's	Women's
1912	63.4	82.2
1920	61.4	73.6
1924	59.0	72.4
1928	58.6	71.0
1932	58.2	66.8
1936	57.6	65.9
1948	57.3	66.3
1952	57.4	66.8
1956	55.4	62.0
1960	55.2	61.2
1964	53.4	59.5
1968	52.2	60.0
1972	51.22	58.59
1976	49.99	55.65
1980	50.40	54.79
1984	49.80	55.92
1988	48.63	54.93
1992	49.02	54.64

Notice that both men's and women's Olympic winning times have been generally decreasing since 1912. Also, the women's winning time has been decreasing faster than the men's. Lines have been fitted to the data. These lines have the following equations:

$$y = -0.176x + 398.204 \text{ (for men)}$$
$$\text{and } y = -0.299x + 647.308 \text{ (for women)}$$

where x is the year and y is the winning time in seconds. The lines intersect near (2025, 41.8). This means that if the winning times continue to decrease at these rates, the women's winning time will be faster than the men's in the Olympic year 2028. The winning times then will each be about 42 seconds.

Finding points of intersection of lines or other curves by working with their equations is called **solving a system.** In this chapter you will learn various ways of solving **linear systems.**

LESSON

11-1

An Introduction to Systems

On your marks. *The 1992 U.S. women's 400-meter freestyle relay team won a gold medal with a record time of 3:39.46 at the Olympic games in Barcelona. Often members of the relay team also participate in the 100-meter freestyle race.*

A **system** is a set of sentences joined by the word "and," which together describe a single situation. The two equations on page 665 describing the Olympic winning times of men and women in the 100-meter freestyle events are an example of a system of equations.

Systems are often signified by a single left-hand brace $\{$ in place of the word "and." So we can write this system as

$$\begin{cases} y = -0.176x + 398.204 \\ y = -0.299x + 647.308. \end{cases}$$

What Is a Solution to a System?

Each sentence in a system is sometimes called a **condition of the system.** Thus, the system above has two conditions.

A **solution to a system** of sentences with two variables is a pair of numbers which satisfies all the conditions of the system. On a graph, each solution of a system is a point of intersection of the graphs of the sentences.

Example 1

Verify that the point (2025, 41.8) is a good estimate of the solution to the system of equations on page 666.

Solution

Substitute $x = 2025$ and $y = 41.8$ into each of

$$y = -0.176x + 398.204$$
$$\text{and } y = -0.299x + 647.308.$$

men: $y = -0.176x + 398.204$

Does $41.8 = -0.176 \cdot 2025 + 398.204$?

$41.8 \approx 41.804$? Yes.

women: $y = -0.299x + 647.308$

Does $41.8 = -0.299 \cdot 2025 + 647.308$?

$41.8 \approx 41.833$? Yes.

Solving Systems by Graphing

You can find the solutions to any system of equations with two variables by graphing each equation and finding the coordinates of the point(s) of intersection of the graphs.

Example 2

Find two numbers whose sum is 22 and whose difference is 8.

Solution

Translate the conditions into a system of two equations.
Let x and y be the two numbers. Then $x + y = 22$ and $x - y = 8$.
Graph each equation and identify the point of intersection.
As shown at the right, the solution is (15, 7).

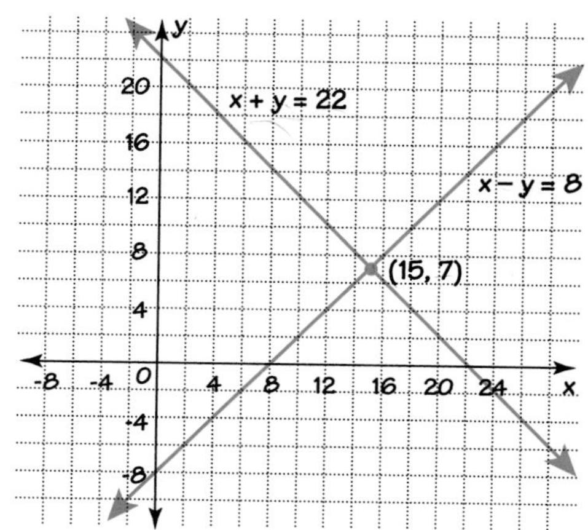

Check

To check that (15, 7) is a solution, $x = 15$ and $y = 7$ must be checked in both conditions.
Does $x + y = 22$? $15 + 7 = 22$? Yes. Does $x - y = 8$? $15 - 7 = 8$? Yes.

In general, there are four ways to write the solution to a system. They are shown below using the solution to the system in Example 2.

as an ordered pair	(15, 7)
as an ordered pair identifying the variables	$(x, y) = (15, 7)$
by naming the variables individually	$x = 15$ and $y = 7$
as a set of ordered pairs	{(15, 7)}

Systems with No Solutions

When the sentences in a system have no solutions in common, we say that there is *no solution* to the system. We cannot write the solution as an ordered pair or by listing the elements. The solution set is the set with no elements { }, or ø.

Example 3

Find all solutions to the system $\begin{cases} y = 2x + 1 \\ y = 2x - 3 \end{cases}$.

Solution

Draw the graph of each equation and look for all intersection points.

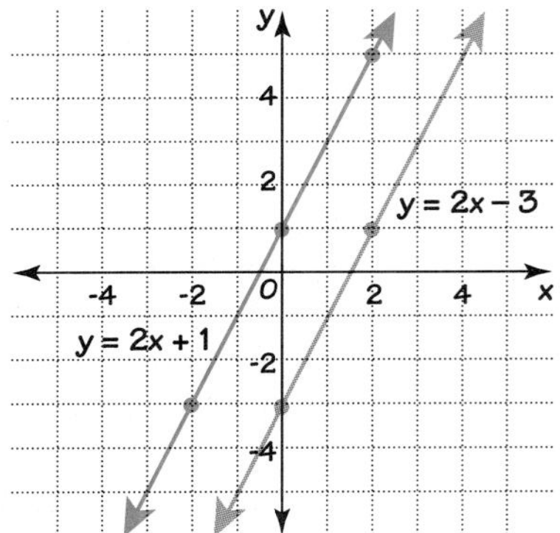

Notice that each line has slope 2, and a different y-intercept. So they are parallel and do not intersect. Thus, there is no pair of numbers that makes both equations true. There is no solution to the system. The solution set is ø.

Check

Twice a number plus 1 could never give the same value as twice the same number minus 3. That is, for all real numbers x, $2x + 1 \neq 2x - 3$.

Graphing can help you find exact solutions to a system, as was shown in Example 2. If solutions do not have integer coordinates, however, it is likely that reading a graph will only give you an estimate. In the following lessons of this chapter, you will learn other techniques to find exact solutions to systems.

QUESTIONS

Covering the Reading

1. a. Define *system.*
   b. Give an example of a system not in this lesson.

2. What does the brace { represent in a system?

3. *True or false.* When a system has two variables, each solution is an ordered pair.

4. What does the solution set to a system represent in a coordinate plane?

5. Refer to the graph at the right.
   a. What system is represented?
   b. What is the solution to the system?
   c. Verify your answer to part **b.**

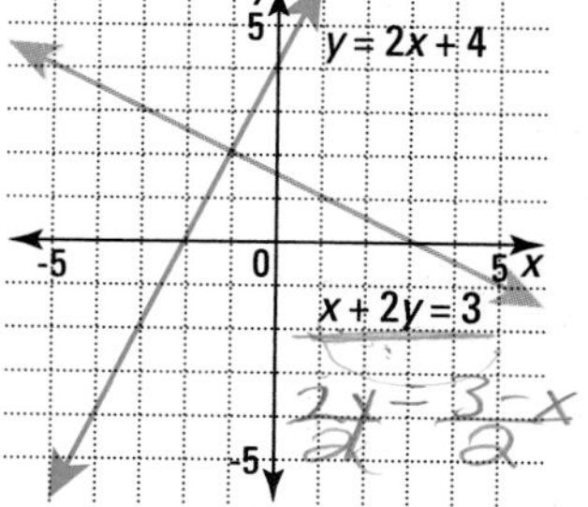

6. a. Verify that the solution to the system $\begin{cases} y = 9x \\ y = 2x - 7 \end{cases}$ is (-1, -9).
   b. Write this solution in two other ways.

7. Is (4, 8) a solution to the following system? How can you be sure?
$\begin{cases} 10x - y = 32 \\ y - x = 4 \end{cases}$

8. The sum of two numbers is 18 and their difference is 8.
   a. If the numbers are x and y, translate the two conditions of the sentence above into two equations.
   b. Graph both of these equations on the same coordinate system.
   c. What are the numbers?

9. Find all solutions to the system $\begin{cases} y = 4x - 2 \\ y = 4x + 5 \end{cases}$.

In 10 and 11, a system is given.
a. Solve each system by graphing.
b. Check your work.

10. $\begin{cases} y = 2x + 1 \\ y = -3x + 6 \end{cases}$

11. $\begin{cases} y = x \\ 2x + 3y = -15 \end{cases}$

Applying the Mathematics

In 12 and 13, a system is given.
a. Graph the system.
b. Find the coordinates of the point(s) of intersection. (Hint: In Question 13, one graph is not a line.)

12. $\begin{cases} x = -5 \\ x - y = 3 \end{cases}$

13. $\begin{cases} y = -\frac{1}{2}x^2 \\ y = \frac{1}{2}x - 1 \end{cases}$

14. Below is a graph of the system $\begin{cases} y = x^2 \\ y = 2^x \end{cases}$.

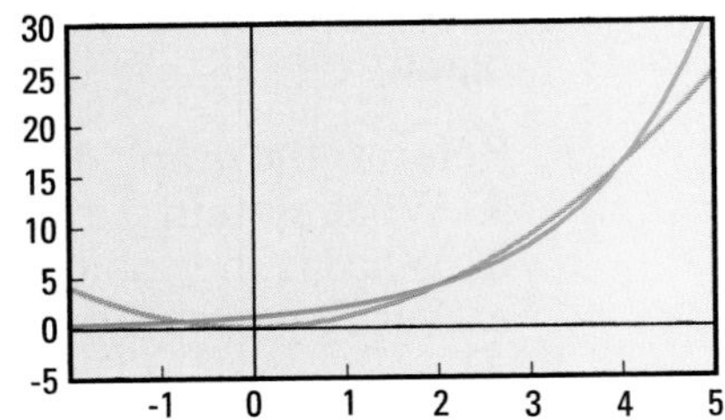

a. Use a calculator or automatic grapher to find three solutions to this system.

b. Check each solution.

15. Below are a table and graph of the winning times in seconds for the Olympic men's and women's 100-meter backstroke events.

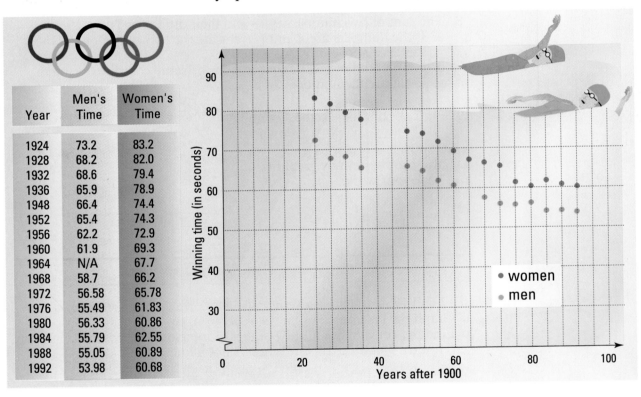

Year	Men's Time	Women's Time
1924	73.2	83.2
1928	68.2	82.0
1932	68.6	79.4
1936	65.9	78.9
1948	66.4	74.4
1952	65.4	74.3
1956	62.2	72.9
1960	61.9	69.3
1964	N/A	67.7
1968	58.7	66.2
1972	56.58	65.78
1976	55.49	61.83
1980	56.33	60.86
1984	55.79	62.55
1988	55.05	60.89
1992	53.98	60.68

Based on these data, do you think the women's winning time will ever equal the men's winning time in the 100-meter backstroke? If yes, estimate the year when this will happen. If no, explain why not.

Review

16. *Skill sequence.* Solve. *(Lessons 3-7, 5-3, 9-5, 10-6)*
  a. $3x + 8 = x - 12$
  b. $3(x + 8) = -4(x - 12)$
  c. $3(x + 8)^2 = -3(x - 12)$

17. a. Evaluate $\frac{y_2 - y_1}{x_2 - x_1}$ where $y_2 = 7$, $y_1 = -1$, $x_2 = 8$, and $x_1 = 10$.
  b. What have you calculated in part **a**? *(Lessons 1-4, 7-2)*

18. Suppose that on a map 1 inch represents 325 miles. The map distance from Los Angeles to Houston is $4\frac{3}{4}$ inches. Suppose you want the actual distance in miles from Los Angeles to Houston.
  a. Write a proportion that will help solve the problem.
  b. Find the distance. *(Lesson 6-8)*

Houston, Texas

19. In 1993, American families owned an average of 1.8 cars per family, and drove an average of 18,600 miles. How many miles was the typical American family car driven that year? *(Lesson 6-2)*

20. Simplify $7\pi \div \left(\frac{2\pi}{3}\right)$. *(Lesson 6-1)*

21. Solve $2x + y = 7$ for y. *(Lesson 5-7)*

22. Eight more than three times a number is two more than six times the number. What is the number? *(Lesson 5-3)*

23. If a country has population P now and the population is increasing by X people per year, what will its population be in Y years? *(Lesson 3-8)*

Exploration

24. Some experts believe that even though the women's swim times are decreasing faster than the men's, it is the ratio of the times that is the key to predictions.
  a. Compute the ratio of the men's time to the women's time for the 100-meter freestyle for each Olympic year.
  b. Graph your results. (Plot Olympic year on the horizontal axis and the ratio of times on the vertical axis.)
  c. What do you think the ratio will be in 2025? Does this agree with the prediction on page 665?

25. Question 13 describes a system in which a line and a parabola intersect in two points. Sketch examples of the following systems.
  a. a line and a parabola with no points of intersection
  b. a line and a parabola with exactly one point of intersection
  c. two parabolas intersecting in two points
  d. two parabolas intersecting in exactly one point

LESSON

11-2

Solving Systems Using Substitution

Taxi! *A system of equations can be used to compare the fees charged by two different taxicab companies. See Example 2.*

Consider the system

$$\begin{cases} y = 5x - 25 \\ y = -8x + 27 \end{cases}$$

graphed below. It appears that the x-coordinate of the point of intersection is 4 and that the y-coordinate is -5.

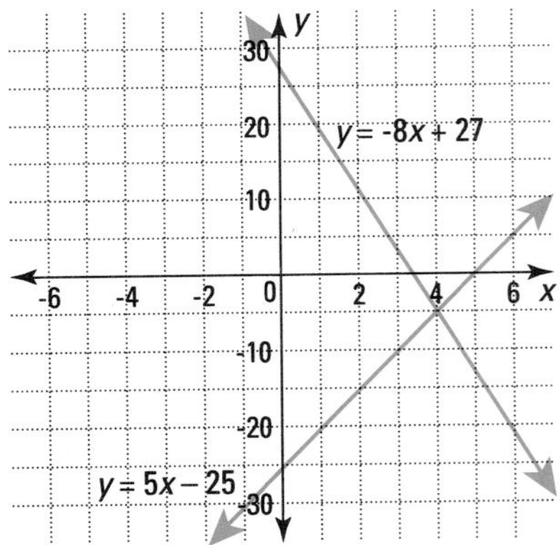

When the coordinates of the intersection point are not integers, it is difficult to find their exact values from a graph. However, you can find the exact coordinates of the point of intersection by using *substitution.* To solve a system using substitution, take the expression equal to y in one equation, and substitute it for y in the other equation. Example 1 illustrates this technique.

Example 1

Solve the system $\begin{cases} y = 5x - 25 \\ y = -8x + 27 \end{cases}$ using substitution.

Solution

Substitute $-8x + 27$ for y in the first equation.

$$-8x + 27 = 5x - 25$$

Now solve for x.

$$27 = 13x - 25$$
$$52 = 13x$$
$$4 = x$$

To find y, substitute 4 for x in either of the original equations. We use the first equation.

$$y = 5x - 25$$
$$y = 5 \cdot 4 - 25 = -5$$

The solution is (4, -5).

Check

The point is on both lines, as substitution shows.
$-5 = 5 \cdot 4 - 25$ and $-5 = -8 \cdot 4 + 27$

Suppose two quantities are increasing or decreasing at different constant rates. Then each quantity can be described by an equation of the form $y = ax + b$. To find out when the quantities are equal, you can solve a system by using substitution. Example 2 illustrates this idea.

Water taxis. *Canals take the place of streets in Venice, Italy. Gondolas—black, flat-bottomed boats—once served as the chief means of transportation. Today, most people use motorboat taxis.*

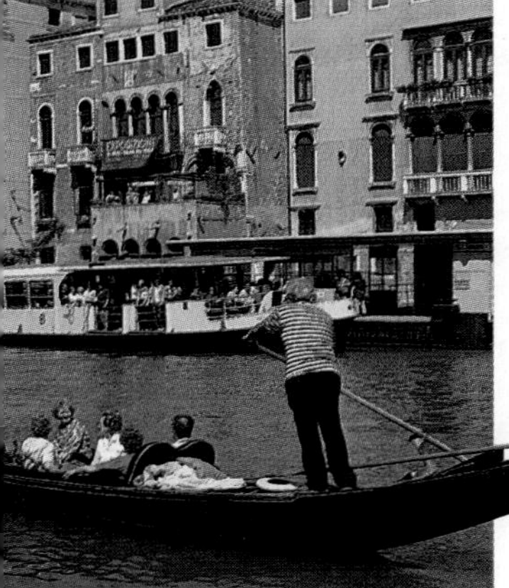

Example 2

A taxi ride in Burford costs \$1.50 plus 20¢ for each $\frac{1}{10}$ mile traveled. In Spotswood, taxi rides cost 90¢ plus 25¢ for each $\frac{1}{10}$ mile traveled. For what distance do rides cost the same?

Solution

Let d = the distance of a cab ride in tenths of a mile.
Let C = the cost of a cab ride of distance d.

In Burford: $C = 1.50 + .20d$
In Spotswood: $C = .90 + .25d$

The rides cost the same when the values of C and d in Burford equal the values in Spotswood. So there is a system to solve. We substitute $1.50 + .20d$ for C in the second equation.

$$1.50 + .20d = .90 + .25d$$

Now solve as usual. Add $-.90$ and $-.20d$ to both sides.

$$.60 = .05d$$
$$d = 12$$

A ride of a distance 12 tenths of a mile, or 1.2 miles, will cost the same in both cities.

Check

Check to see if the cost will be the same for a ride of 12 tenths of a mile.
The cost in Burford is $1.50 + .20 \cdot 12 = 1.50 + 2.40 = 3.90$.
The cost in Spotswood is $.90 + .25 \cdot 12 = .90 + 3.00 = 3.90$.
The cost is $3.90 in each city, so it checks.

QUESTIONS

Covering the Reading

London's black cabs are rated best in the world. Drivers must pass an exam that includes going to the most obscure addresses on a bicycle.

1. *True or false.* Solving a system by substitution only approximates the answer.

In 2 and 3, a system is given. **a.** Use substitution to find the point of intersection of the two lines. **b.** Check your answer.

2. $\begin{cases} y = 3x - 19 \\ y = x + 1 \end{cases}$

3. $\begin{cases} y = 12x + 50 \\ y = 10x - 60 \end{cases}$

In 4–9, refer to Example 2.

4. What would it cost for a 3-mile taxi ride in Burford? (Hint: convert to tenths of a mile.)

5. What would it cost for a 2.5-mile taxi ride in Spotswood?

6. Check the answer to Example 2 by graphing the lines $C = 1.50 + .20d$ and $C = .90 + .25d$ on the same axes. (Let d be the first coordinate and C be the second coordinate.)

7. For what distances are taxi rides more expensive in Burford than in Spotswood?

8. For what distances are taxi rides more expensive in Spotswood than in Burford?

9. Suppose that in Manassas, a taxi ride costs $1.70 plus 15¢ each $\frac{1}{10}$ mile.
a. What does it cost to ride d tenths of a mile?
b. At what distance does a ride in Manassas cost the same as a ride in Spotswood?

Applying the Mathematics

In 10–13, **a.** solve each system; **b.** check your answers.

10. $\begin{cases} y = 21 - x \\ y = 3 + x \end{cases}$

11. $\begin{cases} y = \frac{1}{2}x - 5 \\ y = -\frac{3}{4}x + 10 \end{cases}$

12. $\begin{cases} y = \frac{1}{3}x + 2 \\ y = -3x - 5 \end{cases}$

13. $\begin{cases} y = x \\ y = \frac{1}{4}x^2 \end{cases}$

Phoenix mystery. *This building, called* Mystery Castle, *is located in Phoenix, Arizona.*

14. Jana has \$290 and saves \$5 a week. Dana has \$200 and saves \$8 a week.
   a. After how many weeks will they each have the same amount of money?
   b. How much money will each one have?

15. One plumbing company charges \$45 for the first half-hour of work and \$23 for each additional half-hour. Another company charges \$35 for the first half-hour and then \$28 for each additional half-hour. For how many hours of work will the cost of each company be the same?

16. In 1990 the population of New Orleans, Louisiana, was about 497,000 and was decreasing by about 6,000 people a year. The 1990 population of Phoenix, Arizona, was about 983,000 and was increasing by 20,000 people a year.
   a. If these trends had been going on for some time, how many years before 1990 did each city have the same population?
   b. What was this population?

Review

In 17 and 18, solve by graphing. *(Lesson 11-1)*

17. the system of Question 11 **18.** the system of Question 12

19. **a.** How many solutions does this system have?

$$\begin{cases} y = |x| \\ y = 5 \end{cases}$$

   b. Find the solutions. *(Lessons 9-8, 11-1)*

20. *Skill sequence.* Solve for x. *(Lessons 3-5, 9-1, 9-5)*
   a. $2x - 18 = 0$ **b.** $2x^2 - 18 = 0$ **c.** $2x^2 - 18x = 0$

21. Multiply $2p(p^2 + 3p + 1)$. *(Lesson 10-3)*

22. Simplify $\frac{7m^4n^5}{343m^3n^6}$. *(Lesson 8-7)*

23. Graph all points (x, y) for which $y > -x + 5$. *(Lesson 7-9)*

24. *Skill sequence.* Determine the y-intercept in your head. *(Lesson 7-4)*
   a. $y = 7x - 2$ **b.** $2x + y = 7$ **c.** $7 + y = 2x$

25. What is the cost of x videotapes at \$4 each and y computer disks at \$2 each? *(Lessons 2-4, 3-1)*

Exploration

26. Find the taxi rates where you live or in a nearby community. How do these rates compare to those in Example 2?

27. Make up a question comparing two quantities which are increasing or decreasing, each at its own constant rate (as in Example 2 and Questions 14–16 above). Use substitution to answer your question.

LESSON

11-3

More Uses of Substitution

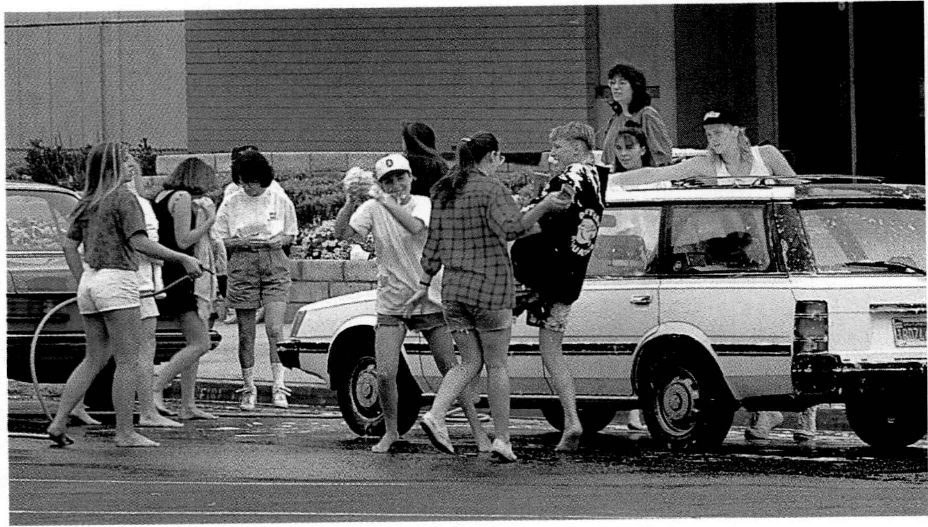

Pass the sponges. *Car washes are a common way for youth organizations to raise money for good causes.*

In the previous lesson, you saw how to use substitution to solve systems of equations when each was solved for y. Substitution may also be used when one or more of the equations is not solved for y.

Example 1

Some Boy Scouts and Girl Scouts sponsored a car wash that made $109. There were twice as many girls as boys working, so a decision was made to give the girls twice as much money. How much did each group receive?

Solution

Translate each condition into an equation. Let B equal the proceeds received by the Boy Scouts and G equal the proceeds received by the Girl Scouts.

The total proceeds equal $109. $B + G = 109$
The girls' proceeds are twice that of the boys'. $G = 2B$

Since $G = 2B$, you can substitute $2B$ for G in the first equation.

$$\begin{aligned} B + G &= 109 \\ B + 2B &= 109 \\ 3B &= 109 \\ B &= 36\tfrac{1}{3} \end{aligned}$$

To find G, substitute $36\frac{1}{3}$ for B in either equation. We use the second equation because it is solved for G.

$$\begin{aligned} G &= 2B \\ &= 2 \cdot 36\tfrac{1}{3} \\ &= 72\tfrac{2}{3} \end{aligned}$$

▶

So the solution is $(B, G) = \left(36\frac{1}{3}, 72\frac{2}{3}\right)$. The Boy Scouts will receive $\$36\frac{1}{3}$, or about \$36.33, and the Girl Scouts will get $\$72\frac{2}{3}$, or about \$72.67.

Check

Are both conditions satisfied? Will the groups receive a total of \$109? Yes, \$36.33 + \$72.67 = \$109. Will the girls get twice as much as the boys? Yes, \$72.67 is about twice as much as \$36.33.

In the equation $G = 2B$ from Example 1, G is given in terms of B. Substitution is usually a good method to use when one variable is given in terms of others. In Example 2, one equation gives y in terms of x.

Example 2

Two lines have equations $3x + 2y = 10$ and $y = 4x + 1$. Where do they intersect?

Solution

Notice that the equation $y = 4x + 1$ is solved for y. Substitute $4x + 1$ for y in the other equation and solve the linear equation that results.

$$3x + 2y = 10$$
$$3x + 2(4x + 1) = 10$$
$$3x + 8x + 2 = 10$$
$$11x + 2 = 10$$
$$11x = 8$$
$$x = \frac{8}{11}$$

Thus $\frac{8}{11}$ is the x-coordinate of the point of intersection. To find the y-coordinate, substitute $\frac{8}{11}$ for x in one of the equations. We use the second equation because it is solved for y.

$$y = 4 \cdot \frac{8}{11} + 1$$
$$= \frac{32}{11} + 1$$
$$= \frac{43}{11}$$

The lines intersect at $\left(\frac{8}{11}, \frac{43}{11}\right)$.

Check

You could graph the lines to see if your solution is reasonable. But to produce an exact check, substitute $\frac{8}{11}$ for x and $\frac{43}{11}$ for y in both equations.

Does $3\left(\frac{8}{11}\right) + 2\left(\frac{43}{11}\right) = 10$?

Does $4\left(\frac{8}{11}\right) + 1 = \left(\frac{43}{11}\right)$?

You should verify that they do.

In Example 3, one equation can easily be solved for a variable.

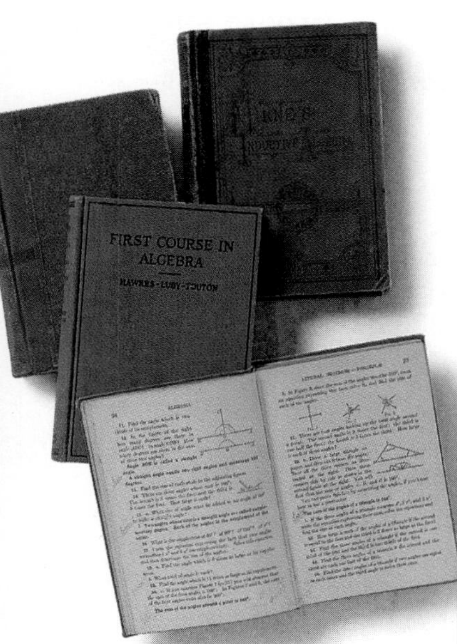

These algebra books were printed between the years 1877 and 1923.

Example 3

(This problem is taken from an 1881 algebra text; the prices are out of date, but the situation is not.) A farmer purchased 100 acres of land for \$2450. For a part of it he paid \$20 an acre, and \$30 an acre for the rest of it. How many acres were there in each part?

Solution

You want to find two amounts, so use two variables.
Let x = the number of acres at \$20/acre,
and y = the number of acres at \$30/acre.

A total of 100 acres was purchased, so

$$x + y = 100.$$

The total cost was \$2450. So

$$20x + 30y = 2450.$$

Solve the system $\begin{cases} x + y = 100 \\ 20x + 30y = 2450 \end{cases}$.

Note that neither equation is solved for a variable. But notice that the first equation can be rewritten as

$$y = 100 - x.$$

Now substitute $100 - x$ for y in the second equation.

$$\begin{aligned} 20x + 30y &= 2450 \\ 20x + 30(100 - x) &= 2450 \\ 20x + 3000 - 30x &= 2450 \\ 3000 - 10x &= 2450 \\ -10x &= -550 \\ x &= 55 \end{aligned}$$

To find y, substitute $x = 55$ in either of the original equations. We use the first equation because it is simpler.

$$\begin{aligned} x + y &= 100 \\ 55 + y &= 100 \\ y &= 45 \end{aligned}$$

The farmer bought 55 acres at \$20/acre, and 45 acres at \$30/acre.

Check

Substitute $x = 55$, $y = 45$ in both equations.
Does $x + y = 55 + 45 = 100$? Yes, it checks.
Does $20x + 30y = 20(55) + 30(45) = 1100 + 1350 = 2450$? Yes, it checks.

QUESTIONS

Covering the Reading

1. Complete the check of Example 2.

In 2–5, **a.** find the point of intersection of the lines without graphing; **b.** check your answer.

2. $\begin{cases} x + y = 14 \\ x = 6y \end{cases}$

3. $\begin{cases} 12x - 5y = 30 \\ y = 2x - 6 \end{cases}$

4. $\begin{cases} y = x - 2 \\ -4x + 7y = 10 \end{cases}$

5. $\begin{cases} 3x + 4y = -15 \\ y = 2x - 3 \end{cases}$

6. Suppose the service club made $180 on a car wash. If the boys were to get three times as much money as the girls, how much would each group receive?

7. a. Solve the system $\begin{cases} B = 7t \\ B - t = 30 \end{cases}$. **b.** Check your answer.

8. Here is another problem from the 1881 algebra book. A farmer bought 100 acres of land, part at $37 and part at $45 an acre, paying for the whole $4220. How much land was there in each part?

Applying the Mathematics

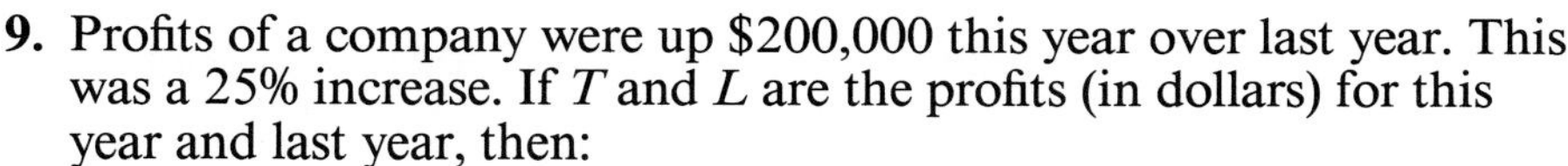

This 1855 painting, Guarding the Corn Fields *by Seth Eastman, depicts Dakota Indians frightening birds away from crops by banging sticks together.*

9. Profits of a company were up $200,000 this year over last year. This was a 25% increase. If T and L are the profits (in dollars) for this year and last year, then:

$$\begin{cases} T = L + 200{,}000 \\ T = 1.25L \end{cases}$$

Find the profits for this year and last year.

10. A will states that John is to get 3 times as much money as Mary. The total amount they will receive is $11,000.
a. Write a system of equations describing this situation.
b. Solve to find the amounts of money John and Mary will get.

11. Solve for A, B, and K. $\begin{cases} A = 40K \\ B = 30K \\ A + B = 1400 \end{cases}$

12. A homemade sealer to use on furniture after it is stained can be made by mixing one part shellac with five parts denatured alcohol. To make a pint (16 fluid ounces) of sealer, how many fluid ounces s of shellac and how many fluid ounces A of denatured alcohol are needed? (Hint: $\frac{A}{s} = \frac{5}{1}$.)

In 13 and 14, solve the system using substitution.

13. $\begin{cases} 4x - 3y = 8 \\ 2x + y = -1 \end{cases}$

14. $\begin{cases} 8a - 1 = 4b \\ 3a = b + 1 \end{cases}$

Full of hot air. *Pictured are balloons at a fiesta in Albuquerque. Hot-air balloons rise because the air inside the bag is warmer, and therefore lighter, than the surrounding air.*

15. Noah got the results of tests on mathematics and verbal achievement. His verbal score is 70 points less than his mathematics score. His total score for the two parts is 1250.
  a. Let v = Noah's verbal score, and m = his mathematics score. Write a system of equations for this situation.
  b. Find Noah's two scores.

Review

16. One hot-air balloon takes off from Albuquerque, New Mexico, and rises at a rate of 120 ft/min. At the same time, another balloon takes off from Santa Fe, NM, and rises at a rate of 75 ft/min. The altitude of Albuquerque is about 4950 ft; the altitude of Santa Fe is about 6950 ft.
  a. When are the two balloons at the same altitude?
  b. What is their altitude at that time? *(Lessons 3-8, 5-2, 5-3, 11-2)*

In 17 and 18, a system is given.
a. Graph or use substitution to find the solution to the system.
b. Check your answer. *(Lessons 11-1, 11-2)*

17. $\begin{cases} y = x + 1 \\ y = -2x + 13 \end{cases}$

18. $\begin{cases} y = 3x - 2 \\ y = 3x + 3 \end{cases}$

In 19 and 20, simplify. *(Lessons 4-5, 10-6)*

19. $(3a - 2b) - (a + b)$

20. $x^2 + y^2$ when $y = x - 1$

21. Graph $y \geq 2x + 2$ on a coordinate plane. *(Lesson 7-9)*

22. Solve $5x + 9y = 7$ for y. *(Lesson 7-4)*

23. If you travel 120 miles in 40 minutes, what is your average speed in miles per hour? *(Lesson 6-2)*

24. Suppose you have n nickels and d dimes and these are the only coins you have. **a.** How many coins do you have? **b.** What is the total value of the coins in cents? *(Lessons 2-4, 3-1)*

Exploration

25. Neighboring cities Lovely (population 35,729) and Elylov (population 74,212) are building a joint conference center that will cost $10 million. How would you suggest they split the cost? Why?

26. A homing pigeon has been clocked doing 94.3 mph in still air.
  a. How fast can it fly *with* the wind if the wind speed is s mph?
  b. How fast can it fly *against* the wind if the wind speed is s mph?
  c. How far can it fly in m minutes in still air?
  d. If it is flying down the highway (where the speed limit is 65 mph) against a 25 mph headwind, would you give it a ticket?

LESSON

11-4

Solving Systems by Addition

Young pilot. *In 1994, 12-year-old Vicki Van Meter, accompanied by her flight instructor Curt Arnspiger, became the youngest female pilot to cross the Atlantic Ocean. Systems of equations can help find plane and wind speeds. See Example 1.*

The numbers $\frac{3}{4}$ and 75% are equal even though they may not look equal. So are $\frac{1}{5}$ and 20%. If you add pairs of these numbers (as seen below), the sums are equal.

$$\frac{3}{4} = 75\%$$

$$\frac{1}{5} = 20\%$$

So $$\frac{3}{4} + \frac{1}{5} = 75\% + 20\%.$$

Simplifying each side, we get $$\frac{19}{20} = 95\%.$$

This is one instance of the following generalization of the Addition Property of Equality.

> **Generalized Addition Property of Equality**
> **For all numbers or expressions *a, b, c,* and *d*:**
> If $a = b$
> and $c = d$,
> then $a + c = b + d$.

The Generalized Addition Property of Equality can be used to solve some systems. Consider this situation:

The sum of two numbers is 63.
Their difference is 12. What are the numbers?

If x and y are the two numbers, with x the larger, we can write the following system.

$$\begin{cases} x + y = 63 \\ x - y = 12 \end{cases}$$

Notice what happens when the left sides are added and the right sides are added.

$$\begin{array}{r} x + y = 63 \\ + \; x - y = 12 \\ \hline 2x + 0 = 75 \end{array}$$

Because y and $-y$ add to 0, the sum is an equation with only one variable. Solve $2x = 75$ as usual.

$$x = 37.5$$

To find y, substitute 37.5 for x in one of the original equations.

$$\begin{aligned} x + y &= 63 \\ 37.5 + y &= 63 \\ y &= 25.5 \end{aligned}$$

The ordered pair (37.5, 25.5) checks in both equations, so the solution to the system $\begin{cases} x + y = 63 \\ x - y = 12 \end{cases}$ is (37.5, 25.5).

Using the Generalized Addition Property of Equality to eliminate one variable from a system is sometimes called the **addition method** for solving a system. The addition method is an efficient way to solve systems when the coefficients of the same variable are opposites.

They can wing it. *These small planes are from the Lion's Club Fly-In Fish Boil on Washington Island, Wisconsin. Planes like these were often used for training pilots during World War II.*

Example 1

A pilot flew a small plane 120 miles from Washington Island, Wisconsin, to Appleton, Wisconsin, in one hour against the wind. The pilot returned to Washington Island in 48 minutes $\left(\frac{48}{60} = \frac{4}{5} \text{ hour}\right)$ with the wind at the plane's back. How fast was the plane flying (without wind)? What was the speed of the wind?

Solution

Let A be the average speed of the airplane without wind and W be the speed of the wind, both in miles per hour. The total speed against the wind is then $A - W$, and the speed with the wind is $A + W$. There are two conditions given on these total speeds.

From Washington Island to Appleton the rate was $\frac{120 \text{ miles}}{1 \text{ hour}}$, or $120 \frac{\text{miles}}{\text{hour}}$.

This was against the wind, so A – W = 120.

From Appleton to Washington Island the rate was $\frac{120 \text{ miles}}{\frac{4}{5} \text{hour}}$, or $150 \frac{\text{miles}}{\text{hour}}$.

This was with the wind, so A + W = 150.

Now solve the system. Since the coefficients of W are opposites (1 and -1), add the equations.

$$\begin{array}{r} A - W = 120 \\ + \quad A + W = 150 \\ \hline 2A = 270 \\ A = 135 \end{array}$$

Substitute 135 for A in either of the original equations. We choose the second equation.

$$135 + W = 150$$
$$W = 15$$

The average speed of the airplane was 135 mph and the speed of the wind was 15 mph.

Check

Refer to the original question. Against the wind, the plane flew at 135 − 15, or 120 mph. With the wind the plane flew at 135 + 15, or 150 mph. At that rate, in 48 minutes the pilot flew $\frac{48}{60}$ hr · 150 $\frac{\text{mi}}{\text{hr}}$ = 120 miles, which checks with the given conditions.

Sometimes the coefficients of the same variable are equal. In this case, use the Multiplication Property of Equality to multiply both sides of one of the equations by -1. This changes all the numbers in that equation to their opposites. Then you can use the addition method to find solutions to the system.

Example 2

Solve this system.

$$\begin{cases} 4x + 13y = 40 \\ 4x + 3y = -40 \end{cases}$$

Solution

We rewrite the equations and number them.

$$4x + 13y = 40 \quad \#1$$
$$4x + 3y = -40 \quad \#2$$

Notice that the coefficients of x in the two equations are equal. Multiply the second equation by -1. Call the resulting equation #3.

$$-4x - 3y = 40 \quad \#3$$

Now use the addition method with the first and third equations.

$$\begin{array}{rl} 4x + 13y = 40 & \#1 \\ + \; -4x - 3y = 40 & \#3 \\ \hline 10y = 80 & \#1 + \#3 \\ y = 8 & \end{array}$$

To find x, substitute 8 for y in one of the original equations. We use the first equation.

$$4x + 13(8) = 40$$
$$4x + 104 = 40$$
$$4x = -64$$
$$x = -16$$

So (x, y) = (-16, 8).

Check

Substitute -16 for x and 8 for y in both equations.

Does $4x + 13y = 4 \cdot -16 + 13 \cdot 8 = 40$? Yes.

Does $4x + 3y = 4 \cdot -16 + 3 \cdot 8 = -40$? Yes.

Alternatively, in Example 2, you could subtract the second equation from the first. This again gives $10y = 80$ and from there you can continue as in the solution. The solution to Example 3 also uses subtraction.

Example 3

A resort hotel offers two weekend specials.

Plan *A:* 3 nights with 6 meals for \$264
Plan *B:* 3 nights with 2 meals for \$218

At these rates, what is the cost of one night's lodging and what is the average cost per meal? (Assume there is no discount for 6 meals.)

Solution

Let N = price of one night's lodging.
Let M = average price of one meal.
From Plan A: $3N + 6M = 264$
From Plan B: $3N + 2M = 218$
The coefficients of *N* are the same; so subtract the second equation from the first equation.

$$4M = 46$$
$$M = 11.5$$

Substitute 11.5 for *M* in either equation. We select the first equation.

$$3N + 6(11.5) = 264$$
$$3N + 69 = 264$$
$$3N = 195$$
$$N = 65$$

Thus, (N, M) = (65, 11.5). One night's lodging costs \$65.00; and an average meal costs \$11.50.

Check

In the Questions, you are asked to check that at these rates the totals for Plans *A* and *B* are correct.

QUESTIONS

Covering the Reading

1. **a.** When is adding equations an appropriate method to solve systems?
   b. What is the goal in adding equations to solve systems?

2. Which property allows you to add corresponding sides of two equations to get a new equation?

In 3 and 4, **a.** solve the system; **b.** check your solution.

3. $\begin{cases} 2x + 8y = 2 \\ -2x - 4y = 6 \end{cases}$

4. $\begin{cases} a + b = 11 \\ a - b = 4 \end{cases}$

5. The sum of two numbers is 90; their difference is 75. Find the numbers.

6. Find two numbers whose sum is -1 and whose difference is 5.

7. In Example 1, suppose it took 50 minutes to fly from Washington Island to Appleton against the wind and 40 minutes for the return flight with the wind. Find the speed of the plane and the speed of the wind in miles per hour under these conditions.

8. When is it useful to multiply an equation by -1 or subtract as a first step in solving a system?

In 9 and 10, solve the system.

9. $\begin{cases} 2x - 3y = 5 \\ 5x - 3y = 11 \end{cases}$

10. $\begin{cases} 2m + n = -5 \\ 2m + 3n = 7 \end{cases}$

11. Check Example 3.

12. A hotel offers the following specials: Plan A is two nights and one meal for \$153. Plan B is 2 nights and 4 meals for \$195. What price is the hotel charging per night and per meal?

Applying the Mathematics

13. As you know, $\frac{3}{4} = 75\%$ and $\frac{1}{5} = 20\%$. Is it true that $\frac{3}{4} - \frac{1}{5} = 55\%$? Justify your answer.

In 14 and 15, solve.

14. $\begin{cases} 2x - 3y = 17 \\ 3y + x = 1 \end{cases}$

15. $\begin{cases} 4z - 5w = 15 \\ 2w + 4z = -6 \end{cases}$

16. Suppose two eggs with bacon cost \$2.70. One egg with bacon costs \$1.80. At these rates, what should bacon alone cost?

17. Five gallons of regular unleaded gas and eight gallons of premium gas cost \$17.15. Five gallons of regular unleaded and two gallons of premium gas cost \$8.75. Find the cost per gallon of each kind of gasoline.

18. The tallest man known to have lived was Robert Wadlow (1918–1940). One of the most famous dwarfs was Charles Sherwood Stratton, alias "General Tom Thumb" (1838–1883). If they stood next to each other, Wadlow would have been 67″ taller. If one had stood on the other's head, they would have stood 12′3″ tall. How tall was each man?

General Tom Thumb. *Charles Sherwood Stratton is shown here with his wife Lavinia Warren. Stratton stopped growing when he was 6 months old. He remained 25 in. (0.6 m) tall until his teens when he began growing again.*

Review

In 19 and 20, solve by using any method. *(Lessons 11-1, 11-2, 11-3)*

19. $\begin{cases} y = 2x - 1 \\ y = 9x + 6 \end{cases}$

20. $\begin{cases} Q = 4z \\ R = -5z \\ 4R + Q = 40 \end{cases}$

21. **a.** Solve $x^2 + 5x - 14 = 0$.
    b. Find the x-intercepts of the graph of $y = x^2 + 5x - 14$. *(Lesson 9-5)*

In 22 and 23, let $P = (-3, -7)$ and $Q = (5, -1)$. *(Lessons 7-2, 9-8)*

22. Find the distance between P and Q.

23. Find the slope of $\overline{PQ}$.

24. Ida is playing with toothpicks. It takes 5 toothpicks to make a pentagon and 6 toothpicks to make a hexagon. She has 100 toothpicks. She wants to make P pentagons and H hexagons.

a. Give three different possible pairs of values of P and H that use all the toothpicks.

b. Write a sentence that expresses the relation between the number of pentagons and hexagons she can make with 100 toothpicks. *(Lesson 7-8)*

All that glitters. *Most women in Sri Lanka own at least one piece of gold jewelry. The jewelry serves as a form of insurance as well as an investment and symbol of wealth. Because of the constant demand for gold, there are goldsmiths throughout the country.*

25. Rewrite $-15 - 8x = y$ in the form $Ax + By = C$. *(Lesson 7-8)*

26. When jewelry is made of 14K gold, $\frac{14}{24}$ of the jewelry is gold and the rest is made of other materials. How many grams of gold are in a 14K gold necklace weighing 30 grams? *(Lesson 6-3)*

27. *Skill sequence.* Simplify. *(Lessons 3-2, 4-5, 5-9)*

a. $100 - (80 - 4p) + 2(p + 5)$ **b.** $\frac{100}{p+2} - \frac{80 - 4p}{p+2} + \frac{2(p+5)}{p+2}$

28. a. How many quarts are there in 5 gallons?

b. How many quarts are there in n gallons? *(Previous course)*

Exploration

29. Subtracting equations is part of a process that can be used to find simple fractions for repeating decimals. For instance, to find a fraction for $.\overline{39}$, first let

$$d = .\overline{39}.$$

Then multiply both sides of the equation by 10^2 because $.\overline{39}$ has a two-digit block that repeats.

$$100d = 39.\overline{39} \quad \#1$$
$$d = .\overline{39} \quad \#2$$

Subtract the second equation from the first.

$$99d = 39 \quad \#1 - \#2$$

Solve for d. $\quad d = \frac{39}{99}$

Simplify the fraction. $\quad d = \frac{13}{33}$

A calculator shows that $\frac{13}{33}$ has a decimal equivalent 0.393939 . . .

a. Use the above process to find a simple fraction equal to $.\overline{81}$

b. Modify the process to find a simple fraction equal to $.\overline{003}$.

c. Find a simple fraction equal to $3.89\overline{5}$.

LESSON

11-5

Solving Systems by Multiplication

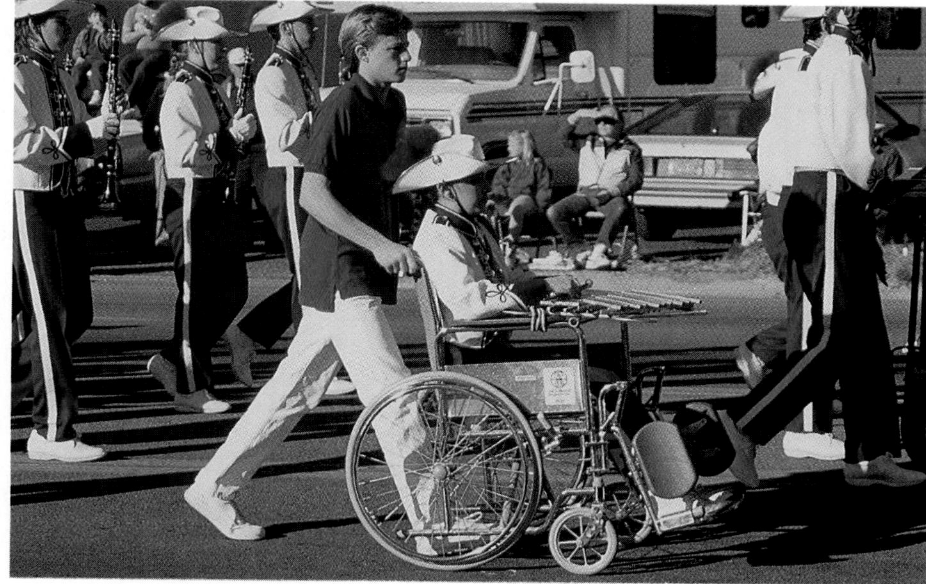

. . . and the band played on. *Musicians in a marching band often arrange themselves into various formations. Many marching bands include drum majors, baton twirlers, and pompon squads. See Example 3.*

Recall that there are two common forms for equations of lines.

Name of form	General form	Sample
standard form	$Ax + By = C$	$3x + 8y = 20$
slope-intercept form	$y = mx + b$	$y = -2x + 1$

The substitution method described in Lessons 11-2 and 11-3 is convenient for solving systems in which one or both equations are in slope-intercept form. The addition method studied in Lesson 11-4 is convenient for solving systems in which both equations are in standard form, and a pair of coefficients are either equal or opposites. However, not all systems fall into one of these two categories.

$$\text{Consider the system } \begin{cases} 5x + 8y = 21 \\ 10x - 3y = -15 \end{cases}.$$

Substitution could be used, but you would first have to solve one of the equations for one of the variables. This is messy! Adding or subtracting the two equations will not result in an equation with just one variable, because neither the x nor the y terms are equal or opposites.

To solve this system, you can use the Multiplication Property of Equality to create an equivalent system of equations. **Equivalent systems** are systems with exactly the same solutions. Notice that if you multiply each side of the first equation by -2, the x-terms of the resulting system have opposite coefficients.

$$\begin{cases} 5x + 8y = 21 \\ 10x - 3y = -15 \end{cases} \rightarrow \begin{cases} -10x - 16y = -42 \\ 10x - 3y = -15 \end{cases}$$

These systems may look different, but they have the same solutions. As Example 1 shows, the system on the right at the bottom of page 687 can be solved using the addition method.

Example 1

Solve the following system.

$$\begin{cases} 5x + 8y = 21 \\ 10x - 3y = -15 \end{cases}$$

Solution 1

Multiply the first equation by -2.

$$\begin{cases} -10x - 16y = -42 \\ 10x - 3y = -15 \end{cases}$$

Add the equations. $-19y = -57$

Solve for y. $y = 3$

Substitute $y = 3$ in one of the original equations to find x.

$$5x + 8y = 21$$
$$5x + 8 \cdot 3 = 21$$
$$5x + 24 = 21$$
$$5x = -3$$
$$x = -\frac{3}{5}$$

So the solution is $(x, y) = \left(-\frac{3}{5}, 3\right)$.

Solution 2

Multiply the second equation by $-\frac{1}{2}$. This also makes the coefficients of x opposites.

$$\begin{cases} 5x + 8y = 21 \\ 10x - 3y = -15 \end{cases} \rightarrow \begin{cases} 5x + 8y = 21 \\ -5x + 1.5y = 7.5 \end{cases}$$

Add. $9.5y = 28.5$

$y = 3$

Proceed as in Solution 1 to find x.

Again $(x, y) = \left(-\frac{3}{5}, 3\right)$.

Check

Does $5 \cdot -\frac{3}{5} + 8 \cdot 3 = 21$? Yes, $-3 + 24 = 21$.

Does $10 \cdot -\frac{3}{5} - 3 \cdot 3 = -15$? Yes, $-6 - 9 = -15$.

Example 1 shows that the solution is the same no matter which equation is multiplied by a number. The goal is to obtain opposite coefficients for one of the variables in the two equations. Then the resulting equations can be added to eliminate that variable. This technique is sometimes called the **multiplication method** for solving a system.

Sometimes it is easier to multiply **each** equation by a number before adding.

Example 2

Solve the system $\begin{cases} 3a + 5b = 8 \\ 2a + 3b = 4.6. \end{cases}$

Solution

You have your choice of eliminating a or b. In either case, you can multiply by a number so that the resulting system has a pair of opposite coefficients. To make the coefficients of a opposites, multiply the first equation by 2 and the second equation by -3.

$$\begin{cases} 6a + 10b = 16 \\ -6a - 9b = -13.8 \end{cases}$$

Now add. $b = 2.2$

Substitute 2.2 for b in the first equation to find the value of a.

$$3a + 5(2.2) = 8$$
$$3a + 11 = 8$$
$$3a = -3$$
$$a = -1$$

Therefore, $(a, b) = (-1, 2.2)$.

Check

Does $3(-1) + 5(2.2) = 8$? Yes, $-3 + 11 = 8$.
Does $2(-1) + 3(2.2) = 4.6$? Yes, $-2 + 6.6 = 4.6$.

Many situations naturally lead to linear equations in standard form. This results in a linear system that can be solved using multiplication.

Example 3

A marching band has 52 musicians (M) and 24 people in the pompon squad (P). They wish to form hexagons and squares like those diagrammed below. Can it be done with no people left over? If so, how many hexagons and how many squares can be made?

Hexagon
pompon person in center

Square
musician in center

It's a great day for a parade. *This pompon squad is marching in Chicago's annual St. Patrick's Day Parade.*

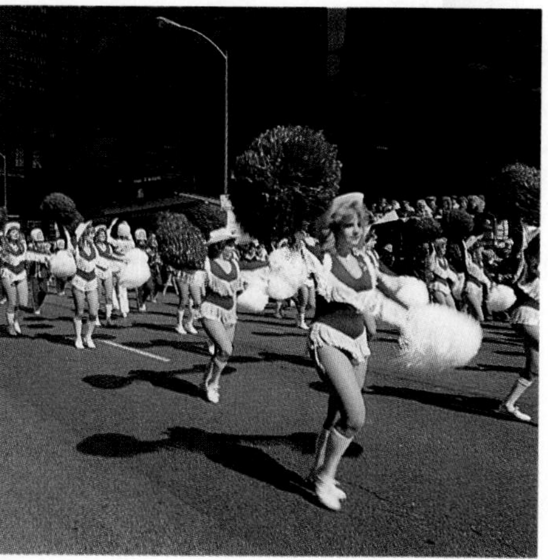

Solution

Consider the formation to include h hexagons and s squares. There are two conditions in the system: one for musicians and one for pompon people.

There are $6 \frac{\text{musicians}}{\text{hexagon}}$ and $1 \frac{\text{musician}}{\text{square}}$, so $6h + s = 52$.

There is $1 \frac{\text{pompon person}}{\text{hexagon}}$ and $4 \frac{\text{pompon people}}{\text{square}}$, so $h + 4s = 24$.

Multiply the first equation by -4 and add the result to the second equation.

$$\begin{aligned} -24h - 4s &= -208 \\ h + 4s &= 24 \\ \hline -23h &= -184 \\ h &= 8 \end{aligned}$$

Substitute 8 for h in one of the original equations. We use the first equation.

$$\begin{aligned} 6 \cdot 8 + s &= 52 \\ 48 + s &= 52 \\ s &= 4 \end{aligned}$$

Since h and s are both positive integers, the formations can be done with no people left over. The band can make 8 hexagons and 4 squares.

Check

Making 8 hexagons would use 48 musicians and 8 pompon people. Making 4 squares would use 4 musicians and 16 pompon people. This setup uses exactly $48 + 4 = 52$ musicians and $8 + 16 = 24$ pompon people.

When a system involves equations that are not in either standard or slope-intercept form, it is wisest to rewrite the equations in one of these forms before proceeding. For example, to solve the system

$$\begin{cases} 5b = 8 - 3a \\ 2a + 3b = 4.6 \end{cases}$$

you could add $3a$ to both sides of the first equation. The result is the system which was solved in Example 2.

QUESTIONS

Covering the Reading

1. Which property allows you to multiply both sides of an equation by a number without affecting the solutions?

2. a. What are equivalent systems?
b. Give an example of two equivalent systems in this lesson.

3. Consider this system. $\begin{cases} 6u - 5v = 2 \\ 12u - 8v = 5 \end{cases}$
   a. If the first equation is multiplied by _?_, then adding the equations will eliminate u.
   b. Solve the system.

4. In Example 2, suppose the first equation is multiplied by 3 and the second equation is multiplied by -5.
   a. What is the resulting system?
   b. Use the system from part **a** to verify the answer to Example 2.

5. Consider this system. $\begin{cases} 3a - 2b = 20 \\ 9a + 4b = 40 \end{cases}$
   a. If the first equation is multiplied by -3, and the result is added to the second equation, what is the resulting equation?
   b. If the first equation is multiplied by 2, and the result is added to the second equation, what is the resulting equation?
   c. Use one of these methods to solve the system.

6. Solve the system $\begin{cases} x + 3y = 19 \\ 2x + y = 3. \end{cases}$
   a. by eliminating x first.
   b. by eliminating y first.

7. Use the hexagon and square formations of Example 3. Will there be an exact fit if the marching band consists of 100 musicians and 42 pompon people? Why or why not?

8. A marching band has 67 musicians and 47 pompon people. They wish to form pentagons and squares like those diagramed below. Will every person have a spot? If so, how many of each formation will be needed?

Canadian bands. *These marchers are in a parade at the Calgary Stampede, an annual exhibition and rodeo in Alberta, Canada. It lasts several days and includes many unusual rodeo events.*

In 9–12, solve the system.

9. $\begin{cases} 5x + y = 30 \\ 3x - 4y = 41 \end{cases}$

10. $\begin{cases} 4a + b = 38 \\ 2a + 3b = 24 \end{cases}$

11. $\begin{cases} 2m - 5n = 0 \\ 6m + n = 0 \end{cases}$

12. $\begin{cases} 6m - 7n = 6 \\ 7m - 8n = 15 \end{cases}$

Applying the Mathematics

In 13 and 14, solve the system by first rewriting each equation in standard form.

13. $\begin{cases} 3x = 4y + 2 \\ 9x - 5y = 7 \end{cases}$

14. $\begin{cases} 3a = 2b + 5 \\ a - 4b = 6 \end{cases}$

15. The sum of two numbers is 45. Three times the first number plus seven times the second is 115. Find the two numbers.

16. A test has m multiple-choice questions and t true-false questions. If multiple choice questions are worth 7 points each and the true-false questions 2 points each, the test will be worth a total of 185 points. If multiple-choice and true-false questions are all worth 4 points apiece, the test will be worth a total of 200 points. Find m and t.

17. A wildlife management station is to care for sick animals. At present they have only birds and deer. The animals being cared for have 25 heads and 74 feet. (No animal is missing a foot!) How many of each type of animal are at the station? Explain your reasoning.

Caretakers. *Pictured is a park ranger from Cumberland Island, Georgia. National park rangers often find injured animals and arrange for their care.*

18. Solve the system $\begin{cases} 4x - 3y = 2x + 5 \\ 8y = 5x - 13 \end{cases}$.

Review

In 19 and 20, solve the system using any method. *(Lessons 11-1, 11-2, 11-3, 11-4)*

19. $\begin{cases} 10x + 5y = 32 \\ 8x + 5y = 10 \end{cases}$

20. $\begin{cases} y = \frac{1}{2}x + 3 \\ y = \frac{1}{3}x - 2 \end{cases}$

21. The two diagrams below illustrate a system of equations.

a. Write an equation for the diagram on the left.
b. Write an equation for the diagram on the right.
c. Solve the system for x and y.
d. Check your work. *(Lessons 3-1, 11-2, 11-4)*

22. Molly has \$400 and saves \$25 a week. Vince has \$1400 and spends \$25 a week.
a. How many weeks from now will they each have the same amount of money?
b. What will this amount be? *(Lesson 11-2)*

23. **Pythagorean triples** are three whole numbers A, B, and C, such that $A^2 + B^2 = C^2$. Here is a program that will generate a Pythagorean triple from two positive integers M and N, where $M > N$. *(Lesson 7-5)*

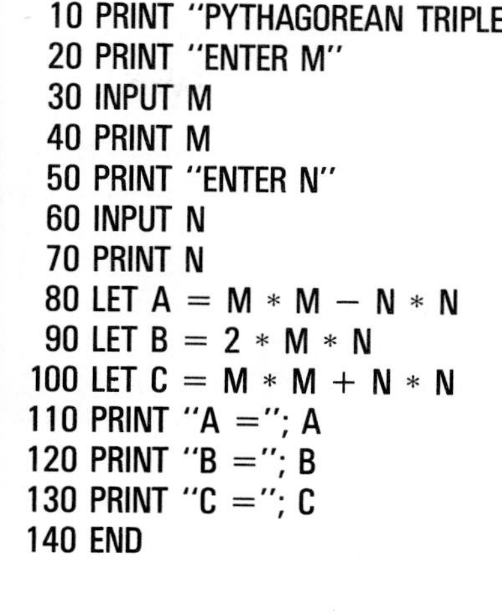

```
 10 PRINT "PYTHAGOREAN TRIPLES"
 20 PRINT "ENTER M"
 30 INPUT M
 40 PRINT M
 50 PRINT "ENTER N"
 60 INPUT N
 70 PRINT N
 80 LET A = M * M - N * N
 90 LET B = 2 * M * N
100 LET C = M * M + N * N
110 PRINT "A ="; A
120 PRINT "B ="; B
130 PRINT "C ="; C
140 END
```

a. What will the program print when $M = 5$ and $N = 3$?
b. What will the program print when $M = 7$ and $N = 1$?
c. Verify that the answers to parts **a** and **b** satisfy the Pythagorean Theorem.

24. *True or false.* *(Lesson 6-4)*
a. Probabilities are numbers from 0 to 1.
b. A probability of 1 means that an event must occur.
c. A relative frequency of -1 cannot occur.

25. In December, 1986, Jeana Yeager and Dick Rutan flew the *Voyager* airplane nonstop around the earth without refueling. The average rate for the 24,987-mile trip was 116 mph. How many days long was this flight? *(Lesson 6-2)*

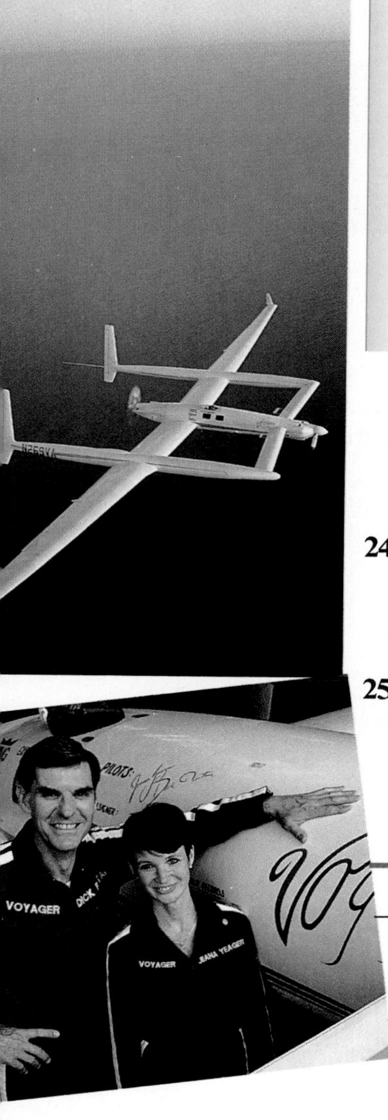

Heavy load. *The fuel for the* Voyager *on its famous flight around the world was stored in the fuselage, wings, and other frame elements. The fuel weighed four times as much as the airplane!*

Exploration

26. If your school has a band with a pompon squad, determine whether the members could fit exactly into the formations of Example 3.

27. Create formations that would require all members of a band consisting of 80 musicians and a 30-member pompon squad.

LESSON

11-6

Systems and Parallel Lines

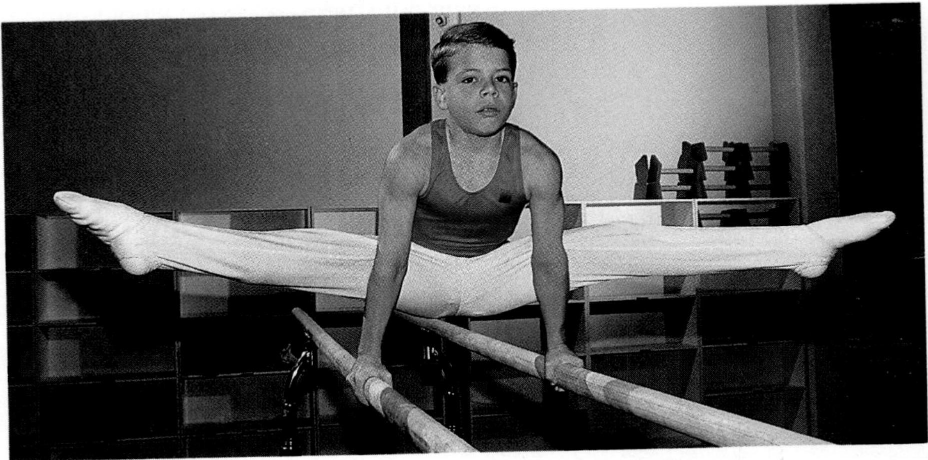

Drawing a parallel. *The parallel bars used in men's gymnastics competitions were invented by Friedrich Jahn. The bars are 1.7 m (5 ft 5 in.) high and are from 42 cm (16.5 in.) to 48 cm (19 in.) apart.*

Recall that parallel lines "go in the same direction." All vertical lines are parallel to each other. So are all horizontal lines. But not all oblique lines are parallel. For oblique lines to be parallel, they must have the same slope.

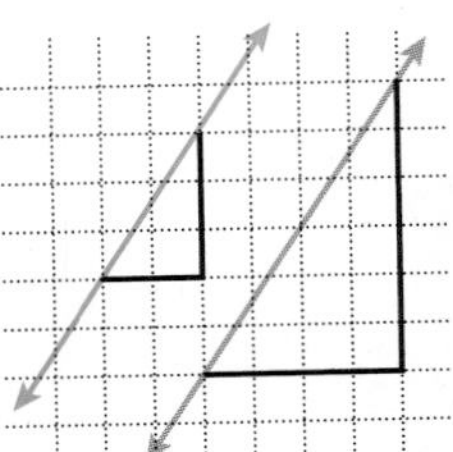

Slopes and Parallel Lines Property
If two lines have the same slope, then they are parallel.

Nonintersecting Parallel Lines

You have learned that when two lines intersect in exactly one point, the coordinates of the point of intersection can be found by solving a system. But what happens when the lines are parallel? Consider this linear system.

$$\begin{cases} 2x + 3y = -6 \\ 4x + 6y = 24 \end{cases}$$

You can solve the system by multiplying the first equation by -2 and adding the result to the second equation.

$$\begin{array}{r} -4x - 6y = 12 \\ + \ 4x + 6y = 24 \\ \hline 0 = 36. \end{array}$$

When you add you get $0 = 36$.

This is impossible! The false statement $0 = 36$ signals that the system has no solution. The graphs have no point of intersection. To check, graph the lines. The graph on page 695 shows that the lines are parallel with no points in common.

As another check, rewrite the equations for the lines in slope-intercept form.

$$\begin{aligned} \text{line } \ell\text{:} \quad 2x + 3y &= -6 \\ 3y &= -2x - 6 \\ y &= -\tfrac{2}{3}x - 2 \\ \text{line } m\text{:} \quad 4x + 6y &= 24 \\ 6y &= -4x + 24 \\ y &= -\tfrac{2}{3}x + 4 \end{aligned}$$

Lines ℓ and m have the same slope, $-\frac{2}{3}$. Thus they are parallel.

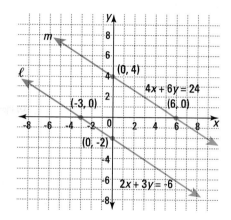

Planely parallel. *Uneven parallel bars are used in women's gymnastic competitions. While the bars are parallel to each other, the plane that contains the bars is not parallel to the floor.*

When an equation with no solution (such as 0 = 36) results from correct applications of the addition and multiplication methods on a system of linear equations, the system has no solutions. That is, there are no pairs of numbers that work in *both* equations. The graph of the system is two parallel nonintersecting lines.

Coincident Lines

Lines that coincide are sometimes called **coincident.** They too are parallel because they go in the same direction and have the same slope.

Example

Solve the system $\begin{cases} y = 1.6x - 6 \\ 16x - 10y = 60 \end{cases}$.

Solution 1

Draw a graph of each equation. As shown below, the graphs are the same line. No matter how much you zoom with an automatic grapher, the two lines are identical. The solution set consists of all ordered pairs on these lines.

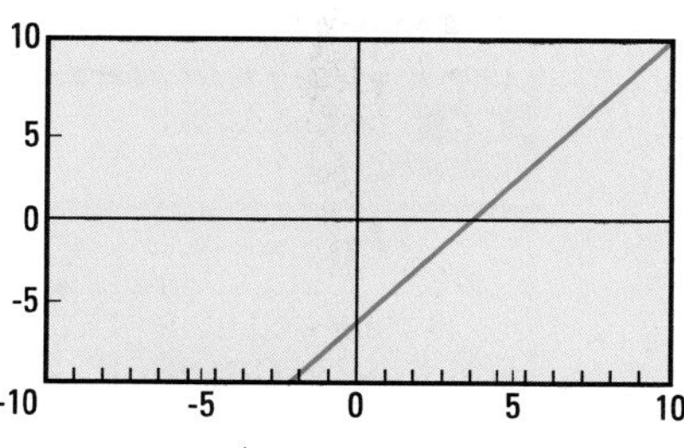

Solution 2

Rewrite the second equation in slope-intercept form.

$$16x - 10y = 60$$

$$-10y = -16x + 60 \quad \text{Add } -16x \text{ to each side.}$$

$$y = 1.6x - 6 \quad \text{Divide each side by } -10.$$

Notice that this equation is identical to the first. So, any ordered pair that is a solution to one equation is also a solution to the other equation.

Solution 3

Substitute $1.6x - 6$ for y in the second equation.

$$16x - 10(1.6x - 6) = 60$$

$$16x - 16x + 60 = 60$$

$$60 = 60$$

The sentence $60 = 60$ is always true. So, any ordered pair that is a solution to one equation is a solution to the other equation.

Check

Find an ordered pair that satisfies $y = 1.6x - 6$, say (0, -6). Check that it also satisfies the second equation. Does $16 \cdot 0 - 10 \cdot -6 = 60$? Yes, $60 = 60$.

When a sentence that is always true ($60 = 60$) occurs from correct work with a system of linear equations, the system has infinitely many solutions.

The table summarizes the ways that two lines in the plane can be related and gives the corresponding solutions for a system of their equations.

Graph of system	Number of solutions to system	Slopes of lines
Two intersecting lines	One (the point of intersection)	Different
Two parallel and nonintersecting lines	Zero	Equal
One line (parallel and coincident lines)	Infinitely many	Equal

QUESTIONS

Covering the Reading

1. What is true about the slopes of parallel lines?

2. Which two lines are parallel?
   (a) $y = 3x + 5$ (b) $y = 2x + 5$
   (c) $y = 3x + 6$ (d) $x = 2y + 5$

3. **a.** Graph the line $y = -2x + 5$.
   b. Draw a line parallel to it through the origin.
   c. What is an equation of the line you drew in part **b**?

4. Give an example of a system with two nonintersecting lines.

5. Give an example of a system with two coincident lines.

In 6 and 7, a system is given. **a.** Determine whether the system includes nonintersecting or coincident lines. **b.** Check by graphing.

6. $\begin{cases} 2x - 3y = 12 \\ 8x - 12y = 12 \end{cases}$

7. $\begin{cases} x - y = 5 \\ y - x = -5 \end{cases}$

8. Consider the system $\begin{cases} y = 2x - 1 \\ 4x - 2y = 2 \end{cases}$.
   a. Give the coordinates of 3 ordered pairs that are solutions of $y = 2x - 1$.
   b. Show that each pair you gave also is a solution of $4x - 2y = 2$.
   c. Use the substitution method to solve this system.

In 9–11, match the description of the graph with the number of solutions to the system.

9. lines intersect in one point — (a) no solution

10. lines nonintersecting — (b) infinitely many solutions

11. lines coincident — (c) one solution

Applying the Mathematics

In 12–15, *multiple choice.* Describe the graph of the system.
(a) two intersecting lines
(b) two parallel non-intersecting lines
(c) one line

12. $\begin{cases} 3u + 2t = 7 \\ 14 - 2t = 6u \end{cases}$

13. $\begin{cases} 6a + 2b = 9 \\ 9a + 3b = 12 \end{cases}$

14. $\begin{cases} 2x - 5y = -3 \\ -4x + 10y = 6 \end{cases}$

15. $\begin{cases} \frac{1}{2}x - \frac{3}{2} = y \\ 2x + y = 3 \end{cases}$

16. Could the given situation have happened? Justify your answer by using a system of equations.

    A pizza parlor sold 39 pizzas and 21 gallons of soda for $396. The next day, at the same prices, they sold 52 pizzas and 28 gallons of soda for $518.

Review

17. Suppose you negotiated a deal to buy 500 square yards of carpet and 500 square yards of linoleum for a total of $9600, which is the price after receiving a 10% discount, or so you've been told. You later learn that another customer bought 60 square yards of the same carpet and 30 square yards of the same linoleum for $1080. Still another customer purchased 50 square yards of the carpet and 10 square yards of the linoleum for $840. If the others paid full price, did you receive the discount you were promised? Explain. *(Lesson 11-5)*

18. A band has 94 musicians (M) and 28 flag bearers (F). They plan to form the following formations:

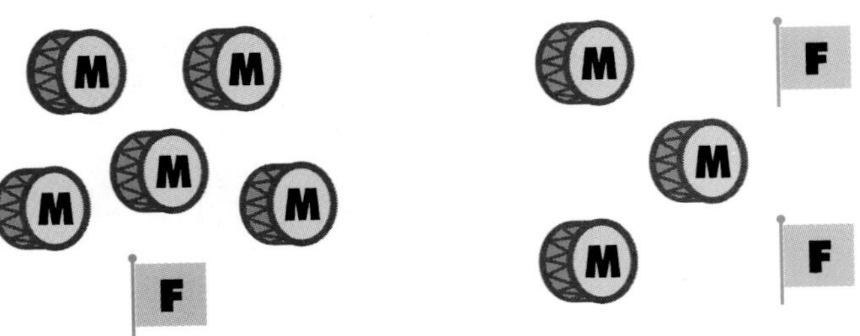

Can all members be fit into these formations? Why or why not? *(Lesson 11-5)*

19. Each diagram at the left represents an equation involving lengths t and u.
a. Write a pair of equations describing these relations.
b. Use either your equations or the diagrams to find the lengths of t and u. Explain your reasoning.
c. Check your work. *(Lesson 11-4)*

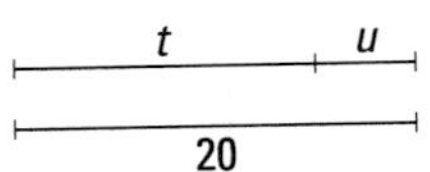

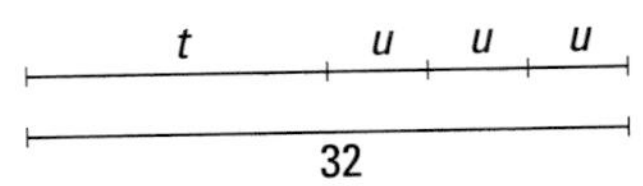

20. Solve by any method. $\begin{cases} 4x + 3y = 7 \\ x - 1.5y = 7 \end{cases}$ *(Lessons 11-1, 11-3, 11-5)*

21. a. Write 1872 as a polynomial with 10 substituted for the variable.
b. Write 1872 in scientific notation. *(Lesson 10-1, Appendix)*

22. Draw the graph of $y \le 2x - 3$. *(Lesson 7-9)*

23. A sweater costs a store owner $15.50. If the profit is to be 40% of the cost, what is the selling price? *(Lessons 3-7, 6-5)*

24. a. Give an equation of the horizontal line through (8, -12).
b. Give an equation of the vertical line through (8, -12). *(Lesson 5-1)*

Exploration

25. Here are some examples of parallel lines, both non-intersecting and coincident. The first system in the first group is from the lesson.

Nonintersecting

$\begin{cases} 2x + 3y = -6 \\ 4x + 6y = 24 \end{cases}$ $\begin{cases} 3x + 6y = -2 \\ 4x + 8y = 4 \end{cases}$

$\begin{cases} -9x - 3y = 5 \\ 6x + 2y = 1 \end{cases}$

Coincident

$\begin{cases} 8x - 7y = 3 \\ 16x - 14y = 6 \end{cases}$ $\begin{cases} -2x + 8y = 30 \\ 5x - 20y = -75 \end{cases}$

$\begin{cases} 12x - 4y = 80 \\ 15x - 5y = 100 \end{cases}$

Look for patterns in the two columns of equations. (Hint: Compare the ratios of the coefficients of x, the coefficients of y, and the constant terms.) Use the patterns you find to explain how to recognize parallel lines directly from their equations.

LESSON

11-7 *Situations Which Always or Never Happen*

Believe it or not. *Lumiere, the talking candelabra from Walt Disney's* Beauty and the Beast, *sings and dances atop a cake. This scene is clearly a situation that never occurs in real life.*

Which job would you take?

Job 1	**Job 2**
Starting wage \$5.60/hour; every 3 months the wage increases \$.10/hour.	Starting wage \$5.50/hour; every 3 months the wage increases \$.10/hour.

Of course, the answer is obvious. Job 1 will always pay better than Job 2. But what happens when this is solved algebraically? Let n = number of 3-month periods worked.

Wage at Job 1	Wage at Job 2
$5.60 + 0.10n$	$5.50 + 0.10n$

When is the pay at Job 1 better than the pay at Job 2? You must solve

$$5.60 + 0.10n > 5.50 + 0.10n.$$

Add $-0.10n$ to each side.

$$5.60 + 0.10n - 0.10n > 5.50 + 0.10n - 0.10n$$
$$5.60 > 5.50$$

As was the case in solving systems representing parallel lines, the variable has disappeared. Since $5.60 > 5.50$ is always true, n can be any real number. Job 1 will always pay a better wage than Job 2, as expected. For any equation or inequality the following generalization is true.

If, in solving a sentence, you get a sentence which is *always* true, then the original sentence is always true.

When does Job 1 pay less than Job 2? To answer this, you could solve:

$$5.60 + 0.10n < 5.50 + 0.10n.$$
$$5.60 + 0.10n - 0.10n < 5.50 + 0.10n - 0.10n$$
$$5.60 < 5.50$$

It is never true that 5.60 is less than 5.50. So Job 1 never pays less than Job 2, something which was obvious from the pay rates. The following generalization is also true.

> **If, in solving a sentence, you get a sentence which is *never* true, then the original sentence is never true.**

Suppose the sentence you are solving has only one variable. If it leads to a sentence (such as $0 = 0$) which is always true, the solution set is the set of all real numbers. When a false statement (such as $0 = 2$) arises, there is no real solution to the original equation or inequality.

Example 1

Solve $5 + 3x = 3(x - 2)$.

Solution

$$5 + 3x = 3x - 6$$
$$5 + 3x + -3x = -3x + 3x + -6$$
$$5 = -6$$

The statement $5 = -6$ is never true, so the original sentence has no solution.

Example 2

Solve $8(2y + 5) < 16y + 60$.

Solution

$$16y + 40 < 16y + 60$$
$$-16y + 16y + 40 < -16y + 16y + 60$$
$$40 < 60$$

This is always true, so the original sentence is true for every possible value of y. Thus, y may be any real number.

Check

Substitute any real number for y, say -3. Is it true that $8(2 \cdot -3 + 5) < 16 \cdot -3 + 60$? Is $8(-1) < -48 + 60$? Yes. $-8 < 12$. It checks.

CAUTION: Here is a problem that looks like the one at the beginning of the lesson but is different.

Example 3

Suppose Town 1 and Town 2 each have populations of 23,000 at present. Town 1's population is growing by 1000 people per year; Town 2 is growing by 1200 per year. When will the population of the towns be the same?

Solution

Let t = number of years from now.

Population of Town 1	Population of Town 2
23,000 + 1000t	23,000 + 1200t

The populations will be the same when

$$23{,}000 + 1000t = 23{,}000 + 1200t.$$
$$23{,}000 = 23{,}000 + 200t$$
$$0 = 200t$$
$$0 = t$$

The solution $t = 0$ means The populations of the towns are the same now.

Notice that having zero for a solution, as in Example 3, is different from having no solution at all.

QUESTIONS

Covering the Reading

Ticket to ride. *These rides are at Knott's Berry Farm in California. Amusement parks were first developed in the United States in the late 1800s.*

1. The Tri-City Amusement Park pays a starting salary of \$6.20 an hour, and each year increases it by \$1.00 an hour. Molly's Supermart starts at \$6.00 an hour, and also increases \$1.00 an hour per year.
a. When does Tri-City pay more?
b. Show how algebra can be used to represent this situation.

2. a. Add $-2x$ to both sides of the sentence $2x + 10 < 2x + 8$. What sentence results?
b. What should you write to describe the solutions to this sentence?

3. a. Add $5y$ to both sides of the sentence $-5y + 9 = 3 - 5y + 6$. What sentence results?
b. What should you write to describe the solution(s) to this sentence?
c. Check your solution(s).

In 4–9, solve.

4. $2(2y - 5) \leq 4y + 6$

5. $3x + 5 = 5 + 3x$

6. $-2m = 3 - 2m$

7. $2A - 10A > 4(1 - 2A)$

8. $\frac{1}{2}x + 6 = 3\left(\frac{1}{4}x + 2\right)$

9. $7 + 4y > 7 - y$

10. The population of Homsburg is about 200,000 and growing at about 5,000 people a year. Prairieville has a population of about 200,000 and is growing at about 4,000 people a year.

a. In y years, what will be the population of Homsburg?

b. In y years what will be the population of Prairieville?

c. When will their populations be the same?

Applying the Mathematics

In 11–13, consider the following information. Apartment A rents for \$375 per month including utilities. Apartment B rents for \$315 per month but the renter must pay \$60 per month for utilities, and a one-time \$25 fee for a credit check.

11. a. What sentence could you solve to find out when apartment A is cheaper?

b. Solve this sentence.

12. a. What sentence could you solve to find out when apartment B is cheaper?

b. Solve this sentence.

13. If you wanted to rent one of these apartments for two years, which one would be cheaper?

In 14–16, make up an example of an equation different from those in this lesson with the given solution.

14. The only solution is $x = 0$.

15. There is no real solution.

16. The equation is true for all real numbers.

Review

17. Without drawing any graphs, explain how you can tell whether the graphs of $11x - 10y = 102$ and $12x - 10y = 101$ are two intersecting lines, one line, or two nonintersecting parallel lines. *(Lesson 11-6)*

In 18 and 19, solve by using any method. *(Lessons 11-1, 11-2, 11-4, 11-6)*

18. $\begin{cases} 4x - 3y = 12 \\ 8x - 6y = 24 \end{cases}$

19. $\begin{cases} y = 5x - 7 \\ y = 7 - 5x \end{cases}$

20. Two hungry football players go through a cafeteria line. One orders 3 slices of lasagna and 3 salads and pays \$8.64. The other orders 4 slices of lasagna and 2 salads, and pays \$9.54. How much would two slices of lasagna and one salad cost? *(Lesson 11-5)*

21. In parts **a-c**, solve the system $\begin{cases} y = -x \\ y = x + 3 \end{cases}$.
   a. by graphing.
   b. by substitution.
   c. by the addition method.
   d. Which method in parts **a-c** do you prefer? Why?
   (Lessons 11-1, 11-2, 11-3)

22. The graph of the equations $x + y = 5$, $x = -1$, and $y = 1$ form a triangle.
   a. Find the vertices of the triangle.
   b. Find the length of each side of the triangle.
   c. Find the area of the triangle. *(Lessons 1-8, 9-9, 11-1)*

Ancient dating. *The Aztec "Calendar Stone" that appears on this stamp was uncovered in Mexico City in 1790. The Aztec calendar consists of 18 months of 20 days each, plus an additional five days called* nemontemi. *The* nemontemi *were considered very unlucky.*

23. Simplify so that no parentheses are needed. *(Lesson 8-8)*
   a. $\left(\frac{2}{3}\right)^2$ **b.** $(5d^2g)^2$ **c.** $\frac{1}{4}\left(\frac{c}{a}\right)^2$

24. Graph the half-plane $x + 2y > 0$. *(Lesson 7-9)*

25. An Aztec calendar is being placed on a rectangular mat that is twice as wide and 1.5 times as high as the calendar. What percent of the mat is taken up by the calendar? *(Lessons 6-1, 6-6)*

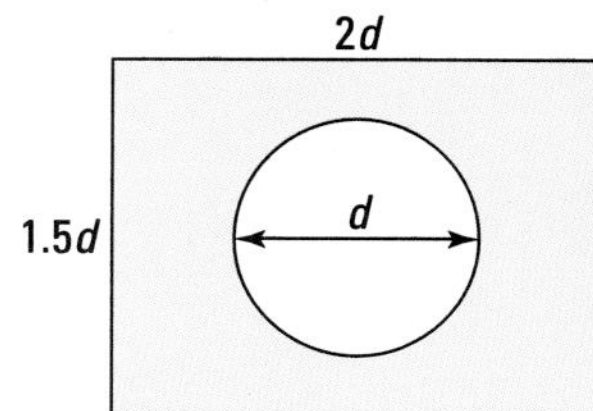

26. Mrs. Chang wants to lease about 2,500 square meters of floor space for a business. As she walked by a set of vacant stores, she wondered if the 4 stores together would meet this requirement. Their widths are given in the floor plan below. How deep must these stores be to give the required area? *(Lessons 2-1, 2-5)*

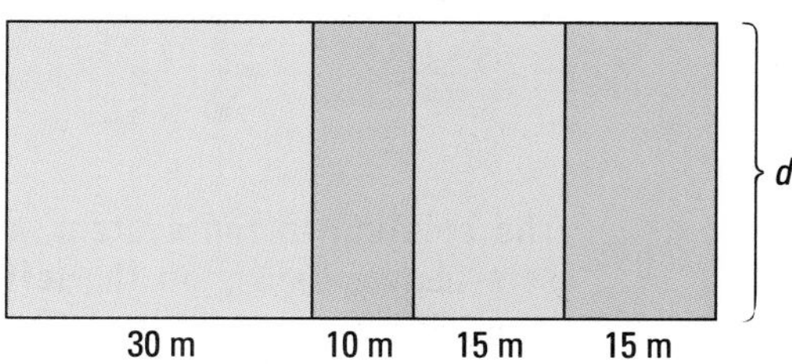

27. Suppose a book has an average of 25 lines per page, w words per line, and 21 pages per chapter. Estimate the number of words per chapter in this book. *(Lesson 2-4)*

Exploration

28. When you solve the equation $ax + b = cx + d$ for x, you may find no solution, exactly one solution, or infinitely many solutions. What must be true about a, b, c, and d to guarantee each of the following?
   a. There is exactly one solution.
   b. There are no solutions.
   c. There are infinitely many solutions.

LESSON

11-8

Systems of Inequalities

In Lesson 7–9, you graphed linear inequalities like $x > -5$ and $y \le 2x + 3$ on a plane. These sentences describe half-planes. Now you will graph regions described by two or more inequalities. Solving a system of inequalities involves finding the common solutions of two or more inequalities. As with systems of equations, in this course we will concentrate on linear sentences.

Example 1

Graph all solutions to the system $\begin{cases} x \ge 0 \\ y \ge 0 \end{cases}$.

Solution

First graph the solution to $x \ge 0$. It is shown on the left below. Then on it superimpose the graph of $y \ge 0$, shown by itself at the right below.

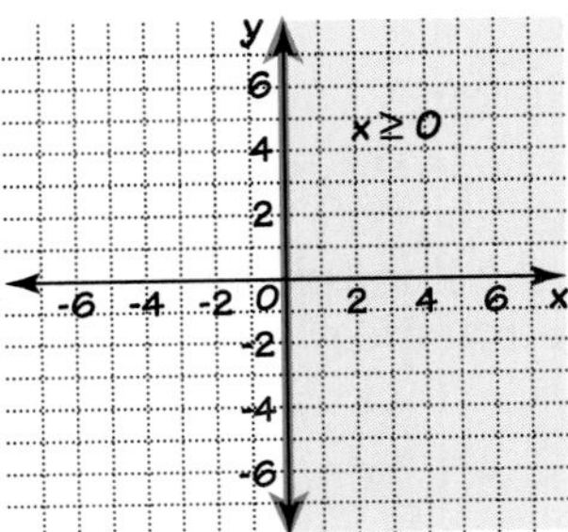

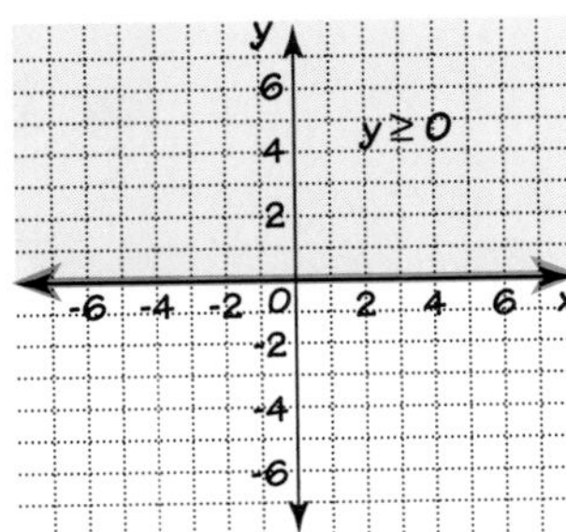

The solution to the system is the set of points common to both of the sets above. Below on the left we show the solutions to the two inequalities superimposed. On the right is the solution to the system. Notice that it is the intersection of the two solution sets above. It consists of the first quadrant and the nonnegative *x*- and *y*-axes.

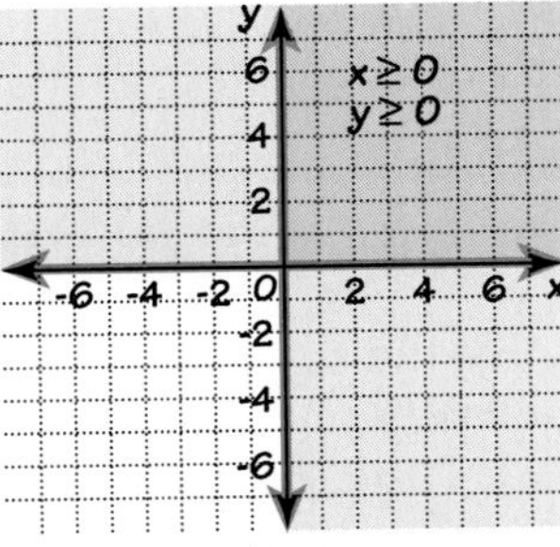

what your paper should look like

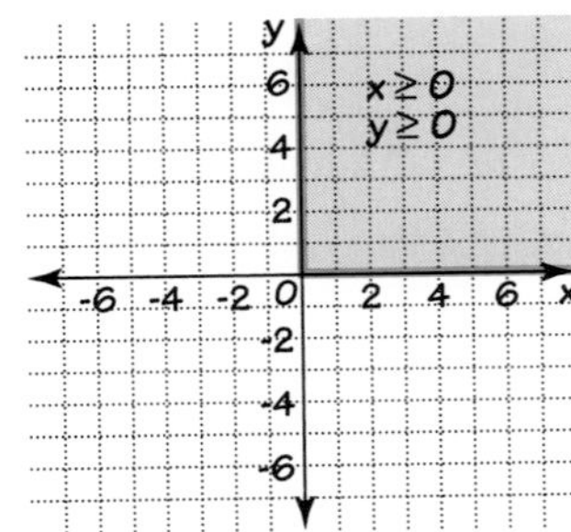

the solution to the system

Recall that in general, the graph of $Ax + By < C$ is a half-plane, and that it lies on one side of the boundary line $Ax + By = C$.

Example 2

Graph all solutions to the system $\begin{cases} y \geq -3x + 2 \\ y < x - 2 \end{cases}$.

Solution

First graph the boundary line $y = -3x + 2$ for the first inequality. The graph of $y \geq -3x + 2$ consists of points on or above the line with equation $y = -3x + 2$.

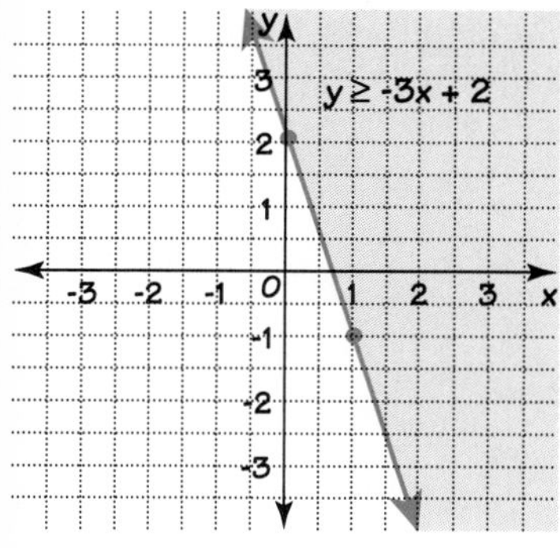

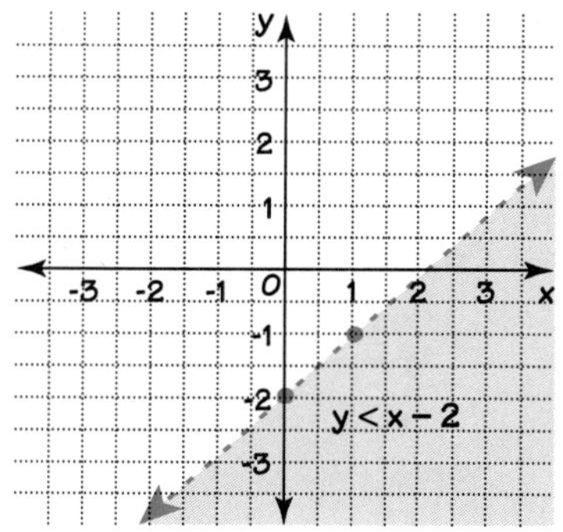

Next, graph the boundary line $y = x - 2$ for the second inequality. The graph of $y < x - 2$ consists of points below the line given by $y = x - 2$. Points on the line are excluded, as shown at the right above.

We have drawn the half-planes on different axes only to make it easier to see them. You should draw them on the same axes. The part of the plane marked with both types of shading is the solution set for the system.

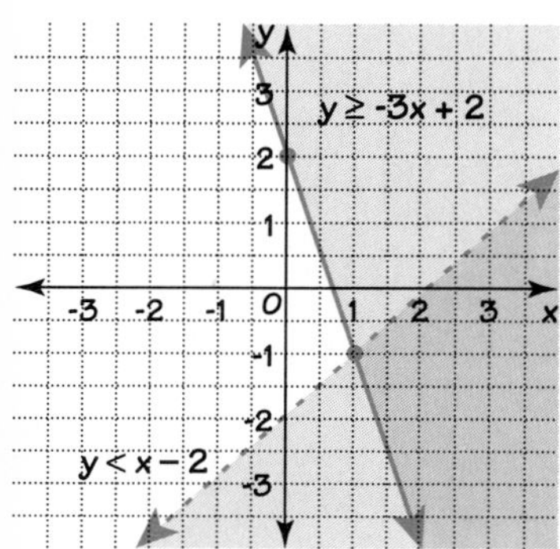

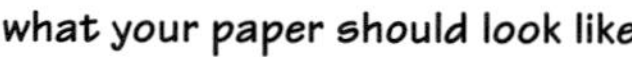
what your paper should look like

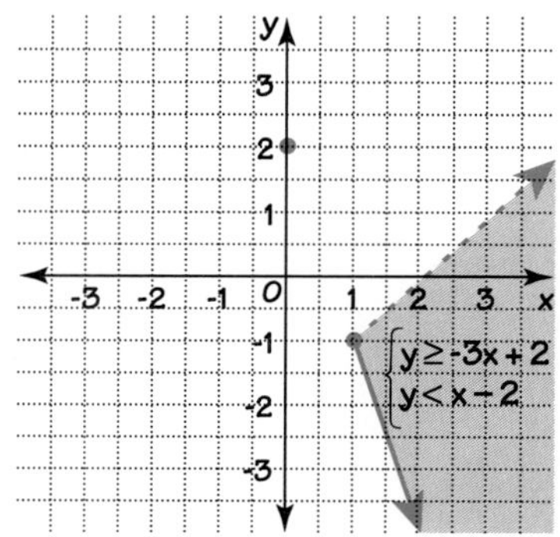

the solution to the system

In Example 2 notice that the point (1, -1) of intersection of the two boundary lines was seen easily from the graph. If an intersection point cannot be found easily from a graph, you can use any of the other techniques learned in this chapter to find its coordinates.

Example 3 shows a system of four linear inequalities.

Example 3

Suppose the sum of two positive numbers x and y is less than 50 and greater than 25. Show all possible values for x and y graphically.

Solution

Because x and y are positive, $x > 0$ and $y > 0$. The desired numbers are the solution to this system of inequalities.

$$\begin{cases} x > 0 \\ y > 0 \\ x + y < 50 \\ x + y > 25 \end{cases}$$

The graph of the solutions to the system is the intersection of the four sets graphed below.

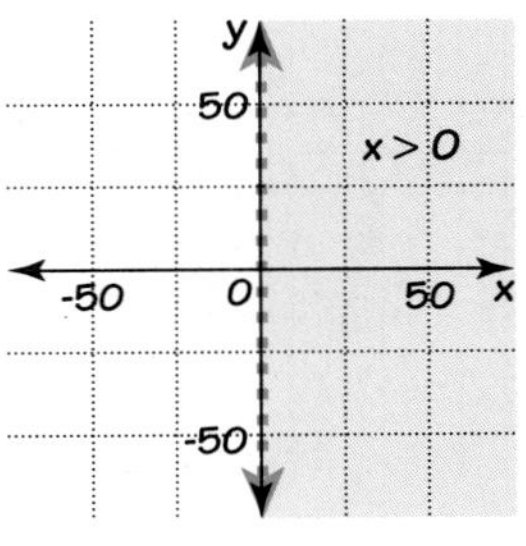

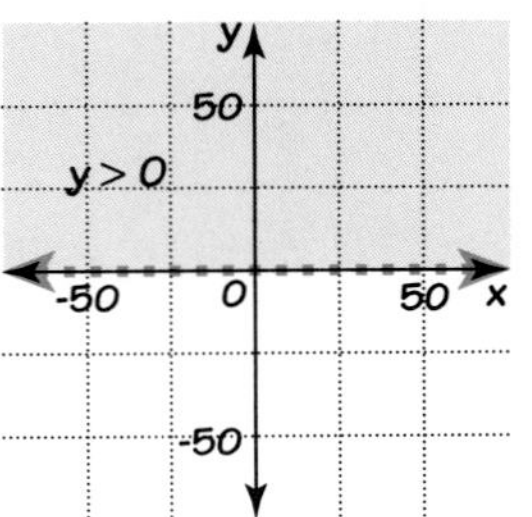

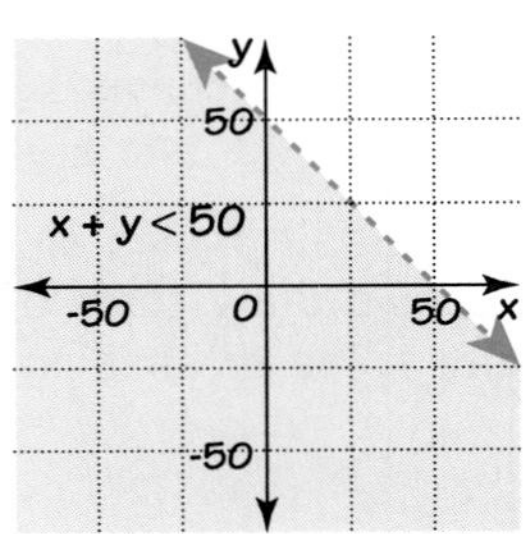

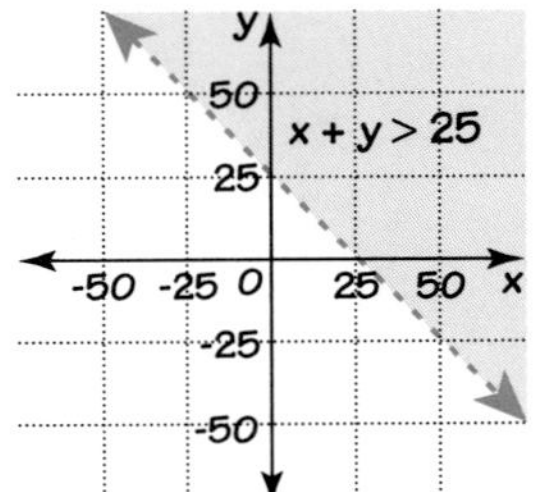

Here is the intersection.

$$\begin{cases} x > 0 \\ y > 0 \\ x + y < 50 \\ x + y > 25 \end{cases}$$

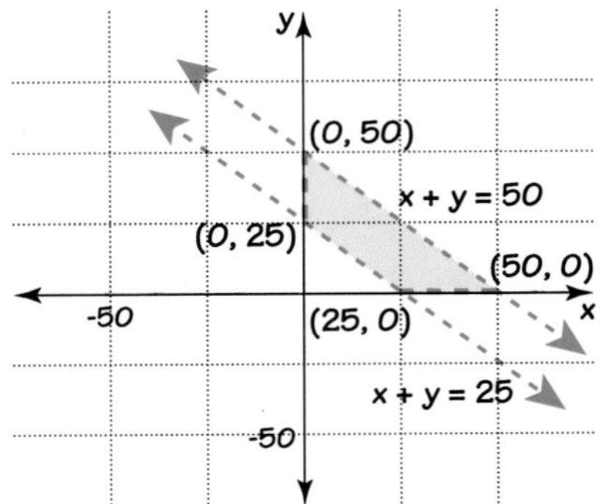

Because there are infinitely many solutions to the system, they cannot be listed. But the graph is easy to describe. It is the interior of the quadrilateral with vertices (0, 50), (50, 0), (25, 0), and (0, 25).

Systems of inequalities arise from many different kinds of situations. Here is one.

Get a kick out of this. *George Blanda, kicker and quarterback, holds the record for scoring the most points, 2002, in the NFL. Blanda played 26 seasons, then retired at age 48 in 1975.*

Example 4

In football, a kicker scores 3 points for a field goal and 1 point for a point after touchdown. Suppose a kicker has scored no more than 20 points in a season. How many field goals and points after touchdowns might the kicker have made?

Solution

You could answer this question using trial-and-error method, but there are a lot of possibilities. It is much easier to show the answers on a graph.

Let f = the number of field goals kicked, and
p = the number of points after touchdowns.

The numbers f and p must be nonnegative integers, so

$$f \geq 0 \text{ and } p \geq 0.$$

Since the total number of points is less than or equal to 20,

$$3f + p \leq 20.$$

Since this is a discrete situation, only points with integer coordinates are possible solutions. These points must be on or above the f-axis, on or to the right of the p-axis, and on or below the line with equation $3f + p = 20$.

The graph below is the solution to this system of

inequalities $\begin{cases} f \geq 0 \\ p \geq 0 \\ 3f + p \leq 20. \end{cases}$

There are 84 points on the graph representing the 84 ways the kicker might have scored no more than 20 points.

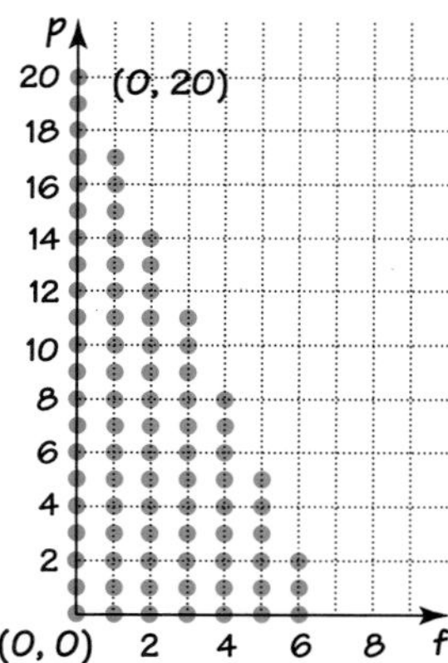

Check

Select a point from the solution set, such as (5, 2), and see if it checks. The point (5, 2) represents the possibility of the kicker making 5 field goals and 2 points after touchdowns. This yields $5 \cdot 3 + 2 = 17$, which is fewer than 20. Also $5 \geq 0$ and $2 \geq 0$. Thus (5, 2) is on the graph of the solution.

If the above situation allowed fractions, the entire triangle would be shaded.

QUESTIONS

Covering the Reading

1. The graph of $\begin{cases} x > 0 \\ y > 0 \end{cases}$ consists of all points in which quadrant?

2. Graph the system $\begin{cases} x < 0 \\ y > 0 \end{cases}$.

In 3–6, refer to Example 2.

3. How is the graph of all solutions to the system
$$\begin{cases} y \ge -3x + 2 \\ y < x - 2 \end{cases}$$
related to the graphs of $y \ge -3x + 2$ and $y < x - 2$?

4. The graph of $y < x - 2$ is a __?__

5. Why does the graph of $y \ge -3x + 2$ include its boundary line?

6. Is (2, 1) a solution to this system? How can you tell?

In 7 and 8, consider the system in the solution to Example 3.
$$\begin{cases} x > 0 \\ y > 0 \\ x + y < 50 \\ x + y > 25 \end{cases}$$

7. The graph of all solutions to this system is the interior of a quadrilateral with what vertices?

8. Name two points that are solutions to the system.

In 9 and 10, refer to Example 4.

9. Give at least three possible combinations of field goals and points after touchdowns that would total exactly 20 points.

10. Suppose the player made at least five field goals. How many possibilities are there then for $3f + p \le 20$?

In 11 and 12, graph the solution set.

11. $\begin{cases} x > 0 \\ y > 0 \\ x + y < 6 \end{cases}$

12. $\begin{cases} x \ge -1 \\ y \le 2 \\ y \ge x - 1 \end{cases}$

Applying the Mathematics

In 13 and 14, describe the shaded region with a system of inequalities.

13.

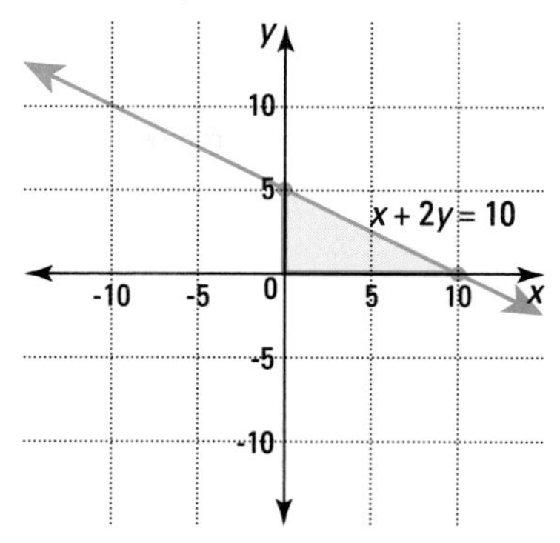

14.

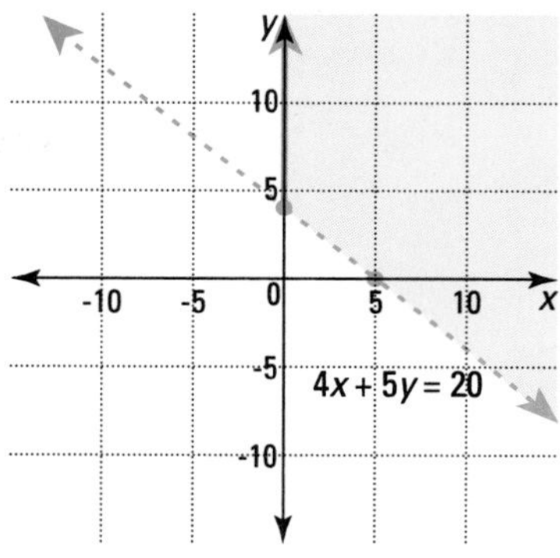

15. Make up a system of inequalities whose solution set is a rectangle and all the points inside it.

16. It takes a good typist about 10 minutes to type a letter of moderate length and about 8 minutes to type a normal double-spaced page.

a. Write a system of inequalities that describes the total number of letters L and pages P a typist can do in an hour or less.

b. Accurately graph the set of points that satisfies the system.

17. A hockey team is scheduled to play 12 games during a season. The coach estimates that the team needs at least 16 points to make the playoffs. A win is worth 2 points and a tie is worth 1 point.

a. Make a graph of all the combinations of wins w and ties t that will get the team into the playoffs.

b. How many ways are there for the team to make the playoffs?

18. An actress is paid $250 per day to understudy a part and $500 per day to perform the role before an audience. During one run, an actress earned between $3000 and $5000 as Maria in *West Side Story.*

a. What is the maximum number of times she might have performed the role of Maria?

b. What is the maximum number of times she might have been an understudy?

c. Graph all possible ways she might have earned her salary.

Action! *Rita Moreno is shown here in the film version of* West Side Story.

Review

19. Solve. *(Lesson 11-7)*

a. $6(3 - 2x) = -12(x - 3)$

b. $6(3 - 2y) = -12\left(y - \frac{3}{2}\right)$

20. Consider the equation $\frac{150(z - 3)}{10(z - 3)} = 15$. *(Lesson 11-7)*

a. What value can z not have?

b. Solve for z.

In 21 and 22, determine whether the lines are parallel and non-intersecting, coincident, or intersecting in only one point. *(Lesson 11-6)*

21. $\begin{cases} 4x - y = 8 \\ 8x - y = 8 \end{cases}$

22. $\begin{cases} y = 3x + 9 \\ 6x - 2y = -27 \end{cases}$

23. A commuter has 27 coins consisting only of dimes and quarters. The total value of the coins is \$5.10. Let d = the number of dimes, and q = the number of quarters. *(Lessons 2-4, 3-1, 11-4)*

a. Write two equations determined by the information above.

b. How many of each coin does the commuter have?

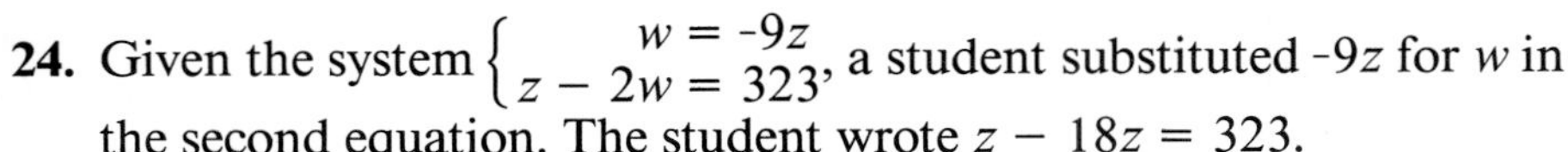

24. Given the system $\begin{cases} w = -9z \\ z - 2w = 323 \end{cases}$, a student substituted $-9z$ for w in the second equation. The student wrote $z - 18z = 323$.

a. Is the student's work correct?

b. If it is correct, finish solving the system. If not, describe what is wrong with it. *(Lessons 11-2, 11-3)*

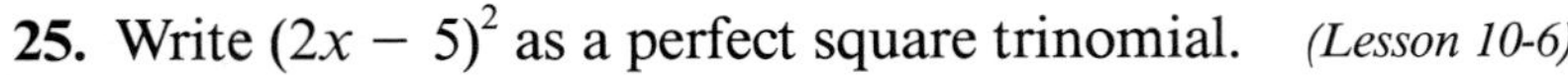

25. Write $(2x - 5)^2$ as a perfect square trinomial. *(Lesson 10-6)*

26. *Skill sequence.* Solve. *(Lessons 1-6, 9-1, 9-5)*

a. $x^2 = 121$

b. $x^2 + 21 = 121$

c. $4x^2 + 21 = 121$

d. $4x^2 + 24x + 157 = 121$

27. Suppose a bank offers an 8.5% annual yield. What would be the amount in an account after 5 years if \$1000 is invested? *(Lesson 8-1)*

Loose change. *These automated ticket machines are in a subway station in Washington, D.C.*

Exploration

28. The graph of the solution to the system of inequalities in Example 3 on page 706 is a special type of quadrilateral called an *isosceles trapezoid.*

a. Look in a dictionary for a definition of isosceles trapezoid.

b. Make up another system of inequalities whose solution set is the interior of an isosceles trapezoid.

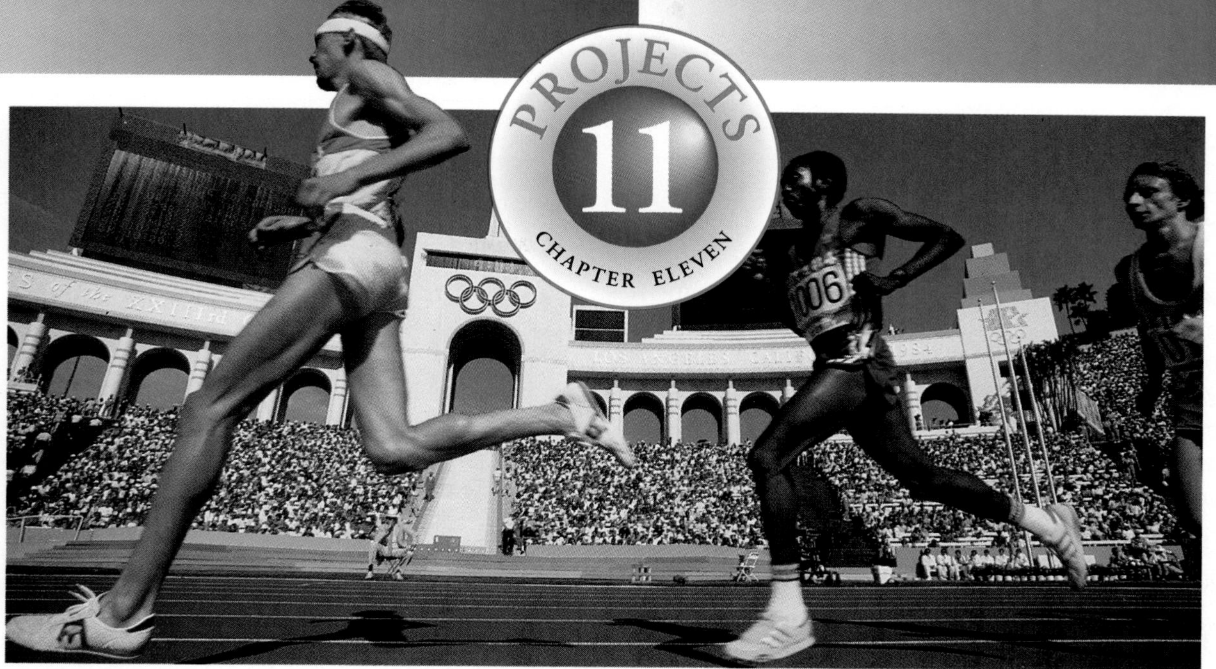

A project presents an opportunity for you to extend your knowledge of a topic related to the material in this chapter. You should allow more time for a project than you do for typical homework questions.

1 Olympic Records

Find the winning times in an Olympic sport in which both men and women participate.

a. Graph the data with the year on the horizontal axis and winning times on the vertical axis.

b. What trends do you observe in the data?

c. Based on the data, does it seem reasonable that in some year in your lifetime the women's winning time will be better than the men's? If so, when do you predict this will happen? Explain your reasoning.

2 Restaurant Prices

Check a restaurant's breakfast menu. Compare the costs of combination meals with the costs of purchasing items separately.

a. How much does the restaurant charge for two eggs and bacon? How much does it charge for one egg and bacon? At these rates, what should be the price of one egg alone? . . . one serving of bacon alone? How do your calculations compare to the restaurant's prices for a "side" order?

b. Investigate the costs of some other combination meals. Are the costs of individual items consistent with the costs of these items if purchased in a combination meal? Explain your reasoning.

(continued)

3 Systems with Equations and Inequalities

Linear equations and inequalities can both be present in a single system. When they are, the solution to the system may be graphed as a line, a ray, a line segment, or a point. The system may also have no solution. Graph the solutions to the following systems, and describe the solution in words.

a. $\begin{cases} y \geq \frac{1}{3}x - 3 \\ y \leq \frac{1}{3}x + 2 \\ y = -\frac{2}{3}x \end{cases}$ **b.** $\begin{cases} 2x + y \leq 5 \\ 4x - 5y \leq 10 \\ 2x + 5y = 5 \end{cases}$

c. Write systems whose solutions are each of the following.
  i. the ray that begins at (3, 5) and contains the point (5, 9)
  ii. the segment with endpoints (-6, 1) and (4, 0)
  iii. the part of the graph of the parabola $y = x^2$ between (-4, 16) and (4, 16)

d. Make up a system of equations and inequalities whose solution is different than those in parts **a-c.**

4 Adding and Subtracting Equations

Select two linear equations and do the following:

a. Graph the two linear equations.
b. Graph the equation found by adding the two selected equations.
c. Graph the equation found by subtracting them.
d. Do multiplications, additions, and subtractions and continue to graph the resulting equations.
e. Do the graphs have any common features? If so, describe them.
f. Verify your calculations starting with a different pair of equations.

5 Life Expectancy

In this century, the number of years of life expected at birth has gradually been increasing over time in most countries of the world, for both men and women of all ethnicities. Almanacs and other reference books contain tables of what has been the life expectancy for many years. Graph the life expectancy of males and females in the United States since 1940 or earlier. Find lines of fit for both males and females. According to your lines, in what year will or did males and females have the same life expectancy? You may wish to consider the life expectancies in other nations, or of people in various ethnic groups.

SUMMARY

A system is a set of sentences which together describe a single situation. The solution set to a system is the set of all solutions common to all of the sentences in the system. A solution to a system of two linear equations is an ordered pair (x, y) that satisfies each equation. Systems of two linear equations may have no solution, one solution, or infinitely many solutions.

One way to solve a system is by graphing. By looking for the intersection point(s) on a graph you can quickly tell if there are any solutions to the system. There are as many solutions as intersection points. Graphing is also a way to describe solutions to systems that have infinitely many solutions. For instance, overlapping half-planes, which arise from systems of linear inequalities, and coincident lines have infinitely many solutions.

However, graphing does not always yield exact solutions. In this chapter, three strategies are presented for finding exact solutions to systems of linear equations. They are substitution, addition, and multiplication. Substitution is a good method to use if at least one equation is given in $y = mx + b$ form. Addition is appropriate if the same term has opposite signs in the two equations in the system. Multiplication is a good method when both equations are in $Ax + By = C$ form. Each method changes the system into an equivalent system whose solutions are the same as those of the original system.

Any kind of situation that leads to a linear equation can lead to a linear system. All that is needed is more than one condition which must be satisfied.

VOCABULARY

Below are the new terms and phrases for this chapter. You should be able to give a general description and a specific example for each.

Lesson 11-1
system
condition of a system
solution to a system
linear systems

Lesson 11-2
substitution

Lesson 11-4
Generalized Addition Property of Equality
addition method for solving a system

Lesson 11-5
equivalent systems
multiplication method for solving a system

Lesson 11-6
coincident lines

Lesson 11-8
system of inequalities

CHAPTER ELEVEN

PROGRESS SELF-TEST

Take this test as you would take a test in class. Then check your work with the solutions in the Selected Answers section in the back of the book.

In 1–4, solve by using any method.

1. $\begin{cases} a - 3b = -8 \\ b = 3a \end{cases}$ **2.** $\begin{cases} p = 5r + 80 \\ p = -7r - 40 \end{cases}$

3. $\begin{cases} m - n = -1 \\ -m + 2n = 4 \end{cases}$ **4.** $\begin{cases} 7x + 3y = 1 \\ 4x - y = 6 \end{cases}$

5. Line ℓ has equation $y = -5x - 15$. Line m has equation $y = x - 3$. Find the point of intersection.

In 6 and 7, determine whether the lines coincide, intersect in one point, or are parallel and nonintersecting.

6. $\begin{cases} 5 = 2A + 7B \\ 10 = 4A + 14B \end{cases}$ **7.** $\begin{cases} y = 2x + 7 \\ y - 2x = 3 \end{cases}$

In 8 and 9, solve the system by graphing.

8. $\begin{cases} y = 3x - 2 \\ x + y = 2 \end{cases}$ **9.** $\begin{cases} x \geq 0 \\ y \geq 0 \\ x + y \leq 20 \\ x + y > 10 \end{cases}$

10. Lisa weighs 4 times as much as her baby sister. Together they weigh 95 pounds. How much does each person weigh?

11. The Reid family went to a restaurant and ordered 3 hamburgers and 4 small salads. Without tax, the bill was \$21.30. At the same restaurant, the Millers ordered 5 hamburgers and 2 small salads. Their bill without tax was \$22.90. What was the cost of a small salad?

12. If the cost per unit is constant, could the following situation have happened? Why or why not? 10 roses and 15 daffodils were sold for \$35. 2 roses and 3 daffodils were sold for \$8.

13. Massachusetts had a 1990 population of about 6,016,000 and was growing at a rate of about 30,000 people each year. Minnesota had a 1990 population of about 4,375,000 and was growing at a rate of about 30,000 people each year. If these rates continue, in how many years after 1990 will these two states have the same population? Explain how you arrived at your answer.

14. Joseph needs to buy at least 3 birthday cards, and he has \$5 to spend. If the big ones cost \$2 and the regular ones cost \$1, show on a graph how many of each kind he might buy.

In 15 and 16, solve and check.

15. $12z + 8 = 12z - 3$ **16.** $-19p < 22 - 19p$

Field of daffodils.

CHAPTER REVIEW

Questions on SPUR Objectives

SPUR stands for **S**kills, **P**roperties, **U**ses, and **R**epresentations. The Chapter Review questions are grouped according to the SPUR Objectives for this chapter.

SKILLS DEAL WITH THE PROCEDURES USED TO GET ANSWERS.

Objective A: *Solve systems using substitution.* *(Lessons 11-2, 11-3)*

1. Find (a, b). $\begin{cases} b = 3a \\ 60 = a + b \end{cases}$

2. Find (x, y). $\begin{cases} x - y = 13 \\ x = 6y - 7 \end{cases}$

3. Solve for p and q. $\begin{cases} p + q = 50 \\ p + 6q = 200 \end{cases}$

4. Solve the system. $\begin{cases} 10y = 20x + 20 \\ 2x + 4y = 29 \end{cases}$

In 5–8, solve the system.

5. $\begin{cases} y = x + 5 \\ 300 + 10x = 35y \end{cases}$

6. $\begin{cases} q = z + 8 \\ 14z + 89 = 13q \end{cases}$

7. $\begin{cases} a = 2b + 3 \\ a = 3b + 20 \end{cases}$

8. $\begin{cases} 16 - 2x = y \\ x + 4 = y \end{cases}$

In 9 and 10, two lines have the given equations. Find the point of intersection, if any.

9. Line ℓ: $y = 7x + 20$
Line m: $y = 3x - 16$

10. Line ℓ: $y = \frac{2}{3}x - \frac{1}{6}$
Line m: $y = \frac{1}{3}x + \frac{1}{3}$

Objective B: *Solve systems by addition.* *(Lesson 11-4)*

In 11–16, solve.

11. $\begin{cases} 3m + b = 11 \\ -4m - b = 11 \end{cases}$

12. $\begin{cases} 6a + 2c = 200 \\ 9a - 2c = 25 \end{cases}$

13. $\begin{cases} 0.6x - 0.4y = 1.1 \\ 0.2x - 0.4y = 2.3 \end{cases}$

14. $\begin{cases} \frac{1}{2}x + 3y = -6 \\ \frac{1}{2}x + y = 2 \end{cases}$

15. $\begin{cases} 4f + g = 15 \\ 3g - 4f = -3 \end{cases}$

16. $\begin{cases} s + \frac{2}{3}t = 3 \\ \frac{2}{3}t - 6s = 10 \end{cases}$

Objective C: *Solve systems by multiplying.* *(Lesson 11-5)*

In 17 and 18, **a.** multiply one of the equations by a number which makes it possible to solve the system by adding. **b.** Solve the system.

17. $\begin{cases} 5x + y = 30 \\ 3x - 4y = 41 \end{cases}$

18. $\begin{cases} 5u + 6v = -295 \\ u - 9v = 400 \end{cases}$

In 19–22, solve the system.

19. $\begin{cases} 3y - 2z = 3 \\ 2y + 5z = 21 \end{cases}$

20. $\begin{cases} 7m - 4n = 0 \\ 9m - 5n = 1 \end{cases}$

21. $\begin{cases} a + b = 3 \\ 5b - 3a = -17 \end{cases}$

22. $\begin{cases} 46 = 2t + u \\ 20 = 8t - 4u \end{cases}$

PROPERTIES DEAL WITH THE PRINCIPLES BEHIND THE MATHEMATICS.

Objective D: *Recognize sentences with no solution, one solution, or all real numbers as solutions.* *(Lesson 11-7)*

In 23–26, solve.

23. $2a + 4 < 2a + 3$

24. $12c < 6(3 + 2c)$

25. $7x - x = 12x$

26. $-10x = 15 - 10x$

27. Is $2k - 7 = 2k$ ever true? Explain why or why not.

28. Is $100d < 100(d - 1)$ ever true? Explain why or why not.

Objective E: *Determine whether a system has no solution, one solution, or infinitely many solutions.* *(Lesson 11-6)*

29. When will the following system have no solution? $\begin{cases} y = mx + b \\ y = mx + c \end{cases}$ Explain how you know this.

30. Tell how you can determine the number of solutions to the system $\begin{cases} r - s = 1 \\ 3s - 3r = -3 \end{cases}$.

In 31–34, determine whether each system describes lines that coincide, intersect, or are parallel and nonintersecting.

31. $\begin{cases} 2x + 4y = 7 \\ 10x + 20y = 35 \end{cases}$

32. $\begin{cases} y - 2x = 5 \\ y = 2x + 4 \end{cases}$

33. $\begin{cases} 6 = m - n \\ -6 = n - m \end{cases}$

34. $\begin{cases} a - 3b = 2 \\ a - 4b = 2 \end{cases}$

35. *Multiple choice.* Two straight lines *cannot* intersect in
(a) exactly one point.
(b) no points.
(c) exactly two points.
(d) infinitely many points.

36. Parallel, nonintersecting lines have the same __?__ but different y-intercepts.

USES DEAL WITH APPLICATIONS OF MATHEMATICS IN REAL SITUATIONS.

Objective F: *Use systems of linear equations to solve real-world problems.*
(Lessons 11-2, 11-3, 11-4, 11-5, 11-6, 11-7)

37. Suppose Joe earned three times as much as Marty during the summer. Together they earned $210. How much did each earn?

38. Renting a car for a day from company C costs $39 plus $.10 a mile. Renting a car for a day from company D costs $22.95 plus $.25 a mile. At what distance does renting the cars cost the same?

39. The starting salary on Job (1) is $7.00 an hour and every 6 months increases $0.50 an hour. For Job (2) the starting salary is $7.20 an hour and every 6 months increases $0.50 an hour. When does Job (2) pay more than Job (1)?

40. From 1980 to 1990, Tucson, Arizona, grew at about 7,500 people a year, to a population of about 405,000. Mesa, Arizona, grew about 13,500 people a year to a population of 290,000. If these rates of increase stay the same, in about how many years will Mesa and Tucson have the same population?

41. In her restaurant Charlene sells 2 eggs and a muffin for $1.80. She sells 1 egg with a muffin for $1.35. At these rates, how much is she charging for the egg and how much for the muffin?

42. A hotel offers two weekend packages. Plan A, which costs $315, gives one person 3 nights lodging and 2 meals. Plan B gives 2 nights lodging and 1 meal and costs $205. At these rates, what is the charge for a room for one night?

43. If the cost per unit is constant, could the given situation have happened? Why or why not? Lydia bought 16 pencils and 5 erasers for $8.00. Then she bought 32 pencils and 10 erasers for $16.00.

44. Tickets to a school play cost $2.50 for students and $4.00 for adults. The school treasurer reported that 850 tickets were sold, and the total revenue was $2395. How many student tickets were sold?

This scene is from Godspell, *performed by students at Anaheim High School in California.*

CHAPTER ELEVEN

Objective G: *Use systems of linear inequalities to solve real-world problems.* *(Lesson 11-8)*

45. Jean has 11 cups of flour on hand, plenty of cookie sheets, but only two cake pans. If it takes 2 cups of flour to make a batch of cookies, and 3 cups to make a cake, what can Jean bake? Make a graph showing all the possibilities.

46. Romeo wants no more than 70 Capulets at the wedding, while Juliet insists that there be no more than 60 Montagues. The hall for the reception is big enough to hold only 100 people. Use a graph to show how many people from each family could attend.

47. Rochelle bought 50 ft of fence for a rectangular pen. She wants to make the pen at least 8 ft long and 6 ft wide.

a. Draw a graph to show all possible dimensions (to the nearest foot) of the pen.

b. At most, how long could the pen be?

48. Rhiann budgeted $5,000 to buy office machines for 8 new typists. A computer costs $1,000 and a typewriter costs $400.

a. Use a graph to show all possible combinations of computers and typewriters Rhiann could buy if she must buy at least 8 machines.

b. At most, how many computers can Rhiann buy if she must provide all 8 new typists with at least one computer or one typewriter?

REPRESENTATIONS DEAL WITH PICTURES, GRAPHS, OR OBJECTS THAT ILLUSTRATE CONCEPTS.

Objective H: *Find solutions to systems of equations by graphing.* *(Lessons 11-1, 11-6)*

In 49–54, solve each system by graphing.

49. $\begin{cases} y = 4x + 6 \\ y = \frac{1}{2}x - 1 \end{cases}$

50. $\begin{cases} y = x - 4 \\ y = -3x \end{cases}$

51. $\begin{cases} y = x + 3 \\ -2x + 3y = 4 \end{cases}$

52. $\begin{cases} 2y - 4x = 1 \\ y = 2x + \frac{1}{2} \end{cases}$

53. $\begin{cases} 3x - 3y = 3 \\ \frac{1}{2}x - \frac{1}{2}y = -1 \end{cases}$

54. $\begin{cases} x + 3y = 5 \\ y = 3x + 5 \end{cases}$

Objective I: *Graphically represent solutions to systems of linear inequalities.* *(Lesson 11-8)*

In 55–58, graph all solutions to the system.

55. $\begin{cases} y \le \frac{1}{2}x + 4 \\ y \ge -x + 1 \end{cases}$

56. $\begin{cases} -x + 2y > 4 \\ x + \frac{1}{2}y < 2 \end{cases}$

57. $\begin{cases} x > 0 \\ y > 0 \\ x + y < 2 \end{cases}$

58. $\begin{cases} x \ge 0 \\ y \ge 0 \\ x + y \le 6 \\ x + y \ge 4 \end{cases}$

In 59 and 60, accurately graph the set of points that satisfies each situation.

59. A small elevator in a building has a capacity of 280 kg. If a child averages 40 kg and an adult 70 kg, how many children *C* and adults *A* can the elevator hold without being overloaded?

60. A person wants to buy *x* pencils at 5¢ each and *y* erasers at 15¢ each and cannot spend more than 60¢. What are the possible values of *x* and *y*?

CHAPTER 12

FACTORING

In March, 1994, six hundred people on five continents using 1,600 computers, having worked over a period of 8 months, completed a task that was thought to be too difficult to do in our day. They found the prime factorization of this 129-digit number:

114,381,625,757,888,867,669,235,779,976,146,612,010,218,296,
721,242,362,562,561,842,935,706,935,245,733,897,830,597,123,
563,958,705,058,989,075,147,599,290,026,879,543,541.

The fact that so many people collaborated and used so many computers for so long on a single task indicates that the task was quite important. Numbers of this size are used as codes to protect information that is held within computers. This information might include a company's recipe for making a food, a military secret, private information about individuals, or code numbers for money accounts. For instance, when a person goes to an automatic teller to obtain money, the person needs to enter a code number. Prime numbers are involved in the process by which this code number is checked to ensure that it goes with the correct account. Because prime numbers are involved in the encoding and decoding, ways of factoring large numbers into primes are studied by some mathematicians.

By breaking the code for this number, the 600 people demonstrated that a 129-digit number was not large enough to be used for coding, and that information thought to be protected might no longer be so safe. Larger numbers are needed.

In this chapter, you will review the prime factorization of positive integers and then extend these ideas to factoring polynomials and other algebraic expressions. These factorizations help in factoring large numbers, simplifying fractions, and solving equations.

LESSON

12-1

Factoring Integers into Primes

Olympic array. *Multiplication helps in calculating the number of band members at the opening ceremonies of the 1984 Olympics in Los Angeles.*

You have probably studied prime numbers in earlier mathematics classes. Here we review some information and perhaps show you some things you have not seen before.

Factors and Multiples

Throughout this lesson, the domain of each variable is the set of positive integers {1, 2, 3, 4, . . . }. When $ab = c$, we say that a and b are **factors** of c. We also say that c is a **multiple** of a, and c is a multiple of b. For instance, since $3 \cdot 15 = 45$,

3 and 15 are factors of 45,
45 is a multiple of 3, and
45 is a multiple of 15.

We also say that 45 **is divisible by** 15. Because for every number n, $n \cdot 1 = n$, every number is a factor of itself and every number is also a multiple of itself.

Suppose two numbers are divisible by the same number. For instance, both 45 and 522 are divisible by 3. We say that 3 is a **common factor** of 45 and 522. The sum 45 + 522, or 567, is also divisible by 3, and so is the difference 522 − 45 = 477. The reason this works for any integers can be deduced by using the Distributive Property.

> Suppose c is a multiple of a. Then there is a number m with $c = ma$.
> If d is also a multiple of a, then there is a number n with $d = na$.
> Add the equations, as you did with systems: $c + d = ma + na$.
> But, by the Distributive Property, $ma + na = (m + n)a$.
> Thus $c + d = (m + n)a$, and so $c + d$ is a multiple of a.

This argument can be repeated for $c - d$.

Common Factor Sum Property
If a is a common factor of b and c, then it is a factor of $b + c$.

As a special case of this property, if a number is divisible by x, then you can repeatedly add x to the number to generate many other numbers that are divisible by x. For instance, 2600 is divisible by 13. Therefore, so is

2613,	2626,	2639,	and so on.
(2600 + 13)	(2600 + 13 + 13)	(2600 + 13 + 13 + 13)	

The same idea works with subtraction. By repeatedly subtracting 13, you can conclude that 2587, 2574, 2561, and so on, are also divisible by 13.

Primes and Composites

Every integer greater than 1 has at least two factors, the number itself and 1. But some integers have only two factors. One such integer is 47. Numbers with this property are the *prime numbers.* A **prime number** is an integer greater than 1 whose only integer factors are itself and 1. Here is a list of the prime numbers less than 50:

2, 3, 5, 7, 11, 13, 17, 19, 23, 29, 31, 37, 41, 43, 47.

Most integers are not prime. They are *composite.* A **composite number** is an integer that has more than two factors. For instance, the number 60 has 12 different factors: 1, 2, 3, 4, 5, 6, 10, 12, 15, 20, 30, and 60. You can use the Area Model for Multiplication to show each factor as the number of dots on a side of a rectangular array containing 60 dots. Here are four of these arrays.

1-by-60 array:

2-by-30 array:

3-by-20 array:

4-by-15 array:

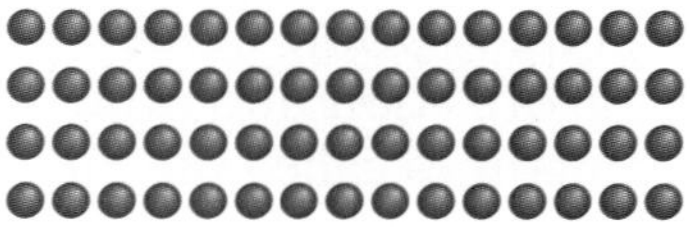

So you can think of a prime number as a number p of dots that cannot be arranged in any array other than a 1-by-p array.

To determine whether or not an integer is prime, you can divide it by the primes less than it. For small numbers, there are not as many divisions needed as you might think.

Example 1

Is 113 a prime number?

Solution

Divide 113 by the primes less than 113. If any quotient is an integer, then 113 has a prime factor and so is not prime.

$$113 \div 2 = 56.5$$
$$113 \div 3 = 37.6\ldots$$
$$113 \div 5 = 22.6$$
$$113 \div 7 = 16.1\ldots$$
$$113 \div 11 = 10.2\ldots$$

We need go no further. For any prime divisor greater than 11, the quotient will be less than the divisor. So it would have been found as a factor earlier. 113 is prime.

Secret codes. *Prime numbers are used to code information. Pictured are PFC Preston Toledo and his cousin, PFC Frank Toledo. During World War II, they transmitted orders in their own Navajo language. This was like a code because the Japanese did not know the language.*

Prime Factorizations

When a number is composite, it can be factored into primes. Here is how this is done. Consider factoring 60. Because 60 is composite, it can be factored into two or more factors, with each factor less than the original number. For instance,

$$60 = 2 \cdot 30.$$

Since 2 is prime, we leave it alone. But 30 is not prime. Pick two factors that multiply to 30 (but don't pick 1 and 30). We pick 5 and 6.

$$60 = 2 \cdot 5 \cdot 6$$

Now 2 and 5 are prime, but 6 is composite. Factor 6.

$$60 = 2 \cdot 5 \cdot 2 \cdot 3$$

This is the *prime factorization* of 60. A **prime factorization** of *n* is the writing of *n* as a product of primes. To put the prime factorization in **standard form,** order the primes and use exponents if a prime is repeated.

$$60 = 2^2 \cdot 3 \cdot 5$$

If you started with different factors of 60, would you get a different final result? No. This was proved by the ancient Greeks, who seem to have been the first people to study prime numbers in a systematic way. Every integer has a unique factorization into primes. This statement is so important that it is sometimes called the Fundamental Theorem of Arithmetic.

Unique Factorization Theorem
Every integer can be represented as a product of primes in exactly one way, disregarding order of the factors.

Example 2

Write the prime factorization of 40,768 in standard form.

Solution

Don't be psyched out by the size of the number! Since 2 is obviously a factor, divide by it.

$$40{,}768 = 2 \cdot 20384$$

Keep dividing by 2 until an odd number results.

$$\begin{aligned}40{,}768 &= 2 \cdot 2 \cdot 10{,}192\\ &= 2 \cdot 2 \cdot 2 \cdot 5096\\ &= 2 \cdot 2 \cdot 2 \cdot 2 \cdot 2548\\ &= 2 \cdot 2 \cdot 2 \cdot 2 \cdot 2 \cdot 1274\\ &= 2 \cdot 2 \cdot 2 \cdot 2 \cdot 2 \cdot 2 \cdot 637\end{aligned}$$

To determine whether 637 is prime, divide it by primes larger than 2.

$637 \div 3 = 212.3\ldots$

637 is obviously not divisible by 5. Do you know why?

$637 \div 7 = 91$. So 637 can be factored.

$$40{,}768 = 2 \cdot 2 \cdot 2 \cdot 2 \cdot 2 \cdot 2 \cdot 7 \cdot 91$$

You may know that $91 = 7 \cdot 13$, or you would try 7 again to find this out.

$$40{,}768 = 2 \cdot 2 \cdot 2 \cdot 2 \cdot 2 \cdot 2 \cdot 7 \cdot 7 \cdot 13$$

This is the prime factorization. Now write it in standard form.

$$40{,}768 = 2^6 \cdot 7^2 \cdot 13$$

By factoring two numbers into primes, you may be able to multiply and divide them more easily.

Example 3

Use the prime factorizations of 40,768 and 294 to do the following:

a. Write the prime factorization of $40{,}768 \cdot 294$.

b. Write $\frac{40{,}768}{294}$ in lowest terms.

Solution

a. From Example 2,

$$40{,}768 = 2^6 \cdot 7^2 \cdot 13.$$

We can find, by the process used in Example 2, that

$$294 = 2 \cdot 3 \cdot 7^2.$$

Consequently, by substitution,

$$40{,}768 \cdot 294 = (2^6 \cdot 7^2 \cdot 13) \cdot (2 \cdot 3 \cdot 7^2).$$

Now use the Product of Powers Property to multiply the powers with the same base.

$$40{,}768 \cdot 294 = 2^7 \cdot 3 \cdot 7^4 \cdot 13$$

b. Use the prime factorizations from part **a**.

$$\frac{40{,}768}{294} = \frac{2^6 \cdot 7^2 \cdot 13}{2 \cdot 3 \cdot 7^2}$$

Now use the Quotient of Powers Property.

$$= \frac{2^5 \cdot 13}{3}$$

$$= \frac{416}{3}$$

Is it prime? *As of September, 1994, the largest known prime number was* $2^{859433} - 1$. *This number is 258,716 digits long! If set in type* this size, *the digits would fill 52 pages of standard notebook paper.*

QUESTIONS

Covering the Reading

1. Because $13 \cdot 17 = 221$, 13 is a _?_ of 221, and 221 is a _?_ of 13.

In 2–5, give four values of the variable that satisfy the sentence.

2. 60 is divisible by b.

3. c is a factor of 35.

4. d is a multiple of 17.

5. e is a common factor of 48 and 60.

6. Explain why 9 is not a prime number.

7. Determine whether or not 133 is a prime number.

8. Draw an array to show that 27 is a composite number.

9. Without calculating, how do you know that $7^2 + 3 \cdot 7^3$ is divisible by 7?

10. Give the prime factorization of 3216 and put it in standard form.

In 11–13, use this information. The prime factorization of 2025 is $3^4 \cdot 5^2$. The prime factorization of 735 is $3 \cdot 5 \cdot 7^2$.

11. What is the prime factorization of $2025 \cdot 735$?

12. Write $\frac{735}{2025}$ in lowest terms.

13. Write $\frac{2025}{735}$ in lowest terms.

14. To what important practical use have prime numbers been put?

Applying the Mathematics

15. List the prime numbers between 50 and 100.

In 16 and 17, find all solutions.

16. e is a multiple of 3 and a factor of 300.

17. f is a factor of 20 and divisible by 20.

18. Give an argument to show that the following is true for all positive integers a, b, and c: If a is a common factor of b and c, then it is a factor of $b - c$.

19. **a.** Explain why the 13-digit number $2^{40} + 332$ must be divisible by 4.
b. Is this number divisible by 8? Why or why not?

Review

20. Draw a picture of the following multiplication using tiles.
$2x(3x + 5) = 6x^2 + 10x$. *(Lesson 10-3)*

21. Find all values of d that satisfy $(d^2)^2 - 25d^2 + 144 = 0$.
(Lessons 5-9, 9-5)

Play ball! *The Cleveland Indians began the 1994 baseball season in a new ballpark,* Jacobs Field. *The Indians won that game for their first opening day victory in five years.*

22. The orbits of Venus and Earth around the Sun are not circles but are reasonably close to circles. Venus averages a distance of about 108 million kilometers from the Sun. The Earth averages about 150 million kilometers from the Sun. Venus takes 225 days to go around the Sun. Which planet travels farther in an Earth year?
(Previous course, Lesson 6-8)

23. The Cleveland Indians won 12 of its first 19 games in 1994. At this rate, how many games would you expect them to win in a 162-game season? *(Lesson 6-8)*

24. A student has scored 75%, 80%, and 90% on three exams and has to take one more exam.
a. What are the highest and lowest mean percents the student can have on the four exams?
b. What are the highest and lowest median percents the student can have?
c. What are the highest and lowest mode percents if there is to be a single mode? *(Previous course)*

25. On page 719, mention was made of a 129-digit number. Which of these numbers has 129 digits? *(Appendix B)*
(a) $2 \cdot 10^{128}$ (b) $2 \cdot 10^{129}$ (c) $2 \cdot 10^{130}$ (d) $(2 \cdot 10)^{64}$

Exploration

26. A statement proved by Pierre Fermat in 1675, known as Fermat's Little Theorem, is that, if p is any prime number and a is any positive integer, then $a^p - a$ is divisible by p. Verify this theorem for five different pairs of values of a and p larger than 2.

LESSON

12-2

Common Monomial Factoring

A tight squeeze. *There are three ways to pack books into a box where all are facing the same way: flat, on end, and on their sides. The books will fit snugly in all three ways only if all the dimensions of the book are common factors of the dimensions of the box.*

When two or more numbers are multiplied, the result is a single number. Factoring is the reverse process. In factoring, one begins with a single number and expresses it as the product of two or more numbers. For instance, the product of 7 and 4 is 28. So, when we factor 28, we get $28 = 7 \cdot 4$. In Lesson 12–1, you used the process of factoring to obtain the prime factorizations of integers. Now we turn our attention to polynomials.

Example 1 shows how to find factors of a monomial.

Example 1

What are the factors of $25x^3$?

Solution

The factors of 25 are 1, 5, and 25. The factors of x^3 are 1, x, x^2, and x^3. The factors of $25x^3$ are the 12 possible products formed from those factors:

$$1,\ 5,\ 25,\ x,\ 5x,\ 25x,\ x^2,\ 5x^2,\ 25x^2,\ x^3,\ 5x^3,\ 25x^3.$$

The **greatest common factor** (GCF) of two or more monomials is found by multiplying the greatest common factor of their coefficients by the greatest common factor of their variables.

Example 2

Find the greatest common factor of $24x^2y$ and $6x$.

Solution

The GCF of 24 and 6 is 6. The GCF of x^2 and x is x. Since the factor y does not appear in all terms, y does not appear in the GCF. So, The GCF of $24x^2y$ and $6x$ is $6 \cdot x$, which is $6x$.

In this book, all the polynomials we factor will have integer coefficients. These are called **polynomials over the integers.** Unless you are told otherwise, you should factor over the integers.

As with integers, the result of factoring a polynomial is called a **factorization.** Here is a factorization of $8x^2 + 12x$.

$$8x^2 + 12x = 2x(4x + 6)$$

Again, as with integers, a factorization with two factors means that a rectangular figure can be formed with the factors as its dimensions. Here is a picture of the above factorization.

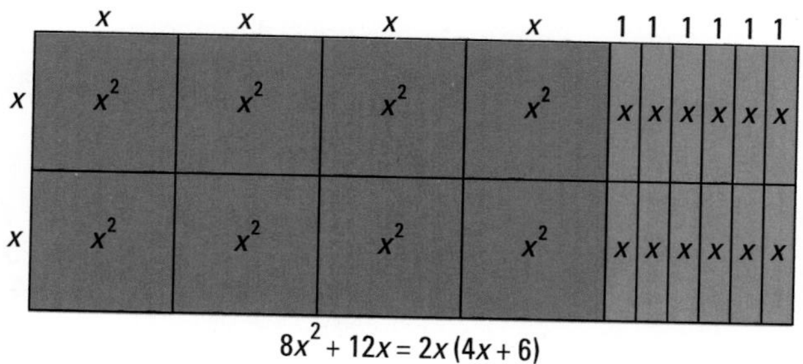

$8x^2 + 12x = 2x(4x + 6)$

Because the terms in $4x + 6$ have the common factor 2, another factorization of $8x^2 + 12x$ is possible.

$$\begin{aligned} 8x^2 + 12x &= 2x(4x + 6) \\ &= 2x \cdot 2(2x + 3) \\ &= 4x(2x + 3) \end{aligned}$$

Monomials such as $4x$, and binomials such as $2x + 3$ which cannot be factored into polynomials of a lower degree are called **prime polynomials.** To factor a polynomial completely means to factor it into prime polynomials. When there are no common numerical factors in the terms of any of the prime polynomials, the result is called a **complete factorization.** The complete factorization of $8x^2 + 12x$ is $4x(2x + 3)$.

The complete factorization of a polynomial has a uniqueness property much like that of the prime factorization of an integer.

Unique Factorization Theorem for Polynomials
The complete factorization of a polynomial is unique, disregarding the order of the factors and multiplication by -1.

Example 3

Factor $24x^2y + 6x$ completely.

Solution

From Example 2, we know that $6x$ is the greatest common factor of $24x^2y$ and of $6x$. So it is a factor of $24x^2y + 6x$. Now find the binomial which when multiplied by $6x$ gives $24x^2y + 6x$.

$$24x^2y + 6x = 6x(\underline{\ ?\ } + \underline{\ ?\ })$$

Divide each term by $6x$ to fill in the factors.

$$\frac{24x^2y}{6x} = 4xy \qquad \frac{6x}{6x} = 1$$

$$24x^2y + 6x = 6x(4xy + 1)$$

Check

Test a special case. Let $x = 3$, $y = 4$. Follow the order of operations.
$24x^2y + 6x = 24 \cdot 3^2 \cdot 4 + 6 \cdot 3 = 864 + 18 = 882.$
$6x(4xy + 1) = 6 \cdot 3(4 \cdot 3 \cdot 4 + 1) = 18 \cdot 49 = 882.$ It checks.

Three or more polynomials can also have common factors.

Example 4

Find the greatest common factor of $-30x^3y$, $20x^4$, and $50x^2y^5$.

Solution

The greatest common factor of -30, 20, and 50 is 10. The GCF of x^3, x^4, and x^2 is x^2. Since the variable y does not appear in all terms, y does not appear in the GCF. The GCF of $-30x^3y$, $20x^4$, and $50x^2y^5$ is $10x^2$.

Notice the pattern from Examples 3 and 4. The GCF of two or more terms includes the GCF of the coefficients of the terms. It also includes any common variable raised to the *lowest* exponent of that variable found in the terms.

Example 5

Factor $-30x^3y + 20x^4 + 50x^2y^5$ completely.

Solution

In Example 4, we found that the GCF of the three terms of this polynomial is $10x^2$. Thus

$$-30x^3y + 20x^4 + 50x^2y^5 = 10x^2(\underline{\ ?\ } + \underline{\ ?\ } + \underline{\ ?\ }).$$

Now divide to find the terms in parentheses. Experts do these steps in their heads.

$$\frac{-30x^3y}{10x^2} = -3xy \qquad \frac{20x^4}{10x^2} = 2x^2 \qquad \frac{50x^2y^5}{10x^2} = 5y^5$$

So $\quad -30x^3y + 20x^4 + 50x^2y^5 = 10x^2(-3xy + 2x^2 + 5y^5).$

Factoring provides a way of simplifying some fractions.

Example 6

Simplify $\frac{5n^2 + 3n}{n}$. (Assume $n \neq 0$.)

Solution 1

Factor the numerator and simplify the fraction.

$$\frac{5n^2 + 3n}{n} = \frac{n(5n + 3)}{n}$$
$$= 5n + 3$$

Solution 2

Separate the given expression into the sum of two fractions and divide.

$$\frac{5n^2 + 3n}{n} = \frac{5n^2}{n} + \frac{3n}{n}$$
$$= 5n + 3$$

QUESTIONS

Covering the Reading

1. List all the factors of $14x^4$.

2. What does the abbreviation GCF represent?

In 3 and 4, find the GCF.

3. $15x^3$ and $10x^2$

4. $30ab^2$ and $24a^2$

5. Picture the factorization $8x^2 + 12x = 4x(2x + 3)$ with algebra tiles.

6. **a.** Factor $3x^2 + 6x$ by first finding the greatest common factor of the terms.
   b. Illustrate the factorization by drawing a rectangle whose sides are the factors.

7. Tell whether or not the polynomial is prime.
   a. $9x^3 + 12x^2$ **b.** $9x^2 + 12x$ **c.** $9x + 12$ **d.** $9 + 12x$

8. In parts **a–c**, complete the products.
   a. $20x^3 + 10x^2 = 5(\underline{\ ?\ } + \underline{\ ?\ })$
   b. $20x^3 + 10x^2 = 10(\underline{\ ?\ } + \underline{\ ?\ })$
   c. $20x^3 + 10x^2 = 10x(\underline{\ ?\ } + \underline{\ ?\ })$
   d. Are any of the products in parts **a–c** a complete factorization of $20x^3 + 10x^2$?

In 9–11, copy and fill in the blanks.

9. $8x^3 + 40x = 8x(\underline{\ ?\ } + \underline{\ ?\ })$

10. $4p^2 - 3p = p(\underline{\ ?\ } - \underline{\ ?\ })$

11. $8a^2b + 4ab^2 = 4ab(\underline{\ ?\ } + \underline{\ ?\ })$

12. Simplify $\frac{12n^2 + 15n}{3n}$.

13. Find the greatest common factor of $6x^2y^2$, $-9xy^3$, and $12x^2y^4$.

In 14–17, factor the polynomial completely.

14. $27b^3 - 27c^3 + 27bc$

15. $23a^3 + a^2$

16. $12a + 16a^2$

17. $14p^3 - 21p^2q$

Applying the Mathematics

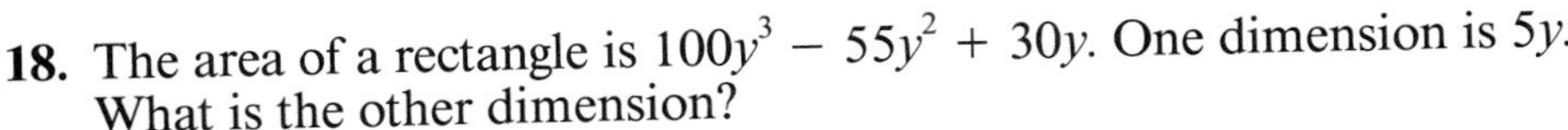

Drums of steel. *Steel drums, originating in the 20th century in Trinidad, West Indies, are made from the unstoppered end and part of the wall of a metal shipping drum.*

18. The area of a rectangle is $100y^3 - 55y^2 + 30y$. One dimension is $5y$. What is the other dimension?

19. Use algebra tiles to draw two different rectangles each with area equal to $12x^2 + 8x$.

20. **a.** Graph $y = 2x^2 - 6x$.
    b. Find an equation of the form $y = (x - 3) \cdot$ __?__ which has the same graph as the one in part **a.**

In 21 and 22, simplify.

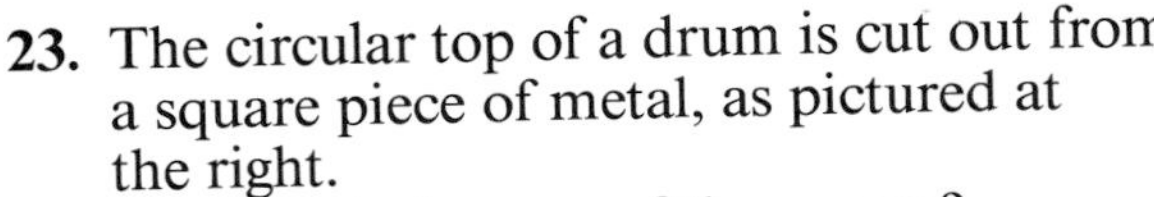

21. $\frac{-3x^2y + 6xy - 9xy^2}{3xy}$

22. $\frac{n^{14} - 3n^{10} + 5n^6 - n^2}{n^2}$

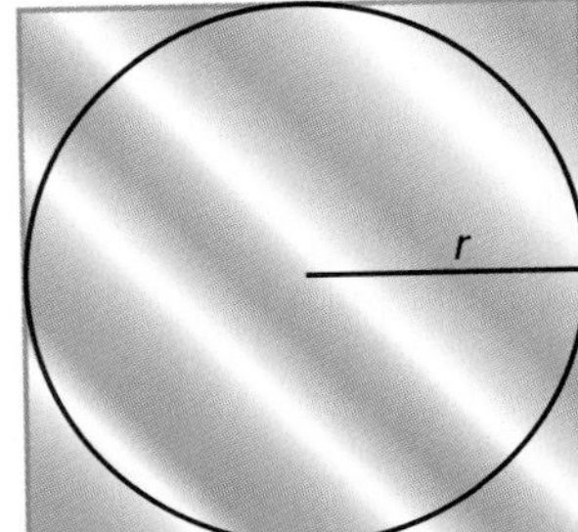

23. The circular top of a drum is cut out from a square piece of metal, as pictured at the right.
    a. What is the area of the square?
    b. What is the area of the circle?
    c. How much metal is not used?
    d. If a dozen drum tops are needed, how much metal will not be used?
    e. If the unused metal in part **d** could be recycled, how many complete drum tops could be made?

Review

24. **a.** Give the prime factorization of 1001.
    b. Give the prime factorization of 91.
    c. Give the prime factorization of $1001 \cdot 91$.
    d. Write $\frac{91}{1001}$ in lowest terms. *(Lesson 12-1)*

25. **a.** Multiply $(x + 14)(x - 11)$.
    b. Solve: $(x + 14)(x - 11) = 0$. *(Lessons 9-5, 10-5)*

26. A person is on a supervised diet. On the third day of the diet, the person weighed 93.4 kg. On the 10th day, the person weighed 91.2 kg.
    a. What was the average rate of change in weight per day?
    b. Express the rate in part **a** in lb/week. *(Lessons 2-4, 7-1)*

27. Simplify $(3a + 5) \cdot a + (3a + 5) \cdot 2$. *(Lesson 3-7)*

28. Solve $3(2x^2 + 2x) - 6(x^2 - 2) = 18$. *(Lessons 3-5, 3-6, 3-7)*

In 29 and 30, find x if $AB = 12$. *(Lesson 3-2)*

29. Points A, C, B on a segment; AC = 3, CB = x

30. Points A, D, E, B on a segment; AD = 5, DE = $x + 1$, EB = 4

Computer room. *One of the early computers, AVIDAC's Arithmetic Unit, contained over 1500 electronic tubes. Its memory unit appears at the left, and its power distribution cabinets are in the background.*

Exploration

31. Computers often have to do millions of calculations, so reducing the number of calculations can save money. The following program asks the computer to evaluate

$$50x^4 + 50x^3 + 50x^2 + 50x + 50$$

for the 20,000 values of x from 1.00001 to 1.20000, increasing by steps of 0.00001. You saw a use for this polynomial in Chapter 10.

```
10 FOR X = 1.00001 TO 1.2 STEP 0.00001
20 LET P = 50 * X ^ 4 + 50 * X ^ 3 + 50 * X ^ 2 +
   50 * X + 50
30 NEXT X
40 PRINT P
50 END
```

a. If, on the average, each operation takes the computer a millionth of a second, how long will it take this program to run?

b. If the expression in line 20 is rewritten in factored form, how long will it take the program to run?

c. If each second of running time costs \$.25, how much will factoring save?

d. Write a problem that leads to the need to evaluate this polynomial.

Computer palm. *Compared to the AVIDAC, any of today's hand-held computers has far more capability and speed.*

Introducing Lesson 12-3

Factoring Trinomials

IN-CLASS ACTIVITY

Materials: algebra tiles
Work in small groups.

Factoring of some polynomials can be shown with algebra tiles.

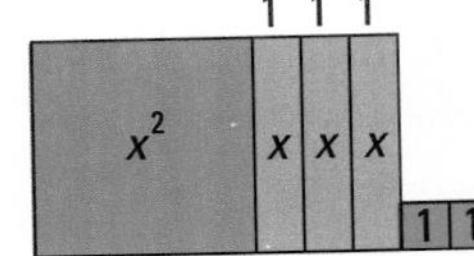

1 Displayed at the right are the algebra tiles for
$$x^2 + 3x + 2.$$

a. Rearrange the tiles to form a rectangle. What are its dimensions?

b. Copy and complete the following: $x^2 + 3x + 2 = (x + \underline{\ ?\ })(x + \underline{\ ?\ })$.

In 2 and 3, follow the idea of Question 1 to factor the expression.

2 $x^2 + 7x + 6$

3 $x^2 + 7x + 12$

4 Displayed at the right is
$$2x^2 + 7x + 3.$$

a. Show how the tiles can be rearranged to form a rectangle with dimensions $x + 3$ by $2x + 1$.

b. Factor $2x^2 + 7x + 3$.

In 5 and 6, follow the idea of Question 4 to factor the expression.

5 $2x^2 + 7x + 6$

6 $3x^2 + 6x + 3$

7 **a.** Display algebra tiles for $4x^2 + 16x + 16$.

b. Arrange the tiles to form a square. What is area of the square? What is the length of a side?

c. Copy and complete the following: $4x^2 + 16x + 16 = (\underline{\ ?\ } + \underline{\ ?\ })^2$.

8 Follow the idea of Question 7 to show that $9x^2 + 12x + 4$ is a perfect square trinomial.

LESSON

12-3

Factoring $x^2 + bx + c$

(x+3) (x−2)(x+9) (x+p)(x+q) (x−5)(x−3) (x+5)(x+2) (x+2) (x−7)(x+3)

The terms in trinomials of the form $x^2 + bx + c$ have no common factors. Yet some of these trinomials can be factored. Notice the pattern which results from the multiplication of the binomials $(x + p)$ and $(x + q)$.

	Square Term		Linear Term		Constant Term
$(x + 5)(x + 2) = x^2 + 2x + 5x + 10 =$	x^2	+	$7x$	+	10
$(x - 5)(x - 3) = x^2 - 3x - 5x + 15 =$	x^2	−	$8x$	+	15
$(x - 2)(x + 9) = x^2 + 9x - 2x - 18 =$	x^2	+	$7x$	−	18
$(x - 7)(x + 3) = x^2 + 3x - 7x - 21 =$	x^2	−	$4x$	−	21
$(x + p)(x + q) = x^2 + qx + px + pq =$	x^2	+	$(p + q)x$	+	pq

The constant term of each trinomial above is the product of the last terms of the binomial factors. The coefficient in the linear term is the sum of the last terms. This pattern suggests a way to factor trinomials in which the leading coefficient is 1.

Example 1

Factor $x^2 + 11x + 18$.

Solution

To factor, we need to identify two binomials, $(x + p)(x + q)$, whose product equals $x^2 + 11x + 18$. We must find p and q, two numbers whose product is 18 and whose sum is 11. Since the product is positive and the sum is positive, both p and q are positive. List the positive factors of 18; then calculate their sums.

Factors of 18	Sum of Factors
1, 18	19
2, 9	11
3, 6	9

The sum of the numbers 2 and 9 is 11. So
$x^2 + 11x + 18$ can be factored into $(x + 9)(x + 2)$.

Check

Factoring can always be checked by multiplication.
$(x + 9)(x + 2) = x^2 + 2x + 9x + 18 = x^2 + 11x + 18$. It checks.

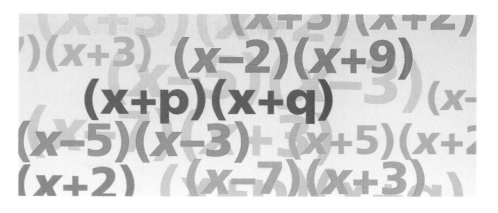

Example 2

Factor $x^2 - x - 12$.

Solution

Think of this trinomial as $x^2 + -1x + -12$.

We need two numbers whose product is -12 and whose sum is -1. Since the product is negative, one of the factors is negative. List the possibilities.

Factors of -12	Sum of Factors
-1, 12	11
-2, 6	4
-3, 4	1
-4, 3	-1
-6, 2	-4
-12, 1	-11

The only two factors of -12 whose sum is -1 are -4 and 3. So

$$x^2 - x - 12 = (x - 4)(x + 3).$$

Check 1

$(x - 4)(x + 3) = x^2 + 3x - 4x - 12 = x^2 - x - 12$. It checks.

Check 2

Graph $y = x^2 - x - 12$ and $y = (x - 4)(x + 3)$. The graphs should be identical. Below we show the output from our automatic grapher.

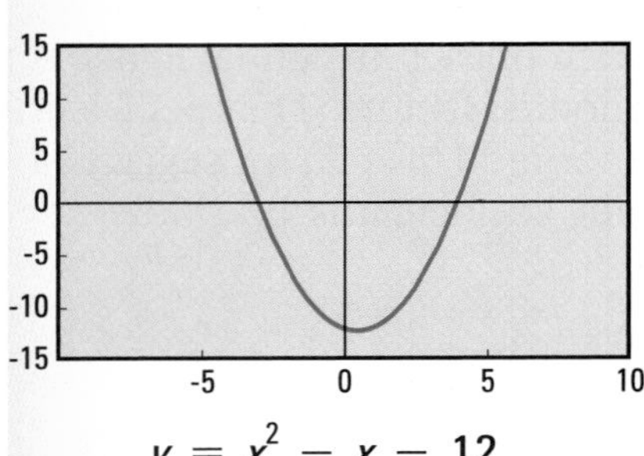

$y = x^2 - x - 12$

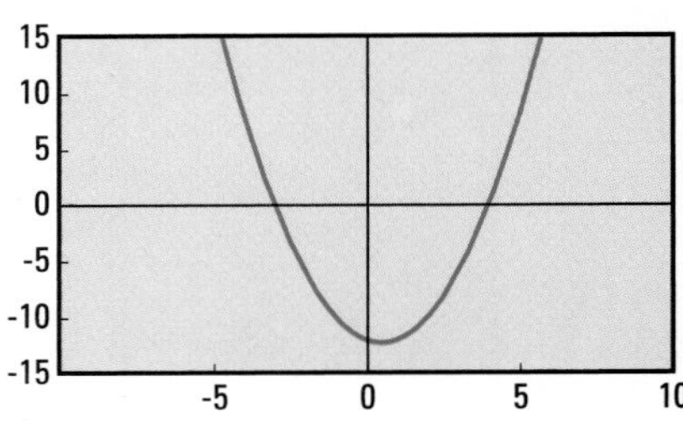

$y = (x - 4)(x + 3)$

It checks.

Example 3

Factor $t^2 - 6t + 9$.

Solution

Think: Which factors of 9 have a sum of -6? Because the product is positive and the sum negative, both numbers are negative. You need only consider negative factors of 9.

Factors of 9	Sum of Factors
-9, -1	-10
-3, -3	-6

So $t^2 - 6t + 9 = (t - 3)(t - 3)$.

Another way to write the answer is as $(t - 3)^2$.

Example 3 shows that $t^2 - 6t + 9$ is a *perfect square trinomial.*

Not all trinomials of the form $x^2 + bx + c$ can be factored into polynomials with integer coefficients. For instance, to factor $x^2 + 8x + 14$ as two binomials, $(x + p)(x + q)$, you would have to find two numbers p and q whose product is 14 and whose sum is 8.

Factors of 14	Sum of Factors
1, 14	15
2, 7	9

None of the pairs of factors have a sum of 8, so $x^2 + 8x + 14$ is not factorable over the integers. It is *prime.*

Binomials that are differences of squares can be factored.

Example 4

Factor $d^2 - 49$.

Solution 1

Both d^2 and 49 are perfect squares. Use the Difference of Squares pattern from Lesson 10-6.

$$d^2 - 49 = d^2 - 7^2 = (d + 7)(d - 7)$$

Solution 2

Think of $d^2 - 49$ as $d^2 + 0d + -49$. Two numbers are needed whose product is -49 and whose sum is 0. They must be opposites, so use 7 and -7.

The factors of $d^2 - 49$ are $(d + 7)(d - 7)$.

Check

Multiply: $(d + 7)(d - 7) = d^2 - 7d + 7d - 49 = d^2 - 49$.

Some trinomials have both monomial and binomial factors. If the terms of a trinomial have a greatest common factor other than 1, you should first factor out the GCF.

Example 5

Factor $3r^3 + 30r^2 + 75r$.

Solution

$3r$ is the GCF of the three terms.

$$3r^3 + 30r^2 + 75r = 3r(r^2 + 10r + 25)$$

To factor $r^2 + 10r + 25$, we need binomials whose constant terms have a product of 25 and a sum of 10.

Factors of 25	Sum of Factors
1, 25	26
5, 5	10

$$r^2 + 10r + 25 = (r + 5)(r + 5).$$

So $\quad 3r^3 + 30r^2 + 75r = 3r(r + 5)(r + 5)$ or $3r(r + 5)^2$. ▶

Check

$$\begin{aligned} 3r(r+5)(r+5) &= (3r^2+15r)(r+5) \\ &= 3r^3+15r^2+15r^2+75r \\ &= 3r^3+30r^2+75r \quad \text{It checks.} \end{aligned}$$

QUESTIONS

Covering the Reading

1. What was the coefficient of the square term in all the trinomials in Examples 1 to 3?

2. a. How can you check that you have found the correct factors of a quadratic trinomial?
b. Perform the check for Example 3.

3. a. In order to factor $x^2 + 8x + 12$, list the possible integer factors of the last term, and their sums.
b. Factor $x^2 + 8x + 12$. **c.** Check your work.

4. Suppose $(x + p)(x + q) = x^2 + bx + c$.
a. What must pq equal? **b.** What must $p + q$ equal?

In 5–10, write the trinomial as the product of two binomials.

5. $x^2 + 7x + 10$ **6.** $p^2 + 14p + 13$ **7.** $t^2 - t - 30$

8. $a^2 - 81$ **9.** $q^2 - 10q + 16$ **10.** $n^2 + 4n - 12$

11. Explain why the trinomial $x^2 + 2x + 3$ cannot be factored over the integers.

12. a. Factor $x^3 - 5x^2 + 6x$ into the product of a monomial and a trinomial.
b. Complete the factoring by finding factors of the trinomial in part **a.**

In 13 and 14, factor completely.

13. $2m^2 + 18m + 36$ **14.** $4y^3 - 20y^2 - 24y$

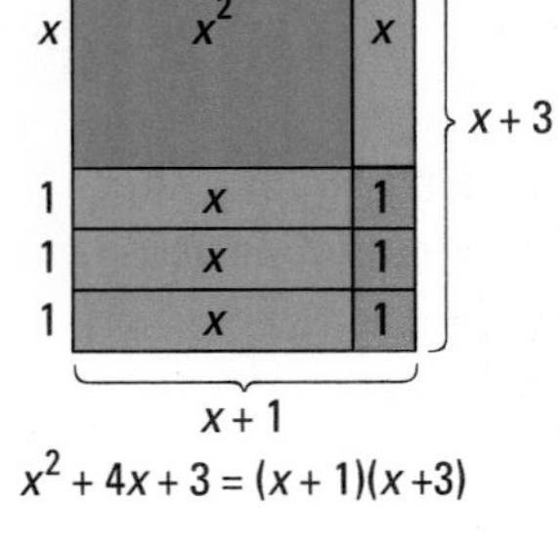

$x^2 + 4x + 3 = (x + 1)(x + 3)$

15. The diagram at the left uses algebra tiles to show the factorization of $x^2 + 4x + 3$. Make a drawing of algebra tiles to show the factorization of $x^2 + 5x + 6$.

Applying the Mathematics

16. a. Find an equation of the form $y = (x + p)(x + q)$ whose graph is identical to the graph of $y = x^2 - 10x + 21$.
b. Check your work by graphing both equations on the same set of axes.

In 17–20, factor.

17. $35 - 12x + x^2$ **18.** $a^2 - 8ab + 7b^2$

19. $40 - 10x^2$

20. $10r^2 + 5r^3 + 5r$

21. Notice that $1{,}032{,}060 = 10^6 + 32 \cdot 10^3 + 60$. Use this information to help find the prime factorization of 1,032,060. (Hint: Begin by letting $x = 10^3$.)

22. Explain how $28^2 - 22^2$ can be calculated in your head.

Ending Digit	Frequency
0	10
1	7
2	2
3	3
4	4
5	3
6	5
7	7
8	2
9	7

Review

23. Phone numbers of businesses often end in 0. But does this apply to state tourism offices? We wondered, so we examined the phone numbers of the in-state tourism offices in the 50 states. At the left are the frequencies for the digits in which they ended. Use a chi-square test to test whether these ending digits depart unusually from what would be expected if last digits occurred at random. Explain your reasoning. *(Lesson 10-7)*

In 24 and 25, multiply. *(Lessons 10-5, 10-6)*

24. $(3x + 1)(2x - 4)$

25. $(8a + 1)(8a - 1)$

26. Solve $x^2 + 5x + 4 = 0$. *(Lesson 9-5)*

In 27 and 28, a thermometer is taken from a room temperature of 70°F to a point outside where it is 45°F. After t minutes, the thermometer reading is $24(1 - 0.04t)^2 + 45$. *(Lesson 9-3)*

27. About what temperature will the thermometer be after 10 minutes?

28. About how long will it take the thermometer to get down to 45°F?

In 29 and 30, solve. *(Lesson 6-8)*

29. $\frac{3}{5} = \frac{x}{15}$

30. $\frac{d}{9} = \frac{5}{d}$

31. The World Wide Wrench Company increased prices by 12%. Their basic wrench now sells for \$9.25. What did it sell for before? *(Lesson 6-5)*

Exploration

32. The fraction $\frac{x^2 - 4x + 3}{x^2 - 9}$ can be simplified when $x \neq 3$ and $x \neq -3$ by the following process:

$$\frac{x^2 - 4x + 3}{x^2 - 9} = \frac{(x - 3)(x - 1)}{(x - 3)(x + 3)} \quad \text{Factoring}$$

$$= \frac{x - 1}{x + 3} \quad \text{Equal Fractions Property}$$

a. Check this answer by letting $x = 2$ in the original and final fractions.

b. Write another fraction with quadratic expressions in its numerator and denominator that can be simplified.

c. Simplify your expression in part **b** and check your work.

LESSON

12-4

Solving Some Quadratic Equations By Factoring

Side out. *Hits (volleys) in volleyball, except when straight down or straight up, follow the path of a parabola.*

Suppose a ball is thrown upward from waist level at a speed of 48 feet per second. If there were no pull of gravity, the distance above the waist after t seconds would be $48t$, and the ball would never come down. Gravity pulls the ball down $16t^2$ feet in t seconds. So the distance d above the waist after t seconds is given by

$$d = 48t - 16t^2.$$

How long after being thrown does the ball return to waist level, where it can be caught? At waist level, the distance d is zero, so to answer this question you need to solve the equation $48t - 16t^2 = 0$.

In Chapter 9 you learned two ways to solve this equation. You can make a graph of $d = 48t - 16t^2$, and find the values of t when $d = 0$. Below is a graph of $y = 48x - 16x^2$ on the window $0 \le x \le 5$, $0 \le y \le 40$.

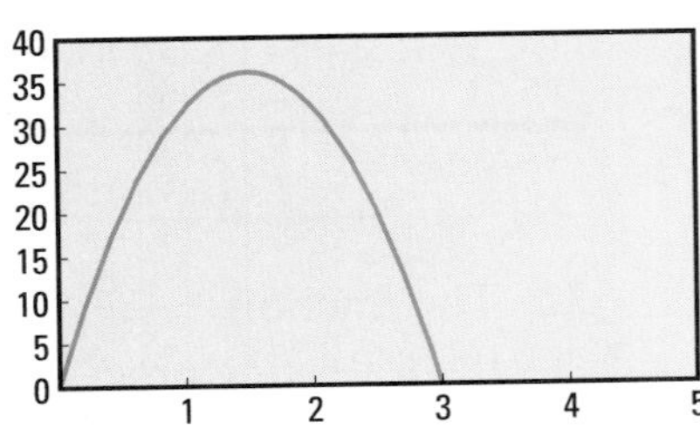

The graph shows that $y = 0$ when $x = 0$ or $x = 3$. So the ball returns to waist level 3 seconds after being thrown upward.

You can also solve the equation $48t - 16t^2 = 0$ by using the Quadratic Formula. It helps to rewrite the equation in the form $-16t^2 + 48t = 0$, so you can see that $a = -16$, $b = 48$, and $c = 0$.

$$\begin{aligned} t &= \frac{-48 \pm \sqrt{48^2 - 4(-16) \cdot 0}}{2(-16)} \\ &= \frac{-48 \pm \sqrt{48^2}}{-32} \\ &= \frac{-48 \pm 48}{-32} \end{aligned}$$

So $t = \frac{-48 + 48}{-32} = 0$ or $t = \frac{-48 - 48}{-32} = \frac{-96}{-32} = 3$

A third way to solve $48t - 16t^2 = 0$ is to use factoring and the *Zero Product Property.*

The Zero Product Property
If the product of two real numbers a and b is 0, then $a = 0$ or $b = 0$.

Numbers may be represented by expressions. So in words, the Zero Product Property is: If the product of two *expressions* is zero, one or the other (or both) must be zero.

Example 1

Refer to the situation described on the preceding page. Use the Zero Product Property to find how long the ball is in the air.

Solution

We need to solve $48t - 16t^2 = 0$.

$t(48 - 16t) = 0$ Factor.

$t = 0$ or $48 - 16t = 0$ Zero Product Property

$t = 0$ or $-16t = -48$

$t = 0$ or $t = \frac{-48}{-16} = 3$

The solution $t = 0$ means the ball was at waist level at the start.
The solution $t = 3$ means that it returned to waist level again after 3 seconds.
It was in the air for 3 seconds.

In Chapter 9 you learned that the Quadratic Formula can be used to solve all quadratic equations. Factoring and the Zero Product Property can be useful tools to solve those quadratic equations which factor over the integers.

Example 2

Solve $r^2 - 11r - 12 = 0$ by factoring.

Solution

$$r^2 - 11r - 12 = 0$$

$$(r - 12)(r + 1) = 0 \qquad \text{Factor the trinomial.}$$

$$r - 12 = 0 \quad \text{or} \quad r + 1 = 0 \qquad \text{Use the Zero Product Property.}$$

$$r = 12 \quad \text{or} \quad r = -1$$

Check

Let $r = 12$: $12^2 - 11 \cdot 12 - 12 = 144 - 132 - 12 = 0$.
Let $r = -1$: $(-1)^2 - 11 \cdot -1 - 12 = 1 + 11 - 12 = 0$.
It checks.

Recall that in order to solve a quadratic equation with the Quadratic Formula, one side of the equation must be 0. This is also true if you wish to solve an equation by factoring.

Checkmate. *Shown is "Josh Waitzkin" from the movie,* Searching for Bobby Fischer, *who aspires to play chess like Bobby Fischer. At age 14, Fischer won his first U.S. National Championship. At 15, he was the youngest player in the world ever to attain the rank of grand master.*

Example 3

In a round-robin chess tournament with n players, $\frac{n^2 - n}{2}$ games are needed. If 55 games were played, how many players were entered in the tournament?

Solution

Here $\frac{n^2 - n}{2} = 55$.

This is a quadratic equation, but it is not in standard form. Multiply both sides by 2 and then subtract 110 from both sides.

$$n^2 - n = 110$$

$$n^2 - n - 110 = 0$$

$$(n - 11)(n + 10) = 0 \qquad \text{Factor.}$$

$$n - 11 = 0 \quad \text{or} \quad n + 10 = 0 \qquad \text{Zero Product Property}$$

$$n = 11 \quad \text{or} \quad n = -10$$

A negative number of players does not make sense in this situation. So 11 players were entered.

Check

Substitute. Does $\frac{11^2 - 11}{2} = 55$? Yes, it does.

The Zero Product Property applies to products of more than two expressions or numbers.

Example 4

For what values of y does $y(2y + 3)(y - 2) = 0$?

Solution

$$y(2y + 3)(y - 2) = 0$$

$y = 0$ or $2y + 3 = 0$ or $y - 2 = 0$ Zero Product Property

$y = -\frac{3}{2}$ or $y = 2$

So y is 0, 2, or $-\frac{3}{2}$.

Check

Let $y = 0$: $0(2 \cdot 0 + 3)(0 - 2) = 0 \cdot 3 \cdot -2 = 0$.

Let $y = 2$: $2(2 \cdot 2 + 3)(2 - 2) = 2 \cdot 7 \cdot 0 = 0$.

Let $y = -\frac{3}{2}$: $-\frac{3}{2}(2 \cdot -\frac{3}{2} + 3)(-\frac{3}{2} - 2) = -\frac{3}{2} \cdot 0 \cdot -\frac{7}{2} = 0$.

Note: to use the Zero Product Property, you must have a product of *zero*. This property does *not* work for equations such as $(x + 3)(x - 2) = 1$ or $y(2y + 3)(y - 2) = -3$. Also, this property does not work for equations like $x^2 + 3x + 50 = 0$, because $x^2 + 3x + 50$ is prime over the integers.

Service charge. *Shown is tennis star Steffi Graf. Many tennis players bounce the ball in preparation for the service toss.*

QUESTIONS

Covering the Reading

1. State the Zero Product Property.

2. a. For what values of k does $(k + 4)(k - 1) = 0$?
b. Check your answers.

3. A ball bounces upward from ground level at 64 feet per second. The distance d above the ground after t seconds is $d = 64t - 16t^2$.
a. Use factoring and the Zero Product Property to determine after how many seconds the ball will hit the ground.
b. Explain how part **a** could be answered by graphing.

In 4–9, solve by factoring.

4. $x^2 + 2x - 3 = 0$

5. $y^2 - 4y - 5 = 0$

6. $x^2 - 12x + 27 = 0$

7. $a^2 + 13a + 36 = 0$

8. $t^2 - 64 = 0$

9. $v^2 + 14v + 49 = 0$

In 10 and 11, use Example 3.

10. Why was the answer -10 not accepted as a solution to the question?

11. If 21 chess games are played, how many players are entered?

12. To solve a quadratic equation by factoring, first make sure the equation is in standard form. Then **a.** __?__ the quadratic expression and apply the **b.** __?__ Property to solve.

13. Why can't the Zero Product Property be used on the equation $(w + 1)(w + 2) = 3$?

In 14 and 15, solve.

14. $(y - 15)(9y - 8) = 0$

15. $r(2r + 5)(r - 6) = 0$

Applying the Mathematics

16. Consider the equation $a^2 - 5a - 50 = 0$
   a. Solve this equation by using the Quadratic Formula.
   b. Solve this equation by using factoring.
   c. Which solution strategy, the one in part **a** or part **b,** do you prefer? Why?

In 17–22, solve by using any method.

17. $y^2 - 4y = 5$

18. $26 + b^2 = -15b$

19. $7p^2 + 7p - 84 = 0$

20. $g^2 - 100 = 0$

21. $2v^2 = 3v$

22. $0 = x^3 - 4x^2 + 4x$

23. The sum of the consecutive even integers from 2 to $2n$ is $n^2 + n$.
   a. Show that the formula is true for $n = 5$.
   b. How many consecutive even integers need to be added in order to reach a sum of 132?

24. If $a \neq 0$, solve $ax^2 - ax = 0$ for x.

25. *Multiple choice.* Which number must be a solution to the equation $ax^2 + bx = 0$, if $a \neq 0$?
   (a) 1 (b) a (c) b (d) 0

White light. *When a beam of white light, which contains all the wavelengths of visible light, passes through a prism, the beam is split up to form the band of colors called a spectrum.*

Review

26. Factor $2x^3 - 18x^2 + 16x$. *(Lesson 12-3)*

In 27 and 28, simplify. *(Lessons 10-5, 12-2)*

27. $\frac{13x^2 - 14x}{x}$ (Assume $x \neq 0$.)

28. $(u - v)^2 - (u^2 - v^2)$

29. Find the area of a square which is $4e + 1$ units on a side. *(Lesson 10-5)*

30. Describe at least three different ways to evaluate the expression $(\sqrt{9} + \sqrt{16})(\sqrt{9} - \sqrt{16})$. *(Lessons 1-6, 9-7, 10-6)*

31. A certain glass allows 90% of the light hitting it to pass through. The fraction y of light passing through x thicknesses of glass is $y = (0.9)^x$.
   a. Draw a graph of this equation for $0 \leq x \leq 8$.
   b. Use the graph to estimate the number of thicknesses you would need to allow only half the light hitting it to pass through.
   (Lesson 8-4)

Diam. (in.)	Vol. (ft^3)
8.3	10.3
8.6	10.3
8.8	10.2
10.5	16.4
10.7	18.8
11.0	15.6
11.0	18.2
11.1	22.6
11.2	19.9
11.3	24.2
11.4	21.0
11.4	21.4
11.7	21.3
12.0	19.1
12.9	22.2
12.9	33.8
13.3	27.4
13.7	25.7
13.8	24.7
14.0	34.5
14.2	31.7
14.5	36.3
16.3	42.6
17.3	55.4
17.5	55.7
17.9	58.3
18.0	51.0
18.0	51.5
20.6	77.0

32. Foresters in the Allegheny National Forest were asked to estimate the volume of timber in the black cherry trees in the forest. They had to do it without cutting all the black cherry trees down. They did cut down 29 trees of varying sizes. They measured the diameter of each tree 4.5 ft above ground level and the volume (in cubic feet) of wood which the tree produced. At the left are their data. A scatterplot is shown below.

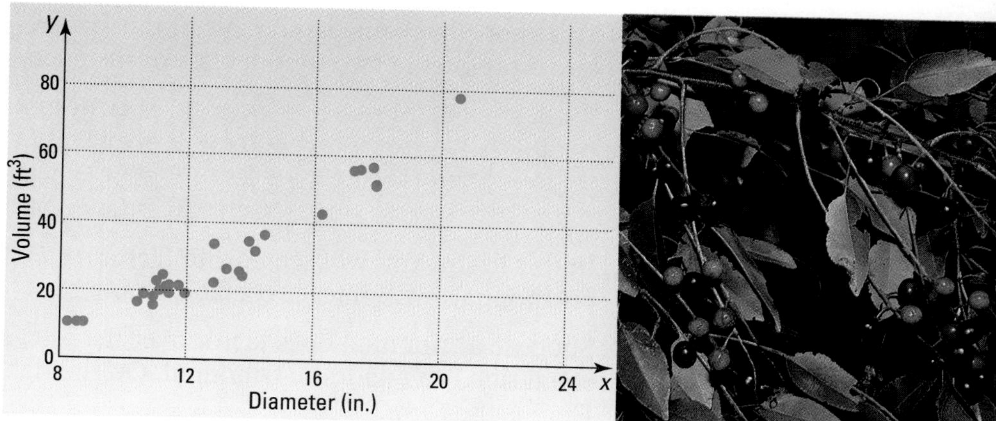

Fruit from a black cherry tree.

a. Fit a line to the data by eye.
b. Find an equation for your line.
c. Estimate the volume of a black cherry tree with diameter 15 inches. *(Lesson 7-7)*

In 33 and 34, calculate the area of the shaded region. *(Lessons 2-1, 4-2)*

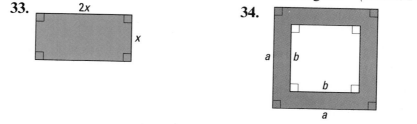

33. $2x$, x

34.

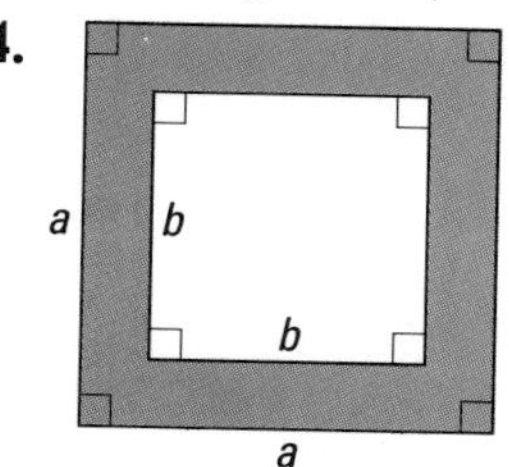

Exploration

35. Some polynomials of degree higher than 2 can be solved by using chunking and the method of this lesson. For example,

$$x^4 - 3x^2 - 4 = (x^2)^2 - 3(x^2) - 4 = (x^2 - 4)(x^2 + 1).$$

Use this idea to solve the following equations.

a. $x^4 - 3x^2 - 4 = 0$ **b.** $m^4 - 13m^2 + 36 = 0$

LESSON

12-5

Factoring $ax^2 + bx + c$

You know that some trinomials of the form $x^2 + bx + c$ can be factored into a product of two binomials. For instance,

$$x^2 - y^2 = (x + y)(x - y)$$
$$x^2 + 6x + 9 = (x + 3)(x + 3)$$
$$x^2 - 2x + 1 = (x - 1)(x - 1)$$
$$x^2 - 6x - 16 = (x + 2)(x - 8).$$

In this lesson you will see how to factor $ax^2 + bx + c$ in the set of polynomials with integer coefficients.

Suppose a trinomial does factor over the integers. On the left side of the equal sign is a quadratic trinomial. On the right side are two binomials. Here is the form.

$$ax^2 + bx + c = (dx + e)(fx + g)$$

The product of d and f, from the first terms of the binomials, is a. The product of e and g, the last terms of the binomials, is c. The task is to find these numbers so that the rest of the multiplication gives b.

Example 1

Factor $2x^2 + 11x + 14$.

Solution

The idea is to rewrite the expression as a product of two binomials.

$$2x^2 + 11x + 14 = (dx + e)(fx + g)$$

So we need to find integers d, e, f, and g.
The coefficient of $2x^2$ is 2, so $df = 2$. Thus one of d or f is 2, and the other is 1. Now you know

$$2x^2 + 11x + 14 = (2x \_\_ e)(x \_\_ g).$$

The product of e and g is 14, so $eg = 14$. Thus, e and g might equal 1 and 14, or 2 and 7, in either order. Try all four possibilities.

Can e and g be 1 and 14?

$$(2x + 1)(x + 14) = 2x^2 + 29x + 14$$
$$(2x + 14)(x + 1) = 2x^2 + 16x + 14$$

No, we want $b = 11$, not 29 or 16.

Can e and g be 2 and 7?

$$(2x + 2)(x + 7) = 2x^2 + 16x + 14$$
$$(2x + 7)(x + 2) = 2x^2 + 11x + 14$$

The last product works.

$$2x^2 + 11x + 14 = (2x + 7)(x + 2)$$

Check 1

Substitute a value for x, say 4. Then does

$$2x^2 + 11x + 14 = (2x + 7)(x + 2)?$$
$$2 \cdot 4^2 + 11 \cdot 4 + 14 = (2 \cdot 4 + 7)(4 + 2)?$$
$$32 + 44 + 14 = 15 \cdot 6?$$

Yes. Each side equals 90.

Check 2

Graph $y = 2x^2 + 11x + 14$ and $y = (2x + 7)(x + 2)$ on the same set of axes. The graphs should be identical. At the right is the output of our automatic grapher. It checks.

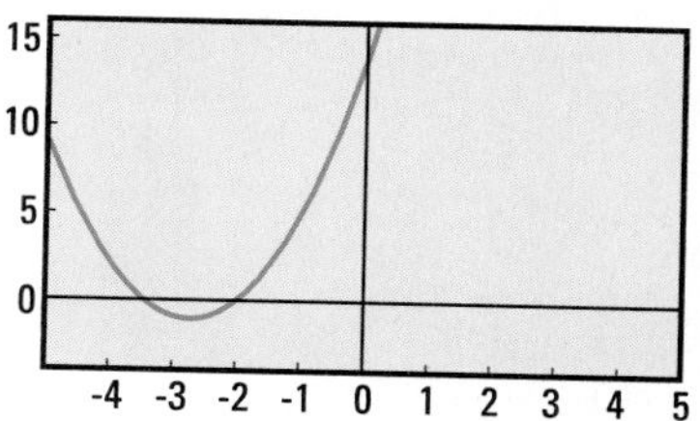

In Example 1, because the coefficient of x^2 is 2 and all terms are positive, there are only a few possible factors. Example 2 has more possibilities, but the idea is still the same.

Example 2

Factor $6y^2 - 7y - 5$.

Solution

First put down the form. $6y^2 - 7y - 5 = (ay + b)(cy + d)$

Now $ac = 6$. Thus either a and c are 3 and 2 or they are 1 and 6. The product $bd = -5$. So b and d are either 1 and -5, or -1 and 5. Here are all the possible factors with $a = 3$ and $c = 2$.

$$(3y + 1)(2y - 5)$$
$$(3y - 1)(2y + 5)$$
$$(3y - 5)(2y + 1)$$
$$(3y + 5)(2y - 1)$$

Here are all the possible factors with $a = 1$ and $c = 6$.

$$(y + 1)(6y - 5)$$
$$(y - 1)(6y + 5)$$
$$(y - 5)(6y + 1)$$
$$(y + 5)(6y - 1)$$

At most, you need to do these eight multiplications. If one of them gives $6y^2 - 7y - 5$, then that is the correct factorization.

We show all eight multiplications. Notice that the third one is correct.

$$(3y + 1)(2y - 5) = 6y^2 - 13y - 5$$
$$(3y - 1)(2y + 5) = 6y^2 + 13y - 5$$
$$(3y - 5)(2y + 1) = 6y^2 - 7y - 5$$
$$(3y + 5)(2y - 1) = 6y^2 + 7y - 5$$
$$(y + 1)(6y - 5) = 6y^2 + y - 5$$
$$(y - 1)(6y + 5) = 6y^2 - y - 5$$
$$(y - 5)(6y + 1) = 6y^2 - 29y - 5$$
$$(y + 5)(6y - 1) = 6y^2 + 29y - 5$$

So $6y^2 - 7y - 5 = (3y - 5)(2y + 1)$.

In Example 2, notice that each choice of factors gives a product that differs only in the coefficient of y (the middle term). If the original problem were to factor $6y^2 - 100y - 5$, the products listed show that no factors with integer coefficients will work. The quadratic $6y^2 - 100y - 5$ is prime in the set of polynomials with integer coefficients.

Factoring quadratic trinomials using trial and error is a skill many people learn to do in their heads. But it takes practice. This skill can save you time in solving some quadratic equations or in recognizing equivalent forms of expressions.

If at first you don't succeed . . . *To find your way out of the* Maze, *displayed at the Museum of Science and Industry in Chicago, you would probably use trial and error, just as you might do in factoring.*

Example 3

Solve $6y^2 - 7y - 5 = 0$.

Solution

Use the factorization of $6y^2 - 7y - 5$ found in Example 2.

$$6y^2 - 7y - 5 = 0$$
$$(3y - 5)(2y + 1) = 0$$

So

$$3y - 5 = 0 \quad \text{or} \quad 2y + 1 = 0$$
$$y = \frac{5}{3} \quad \text{or} \quad y = -\frac{1}{2}$$

QUESTIONS

Covering the Reading

1. Perform the multiplications in **a–d.**
 a. $(5x + 3)(x + 7)$ **b.** $(5x + 7)(x + 3)$
 c. $(5x + 1)(x + 21)$ **d.** $(5x - 1)(x - 21)$
 e. Explain how these multiplications are related to factoring $5x^2 + 26x + 21$.

2. Suppose $ax^2 + bx + c = (dx + e)(fx + g)$ for all values of x.
 a. The product of d and f is __?__.
 b. The product of __?__ and __?__ is c.

3. Give two ways to check the factorization of a quadratic trinomial.

In 4–9, factor the trinomial, if possible.

4. $2x^2 + 7x + 5$

5. $7x^2 - 36x + 5$

6. $y^2 + 10y + 9$

7. $4x^2 - 12x - 7$

8. $3x^2 + 11x - 4$

9. $10k^2 - 23k + 12$

10. Check the solutions of Example 3 by substitution.

11. Solve $2x^2 + 11x + 14 = 0$ by factoring.

Applying the Mathematics

12. Without graphing, what can you say about the graphs of $y = (2x - 5)(x + 1)$ and $y = 2x^2 - 3x - 5$? Justify your answer.

13. Find k if $8x^2 + 10x - 25 = (2x + k)(4x - k)$.

14. a. Solve the equation $6t^2 - 5t - 21 = 0$
  i. by using the Quadratic Formula.
  ii. by factoring.
b. Which method in part **a** do you prefer? Why?

15. a. Find the x-intercepts of the graph of $y = 2x^2 + x - 36$
  i. by making a graph and estimating the coordinates.
  ii. by factoring.
b. Which method in part **a** do you prefer? Why?

16. a. Factor $20x^3 - 70x^2 + 60x$ into the product of a monomial and a trinomial.
b. Give the complete factorization of $20x^3 - 70x^2 + 60x$.

In 17 and 18, find the complete factorization.

17. $9p^4 + 12p^3 + 4p^2$

18. $2x^2 - 5xy + 3y^2$

Review

19. State the Zero Product Property. *(Lesson 12-4)*

In 20–23, solve by using any method. *(Lessons 9-5, 12-4)*

20. $x^2 - 8x + 7 = 0$

21. $y^2 + 9y + 7 = 0$

22. $d^2 + 16 = 8d$

23. $100v^2 - 400 = 0$

24. a. Factor $x^2 + 8x + 7$.
b. Use the factoring of part **a** to find the prime factorization of 187. *(Lessons 12-1, 12-3)*

25. Rewrite $x^4 - 81$ as the product of
a. two binomials.
b. three binomials. *(Lessons 10-6, 12-3)*

26. Which holds more?
a. a cube with edges of length 4, or a rectangular box with dimensions 3 by 4 by 5
b. a cube with edges of length e, or a box with dimensions $e - 1$ by e by $e + 1$. *(Lessons 2-1, 10-6)*

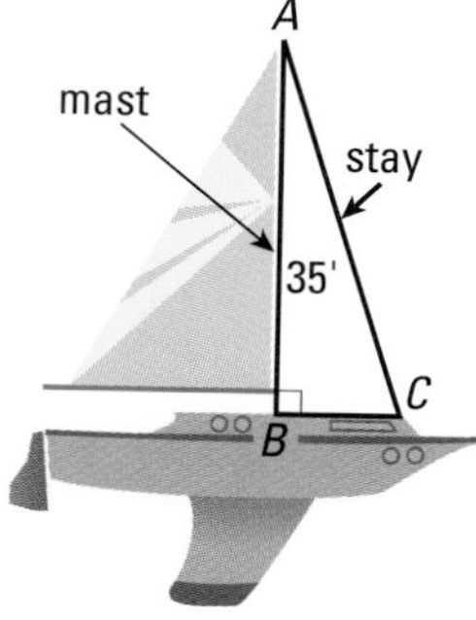

In 27 and 28, a 35-foot mast on a sailboat is strengthened by a wire called a *stay,* as shown at the left.

27. If the stay is attached to the boat 12 feet from B, the base of the mast, how long is the stay?

28. If the stay is 50 ft long, how far from the base of the mast is it attached? *(Lessons 1-8, 9-7)*

29. **a.** Here are 3 instances of a pattern. Describe the general pattern using two variables, x and y. *(Lessons 1-7, 10-6)*

$$(72 + 100)(36 - 50) = 2(36^2 - 50^2)$$
$$(30 + 20)(15 - 10) = 2(15^2 - 10^2)$$
$$(2 + 1)(1 - 0.5) = 2(1^2 - 0.5^2)$$

b. Does your general pattern hold for all real values of x and y? Justify your answer.

30. *Multiple choice.* Which equation has the graph at the left? *(Lesson 9-1)*

(a) $y = \frac{1}{2}x^2$ (b) $y = 2x^2$

(c) $y = -2x^2$ (d) $y = -\frac{1}{2}x^2$

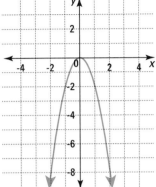

In 31 and 32, give the slope and y-intercept for each line. *(Lesson 7-4)*

31. $y = \frac{1}{2}x$ **32.** $8x - 5y = 1$

33. If "One picture is worth a thousand words," what is the worth of 325 words? *(Lesson 6-8)*

One picture is worth a thousand words.

Exploration

34. The polynomial $x^3 + 6x^2 + 11x + 6$ has the factorization over the integers

$$(x + a)(x + b)(x + c).$$

Find a, b, and c. (Hint: What is abc?)

LESSON

12-6

How Was the Quadratic Formula Found?

Heavy notebooks. *Pictured is a 14th century* B.C. *Babylonian clay tablet showing a seal dedicated to King Kidin-Marduk. Much of the ancient Babylonian work in mathematics was recorded on clay tablets.*

What Problem First Led to Quadratics?

Our knowledge of ancient civilizations is based only on what survives today. The earliest known problems that led to quadratic equations are on Babylonian tablets dating from 1700 B.C. These problems are to find two numbers x and y that satisfy the system

$$\begin{aligned} x + y &= b \\ xy &= c. \end{aligned}$$

This suggests that the Babylonian scribes were interested in finding the length x and width y of a rectangle with a given area c and a given perimeter $2b$. The historian Victor Katz suggests that maybe there were some people who believed that if you knew the area of a rectangle, then you knew its perimeter. In solving these problems, the Babylonian scribes may have been trying to show that there are many rectangles with different dimensions and the same area.

Example

Find the dimensions of a rectangular field whose perimeter is 200 meters and whose area is 2475 square meters.

Solution

Let L and W be the length and width of this rectangle.

Then
$$\begin{aligned} 2L + 2W &= 200 \\ LW &= 2475. \end{aligned}$$

This system can be solved by substitution. First solve the first equation for *W*.

$$\begin{aligned} 2W &= 200 - 2L \\ W &= 100 - L \end{aligned}$$

Now, substitute $100 - L$ for W in the second equation.

$$L(100 - L) = 2475$$

This is a quadratic equation and so it can be solved.

Activity 1

Solve the quadratic equation of the Example using either the Quadratic Formula or factoring.

How Were the First Quadratics Solved?

The Babylonians, like the Greeks who came after them, used a geometric approach to solve such problems. Using today's algebraic language and notation, here is what they did. It is a very sneaky way to solve this sort of problem. Refer back to the Example.

Since $L + W = 100$, the average of L and W is 50. This means that L is as much bigger than 50 as W is smaller than 50. So let $L = 50 + x$ and $W = 50 - x$. Substitute these values into the second equation.

$$(50 + x)(50 - x) = 2475$$
$$2500 - x^2 = 2475$$
$$x^2 = 25$$

So $x = 5$ or $x = -5$.
$L = 50 + x$, so $L = 50 + 5 = 55$ or $L = 50 + -5 = 45$.
$W = 50 - x$, so $W = 50 - 5 = 45$ or $W = 50 - -5 = 55$.

Either solution tells us that the rectangle has dimensions 45 meters by 55 meters.

This stamp from the former Soviet Union honors al-Khwarizmi.

Activity 2

Use the Babylonian method to find two numbers whose sum is 64 and whose product is 903. (Hint: Let one of the numbers be $32 + x$, the other $32 - x$.)

Notice what the Babylonians did. They took a complicated quadratic equation and, with a clever substitution, reduced it to an equation of the form $x^2 = k$. That equation is easy to solve. Then they substituted back.

The work of the Babylonian scribes was lost for many years. In 825 A.D., about 2500 years after the Babylonian tablets were carved, the general method that is similar to today's Quadratic Formula was written down in words by the Arab mathematician al-Khwarizmi in a book entitled *Hisab al-jabr w'al muqabala.* Al-Khwarizmi's techniques were more general than those of the Babylonians. He gave a method to solve any equation of the form $ax^2 + bx = c$, where a, b, and c are positive numbers. His book was very influential. From the second word in that title comes our modern word "algebra." From his name comes our word "algorithm."

How Do We Know the Quadratic Formula Is True?

Neither the Babylonians nor al-Khwarizmi worked with an equation of the form $ax^2 + bx + c = 0$, because they considered only positive numbers, and if a, b, and c are positive, the equation has no positive solutions.

Not until the 1700s was the general solution to the quadratic given as you have learned it in this book. Now we show why the formula works.

Examine the argument below closely. See how each equation follows from the preceding equation. The idea is quite similar to the one used by the Babylonians, but a little more general. We work with the equation $ax^2 + bx + c = 0$ until the left side is a perfect square. Then the equation has the form $t^2 = k$, which you know how to solve for t.

The goal is to solve $ax^2 + bx + c = 0$.

Step 1: Multiply both sides by $4a$. This makes the first term of the expression on the left equal to $4a^2x^2$, the square of $2ax$.

$$4a^2x^2 + 4abx + 4ac = 0$$

Step 2: When the quantity $2ax + b$ is squared, it equals $4a^2x^2 + 4abx + b^2$. The left and center terms of this trinomial match what is in the equation. We add b^2 to both sides of the equation to get all three terms into our equation.

$$4a^2x^2 + 4abx + b^2 + 4ac = b^2$$

Step 3: The first three terms are the square of $2ax + b$.

$$(2ax + b)^2 + 4ac = b^2$$

Step 4: Add $-4ac$ to both sides.

$$(2ax + b)^2 = b^2 - 4ac$$

Step 5: Now the equation has the form $t^2 = k$, with $t = 2ax + b$ and $k = b^2 - 4ac$. This is where the discriminant $b^2 - 4ac$ becomes so important. If $b^2 - 4ac \geq 0$, then there are real solutions. They are found by taking the square roots of both sides.

$$2ax + b = \pm\sqrt{b^2 - 4ac}$$

Step 6: It's beginning to look like the formula. Now add $-b$ to each side.

$$2ax = -b \pm \sqrt{b^2 - 4ac}$$

Step 7: Divide both sides by $2a$ to obtain the Quadratic Formula.

$$x = \frac{-b \pm \sqrt{b^2 - 4ac}}{2a}$$

What if $b^2 - 4ac < 0$? Then the quadratic equation has no real number solutions. The formula still works, but you have to take square roots of negative numbers to get solutions. You will study these non-real solutions in a later course.

QUESTIONS

Covering the Reading

1. *Multiple choice.* The earliest known problems that led to the solving of quadratic equations were considered about how many years ago?
(a) 1175 (b) 1700 (c) 2500 (d) 3700

2. In what civilization do quadratic equations first seem to have been considered and solved?

3. Show your work in Activity 1.

4. Show your work in Activity 2.

5. Suppose a rectangular room has a floor area of 60 square yards. Give two different lengths and widths that this floor might have.

In 6 and 7, suppose a rectangular room has a floor area of 60 square yards and that the perimeter of its floor is 34 yards.

6. Find its length and width by solving a quadratic equation using the Quadratic Formula or factoring.

7. Find its length and width using a more ancient method.

8. Find two numbers whose sum is 12 and whose product is 9.

9. What is the significance of the work of al-Khwarizmi in the history of the Quadratic Formula?

10. Here are steps in the derivation of the Quadratic Formula. Tell what was done to get each step.

$$ax^2 + bx + c = 0.$$

a. $4a^2x^2 + 4abx + 4ac = 0$

b. $4a^2x^2 + 4abx + 4ac + b^2 = b^2$

c. $4a^2x^2 + 4abx + b^2 = b^2 - 4ac$

d. $(2ax + b)^2 = b^2 - 4ac$

e. $2ax + b = \pm\sqrt{b^2 - 4ac}$

f. $2ax = -b \pm \sqrt{b^2 - 4ac}$

g. $x = \frac{-b \pm \sqrt{b^2 - 4ac}}{2a}$

11. When the discriminant of a quadratic equation is negative, how many real solutions does the equation have?

Applying the Mathematics

12. Solve the equation $6x^2 - 5x - 1 = 0$ by following the steps in the derivation of the Quadratic Formula.
(Hint: The first step is to multiply both sides by $4 \cdot 6$, or 24.)

13. Explain why there are no real numbers x and y whose sum is 15 and whose product is 60.

14. In a Chinese text that is thousands of years old, the following problem is given: The height of a door is 6.8 more than its width. The distance between its corners is 10. Find the height and width of the door.

Review

In 15–17, factor. *(Lessons 12-2, 12-5)*

15. $8abc - 4ab^2 + 12abd$ **16.** $4x^2 - 9y^2$ **17.** $20m^2 + 20 - 41m$

18. Solve: $12x^3 + 20x^2 + 3x = 0$. *(Lessons 12-2, 12-5)*

19. Refer to the cartoon below.

Use the prime factorization of one million to determine all possible whole numbers of years and amounts over which $1,000,000 could be spread and still give the winner of the lottery a whole number of dollars. *(Lesson 12-1)*

20. According to the National Center for Health Statistics, the following percentages of the adult population smoked in particular years. *(Lesson 7-7)*

Year	Men	Women
1965	52	34
1979	38	30
1983	35	30
1991	28	24

Assume the trends continue.

a. Use any method to predict what percentage of adult men will be smokers in the year 2000.

b. Use any method to predict when the same percentage of adult men and women will smoke.

c. Use any method to predict when no men or women will smoke.

Exploration

21. Solve a few quadratic equations of your own choosing. If the solutions are fractions, keep them that way. Find the sum of the two solutions to each equation. Then find the product of the two solutions. In the late 1500s, Viète discovered how the sum and product of the solutions are related to the coefficients *a, b,* and *c* of the equation. Try to rediscover what Viète discovered.

LESSON

12-7

Rational Numbers and Irrational Numbers

Every number that can be represented as a decimal is a real number. All the real numbers are either rational or irrational. In this lesson you will learn to distinguish between these two types of real numbers.

What Are Rational Numbers?

Recall that a **simple fraction** is a fraction with integers in its numerator and denominator. For instance, $\frac{2}{3}$, $\frac{5488}{212}$, $\frac{-7}{-2}$, and $\frac{-43}{1}$ are simple fractions.

Some numbers are not written as simple fractions, but *equal* simple fractions. Any mixed number equals a simple fraction. For example, $3\frac{2}{7} = \frac{23}{7}$. Also, any integer equals a simple fraction. For example, $-10 = \frac{-10}{1}$. And any finite decimal equals a simple fraction. For instance, $3.078 = 3\frac{78}{1000} = \frac{3078}{1000}$. All these numbers are *rational numbers.* A **rational number** is a number that is a simple fraction or equals a simple fraction. All repeating decimals are also rational numbers.

Example

Show that $5.7\overline{62}$ is a rational number.

Solution

Let $\quad x = 5.7\overline{62} = 5.7626262626262\ldots$

Multiply both sides by 10^n, where n is the number of digits in the repetend. Here there are two digits in the repetend, so we multiply by 10^2, or 100.

$$100x = 576.2\overline{62} = 576.2626262626262\ldots$$

Subtract the top equation from the bottom equation. The key idea here is that the result is no longer an infinite repeating decimal; in this case, after the first decimal place the repeating parts subtract to zero.

$$100x = 576.2\overline{62}$$
$$-\quad x = -\ 5.7\overline{62}$$
$$99x = 570.5$$

Divide both sides by 99. $x = \frac{570.5}{99} = \frac{5705}{990} = \frac{1141}{198}$

Since x can be represented as a simple fraction, x is a rational number.

Rational numbers have nice properties. They can be added, subtracted, multiplied, and divided, and give answers that are rational numbers. They can always be written as terminating or repeating decimals.

What Are Irrational Numbers?

The ancient Greeks seem to have been the first to discover that there are numbers that are not rational numbers. They called them *irrational.* An **irrational number** is a real number that is not a rational number. Some of the most commonly found irrational numbers in mathematics are the square roots of integers that are not perfect squares. That is, numbers like $\sqrt{2}$, $\sqrt{3}$, $\sqrt{5}$, $\sqrt{6}$, $\sqrt{7}$, $\sqrt{8}$, $\sqrt{10}$, and so on, are irrational. As you know, these numbers arise from situations involving right triangles. At the left is a figure that you saw in Chapter 1.

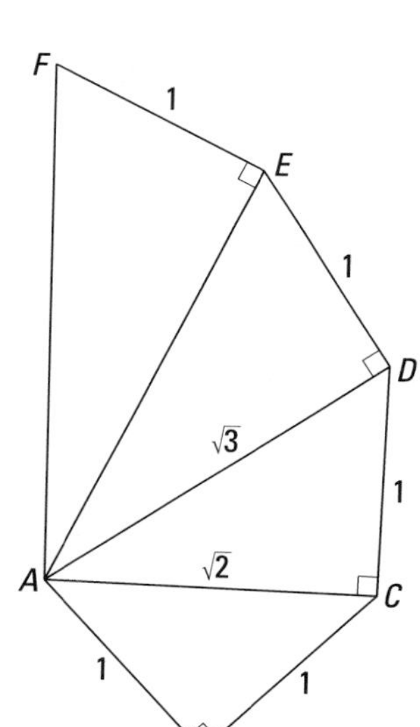

$\triangle ABC$ is a right triangle with legs of 1 and 1. By the Pythagorean Theorem,

$$AC^2 = AB^2 + BC^2$$
$$= 1^2 + 1^2$$
$$= 2.$$

So $$AC = \sqrt{2}.$$

$\triangle ACD$ is drawn with leg $\overline{AC}$, and another leg $CD = 1$. So, by the Pythagorean Theorem,

$$AD^2 = AC^2 + CD^2$$
$$= (\sqrt{2})^2 + 1^2$$
$$= 2 + 1$$
$$= 3.$$

So $$AD = \sqrt{3}.$$

You are asked to find other lengths in this figure in the Questions.

Prime Factorizations of Perfect Squares

Consider the square of 28. The prime factorization of 28 is $2 \cdot 2 \cdot 7$. So the prime factorization of 28^2, or 784, is $(2 \cdot 2 \cdot 7) \cdot (2 \cdot 2 \cdot 7)$. Notice that 28 is a product of 3 primes, while its square is a product of 6 prime numbers.

Activity

Pick a perfect square other than 784. Find its prime factorization. Verify that the number is the product of an even number of primes.

In general, if a number is the product of n primes, then we can write the number as $x_1 \cdot x_2 \cdot x_3 \cdot \ldots \cdot x_n$. Its square is

$$(x_1 \cdot x_2 \cdot x_3 \cdot \ldots \cdot x_n) \cdot (x_1 \cdot x_2 \cdot x_3 \cdot \ldots \cdot x_n),$$

which is a product of $2n$ primes. Since n is an integer, $2n$ is an even number. The important conclusion here is: *If a number is a perfect square, then it is the product of an even number of primes.*

How Do We Know That Certain Numbers Are Irrational?

To show that $\sqrt{7}$ is an irrational number, we first show that there cannot be two perfect squares a^2 and b^2 with $a^2 = 7b^2$. (The same process works with the square root of every prime.) That is, no perfect square can be exactly 7 times another perfect square. (You can come very close; 64 is one more than 7 times 9.)

Here is the argument: If the numbers $7b^2$ and a^2 were the same number, the Unique Factorization Theorem says they must have the same prime factorization. The number a^2 is the product of an even number of primes, because a^2 is a perfect square. Since b^2 is a perfect square, it is also the product of an even number of primes. But $7b^2$ has one more prime factor (the number 7) than b^2 has. So $7b^2$ is the product of an odd number of primes. Therefore, a^2 and $7b^2$ cannot be the same number.

Now work backwards. Since there are no integers a and b with $7b^2 = a^2$, then there cannot be integers a and b with $7 = \frac{a^2}{b^2}$, and so there are no integers a and b with $\sqrt{7} = \frac{a}{b}$. This means that $\sqrt{7}$ cannot be written as a simple fraction, and so it is irrational.

Arguments like this one can be used to prove the following theorem.

Theorem
If a positive integer n is not a perfect square, then $\sqrt{n}$ is irrational.

Today we know that there are many irrational numbers. For instance, every number that has a decimal expansion that does not end or repeat is irrational. Among the irrational numbers is the famous number π. But the argument to show that π is irrational is far more difficult than the argument used above for some square roots of integers. It requires advanced mathematics, and was first given by the German mathematician Johann Lambert in 1767, over 2000 years after the Greeks had first discovered that some numbers were irrational.

There is a practical reason for knowing whether a number is rational or irrational. When a number is rational, arithmetic can be done with it rather easily. Just write the number as a simple fraction and work as you do with fractions. But if a number is irrational, then it is generally more difficult to do arithmetic with it.

QUESTIONS

Covering the Reading

In 1–3, give an example.

1. a simple fraction

2. a fraction that is not a simple fraction

3. a rational number

In 4–6, write the number as a simple fraction.

4. 99.44 **5.** $.\overline{15}$ **6.** $14.8\overline{3}$

7. Take $\overline{MN}$ drawn at the left as having a length of 1 unit. Use it to draw a segment with length $\sqrt{2}$ units.

8. Use the drawing on page 755.
a. Find *AE*.
b. Determine whether *AE* is rational or irrational.
c. Find *AF*.
d. Determine whether *AF* is rational or irrational.

9. Refer to the Activity on page 755.
a. What perfect square did you use for the Activity?
b. Give its prime factorization.
c. How many prime factors are in the factorization?

10. **a.** How many prime factors are in the factorization of 50^2?
b. How many prime factors are in the factorization of $13 \cdot 50^2$?
c. Why does the answer to part **b** guarantee that $\sqrt{13 \cdot 50^2}$ is irrational?

In 11–13, tell whether the number is a rational number, an irrational number, or neither.

11. π **12.** -45 **13.** $\sqrt{64}$

14. Who first discovered that there were numbers that were not rational numbers?

Applying the Mathematics

15. Is 0 a rational number? Why or why not?

16. Is it possible for two irrational numbers to have a sum that is a rational number? Explain why or why not.

17. A square card table has a side of length 30″. Find the length of its diagonal and tell whether the length is rational or irrational.

18. If a circular table has a diameter of 120 cm, is its circumference rational or irrational?

19. Determine whether the solutions to the equation $x^2 - 6x - 1 = 0$ are rational or irrational.

Review

20. Find two numbers whose sum is 562 and whose product is 74,865. *(Lesson 12-6)*

21. Suppose $12x^2 + 7xy - 12y^2 = (ax + b)(cx + d)$.

a. Find the value of $ad + bc$.

b. Find a, b, c, and d. *(Lesson 12-5)*

Still waters run deep. *Pictured is a water well in Egypt.*

22. Factor $9 + 6x + x^2$. *(Lesson 12-3)*

23. The sector at the right is a quarter of a circle. Write a formula for the perimeter p of this sector in factored form. *(Previous course, Lesson 12-2)*

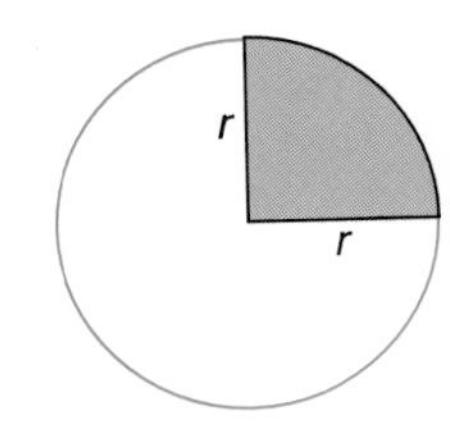

24. Find all values of x that satisfy the equation $bx^2 + cx + a = 0$, when $b \neq 0$. *(Lesson 9-5)*

25. With a stopwatch and a stone, you can estimate the depth of a well. If the stone takes 2.4 seconds to reach bottom, how deep is the well? *(Lesson 9-4)*

Exploration

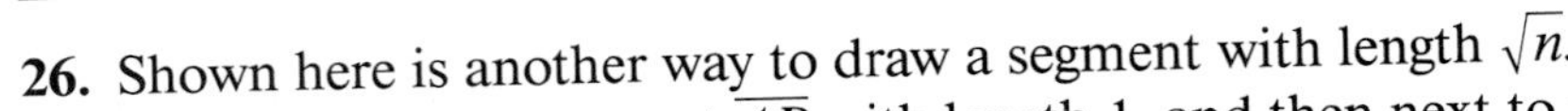

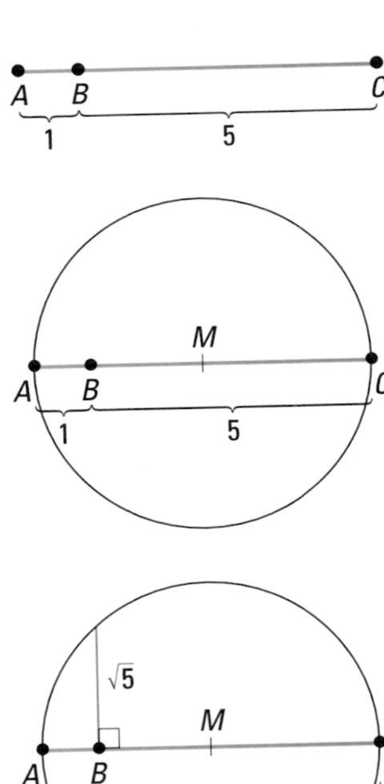

26. Shown here is another way to draw a segment with length $\sqrt{n}$.

Step 1: Draw a segment $\overline{AB}$ with length 1, and then next to it a segment $\overline{BC}$ with length n. (In the drawing at the left, $n = 5$.)

Step 2: Find the midpoint M of segment $\overline{AC}$. Draw the circle with center M that contains A and C. ($\overline{AC}$ will be a diameter of this circle.)

Step 3: Draw a segment perpendicular to $\overline{AC}$ from B to the circle. This segment has length $\sqrt{n}$. (In our drawing it should have length $\sqrt{5}$.)

a. Try this algorithm to draw a segment with length $\sqrt{6}$.

b. Measure the segment you get.

c. How close is its length to $\sqrt{6}$?

LESSON

12-8

Which Quadratic Expressions Are Factorable?

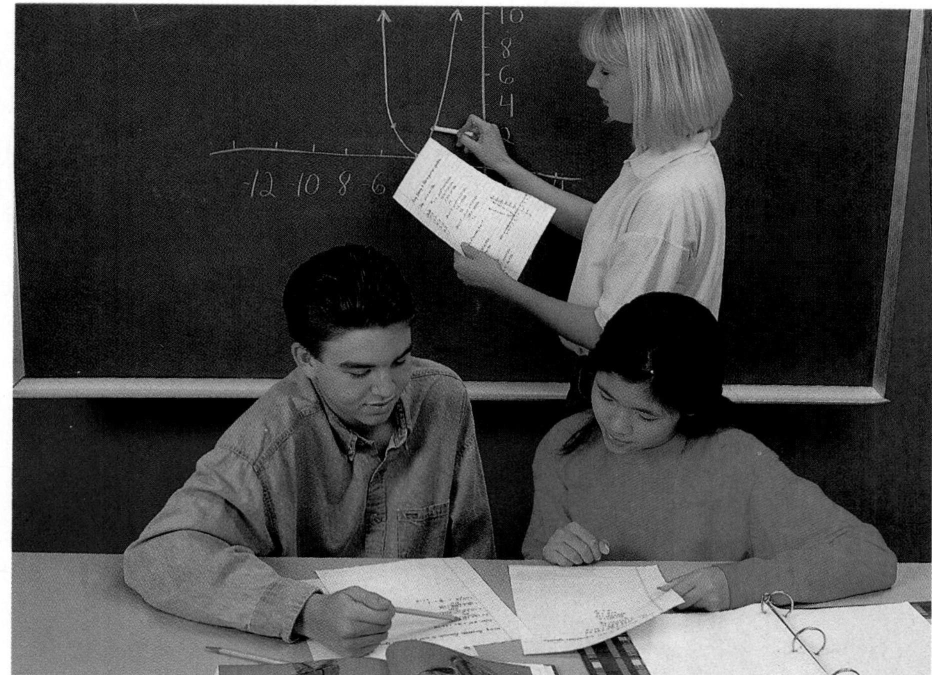

Factoring in what to do. *When solving a quadratic equation, you may try any of three methods: graphing, using the Quadratic Formula, or applying the Zero Product Property.*

This lesson connects three topics you have seen in this chapter: factoring, irrational numbers, and solutions to quadratic equations. These topics seem quite different. Their relationship to each other is an example of how what you learn in one part of mathematics is often useful in another part.

You have studied three ways to find the real-number values of x that satisfy $ax^2 + bx + c = 0$. (1) You can graph $y = ax^2 + bx + c$ and look for its x-intercepts. (2) You can use the Quadratic Formula. (3) You can factor $ax^2 + bx + c$ and use the Zero Product Property. The first two ways can always be done. But you know that it is not always possible to factor $ax^2 + bx + c$ over the integers. So it is useful to know when it is possible.

A Quadratic Equation with Rational Solutions

Consider the equation $2x^2 + 5x - 12 = 0$.

By the Quadratic Formula,

Step 1: $x = \frac{-5 \pm \sqrt{25 - 4 \cdot 2 \cdot (-12)}}{4}$

Step 2: $= \frac{-5 \pm \sqrt{121}}{4}$

Step 3: $= \frac{-5 \pm 11}{4}$

Step 4: So, $x = \frac{-5 + 11}{4} = \frac{3}{2}$, or $x = \frac{-5 - 11}{4} = -4$.

Notice that there are no radical signs in the solutions in Step 4. This is because the number 121 under the square root sign is a perfect square (Step 2). So, after the square root is calculated (Step 3), one integer is divided by another, and the solutions are rational numbers.

In general, for polynomials over the integers: If the discriminant $b^2 - 4ac$ of the quadratic equation $ax^2 + bx + c = 0$ is a perfect square, then the solutions are rational. Otherwise, the square root will not equal an integer, it will remain in the solutions, and the solutions will be irrational. These results can be summarized in one sentence. When a, b, and c are integers, the solutions to $ax^2 + bx + c = 0$ are rational numbers if and only if $b^2 - 4ac$ is a perfect square.

Now we connect this with factoring. Consider the same equation as before.

$$2x^2 + 5x - 12 = 0$$

Factor the left side.

$$(2x - 3)(x + 4) = 0.$$

Use the Zero Product Property.

$$2x - 3 = 0 \quad \text{or} \quad x + 4 = 0.$$

So, as before, $x = \frac{3}{2}$ or $x = -4$.

Relating the Solutions to $ax^2 + bx + c = 0$ to the Factorization of $ax^2 + bx + c$

Recall that polynomials can be factored in only one way, except for order and for real number multiples. Now you can see why this is true for quadratics. If $ax^2 + bx + c$ could be factored in more than one way, then you would wind up with different solutions to the equation $ax^2 + bx + c = 0$. And we already know there can be at most two solutions; they can be found by using the Quadratic Formula.

Notice how the solutions are related to the factors. When $(dx + e)$ is one of the factors of the left side, then we set $dx + e = 0$ and solve for x to obtain $x = \frac{-e}{d}$. Because d and e are integers, $\frac{-e}{d}$ is a simple fraction. This tells us: If a, b, c are integers and $ax^2 + bx + c$ is factorable over the integers, then each solution to $ax^2 + bx + c = 0$ is a rational number.

The argument can be reversed. Suppose each solution to $ax^2 + bx + c = 0$ is a rational number. Then the solutions can be written as $\frac{-e}{d}$ and $\frac{-g}{f}$, where d, e, f, and g are integers. Because $x = \frac{-e}{d}$ or $x = \frac{-g}{f}$, clearing the equations of fractions gives,

$$dx = -e \quad \text{or} \quad fx = -g.$$

Now, adding to obtain 0 on the right side of each equation, we get

$$dx + e = 0 \quad \text{or} \quad fx + g = 0.$$

Consequently,

$$(dx + e)(fx + g) = 0.$$

This must be a factorization of $ax^2 + bx + c$, because all factorizations give the same solutions. Consequently, when a, b, and c are integers, if there are rational solutions to $ax^2 + bx + c = 0$, then $ax^2 + bx + c$ is factorable.

All of these arguments together show the importance of the discriminant $b^2 - 4ac$. You need only examine it to determine whether a quadratic expression is factorable.

Discriminant Theorem

When a, b, and c are integers, with $a \neq 0$, either all three of the following conditions happen at exactly the same time, or none of these conditions happens.

(1) The solutions to $ax^2 + bx + c = 0$ are rational numbers.

(2) $b^2 - 4ac$ is a perfect square.

(3) $ax^2 + bx + c$ is factorable over the set of polynomials with integer coefficients.

Example 1

Is $5x + 1 + 4x^2$ factorable into polynomials with integer coefficients?

Solution

First rewrite this expression in the standard form of a polynomial: $4x^2 + 5x + 1$. Thus $a = 4$, $b = 5$, and $c = 1$, so $b^2 - 4ac = 9$. Since 9 is a perfect square, the expression is factorable.

Activity

Verify Example 1 by finding the factorization of $5x + 1 + 4x^2$.

Caution: The phrase "with integer coefficients" is necessary in Example 1, because every quadratic expression is factorable if noninteger coefficients are allowed. For instance, the quadratic expression $x^2 - 12 = x^2 - (\sqrt{12})^2 = (x - \sqrt{12})(x + \sqrt{12})$. The polynomial $x^2 - 12$ is prime over the set of polynomials with integer coefficients, but not over the set of all polynomials.

Applying the Discriminant Theorem

Knowing whether an expression is factorable can help determine what methods are available to solve an equation.

Example 2

Solve $m^2 - 31m - 24 = 0$ by any method.

Solution

Since the coefficient of m^2 is 1, it is reasonable to try to factor the left side. But first evaluate $b^2 - 4ac$ to see whether this is possible.
In this equation, $a = 1$, $b = -31$, and $c = -24$. So

$$b^2 - 4ac = (-31)^2 - 4 \cdot 1 \cdot (-24) = 1057.$$

This is not a perfect square, so the equation does not factor over the integers. So use the Quadratic Formula. This is easier to do because the discriminant has already been calculated.

$$x = \frac{31 \pm \sqrt{1057}}{2}$$

Check

$\frac{31 \pm \sqrt{1057}}{2} \approx 15.5 \pm 16.3 \approx 31.8$ or -0.8.

Substitute 31.8 for x in the original equation. Is $(31.8)^2 - 31 \cdot 31.8 - 24 = 0$? The left side equals 1.44, which is very close to 0 given the size of the numbers. You should check the second solution.

You may wonder how often quadratic expressions are factorable. Example 3 asks this for an application of quadratics. Recall that, neglecting such factors as air resistance, an object shot into the air at an upward velocity of v feet per second will, after t seconds, be at a height given by $h = vt - 16t^2$.

Example 3

At what upward integer velocities from 80 to 100 feet per second can an object be shot and reach a height of 100 feet at a time that is a rational number of seconds?

Solution

Begin with equation $h = vt - 16t^2$. Here $h = 100$, so $100 = vt - 16t^2$. The question asks for the integer values of v so that t is rational. To find these values, put the equation into standard form.

$$16t^2 - vt + 100 = 0$$

Use the Discriminant Theorem. If the discriminant is a perfect square, then t will be rational. Calculate the discriminant.

$$b^2 - 4ac = v^2 - 4 \cdot 16 \cdot (100) = v^2 - 6400$$

►

	A	B	C
1	V	V*V−6400	SQRT(V*V−6400)
2	80	0	0
3	81	161	12.68857754
4	82	324	18
5	83	489	22.11334439
6	84	656	25.61249695
7	85	825	28.72281323
8	86	996	31.55946768
9	87	1169	34.19064199
10	88	1344	36.66060556
11	89	1521	39
12	90	1700	41.23105626
13	91	1881	43.37049688
14	92	2064	45.43126677
15	93	2249	47.42362281
16	94	2436	49.35585072
17	95	2625	51.23475383
18	96	2816	53.06599665
19	97	3009	54.85435261
20	98	3204	56.60388679
21	99	3401	58.31809325
22	100	3600	60

Now the problem requires substituting the numbers 80 to 100 for v and checking whether $v^2 - 6400$ is a perfect square for each value. This is tedious without some technology. However, with a spreadsheet, it is not difficult. At the left is a spreadsheet that shows the calculations. The top row of each column gives the formula for that column. For instance, the formulas for the fourth row, which starts with 82, are = A4*A4 − 6400 in cell B4, and = SQRT(B4) in cell C4.
At velocities of 80, 82, 89, and 100 feet per second the object will reach 100 feet at a rational number of seconds.

QUESTIONS

Covering the Reading

In 1–4, a quadratic expression is given. If possible, factor the expression into polynomials with integer coefficients. If this is not possible, explain why not.

1. $7x^2 - 11x - 4$

2. $y^2 - 18$

3. $9n^2 - 12n + 4$

4. $3 + m^2 + 2m$

5. What factorization did you get as a result of the Activity of this lesson?

6. Consider the equation $ax^2 + bx + c = 0$ when a, b, and c are integers. If $b^2 - 4ac$ is a perfect square, explain why x is rational.

7. Give an example of a quadratic expression that can only be factored if noninteger coefficients are allowed.

In 8 and 9, refer to Example 3. At what times will the ball reach a height of 100 feet if

8. the ball is shot into the air with an upward velocity of 89 feet per second?

9. the ball is thrown into the air with an upward velocity of 88 feet per second?

10. What in this lesson tells you that the solutions to the quadratic equation in Question 9 are irrational?

Applying the Mathematics

11. Find a value of k such that $4x^2 + kx - 3$ is factorable.

12. Suppose a, b, and c are integers. When will the x-intercepts of $y = ax^2 + bx + c$ be rational numbers?

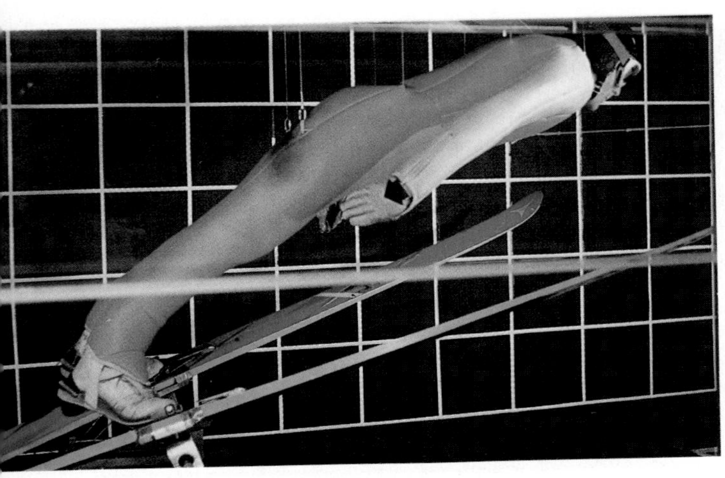

This skier is in a wind tunnel at the Calspan Corporation in Rochester, New York. The wind tunnel is used to find ways to streamline bodies in order to minimize air resistance.

13. Try a version of Example 3 using metric units. Neglecting such factors as air resistance, an object shot into the air at an upward velocity of v meters per second will, after t seconds, be at a height $h = vt - 4.9t^2$. At what upward integer velocities from 25 to 40 meters per second can an object be shot and reach a height of 30 meters at a time that is a rational number of seconds?

14. **a.** By multiplying, verify that $(x + 3 + \sqrt{2})(x + 3 - \sqrt{2}) = x^2 + 6x + 7$.
b. Verify that the discriminant of the expression $x^2 + 6x + 7$ is not a perfect square.
c. Part **a** indicates that $x^2 + 6x + 7$ is factorable, and yet its discriminant is not a perfect square. Does this situation contradict the Discriminant Theorem?

Review

15. Tell whether the number is rational or irrational. *(Lesson 12-7)*
a. $\sqrt{25}$ **b.** $\sqrt{26}$ **c.** $\sqrt{27}$

16. **a.** Is $\frac{\sqrt{3}}{\sqrt{12}}$ rational or irrational?
b. Explain why or why not. *(Lesson 12-7)*

17. The winning percentage of a high school tennis team was $.58\overline{3}$. If the team had fewer than 20 matches, how many matches did they have and how many did they win? *(Lesson 12-7)*

18. Roseanne was surprised to learn that a farm with an area of 1 square mile had a perimeter of 4.5 miles. Explain to Roseanne how this is possible. *(Lesson 12-6)*

19. The surface area *S.A.* of a cylinder with radius r and height h is given by the formula $S.A. = 2\pi r^2 + 2\pi rh$.
a. Factor the right side of this formula into prime factors.
b. Calculate the surface area of a cylinder 10 cm in diameter and 8 cm high using either the given formula or its factored form. Which form do you think is easier? *(Lessons 1-5, 12-2)*

20. Find an integer that has exactly five integer factors. (The factors need not be prime.) *(Lesson 12-1)*

Exploration

21. Refer to Question 11. Find all values of k such that $4x^2 + kx - 3$ is factorable. Explain how you know that you have found all values.

PROJECTS 12 CHAPTER TWELVE

A project presents an opportunity for you to extend your knowledge of a topic related to the material in this chapter. You should allow more time for a project than you do for typical homework questions.

1 What Percentage of Some Simple Quadratics Are Factorable?

Use a spreadsheet or computer program to consider the discriminant of all quadratics of the form $x^2 + bx + c$, where b and c are integers from -10 to 10. Use this information to determine what percentage of these quadratics are factorable over the integers. If it is easy for you to do so, you might extend the ranges of b and c. When the absolute values of b and c are larger, is it more or less likely that the quadratic will be factorable?

2 Perfect, Abundant, and Deficient Numbers

For a given positive integer, consider the sum of all its factors that are less than the given integer. For instance, for the number 10, this sum is 8 because $1 + 2 + 5 = 8$. Positive integers are classified as perfect, abundant, and deficient according to whether this sum is equal to, is greater than, or is less than the integer itself. The number 10 is deficient because $8 < 10$. Classify each number from 1 to 100 as perfect, abundant, or deficient. Find patterns in the numbers that fall into these categories. What numbers are certain to be abundant? Which are certain to be deficient?

(continued)

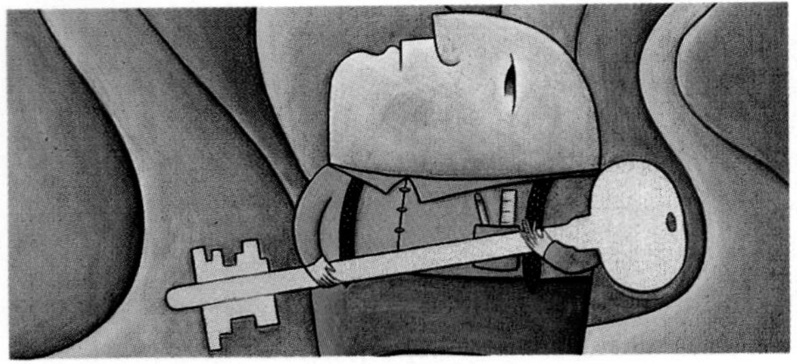

3 Public-key Cryptography

The use of codes based on prime numbers to protect information is called public-key cryptography. Do research in your library and write an essay describing how public-key cryptography works.

4 Infinite Repeating Continued Fractions

Consider this sequence of complex fractions.

$$\frac{1}{2}, \frac{1}{2 + \frac{1}{2}}, \frac{1}{2 + \frac{1}{2 + \frac{1}{2}}}, \frac{1}{2 + \frac{1}{2 + \frac{1}{2 + \frac{1}{2}}}}, \ldots$$

a. Calculate the values of the first five terms of this sequence. (Four terms are shown.)

b. As you calculate more and more terms of this sequence, the sequence approaches the value of x, where

$$x = \frac{1}{2 + \frac{1}{2 + \frac{1}{2 + \frac{1}{2 + \frac{1}{2 + \ldots}}}}}.$$

Then $x = \frac{1}{2 + x}$.

Solve this equation to find x.

c. Replace the 2s by 3s and repeat parts **a** and **b.**

d. If you can, generalize what you have found.

5 Packing Boxes

A rectangular box containing a stapler is 8″ long, 2″ wide, and 3″ high. The manufacturer wants to pack 144 (a *gross*) of these boxes in a crate for shipping to stores.

a. One crate that will hold 144 boxes without any space left over is to have 4 layers of 36 boxes, each layer having 6 boxes in each row and column. Find the dimensions of other crates that will hold 144 of these boxes without any space left over.

b. Which crate of those possible has the least surface area?

c. Which crate of those possible do you think the manufacturer should choose for packing, and why?

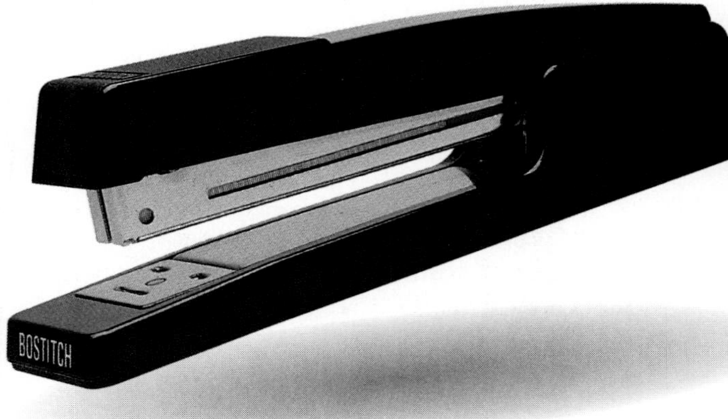

6 Factors and Graphs

a. Multiply $x - 1$ by a binomial of the form $ax + b$, where a and b are integers of your own choosing. Then graph the equation $y = (x - 1)(ax + b)$. Next, repeat this procedure three times. What do the graphs have in common? If you can, explain why they have this commonality.

b. Repeat part **a** but with some factor other than $x - 1$. What do the four new graphs have in common?

c. If you can, generalize the results you find in parts **a** and **b.**

SUMMARY

There are many similarities between the factoring of integers and the factoring of polynomials over the integers. A prime number is an integer greater than 1 that has exactly two integer factors, itself and 1. Every integer can be factored into primes in exactly one way, except for order. A prime polynomial is a polynomial that cannot be factored into polynomials of lower degree. The complete factorization of a polynomial is unique, except for the order of the factors. Prime factorizations are used nowadays in the construction of codes to protect information.

If an integer is a perfect square, then it is the product of an even number of primes. From this fact it can be deduced that certain square roots cannot be written as simple fractions. Consequently, such square roots are irrational numbers. More generally, any number that cannot be written as a finite or an infinitely repeating decimal is an irrational number.

If each of two integers has a common factor, then so does their sum, and the common factor can be factored out using the Distributive Property $ab + ac = a(b + c)$. Similarly, if each of the terms of a polynomial has a common monomial factor, then so does their sum, and it can be factored out using the Distributive Property.

Factoring the general quadratic trinomial $ax^2 + bx + c$ is more difficult, and may require trial and error procedures. However, when $a = 1$, then the quadratic is relatively easy to factor. Specifically, $x^2 + bx + c$ can be factored into $(x + p)(x + q)$ provided there are integers p and q such that $p + q = b$ and $pq = c$. In general, $ax^2 + bx + c$ can be factored over the integers if and only if its discriminant, $b^2 - 4ac$, is a perfect square.

The Zero Product Property states that if the product of two or more factors is zero, then at least one of the factors must be zero. Thus, if $ax^2 + bx + c = 0$ and $ax^2 + bx + c$ can be factored, then the solutions of the equation can be found quickly by setting each factor equal to zero and solving these simpler equations. This is one of the major uses for factoring, even though most quadratic expressions do not factor over the integers.

About 3700 years ago, Babylonian scribes showed how to solve certain quadratics. They used what we today call substitution to convert a quadratic equation into one of the form $x^2 = k$. In the 18th century, a similar idea was used to derive the Quadratic Formula we know today.

VOCABULARY

Below are the new terms and phrases for this chapter. You should be able to give a general description and a specific example for each.

Lesson 12-1
factor, multiple
is divisible by
common factor
Common Factor Sum Property
prime number
composite number
prime factorization
standard form of a factorization
Unique Factorization Theorem

Lesson 12-2
greatest common factor
polynomial over the integers
factorization of a polynomial
prime polynomial
complete factorization
Unique Factorization Theorem for Polynomials

Lesson 12-4
Zero Product Property

Lesson 12-7
simple fraction
rational number
irrational number

Lesson 12-8
Discriminant Theorem

PROGRESS SELF-TEST

Take this test as you would take a test in class. You will need a calculator. Then check your work with the solutions in the Selected Answers section in the back of the book.

1. Write the prime factorization of 300 in standard form.

2. Explain how you know that the number $6^{1000} + 36$ is divisible by 3.

3. Give the greatest common factor of $15a^2b^3$, $30a^2b$, and $25a^3b$.

4. Simplify $\frac{8c^2 + 4c}{c}$.

In 5–10, factor completely over the integers.

5. $12m - 2m^3$

6. $500x^2y + 100xy + 50y$

7. $z^2 - 81$

8. $k^2 - 9k + 14$

9. $3y^2 - 17y - 6$

10. $4x^2 - 20xy + 25y^2$

11. *Multiple choice.* Which of the following can be factored over the integers?
(a) $x^2 + 7x + 12$
(b) $x^2 + 7x - 12$
(c) $x^2 + 12x - 7$
(d) $x^2 - 12x - 7$

12. Explain how you can determine whether or not $ax^2 + bx + c$ can be factored over the integers.

13. Show the factorization

$$2x^2 + 11x + 15 = (x + 3)(2x + 5)$$

using areas of rectangles.

In 14–17, solve.

14. $(q - 7)^2 = 0$

15. $d^2 - 20 = d$

16. $(2a - 5)(3a + 1) = 0$

17. $x^3 + 6x^2 = 7x$

18. Explain why 3.54 is a rational number.

19. How many prime factors does the number 26^2 have?

20. Give an example of an irrational number between 10 and 11.

21. Is the larger solution to

$$5x^2 - 18x - 18 = 0$$

rational or irrational?

22. A square frame is d ft on a side. The artwork it holds is 1 ft shorter and 2 ft narrower than the frame. If the area of the artwork is 12 sq ft, how big is the frame?

23. A tennis ball bounces up from ground level at 8 meters per second. An equation that estimates the distance d above ground (in meters) after t seconds is $d = 8t - 5t^2$. After how many seconds will the ball return to the ground?

24. Find two numbers whose product is 15.51 and whose sum is 8.

CHAPTER REVIEW

Questions on SPUR Objectives

SPUR stands for **S**kills, **P**roperties, **U**ses, and **R**epresentations. The Chapter Review questions are grouped according to the SPUR Objectives for this chapter.

SKILLS DEAL WITH THE PROCEDURES USED TO GET ANSWERS.

Objective A: *Factor positive integers into primes.* *(Lesson 12-1)*

In 1–4, write the prime factorization of the given integer in standard form.

1. 175 **2.** 8888

3. $441 \cdot 9$ **4.** $1024 + 512$

Objective B: *Find common monomial factors of polynomials.* *(Lesson 12-2)*

5. Copy and complete: $7x^4 + 49x = 7x(\underline{\ ?\ } + \underline{\ ?\ })$.

6. Find the greatest common factor of $8a^2$ and $12a$.

7. Find the greatest common factor of $27x^2y$, $12x^2y^2$, and $3x^3y$.

8. Find the greatest common factor of $20ay^3$, $-15y^4$, and $35a^2y^6$.

In 9 and 10, factor.

9. $14m^4 + m^2$

10. $18b^3 - 21ab + 3b$

In 11 and 12, simplify, assuming the denominator is not 0.

11. $\frac{6z^3 - z}{z}$ **12.** $\frac{14x^2 + 12x}{2x}$

Objective C: *Factor quadratic expressions.* *(Lessons 12-3, 12-5)*

In 13–18, factor if possible.

13. $x^2 + 7x + 6$ **14.** $p^2 + 9p - 10$

15. $r^2 - 10r + 28$ **16.** $x^2 - 1$

17. $4L^3 + 28L^2 + 48L$ **18.** $d^2 - 8d - 20$

In 19 and 20, *multiple choice.*

19. $11a^2 + 26a - 21 =$

(a) $(11a - 7)(a - 3)$
(b) $(11a + 7)(a - 3)$
(c) $(11a - 7)(a + 3)$
(d) $(11a + 7)(a + 3)$

20. $24x^2 - 83x + 10 =$

(a) $(8x + 1)(3x + 10)$
(b) $(8x - 1)(3x + 10)$
(c) $(8x - 1)(3x - 10)$
(d) $(8x + 1)(3x - 10)$

In 21–24, factor.

21. $3y^2 + 2xy - 8x^2$

22. $10a^2 - 19a + 7$

23. $12m^3 + 117m^2 + 81m$

24. $-3 - 2k + 8k^2$

In 25 and 26, write each perfect square trinomial as the square of a binomial.

25. $m^2 + 16m + 64$ **26.** $9a^2 - 24ab + 16b^2$

In 27–30, write each difference of squares as the product of two binomials.

27. $a^2 - 4$ **28.** $b^2 - 81m^2$

29. $4x^2 - 1$ **30.** $25t^2 - 25$

Objective D: *Solve quadratic equations by factoring.* *(Lessons 12-4, 12-5)*

In 31–40, solve by factoring.

31. $x^2 - 2x = 0$ **32.** $z^2 + 7z = -12$

33. $y^2 - 2y - 3 = 0$ **34.** $2r^2 - 10r + 12 = 0$

35. $b^2 - 48 = 2b$ **36.** $k^2 = 9k - 14$

37. $0 = m^2 - 16$ **38.** $9w^2 + 12w = 0$

39. $6y^2 + y - 2 = 0$ **40.** $0 = 16m^2 - 8m + 1$

CHAPTER TWELVE

PROPERTIES DEAL WITH THE PRINCIPLES BEHIND THE MATHEMATICS.

Objective E: *Apply the definitions and properties of primes and factors.* *(Lesson 12-1)*

41. Is 203 prime? Explain why or why not.

42. Is 311 prime? Explain why or why not.

43. Give the number of factors in the prime factorization of 38^2.

44. Indicate why the number $3^{40} + 3^{39} + 3^{38}$ could not be prime.

Objective F: *Recognize and use the Zero Product Property.* *(Lesson 12-4)*

45. What is the Zero Product Property?

In 46–48, why can't the Zero Product Property be used on the given equation?

46. $(x + 3)(x + 4) = 5$

47. $(x + 3) + (x - 4) = 0$

48. $(x + 3)^2 = 25$

In 49–52, solve.

49. $5q(2q - 7) = 0$

50. $(m - 3)(m - 1) = 0$

51. $(2w - 3)(3w + 5) = 0$

52. $(y - 3)(2y - 1)(2y + 1) = 0$

Objective G: *Determine whether a quadratic polynomial can be factored over the integers.* *(Lessons 12-3, 12-8)*

53. *Multiple choice.* Which polynomial can be factored over the integers?
(a) $x^2 - 11$ (b) $x^2 - 121$
(c) $x^2 + 121$ (d) $x^2 + 112$

54. Suppose *m, n,* and *p* are integers. When will the quadratic expression $mx^2 + nx + p$ be factorable over the integers?

In 55–57, use the discriminant to determine whether the expression can be factored over the integers.

55. $x^2 + 7x - 13$

56. $x^2 + 7x - 60$

57. $3r^2 + 2r - 21$

58. In attempting to factor $x^2 - 16x + 20$, Rachel made a list of pairs of factors of 20 and checked the sum of each pair.

Factors of 20	Sums of factors
-1, -20	-21
-2, -10	-12
-4, -5	-9

From this list she deduced that $x^2 - 16x + 20$ was not factorable over the integers. Determine whether she was right or wrong. Explain your answer.

59. Find two integer factors of 24 whose sum is 10. What does this tell you about $x^2 + 10x + 24$?

Objective H: *Apply the definitions and properties of rational and irrational numbers.* *(Lesson 12-7)*

In 60–67, tell whether the number is rational or irrational.

60. $\sqrt{50}$

61. $\sqrt{100}$

62. $2 + \sqrt{2}$

63. $\frac{\sqrt{12}}{3}$

64. π

65. -100

66. 3.14

67. $\frac{2}{87}$

68. Show that $5.8\overline{7}$ is a rational number by finding a simple fraction equal to it.

69. Show that $0.\overline{428}$ is a rational number by finding a simple fraction equal to it.

USES DEAL WITH APPLICATIONS OF MATHEMATICS IN REAL SITUATIONS.

Objective I: *Solve quadratic equations in real situations.* *(Lessons 12-4, 12-6)*

70. A circular swimming pool with radius r feet will have a seating area 6 feet in width about the pool.

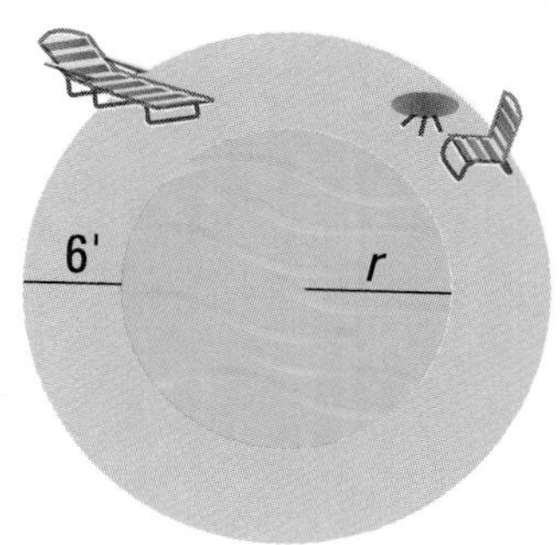

The total area of the pool and seating area is to be 256π square ft. What is the radius of the pool?

71. The area of a rectangular picture is 90 square inches. The length is 4 inches greater than the width. What are the dimensions of the picture?

In 72 and 73, use this information. When a golf ball is hit with an upward velocity of 80 feet per second, an equation that gives its height (in feet) above the ground after t seconds is $h = 80t - 16t^2$.

72. How long will the golf ball be in the air?

73. After how many seconds will the golf ball be 96 feet high?

74. The area of a rectangular field is 44,800 square feet. The perimeter of the field is 1640 feet. What are the dimensions of the field?

75. The area of a rectangular poster is 10,350 cm^2. The perimeter of the poster is 410 cm. What are the dimensions of the poster?

REPRESENTATIONS DEAL WITH PICTURES, GRAPHS, OR OBJECTS THAT ILLUSTRATE CONCEPTS.

Objective J: *Represent quadratic expressions and their factorizations with areas.* *(Lessons 12-2, 12-3)*

In 76 and 77,

a. Write the area of the figure as a polynomial.

b. Write the area in factored form.

76.

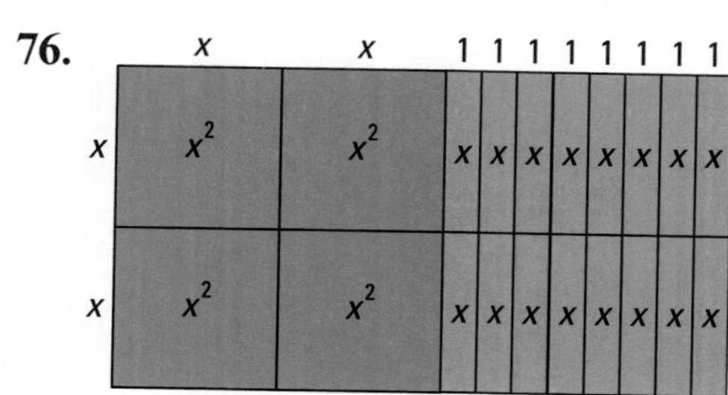

77.

	x	x	1	1	1	1
x	x^2	x^2	x	x	x	x
1	x	x	1	1	1	1
1	x	x	1	1	1	1

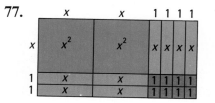

78. Show that $x^2 + 7x + 6$ can be factored by arranging tiles representing the polynomial in a rectangle. Sketch your arrangement.

79. A square has an area of $9a^2 + 30ab + 25b^2$. What is the length of a side of the square?

CHAPTER 13

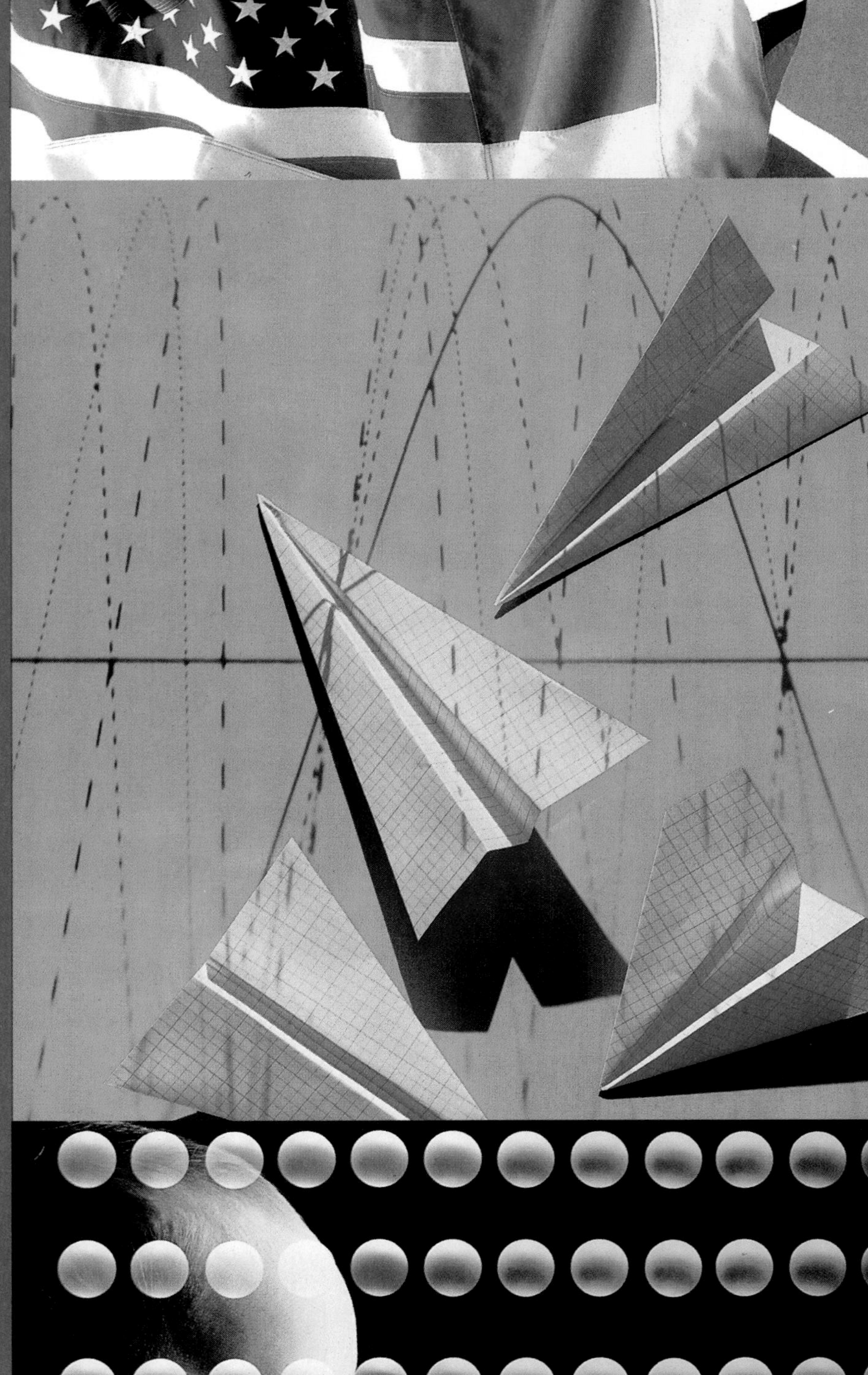

FUNCTIONS

In earlier chapters, you studied the graphs of many equations. Here are two of them.

Value of Investment at 6% Annual Yield

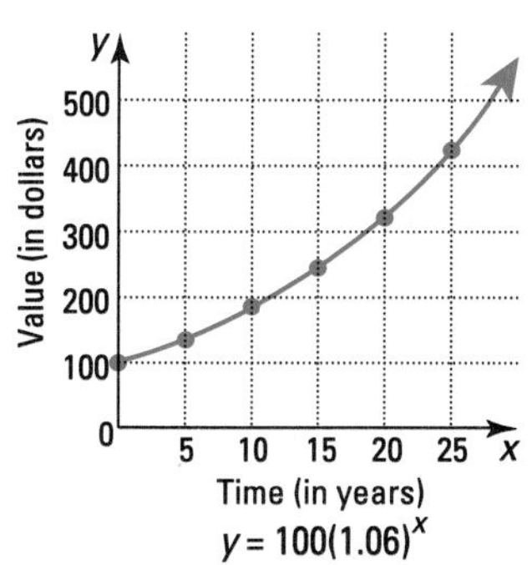

Fahrenheit-Celsius Temperatures

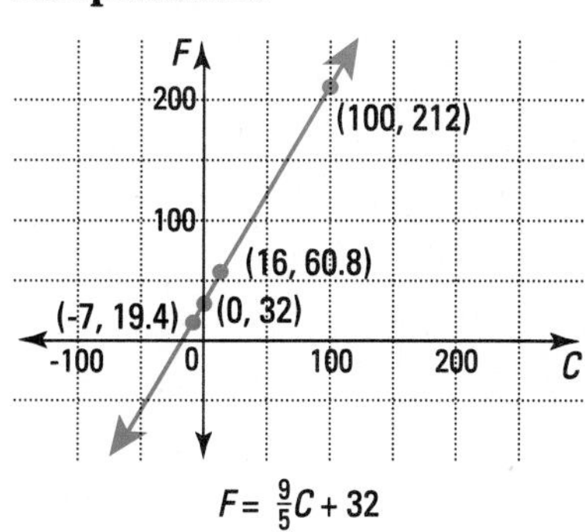

Despite their differences, these graphs describe situations that have several things in common. In each situation there are two variables, and every value of the first variable determines exactly one value of the second variable. In the investment situation, the length of time that the money has been invested determines the value of the investment. In the Fahrenheit-Celsius situation, the Celsius temperature determines the Fahrenheit temperature. When a first variable determines a second, we call the relationship between the two variables a *function.*

Functions are found in every branch of mathematics and its applications. As a result, there are many ways to describe functions: by graphs; by equations; by lists of ordered pairs; by rules written in words. The analysis of functions is extremely important in mathematics, and entire courses are often devoted to them. In this chapter, you will use the language of functions to review many ideas you have seen in earlier chapters. You will also encounter some important functions you have not seen before.

LESSON

13-1

What Is a Function?

A function is a relationship between two variables in which the first variable determines the second variable. Here is a function you first saw in Chapter 9 while studying quadratic equations. In this function, the horizontal distance of the La Quebrada cliff diver from the cliff determines the height of the diver above the water. This function can be described nicely by a graph or by an equation. The table gives some, but not all, values of the function. The description in words is quite long.

La Quebrada Cliff-Diver Function

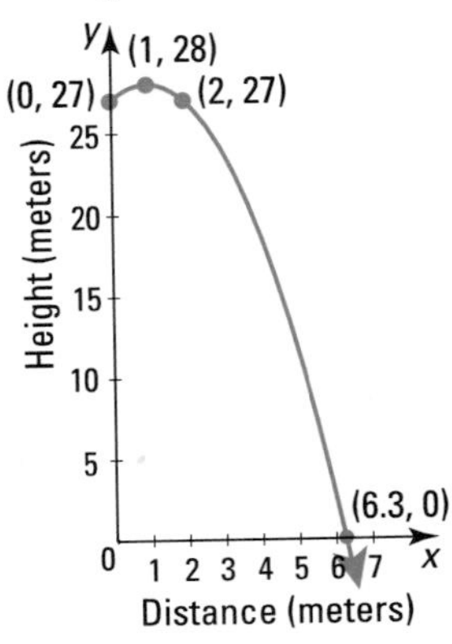

Graph: (at the left)

Equation: $y = -x^2 + 2x + 27$

Table:

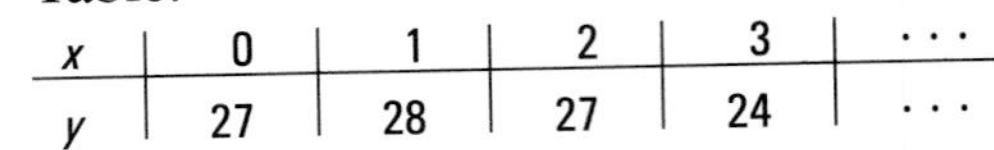

x	0	1	2	3	. . .
y	27	28	27	24	. . .

Words:
The height of the diver is 27 meters plus twice the diver's horizontal distance from the cliff less the square of that distance.

Two Definitions of Function

In general, you can think of functions as special kinds of correspondences or as special sets of ordered pairs. A correspondence between x and y is shown in the table of the La Quebrada cliff-diver function. A **function** is a correspondence between two variables in which each value of the first variable corresponds to *exactly one* value of the second variable.

The graph shows the function as a set of ordered pairs. A **function** is a set of ordered pairs in which each first coordinate appears with *exactly one* second coordinate. That is, once you know the value of the first variable (often called x), then there is only one value for the second variable (often called y). The value of the second variable is called a **value of the function.**

Example 1

Does the equation $3x + 4y = 12$ describe a function? Why or why not?

Solution 1

Solve the equation for y.

$$4y = -3x + 12$$
$$y = -\frac{3}{4}x + 3$$

Because each value of x corresponds to just one value of y, the equation describes a function.

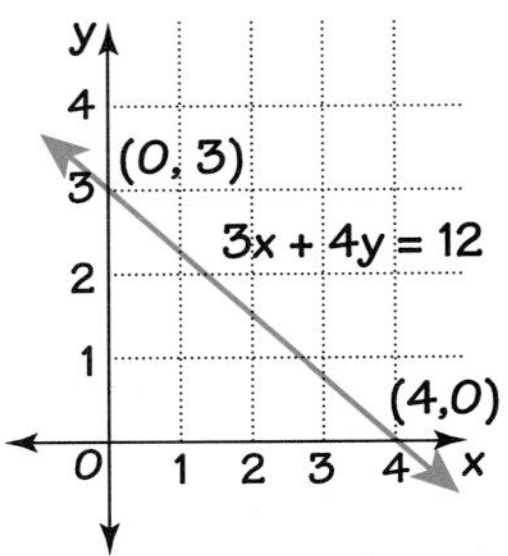

Solution 2

Graph $3x + 4y = 12$. Because no two points have the same first coordinate, the graph describes a function.

What Are Some Types of Functions?

The function of Example 1 is a **linear function.** Linear functions have equations of the form $y = mx + b$. Their graphs are lines. A **quadratic function** has an equation of the form $y = ax^2 + bx + c$. Its graph is a parabola. The cliff-diver graph on page 774 is an example of a quadratic function. An **exponential function** has an equation of the form $y = ab^x$. The investment graph on page 773 is an example of an exponential function.

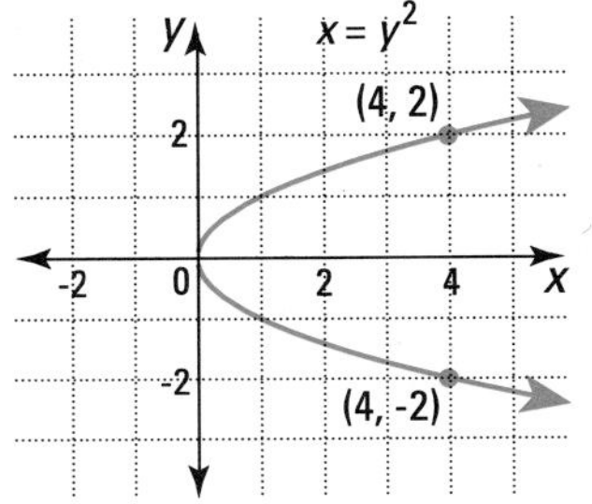

A Relation That Is Not a Function

At the left is a graph of $x = y^2$. This equation does *not* describe a function. When $x = y^2$, the value $x = 4$ corresponds to two different values for y, 2 and -2. Since the points (4, 2) and (4, -2) are both on the graph, the set of ordered pairs satisfying $x = y^2$ is not a function.

By solving $x = y^2$ for y, you can tell without graphing that this equation is not a function. $x = y^2$ implies $y = \pm\sqrt{x}$. Because in $y = \pm\sqrt{x}$ every positive value of x corresponds to two values of y, the equation $y^2 = x$ does not describe a function.

Located at the southern end of Moscow's Red Square is St. Basil's Cathedral. Each of its 10 domes differs in design and color.

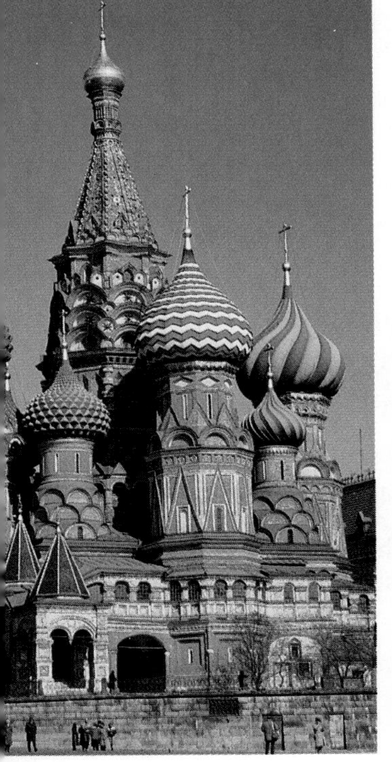

Example 2

This scatterplot from page 458 gives latitudes and April mean high temperatures (°F) for 10 selected cities. Does it describe a function? Why or why not?

Solution

No. The latitudes for Copenhagen and Moscow are the same, but the temperatures are different. These ordered pairs have the same first coordinate but different second coordinates. This scatterplot does not describe a function.

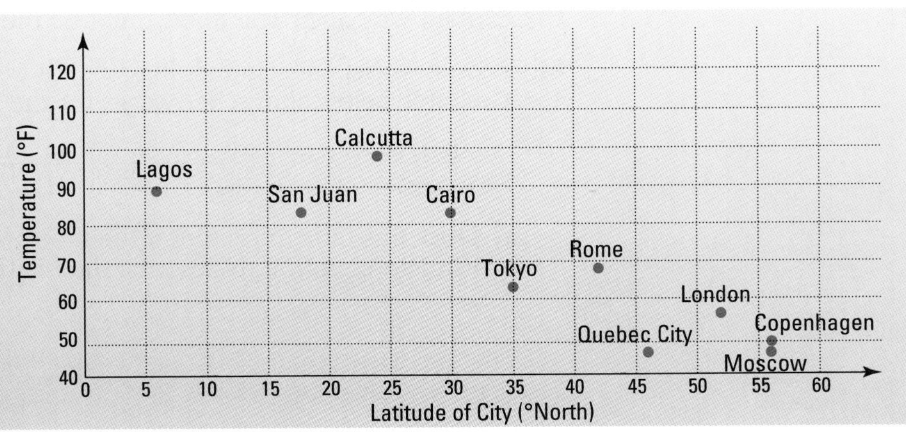

QUESTIONS

Covering the Reading

1. Name four ways in which a function can be described.
2. Define *function* using the definition you prefer.
3. In your own words explain why $x = y^2$ is not an equation for a function.

In 4 and 5, a relation and a first coordinate in the relation are given. **a.** Find all corresponding second coordinates. **b.** Is the relation also a function?

4. cliff-diver graph on page 774; $x = 4$
5. latitude-high temperature relation on page 775; latitude = 56
6. Does the graph of $x - 3y = 6$ describe a function? Why or why not?

In 7 and 8, give an equation for the type of function.

7. exponential
8. quadratic

Applying the Mathematics

9. *Multiple choice.* Which set of ordered pairs is *not* a function?
   (a) $\{(0, 0), (1, 1), (2, 2)\}$
   (b) $\{(3, 5), (5, 3), (4, 4)\}$
   (c) $\{(0, 0), (1, 0), (0, 1)\}$
   (d) $\left\{\left(\frac{1}{2}, 1\right), (\sqrt{7}, \sqrt{8}), \left(6, \frac{-9}{23}\right)\right\}$

In 10–12, does the graph represent a function?

10. 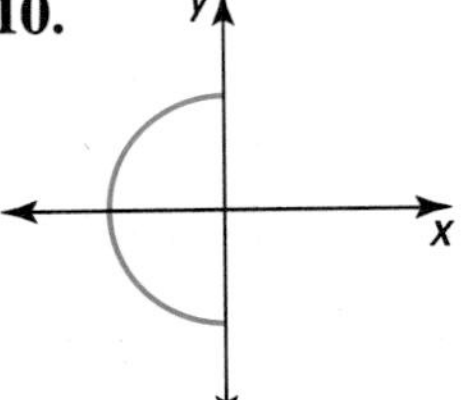

11. 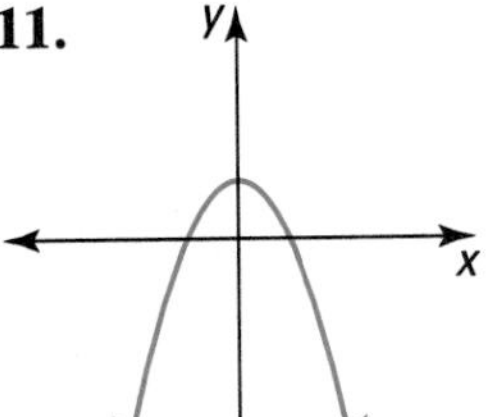

12. 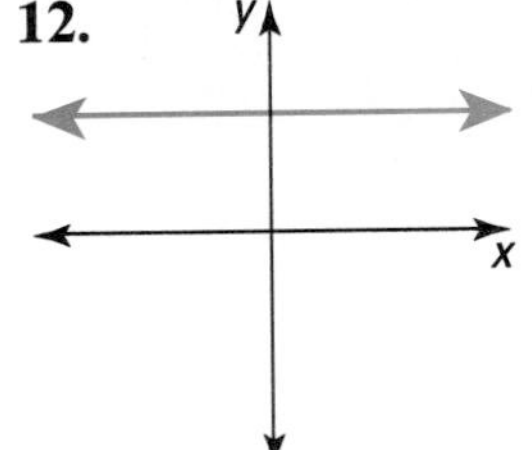

13. Explain why a vertical line cannot be the graph of a function.
14. Rennie works part-time doing jobs for people in her neighborhood. The table below shows her wages for the past six weeks.

h = hours worked	12	10	18	9	4	16
w = wages earned ($)	54	45	81	40.50	18	72

   a. Does this table represent a function? Why or why not?
   b. Find an equation to describe the relation between w and h.

In 15 and 16, an equation for a function is given. **a.** Identify three points of the function. **b.** Graph the function.

15. $y = \frac{1}{2}x^2 + 3$

16. $x = y + 400$

In 17 and 18, a written rule for a function is given. **a.** Translate the written description into an equation. **b.** Name three ordered pairs of the function.

17. The cost of wrapping and mailing a package is fifty cents a pound plus a three-dollar handling charge.

18. To find the volume of a sphere, multiply the cube of the radius by $\frac{4}{3}\pi$.

Review

This scene is from Anaheim High School's production of The King and I.

19. The Drama Club sold 520 tickets to the school musical. Adult tickets cost $5 and student tickets cost $3. If club members sold $1840 in tickets, how many of each kind did they sell? *(Lesson 11-5)*

20. Write $2^{-3} + 4^{-3}$ as a simple fraction. *(Lesson 8-6)*

21. *True or false.* The slope of the line through (x_1, y_1) and (x_2, y_2) is the opposite of the slope of the line through (x_2, y_2) and (x_1, y_1). *(Lesson 7-2)*

22. A class of 24 students contains 3% of all the students in the school. How many students are in the school? *(Lesson 6-5)*

23. A case contains *c* cartons. Each carton contains *b* boxes. Each box has 100 paper clips. How many paper clips are in the case? *(Lesson 2-4)*

24. When $m > n > 0$, which is larger, $\frac{1}{m}$ or $\frac{1}{n}$? *(Previous course)*

Exploration

25. A function contains the ordered pairs (1, 1) and (2, 4).
- **a.** Find a possible linear equation describing this function.
- **b.** Find a possible quadratic equation describing this function.
- **c.** Find a third possible equation describing this function that is not equivalent to those in **a** and **b.**

LESSON

13-2

Function Notation

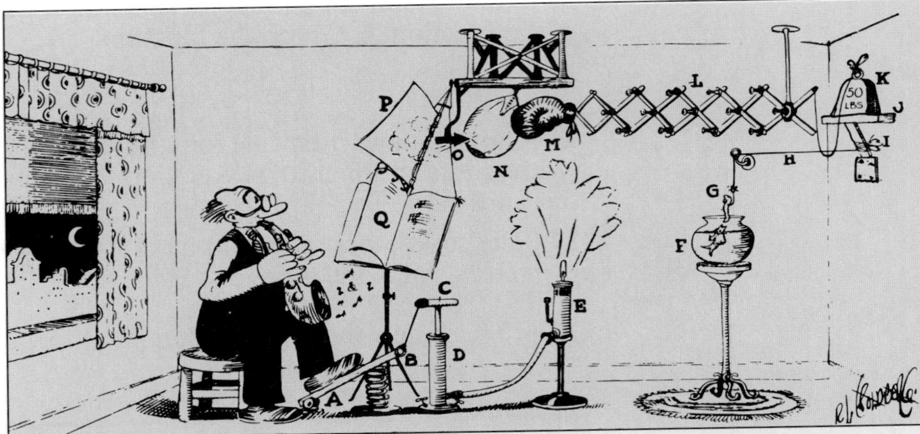

Press left foot (**A**) on pedal (**B**) which pulls down handle (**C**) on tire pump (**D**). Pressure of air blows whistle (**E**)—goldfish (**F**) believes this is dinner signal and starts feeding on worm (**G**). The pull on string (**H**) releases brace (**I**), dropping shelf (**J**), leaving weight (**K**) without support. Naturally, hatrack (**L**) is suddenly extended and boxing glove (**M**) hits punching bag (**N**) which, in turn, is punctured by spike (**O**). Escaping air blows against sail (**P**) which is attached to page of music (**Q**), which turns gently and makes way for the next outburst of sweet or sour melody.

A function machine. *This 1929 cartoon entitled* Automatic Sheet Music Turner, *is by Rube Goldberg who specialized in drawing absurdly connected machines. Here the musician presses a pedal and a page is turned. With a function, a number or other thing is input and out comes a value of the function.*

What Is Function Notation?

Ordered pairs in functions need not be numbers. You are familiar with the abbreviation $P(E)$, read "the probability of E," or even shorter, "P of E." In $P(E)$, the letter E names an event and $P(E)$ names the probability of that event. An event can have only one probability, so any set of events and their probabilities is a function.

This kind of abbreviation is used for all functions. For instance, we can use the shorthand $s(x)$, read "s of x," for the *square of* x. Then for each number x, the value of the function $s(x)$ is its square. The function is named s and called the *squaring function.* In the abbreviation $s(x)$, as in $P(E)$, the parentheses do *not* mean multiplication. The abbreviation $s(x)$ stands for the square of x.

The most common letter used to name a function is f. For example, $f(x) = -5x + 40$ represents the linear function f with slope -5 and y-intercept 40. It is read "f of x equals negative 5 times x plus 40." We call this description **function notation.** The $f(x)$ function notation was first used by the great mathematician Leonhard Euler (pronounced "oiler") in the 1700s.

Example 1

Suppose $s(x) = x^2$.

a. Evaluate $s(3)$.

b. Give the value of $s(-3)$.

Solution

a. $s(3)$ means "the square of 3." Substitute 3 for x in the formula $s(x) = x^2$. $s(3) = 3^2 = 9$

b. Substitute -3 for x in the formula $s(x) = x^2$. $s(-3) = (-3)^2 = 9$

Here are some values of the squaring function and a graph of $s(x) = x^2$. You are familiar with this function. In previous lessons it was described by $y = x^2$. Its graph is a parabola.

x	$s(x)$
-4	16
-3	9
0	0
1	1
1.7	2.89
2	4
3	9
4	16

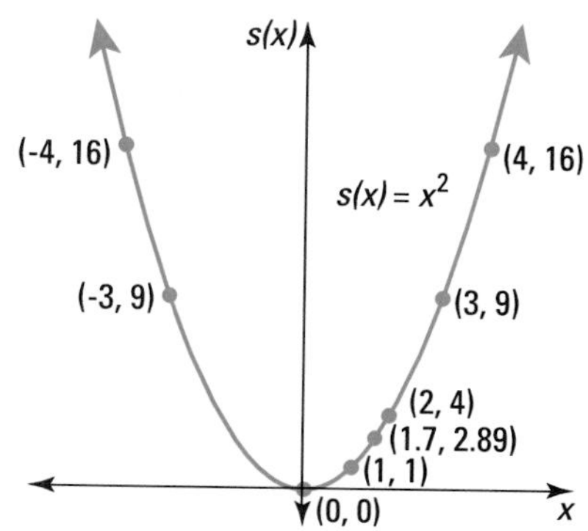

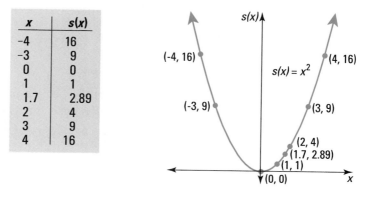

Function Notation in Calculators and Computers

Calculator and computer programs take advantage of function notation. Recall that in BASIC,

SQR(x)	means	the square *root* of x.
ABS(x)	means	the absolute value of x.

In this way, BASIC uses function notation. The names of the functions are SQR and ABS. Many automatic graphers use function notation.

Example 2

Evaluate the following.

a. ABS(-3.4)

b. SQR(4 + 9)

Solution

a. ABS(-3.4) = $|-3.4| = 3.4$

b. Work within parentheses before applying the function.
SQR(4 + 9) = $\sqrt{4 + 9} = \sqrt{13} \approx 3.6$

Function notation is particularly useful when one variable determines two or more values. The function notation helps you distinguish between the values.

Example 3

Khalil is offered two jobs as a waiter. His earnings at each restaurant depend on the total value x of the meals he serves. He thinks he can make 15% in tips at the Comfy Cafe but only 10% at the Dreamy Diner. He estimates his weekly wages at the Comfy Cafe and the Dreamy Diner would be given by these formulas, where 160 and 200 represent base pay.

Comfy Cafe: $c(x) = 160 + 0.15x$

Dreamy Diner: $d(x) = 200 + 0.10x$.

a. Evaluate $c(500)$ and $d(500)$, and tell what the values represent.

b. For what values of x is $c(x) > d(x)$? Explain what your answer means in words.

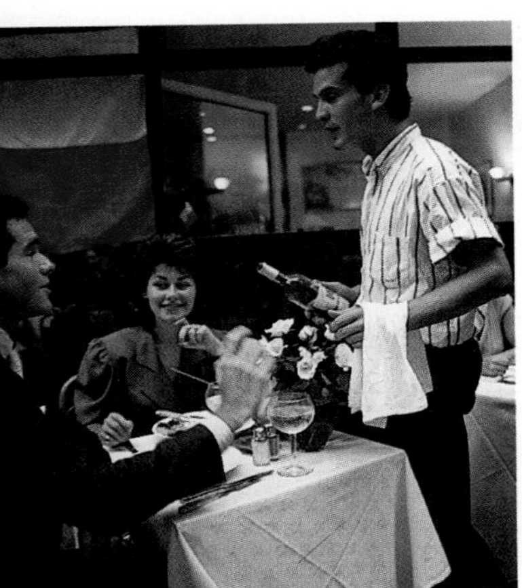

Check, please. *Waiters usually receive an hourly or weekly wage in addition to tips.*

Solution

a. To find $c(500)$, substitute $x = 500$ into $160 + 0.15x$.

$$c(500) = 160 + 0.15(500) = 160 + 75 = 235$$

Similarly,

$$d(500) = 200 + 0.1(500) = 200 + 50 = 250.$$

If Khalil serves \$500 worth of meals in one week, he will earn \$235 at the Comfy Cafe and \$250 at the Dreamy Diner.

b. To solve $c(x) > d(x)$, use their formulas and substitute.

$$c(x) > d(x)$$
$$160 + 0.15x > 200 + 0.10x$$
$$0.15x > 40 + 0.10x$$
$$0.05x > 40$$
$$x > 800$$

If he serves more than \$800 worth of meals each week, Khalil will earn more at the Comfy Cafe than at the Dreamy Diner.

Check

b. Draw the graphs of $c(x) = 160 + 0.15x$ and $d(x) = 200 + 0.10x$ on the same set of axes. Identify the y-coordinates of the points where the graph for Comfy Cafe's wages is higher than the graph of Dreamy Diner's wages.

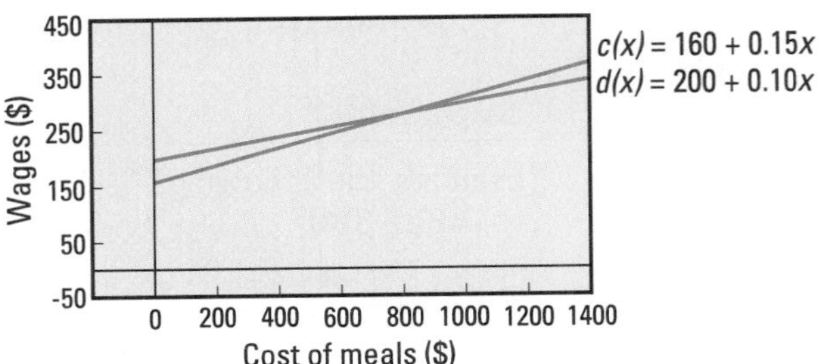

The graph shows that when $x > 800$, $c(x)$ is higher than $d(x)$.

Advantages of Function Notation

In Example 3, if we had used $y = 160 + .15x$ and $y = 200 + .10x$, it would have been more difficult to remember which equation stood for which restaurant. By using a different letter for each function, c and d, the first letters of *Comfy* and *Dreamy*, the functions are easier to distinguish. Another advantage of function notation is that it is shorter than a verbal description. For instance,

$d(500) > c(500)$ means: Khalil's earnings at Dreamy Diner will be greater than his earnings at Comfy Cafe if he serves \$500 worth of meals during the week.

QUESTIONS

Covering the Reading

In 1–3, write out how each symbol is read.

1. $P(E)$ **2.** SQR(x) **3.** $f(x) = 100 - x$

In 4–6, let $s(x) = x^2$. Give the value of:

4. $s(8)$ **5.** $s(-8)$ **6.** $s\left(\frac{2}{5}\right)$

In 7–9, evaluate.

7. SQR(40) **8.** ABS(-2.5) **9.** SQR(9) – ABS(-9)

In 10–12, an equation for a function is given. Find $f(-2)$.

10. $f(x) = 4x$ **11.** $f(x) = x^4$ **12.** $f(x) = 4^x$

In 13 and 14, refer to Example 3.

13. a. Evaluate $c(650)$.
b. What does $c(650)$ mean for Khalil?

14. a. Which is larger, $c(1000)$ or $d(1000)$?
b. What does the answer to part **a** mean for Khalil?

Car shopping.
81.3 million American households owned at least one motor vehicle in 1990.

Applying the Mathematics

15. Peggy has to choose between buying two autos, one new and one used. Including the cost of gas, maintenance, and insurance, she figured the cost of owning each as

new: $n(t) = 11{,}300 + 160t$
used: $u(t) = 6{,}500 + 200t$,

where t is the number of months she owns the auto.

a. What is meant by the sentence $n(24) > u(24)$?
b. How long would Peggy have to keep the new auto for it to be a better deal than the used one?

In 16 and 17, an equation for a function is given. **a.** Graph the function. **b.** Identify its x- and y-intercepts.

16. $f(x) = -5x + 40$

17. $f(x) = 5x - x^2$

18. Let $c(n) = n^3$.
a. Calculate $c(1)$, $c(2)$, $c(3)$, $c(4)$, and $c(5)$.
b. What might be an appropriate name for c?

19. Let $s(p)$ = the number of sisters of a person p. Let $b(p)$ = the number of brothers of a person p.
a. If you are the person p, give the values of $s(p)$ and $b(p)$.
b. What does $s(p) + b(p) + 1$ stand for?

20. A computer purchased for \$3200 is estimated to depreciate at a rate of 25% per year. That is, its worth $W(t)$ after t years is given by $W(t) = 3200(0.75)^t$.
a. Evaluate $W(3)$ and explain what the value means.
b. Graph the function for $0 \le t \le 6$.
c. Use your graph to solve $W(t) < 1000$, and explain what your answer means.

21. Suppose $L(x) = 17x + 10$.
a. Calculate $L(5)$.
b. Calculate $L(2)$.
c. Calculate $\frac{L(5) - L(2)}{5 - 2}$.
d. What have you calculated in part **c?**

Review

In 22 and 23, **a.** graph the equation; **b.** tell whether the graph represents a function. *(Lessons 13-1, 9-8)*

22. $y = |x|$

23. $x = |y|$

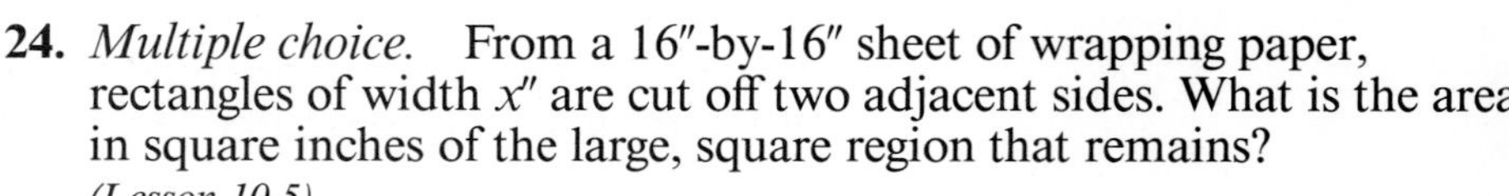

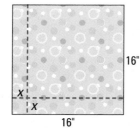

24. *Multiple choice.* From a 16″-by-16″ sheet of wrapping paper, rectangles of width x″ are cut off two adjacent sides. What is the area in square inches of the large, square region that remains? *(Lesson 10-5)*

(a) $16 - x$ (b) $256 + x^2$ (c) $256 - x^2$ (d) $256 - 32x + x^2$

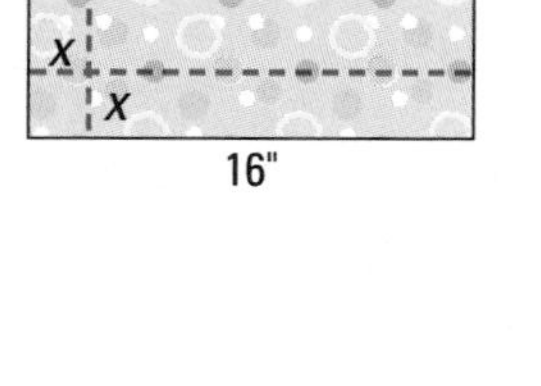

In 25–27, simplify. *(Lessons 9-7, 8-6, 1-6)*

25. $6 \cdot 3^{-2}$

26. $(\sqrt{7})^2$

27. $\frac{6 \pm \sqrt{24}}{2}$

28. *Skill sequence.* Write as a single fraction. *(Lesson 3-9)*
a. $3 + \frac{2}{5}$ **b.** $3 + \frac{7}{5}$ **c.** $3 + \frac{k}{5}$

29. In June 1994, the U.S. Postal Service put out an advisory containing crime-prevention tips. Among these was a warning against participation in chain letters that guarantee money with one small investment. Chain letters can violate federal mail fraud laws. Here is what the advisory said: "A typical scheme may require you to mail the chain letter, along with a specified amount of money to six people, each of whom must then mail letters to six more people, and so on. But looking at the chart, you can see that more participants are required than there are people in the entire world!"

Number of Mailings	Number of Participants
1	6
2	36
3	216
4	1,296
5	7,776
6	46,656
7	279,936
8	1,679,616
9	10,077,696
10	60,466,176
11	362,797,056
12	2,176,782,336
13	13,060,694,016

The table suggests the name sometimes given to this chain letter idea: a *pyramid scheme.*

a. The table defines a function because the number of mailings m determines the number of participants p. Give an equation relating m and p that describes the function.

b. What kind of function is this?

c. For what number of mailings does the number of participants first surpass the U.S. population?

d. For what number of mailings does the number of participants first surpass the world population? *(Lessons 13-1, 8-2)*

Exploration

30. Let $f(x) = \frac{12}{x - a}$. Use an automatic grapher to graph the function f from $x = -5$ to $x = 5$ when $a = 0$, $a = 1$, $a = 2$, and $a = 3$. What do the graphs have in common? How are they different?

LESSON

13-3

Absolute Value Functions

It's about time. *The line drawn in the street is the Greenwich Meridian. This line has longitude 0° and is the starting point for the world's 24 time zones. See the Example on page 785 concerning people's perception of time.*

The Function $f(x) = |x|$

The function with equation

$$f(x) = |x|,$$

or $f(x) =$ ABS(x), is the simplest example of an *absolute value function.* By substituting, you can find ordered pairs for this function.

x	$f(x) = \lvert x \rvert$	ordered pair
2	$f(2) = \lvert 2 \rvert = 2$	(2, 2)
-8	$f(-8) = \lvert -8 \rvert = 8$	(-8, 8)
0	$f(0) = \lvert 0 \rvert = 0$	(0, 0)

Here is a graph of $f(x) = |x|$. You saw this graph as $y = |x|$ in Lesson 9-8.

In general, when x is positive, $f(x) = x$, so the graph is part of the line $y = x$. When x is negative, then $f(x) = -x$, and the graph is part of the line $y = -x$. The result is that the graph of the function is an angle. The angle has vertex at the origin (0, 0) and has measure 90°. It is a right angle.

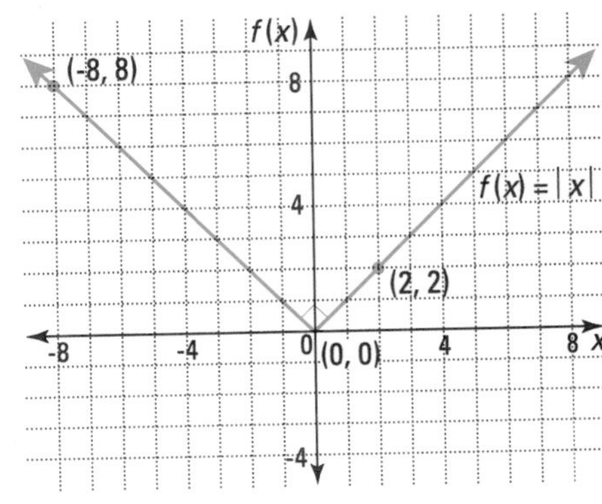

Some Uses of Absolute Value Functions

One place absolute value functions are used is in the study of error.

Example

A psychologist was studying people's perceptions of time. Some people were asked to estimate the length of a minute. With each person, the psychologist rang a bell. The person waited until he or she thought a minute was up, and then rang the bell again. The estimate x(in seconds) has the error $|60 - x|$. Graph the function with equation

$$f(x) = |60 - x|.$$

Solution

Since x represents time in seconds, $x \geq 0$. Draw axes with units of 10. Make a table.

x	error f(x) = |60 − x|
0	60
40	20
50	10
60	0
70	10
100	40

These and other points are plotted below. You can see that the graph again is a right angle, but its vertex is at (60, 0).

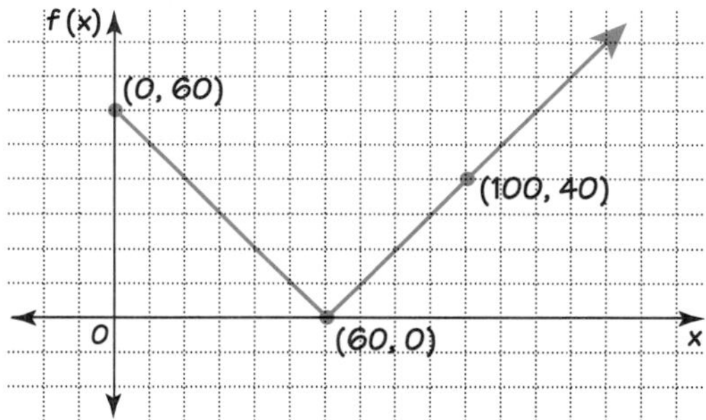

The symmetry of the graph to the vertical line $x = 60$ shows that the error has the same values on either side of 60. The slopes of -1 and 1 for the graph are due to the fact that error increases by 1 as the estimate differs by 1 more from 60.

Another reason for absolute value functions is that they help to explain some complicated situations. They are particularly useful in some situations involving distance.

Suppose that in a test flight, a plane going due east crosses a checkpoint. It flies east at 600 km/h for 2 hours and then returns flying due west.

We let $d(t)$ = the plane's distance from the checkpoint t hours after crossing it.

When $t = 0$, the plane is at the checkpoint. So $d(0) = 0$.
When $t = 1$, the plane is 600 km east. So $d(1) = 600$.
When $t = 2$, the plane is 1200 km east. So $d(2) = 1200$.

All this time the plane has been going at a constant rate. Then it turns back.

When $t = 3$, the plane is again 600 km east. So $d(3) = 600$.
When $t = 4$, the plane crosses the checkpoint again. So $d(4) = 0$.

The graph below results from graphing these and other ordered pairs. Notice how the graph describes the situation.

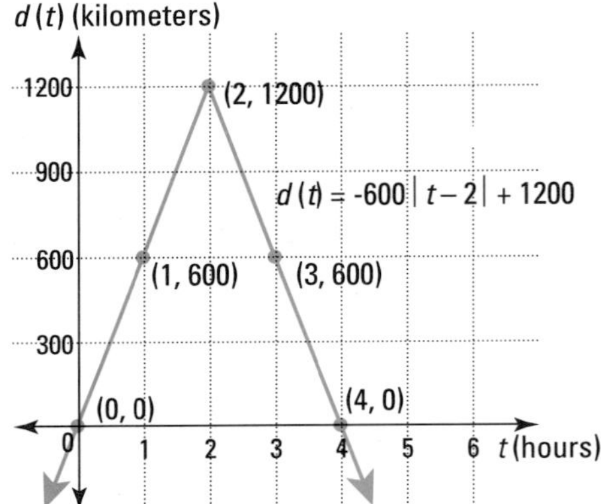

Because the graph is an angle, you should expect a formula for $d(t)$ to involve the absolute value function. And it does.

$$d(t) = -600|t - 2| + 1200.$$

Activity

Check that the formula $d(t) = -600|t - 2| + 1200$ does work for the 5 values we found and for 3 other values of t.

Without knowing the graph of the absolute value function $f(x) = |x|$, you probably would never think that the test-flight situation could involve absolute value. The formula

$$d(t) = -600|t - 2| + 1200$$

is a special case of the **general absolute value function**

$$f(x) = a|x - h| + k,$$

where $a = -600$, $h = 2$, and $k = 1200$. In the Questions, you are asked to explore what the graph of this function looks like for various values of a, h, and k. In your later study of mathematics, you will learn how to determine such formulas.

Earlier in this book, you would have seen the formula $d = -600|t - 2| + 1200$ without function notation. Using the $d(t)$ function notation makes it clear that the value of d depends on t. There may be other quantities that depend on the time. We could write

$a(t)$ = altitude of the plane t hours after crossing the checkpoint,

$f(t)$ = amount of fuel used t hours after crossing the checkpoint, and so on.

Using $a(t)$ and $f(t)$ makes it clear that the altitude and fuel used by the plane depends on time.

Flight of the Eagle. *On January 21, 1987, Lois McCallin pedaled this plane, the* Eagle, *10 miles in 37 min 38 sec. This was the longest such flight by a woman. Until 1988 the* Eagle *held the distance record—37.3 miles—for human-powered flight.*

QUESTIONS

Covering the Reading

In 1–4, let $f(x) = |x|$. Calculate.

1. $f(-3)$ **2.** $f(2)$ **3.** $f\left(-\frac{3}{4}\right)$ **4.** $f(0)$

5. The graph of an absolute value function is _?_.

6. Name two reasons for studying absolute value functions.

In 7–9, let f be the function of the Example.

7. If $x = 90$, $f(x) =$ _?_. **8.** The graph of f has vertex _?_.

9. $f(x)$ stands for the absolute difference between the actual and estimated values of _?_.

In 10–13, let $d(t) = -600|t - 2| + 1200$.

10. Calculate $d(0)$, $d(1)$, $d(2)$, and $d(3)$.

11. Describe a situation that can lead to the function d.

12. The function d contains (1.5, 900). What does this point represent?

13. The function d contains (5, -600). What could this point represent?

Applying the Mathematics

In 14 and 15, graph the function with the given equation.

14. $f(x) =$ ABS$(3x)$ **15.** $y = |x - 10| + 7$

In 16 and 17, suppose you start at the goal line of a football field and walk to the other goal line. After you have walked w yards, you will be on the y yard line.

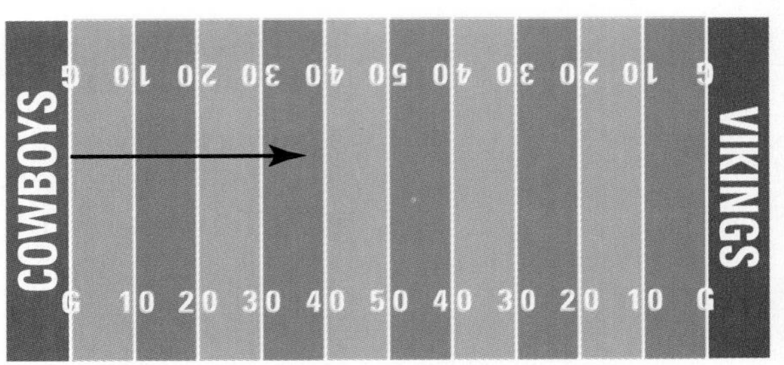

16. *Multiple choice.* Which equation relates w and y?
(a) $y = w$ (b) $y = |w|$
(c) $y = |50 - w| + w$ (d) $y = -|w - 50| + 50$

17. Let f be the function relating w and y. Graph f.

Review

18. Let $f(x) = \frac{2}{3}x + 5$. *(Lesson 13-2)*
a. Calculate $f(120)$. **b.** Calculate $f(-120)$.
c. Describe the graph of f.

19. Let $A(x)$ = the April mean high temperature for city x, as shown in Lesson 13-1. What is A(Moscow)? *(Lesson 13-1)*

20. *Skill sequence.* Factor. *(Lesson 12-3)*
a. $x^2 - 16$ **b.** $y^2 - 6y - 16$ **c.** $a^2 - 6ab - 16b^2$

21. What value(s) can v not have in the expression $\frac{(v - 1)(v - 3)}{(v - 2)(v - 4)}$? *(Lessons 12-4, 6-2)*

22. If 10 pencils and 7 erasers cost \$4.23 and 3 pencils and 1 eraser cost \$0.95, what is the cost of two erasers? *(Lesson 11-5)*

23. *Skill sequence.* Simplify. *(Lesson 3-9)*
a. $x + \frac{x}{2}$ **b.** $\frac{x}{3} + \frac{x}{2}$ **c.** $\frac{x}{3} + \frac{y}{2}$

Exploration

In 24 and 25, consider absolute value functions of the form $f(x) = a|x - h|$.

24. Fix $a = 1$. Then vary the value of h, choosing any numbers you wish. For instance, if you let $h = 3$, then $f(x) = |x - 3|$.
a. Graph the function f for four different values of h.
b. How does the value of h affect the graphs?

25. Fix $h = 2$. Now vary the value of a, choosing any numbers you wish. For instance, if you let $a = 4$, then $f(x) = 4|x - 2|$.
a. Graph the function f for four different values of a.
b. How does the value of a affect the graphs?

LESSON

13-4 Domain and Range

Peacemakers. *Pictured is part of the United Nations headquarters in New York City. The primary objective of the U.N. is the maintenance of international peace and security.*

What Are the Domain and the Range of a Function?

Every function can be thought of as a set of ordered pairs. The set of first coordinates of these pairs is called the **domain** of the function. The set of second coordinates of these pairs is called the **range** of the function. If the function is a finite set of ordered pairs, then you can list the elements of the domain and the range.

Example 1

The set of ordered pairs {(1945, 51), (1965, 117), (1985, 159), (1993, 184)} associates a year with the number of members of the United Nations that year. Give the domain and the range of the function.

Solution

Only four points are given for this function.
The domain is the set of first coordinates {1945, 1965, 1985, 1993}.
The range is the set of second coordinates {51, 117, 159, 184}.

When a function is expressed as an equation or other rule involving x and y, or x and $f(x)$, then the domain is the replacement set for x, and the range is the replacement set for y or $f(x)$. When no domain is given, the domain is assumed to be the largest set possible in the situation.

To find the range, think about what y-values (or $f(x)$-values) are possible, or make a graph. If the problem involves a real use, you will also have to think about the situation.

Functions with the Same Rule but with Different Domains and Ranges

Here are three linear functions that have the same equation, but have different domains and ranges because of the different situations.

Let $h(x)$ be the number of children in x sets of twins. Then $h(x) = 2x$.

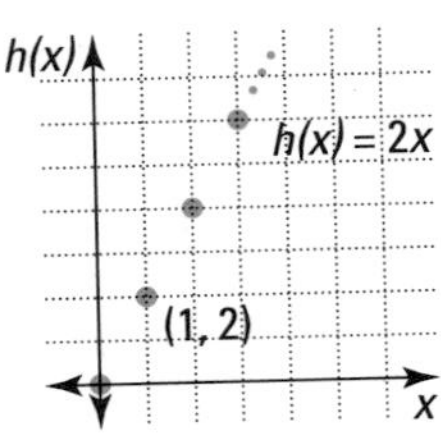

domain = set of all nonnegative integers
range = set of nonnegative even integers

Let $g(x)$ be the length of two ribbons if the length of one ribbon is x. Then $g(x) = 2x$.

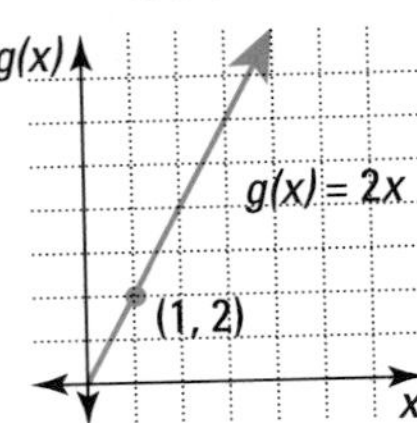

domain = set of nonnegative real numbers
range = set of nonnegative real numbers

Let $f(x)$ be twice x. Then $f(x) = 2x$.

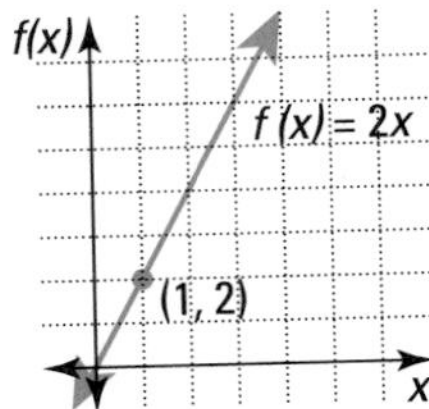

domain = set of all real numbers
range = set of all real numbers

Seeing double. *This Twinsburg school bus certainly lives up to its name!*

Functions with Restricted Range

Absolute value functions and quadratic functions allow any real number to be in their domains. However, not all real numbers are in their ranges.

Example 2

Let $f(x) = |x|$. What are the domain and the range of f?

Solution 1

The domain is the set of possible values of x. x can be any real number. So, **the domain is the set of all real numbers.**
The range is the set of possible values of $f(x)$. $f(x) = |x|$ may equal any nonnegative real number. Thus **the range is the set of nonnegative real numbers.**

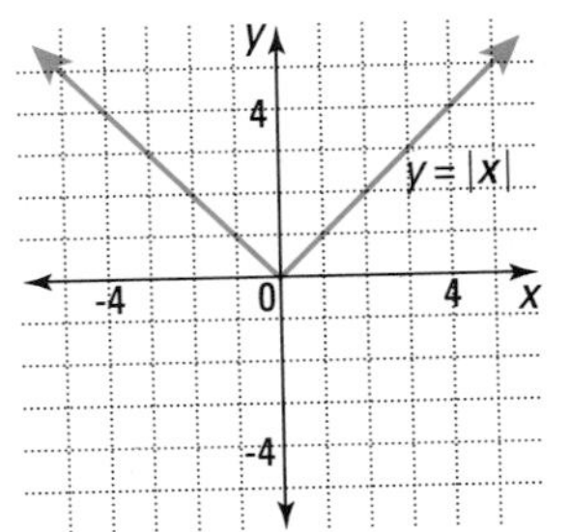

Solution 2

Examine the graph of $y = |x|$. This is the same as $f(x) = |x|$.

The graph shows that any value of x is possible, but only nonnegative values of y are possible. This leads to the same answers as in Solution 1.

Example 3

Find the domain and the range of the function $f(x) = -x^2 + 6x - 4$.

Solution

Any value can be substituted for x, so **the domain = the set of all real numbers.**

Find the range by examining its graph. In the graph, y could be any number less than or equal to 5. So **the range is the set of real numbers less than or equal to 5, or $y \leq 5$.**

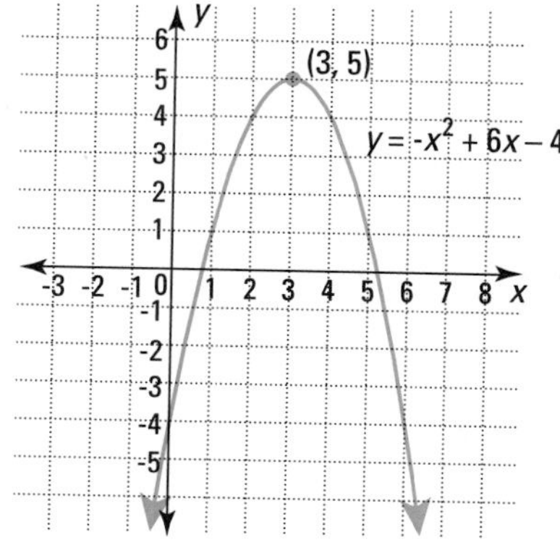

Functions with Restricted Domain

In functions with rules that have variables in the denominator or square roots with variables, the domain sometimes cannot contain particular values.

Example 4

What is the domain of the function f with rule $f(x) = \frac{x+1}{x-2}$?

Solution

The domain is the set of allowable values for x. The numerator can be any number, so it can be ignored. However, the denominator cannot be 0. Therefore, **the domain is the set of all real numbers except 2.**

Example 5

Determine the domain and the range of the function f with equation $f(x) = \sqrt{x} + 4$.

Solution

You can take the square root only of a nonnegative number. **The domain is the set of nonnegative real numbers.** The number $\sqrt{x}$ can be any positive number or zero. Thus $\sqrt{x} + 4$ can be any number greater than or equal to 4. **The range is the set of numbers greater than or equal to 4.**

Check

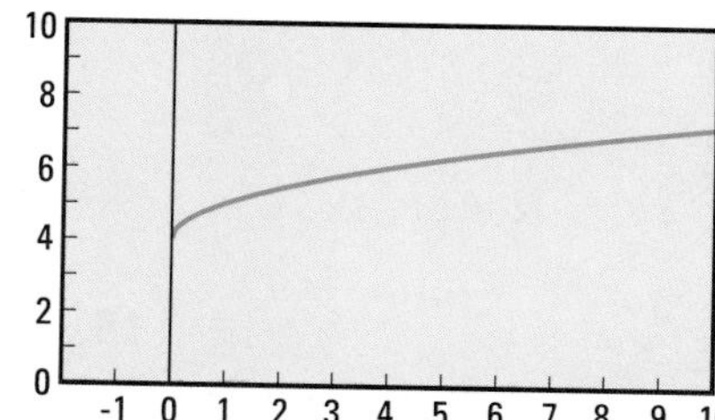

A graph of $f(x) = \sqrt{x} + 4$ with an automatic grapher is shown at the right. Notice that it contains no points for negative values of x or values of $f(x)$ less than 4. This verifies the answer.

QUESTIONS

Covering the Reading

1. Define *domain* of a function. **2.** Define *range* of a function.

3. Refer to Example 1. How many more members did the United Nations have in 1993 than when it was founded in 1945?

In 4–6, a function is given. State its domain and its range.

4. {(1, 2), (3, 4), (5, 7)}

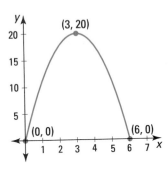

5. the function with equation $f(x) = 4x$

6. the function whose entire graph is shown at the left

(3, 20)
(0, 0)
(6, 0)

7. When no domain is given for a function, what can you assume about the domain?

8. Let $t(x)$ = the price of four tires if one costs x dollars. Let $b(x)$ = the number of people in x foursomes for bridge.
a. Give a formula for t. **b.** Give a formula for b.
c. How do t and b differ?

In 9–12, an equation for a function is given. **a.** Determine the domain of the function. **b.** Determine its range.

9. $f(x) = 3x + 1$

10. $s(x) = |x| + 100$

11. $h(x) = 2x^2 - 3$

12. $f(x) = \sqrt{x} - 5$

13. What number is not in the domain of the function g with rule $g(x) = \frac{x-2}{x-3}$?

Applying the Mathematics

14. Let $n(y)$ equal the number of daily newspapers in circulation in the U.S. in year y. The table below gives some values of this function.

year	1970	1975	1980	1985	1990
number of newspapers	1748	1756	1745	1676	1611

a. Plot these ordered pairs.
b. What trend do you notice in the plot?
c. What is the largest value of the domain listed in the table?
d. What is the largest value of the range listed in the table?
e. Suppose this function is defined for the U.S. from the year of the Declaration of Independence until the present. State the domain of n.

15. a. Graph a function with domain {-1, 2, 3} and range {5, 8, 0}.
b. How many such functions are possible?

16. Sketch a graph or give an equation for a function that has all real numbers for its domain and all real numbers less than or equal to -1 for its range.

In 17–19, state the domain and the range of the function.

17. the function with equation $y = 2^x$ graphed below at the left

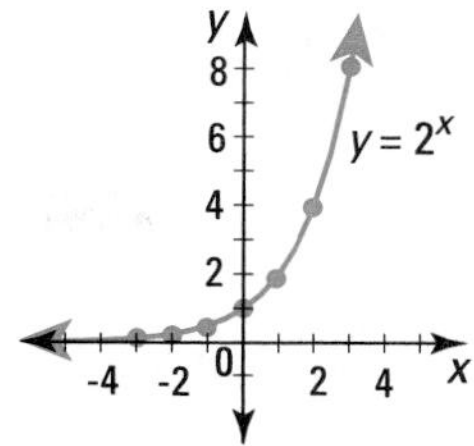

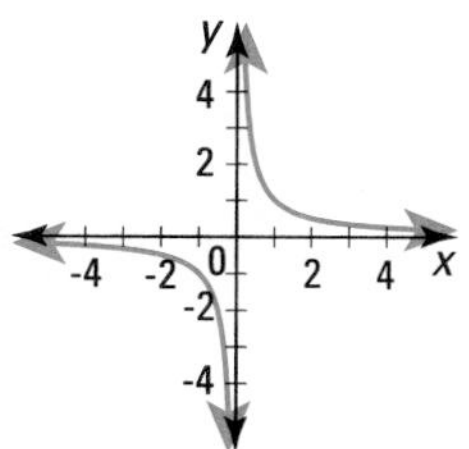

18. the function with equation $y = \frac{1}{x}$ graphed above at the right

19. the investment function described on page 773

Review

20. Let $f(x) = (x - 1)(x + 2)$. Calculate each value. *(Lesson 13-2)*

a. $f(1)$ **b.** $f(2)$ **c.** $f\left(\frac{7}{3}\right)$

In 21 and 22, tell whether or not the graph represents a function. *(Lesson 13-1)*

21.

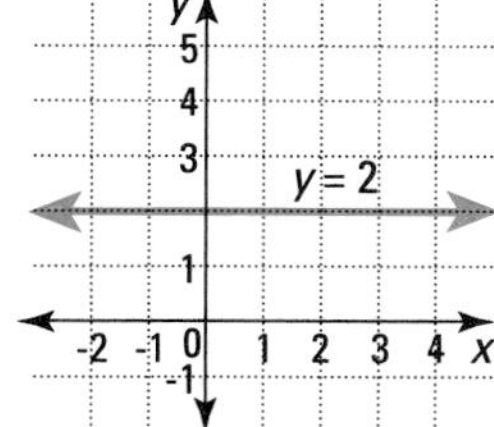

22.

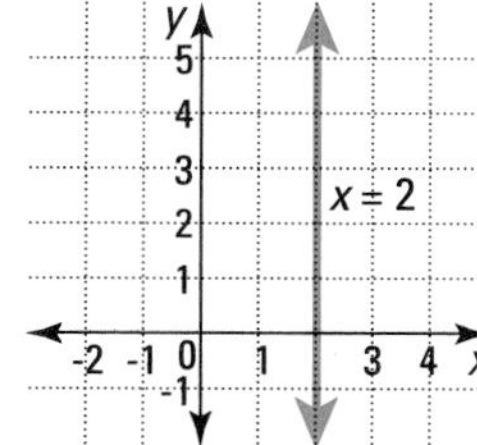

Havana, Cuba. *Pictured is the Vedado District in Havana, the capital of Cuba.*

23. The price of a hat was increased by $\frac{1}{3}$ to the new price of \$10.00.
a. What equation can be solved to find the former price?
b. What was that price? *(Lesson 6-7)*

24. What is the probability that one toss of a fair die does not show a 6? *(Lesson 6-4)*

25. In 1991, Cuba had a population of about 10.7 million living on about 44,218 square miles of land. Mexico had a population of 90.0 million living on 761,604 square miles. **a.** Which country was more densely populated? **b.** How can you tell? *(Lesson 6-2)*

26. Solve for x: $y = 2x + 6$. *(Lesson 3-5)*

Exploration

27. Consider the absolute value functions of the form $f(x) = |x| + k$.
a. How does the value of k affect the graph of f? (You may need to graph f for different values of k.)
b. Describe the range of f in terms of k.

Introducing Lesson 13-5

Results of Tossing Two Dice

IN-CLASS

ACTIVITY

Materials: Dice, graph paper, and a calculator
Work in pairs.

1 **a.** Each pair should toss two dice at least 50 times and record the sum after each toss.
b. Tally the number of times each sum from 2 to 12 comes up, and make a table of frequencies. After you have tallied the frequency of each sum, calculate its relative frequency. For instance, if you toss the dice 50 times and get a sum of 2 three times, your table should have the following entries.

sum	frequency	relative frequency
2		$\frac{3}{50} = .06$
⋮	⋮	⋮
12		
total	50	

2 All the pairs in the class should combine their data in one table as in Step 1. Graph the function whose ordered pairs are (sum, relative frequency of that sum).

3 To calculate the probabilities, answer the following.
a. When one fair die is tossed, how many outcomes are possible? What are they?
b. When two fair dice are tossed, how many outcomes are possible? List them all.
c. How many of these outcomes give a sum of 2? What is the probability of getting a sum of 2 when two dice are tossed?
d. How many of these outcomes give a sum of 5? What is the probability of getting a sum of 5 when two dice are tossed?
e. Copy and complete this table of probabilities.

sum	probability
2	.
.	.
.	.
12	.

4 ***Draw conclusions.*** Do your relative frequencies approximate the probabilities? Explain why or why not.

LESSON

13-5 Probability Functions

A Probability Function for Two Dice

In many board games, two dice are tossed and the sum of the numbers that appear is used to make a move. Since the outcome of the game depends on landing or not landing on particular spaces, it is helpful to know the probability of obtaining each sum. The following diagram shows the 36 possibilities for two fair dice.

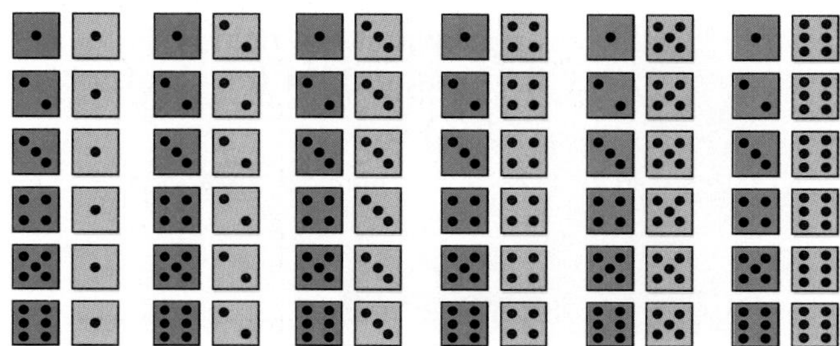

If the dice are fair, then each of the 36 outcomes has a probability of $\frac{1}{36}$. Let $P(n)$ = the probability of getting a sum of n. The domain of P is the set of possible values for n, namely {2, 3, 4, 5, 6, 7, 8, 9, 10, 11, 12}. By counting, you can find the values of $P(n)$ given in the table below. The range of the function P is thus $\left\{\frac{1}{36}, \frac{2}{36}, \frac{3}{36}, \frac{4}{36}, \frac{5}{36}, \frac{6}{36}\right\}$. This probability function is graphed below.

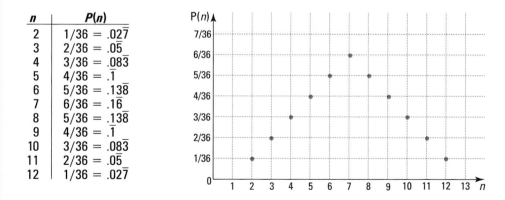

n	$P(n)$
2	$1/36 = .0\overline{27}$
3	$2/36 = .0\overline{5}$
4	$3/36 = .08\overline{3}$
5	$4/36 = .\overline{1}$
6	$5/36 = .13\overline{8}$
7	$6/36 = .1\overline{6}$
8	$5/36 = .13\overline{8}$
9	$4/36 = .\overline{1}$
10	$3/36 = .08\overline{3}$
11	$2/36 = .0\overline{5}$
12	$1/36 = .0\overline{27}$

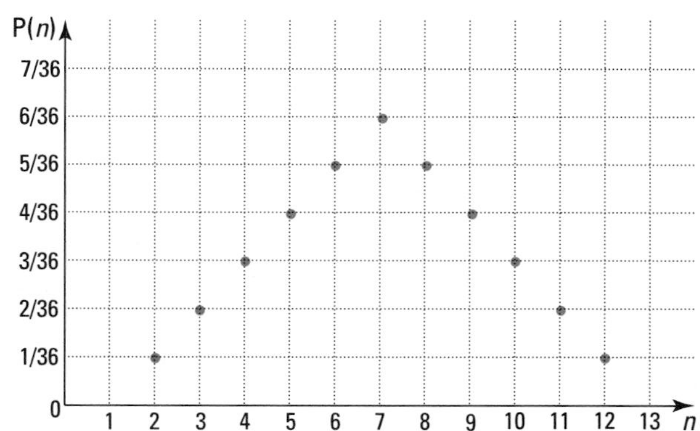

Your table and graph in Step 2 from the In-class Activity on page 794 should approximate these. Notice that the graph is part of an angle.

In general, a **probability function** is a function whose domain is a set of outcomes in a situation, and in which each ordered pair of the function contains an outcome and its probability.

Other Probability Functions

Example 1

Consider the spinner at the right. Assume all regions have the same probability of the spinner landing in them. Let $P(n)$ = the probability of the spinner landing in region n.

a. Graph the function P.

b. Give an equation for the function.

Solution

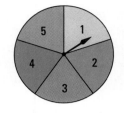

a. Since there are five regions, each with the same probability, each has probability $\frac{1}{5}$. So $P(1) = \frac{1}{5}$, $P(2) = \frac{1}{5}$, and so on. The graph is shown at the right.

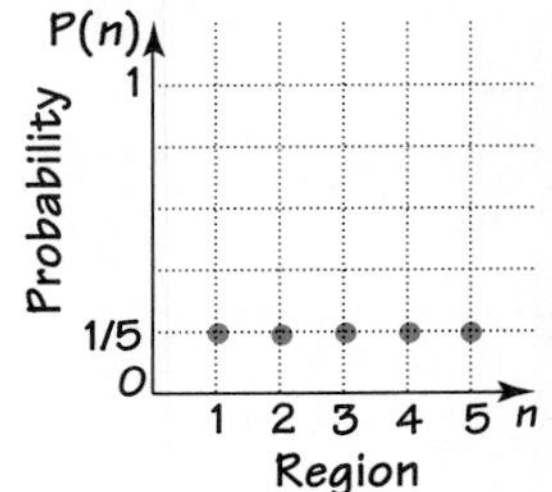

b. An equation for the function is $P(n) = \frac{1}{5}$.

Example 2

Three fair coins—a quarter, a dime, and nickel—are tossed. Let $P(h)$ = the probability of tossing exactly h heads.

a. Make a table of values for the function P.

b. Give the domain of P.

c. Give the range of P.

Solution

a. From the Multiplication Counting Principle you know there are $2 \cdot 2 \cdot 2 = 8$ possible ways the three coins could come up. List them and count the number of heads for each outcome.

quarter	dime	nickel	no. of heads
H	H	H	3
H	H	T	2
H	T	H	2
H	T	T	1
T	H	H	2
T	H	T	1
T	T	H	1
T	T	T	0

Since the coins are fair, all of the outcomes are equally likely. There are eight outcomes, so each has probability $\frac{1}{8}$. Now make a table showing the number of heads and $P(h)$ the probability that h heads occur.

h	0	1	2	3
P(h)	$\frac{1}{8}$	$\frac{3}{8}$	$\frac{3}{8}$	$\frac{1}{8}$

Notice that getting one or two heads is more likely than getting either zero or three heads.

b. The domain is the set of h-values. domain = $\{0, 1, 2, 3\}$

c. The range is the set of probability values. range = $\left\{\frac{1}{8}, \frac{3}{8}\right\}$

Relative frequencies. *Although more boys are born each year than girls, the female population eventually exceeds the male population at around age 31—and continues in that direction thereafter.*

Probability and Relative Frequency Functions

Below are graphs for two functions related to the births of boys and girls. There were about 2,129,000 boys and 2,029,000 girls born in the U.S. in 1990. The function below left assumes that the two events—birth of a girl and birth of a boy—are equally likely. The function below right gives the relative frequencies of boy and girl births in 1990. A basic question for statisticians is: "Could you expect relative frequencies like those below right if the probabilities below at the left are true?"

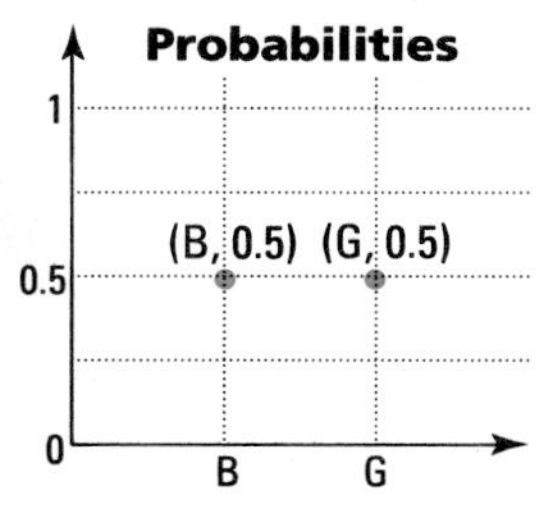

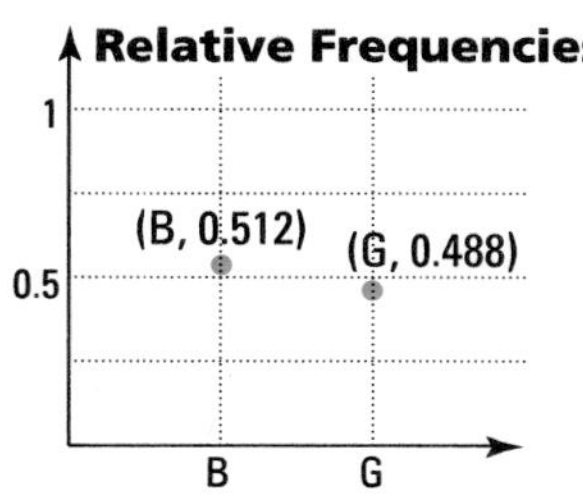

The answer in this case (which requires more advanced mathematics to prove) is "No." If the probabilities were equal, these relative frequencies would be unlikely. Thus, a baby is more likely to be a boy than a girl. However, the ratio of males to females changes with age. The U.S. population contains more women than men because women live longer.

QUESTIONS

Covering the Reading

1. Explain what a probability function is.

In 2–5, let $P(n)$ = the probability of getting a sum of n when two fair dice are tossed.

2. Evaluate $P(3)$.

3. If $P(n) = \frac{5}{36}$, then n = __?__ or __?__.

4. Describe the shape of the graph of P.

5. Does the graph of P have an axis of symmetry? If so, what is an equation for that line?

6. Consider the spinner at the right. Assume the spinner has the same probability of landing in each region. Let $P(n)$ = the probability of the spinner landing in region n.

a. Graph the function P.
b. Find an equation for the function.
c. What is the domain of P?
d. What is the range of P?

In 7 and 8, refer to Example 2.

7. *True or false.* $P(0) = P(3)$.

8. Explain why $P(1) = \frac{3}{8}$.

In 9–11, *true or false.*

9. In 1990, more boys than girls were born in the U.S.

10. In 1990, more men than women were living in the U.S.

11. In births in the U.S., the relative frequency that a baby is a boy is $\frac{1}{2}$.

Applying the Mathematics

12. At the right is graphed a probability function for a weighted (unfair) 6-sided die.

a. $P(3) =$ ___?___

b. $P(\text{a number greater than } 3) =$ ___?___

c. Describe how this graph differs from the probability function for a fair die.

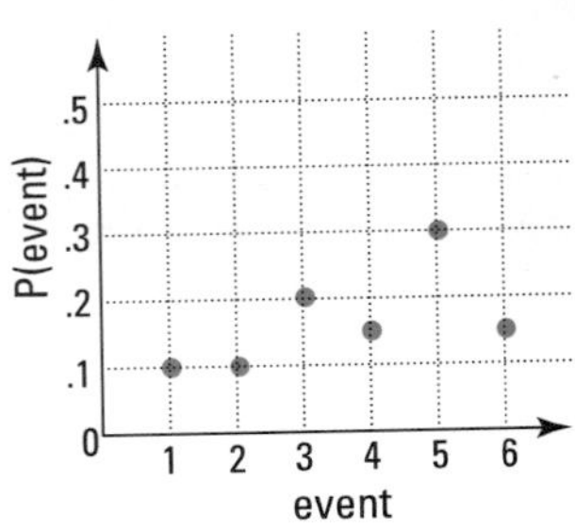

13. Why is it impossible for the number 2 to be in the range of a probability function?

14. Suppose that when two brown-eyed people have a child, the probability that the child is brown-eyed is $\frac{3}{4}$.

a. What is the probability that they have a child who is not brown-eyed?

b. Graph the probability function suggested by this situation.

c. What is the range of this function?

15. A letter is mailed Saturday with the following probabilities.
$P(\text{it arrives Monday}) = \frac{1}{2}$. $P(\text{it arrives Tuesday}) = \left(\frac{1}{2}\right)^2$.
$P(\text{it arrives Wednesday}) = \left(\frac{1}{2}\right)^3$. $P(\text{it arrives Thursday}) = \left(\frac{1}{2}\right)^4$.
$P(\text{it arrives Friday}) = \left(\frac{1}{2}\right)^5$.

a. Calculate $P(\text{it does not arrive by Friday})$.

b. Graph an appropriate probability function.

16. Make up a probability function for which the only two elements in the range are 0 and 1.

Where do they put large packages? *This is one of the smallest post office buildings in the United States. It is located in Ochopee, Florida.*

Gone fishin'. *Pictured are fish being processed at a plant in Newport, Oregon.*

Review

17. Refer to the table and graph below showing domestic catch of fish for human food between 1960 and 1990. Let $f(x)$ = the amount (in millions of pounds) of fish caught in year x. *(Lessons 13-2, 7-1)*

x	f(x)
1960	2498
1965	2587
1970	2537
1975	2465
1980	3654
1985	3294
1990	7041

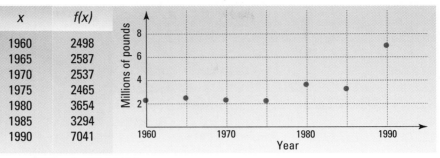

a. What is $f(1980)$?
b. Evaluate $f(1980) - f(1985)$ and tell what the answer represents.
c. What is the domain of f?
d. Between 1980 and 1985, tell whether the rate of change of fish caught was positive, negative, or zero.
e. In which five-year period was the rate of change of fish caught the lowest?

18. Does $y < -3x + 1$ describe a function? Why or why not? *(Lesson 13-1)*

19. a. Factor $2x^2 + 3x - 20$.
b. Find a value of x for which $2x^2 + 3x - 20$ is a prime number. *(Lessons 12-5, 12-1)*

20. Do this problem in your head. Since one thousand times one thousand equals one million, $1005 \cdot 995 =$ __?__. *(Lesson 10-6)*

21. Simplify $\sqrt{12} + \sqrt{3}$. *(Lesson 9-7)*

In 22 and 23, suppose the two triangles below are similar with corresponding sides parallel. *(Lessons 6-9, 4-7)*

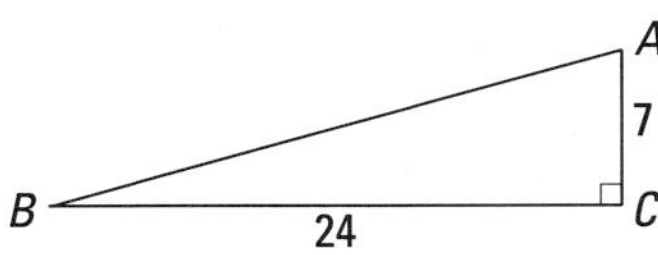

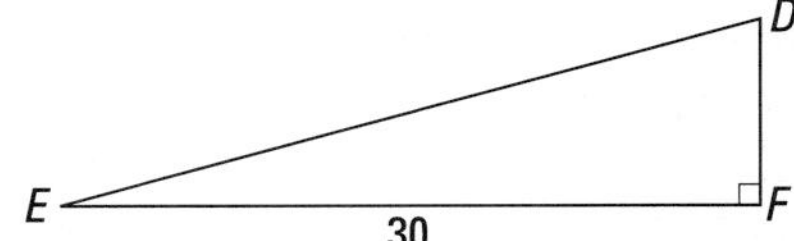

22. Find the missing lengths.

23. If $m\angle B \approx 16°$, then $m\angle D \approx$ __?__

Exploration

24. Consider the function $f(x) = -\frac{1}{36}|x - 7| + \frac{1}{6}$.
a. Use an automatic grapher to plot $f(x)$ for $2 \le x \le 12$.
b. What is $f(2)$?
c. What is $f(7)$?
d. How is this graph related to one of the probability functions in this lesson?

LESSON

13-6 Polynomial Functions

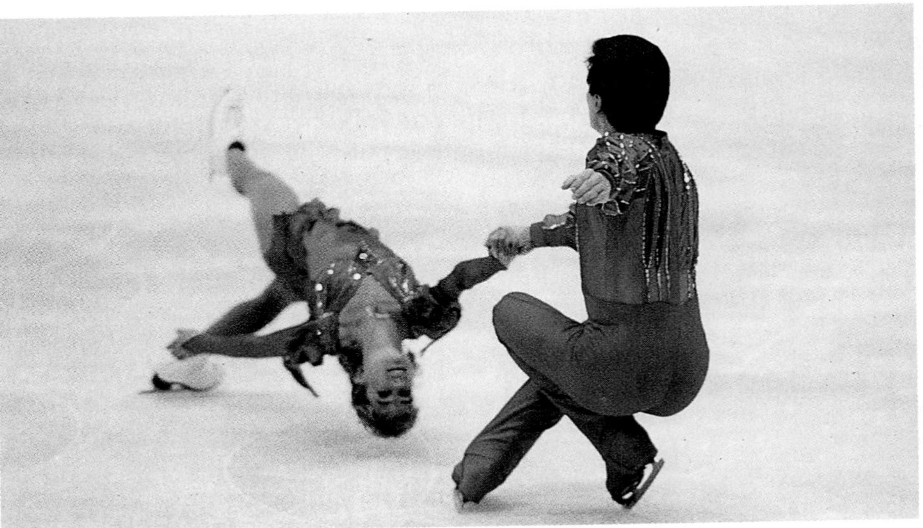

One good turn deserves another. *These pairs skaters are performing a "death spiral," in which both people rotate 360° around the fixed back toe of the male skater.*

The Cubing Function

As you know, some types of equations have graphs with predictable shapes. Graphs of all equations of the form $y = mx + b$ are lines with slope m and y-intercept b. Therefore, the function f with $f(x) = mx + b$ is called a *linear function.* Graphs of quadratic functions of the form $f(x) = ax^2 + bx + c$ are parabolas. In contrast, not all polynomial equations of degree 3 have the same shape. Consider the simplest cubic equation $y = x^3$. This equation defines the **cubing function** with equation $f(x) = x^3$.

Example 1

Graph $f(x) = x^3$ by plotting points where x goes from -2 to 2 in steps of 0.5.

Solution

Make a table of values. Connect them with a smooth curve. A table and graph are shown below.

x	y = f(x)
-2	-8
-1.5	-3.375
-1	-1
-0.5	-0.125
0	0
0.5	0.125
1	1
1.5	3.375
2	8

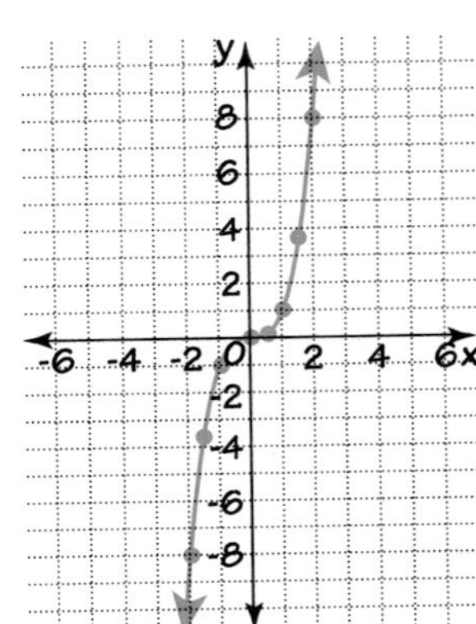

To explore the shape of the graph of $y = x^3$, we can use an automatic grapher. Below are two views of this curve on different windows.

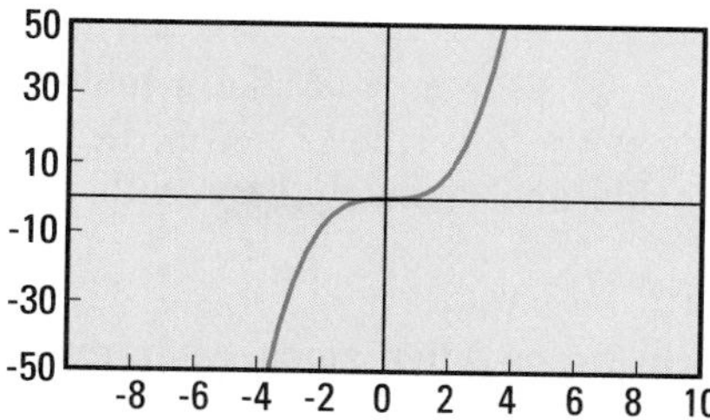

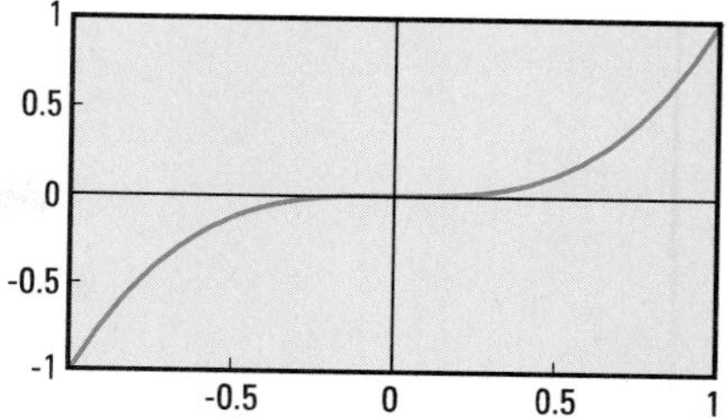

The window on the left above shows that when $|x|$ is moderately large, $|x^3|$ is very large. The one on the right shows that near the origin the graph of $y = x^3$ is fairly flat.

Activity

Trace one of the graphs of $y = x^3$ onto a piece of paper. Trace the axes also. Hold the paper in place on your book with a pencil point at the origin. Rotate the paper

a. 90° clockwise; **b.** 180° clockwise.

After each rotation, describe how the graph on the paper is related to the graph in your book.

You are asked for your answer to part **a** in the Questions. For part **b,** you should find that the graph on your paper coincides exactly with that in your book. Because a half-turn makes these two graphs coincide, we say that the graph of $y = x^3$ has **180° rotation symmetry** around the origin. In general, if you can turn a figure 180° around some point so that the figure coincides with itself, that point is called a **center of symmetry.** Notice that the graph does *not* have a *line* of symmetry.

Spin cycle. *Pictured are folk dancers in Mexico City. Many dances include examples of rotations and rotational symmetry.*

Another Cubic Polynomial Function

Example 2

Consider $g(x) = x^3 - x + 1$. Make a table of values satisfying the equation, using x = -2, -1.5, -1, -0.5, 0, 0.5, 1, 1.5, and 2. Plot the points in the table and connect them with a smooth curve.

Solution

Be careful with the negative signs when evaluating the expression. For instance, when $x = -2$, $x^3 - x + 1 = (-2)^3 - (-2) + 1 = -8 + 2 + 1 = -5$.

Below are the table and graph.

x	$g(x) = x^3 - x + 1$
-2	-5
-1.5	-.875
-1	1
-0.5	1.375
0	1
0.5	.625
1	1
1.5	2.875
2	7

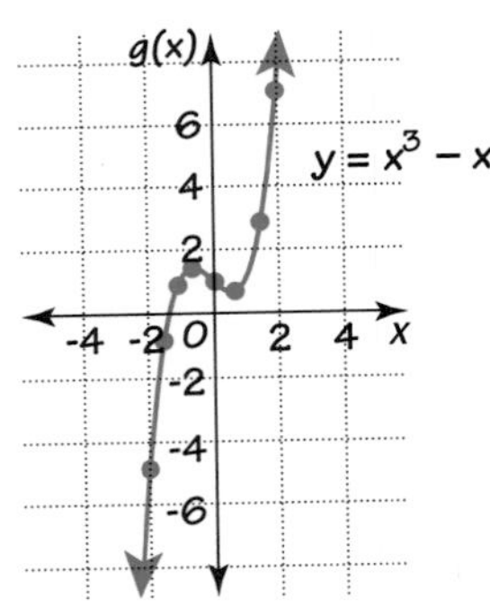

Like the graph in Example 1, the graph of $g(x) = x^3 - x + 1$ increases very quickly as x gets large, and decreases very quickly when x gets very small (that is, "very negative").

The table and graph show that there is an x-intercept between -2 and -1. To estimate the x-intercept, you can use an automatic grapher to draw the graph of $g(x) = x^3 - x + 1$. On the window $-1.5 \leq x \leq 1.5$, $-1 \leq y \leq 2$, we trace and find the x-intercept to be about -1.3.

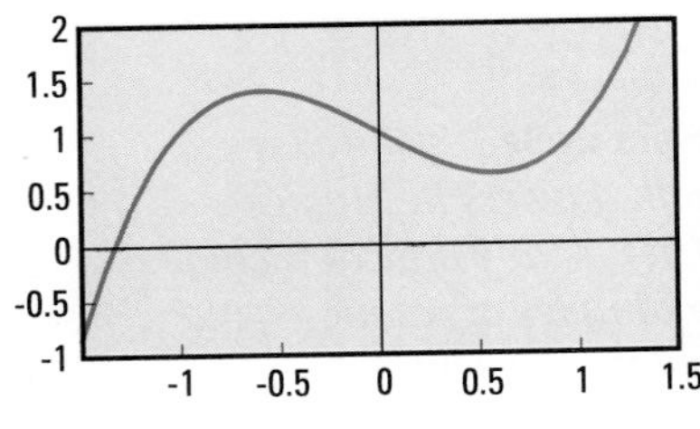

Like the graph of $y = x^3$, the graph of $y = x^3 - x + 1$ also has a point of symmetry. (In the Questions you are asked to find it.) In a later course you will see that algebra can be used to show that the graph of every polynomial function of degree 3 has 180° rotation symmetry.

Higher-Degree Polynomial Functions

Making graphs by hand of most higher-degree polynomial functions involves a great deal of computation. For all but the simplest cases, we recommend that you use an automatic grapher. Unless you are looking only for a certain point or at a certain region on the graph, it is a good idea to use a window that allows you to see all the intercepts, and all the "peaks and valleys" polynomial graphs can have.

It is helpful to know that every graph of a polynomial function of degree n has exactly one y-intercept and no more than n x-intercepts. Aside from the above, there are very few "rules" about polynomial graphs.

Example 3

a. Draw a graph of $y = x^4 - 7x^3 - 9x^2 + 63x$.
b. Identify all intercepts and show the behavior of the graph between those intercepts.

Solution

a. Calculate a few values to get some idea of what window to use.

When $x = -2, y = (-2)^4 - 7(-2)^3 - 9(-2)^2 + 63(-2) = -90.$
When $x = 0, y = 0^4 - 7(0)^3 - 9(0)^2 + 63(0) = 0.$
When $x = 2, y = (2)^4 - 7(2)^3 - 9(2)^2 + 63(2) = 50.$

We use a window of $-15 \le x \le 15$ and $-100 \le y \le 100$. The graph is below on the left.

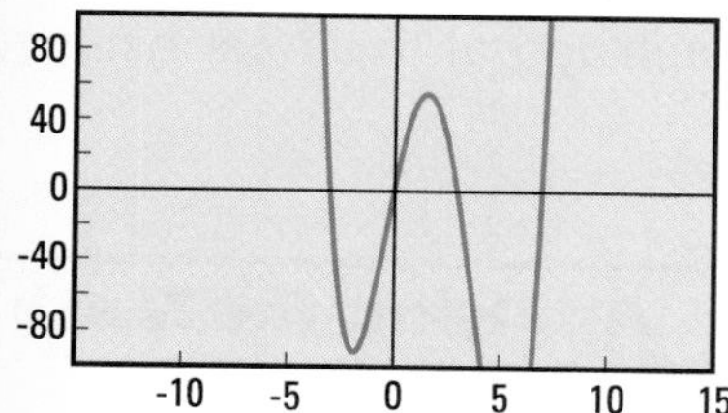
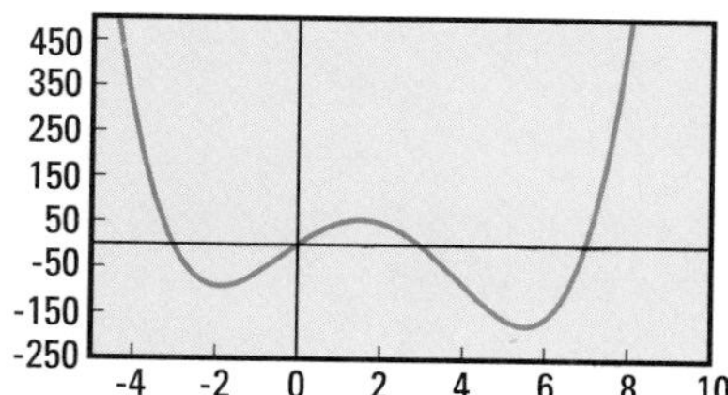

This graph shows the y-intercept and 4 x-intercepts, the maximum possible for a 4th-degree polynomial function. So there are no other intercepts. We need to increase the range of y-values to see where the peaks and valleys occur. The sketch above on the right, using the window $-5 \le x \le 10$, $-250 \le y \le 500$, captures all the essential features of the graph.

b. The x-intercepts are -3, 0, 3, and 7; the y-intercept is 0.

Check

b. The y-intercept was checked as part of the solution. To check the x-intercepts, calculate y when $x = -3, 3$, and 7.

$$\text{When } x = -3,\ y = (-3)^4 - 7(-3)^3 - 9(-3)^2 + 63(-3) = 81 + 189 - 81 - 189 = 0.$$

You are asked to check the other two x-intercepts in the Questions.

QUESTIONS

Covering the Reading

1. How many x-intercepts does the graph of $f(x) = x^3$ have?
2. *True or false.* The graph of the cubing function has a line of symmetry.
3. What does it mean to say that a graph has 180° rotation symmetry?
4. Copy one of the graphs of $f(x) = x^3$ from the book. On the same set of axes, draw the image you got by doing part **a** of the Activity on page 801.

In 5 and 6, refer to Example 2.

5. Check by substitution that -1.3 is close to an x-intercept of $y = x^3 - x + 1$.

6. Use tracing to estimate the coordinates of the point of symmetry of the graph.

7. *True or false.* The graph of every cubic polynomial function has a point of symmetry.

8. The graph of a polynomial equation of degree n has __?__ y-intercept(s) and __?__ x-intercept(s).

In 9 and 10, refer to Example 3.

9. Verify that $x = 3$ and $x = 7$ are x-intercepts of the graph.

10. Does the graph appear to have a symmetry point? If so, what are its coordinates?

Applying the Mathematics

In 11–13, an equation for a polynomial function is given.

a. Make a table of x- and y-values for x-values -2, -1.5, -1, -0.5, 0, 0.5, 1, 1.5, and 2.
b. Graph the values.
c. Describe the graphs.

11. $y = x^3 + 3$ **12.** $y = 2x^3$ **13.** $y = -x^3$

14. A spreadsheet was used to compute the value of $f(x) = x^3 + 3x^2 - 10x$ for values of x from -6 to 4.
a. Copy and complete the table.
b. Graph the function f for $-6 \le x \le 4$.
c. Identify all x-intercepts.

	A	B
1	X	VALUE
2	-6	
3	-5	
4	-4	24
5	-3	30
6	-2	24
7	-1	
8	0	
9	1	
10	2	0
11	3	24
12	4	72

15. You will need an automatic grapher. In Example 2 of Lesson 10-2 you learned that the polynomial $50x^3 + 60x^2 + 70x + 80$ represents the amount of Cole's savings after three years when invested at a scale factor x. Let $A(x) = 50x^3 + 60x^2 + 70x + 80$.
a. What is a reasonable domain for the function A?
b. Graph A on this domain.
c. Use the trace feature to find the amount Cole would have if he invested at a 10% annual yield.
d. At what annual yield would Cole have to invest in order to accumulate \$400 in 3 years?

Review

16. Assume a coin is fair. Then if it is tossed twice, there are four possible arrangements of heads and tails: HH, HT, TT, and TH. Let $P(n)$ be the probability that there are n heads in 2 tosses of the coin.

a. What is the domain of the function P?

b. Give all the ordered pairs in this function. *(Lessons 13-5, 13-4)*

In 17–19, an open soup can has radius r and height h. Its volume $V = \pi r^2 h$, and its surface area $S = \pi r^2 + 2\pi rh$.

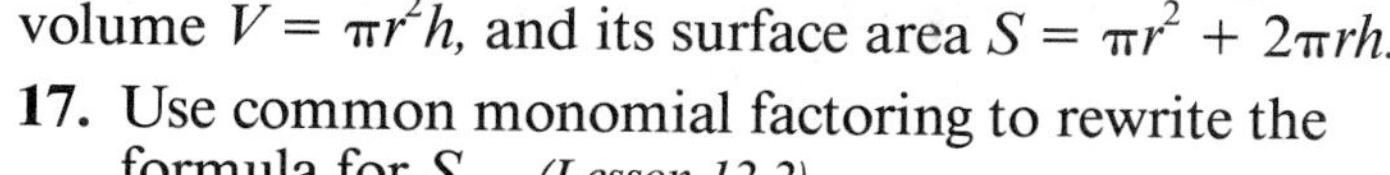

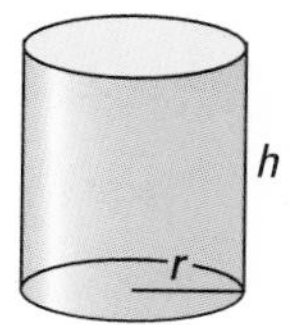

17. Use common monomial factoring to rewrite the formula for S. *(Lesson 12-2)*

18. Find each of the following. *(Lesson 10-1)*

a. the degree of V

b. the degree of S

19. If the can has a diameter of 8 cm and a height of 12 cm, about how many milliliters of soup can it hold? (Remember that 1 liter = 1000 cm^3.) *(Lesson 1-5, Previous course)*

20. Solve by graphing. $\begin{cases} y = |x| \\ y = \frac{1}{2}x^2 \end{cases}$ *(Lesson 11-1)*

In 21–26, solve. *(Lessons 9-8, 9-5, 6-8, 5-6, 3-5)*

21. $|3y - 6| = 2$

22. $100B^2 + 100B - 100 = 0$

23. $\frac{A}{0.2} = \frac{10}{A}$

24. $2y + 14 > 5y - 19$

25. $14.7 = 7x + 21$

26. $\sqrt{C} = 400$

Exploration

27. Explore with an automatic grapher.

a. Find an equation for a polynomial function with a graph shaped like the following.

i.

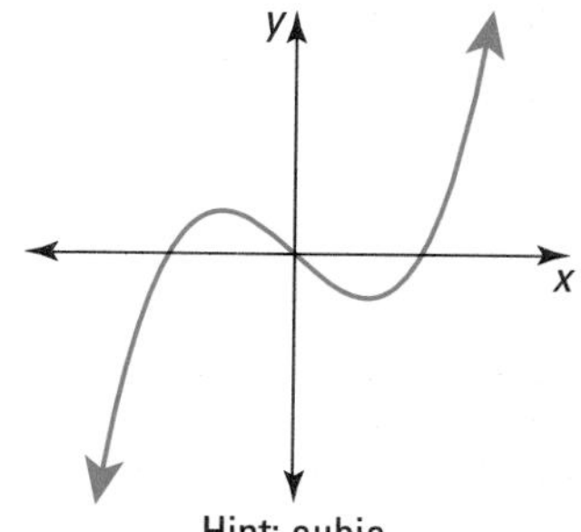

ii.

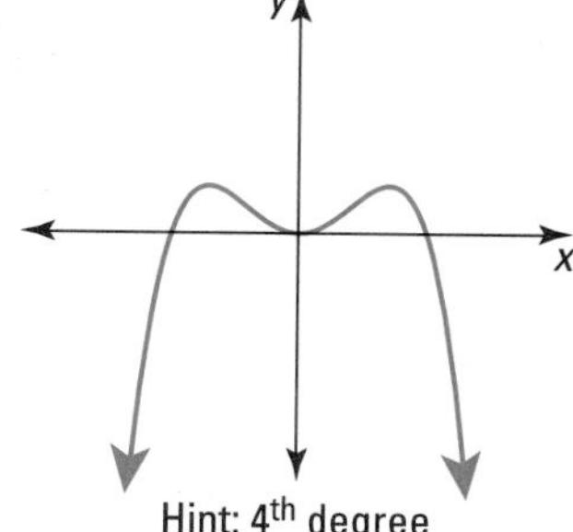

b. Draw the graph of some 5th degree polynomial function.

Introducing Lesson 13-7

Ratios of Sides in Right Triangles

IN-CLASS

ACTIVITY

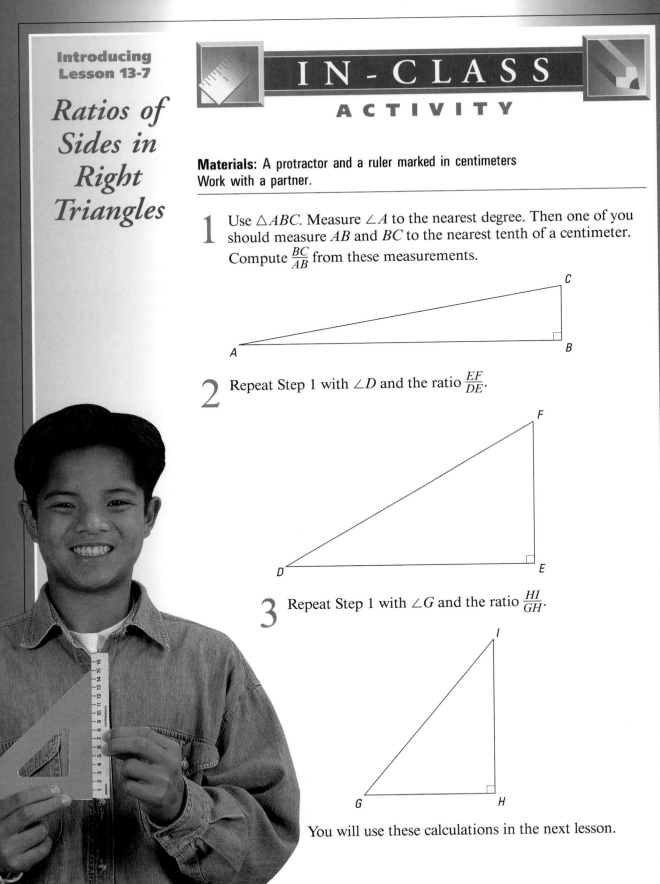

Materials: A protractor and a ruler marked in centimeters
Work with a partner.

1 Use $\triangle ABC$. Measure $\angle A$ to the nearest degree. Then one of you should measure AB and BC to the nearest tenth of a centimeter. Compute $\frac{BC}{AB}$ from these measurements.

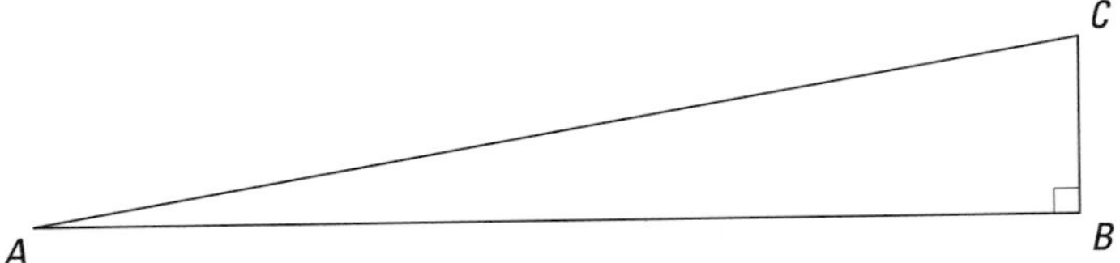

2 Repeat Step 1 with $\angle D$ and the ratio $\frac{EF}{DE}$.

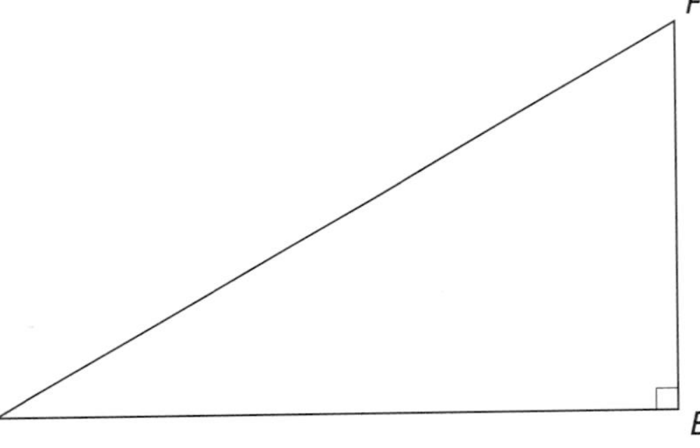

3 Repeat Step 1 with $\angle G$ and the ratio $\frac{HI}{GH}$.

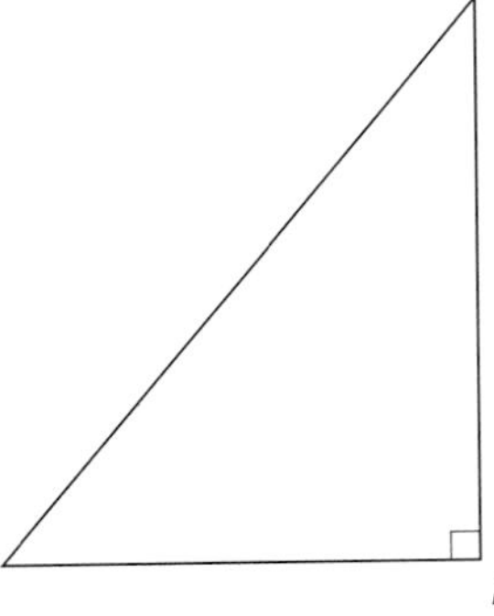

You will use these calculations in the next lesson.

LESSON

13-7 *The Tangent Function*

Tree top. *To determine the height of this Giant Sequoia tree, a person could use the tangent function. See the Example on pages 808–809.*

Calculators and computers have built-in functions. Sometimes these functions can be accessed by pressing buttons. At other times they may be found by highlighting an item on a menu. In this lesson, we consider the tan key. It gives the values of the **tangent function.** In the Activity preceding this lesson, you calculated three values of this function.

The Tangent of an Angle

Consider right triangle *ABC* below, with legs of lengths 3 and 4. $\angle C$ is a right angle and is thus a 90° angle. Since the sum of the measures of the three angles of a triangle is 180°, $m\angle A + m\angle B = 90°$.

The tangent of angle *A* in a right triangle is defined as a particular ratio of legs in the triangle. Here $\overline{BC}$ is the **leg opposite** $\angle A$, and $\overline{AC}$ is the **leg adjacent** to $\angle A$.

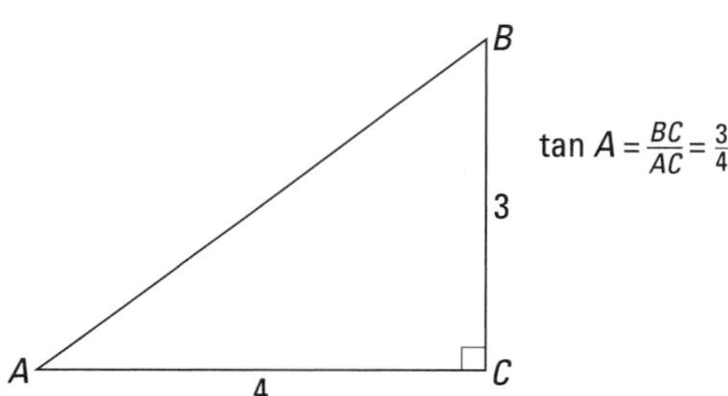

The Tangent of an Angle
In a right triangle with acute angle *A*, the **tangent of $\angle A$,** abbreviated **tan *A*,** equals the ratio:

$$\frac{\text{length of the leg opposite angle } A}{\text{length of the leg adjacent to angle } A}.$$

Most scientific calculators have a tan key. This key gives values of the tangent of $\angle A$ when you specify the measure of $\angle A$. Make sure your calculator is set to handle degrees when you do this lesson.

The key sequence that finds tangents varies by calculator. Here are some possible sequences for tan 30°.

30 tan, tan 30, 30 tan =, tan 30 =

Your calculator may use enter for =. You should see the display 0.5773503 or something close. If you see -6.4053312, then your calculator is not set to degrees, but is using another unit, the *radian*. If you see 0.5095254, then your calculator is using a third unit, called the *grad*.

Activity

Determine the key sequence that will give you tan 30° on your calculator. The values you obtained in the In-class Activity should be in agreement with this table of some values of the tangent function, rounded to three decimal places. (All the tangents in this table are irrational, so none is a finite or repeating decimal.) Below is a graph of this function for values of x between 0° and 90°.

x	tan x
10°	0.176
20°	0.364
30°	0.577
40°	0.839
50°	1.192
60°	1.732
70°	2.747
80°	5.671

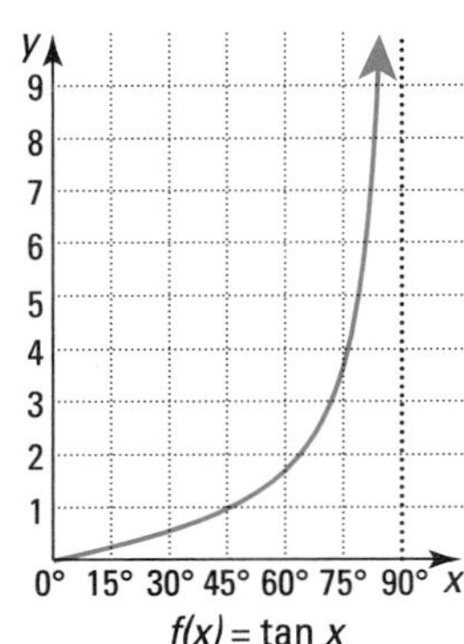

Using the Tangent Function

The tangent function can be used to estimate inaccessible heights.

Example

Nancy wants to estimate the height of a tree. As shown at the left she has to look up 50° to see the top of a tree 5 meters away. If her eyes are 1.5 meters above the ground, about how tall is the tree?

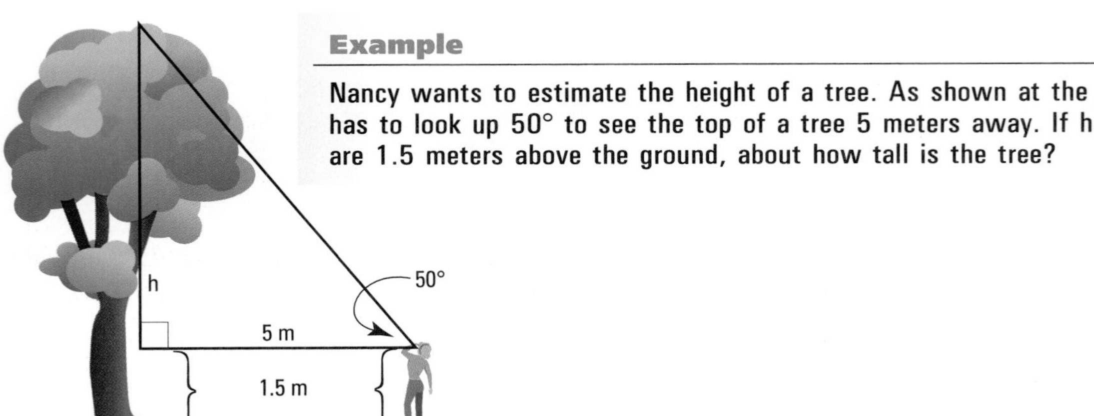

Solution

If h is the height of the tree above eye level, then the height of the tree is $h + 1.5$. First use the triangle to find h and then add 1.5. To find h, note that

$$\tan 50° = \frac{h}{5}.$$

Use your calculator to evaluate tan 50°.

$$\tan 50° \approx 1.19$$

Substitute. $$\frac{h}{5} \approx 1.19$$

$$h \approx 5 \cdot 1.19$$

$$h \approx 6.0$$

So the full height of the tree is about $h + 1.5 \approx 6.0 + 1.5 = 7.5$ m.

You may not realize it, but you have already calculated tangents on the coordinate plane. Tangents of angles are related to equations of lines.

Consider the line $y = \frac{2}{3}x - 2$. This is a line with a slope of $\frac{2}{3}$. It crosses the x-axis at (3, 0) and also passes through $\left(4, \frac{2}{3}\right)$.

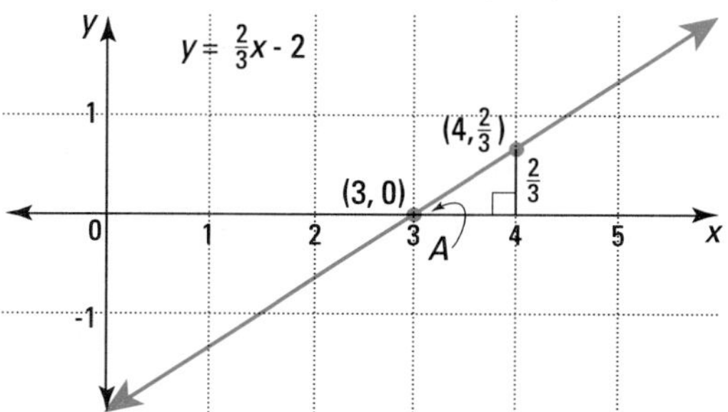

Consider the acute angle formed by the line and the positive ray of the x-axis. From the graph you can see that

$$\tan A = \frac{\frac{2}{3}}{1} = \frac{2}{3}$$

This is the slope of the line!

If A is the acute angle formed by the upper half of the oblique line $y = mx + b$ and the positive ray of the x-axis, then

$$\tan A = m.$$

The tangent function is quite a function; it combines the concepts of slope, graphing, angles, ratios, and triangles. In later courses you will study the tangent function with a larger domain and learn of other applications.

QUESTIONS

Covering the Reading

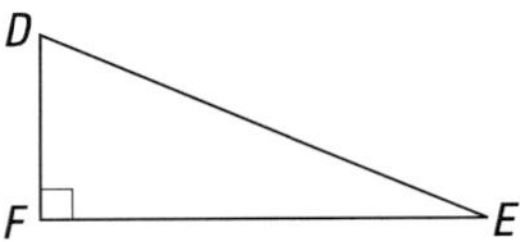

1. Use right triangle *DEF* at the right.
   a. Name the side opposite $\angle E$.
   b. Name the side adjacent to $\angle E$.
   c. What ratio equals tan *E*?

In 2–4, give the values you found from the In-class Activity on page 806.

2. $m\angle A$ and tan *A*
3. $m\angle D$ and tan *D*
4. $m\angle G$ and tan *G*

In 5 and 6, use your calculator. Give a three-place decimal approximation.

5. tan 57°
6. tan 3°

7. Refer to $\triangle KLM$ at the right. Find tan *K*.

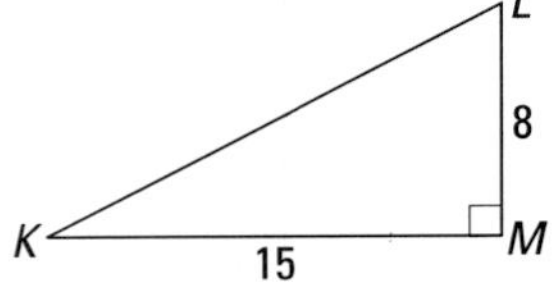

8. a. Measure angle *F* on page 806.
   b. Estimate tan *F* by dividing the lengths of two sides of $\triangle DEF$.
   c. Estimate tan *F* by using a calculator.
   d. How close are the values you found in parts **b** and **c?**

9. Lester had to look up 65° to see the top of a tree 6 meters away. If his eyes are 1.7 meters above the ground, how tall is the tree?

10. What is the relationship between the tangent function and the slope of a line $y = mx + b$?

11. Consider the line with equation $y = \frac{6}{5}x + 12$.
    a. What is the tangent of the acute angle formed by this line and the positive ray of the *x*-axis?
    b. Use the table on page 808 to give the approximate measure of that angle.

Applying the Mathematics

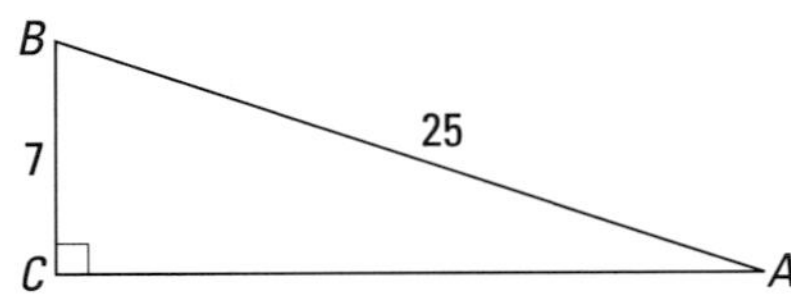

12. Refer to $\triangle ABC$ at the right.
    a. Find *AC*.
    b. Find tan *A*.
    c. Find tan *B*.

13. A meter stick casts a shadow 0.6 meter long. Use the table in this lesson to estimate the measure of the angle at which the sun appears above the horizon, to the nearest 10°. (This is called the *angle of elevation* of the sun.)

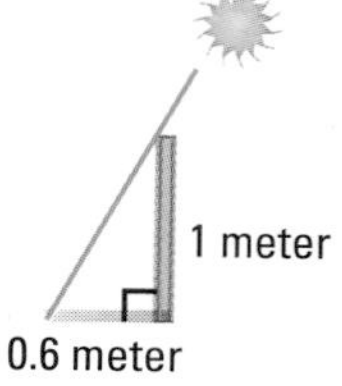

14. A line goes through the origin and the upper half makes an angle of 140° with the positive ray of the x-axis.
  a. Find the slope of this line.
  b. Find an equation for this line.

15. To the nearest 10°, find the measure of the angle formed by the upper half of the line $y = 4x - 3$ and the positive ray of the x-axis. Use the table on page 808.

Review

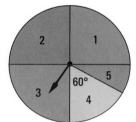

16. In the spinner at the left, the two diameters are perpendicular and the central angle of sector 4 has a measure of 60°.
  a. What is the measure of the angle in sector 5?
  b. Give P(landing in sector n) for $n = 1, 2, 3, 4$, and 5.
  c. Graph the probability function P. *(Lessons 13-5, 6-6)*

17. What number is not in the domain of the function f if $f(x) = \frac{x - 1}{2x + 4}$? *(Lesson 13-4)*

18. a. If $y = 3(x + 1)^2 - 4$, what is the smallest possible value of y?
  b. If $y = 3|x + 1| - 4$, what is the smallest possible value of y?
  c. If $y = 3\sqrt{x + 1} - 4$, what is the smallest possible value of y? *(Lessons 13-3, 13-2)*

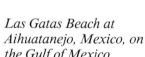

Las Gatas Beach at Aihuatanejo, Mexico, on the Gulf of Mexico.

19. Solve for x: $ax^2 + bx + c = 0$. *(Lesson 9-5)*

20. If $\frac{(a^{11})^{12} \cdot a^{13}}{a^{14}} = a^t$, what is the value of t? *(Lessons 8-7, 8-5)*

21. In right triangle DEF, $\angle D$ is a right angle. If $m\angle E$ is four times $m\angle F$, find $m\angle E$. *(Lesson 4-7)*

Exploration

22. When $x = \frac{\pi A}{180°}$, $\tan A \approx \frac{2x^5 + 5x^3 + 15x}{15}$.
  a. Let $A = 10°$. How close is the polynomial approximation to the calculator value of $\tan A$?
  b. Repeat part **a** when $A = 70°$.

23. What became of the man who sat on a beach along the Gulf of Mexico?

LESSON

13-8

Functions on Calculators and Computers

Purely musical. *The music played by a violinist or other musician is based on combinations of pure tones. The graph of a pure tone is a sine wave, like the one shown here.*

Some Familiar Calculator Functions

In Lesson 13-7, you studied a function whose values are given or estimated by the [tan] key on a calculator. Other functions are pre-programmed into calculators. They will give you function values when you enter a value from their domains. The calculator will indicate an error message when you enter a value not in the domain. Here are some familiar keys, the functions they define, and their domains.

Key	Function	Domain	Error if:
the square root key [$\sqrt{\ }$]	SQR(x) = $\sqrt{x}$	set of nonnegative reals	$x < 0$
the factorial function key [!]	FACT(n) = $n!$	set of nonnegative integers	x is not an integer or $x < 0$
reciprocal function key [1/x]	$f(x) = \frac{1}{x}$	set of nonzero reals	$x = 0$
the squaring key [x^2]	$s(x) = x^2$	set of all reals	none

You have learned some applications of these functions earlier in this book. In this lesson, we introduce you to the meaning of some of the other function keys on a calculator. These keys represent functions that are built into virtually every computer language as well.

Trigonometric Functions

Two functions related to the tangent function are the **sine** and **cosine functions,** whose values are found by the `sin` and `cos` keys on your calculator. Like the tangent function value, these values are ratios of sides in right triangles. Again, consider a right triangle *ABC*. The sine and the cosine of $\angle A$ are as follows.

$$\sin A = \frac{\text{length of the leg opposite angle } A}{\text{length of the hypotenuse}}$$

$$\cos A = \frac{\text{length of leg adjacent to angle } A}{\text{length of the hypotenuse}}$$

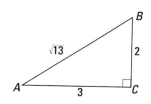

So in this case, $\sin A = \frac{2}{\sqrt{13}}$ and $\cos A = \frac{3}{\sqrt{13}}$. For $\angle B$, the values of sine and cosine are interchanged: $\sin B = \frac{3}{\sqrt{13}}$ and $\cos B = \frac{2}{\sqrt{13}}$.

It is possible to define these functions and the tangent function for angles of any measure. The graph of the sine or cosine function may surprise you. Below is the graph of $y = \sin x$.

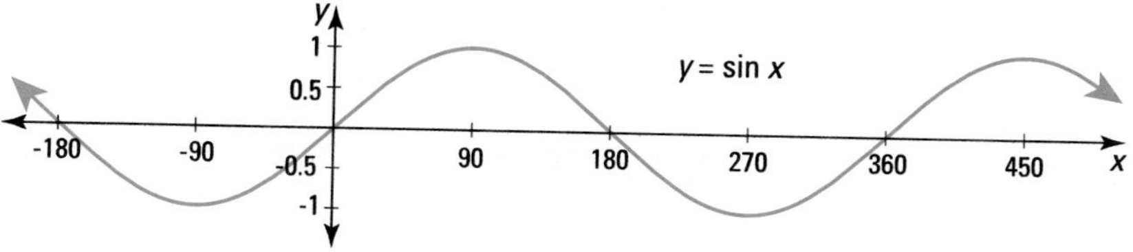

The curve is called a *sinusoidal curve,* or *sinusoid,* and it has the same shape as sound waves and radio waves. The sine, cosine, and tangent functions are part of a branch of mathematics called **trigonometry.** These functions are so important that some high schools offer a full course in trigonometry devoted to studying them and their applications. In BASIC, values of the tangent, sine, and cosine functions are denoted by TAN(x), SIN(x), and COS(x), respectively.

The Common Logarithm Function

Another function built into almost all scientific calculators is the **common logarithm function,** activated by the `log` key. This key defines a function $y = \log x$, read "*y* equals the common logarithm of *x*." The common logarithm of a number is the power to which 10 must be raised to equal that number. So, since 1 million $= 10^6$, $\log(1000000) = 6$. Logarithms provide a way to deal easily with very large or very small numbers. A graph of the common logarithm function is given on the next page.

The part of this graph above the x-axis pictures the kind of growth often found in learning. At first, one learns an idea quickly, so the curve increases quickly. But after a while it is more difficult to improve one's performance, so the curve increases more slowly.

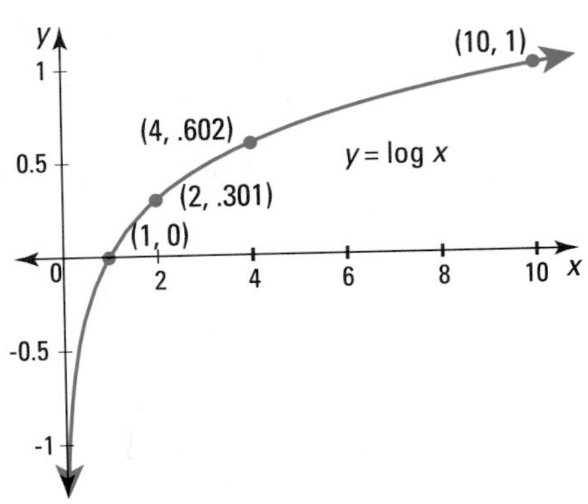

Learning curve. *When learning ballet techniques, the beginner often masters the basic movements quickly. As the dancer progresses, the techniques usually take longer to master, as suggested by the graph at the right.*

Most scientific calculators and computers have other built-in functions. These functions would not be there unless many people needed to get values of that function. Calculators and computers have made it possible for people to obtain values of these functions more easily than most people ever imagined. The algebra that you have studied this year gives you the background to understand these functions and to deal with them.

QUESTIONS

Covering the Reading

In 1–6, use a calculator to approximate each value to the nearest thousandth.

1. tan 11° **2.** sin 45° **3.** cos 47°

4. $\log(10^7)$ **5.** $(-3.489)^2$ **6.** $\sqrt{0.5}$

In 7–9, consider the [√], [!], [1/x], and [x^2] function keys on your calculator. Which produce error messages when the given number is entered?

7. 3.5 **8.** -4 **9.** 0

10. Which function has a graph that is the shape of a sound wave?

11. Which function has a graph that is sometimes used to model learning?

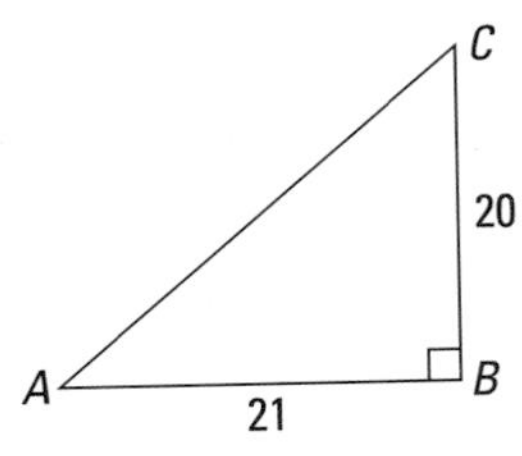

12. Find each number related to $\triangle ABC$ at the left.
a. AC **b.** sin A **c.** cos A **d.** tan A

In 13–15, refer to the graph of $y = \sin x$ in this lesson.
a. Estimate the value from the graph.
b. Use a calculator to check your estimate.

13. sin 90° **14.** sin 360° **15.** sin(-70°)

Applying the Mathematics

16. **a.** Make a table of values for $y = \cos x$ for values of x from 0° to 360° in increments of 15°.
b. Carefully graph this function.
c. What graph in this lesson does the graph of $y = \cos x$ most resemble?

17. Many computers use the name LOG to refer to a logarithm function different from the one in the lesson.
a. Run this program or use your calculator to determine what is printed.

```
10 PRINT "X", "LOG X"
20 FOR X = 1 TO 10
30 Y = LOG(X)
40 PRINT X, Y
50 NEXT X
60 END
```

b. Graph the ordered pairs that are printed.
c. Is your graph like the one in the lesson, or is it different? If it is different, how does it differ?

18. **a.** Graph $y = \tan x$ on an automatic grapher with x set for degrees. Use the graph to estimate tan(-10°).
b. Verify the value of tan(-10°) on your calculator.

19. Graph the reciprocal function $f(x) = \frac{1}{x}$.

20. What curve is the graph of the squaring function, $s(x) = x^2$?

Review

21. In $\triangle ABC$ at the right, find tan A to the nearest tenth. *(Lesson 13-7)*

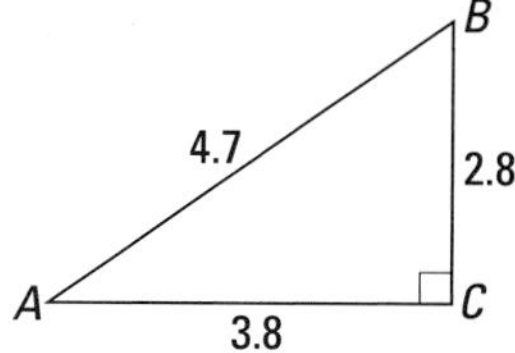

22. Joan had to look up 40° to see the top of a flagpole 20 feet away. The situation is pictured at the left. If her eyes are 5 feet above the ground, how tall is the flagpole? *(Lesson 13-7)*

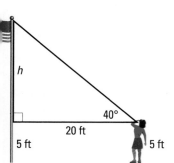

23. **a.** In a toss of two fair dice, what sum is most likely to appear?
b. What is its probability of occurring? *(Lesson 13-5)*

In 24 and 25, $S(x) = x + \frac{x^2}{20}$ gives the number of feet a car traveling at x miles per hour will take to stop. $B(x) = \frac{x^2}{20}$ gives the number of feet traveled after brakes are applied.

The right to vote. *Shown is a drawing of a meeting of the National Women's Suffrage Association from the late 1800s. The Association argued that women should have the right to vote. It wasn't until 1920, when the 19th Amendment was ratified, that women throughout the country obtained the right to vote.*

24. About how many feet does it take to stop at 40 mph? *(Lesson 13-2)*

25. If skid marks in an accident are 100 feet long, about how fast was the car traveling? *(Lesson 13-2)*

26. A rectangular box has dimensions p by $p + 5$ by $p - 1$. *(Lessons 10-1, 10-5)*
 a. Find its volume.
 b. What is the degree of the polynomial in part **a**?

27. Here are the total numbers of votes (to the nearest million) cast for all of the major candidates in the presidential elections since 1940.

Year	All Votes Cast for Major Candidates (millions)	Winner
1940	50	Franklin D. Roosevelt
1944	48	Franklin D. Roosevelt
1948	49	Harry S. Truman
1952	61	Dwight D. Eisenhower
1956	62	Dwight D. Eisenhower
1960	68	John F. Kennedy
1964	70	Lyndon B. Johnson
1968	73	Richard M. Nixon
1972	77	Richard M. Nixon
1976	81	Jimmy Carter
1980	86	Ronald Reagan
1984	92	Ronald Reagan
1988	91	George Bush
1992	104	Bill Clinton

 a. Graph the ordered pairs (year, number of votes).
 b. Use the graph to predict how many votes will be cast for major candidates in the presidential election of 2000. *(Lessons 7-7, 3-3)*

28. The explorers were 13 km from home base at 2 P.M. and 10 km from home base at 3:30 P.M. At this rate, when will they reach home? *(Lesson 7-6)*

29. If you read 17 pages of a 300-page novel in 45 minutes, about how long will it take you to read the entire novel? *(Lesson 6-8)*

In 30–32, solve. *(Lessons 6-8, 5-6, 1-6)*

30. $\frac{m}{2} = \frac{m + 36}{11}$

31. $3x + 9 > x$

32. $\sqrt{v - 6} = 4$

Exploration

33. List all the function keys of a scientific calculator to which you have access. Separate those you have studied from those you have not. Identify at least one situation in which each function you have studied might be used.

A project presents an opportunity for you to extend your knowledge of a topic related to the material of this chapter. You should allow more time for a project than you do for typical homework questions.

1 Tossing Three Dice

Imagine tossing three fair dice.

a. What are the possible outcomes for the sum of the three numbers that appear?

b. If each die is fair, what is the probability of each possible sum?

c. What is the most likely sum? What is the least likely sum?

d. Graph the ordered pairs (sum of x, probability of that sum).

e. Describe the graph.

f. What is the probability of an even sum? What is the probability of an odd sum?

2 Polynomial Functions of Degree 4

With an automatic grapher, explore the possible shapes of functions of the form $P(x) = ax^4 + bx^3 + cx^2 + dx + e$, for various values of a, b, c, d, and e. Find and describe as many shapes as you can.

3 Absolute Value Functions

a. Graph $f(x) = 3|x - 5| + 4$.

b. Discuss the effects of the values of a, b, and c on the graphs of the absolute value functions with equations of the form $y = a|x - b| + c$. (Hint: Let $b = 0$ and $c = 0$ to find the effects of a. Let $a = 1$ and $c = 0$ to find the effects of b. Let $a = 1$ and $b = 0$ to find the effects of c.)

4 Reciprocal Functions

When $f(x)$ is the value of a function, then $\frac{1}{f(x)}$ is the corresponding value of its *reciprocal function.*

a. On the same pair of axes, graph the function with equation $y = x$ and its reciprocal, the function with equation $y = \frac{1}{x}$.

b. On another pair of axes, graph the function $y = x^2$ and its reciprocal.

c. On a third pair of axes, graph the function $f(x) = 2^x$ and its reciprocal.

d. Pick two other functions and graph them and their reciprocals.

e. Describe any general patterns that you find.

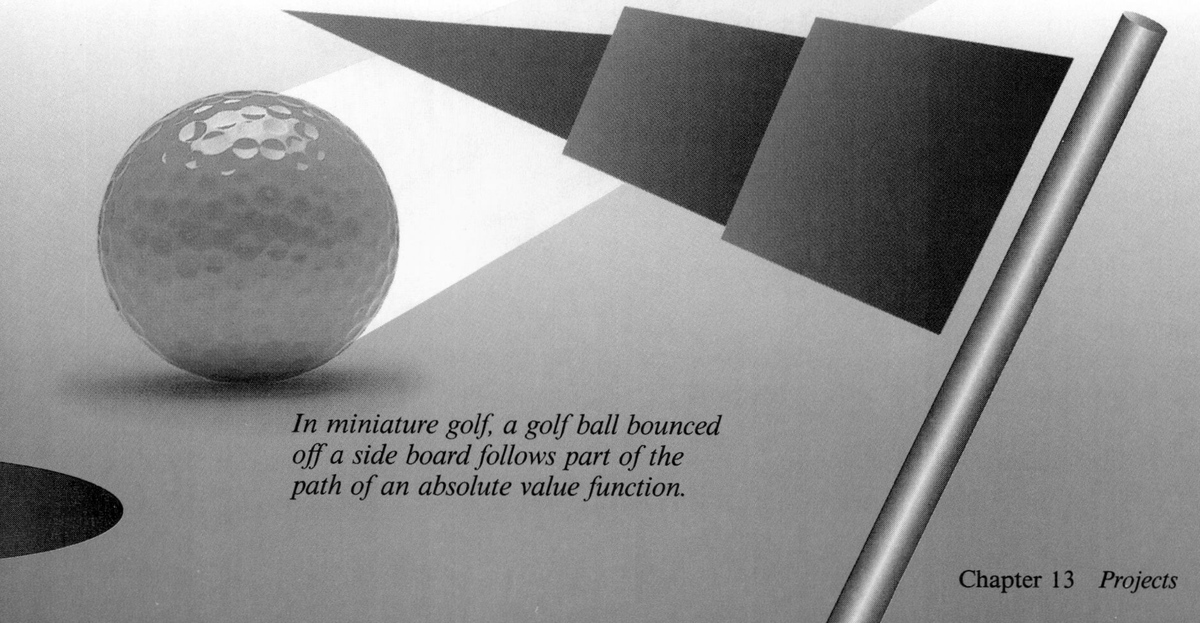

In miniature golf, a golf ball bounced off a side board follows part of the path of an absolute value function.

(continued)

5 Birthday Cubics

Consider the equation $y = ax^3 - bx^2 + cx - d$, and the following code. Let a and b be the last two digits of your birth year. Let c and d be your 2-digit birth month, where January = 01 and December = 12. For instance, if you were born in April 1982, your code is $a = 8$, $b = 2$, $c = 0$, $d = 4$, and your birthday cubic is $y = 8x^3 - 2x^2 - 4$.

a. Graph your birthday cubic. Describe its intercepts and its peaks and valleys.

b. What is true about the graph of all people who share your birth month?

c. All graphs of cubic polynomials have one of the following shapes.

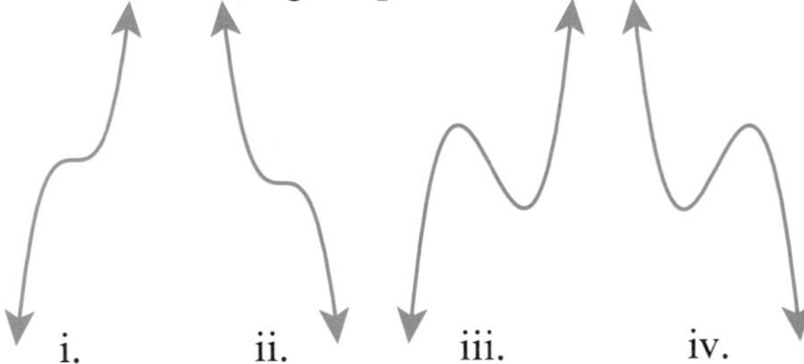

Which shape is your birthday cubic?

d. Make a collage or poster with the birthday cubics of your family or friends. Group them in some way you find interesting. Describe your results.

e. Graph the opposite of your birthday cubic. (The opposite of $f(x) = 8x^3 - 2x^2 - 4$ is $g(x) = -8x^3 + 2x^2 + 4$.) How is its graph related to your birthday cubic?

6 Functions in the Real World

Make a list of at least ten real-life situations in which the value of one variable determines the value of another. Give the domain and the range of each variable. Pick at least two of the situations to be of the type that you can graph, and show the graphs.

7 Functions of Two Variables

Recall the formula $p = ns$ for the perimeter of a regular polygon with n sides in which each side has length s. In this case, p is a function of the two variables n and s. We could write $p = f(n, s) = ns$. There are computer programs that can graph functions of two variables in three dimensions. Find such a program and graph this function and some other functions of your own choosing. Print out the graphs and arrange them in a nice display.

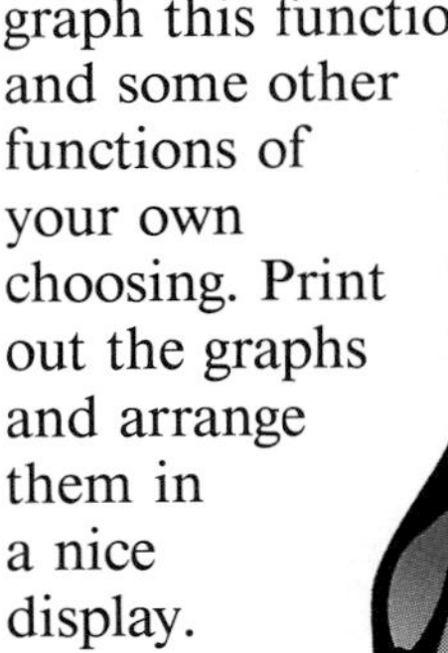

SUMMARY

A function is a set of ordered pairs in which each first coordinate appears with exactly one second coordinate. A function may be described by a graph, a written rule, a list of pairs, or an equation. The key idea in functions is that knowing the first coordinate of a pair is enough to determine the second coordinate. Thus, functions exist whenever one variable determines another.

If a function f contains the ordered pair (a, b), then we write $f(a) = b$. We say that b is the value of the function at a. The set of possible values of a is the domain of the function. The set of possible values of b is the range of the function. If a and b are real numbers, then the function can be graphed on the coordinate plane and values of the function can be approximated by reading the graph. An automatic grapher saves time when investigating complicated functions or multiple functions. Though convenient, it is not necessary to use $f(x)$ notation for functions; y is often used to stand for the second coordinate. Many of the graphs you studied in earlier chapters describe functions.

Equation	Graph	Type of Function
$f(x) = mx + b$	line	linear
$f(x) = ax^2 + bx + c$	parabola	quadratic
$f(x) = ab^x$	exponential curve	exponential
$f(x) = \|x\|$	angle	absolute value
$f(x) = ax^n + bx^{n-1} + \ldots + d$	(varied)	polynomial

In a probability function the domain is a set of outcomes in a situation and the range is the set of probabilities of these outcomes.

An important use of calculators is to evaluate functions at various values of their domains. Trigonometric functions, such as tangent, sine, and cosine, have their own keys on most calculators: tan, sin, cos. The tangent function can be used to determine lengths of sides and measures of angles in right triangles. Other functions on your calculator may include the squaring function x^2, factorial function !, and common logarithmic function log. Computer languages often build in functions such as SQR(X) and ABS(X) in BASIC.

VOCABULARY

Below are the most important terms and phrases for this chapter. You should be able to give a general description and a specific example of each.

Lesson 13-1
function
value of a function
linear function
quadratic function
exponential function
polynomial function

Lesson 13-2
$f(x)$ notation, function notation
squaring function

Lesson 13-3
absolute value function

Lesson 13-4
domain of a function
range of a function

Lesson 13-5
probability function

Lesson 13-6
cubing function
rotation symmetry, center of symmetry

Lesson 13-7
tangent function
tangent of an angle, tan A tan
leg opposite, leg adjacent

Lesson 13-8
sine function, sin A
cosine function, cos A
sinusoidal curve, sinusoid
sin, cos, log
trigonometry
common logarithm function

CHAPTER THIRTEEN

PROGRESS SELF-TEST

Take this test as you would take a test in class. You will need graph paper and a calculator. Then check your work with the solutions in the Selected Answers section in the back of the book.

1. If $f(x) = 3x + 5$, then $f(2) =$ ___?___.
2. Explain in your own words what a function is.
3. Estimate tan 82° to the nearest thousandth.
4. Give the value of sin 30°.
5. What is the tangent of the acute angle formed by the line $y = 4x - 2$ and the x-axis?
6. Give an example of a quadratic function.
7. Explain why the equation $x = |y|$ does not describe a function.
8. If the set $\{(10, 4), (x, 5), (30, 6)\}$ is a function, what values can x not have?

In 9 and 10, tell whether the graph represents a function. If so, give its domain and range. If not, tell why not.

9. **10.**

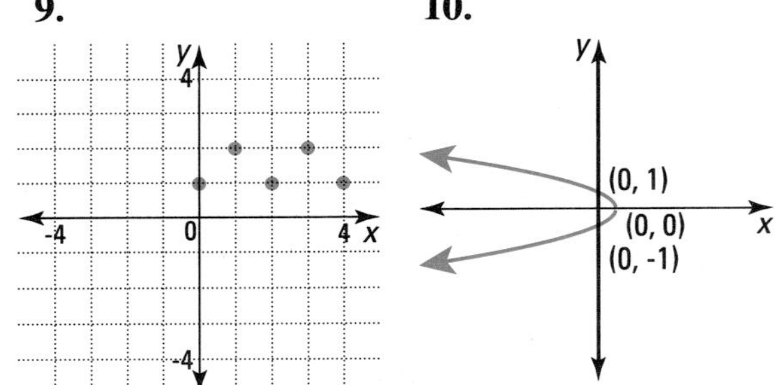

11. In the tossing of two fair dice, what is the probability of obtaining a sum of 10?

In 12 and 13, assume each of ten regions in the circle at the right has the same probability that the spinner will land on it. Let $P(n)$ = the probability of landing in region n.

12. Calculate $P(2)$.
13. Graph the function P.

14. Suppose you are making double decker hamburger sandwiches. Let n = the number of sandwiches to be made. If $b(n)$ = (number of pieces of bread) and $h(n)$ = (number of hamburger patties), when is $b(n) > h(n)$?

In 15 and 16, graph the function for values of x between -4 and 6.

15. $f(x) = 2|x - 3|$ **16.** $g(x) = 10 - x^2$

17. Kristin is having carpet installed. Fabulous Floor's price is determined by the function $f(x) = 11.75x + 300$ and Rapid Rug's price is determined by the function $r(x) = 14.50x$, where x is the number of square yards to be installed. If she needs 220 yards, where should she buy her carpet, and how much will it cost?
18. Use the spreadsheet information to the right to graph $y = x^3 - x^2 - 2x$.

	A	B
1	x	y
2	-2	-8
3	-1.5	-2.625
4	-1	0
5	-.5	.625
6	0	0
7	.5	-1.125
8	1	-2
9	1.5	-1.875
10	2	0
11	2.5	4.375

19. Graph the cubing function and give its domain and range.

CHAPTER REVIEW

Questions on SPUR Objectives

SPUR stands for **S**kills, **P**roperties, **U**ses, and **R**epresentations. The Chapter Review questions are grouped according to the SPUR Objectives for this chapter.

SKILLS DEAL WITH THE PROCEDURES USED TO GET ANSWERS.

Objective A: *Evaluate functions and solve equations involving function notation.* *(Lessons 13-2, 13-3)*

In 1–4, $f(x) = x^2 - 3x + 8$. Calculate.

1. $f(2)$ **2.** $f(3)$ **3.** $f(-7)$ **4.** $f(0)$

5. If $A(t) = 2|t - 5|$, calculate $A(1)$.

6. If $g(n) = 2^n$, calculate $g(3) + g(4)$.

7. If $f(x) = -x$, what is $f(-1.5)$?

8. If $h(t) = 64t - 16t^2$, find $h(4)$.

9. If $f(x) = |x + 3|$, solve $f(x) = 5$.

10. If $A(r) = \pi r^2$, what is r if $A(r) = 18\pi$?

Objective B: *Use function keys on a calculator.* *(Lessons 13-7, 13-8)*

In 11–20, approximate answers to the nearest thousandth.

11. $\frac{1}{12.5}$ **12.** 10!

13. $\sqrt{11469}$ **14.** 0.8^{-3}

15. tan 30° **16.** SQR(6.5)

17. sin 82.4° **18.** log 5

19. ABS(16 − 20) **20.** tan 89.5°

PROPERTIES DEAL WITH THE PRINCIPLES BEHIND THE MATHEMATICS.

Objective C: *Determine whether a set of ordered pairs is a function.* *(Lesson 13-1)*

In 21–26, tell whether or not the sentence determines a function.

21. $y = 300(1.04)^x$ **22.** $x^2 = y$

23. $3x - 5y = 7$ **24.** $y = \tan x$

25. $y^2 = x$ **26.** $y < 3$

In 27–30, tell whether or not the set of ordered pairs is a function.

27. {(0, 1), (1, 2), (2, 3), (3, 4)}

28. {(1, 8), (1, 9), (1, 10), (1, 11)}

29. the set of pairs (students, age) for students in your class

30. the set of pairs (day of the week, date), for the days in this month

Objective D: *Find the domain and the range of a function from its formula, graph, or rule.* *(Lesson 13-4)*

31. If the domain of a function is not given, what should you assume?

32. *True or false.*

a. If $m \neq 0$, the range of the function $f(x) = mx + b$ is the set of all real numbers.

b. If $a \neq 0$, the range of the function $g(x) = ax^2$ is the set of all nonnegative real numbers.

33. *Multiple choice.* The domain of a function is {1, 2, 3}. The range is {4, 5, 6}. Which of these could *not* be a rule for the function?

(a) $y = x + 3$ (b) $y = 7 - x$

(c) $y = x - 3$ (d) $y = |x| + 3$

34. What is the range of the function $A(x) = |x - 2|$?

In 35 and 36, determine the domain and range of the function from its graph.

35.

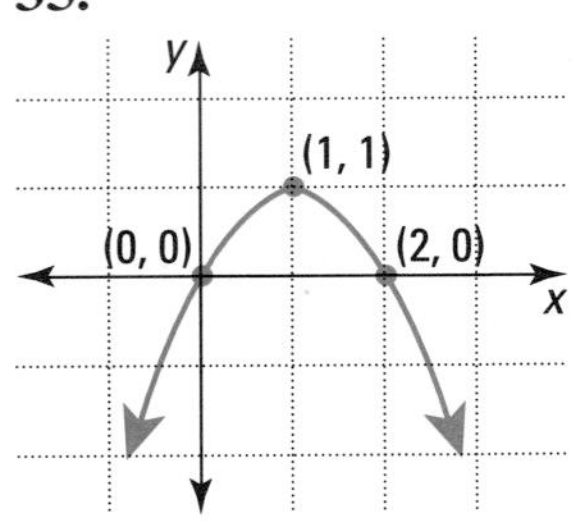

36.

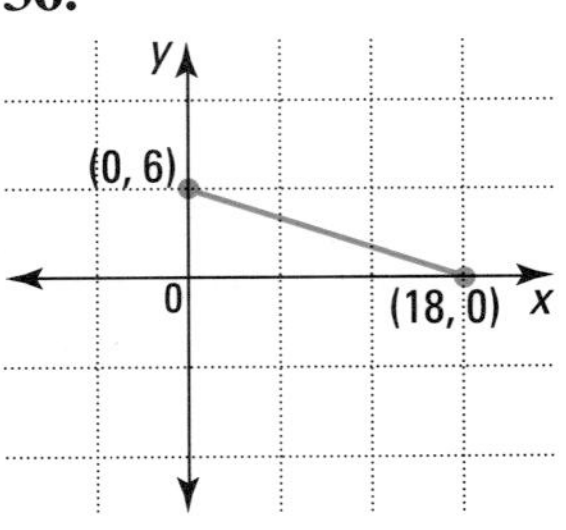

CHAPTER THIRTEEN

37. What is the largest possible value in the range of the function f, where $f(x) = -x^2 + 10$?

38. What is the smallest possible value in the range of the function A with equation $A(n) = 5|n - 3| - 9$?

39. What is the domain of $d(x) = \frac{6x + 2}{2x - 6}$?

40. What is the range of $r(x) = \frac{1}{2}\sqrt{r}$?

USES DEAL WITH APPLICATIONS OF MATHEMATICS IN REAL SITUATIONS.

Objective E: *Use function notation and language in real situations.* *(Lessons 13-2, 13-3, 13-4)*

41. Let $N(t)$ = the number of chirps of a cricket in a minute at a temperature $t°$ Fahrenheit. If $N(t) = \frac{1}{4}t + 37$, for what value of t is $N(t) = 60$?

42. Give the domain and range of this population function for Los Angeles, California.
{(1850, 1610), (1900, 102479), (1950, 1970358), (1960, 2479015), (1970, 2811801), (1980, 2966850), (1990, 3485398)}.

In 43 and 44, Bonita is comparing two health clubs. Daily Workout charges \$100 to join, and \$3.50 per visit. George's Gym charges \$250 to join and \$1 per visit.

43. If $D(v)$ is a function for the cost of joining Daily Workout and making v visits, and $D(v) = 3.5v + 100$, write a function $G(v)$ for joining and making v visits to George's Gym.

44. For how many visits is $G(v) < D(v)$?

45. S is the function that relates the time of day to the number of shoppers in MacGregor's Mart. S(time) = number of shoppers.

a. What is the domain of S?

b. What is the range of S?

46. A jar contains 437 jelly beans. You guess that there are G beans in the jar. Let $f(G)$ be the error in your guess. Find a formula for $f(G)$.

Objective F: *Determine values of probability functions.* *(Lesson 13-5)*

47. If it is equally likely that the spinner below will land in any direction, find each probability.

a. $P(1)$ **b.** $P(2)$ **c.** $P(3)$

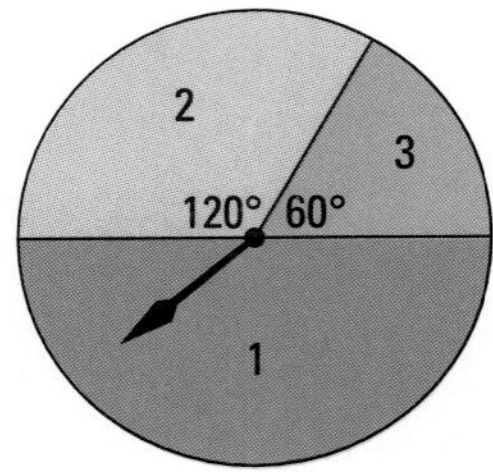

48. What is the probability of tossing a sum of 12 with two fair dice?

49. If you guess on three multiple-choice questions with four choices each, the probability that you will get exactly n correct is $\frac{3!}{n!(3-n)!} \cdot \left(\frac{1}{4}\right)^n \cdot \left(\frac{3}{4}\right)^{3-n}$. Calculate the probability that you will get exactly 2 questions correct.

50. A letter is mailed. Suppose P(the letter arrives the next day) = 0.75. What is P(the letter does not arrive the next day)?

Objective G: *Find lengths of sides or tangents of angles in right triangles using the tangent function.* (Lesson 13-7)

51. Find tan A in the triangle at the right, to the nearest thousandth.

80 m

92 m

A

52. On a field trip, a girl whose eyes are 150 cm above the ground sights a nest high on a pole 25 meters away. If she has to raise her eyes 70° to see the nest, how high is the nest?

53. What is the slope of line ℓ graphed below?

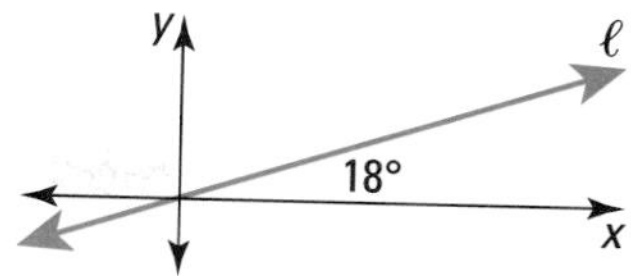

54. A line has equation $4x - 3y = 2$. What is the tangent of the acute angle this line makes with the x-axis?

REPRESENTATIONS DEAL WITH PICTURES, GRAPHS, OR OBJECTS THAT ILLUSTRATE CONCEPTS.

Objective H: *Determine whether or not a graph represents a function.* (Lesson 13-1)

In 55–60, tell whether or not the set of ordered pairs graphed is a function.

55.

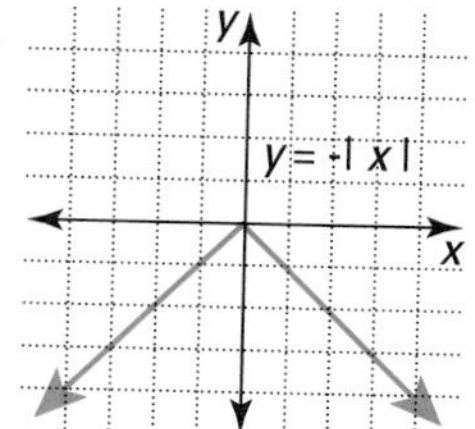

56.

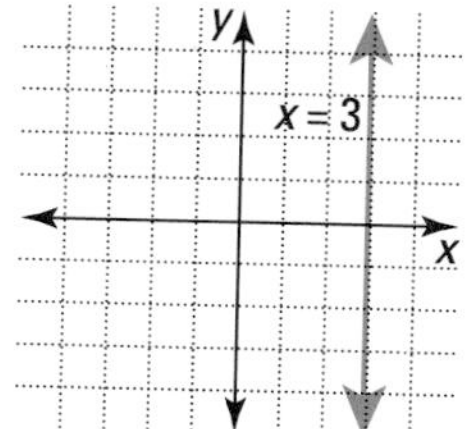

57.

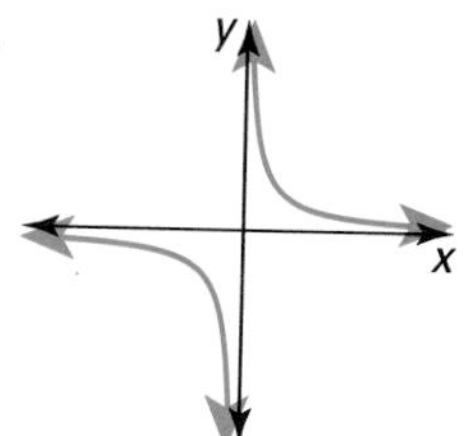

58.

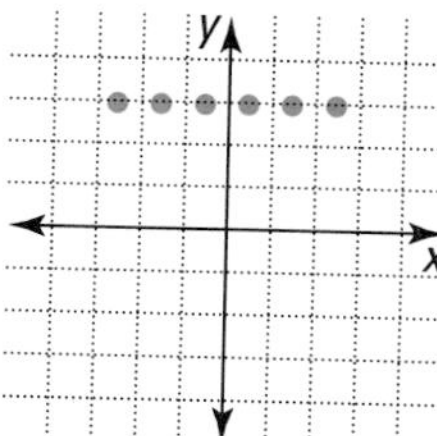

59.

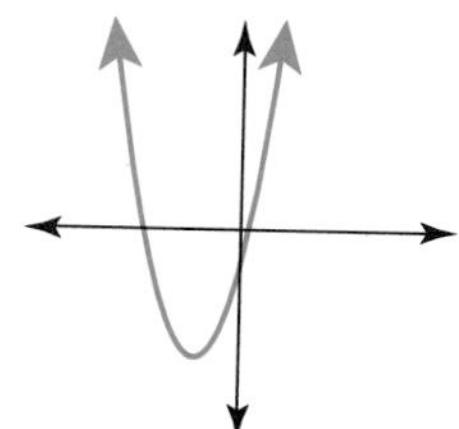

60.

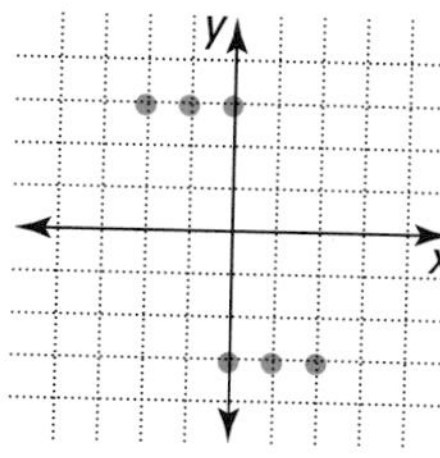

Objective I: *Graph functions.* (Lessons 13-1, 13-2, 13-3, 13-5)

In 61–64, graph each function over the domain $-5 \le x \le 5$.

61. $f(x) = 3|2x + 1|$

62. $g(t) = t^2 - 10$

63. $y = \frac{1}{5}x$

64. $f(n) = 2^n$, n an integer

65. A weighted die has the following probabilities of landing on its sides. $P(1) = 0.12$; $P(2) = 0.19$; $P(3) = 0.09$; $P(4) = 0.21$; $P(5) = 0.15$. Find $P(6)$ and graph the probability function.

66. Graph the probability function for a fair die.

67. Graph the probability function for the number of heads that appear when two fair coins are tossed.

Objective J: *Graph polynomial functions.* (Lesson 13-6)

68. Graph $y = x^3 - 4x^2 + 5$ from $x = -4$ to $x = 4$.

69. Graph $y = x^4 - 5x^2$ from $x = -3$ to $x = 3$.

70. *Multiple choice.* The graph of which equation is symmetric about the origin?
(a) $y = x^2$ (b) $y = x^3$
(c) $y = x^4$ (d) $y = -x^2$

71. a. Use the table at the right to plot the graph of the polynomial function $P(x) = 2x^3 - x^2 - 6x$.

b. Why can you be sure that all the x-intercepts are listed in the table?

x	value
-3	-45
-2	-8
-1.5	0
-1	3
-.5	2.5
0	0
.5	-3
1	-5
1.5	-4.5
2	0
3	27

APPENDIX A

Scientific Calculators

You should be using a scientific calculator throughout this book, so it is important for you to know how to use one. As you read, you should use your calculator to do all the calculations described. Some of the problems are very easy. They were selected so that you can check whether your calculator does the computations in the proper order. Your scientific calculator should follow the order of operations used in algebra.

Suppose you want to use your calculator to find 3 + 4. Here is one way to do it:

	Display shows
Press 3	3
Now press [+]	3
Now press 4	4
Now press [=]	7

Pressing calculator keys is called **entering** or **keying in.** The set of instructions in the left column is called the **key sequence** for this problem. We write the key sequence for this problem using boxes for everything pressed except the numbers.

3 [+] 4 [=]

Sometimes we put what you would see in the calculator display underneath the key presses.

Key sequence:	3	[+]	4	[=]
Display:	3	3	4	7

Some calculators do not have an equal sign, but have a key that enters the calculation [ENTER] or executes it [EXE]. Key sequences for finding 3 + 4 on these calculators are:

3 [+] 4 [ENTER]   3 [+] 4 [EXE]

Next consider 12 + 3 · 5. In the algebraic order of operations, multiplication is performed before addition. Perform the key sequence below on your calculator. See what your calculator does.

Key sequence:	12	[+]	3	[×]	5	[=]
Display:	12	12	3	3	5	27

Different calculators may give different answers even when the same buttons are pushed. If you have a calculator appropriate for algebra, the calculator displayed 27. If your calculator gave you the answer 75, then it has done the addition first and does not follow the algebraic order of operations. Using such a calculator with this book may be confusing.

Example 1

Evaluate $ay + bz$ when $a = 0.05$, $y = 2000$, $b = 0.06$ and $z = 9000$. (This is the total interest in a year if $2000 is earning 5% and $9000 is earning 6%.)

Solution

Key sequence: a [×] y [+] b [×] z [=]

Substitute in the key sequence:

Key sequence:	0.05	[×]	2000	[+]	0.06	[×]	9000	[=]
Display	0.05	0.05	2000	100	0.06	0.06	9000	640

The total interest is $640.

Most scientific calculators have parentheses keys, [( ] and [ )]. To use them just enter the parentheses when they appear in the problem. You may need to use the [×] key every time you do a multiplication, even if × is not in the expression.

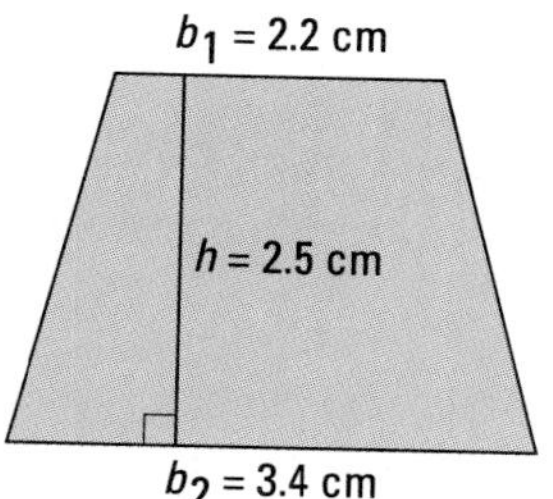

Example 2

Use the formula $A = 0.5h(b_1 + b_2)$ to calculate the area of the trapezoid at the left.

Solution

Remember that $0.5h(b_1 + b_2)$ means $0.5 \cdot h \cdot (b_1 + b_2)$.

Key sequence:	0.5	[×]	h	[×]	[( ]	b_1	[+]	b_2	[ )]	[=]
Substitute:	0.5	[×]	2.5	[×]	[( ]	2.2	[+]	3.4	[ )]	[=]
Display:	0.5	0.5	2.5	1.25	1.25	2.2	2.2	3.4	5.6	7.

The area of the trapezoid is 7 square centimeters.

Some frequently used numbers have special keys on the calculator.

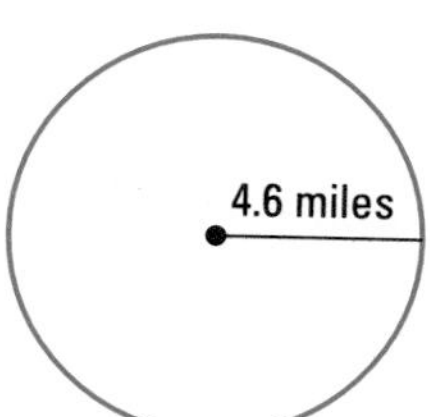

Example 3

Find the circumference of the circle at the left.

Solution

The circumference is the distance around the circle, and is calculated using the formula $C = 2\pi r$, where C = circumference and r = radius. Use the π key.

Key sequence:	2	[×]	[π]	[×]	r	[=]
Substitute:	2	[×]	[π]	[×]	4.6	[=]

Rounding to the nearest tenth, the circumference is 28.9 miles.

As a decimal, $\pi = 3.141592653\ldots$ and the decimal is unending. Since it is impossible to list all the digits, the calculator rounds the decimal. Some calculators, like the one in Example 3, round to the nearest value that can be displayed. Some calculators truncate or round down. If the calculator in the example had truncated, it would have displayed 3.1415926 instead of 3.1415927 for π.

On some calculators you must press two keys to display π. If a small π is written above a key, two keys are probably needed. Then you should press INV, 2nd, or F before pressing the key below π.

Negative numbers can be entered in your calculator. On many calculators this is done with a plus-minus key +/– or ±. Enter -19.

Key sequence:	19	+/–
Display:	19	-19

If your scientific calculator has an opposite key (–), you can enter a negative number in the same order as you write it.

Key sequence:	(–)	19
Display:	–	-19

You will use powers of numbers throughout this book. The scientific calculator has a key y^x (or x^y or ^) used to raise numbers to powers.

The key sequence for 3^4 is	3	y^x	4	=
You should see displayed	3	3	4	81.

This display shows that $3^4 = 81$.

Example 4

A formula for the volume of a sphere is $V = \frac{4\pi r^3}{3}$, where r is the radius. The radius of the moon is about 1080 miles. Estimate the volume of the moon.

Solution

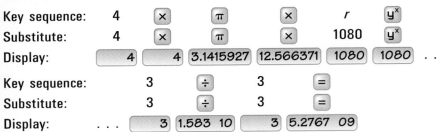

Key sequence:	4	×	π	×	r	y^x
Substitute:	4	×	π	×	1080	y^x
Display:	4	4	3.1415927	12.566371	1080	1080 . . .

Key sequence:	3	÷	3	=
Substitute:	3	÷	3	=
Display: . . .	3	1.583 10	3	5.2767 09

The display shows the answer in scientific notation. If you do not understand scientific notation, read Appendix B.

The volume of the moon is about $5.28 \cdot 10^9$ cubic miles.

Note: You may be unable to use a negative number as a base on your calculator. Try the key sequence 2 [±] [y^x] 5 to evaluate $(-2)^5$. The answer should be -32. However, some calculators will give you an error message. You can, however, use negative *exponents* on scientific calculators.

QUESTIONS

Covering the Reading

1. What is meant by the phrase "keying in"?
2. To calculate $28.5 \cdot 32.7 + 14.8$, what key sequence can you use?
3. Consider the key sequence 13.4 [−] 15 [÷] 3 [=]. What arithmetic problem does this represent?
4. **a.** To evaluate $ab - c$ on a calculator, what key sequence should you use?
   b. Evaluate $297 \cdot 493 - 74{,}212$.
5. Estimate 26π to the nearest thousandth.
6. What number does the key sequence 104 [±] yield?
7. **a.** Write a key sequence for entering -104 divided by -8 on your calculator.
   b. Calculate -104 divided by -8 on your calculator.
8. Calculate the area of the trapezoid below.

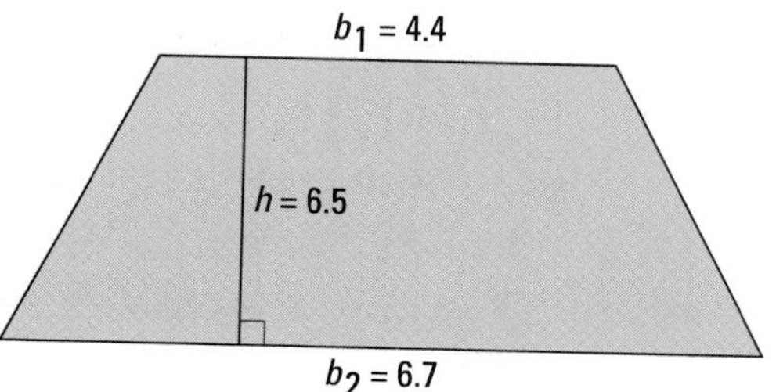

9. Find the circumference of a circle with radius 6.7 inches.
10. Which is greater, $\pi \cdot \pi$ or 10?
11. What expression is evaluated by 5 [y^x] 2 [=]?
12. A softball has a radius of about 1.92 in. What is its volume?
13. What kinds of numbers may not be allowed as bases when you use the [y^x] key on some calculators?

Applying the Mathematics

14. Use your calculator to help find the surface area $2LH + 2HW + 2LW$ of the box below.

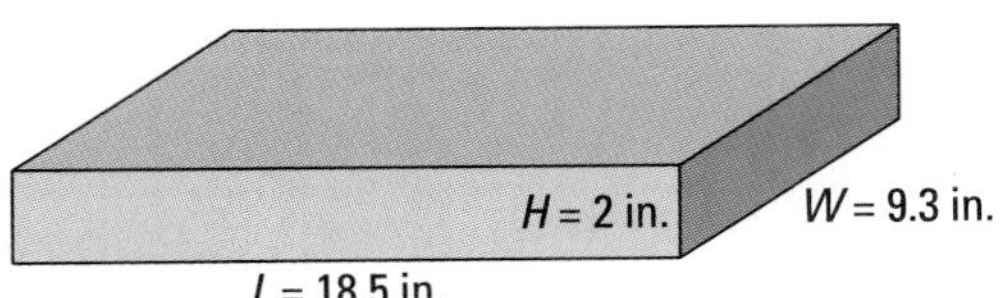

15. Remember that $\frac{2}{3} = 2 \div 3$.

a. What decimal for $\frac{2}{3}$ is given by your calculator?

b. Does your calculator *truncate* or *round to the nearest*?

16. Order $\frac{3}{5}$, $\frac{4}{7}$, and $\frac{5}{9}$ from smallest to largest.

17. Use the clues to find the mystery number y.

Clue 1: y will be on the display if you alternately press 2 and [×] again and again. . . .

Clue 2: $y > 20$.

Clue 3: $y < 40$.

18. $A = \pi r^2$ is a formula for the area A of a circle with radius r. Find the area of the circle in Example 3.

19. What is the total interest in a year if \$350 is earning 5% and \$2000 is earning 8%? (Hint: use Example 1.)

20. To multiply the sum of 2.08 and 5.76 by 2.24, what key sequence can you use?

APPENDIX B

Scientific Notation

The first three columns in the chart below show three ways to represent integer powers of ten: in exponential notation, with word names, and as decimals. The fourth column describes a distance or length in meters. For example, the top row tells that Mercury is about ten billion meters from the sun.

Integer Powers of Ten

Exponential Notation	Word Name	Decimal	Something near this length in meters
10^{10}	ten billion	10,000,000,000	distance of Mercury from Sun
10^{9}	billion	1,000,000,000	radius of Sun
10^{8}	hundred million	100,000,000	diameter of Jupiter
10^{7}	ten million	10,000,000	radius of Earth
10^{6}	million	1,000,000	radius of Moon
10^{5}	hundred thousand	100,000	length of Lake Erie
10^{4}	ten thousand	10,000	average width of Grand Canyon
10^{3}	thousand	1,000	5 long city blocks
10^{2}	hundred	100	length of a football field
10^{1}	ten	10	height of shade tree
10^{0}	one	1	height of waist
10^{-1}	tenth	0.1	width of hand
10^{-2}	hundredth	0.01	diameter of pencil
10^{-3}	thousandth	0.001	thickness of window pane
10^{-4}	ten-thousandth	0.000 1	thickness of paper
10^{-5}	hundred-thousandth	0.000 01	diameter of red blood corpuscle
10^{-6}	millionth	0.000 001	mean distance between successive collisions of molecules in air
10^{-7}	ten-millionth	0.000 000 1	thickness of thinnest soap bubble with colors
10^{-8}	hundred-millionth	0.000 000 01	mean distance between molecules in a liquid
10^{-9}	billionth	0.000 000 001	size of air molecule
10^{-10}	ten-billionth	0.000 000 000 1	mean distance between molecules in a crystal

You probably know the quick way to multiply by 10, 100, 1000, and so on. Just move the decimal point as many places to the right as there are zeros.

$$84.3 \cdot 100 = 8430 \qquad 84.3 \cdot 10{,}000 = 843{,}000$$

It is just as quick to multiply by these numbers when they are written as powers.

$$489.76 \cdot 10^2 = 48{,}976 \qquad 489.76 \cdot 10^4 = 4{,}897{,}600$$

The general pattern is as follows.

To multiply by 10 raised to a positive power, move the decimal point to the *right* as many places as indicated by the exponent.

The patterns in the chart on the previous page also help to explain powers of 10 where the exponent is negative. Each row describes a number that is $\frac{1}{10}$ of the number in the row above it. So 10^0 is $\frac{1}{10}$ of 10^1.

$$10^0 = \frac{1}{10} \cdot 10 = 1$$

To see the meaning of 10^{-1}, think: 10^{-1} is $\frac{1}{10}$ of 10^0 (which equals 1).

$$10^{-1} = \frac{1}{10} \cdot 1 = \frac{1}{10} = .1$$

Remember that to multiply a decimal by 0.1, just move the decimal point one unit to the left. Since $10^{-1} = 0.1$, to multiply by 10^{-1}, just move the decimal point one unit to the left.

$$435.86 \cdot 10^{-1} = 43.586$$

To multiply a decimal by 0.01, or $\frac{1}{100}$, move the decimal point two units to the left. Since $10^{-2} = 0.01$, the same goes for multiplying by 10^{-2}.

$$435.86 \cdot 10^{-2} = 4.3586$$

The following pattern emerges.

To multiply by 10 raised to a negative power, move the decimal point to the *left* as many places as indicated by the exponent.

Example 1

Write $68.5 \cdot 10^{-6}$ as a decimal.

Solution

To multiply by 10^{-6}, move the decimal point six places to the left. So $68.5 \cdot 10^{-6} = 0.0000685$.

The names of the negative powers are very similar to those for the positive powers. For instance, 1 billion = 10^9 and 1 billionth = 10^{-9}.

Example 2

Write 8 billionths as a decimal.

Solution

8 billionths = $8 \cdot 10^{-9} = 0.000000008$

Most calculators can display only the first 8, 9, or 10 digits of a number. This presents a problem if you need to key in a large number like 455,000,000,000 or a small number like 0.00000000271. However, powers of 10 can be used to rewrite these numbers in **scientific notation.**

$$455{,}000{,}000{,}000 = 4.55 \cdot 10^{11}$$
$$0.00000000271 = 2.71 \cdot 10^{-9}$$

Definition: In scientific notation, a number is represented as $x \cdot 10^n$, where $1 \le x < 10$ and n is an integer.

Scientific calculators can display numbers in scientific notation. The display for $4.55 \cdot 10^{11}$ will usually look like one of these shown here.

4.55 E 11   4.55 11   4.55 x10 11

The display for $2.71 \cdot 10^{-9}$ is usually one of these

2.71 E -09   2.71 -09   2.71 x10 -09

Numbers written in scientific notation are entered into a calculator using the EXP or EE key. For instance, to enter $6.0225 \cdot 10^{23}$ (known as Avogadro's number), key in

6.0225 EE 23.

You should see this display.

6.0225 23

In general, to enter $x \cdot 10^n$, key in x EE n.

Example 3

The total number of hands possible in the card game bridge is about 635,000,000,000. Write this number in scientific notation.

Solution

Move the decimal point to get a number between 1 and 10. In this case the number is 6.35. This tells you the answer will be:

$$6.35 \cdot 10^{\text{exponent}}.$$

The exponent of 10 is the number of places you must move the decimal point in 6.35 in order to get 635,000,000,000. You must move it 11 places to the right, so the answer is $6.35 \cdot 10^{11}$.

Example 4

The charge of an electron is 0.00000000048 electrostatic units. Put this number in scientific notation.

Solution

Move the decimal point to get a number between 1 and 10. The result is 4.8. To find the power of 10, count the number of places you must move the decimal to change 4.8 to 0.00000000048. The move is 10 places to the left, so the charge of the electron is $4.8 \cdot 10^{-10}$ *electrostatic units.*

Example 5

Enter 0.00000000123 into a calculator.

Solution

Rewrite the number in scientific notation.

$0.00000000123 = 1.23 \cdot 10^{-9}$.

Key in 1.23 EE 9 +/− or 1.23 EE 9 (−).

QUESTIONS

Covering the Reading

1. Write one million as a power of ten.

2. Write 1 billionth as a power of 10.

In 3–5, write as a decimal.

3. 10^{-4} **4.** $28.5 \cdot 10^{7}$ **5.** 10^{0}

6. To multiply by a negative power of 10, move the decimal point to the __?__ as many places as indicated by the __?__.

7. Write $2.46 \cdot 10^{-8}$ as a decimal.

8. Why is $38.25 \cdot 10^{-2}$ not in scientific notation?

9. Suppose $x \cdot 10^{y}$ is in scientific notation.
a. What is the domain of x? **b.** What is the domain of y?

In 10–14, rewrite the number in scientific notation.

10. 5,020,000,000,000,000,000,000,000,000 tons, the mass of Sirius, the brightest star

11. 0.0009 meters, the approximate width of a human hair

12. 763,000 **13.** 0.00000328 **14.** 754.9876

15. One computer can do an arithmetic problem in $2.4 \cdot 10^{-9}$ seconds. What key sequence can you use to display this number on your calculator?

Applying the Mathematics

In 16 and 17, write in scientific notation.

16. 645 billion

17. 27.2 million

In 18–21, use the graph below. Write the estimated world population in the given year: **a.** as a decimal; **b.** in scientific notation.

18. 10,000 B.C.

19. 1 A.D.

20. 1700

21. 1970

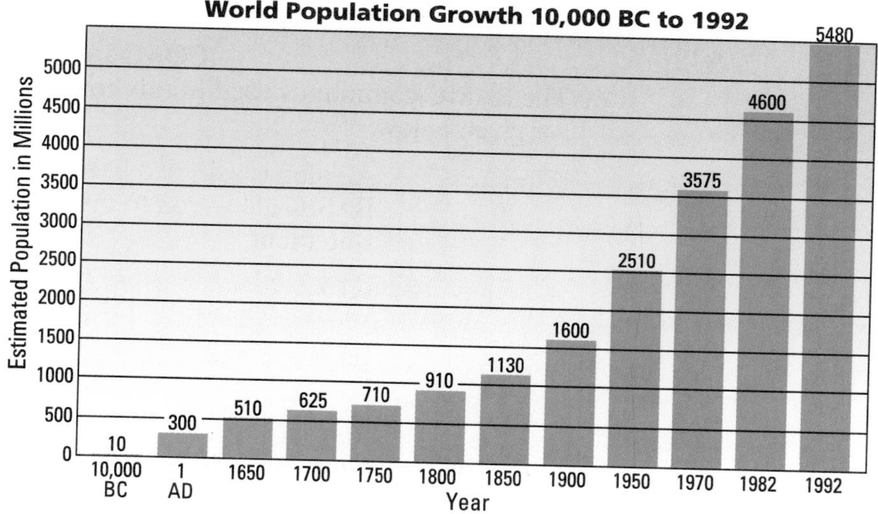

22. How can you enter the world population in 1992 into your calculator?

23. How many digits are in $1.7 \cdot 10^{100}$?

In 24–26, write the number in scientific notation.

24. 0.00002

25. 0.0000000569

26. $400.\dot{0}07$

In 27–29, write as a decimal.

27. $3.921 \cdot 10^{5}$

28. $3.921 \cdot 10^{-5}$

29. $8.6 \cdot 10^{-2}$

Exploration

30. **a.** What is the largest number you can display on your calculator?
b. What is the smallest number you can display? (Use scientific notation and consider negative numbers.)
c. Find out what key sequence you could use to enter -5×10^{-7} in your calculator.

APPENDIX C

BASIC

In BASIC (Beginner's All Purpose Symbolic Instruction Code), the arithmetic symbols are: $+$ (for addition), $-$ (for subtraction), $*$ (for multiplication), / (for division), and ^ (for powering). In some versions of BASIC, $\uparrow$ is used for powering. The computer evaluates expressions according to the usual order of operations. Parentheses ( ) may be used. The comparison symbols $=$, $>$, $<$ are also used in the standard way, but BASIC uses $<=$ instead of $\leq$, $>=$ instead of $\geq$, and $<>$ instead of $\neq$.

Variables are represented by letters or letters in combination with digits. Consult the manual for your version of BASIC for restrictions on the length or other aspects of variable names. Examples of variable names allowed in most versions are N, X1, and AREA.

COMMANDS

The BASIC commands used in this course and examples of their uses are given below.

LET . . . A value is assigned to a given variable. Some versions of BASIC allow you to omit the word LET in the assignment statement.

LET X = 5	The number 5 is stored in a memory location called X.
LET N = N + 2	The value in the memory location called N is increased by 2 and the result is stored in the location N.

PRINT . . . The computer prints on the screen what follows the PRINT command. If what follows is not in quotes, it is a constant or variable, and the computer will print the value of that constant or variable. If what follows is in quotes, the computer prints exactly what is in quotes.

PRINT X	The computer prints the number stored in memory location X.
PRINT "X-VALUES"	The computer prints the phrase X-VALUES.

INPUT . . . The computer asks the user to give a value to the variable named and stores that value.

INPUT X	When the program is run, the computer will prompt you to give X a value by printing a question mark, and then will store that value in memory location X.
INPUT "HOW OLD?";AGE	The computer prints HOW OLD? and stores your response in memory location AGE. (Note: Some computers will not print the question mark.)

REM . . . REM stands for *remark.* This command allows remarks to be inserted in a program. These may describe what the variables represent, what the program does, or how the program works. REM statements are often used in long complex programs or programs other people will use.

```
REM PYTHAGOREAN THEOREM
```

A statement that begins with REM has no effect when the program is run.

FOR . . .
NEXT . . .
STEP . . . FOR and NEXT are used when a set of instructions must be performed more than once, a process which is called a *loop.* The FOR command assigns a beginning and ending value to a variable. The first time through the loop, the variable has the beginning value in the FOR command. When the computer hits the line reading NEXT, the value of the variable is increased by the amount indicated by STEP. The commands between FOR and NEXT are then repeated. When the value of the incremented variable is larger than the ending value in the FOR command, the computer leaves the loop and executes the rest of the program. If STEP is not written, the computer increases the variable by 1 each time through the loop.

```
10 FOR N = 3 TO 6 STEP 2
20 PRINT N
30 NEXT N
40 END
```

The computer assigns 3 to N and then prints the value of N. On reaching NEXT, the computer increases N by 2 (the STEP amount) and prints 5. The next N would be 7 which is too large. The computer executes the command after NEXT, ending the program.

IF . . . THEN . . . The computer performs the consequent (the THEN part) only if the antecedent (the IF part) is true. When the antecedent is false, the computer *ignores* the consequent and goes directly to the next line of the program.

```
IF X > 100 THEN END
PRINT X
```

If the X value is less than or equal to 100, the computer ignores END, goes to the next line, and prints the value stored in X. If the X value is greater than 100, the computer stops and the value stored in X is not printed.

GOTO . . .	The computer goes to whatever line of the program is indicated. GOTO statements are generally avoided because they interrupt program flow and make programs hard to interpret.	
	GOTO 70	The computer goes to line 70 and executes that command.
END . . .	The computer stops running the program. No program should have more than one END statement.	

FUNCTIONS

The following built-in functions and many others are available in most versions of BASIC. Each function name must be followed by a variable or a constant enclosed in parentheses.

ABS	The absolute value of the number that follows is calculated.	
	LET X = ABS(-10)	The computer calculates $\|-10\| = 10$ and assigns the value 10 to memory location X.
SQR	The square *root* of the number that follows is calculated.	
	C = SQR(A*A + B*B)	The computer calculates $\sqrt{A^2 + B^2}$ using the values stored in A and B and stores the result in C.

PROGRAMS

A program is a set of instructions to the computer. In most versions of BASIC, every step in the program must begin with a line number. We usually start numbering at 10 and count by ten, so intermediate steps can be added later. The computer reads and executes a BASIC program in order of the line numbers. It will not go back to a previous line unless told to do so.

To enter a new program, type NEW, and then type the lines of the program. At the end of each line press the key named RETURN or ENTER. You may enter the lines in any order. The computer will keep track of them in numerical order. If you type LIST, the program currently in the computer's memory will be printed on the screen. To change a line, retype the line number and the complete line as you now want it.

To run a new program after it has been entered, type RUN, and then press the RETURN or ENTER key.

Programs can be saved on disk. Consult your manual on how to do this for your version of BASIC. To run a program already saved on disk you must know the exact name of the program including any spaces or punctuation. To run a program called TABLE SOLVE, type RUN "TABLE SOLVE" and press the RETURN or ENTER key.

The following program illustrates many of the commands used in this course.

10 PRINT "A DIVIDING SEQUENCE"	The computer prints A DIVIDING SEQUENCE.
20 INPUT "NUMBER PLEASE?";X	The computer prints NUMBER PLEASE? and waits for you to enter a number. You must give a value to store in the location X. Suppose you use 20. X now contains 20.
30 LET Y = 2	2 is stored in location Y.
40 FOR Z = -5 TO 4	Z is given the value -5. Each time through the loop, the value of Z will be increased by 1.
50 IF Z = 0 THEN GOTO 70	When Z = 0 the computer goes directly to line 70. When Z ≠ 0 the computer executes line 60.
60 PRINT X" TIMES "Y " DIVIDED BY "Z" = " (X*Y)/Z	On the first pass through the loop, the computer prints -8 because (20 · 2)/(-5) = -8.
70 NEXT Z	The value in Z is increased by 1 to -4 and the computer goes back to line 50.
80 END	After going through the FOR . . . NEXT . . . loop with Z = 4, the computer stops.

The output of this program is:

```
A DIVIDING SEQUENCE
NUMBER PLEASE? 20
20  TIMES 2 DIVIDED BY -5 = -8
20  TIMES 2 DIVIDED BY -4 = -10
20  TIMES 2 DIVIDED BY -3 = -13.3333
20  TIMES 2 DIVIDED BY -2 = -20
20  TIMES 2 DIVIDED BY -1 = -40
20  TIMES 2 DIVIDED BY 1 = 40
20  TIMES 2 DIVIDED BY 2 = 20
20  TIMES 2 DIVIDED BY 3 = 13.3333
20  TIMES 2 DIVIDED BY 4 = 10
```

GETTING STARTED (pp. 1–3)

7. a. 3 **b.** scientific calculators, scientific notation, BASIC **9.** page 841 in the back of the book. The answers to the odd-numbered questions in the sections Applying the Mathematics and the Review are given. **11.** Skills, Properties, Uses, Representations **13.** Viète, a French lawyer of the late 16th century, invented the use of letters to describe arithmetic patterns.

LESSON 1-1 (pp. 6–10)

19. $\frac{2}{3} < \frac{7}{10} < \frac{3}{4}$ **21.** 2 and 4 **23. a.** $y < 1990$ **b.** $p > 2.1$ **25.** $-4 + 7 = 3$ **27.** -4 **29.** Sample: 0.3333 **31.** 3

LESSON 1-2 (pp. 11–16)

21. a. closed **b.** x is greater than or equal to zero and less than or equal to ten. **c.** $0 \le x \le 10$ **23. a.** neither open nor closed **b.** z is greater than or equal to negative ten and less than negative four. **c.** $-10 \le z < -4$ **25.** {6, 9} **27.** $y > -15$ **29.** $x = 256$ **31.** -15 **33.** -350 **35.** $350

LESSON 1-3 (pp. 17–22)

13. a. {3, 9} **b.** {1, 3, 5, 6, 7, 9} **15. See below. 17. See below.** **19.** True **21.** True **23.** 4; 8; 12; Sample pattern: $-4 \cdot n = -(4 \cdot n)$

15.

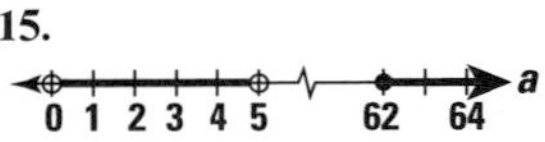

17.

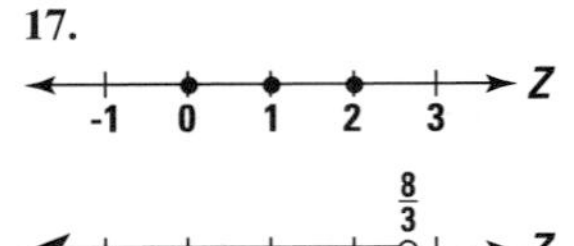

LESSON 1-4 (pp. 23–26)

17. 0 **19.** ≈ 10.9 mph **21.** {2} **23. a.** $d > 1800$ **b. See below.** **25.** $1\frac{11}{16}$ yd **27. a.** 8 sq units **b.** 4 complete squares plus 8 half-squares equals 8.

23. b. d: 1600 1800 2000

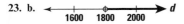

LESSON 1-5 (pp. 27–30)

9. 23 cm **11.** 87°F **13. a.** line 40: L * W; line 50: 2 * L + 2 * W **b.** area = 1995; perimeter = 181 **15. a.** + **b.** / **c.** * **d.** ^ **17.** $q < -36$, $q > 12$ **19. a.** 0 **b.** Sample: -1 **c.** Sample: $\frac{1}{2}$ **21.** $1\frac{3}{4}$ cups

LESSON 1-6 (pp. 31–36)

19. $\sqrt{25} = 5$ and $\sqrt{36} = 6$, so $5 < \sqrt{32} < 6$ **21. a.** 18 **b.** 6 **c.** 12 **23.** 2.5 seconds **25. a.** $A = \$1030$ **b.** $A = \$1045$ **27.** 3

29. See below. **31. a.** positive real numbers **b.** $2 \le w \le 6$ **c. See below.** **33. a.** 2 **b.** $\frac{10}{9}$ **c.** $\frac{38}{39}$

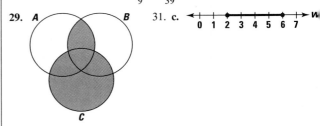

29.

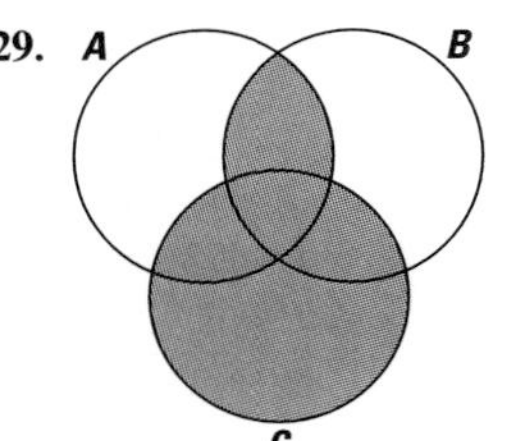

31. c. w: 0 1 2 3 4 5 6 7

LESSON 1-7 (pp. 37–43)

13. a. L heads of lettuce and T tomatoes cost $L \cdot 89¢ + T \cdot 24¢$. **b.** Sample: 10 heads of lettuce and 4 tomatoes cost $10 \cdot 89¢ + 4 \cdot 24¢$. **c.** $C = .89L + .24T$ **15. a.** Sample: $10(11 - 0) = 10 \cdot 11 - 0$, true $110 = 110$, $10(4 - 4) = 10 \cdot 4 - 4$, false $0 \ne 3$ $10(0 - 10) = 10 \cdot 0 - 10$, false $-100 \ne -10$ **b.** No; counterexample can be found. **17. a.** $n \cdot n > n$ **b.** Sample: $-8 \cdot -8 > -8$ **c.** Sample: $0 \cdot 0 = 0$ **d.** Sample: $\frac{1}{2} \cdot \frac{1}{2} < \frac{1}{2}$ **19. a.** ⁝⁝⁝⁝⁝⁝ **b.** Sample: The first design has 3 pennies, the second design has $3 \cdot 2$ pennies, the third design has $3 \cdot 3$ pennies, and so on. **c.** $20 \cdot 3$ or 60 pennies **21.** 20 **23. a.** ≈ 54.8 feet **b.** Sample: 60 feet by 50 feet **25.** 176 pounds **27. a.** Sample: 4 **b.** Sample: **c.** 3 **29.** $1.245 \le d \le 1.255$

LESSON 1-8 (pp. 45–51)

13. a. 6 km **b.** ≈ 4.5 km **c.** ≈ 1.5 km **15. a.** $\sqrt{12} \approx 3.5$ cm **b.** See students' drawings. The hypotenuse ≈ 3.5 cm **c.** Answers will vary. Sample: Both answers are close to 3.5 cm. **17.** > **19.** 41.4 meters **21.** $3\frac{1}{2}$ **23.** $\frac{5}{6}$ **25.** ≈ $4.50

LESSON 1-9 (pp. 53–59)

11. a.

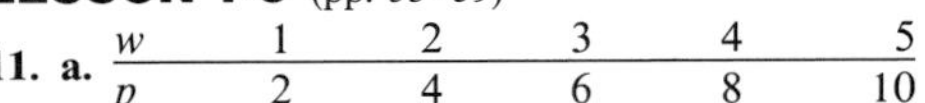

w	1	2	3	4	5
p	2	4	6	8	10

b. The total number of panes in a design is equal to twice the number of panes in one row. **c.** $p = 2w$ **13. a.** 2 **b.** 4

c.

n	1	2	3	4
t	2	4	8	16

d. $t = 2^n$ **e.** 512 **15.** (b) **17.** $\sqrt{45}$ **19. a.** $37.50 **b.** $5.00a + 1.50c$ **21.** ≈ 5.7 kilograms **23. a.** 0 **b.** -6 **c. See below** **d.** No. Between any two real numbers, there are always an infinite number of real numbers.

23. c.

CHAPTER 1 PROGRESS SELF-TEST (pp. 63–64)

1. When $a = 3$ and $b = 5$, $2(a + 3b) = 2(3 + 3 \cdot 5) = 2\ (18) = 36$ **2.** When $n = 4$, $5 \cdot 6^n = 5 \cdot 6^4 = 5 \cdot 1296 = 6480$ **3.** When $p = 5$ and $t = 2$, $\frac{p + t^2}{p - t} = \frac{5 + 2^2}{5 - 2} = \frac{9}{3} = 3$ **4.** $(\sqrt{50})^2 = \sqrt{50} \cdot \sqrt{50} = 50$ **5.** 10 * 3 ^ 2 + 5 = 10 * 9 + 5 = 90 + 5 = 95 **6.** By calculator, $3\sqrt{42} = 19.442221 \approx 19.4$ **7. a.** {2, 3, 4, 6, 8, 9, 10, 12, 14, 15, 16} **b.** {6, 12} **8.** S > 25 **9.** (a) **10.** $4 \cdot 2 + 7 \ne 2 \cdot 2 + 23$; $4 \cdot 5 + 7 \ne 2 \cdot 5 + 23$; $4 \cdot 8 + 7 = 2 \cdot 8 + 23$; so 8 is a solution. **11.** Samples: 6, 7, 7.9 **12.** $C = 23(4 - 1) + 29$, $C = 98¢$

13. $A = 3.14159 \cdot 3^2 = 28.27431\ m^2 \approx 28\ m^2$ **14.** False, $\sqrt{100} + \sqrt{36} = 10 + 6 = 16 \ne \sqrt{136}$ **15.** 13 **16.** Sample: n tick[ets] cost $\$3.50 \cdot n$. **17.** Sample: $\frac{3}{5} - \frac{2}{5} = \frac{3 - 2}{5}$; $\frac{12}{5} - \frac{2}{5} = \frac{12 - 2}{5}$; $\frac{1.9}{5} - \frac{6.13}{5} = \frac{1.9 - 6.13}{5}$ **18.** $y = 8x$ **19. See below.** **20. a.** $x \ge 1$[…] **b.** Answers will vary. **21. See below.** **22.** $\ell^2 = 45^2 + 10^2$, $\ell^2 = 2025 + 100$, $\ell^2 = 2125$, $\ell \approx 46.1$ ft **23.** $\sqrt{193600} = 440$ yar[ds] **24.** 25 in.

19.

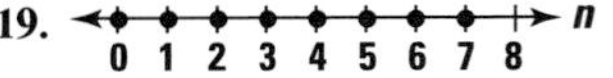

21. x: -4 -3 -2 -1 0 1 2 3 4 5 6

The chart below keys the **Progress Self-Test** questions to the objectives in the **Chapter Review** on pages 65–68. This will enable you to locate those **Chapter Review** questions that correspond to questions missed on the **Progress Self-Test.** The lesson where the material is covered is also indicated on the chart.

Question	**1**	**2**	**3**	**4**	**5**	**6**	**7**	**8**	**9**	**10**	**11**	**12**	**13**	**14**
Objective	C	C	C	F	C	D	B	I	I	A	A	J	J	D
Lesson	1-4	1-4	1-4	1-6	1-4	1-6	1-3	1-2	1-2	1-1	1-1	1-5	1-5	1-6
Question	**15**	**16**	**17**	**18**	**19**	**20**	**21**	**22**	**23**	**24**				
Objective	E	H	G	H	L	L	L	K	K	D				
Lesson	1-2	1-7	1-7	1-9	1-2	1-2	1-2	1-8	1-8	1-6				

CHAPTER 1 REVIEW (pp. 65–68)

1. 4 **3.** Sample: −5, −4.2, −3 **5.** 3, −3 **7. a.** {15, 25} **b.** {10, 11, 15, 19, 20, 23, 25, 30} **9. a.** {−1, 0, 1, 2, 3, . . .} **b.** {0, 1, 2} **11. a.** $8\frac{3}{5}$ **b.** 11 **13.** 529 **15.** 576 **17.** 3 **19.** 9 **21.** 30 **23.** False; $5 + 2 \neq \sqrt{29}$ **25.** 4 and 5 **27.** 14.107 **29.** 50 **31. a.** empty or null set **b.** Sample: the set of integers between −3 and −2 **33.** 7 **35.** 39 **37.** Sample: 2 + 2 = 2 · 2; −3 + −3 = 2 · −3; 4.9 + 4.9 = 2 · 4.9 **39.** Sample: 9 = 4.5 · 2; 36 = 4.5 · 8 **41.** Sample: n sheep have $n \cdot 4$ legs. **43. a.** 21,000 people **b.** 2100y people **45.** $y = 4^x$ **47.** (c) **49.** (d) **51.** 25% **53.** $399 **55.** 10 ft **57.** ≈ 1.78 ft or about 2 ft **59. See right. 61. See right. 63. See right.** **65.** (b) **67. a.** $n \geq 18$ **b.** Sample: A U.S. citizen may vote when he or she is at least 18 years old.

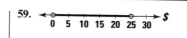

59. 0 5 10 15 20 25 30 $

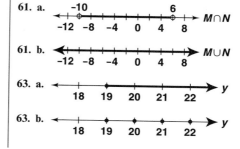

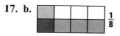

61. a. −10 6 M∩N
−12 −8 −4 0 4 8

61. b. M∪N
−12 −8 −4 0 4 8

63. a. 18 19 20 21 22 y

63. b. 18 19 20 21 22 y

REFRESHER (p. 69)

1. 15.087 **3.** 0.00666 **5.** 20 **7.** $\frac{1}{6}$ **9.** $2\frac{21}{32}$ **11.** $180 **13.** 27,000 cartons **15.** 402 miles **17.** −6 **19.** −24 **21.** 0 **23.** 3540 **25.** $x = 4$ **27.** $z = \frac{1}{2}$ **29.** $a = \frac{3}{25}$ **31.** $c = \frac{1}{2}$ **33.** 180 in² **35.** 8 m² **37.** 96 cm³ **39.** $\frac{1}{8}$ ft³ or 0.125 ft³

LESSON 2-1 (pp. 72–78)

13. a. No **b.** Yes **c.** No **d.** Sample: Washing your hair followed by drying your hair; not commutative. **15.** $240x^2$ **17. a.** kn **b.** $k + k + n + n$ or $2k + 2n$ **19.** 576 m² **21.** 180 in³ **23.** Sample: Area is a measure of the amount of 2-dimensional surface; volume is a measure of 3-dimensional space. The amount of space in a box is its volume; the size of one of its sides is an area. **25.** $2L$ **27.** $\frac{3}{10}$

LESSON 2-2 (pp. 79–83)

17. (c) **19. a.** reciprocals **b.** 200 · 0.005 = 1 **21. a.** reciprocals **b.** $1.5 = \frac{3}{2}$; $\frac{3}{2} \cdot \frac{2}{3} = 1$ **23.** $\frac{5}{2}$ or $2\frac{1}{2}$ times **25.** 0 **27.** $\frac{q}{p}$ **29. a.** 3080 ft² **b.** 513 people **31. a.** $10s^3$ **b.** 10 **c. See below.**

31. c. Sample: The rectangle is two cubes deep and 5 cubes high, for a total of 10 cubes.

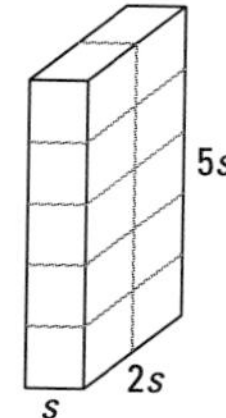

LESSON 2-3 (pp. 85–90)

17. a. One area is $\frac{1}{8}$ of the other. **b. See below.** **19. a.** 5 **b.** 9 **c.** a **d.** a **21.** (c) **23.** 1 **25.** $\frac{21xy}{5}$ **27.** Sample: $\frac{12x}{y^2} \cdot \frac{x}{5y}$ **29.** 280 **31.** 3025 ft²

17. b. $\frac{1}{8}$

LESSON 2-4 (pp. 91–95)

11. a. 2880 ounces **b.** 288c ounces **13.** Size 32. $80 \text{ cm} \cdot \frac{1 \text{ in.}}{2.54 \text{ cm}} \approx 31.5$ in. The size closest to 80 cm is 32 in. **15. a.** $2k \frac{\text{dishes}}{\text{minute}}$ **b.** $\frac{1}{2k} \frac{\text{minutes}}{\text{dish}}$ **17.** 5 days **19.** Sample: If you can read magazines at a rate of 120 minutes per magazine and each magazine has 40 pages, what is your speed in minutes per page? **21.** (d) **23.** Sample: $\frac{8}{15} \cdot \frac{x}{y} = \frac{8x}{15y}$ **25. a.** 0 **b.** Multiplication Property of Zero **27. a.** x^3 **b.** Sample $3x \cdot x \cdot \frac{x}{3}$; $16x \cdot \frac{x}{4} \cdot \frac{x}{4}$

LESSON 2-5 (pp. 96–101)

17. $9x^2$ **19.** $-32a^5$ **21. a.** positive **b.** negative **c.** positive **d.** positive **e.** zero **f.** negative **23.** No **25.** Yes **27.** ≈ .14 km **29.** $\frac{9}{7}a$

LESSON 2-6 (pp. 102–108)
17. a. 30.48 cm **b.** Multiply both sides by 12 to get $12 \cdot 1$ in. $=$ $12 \cdot 2.54$ cm. **19.** $x = 0.13$; $6.5 = 5(1.3)$; $6.5 = 6.5$ **21.** 8 cm **23.** $a = \frac{F}{m}$ **25. a.** positive **b.** negative **c.** zero **27.** 4.5 yd $\cdot$ $36 \frac{\text{in.}}{\text{yd}} \cdot 2.54 \frac{\text{cm}}{\text{in.}} = 411.48$ cm **29.** *DP* **31. a.** 660 ft^2 **b.** Sample: $14 \cdot 6 + 24 \cdot 10 + 14 \cdot 24 = 660$; and $32 \cdot 30 - 14 \cdot 8 - 8 \cdot 10 - 18 \cdot 6 = 660$ square feet **c.** 1320 ft^2 **d.** about 25.7 ft × 25.7 ft

LESSON 2-7 (pp. 109–113)
13. *N* would be 0. **15.** Sample: You are traveling 13 mph. How long does it take you to go 0 miles? **17.** $x = -15$; $(6 - 7)(-15) = (-1)(-15) = 15$ **19. a.** -3; $-4(-3) = 12$ **b.** $-\frac{1}{3}$; $12(-\frac{1}{3}) = \frac{-12}{3} = -4$ **21.** $\frac{10}{3}$ **23.** $\frac{1}{12}$ **25.** Sample: $8x = c$ **27.** 15 hr **29.** 7608 m^2 **31. a.** Yes **b.** Yes **c.** Yes **d.** No

LESSON 2-8 (pp. 114–118)
17. more than 31, or at least 32 rows **19.** $m > -8$ **21.** $x \geq 384$ **23.** $x < 272$ **25. a.** $n = 4$ **b.** no solution **c.** $n = -2$ **d.** $n = -\frac{1}{2}$ **27.** -10 **29.** $\frac{x^2}{15}$ **31.** $A = \frac{1}{2}s^2$

LESSON 2-9 (pp. 119–124)
11. See below. **13. a.** *AJ, AK, AL, BJ, BK, BL, CJ, CK, CL, DJ, DK, DL* **b.** $\frac{1}{12}$ **15.** Sample: A girl can choose from 2 jackets, 3 skirts and 5 blouses. How many 3-piece outfits are possible? **17. a.** 2 **b.** 4 **c.** 8 **d.** 1024 **e.** 2^n **19.** $k = -\frac{1}{25}$ **21.** $n = 1200$ **23. a.** no solution **b.** 0 **c.** $x < 0$ **d.** $x > 0$ **25.** $x = -102$ **27.** $2x + 48$ **29. a.** $-\frac{1}{8}$ **b.** $\frac{5}{2}$ **c.** $-\frac{1}{2}$

11.
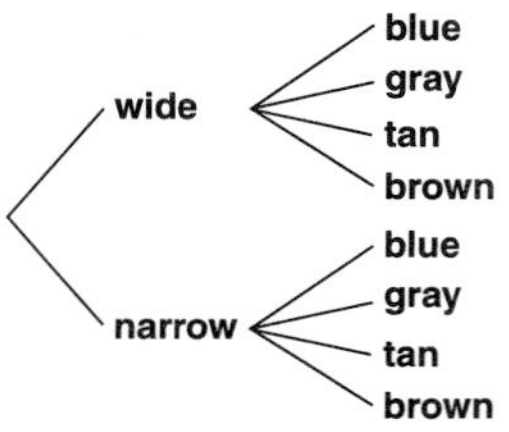

LESSON 2-10 (pp. 125–130)
15. 20! or about $2.4 \cdot 10^{18}$ **17. a.** True **b.** True **c.** $n! = n \cdot (n - 1)!$ **19.** Sample: $100^{100} = 100 \cdot 100 \cdot 100 \ldots \cdot 100$ whereas $100! = 100 \cdot 99 \cdot 98 \cdot \ldots \cdot 1$, so you get a bigger product from 100^{100}. **21.** $\frac{1}{1600}$ **23. a.** Sample: Adam's total earnings for 10 weeks was \$723. How much did he earn each week? **b.** \$72.30 **25.** $y < \frac{10}{3}$ **27.** Multiplication Property of Equality **29.** $6.912 \cdot 10^8$ or 691,200,000 **31.** Sample: $3 \cdot 4 = 4 \cdot 3$ **33. a.** Sample: $2 + 5 > 2$ **b.** Sample: $2 + -1 > 2$ is not true. **c.** positive numbers

CHAPTER 2 PROGRESS SELF-TEST (pp. 134–135)
1. $\frac{22!}{20!} = \frac{22 \cdot 21 \cdot 20!}{20!} = 22 \cdot 21 = 462$ **2.** False because $(-5)^{10} = 9{,}765{,}625 \neq -9{,}765{,}625 = -5^{10}$ **3.** $\frac{20x}{3y} \cdot \frac{5}{4x} = \frac{5}{3y} \cdot \frac{5}{1} = \frac{25}{3y}$ **4.** $\frac{4}{x^2} \cdot \frac{11}{2x} = \frac{2 \cdot 11}{x^3} = \frac{22}{x^3}$ **5.** $-5a \cdot \frac{a}{5} = -a \cdot a = -a^2$ **6.** Sample: $7 \cdot 5 = 5 \cdot 7 = 35$ **7.** $\frac{1}{50} \cdot 50x = \frac{1}{50} \cdot 10$; $x = \frac{10}{50}$; $x = \frac{1}{5}$ **8.** $4 \cdot \frac{1}{4}k = 4 \cdot (-24)$; $k = -96$ **9.** $\frac{1}{3} \cdot 15 \leq \frac{1}{3} \cdot 3m$; $5 \leq m$ **10.** $-y \leq -2$; $y \geq 2$ **11. a.** $-\frac{1}{2} \cdot -2n > -\frac{1}{2} \cdot 18$; $n > -9$ **b.** See right. **12.** $-48 = -\frac{4}{3}n$; $-\frac{3}{4} \cdot -48 = -\frac{3}{4} \cdot -\frac{4}{3}n$; $36 = n$. Check: Does $-48 = -\frac{4}{3}(36)$? Yes, $-48 = -4 \cdot 12$. **13.** $-\frac{n}{3}$ **14.** $\frac{1}{3.2}$ or 0.3125 or $\frac{5}{16}$ **15.** $a \cdot -\frac{1}{a} = -1$ **16.** 8 inches $\cdot\ 2.54 \frac{\text{cm}}{\text{in.}} = 20.32$ cm

17. 24000 people $\cdot\ 0.57 \frac{\text{car}}{\text{person}} \cdot 15 \frac{\text{dollars}}{\text{car}} = 205{,}200$ dollars **18.** Volume $= 4n \cdot 8n \cdot 1.5n = 48n^3$ **19.** $55t = 300$; $\frac{1}{55} \cdot 55t = \frac{1}{55} \cdot 300$; $t \approx 5.45$ hours $\approx$ 5 hours, 27 minutes **20.** $80^2 - 15^2 = 6400 - 225 = 6175$ ft^2 **21. a.** positive **b.** negative × negative = positive **22. a.** negative **b.** odd power of negative number **23. a.** $\frac{5}{6}x, \frac{3}{4}y$ **b.** $\frac{5}{8}xy$ **24. a.** Sample: $0 \cdot x = 48$ **b.** Sample: For any value of x, $0 \cdot x = 0$. So $0 \cdot x$ cannot equal 48. **25. a.** $40r = 600$ **b.** $\frac{1}{40} \cdot 40r = \frac{1}{40} \cdot 600$; $r = 15$ 15 rows **26.** $7 \cdot 5 \cdot 3 = 105$ programs **27.** $5! = 5 \cdot 4 \cdot 3 \cdot 2 \cdot 1 = 120$ **28.** $2^{25} = 33{,}554{,}432$

11. b.
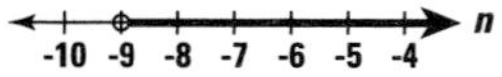

The chart below keys the **Progress Self-Test** questions to the objectives in the **Chapter Review** on pages 136–139. This will enable you to locate those **Chapter Review** questions that correspond to questions missed on the **Progress Self-Test.** The lesson where the material is covered is also indicated on the chart.

Question	**1**	**2**	**3**	**4**	**5**	**6**	**7**	**8**	**9**	**10**	**11**	**12**
Objective	E	B	A	A	A	F	C	C	D	D	D	C
Lesson	2-10	2-5	2-3	2-3	2-3	2-1	2-6	2-6	2-8	2-8	2-8	2-7
Question	**13**	**14**	**15**	**16**	**17**	**18**	**19**	**20**	**21**	**22**	**23**	**24**
Objective	F	F	F	H	H	H	G	G	B	B	J	F
Lesson	2-2	2-2	2-2	2-4	2-6	2-4	2-1	2-1	2-5	2-5	2-1	2-7
Question	**25**	**26**	**27**	**28**								
Objective	G	I	I	I								
Lesson	2-8	2-9	2-10	2-9								

CHAPTER 2 REVIEW (pp. 136–139)

1. $\frac{27}{40}$ **3.** $\frac{x}{y}$ **5.** $3n$ **7.** $\frac{r}{3}$ **9.** 270 **11.** -8 **13.** negative **15. a.** 36 **b.** −36 **c.** −216 **d.** −216 **17.** positive **19.** $m = 150$ **21.** $h = -5$ **23.** $n = -27$ **25.** $g = \frac{d}{c}$ **27.** Sample: $0 \cdot h = 13$ **29.** $m \le 2$ **31.** $u < -2$ **33. a.** $g \le -10$ **b. See right.** **35.** 30 **37.** $\approx 1.3077 \cdot 10^{12}$ **39. a.** False **b.** $\frac{10!}{10} = \frac{10 \cdot 9!}{10} = 9! \ne 1 = 1!$ **41.** 1,030,200 **43. a.** $2200x$ **b.** Commutative and Associative Properties of Multiplication **45.** $-\frac{1}{2}$ **47.** $\frac{4}{3x}$ **49.** Multiplication Property of Equality **51.** opposite **53. a.** $\frac{1}{2}\ell$ by $\frac{2}{3}w$ **b.** The Banjerils' garden is $\frac{1}{3}$ the size of the Cohens' garden. **55.** 31.25 feet **57.** 720; The volume of the solid is $720s^3$ and the volume of the cube is s^3. So $\frac{720s^3}{s^3} = 720$. **59.** $0.033 or $3\frac{1}{3}$¢ **61.** $225 **63.** 8.92 hours or 8 hours 55 minutes **65. a.** $\frac{1}{2}$ second per revolution **b.** It takes a pirouetting ballet dancer $\frac{1}{2}$ second to make one revolution. **67.** $24^3 = 13{,}824$ **69.** $18! \approx 6.4 \times 10^{15}$ **71.** fs **73.** Commutative Property of Multiplication **75. a. See below.** **b.** $\frac{3}{8}s^2$ **77.** $24k^3$ **79.** 2600

33. b.

75. a.

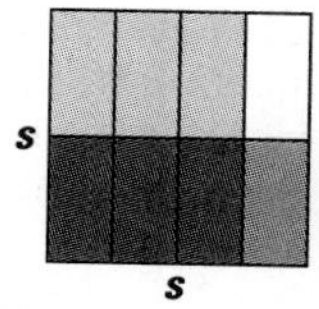

REFRESHER (p. 141)

1. 7.8 **3.** 11.0239 **5.** 9 **7.** 18% **9.** 11.03 cm **11.** 3′ **13.** 158 oz or 9 lb 14 oz **15.** 24 **17.** 1 **19.** −4 **21.** 4 + −7 **23-31. See right.** **33.** $z = 31$ **35.** $m = 5$ **37.** $s = 299$

23-31.

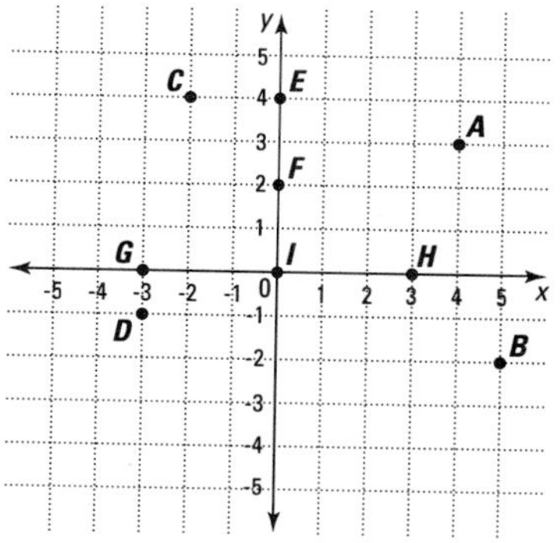

LESSON 3-1 (pp. 142–148)

13. Sample: $5 + 3a = 3a + 5$ **15.** $x° + -3° + 5°$ **17. a.** $A + 3$ **b.** $A + -4$ **19.** $50 + x = 54$ **21.** $M + 4.25 < 10.50$ **23.** $a + b + 5$ **25. a.** 1 **b.** Sample: When you multiply a number by 1, the product is the same as (or identical to) the number. **27.** 13 **29. a.** 12.6 **b.** 0 **c.** 0

LESSON 3-2 (pp. 149–154)

19. 8 cm **21. a.** 3.14 **b.** Opposite of Opposites or the Op-Op Property **23.** $f = 13.05$; $15.2 = 13.05 + 2.15$ **25. a.** $17 + C = -12$ **b.** $C = -29$ **c.** $17 + -29 = -12$ **27.** $40 - n + 2n$ records **29.** (b) **31.** Sample: a line that neither goes up nor down; it is level like the horizon. **33.** $A = (1, 4)$; $B = (5, 2)$; $C = (-2, 2)$; $D = (-5, 0)$; $E = (0, -2)$; $F = (-5, -3)$; $G = (3, -3)$; $H = (3, 0)$

LESSON 3-3 (pp. 155–162)

11. a. L **b.** J **c.** American: The point for the average American student falls in the middle of the scatterplot. **13. a. See right.** **b.** Sample: For every 10° drop in temperature, the wind chill drops about 12°. **15.** (c) **17.** about $1\frac{3}{4}$ hr **19.** (b) **21. a.** −17.3 **b.** Property of Opposites **23. a.** $n = 9$ **b.** $n = 29$ **c.** $n = -29$ **d.** $n = \frac{-10}{19}$ **25.** $-2 + p + -q$ **27.** 9 minutes **29. a.** −13 **b.** 12 **c.** −32

13. a.

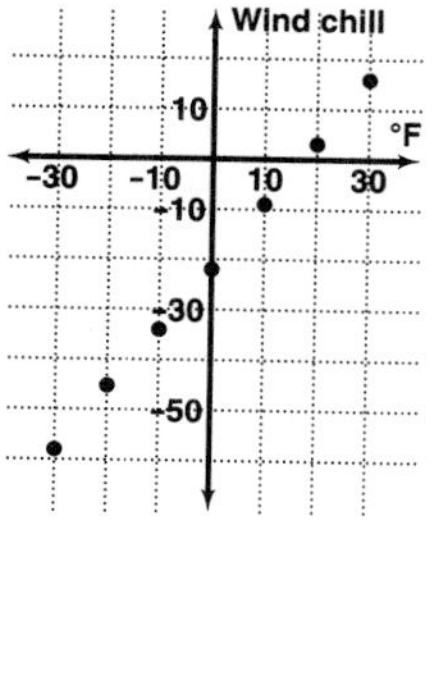

LESSON 3-4 (pp. 163–169)

13. a. (−1, −9) **b. See below.** **15.** 8 right, 6 down **17. a.** 9, 8 **b.** 9, 8 **19.** (4, −10) **21.** 6 minutes **23.** x **25.** −$1.8 million **27. a.** Commutative Property of Addition **b.** $x + y$ **29. a.** 9.6 **b.** 5.9 **c.** a; 9.6 > 5.9

13. b.

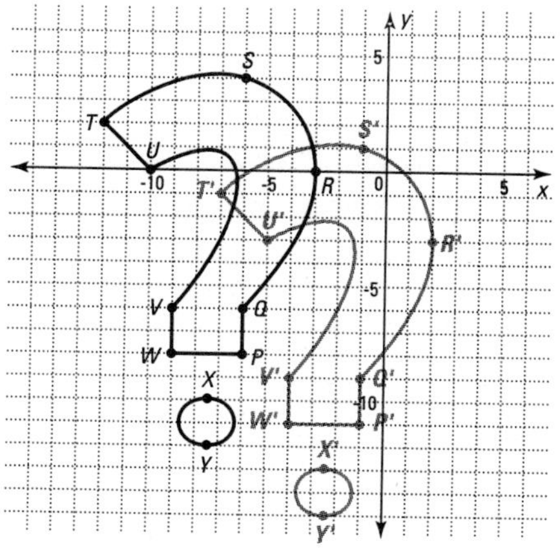

LESSON 3-5 (pp. 170–175)

15. a. $\$4347.59 + \$752.85 + {-\$550.00} + x = \$4574.14$ **b.** $x = \$23.70$; Total interest for June was \$23.70. **17.** $x = 7\frac{1}{4}$; $3\frac{1}{4} + 7\frac{1}{4} = 10\frac{1}{2}$ **19.** See below. **21. a.** 2 right; 4 down **b.** $D' = (2, -4)$; $E' = (3, 0)$ **23. a.** \$3.00 **b.** $.50n = c$ **25. a.** .2 **b.** .02 **c.** 1.02 **d.** 1.20 **27.** B **29.** Yes, several of the higher priced ice creams have a high rating.

19.

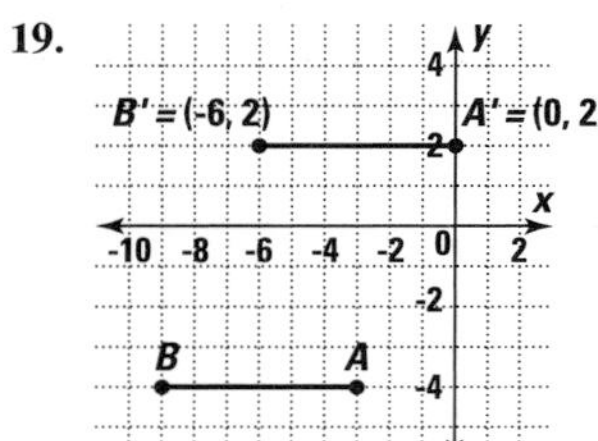

LESSON 3-6 (pp. 177–182)

17. $F - .3F$ or $.7F$ **19.** 0 **21.** 9 camels **23. a.** $3w + 2 = 8$ **b.** $w = 2$ oz **25.** Sample: First add $-b$ to both sides; then multiply both sides by $\frac{1}{a}$. To solve $3x + 5 = 11$, add -5 to both sides to get $3x = 6$; then multiply both sides by $\frac{1}{3}$ to get $x = 2$. **27. a.** -17 **b.** $\frac{1}{17}$ **29.** The minus sign indicates the Death Valley area is 282 feet below sea level. **31.** 900 minutes or 15 hours

LESSON 3-7 (pp. 183–187)

21. 5,999,994 **23. a.** $60 = \frac{1}{2} \cdot 6(5 + b_2)$ **b.** $b_2 = 15$ cm **25.** 123 miles **27.** $2a + 6b + -2c$ **29. a.** $m < 12$; $5 \cdot 12 + 2 \cdot 12 = 84$ and $5 \cdot 10 + 2 \cdot 10 < 84$ **b.** Yes, -5 is less than 12. **31.** 12

LESSON 3-8 (pp. 188–194)

11. 3; 5; 7; 9 **13. a.** $t = 1 + 2n$ **b.** 11

15. a.

Minutes	Cost
100	65.95
200	65.95
300	86.95
400	107.95
500	128.95

b. $c = 65.95 + .21(m - 200)$ **c.** $m \geq 200$

17. Sample:

```
10 PRINT "GALLONS OF GASOLINE CONSUMED"
20 PRINT "DAYS", "GALLONS"
30 FOR N = 1 TO 20
40 LET G = 7500*N
50 PRINT N, G
60 NEXT N
70 END
```

19. $25(\$1.99) = 25(\$2 - 1¢) = 25 \cdot \$2 - 25 \cdot 1¢ = \$50 - 25¢ =$ \$49.75 **21.** $x = \frac{1}{2}$ **23.** $x = 10$ **25.** Quadrant II **27. a.** $z < 77$ **b.** See below.

27. b. 76 77 78 z

LESSON 3-9 (pp. 195–199)

15. $-\frac{1}{x}$ **17.** $-\frac{3}{2x}$ **19. a.** $2W + 2L$ **b.** $L + W$ **21. a.** Samples: After the first slide at 75¢, the cost is 50¢ per slide. The cost is always divisible by 25¢. **b.** \$4.25; $75 + 50(n - 1) = c$; $75 + 50(8 - 1) = 425¢$ **c.** 15

23. a.

n	1	2	3	4
t	4	7	10	13

b. Sample: t increases by 3 for every increase of 1 in n. **c.** 181 toothpicks **d.** $n = 100$; Sample: How many squares are formed by 301 toothpicks arranged side by side in one row? **25.** $n = -1$; Does $8 = 2(-1 + 3) + 4(5 \cdot -1 + 6)$? Yes, it checks. **27.** $u + -d$ **29.** $11 \cdot 10 \cdot 9 \cdot 8 \cdot 7 \cdot 6 \cdot 5 \cdot 4 \cdot 3 \cdot 2 \cdot 1$ **31.** more than $27\frac{7}{9}$ hours **33.** \$18,777,777.78 per year

LESSON 3-10 (pp. 200–204)

9. $n + n + 1 + n + 2 > 79$; 26, 27, 28 **11.** $y \leq 1$ **13.** $x > -10$ **15.** Celsius temperatures greater than 37°C **17. a.** \$9.45 **b.** \$10 **19.** $2a$ **21.** $\frac{-x}{15}$ **23. a.** 4 **b.** 10 **c.** $2(n + 1)$ or $2n + 2$ **25.** $y = 4$. Does $6(4 \cdot 4 + -1) - 2 \cdot 4 = 82$? $6 \cdot 15 - 8 = 90 - 8 = 82$. Yes, it checks. **27.** about 45 hours **29. a.** 75^{75} **b.** Sample: each factor in 75! is less than or equal to 75, but $75^{75} = 75 \cdot 75 \cdot 75 \cdot \ldots \cdot 7$ So $75^{75} > 75!$

CHAPTER 3 PROGRESS SELF-TEST (pp. 208–209)

1. $m + 3m = 1m + 3m = (1 + 3)m = 4m$ **2.** $\frac{5}{2}(4v + 100 + 2) = \frac{5}{2} \cdot 4v + \frac{5}{2} \cdot 100 + \frac{5}{2} \cdot w = \frac{20v}{2} + \frac{500}{2} + \frac{5w}{2} = 10v + 250 + \frac{5w}{2}$ **3.** $-9k + 3(k + 3) = -9k + 3k + 9 = (-9 + 3)k + 9 = -6k + 9$ **4.** $(x + 5 + x) + (-8 + -x) = (x + x + -x) + (5 + -8) = x + -3$ **5.** $-(-(-p)) = -p$ **6.** $\frac{2}{n} + \frac{5}{n} + \frac{-3}{n} = \frac{2 + 5 + -3}{n} = \frac{4}{n}$ **7.** $\frac{3x}{2} + \frac{5x}{3} = \frac{9x}{6} + \frac{10x}{6} = \frac{19x}{6}$ **8.** $8r = 60$; $r = \frac{60}{8} = 7.5$ **9.** $5q + 3 = -12$; $5q = -15$; $q = \frac{-15}{5} = -3$ **10.** $3x + 6 + 100 = 54$; $3x + 106 = 54$; $3x = -52$; $x = \frac{-52}{3} = -17\frac{1}{3}$ **11.** $85 = x + 2 \cdot 3x + 2 \cdot 4$; $85 = x + 6x + 8$; $85 = 7x + 8$; $85 + -8 = 7x + 8 + -8$; $77 = 7x$; $11 = x$ **12.** $30v > 33$; $v > \frac{33}{30}$; $v > \frac{11}{10}$ See right. **13.** Addition Property of Equality **14.** Adding Like Terms form of the Distributive Property **15.** It is not an element because $-100 + 87 = -13$ and $15 > -13$ **16.** $\$137.25 + 2.50w$ **17.** $6 \cdot \$2.99 = 6(\$3 - 1¢) = 6 \cdot \$3.00 - 6 \cdot 1¢ = \$18 - 6¢ = \$17.94$ **18.** $50 = \frac{9}{5}C + 32$; $18 = \frac{9}{5}C$; $\frac{5}{9} \cdot 18 = C$; $10 = C$; 10°C **19.** Let X stand for Jill's share. Juana receives $2X$. $X + 2X = \$58.50$; $3X = \$58.50$; $X = \$19.50$; Jill receives \$19.50. **20. a.** $(a + b)c$; $ac + bc$ **b.** The total area is also the sum of the areas of the two smaller rectangles $(ac + bc)$. So $(a + b)c = ac + bc$. **21.** $(5 + -4, -2 + 5) = (1, 3)$ **22.** $(-4 + x, 7 + y) = (6, 5)$; $x = 10$; $y = -2$; So $B' = (-5 + 10, -2 + -2) = (5, -4)$ **23. a.** Although deaths do not decrease every year, the likelihood of being killed by a tornado has generally decreased during the last 70 years. **b.** Sample: Better health care makes it possible for more victims of natural disaster to survive. **24.** See below.

12.

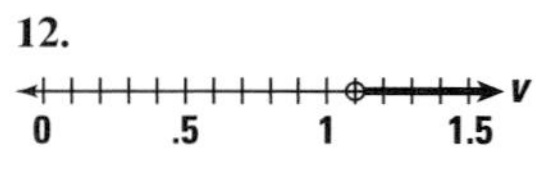

24.

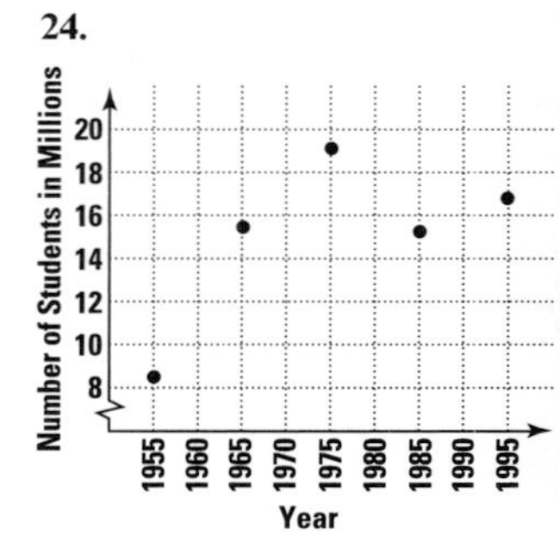

The chart below keys the **Progress Self-Test** questions to the objectives in the **Chapter Review** on pages 210–213. This will enable you to locate those **Chapter Review** questions that correspond to questions missed on the **Progress Self-Test.** The lesson where the material is covered is also indicated on the chart.

Question	**1**	**2**	**3**	**4**	**5**	**6**	**7**	**8**	**9**	**10**	**11**	**12**	**13**
Objective	A	A	A	A	E	C	C	B	B	B	B	D	E
Lesson	3-6	3-7	3-7	3-6	3-2	3-9	3-9	3-5	3-6	3-7	3-7	3-10	3-2
Question	**14**	**15**	**16**	**17**	**18**	**19**	**20**	**21**	**22**	**23**	**24**		
Objective	E	D	H	F	G	G	K	J	J	I	I		
Lesson	3-6	3-10	3-8	3-7	3-5	3-6	3-6	3-4	3-4	3-3	3-3		

CHAPTER 3 REVIEW (pp. 210–213)

1. $15x$ **3.** $\frac{3}{2}c$ **5.** $53x + 46$ **7.** $t = -0.6; 2.5 = -0.6 + 3.1; 2.5 = 2.5$ **9.** $n = 7; (3 + 7) + -11 = -5 + 4; -1 = -1$ **11.** $n = 3;$ $4 \cdot 3 + 3 = 15; 15 = 15$ **13.** $x = \frac{3}{2}; \frac{2}{3} \cdot \frac{3}{2} + 14 = 15; 15 = 15$ **15.** $r = 50; 17 \cdot 50 + 12 + 9 \cdot 50 = 1312; 1312 = 1312$ **17.** $x = 3; 2 \cdot 3 + 3(1 + 3) = 18; 18 = 18$ **19.** $\frac{x + y}{3}$ **21.** x **23.** $\frac{-13x}{10}$ **25.** $x < 95$; Check: $2 \cdot 95 + 11 = 190 + 11 = 201$ and $2 \cdot 10 + 11 = 20 + 11 = 31 < 201$. **27.** $x > 1$; Check: $-2 + (5 + 1) = -2 + 6 = 4$ and $-2 + (5 + 10) = -2 + 15 = 13 > 4$. **29.** $g < \frac{1}{9}$; Check: $-16 \cdot \frac{1}{9} + 7 \cdot \frac{1}{9} + 5 = -\frac{9}{9} + 5 = -1 + 5 = 4$ and $4 < -16 \cdot 0 + 7 \cdot 0 + 5; 4 < 5$. **31.** Commutative Property of Addition **33.** Opposite of the Opposite Prop. **35.** Adding Like Terms form of the Distributive Prop. **37.** $x + -21 = 0$ **39.** $21.28; 7($3.00 + 4¢) = $21.00 + 28¢ = $21.28 **41.** 285; $3(100 - 5) = 300 - 15 = 285$ **43.** 24° **45.** $w + c$ billion pounds **47.** at least $9 **49.** children $15,000 grandchild $7,500 **51. a.** 6, 25 **b.** $y = 3x + 7$ **53. a.** $3.70 **b.** $8.20 **c.** $2.95 + .75(n - 3) when $n > 3$ **55.** halfway up **57.** Ohio **59.** 1950–1960 **61.** $(-38, 56)$ **63.** The image of (x, y) is $(x - 1, y + 8)$. **65.** **See below.** **67. a.** $13 = 5W + 8$ **b.** 1 kg **69.** $ad + bd + cd$; $(a + b + c)d$ **71.** **See below.** **73.** **See below.**

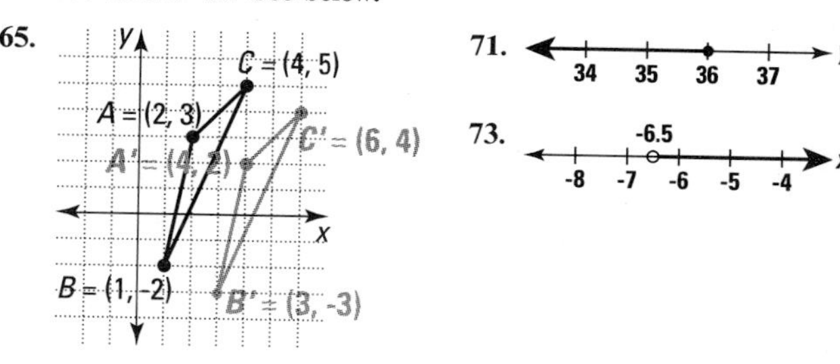

REFRESHER (p. 213)

1. $5\frac{1}{3}$ **3.** 5.65 **5.** 96% **7.** −160 **9.** −8 **11.** 199 **13.** $x = 51$ **15.** $w = 113$ **17.** (b) **19.** 74° **21.** **See right.**

21. Sample:

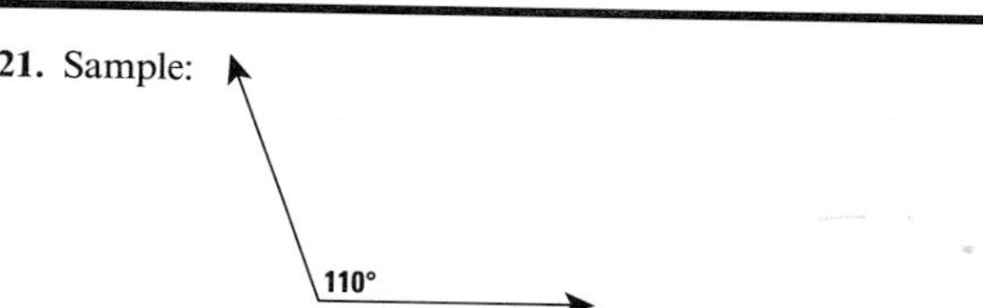

LESSON 4-1 (pp. 216–220)

15. −22 **17.** 0 **19. a.** $-4 + -3 + -3 + 5$ **b.** $-4 - 3 - 3 + 5$ **c.** −5 lb **21.** $x = 1$ **23.** $1.2n + 3$ **25.** $y = 0.1$ **27.** 22.485

LESSON 4-2 (pp. 221–226)

9. $4x$ **11. a.** Bernie **b.** 7 years **13. a.** $.80R$ **b.** $.80R - .10(.80R) = .72R$ **c.** $.72R + .03(.72R) = .7416R$ **15. a.** 47° **b.** $a - b$ or $b - a$ **17.** 0 **19.** $-p + q$ **21. a.** −10 **b.** $-6 + n$ $(n > 0)$ **c.** Sample: Point F could be located anywhere to the right of A. **23.** $x = -1$ **25.** $x > 45.8$ **27. a.** $81 - 16 = 13 \cdot 5$; $961 - 841 = 60 \cdot 2$; $12.25 - 6.25 = 6 \cdot 1$ **b.** $a^2 - b^2 = (a + b)(a - b)$ **c.** Sample: $3^2 - 2^2 = (3 + 2)(3 - 2)$; True

LESSON 4-3 (pp. 227–231)

13. a. $10n - 3 = 84$ **b.** 8.7 **15.** 2044 **17.** $p = -11.64$ **19.** $t > 24$ **21.** $y - 50$ **23. a.** $22.00 **b.** $19.80 **c.** The amount subtracted from 22, 10% of 22, is larger than the amount added to 20, 10% of 20. So the net result is less than 20. **25.** (a) **27.** (a) **29.** 3 **31.** $-3x + 5$

LESSON 4-4 (pp. 232–239)

9. 86 **11.** Marcel's mean **13.** = (J5 + K5 + J6 + K6)/4 **15.** No, because $\sqrt{16 + 25} = \sqrt{41}$, which does not equal 9. **17. a.** $12.00 **b.** Sample: In cell A1 enter the label "Number of words." In cell B1 enter the label "Cost of ad." In cell A2 enter 25, and in cell A3 enter the formula = A2 + 1. Replicate this formula in cells A4 through A77. Enter the formula = 5.00 + .50*(A2 − 25) in cell B2. Finally, replicate the formula from cell B2 in cells B3 through B77. **19.** $x = 66$ **21.** $z > 0$ **23.** $2499.8 \approx 2500$ ft

LESSON 4-5 (pp. 240–245)

17. $-a - 2b + c$ **19.** $(90 - 4f - p)°$ or $(90 - (4f + p))°$ **21.** $t = 1$ **23.** 12 inches **25.** 1.21 D dollars **27. a.** **See below.** **b.** The points lie along the same line. **c.** $y = x + 4$ **29.** TABLE OF (X, Y) VALUES

X VALUE	Y VALUE
0	12
1	11
2	10
3	9
4	8
5	7

27. a.

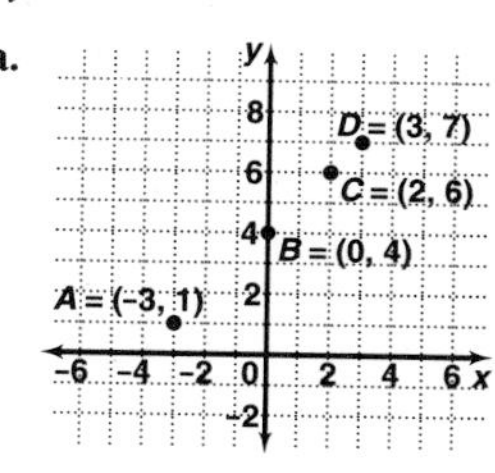

LESSON 4-6 (pp. 246–252)

9. a.

Sal	Al
1	7
2	6
3	5
4	4
5	3
6	2
7	1

b. 3 **11.** (c), (d)

13. a.

J	M
0	10
1	9
2	8
3	7
4	6
5	5
6	4
7	3
8	2
9	1
10	0

b. See below. **15.** $^-3a - 12$ **17. a.** B2 = 55; B3 = 82.5; B7 = 192.5 **b.** = A6 * 55 **19.** $y = \frac{18}{5}$ **21.** \$9.50 **23. a.** 1000 **b.** 2000 **c.** 125 **25.** 60°

13. b.

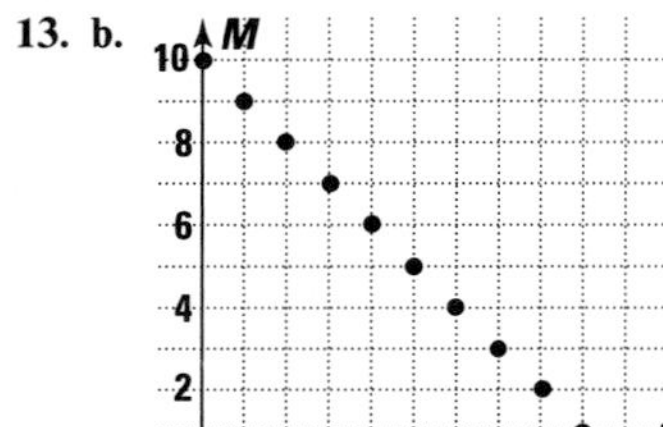

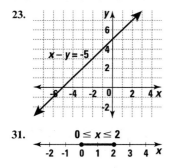

LESSON 4-7 (pp. 253–258)

15. $m\angle B = 90 - x$ **17.** Sample: The angle formed by the jumper's legs and the back part of the skis measures 165°. **19.** 64°, 64°, 52° **21.** $m\angle C = 180 - m\angle A - m\angle B$ **23. See below.** **25.** $n = -\frac{1}{2}$ **27.** $-\frac{7}{12}x$

29. a.

time	charge
$\frac{1}{2}$ hr	\$70
1 hr	\$90
$1\frac{1}{2}$ hr	\$110
2 hr	\$130
$2\frac{1}{2}$ hr	\$150
3 hr	\$170

b. $c = 50 + 20n$ **31. See below.**

23.

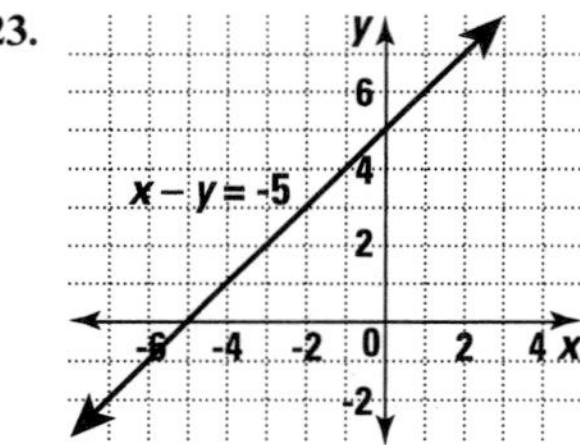

31. $0 \le x \le 2$

-2 -1 0 1 2 3 4 x

LESSON 4-8 (pp. 260–266)

9. between 0 in. and 6 in. **11.** because 1 + 2 is not greater than 4 **13.** 2.6 light-years ≤ m ≤ 20 light-years **15.** Sample: The flying distance will be less because you can fly on a straight line from one city to another, while on land the Triangle Inequality may be applied many times between the cities. **17.** $25\frac{5}{7}°$, $51\frac{3}{7}°$, $102\frac{6}{7}°$ **19. See below.** **21.** $^-3u + {}^-3v^2$ **23.** (7, 12)

19.

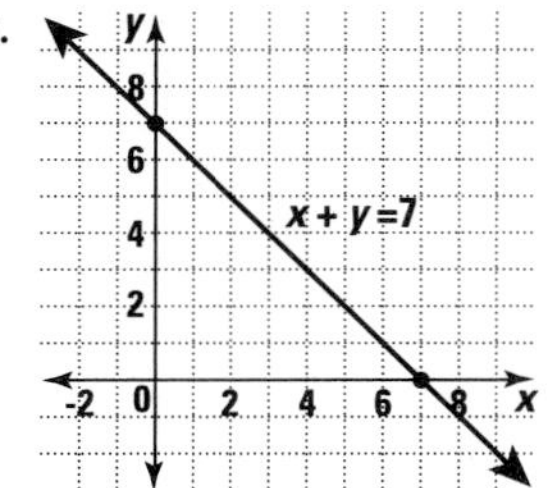

LESSON 4-9 (pp. 267–273)

11. a.

w	b
0	30
1	25
2	20
3	15
4	10

b. $b = 30 - 5w$ **c. See below.** **d.** week 6

13. a. Sample:

x	y
0	10
1	10.5
2	11
3	11.5
4	12

b. the set of nonnegative real numbers **c. See below.** **d.** 20 years **15. a., b. See below.** **c.** (0, 0) **d.** Sample: The lines are reflection images of each other over the y-axis. **17.** 2 points **19.** 7 < length < 47 **21.** 40°, 60°, 80° **23.** (d) **25.** $-\frac{9x}{35}$ **27. a.** 0.25 **b.** 2.5 **c.** 25 **d.** 250,000

11. c.

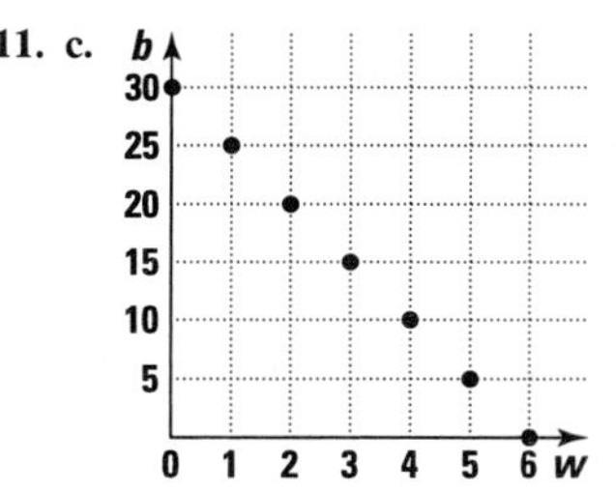

13. c.

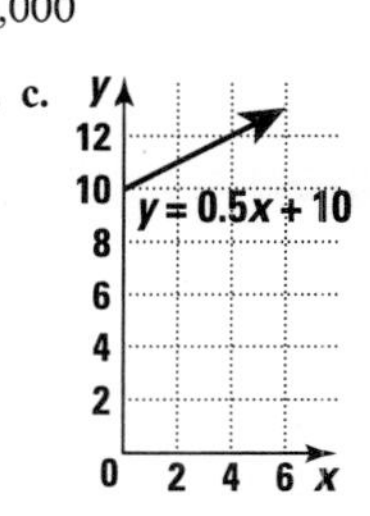

15. a., b.

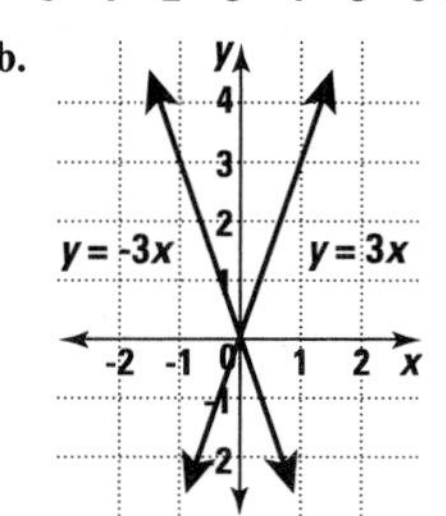

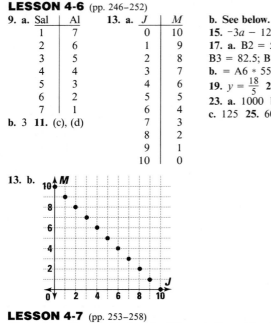

CHAPTER 4 PROGRESS SELF-TEST (p. 277)

1. subtracting 7 **2.** Apply the Comparison Model for Subtraction. $D - 53$ **3.** $9p + -7q + 14z$ **4.** $-n - 16 + 12 = -n - 4$ **5.** $-8x - (2x - x) = -8x - 2x + x = -9x$ **6.** $-2b + 6$ **7.** $\frac{m}{2} - \frac{7m}{2} = \frac{-6m}{2} = -3m$ **8.** $S - .20S = (1 - .20)S = .80S$ **9.** $5n = 60$; $n = 12$ **10.** $2 < -3p$; $-\frac{2}{3} > p$ or $p < -\frac{2}{3}$ **11.** $\frac{3}{4} - \frac{1}{4}m = 12$; $-\frac{1}{4}m = 11\frac{1}{4}$; $-\frac{1}{4}m = \frac{45}{4}$; $-4 \cdot -\frac{1}{4}m = -4 \cdot \frac{45}{4}$; $m = -45$ **12.** $201 = 15f - 6 - 12f$; $207 = 3f$; $f = 69$ **13.** $50 = \frac{5}{9} \cdot (F - 32)$; $50 = \frac{5}{9}F - \frac{160}{9}$; $\frac{450}{9} + \frac{160}{9} = \frac{5}{9}F$; $\frac{9}{5} \cdot \frac{610}{9} = F$; $F = 122°$F **14.** Solve the sentence: $1100 - 6x > 350$; $-6x > -750$; $x < 125$ minutes

15.

Yes	9	8	7	6	5	4	3	2	1	0
No	0	1	2	3	4	5	6	7	8	9

b. See right.

16. a.

x	-1	0	1	2	3
y	-5	-3	-1	1	3

b. See right.

17. Use the Triangle Inequality. $786 + 582 \geq$ distance from LA to SA; so greatest possible distance is 1368 miles. **18. a.** $180 - 18 = 162°$ **b.** $90 - 18 = 72°$ **19.** $x + y = 90$ **See right.** **20.** $n + 2n + 2n - 4 = 180$; $5n - 4 = 180$; $5n = 184$;

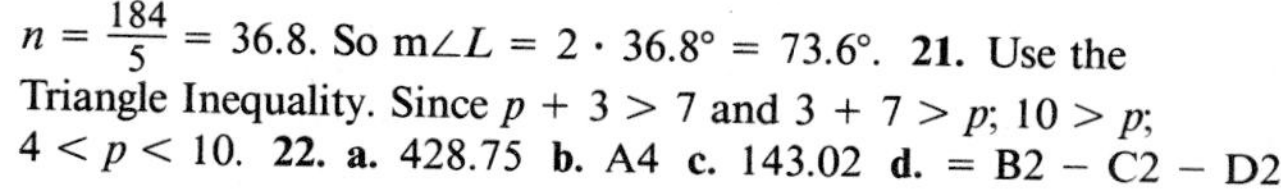

$n = \frac{184}{5} = 36.8$. So $m\angle L = 2 \cdot 36.8° = 73.6°$. **21.** Use the Triangle Inequality. Since $p + 3 > 7$ and $3 + 7 > p$; $10 > p$; $4 < p < 10$. **22. a.** 428.75 **b.** A4 **c.** 143.02 **d.** = B2 − C2 − D2

15. b.

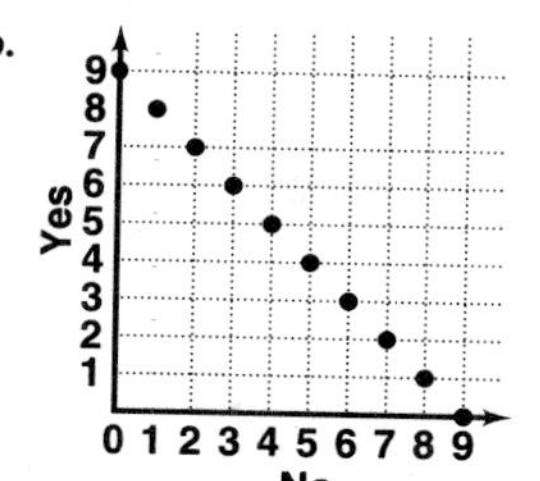

16. b.

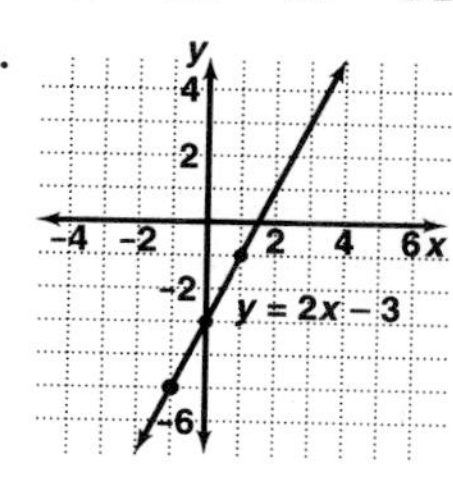

19.

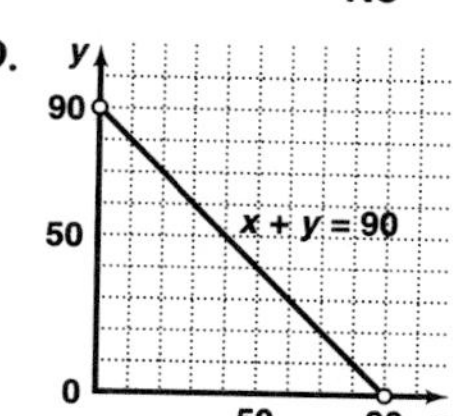

The chart below keys the **Progress Self-Test** questions to the objectives in the **Chapter Review** on pages 278–281. This will enable you to locate those **Chapter Review** questions that correspond to questions missed on the **Progress Self-Test.** The lesson where the material is covered is also indicated on the chart.

Question	**1**	**2**	**3**	**4**	**5**	**6**	**7**	**8**	**9**	**10**
Objective	E	E	A	A	D	D	A	H	B	C
Lesson	4-1	4-2	4-1	4-1	4-5	4-5	4-1	4-2	4-3	4-3
Question	**11**	**12**	**13**	**14**	**15**	**16**	**17**	**18**	**19**	**20**
Objective	B	D	I	I	L	L	J	F	L	F
Lesson	4-3	4-5	4-3	4-3	4-6	4-6	4-8	4-7	4-9	4-7
Question	**21**	**22**								
Objective	G	K								
Lesson	4-8	4-4								

CHAPTER 4 REVIEW (pp. 278–281)

1. $4x$ **3.** $-3a$ **5.** $-\frac{4c}{3} - 2$ **7.** $-3z^3 + 1$ **9.** $x = 45$ **11.** $y = -\frac{1}{2}$ **13.** $n = 5$ **15.** $m = -5$ **17.** $a = \frac{3}{2}$ **19.** $x < 106$ **21.** $y < -13$ **23.** $-4a - 7$ **25.** $2 - z$ **27.** $-3a - 16$ **29.** $p = -8$ **31.** $x > 10$ **33.** $x = \frac{45}{2}$ **35.** $x + -y + z$ **37.** True **39. a.** 73° **b.** 163° **41.** True. The supplement of an angle with measure $x°$ is $180 - x$. The complement is $90 - x$. $180 - x$ is greater than $90 - x$. **43.** 73° **45.** 40°, 100° or 70°, 70° **47.** No; $3 + 5 < 16$ **49.** $15 > m$, $m + 8 > 7$, $m + 7 > 8$ **51.** $3 < y < 25$ **53.** $S = E - P$ **55.** $2500 - F$ **57.** $D - 5$ **59. a.** $.3V$ **b.** $.7V$ **61.** 51 wks; Solve $750 + 15w > 1500$. **63.** 13 days **65.** 1233 mi **67. a.** = A6^2 + A6 **b.** 6 **c.** B6 would change to 380. **69. a.** D3 = 740; D4 = 586; D5 = 1118; D6 = 1180; D7 = 520 **b.** = B4 * 6 + C4 * 4 **c.** Aug. 30 **71.** **See right.** **73.** **See right.** **75.** **See right.** **77.** **See right.**

71.

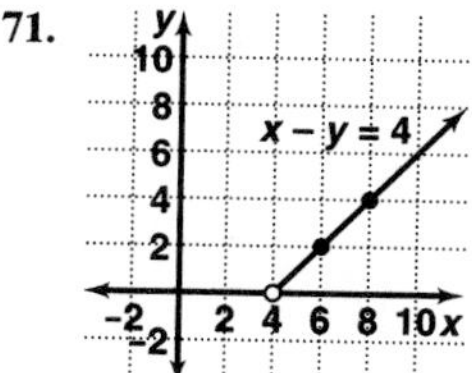

73.

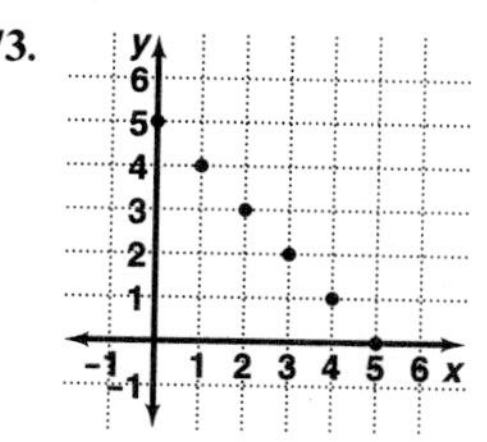

75.

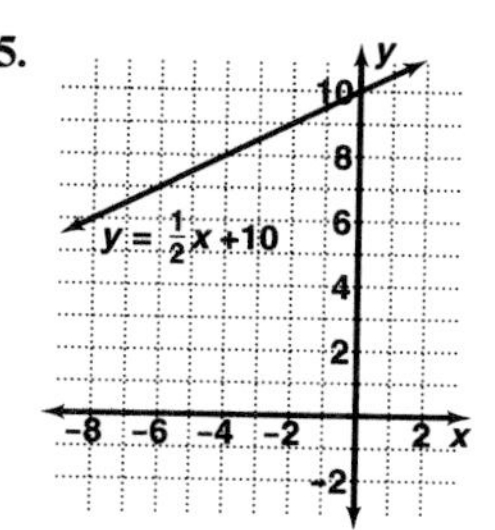

77.

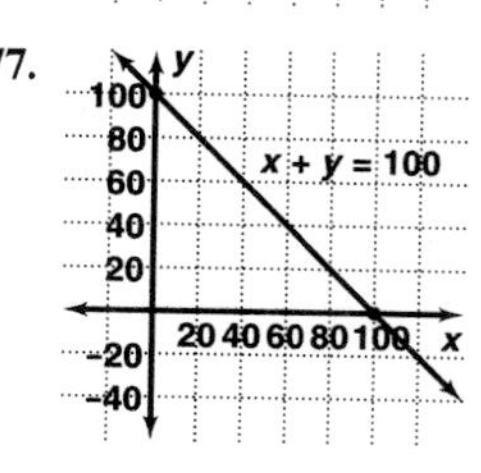

LESSON 5-1 (pp. 284–290)

15. a. **See below.** **b.** Let x = the date and y = the cost (in cents). From 2/17/85 to 4/2/88, $y = 22$. From 4/3/88 to 2/2/91, $y = 25$. From 2/3/91 to the present, $y = 29$. **17.** $x = -6$ **19. a.** $y = -13$ **b.** $x = 7$ **21. a.** $y = 35x + 25$ **b., c. See below.** **d.** 6 hr **e.** $35x + 25 \le 250$; $x \le 6.43$; Ron could work for 6 whole hours. **23.** $.80d$ **25.** $9.8(25) + 14.2(25)$ or $(9.8 + 14.2)25$ **27.** (a) **29. a.** Any real number multiplied by 1 equals that same real number. **b.** Sample: $1 \cdot 5 = 5$

15. a.

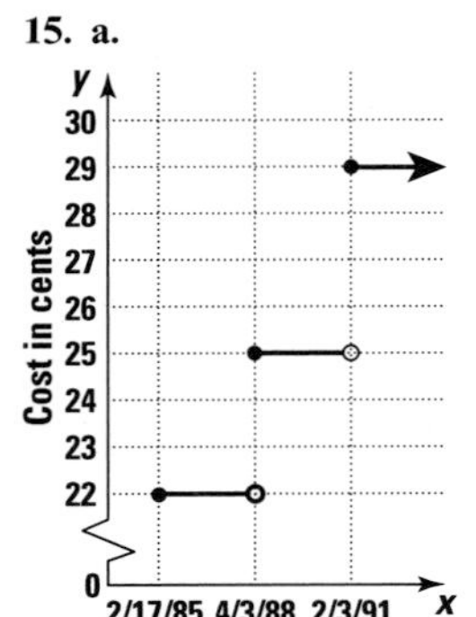

21. b, c

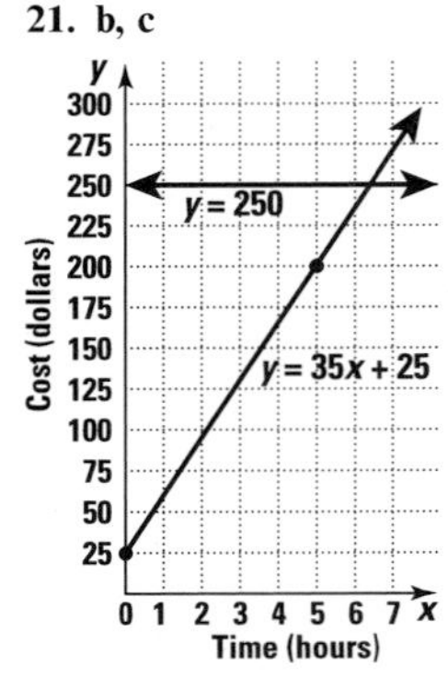

LESSON 5-2 (pp. 291–296)

9. a. $m = 20 + 6w$ **b.** $m = 150 - 4w$

c.

week	1	2	3	4	5	6	7	8	9	10	11	12	13
Kim	26	32	38	44	50	56	62	68	74	80	86	92	98
Jenny	146	142	138	134	130	126	122	118	114	110	106	102	98

After 13 weeks, Kim and Jenny will have the same amount of money. **11.** Sample: $(-3, -6)$, $(0, -6)$, $(17, -6)$ **13. a. See below.** **b.** $(2, 7)$ **c.** $(5, 7)$ **d.** 9 sq units **15. a. See below.** **b.** $(9, 1)$ **17.** $m < 5$ **19.** $x + y - 3(z + w) = x + y - 3z - 3w$. Sample: multiplication by -3 was not distributed over the w term. **21.** 37 **23.** $3.46 \cdot 10^{12}$ **25.** 3

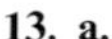

13. a.

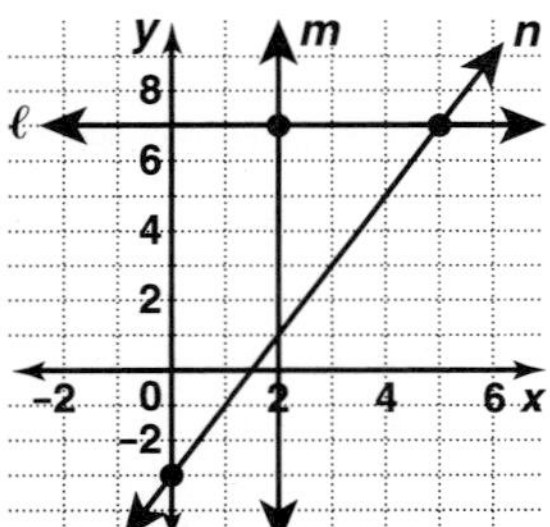

15. a.

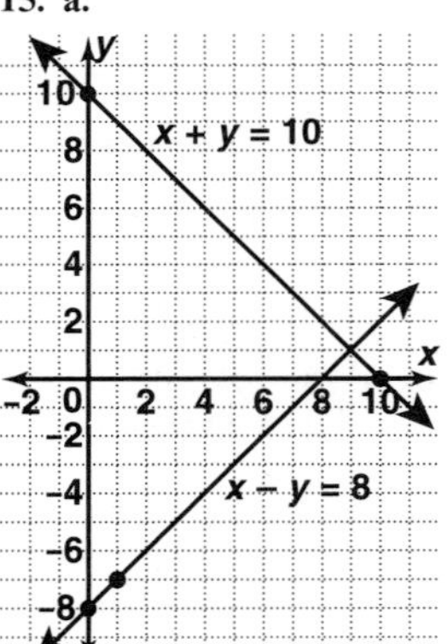

LESSON 5-3 (pp. 297–303)

17. $d = -\frac{1}{2}$ **19.** $y = 6$ **21. a.** $54.64 - 0.33x$ **b.** $49.36 - 0.18x$ **c.** after about 35 years at the Olympics in the year 2028

23. a.

hour	Lamont's distance	Chris's distance
0	24	0
1	33	13
2	42	26
3	51	39
4	60	52
5	69	65
6	78	78

b. 6 hours **25. See below.** **27. a.** $3(2 + x)$ **b.** $(x + 2)3$ **c.** $3x + 6$ **29. a.** Yes **b.** No **c.** Yes

25.

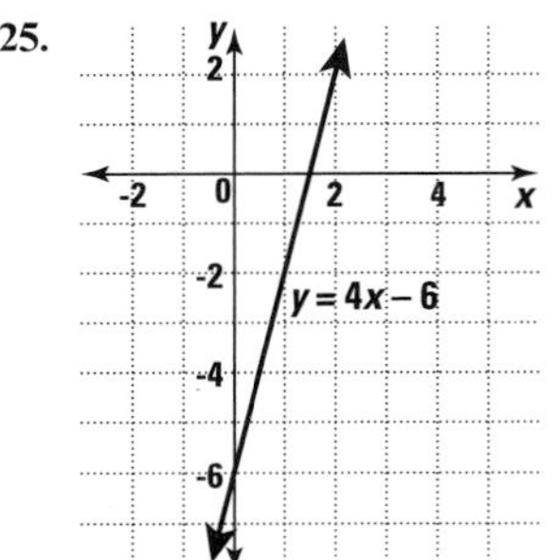

LESSON 5-4 (pp. 304–309)

9. a. See below. At about 9,000 copies costs are equal. **b.** $200 + 0.015x = 70 + 0.03x$; $x \approx 8{,}667$; Costs are equal for about 8,667 copies. **11.** Using a table, you get 312.5 pu.

x = distance dog travels	y = distance hare is ahead
0	50
125	30
250	10
312.5	0

Using a graph, you may approximate the answer. Graph the line from (0, 50) through (125,30). When the distance (y) the hare is ahead reaches zero, the x-value tells the total distance traveled by the dog. **13.** $r = 10$ **15.** $m = \frac{1}{2}$ **17. a. See below.** **b.** $(4, -3)$

19. SQUARE ROOTS AND SQUARES

1	1	1
1.41421356	2	4
1.73205081	3	9
2	4	16
2.23606789	5	25
2.44948974	6	36

9. a.

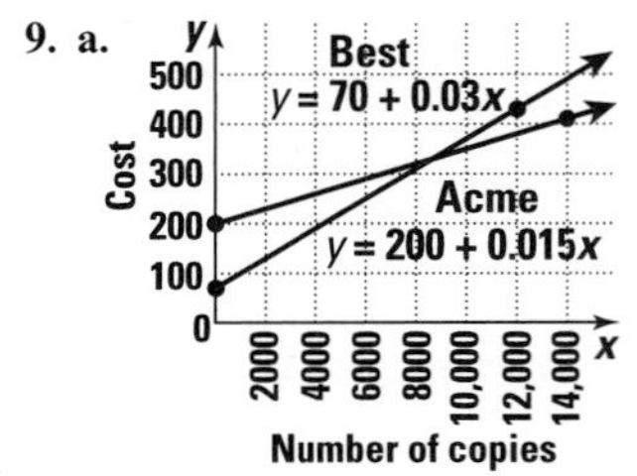

17. a.

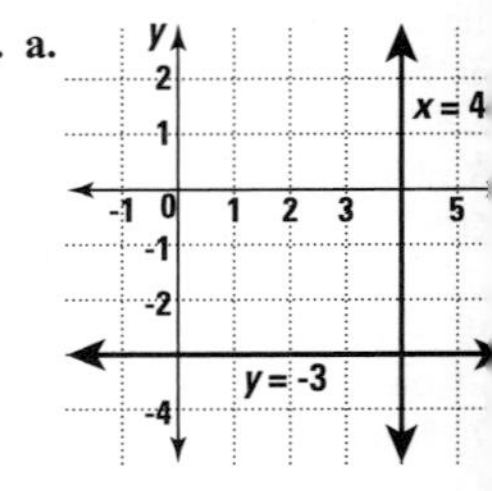

LESSON 5-5 (pp. 310–316)

13. a. See below. **b.** Sample: All lines passing through (0, 0). Some are steeper than others. The higher the coefficient of x, the steeper the graph is. **c.** It will pass through the origin and be steeper than the other graphs. **15. a.** See below. **b.** Sample: The graphs are not equal; one is a line and one is a curve. **17.** $d = 17$ **19.** (c) **21.** $x = 9$ **23. a.** $x - s - 2m$ **b.** $x - (s + 2m)$ **25.** (b) **27. a.** $x = 2$ **b.** $y = 4$ **c.** $z = 16$

13. a.

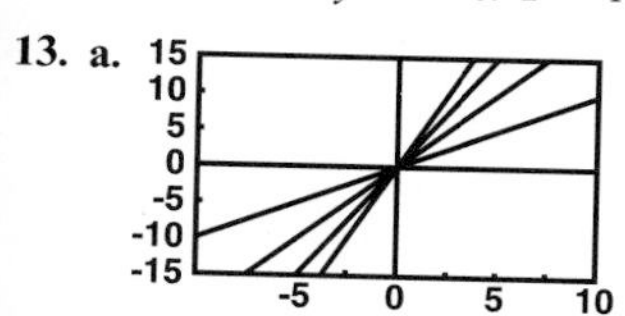

15. a.

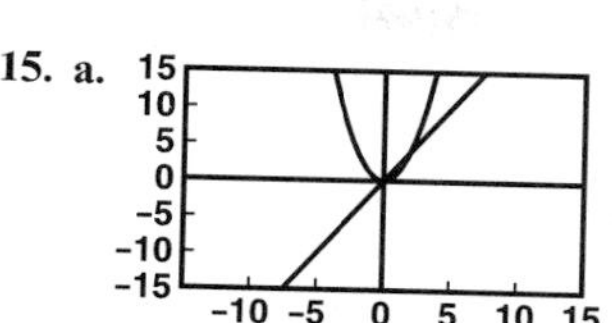

LESSON 5-6 (pp. 317–321)

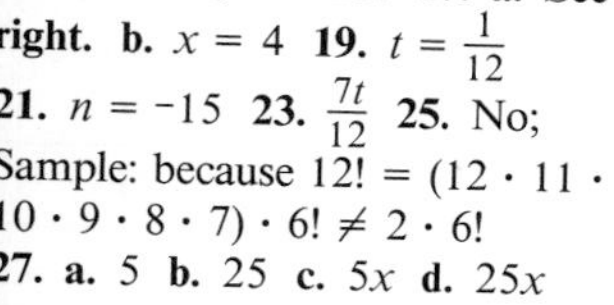

13. $x > -\frac{1}{13}$ **15.** ads more than 25 words; Sample: $2.00 + 0.08x < 1.50 + 0.10x$, $0.5 < 0.02x$, $x > 25$. **17. a.** See right. **b.** $x = 4$ **19.** $t = \frac{1}{12}$ **21.** $n = -15$ **23.** $\frac{7t}{12}$ **25.** No; Sample: because $12! = (12 \cdot 11 \cdot 10 \cdot 9 \cdot 8 \cdot 7) \cdot 6! \neq 2 \cdot 6!$ **27. a.** 5 **b.** 25 **c.** $5x$ **d.** $25x$

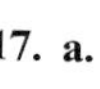

17. a.

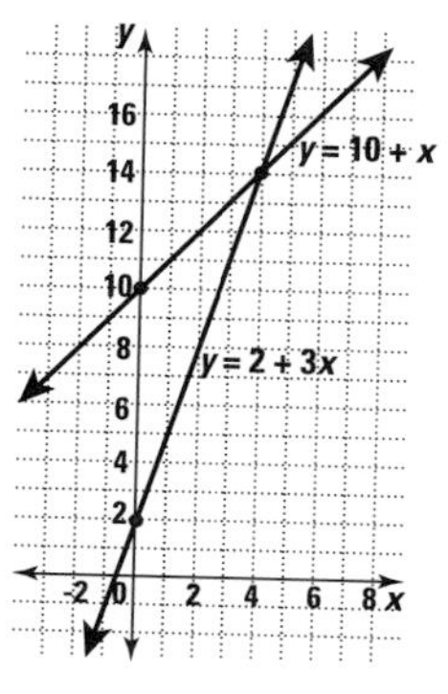

LESSON 5-7 (pp. 322–326)

13. a. $\pi = \frac{C}{d}$ **b.** Sample: If you know the circumference and diameter of a circle, you can substitute those values in the equation to find an approximation for π. **c.** Answers may vary. **15. a.** $D = \frac{9}{10}G$ **b.** 90° **17. a.** −40° **b.** Solve $C = \frac{9}{5}C + 32$ for C, or solve $F = \frac{5}{9}(F - 32)$ for F. **19.** $x \le \frac{6}{5}$ **21.** See above right. **23.** 1994; $8{,}487{,}000 + 395{,}000x > 9{,}170{,}000 - 55{,}000x$; $x > 1.52$ years **25. a.** 40 dots **b.** $4n$ dots

21.

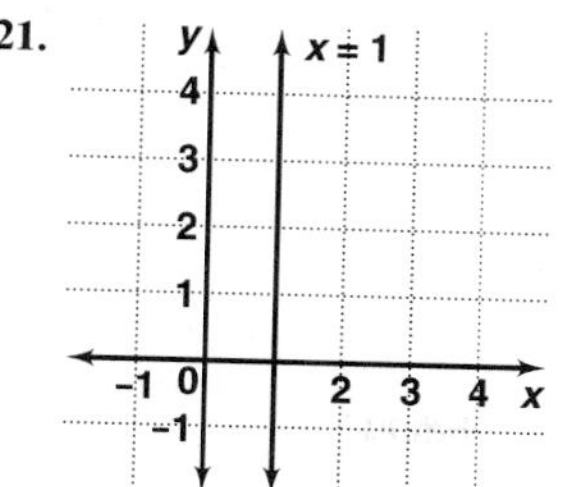

LESSON 5-8 (pp. 327–332)

15. a. \$40,640 **b.** \$15,240 **17.** No. The term $\frac{1}{1000}x$ would appear in the equation, which would not simplify the solution. **19.** $x = 8$ **21.** $y > -15$ **23.** $v = \frac{1}{2}g - \frac{d}{t}$ **25. a.** A **b.** \$2,000

27. a.

years from now	Town A pop.	Town B pop.
0	25000	35500
1	26200	35200
2	27400	34900
3	28600	34600
4	29800	34300
5	31000	34000
6	32200	33700
7	33400	33400
8	34600	33100

b. about 7 years from now **29.** $p = 100$ or $p = -100$

LESSON 5-9 (pp. 333–337)

13. $-2\sqrt{a}$ **15.** $5x^2 - 10y$ **17.** 3 **19.** 52 **21.** 12.5 **23.** $p = \frac{20}{3}$ or $p = -10$ **25.** Multiplication Property of Equality **27.** $x = \frac{1}{2}$ **29.** \$480 **31. a.** $y = -\frac{2}{5}x + 2$ **b.** See below. **33.** Sample: $44 + 31 = 31 + 44$ **35.** 6 units

31. b.

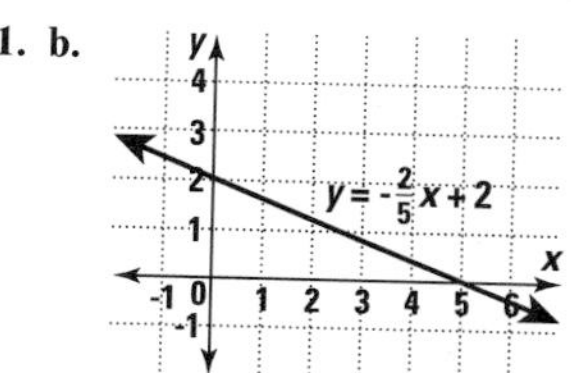

CHAPTER 5 PROGRESS SELF-TEST (pp. 341–342)

. **a.** $3y$ or $-5y$; each gives an equation with a variable on only one de **b.** Addition Property of Equality **2.** 15 (or any other ommon multiple of 3 and 5) **3.** $4x - 3 = 3x + 14$, $x - 3 = 14$, $= 17$ **4.** $3.9z - 56.9 = 6.1 - 4.7z$, $39z - 569 = 61 - 47z$, $5z = 630$, $z \approx 7.33$ **5.** $5n \ge 2n + 12$, $3n \ge 12$, $n \ge 4$ $5(10 - y) = 6(y + 1)$, $5 \cdot 10 - 5y = 6y + 6 \cdot 1$, $50 = 11y + 6$, $4 = 11y$, $4 = y$ **7.** $-5a + 6 < -11a + 24$, $6a + 6 < 24$, $6a < 18$, < 3 **8.** $\frac{1}{2}m - \frac{3}{4} = \frac{2}{3}$, $12 \cdot \frac{1}{2}m - 12 \cdot \frac{3}{4} = 12 \cdot \frac{2}{3}$, $6m - 9 = 8$, $n = 17$, $m = \frac{17}{6}$ **9.** $5000 - 4000v = 11000v + 680000$, $\frac{1}{1000} \cdot$ 00 $- \frac{1}{1000} \cdot 4000v = \frac{1}{1000} \cdot 11000v + \frac{1}{1000} \cdot 680000$, $5 - 4v =$ $v + 680$, $-15v = 675$, $v = -45$ **10.** If $4y = 2.6$, then $20y = 5 \cdot$ 6 or 13. So $20y + 3 = 13 + 3 = 16$ **11.** Use chunking. $\frac{4}{t+7} + \frac{5}{t+7} = \frac{4+5}{t+7} = \frac{9}{t+7}$ **12.** Use chunking. $8(x^2 - 5) + 3(x^2 - 5) =$ $(x^2 - 5) = 11x^2 - 11 \cdot 5 = 11x^2 - 55$ **13.** Use chunking. $= 7^2$ or $(-7)^2$ so $n + 3 = 7$ or $n + 3 = -7$. Hence $n = 4$ or $n = -10$. **14.** $3x + 5y = 15$, $5y = -3x + 15$, $\frac{1}{5} \cdot 5y = \frac{1}{5}(-3x + 15)$, $y = \frac{1}{5} \cdot -3x + \frac{1}{5} \cdot 15$, $y = -\frac{3}{5}x + 3$ **15.** $C = np$, $\frac{1}{n} \cdot C = np \cdot \frac{1}{n}$, $\frac{C}{n} = p$ **16.** See p. 848. **17.** $4x + 6x - 6 + 24 = 3x + 3x + 3x + 3x$, $10x + 18 = 12x$, $18 = 2x$, $9 = x$, Triangle sides are 24, 36, and 48. All sides of the square are 27. **18. a.** See p. 848. After 12 months, the graph of the younger child is higher than the graph of the older child. **b.** Let m = number of months, $1200 + 50m > 1500 + 25m$, $50m > 300 + 25m$, $25m > 300$, $m > 12$

19. a.

Monthly Sales	Sun Fashions Total Salary	Today's Outerwear Total Salary
\$0	\$400	**\$750**
\$5000	**\$1000**	**\$1250**
\$10000	**\$1600**	\$1750
\$15000	**\$2200**	**\$2250**
\$20000	**\$2800**	**\$2750**
\$25000	\$3400	**\$3250**

b. sales $\ge$ \$20,000 **c.** Today's Outerwear would pay more for sales less than or equal to \$15,000.

20. a. **See right. b.** Let v = value of home. $v = 80000 + 3500t$ **21. a.** for more than seven pictures **b.** for fewer than seven pictures **c.** for exactly seven pictures **22. a.** Sample: **See right. b.** $x = -14$ **c.** The y-coordinate increases.

16.

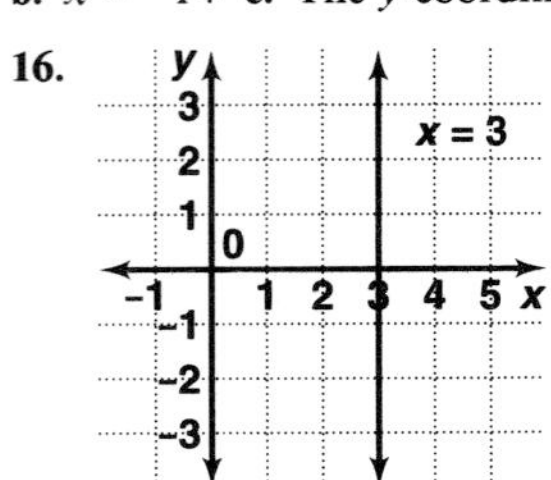

18. a.

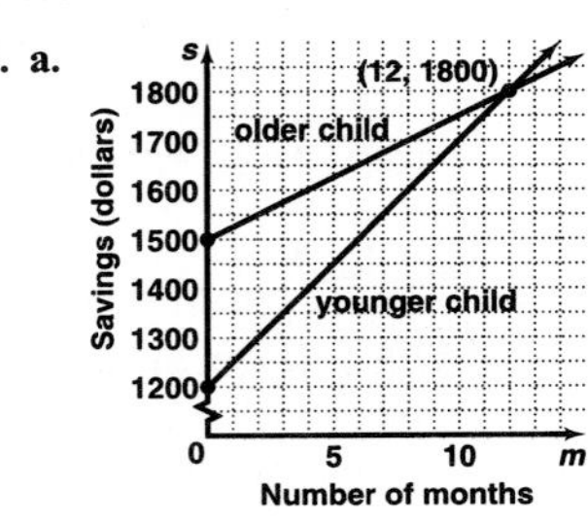

20. a.

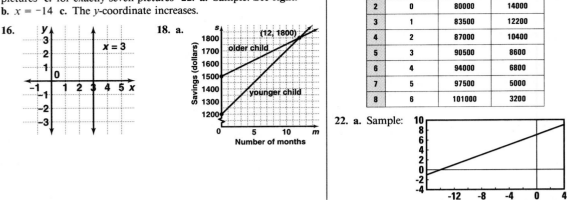

	A	B	C
1	yrs from now	house value	car value
2	0	80000	14000
3	1	83500	12200
4	2	87000	10400
5	3	90500	8600
6	4	94000	6800
7	5	97500	5000
8	6	101000	3200

22. a. Sample:

10 8 6 4 2 0 -2 -4

-12 -8 -4 0 4

The chart below keys the **Progress Self-Test** questions to the objectives in the **Chapter Review** on pages 343–346. This will enable you to locate those **Chapter Review** questions that correspond to questions you missed on the **Progress Self-Test.** The lesson where the material is covered is also indicated on the chart.

Question	**1**	**2**	**3**	**4**	**5**	**6**	**7**	**8**	**9**	**10**
Objective	E	E	A	A	B	A	B	A	A	C
Lesson	5-3	5-8	5-3	5-8	5-6	5-3	5-6	5-8	5-8	5-9
Question	**11**	**12**	**13**	**14**	**15**	**16**	**17**	**18**	**19**	**20**
Objective	C	C	C	D	D	H	F	I, F	G	G
Lesson	5-9	5-9	5-9	5-7	5-7	5-1	5-3	5-4, 5-6	5-2	5-2
Question	**21**	**22**								
Objective	I	J								
Lesson	5-4	5-5								

CHAPTER 5 REVIEW (pp. 343–346)

1. $A = 2$ **3.** $a = \frac{3}{2}$ **5.** $x = \frac{117}{43} \approx 2.72$ **7.** $x = -4$ **9.** 3 **11.** 45 **13.** $h \leq 145$ **15.** $z > 19$ **17. a.** $x \leq -\frac{5}{2}$ **b. See p. 849.** **19.** $n < -\frac{7}{2}$ **21.** $x > -4$ **23.** 67.6 **25.** $3x - 21$ **27.** $\frac{x + y}{z}$ **29.** 6, −10 **31.** 10, −10 **33.** $b = \frac{2A}{h}$ **35.** $w = \frac{1}{2}P - \ell$ **37.** $x = yz$ **39.** $y = -\frac{5}{4}x + 5$ **41. a.** $x = -1$ **b.** $x = -1$ **c.** They are equal. **43. a.** multiply **b.** 16 **c.** distributive **d.** Add $-2x$ **e.** Add 80 **45.** Sample: Multiply by 12; $3 - 24x = 10x + 108$. **47.** by $\frac{1}{100}$; $48t - 1200 = 36t$. **49. a.** Kate will have $1500 + 45n$. Melissa will have $2000 + 20n$ **b.** 20 **51.** $2\frac{2}{3}$ years **53.** after 312 days **55.** 12 gallons
57. a. Charges

Number of CDs	First club	Second Club
2	**$33**	**$30**
4	**$51**	**$49**
6	**$69**	**$68**
8	**$87**	**$87**
10	**$105**	**$106**

b. 8 **c.** more than 8 CDs **d.** less than 8 CDs **59. a., d. See right.** **b.** In B5; = 10200 + A5 * 100 In C5; = 6750 + A5 * 500 **c.** in the 9th year **e.** This year the soybean crop is worth about twice the corn crop's value. Ten years from now it will be worth about three times as much. **61. See page 849.** **63.** True **65.** $x = 5$. **67.** See page 849. **69. See page 849. a.** $x = 8.48$ **b.** $x < 8.48$ **c.** $x > 8.48$ **71.** $-5 \leq x \leq 15$; $-8 \leq y \leq 12$ **73. See page 849.** $x = -4$ when $y = 0$. **75. a. See page 849. b.** $x = 4$

59. a., d.

	A	B	C	D	E
1	yrs from now	bu. corn	bu. soybeans	corn value	Soybean value
2	0	10200	6750	$26520	$51975
3	1	10300	7250	$26780	$55825
4	2	10400	7750	$27040	$59675
5	3	10500	8250	$27300	$63525
6	4	10600	8750	$27560	$67375
7	5	10700	9250	$27820	$71225
8	6	10800	9750	$28080	$75075
9	7	10900	10250	$28340	$78925
10	8	11000	10750	$28600	$82775
11	9	11100	11250	$28860	$86625
12	10	11200	11750	$29120	$90475

17. b.

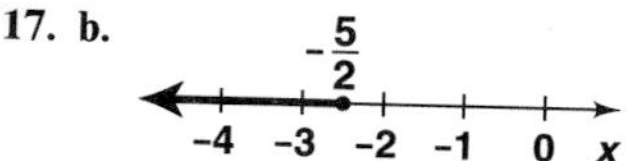

61.

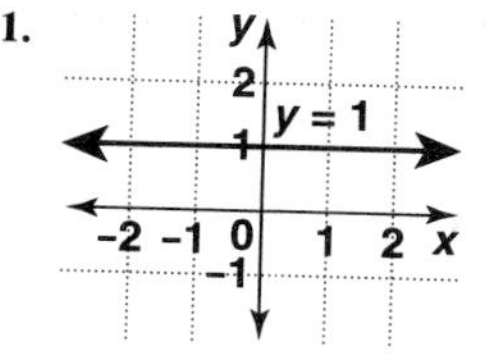

73. Sample:

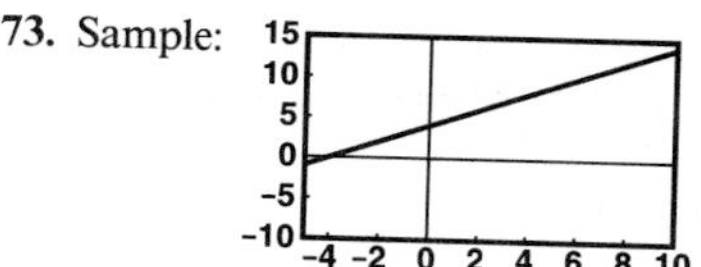

67.

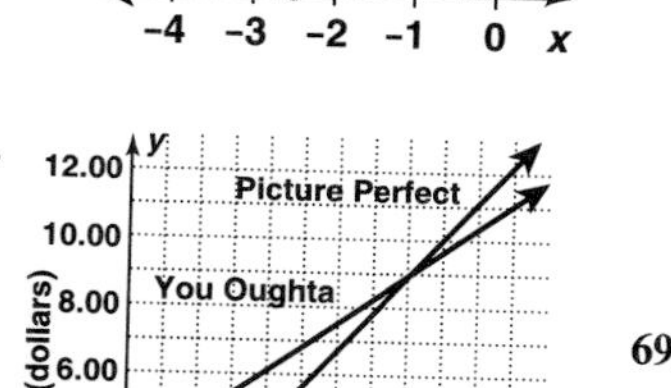

69.

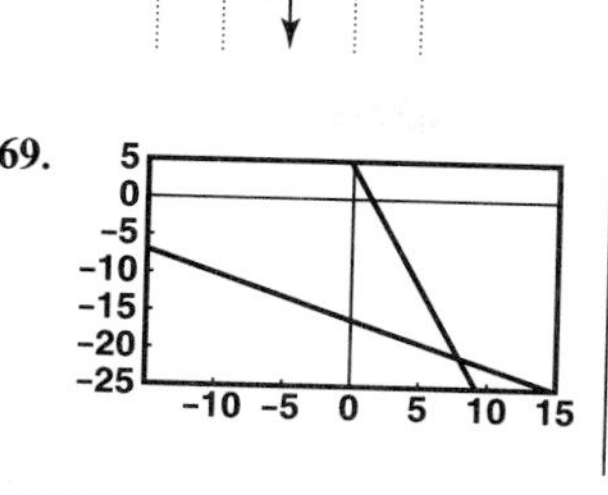

75.

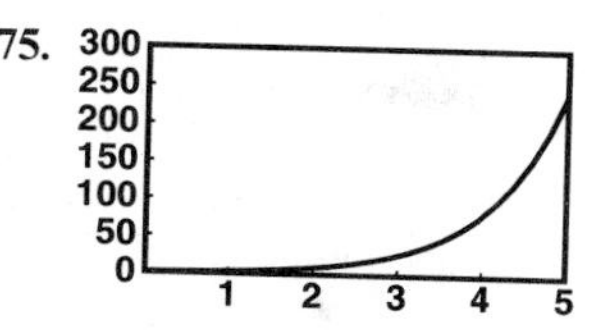

REFRESHER (p. 347)

1. a. 3 **b.** 21 **c.** 7 **3. a.** 7 **b.** 56 **c.** 8 **5. a.** 0.4 **b.** 2.5 **7. a.** 1.25 or $\frac{5}{4}$ **b.** 0.8 or $\frac{4}{5}$ **9.** 1600 **11.** 0.2 **13.** $\frac{4}{5}$ **15.** $1\frac{7}{8}$ **17.** 2.5 m **19.** 0.24 lb **21.** −8 **23.** −0.025 **25.** −100 **27.** 0.025; 2.5% **29.** 0.14 **31.** 6.47 **33.** 27% **35. a.** 0.3 **b.** $\frac{3}{10}$ **37. a.** 3 **b.** $\frac{300}{100}$ **39. a.** 0.0003 **b.** $\frac{3}{10000}$ **41.** 240 **43.** 2942 voters **45.** 0 **47.** 21.75 **49.** ≈ 78.54 cm^2

LESSON 6-1 (pp. 350–355)

21. a. $200 \div \frac{1}{4}$ **b.** 800 **23.** $\frac{1}{6}$ **25.** $\frac{4y^2}{21}$ **27. a.** A positive number divided by a positive number is positive. **b.** A negative number divided by a negative number is positive. **c.** A negative number divided by a positive number is negative. **d.** A positive number divided by a negative number is negative. **29. a.** $3B + 2 = 10$ **b.** $2\frac{2}{3}$ kg **31.** 5 **33.** 4.5 m^2

LESSON 6-2 (pp. 356–361)

15. They had enough money for $2\frac{1}{2}$ weeks. **17. a.** 8.9¢ **b.** 13¢ **c.** 20-ounce can **19. a.** $\frac{1}{4}$ mile per minute **b.** 15 mph **21.** John Olerud **23. a.** An error message because division by 0 is undefined. **b.** 0, because $\frac{0}{100} = 0$. **25.** $\frac{1}{2}$ **27.** 12 **29. a.** $180 - a$ **b.** No. If a supplement were less than 90°, then the angle would be obtuse and not have a complement. **c.** Yes. When an angle is 60°, its complement is 30° and its supplement is 120°. **31. a.** $x < -8$ **b.** $x < -4$ **c.** $x < -\frac{4}{3}$ **d.** $x > 1$

LESSON 6-3 (pp. 362–367)

13. 182 **15.** reciprocal **17.** bx **19. a.** $\frac{6}{4} = \frac{3}{2}$ **b.** $\frac{9}{4}$ **c.** 2.25 **21.** 1.69 meters per second **23.** $\frac{3}{2}$ **25.** 48 **27.** 81.8

LESSON 6-4 (pp. 370–375)

15. $\frac{5}{26}$ **17. a.** $P(X)$ because it has the largest probability. **b.** No; relative frequency can vary from experiment to experiment. **19.** $\frac{50}{50} = 1$ **21.** (d) **23.** 9.6 quarts orange juice, 14.4 quarts ginger ale **25.** 64 **27.** $m = \frac{16}{3}$ **29. a.** $b - .25b = .75b$ **b.** $b + .04b = 1.04b$ **c.** $.75b + .04(.75b) = .78b$

LESSON 6-5 (pp. 376–380)

11. 25% **13.** 658 **15.** 51% **17.** 35% **19.** 0.57 **21. a.** $\frac{7}{15}$ **b.** $\frac{15}{7}$ **23. a.** $245 **b.** rate **25.** 16; −4 **27.** 130.5 **29.** 33%

LESSON 6-6 (pp. 381–386)

11. $\frac{ab}{pq}$ **13. a.** 0.60 **b.** 0.49 **c.** 0.40 **15.** See below. **17. a.** $\frac{c}{s}$ **b.** $\frac{d}{n + c + d + s + x}$ **c.** $\frac{n}{n + c + d + s + x}$ **19. a.** 10.152 for the 100m; 10.142 for the 200m **b.** Burrell is about 0.01 m/sec faster. **21.** $\frac{20}{s}$ **23.** −16 **25.** See below. **27.** 23,040

15.

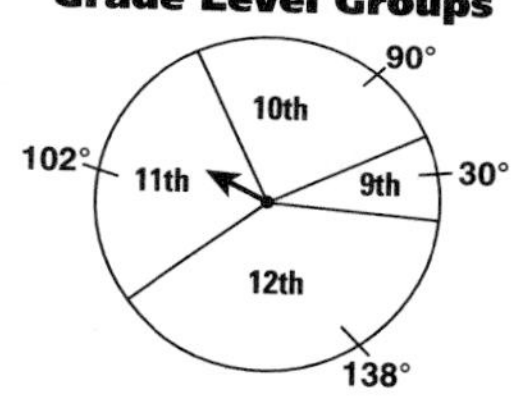

25.

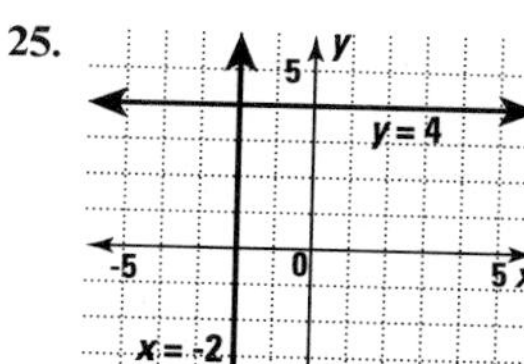

LESSON 6-7 (pp. 387–393)

15. 2.86 kg **17.** (1.5, −7) **19. a.** 8 cm^2; 4.5 cm^2; 3.2768 cm^2; 11.52 cm^2 **b.** False; The area of the image is k^2 the area of the preimage. **21. a.** 78.5% **b.** 0.785 **23.** 37.5% **25. a.** $\frac{1}{86} \approx 0.012$ **b.** $\frac{1}{86}$ **27.** Answers will vary.

LESSON 6-8 (pp. 394–400)

17. 218.5 cm **19.** 0.005 face **21. a.** $\pm\sqrt{22}$ **b.** 4.69; −4.69 **23.** No **25.** No **27.** Yes; $\frac{-15}{-20}$. **29. a.** See right. **b.** 4 **c.** 12 **d.** The perimeter of the image is three times as large as the perimeter of the preimage. **31.** 75% **33.** −5

29. a.

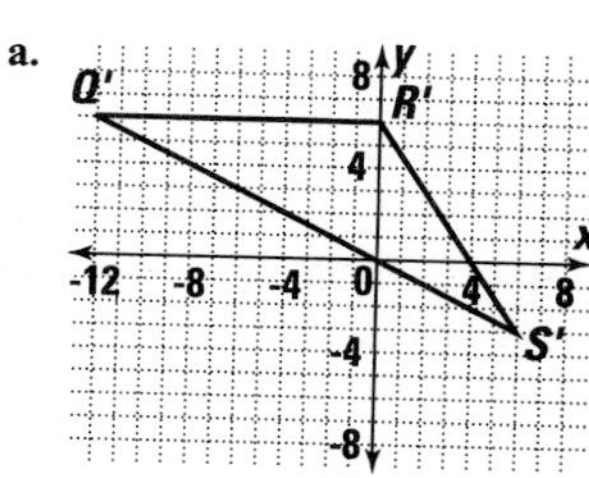

LESSON 6-9 (pp. 402–407)

9. a. $\frac{t}{p}; \frac{w}{k}; \frac{e}{r}$ **b.** $\frac{k}{w}; \frac{r}{e}; \frac{a}{s}$ **11. a.** Drawings will vary. **b.** Sample: $\frac{6}{t} = \frac{10}{25}$ **c.** 15 ft **13. a.** $\frac{8.7 \text{ cm}}{9 \text{ m}}$ **b.** Sample: $\frac{8.7}{9} = \frac{5.5}{x}$ **c.** 5.7 m **15.** $\frac{40}{3}$ **17. a.** $\frac{9}{25}$ **b.** 15; −15 **19.** $\frac{1}{38}$ **21.** 19% **23.** 6 **25.** −3 **27. a.** 15 **b.** $\frac{23}{8}$ **c.** −15 **d.** 184

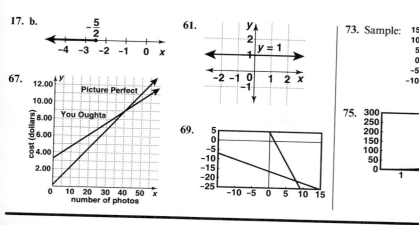

CHAPTER 6 PROGRESS SELF-TEST (pp. 411)

1. $15 \cdot -\frac{2}{3} = -10$ **2.** $\frac{x}{9} \cdot \frac{3}{2} = \frac{3x}{18} = \frac{x}{6}$ **3.** $\frac{2b}{3} \cdot \frac{3}{b} = \frac{6b}{3b} = 2$ **4.** $23y = 22$, $y = \frac{22}{23}$ **5.** $b \cdot b = 25 \cdot 49$, $b^2 = 1225$, $b = 35$ or $b = -35$ **6.** $8(4g - 3) = 26g$, $32g - 24 = 26g$, $6g = 24$, $g = 4$ **7.** $0.14 \cdot b = 60$, $b \approx 428.6$ **8.** $\frac{1}{2} = x \cdot \frac{4}{5}$; $x = \frac{1}{2} \cdot \frac{5}{4} = \frac{5}{8}$; $x =$ 62.5% **9.** $P(5 \text{ or } 6) = \frac{2}{6} = \frac{1}{3}$ **10.** −1, because $-1 + 1 = 0$, therefore the fraction is undefined. **11.** 3; x **12. a.** $36 = x \cdot 30$; $x = \frac{36}{30} = 1.2 = 120\%$ **b.** 20 **13.** The number of animals is $d + c$. So the ratio of the number of dogs to the number of animals is $\frac{d}{c + d}$. **14. a.** $\frac{9}{4}, \frac{4}{9}$ **b.** $\frac{z}{12} = \frac{4}{9}$; $9z = 48$; $z = \frac{48}{9} = 5\frac{1}{3} = 5.\overline{3}$ **15.** $295; $0.8(x) = 236$, $x = 236 \div 0.8$, $x = 295$ **16.** p pages in $7y$ min; the same number of pages are read in less time. **17.** 0.09; $\frac{36}{400} = \frac{\text{no. with Alzheimer's}}{\text{no. surveyed}}$; 9% **18.** $\frac{30°}{360°} = \frac{1}{12}$ **19.** $\frac{280}{12} = \frac{x}{14}$; $12x = 3920$; $x = 326.\overline{6}$. To the nearest mile, this is 327 miles.

20. a., b. See below. **21. a.** $\left(\frac{2}{3} \cdot -6, \frac{2}{3} \cdot 4\right) = \left(-4, \frac{8}{3}\right)$ **b.** contraction **22.** $\frac{x}{7} = \frac{x + 2}{10}$, $10x = 7x + 14$, $3x = 14$, $x = \frac{14}{3} = 4\frac{2}{3} =$ width; $x + 2 = \frac{20}{3} = 6\frac{2}{3} =$ length **23.** $\frac{6 \cdot 8}{14 \cdot 16} = \frac{48}{224} = \frac{3}{14} = \approx 0.21$

20. a., b.

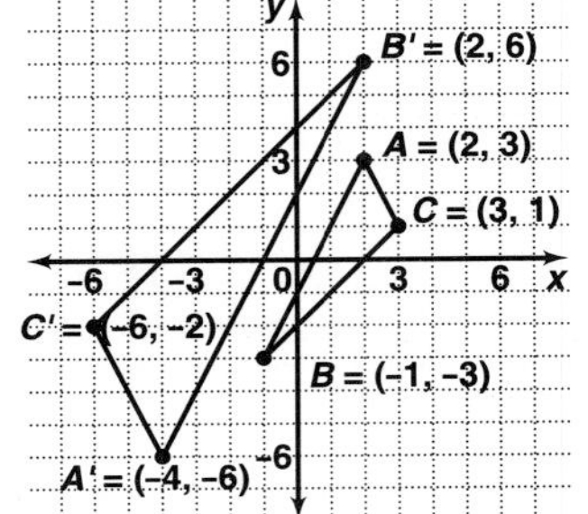

The chart below keys the **Progress Self-Test** questions to the objectives in the **Chapter Review** on pages 412–415. This will enable you to locate those **Chapter Review** questions that correspond to questions you missed on the **Progress Self-Test.** The lesson where the material is covered is also indicated on the chart.

Question	**1**	**2**	**3**	**4**	**5**	**6**	**7**	**8**	**9**	**10**
Objective	A	A	A	C	C	C	B	B	G	A
Lesson	6-1	6-1	6-1	6-8	6-8	6-8	6-5	6-5	6-4	6-1
Question	**11**	**12**	**13**	**14**	**15**	**16**	**17**	**18**	**19**	**20**
Objective	D	H	F	L	H	E	G	J	I	K
Lesson	6-8	6-5	6-3	6-9	6-5	6-2	6-4	6-6	6-8	6-7
Question	**21**	**22**	**23**							
Objective	K	L	J							
Lesson	6-7	6-9	6-6							

CHAPTER 6 REVIEW (pp. 412–415)

1. −125 **3.** $\frac{4}{3}$ **5.** $-\frac{8}{5}$ **7.** 240 **9.** −4 **11.** 24 **13.** 200% **15.** 200 **17.** 156 **19.** 30; −30 **21.** −3 **23. a.** 8 and 15 **b.** 5 and 24 **25.** $\frac{3}{2}$ **27. a.** 11 mph **b.** $\frac{1}{11}$ hour per mile **29.** $\frac{1}{6}$ of an hour or 10 min **31.** Sample: How much did Tony spend per day? $\frac{400}{d}$ dollars per day **33. a.** 46 oz: ≈ 3.8¢ per oz; 6 oz: ≈ 4.7¢ per oz **b.** 46-ounce can **35.** 1.7 **37.** 16.7% **39.** $\frac{x}{x + y}$ **41.** $\frac{9}{11}$ **43.** $\frac{13}{52} = \frac{1}{4}$ **45.** $\frac{999{,}999}{1{,}000{,}000}$ **47.** $\frac{29}{50} = 0.58 = 58\%$ **49.** $\frac{1}{5}$ **51.** 80% **53.** 5.9 lb **55.** $9.93 **57.** $414.61 **59.** $105 **61.** 193 **63.** (b) **65.** $\frac{7}{16}$ **67. a.** 0.049 **b.** 0.589 **69.** $\frac{1}{6} = 0.167$ **71.** (6, 12) **73. a., b.** See right. **c.** contraction **75.** See right. **77.** (c) **79.** No; Ratios of sides are not equal. **81. a.** 5.5 **b.** 30 **83. a.** Sketches will vary. **b.** $\frac{x}{9} = \frac{3}{n}$, $x = \frac{27}{n}$

73. a., b.

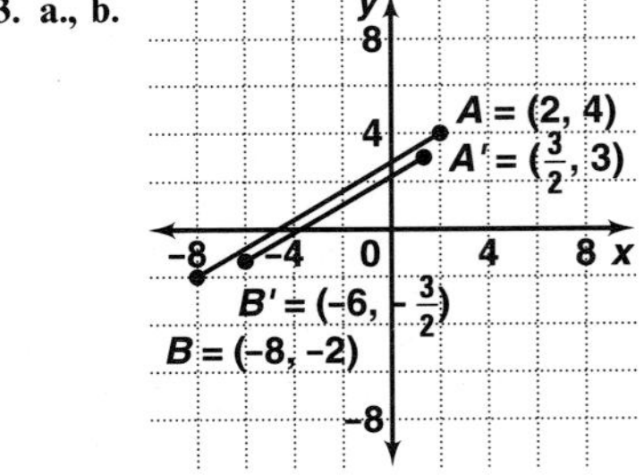

75.

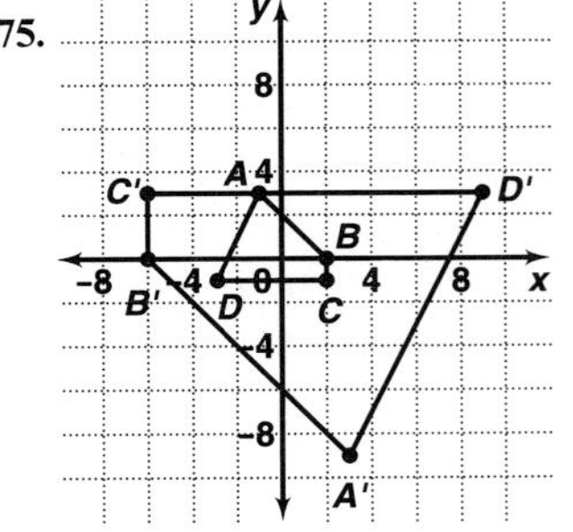

LESSON 7-1 (pp. 418–424)

15. 0 meters per second **17.** positive **19.** −0.1 inch per year **21.** $\frac{y}{4}$ meters per minute **23. a.** $\frac{1}{a}$ **b.** $\frac{1}{3 + 4x}$ **c.** $x = -\frac{4}{5}$ **25. a.** $y = 5.25$ **b.** $y = \frac{3}{4}x - 3$ **c.** $y = \frac{a}{4}x - 3$ **27.** 10,000 + 1,000x **29.** $-\frac{1}{2}$

LESSON 7-2 (pp. 425–431)

13. *C*; *D* **15. a.** *r* **b.** *q* **c.** *p* **17.** $y = 3$

19. a.

	A	B	C
1	*x*	*y*	rate of change
2	0	1	
3	4	6	1.25
4	8	7	0.25
5	12	11	1

b. No; the rate of change is not constant. **21. a.** Zero is in the denominator; division by zero is impossible. **b.** No **23. a.** 1980–1985 **b.** 1960–1965 and 1965–1970 **c.** $\frac{1}{2}$ cent per year **d.** The cost of stamps does not gradually rise over each 5-year period. For example, the cost was never 4.3 cents. **25. See below.** **27.** No. If you substitute 7 for x in the equation $y = 2x - 5$, then $y = 9$. $y \neq 6$. **29.** $-350 + 5x$

25.

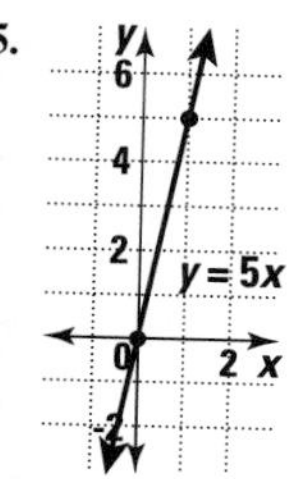

LESSON 7-3 (pp. 432–438)

7.

	A	B
1	*x*	*y*
2	6	10
3	7	2
4	8	−6
5	9	−14

. 0.79 **21. a.,b. See below.** **c.** Sample: Lines with positive slopes ant upward from left to right. Lines with negative slopes slant ownward from left to right. **23.** −2 **25. a.** $y = .25x + 39$ **b. See** low. **c.** $\frac{1}{4}$ **27. a.** 1987–1988 **b.** 1987–1988 **29.** 40 **31. a.** $1.20 2$x$ dollars

a.,b.

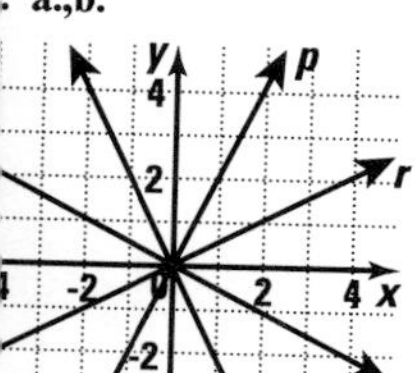

25. b.

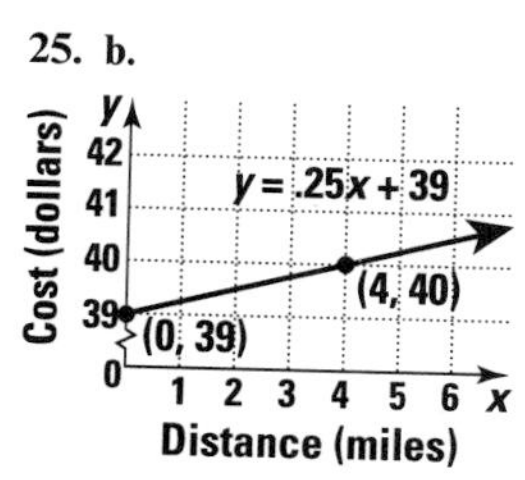

LESSON 7-4 (pp. 439–444)

13. a. ii **b.** 4 **c.** 100 **15. a.** iii **b.** 4 **c.** −100 **17. a.** $y = .50x + 8$ **b. See below.** **19. a.** *q* **b.** *p* **c.** *n* **d.** *r* **21. a. See below.** **b.** a line that passes through (0, 5) with slope *m* **c.** Yes; **See below.** **23.** ≈.09 **25.** $\frac{-10}{3}$ **27.** $z = 2$

17. b.

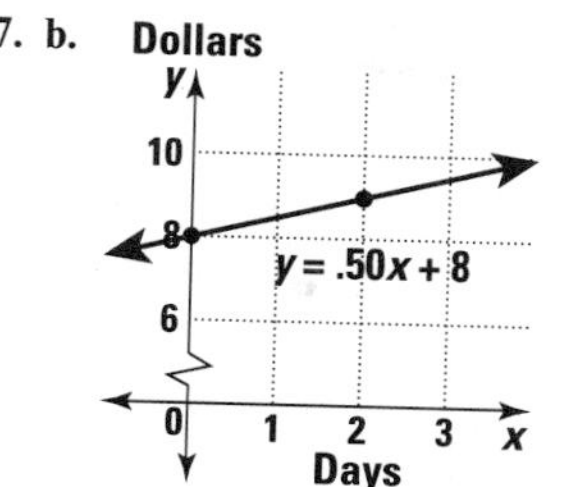

21. a.

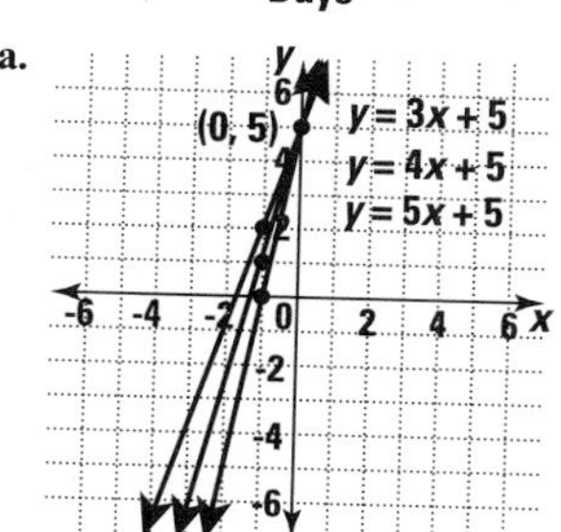

21. c.

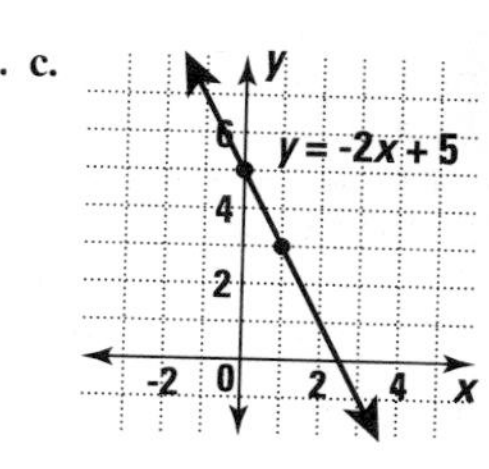

LESSON 7-5 (pp. 445–449)

9. a. $y = 5200x - 10{,}121{,}700$ **b.** 288,700 **c.** 2042 **11. a.** $y = 3x - 7$ **b.** (0, −7) **c.** (4, 5) **d.** $5 = 3(4) - 7$? $5 = 12 - 7$? $5 = 5$? Yes, it checks. **13. a.** −3; (14, 68) **b.** $y = -3x + 110$ **15. a.** *p* **b.** *n* **c.** *q* **17. See below.** **19.** No, (1, 3) and (−3, −5) lie on a line with slope 2, while (1, 3) and (3, 6) lie on a line with slope $\frac{3}{2}$. **21. a.** *A* **b.** *C* **23.** 455 adults **25. a.** $n = \sqrt{24}$ or $n = -\sqrt{24}$ **b.** $n = 576$ **c.** $n = 4$

17.

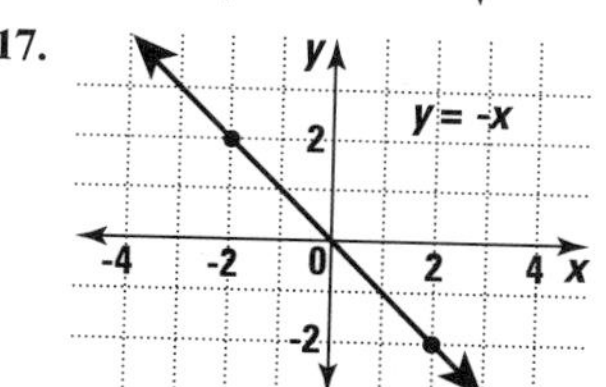

LESSON 7-6 (pp. 450–455)

11. $y = -\frac{5}{6}x + 5$ **13. a.** $C = \frac{5}{9}F - \frac{160}{9}$ **b.** ≈ 65.6°C **c.** 302°F **15.** $S = 180n - 360$ **17.** $y = 7x - 42$ **19.** $a = 10$ **21. a.** 8 or −8 **b.** −1 or 15 **c.** $-\frac{1}{3}$ or 5

LESSON 7-7 (pp. 458–461)

11. No. Negative values of x would result in values of y that are always greater than 107.6°F. **13.** Yes **15.** No **17.** $y = -\frac{7}{4}x + 7$ **19. a.** $-\frac{3}{5}$; $\frac{2}{5}$ **b. See right.** **21. a.** 0 **b.** −2 **23.** about 91 points **25.** $2a + 4c$

19. b.

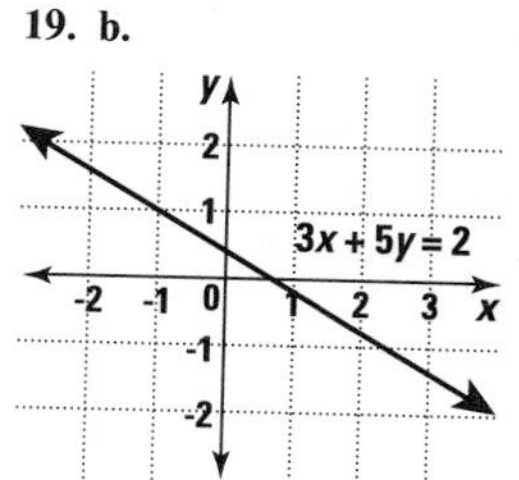

SELECTED ANSWERS

LESSON 7-8 (pp. 463–468)

15. a. $2x + 4y = 100$ **b.** Samples: (2, 24); (10, 20); (30, 10) **17. a.** $y = -\frac{A}{B}x + \frac{C}{B}$ **b.** slope $= -\frac{A}{B}$; y-intercept $= \frac{C}{B}$ **19. a. See below. b.** $y = .38x - 686$ **c.** ≈ 74 meters **d.** Sample: The line is not exact, or there may be a threshold distance beyond which it is physically impossible to throw. **21. a.** No; $\frac{0 - (-2)}{10 - 8} = \frac{2}{2} = 1 \neq 2.$ **b.** Yes; $\frac{18 - (-2)}{18 - 8} = \frac{20}{10} = 2.$ **23. See below. 25.** Yes

19. a.

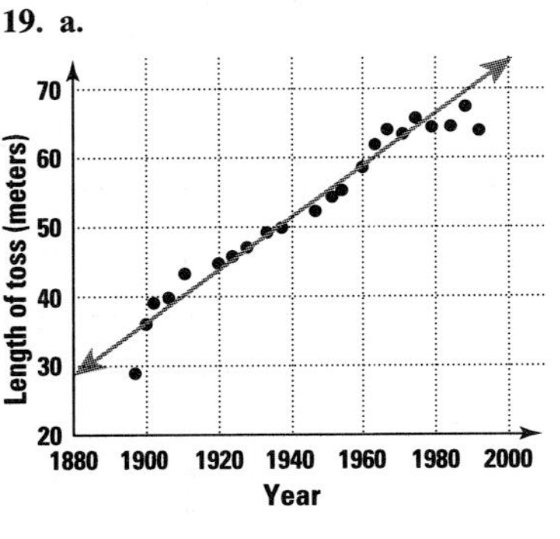

23.

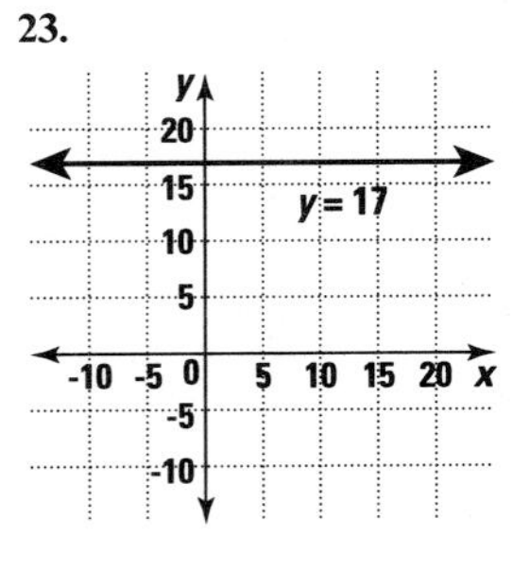

LESSON 7-9 (pp. 469–475)

13. a. $2W + T \geq 20$ **b. See below. 15. a.** 6 **b. See below. 17. a.** Sample: Average annual snowfall is linearly related to the latitude. **b.** Sample: $y = 5x - 172$ **c.** Sample: 8 inches **d.** Sample: altitude, geographic position **19.** $-\frac{1}{3}$ **21. a.** 1.44% **b.** 1.26% **c.** No **23. a.** 0.1 **b.** .03 **c.** .00001

13. b.

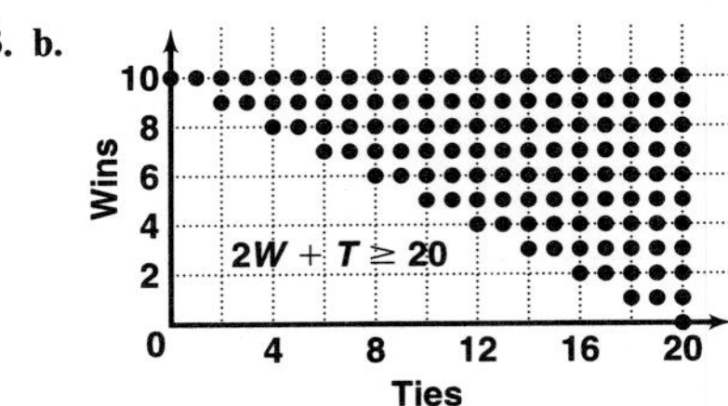

15. b.

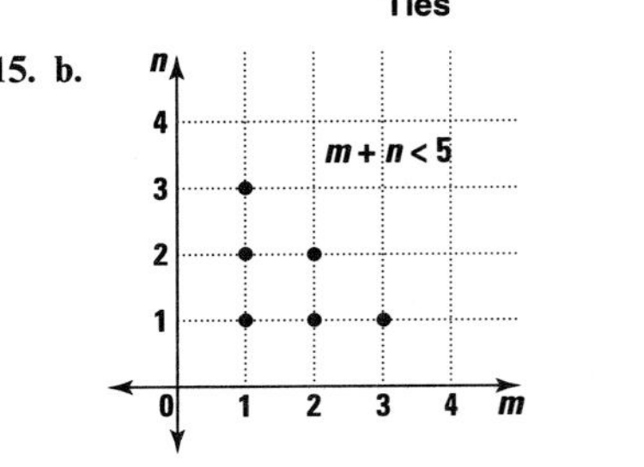

CHAPTER 7 PROGRESS SELF-TEST (p. 479)

1. y-intercept $= 5$ **2.** x-intercept $= 2$ **3.** $\frac{0 - 5}{2 - 0} = \frac{-5}{2}$ **4.** $\frac{1 - (-5)}{-2 - 4} = \frac{6}{-6} = -1$; $\frac{-20 - 1}{20 - (-2)} = \frac{-21}{22}$. No; the slope of the line between the first two points (−1) is not the same as the slope of the line between the last two points $\left(-\frac{21}{22}\right)$. **5.** slope $= -4$; y-intercept $= 8$ **6.** Rewrite as $y = \frac{-5}{2}x + \frac{1}{2}$, so the slope is $\frac{-5}{2}$ and the y-intercept is $\frac{1}{2}$. **7.** $y = \frac{3}{4}x + 13$ **8.** $-5x + y = -2$; $A = -5$, $B = 1$, $C = -2$ **9.** Slope is 0. **10.** up $\frac{3}{5}$ unit **11.** Do a quick estimate. Between 1942 and 1943 the increase was from about 3 million to 7 million, or an increase of 4 million personnel. From 1943 to 1944 the increase was a little under 1 million. From 1944 to 1945 the increase was about 270,000. From 1945 to 1946 there was a decrease. So the greatest increase was between 1942 and 1943. **12.** $\frac{1{,}889{,}690 - 3{,}074{,}184}{1946 - 1942} = -296{,}123.5$ personnel per year **13.** $2x + y = 67$ **14. a. See right. b.** Substitute $m = -2$ and $(x, y) = (-5, 6)$ in the equation $y = mx + b$ to solve for b. Since $b = -4$, the equation of the line is $y = -2x - 4$. **15.** First find the rate of increase of weight; $\frac{50 - 43}{14 - 12} = \frac{7}{2}$ kg per year. Next, substitute $m = \frac{7}{2}$ and $(x, y) = (14, 50)$ in $y = mx + b$ and solve for b; $50 = \frac{7}{2} \cdot 14 + b$, $50 = 49 + b$, $b = 1$. Therefore, the equation of the line is $y = \frac{7}{2}x + 1$. **16. See right. 17. See right. 18.** First graph $y = x + 1$. It has y-intercept 1 and slope 1. Then test (0, 0) in $y < x + 1$; $0 < 0 + 1$, $0 < 1$. So (0, 0) is a solution and the region below the line is shaded. **See right. 19.** c is the only line with negative slope and positive y-intercept. **20. a.** $\overline{AB}$, since it increases from left to right. **b.** $\overline{DE}$, since it is vertical. **21.** Sample: **a.** (60, 70); (120, 140) **See right. b.** Answers will vary. Sample: $\frac{140 - 70}{120 - 60} = \frac{7}{6}$ **c.** Answers will vary. Sample: Substitute (60, 70) in $y = \frac{7}{6}x + b$ to solve for b. Since $b = 0$, an equation of the line is $y = \frac{7}{6}x$. **d.** Answers will vary. Sample: In the equation $y = \frac{7}{6}x$, substitute 100 for x. $y = \frac{7}{6} \cdot 100 \approx 117$; about 117 feet

14. a.

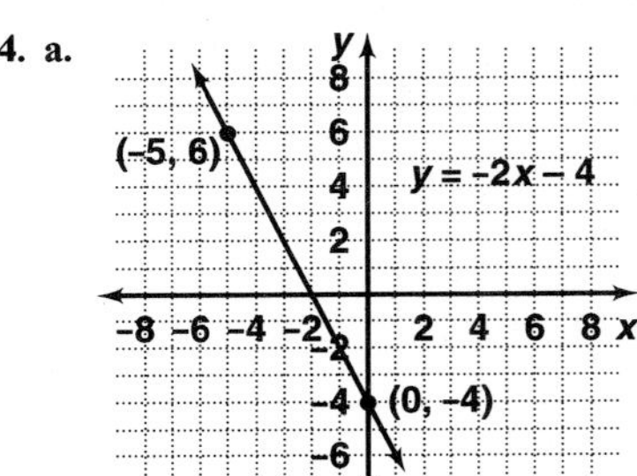

16.

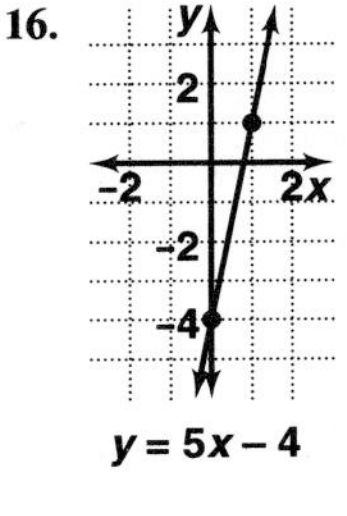

17.

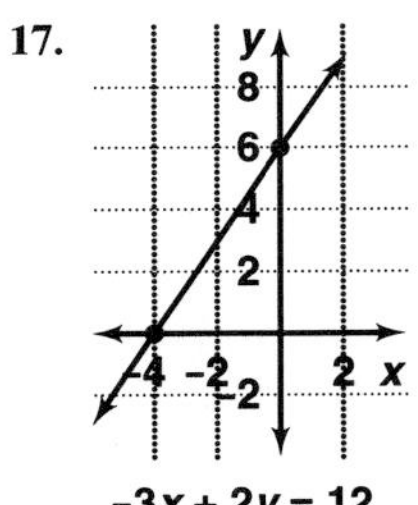

18.

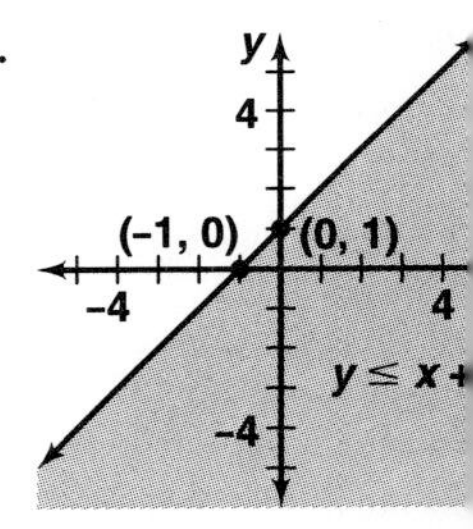

21. a.

Length and wingspan of 2- and 3-engine jets

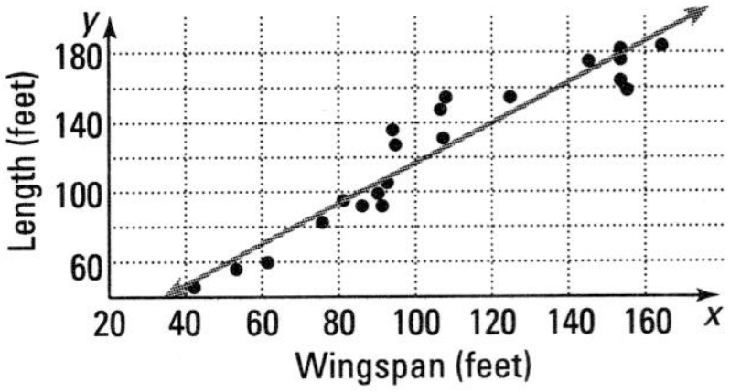

Length (feet): 60, 100, 140, 180

Wingspan (feet): 20, 40, 60, 80, 100, 120, 140, 160

The chart below keys the **Progress Self-Test** questions to the objectives in the **Chapter Review** on pages 480–483. This will enable you to locate those **Chapter Review** questions that correspond to questions missed on the **Progress Self-Test.** The lesson where the material is covered is also indicated on the chart.

Question	**1**	**2**	**3**	**4**	**5**	**6**	**7**	**8**	**9**	**10**
Objective	Voc.	Voc.	A	D	C	C	B	C	D	D
Lesson	7-4	7-5	7-2	7-2	7-4	7-4	7-4	7-8	7-3	7-3
Question	**11**	**12**	**13**	**14**	**15**	**16**	**17**	**18**	**19**	**20**
Objective	E	E	F	B, H	F	H	H	I	H	D
Lesson	7-1	7-1	7-8	7-5	7-6	7-4	7-8	7-9	7-4	7-3
Question	**21**									
Objective	G									
Lesson	7-7									

CHAPTER 7 REVIEW (pp. 480–483)

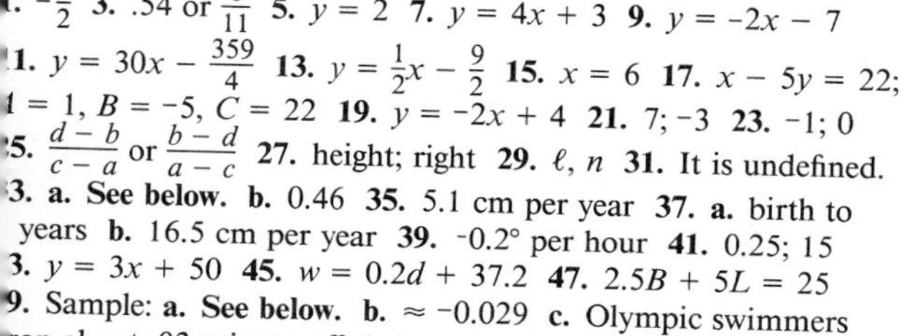

1. $-\frac{1}{2}$ **3.** $.\overline{54}$ or $\frac{6}{11}$ **5.** $y = 2$ **7.** $y = 4x + 3$ **9.** $y = -2x - 7$ **11.** $y = 30x - \frac{359}{4}$ **13.** $y = \frac{1}{2}x - \frac{9}{2}$ **15.** $x = 6$ **17.** $x - 5y = 22$; $A = 1$, $B = -5$, $C = 22$ **19.** $y = -2x + 4$ **21.** 7; −3 **23.** −1; 0 **25.** $\frac{d-b}{c-a}$ or $\frac{b-d}{a-c}$ **27.** height; right **29.** ℓ, n **31.** It is undefined. **33. a.** See below. **b.** 0.46 **35.** 5.1 cm per year **37. a.** birth to 4 years **b.** 16.5 cm per year **39.** -0.2° per hour **41.** 0.25; 15 **43.** $y = 3x + 50$ **45.** $w = 0.2d + 37.2$ **47.** $2.5B + 5L = 25$ **49.** Sample: **a.** See below. **b.** ≈ -0.029 **c.** Olympic swimmers drop about .03 minute off their racing time each year. **d.** $y = -0.029x + 61.7$ **e.** 3.82 min or 3:49.2 min **51.** See below. **53.** See below. **55.** See right. **57.** See right. **59.** half-planes **61.** See right. **63.** See right. **65.** See right. **67.** See right.

33. a.

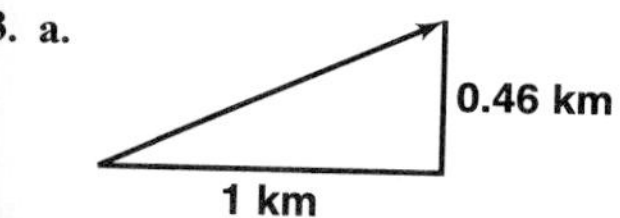

49. a.

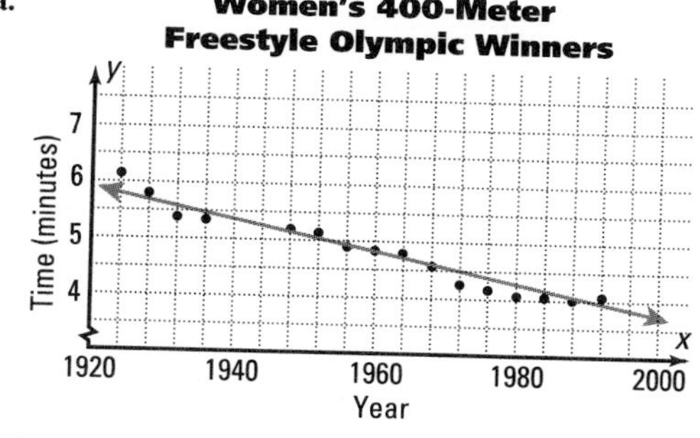

51.

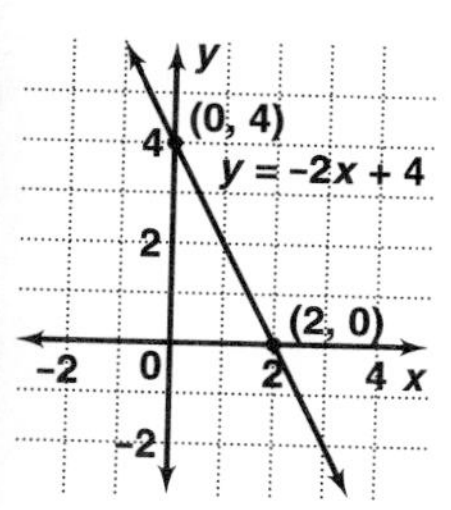

53.

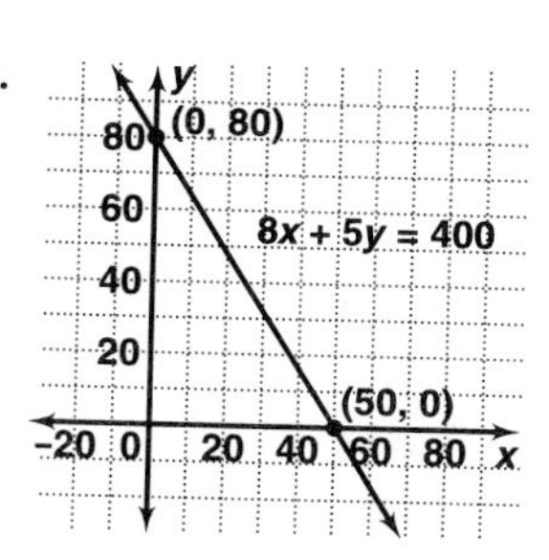

55.

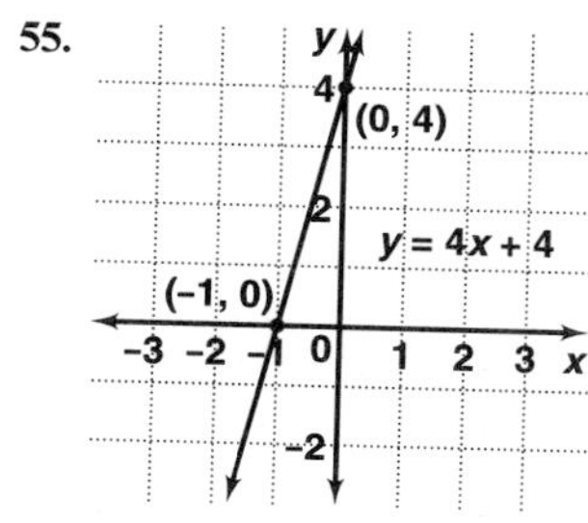

57.

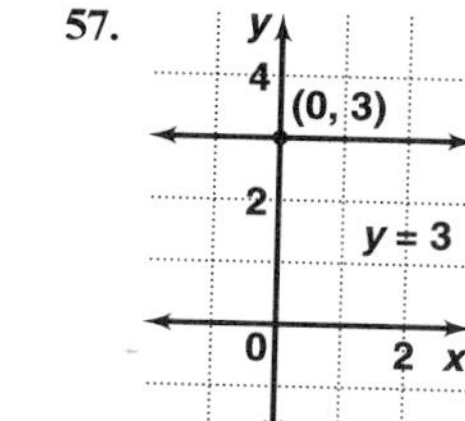

61.

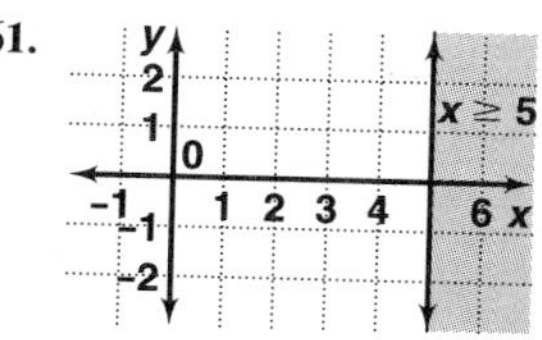

63.

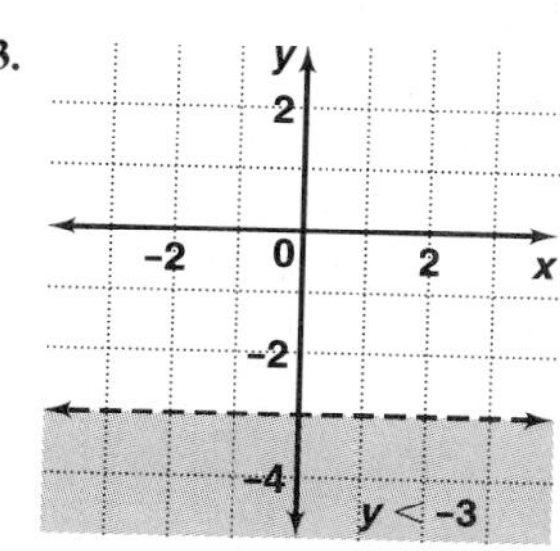

65.

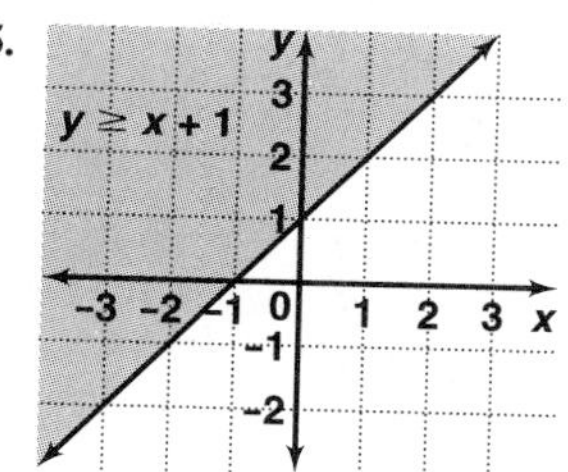

67.

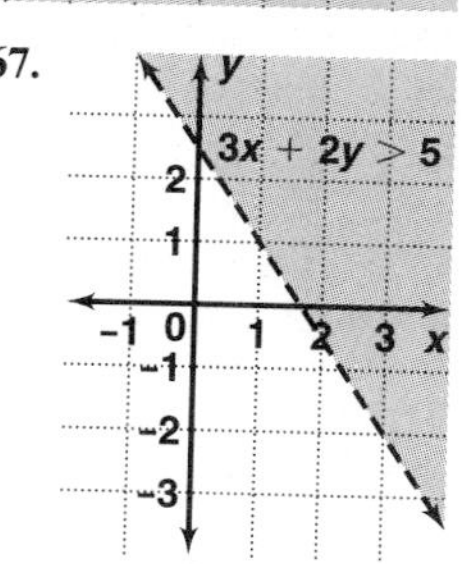

LESSON 8-1 (pp. 486–491)

15. (a); $P(1.06)^5 > P(1.1)^3$ **17. a.** = 1000*1.015^A4–1000*1.015^A3 **b.** 15.45 **19. a.** $T = 2W + 7$ **b. See below.** **21.** $\frac{3}{13}$ **23. a.** $36n$ **b.** $36n - 84$ **c.** $-36n + 84$ **d.** $3n^2 - 7n$ **25.** $-\frac{1}{8}$ **27.** 6

19. b.

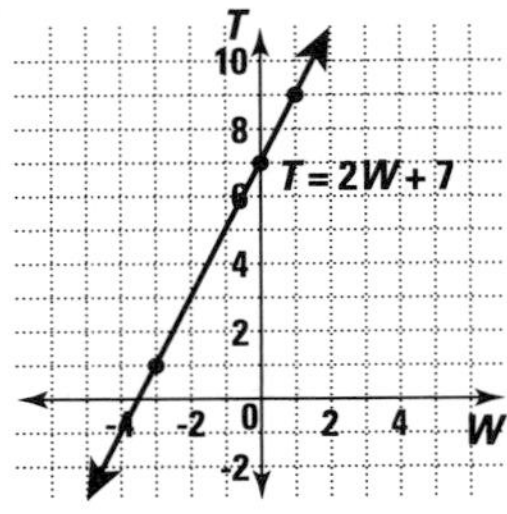

LESSON 8-2 (pp. 492–497)

15. a. 20 minutes **b.** 6 **c.** 1,458,000 **d.** 1,062,882,000 **17.** about 5.14 trillion dollars **19.** 343 **21.** $\frac{8}{9}$ **23.** $\frac{3}{26}$ **25.** $14n + 44$ **27.** $\frac{1}{128}$

LESSON 8-3 (pp. 498–504)

11. a. linear **b.** line **13. a.** exponential **b.** curve

15. a.

	A	B	C
1	years from now	constant increase	exponential
2	0	2520	2520
3	1	2640	2646
4	2	2760	2778
5	3	2880	2917
6	4	3000	3063
7	5	3120	3216

b. There are 96 more students if the growth is exponential. **c.** 4,320 students **d.** 5,239 students **17.** (c) **19.** (b) **21. a.** 100*1.06^YEAR **b.** 20 320.7135 **c.** 30 FOR YEAR = 1 TO 100 **d.** 100 33930.2084 **23. a. See below. b.** Sample: (1, 41), (6, 19) **c.** Sample: $y = -4.4x + 45.4$ **d.** about 12.4 in. **25.** $6a^3 + 12a^2 - 2a$ **27. a.** $\frac{9}{x}$ **b.** $\frac{13}{2y}$ **c.** $\frac{8z + 5}{2z}$ **29.** −13,824

23. a.

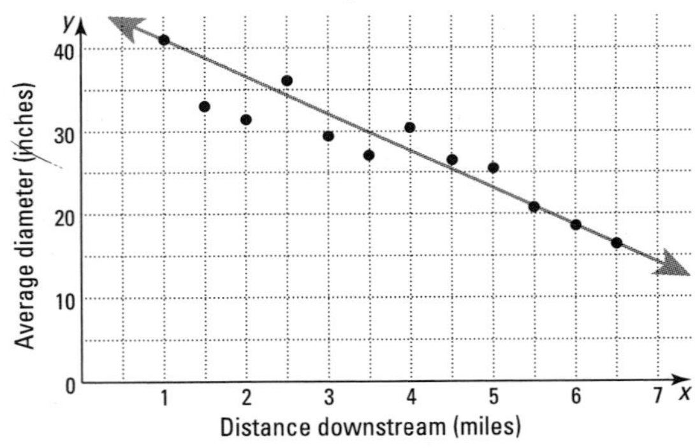

LESSON 8-4 (pp. 505–509)

13. True **15. a.** A decays exponentially; the growth factor is $\frac{1}{2}$. **b.** t grows exponentially; the growth factor is 2. **17. a.** 5832 **b.** 1.033174×10^{-2} **c.** $P = 100{,}000$; $X = 1.02$; $N = 10$ **19.** $k = 30(1.05)^n$ **21.** about 6,230,000 **23.** 1 **25. a.** $\frac{1}{2}y^3$ **b.** Sample: $\frac{1}{2}y$, y, y

LESSON 8-5 (pp. 510–514)

17. $a = 7$ **19.** Samples: x^1, x^6; x^3, x^4; x^0, x^7 **21. a.** $P \cdot 3^5$ **b.** 12 **23.** $2x^7$ **25.** x^{14} **27.** $y^8 - y^3$ **29. See below. 31.** Sample: the population T of a city of 5000 that is growing at a rate of 3.5% pe year, n years from now **33.** \$4.49/lb **35. a.** −1; 1; −1; 1; −1; 1; −1; 1 **b.** 1 **37.** 3.6×10^{-4}

29.

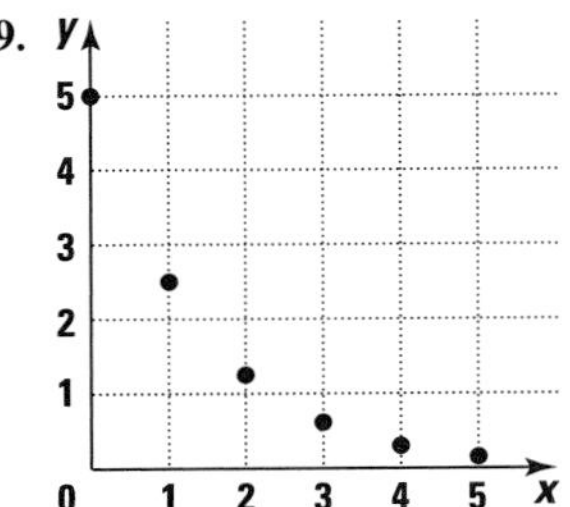

LESSON 8-6 (pp. 515–520)

19. a. See below. b. The y-coordinate approaches zero. **21.** $z = -2$; Check answers using a calculator. **23.** t^{-6} **25. a.** ≈ 5.82 billion **b.** ≈ 4.60 billion **27.** $12x^3$ **29.** $22c^7$ **31.** (c **33.** $y = -5x + 9$ **35. a.** $28a - 8$ **b.** $13a - 11$ **c.** $a = 2$

19. a.

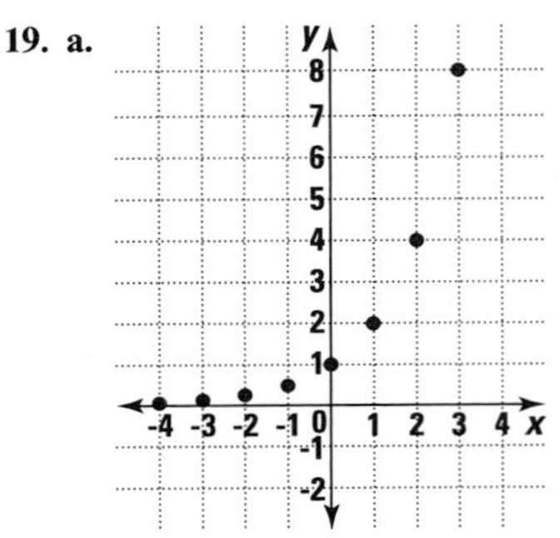

LESSON 8-7 (pp. 521–526)

15. $\frac{6 \cdot 10^9}{2.56 \cdot 10^8} \approx 2.34 \cdot 10^1 \approx 23 \frac{\text{pounds}}{\text{person}}$ **17.** 3^6 **19.** x^n **21.** 1 **23.** $5p$ **25.** about 0.40 $\frac{\text{people}}{\text{km}^2}$ **27.** 4^{x+y} **29.** $-2y^{16}$ **31.** x^{15}; $3^5 \cdot 3^5 \cdot 3^5 = 243 \cdot 243 \cdot 243 = 14{,}348{,}907$; $3^{15} = 14{,}348{,}907$. **33. a.** slope = −2, y-intercept = 54; The elevator descends at a ra of 2 floors per second and started on the 54th floor. **b. See below** **35.** 1000 times as much

33. b.

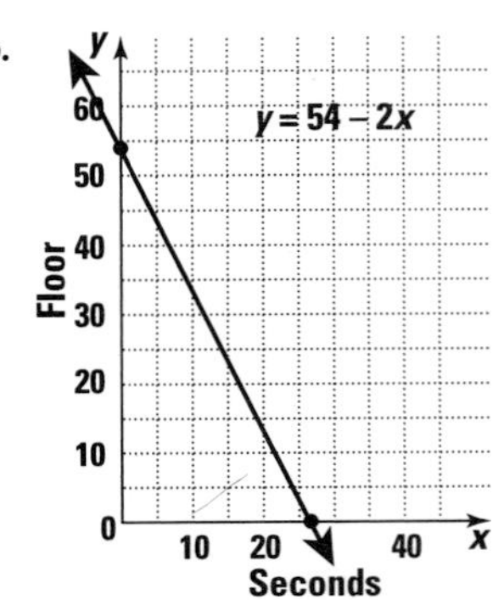

LESSON 8-8 (pp. 527–532)

17. $18x^2$ **19.** $\frac{u^t}{3^t}$ **21.** $\frac{16z^5}{2401}$ **23. a.** $\left(\frac{2}{3}\right)^5 = \frac{32}{243} \approx 0.13$ **b.** Example **25.** (a) **27.** 1728 **29.** $\frac{1}{5}$ **31.** k^3 **33.** v^{-6} **35. a.** $2x^5$ **b.** x^9 **c.** x^{10} **37. a.** \$6.00 − \$.12 = \$5.88 **b.** \$60.00 + \$.20 = \$60.20 **c.** \$6.

LESSON 8-9 (pp. 533–538)

11. Sample: Let $x = 3$. Then $(2x)^3 = (2 \cdot 3)^3 = 6^3 = 216$ and $2x^3 = 2 \cdot 3^3 = 2 \cdot 27 = 54$. **13.** Power of a Product Property **15.** Negative Exponent Property **17.** Sample: Let $a = 2$ and $b = 3$. Then $2^2 + 3^2 = 13$ and $(2 + 3)^2 = 5^2 = 25$. The counterexample shows that the pattern is not true. **19.** $\frac{36}{625x^4}$ **21.** $\frac{25a^9}{2b^3}$ **23.** $2n^4$ **25.** $a = 20$ **27.** 2^{1492} **29. a.** See right. **b.** See right. **c.** Sample: Of two accounts, each starting with \$100, which has more money in it at the end of 10 years, one that earns 6% compound interest or one that has \$6 deposited in it each year? Answer: the one that earns 6% compound interest. **31.** (a); $5m^2$

29. a.

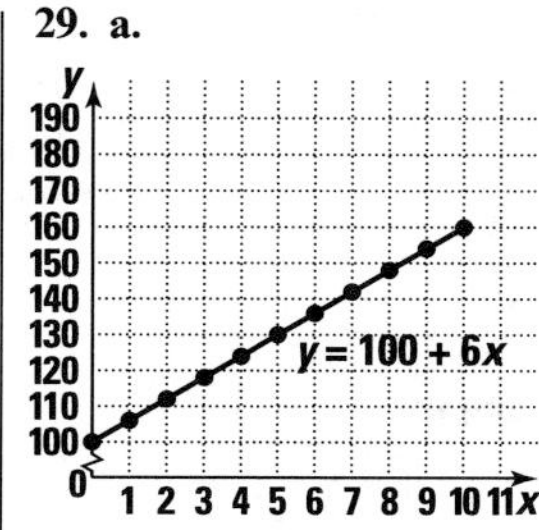

29. b.

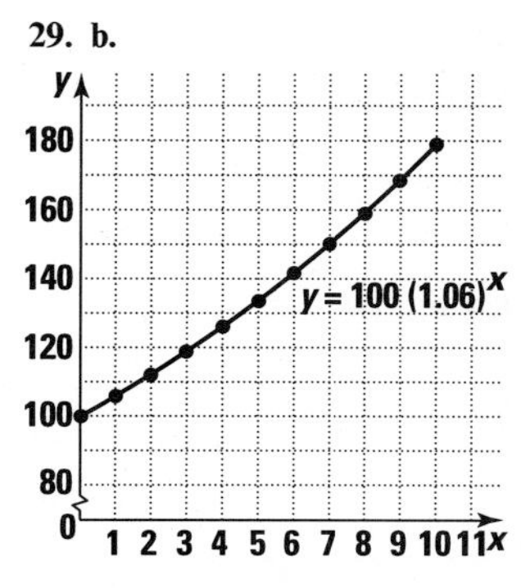

CHAPTER 8 PROGRESS SELF-TEST (p. 542)

1. a. $\frac{4^{12}}{4^6} = 4^{12-6} = 4^6 = 4096$; Apply the Quotient of Powers Property. **b.** $\frac{4^{12}}{4^6} = \frac{16777216}{4096} = 4096$ **2.** $\frac{5 \cdot 10^{20}}{5 \cdot 10^{10}} = \frac{10^{20}}{10^{10}} = 10^{20-10} = 10^{10} = 10{,}000{,}000{,}000$ **3.** $(8)^{-5} = \frac{1}{8^5} = \frac{1}{32768}$ **4.** $b^7 \cdot b^{11} = b^{7+11} = b^{18}$ **5.** $(5y^4)^3 = 5^3(y^4)^3 = 125y^{4 \cdot 3} = 125y^{12}$ **6.** $\frac{3z^6}{12z^4} = \frac{z^6}{4z^4} = \frac{1}{4}(z^{6-4}) = \frac{z^2}{4}$ **7.** $(y^{10})^4 = y^{10 \cdot 4} = y^{40}$ **8.** $\left(\frac{3}{x}\right)^2 \cdot \left(\frac{x}{3}\right)^4 = \frac{3^2}{x^2} \cdot \frac{x^4}{3^4} = \frac{x^2}{3^2} = \frac{x^2}{9}$ **9.** $\frac{48a^3b^7}{12a^4b} = \frac{4b^6}{a} = 4a^{-1}b^6$ **10.** 6 **11.** Sample: $3 \cdot 2^2 = 3 \cdot 4 = 12$ and $(3 \cdot 2)^2 = 6^2 = 36$; $12 \neq 36$. **12.** Product of Powers Property **13.** $\$6500(1.05)^5 \approx \$8295.83$ **14.** $1900(1.058)^3 \approx 2250.15$; interest earned = 2250.15 − 1900 = \$350.15 **15.** $135{,}000(1.03)^5 \approx 157{,}000$ **16.** $135000(1.03)^{-2} \approx 127{,}000$ **17. a.** exponential **b.** not exponential **c.** not exponential **d.** exponential **18.** See right. **19.** $(1.30)^3 = 2.197$ times as large **20.** $V = \frac{4}{3}\pi(6.96 \cdot 10^6)^3 \approx 1.41 \cdot 10^{21}$ km^3

18.

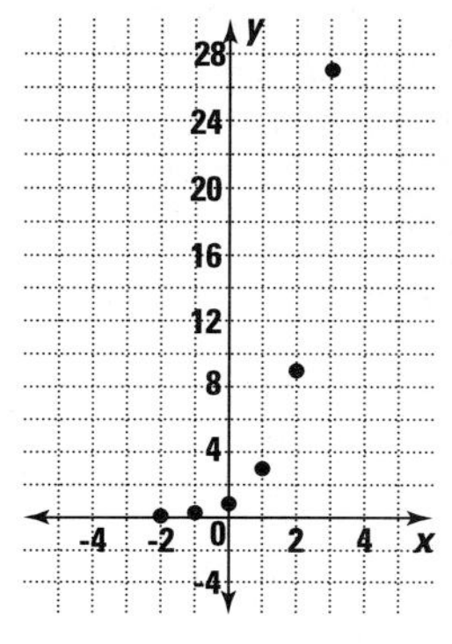

The chart below keys the **Progress Self-Test** questions to the objectives in the **Chapter Review** on pages 543–545. This will enable you to locate those **Chapter Review** questions that correspond to questions missed on the **Progress Self-Test.** The lesson where the material is covered is also indicated on the chart.

Question	**1**	**2**	**3**	**4**	**5**	**6**	**7**	**8**	**9**	**10**
Objective	A	A	A	B	C	B	B	C	B	A
Lesson	8-7	8-7	8-6	8-5	8-8	8-7	8-5	8-8	8-7	8-2
Question	**11**	**12**	**13**	**14**	**15**	**16**	**17**	**18**	**19**	**20**
Objective	D	E	F	F	G	G	I	I	H	H
Lesson	8-9	8-5	8-1	8-1	8-2	8-6	8-4	8-6	8-3	8-8

CHAPTER 8 REVIEW (pp. 543–545)

a. 81 **b.** −81 **c.** 81 **3.** 4 **5.** 8 **7.** $\frac{1}{125}$ **9.** $\frac{8}{343}$ **11.** 81 **13.** x^{11} **5.** x^3y^{12} **17.** n^{13} **19.** $\frac{c}{a}$ **21.** $28x^{15}$ **23.** $5^{-1}m^4$ **25.** 1000 **27.** $\frac{x}{y^2}$ **.** $\frac{x^3}{y^3}$ **31.** $1024x^5$ **33.** $\frac{32}{n^5}$ **35.** $-27n^3$ **37.** $\frac{4k^3}{27}$ **39.** $32x^2$ **. a.** True **b.** True **c.** False **d.** False **43.** Sample: $(1 + 1)^3 = \;= 8$; $1^3 + 1^3 = 1 + 1 = 2$. **45.** Power of a Product Property **.** Zero Exponent Property **49.** Power of a Quotient Property **.** Negative Exponent Property **53.** Samples: $\left(\frac{x^3}{x}\right)^8 = (x^{3-1})^8$; $)^8 = x^{16}$; $\left(\frac{x^3}{x}\right)^8 = \frac{x^{3 \cdot 8}}{x^{1 \cdot 8}} = \frac{x^{24}}{x^8} = x^{24-8} = x^{16}$ **55.** \$2952.33 **.** \$1348.32 **59.** \$9.02 per hour **61. a.** 128,000 **b.** After 4 hours ere will be 128,000 bacteria. **63.** $P = 1{,}500{,}000 \cdot (0.97)^n$ **.** 1,500,000; the population now **67. a.** ≈ 459 **b.** The death e three years earlier (1977) was about 459 per 100,000 people. **.** $(0.9)^x$ **71.** 5 billion cubic miles **73.** $\left(\frac{1}{3}\right)^4 = \left(\frac{1}{81}\right)$ **.** exponential **77.** linear

79.

x	y
−3	15.625
−2	6.25
−1	2.5
0	1
1	0.4
2	0.16
3	0.064

See right.

81. $y = 5 \cdot (1.04)^x$; because if the growth factor g is greater than one, exponential growth always overtakes constant increase.

79.

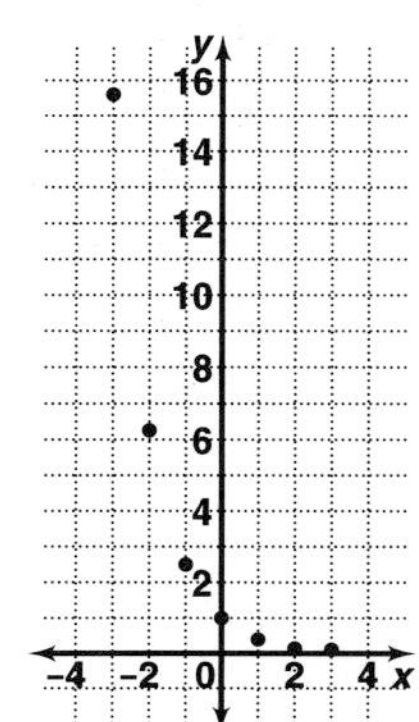

LESSON 9-1 (pp. 548–553)

15. about 1.46 seconds **17.** 11; −13 **19.** x^7y^4 **21.** $\frac{8a^3}{125}$ **23.** It is at least 5 km and at most 11 km since $8 - 3 = 5$ and $8 + 3 = 11$. **25.** $7500 - 1800 - k$ or $5700 - k$

LESSON 9-2 (pp. 554–560)

15. (c) **17.** (a) **19.** (b); Sample: if $x = 1$, $y = 1 - 6 + 8 = 3$. Graph (b) contains this point. **21.** He must change the y-values in his table to their opposites, and turn his graph upside down. **23.** $\sqrt{640} \approx 25.3$ or $-\sqrt{640} \approx -25.3$ **25.** $\frac{c + d + r}{(c - 3) + d + r}$

LESSON 9-3 (pp. 562–566)

11. a. **See below.** **b.** Sample: All the parabolas look alike in the window. They have different vertices and are translation images of each other. **c.** It will be congruent to those in part **a.** It will be a translation image of them and have vertex (0, 6). **d.** **See below.** **13. a.** **See below.** **b.** (−1.25, 1.25), (4, 17) **15. a.** **See below.** **b.** 1 **17.** $x = 7$ or $x = -7$ **19. a.** 4 miles **b.** for distances less than 4 miles; $.45 + 1.20x < 1.25 + 1.00x$; $x < 4$ **21. a.** 420 **b.** 456 **c.** ≈ 21.4

11. a.

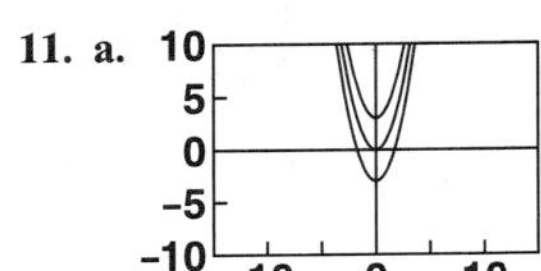

11. d.

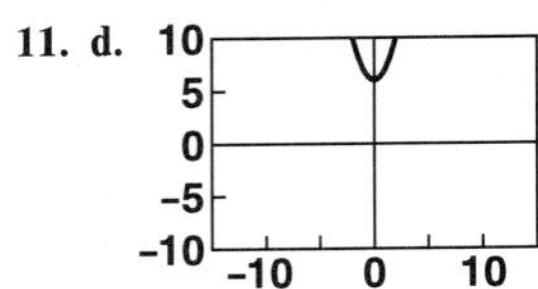

13. a.

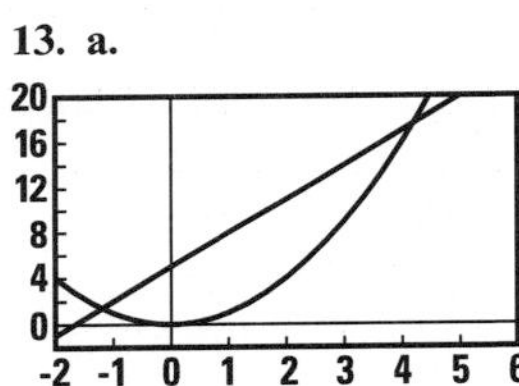

15. a.

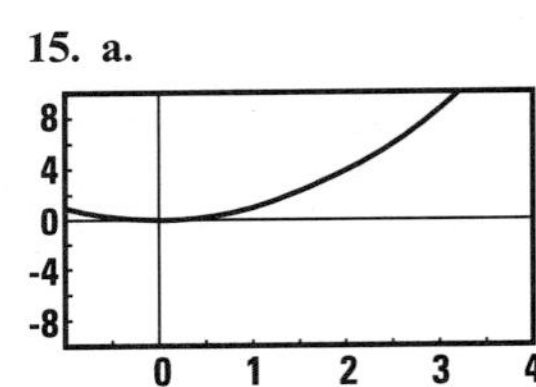

LESSON 9-4 (pp. 567–572)

13. 25 meters **15.** 35 meters **17.** 10.125 m above the surface of the water

19. a.

x	y
−3	$\frac{9}{2}$
−2	2
−1	$\frac{1}{2}$
0	0
1	$\frac{1}{2}$
2	2
3	$\frac{9}{2}$

b. **See right.** **21.** $A = (8, 4)$; $B = (9, 7)$ **23.** (c) **25.** (−3, −5) **27.** ≈3.4

19. b.

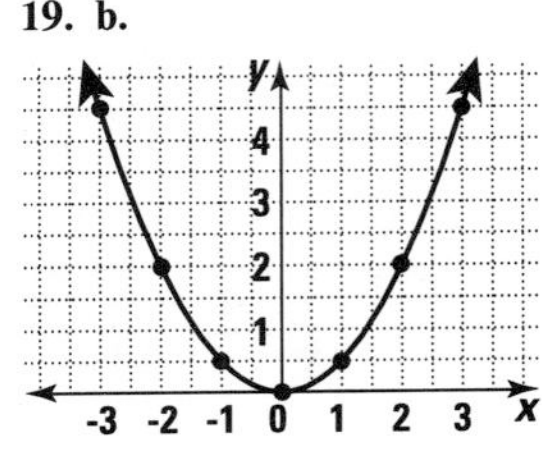

LESSON 9-5 (pp. 573–578)

15. a. 2.5; 5.0 **b.; c.** **See right.** **17.** 5.1 meters above the cliff **19.** 10 ft **21.** ≈ 13.5 ft **23.** (d) **25.** 540

15. b.; c.

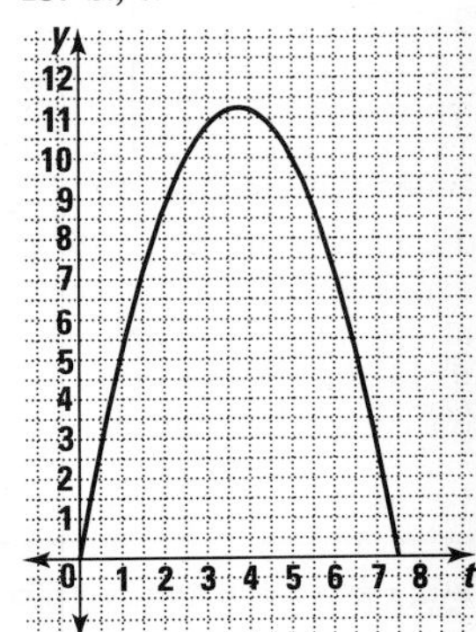

LESSON 9-6 (pp. 579–585)

11. a. 5 meters **b.** 0.4 seconds and 1.6 seconds **c.** ≈ 2.4 seconds **13.** 9 **15. a.** $0 = -16t^2 + 28t - 12$ **b.** positive **c.** 2 **17.** 12 feet **19.** down **21.** Sample: $y = -x^2$ **23. a.** $a < \frac{1}{3}$ **b.** $b > \frac{1}{3}$ **c.** $c > 0$ **25.** 6435

LESSON 9-7 (pp. 586–592)

15. $t = 3\sqrt{3}$ **17.** 6 **19.** 55 **21.** $a = \pm 2\sqrt{3}$; $(2(2\sqrt{3}))^2 = (4\sqrt{3})^2 = 16 \cdot 3 = 48$; $(2(-2\sqrt{3}))^2 = (-4\sqrt{3})^2 = 16 \cdot 3 = 48$ **23. a.** $5\sqrt{3}$ **b.** $2\sqrt{3}$ **c.** $7\sqrt{3}$ **25.** Sample: $w\sqrt{20}$, $2w\sqrt{5}$ **27.** It equals zero. **29.** 49 meters **31.** 2 **33.** 1 **35.** $0.25q + 0.10d \geq 5.20$

LESSON 9-8 (pp. 593–598)

23. 0; no real solutions **25.** 2; $q = -31$ or $q = 31$ **27.** −10; 4 **29. a.** $x - y$ **b.** $y - x$ **c.** $|y - x|$ or $|x - y|$ **31.** $\sqrt{7}$ **33.** $4 \pm 3\sqrt{2}$ **35. a.** $10\sqrt{2}$ ft **b.** 14.1 ft **c.** $10\sqrt{5} \approx 22.4$ ft **37.** $x = 4 + 3\sqrt{2}$ or $x = 4 - 3\sqrt{2}$

LESSON 9-9 (pp. 599–604)

15. 7 km **17.** $\sqrt{5} \approx 2.2$ km **19.** $\sqrt{|a|^2 + |b|^2}$ **21.** $JK = 10$, $KL = 17$, $JL = 21$ **23.** 29.732137 **25.** $x = 3.5$ or $x = -3.5$ **27. a.** $t = \pm 3$ **b.** $t = 81$ **c.** $t = 9$ or $t = -9$ **29.** 2; the line $y = 1$ intersects the graph twice. **31. a.** $\frac{-9 \pm \sqrt{101}}{2} \approx 0.52$ or -9.52 **b.** **See below.** **33.** $12\frac{2}{3}$

31. b.

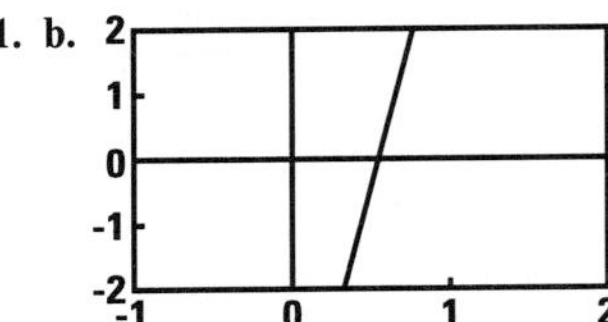

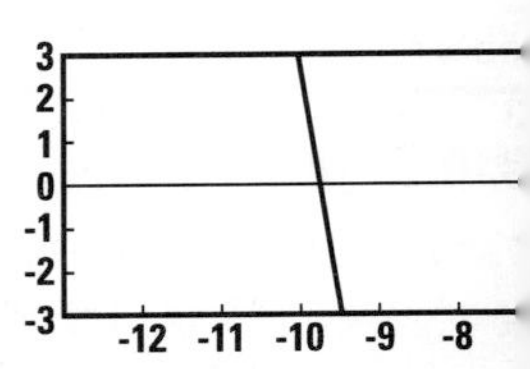

CHAPTER 9 PROGRESS SELF-TEST (p. 609)

1. $\frac{-(-9) \pm \sqrt{(-9)^2 - 4(1)(20)}}{2(1)} = \frac{9 \pm \sqrt{81 - 80}}{2} = \frac{9 \pm 1}{2} = 5$ or 4 **2.** $5y^2 - 3y - 11 = 0$, $\frac{-(-3) \pm \sqrt{(-3)^2 - 4(5)(-11)}}{2(5)} = \frac{3 \pm \sqrt{9 + 220}}{10} = \frac{3 \pm \sqrt{229}}{10}$; $y \approx 1.81$ or $y \approx -1.21$ **3.** no real solutions **4.** $z^2 - 16y + 64 = 0$, $\frac{-(-16) \pm \sqrt{(-16)^2 - 4(1)(64)}}{2(1)} = \frac{16 \pm \sqrt{256 - 256}}{2} = \frac{16}{2} = 8$ **5.** 2 **6.** (a) because as a gets larger than 1 in $y = ax^2$, the graph gets narrower.

7. a.

x	y
−3	−18
−2	−8
−1	−2
0	0
1	−2
2	−8
3	−18

b. See p. 857.

8. a.

x	y
−3	24
−2	15
−1	8
0	3
1	0
2	−1
3	0

b. See p. 857.

9. False. If $a > 0$, the graph opens up. **10.** (2, −2) **11.** (1, 0); (3,

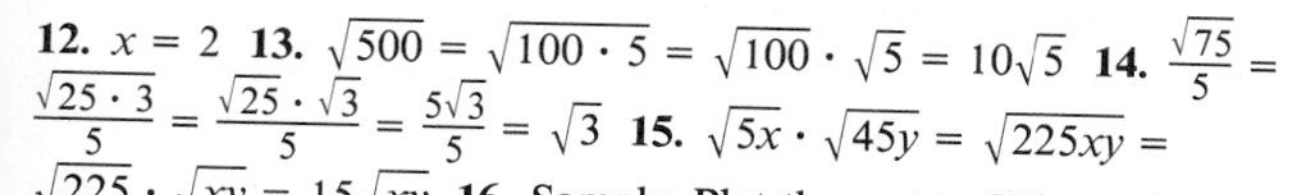
12. $x = 2$ **13.** $\sqrt{500} = \sqrt{100 \cdot 5} = \sqrt{100} \cdot \sqrt{5} = 10\sqrt{5}$ **14.** $\frac{\sqrt{75}}{5} = \frac{\sqrt{25 \cdot 3}}{5} = \frac{\sqrt{25} \cdot \sqrt{3}}{5} = \frac{5\sqrt{3}}{5} = \sqrt{3}$ **15.** $\sqrt{5x} \cdot \sqrt{45y} = \sqrt{225xy} = \sqrt{225} \cdot \sqrt{xy} = 15\sqrt{xy}$ **16.** Sample: Plot the vertex. When a is negative, the graph opens down. You can't tell how narrow or broad the parabola is without more information. **See right.** **17.** (7, −6) **18.** True **19.** $\sqrt{|2 - 7|^2 + |-6 - 4|^2} = \sqrt{|-5|^2 + |-10|^2} = \sqrt{25 + 100} = \sqrt{125} = \sqrt{25 \cdot 5} = \sqrt{25} \cdot \sqrt{5} = 5\sqrt{5}$ **20.** $\sqrt{|3 - x|^2 + |-2 - y|^2}$ **21.** Since $h = 0$ at ground-level, solve $0 = -16t^2 + 21t + 40$. $t \approx 2.4$ seconds **22.** Solve $43 = -16t^2 + 21t + 40$. $t = 0.2$ or 1.1 seconds **23.** The highest point is the vertex located at 10 yards. When $x = 10$, $h = -0.07 \cdot 10^2 + 1.4 \cdot 10 + 5 = 12$. So the vertex is (10, 12). **See right.** **24.** Melody is $20 - 2 = 18$ yards from Harry. So, the ball is $-0.07(18)^2 + 1.4(18) + 5 = 7.52$ ft high when it passes Melody. **25.** $\frac{1}{3}x^2 = 10$ $x^2 = 30$; $x \approx 5.48$ or -5.48 **26.** $x = -57$ or $x = 57$ **27.** $3 - n = 0.5$ or $3 - n = -0.5$; $-n = -2.5$ or $-n = -3.5$; $n = 2.5$ or $n = 3.5$

7. b.

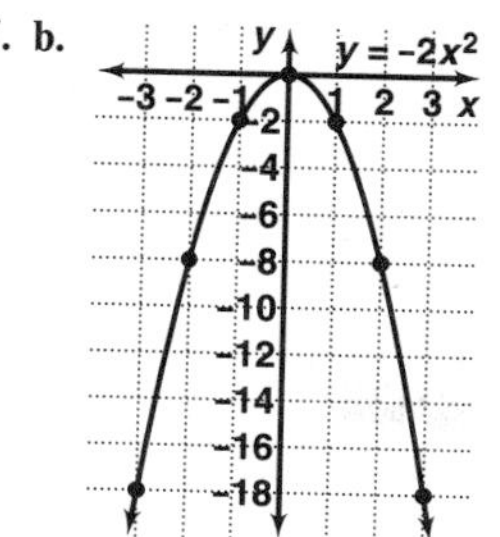

8. b.

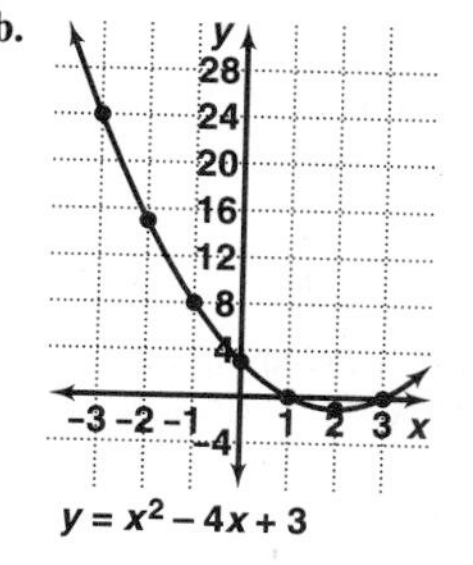

16.

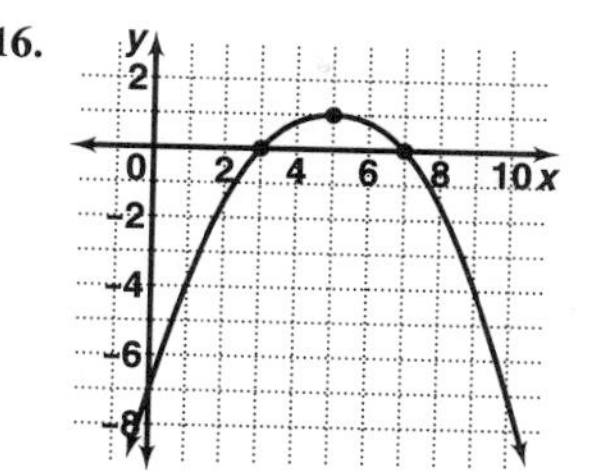

23.

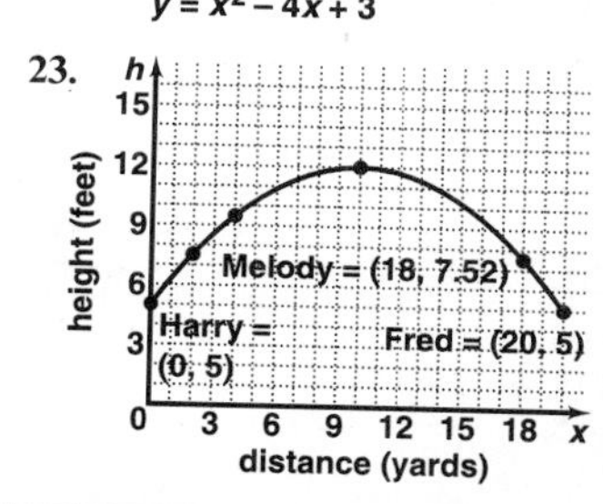

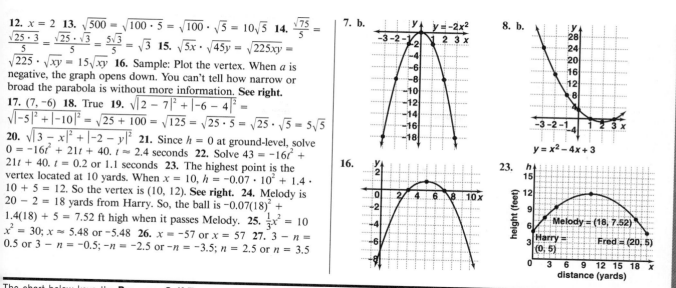

The chart below keys the **Progress Self-Test** questions to the objectives in the **Chapter Review** on pages 610–613. This will enable you to locate those **Chapter Review** questions that correspond to questions missed on the **Progress Self-Test.** The lesson where the material is covered is also indicated on the chart.

Question	**1**	**2**	**3**	**4**	**5**	**6**	**7**	**8**	**9**	**10**
Objective	A	A	A	A	D	F	F	F	F	F
Lesson	9-5	9-5	9-5	9-5	9-6	9-3	9-1	9-2	9-2	9-3
Question	**11**	**12**	**13**	**14**	**15**	**16**	**17**	**18**	**19**	**20**
Objective	F	F	B	B	B	F	G	G	G	G
Lesson	9-3	9-3	9-7	9-7	9-7	9-3	9-9	9-9	9-9	9-9
Question	**21**	**22**	**23**	**24**	**25**	**26**	**27**			
Objective	E	E	E	E	A	C	C			
Lesson	9-4	9-4	9-4	9-4	9-1	9-8	9-8			

CHAPTER 9 REVIEW (pp. 610–613)

1. −5, 5 **3.** −7, 7 **5.** $-\frac{5}{2}, \frac{4}{3}$ **7.** 7 **9.** −0.27, 7.27 **11.** no real solutions **13.** $\frac{5 \pm \sqrt{13}}{2}$, ≈ 4.30 or ≈ 0.70 **15.** 14 **17.** $20\sqrt{2}$ **19.** $10\sqrt{5}$ **21.** $18\sqrt{2}$ **23.** $3 \pm 3\sqrt{6}$ **25.** $x\sqrt{5}$ **27.** 17 **29.** 43 **31.** 6 **33.** −5 **35.** −16; 16 **37.** −7; 7 **39.** 5; 15 **41.** $\frac{-b \pm \sqrt{b^2 - 4ac}}{2a}$ **43.** True **45.** $D = -23$; no real solutions **47.** $D = -1199$; no real solutions **49. a.** 576 ft **b.** ≈ 11.2 seconds **51. a.** See right. **b.** 2.5 and 10 seconds $\frac{-10 \pm \sqrt{10^2 - 4(-0.8)(-20)}}{2(-0.8)} = \frac{-10 \pm \sqrt{100 - 64}}{-1.6} = \frac{-10 \pm 6}{-1.6} =$ 2.5 or 10 **53. a.** ≈ 48 ft **b.** ≈ 0.65 or 3.35 seconds **c.** 4 **d.** Sample: What is the maximum height the ball will reach? **55. a.** (10, 13) **b.** $x = 10$ **c.** $A = (7, 4)$, $B = (8, 9)$, $C = (9, 12)$ **57.** False **59.** $ax^2 + bx + c = 0$

61. a.

x	y
−3	27
−2	12
−1	3
0	0
1	3
2	12
3	27

61. b. See right.

63. a.

x	y
−5	6
−4	2
−3	0
−2.5	−.25
−2	0
−1	2
0	6

63. b. See right.

65. (b) **67. a.** See below. **b.** (20, 16) **c.** maximum **69.** Sample: $-5 \le x \le 15$ and $-10 \le y \le 30$ **71.** 31 **73.** −8, 2 **75.** 5 **77.** 9 **79.** $10\sqrt{26} \approx 50.99$ **81.** 17 **83. a.** See below. **b.** $7\sqrt{2} \approx 9.9$ miles

51. a. **61. b.** **63. b.**

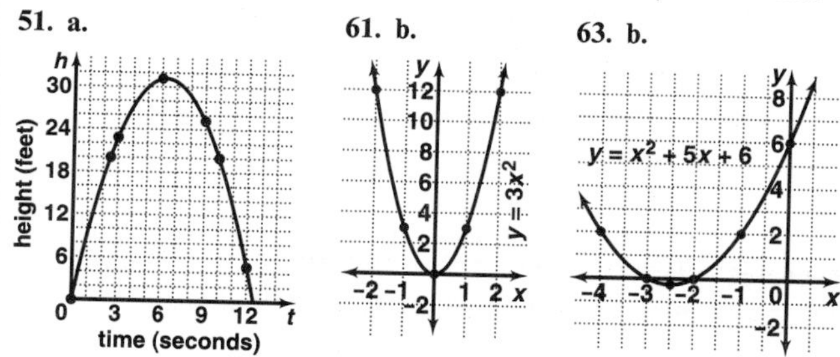

67. b.

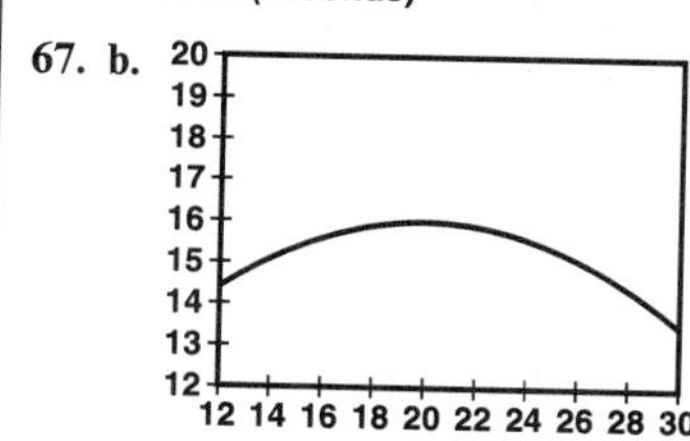

83. a.

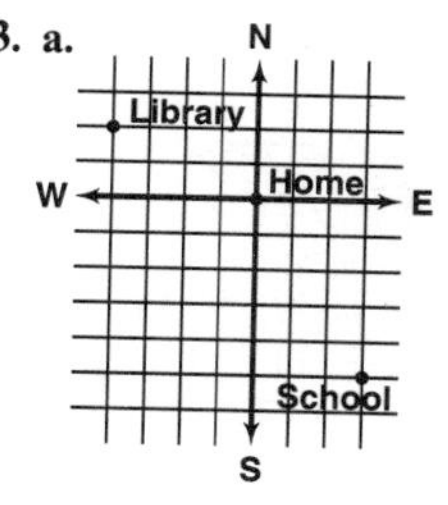

LESSON 10-1 (pp. 616–620)

21. a. $4x$ **b.** 1 **23. a.** $-30n^4$ **b.** 4 **25. a.** $64a^6b^6$ **b.** 12 **27. a.** See below. **b.** See below. **29. a.** $2 \cdot 10^2 + 4 \cdot 10 + 6$; $1 \cdot 10^3 + 3 \cdot 10 + 2$ **b.** The sum is $1 \cdot 10^3 + 2 \cdot 10^2 + 7 \cdot 10 + 8 = 1278$; $246 + 1032 = 1278$; yes the sums are equal. **31. a.** See below. **b.** See below. **33.** See below. **35. a.** $h = 5$ **b.** $r = 5\sqrt{10}$

27. a.

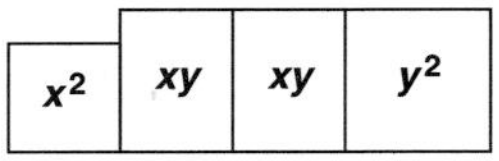

27. b.

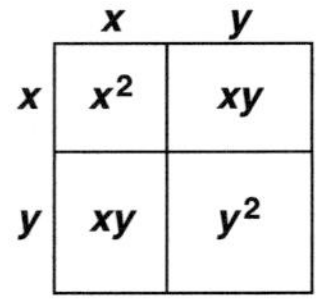

31. a.

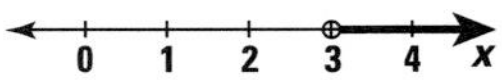

31. b.

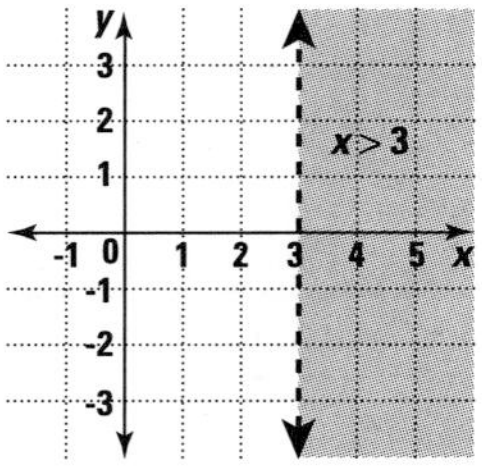

33.

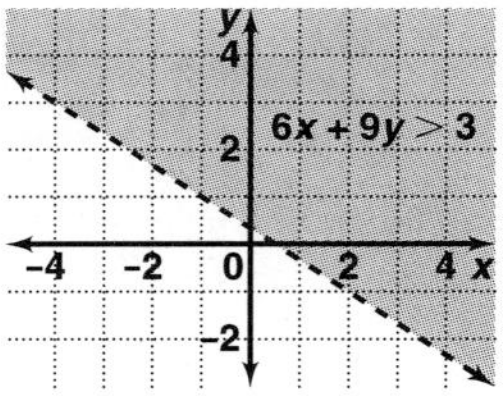

LESSON 10-2 (pp. 621–626)

9. \$100 **11.** \$480.18 **13.** $16y^2 + y - 17$ **15.** $2w^2 - w + 11$ **17.** $-\frac{3}{2}, -\frac{2}{3}$ **19.** $7y^2 - 47$ **21.** Sample: area of a circle with radius r; degree 2 **23.** Sample: amount saved if \$16 was invested at some rate t two years ago, and then \$48 was added to the account one year ago; degree 2 **25.** 61.4 ft **27.** $-\frac{b}{a}$ **29.** (c) **31.** .5 and 1.5

LESSON 10-3 (pp. 627–632)

15. $w(5w - 1)$; $5w^2 - w$ **17.** $8x^2 + 2x$ **19.** $6a^3b^3c^2$ **21.** $ay^2 + ya^2$ **23.** $6n^2$ **25. a.** No. **b.** $2x^{-3}$ cannot be written as a product of variables with nonnegative exponents. **27.** Sample: $-4x^7 + 5x^5 + 1$ **29.** $\frac{3 + 4v}{2v}$ **31.** $y = 3x + 5$ **33. a.** 2 meters per second **b.** Sample: snake

LESSON 10-4 (pp. 633–637)

11. $2n^3 + 7n^2 - 19n - 60$ **13.** $40x - 240$ **15.** $m^2 - 4n^2 - 9p^2 - 16q^2 - 12np - 16nq - 24pq$ **17.** Plan B; At the end of 10 years, Plan B is worth \$1437.86 and Plan A is worth \$1397.16. **19. a.** $10\sqrt{2}$ **b.** Sample: What is the length of the diagonal of a square with side 10? **21.** $\frac{7a^2}{3b}$ **23.** $y = 81$

LESSON 10-5 (pp. 639–645)

15. $9x^2 + 24x + 16$ **17.** $9y^2 + 12y + 4$

19. a.

x	−4	−3	−2	−1	0	1	2	3
y	14	6	0	−4	−6	−6	−4	0

b. See right. **c.** They are the same; the equation in part **b** is the expansion of the equation in part **a.** **21.** $x^3 + 3x^2 + 2x$ **23.** 2, 3 **25.** $a + b - c + d$ **27. a.** −106 feet **b.** The rocket hit the ground before seven seconds elapsed. **29.** $y = \frac{1}{2}x + \frac{17}{2}$ **31.** $x = 512$

19.

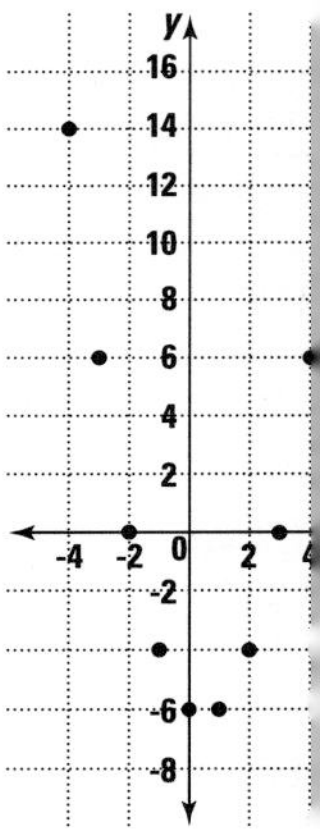

LESSON 10-6 (pp. 646–650)

19. Sample: $(1 + 2)^2 = 3^2 = 9$; $1^2 + 2^2 = 1 + 4 = 5$. So, $(1 + 2)^2 \neq 1^2 + 2^2$. **21.** $324 + 72y + 4y^2$ **23.** $x^2 - 11$ **25. a.** $s^3 + 10s^2 + 25s$ **b.** $s^3 - 2s^2$ **c.** $2s^3 + 8s^2 + 25s$ **d.** Does $2 \cdot 4^3 + 8 \cdot 4^2 + 25 \cdot 4 = 9 \cdot 9 \cdot 4 + 4 \cdot 4 \cdot 2$? Yes, $356 = 356$. **27.** $2c^2 + 3cd - 35d^2$ **29. a.** $x = 0$ **b.** $x < 0$ **31.** 61, 62, 63, 64

LESSON 10-7 (pp. 651–656)

9. Chi-square value ≈ 3.58. Such a value would occur over 10% the time. This is not enough evidence to support the view that earthquakes occur more in certain seasons. **11.** Chi-square value ≈ 5.52. Such a value would occur over 5% of the time. This is n enough evidence to say that more runs are scored in one part of game. **13. a.** $9a^2 - 6ab + b^2$ **b.** $9a^2 + 6ab + b^2$ **c.** $9a^2 - b^2$ **15. a.** $8p(4p + 2)$ **b.** $3p(p + 1)$ **c.** $8p(4p + 2) - 3p(p + 1) = 29p^2 + 13p$ **17.** $12y^4 + 2y^3 - 10y^2 + y + 7$ **19.** 98 teams

CHAPTER 10 PROGRESS SELF-TEST (p. 660)

1. $4x^2 - 7x + 9x^2 - 12 - 11 = 4x^2 + 9x^2 - 7x - 23 = 13x^2 - 7x - 23$ **2.** 2 **3.** trinomial **4.** $4(3v^2 - 9 + 2v) = 4 \cdot 3v^2 - 4 \cdot 9 + 4 \cdot 2v = 12v^2 - 36 + 8v = 12v^2 + 8v - 36$ **5.** $-5z(z^2 - 7z + 8) = -5z \cdot z^2 - (-5z) \cdot 7z + -5z \cdot 8 = -5z^3 + 35z^2 - 40z$ **6.** $(3x - 8)(3x + 8) = 9x^2 + 24x - 24x - 64 = 9x^2 - 64$ **7.** $(4y - 2)(3y - 16) = 12y^2 - 64y - 6y + 32 = 12y^2 - 70y + 32$ **8.** $(d - 12)^2 = (d - 12)(d - 12) = d^2 - 12d - 12d + 12^2 = d^2 - 24d + 144$ **9.** $(x - 3)(x^2 - 6x + 9) = x^3 - 6x^2 + 9x - 3x^2 + 18x - 27 = x^3 - 9x^2 + 27x - 27$ **10.** $(3x^2 - 10x) + (15x^3 - 7x^2 + x - 1) = 15x^3 + 3x^2 - 7x^2 - 10x + x - 1 = 15x^3 - 4x^2 - 9x - 1$ **11.** $8t^3 + t^2 - 7t + 1 - (5t^3 - 7t^2) = 8t^3 - 5t^3 + t^2 + 7t^2 - 7t + 1 = 3t^3 + 8t^2 - 7t + 1$ **12.** $(x + y + 5)(a + b + 2) = ax + bx + 2x + ay + by + 2y + 5a + 5b + 10$ **13.** $(4x + 1)(3x + 2) - x(x + 10) = 12x^2 + 11x + 2 - x^2 - 10x = 11x^2 + x + 2$ **14.** See right.

15. $(30 - 1)(30 + 1) = 900 - 1 = 899$ **16.** $2 \cdot 10^4 + 6 \cdot 10^3 + 3 \cdot 10^2 + 8 \cdot 10 + 4$ **17.** $80x^2 + 60x + 90$ **18.** $80 \cdot 1.04^2 + 60 \cdot 1.04 + 90 = \$238.93$ **19.** $\frac{861 + 748 + 812 + 939}{4} = 840$ **20.** $\frac{(861 - 840)^2}{840} + \frac{(748 - 840)^2}{840} + \frac{(812 - 840)^2}{840} + \frac{(939 - 840)^2}{840} = \frac{441 + 8464 + 784 + 9801}{840} = \frac{19490}{840} \approx 23.2$ **21.** The number of eve n is 4; $n - 1$ is 3. So look at the 3rd row of the Critical Chi-Squa Values table. The chi-square statistic 23.2 is greater than the criti value 16.3 that would occur with probability .001. So it is very unlikely that the sophomores were being given fewer lines in the newspaper simply by chance. **22.** $(2y)(5y - 3)(y + 9) = 2y(5y^2 + 45y - 3y - 27) = 2y(5y^2 + 42y - 27) = 10y^3 + 84y^2 - 54y$

14.

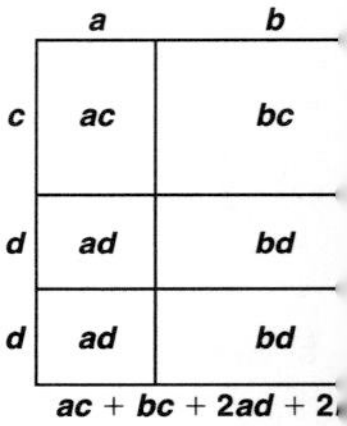

The chart below keys the **Progress Self-Test** questions to the objectives in the **Chapter Review** on pages 661–663. This will enable you to locate those **Chapter Review** questions that correspond to questions missed on the **Progress Self-Test.** The lesson where the material is covered is also indicated on the chart.

Question	**1**	**2**	**3**	**4**	**5**	**6**	**7**	**8**	**9**	**10**
Objective	Voc.	E	E	C	C	C	C	D	B	A
Lesson	10-1, 10-2	10-1	10-1	10-3	10-3	10-6	10-5	10-6	10-4	10-2
Question	**11**	**12**	**13**	**14**	**15**	**16**	**17**	**18**	**19**	**20**
Objective	A	B	I	I	G	G	C	F	H	H
Lesson	10-2	10-4	10-3, 10-5	10-5	10-2	10-2	10-6	10-1	10-7	10-7
Question	**21**	**22**								
Objective	H	I								
Lesson	10-7	10-4								

CHAPTER 10 REVIEW (pp. 661–663)

1. $7x^2 + 4x + 1$; 2 **3.** $3.9x^2 + 1.7x + 19$; 2 **5.** $-k^2 + k - 5$ **7.** $ac + ad + a + bc + bd + b + c + d + 1$ **9.** $y^3 - y^2 - y + 1$ **11.** $ax + bx + 3x + a + b + 3$ **13.** $3k^3 + 12k^2 - 3k$ **15.** $8x^2 + 10x - 36$ **17.** $y^2 - 12y - 13$ **19.** $a^2 - 225$ **21.** $-4z^2 - 5z - 1$ **23.** $d^2 - 2d + 1$ **25.** $48x^2 + 120x + 75$ **27.** $x^3 + 2x^2 + x$ **29.** (a) **31.** (a), (c), (d) **33.** Sample: z^4 **35.** 30,200,901 **37.** $9 \cdot 10^4 +$ $\cdot 10^3 + 1 \cdot 10^2 + 3$ **39. a.** $250x^4 + 250x^3 + 250x^2 + 250x + 250$ **b.** \$811.60 **41. a.** 50 **b.** 1.44 **c.** No; the chi-square value 1.44 is less than the critical value 3.84 that would occur with probability .05. **43. a.** 65° **b.** ≈ 7.11 **c.** Yes. The chi-square value 7.11 is smaller than the value 19.7 that would occur with probability .05. This is not a high enough chi-square value to support a claim that the temperatures are different throughout the year. **45. a.** $2x^2 + 8x + 8$ **b.** $(x + 2)(2x + 4)$ **47. a.** $xy + 3y + 2x + 6$ **b.** $(x + 3)(y + 2)$ **c.** Yes **49.** $10x^4 + 10x^3 + 9x^2 - 2x - 1$ **51.** $21x^2 - \pi x^2$ **53.** $x^3 - x$

LESSON 11-1 (pp. 666–671)

13. a., b. (−2, −2); (1, −.5) **See right. 15.** Sample: Yes. Since the two lines are not parallel and the women's times are decreasing faster than the men's times, the times will be equal in 2044. **17. a.** −4 the slope of the line that passes through the points (0, −1) and (8, 7) ≈ 10,333 miles $y = 7 - 2x$ **23.** $P + XY$

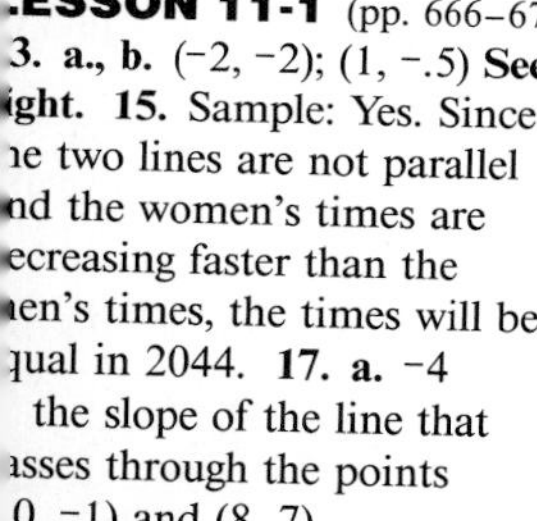

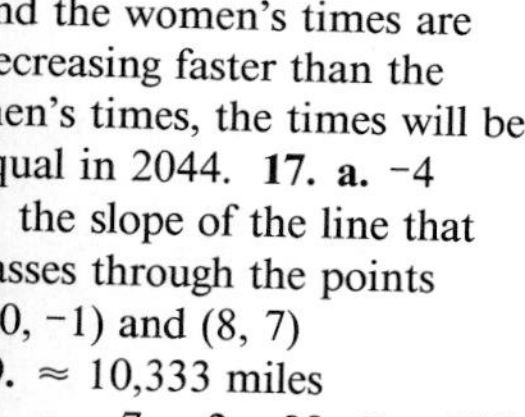

13. a.

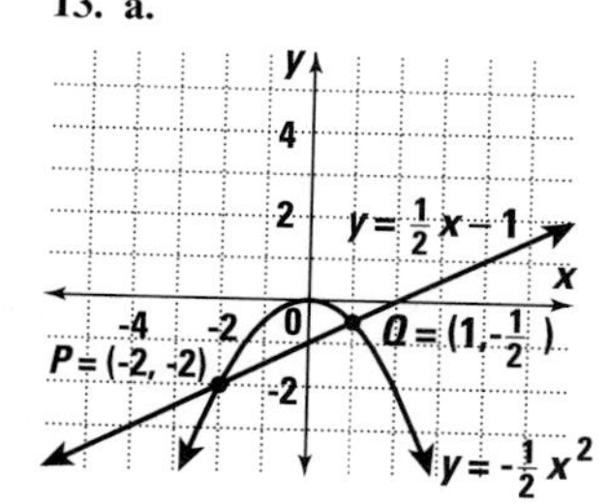

LESSON 11-2 (pp. 672–675)

a. (12, 1) **b.** Does $\frac{1}{2}(12) - 5 = 1$? Yes. Does $-\frac{3}{4}(12) + 10 = 1$? Yes. **13. a.** (0, 0), (4, 4) **b.** Does 0 = 0? Yes. Does $0 = \frac{1}{4}(0)^2$? Yes. Does 4 = 4? Yes. Does $4 = \frac{1}{4}(4)^2$? Yes. **15.** 1.5 hours **17. See below.** **a.** 2 **b.** (5, 5), (−5, 5) **21.** $2p^3 + 6p^2 + 2p$ **23. See below.** $4x + 2y$ dollars

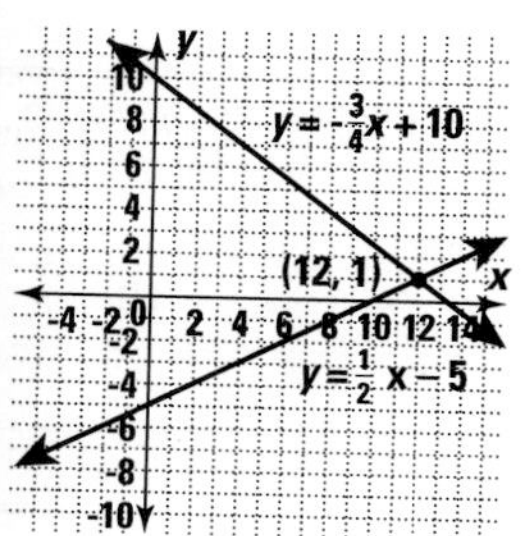

23.

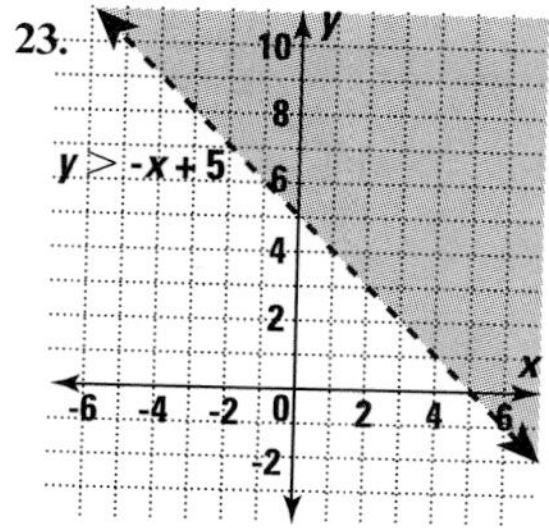

LESSON 11-3 (pp. 676–680)

9. L = \$800,000 and T = \$1,000,000 **11.** $A = 800$, $B = 600$, $K = 20$ **13.** $\left(\frac{1}{2}, -2\right)$ **15. a.** $m - 70 = v$; $m + v = 1250$ **b.** $m = 660$; $v = 590$ **17. a.** (4, 5) **See below. b.** Does 5 = 4 + 1? Yes. Does 5 = −2(4) + 13 = −8 + 13 = 5? Yes, it checks. **19.** $2a - 3b$ **21. See below. 23.** 180 mph

17. a.

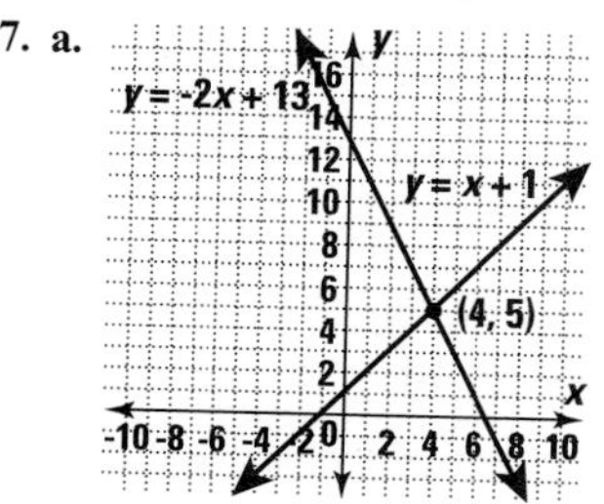

21.

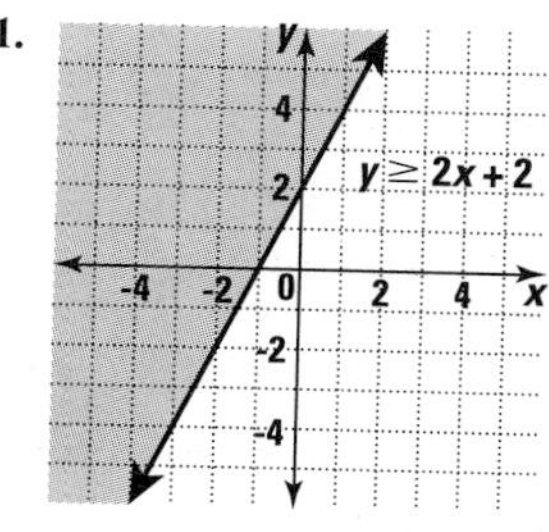

LESSON 11-4 (pp. 681–686)

13. Yes; by the Generalized Addition Property of Equality, $\frac{3}{4} - \frac{1}{5} = \frac{3}{4} + -\frac{1}{5}$ = 75% + −20% = 75% − 20% = 55%. **15.** $z = 0$; $w = -3$ **17.** regular, \$1.19 per gallon; premium, \$1.40 per gallon **19.** (−1, −3) **21. a.** $x = -7$ or $x = 2$ **b.** −7; 2 **23.** $\frac{3}{4}$ **25.** $8x + y = -15$ or $-8x - y = 15$ **27. a.** $30 + 6p$ **b.** $\frac{30 + 6p}{p + 2}$

LESSON 11-5 (pp. 687–693)

13. $\begin{cases} 3x - 4y = 2 \\ 9x - 5y = 7 \end{cases}$; $\left(\frac{6}{7}, \frac{1}{7}\right)$ **15.** (50, −5) **17.** 13 birds and 12 deer; sample reasoning: Let b = number of birds and d = number of deer. Since all animals have one head, $b + d = 25$. Since birds have two feet and deer have four feet, $2b + 4d = 74$. Solving the system gives $b = 13$ and $d = 12$. **19.** $\left(11, -\frac{78}{5}\right)$ **21. a.** $3x + y = 29$ **b.** $3x = y + 19$ **c.** (8, 5) **d.** Does 3(8) + 5 = 29? Does 24 + 5 = 29? Yes, it checks. Does 3(8) = 24 = 5 + 19? Yes, it checks.

23. a. PYTHAGOREAN TRIPLES
ENTER M
5
ENTER N
3
A = 16
B = 30
C = 34

b. PYTHAGOREAN TRIPLES
ENTER M
7
ENTER N
1
A = 48
B = 14
C = 50

c. Does $16^2 + 30^2 = 34^2$? Does $256 + 900 = 1156$? Yes. Does $48^2 + 14^2 = 50^2$? Does $2304 + 196 = 2500$? Yes. **25.** about 9 days

LESSON 11-6 (pp. 694–698)

13. (b) **15.** (a) **17.** No, with a 10% discount, you should have paid only $9000. The full price would have been $10,000. You paid only $400 less, getting a 4% discount. **19. a.** $\begin{cases} t + u = 20 \\ t + 3u = 32 \end{cases}$ **b.** Sample: Substitute $t = 20 - u$ in the equation $t + 3u = 32$ to get $20 - u + 3u = 32$. Solve this equation for u to get $u = 6$. Substitute $u = 6$ in $t = 20 - u$ to get $t = 14$. **c.** Does $14 + 6 = 20$? Yes. Does $14 + 3(6) = 32$? Yes. **21. a.** $10^3 + 8 \cdot 10^2 + 7 \cdot 10 + 2$ **b.** $1.872 \cdot 10^3$ **23.** $21.70

LESSON 11-7 (pp. 699–703)

11. a. $375m < 315m + 60m + 25$ **b.** $0 < 25$; m can be any number of months. **13.** A **15.** Sample: $y - 2 = y - 1$ **17.** two intersecting lines because the slopes, $\frac{11}{10}$ and $\frac{12}{10}$, are not equal **19.** $\left(\frac{7}{5}, 0\right)$ **21. a.** See above right. **b.** $-x = x + 3$; $-2x = 3$; $x = -1.5$, $y = 1.5$ **c.** $2y = 3$; $y = 1.5$, $x = -1.5$ **d.** Sample: The addition method requires the fewest steps. **23. a.** $\frac{4}{9}$ **b.** $25d^4g^2$ **c.** $\frac{c^2}{4a^2}$ **25.** about 26% **27.** $525w$

21. a.

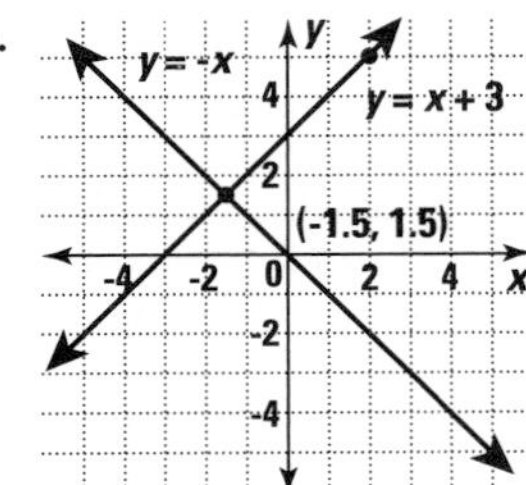

LESSON 11-8 (pp. 704–710)

13. $x \geq 0$, $y \geq 0$, $x + 2y \leq 10$ **15.** Sample: $\begin{cases} x \geq -1 \\ x \leq 1 \\ y \geq -2 \\ y \leq 2 \end{cases}$ **17. a.** See right. **b.** 25 **19. a.** no solution **b.** y may be any real number. **21.** intersecting **23. a.** $\begin{cases} d + q = 27 \\ 0.1d + 0.25q = 5.10 \end{cases}$

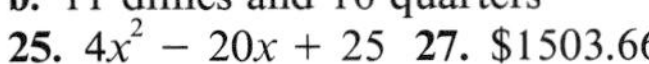

b. 11 dimes and 16 quarters **25.** $4x^2 - 20x + 25$ **27.** $1503.66

17.

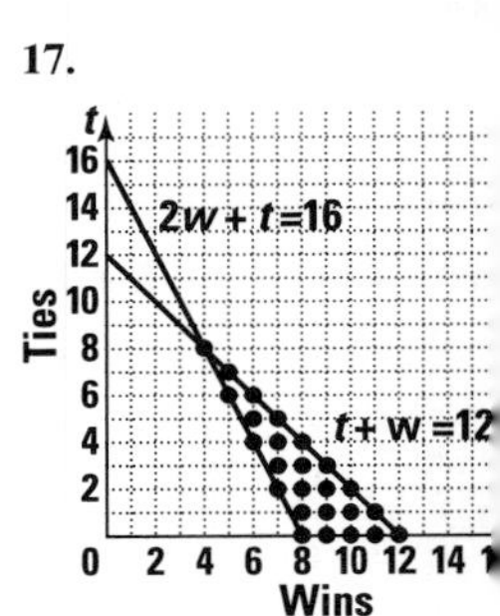

CHAPTER 11 PROGRESS SELF-TEST (p. 714)

1. Sample: Substitute $3a$ for b in the first equation and solve for a. $a - 3(3a) = a - 9a = -8a = -8a = -8$; $a = 1$, $b = 3 \cdot 1 = 3$; $a = 1$; $b = 3$ **2.** Sample: Substitute $-7r - 40$ for p in first equation. $5r + 80 = -7r - 40$; $120 = -12r$; $r = -10$; $p = 5 \cdot -10 + 80 = 30$; $p = 30$; $r = -10$ **3.** Sample: Add equations. $n = 3$; $m - 3 = -1$; $m = 2$; $m = 2$; $n = 3$ **4.** Sample: Multiply second equation by 3 and add. $3 \cdot 4x - 3 \cdot y = 3 \cdot 6$; $12x - 3y = 18$; $19x = 19$; $x = 1$; $4 \cdot 1 - y = 6$; $4 - y = 6$; $-2 = y$; $x = 1$; $y = -2$ **5.** $-5x - 15 = x - 3$; $-6x = 12$; $x = -2$; $y = -2 - 3 = -5$; $(x, y) = (-2, -5)$ **6.** $2 \cdot 5 = 2 \cdot 2A + 2 \cdot 7B$; $10 = 4A + 14B$; $0 = 0$; the lines coincide. **7.** $2x + 7 - 2x = 3$; $7 \neq 3$; the lines are parallel. **8.** See right. **9.** See right. **10.** Let L = Lisa's weight and S = sister's weight. $L = 4S$; $L + S = 95$; $4S + S = 95$; $5S = 95$; $S = 19$; $L = 4 \cdot 19 = 76$; Lisa weighs 76 pounds and the baby weighs 19 pounds. **11.** Let h = price of hamburgers and s = price of salads. $3h + 4s = 21.30$; $5h + 2s = 22.90$; $10h + 4s = 45.80$; $7h = 24.50$; $h = 3.50$; $5 \cdot 3.50 + 2s = 22.90$; $2s = 5.40$; $s = 2.70$. A small salad costs $2.70. **12.** No, let r = price of a rose and d = price of a daffodil. $5(2r + 3d) = 5 \cdot 8$; $10r + 15d = 40$; $35 \neq 40$. **13.** They will never have the same population, because $6{,}016{,}000 - 30{,}000y = 4{,}375{,}000 + 30{,}000y$ has no solution. The lines are parallel. **14.** See below. **15.** $12z + 8 - 12z = -3$; $8 = -3$; z has no solutions. **16.** $0 < 22 - 19p + 19p$; $0 < 22$; p may be any real number.

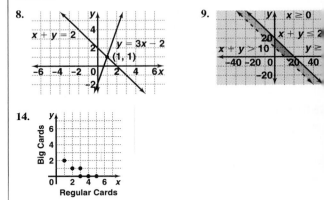

8.

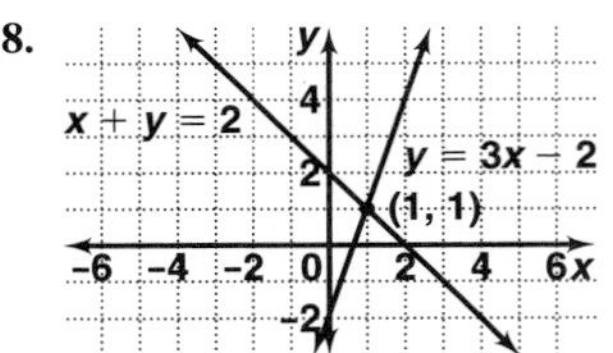

9.

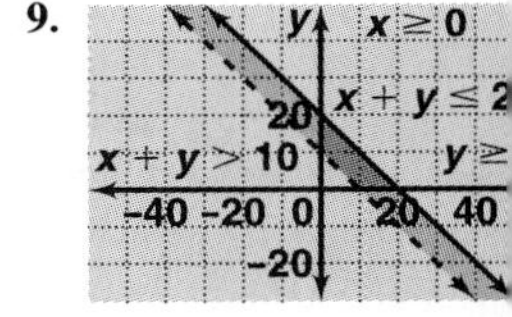

14.

Big Cards

Regular Cards

The chart below keys the **Progress Self-Test** questions to the objectives in the **Chapter Review** on pages 715–717. This will enable you to locate those **Chapter Review** questions that correspond to questions missed on the **Progress Self-Test.** The lesson where the material is covered is also indicated on the chart.

Question	**1**	**2**	**3**	**4**	**5**	**6**	**7**	**8**	**9**	**10**
Objective	A	A	B	C	A	E	E	H	I	F
Lesson	11-3	11-2	11-4	11-5	11-2	11-6	11-6	11-1	11-8	11-3
Question	**11**	**12**	**13**	**14**	**15**	**16**	**17**	**18**	**19**	**20**
Objective	F	F	F	G, I	D	D				
Lesson	11-5	11-6	11-7	11-8	11-7	11-7				

HAPTER 11 REVIEW (pp. 715–717)

(15, 45) **3.** $p = 20$, $q = 30$ **5.** $(x, y) = (5, 10)$ **7.** $(a, b) =$ 31, −17) **9.** (−9, −43) **11.** $(m, b) = (-22, 77)$ **13.** $(x, y) =$ 3, −7.25) **15.** $(f, g) = (3, 3)$ **17. a.** Sample: Multiply equation by 4 to give $20x + 4y = 120$. **b.** $(x, y) = (7, -5)$ **19.** $(y, z) =$ 3) **21.** $(a, b) = (4, -1)$ **23.** no solutions **25.** $x = 0$ (one ution) **27.** No. Subtract $2k$. $-7 = 0$ is never true; so the original ntence is never true. **29.** No solutions when $b \neq c$. If the slopes equal but the y-intercepts are different, the lines are parallel and system has no solutions. **31.** coincide **33.** coincide **35.** (c) Marty, \$52.50; Joe, \$157.50. **39.** always **41.** egg, \$0.45; iffin, \$0.90 **43.** Yes; $16p + 5e = 8$ is equivalent to $32p + 10e =$ There are infinitely many solutions. **45. See below. 47. a. See low. b.** 19 ft **49. See right. 51. See right. 53.** parallel, no utions **See right. 55. See right. 57. See right. 59. See right.**

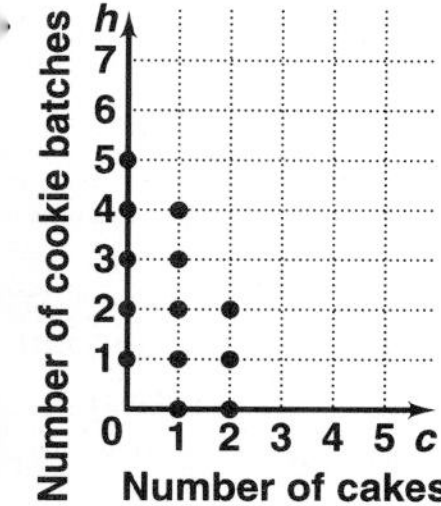

47. a.

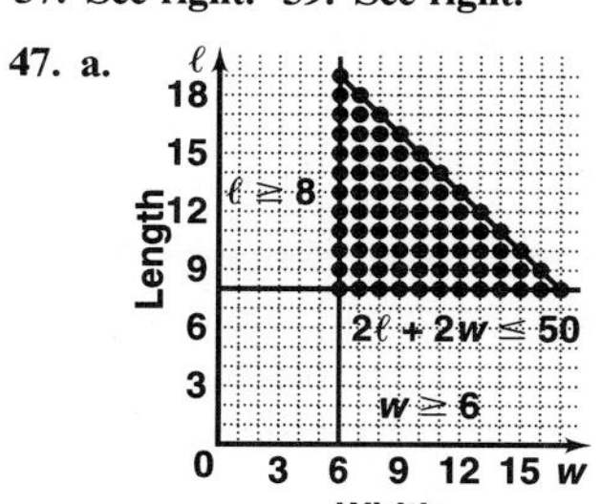

49.

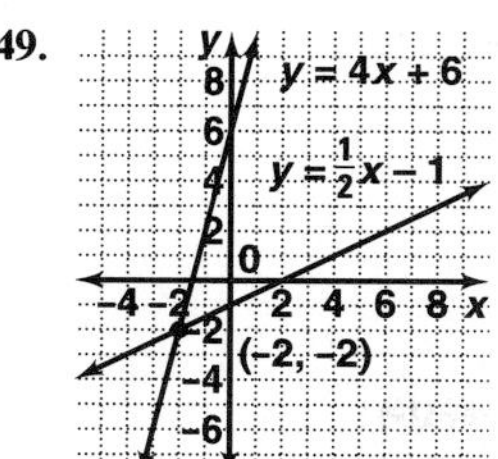

51.

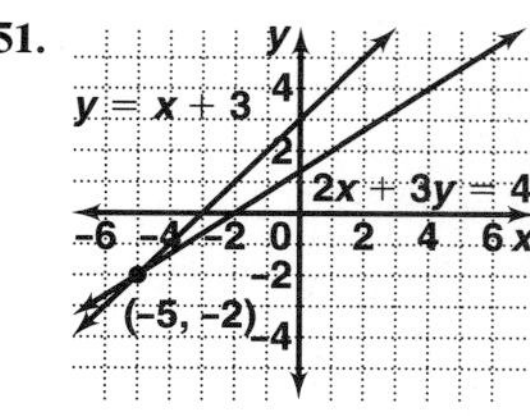

53.

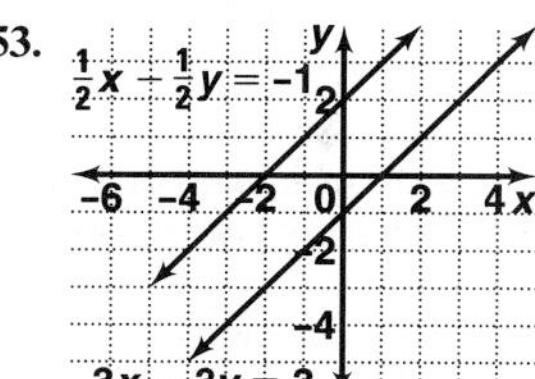

55.

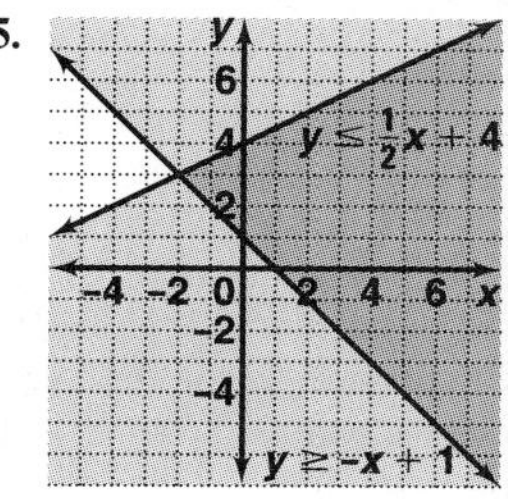

57.

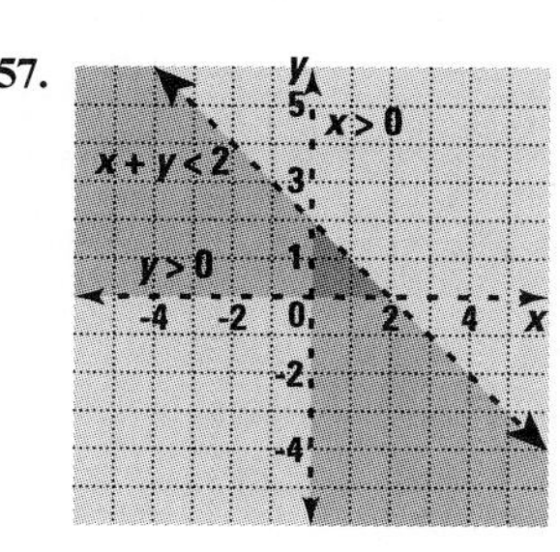

59.

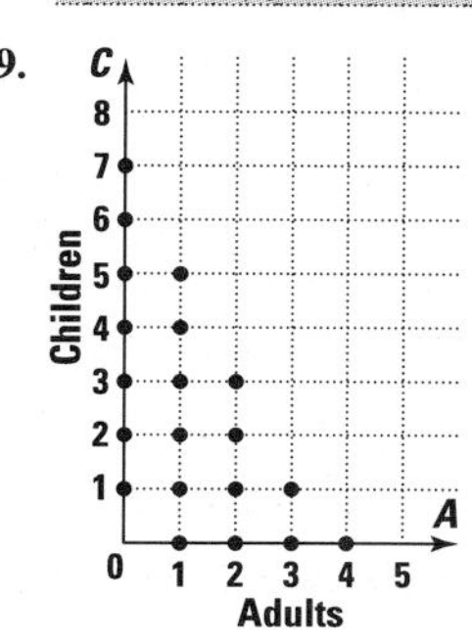

ESSON 12-1 (pp. 718–725)

53, 59, 61, 67, 71, 73, 79, 83, 89, 97 **17.** 20 **19. a.** Because $= 4 \cdot 2^{38}$ and $332 = 4 \cdot 83$, then 4 is a common factor of 2^{40} d 332. Using the Common Factor Sum Property, we know 4 is o a factor of $2^{40} + 332$. **b.** No; 332 is not divisible by 8. $d = 3; -3; 4; -4$ **23.** about 102 games **25.** (a)

ESSON 12-2 (pp. 726–732)

See below. 21. $-x + 2 - 3y$ **23. a.** $4r^2$ **b.** πr^2 **c.** $(4 - \pi)r^2$ $(48 - 12\pi)r^2$ **e.** 3 **25. a.** $x^2 + 3x - 154$ **b.** $x = 11$ or $x = -14$ $3a^2 + 11a + 10$ **29.** $x = 9$

Samples:

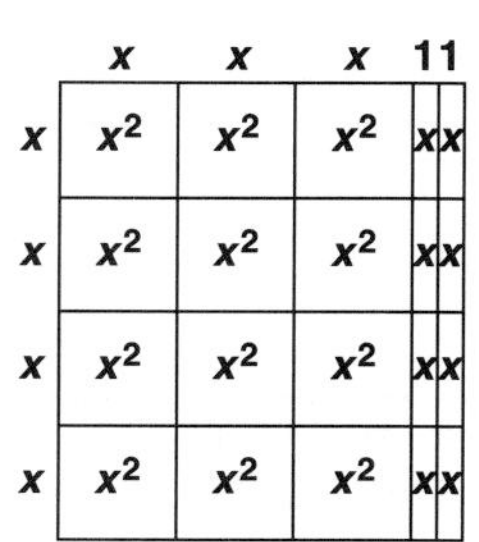

	x	x	x	x	x	x	1111
x	x^2	x^2	x^2	x^2	x^2	x^2	xxxx
x	x^2	x^2	x^2	x^2	x^2	x^2	xxxx

LESSON 12-3 (pp. 733–737)

17. $(7 - x)(5 - x)$ **19.** $10(2 + x)(2 - x)$ **21.** $1{,}032{,}060 = 10^6 + 32 \cdot 10^3 + 60 = (10^3 + 2)(10^3 + 30) = 1002 \cdot 1030 = 2 \cdot 501 \cdot 2 \cdot 515 = 2^2 \cdot 3 \cdot 167 \cdot 5 \cdot 103 = 2^2 \cdot 3 \cdot 5 \cdot 103 \cdot 167$ **23.** The chi-square value is 12.8. A chi-square value this size would be expected more than 10% of the time, so this is not an unusual distribution of digits. **25.** $64a^2 - 1$ **27.** 53.64° **29.** $x = 9$ **31.** \$8.26

LESSON 12-4 (pp. 738–743)

17. $y = 5$ or $y = -1$ **19.** $p = 3$ or $p = -4$ **21.** $v = 0$ or $v = \frac{3}{2}$ **23. a.** $5^2 + 5 = 30$; $2 + 4 + 6 + 8 + 10 = 30$ **b.** 11 **25.** (d) **27.** $13x - 14$ **29.** $16e^2 + 8e + 1$ square units **31. a. See right. b.** about 6.6 thicknesses **33.** $2x^2$

31. a.

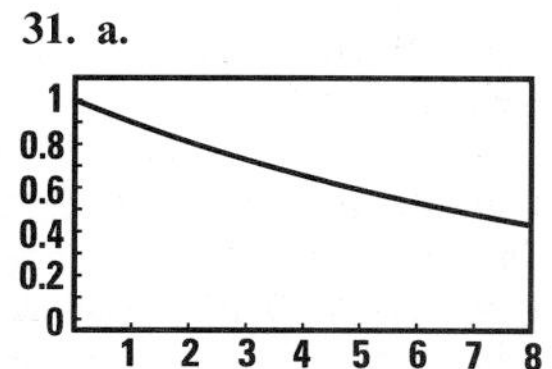

LESSON 12-5 (pp. 744–748)

13. $k = 5$ **15. a. i. See p. 862. ii.** $(x - 4)(2x + 9) = 0$; $x = 4$ or $x = \frac{-9}{2}$ **b.** Sample: Factoring, because it is more precise **17.** $p^2(3p + 2)^2$ **19.** If the product of two real numbers a and b is zero, then $a = 0$ or $b = 0$. **21.** $y \approx -8.14$ or $y \approx -0.86$ **23.** $v = 2$ or $v = -2$ **25. a.** $(x^2 + 9)(x^2 - 9)$ **b.** $(x^2 + 9)(x + 3)(x - 3)$ **27.** 37 ft

29. a. $(2x + 2y)(x - y) = 2(x^2 - y^2)$ **b.** Yes; $(2x + 2y)(x - y) = 2x^2 - 2xy + 2xy - 2y^2 = 2x^2 - 2y^2 = 2(x^2 - y^2)$ **31.** slope $= \frac{1}{2}$, y-intercept $= 0$ **33.** $\frac{13}{40}$, or .325 of a picture

15. a.

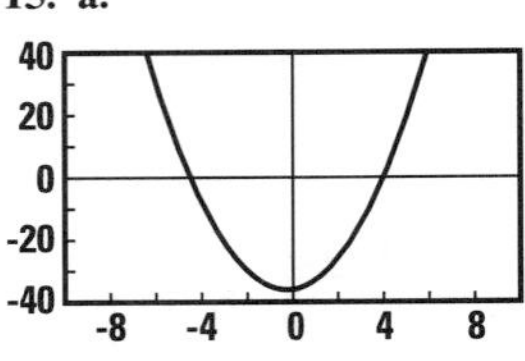

LESSON 12-6 (pp. 749–753)
13. The discriminant is negative. **15.** $4ab(2c - b + 3d)$ **17.** $(4m - 5)(5m - 4)$ **19.** Each factor gives a possibility. Here we organize by the powers of 2, and then by multiplying each by a power of 5. 1 year and $1,000,000 (per year); 2 years and $500,000; 4 years and $250,000; 8 years and $125,000; 16 years and $62,500; 32 years and $31,250; 64 years and $15,625; 5 years and $200,000; 25 years and $40,000; 125 years and $8,000; 625 years and $1600; 3,125 years and $320; 15,625 years and $64; 10 years and $100,000; 50 years and $20,000; 250 years and $4,000; 1,250 years and $800; 6,250 years and $160; 31,250 years and $32; 20 years and $50,000; 100 years and $10,000; 500 years and $2,000; 2,500 years and $400; 12,500 years and $80; 62,500 years and $16; 40 years and $25,000; 200 years and $5,000; 1,000 years and $1,000; 5,000 yea and $200; 25,000 years and $40; 125,000 years and $8; 80 years and $12,500; 400 years and $2,500; 2,000 years and $500; 10,000 years and $100; 50,000 years and $20; 250,000 years and $4; 160 years and $6,250; 800 years and $1,250; 4,000 years and $250; 20,000 years and $50; 100,000 years and $10; 500,000 years and $2; 320 years and $3,125; 1,600 years and $625; 8,000 years and $125; 40,000 years and $25; 200,000 years and $5; 1,000,000 yea and $1

LESSON 12-7 (pp. 754–758)
15. Yes, because it equals a simple fraction. **17.** $30\sqrt{2}$ in.; irratio **19.** irrational **21. a.** $7y$ **b.** $a = 3, b = 4y, c = 4$, and $d = -3y$ or $a = 4, b = -3y, c = 3$, and $d = 4y$ **23.** $\left(2 + \frac{\pi}{2}\right)r$ **25.** 92.16 ft

LESSON 12-8 (pp. 759–764)
11. Sample: $k = 4$ **13.** only at 28 meters per second **15. a.** rational **b.** irrational **c.** irrational **17.** 12; 7 **19. a.** $2\pi r(r + h)$ **b.** ≈ 408.4 cm^2; the factored form

CHAPTER 12 PROGRESS SELF-TEST (p. 768)

1. $300 = 2^2 \cdot 3 \cdot 5^2$ **2.** Both 6^{1000} and 36 are divisible by 3, so by the Common Factor Sum Property, $6^{1000} + 36$ is divisible by 3. **3.** $15a^2b^3 = 3 \cdot 5a^2b^3$, $30a^2b = 6 \cdot 5a^2b$, $25a^3b = 5 \cdot 5a^3b$. So, $5a^2b$ is the greatest common factor. **4.** $\frac{8c^2 + 4c}{c} = \frac{4c(2c + 1)}{c} = 4(2c + 1) = 8c + 4$ **5.** $12m - 2m^3 = 2m(6 - m^2)$ **6.** $500x^2y + 100xy + 50y = 50y(10x^2 + 2x + 1)$ **7.** $z^2 - 81 = (z + 9)(z - 9)$ **8.** Factors of 14 = −1, −14 and −2, −7; −2 + −7 = −9; $k^2 - 9k + 14 = (k - 2)(k - 7)$ **9.** $3y^2 - 17y - 6 = (3y + 1)(y - 6)$ **10.** $4x^2 - 20xy + 25y^2 = (2x - 5y)^2$ **11.** The discriminants are: **(a)** $7^2 - 4 \cdot 1 \cdot 12 = 1$ **(b)** $7^2 - 4 \cdot 1 \cdot (-12) = 97$ **(c)** $12^2 - 4 \cdot 1 \cdot (-7) = 172$ **(d)** $(-12)^2 - 4 \cdot 1 \cdot (-7) = 172$. Equation **(a)** is the only one with a discriminant which is a perfect square. So it is the only one which can be factored over the integers. **12.** Calculate the discriminant ($b^2 - 4ac$). If it is a perfect square, the trinomial is factorable over the integers. **13.** One side of the rectangle has length $2x + 5$, the other side has length $x + 3$. **See right.** **14.** $(q - 7)^2 = 0$; $(q - 7)^2 = (q - 7)(q - 7)$; by the Zero Product Theorem $q - 7 = 0$; $q = 7$ **15.** $d^2 - 20 - d = 0$; $(d - 5)(d + 4) = 0$; by the Zero Product Theorem $d = 5$ or $d = -4$. **16.** By the Zero Product Theorem $2a - 5 = 0$ or $3a + 1 = 0$; $a = \frac{5}{2}$ or $a = -\frac{1}{3}$ **17.** $x^3 + 6x^2 - 7x = x(x^2 + 6x - 7) = x(x + 7)(x - 1) = 0$; by the Zero Product Theorem $x = 0$ or $x =$ or $x = 1$. **18.** Because 3.54 can be written as the simple fraction $\frac{354}{100}$. **19.** 26 has the prime factors 13 and 2. $(26)^2$ has twice the number of prime factors, or 4. **20.** $10^2 = 100$ and $11^2 = 121$. Ar integer between 100 and 121 will not be a perfect square; its squa root will therefore be an irrational number between 10 and 11. Sample: $\sqrt{105}$ **21.** Because $b^2 - 4ac = 18^2 + 4 \cdot 5 \cdot 18 = 684$ is r a perfect square, both solutions are irrational. **22.** $(d - 1)(d - 2)$ 12; $d^2 - 3d + 2 = 12$; $d^2 - 3d - 10 = 0$; $(d - 5)(d + 2) = 0$; $d = 5$ or $d = -2$. The frame is 5 ft by 5 ft. **23.** When the ball returns to the ground, $d = 0$. $0 = 8t - 5t^2 = t(8 - 5t)$ when $t =$ and $t = \frac{8}{5}$. So, the ball returns to the ground after $\frac{8}{5} = 1.6$ second **24.** $\begin{cases} x + y = 8 \\ xy = 15.51 \end{cases}$; $x(8 - x) = 15.51$; $-x^2 + 8x - 15.51 = 0$
$x^2 - 8x + 15.51 = 0$;
$x = \frac{8 \pm \sqrt{8^2 - 4 \cdot 15.51}}{2} = \frac{8 \pm 1.4}{2}$.
When $x = \frac{8 + 1.4}{2} = 4.7$, $y = 8 - x = 3.3$. When $x = \frac{8 - 1.4}{2} = 3.3$, $y = 8 - x = 4.7$. So, the two numbers are 3.3 and 4.7.

13.

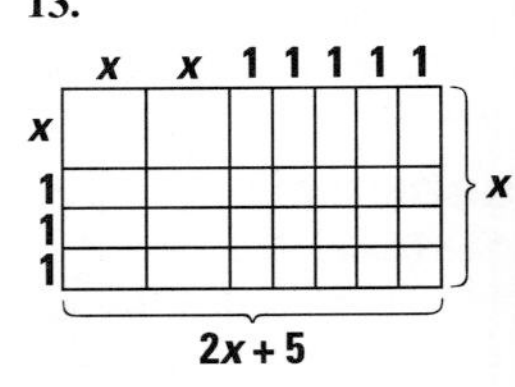

The chart below keys the **Progress Self-Test** questions to the objectives in the **Chapter Review** on pages 769–771. This will enable you to locate those **Chapter Review** questions that correspond to questions missed on the **Progress Self-Test.** The lesson where the material is covered is also indicated on the chart.

Question	**1**	**2**	**3**	**4**	**5**	**6**	**7**	**8**	**9**	**10**
Objective	A	E	B	B	B	B	C	C	C	C
Lesson	12-1	12-1	12-2	12-2	12-2	12-2	12-3	12-3	12-5	12-5
Question	**11**	**12**	**13**	**14**	**15**	**16**	**17**	**18**	**19**	**20**
Objective	G	G	J	F	D	F	D	H	E	H
Lesson	12-8	12-8	12-3	12-4	12-4	12-4	12-4	12-7	12-7	12-7
Question	**21**	**22**	**23**	**24**						
Objective	G	I	I	I						
Lesson	12-8	12-4	12-4	12-6						

CHAPTER 12 REVIEW (pp. 769–771)

1. $5^2 \cdot 7$ **3.** $3^4 \cdot 7^2$ **5.** x^3; 7 **7.** $3x^2y$ **9.** $m^2(14m^2 + 1)$ **11.** $6z^2 - 1$ **13.** $(x + 6)(x + 1)$ **15.** not factorable **17.** $4L(L + 3)(L + 4)$ **19.** (c) **21.** $(3y - 4x)(y + 2x)$ **23.** $3m(4m + 3)(m + 9)$ **25.** $(m + 8)^2$ **27.** $(a + 2)(a - 2)$ **29.** $(2x + 1)(2x - 1)$ **31.** 0 or 2 **33.** -1 or 3 **35.** -6 or 8 **37.** -4 or 4 **39.** $\frac{1}{2}$ or $-\frac{2}{3}$ **41.** No; $203 = 7 \cdot 29$ **43.** 4 **45.** For any two numbers a and b, if $ab = 0$, then $a = 0$ or $b = 0$. **47.** There is no product. **49.** 0 or $\frac{7}{2}$ **51.** $\frac{3}{2}$ or $-\frac{5}{3}$ **53.** (b) **55.** No **57.** Yes **59.** 6 and 4; it is factorable over the integers. **61.** rational **63.** irrational **65.** rational **67.** rational **69.** $\frac{428}{999}$ **71.** $\approx$ 7.7 in. by 11.7 in. **73.** after 2 and 3 seconds **75.** 115 cm by 90 cm **77. a.** $2x^2 + 8x + 8$ **b.** $(x + 2)(2x + 4)$ or $2(x + 2)^2$ **79.** $3a + 5b$

LESSON 13-1 (pp. 774–777)

9. (c) **11.** Yes **13.** One x value is paired with infinitely many y values. **15. a.** Sample: $(-2, 5)$, $(0, 3)$, $(2, 5)$ **b. See below.** **17. a.** Let cost in dollars = y and weight in pounds = x. $y = 0.50x + 3.00$ **b.** Sample: (0, 3.00), (1, 3.50), (2, 4.00) **19.** 140 adult tickets and 380 student tickets **21.** False **23.** $100bc$

15. b.

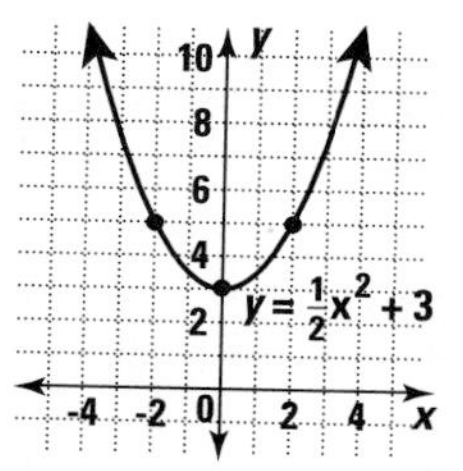

LESSON 13-2 (pp. 778–783)

15. a. After 24 months, the cost of owning the new car is greater than the cost of owning the used car. **b.** more than 120 months **17. a. See below.** **b.** 0, 5; 0 **19. a.** Answers will vary. **b.** total number of children in p's family **21. a.** 95 **b.** 44 **c.** 17 **d.** the slope of the graph of L **23. a. See below.** **b.** No **25.** $\frac{2}{3}$ **27.** $3 \pm \sqrt{6}$ **29. a.** $p = 6^m$ **b.** exponential **c.** 11 **d.** 13

17. a.

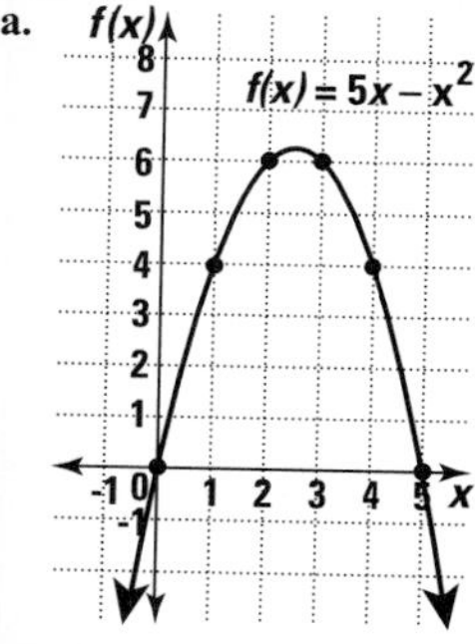

23. a.

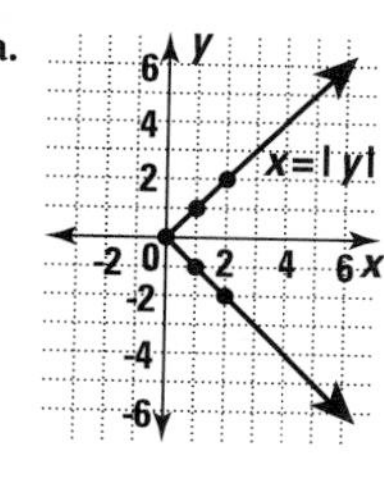

LESSON 13-3 (pp. 784–788)

15. See below. **17. See below.** **19.** 47°F **21.** 2 and 4 **23. a.** $\frac{3x}{2}$ **b.** $\frac{5x}{6}$ **c.** $\frac{2x + 3y}{6}$

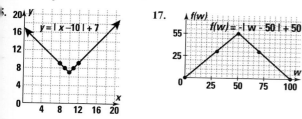

15.

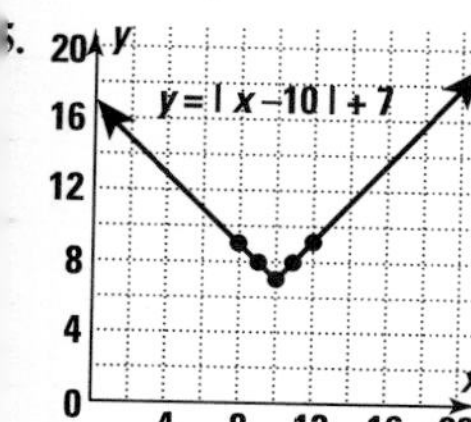

17.

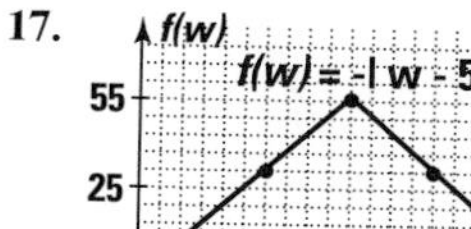

LESSON 13-4 (pp. 789–793)

15. a. See right. **b.** 6 **17.** domain: set of all real numbers; range: set of all real numbers > 0 **19.** domain: set of all real numbers ≥ 0; range: set of all real numbers ≥ 100 **21.** Yes **23. a.** $\left(1 + \frac{1}{3}\right)x = 10.00$ **b.** $7.50 **25. a.** Cuba **b.** $\frac{10{,}700{,}000}{44{,}218} > \frac{90{,}000{,}000}{761{,}604}$ or $242 > 118$

15. a.

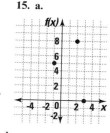

f(x)
8
6
4
2
-4 -2 0 2 4 x
-2

LESSON 13-5 (pp. 795–799)

13. Sample: Probabilities must be from 0 to 1. **15. a.** $\frac{1}{32}$ **b. See right.** **17. a.** 3654 **b.** 360 million pounds more fish were caught in 1980 than in 1985. **c.** {1960, 1965, 1970, 1975, 1980, 1985, 1990} **d.** negative **e.** between 1965 and 1970 **19. a.** $(2x - 5)(x + 4)$ **b.** $x = 3$ **21.** $3\sqrt{3}$ **23.** 74°

15. b.

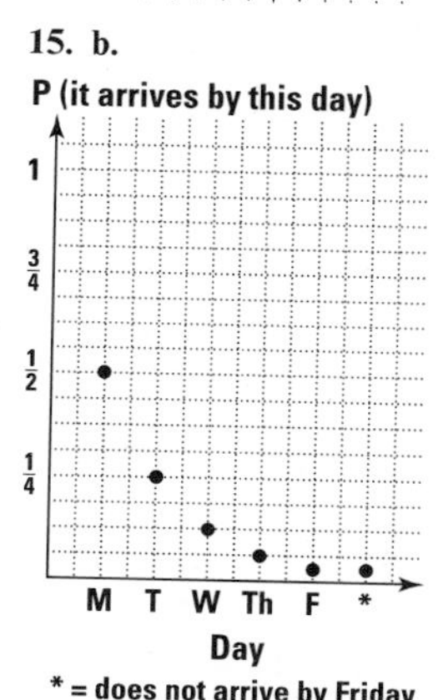

LESSON 13-6 (pp. 800–805)

11. a.

x	$y = x^3 + 3$
−2	−5
−1.5	−0.375
−1	2
−0.5	2.875
0	3
0.5	3.125
1	4
1.5	6.375
2	11

b. See page 864. **c.** This curve has a y-intercept of 3 and an x-intercept of ≈ -1.4 and does not have an axis of symmetry, but it does have a point of rotation symmetry at (0, 3).

13. a.

x	$y = -x^3$
−2	8
−1.5	3.375
−1	1
−0.5	0.125
0	0
0.5	−0.125
1	−1
1.5	−3.375
2	−8

b. See page 864. **c.** This curve has its one intercept and its point of rotation symmetry at the origin. It does not have an axis of symmetry. **15. a.** Sample: the set of numbers x with $1 \le x \le 2$. **b. See page 864.** **c.** about $298 **d.** 33% **17.** $S = \pi r(r + 2h)$ **19.** about 603 milliliters **21.** $y = \frac{4}{3}$ or $\frac{8}{3}$ **23.** $A = \pm\sqrt{2}$ **25.** $x = -.9$

11. b.

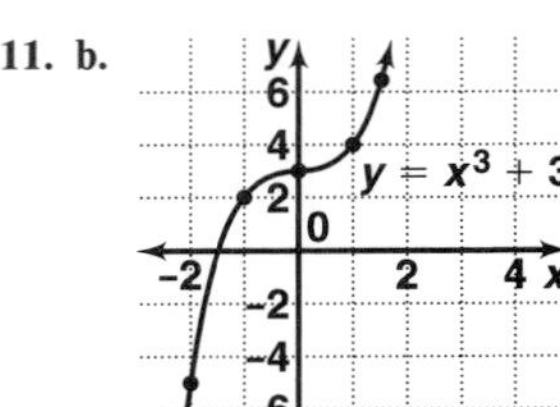

13. b.

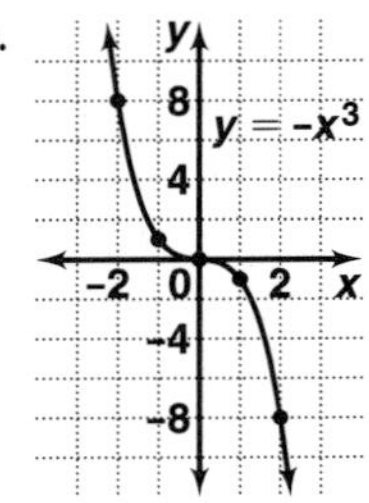

15. b.

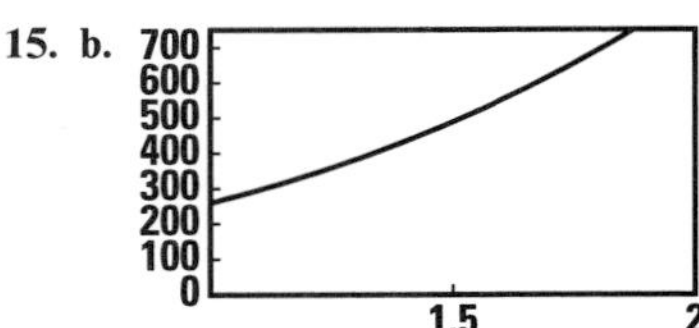

LESSON 13-7 (pp. 807–811)

13. ≈ 60° **15.** 80° **17.** −2 **19.** $x = \frac{-b \pm \sqrt{b^2 - 4ac}}{2a}$ **21.** 72°

LESSON 13-8 (pp. 812–816)

17. a. It depends on how your BASIC defines LOG(X).

X	LOG X	or	LOG X
1	0		0
2	0.30103		.69315
3	0.47712		1.0986
4	0.60206		1.3863
5	0.69897		1.6094
6	0.77815		1.7918
7	0.84510		1.9459
8	0.90309		2.0794
9	0.95424		2.1972
10	1		2.3026

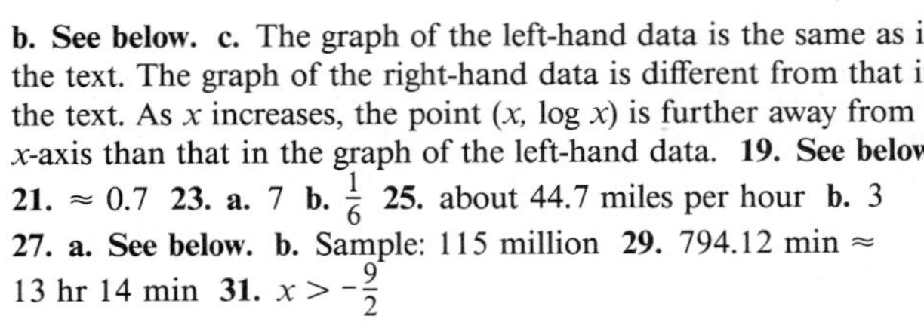

b. See below. **c.** The graph of the left-hand data is the same as in the text. The graph of the right-hand data is different from that in the text. As x increases, the point (x, log x) is further away from the x-axis than that in the graph of the left-hand data. **19.** See below. **21.** ≈ 0.7 **23. a.** 7 **b.** $\frac{1}{6}$ **25.** about 44.7 miles per hour **b.** 3 **27. a.** See below. **b.** Sample: 115 million **29.** 794.12 min ≈ 13 hr 14 min **31.** $x > -\frac{9}{2}$

17. b.

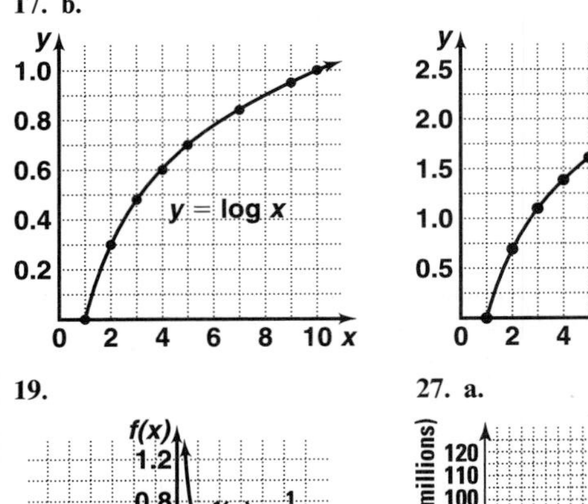

19.

$f(x) = \frac{1}{x}$

27. a.

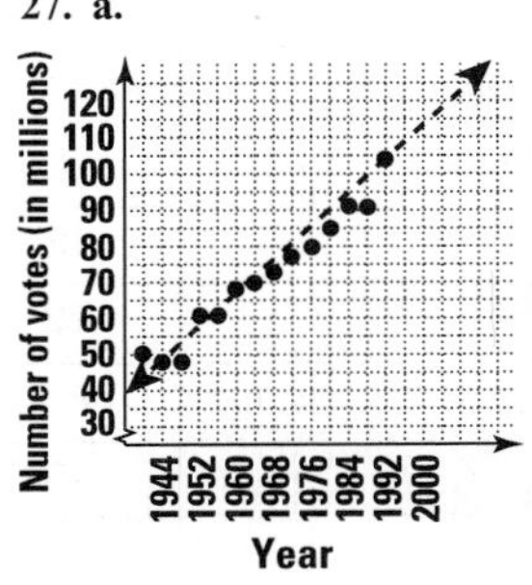

CHAPTER 13 PROGRESS SELF-TEST (p. 820)

1. $f(2) = 3 \cdot 2 + 5 = 11$ **2.** Sample: A function is a set of ordered pairs for which every first coordinate is paired with exactly one second coordinate. **3.** 7.115 **4.** 0.5 **5.** It's the slope of the line: 4. **6.** Sample: $f(x) = x^2 + 2x + 1$ **7.** $x = |y|$ is not a function because when $x > 0$, each x-value is paired with two y-values. For instance, if $x = 3$, $y = 3$ or $y = -3$. **8.** x cannot equal 10 or 30. $x \neq 30$, $x \neq 10$ **9.** Yes; domain = {0, 1, 2, 3, 4}, range = {1, 2}. **10.** No; (0, 1) and (0, −1) are both in the relation. When $x < 1$, every x-value is paired with two y-values. **11.** There are three ways of tossing a sum of 10: (6, 4), (5, 5), (4, 6). So $P(10) = \frac{3}{36} = \frac{1}{12}$. **12.** $P(2) = \frac{1}{10}$ **13.** See right. **14.** $b(n) = 3n$ and $h(n) = 2n$; $3n > 2n$; $3n - 2n > 0$; $n > 0$, $b(n) > h(n)$ for any positive number of sandwiches. **15.** See right. **16.** See page 865. **17.** $f(220) = 11.75 \cdot 220 + 300 = 2885$. $r(220) = 14.50 \cdot 220 = 3190$. Fabulous Floor's carpet costs $2885, which is cheaper than Rapid Rug's price. Therefore, she should buy her carpet from Fabulous Floor. **18.** See page 865. **19.** See page 865. The cubing function is $f(x) = x^3$. domain = the set of all real numbers; range = the set of all real numbers

13.

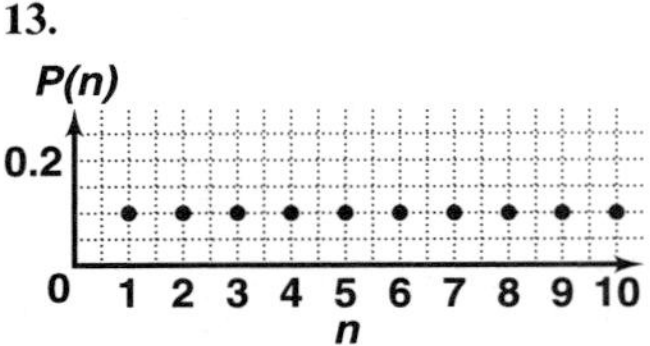

15.

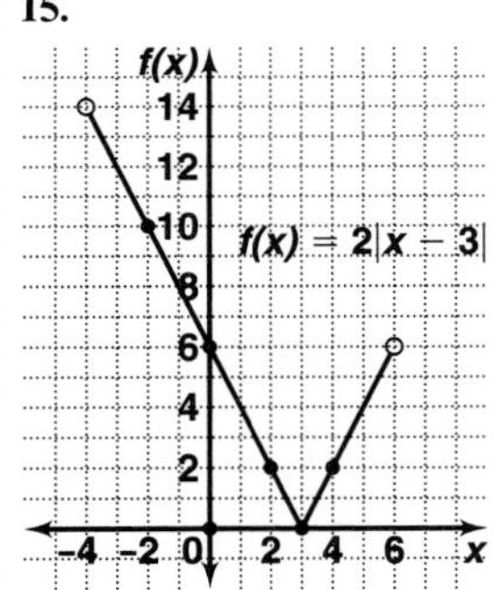

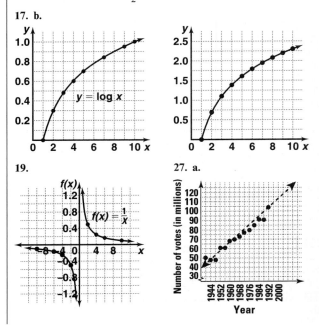

16.

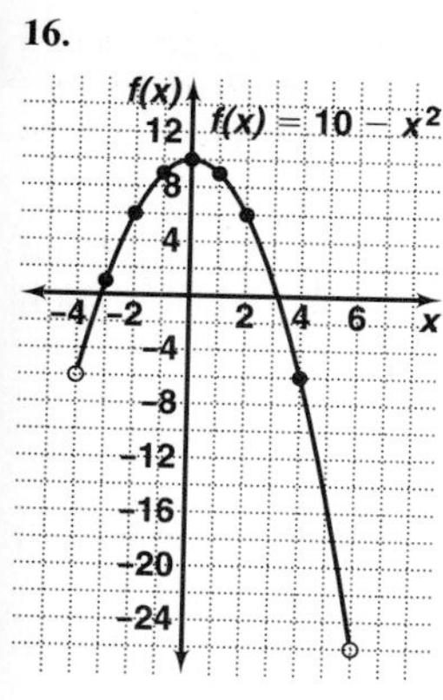

18.

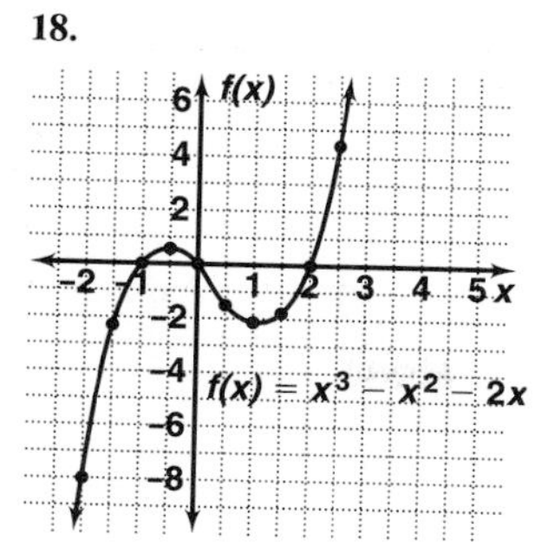

19.

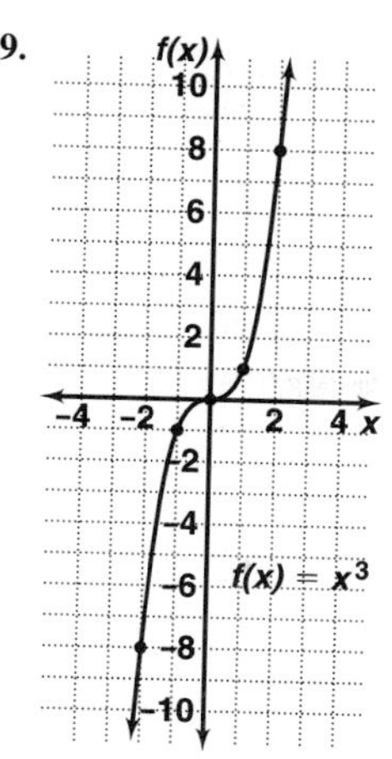

The chart below keys the **Progress Self-Test** questions to the objectives in the **Chapter Review** on pages 821–823. This will enable you to locate those **Chapter Review** questions that correspond to questions you missed on the **Progress Self-Test.** The lesson where the material is covered is also indicated on the chart.

Question	**1**	**2**	**3**	**4**	**5**	**6**	**7**	**8**	**9**	**10**
Objective	A	C	B	B	C	Voc.	C	C	D	H
Lesson	13-2	13-1	13-7	13-8	13-7	13-2	13-1	13-1	13-4	13-1
Question	**11**	**12**	**13**	**14**	**15**	**16**	**17**	**18**	**19**	
Objective	F	F	I	E	I	I	E	J	J	
Lesson	13-5	13-5	13-5	13-3	13-3	13-2	13-2	13-6	13-6	

CHAPTER 13 REVIEW (pp. 821–823)

1. 6 **3.** 78 **5.** 8 **7.** 1.5 **9.** $x = 2$ or $x = -8$ **11.** 0.080 **13.** 107.093 **15.** 0.577 **17.** 0.991 **19.** 4 **21.** Yes **23.** Yes **25.** No **27.** Yes **29.** Yes **31.** The domain is the largest set for which the function makes sense. **33.** (c) **35.** set of all real numbers; set of real numbers ≤ 1 **37.** 10 **39.** set of all real numbers except 3 **41.** $t = 92$ **43.** $G(v) = v + 250$ **45. a.** set of times of the day the Mart is open **b.** Sample: If the largest number of people MacGregor's Mart can hold is 2000, then the range = set of whole numbers ≤ 2000. **47. a.** $\frac{1}{2}$ **b.** $\frac{1}{3}$ **c.** $\frac{1}{6}$ **49.** $\frac{9}{64}$ **51.** .870 **53.** ≈ 0.325 **55.** Yes **57.** Yes **59.** Yes **61. See below.** **63. See below.** **65. See right.** **67. See right.** **69. See right.** **71. a. See right. b.** Since the degree of $P(x)$ is 3, it has at most 3 x-intercepts.

61.

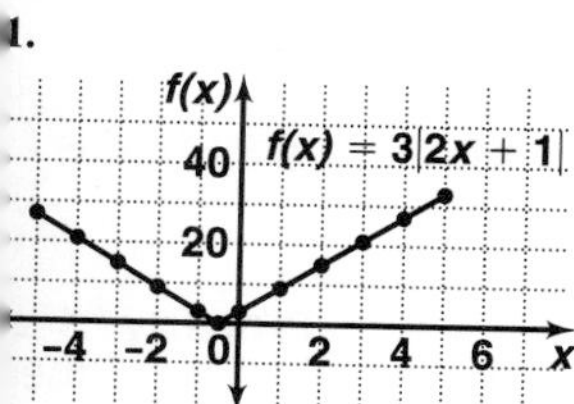

63.

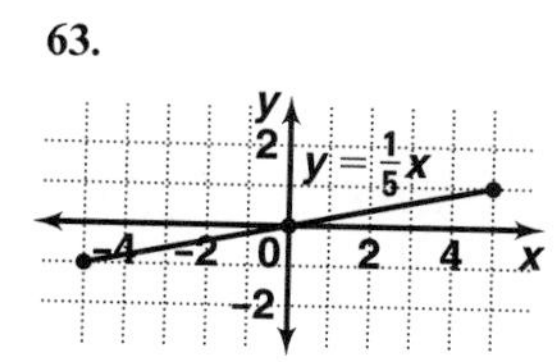

65.

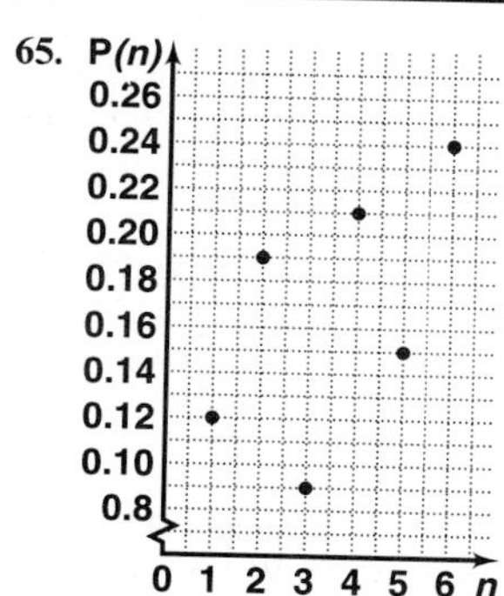

67.

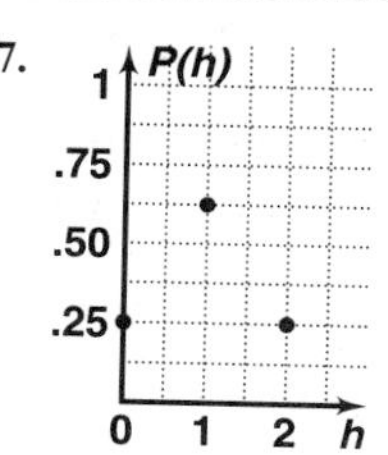

69.

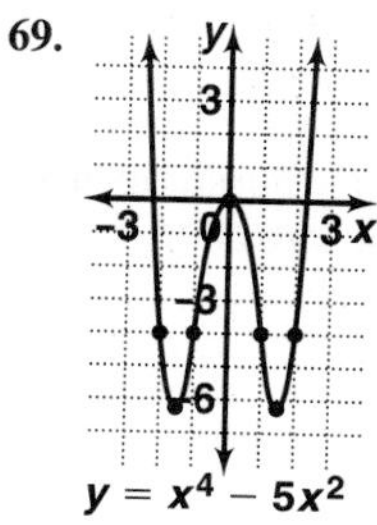

71. a.

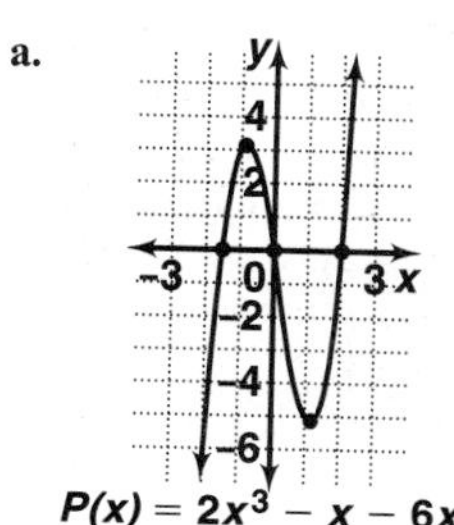

ABS(x) A BASIC function that gives the absolute value of *x*. (784, 836)

absolute value function A function with an equation of the form $f(x) = a|x - b| + c$. (784)

absolute value If $n < 0$, then the absolute value of *n* equals $-n$; if $n \geq 0$, then the absolute value of *n* is *n*. The absolute value of a number is its distance on a number line from the point with coordinate 0. (593)

Absolute Value-Square Root Property For all real numbers *x*, $\sqrt{x^2} = |x|$. (596)

acute angle An angle with measure between 0° and 90°.

Adding Like Terms Property See *Distributive Property: Adding or Subtracting Like Terms.*

addition method for solving a system The method of adding the sides of two equations to yield a third equation which contains solutions to the system. (682)

Addition Property of Equality For all real numbers *a, b,* and *c*: if $a = b$, then $a + c = b + c$. (151)

Addition Property of Inequality For all real numbers *a, b,* and *c*: if $a < b$, then $a + c < b + c$. (200)

additive identity The number zero. (149)

Additive Identity Property For any real number *a*: $a + 0 = a$. (149)

additive inverse The additive inverse of any real number *x* is $-x$. Also called *opposite.* (149)

Algebraic Definition of Division For any real numbers *a* and *b*, $b \neq 0$: $a \div b = a \cdot \frac{1}{b}$. (350)

Algebraic Definition of Subtraction For all real numbers *a* and *b*: $a - b = a + -b$. (216)

algebraic expression An expression that includes one or more variables. (23)

algebraic fraction A fraction which has variables in its numerator or denominator. (86)

algorithm A finite step-by-step recipe or procedure. (640)

annual yield The percent the money in an account earns per year. (486)

array See *rectangular array.* (74)

Area Model for Multiplication (discrete version) The number of elements in a rectangular array with *x* rows and *y* columns is *xy*. (74)

Area Model for Multiplication The area of a rectangle with length ℓ and width *w* is ℓw. (72)

"as the crow flies" The straight line distance between two points. (599)

Associative Property of Addition For any real numbers *a, b,* and *c*: $(a + b) + c = a + (b + c)$. (145)

Associative Property of Multiplication For any real numbers *a, b,* and *c*: $(ab)c = a(bc)$. (75)

automatic grapher A graphing calculator or a computer graphing program that enables equations to be graphed on a coordinate plane. (310)

average See *mean.*

axes The perpendicular number lines in a coordinate graph from which the coordinates of points are determined. (155)

axis of symmetry The line over which a figure coincides with its reflection image. (548)

bar graph A way of displaying data using rectangles or bars with lengths corresponding to the data. (142)

base The number *x* in the power x^n. (486)

BASIC A computer language, short for Beginner's All-purpose Symbolic Instruction Code. (24, 834)

binomial A polynomial with two terms. (617)

boundary point A point that separates solutions from nonsolutions on a number line. (469)

box See *rectangular solid*. (74)

cell The location or box formed by the intersection of a row and a column in a spreadsheet. (233)

Celsius scale The temperature scale in which 0° is the freezing poi of water and 100° is the boiling poi Also called the centigrade scale. (322)

center of symmetry A point about which a figure with rotation symmetry can be rotated 180° to coincide with itself. (801)

centigrade scale See *Celsius sca* (322)

changing the sense (direction) of a inequality Changing from $<$ to $>$, or from $\leq$ to $\geq$, or vice-versa. (116)

Chi-Square Statistic A number calculated from data used to determine whether the difference in two frequency distributions is great than that expected by chance. (652)

chunking The process of groupin several bits of information into a single piece of information. In algebra, viewing an entire algebraic expression as one variable. (333)

clearing fractions Multiplying eac side of an equation by a constant t get an equivalent equation without fractions as coefficients. (327)

closed interval An interval that includes its endpoints. (14)

coefficient A number multiplied by a variable or variables. In the term $-6x$, -6 is the coefficient of *x*. (178)

coincident lines Two lines that contain the same points. (695)

column A vertical list in a table, rectangular array, or spreadsheet. (233)

common denominator The same denominator for two or more fractions. (720)

common factor A number that is a factor of two or more given numbers. (721)

Common Factor Sum Property If a is a common factor of *b* and *c*, the it is a factor of $b + c$. (721)

common logarithm function The function with equation $y = \log x$. (813)

common monomial factoring Isolating a common factor from eac term of a polynomial. (726)

ommutative Property of Addition or any real numbers a and b: $+ b = b + a$. (144)

ommutative Property of lultiplication For any real umbers a and b: $ab = ba$. (73)

omparison Model for Subtraction he quantity $x - y$ tells how much uantity x differs from the quantity y. 23)

omplementary angles Two angles hose measures have a sum of 90°. lso called *complements*. (256)

omplementary events Two events hich have no elements in common nd whose union is the set of all ossible outcomes. (372)

omplete factorization A ctorization into prime polynomials which there are no common umerical factors left in the terms. 27)

omplex fraction A fraction whose umerator and/or denominator ontains a fraction. (351)

omposite number A positive teger that has one or more positive teger factors other than 1 and self. (721)

ompound interest A form of terest payment in which the terest is placed back into the ccount so that it too earns interest. 87)

ompound Interest Formula $= P(1 + i)^n$ where T is the total fter n years if a principal P earns an nnual yield of i. (488)

ondition of the system A entence in a system. (666)

onstant decrease A situation in hich a positive number is epeatedly subtracted. (425)

onstant increase A situation in hich a positive number is epeatedly added. (498)

onstant term A term in a olynomial without a variable. (574)

onstant difference A situation in hich the difference of two xpressions is a constant. (247)

onstant sum A situation in which e sum of two expressions is a onstant. (247)

ontinuous A situation in which umbers between any two numbers ave meaning. (158)

contraction A size change in which the factor k is nonzero and between -1 and 1. (388)

coordinate graph A graph displaying points as ordered pairs of numbers. (155)

coordinates The numbers identified with a point in the coordinate plane. (156)

coordinate plane A plane in which every point can be identified by two numbers. (156)

cosine function A function defined by $y = \cos x$. (813)

cosine of an angle (cos *A*) The ratio of sides in a right triangle given by $\cos A = \frac{\text{length of leg adjacent}}{\text{length of hypotenuse}}$. (813)

counterexample An example for which a pattern is false. (39)

cubic unit The basic unit of volume. (74)

cubing function The function defined by $y = x^3$. (800)

cursor An arrow or pixel which may be moved along a graph on an automatic grapher. (312)

default window The window preprogrammed for use by an automatic grapher when no other window is specified. (311)

degree of a monomial The sum of the exponents of the variables in the monomial. (616)

degree of a polynomial The highest degree of any of its terms after the polynomial has been simplified. (617)

degrees of freedom The number of events minus one, used in the Chi-Square Statistic. (653)

deviation The absolute value of the difference between an expected number and an actual observed number. (652)

difference of squares An expression of the form $x^2 - y^2$. (648)

Difference of Two Squares Pattern $(a + b)(a - b) = a^2 - b^2$. The product of the sum and the difference of two numbers equals the difference of squares of the numbers. (648)

dimensions The number of rows and columns of an array. The lengths of sides of a rectangle or a rectangular solid. (74)

discount The percent by which the original price of an item is lowered. (222)

discrete A situation in which some numbers between given numbers do not have meaning. (13)

discrete set A set of objects that can be counted. (74)

discriminant In the quadratic equation $ax^2 + bx + c = 0$, $b^2 - 4ac$. (582)

Discriminant Property Suppose $ax^2 + bx + c = 0$ and a, b, and c are real numbers with $a \neq 0$. Let $D = b^2 - 4ac$. Then when $D > 0$, the equation has exactly two real solutions. When $D = 0$, the equation has exactly one real solution. When $D < 0$, the equation has no real solutions. (582)

Discriminant Theorem When a, b, and c are integers, with $a \neq 0$, three conditions happen at exactly the same time:
1. The solutions to $ax^2 + bx + c = 0$ are rational numbers.
2. $b^2 - 4ac$ is a perfect square.
3. $ax^2 + bx + c$ is factorable over the set of polynomials with integer coefficients. (761)

Distance Formula in the Coordinate Plane The distance AB between points $A = (x_1, y_1)$ and $B = (x_2, y_2)$ is $AB = \sqrt{(x_2 - x_1)^2 + (y_2 - y_1)^2}$. Also called *Pythagorean Distance Formula*. (601)

Distance Formula on a Number Line If two points on a line have coordinates x_1 and x_2, the distance between them is $|x_1 - x_2|$. (595)

Distributive Property: Adding Fractions For all real numbers a, b, and c, with $c \neq 0$, $\frac{a}{c} + \frac{b}{c} = \frac{a+b}{c}$. (195)

Distributive Property: Adding or Subtracting Like Terms For all real numbers a, b, and c: $ac + bc = (a + b)c$ and $ac - bc = (a - b)c$. (177)

Distributive Property: Removing Parentheses For all real numbers a, b, and c: $c(a + b) = ca + cb$ and $c(a - b) = ca - cb$. (183)

GLOSSARY

domain The values which may be meaningfully substituted for a variable. (12)

domain of a function The set of possible replacements for the first variable in a function. (789)

edge A line in a plane that separates solutions from nonsolutions. (470)

element An object in a set. Also called *member*. (11)

empty set A set which has no elements in it. (19)

END A BASIC command which stops a program. (30, 836)

endpoints The smallest and largest numbers in an interval. The points *A* and *B* in the segment $\overline{AB}$. (14)

Equal Fractions Property For all numbers *a, b,* and *k*, if $k \neq 0$ and $b \neq 0$, then $\frac{a}{b} = \frac{ak}{bk}$. (87)

equal sets Two sets that have the same elements. (11)

equally likely outcomes Outcomes in a situation where the likelihood of each outcome is assumed to be the same. (381)

equation A sentence with an equal sign. (6)

equivalent formulas Formulas in which every set of values that satisfies one formula also satisfies the other. (323)

equivalent systems Systems with exactly the same solutions. (687)

evaluating an expression Finding the numerical value of an expression. (23)

event A set of possible outcomes. (371)

expanding a power of a polynomial Writing the power of a polynomial as a single polynomial. (646)

expansion A size change in which the size change factor *k* is greater than 1 or less than -1. (388)

expected number The mean frequency of a given event that is predicted by probability. (651)

exponent The number *n* in the power x^n. (486)

exponential decay A situation in which the original amount is repeatedly multiplied by a growth factor between zero and one. (505)

exponential function A function with an equation of the form $y = ab^x$. (775)

exponential growth A situation in which the original amount is repeatedly multiplied by a growth factor greater than one. (493)

exponential curves Graphs of equations of the form $y = b \cdot g^x$, where $b \neq 0$, $g > 0$, and $g \neq 1$. (493)

Extended Distributive Property To multiply two sums, multiply each term in the first sum by each term in the second sum. (634)

extremes The numbers *a* and *d* in the proportion $\frac{a}{b} = \frac{c}{d}$. (394)

f(x) The value of the function *f* at *x*. (778)

factorial The product of the integers from 1 to a given number. $n! = 1 \cdot 2 \cdot \ldots \cdot (n - 1) \cdot n$. (126)

factoring The process of expressing a given number (or expression) as the product of two or more numbers (or expressions). (726)

factorization The result of factoring a number or expression. (727)

factors Numbers or expressions whose product is a given number or expression. If $ab = c$, then *a* and *b* are factors of *c*. (720)

Fahrenheit scale A temperature scale in which 32° is the freezing point and 212° is the boiling point. (322)

fitting a line to data Finding a line that closely describes data points which themselves may not all lie on a line. (458)

FOIL algorithm A method for multiplying two binomials; the sum of the product of the First terms, plus the product of the Outside terms, plus the product of the Inside terms, plus the product of the Last terms: $(a + b)(c + d) = ac + ad + bc + bd$. (640)

FOR/NEXT loop A sequence of steps in a BASIC program which enables a procedure to be repeated a certain number of times. (191, 835)

formula A sentence in which one variable is given in terms of other variables and numbers. (27)

frequency The number of times an event occurs. (370)

function A set of ordered pairs in which each first coordinate appears with exactly one second coordinate. (774)

function key on a calculator A key which produces the value of a function when a value in the domain is entered. (812)

function notation Notation to indicate a function, such as *f*(*x*), and read as "*f* of *x*". (778)

general form of a linear equation An equation of the form $ax + b = cx + d$, where $a \neq 0$. (299)

general form of a quadratic equation A quadratic equation in which one side is 0 and the other side is arranged in descending order of exponents: $ax^2 + bx + c = 0$, where $a \neq 0$. (573)

Generalized Addition Property of Equality For all numbers or expressions *a, b, c,* and *d*: if $a = b$ and $c = d$, then $a + c = b + d$. (681)

GOTO A BASIC command which tells the computer to go to the indicated program line number. (836)

greatest common factor for integers The greatest integer that is a common factor of two or more integers. (726)

greatest common factor for monomials The product of the greatest common factor of their coefficients and the greatest common factor of their variables. (726)

growth factor In exponential growth or decay, the nonzero number which is repeatedly multiplied by the original amount. (493)

growth model for powering When an amount is multiplied by *g*, the growth factor in each of *x* time periods, then after the *x* periods, the original amount will be multiplied by g^x. (493)

half-plane In a plane, the region on either side of a line. (470)

hard copy A paper copy printed by a computer or calculator of a graph or other information on a screen. (313)

horizontal line A line with an equation of the form $y = k$, where k is a fixed real number. (285)

hypotenuse The longest side of a right triangle. (45)

IF . . . THEN A BASIC command which tells the computer to perform the THEN part only if the IF part is true. (111, 835)

image The final figure resulting from a transformation. (164)

inequality A sentence with one of the following signs: "$\neq$", "$<$", "$>$", "$\leq$", or "$\geq$". (7)

INPUT A BASIC statement that makes the computer pause and wait for a value of the variable to be entered. (30, 834)

instance An example for which a pattern is true. (37)

integers The whole numbers and their opposites. (12)

interest The money a bank pays on the principal in an account. (486)

intersection of sets The set of elements in both set A and set B, written $A \cap B$. (17)

interval The set of numbers between two numbers a and b, possibly including a or b. (14)

irrational number A real number that is not rational. A number that can be written as a nonrepeating infinite decimal. (755)

latitude A measure of the distance of a place on Earth north or south of the equator, given in degrees. (456)

leg of a right triangle One of the sides forming the right angle of a triangle. (45)

LET A BASIC command which assigns a value to a given variable. (30, 834)

like terms Two or more terms in which the variables and corresponding exponents are the same. (178)

linear equation An equation in which the variable or variables are all to the first power and none multiply each other. (267)

linear expression An expression in which all variables are to the first power. (190)

linear function A function with an equation of the form $y = mx + b$. (775)

linear inequality A linear sentence with an inequality symbol. (317, 472)

linear polynomial A polynomial of degree one. (618)

linear sentence A sentence in which the variable or variables are all to the first power and none multiply each other. (618)

linear system A system of equations, each of degree one. (665)

log (x) The common logarithm of x. (813)

loop Repetition of a set of instructions in a computer program. (191, 835)

magnitude of a size change See *size change factor*. (388)

mark-up A percent by which the original price of an item is raised. (222)

maximum The greatest value in a set of numbers; the highest point on a graph. (223, 550)

mean The sum of the numbers in a collection divided by the number of numbers in the collection. Also called *average*. (725)

means The numbers b and c in the proportion $\frac{a}{b} = \frac{c}{d}$. (394)

Means-Extremes Property For all real numbers a, b, c, and d (b and d nonzero): if $\frac{a}{b} = \frac{c}{d}$, then $ad = bc$. (395)

median In a collection consisting of an odd number of numbers in numerical order, the middle number. In a collection of an even number of numbers arranged in numerical order, the average of the two middle terms. (725)

member An object in a set. Also called *element*. (11)

minimum The smallest value in a set of numbers; the lowest point on a graph. (223, 550)

mirror image The reflection image of a figure. (548)

mode The object(s) in a collection that appear(s) most often. (725)

model A general pattern for an operation that includes many of the uses of the operation. (72)

monomial A polynomial with one term. An expression that can be written as a real number, a variable, or a product of a real number and one or more variables with non-negative exponents. (616)

multiple of a number *n* A number that has n as a factor. (720)

Multiplication Counting Principle If one choice can be made in m ways and a second choice can be made in n ways, then there are mn ways of making the choices in order. (119)

Multiplication of Positive and Negative Numbers The product of an odd number of negative numbers is negative, and the product of an even number of negative numbers is positive. (98)

Multiplication Property of -1 For any real number a: $a \cdot -1 = -1 \cdot a = -a$. (97)

Multiplication Property of Equality For all real numbers a, b, and c: if $a = b$, then $ca = cb$. (102)

Multiplication Property of Inequality If $x < y$ and a is positive, then $ax < ay$. If $x < y$ and a is negative, then $ax > ay$. (114, 116)

Multiplication Property of Zero For any real number a: $a \cdot 0 = 0 \cdot a = 0$. (81)

Multiplicative Identity Property of One For any real number a: $a \cdot 1 = 1 \cdot a = a$. (79)

multiplicative identity The number 1. (79)

multiplicative inverse The multiplicative inverse of a nonzero number n is $\frac{1}{n}$. Also called *reciprocal*. (79)

Multiplying Fractions Property For all real numbers a and c, and all nonzero b and d: $\frac{a}{b} \cdot \frac{c}{d} = \frac{ac}{bd}$. (86)

Multiplying Positive and Negative Numbers If two numbers have the same sign, their product is positive. If the two numbers have different signs, their product is negative. (97)

multiplying through The process of multiplying each side of an equation by a common multiple of the denominators to result in an equation for which all coefficients are integers. (327)

n factorial ($n!$) The product of the integers from 1 to n. (126)

Negative Exponent Property For all n and all nonzero b, $b^{-n} = \frac{1}{b^n}$, the reciprocal of b^n. (516)

nth power The nth power of a number x is the number x^n. (486)

null set A set which has no elements in it. Also called *empty set.* (19)

numerical expression An expression which includes numbers and operations and no variables. (23)

oblique line A line which is neither horizontal nor vertical. (465)

obtuse angle An angle with measure greater than 90° and less than 180°.

open interval An interval that does not include its endpoints. (14)

open sentence A sentence that contains at least one variable. (7)

Opposite of a Difference Property For all real numbers a and b, $-(a - b) = -a + b$. (241)

Opposite of a Sum Property For all real numbers a and b, $-(a + b) = -a + -b = -a - b$. (240)

Opposite of Opposites Property (Op-op Property) For any real number a: $-(-a) = a$. (150)

opposite The opposite of any real number x is $-x$. Also called *additive inverse*. (97)

order of operations The correct order of evaluating numerical expressions: first, work inside parentheses, then do powers. Then do multiplications or divisions, from left to right. Then do additions or subtractions, from left to right. (23)

origin The point (0, 0) on a coordinate plane. (155)

outcome A result of an experiment. (371)

parabola The curve that is the graph of an equation of the form $y = ax^2 + bx + c$, where $a \neq 0$. (548)

pattern A general idea for which there are many instances. (37)

P(E) The probability of event E or "P of E." (371)

percent (%), times $\frac{1}{100}$, or "per 100." (376)

percent of discount The ratio of the discount to the original price. (363)

percent of tax The ratio of tax to the selling price. (363)

perfect square A number which is the square of a whole number. (32)

perfect square trinomial A trinomial which is the square of a binomial. (647)

Perfect Square Patterns $(a + b)^2 = a^2 + 2ab + b^2$ and $(a - b)^2 = a^2 - 2ab + b^2$. (647)

permutation An arrangement of letters, names, or objects. (126)

Permutation Theorem There are $n!$ possible permutations of n different objects, when each object is used exactly once. (126)

plus or minus symbol (±) A symbol which shows that a calculation should be done twice, once by adding and once by subtracting. (574)

polynomial An algebraic expression that is either a monomial or a sum of monomials. (617)

polynomial function A function whose range values are given by a polynomial. (800)

polynomial in the variable x An expression of the form $a_nx^n + a_{n-1}x^{n-1} + \ldots a_1x + a_0$, where $a_0, a_1, \ldots, a_n$ are real numbers. (617)

polynomial over the integers A polynomial with integer coefficients. (727)

population density The number of people per unit of area. (360)

power An expression written in the form x^n. (486)

Power of a Power Property For all m and n, and all nonzero b, $(b^m)^n = b^{mn}$. (512)

Power of a Product Property For all n, and all nonzero a and b, $(ab)^n = a^n \cdot b^n$. (527)

Power of a Quotient Property For all n, and all nonzero a and b, $\left(\frac{a}{b}\right)^n = \frac{a^n}{b^n}$. (529)

preimage The original figure before a transformation takes place. (164)

prime factorization The writing of a number as a product of primes. (722)

prime number An integer greater than 1 whose only integer factors are itself and 1. (721)

prime polynomial A polynomial which cannot be factored into polynomials of a lower degree. (727)

principal Money deposited in an account. (486)

PRINT A BASIC command which tells the computer to print what follows the command. (30, 834)

Probability Formula for Geometric Regions If all points occur randomly in a region, then the probability P of an event is given by $P = \frac{\text{measure of region for event}}{\text{measure of entire region}}$, where the measure may be length, area, etc. (382)

probability function A function that maps a set of outcomes onto their probabilities. (795)

probability of an event A number from 0 to 1 that measures the likelihood that an event will occur. (371)

Product of Powers Property For all m and n, and all nonzero b, $b^m \cdot b^n = b^{m+n}$. (510)

Product of Square Roots Property For all nonnegative real numbers a and b, $\sqrt{a} \cdot \sqrt{b} = \sqrt{ab}$. (587)

projectile An object that is thrown, dropped, or shot by an external force and continues to move on its own. (567)

Property of Opposites For any real number a: $a + -a = 0$. (149)

Property of Reciprocals For any nonzero real number a: $a \cdot \frac{1}{a} = \frac{1}{a} \cdot a = 1$. (79)

proportion A statement that two fractions are equal. Any equation of the form $\frac{a}{b} = \frac{c}{d}$. (394)

Putting-Together Model for Addition If a quantity x is put together with a quantity y with the same units and if there is no overlap, then the result is the quantity $x + y$. (143)

Pythagorean Distance Formula See *Distance Formula in the Coordinate Plane.*

Pythagorean Theorem In a right triangle with legs a and b and hypotenuse c, $a^2 + b^2 = c^2$. (46)

quadrant One of the four regions of the coordinate plane formed by the x-axis and y-axis. (163)

quadratic equation An equation that can be written in the form $ax^2 + bx + c = 0$. (573)

Quadratic Formula If $a \neq 0$ and $ax^2 + bx + c = 0$, then $x = \frac{-b \pm \sqrt{b^2 - 4ac}}{2a}$. (574)

quadratic function A function with an equation of the form $y = ax^2 + bx + c$ or $y = \frac{k}{x}$. (775)

quadratic polynomial A polynomial of degree two. (618)

Quotient of Powers Property For all m and n, and all nonzero b, $\frac{b^m}{b^n} = b^{m-n}$. (521)

radical sign ($\sqrt{\ }$) The symbol for square root. (31)

random outcomes Outcomes in a situation where each outcome is assumed to have the same probability. (372)

range The length of an interval. The maximum value minus the minimum value. (223)

range of a function The set of possible values of a function. (789)

Rate Factor Model for Multiplication When a rate is multiplied by another quantity, the unit of the product is the product of units. Units are multiplied as though they were fractions. The product has meaning when the units have meaning. (92)

Rate Model for Division If a and b are quantities with different units, then $\frac{a}{b}$ is the amount of quantity a per quantity b. (356)

rate of change The rate of change between points (x_1, y_1) and (x_2, y_2) is $\frac{y_2 - y_1}{x_2 - x_1}$. (419)

ratio A quotient of quantities with the same units. (363)

Ratio Model for Division Let a and b be quantities with the same units. Then the ratio $\frac{a}{b}$ compares a to b. (363)

ratio of similitude The ratio of corresponding sides of two similar figures. (402)

rational number A number that can be written as the ratio of two integers. (754)

real numbers Numbers which can be represented as finite or infinite decimals. (12)

reciprocal The reciprocal of a nonzero number n is $\frac{1}{n}$. Also called *multiplicative inverse*. (79)

Reciprocal of a Fraction Property If $a \neq 0$ and $b \neq 0$ the reciprocal of $\frac{a}{b}$ is $\frac{b}{a}$. (80)

reciprocal rates Two rates in which the quantities are compared in reverse order. (93)

rectangular array A two-dimensional display of numbers or symbols arranged in rows and columns. (74)

rectangular solid A 3-dimensional figure with 6 rectangular faces. (74)

reflection symmetry The property held by a figure that coincides with its image under a reflection over a line. Also called *symmetry with respect to a line*. (548)

relation A set of ordered pairs. (775)

relative frequency The ratio of the number of times an event occurred to the total number of possible occurrences. (370)

Relative Frequency of an Event Suppose a particular event has occurred with a frequency of f times in a total of T opportunities for it to happen. Then the relative frequency of the event is $\frac{f}{T}$. (370)

REM A BASIC statement for a remark or explanation that will be ignored by the computer. (835)

Removing Parentheses Property See *Distributive Property*. (183)

Repeated Multiplication Model for Powering When n is a positive integer, $x^n = x \cdot x \cdot \ldots \cdot x$ where there are n factors of x. (486)

replication The process of copying a formula in a spreadsheet in which the cell references in the original formula are adjusted for new positions in the spreadsheet. (235)

right angle An angle with measure 90°.

rotation symmetry A property held by some figures where a rotation of some amount other than 360° results in an image which coincides with the original image. (801)

row A horizontal list in a table, rectangular array, or spreadsheet. (233)

scatterplot A two-dimensional coordinate graph of individual points. (156)

scientific notation A number represented as $x \cdot 10^n$, where $1 \leq x < 10$ and n is an integer. (829–833)

sentence Two algebraic expressions connected by "=", "≠", "<", ">", "≤", "≥", or "≈". (6)

sequence A set of numbers or objects in a specific order. (188)

set A collection of objects called elements. (11)

similar figures Two or more figures that have the same shape. (402)

simple fraction A numerical expression of the form $\frac{a}{b}$, where a and b are integers. (754)

simplifying radicals Rewriting a radical with a smaller integer under the radical sign. (588)

sine function The function defined by $y = \sin x$. (813)

sine of an angle (sin A) In a right triangle, $\sin A = \frac{\text{length of the leg opposite}}{\text{length of the hypotenuse}}$. (813)

sinusoidal curve The curve that is the graph of a sine or cosine function. (813)

size change A transformation in which the image of (x,y) is (kx, ky). (387)

size change factor The number k in the transformation in which the image of (x, y) is (kx, ky). Also called *magnitude*. (387)

Size Change Model for Multiplication If a quantity x is multiplied by a size change factor k, $k \neq 0$, then the resulting quantity is kx. (388)

Slide Model for Addition If a slide x is followed by a slide y, the result is the slide $x + y$. (143)

Slope and Parallel Lines Property If two lines have the same slope, then they are parallel. (694)

slope The rate of change between points on a line. The amount of change in the height of the line as you go 1 unit to the right. The slope of the line through (x_1, y_1) and (x_2, y_2) is $\frac{y_2 - y_1}{x_2 - x_1}$. (432)

slope-intercept form An equation of a line in the form $y = mx + b$, where m is the slope and b is the y-intercept. (439)

Slope-Intercept Property The line with equation $y = mx + b$ has slope m and y-intercept b. (440)

solution A replacement of the variable(s) in a sentence that makes the sentence true. (7)

solution set of an open sentence The set of numbers from the domain that are solutions. (12)

solution set to a system The intersection of the solution sets for each of the sentences in the system. (666)

spreadsheet A computer program in which data are presented in a table and calculations upon entries in the table can be made. The table itself. (232)

SQR(X) A BASIC function that gives the square root of x. (812, 836)

Square of the Square Root Property For any nonnegative number n, $\sqrt{n} \cdot \sqrt{n} = n$. (33)

square root If $A = s^2$, then s is a square root of A. (31)

square unit The basic unit for area. (72)

squaring function A function defined by $y = x^2$. (778)

stacked bar graph A display of data using rectangles or bars stacked on top of each other. (143)

standard form for an equation of a line An equation of the form $Ax + By = C$, where not both A and B are zero. (464)

standard form of a prime factorization The form of a factorization where the factors are primes in increasing order and where exponents are used if primes are repeated. (722)

standard form of a polynomial A polynomial written with the terms in descending order of the exponents of its terms, with the largest exponent first. (618)

standard form of a quadratic equation An equation written in the form $ax^2 + bx + c = 0$, where $a \neq 0$. (576)

STEP A BASIC command that tells the computer how much to add to the counter each time through a FOR/NEXT loop. (556, 835)

stopping distance The length of time for a car to slow down from the instant the brake is applied until the car is no longer moving. (554)

substitution method for solving a system A method in which one variable is written in terms of other variables, and then this expression is used in place of the original variable in subsequent equations. (672)

supplementary angles Two angles whose measures have a sum of 180°. Also called *supplements*. (254)

symmetric Having some symmetry. See *reflection symmetry* and *rotation symmetry*. (548)

symmetry with respect to a line See *reflection symmetry*.

system A set of conditions separated by the word *and*. (666)

system of equations A system in which the conditions are equations. (666)

system of inequalities A system in which the conditions are inequalities. (704)

Take-Away Model for Subtraction If a quantity y is taken away from an original quantity x, the quantity left is $x - y$. (221)

tangent function A function defined by $y = \tan x$. (807)

tangent of an angle (tan A) The ratio of sides given by $\tan A = \frac{\text{length of leg opposite}}{\text{length of leg adjacent}}$. (808)

term A number, a variable, or a product of numbers and variables. (178)

testing a special case A strategy for determining whether a pattern is true by trying out specific instances. (535)

theorem A property that has been proved to be true. (46)

Third Side Property If x and y are the lengths of two sides of a triangle, and $x > y$, then the length z of the third side must satisfy the inequality $x - y < z < x + y$. (263)

tick marks Marks on the x and y axes of a graph to show distance. (312)

tolerance The specific amount that manufactured parts are allowed to vary from an accepted standard size. (598)

trace An option on an automatic grapher that allows the user to move a cursor along the graph while displaying the coordinates of the point the cursor indicates. (312)

translation A two-dimensional slide. (164)

tree-diagram A tree-like way of organizing the possibilities of choices in a situation. (120)

Triangle Inequality The sum of the lengths of two sides of any triangle is greater than the length of the third side. (261)

Triangle-Sum Theorem In any triangle with angle measures a, b, and c: $a + b + c = 180$. (255)

trigonometry The study of the trigonometric functions sine, cosine, and tangent, and their properties. (813)

trinomial A polynomial with three terms. (617)

two dimensional slide A transformation in which the image of (x, y) is $(x + h, y + k)$. (164)

undefined slope The situation regarding the slope of a vertical line, which does not exist. (435)

union of sets The set of elements in either set A or set B, written $A \cup B$. (18)

Unique Factorization Theorem Every integer can be represented as a product of primes in exactly one way, disregarding order of the factors. (722)

Unique Factorization Theorem for Polynomials Every polynomial can be represented as a product of prime polynomials in exactly one way, disregarding order and real number multiples. (727)

value of a function The value of the second variable (often called y) in a function for a given value of the first variable. (774)

variable A letter or other symbol that can be replaced by a number (or other object). (6)

Venn diagram A diagram used to show relationships among sets. (18)

vertex The intersection of a parabola with its axis of symmetry. (548)

vertical line A line with an equation of the form $x = h$, where h is a fixed real number. (286)

volume The space contained by a three-dimensional figure. The volume of a rectangular solid is the product of its dimensions. (74)

whole numbers The set of numbers {0, 1, 2, 3, . . . }. (12)

window The part of the coordinate grid that is shown on an automatic grapher. (310)

x-axis The horizontal axis in a coordinate graph. (163)

x-coordinate The first coordinate of a point. (163)

x-intercept The x-coordinate of a point where a graph crosses the x-axis. (446)

y-axis The vertical axis in a coordinate graph. (163)

y-coordinate The second coordinate of a point. (163)

y-intercept The y-coordinate of a point where a graph crosses the y-axis. (439)

Zero Exponent Property If g is any nonzero real number, then $g^0 = 1$. (493)

Zero Product Property For any real numbers a and b, if $ab = 0$, then $a = 0$ or $b = 0$. (739)

zoom A feature on an automatic grapher that allows the user to see a graph on a window of different dimensions without having to input the dimensions. (313)

MATHEMATICAL SYMBOLS

$=$	is equal to
$\neq$	is not equal to
$<$	is less than
$\leq$	is less than or equal to
$\approx$	is approximately equal to
$>$	is greater than
$\geq$	is greater than or equal to
$\pm$	plus or minus
π	Greek letter pi; $= 3.141592...$ or $\approx \frac{22}{7}$
A'	image of point A
$\overleftrightarrow{AB}$	line through A and B
$\overrightarrow{AB}$	ray starting at A and containing B
$\overline{AB}$	segment with endpoints A and B
AB	length of segment from A to B
$\angle ABC$	angle ABC
$m\angle ABC$	measure of angle ABC
$\triangle ABC$	triangle ABC
$\{\ldots\}$	the symbol used for a set
$\emptyset, \{\ \}$	the empty or null set
$A \cap B$	the intersection of sets A and B
$A \cup B$	the union of sets A and B
W	the set of whole numbers
I	the set of integers
R	the set of real numbers
⌐	symbol for 90° angle
%	percent
$\sqrt{}$	square root symbol; radical sign
$\sqrt{n}$	positive square root of n
$\lvert x \rvert$	absolute value of x
^-x	opposite of x
$n°$	n degrees
$n!$	n factorial
$f(x)$	function notation "f of x"; the second coordinates of the points of a function
(x, y)	ordered pair x, y
$N(E)$	the number of elements in set E
$P(E)$	the probability of an event E
$P(A$ and $B)$	the probability that A and B occur
$\tan A$	tangent of $\angle A$
$\sin A$	sine of $\angle A$
$\cos A$	cosine of $\angle A$
ABS(X)	in BASIC, the absolute value of X
SQR(X)	in BASIC, the square root of X
X * X	in BASIC, $X \cdot X$
X ^ Y	in BASIC, X^Y
1/x	calculator reciprocal key
y^x	calculator powering key
x^2	calculator squaring function key
√	calculator square root function key
x!	calculator factorial function key
tan	calculator tangent function key
sin	calculator sine function key
cos	calculator cosine function key
log	calculator logarithm function key
INV, 2nd, or F	calculator second function key
EE or EXP	calculator scientific notation key

Acknowledgments

Unless otherwise acknowledged, all photographs are the property of Scott, Foresman and Company. Page abbreviations are as follows: (T) top, (C) center, (B) bottom, (L) left, (R) right.

COVER & TITLE PAGE Steven Hunt (c)1994 **vi(L)** Stephen Studd/Tony Stone Images **vi(R)** George Hall/Check Six **vii** Uniphoto **viii** Index Stock International **ix** Nadia Mackenzie/Tony Stone Images **x(R)** Pete McArthur/Tony Stone Images **x(L)** West Light **3** AP/Wide World **4-5T** Profiles West **4C** Stephen Studd/Tony Stone Images **4BL** Backgrounds/West Light **4BR** Michael Mazzeo/The Stock Market **5** Ed Manowicz/Tony Stone Images **6** Tom Ives **8** Clive Brunskill/ALLSPORT USA **9** PhotoFest **11** B.Markel/Gamma-Liaison **13** Bob Daemmrich/The Image Works **15** Louis Psihoyos/Matrix **16** Hank Ketcham **17** Clearwater Florida Fire and Police Departments of Public Safety **20** Tony Freeman/Photo Edit **21** Paul Conklin **22** Rita Boseruf **26T** Library of Congress **26C&B** Courtesy United Air Lines **27** Michael Newman/Photo Edit **28** Robinson/ANIMALS ANIMALS **29B** Zig Leszczynski/ANIMALS ANIMALS **30** California Institute of Technology **31** Sidney Harris **35** Milt & Joan Mann/Cameramann International, Ltd. **37** Lawrence Migdale **38** Dr. Duane de Temple **39** Bob Daemmrich/Tony Stone Images **41** Photo: Bill Hogan/Copyrighted, Chicago Tribune Company, all rights reserved, **45** The Vatican/Art Resource, New York **53** L.Rorke/The Image Works **55** David Spangler **59** NASA **60B** Telegraph Color Library/FPG **60T** Eddie Adams/Leo de Wys **70-71T** Tony Hallas/SPL/Photo Researchers **70C** George Hall/Check Six **70-71B** David Lawrence/Panoramic Stock Images **71C** Steven E.Sutton/Duomo Photography Inc. **72** Robert Frerck/Odyssey Productions, Chicago **74-75** Tony Stone Images **76** NASA **79** Martha Swope **82** Milt & Joan Mann/Cameramann International, Ltd. **85** AP/Wide World **89** Christopher Morris/Black Star **91** Focus On Sports **92** Milt & Joan Mann/Cameramann International, Ltd. **93** Bob Daemmrich/Stock Boston **94** John Elk III/Stock Boston **95** David Falconer/David R. Frazier Photolibrary **96** JPL/NASA **99B** Grant Heilman/Grant Heilman Photography **101** Johnny Johnson/ANIMALS ANIMALS **104** Pasley/Stock Boston **106** Don DuBroff Photo **107T** Mary Kate Denny/Photo Edit **107B** Conte/ANIMALS ANIMALS **108** Michael Newman/Photo Edit **109** Milt & Joan Mann/Cameramann International, Ltd. **112** Robert Frerck/Tony Stone Images **113** Robert Frerck/Odyssey Productions, Chicago **117** Milt & Joan Mann/Cameramann International, Ltd. **118** Scala/Art Resource, New York **119** Alex S.MacLean/Landslides **120** Bob Daemmrich **122** Milt & Joan Mann/Cameramann International, Ltd. **125** Patrick Ward/Stock Boston **126** AP/Wide World **129** Chip Henderson/Tony Stone Images **131T** Julian Baum/SPL/Photo Researchers **131B** Ken Korsh/FPG **132T** Charly Franklin/FPG **132C** Scott Spiker/The Stock Shop **132B** Donovan Reese/Tony Stone Images **134** Ron Thomas/FPG **135** Bob Daemmrich/Stock Boston **140B** Telegraph Colour Library/FPG **140-141(TR)** Imtek Imagineering/Masterfile **140TL** Pelton & Associates/West Light **140-141C** G.Biss/Masterfile **141B** Uniphoto **143** Milt & Joan Mann/Cameramann International, Ltd. **144** Mark Burnett/Photo Edit **146** William Johnson/Stock Boston **149** Kevin Syms/David R. Frazier Photolibrary **150** Robert W.Ginn/Photo Edit **155** Jeff Greenberg/dMRp/Photo Edit **156** Jeff Greenberg/dMRp/Photo Edit **157** Rick Maiman/Sygma **160** David R. Frazier Photolibrary **165** Andy Hayt 1994 **167** Lee Boltin **169** Mark Twain National Forest/U.S.Forestry Service **172** J.C.Stevenson/ANIMALS ANIMALS **174** William Johnson/Stock Boston **177** Milt & Joan Mann/Cameramann International, Ltd. **179** Milt & Joan Mann/Cameramann International, Ltd. **182** Janice Rubin/Black Star **184** Mary Kate Denny/Photo Edit **185** Milt & Joan Mann/Cameramann International, Ltd. **187** Elk/Bruce Coleman Inc. **188** Scott Camazine/Photo Researchers **190** Art Pahlke **195** FPG **196** Reuters/Bettmann **197** Milt & Joan Mann/Cameramann International, Ltd. **198** Freeman/Grishaber/Photo Edit **199** Ralph Nelson, Jr./PhotoFest **200ALL** Carol Zacny **201** Leslye Borden/Photo Edit **205C** Ken Reid/FPG **205BL** Bruce Bishop/PhotoFile **205BR** Marc Chamberlain/Tony Stone Images **206T** John Terence Turner/FPG **206B** The name Cuisenaire and the color sequences of the rods, squares, and cubes are registered trademarks of the Cuisenaire Company of America, Inc. **214-215T** Steven Curtis **214CL** Gordon Wilts/Adventure Photo **214CR** C.Moore/West Light **214-215B** Gerry Ellis Nature Photography **215C** Perry Conway/The Stock Broker **216** Jan Kanter **219** Jim Pickerell/Stock Boston **220B** Museum of Modern Art/Film Stills Archive **221** Naoki Okamoto/The Stock Market **222** Photo Edit **223** Mike Penney/David R. Frazier Photolibrary **224** James Blank/Bruce Coleman Inc. **225** Beryl Goldberg **226** Independence National Historical Park Collection **227** JPL/NASA **228-229** Don W.Fawcett/Visuals Unlimited **230** Sygma **232** Karen Usiskin **233** David Young-Wolff/Photo Edit **234** The Stock Market **240** Reproduced from the Story of the Great American West ©1977 The Readers Digest Association, Inc. Used by permission. Artist: David K.Stone **243** Milt & Joan Mann/Cameramann International, Ltd. **245** David R. Frazier Photolibrary **246** Milt & Joan Mann/Cameramann International, Ltd. **249** Tony Freeman/Photo Edit **250** Max Gibbs/Oxford Scientific Films/ANIMALS ANIMALS **251** Tony Stone Images **253** Milt & Joan Mann/Cameramann International, Ltd. **255** Sygma **257** Focus On Sports **260** Copyright, National Geographic Society **261** Stacy Pick/Stock Boston **268** Milt & Joan Mann/Cameramann International, Ltd. **272** Rhodes/Earth Scenes **273** Milt & Joan Mann/Cameramann International, Ltd. **274** Adamsmith Productions/ West Light **275L** Telegraph Color Library/FPG **282-283T** The Stock Market **282C** Charly Franklin/FPG **282BL** Ron Watts/West Light **282-283B** Jean Miele/The Stock Market **283C** Jeff Schultz/Leo de Wys **284** M.Richards/Photo Edit **285** Phil McCarten/Photo Edit **290** Focus On Sports **291** Milt & Joan Mann/Cameramann International, Ltd. **293** Milt & Joan Mann/Cameramann International, Ltd. **294** Carol Zacny **297** Jerry Wachter/Focus On Sports **300** Bob Daemmrich/Stock Boston **302** Focus On Sports **304** Arthur Rackham Illustration **309** Courtesy General Electric Corp. **315** Michael Newman/Photo Edit **317** Stephen McBrady/Photo Edit **318T** John Stern/Earth Scenes **318B** Donald Specker/Earth Scenes **321** John Neubauer/Photo Edit **322** Scott Zapel **325** Danny Daniels/AlaskaStock Images **326T** Gail McCann/Photo Researchers **326B** Jerry Cooke **327** Focus On Sports **328** Milt & Joan Mann/Cameramann International, Ltd. **330** Joanne K.Peterson **331** Nick Sapiena/Stock Boston **337** Milt & Joan Mann/Cameramann International, Ltd. **338T** C.Brewer/H. Armstrong Roberts **338C** Dave Reede/First Light **338-339B** Alan Briere/Natural Selection **339T** Chris Springman/PhotoFile **339** Charly Franklin/FPG **348-349T** R.Gage/FPG **348CL** Robert George Young/Masterfile **348CR** Charly Franklin/FPG **348-349B** Richard Fukubara/West Light **349C** Index Stock International **350** David R. Austen/Stock Boston **353** Robert Torre/Tony Stone Images **355** Yoram Lehmann/Peter Arnold, Inc. **356** Jim Schwabel/Southern Stock Photos **357** James Blank/Southern Stock Photos **359** William Johnson/Stock Boston **360** B.P.Wolff/Photo Researchers **362** Michael Newman/Photo Edit **364** Courtesy Levi Strauss & Company, San Francisco, CA. **366** City Art Museum of St.Louis **367** Focus On Sports **370** Barbara Campbell/Gamma-Liaison **371** NASA **372B** Timothy White/ABC News **376** Stephen Ferry/Gamma-Liaison **377** Michael Newman/Photo Edit **379** Ric Patzke **380** Zig Leszczynski/Earth Scenes **385** Focus On Sports **386T** AP/Wide World **387** Everett Collection **390** Milt & Joan Mann/Cameramann International, Ltd. **391** David Spangler **393** Everett Collection **398** Joseph F.Viesti/Viesti Associates **399** Barry Iverson/Woodfin Camp & Associates **407** Milt & Joan Mann/Cameramann International, Ltd. **408** Telegraph Colour Library/FPG **409T** Tony Garcia/Tony Stone Images **409BR** Tecmap/West Light **416T** Index Stock International **416C** Richard Laird/FPG **416BL** Penny Tweedie/Tony Stone Images **416BR** R.Ian Lloyd/West Light **417T** Deuter/Zefa/H. Armstrong Roberts **418** Robert E. Daemmrich/Tony Stone Images **420** Willard Luce/ANIMALS ANIMALS **422** M.A.Chappell/ANIMALS ANIMALS **423** Charles Gupton/Stock Boston **425** Mark M. Lawrence/The Stock Market **432** James Blank/Stock Boston **434** David Spangler **437** Patricia Woeber **439** Owaki/Kulla/The Stock Market **442** Tony Freeman/Photo Edit **444** Katoomba Scenic Railway, New South Wales, Australia **447** Milt & Joan Mann/Cameramann International, Ltd. **448** Fritz Prenzel/ANIMALS ANIMALS **449** Robert Frerck/Odyssey Productions, Chicago **454** Jim Merli/Visuals Unlimited **457** Imtek Imagineering-1/Masterfile **458** Boisvieux/Photo Researchers **459** Robert Frerck/Woodfin Camp & Associates **460T** Robert Frerck/Odyssey Productions, Chicago **460B** Delip Mehta/Woodfin Camp & Associates **463** Robert E. Daemmrich/The Image Works **465** Felicia Martinez/Photo Edit **469** Robert Rathe/Stock Boston **472** Robert E. Daemmrich/The Image Works **473** Focus On Sports **476T** AP/Wide World **476BL** C.Ursillo/H. Armstrong Roberts **476R** L.Powers/H. Armstrong Roberts **477C** Gregory Heisler/The Image Bank **482** Historical Pictures/Stock Montage, Inc. **483** Mike Andrews/Earth Scenes **484T** Chris Michaels/FPG **484CL** SuperStock, Inc. **484CR** Ralph Mercer/Tony Stone Images **484B** Joe Riley/Folio **485B** Mark Tomalty/Masterfile **488ALL** Scott Zapel **491** Murray Alcosser/The Image Bank **492T** Oxford Scientific Films/ANIMALS ANIMALS **492B** Leonard Lee Rue III/ANIMALS ANIMALS **493ALL** Leonard Lee Rue III/ANIMALS ANIMALS **496** Benn Mitchell/The Image Bank **498** Everett Collection **500** PhotoFest **502** Mary Kate Denny/Photo Edit **505** Carol Zacny **506** John Elk III/Stock Boston **509** Leo Touchet/Woodfin Camp & Associates **511** Dr. Kari Lounatmaa/SPL/Photo Researchers **513** The National Archives **515** Nuridsany et Perennou/Photo Researchers **519** MGM/Photo:/The Kobal Collection **522T** Smithsonian Institution **522-523** U. S. Bureau of Printing and Engraving **525** Felicia Martinez/Photo Edit **526** Martin Rogers/Stock Boston **527** NASA **528** Copyright the British Museum **530** NASA **531** Cara Moore/The Image Bank **537** D.Woo/Stock Boston **538** Mark C. Burnett/Photo Researchers **539** Imtek Imagineering/ Masterfile **540** Telegraph Colour Library/FPG **546-547T** R.Krubner/H. Armstrong Roberts **546C** Nadia Mackenzie/Tony Stone Images **546B** Tom Sanders/Adventure Photo **547C** R.Faris/West Light **547B** SuperStock, Inc. **550-551** David Young-Wolff/Photo Edit **558** Marco Corsetti/FPG **560** Robert Pearcy/ANIMALS ANIMALS **562** Jan Kanter **567** Eric Meola/The Image Bank **568** Focus On Sports **569** Focu On Sports **570** Shaun Botterill/ALLSPORT USA **571** Focus On Sports **573** Travelpix/FPG **576** Ken Cole/Earth Scenes **579** Tom Nebbia/The Stock Market **583** Richard Hutchings/Photo Edit **584T** Tony Freeman/Photo Edit **584B** Dennis MacDonald/Photo Edit **586** Brett Froomer/The Image Bank **591** NASA **593** Oliver Strewe/Tony Stone Images **598** Kaluzny/Thatcher/Tony Stone Images **605T** Mason Morfit/FPG **605B** Erich Lessing/Art Resource, New York **606L** Dennis Hallinan/FPG **606R** Tim Davis/AllStock Inc. **611** Bruno Brokken/ALLSPORT USA **613** Tom Stewart/The Stock Market **614TL** Charles O'Rear/West Light **614-615T** Steven Hunt/The Image Bank **614-615C** Telegraph Colour Library/FPG **614B** Telegraph Colour Library/FPG **615** Arthur Tilley/FPG **619** THE STUDIO, 1977, Jacob Lawrence, Gift of Gull Industries John H. and A H. Hauberg, Links, Seattle and by exchange from the estate of Mark Tobey. Photo: Paul Macapia **620** Goltzer/The Stock Market **622-623** Brett Froomer/The Image Bank **625** David R. Frazier/Tony Stone Images **631** Photo: Milbert Orlando Brown/Copyrighted, Chicago Tribune Company, all rights reserved **633** COMPOSITION WITH RED, BLUE, YELLOW & BLACK, 1921 - P Mondrian. Collection Haags Gemeentemuseum, The Hague. **642** Myrleen Ferguson Cate/Photo Edit **644** Therese Smith **645** Mark Antman/The Image Works **650** John Running **651** Milt Joan Mann/Cameramann International, Ltd. **654T** Courtesy The White House Collection **654C** Courtesy The White House Collection **654B** Copyright by the White House Historical Associatic **655** AP/Wide World **658L** Gwendolen Cates/Sygma **658CL** Aloma/Shooting Star **658C** Julie Dennis/SS/Shooting Star **658CR** Stephen Begleiter/Shooting Star **658R** Terry O'Neill/Sygma **658BG** Pelton & Associates, Inc./West Light **664T** Chuck O'Rear/West Light **664C** The Stock Shop **664B** Mike Fizer/Check Six **665C** Mark MacLaren **665B** Gary Conner/Photo Edit **66** Focus On Sports **671** Bob Thomason/Tony Stone Images **672** John Madere/The Stock Mark **673** Fay Torresyap/Stock Boston **674** Paul Steel/The Stock Market **675** Mark E. Gibson/The Stock Market **676** Spunbarg/Photo Edit **679** James J. Hill. Reference Library St. Paul, MN. Pechter/The Stock Market **681** Reuters/Bettmann **682** Photo by Gail Toerpe Publishers-Washington Island Observer Newspaper **685** The Museum of the City of New York **687** Ton Freeman/Photo Edit **689** Milt & Joan Mann/Cameramann International, Ltd. **691** Milt & Joar Mann/Cameramann International, Ltd. **692** Fred Whitehead/ANIMALS ANIMALS **693** Mark Greenberg/Visions **693B** Mark Greenberg/Visions **694** Photo Edit **695** Focus On Sports Tony Freeman/Photo Edit **699** Walt Disney Collection/Everett Collection **701** Bob Torrez/To Stone Images **707** Focus On Sports **709** The Kobal Collection **710** Paul Conklin **711T** FF **711C** McFarland/SuperStock, Inc. **712T** L.Powers/H. Armstrong Roberts **712C** D.E.Cox/Ton Stone Images **712B** Denis Scott/FPG **714** Vince Streano/Tony Stone Images **716** Tony Freeman/Photo Edit **718T** Jook Leung/FPG **718-719C** West Light **718BL** Larry Lee/West L **718-719BR** Ralph Mercer/Tony Stone Images **719T** West Light **720** Pete Saloutos/The Stoc Market **722** U.S. Department of Defense **725** Michael Yelman **730** Roy Morsch/The Stoc Market **731** Argonne National Lab **738** Jan Kanter **740** Everett Collection **741** Focus O Sports **742** David Parker/SPL/Photo Researchers **743** Fred Whitehead/Earth Scenes **746** & Joan Mann/Cameramann International, Ltd. **748** Susan Copen Oken/Dot **749** Bettmann Archive **753** MISS PEACH/Creators Syndicate **754** Sidney Harris **757** Carol Zacny **758** Owen Franken/Stock Boston **764** Calspan Corporation **765L** Jon Riley/Tony Stone Images **765C** Ralph Mercer/Tony Stone Images **765R** Karageorge/H. Armstrong Roberts **766T** Ro Villa/Stock Illustration Source, Inc. **772-773T** Index Stock International **772C** Pete McArthur/Tony Stone Images **772-773B** Kazu Studio Ltd/FPG **773C** Telegraph Colour Library/FPG **775** Peter Sidebotham/Tony Stone Images **777** Tony Freeman/Photo Edit **77** Reprinted with Special permission of King Features Syndicate **780** Claudia Parks/The Stock Market **784** Judith Canty/Stock Boston **787** Mark Richards/Photo Edit **789** Woodfin Car Associates **790** Jean-Claude Figenwald **793** Rob Crandall/Stock Boston **797** Frank Siteman/Stock Boston **798** Gay Bumgarner/Tony Stone Images **799** Milt & Joan Mann/Cameramann International, Ltd. **800** Focus On Sports **801** Tom & Michelle Grimm/T Stone Images **807** John Gerlach/Earth Scenes **811** Mark Lewis/Tony Stone Images **812** Young-Wolff/Photo Edit **814** Paul Conklin/Photo Edit **816** Culver Pictures **818TL** Elizabe Simpson/FPG **818TR** The Wood River Gallery **818CL** D.Degnan/H. Armstrong Roberts **81** Stock Illustration Source, Inc. **818B** Arthur Tilley/FPG